Advanced Structured Materials

Volume 238

Common engineering materials are reaching their limits in many applications, and new developments are required to meet the increasing demands on engineering materials. The performance of materials can be improved by combining different materials to achieve better properties than with a single constituent, or by shaping the material or constituents into a specific structure. The interaction between material and structure can occur at different length scales, such as the micro, meso, or macro scale, and offers potential applications in very different fields.

This book series addresses the fundamental relationships between materials and their structure on overall properties (e.g., mechanical, thermal, chemical, electrical, or magnetic properties, etc.). Experimental data and procedures are presented, as well as methods for modeling structures and materials using numerical and analytical approaches. In addition, the series shows how these materials engineering and design processes are implemented and how new technologies can be used to optimize materials and processes.

Advanced Structured Materials is indexed in Google Scholar and Scopus.

Jürgen Fuhrmann · Dietmar Hömberg ·
Wolfgang H. Müller · Wolf Weiss
Editors

Advances in Continuum Physics

In Memoriam Wolfgang Dreyer

Springer

Editors
Jürgen Fuhrmann
WIAS
Berlin, Germany

Dietmar Hömberg
WIAS
Berlin, Germany

Wolfgang H. Müller
Institut für Mechanik
Technische Universität Berlin
Berlin, Germany

Wolf Weiss
WIAS
Berlin, Germany

ISSN 1869-8433 ISSN 1869-8441 (electronic)
Advanced Structured Materials
ISBN 978-3-031-93917-4 ISBN 978-3-031-93918-1 (eBook)
https://doi.org/10.1007/978-3-031-93918-1

This Springer imprint is published by the registered company Springer Nature Switzerland AG
The registered company address is: Gewerbestrasse 11, 6330 Cham, Switzerland

Dedicated to the memory of a creative scientific mind, Wolfgang Dreyer

Preface

Immediately after the unexpected death of Wolfgang Dreyer, invitation letters for a special memorial volume were sent to his many friends and colleagues—scientists working in the fields of Continuum and Mathematical Physics.

The more than 20 papers compiled in this book are the result of this call. They can be categorized into

- general concepts of continuum physics
- applications of continuum physics
- kinetic theory of gases

Of particular interest is an article on Wolfgang Dreyer's eventful life, which also contains a full list of his publications.

We regret to inform that one of the original group of Editors, Prof. Michael Herrmann, Universität Braunschweig, died unexpectedly before the book project got started. This volume is also dedicated to his memory.

The editors would like to thank all contributors for their dedicated work. Moreover, we gratefully acknowledge Prof. Dr. hc. mult. Holm Altenbach and Dr. Christoph Baumann (Springer Publisher) for their support of the book project.

Berlin, Germany
April 2025

Jürgen Fuhrmann
Dietmar Hömberg
Wolfgang H. Müller
Wolf Weiss

About Wolfgang Dreyer

Wolfgang Dreyer
(1950–2024)

Wolfgang Dreyer was born in Bielefeld on November 11, 1950. He passed away on May 12, 2024, in Gülpe (Germany). He retired from the Weierstraß-Institut für Angewandte Analysis und Stochastik (WIAS) in Berlin in 2015 where he had been the Head of the Group Thermodynamische Modellierung und Analyse von Phasenübergängen for many years.

Mathematical Physics was the paramount topic of his professional career. This lifelong endeavor started with his diploma thesis in 1979 supported by Prof. Ingo Müller at the Universität Paderborn. It continued in his doctoral and habilitation theses at the Technische Universität Berlin in 1983 and 1990, respectively—also accompanied by Prof. Ingo Müller.

Wolfgang Dreyer was the leading author of many important, internationally renowned, ground-breaking papers on thermodynamics, continuum mechanics and the kinetic theory of gases.

He will remain in the memories of his many friends and colleagues in the Scientific Community.

Berlin, Germany
April 2025

Jürgen Fuhrmann
Dietmar Hömberg
Wolfgang H. Müller
Wolf Weiss

Contents

Chapter 1
Wolfgang Dreyer—The Life of a Free-Spirited Scientist

Wolfgang H. Müller, Wolf Weiss, and Wolfgang Brocks

Epitaph
... Sie zu lieben, mein Freund, ist eine hohe Kunst, die nicht jeder fertig bringt – ein komplizierter Solo-Tanz – Leben Sie wohl.

Agnes E. Meyer about Thomas Mann, April 7th, 1941

Abstract In this article we present some information about the life and the scientific work of the late Prof. Wolfgang Dreyer. It contains various details regarding his personal and scientific life, his education, the academic institutions he was affiliated with, the principal direction of his research, his Ph.D. students, and a (hopefully) complete list of his scientific publications, which is compiled here for the first time.

1.1 Beginnings and Scope of Scientific Work of Wolfgang Dreyer

Telling the life story of someone like Wolfgang Dreyer (Fig. 1.1) is anything but easy due to the complexity of his character. We feel like Agnes E. Meyer in her attempt to get closer to Thomas Mann (who, by the way, was highly appreciated by Wolfgang Dreyer): his contradictory nature is actually expressed in everything he did. He was always good to animals and small children but all too often the fright and the nemesis

W. H. Müller (✉)
Technische Universität Berlin, Institut für Mechanik, Fachgebiet Kontinuumsmechanik und Materialtheorie, Berlin, Germany
e-mail: wolfgang.h.mueller@tu-berlin.de

W. Weiss
Weierstrass-Institut für Angewandte Analysis und Stochastik, Berlin, Germany
e-mail: wolf.weiss@alumni.tu-berlin.de

W. Brocks
Helmholtz-Zentrum Geesthacht, Geesthacht, Germany
e-mail: WBrocks@kabelmail.de

J. Fuhrmann et al. (eds.), *Advances in Continuum Physics*, Advanced Structured Materials 238, https://doi.org/10.1007/978-3-031-93918-1_1

Fig. 1.1 Wolfgang Dreyer 1950–2024 (courtesy of Thorsten Köhler)

of his colleagues. Witty and sharp-tongued, he immediately noticed even the smallest scientific flaws and did not hesitate to search for errors with relish, expose them and announce them in public with loud aplomb. So he took great pride in being called the Donald Trump of science by his enemies, and we remember hearing him say that nothing invigorated him more than a good argument and the subsequent destruction of his opponents—this was as true in the neighborly dispute at his house in Lemgo as it was in scientific discourse. On the other hand, he could be overly sensitive when he was criticized himself.

But let us start at the beginning. Wolfgang Dreyer was born on November 10, 1950 in Bielefeld and grew up in the small East Westphalian town of Lemgo. Highly talented and promising, he attended elementary school there from 1956 to 1961. He started making radios at an early age and his teachers were amazed at the devices he soldered together even as a young schoolchild. Then he transferred to the local Engelbert-Kämpfer-Gymnasium, suddenly puberty struck, and he turned into a rebellious *enfant terrible*—against his parents in particular and all authority in general (the latter characteristic remained with him all his life until his death). His *curriculum vitae* therefore succinctly states that he left the Gymnasium in 1965 *without graduating*.

And so began Wolfgang Dreyer's *anabasis*: in their powerlessness and probably with the hope that their son would improve through punishment, he was forced into an apprenticeship in mechanical engineering at the Heinrich Brandt company in Lemgo.

Telling the stories of the tortures inflicted by his colleagues was a constant trauma of his, and it made us, the listeners, realize the perfidies that humans are capable of.

Be that as it may, the fact is that he completed his apprenticeship in 1968 with the *Gesellenprüfung* and, after a further four-month intermezzo at the aforementioned company, switched to a technical college in Lemgo and obtained the *Fachhochschulreife* in 1971. This enabled him to study *Physikalische Technik* at the University of Applied Sciences in Hagen, Iserlohn department, which he completed in 1975 with a thesis on nuclear radiation measurement.

He then went on to study physics at the University of Paderborn, where he completed his diploma thesis on the thermodynamics and kinetic theory of rubber membranes under the support of Professor Ingo Müller in 1979. In fact, this turned out to be the next fateful encounter of his life.

He then went to the Technische Universität in Berlin as a research assistant with Ingo Müller and received his doctorate with distinction in 1983. The title of his dissertation was *Zur Thermodynamik von Helium II Superfluides Helium mit und ohne Wirbellinien als binäre Mischung*. At the same institution he finished his habilitation thesis in 1990 entitled *Molekulare Erweiterte Thermodynamik realer Gase—Hyperbolizität und Entropie* and obtained the *venia legendi* for continuum physics and statistical mechanics. He then continued to work as a scientist and a *Privatdozent* in Ingo Müller's group until 1992. His fields of research at this time were mainly Extended Thermodynamics and low-temperature physics (with Peter Strehlow at the Physikalisch Technische Bundesanstalt). Moreover, from 1993 until 1997 he was working as a scientist at the Bundesanstalt für Materialforschung und -Prüfung (BAM), where he was engaged in various solid mechanics research projects, such as the development of material models for single-crystal multiphase high-temperature materials and the investigation of micromechanical material models for their suitability of describing the development of damage in porous metals.

It is worth mentioning that Wolfgang Dreyer never applied for scientific positions. On the contrary, he was "called to the rescue" by the institutions as he used to emphasize. And indeed, the ultimate call to his most favorite research institute came in 1997, when he became first Deputy Head of the research group Continuum Mechanics at the Weierstraß-Institut für Angewandte Analysis und Stochastik (WIAS) in Berlin. Finally, after the retirement of Professor Krzysztof Wilmański, he was appointed Head of that group and became an Adjoint Professor of Continuum Mechanics at the Technische Universität Berlin in 2008. He stayed at the WIAS until 2015 and it is fair to say that he added considerably to its reputation in the field of mathematical physics. Let us explore some of his contributions there.

Above all, the position at WIAS was much more to his liking: the work at BAM had simply been too applied and mundane. Now he was free to engage in whatever research *he* he wanted to do. Particularly noteworthy are his contributions to phase field modeling, molecular dynamics and statistical mechanics of multi-phase alloys, or initial and boundary value problems of hyperbolic systems.

In his last active years, he made significant, internationally renowned contributions to the model-based understanding of the functioning of electrochemical systems. These include a fundamentally new model for the intercalation of lithium in

Fig. 1.2 Wolfgang Dreyer on his premises in Gülpe

the nanoparticles of the cathodes of lithium-ion batteries and the thermodynamically consistent formulation of the Nernst-Planck-Poisson system for ion transport in electrolytes, including pressure.

He was also heavily involved in the training of young scientists, whereby the border areas and bridges between the various individual sciences were particularly important to him. His students will always remember the many fruitful and often extremely passionate discussions about mathematical modeling of the real world. From 2007 to 2015, Wolfgang Dreyer was actively engaged in teaching at the Institute of Mechanics of Faculty V in the field of continuum physics and materials theory. He taught students of mechanical engineering, transportation and civil engineering as well as physical engineering. The course of the latter students was particularly important to him, as he always knew how to arouse understanding and interest with great didactic skill. His inspiring lectures and talks are legend.

Now a word of advice. There was an easy way to prevent or even to stop the wrath of Archilles, which unfortunately was never followed by Wolfgang Dreyer's scientific opponents, most likely because they were too caught up in their own importance: you simply had to ask and let him explain. He loved to do that. But beware, comments of dissent or even defiance were not allowed. Listening to the words and explanations of the Master had to be enough of a reward. Maybe an occasional add-on question or admission of one's own intellectual deficiencies were allowed. And if you did that you could learn—a lot.

Of course he also always devoted his free time to science: Wolfgang Dreyer lived in the darkest place in Germany, in Gülpe, see Fig. 1.2. It was therefore only natural for him to become involved in astronomy, not only in theory, but also in practice

Fig. 1.3 Wolfgang Dreyer with one of his many Ph.D. students (Thomas Böhme)

with a telescope. He reported on his sightings and interesting additional information in lectures to the stargazers in the "Sternenpark Westhavelland."

Now, is that all there is? Was he "just" an extremely competent physicist and gifted scientist? No! It was mentioned at the beginning that his character was always ambivalent and good for surprises. Thus it is not too astonishing to hear that Wolfgang Dreyer was a friend of all arts. He loved music. In fact, he learned to play the guitar and was a passionate drummer. In his youth he sang and played songs by Johnny Cash, among others. Later he turned to opera, especially Richard Wagner, and attended the Wagner Festival in Bayreuth on a regular basis. He collected, loved and possessed many paintings and prints by Stassen and Turner. He enjoyed Germanic mythology and watching movies, in particular SciFi. He abhorred sports.

Wolfgang Dreyer will be remembered by all who knew him and missed by his friends (Fig. 1.3), but not necessarily by his enemies.

He is survived by his wife Ella Dreyer and his two sons, for whom he always had an open ear and a big heart.

1.2 List of Publications

Books and Monographs

1. Dreyer, W. *Zur Thermodynamik von Helium II: superfluides Helium, mit und ohne Wirbellinien als binäre Mischung* (Dissertation TU Berlin, 1983).

Peer-Reviewed Journals

2. Dreyer, W. On thermodynamics and kinetic theory of ideal rubber membranes. *Journal of Applied Mechanics* **48** (1981).
3. Dreyer, W., Müller, I. & Strehlow, P. A study of equilibria of interconnected balloons. *The Quarterly Journal of Mechanics and Applied Mathematics* **35,** 419–440 (1982).
4. Dreyer, W. & Weiss, W. The classical limit of relativistic extended thermodynamics. *Annales de l'IHP Physique théorique* **45,** 401–418 (1986).
5. Dreyer, W. Maximisation of the entropy in non-equilibrium. *Journal of Physics A: Mathematical and General* **20,** 6505 (1987).
6. Dreyer, W. & Seelecke, S. Entropy and causality as criteria for the existence of shock waves in low temperature heat conduction. *Continuum Mechanics and Thermodynamics* **4,** 23–36 (1992).
7. Dreyer, W. & Struchtrup, H. Heat pulse experiments revisited. *Continuum Mechanics and Thermodynamics* **5,** 3–50 (1993).
8. Strehlow, P. & Dreyer, W. Heat propagation in glasses. *Physica B: Condensed Matter* **194,** 485–486 (1994).
9. Dreyer, W. & Kunik, M. Maximum entropy principle revisited. *Continuum Mechanics and Thermodynamics* **10,** 331–347 (1998).
10. Dreyer, W. & Kunik, M. Reflections of Eulerian shock waves at moving adiabatic boundaries. *Monte Carlo Methods and Applications,* 231–252 (1998).
11. Dreyer, W., Kunik, M., Sabelfeld, K., Simonov, N. & Wilmanski, K. Iterative procedure for multidimensional Euler equations. *Monte Carlo Methods and Applications,* 253–272 (1998).
12. Dreyer, W. & Kunik, M. Initial and boundary value problems of hyperbolic heat conduction. *Continuum Mechanics and Thermodynamics* **11,** 227–245 (1999).
13. Dreyer, W. & Müller, W. Computer modeling of micromorphological change by phase field models: applications to metals and ceramics. *Journal of the Australasian ceramic society* **36** (2000).
14. Dreyer, W. & Kunik, M. Cold, thermal and oscillator closure of the atomic chain. *Journal of Physics A: Mathematical and General* **33,** 2097 (2000).
15. Dreyer, W. & Mueller, W. H. A study of the coarsening in tin/lead solders. *International Journal of solids and structures* **37,** 3841–3871 (2000).
16. Dreyer, W., Müller, W. H. & Weiss, W. Tales of thermodynamics and obscure applications of the second law. *Continuum Mechanics and Thermodynamics* **12,** 151–184 (2000).
17. Dreyer, W., Junk, M. & Kunik, M. On the approximation of the Fokker-Planck equation by moment systems. *Nonlinearity* **14,** 881 (2001).
18. Dreyer, W. & Mueller, W. H. Modeling diffusional coarsening in eutectic tin/lead solders: a quantitative approach. *International Journal of Solids and Structures* **38,** 1433–1458 (2001).

19. Bonetti, E., Colli, P., Dreyer, W., Gilardi, G., Schimperna, G. & Sprekels, J. On a model for phase separation in binary alloys driven by mechanical effects. *Physica D: Nonlinear Phenomena* **165,** 48–65 (2002).
20. Brown, C. M., Dreyer, W. & Müller, W. H. Discrete Fourier transforms and their application to stress—strain problems in composite mechanics: a convergence study. *Proceedings of the Royal Society of London. Series A: Mathematical, Physical and Engineering Sciences* **458,** 1967–1987 (2002).
21. Dreyer, W. & Müller, W. H. Toward quantitative modeling of morphology changes in solids with phase field theories: Atomistic arguments for the determination of higher gradient coefficients. *Key Engineering Materials* **240,** 901–914 (2002).
22. Herrmann, M., Dreyer, W. & Kunik, M. Kinetic schemes and initial boundary value problems for the Euler system. *Transport Theory and Statistical Physics* **31,** 1–33. https://doi.org/10.1081/TT-120003009 (2002).
23. Bonetti, E., Dreyer, W. & Schimperna, G. Global solutions to a generalized Cahn-Hilliard equation with viscosity. *Advances in Differential Equations* **8** (2003).
24. Dreyer, W. & Duderstadt, F. Towards the thermodynamic modeling of nucleation and growth of liquid droplets in single crystals. *Free boundary problems,* 113–130 (2004).
25. Dreyer, W., Herrmann, M. & Kunik, M. Kinetic solutions of the Boltzmann-Peierls equation and its moment systems. *Continuum Mechanics and Thermodynamics* **16,** 453–469 (2004).
26. Dreyer, W. & Qamar, S. Kinetic flux-vector splitting schemes for the hyperbolic heat conduction. *Journal of Computational Physics* **198,** 403–423 (2004).
27. Dreyer, W., Herrmann, M., Kunik, M. & Qamar, S. Kinetic Schemes for Selected Initial and Boundary Value Problems. *Analysis and Numerics for Conservation Laws* (Jan. 2005).
28. Dreyer, W., Herrmann, M. & Mielke, A. Micro–macro transition in the atomic chain via Whitham's modulation equation. *Nonlinearity* **19,** 471 (2005).
29. Dreyer, W. & Qamar, S. Second order accurate explicit finite volume schemes for the solution of Boltzmann-Peierls equation. *ZAMM-Journal of Applied Mathematics and Mechanics/Zeitschrift für Angewandte Mathematik und Mechanik: Applied Mathematics and Mechanics* **85,** 4–22 (2005).
30. Dreyer, W. & Wagner, B. A. Sharp-interface model for eutectic alloys. Part I: Concentration dependent surface tension. *Interfaces and Free Boundaries* **7,** 199–227 (2005).
31. Dreyer, W. & Duderstadt, F. On the Becker/Döring theory of nucleation of liquid droplets in solids. *Journal of statistical physics* **123,** 55–87 (2006).
32. Dreyer, W., Duderstadt, F., Eichler, S. & Jurisch, M. Stress analysis and bending tests for GaAs wafers. *Microelectronics Reliability* **46,** 822–835 (2006).
33. Böhme, T., Dreyer, W. & Müller, W. H. Determination of stiffness and higher gradient coefficients by means of the embedded-atom method: An approach for binary alloys. *Continuum Mechanics and Thermodynamics* **18,** 411–441 (2007).

34. Dreyer, W., Duderstadt, F. & Naldzhieva, M. Thermodynamics and kinetic theory of nucleation and the evolution of liquid precipitates in gallium arsenide wafer. *Journal of crystal growth* **303,** 18–22 (2007).
35. Dreyer, W. & Duderstadt, F. On the modelling of semi-insulating GaAs including surface tension and bulk stresses. *Proceedings of the Royal Society A: Mathematical, Physical and Engineering Sciences* **464,** 2693–2720 (2008).
36. Dreyer, W. & Herrmann, M. Numerical experiments on the modulation theory for the nonlinear atomic chain. *Physica D: Nonlinear Phenomena* **237,** 255–282 (2008).
37. Dreyer, W., Jamnik, J., Guhlke, C., Huth, R., Moškon, J. & Gaberšček, M. The thermodynamic origin of hysteresis in insertion batteries. *Nature materials* **9,** 448–453 (2010).
38. Dreyer, W. & Kraus, C. On the van der Waals–Cahn–Hilliard phase-field model and its equilibria conditions in the sharp interface limit. *Proceedings of the Royal Society of Edinburgh Section A: Mathematics* **140,** 1161–1186 (2010).
39. Guhlke, C. & Dreyer, W. Hysteresis due to non-monotone material behaviour inside many particle systems. *Verhandlungen der Deutschen Physikalischen Gesellschaft* (2010).
40. Dreyer, W., Gaberšček, M., Guhlke, C., Huth, R. & Jamnik, J. Phase transition in a rechargeable lithium battery. *European Journal of Applied Mathematics* **22,** 267–290 (2011).
41. Dreyer, W., Gaberšček, M. & Jamnik, J. Phase transition in a rechargeable lithium battery. *European Journal of Applied Mathematics* **22,** 267–290 (2011).
42. Dreyer, W., Guhlke, C. & Herrmann, M. Hysteresis and phase transition in many-particle storage systems. *Continuum Mechanics and Thermodynamics* **23,** 211–231 (2011).
43. Dreyer, W., Guhlke, C. & Huth, R. The behavior of a many-particle electrode in a lithium-ion battery. *Physica D: Nonlinear Phenomena* **240,** 1008–1019 (2011).
44. Dreyer, W., Druet, P.-É., Klein, O. & Sprekels, J. Mathematical modeling of Czochralski type growth processes for semiconductor bulk single crystals. *Milan Journal of Mathematics* **80,** 311–332 (2012).
45. Dreyer, W., Duderstadt, F., Hantke, M. & Warnecke, G. Bubbles in liquids with phase transition: Part 1. On phase change of a single vapor bubble in liquid water. *Continuum Mechanics and Thermodynamics* **24,** 461–483 (2012).
46. Dreyer, W., Giesselmann, J., Kraus, C. & Rohde, C. Asymptotic analysis for Korteweg models. *Interfaces and Free Boundaries* **14,** 105–143 (2012).
47. Dreyer, W., Guhlke, C. & Müller, R. Overcoming the shortcomings of the Nernst–Planck model. *Physical Chemistry Chemical Physics* **15,** 7075–7086 (2013).
48. Dreyer, W., Guhlke, C. & Müller, R. Simulation of Electrolytes in the Modeling of Batteries. *PAMM* **13,** 497–498 (2013).
49. Hantke, M., Dreyer, W. & Warnecke, G. Exact solutions to the Riemann problem for compressible isothermal Euler equations for two-phase flows with and without phase transition. *Quarterly of Applied Mathematics* **71,** 509–540 (2013).

50. Aki, G. L., Dreyer, W., Giesselmann, J. & Kraus, C. A quasi-incompressible diffuse interface model with phase transition. *Mathematical Models and Methods in Applied Sciences* **24,** 827–861 (2014).
51. Dreyer, W., Giesselmann, J. & Kraus, C. A compressible mixture model with phase transition. *Physica D: Nonlinear Phenomena* **273,** 1–13 (2014).
52. Dreyer, W., Guhlke, C. & Landstorfer, M. A mixture theory of electrolytes containing solvation effects. *Electrochemistry communications* **43,** 75–78 (2014).
53. Dreyer, W., Hantke, M. & Warnecke, G. Bubbles in liquids with phase transition—part 2: on balance laws for mixture theories of disperse vapor bubbles in liquid with phase change. *Continuum Mechanics and Thermodynamics* **26,** 521–549 (2014).
54. Bothe, D. & Dreyer, W. Continuum thermodynamics of chemically reacting fluid mixtures. *Acta Mechanica* **226,** 1757–1805 (2015).
55. Dreyer, W., Guhlke, C. & Müller, R. Modeling of electrochemical double layers in thermodynamic non-equilibrium. *Physical Chemistry Chemical Physics* **17,** 27176–27194 (2015).
56. Dreyer, W., Huth, R., Mielke, A., Rehberg, J. & Winkler, M. Global existence for a nonlocal and nonlinear Fokker–Planck equation. *Zeitschrift für angewandte Mathematik und Physik* **66,** 293–315 (2015).
57. Dreyer, W., Huth, R., Mielke, A., Rehberg, J. & Winkler, M. Global existence for a nonlocal and nonlinear Fokker–Planck equation. *Z. Angew. Math. Phys.* **66,** 293–315 (2015).
58. Dreyer, W., Guhlke, C. & Müller, R. A new perspective on the electron transfer: recovering the Butler–Volmer equation in non-equilibrium thermodynamics. *Physical Chemistry Chemical Physics* **18,** 24966–24983 (2016).
59. Landstorfer, M., Guhlke, C. & Dreyer, W. Theory and structure of the metal-electrolyte interface incorporating adsorption and solvation effects. *Electrochimica Acta* **201,** 187–219 (2016).
60. Dreyer, W. & Guhlke, C. Sharp limit of the viscous Cahn–Hilliard equation and thermodynamic consistency. *Continuum Mechanics and Thermodynamics* **29,** 913–934 (2017).
61. Dreyer, W., Guhlke, C., Landstorfer, M. & Müller, R. New insights on the interfacial tension of electrochemical interfaces and the Lippmann equation. *European Journal of Applied Mathematics* **29,** 708–753 (2018).
62. Dreyer, W., Friz, P. K., Gajewski, P., Guhlke, C. & Maurelli, M. Stochastic many-particle model for LFP electrodes. *Continuum Mechanics and Thermodynamics* (2018).
63. Dreyer, W., Guhlke, C. & Müller, R. Bulk-surface electrothermodynamics and applications to electrochemistry. *Entropy* **20,** 939 (2018).
64. Guhlke, C., Gajewski, P., Maurelli, M., Friz, P. K. & Dreyer, W. Stochastic many-particle model for LFP electrodes. *Continuum Mechanics and Thermodynamics* **30,** 593–628 (2018).
65. Dreyer, W., Guhlke, C. & Müller, R. The impact of solvation and dissociation on the transport parameters of liquid electrolytes: Continuum modeling and

numerical study. *The European Physical Journal Special Topics* **227,** 2515–2538 (2019).
66. Dreyer, W., Druet, P.-É., Gajewski, P. & Guhlke, C. Analysis of improved Nernst-Planck-Poisson models of compressible isothermal electrolytes. *Zeitschrift für angewandte Mathematik und Physik* **71** (2020).
67. Coghi, M., Dreyer, W., Friz, P. K., Gajewski, P., Guhlke, C. & Maurelli, M. A McKean–Vlasov SDE and Particle System with Interaction from Reflecting Boundaries. *SIAM Journal on Mathematical Analysis* **54,** 2251–2294 (2022).
68. Bothe, D., Dreyer, W. & Druet, P.-É. Multicomponent incompressible fluids—An asymptotic study. *ZAMM-Journal of Applied Mathematics and Mechanics/Zeitschrift für Angewandte Mathematik und Mechanik* **103,** e202100174 (2023).

Conference Proceedings

69. Dreyer, W., Müller, I. & Strehlow, P. *Rubber Balloons - Taken Seriously* in *Proceedings 4th Symposium on Mechanics and Mathematics, 1981* (1981).
70. Dreyer, W. & Müller, I. *Extended Relativistic Thermodynamics of Gases* in *Proceedings of the Workshop on Mathematical Aspects of Fluids and Plasma Dynamics, Trieste* (1984).
71. Dreyer, W. *Superfluid Helium as a Binary Mixture* in *Proceedings of the Symposium on Macroscopic Theories of Superfluids* (Academia Nazionale Dei Lincei, Rom, 1986).
72. Dreyer, W. *Maximisation of entropy for relativistic and degenerate gases in non-equilibrium* in *Kinetic theory and extended thermodynamics* (eds Müller, I. & Ruggeri, T.) (Pigatora, 1987), 107–123.
73. Strehlow, P., Dreyer, W. & Meissner, M. *Kinetic Theory of Phonons in Glasses* in *Phonon Scattering in Condensed Matter VII: Proceedings of the Seventh International Conference, Cornell University, Ithaca, New York, August 3–7, 1992* (1993), 229–230.
74. Dreyer, W. & Olschewski, J. *Order-disorder transitions under load in single crystal superalloys: theory* in (1994), 419–424.
75. Dreyer, W. *Stability of Single Crystal Superalloys* in (Delft University Press, 1995), 419–424.
76. Müller, W., Albrecht, H., Dreyer, W., Hauck, T., Jendrny, J. & Tilgner, R. *Modeling the lifetime of microelectronic components: A challenge for material science and computational engineering* in *Proc. Materials Mechanics, Fracture Mechanics, Micro Mechanics.* (1999).
77. Müller, W. & Dreyer, W. *Phase field models and prediction of micromorphological changes in alloys I–II* in *Proceedings of the Fourth International Congress on Industrial and Applied Mathematics, Edinburgh, 1999* (2000), p. 311.

78. Dreyer, W., Herrmann, M. & Rademacher, J. D. *Wave trains, solitons and modulation theory in FPU chains* in *Analysis, Modeling and Simulation of Multiscale Problems* (Springer, 2006).
79. Aki, G., Daube, J., Dreyer, W., Giesselmann, J., Kränkel, M. & Kraus, C. *A diffuse interface model for quasi–incompressible flows: Sharp interface limits and numerics* in *ESAIM: Proceedings* **38** (2012), 54–77.
80. Landstorfer, M., Dreyer, W. & Guhlke, C. *A New Picture of the Electrochemical Double Layer Based on Modern Continuum Thermodynamics* in *Electrochemical Society Meeting Abstracts 226* (2014), 1359–1359.

Reports

81. Dreyer, W. *Development of microstructure based viscoplastic models for an advanced design of single crystal hot section components* (1994), 1–17.

Miscellaneous

82. Dreyer, W. & Strehlow, P. *Quantenthermodynamik und ihre Bedeutung für thermische Präzisionsmessungen* 1993.
83. Dreyer, W. & Olschewski, J. *A Microscopic Model for the Simulation of the Thermomechanical Behaviour of Single Crystal Nickelbased Superalloys* 1995.
84. Dreyer, W. & Guckel, R. *Micro-macro transitions by interpolation, smoothing, averaging and scaling of particle trajectories* 2002.
85. Dreyer, W. & Herrmann, M. *A simple but rigorous micro-macro transition* 2002.
86. Dreyer, W. *On jump conditions at phase boundaries for ordered and disordered phases* 2005.
87. Dreyer, W. & Herrmann, M. *On the approximation of periodic traveling waves for the nonlinear atomic chain* 2005.
88. Dreyer, W. & Herrmann, M. *Towards rigorous micro-macro transitions: The microscopic oscillator motion* 2005.
89. Dreyer, W., Duderstadt, F. & Qamar, S. *Diffusion in the vicinity of an evolving spherical arsenic droplet* 2006.
90. Böhme, T., Dreyer, W., Duderstadt, F. & Müller, W. H. *A higher gradient theory of mixtures for multi-component materials with numerical examples for binary alloys* 2007.
91. Dreyer, W., Gaberšček, M. & Jamnik, J. *Phase transition and hysteresis in a rechargeable lithium battery* 2007.
92. Dreyer, W. & Kraus, C. *The equilibria of vapour-liquid systems revisited* 2007.
93. Dreyer, W., Duderstadt, F., Eichler, S. & Naldzhieva, M. *On unwanted nucleation phenomena at the wall of a VGF chamber* 2008.

94. Dreyer, W., Duderstadt, F. & Kimmerle, S.-J. *A study on the eigenstrain problem in solid mixtures* 2008.
95. Dreyer, W., Guhlke, C. & Huth, R. *Hysteresis in the context of hydrogen storage and lithium-ion batteries* 2009.
96. Dreyer, W., Hömberg, D. & Petzold, T. *A model for the austenite-ferrite phase transition in steel including misfit stress* 2009.
97. Dreyer, W. & Kimmerle, S.-J. *Mean field diffusion models for precipitation in crystalline GaAs including surface tension and bulk stresses* 2009.
98. Dreyer, W., Giesselmann, J. & Kraus, C. *Modeling of compressible electrolytes with phase transition* 2014.
99. Dreyer, W., Druet, P.-É., Gajewski, P. & Guhlke, C. *Existence of weak solutions for improved Nernst-Planck-Poisson models of compressible reacting electrolytes* 2016.

Chapter 2
An Eulerian Formulation for Dissipative Materials Using Lie Derivatives and GENERIC

Alexander Mielke

In memory of Wolfgang Dreyer

Abstract We recall the systematic formulation of Eulerian mechanics in terms of Lie derivatives along the vector field of the material points. Using the abstract properties of Lie derivatives we show that the transport via Lie derivatives generates in a natural way a Poisson structure on the chosen phase space. The evolution equations for thermo-viscoelastic-viscoplastic materials in the Eulerian setting are formulated in the abstract framework of GENERIC (General Equations for Non-Equilibrium Reversible-Irreversible Coupling). The equations may not be new, but the systematic splitting between reversible Hamiltonian and dissipative effects allows us to see the equations in a new light that is especially useful for future generalizing of the system, e.g. for adding new effects like reactive species.

2.1 Introduction

Motivated by the related work (Mielke and Roubíček 2025) we reconsider the theory of Lie derivatives for the formulation of dissipative continuum-mechanical systems in the Eulerian setting. For this, we provide in Sect. 2.2 an introduction to the theory of Lie derivatives, first in the differential geometrical setting involving multi-linear forms and then for vectors, co-vectors, and operators as they are used in continuum mechanics. The main point here is to gain an understanding that the usage of such objects may have different interpretations in differential geometry and hence need a suitable Lie derivative. It is well-known that there are different Lie derivatives for stress tensors (stress rates) but the issue is already relevant to the distinction of extensive and intensive field variables and for vectors and co-vectors such as the momentum.

A. Mielke (✉)
Weierstrass-Institut für Angewandte Analysis und Stochastik, Berlin, Germany
e-mail: alexander.mielke@wias-berlin.de

J. Fuhrmann et al. (eds.), *Advances in Continuum Physics*, Advanced Structured Materials 238, https://doi.org/10.1007/978-3-031-93918-1_2

In Sect. 2.3 we first recall the GENERIC framework, where the acronym GENERIC stands for *General Equation for Non-Equilibrium Reversible-Irreversible Coupling* and was introduced in Grmela and Öttinger (1997). However, this class of models originates in Grmela (1984, 1985) and in the metriplectic theory developed in Morrison (1984, 1986), cf. the survey (Morrison 2009). Over the last decade, GENERIC has proved to be a versatile modeling tool for various complex coupled models for fluids and solids, see, e.g., Lebon et al. (2008), Mielke (2011), Hütter and Svendsen (2012), Duong et al. (2013), Pavelka et al. (2018, 2020), Betsch and Schiebl (2019), Lasarzik (2021), Pelech et al. (2022), Zafferi et al. (2023) and the references therein. We also refer to Mielke (2015), Kantner et al. (2019) for applications in semiconductor and quantum devices. In Mielke et al. (2024) it is shown that a dissipative GENERIC system can be rigorously derived from an infinite-dimensional (non-dissipative) Hamiltonian systems.

In Sect. 2.3.2 we start to develop continuum mechanics system at finite strain in a systematic way that is compatible with Lie derivatives and GENERIC. In particular, we consider elastoplastic materials based on the multiplicative split $\boldsymbol{F} = \boldsymbol{F}_{\mathrm{e}}\boldsymbol{F}_{\mathrm{p}}$ where $\boldsymbol{F}$ remains related with the velocity field $\boldsymbol{v}$ through the kinematic relation

$$\partial_t \boldsymbol{F} = -\mathfrak{L}_{\boldsymbol{v}}^{\mathrm{ve}} \boldsymbol{F} := -\boldsymbol{v}\cdot\nabla \boldsymbol{F} + (\nabla \boldsymbol{v})\boldsymbol{F},$$

which is indeed given by a suitable Lie derivative, see Lemma 2.1.

In addition to the momentum $\boldsymbol{\pi} = \rho \boldsymbol{v}$ (with $\rho = \rho_{\mathrm{ref}}/\det \boldsymbol{F}$) and $\boldsymbol{F}$, we use the plastic distorting $\boldsymbol{F}_{\mathrm{p}} \in \mathrm{GL}(\mathbb{R}^d)$ and a scalar thermodynamical field variable τ for form the state variable $q = (\boldsymbol{\pi}, \boldsymbol{F}, \boldsymbol{F}_{\mathrm{p}}, \tau)$. The variable τ can be chosen quite general, e.g.m as density of internal energy, enthalpy, or entropy or as the temperature.

In Sect. 2.3.3 we discuss how the total energy $\mathcal{E}(q)$ and the total entropy $\mathcal{S}(q)$ as main objects in the GENERIC framework generate thermodynamical driving forces that generate the evolution equations

$$\partial_t q = \mathbb{J}(q)\mathrm{D}\mathcal{E}(q) + \partial_\xi \mathcal{R}^*\big(q, \mathrm{D}\mathcal{S}(q)\big),$$

which include the balance laws of mechanics as well as the constitutive relations. Here $\mathbb{J}$ is the Poisson operator, and $\mathcal{R}^*$ is the dual dissipation potential, see Sect. 2.3.1. All the material properties for the dissipative mechanisms are contained in $\mathcal{R}^*$ and the mapping $\xi \mapsto \boldsymbol{j} := \partial_\xi \mathcal{R}^*(q, \xi)$ is called the abstract kinetic relation between the vector ξ of all thermodynamic driving forces and the vector $\boldsymbol{j}$ of all corresponding fluxes.

The main theoretical result of this paper is Theorem 2.3 that shows that there is a straightforward way for the construction of a Poisson operator $q \mapsto \mathbb{J}(q)$ for Eulerian mechanics based on the general properties for Lie derivatives. It is usually difficult to show that $\mathbb{J}$ satisfies the so-called Jacobi identity (see (2.3.1)), but exploiting the calculation rules for general Lie derivatives (in particular the commutator rule (2.2.3)) they can be established without too much efforts.

In Sect. 2.3 we then follow the full program of the GENERIC framework and derive our Eulerian model for thermo-visco-elastoplasticity using the multiplicative split $\boldsymbol{F} = \boldsymbol{F}_{\mathrm{e}}\boldsymbol{F}_{\mathrm{p}}$. One of the advantages of this framework is that we are able to derive the reversible Hamiltonian part and the irreversible dissipative parts independently, namely,

$$\partial_t q = \boldsymbol{V}_{\mathrm{Ham}}(q) + \boldsymbol{V}_{\mathrm{diss}}(q) \quad \text{with } \boldsymbol{V}_{\mathrm{Ham}}(q) := \mathbb{J}(q)\mathrm{D}\mathcal{E}(q) \text{ and } \boldsymbol{V}_{\mathrm{diss}}(q) := \partial_\xi \mathcal{R}^*\big(q, \mathrm{D}\mathcal{S}(q)\big).$$

This allows us to study the different physical effects separately, thus providing a better overview of the physical principles.

Our final system can be written in terms of the variables $q = (\boldsymbol{\pi}, \mathsf{w})$ where $\boldsymbol{\pi} = \rho\boldsymbol{v}$ is the momentum and $\mathsf{w} = (\boldsymbol{F}, \boldsymbol{F}_{\mathrm{p}}, \tau)$, and it takes the form

$$\partial_t \boldsymbol{\pi} = -\mathfrak{L}_{\boldsymbol{v}}^{\mathrm{mo}} \boldsymbol{\pi} + \operatorname{div}\big(\boldsymbol{\Sigma}_{\mathrm{Cauchy}}(\mathsf{w}) + \mathbb{D}_{\mathrm{visc}}(\mathsf{w})\boldsymbol{D}(\boldsymbol{v})\big), \tag{2.1.1a}$$

$$\partial_t \boldsymbol{F} = -\mathfrak{L}_{\boldsymbol{v}}^{\mathrm{ve}} \boldsymbol{F}, \tag{2.1.1b}$$

$$\partial_t \boldsymbol{F}_{\mathrm{p}} = -\mathfrak{L}_{\boldsymbol{v}}^{\mathrm{in}} \boldsymbol{F}_{\mathrm{p}} + \boldsymbol{F}_{\mathrm{p}} \boldsymbol{L}_{\mathrm{vi.pl}}(\mathsf{w}), \tag{2.1.1c}$$

$$\partial_t \tau = j_{\mathrm{Ham}}^S(\mathsf{w}) + j_{\mathrm{diss}}^E(\mathsf{w}) - \frac{1}{\partial_\tau E(\mathsf{w})} \operatorname{div}\Big(\mathbb{K}_{\mathrm{heat}}(\mathsf{q})\nabla \frac{1}{\Theta(\mathsf{w})}\Big). \tag{2.1.1d}$$

The first equation is the momentum balance including the Cauchy stress tensor and the viscoelastic term with strain-rate tensor $\boldsymbol{D}(\boldsymbol{v}) = \frac{1}{2}\big(\nabla\boldsymbol{v} + (\nabla\boldsymbol{v})^*\big)$. The second equation states that the deformation gradient $\boldsymbol{F}$ is simply transported by a (suitable) Lie derivative, and the third equation contains the viscoplastic flow rule encoded by $\boldsymbol{L}_{\mathrm{vi.pl}}$. The last equation is a scalar thermodynamic equation which reduces (i) to the energy balance if τ is chosen as the density of the internal energy, (ii) to the heat equation for the choice $\tau = \theta =$temperature, or (iii) to the entropy (im)balance if we choose τ as entropy density. We refer to Sect. 2.4.3 for the explanation of all symbols.

At this stage, we want to highlight that the first three equations clearly show the relevance of the Lie derivative $\mathfrak{L}_{\boldsymbol{v}}$, while in the fourth equation, it will only appear, if one chooses τ as an intensive or an extensive variable. Here we have added the superscripts "mo", "ve", "in" for "momentum", "vector", and "internal variable" to the Lie derivative $\mathfrak{L}_{\boldsymbol{v}}$ to indicate that the Lie derivatives depend on the tensorial nature of the variables $\boldsymbol{\pi}$, $\boldsymbol{F}$, and $\boldsymbol{F}_{\mathrm{p}}$, respectively.

Section 2.5 concludes the paper with a discussion of the developed theory, in particular on the relevance of the proper usage of Lie derivatives and the GENERIC framework.

2.2 Lie Derivatives

The theory of Lie derivatives is a well-known tool in differential geometry as well as in Eulerian fluid and solid mechanics. It is relevant when tensors (functions, vectors, forces, densities, etc.) are defined over a manifold and there is a vector field $\boldsymbol{v}$ on this manifold, see e.g. Yano (1955), (Chandrasekhar 1983, Chap. 1), (Marsden and

Hughes 1983, Sect. 1.6), and (Marsden and Ratiu 1999, Sect. 4.3). See also (Marsden and Hughes 1983, footnote 26) for historical remarks concerning the first applications of Lie derivatives in continuum mechanics.

2.2.1 Differential Geometric Approach

In our case the manifold is the Eulerian space $\mathbb{R}^d$ or a subset S thereof, called the body in Eulerian description. The tangent and cotangent spaces are $\mathrm{T}_\mathsf{x}\mathsf{S} = \mathbb{R}^d$ and $\mathrm{T}^*_\mathsf{x}\mathsf{S} = \mathbb{R}^d_*$, respectively. A vector field $\boldsymbol{v}$ on S means that $\boldsymbol{v}(\mathsf{x}) \in \mathrm{T}_\mathsf{x}\mathsf{S}$ for all $\mathsf{x} \in \mathsf{S}$, and a co-vector field $\boldsymbol{\alpha}$ satisfies $\boldsymbol{\alpha}(\mathsf{x}) \in \mathrm{T}^*_\mathsf{x}\mathsf{S}$. In the differential geometric setting, a tensor $\boldsymbol{A}$ of type $(i_\circ, j_\circ)$ on S, written as $\boldsymbol{A} \in \mathfrak{T}^{i_\circ}_{j_\circ}(\mathsf{S})$, means that $\boldsymbol{A}(\mathsf{x})$ is a multi-linear mapping from $(\mathrm{T}_\mathsf{x}\mathsf{S})^{i_\circ} \times (\mathrm{T}^*_\mathsf{x}\mathsf{S})^{j_\circ}$, i.e., linear in each of the $i_\circ{+}j_\circ$ arguments. By the dual pairing

$$(\boldsymbol{v}, \boldsymbol{\alpha}) \mapsto \langle \boldsymbol{\alpha}, \boldsymbol{v}\rangle_\mathsf{S} : x \mapsto \langle \boldsymbol{\alpha}(\mathsf{x}), \boldsymbol{v}(\mathsf{x})\rangle_{\mathrm{T}_\mathsf{x}\mathsf{S}}$$

we can identify $\boldsymbol{v}$ with a tensor in $\mathfrak{T}^0_1(\mathsf{S})$ and $\boldsymbol{\alpha}$ with a tensor in $\mathfrak{T}^1_0(\mathsf{S})$. The anti-symmetric tensors in $\mathfrak{T}^0_{j_\circ}(\mathsf{S})$ are the differentials forms denoted by the set $\Lambda_{j_0}(\mathsf{S})$.

For a general vector field $\boldsymbol{v}$ and a tensor field $\boldsymbol{A}$, the Lie derivative is defined by taking the derivative of $\boldsymbol{A}$ along the flow of $\boldsymbol{v}$. Throughout this section we assume that all objects like $\boldsymbol{v}$ and $\boldsymbol{A}$ are independent of time. First observe, that $\boldsymbol{v}$ defines a flow map $s \mapsto \boldsymbol{\Psi}(s, \cdot) : \mathsf{S} \to \mathbb{R}^d$ such that the following ODE is satisfied:

$$\frac{\mathrm{d}}{\mathrm{d}s}\boldsymbol{\Psi}(s, \mathsf{x}) = \boldsymbol{v}(\boldsymbol{\Psi}(s, \mathsf{x})) \text{ for } s \in \,]-\delta, \delta[, \qquad \boldsymbol{\Psi}(0, \mathsf{x}) = \mathsf{x}.$$

The Lie derivative of $\boldsymbol{A}$ along $\boldsymbol{v}$ is now defined via

$$\mathfrak{L}_v \boldsymbol{A} = \Big(\frac{\mathrm{d}}{\mathrm{d}s}\boldsymbol{\Psi}(s, \cdot)^* \boldsymbol{A}\Big)\Big|_{s=0}, \tag{2.2.1}$$

where the pull-back $\boldsymbol{\Phi}^*\boldsymbol{A}$ of $\boldsymbol{A}$ by $\boldsymbol{\Phi}$ is defined via

$$\begin{aligned}&\boldsymbol{\Phi}^* \boldsymbol{A}(x)\big[\boldsymbol{v}_1, \dots, \boldsymbol{v}_{i_\circ}; \boldsymbol{\alpha}_1, \dots, \boldsymbol{\alpha}_{j_\circ}\big]\\ &= \boldsymbol{A}(\boldsymbol{\Phi}(x))\big[\mathrm{D}\boldsymbol{\Phi}\boldsymbol{v}_1, \dots, \mathrm{D}\boldsymbol{\Phi}\boldsymbol{v}_{i_\circ}; \big(\mathrm{D}\boldsymbol{\Phi}\big)^{-*}\boldsymbol{\alpha}_1, \dots, \big(\mathrm{D}\boldsymbol{\Phi}\big)^{-*}\boldsymbol{\alpha}_{j_\circ}\big],\end{aligned}$$

where $\mathrm{D}\boldsymbol{\Phi}$ is evaluated at $x \in \mathcal{B}$, see Appendix 2.6. In particular, $\mathfrak{L}_v$ is again a tensor of the same order as $\boldsymbol{A}$.

Because we have $\boldsymbol{\Psi}(s, \mathsf{x}) = \mathsf{x} + s\boldsymbol{v}(\mathsf{x}) + O(s^2)$ we find $\mathrm{D}\boldsymbol{\Psi}(s, \mathsf{x}) = \boldsymbol{I} + s\nabla\boldsymbol{v}(\mathsf{x}) + O(s^2)$ and conclude

$$\mathfrak{L}_v A\big[v_1, .., v_{i_\circ}; \alpha_1, .., \alpha_{j_\circ}\big] = \big(v\cdot\nabla A\big)\big[v_1, ..., v_{i_\circ}; \alpha_1, ..., \alpha_{j_\circ}\big] + \sum_{i=1}^{i_\circ} A\big[v_1, ..., (\nabla v)v_i, ..., v_{i_\circ}; \alpha_1, ..., \alpha_{j_\circ}\big] - \sum_{j=1}^{j_\circ} A\big[v_1, ..., v_{i_\circ}; \alpha_1, ..., (\nabla v)^*\alpha_j, ..., \alpha_{j_\circ}\big]. \tag{2.2.2}$$

One of the fundamental properties of Lie derivatives is the commutator relation

$$\mathfrak{L}_v\big(\mathfrak{L}_w T\big) - \mathfrak{L}_w\big(\mathfrak{L}_v T\big) = \mathfrak{L}_{[\![v,w]\!]} T, \tag{2.2.3}$$

where the commutator between vector fields is given by

$$[\![v, w]\!] := (\nabla w)v - (\nabla v)w = (v\cdot\nabla)w - (w\cdot\nabla)v = \mathfrak{L}_v w = -\mathfrak{L}_w v.$$

It is easily checked that we have the Jacobi identity for vector fields, viz.,

$$\big[\![v_1, [\![v_2, v_3]\!]\big]\!] + \big[\![v_2, [\![v_3, v_1]\!]\big]\!] + \big[\![v_3, [\![v_1, v_2]\!]\big]\!] = 0 \quad \text{for all } v_1, v_2, v_3 \in \mathfrak{T}^1(\mathsf{S}). \tag{2.2.4}$$

The identities (2.2.3) and (2.2.4) will be extremely useful for showing that the operator $\mathbb{J}$ to be constructed satisfies the Jacobi identity, and hence is a Poisson structure. It is not easy to find explicit statements of (2.2.3) in the literature, but it is an easy consequence of its validity for functions, vectors, and co-vectors (1-forms) and of the well-known derivation rule $\mathfrak{L}_v(T\otimes S) = (\mathfrak{L}_v T)\otimes S + T\otimes(\mathfrak{L}_v S)$ for tensor products by doing induction over the rank of the tensors (see, e.g., condition "(DO1)" in Abraham and Marsden (1978, Ch. 2)).

Here the tensor product of $A \in \mathfrak{T}^{i_\circ}_{j_\circ}(\mathsf{S})$ and $A \in \mathfrak{T}^{n_\circ}_{m_\circ}(\mathsf{S})$ is the tensor $A\otimes B \in \mathfrak{T}^{i_\circ+n_\circ}_{j_\circ+m_\circ}(\mathsf{S})$ defined via

$$A\otimes B[v_1, \ldots, v_{i_\circ+n_\circ}; \alpha_1, \ldots, \alpha_{j_\circ+m_\circ}] := A[v_1, \ldots, v_{i_\circ}; \alpha_1, \ldots, \alpha_{j_\circ}] \cdot B[v_{i_\circ+1}, \ldots, v_{i_\circ+n_\circ}; \alpha_{j_\circ+1}, \ldots, \alpha_{j_\circ+m_\circ}].$$

Inner products can be defined by tensor products and subsequent contraction of indices. For $A \in \mathfrak{T}^{i_\circ}_{j_\circ}(\mathsf{S})$ and $n, m \in \mathbb{N}$ with $1 \le n \le i_\circ$ and $1 \le m \le j_\circ$ we can contract the nth vector slot with the m co-vector slot to obtain $\mathsf{C}^n_m A \in \mathfrak{T}^{i_\circ-1}_{j_\circ-1}(\mathsf{S})$ defined via

$$\mathsf{C}^n_m A\big[v_1, \ldots, v_{i_\circ-1}; \alpha_1, \ldots, \alpha_{j_\circ-1}\big] := \sum_{k=1}^{d} A\big[v_1, .., v_{k-1}, e_k, v_k, .., v_{i_\circ-1}; \alpha_1, .., \alpha_{m-1}, \varepsilon_k, \alpha_m, \ldots, \alpha_{j_\circ-1}\big],$$

where $\{e_1, \ldots, e_d\}$ is an arbitrary basis in $\mathrm{T_x S}$ and $\varepsilon_k \in \mathrm{T_x^* S}$ satisfy $\langle \varepsilon_k, e_l \rangle = \delta_{k,l}$ (the Kronecker symbol) for $k, l \in \{1, \ldots, d\}$.

2.2.2 Differential Forms

The special inner product with a vector field w is denoted by $i_w : \mathfrak{T}^{i_\circ+1}_{j_\circ}(\mathrm{S}) \to \mathfrak{T}^{i_\circ}_{j_\circ}(\mathrm{S})$ and is given by

$$i_w A\big[v_1, .., v_{i_\circ}; \alpha_1, .., \alpha_{j_\circ}\big] = A\big[w, v_1, .., v_{i_\circ}; \alpha_1, .., \alpha_{j_\circ}\big],$$

i.e., by simple insertion into the first vector slot. Identifying w with the tensor $B[\alpha] = \langle \alpha, w \rangle$ we have the relation $i_w A = \mathrm{C}_1^1\big(A \otimes B\big)$. It is not difficult to see that Lie derivatives commute with contraction, hence we also have the product rule for inner products:

$$\mathfrak{L}_v \mathrm{C}^n_m A = C^n_m \mathfrak{L}_v A \quad \text{and} \quad \mathfrak{L}_v(i_w A) = i_w(\mathfrak{L}_v A) + i_{\mathfrak{L}_v w} A.$$

An important subclass of tensor is given by the differential forms that appear as k-forms for $k \in \{0, \ldots, d\}$. They are given as tensors $\beta \in \Lambda_m(S) \subset \mathfrak{T}^0_k(\mathrm{S})$, where Λ_m denotes the subset of tensors that are anti-symmetric, i.e., interchanging any two arguments changes the result by a factor (-1). One can define the differential $d : \Lambda_k(S) \to \Lambda_{k+1}(\mathrm{S})$ and has $i_w : \Lambda_{k+1}(S) \to \Lambda_k(S)$. For differential forms β one has the identities

$$\mathfrak{L}_v \beta = i_v d\beta + d(i_v \beta) \quad \text{and} \quad \mathfrak{L}_v(d\beta) = d\big(\mathfrak{L}_v \beta\big),$$

where the first one is called "Cartan's magic formula" (also known as Cartan's first structural equation) and the second follows easily from the first when using $dd\gamma = 0$ for all differential forms γ.

The wedge product $\wedge$ maps $\Lambda_k(\mathrm{S}) \times \Lambda_l(\mathrm{S})$ into $\Lambda_{k+l}(\mathrm{S})$ by $\beta \wedge \gamma = \mathrm{anti}(\beta \otimes \gamma)$ and satisfies

$$\mathfrak{L}_v(\beta \wedge \gamma) = \big(\mathfrak{L}_v \beta\big) \wedge \gamma + \beta \wedge \big(\mathfrak{L}_v \gamma\big).$$

2.2.3 Tensors with Values in a Linear Space

The above theory can easily generalized to the case that the tensors $A(x)$ do not map into $\mathbb{R}$ but into a general linear space U_x, i.e.,

$$A(x) : (\mathrm{T_x S})^{i_\circ} \times (\mathrm{T_x^* S})^{j_\circ} \to U_\mathrm{x}$$

is still multi-linear in $(\boldsymbol{v}_1, \ldots, \boldsymbol{v}_{i_\circ}; \boldsymbol{\alpha}_1, \ldots, \boldsymbol{\alpha}_{j_\circ})$. We will then speak of $\boldsymbol{U}$-valued tensors and remark that $\boldsymbol{U}_\mathsf{x}$ has to be independent of $\mathrm{T}_\mathsf{x}\mathsf{S}$ and $\mathrm{T}^*_\mathsf{x}\mathsf{S}$, which means that the vector field $\boldsymbol{v}$ is not moving the objects in $\boldsymbol{U}_\mathsf{x}$.

2.2.4 *Lie Derivatives of Vectors, Operators, and Volume Forms*

A proper usage of differential geometric concepts in continuum mechanics is not standard, but we refer to Segev (2023) for a careful and detailed approach in this direction. Traditionally, it is more common to use vector, co-vector, and matrices, called tensors there. To distinguish these notions we use the name "operator" for matrices acting as linear mappings between linear spaces.

Even for scalar-valued functions, the Lie derivatives are not simple, because functions in continuum mechanics occur in different instances. First, we have *intensive* and *extensive* functions, which have the differential geometric interpretation as 0-forms (i.e., $f \in \Lambda_0(\mathsf{S})$) and volume or d-form (i.e., $\boldsymbol{\rho} \in \Lambda_d(\mathsf{S})$), respectively. In the latter case identify a function ρ with the volume form $\boldsymbol{\rho}[\boldsymbol{v}_1, \ldots, \boldsymbol{v}_d] = \rho \det\big(\boldsymbol{v}_1|..|\boldsymbol{v}_d\big)$, and we obtain

$$\mathfrak{L}_v f = \boldsymbol{v}\cdot\nabla f \quad \text{and} \quad \mathfrak{L}_v \rho = \operatorname{div}(\rho \boldsymbol{v}). \tag{2.2.5}$$

In continuum mechanics, the typical extensive variables are mass density ρ, internal-energy density e, entropy density s, or any one-homogeneous function of those. The typical intensive variables are concentrations, temperature $\theta = \mathrm{d}e/\mathrm{d}s$, velocity, pressure, or chemical potentials. For general functions, not falling into the classes of intensive or extensive functions, the Lie derivative has to be obtained by representing it via intensive and extensive variables and then applying the chain rule and the appropriate Lie derivatives.

A vector field $\boldsymbol{w}$ and a co-vector field $\boldsymbol{\alpha}$ can be identified with tensors $\boldsymbol{A}_w \in \mathfrak{T}^0_1(\mathsf{S})$ and $\boldsymbol{B}_\beta \in \mathfrak{T}^1_0(\mathsf{S})$, respectively, by

$$\boldsymbol{A}_w[\boldsymbol{\alpha}] = \langle \boldsymbol{\alpha}, \boldsymbol{w} \rangle \quad \text{and} \quad \boldsymbol{B}_\beta[\boldsymbol{v}] = \langle \boldsymbol{\beta}, \boldsymbol{v} \rangle.$$

Using this, we can derive the form of the Lie derivatives from the tensor Lie derivatives, namely,

$$\mathfrak{L}_v \boldsymbol{w} = \boldsymbol{v}\cdot\nabla \boldsymbol{w} - (\nabla \boldsymbol{v})\boldsymbol{w} = [\![\boldsymbol{v}, \boldsymbol{w}]\!] \quad \text{and} \quad \mathfrak{L}_v \boldsymbol{\beta} = \boldsymbol{v}\cdot\nabla \boldsymbol{\beta} + (\nabla \boldsymbol{v})^* \boldsymbol{\beta}. \tag{2.2.6}$$

Moreover, in continuum mechanics the momentum vector $\boldsymbol{\pi} = \rho I_\mathrm{R} \boldsymbol{v}$ (where $I_\mathrm{R} : \mathrm{T}_\mathsf{x}\mathsf{S} \to \mathrm{T}^*_\mathsf{x}\mathsf{S}$) is the Riesz isomorphism) is a co-vector but its correct interpretation is as a $(d-1)$-form (by the Hodge star operator mapping $\Lambda_k(\mathsf{S})$ into $\Lambda_{d-k}(\mathsf{S})$), namely,

$$\boldsymbol{\pi} = \rho I_{\mathrm{R}} \boldsymbol{v} = \boldsymbol{i}_{\boldsymbol{v}} \boldsymbol{\rho}. \tag{2.2.7a}$$

This leads to the Lie derivative

$$\mathfrak{L}_{\boldsymbol{v}} \boldsymbol{\pi} = \mathrm{div}(\boldsymbol{\pi} \otimes \boldsymbol{v}) + (\nabla \boldsymbol{v})^* \boldsymbol{\pi}, \tag{2.2.7b}$$

see (Mielke and Roubíček 2025, Prop. 4.2). A "vector" may also be considered as a tensor with values in the linear $\mathbb{R}^d$ which is not connected to $\mathsf{T_x S}$. Such a case occurs in (Benešová et al. 2024, Eq. (2.3)), where the magnetization vector $\boldsymbol{M}(t, \mathsf{x}) \in \mathbb{R}^3$ has the Lie derivative $\mathfrak{L}_{\boldsymbol{v}}^{\mathrm{mag}} \boldsymbol{M} = \boldsymbol{v} \cdot \nabla M + (\mathrm{div}\, \boldsymbol{v}) \boldsymbol{M} = \mathrm{div}\big(\boldsymbol{M} \otimes \boldsymbol{v}\big)$. The proper tensorial interpretation of "flux fields" is reported in Segev (2023, p. 71):

> A flux field, such as the heat flux field, should be regarded as a 2-form in the three-dimensional body rather than as a vector field. As a 2-form, a heat flux field may be considered independently of the configuration of the body in space.

We have four types of operators, namely,

$$\mathbb{B} : \mathsf{T_x S} \to \mathsf{T_x S}, \quad \mathbb{C} : \mathsf{T_x S} \to \mathsf{T_x^* S}, \quad \mathbb{D} : \mathsf{T_x^* S} \to \mathsf{T_x^* S}, \quad \mathbb{E} : \mathsf{T_x^* S} \to \mathsf{T_x S}.$$

They can be identified with tensors $\boldsymbol{B} \in \mathfrak{T}_1^1$, $\boldsymbol{C} \in \mathfrak{T}_0^2$, $\boldsymbol{D} \in \mathfrak{T}_1^1$, and $\boldsymbol{E} \in \mathfrak{T}_2^0$ via

$$\begin{aligned} \boldsymbol{B}[\boldsymbol{v}, \boldsymbol{\alpha}] &= \langle \boldsymbol{\alpha}, \mathbb{B}\boldsymbol{v} \rangle, \quad & \boldsymbol{C}[\boldsymbol{v}_1, \boldsymbol{v}_2] &= \langle \mathbb{C}\boldsymbol{v}_1, \boldsymbol{v}_2 \rangle, \\ \boldsymbol{D}[\boldsymbol{v}, \boldsymbol{\alpha}] &= \langle \mathbb{D}\boldsymbol{\alpha}, \boldsymbol{v} \rangle, \quad & \boldsymbol{E}[\boldsymbol{\alpha}_1, \boldsymbol{\alpha}_2] &= \langle \boldsymbol{\alpha}_1, \mathbb{E}\boldsymbol{\alpha}_2 \rangle. \end{aligned}$$

Thus, the corresponding Lie derivatives can be calculated by those for $\boldsymbol{B}$ to $\boldsymbol{E}$ and we obtain

$$\begin{aligned} \mathfrak{L}_{\boldsymbol{v}} \mathbb{B} &= \boldsymbol{v} \cdot \nabla \mathbb{B} + \mathbb{B}(\nabla \boldsymbol{v}) - (\nabla \boldsymbol{v}) \mathbb{B}, \qquad & \mathfrak{L}_{\boldsymbol{v}} \mathbb{C} &= \boldsymbol{v} \cdot \nabla \mathbb{C} + \mathbb{C}(\nabla \boldsymbol{v}) + (\nabla \boldsymbol{v})^* \mathbb{C}, \\ \mathfrak{L}_{\boldsymbol{v}} \mathbb{D} &= \boldsymbol{v} \cdot \nabla \mathbb{C} - \mathbb{D}(\nabla \boldsymbol{v})^* + (\nabla \boldsymbol{v})^* \mathbb{D}, \quad & \mathfrak{L}_{\boldsymbol{v}} \mathbb{E} &= \boldsymbol{v} \cdot \nabla \mathbb{E} - \mathbb{E}(\nabla \boldsymbol{v})^* - (\nabla \boldsymbol{v}) \mathbb{E}. \end{aligned} \tag{2.2.8}$$

Note that only $\mathbb{C}$ and $\mathbb{E}$ can be symmetric tensors and that in this case the two signs of the second and third terms are the same and one times we have $\nabla \boldsymbol{v}$ and the other time $(\nabla \boldsymbol{v})^*$. See also (Marsden and Hughes 1983, Ch. 1 Box 6.1).

As an example consider the Euclidean tensor $\boldsymbol{C}_{\mathrm{Euc}} \in \mathfrak{T}_0^2$ with $\boldsymbol{C}_{\mathrm{Euc}}[\boldsymbol{v}_1, \boldsymbol{v}_2] = (\boldsymbol{v}_1 | \boldsymbol{v}_2)_{\mathrm{Euc}}$ (the Euclidean scalar product). The associated operator $\mathbb{C}_{\mathrm{Euc}} = \mathbb{I}_{\mathrm{Riesz}} : \mathsf{T_x}\mathbb{R}^3 \to \mathsf{T_x^*}\mathbb{R}^3$ is the Riesz isomorphism. The Lie derivative provides twice the classical strain-rate tensor

$$\mathfrak{L}_{\boldsymbol{v}} \mathbb{C}_{\mathrm{Euc}} = 2\, \boldsymbol{D}(\boldsymbol{v}) \text{ with } \boldsymbol{D}(\boldsymbol{v}) := \frac{1}{2}\big(\mathbb{I}_{\mathrm{Riesz}} \nabla \boldsymbol{v} + (\nabla \boldsymbol{v})^* \mathbb{I}_{\mathrm{Riesz}}\big) \in \mathrm{Lin}_{\mathrm{sym}}\big(\mathbb{R}^3; \mathbb{R}^3_*\big). \tag{2.2.9}$$

There are various objective stress rates which arise from the fact that there are various stress tensors which have different tensorial properties. Similar to the construction of the momentum as a $(d-1)$ form (2.2.7) one also has to consider the Cauchy stress tensors $\boldsymbol{\Sigma}_{\mathrm{Cauchy}}$ as an extensive version, and its Lie derivative is the so-called Truesdell stress rate

$$\mathfrak{L}_v \boldsymbol{\Sigma}_{\text{Cauchy}} = \boldsymbol{v}\boldsymbol{\cdot}\nabla\boldsymbol{\Sigma}_{\text{Cauchy}} + (\operatorname{div}\boldsymbol{v})\boldsymbol{\Sigma}_{\text{Cauchy}} - \boldsymbol{\Sigma}_{\text{Cauchy}}(\nabla\boldsymbol{v})^* - (\nabla\boldsymbol{v})\boldsymbol{\Sigma}_{\text{Cauchy}}.$$

However, the Kirchhoff stress tensor $\boldsymbol{T}_{\text{Kir}} = \det \boldsymbol{F}\, \boldsymbol{\Sigma}_{\text{Cauchy}} = \frac{\rho_{\text{ref}}}{\rho}\boldsymbol{\Sigma}_{\text{Cauchy}}$ (see Sect. 2.3.2 for the last identity) has "intensive" properties, because it is the quotient of two extensive objects. Thus, $\boldsymbol{T}_{\text{Kir}}$ has to be interpreted as element of $\mathfrak{T}^0_2$ like $\boldsymbol{E}$ in (2.2.8), and its Lie derivative is the upper convected or Oldroyd stress rate

$$\mathfrak{L}_v \boldsymbol{T}_{\text{Kir}} = \boldsymbol{v}\boldsymbol{\cdot}\nabla\boldsymbol{T}_{\text{Kir}} - \boldsymbol{T}_{\text{Kir}}(\nabla\boldsymbol{v})^* - (\nabla\boldsymbol{v})\boldsymbol{T}_{\text{Kir}}.$$

Here, we do not dwell on this subject any further, but refer to (Marsden and Hughes 1983, Ch. 1 Box 6.1) for this and to Fiala (2008) for a way to write the Zaremba-Jaumann derivative as a Lie derivative.

2.3 GENERIC Structures and Eulerian Mechanics

The acronym GENERIC stands for

General Equation for Non-Equilibrium Reversible-Irreversible Coupling

and was introduced in Grmela and Öttinger (1997), but the class of models appears already in Grmela (1984, 1985) and as metriplectic systems in Morrison (1984, 1986), cf. the survey Morrison (2009). Over the last decade, the GENERIC framework has proved to be a versatile modeling tool for various complex coupled models for fluids and solids, see, e.g., Lebon et al. (2008), Mielke (2011), Hütter and Svendsen (2012), Duong et al. (2013), Pavelka et al. (2018), Betsch and Schiebl (2019), Lasarzik (2021), Pelech et al. (2022), Zafferi et al. (2023), Mielke and Roubíček (2025) and the references therein.

2.3.1 *Setup of GENERIC*

We consider states q in a state space $\boldsymbol{Q}$ which is either a flat space or a smooth manifold. A GENERIC system is a quintuple $(\boldsymbol{Q}, \mathcal{E}, \mathcal{S}, \mathbb{J}, \mathbb{K})$, where the energy $\mathcal{E}$ and the entropy $\mathcal{S}$ are differentiable functions on $\boldsymbol{Q}$ with differentials $\mathrm{D}\mathcal{E}(q)$ and $\mathrm{D}\mathcal{S}(q)$ lie $\mathrm{T}_q\boldsymbol{Q}$. The geometric structures are the Poisson operator $\mathbb{J}$ for Hamiltonian dynamics and a (dual) dissipation potential $\mathcal{R}^*$, where $\mathbb{J}(q)$ maps $\mathrm{T}^*\boldsymbol{Q}$ to $\mathrm{T}\boldsymbol{Q}$ and $\mathcal{R}^*(q,\cdot) : \mathrm{T}^*_q\boldsymbol{Q} \to [0,\infty]$ is a convex functional with $\mathcal{R}^*(q,0) = 0$. In many cases $\mathcal{R}^*(q,\cdot)$ is a quadratic functional given in terms of a symmetric, positive semi-definite operator $\mathbb{K}(q)$, namely, $\mathcal{R}^*(q,\xi) = \frac{1}{2}\langle\xi, \mathbb{K}(q)\xi\rangle$. Such operators $\mathbb{K}(q)$ are often called "Onsager operators" because of Onsager's fundamental reciprocal relations in Onsager (1931), earning him the Nobel Prize in Chemistry in 1968.

The evolution equation then takes the form

$$\frac{\partial q}{\partial t} = \mathbb{J}(q)\mathrm{D}\mathcal{E}(q) + \partial_\xi \mathcal{R}^*\big(q, \mathrm{D}\mathcal{S}(q)\big),$$

where $\partial_\xi \mathcal{R}^*(q,\cdot)$ denotes the convex subdifferential of $\mathcal{R}^*(q,\cdot)$, which is possibly set-valued (e.g., in plasticity).

The Poisson operator is defined by being skew-symmetric and satisfying the Jacobi identity, i.e.,

$$\big\langle \zeta_1, \mathrm{D}\mathbb{J}(q)[\mathbb{J}(q)\zeta_2]\zeta_3\big\rangle + \text{cycl.perm.} \equiv 0 \quad \text{for all } \zeta_1, \zeta_2, \zeta_3 \in \mathrm{T}^*_q \boldsymbol{Q}. \tag{2.3.1}$$

The main condition for GENERIC is the so-called *non-interaction condition*, namely,

$$\text{(a)}\ \mathbb{J}(q)\mathrm{D}\mathcal{S}(q) \equiv 0 \quad \text{and} \quad \text{(b)}\ \mathcal{R}^*(q, \lambda\mathrm{D}\mathcal{E}(q)) \equiv 0 \ \text{ for all } \lambda \in \mathbb{R}. \tag{2.3.2}$$

By convexity, the latter condition implies $\mathcal{R}^*(q, \zeta{+}\lambda\mathrm{D}\mathcal{E}(q)) = \mathcal{R}^*(q,\zeta)$ for all $(q,\zeta) \in \mathrm{T}^*\boldsymbol{Q}$ and $\lambda \in \mathbb{R}$, and as a consequence we have $\big\langle \mathrm{D}\mathcal{E}(q)), \partial_\xi \mathcal{R}^*(q,\xi)\big\rangle = 0$ for all ξ.

Using the chain rule, a simple consequence of Condition (2.3.2.b) (and $\mathbb{J} = -\mathbb{J}^*$) is the conservation of energy along solutions, i.e., $\frac{\mathrm{d}}{\mathrm{d}t}\mathcal{E}(q(t)) = 0$. Condition (2.3.2.a) (together with convexity of $\mathcal{R}^*(q,\cdot)$) implies entropy increase: $\frac{\mathrm{d}}{\mathrm{d}t}\mathcal{S}(q(t)) = \big\langle \mathrm{D}\mathcal{S}(q), \partial_\xi \mathcal{R}^*\big(q, \mathrm{D}\mathcal{S}(q)\big)\big\rangle \geq 0$, i.e., the second law of thermodynamics is automatically satisfied for GENERIC systems.

However, we emphasize that (2.3.2) is much stronger than energy conservation and entropy increase. We refer to Öttinger (2005) and (Mielke 2011, Sect. 2.2) for further properties of GENERIC systems, in particular concerning additional conservation laws and the maximum entropy principle providing thermal equilibrium states when maximizing $\mathcal{S}(q)$ subject to the conserved quantities. We also refer to Muschik et al. (2000) for a discussion on the differences between GENERIC and other thermodynamical approaches.

An efficient way of constructing GENERIC systems is using the "special form" of GENERIC systems as described in Öttinger (2005, Sect. 2.3.2) (following Edwards (1998)) and Mielke (2011, Sect. 2.4+4.3). Having a suitable "simpler" Poisson structure $\mathbb{J}_{\text{simple}}$ and a "simpler" dual dissipation potential $\mathcal{R}^*_{\text{simple}}$, one can define more complex structures via

$$\mathbb{J}(q) = M_\mathcal{S}(q)\mathbb{J}_{\text{simple}}(\Phi(q))M_\mathcal{S}(q)^* \ \text{ and } \ \mathcal{R}^*(q,\xi) = \mathcal{R}^*_{\text{simple}}\big(q, M_\mathcal{E}(q)^*\xi\big) \tag{2.3.3a}$$

where the following additional conditions have to be met:

$$M_\mathcal{S}(q)^*\mathrm{D}\mathcal{S}(q) = f_*, \quad \mathbb{J}_{\text{simple}}(q)f_* = 0, \quad M_\mathcal{S}(q) = \big(\mathrm{D}\Phi(q)\big)^{-1}, \tag{2.3.3b}$$

$$M_\mathcal{E}(q)^*\mathrm{D}\mathcal{E}(q) = g_*, \quad \mathcal{R}^*_{\text{simple}}(q, \lambda g_*) = 0 \text{ for all } \lambda \in \mathbb{R}, \tag{2.3.3c}$$

where f_* and g_* are fixed vectors.

The typical application occurs in continuum systems where $q = (w, \tau)$ and τ is a scalar thermodynamical variable such as temperature θ, internal energy density u, or the entropy density s. Then, we set $\Phi(w, \tau) = (w, S(w, \tau))^\top$ and obtain

$$M_S(w,\tau) = \begin{pmatrix} I & 0 \\ \partial_w S(w,\tau) & \partial_\tau S(w,\tau) \end{pmatrix}^{-1} = \begin{pmatrix} I & 0 \\ -\frac{\langle \partial_w S(w,\tau), \square\rangle}{\partial_\tau S(w,\tau)} & \frac{1}{\partial_\tau S(w,\tau)} \end{pmatrix}.$$

Here and below the symbol "$\square$" indicates at which position the corresponding argument from matrix multiplication has to be placed. In an example this means

$$\begin{pmatrix} a\,\nabla\square & \mathrm{e}^{\square\tau}V \\ \square\nabla\phi & M\,\nabla\square \end{pmatrix}\begin{pmatrix}\xi \\ \eta\end{pmatrix} = \begin{pmatrix} a\nabla\xi + \mathrm{e}^{\eta\tau}V \\ \xi\nabla\phi + M\,\nabla\eta \end{pmatrix}.$$

With the similar construction for $\mathcal{E}$ we obtain

$$M_S(w,\tau)^* = \begin{pmatrix} I & -\frac{\square}{\partial_\tau S(w,\tau)}\partial_w S(w,\tau) \\ 0 & \frac{1}{\partial_\tau S(w,\tau)} \end{pmatrix} = \begin{pmatrix} I & 0 \\ 0 & 1 \end{pmatrix} + \begin{pmatrix} -\partial_w S \\ 1-\partial_\tau S \end{pmatrix} \otimes \begin{pmatrix} 0 \\ \frac{1}{\partial_\tau S} \end{pmatrix} \text{ and}$$

$$M_{\mathcal{E}}(w,\tau)^* = \begin{pmatrix} I & -\frac{\square}{\partial_\tau E(w,\tau)}\partial_w E(w,\tau) \\ 0 & \frac{1}{\partial_\tau E(w,\tau)} \end{pmatrix} = \begin{pmatrix} I & 0 \\ 0 & 1 \end{pmatrix} + \begin{pmatrix} -\partial_w E \\ 1-\partial_\tau E \end{pmatrix} \otimes \begin{pmatrix} 0 \\ \frac{1}{\partial_\tau E} \end{pmatrix}.$$

Clearly, we have (2.3.3b) with $f_* = g_* = (0, 1)^\top$. The remaining relation (2.3.3c) will follow from the classical balance equations via $\nabla 1 \equiv 0$.

Looking at M_S the best choice for τ is the entropy density $\tau = s$ such that $S(w, s) = s$, because this gives $M_S(w, s)^* = \mathbb{I}$. However, this choice is not optimal for $M_{\mathcal{E}}$. In engineering often $\tau = \theta$ is chosen, which makes both matrices M_S and $M_{\mathcal{E}}$ nontrivial.

2.3.2 *Eulerian Thermo-elastoplasticity and Its Kinematics*

In Lagrangian mechanics a body is described on a reference domain $\mathrm{M} \subset \mathbb{R}^d$ with material points $\mathrm{x} \in \mathrm{M}$. The time-dependent deformation is denoted by $\mathsf{x} = y(t, \mathrm{x})$, where $\mathsf{x} \in \mathsf{S}$ is the spatial coordinate. Denoting the inverse mapping (also called return mapping) by $\mathrm{x} = \mathrm{Y}(t, \mathsf{x})$ the Eulerian velocity $\boldsymbol{v}$ and the Eulerian deformation gradient $\boldsymbol{F}$ are given by

$$\boldsymbol{v}(t,\mathsf{x}) = \frac{\partial}{\partial t}\boldsymbol{y}\big(t, \mathrm{Y}(t,\mathsf{x})\big) \quad\text{and}\quad \boldsymbol{F}(t,\mathsf{x}) = \nabla_{\mathrm{x}}\boldsymbol{y}\big(t, \mathrm{Y}(t,\mathsf{x})\big).$$

Note that $\boldsymbol{v}$ is a vector field on $\mathsf{S} \subset \mathbb{R}^d$, whereas $\boldsymbol{F}(t, x)$ is a two-point tensor mapping $\mathrm{T}_{\mathrm{Y}(t,\mathsf{x})}\mathrm{M}$ to $\mathrm{T}_{\mathsf{x}}\mathsf{S}$. Thus, we have to understand $\boldsymbol{F}$ as a $\boldsymbol{U}$-valued 1-form on $\mathcal{S}$, where $\boldsymbol{U}_{\mathsf{x}} = \mathrm{T}_{\mathrm{Y}(t,\mathsf{x})}\mathrm{M}$. The important kinematic relation is that $\boldsymbol{F}$ is directly transported by its Lie derivative. The same happens to the Eulerian density $\rho(t,\mathsf{x}) = \rho_{\mathrm{ref}}/\det\big(\boldsymbol{F}(t,\mathsf{x})\big)$, where the referential density ρ_{ref} is assumed to be constant.

These results are well-known, and we state them to highlight their Lie derivative aspect.

Lemma 2.1 (Kinematic relation) *Considering $\boldsymbol{v}$, $\boldsymbol{F}$, and ρ as above we have the relations (see* (2.2.5) *and* (2.2.6)*)*

$$\partial_t \rho = -\mathfrak{L}_{\boldsymbol{v}} \rho = -\operatorname{div}(\rho \boldsymbol{v}) \quad \textit{and} \quad \partial_t \boldsymbol{F} = -\mathfrak{L}_{\boldsymbol{v}} \boldsymbol{F} = -\boldsymbol{v}\cdot\nabla \boldsymbol{F} + (\nabla \boldsymbol{v})\boldsymbol{F}.$$

In finite-strain elastoplasticity, the deformation tensor $\boldsymbol{F}$ is multiplicatively decomposed by the so-called Kröner-Lee-Liu form (cf. Kröner (1960); Lee and Liu (1967))

$$\boldsymbol{F}(t,\mathsf{x}) = \boldsymbol{F}_{\mathrm{e}}(t,\mathsf{x})\,\boldsymbol{F}_{\mathrm{p}}(t,\mathsf{x}), \tag{2.3.4}$$

where $\boldsymbol{F}_{\mathrm{p}}(t,\mathsf{x})$ maps $\mathrm{T}_{\mathrm{Y}(t,\mathsf{x})}\mathcal{B}$ into itself, so that the Lie derivatives are

$$\mathfrak{L}_{\boldsymbol{v}} \boldsymbol{F}_{\mathrm{e}} = -\boldsymbol{v}\cdot\nabla \boldsymbol{F}_{\mathrm{e}} + (\nabla \boldsymbol{v})\boldsymbol{F}_{\mathrm{e}} \quad \text{and} \quad \mathfrak{L}_{\boldsymbol{v}} \boldsymbol{F}_{\mathrm{p}} = -\boldsymbol{v}\cdot\nabla \boldsymbol{F}_{\mathrm{p}}.$$

Inserting these relations into the kinematic relation for $\partial_t \boldsymbol{F}$ we obtain an additive kinematic relation between the elastic and the plastic strain rates in the form

$$\boldsymbol{F}_{\mathrm{e}}^{-1}\big(\partial_t \boldsymbol{F}_{\mathrm{e}} + \mathfrak{L}_{\boldsymbol{v}} \boldsymbol{F}_{\mathrm{e}}\big) + \big(\partial_t \boldsymbol{F}_{\mathrm{p}} + \mathfrak{L}_{\boldsymbol{v}} \boldsymbol{F}_{\mathrm{p}}\big)\boldsymbol{F}_{\mathrm{p}}^{-1} = 0, \tag{2.3.5}$$

which is the counterpart to the often-used split in the Lagrangian setting, see, e.g., Lee (1969), Besseling and Van der Giessen (1994), Gurtin and Anand (2005).

We will describe our full model for thermo-elastoplasticity by the state vector

$$q = \big(\boldsymbol{\pi}, \boldsymbol{F}, \boldsymbol{F}_{\mathrm{p}}, \tau\big)^{\top},$$

where $\boldsymbol{\pi} = \rho \boldsymbol{v}$ is the momentum and τ is a scalar-valued thermodynamical field variable such as internal energy, enthalpy, entropy, or temperature. At this stage the structure remains more transparent, when τ is staying more general, see Mielke (2011), Betsch and Schiebl (2019).

The class of models considered here can also be described by a smaller set of variables, namely, $(\boldsymbol{\pi}, \boldsymbol{F}_{\mathrm{e}}, \tau)$, see Mielke and Roubíček (2025), where then (2.3.5) is rewritten as $\partial_t \boldsymbol{F}_{\mathrm{e}} = -\mathfrak{L}_{\boldsymbol{v}} \boldsymbol{F}_{\mathrm{e}} - \boldsymbol{F}_{\mathrm{e}} \boldsymbol{L}_{\mathrm{p}}$ and $\boldsymbol{L}_{\mathrm{p}}$ is modeled accordingly. We hope that the present formulation is clearer, as it treats $\boldsymbol{F}_{\mathrm{p}}$ as an integral part of the state of the system.

2.3.3 Energy, Entropy, and Driving Forces

The total energy is given as the sum of the kinetic energy and the stored energy in the form

$$\mathcal{E}(\boldsymbol{\pi}, \boldsymbol{F}, \boldsymbol{F}_{\mathrm{p}}, \tau) = \int_{\mathcal{S}} \Big(\frac{\det \boldsymbol{F}}{2\rho_{\mathrm{ref}}} |\boldsymbol{\pi}|^2 + E(\boldsymbol{F}, \boldsymbol{F}_{\mathrm{p}}, \tau) \Big) \mathrm{d}x. \tag{2.3.6}$$

Here we have written E as a general function of $\boldsymbol{F}$, $\boldsymbol{F}_{\mathrm{p}}$, and τ, however, often it is assumed that E depends on $\boldsymbol{F}$ only through $\boldsymbol{F}_{\mathrm{e}} = \boldsymbol{F}\boldsymbol{F}_{\mathrm{p}}^{-1}$ which is a consequence of the Kröner-Lee-Liu decomposition (2.3.4). A typical choice would be $E(\boldsymbol{F}, \boldsymbol{F}_{\mathrm{p}}, \tau) = W(\boldsymbol{F}\boldsymbol{F}_{\mathrm{p}}^{-1}, \tau) + H(\boldsymbol{F}_{\mathrm{p}}, \tau)$, where the latter term can be used to describe (kinematic) hardening effects. At this stage the structure remains clearer, if we stay with the general function $E(\boldsymbol{F}, \boldsymbol{F}_{\mathrm{p}}, \tau)$, but later on, we will specify to the case $E(\boldsymbol{F}, \boldsymbol{F}_{\mathrm{p}}, \tau) = \widetilde{E}(\boldsymbol{F}\boldsymbol{F}_{\mathrm{p}}^{-1}, \boldsymbol{F}_{\mathrm{p}}, \tau)$.

Similarly, the entropy is given in the form

$$\mathcal{S}(\boldsymbol{F}, \boldsymbol{F}_{\mathrm{p}}, \tau) = \int_{\mathsf{S}} S(\boldsymbol{F}, \boldsymbol{F}_{\mathrm{p}}, \tau) \, \mathrm{d}\mathsf{x}, \tag{2.3.7}$$

where no dependence on the momentum $\boldsymbol{\pi}$ is present because of Galilean invariance. Here S is assume to be strictly increasing in τ, and the temperature is defined via

$$\theta = \Theta(\boldsymbol{F}, \boldsymbol{F}_{\mathrm{p}}, \tau) := \frac{\partial_\tau E(\boldsymbol{F}, \boldsymbol{F}_{\mathrm{p}}, \tau)}{\partial_\tau S(\boldsymbol{F}, \boldsymbol{F}_{\mathrm{p}}, \tau)} \quad \Longleftrightarrow \quad \tau = \widehat{\tau}(\boldsymbol{F}, \boldsymbol{F}_{\mathrm{p}}, \theta).$$

Note that E and S are densities with respect to the spatial domain $\mathsf{S} \subset \mathbb{R}^3$ and that they must be frame indifferent, i.e., $E(\boldsymbol{Q}\boldsymbol{F}, \boldsymbol{F}_{\mathrm{p}}, \tau) = E(\boldsymbol{F}, \boldsymbol{F}_{\mathrm{p}}, \tau)$ and $S(\boldsymbol{Q}\boldsymbol{F}, \boldsymbol{F}_{\mathrm{p}}, \tau) = S(\boldsymbol{F}, \boldsymbol{F}_{\mathrm{p}}, \tau)$ for all $\boldsymbol{Q} \in \mathrm{SO}(\mathbb{R}^d)$, which implies that $\partial_{\boldsymbol{F}} E(\boldsymbol{F}, \boldsymbol{F}_{\mathrm{p}}, \tau)\boldsymbol{F}^*$ and $\partial_{\boldsymbol{F}} S(\boldsymbol{F}, \boldsymbol{F}_{\mathrm{p}}, \tau)\boldsymbol{F}^*$ are symmetric. Here $\boldsymbol{F}^*$ is the adjoint operator to $\boldsymbol{F} : \mathrm{T_xM} \to \mathrm{T}_{\mathsf{x}}\mathsf{S}$ with $\mathsf{x} = \boldsymbol{y}(t, \mathrm{x})$, i.e., $\boldsymbol{F}^* : \mathrm{T}^*_{\mathsf{x}}\mathsf{S} \to \mathrm{T}^*_{\mathrm{x}}\mathrm{M}$, and we will also use $\boldsymbol{F}^{-*} = (\boldsymbol{F}^*)^{-1} = (\boldsymbol{F}^{-1})^* : \mathrm{T}^*_{\mathrm{x}}\mathrm{M} \to \mathrm{T}^*_{\mathsf{x}}\mathsf{S}$.

As GENERIC structures suggest, the main driving forces for the reversible (Hamiltonian) dynamics are $\mathrm{D}\mathcal{E}(q)$ and those for the dissipative dynamics are $\mathrm{D}\mathcal{S}(q)$:

$$\mathrm{D}\mathcal{E}(q) = \begin{pmatrix} \boldsymbol{v} \\ \partial_{\boldsymbol{F}} E(..) + \frac{\rho|\boldsymbol{v}|^2}{2} \boldsymbol{F}^{-*} \\ \partial_{\boldsymbol{F}_{\mathrm{p}}} E(..) \\ \partial_\tau E(..) \end{pmatrix} \quad \text{and} \quad \mathrm{D}\mathcal{S}(q) = \begin{pmatrix} 0 \\ \partial_{\boldsymbol{F}} S(..) \\ \partial_{\boldsymbol{F}_{\mathrm{p}}} S(..) \\ \partial_\tau S(..) \end{pmatrix}.$$

However, in light of the special GENERIC structure introduced at the end of Sect. 2.3.1, it will be more natural to look at the combined generalized driving forces $M_{\mathcal{S}}(q)\mathrm{D}\mathcal{E}(q)$ and $M_{\mathcal{E}}(q)\mathrm{D}\mathcal{S}(q)$.

Proposition 2.2 *With* $e_\tau = (0, 0, 0, 1)^\top$ *we have the relations*

$$M_{\mathcal{S}}(q)^*\mathrm{D}\mathcal{E}(q) = \begin{pmatrix} \boldsymbol{v} \\ \boldsymbol{\Sigma}_{\mathrm{e}}^F + \frac{\rho|\boldsymbol{v}|^2}{2}\boldsymbol{F}^{-*} \\ \boldsymbol{\Sigma}_{\mathrm{p}}^F \\ \Theta \end{pmatrix} \quad \textit{and}$$
$$M_{\mathcal{E}}(q)^*\mathrm{D}\mathcal{S}(q) = \frac{-1}{\Theta(q)}M_{\mathcal{S}}(q)^*\mathrm{D}\mathcal{E}(q) + \frac{1{+}\Theta}{\Theta}e_\tau, \tag{2.3.8}$$

where $\boldsymbol{\Sigma}_{\mathrm{e}}^F = \partial_{\boldsymbol{F}} E(..){-}\Theta(..)\partial_{\boldsymbol{F}} S(..)$ *and* $\boldsymbol{\Sigma}_{\mathrm{p}}^F = \partial_{\boldsymbol{F}_{\mathrm{p}}} E(..){-}\Theta(..)\partial_{\boldsymbol{F}_{\mathrm{p}}} S(..)$.

Here $\boldsymbol{\Sigma}_{\mathrm{e}}^F$ *and* $\boldsymbol{\Sigma}_{\mathrm{p}}^F$ *are the derivatives of the free energy* $F(q) = E(q){-}\Theta(q)S(q)$ *at fixed temperature with respect to* $\boldsymbol{F}$ *and* $\boldsymbol{F}_{\mathrm{p}}$*, respectively. Inverting the relation* $\theta = \Theta(w,\tau)$ *into* $\tau = \widehat{\tau}(w,\theta)$ *and setting* $\Psi(w,\theta) := F\big(w,\widehat{\tau}(w,\theta)\big)$*, this means*

$$\boldsymbol{\Sigma}_{\mathrm{e}}^F = \partial_{\boldsymbol{F}}\Psi(\boldsymbol{F},\boldsymbol{F}_{\mathrm{p}},\theta)\big|_{\theta=\Theta(\boldsymbol{F},\boldsymbol{F}_{\mathrm{p}},\tau)} \quad \textit{and} \quad \boldsymbol{\Sigma}_{\mathrm{p}}^F = \partial_{\boldsymbol{F}_{\mathrm{p}}}\Psi(\boldsymbol{F},\boldsymbol{F}_{\mathrm{p}},\theta)\big|_{\theta=\Theta(\boldsymbol{F},\boldsymbol{F}_{\mathrm{p}},\tau)}.$$

Proof The transformation matrices in Sect. 2.3.1 read

$$M_{\mathcal{S}}(q)^* = I - \tfrac{1}{\partial_\tau S}\big(\mathrm{D}\mathcal{S}(q){-}e_\tau\big)\otimes e_\tau \text{ and } M_{\mathcal{E}}(q)^* = I - \tfrac{1}{\partial_\tau E}\big(\mathrm{D}\mathcal{E}(q){-}e_\tau\big)\otimes e_\tau. \tag{2.3.9}$$

Thus, the generalized driving forces take the form

$$M_{\mathcal{S}}(q)^*\mathrm{D}\mathcal{E}(q) = \mathrm{D}\mathcal{E}(q) - \Theta\mathrm{D}\mathcal{S}(q) + \Theta e_\tau \quad \text{and}$$
$$M_{\mathcal{E}}(q)^*\mathrm{D}\mathcal{S}(q) = \mathrm{D}\mathcal{S}(q) - \frac{1}{\Theta}\mathrm{D}\mathcal{E}(q) + \tfrac{1}{\Theta}e_\tau,$$

which means that the two driving forces, except for the last scalar component, are the same up to constant $-\Theta(..)$. Hence, the second statement in (2.3.8) is shown.

The first relation in (2.3.8) follows simply by using (2.3.9), the formulas for $\mathrm{D}\mathcal{E}$ and $\mathrm{D}\mathcal{S}$, and the definition of Θ.

The last statement about the free energy is established in Mielke (2011, Eq. (2.13)). □

In fact, for the dissipative driving forces, it will be more useful to generalize $M_{\mathcal{E}}(q)^*$ to the better adapted operator $N_{\mathcal{E}}$. We emphasize that the conditions (2.3.3b) for $M_{\mathcal{S}}$ are much more restrictive, because the third condition asks $M_{\mathcal{S}}$ to be the inverse of a derivative. In contrast, the condition (2.3.3c) for $M_{\mathcal{E}}$ are simpler, e.g., $M_{\mathcal{E}} \equiv 0$ would be allowed (but not really useful). In the following we will use the operator

$$N_{\mathcal{E}}(q)^* = \begin{pmatrix} \boldsymbol{D}(\square)\ 0\ 0 & \frac{-\square}{\partial_\tau E}\boldsymbol{D}(\boldsymbol{v}) \\ 0 \quad 0\,\square & \frac{-\square}{\partial_\tau E}\partial_{\boldsymbol{F}_{\mathrm{p}}}E \\ 0 \quad 0\ 0 & \square/\partial_\tau E \end{pmatrix}, \text{ i.e. } N_{\mathcal{E}}(q)^*\begin{pmatrix} \boldsymbol{w} \\ \boldsymbol{\xi} \\ \boldsymbol{\eta} \\ \kappa \end{pmatrix} = \begin{pmatrix} \boldsymbol{D}(\boldsymbol{w}){-}\frac{\kappa}{\partial_\tau E}\boldsymbol{D}(\boldsymbol{v}) \\ \boldsymbol{\eta} - \frac{\kappa}{\partial_\tau E}\partial_{\boldsymbol{F}_{\mathrm{p}}}E \\ \frac{\kappa}{\partial_\tau E} \end{pmatrix}. \tag{2.3.10}$$

By construction, we have $N_{\mathcal{E}}(q)^*\mathrm{D}\mathcal{E}(u) = (0,0,1)^\top$ and find the adapted driving forces

$$\eta = \begin{pmatrix} \eta_{\rm m} \\ \eta_{\rm p} \\ \eta_{\rm t} \end{pmatrix} = N_{\mathcal{E}}(q)^* {\rm D}\mathcal{S}(q) = \begin{pmatrix} -\frac{1}{\Theta} \boldsymbol{D}(\boldsymbol{v}) \\ \frac{1}{\Theta}\big(\partial_{\boldsymbol{F}_{\rm p}} S - \Theta \partial_{\boldsymbol{F}_{\rm p}} E\big) \\ 1/\Theta \end{pmatrix} = \frac{1}{\Theta} \begin{pmatrix} -\boldsymbol{D}(\boldsymbol{v}) \\ -\boldsymbol{\Sigma}_{\rm p}^{F} \\ 1 \end{pmatrix}. \tag{2.3.11}$$

The major reason for choosing $N_{\mathcal{E}}(q)^*$ is that the arising driving forces are Galilean invariant, in contrast to $M_{\mathcal{E}}(q)^*{\rm D}\mathcal{S}(q)$ which involves $\boldsymbol{v}$ and $\rho|\boldsymbol{v}|^2$. Thus, we will have more flexibility in making physically reasonable choices for $\mathcal{R}^*_{\rm simple}$. Note also that $N_{\mathcal{E}}(q)^*$ is not quadratic and hence not invertible (in contrast to $M_{\mathcal{S}}(q)^*$). This reflects the fact that the equation of $\boldsymbol{F}$ is purely kinematic and thus will not include any dissipative effects.

2.3.4 The Poisson Structure for the Hamiltonian Part

The main result of the paper states that skew-symmetric operators $\mathbb{J}$ build with Lie derivatives automatically satisfy the Jacobi identity and hence qualify as Poisson operators for Eulerian continuum mechanics.

Theorem 2.3 (Jacobi identity via Lie derivatives) *We consider the case that the state z is given by $z = (\boldsymbol{\pi}, \boldsymbol{X})$ where $\boldsymbol{\pi}$ is the momentum (considered as a $(d-1)$ form as in* (2.2.7)*) and a collection of variables comprised into a tuple $\boldsymbol{X}$. We define the state-dependent skew-symmetric operator $\mathbb{J}(\boldsymbol{\pi}, \boldsymbol{X})$ via*

$$\mathbb{J}(\boldsymbol{\pi}, \boldsymbol{X})\begin{pmatrix} \boldsymbol{v} \\ \boldsymbol{\xi} \end{pmatrix} = \begin{pmatrix} -\mathfrak{L}_{\Box}\boldsymbol{\pi} & \mathbb{B}(\boldsymbol{X})\Box \\ -\mathfrak{L}_{\Box}\boldsymbol{X} & 0 \end{pmatrix}\begin{pmatrix} \boldsymbol{v} \\ \boldsymbol{\xi} \end{pmatrix} = \begin{pmatrix} -\mathfrak{L}_{\boldsymbol{v}}\boldsymbol{\pi} + \mathbb{B}(\boldsymbol{X})\boldsymbol{\xi} \\ -\mathfrak{L}_{\boldsymbol{v}}\boldsymbol{X} \end{pmatrix},$$

where $\mathbb{B}(\boldsymbol{X})$ is defined by skew-symmetry, namely, $\langle \boldsymbol{v}, \mathbb{B}(\boldsymbol{X})\boldsymbol{\xi}\rangle = \langle \boldsymbol{\xi}, \mathfrak{L}_{\boldsymbol{v}}\boldsymbol{X}\rangle$.

Then, $\mathbb{J}$ satisfies the Jacobi identity (2.3.1)*, and hence provides a Poisson structure.*

Proof The proof relies on the rules for Lie derivatives, in particular the commutator rule (2.2.3), and the simple observation that the mapping $(\boldsymbol{\pi}, \boldsymbol{X}) \mapsto \mathbb{J}(\boldsymbol{\pi}, \boldsymbol{X})$ is linear. A special case of our result was already obtained in Mielke and Roubíček (2025, Prop. A.1); however, the present proof is significantly shorter and still self-contained.

We consider one of the three terms in the Jacobi identity and simplify it systematically using ${\rm D}\mathbb{J}(\boldsymbol{\pi}, \boldsymbol{X})[\boldsymbol{\pi}_2, \boldsymbol{X}_2] = \mathbb{J}(\boldsymbol{\pi}_2, \boldsymbol{X}_2)$ (because of linearity), where $(\boldsymbol{\pi}_2, \boldsymbol{X}_2)^\top = \mathbb{J}(\boldsymbol{\pi}, \boldsymbol{X})(\boldsymbol{v}_2, \boldsymbol{\xi}_2)^\top$. We obtain

$$\begin{aligned} \mathsf{T}_{1,2,3} &:= \Big\langle \big(\begin{smallmatrix} \boldsymbol{v}_1 \\ \boldsymbol{\xi}_1 \end{smallmatrix}\big), {\rm D}\mathbb{J}(\boldsymbol{\pi}, \boldsymbol{X})[\boldsymbol{\pi}_2, \boldsymbol{X}_2]\big(\begin{smallmatrix} \boldsymbol{v}_3 \\ \boldsymbol{\xi}_3 \end{smallmatrix}\big)\Big\rangle = \langle \boldsymbol{v}_1, -\mathfrak{L}_{\boldsymbol{v}_3}\boldsymbol{\pi}_2 + \mathbb{B}(\boldsymbol{X}_2)\boldsymbol{\xi}_3\rangle - \langle \boldsymbol{\xi}_1, \mathfrak{L}_{\boldsymbol{v}_3}\boldsymbol{X}_2\rangle \\ &= \langle \mathfrak{L}_{\boldsymbol{v}_3}\boldsymbol{v}_1, \boldsymbol{\pi}_2\rangle + \langle \boldsymbol{\xi}_3, \mathfrak{L}_{\boldsymbol{v}_1}\boldsymbol{X}_2\rangle - \langle \boldsymbol{\xi}_1, \mathfrak{L}_{\boldsymbol{v}_3}\boldsymbol{X}_2\rangle \end{aligned}$$

Inserting the definitions of $\boldsymbol{\pi}_2 = -\mathfrak{L}_{\boldsymbol{v}_2}\boldsymbol{\pi} + \mathbb{B}(\boldsymbol{X})\boldsymbol{\xi}_2$ and $\boldsymbol{X}_2 = -\mathfrak{L}_{\boldsymbol{v}_2}\boldsymbol{X}$ and using the definition of $\mathbb{B}(\boldsymbol{X})$ we obtain

$$\mathsf{T}_{1,2,3} = -\langle [\![\boldsymbol{v}_3, \boldsymbol{v}_1]\!], \mathfrak{L}_{\boldsymbol{v}_2}\boldsymbol{\pi}\rangle + \langle \boldsymbol{\xi}_2, \mathfrak{L}_{[\![\boldsymbol{v}_3, \boldsymbol{v}_1]\!]}\boldsymbol{X}\rangle - \langle \boldsymbol{\xi}_3, \mathfrak{L}_{\boldsymbol{v}_1}\mathfrak{L}_{\boldsymbol{v}_2}\boldsymbol{X}\rangle + \langle \boldsymbol{\xi}_1, \mathfrak{L}_{\boldsymbol{v}_3}\mathfrak{L}_{\boldsymbol{v}_2}\boldsymbol{X}\rangle.$$

The first term can be rewritten as $\langle \mathfrak{L}_{v_2}[\![v_3, v_1]\!], \pi\rangle = \big\langle \big[\!\big[v_2, [\![v_3, v_1]\!]\big]\!\big], \pi\big\rangle$. Hence adding the corresponding cyclic permutations, we can exploit the Jacobi identity for vector fields in (2.2.4) and find that the contribution of the terms linear in π in $\mathsf{C} := \mathsf{T}_{1,2,3} + \mathsf{T}_{2,3,1} + \mathsf{T}_{3,1,2}$ cancel each other.

Similarly, we can look at all terms in C involving ξ_1 and find

$$\langle \xi_1, \mathfrak{L}_{v_3}\mathfrak{L}_{v_2} X\rangle - \langle \xi_1, \mathfrak{L}_{v_2}\mathfrak{L}_{v_3} X\rangle + \langle \xi_1, \mathfrak{L}_{[\![v_2,v_3]\!]} X\rangle = 0$$

by using the commutator rule (2.2.3). The same holds for the terms involving ξ_2 and ξ_3, and hence the Jacobi identity for $\mathbb{J}$ is established. □

In the sense of the special form of GENERIC we define $\mathbb{J}_{\text{simple}}$ by using the special choice $\tau = s$, namely, the entropy density. The reason is that we know that the Hamiltonian dynamics does not change the entropy. Hence, we know that the evolution should be the simple transport along the Eulerian vector field v as an extensive variable. Note also that F_p is an intensive variable that is U-valued with $F_\text{p}(t, x) \in U_x = \text{GL}(\text{T}_x\text{M})$. Denoting the Lie derivatives for π, F, F_p, and s by $\mathfrak{L}^{\text{mo}}_v$, $\mathfrak{L}^{\text{ve}}_v$, $\mathfrak{L}^{\text{in}}_v$, and $\mathfrak{L}^{\text{ex}}_v$, respectively, we set

$$\mathbb{J}_{\text{simple}}(\pi, F, F_\text{p}, s) = \begin{pmatrix} -\mathfrak{L}^{\text{mo}}_{\square}\pi & \mathbb{B}^{\text{ve}}(F)\square & \mathbb{B}^{\text{in}}(F_\text{p})\square & \mathbb{B}^{\text{ex}}(s)\square \\ -\mathfrak{L}^{\text{ve}}_{\square}F & 0 & 0 & 0 \\ -\mathfrak{L}^{\text{in}}_{\square}F_\text{p} & 0 & 0 & 0 \\ -\mathfrak{L}^{\text{ex}}_{\square}s & 0 & 0 & 0 \end{pmatrix}$$

where the operators in the first row are obtained by skew-symmetry from the first column:

$$\mathbb{B}^{\text{ve}}(F)\Xi_\text{e} = \nabla F{:}\Xi_\text{e} + \text{div}\big(\Xi_\text{e} F^*\big), \quad \mathbb{B}^{\text{in}}(F_\text{p})\Xi_\text{p} = \nabla F_\text{p}{:}\Xi_\text{p}, \quad \mathbb{B}^{\text{ex}}(s)\xi = -s\nabla\xi.$$

The full Poisson structure $\mathbb{J}(\pi, F, F_\text{p}, \tau)$ is now obtained by using the transformation $s = S(F, F_\text{p}, \tau)$ inducing the transformation operator M_S resulting in

$$\mathbb{J}(\pi, F, F_\text{p}, \tau) = M_S(\pi, F, F_\text{p}, \tau)\mathbb{J}_{\text{simple}}\big(\pi, F, F_\text{p}, S(F, F_\text{p}, \tau)\big)M_S(\pi, F, F_\text{p}, \tau)^*.$$

From Theorem 2.3 we know that $\mathbb{J}_{\text{simple}}$ is a Poisson structure, hence the transformation shows that $\mathbb{J}$ is again a Poisson structure.

2.4 The Eulerian form of Thermo-visco-elastoplasticity

We now collect the evolutionary equations for thermo-visco-elastoplasticity at finite strain from the GENERIC form. Abbreviating $q = (\pi, F, F_\text{p}, \tau)$ we have

$$\partial_t q = V_{\text{Ham}}(q) + V_{\text{diss}}(q)$$

$$\text{with} \quad \boldsymbol{V}_{\text{Ham}}(q) = \mathbb{J}(q)\mathrm{D}\mathcal{E}(q) \quad \text{and} \quad \boldsymbol{V}_{\text{diss}}(q) = N_{\mathcal{E}}(q)\partial_\xi \mathcal{R}^*_{\text{simple}}\big(q, N_{\mathcal{E}}(q)^*\mathrm{D}\mathcal{S}(q)\big).$$

One of the advantages of the GENERIC framework is that we can derive the Hamiltonian part and the dissipative part completely independently, thus separating the two quite different physical phenomena.

2.4.1 The Hamiltonian Part of the Model

We now discuss the terms arising from of the Hamiltonian part of the dynamics, namely, $\frac{\partial q}{\partial t} = \boldsymbol{V}_{\text{Ham}}(q) = \mathbb{J}(q)\mathrm{D}\mathcal{E}(q)$. Using the form of $\mathcal{E}$ and the definition of $\mathbb{J}$ we obtain

$$\boldsymbol{V}_{\text{Ham}}(q) = \mathbb{J}(q)\mathrm{D}\mathcal{E}(q) = \begin{pmatrix} -\operatorname{div}\big(\rho\boldsymbol{v}\otimes\boldsymbol{v}\big) + \operatorname{div}\boldsymbol{\Sigma}_{\text{Cauchy}} \\ -\boldsymbol{v}\cdot\nabla\boldsymbol{F} + (\nabla\boldsymbol{v})\,\boldsymbol{F} \\ -\boldsymbol{v}\cdot\nabla\boldsymbol{F}_{\text{p}} \\ -\boldsymbol{v}\cdot\nabla\tau - \dfrac{S(..)\operatorname{div}\boldsymbol{v} + \partial_{\boldsymbol{F}}S(..)\boldsymbol{F}^*{:}\boldsymbol{D}(\boldsymbol{v})}{\partial_\tau S(..)} \end{pmatrix}, \tag{2.4.1}$$

where the Cauchy stress tensor $\boldsymbol{\Sigma}_{\text{Cauchy}}$ is defined via the free-energy density $\Psi = E - \Theta S$:

$$\boldsymbol{\Sigma}_{\text{Cauchy}} = \boldsymbol{\Sigma}_{\text{e}}^F\boldsymbol{F}^* + \Psi\,\mathbb{I} \quad \text{with} \quad \boldsymbol{\Sigma}_{\text{e}}^F = \partial_{\boldsymbol{F}}E(\boldsymbol{F}, \boldsymbol{F}_{\text{p}}, \tau) - \Theta(\boldsymbol{F}, \boldsymbol{F}_{\text{p}}, \tau)\partial_{\boldsymbol{F}}S(\boldsymbol{F}, \boldsymbol{F}_{\text{p}}, \tau).$$

A simplified version of (2.4.1) was derived in in Hütter and Tervoort (2008, Sect. 3), where $\boldsymbol{F}_{\text{p}}$ is neglected and $\tau = \theta$ is chosen.

The rest of this subsection will explain the form of the four components of $\boldsymbol{V}_{\text{Ham}}(q)$ in detail. For this it is important to keep the adapted reversible driving forces $M_{\mathcal{S}}(q)^*\mathrm{D}\mathcal{E}(q)$ from (2.3.8) in mind, where the first component remains $\boldsymbol{v}$. This vector will be applied to $M_{\mathcal{S}}(q)\mathbb{J}_{\text{simple}}(\Phi(q))$ to yield $\boldsymbol{V}_{\text{Ham}}(q) =$

$$\begin{pmatrix} I & 0 & 0 & 0 \\ 0 & I & 0 & 0 \\ 0 & 0 & I & 0 \\ -\frac{\partial_{\boldsymbol{v}} S}{\partial_\tau S} & -\frac{\partial_{\boldsymbol{F}} S}{\partial_\tau S} & -\frac{\partial_{\boldsymbol{F}_{\text{p}}} S}{\partial_\tau S} & \frac{1}{\partial_\tau S} \end{pmatrix} \begin{pmatrix} -\mathfrak{L}^{\text{mo}}_{\square}\boldsymbol{\pi} & \mathbb{B}^{\text{ve}}(\boldsymbol{F}) & \mathbb{B}^{\text{in}}(\boldsymbol{F}_{\text{p}}) & \mathbb{B}^{\text{cx}}(S) \\ -\mathfrak{L}^{\text{ve}}_{\square}\boldsymbol{F} & 0 & 0 & 0 \\ -\mathfrak{L}^{\text{in}}_{\square}\boldsymbol{F}_{\text{p}} & 0 & 0 & 0 \\ -\mathfrak{L}^{\text{ex}}_{\square}S & 0 & 0 & 0 \end{pmatrix} \begin{pmatrix} \boldsymbol{v} \\ \boldsymbol{\Sigma}_{\text{e}}^F + \frac{\rho|\boldsymbol{v}|^2}{2}\boldsymbol{F}^{-*} \\ \boldsymbol{\Sigma}_{\text{p}}^F \\ \Theta \end{pmatrix}$$

Note that the first matrix $M_{\mathcal{S}}(q)$ does not change the first three components; hence, the second and third components of $\boldsymbol{V}_{\text{Ham}}(q)$ are simply given by $-\mathfrak{L}^{\text{vec}}_{\boldsymbol{v}}\boldsymbol{F}$ and $-\mathfrak{L}^{\text{int}}_{\boldsymbol{v}}\boldsymbol{F}_{\text{p}}$.

For the fourth component we can take advantage from $\partial_{\boldsymbol{v}}S \equiv 0$; hence it is a linear combination of the three Lie derivatives, namely,

$$\frac{1}{\partial_\tau S}\Big(\partial_{\boldsymbol{F}}S{:}\mathfrak{L}^{\text{ve}}_{\boldsymbol{v}}\boldsymbol{F} + \partial_{\boldsymbol{F}_{\text{p}}}S{:}\mathfrak{L}^{\text{in}}_{\boldsymbol{v}}\boldsymbol{F}_{\text{p}} - \mathfrak{L}^{\text{ex}}_{\boldsymbol{v}}S\Big) = -\boldsymbol{v}\cdot\nabla\tau - \frac{S(..)\operatorname{div}\boldsymbol{v} + \partial_{\boldsymbol{F}}S(..)\boldsymbol{F}^*{:}\boldsymbol{D}(\boldsymbol{v})}{\partial_\tau S(..)}.$$

For the last identity, we use that the terms involving $\boldsymbol{v}\cdot\nabla\boldsymbol{F}$ and $\boldsymbol{v}\cdot\nabla\boldsymbol{F}_{\mathrm{p}}$ arising from $\mathfrak{L}^{\mathrm{ve}}_{\boldsymbol{v}}\boldsymbol{F}$ and $\mathfrak{L}^{\mathrm{in}}_{\boldsymbol{v}}\boldsymbol{F}_{\mathrm{p}}$, respectively, cancel with those arising from $\mathfrak{L}^{\mathrm{ex}}_{\boldsymbol{v}}S = \mathrm{div}(S(..)\boldsymbol{v})$. The last term in the above relation equals $\partial_{\boldsymbol{F}}S(..)\mathbf{:}\big((\nabla\boldsymbol{v})\boldsymbol{F}^*\big)$ because of the symmetry of $\partial_{\boldsymbol{F}}S(..)\boldsymbol{F}^*$.

Thus, it remains to establish the simple representation of the first component of $V_{\mathrm{Ham}}(q)$ in (2.4.1). For this, we first calculate the terms involving $\boldsymbol{v}$ (see also Zafferi et al. (2023, Eq. (4.6))), namely,

$$\begin{aligned}- \mathfrak{L}_{\boldsymbol{v}}\boldsymbol{\pi} + \mathbb{B}^{\mathrm{ve}}(\boldsymbol{F})\tfrac{\rho|\boldsymbol{v}|^2}{2}\boldsymbol{F}^{-*} &= -\,\mathrm{div}(\boldsymbol{\pi}\otimes\boldsymbol{v}) - (\nabla\boldsymbol{v})^*\boldsymbol{\pi} + \frac{\rho|\boldsymbol{v}|^2}{2}\boldsymbol{F}^{-*}\mathbf{:}\nabla\boldsymbol{F} + \mathrm{div}\Big(\frac{\rho|\boldsymbol{v}|^2}{2}\mathbb{I}\Big)\\ &= -\,\mathrm{div}(\rho\boldsymbol{v}\otimes\boldsymbol{v}) - \rho\,\nabla\big(\tfrac{|\boldsymbol{v}|^2}{2}\big) + \tfrac{\rho|\boldsymbol{v}|^2}{2\det\boldsymbol{F}}\nabla\det\boldsymbol{F} + \nabla\big(\rho\tfrac{|\boldsymbol{v}|^2}{2}\big) = -\,\mathrm{div}(\rho\boldsymbol{v}\otimes\boldsymbol{v}),\end{aligned}$$

because $\rho = \rho_{\mathrm{ref}}/\det\boldsymbol{F}$ implies $\nabla\rho = -(\rho/\det\boldsymbol{F})\nabla\det\boldsymbol{F}$, where ρ_{ref} is a constant.

To simplify the remaining terms in the first component we first use $\Psi = E - \Theta S$ to obtain

$$\begin{aligned}\nabla\Psi &= \nabla E - \Theta\nabla S - S\nabla\Theta\\ &= \nabla\boldsymbol{F}\mathbf{:}(\partial_{\boldsymbol{F}}E - \Theta\partial_{\boldsymbol{F}}S) + \nabla\boldsymbol{F}_{\mathrm{p}}\mathbf{:}(\partial_{\boldsymbol{F}_{\mathrm{p}}}E - \Theta\partial_{\boldsymbol{F}_{\mathrm{p}}}S) + \underbrace{(\partial_\tau E - \Theta\partial_\tau S)}_{=0}\nabla\tau - S\nabla\Theta\\ &= \nabla\boldsymbol{F}\mathbf{:}\boldsymbol{\Sigma}^F_{\mathrm{e}} + \nabla\boldsymbol{F}_{\mathrm{p}}\mathbf{:}\boldsymbol{\Sigma}^F_{\mathrm{p}} - S\nabla\Theta,\end{aligned}$$

where we recall $\boldsymbol{\Sigma}^F_{\mathrm{e,p}}$ from Proposition 2.2. Hence, the remaining terms are

$$\begin{aligned}&\mathbb{B}^{\mathrm{ve}}(\boldsymbol{F})\boldsymbol{\Sigma}^F_{\mathrm{e}} + \mathbb{B}^{\mathrm{in}}(\boldsymbol{F}_{\mathrm{p}})\boldsymbol{\Sigma}^F_{\mathrm{p}} + \mathbb{B}^{\mathrm{ex}}(S)\Theta\\ &= \mathrm{div}(\boldsymbol{\Sigma}^F_{\mathrm{e}}\boldsymbol{F}^*) + \nabla\boldsymbol{F}\mathbf{:}\boldsymbol{\Sigma}^F_{\mathrm{e}} + \nabla\boldsymbol{F}_{\mathrm{p}}\mathbf{:}\boldsymbol{\Sigma}^F_{\mathrm{p}} - S\nabla\Theta = \mathrm{div}\big(\boldsymbol{\Sigma}^F_{\mathrm{e}}\boldsymbol{F}^* + \Psi I\big).\end{aligned}$$

Hence, also the desired form of the first component in (2.4.1) is established.

2.4.2 *The Dissipative Part of Eulerian Thermo-elastoplasticity*

The modeling of the dissipative effects is considerably simpler than that of the reversible Hamiltonian part because there is much more freedom in choosing dissipation potentials than in choosing Poisson structures. One of the main advantages of GENERIC is indeed the very structured modeling of dissipative processes allowing for general couplings. Using the subdifferential of the dual dissipation potential will automatically satisfy the so-called Onsager symmetries. In the quadratic case $\mathcal{R}^*(q,\xi) = \frac{1}{2}\langle\xi, \mathbb{A}(q)\xi\rangle$ this is seen by the fact that $\partial_\xi\mathcal{R}^*(q,\xi) = \mathbb{K}(q)\xi$ with $\mathbb{K}(q) = \frac{1}{2}\big(\mathbb{A}(q){+}\mathbb{A}(q)\big) = \mathbb{K}(q)^*$, which is Onsager's reciprocal relation. Note that $\mathbb{K}(q) = \mathrm{D}^2\mathcal{R}^*(q,0)$ is the symmetric Hessian of $\mathcal{R}^*(q,\cdot)$.

Using the special form of GENERIC as described in (2.3.3) we can construct suitable nonlinear dissipation potentials $\mathcal{R}^*$ (or linear Onsager operators $\mathbb{K}$) by collecting the building blocks of the dissipative effects and combining them with a nontrivial operator $N_\mathcal{E}$ in the form

$$\mathcal{R}^*(q,\boldsymbol{\zeta}) = \mathcal{R}^*_{\text{simple}}(q, N_\mathcal{E}(q)^*\boldsymbol{\zeta}) \quad \text{or} \quad \mathbb{K}(q) = N_\mathcal{E}(q)\mathbb{K}_{\text{simple}}(q)N_\mathcal{E}(q)^*.$$

This strategy is propagated in Öttinger (2005, Sect. 2.3.2) and follows Edwards (1998).

In our model we can have three dissipative processes:
(A) viscoelastic dissipation induced by $\boldsymbol{D}(\boldsymbol{v}) = \frac{1}{2}\big(\nabla\boldsymbol{v}+(\nabla\boldsymbol{v})^\top\big)$,
(B) plastic dissipation induced by $\partial_t \boldsymbol{F}_\text{p}$,
(C) heat flow induced by $\nabla(1/\theta)$.
Thus, we can construct a suitable dual dissipation potential in the additive form

$$\mathcal{R}^*_{\text{simple}} = \mathcal{R}^*_\text{A} + \mathcal{R}^*_\text{B} + \mathcal{R}^*_\text{C}.$$

However, we emphasize that this simplistic assumption is by far not necessary. Of course, it is possible to construct much more general thermodynamically consistent models where there is a strong interaction of the different dissipation mechanics, but to keep simplicity and clarity we will restrict our approach to the case of a simple block structure.

The advantage of using an operator $N_\mathcal{E}$ is three-fold. First, it is used to guarantee the second non-interaction condition by asking

$$N_\mathcal{E}(q)^*\mathrm{D}\mathcal{E}(q) = e_\tau = (0, ..., 0, 1)^\top \quad \text{and} \quad \mathcal{R}^*_{\text{simple}}\big(q, \lambda\, e_\tau\big) = 0 \ \text{ for all } \lambda \in \mathbb{R}. \tag{2.4.2}$$

Here $\lambda \in \mathbb{R}$ stand for the constant (reciprocal of the) temperature $1/\theta$, which does not generate any dissipation.

The second advantage is the fundamental observation in Mielke (2011, Sect. 4.3) that the dissipative (a.k.a. irreversible) driving forces are now given by the generalized driving forces

$$\boldsymbol{\eta} = N_\mathcal{E}(q)^*\mathrm{D}\mathcal{S}(q)\,,$$

which contains important thermodynamical information by identifying the correct combinations of $\mathrm{D}\mathcal{S}$ and $\mathrm{D}\mathcal{E}$ governing dissipative processes and satisfying Galilcan invariance.

Finally, the operator $N_\mathcal{E}$ acting from the left on $\partial\mathcal{R}^*_{\text{simple}}$ adjusts the dissipative terms in such a way that energy conservation holds for

$$\partial_\xi\mathcal{R}^*(q, \mathrm{D}\mathcal{S}(q)) = N_\mathcal{E}(q)\,\partial\mathcal{R}^*_{\text{simple}}\big(q, N_\mathcal{E}(q)^*\mathrm{D}\mathcal{S}(q)\big).$$

Moreover, this form involving $N_\mathcal{E}$ and $N_\mathcal{E}^*$ guarantees the Onsager symmetries: if $\mathcal{R}^*_{\text{simple}}(q,\boldsymbol{\eta}) = \frac{1}{2}\langle\boldsymbol{\eta}, \mathbb{K}_{\text{simple}}(q)\boldsymbol{\eta}\rangle$ is quadratic, then $\mathcal{R}^*$ is still quadratic with

$\mathcal{R}^*(q,\boldsymbol{\xi}) = \frac{1}{2}\langle \boldsymbol{\xi}, \mathbb{K}(q)\boldsymbol{\xi}\rangle$, where $\mathbb{K}(q) = N_\mathcal{E}(q)\mathbb{K}_{\text{simple}}(q)N_\mathcal{E}(q)^* = \mathbb{K}(q)^*$ is again symmetric.

We recall our special choice for $N_\mathcal{E}(q)^*$ from (2.3.10) and find the adjoint

$$N_\mathcal{E}(q) = \begin{pmatrix} -\operatorname{div}(\Box) & 0 & 0 \\ 0 & 0 & 0 \\ 0 & I & 0 \\ \frac{-1}{\partial_\tau E}\boldsymbol{D}(\boldsymbol{v}):\Box & \frac{-1}{\partial_\tau E}\partial_{F_\text{p}}E:\Box & \frac{1}{\partial_\tau E} \end{pmatrix}.$$

The three 0's in the second row indicate that dissipative forces cannot change the kinematic relation of $\boldsymbol{F}$.

The full dual dissipation potential takes the form

$$\mathcal{R}^*(q,\boldsymbol{\zeta}) = \mathcal{R}^*_{\text{simple}}(q, N_\mathcal{E}(q)^*\boldsymbol{\zeta}),$$

and now assume that $\mathcal{R}^*_{\text{simple}}$ has a block structure

$$\mathcal{R}^*_{\text{simple}}(q,\boldsymbol{\eta}) = \mathcal{R}^*_{\text{vi.el}}(q,\boldsymbol{\eta}_\text{m}) + \mathcal{R}^*_{\text{vi.pl}}(q,\boldsymbol{\eta}_\text{p}) + \mathcal{R}^*_{\text{heat}}(q,\eta_\text{t}).$$

Yet, we hasten to say that this is a simplification that is not necessary at all; in fact, it is one of the big advantages of the GENERIC framework that it can easily handle coupling phenomena between different effects.

Even for these three blocks we only write the simplest forms and leave the study of more general dissipation potentials for future work.

$$\mathcal{R}^*_{\text{vi.el}}(q,\boldsymbol{\eta}_\text{m}) = \int_\text{S} \frac{\Theta}{2}\boldsymbol{\eta}_\text{m}:\mathbb{D}_{\text{vi.el}}(q)\boldsymbol{\eta}_\text{m}\,\mathrm{dx}, \tag{2.4.3a}$$

$$\mathcal{R}^*_{\text{vi.pl}}(q,\boldsymbol{\eta}_\text{p}) = \int_\text{S} R^*_{\text{vi.pl}}\big(q,\Theta\boldsymbol{F}_\text{p}^*\boldsymbol{\eta}_\text{p}\big)\,\mathrm{dx}, \tag{2.4.3b}$$

$$\mathcal{R}^*_{\text{heat}}(q,\eta_\text{t}) = \int_\text{S} \frac{1}{2}\nabla\eta_\text{t}\cdot\mathbb{K}_{\text{heat}}(q)\nabla\eta_\text{t}\,\mathrm{dx}. \tag{2.4.3c}$$

Here the pre-multiplication of $\boldsymbol{\eta}_\text{p}$ by $\boldsymbol{F}_\text{p}^*$ will generate the so-called plastic indifference, see Mielke (2003, Sect. 3.1) and below.

With these choices the dissipative (irreversible) part $\boldsymbol{V}_\text{diss}(q) = \partial_\zeta\mathcal{R}^*\big(q,\mathrm{D}\mathcal{S}(q)\big)$ of the evolution takes the form

$$\boldsymbol{V}_\text{diss}(q) = N_\mathcal{E}(q)\partial_\eta\mathcal{R}^*_{\text{simple}}\big(q, N_\mathcal{E}(q)^*\mathrm{D}\mathcal{S}(q)\big) = \begin{pmatrix} -\operatorname{div}\big(\partial_{\eta_\text{m}}\mathcal{R}^*_{\text{vi.el}}(q,\boldsymbol{\eta}_\text{m})\big) \\ 0 \\ \partial_{\eta_\text{p}}\mathcal{R}^*_{\text{vi.pl}}(q,\boldsymbol{\eta}_\text{p}) \\ \frac{1}{\partial_\tau E}\,j_\text{ener} \end{pmatrix}$$

$$\text{with } j_\text{ener} = -\partial_{\eta_\text{m}}\mathcal{R}^*_{\text{vi.el}}(..):\boldsymbol{D}(\boldsymbol{v}) - \partial_{\eta_\text{p}}\mathcal{R}^*_{\text{vi.pl}}(..):\partial_{F_\text{p}}E(q) + \partial_{\eta_\text{t}}\mathcal{R}^*_{\text{heat}}(q,\eta_\text{t}),$$

Inserting $\boldsymbol{\eta} = N_{\mathcal{E}}(q)^* \mathrm{D}\mathcal{S}(q)$ from (2.3.11) and the choices (2.4.3) we arrive at

$$\boldsymbol{V}_{\text{diss}}(q) = \begin{pmatrix} \operatorname{div}\big(\mathbb{D}_{\text{visc}}(q)\boldsymbol{D}\big) \\ 0 \\ \boldsymbol{F}_{\text{p}}\boldsymbol{L}_{\text{vi.pl}}(q) \\ \frac{1}{\partial_\tau E}\Big(\boldsymbol{D}{:}\mathbb{D}_{\text{visc}}\boldsymbol{D} + (\boldsymbol{F}_{\text{p}}\boldsymbol{L}_{\text{vi.pl}})(q){:}\partial_{\boldsymbol{F}_{\text{p}}}E(q) - \operatorname{div}\big(\mathbb{K}_{\text{heat}}\nabla\frac{1}{\Theta}\big)\Big) \end{pmatrix}$$

with strain-rate $\boldsymbol{D} = \boldsymbol{D}(\boldsymbol{v}) = \frac{1}{2}\big(\mathbb{I}_{\text{Riesz}}\nabla\boldsymbol{v}+(\nabla\boldsymbol{v})^*\mathbb{I}_{\text{Riesz}}\big)$ (cf. (2.2.9)) and plastic rate $\boldsymbol{L}_{\text{vi.pl}}(q) = \Theta(q)\partial_{\boldsymbol{\xi}_{\text{p}}} R^*_{\text{vi.pl}}\big(q, -\boldsymbol{F}_{\text{p}}^*\boldsymbol{\Sigma}_{\text{p}}^F\big)$, where $\boldsymbol{\Sigma}_{\text{p}}^F = \partial_{\boldsymbol{F}_{\text{p}}}E(q) - \Theta(q)\partial_{\boldsymbol{F}_{\text{p}}}S(q)$.

2.4.3 *The Eulerian Model for Thermo-visco-elastoplasticity*

We can now assemble the whole continuum model for non-isothermal finite-strain visco-elastoplasticity as obtained from the GENERIC framework:

$$\dot{q} = \boldsymbol{V}_{\text{Ham}}(q) + \boldsymbol{V}_{\text{diss}}(q) = \mathbb{J}(q)\mathrm{D}\mathcal{E}(q) + \partial_\zeta\mathcal{R}^*\big(q, \mathrm{D}\mathcal{S}(q)\big).$$

Combining the results from Sects. 2.4.1 and 2.4.2, this leads to the following system for $(\boldsymbol{v}, \mathsf{q})$ with $\mathsf{q} = (\boldsymbol{F}, \boldsymbol{F}_{\text{p}}, \tau)$:

$$\partial_t(\rho\boldsymbol{v}) = -\operatorname{div}\big(\rho\boldsymbol{v}\otimes\boldsymbol{v}\big) + \operatorname{div}\big(\boldsymbol{\Sigma}_{\text{Cauchy}}(\mathsf{q}) + \mathbb{D}_{\text{visc}}(\mathsf{q})\boldsymbol{D}\big), \tag{2.4.4a}$$

$$\partial_t\boldsymbol{F} = -\mathfrak{L}_{\boldsymbol{v}}\boldsymbol{F} = -(\boldsymbol{v}\cdot\nabla)\boldsymbol{F} + (\nabla\boldsymbol{v})\boldsymbol{F}, \tag{2.4.4b}$$

$$\partial_t\boldsymbol{F}_{\text{p}} = -\boldsymbol{v}\cdot\nabla\boldsymbol{F}_{\text{p}} + \boldsymbol{F}_{\text{p}}\boldsymbol{L}_{\text{vi.pl}}(\mathsf{q}), \tag{2.4.4c}$$

$$\partial_t\tau = j^S_{\text{Ham}}(\mathsf{q}) + j^E_{\text{diss}}(\mathsf{q}) - \frac{1}{\partial_\tau E(\mathsf{q})}\operatorname{div}\Big(\mathbb{K}_{\text{heat}}(\mathsf{q})\nabla\frac{1}{\Theta(\mathsf{q})}\Big), \tag{2.4.4d}$$

where $\rho = \rho_{\text{ref}}/\det\boldsymbol{F}$ and where we have

$$\boldsymbol{\Sigma}_{\text{Cauchy}}(\mathsf{q}) = \big(\partial_{\boldsymbol{F}}E(\mathsf{q}) - \Theta(\mathsf{q})\partial_{\boldsymbol{F}}S(\mathsf{q})\big)\boldsymbol{F}^* + \big(E(\mathsf{q}) - \Theta(\mathsf{q})S(\mathsf{q})\big)\mathbb{I}, \tag{2.4.5a}$$

$$\boldsymbol{L}_{\text{vi.pl}}(\mathsf{q}) - \Theta(\mathsf{q})\partial_{\boldsymbol{\xi}_{\text{p}}}R^*_{\text{vi.pl}}\big(\mathsf{q}, -\boldsymbol{F}_{\text{p}}^*\boldsymbol{\Sigma}_{\text{p}}^F\big) \text{ with } \boldsymbol{\Sigma}_{\text{p}}^F = \partial_{\boldsymbol{F}_{\text{p}}}E(\mathsf{q}) - \Theta(\mathsf{q})\partial_{\boldsymbol{F}_{\text{p}}}S(\mathsf{q}), \tag{2.4.5b}$$

$$j^S_{\text{Ham}}(\mathsf{q}) = -\boldsymbol{v}\cdot\nabla\tau - \frac{1}{\partial_\tau S(\mathsf{q})}\Big(S(\mathsf{q})\operatorname{div}\boldsymbol{v} + \partial_{\boldsymbol{F}}S(\mathsf{q})\boldsymbol{F}^*{:}\boldsymbol{D}(\boldsymbol{v})\Big), \tag{2.4.5c}$$

$$j^E_{\text{diss}}(\mathsf{q}) = \frac{1}{\partial_\tau E(\mathsf{q})}\Big(\boldsymbol{D}(\boldsymbol{v}){:}\mathbb{D}_{\text{visc}}(\mathsf{q})\boldsymbol{D}(\boldsymbol{v}) + \big(\boldsymbol{F}_{\text{p}}\boldsymbol{L}_{\text{vi.pl}}(\mathsf{q})\big){:}\partial_{\boldsymbol{F}_{\text{p}}}E(\mathsf{q})\Big). \tag{2.4.5d}$$

Specifying the choice of τ either to $\tau = e$, the density of the internal energy giving $E(\boldsymbol{F}, \boldsymbol{F}_{\text{p}}, e) = e$, or to $\tau = s$, the density of the entropy giving $S(\boldsymbol{F}, \boldsymbol{F}_{\text{p}}, s) = s$, reveals more structure to the terms $j^S_{\text{Ham}}(\mathsf{q}) + j^E_{\text{diss}}(\mathsf{q})$. In the first case we can read of the heating contribution through mechanical processes, and in the second case we see the entropy production through plasticity. We refer to Mielke and Roubíček (2025, Sect. 4.6) for a more elaborate discussion.

2.5 Discussion and Outlook

In this paper we have brought together the theory of Lie derivatives and the GENERIC framework to provide the right modeling tools for Eulerian mechanics of solid materials.

The GENERIC framework takes care that the model is thermodynamically consistent in the sense that energy is conserved (first law of thermodynamics) and that the entropy is non-decreasing (second law of thermodynamics). However, the GENERIC framework enforces an additional structural condition on the reversible/conservative part and on the irreversible/dissipative part of the system. The reversible part has to be strictly Hamiltonian which is more restrictive than being energy and entropy preserving. For this, it would be sufficient to have the skew-symmetry $\mathbb{J}(q)^* = -\mathbb{J}(q)$ and the non-interaction condition $\mathbb{J}(q)\mathrm{D}\mathcal{S}(q) \equiv 0$, but GENERIC imposes the Jacobi identity (2.3.1). The irreversible part is restricted by enforcing that the kinetic relations for the dissipative processes derive from the (sub)-differential of a dissipation potential, which automatically enforces Onsager's symmetry relations.

Thus, GENERIC systems form a subclass of all thermodynamically consistent systems. The additional structural conditions may be restrictive but appear to be valid for many continuum models. In particular, in situations where complex coupled models are to be developed, the framework provides easy ways for the derivation of good coupling terms between different variables; see, e.g., Kantner et al. (2019, Sect. 3.4+4.6), where a reaction-diffusion system for charge carriers is coupled to a quantum mechanical system. We expect that the GENERIC framework will also be very useful in solid mechanics when in addition to deformation and temperature also internal variables such as a plastic distortion, phase indicators, dislocation densities, or magnetization have to be modeled, see, e.g., Gladkov et al. (2016), Benešová et al. (2024).

This approach was used already very successfully in Mielke (2011), Hütter and Svendsen (2012), Gladkov et al. (2016) for models based on the Lagrangian description, where a referential body M is used for describing all the fields. While the Eulerian perspective working in the spatial domain $\mathsf{S} \subset \mathbb{R}^d$ is commonly used in fluid mechanics, see, e.g., Grmela and Öttinger (1997), Öttinger (2005), Zafferi et al. (2023), the combination of GENERIC, Eulerian description, and solid mechanics was brought together only recently in Mielke and Roubíček (2025). One of the reasons for using the Eulerian description for solids is the growing mathematical interest in geophysical flows in the upper mantle of our planet Earth, see, e.g., Roubíček and Tomassetti (2021), Roubíček (2023), Mielke et al. (2023). Depending on the time scale under consideration the solid rock behaves visco-elastically or visco-plastically, but there are many other variables to be taken into account, like concentrations of different chemicals, water contents, temperature, or aging variables, see, e.g., Gerya and Yuen (2007), Herrendörfer et al. (2018), Sandiford et al. (2021).

The Eulerian description of solids is intrinsically linked to the notion of Lie derivatives, because the variables are transported with the velocity $\boldsymbol{v}$ of the points in the moving and deforming body. This observation from Mielke and Roubíček (2025) served as a motivation to provide a short and self-contained introduction to Lie derivatives with a focus on the applications in continuum thermodynamics in the present paper. As an major outcome we found Theorem 2.3, which states that in the Eulerian setting the Jacobi identity (2.3.1) for Poisson operators $\mathbb{J}$ is intimately related to the commutator relation (2.2.3) for Lie derivatives. The result is now more general and the proof is shorter and more direct as for Mielke and Roubíček (2025, Thm. 4.1).

2.6 Pull-Backs and Push-Forwards in Differential Geometry

A diffeomorphism $\boldsymbol{\Phi} : M \to N$ defines pull-backs $\boldsymbol{\Phi}^*$ and push-forwards $\boldsymbol{\Phi}_*$ as follows:

0. Functions:

$$\big(\boldsymbol{\Phi}^* g\big)(m) = g(\boldsymbol{\Phi}(m)) \quad \text{and} \quad \big(\boldsymbol{\Phi}_* f\big)(n) = f(\boldsymbol{\Phi}^{-1}(n)).$$

1. Vector fields:

$$\begin{aligned}\big(\boldsymbol{\Phi}^* \boldsymbol{w}\big)(m) &= \big(\mathrm{D}\boldsymbol{\Phi}(\boldsymbol{\Phi}(m))\big)^{-1}\boldsymbol{w}(\boldsymbol{\Phi}(m)) = \big(\mathrm{D}\boldsymbol{\Phi}(m)\big)^{-1}\boldsymbol{w}(\boldsymbol{\Phi}(m)) \in \mathrm{T}_m M\\ \big(\boldsymbol{\Phi}_* \boldsymbol{v}\big)(n) &= \mathrm{D}\boldsymbol{\Phi}(\boldsymbol{\Phi}^{-1}(n))\boldsymbol{v}(\boldsymbol{\Phi}^{-1}(n)) = \mathrm{D}(\boldsymbol{\Phi}^{-1})(n)\boldsymbol{v}(\boldsymbol{\Phi}^{-1}(n)) \in \mathrm{T}_n N\end{aligned}$$

1*. one-forms:

$$\begin{aligned}\big(\boldsymbol{\Phi}^* \boldsymbol{\beta}\big)(m) &= \mathrm{D}\boldsymbol{\Phi}(m)^* \boldsymbol{\beta}(\boldsymbol{\Phi}(m)) \in \mathrm{T}^*_m M\\ \big(\boldsymbol{\Phi}_* \boldsymbol{\alpha}\big)(n) &= \big(\mathrm{D}\boldsymbol{\Phi}(\boldsymbol{\Phi}^{-1}(m))\big)^{-*}\boldsymbol{\alpha}(\boldsymbol{\Phi}^{-1}(n)) \in \mathrm{T}_n N\end{aligned}$$

Here for a matrix H we denote by H^{-*} the transposed of the inverse, namely, $H^{-*} = \big(H^{-1}\big)^* = (H^*)^{-1}$. Pull-backs and push-forwards are consistent with dual pairings, namely,

$$\begin{aligned}&\boldsymbol{\Phi}^*\big(\langle \boldsymbol{\beta}, \boldsymbol{w}\rangle_N\big)(m) = \big\langle \boldsymbol{\Phi}^*\boldsymbol{\beta}(m), \boldsymbol{\Phi}^*\boldsymbol{w}(m)\big\rangle_{\mathrm{T}_m M}\\ &\boldsymbol{\Phi}_*\big(\langle \boldsymbol{\alpha}, \boldsymbol{v}\rangle_M\big)(n) = \big\langle \boldsymbol{\Phi}_*\boldsymbol{\alpha}(\phi^{-1}(n)), \boldsymbol{\Phi}_*\boldsymbol{v}(\phi^{-1}(n))\big\rangle_{\mathrm{T}_n N},\\ &\langle \boldsymbol{\Phi}^*\boldsymbol{\beta}, \boldsymbol{v}\rangle_M = \langle \boldsymbol{\beta}, \boldsymbol{\Phi}_*\boldsymbol{v}\rangle_N \quad \text{and} \quad \langle \boldsymbol{\alpha}, \boldsymbol{\Phi}^*\boldsymbol{w}\rangle_M = \langle \boldsymbol{\Phi}_*\boldsymbol{\alpha}, \boldsymbol{w}\rangle_N.\end{aligned}$$

General tensors $\boldsymbol{A} \in \mathfrak{T}^{i_\circ}_{j_\circ}(M)$ with $\boldsymbol{A}(m) \in \mathrm{ML}\big((\mathrm{T}_m M)^{i_\circ} \times (\mathrm{T}_m M)^{j_\circ}\big)$ and $\boldsymbol{B} \in \mathfrak{T}^{i_\circ}_{j_\circ}(N)$ with $\boldsymbol{B}(n) \in \mathrm{ML}\big((\mathrm{T}_n N)^{i_\circ} \times (\mathrm{T}_n N)^{j_\circ}\big)$:

$$\boldsymbol{\Phi}^* \boldsymbol{B}(m)[\boldsymbol{v}_1, .., \boldsymbol{v}_{i_\circ}, \boldsymbol{\alpha}_1, .., \boldsymbol{\alpha}_{j_0}] = \boldsymbol{B}(n)\big[\boldsymbol{\Phi}_* \boldsymbol{v}_1, .., \boldsymbol{\Phi}_* \boldsymbol{v}_{i_\circ}, \boldsymbol{\Phi}_* \boldsymbol{\alpha}_1, .., \boldsymbol{\Phi}_* \boldsymbol{\alpha}_{j_\circ}\big]\Big|_{n=\boldsymbol{\Phi}(m)},$$

$$\boldsymbol{\Phi}_* \boldsymbol{A}(n)[\boldsymbol{w}_1, .., \boldsymbol{w}_{i_\circ}, \boldsymbol{\beta}_1, .., \boldsymbol{\beta}_{j_0}] = \boldsymbol{A}(m)\big[\boldsymbol{\Phi}^* \boldsymbol{w}_1, .., \boldsymbol{\Phi}^* \boldsymbol{w}_{i_\circ}, \boldsymbol{\Phi}^* \boldsymbol{\beta}_1, .., \boldsymbol{\Phi}^* \boldsymbol{\beta}_{j_\circ}\big]\Big|_{m=\boldsymbol{\Phi}^{-1}(n)}.$$

Acknowledgements This research has been partially funded by Deutsche Forschungsgemeinschaft (DFG) through the Berlin Mathematics Research Center MATH+ (EXC-2046/1, project ID: 390685689) subproject "DistFell". The author is grateful for stimulating discussions with Tomáš Roubíček.

References

Abraham R, Marsden JE (1978) Foundations of mechanics. Benjamin/Cummings Publishing Co. Inc. Advanced Book Program, Reading, Mass., 1978, Second edition, revised and enlarged, With the assistance of Tudor Raţiu and Richard Cushman

Betsch P, Schiebl M (2019) Energy-momentum-entropy consistent numerical methods for large-strain thermoelasticity relying on the GENERIC formalism. Int J Numer Meth Eng 119(12):1216–1244

Besseling J, Van der Giessen E (1994) Mathematical modelling of inelastic deformation. Chapman & Hall/Springer

Benešová B, Nečasová Š, Scherz J, Schlömerkemper A (2024) A variational approach to the modeling of compressible magnetoelastic materials. Preprint (2024), arXiv:2410.15196

Chandrasekhar S (1983) The mathematical theory of black holes. Oxford University Press

Duong MH, Peletier MA, Zimmer J (2013) GENERIC formalism of a Vlasov-Fokker-Planck equation and connection to large-deviation principles. Nonlinearity 26(11):2951–2971

Edwards BJ (1998) An analysis of single and double generator thermodynamics formalisms for the macroscopic description of complex fluids. J Non-Equilib Thermodyn 23(4):301–333

Fiala Z (2008) Geometry of finite deformations, linearizations, and incremental deformations under initial stress/strain. In: Proceedings of engineering mechanics 2008 (held in Svratka, Czech Republic, 12–15 May 2008), pp 1–20

Gerya T, Yuen DA (2007) Robust characteristics method for modelling multiphase visco-elasto-plastic thermo-mechanical problems. Phys Earth Plan Inter 163(1–4):83–105

Gladkov S, Kochmann J, Reese S, Hütter M, Svendsen B (2016) Thermodynamic model formulations for inhomogeneous solids with application to non-isothermal phase field modeling. J Non-Equil Thermodyn 42(2):131–139

Grmela M, Öttinger HC (1997) Dynamics and thermodynamics of complex fluids. I. Development of a general formalism. II. Illustrations of a general formalism. Phys Rev E (3) 56(6):6620–6655

Grmela M (1984) Particle and bracket formulations of kinetic equations. In: Fluids and plasmas. Geometry and Dynamics (Boulder, Colo., 1983), Contemp Math 28:125–132

Grmela M (1985) Bracket formulation of dissipative time evolution equations. Phys Lett A 111(1–2):36–40

Gurtin M, Anand L (2005) The decomposition $\mathbf{F} = \mathbf{F}^{\mathrm{e}}\mathbf{F}^{\mathrm{p}}$, material symmetry, and plastic irrotationality for solids that are isotropic-viscoplastic or amorphous. Int J Plast 21:1686–1719

Herrendörfer R, Gerya T, Van Dinther Y (2018) An invariant rate- and state-dependent friction formulation for viscoelastoplastic earthquake cycle simulations. J Geophys Res Solid Earth 123:5018–5051

Hütter M, Svendsen B (2012) Thermodynamic model formulation for viscoplastic solids as general equations for non-equilibrium reversible-irreversible coupling. Contin Mech Thermodyn 24(3):211–227

Hütter M, Tervoort TA (2008) Finite anisotropic elasticity and material frame indifference from a nonequilibrium thermodynamics perspective. J Non-Newtonian Fluid Mech 152:45–52

Kantner M, Mielke A, Mittnenzweig M, Rotundo N (2019) Mathematical modeling of semiconductors: from quantum mechanics to devices. In: Rodrigues J, Hintermüller M (eds) Topics in applied analysis and optimisation, CIM series in mathematical sciences. Springer, pp 269–293

Kröner E (1960) Allgemeine Kontinuumstheorie der Versetzungen und Eigenspannungen. Arch Rational Mech Anal 4(1960):273–334

Lasarzik R (2021) Analysis of a thermodynamically consistent Navier-Stokes-Cahn-Hilliard model. Nonl Anal 213:112526/1–33

Lee EH (1969) Elastic-plastic deformation at finite strains. J Appl Mech 36:1–6

Lee EH, Liu DT (1967) Finite-strain elastic-plastic theory with application to plain-wave analysis. J Appl Phys 38:19–27

Lebon G, Jou D, Casas-Vázquez J (2008) Understanding non-equilibrium thermodynamics. Springer

Marsden J, Hughes TJ (1983) Mathematical foundations of elasticity. Dover Publications Inc., New York, Corrected reprint of the 1983 original

Marsden JE, Ratiu TS (1999) Introduction to mechanics and symmetry, 2nd edn, Texts in Applied Mathematics, vol 17. Springer, New York

Muschik W, Gümbel S, Kröger M, Öttinger H (2000) A simple example for comparing generic with rational non-equilibrium thermodynamics. XXPhys A 285:448–466

Mielke A (2003) Energetic formulation of multiplicative elasto-plasticity using dissipation distances. Contin Mech Thermodyn 15:351–382

Mielke A (2011) Formulation of thermoelastic dissipative material behavior using GENERIC. Contin Mech Thermodyn 23(3):233–256

Mielke A (2015) On thermodynamical couplings of quantum mechanics and macroscopic systems. In: Exner P, König W, Neidhardt H (eds) Mathematical results in quantum mechanics (Singapore). World Scientific, Proceedings of the QMath12 Conference, pp 331–348

Mielke A, Roubíček T (2025) A general thermodynamical model for finitely-strained continuum with inelasticity and diffusion, its GENERIC derivation in Eulerian formulation, and some application. Z angew Math Phys (ZAMP) 76(11):1–28

Morrison PJ (1984) Bracket formulation for irreversible classical fields. Phys Lett A 100(8):423–427

Morrison PJ (1986) A paradigm for joined Hamiltonian and dissipative systems. Phys D 18(1–3):410–419

Morrison PJ (2009) Thoughts on brackets and dissipation: old and new. J Phys Conf Ser 169:012006, 12

Mielke A, Peletier MA, Zimmer J (2024) Deriving a GENERIC system from a Hamiltonian system. Submitted (2024), WIAS preprint 3108, arXiv:2404.09284

Mielke A, Roubíček T, Stefanelli U (2023) A model of gravitational differentiation of compressible self-gravitating planets. Submitted, arXiv:2305.06232, WIAS preprint 3015

Onsager L (1931) Reciprocal relations in irreversible processes, I+II. Phys Rev 37:405–426, (part II, 38:2265–2279)

Öttinger HC (2005) Beyond equilibrium thermodynamics. Wiley, New Jersey

Pavelka M, Klika V, Grmela M (2018) Multiscale thermo-dynamics. Introduction to GENERIC. De Gruyter

Pavelka M, Peshkov I, Klika V (2020) On Hamiltonian continuum mechanics. Physica D 408:132510/1–19

Pelech P, Tůma K, Pavelka M, Šípka M, Sýkora M (2022) On compatibility of the natural configuration framework with general equation for non-equilibrium reversible-irreversible coupling (GENERIC): derivation of anisotropic rate-type models. J Non-Newtonian Fluid Mech 305, Art. no. 104808

Roubíček T, Tomassetti G (2021) A convective model for poro-elastodynamics with damage and fluid flow towards Earth lithosphere modelling. Contin Mech Thermodyn 33:2345–2361

Roubíček T (2023) A thermodynamical model for paleomagnetism in Earth's crust. Math Mechanics Solids 28:1063–1090

Sandiford D, Brune S, Glerum A, Naliboff J, Whittaker JM (2021) Kinematics of footwall exhumation at oceanic detachment faults: Solid-block rotation and apparent unbending. Geochem Geophys Geosyst 22:e2021GC009681 1–12

Segev R (2023) Foundations of geometric continuum mechanics. Birkhäuser

Yano K (1955) The theory of Lie derivatives and its applications. North-Holland Publ. Co., Amsterdam

Zafferi A, Peschka D, Thomas M (2023) GENERIC framework for reactive fluid flows. Z angew Math Mech (ZAMM) 103:e202100254 70 pp

Chapter 3
Reference Map Approach to Eulerian Thermomechanics Using GENERIC

Dirk Peschka, Marita Thomas, and Andrea Zafferi

In memory of Wolfgang Dreyer, who we will always remember for his passion for thermodynamics and his love for a good argument.

Abstract An Eulerian GENERIC model for thermo-viscoelastic materials with diffusive components is derived based on a transformation framework that maps a Lagrangian formulation to corresponding Eulerian coordinates. The key quantity describing the deformation in Eulerian coordinates is the inverse of the deformation, i.e., the *reference map*. The Eulerian model is formally constructed, and by reducing the GENERIC system to a damped Hamiltonian system, the isothermal limit is derived. A structure-preserving weak formulation is developed. As an example, the coupling of finite strain viscoelasticity and diffusion in a multiphase system governed by Lagrangian indicator functions is demonstrated.

3.1 Introduction

There is a plethora of thermomechanical models with diffusion that are relevant for numerous applications in the natural sciences. Among them, one can distinguish models for fluids or solids that may be either compressible or incompressible, cf.

D. Peschka
Weierstrass Institut, Berlin, Germany
e-mail: peschka@wias-berlin.de

M. Thomas · A. Zafferi (✉)
Freie Universität Berlin, Berlin, Germany
e-mail: andrea.zafferi@fu-berlin.de

M. Thomas
e-mail: marita.thomas@fu-berlin.de

M. Thomas
Weierstrass Institut, Berlin, Germany

J. Fuhrmann et al. (eds.), *Advances in Continuum Physics*, Advanced Structured Materials 238, https://doi.org/10.1007/978-3-031-93918-1_3

e.g., Noll and Coleman (1974), Green and Naghdi (1934), Lions (1996a, b), Müller and Müller (2009), Mielke (2011b). These models can include diffusive or reactive components (Biot 1984; Datta and Vilekar 2010; Bothe and Dreyer 2015; Gay-Balmaz and Yoshimura 2017) or even incorporate multiple phases, featuring phase transitions or phase separation (Anand 2012; Alt 2009; Ding et al. 2007; Abels et al. 2012). For solids, complex viscoelastic effects are relevant for hydrogels (Mao et al. 2017), in soft matter physics (Doi 2011) as well as in geomechanics (Mielke and Roubíček 2025).

The wide applicability of these models, which all have a similar structure and thermodynamic basis, necessitates a unified thermomechanical framework that allows for large deformations. While thermo-viscoelastic models with large deformations are naturally formulated in the Lagrangian frame (Noll and Coleman 1974; Marsden and Hughes 1994; Mielke and Roubíček 2020; Ciarlet 1988; Badal et al. 2023), this can be impractical for certain cases, e.g., when coupled with other effects that act in the current configuration, such as diffusion or fluid flow. In such situations, an Eulerian formulation might be more suitable (Roubíček and Stefanelli 2023). In a Lagrangian frame, compressible fluids and solids share a canonical skew-symmetric thermomechanical structure with irreversible dissipation but differ in their energy dependence on mechanical deformation, i.e., on the tensorial deformation gradient for hyperelastic solids and on the scalar volume for fluids. In contrast, in an Eulerian description, fluids and solids already differ in their choice of unknowns. Besides momentum as a state variable, fluid models typically track the volume via the Eulerian mass density, whereas solids evolve deformation via a tensor related to stress or strain (Oldroyd 1950; Roubíček 2024) transported by suitable Lie derivatives (Marsden and Hughes 1994). An alternative approach employs evolution equations for the vectorial *reference map* (Kamrin et al. 2012; Rycroft et al. 2020), i.e., the inverse of the deformation, which is therefore sometimes also called in literature the *inverse* map (Cotter et al. 2007; Richter 2013), *return* map (Wei and Min 2023), or *back-to-labels* map (Constantin 2001). These approaches can be especially useful when dealing with *Fluid-Structure Interaction* (FSI) problems (Richter 2013; Rycroft et al. 2020). In Aland and Mokbel (2021), the authors implicitly also employ the reference map technique to evolve Eulerian displacements in a fluid-structure interaction problem involving capillary interfaces. This requires a careful and systematic consideration of the transformations between the Lagrangian reference configuration and the Eulerian deformed configuration in a thermodynamic framework.

In his book *Understanding Non-Equilibrium Thermodynamics*, Lebon presents a clear overview of the various schools that have evolved to study systems far from equilibrium (Lebon et al. 2008). There is no uniquely correct way to describe these processes, and different approaches have emerged based on distinct ideas of how to construct such theories. For example, classical irreversible thermodynamics is built on conservation laws and constitutive relations, while rational thermodynamics, pioneered by Truesdell Truesdell (1984), Coleman and Noll (1974), follows a more systematic method to ensure consistency with the second law of thermodynamics. For fluids and mixtures, many well-developed theoretical extensions offer thermodynamically consistent formulations of complex systems, e.g., see *Thermo-Fluid*

Dynamics of Two-Phase Flow by Ishii and Hibiki (2010), *Thermodynamics of flowing systems: with internal microstructure* by Beris and Edwards (1994), as well as the treatments by deGroot and Mazur in (2013). We also highlight the work *Continuum thermodynamics of chemically reacting fluid mixtures* by Dreyer and Bothe, who present a general model for diffusive and reactive mixtures and flows, along with an excellent overview of the different thermodynamic schools (Bothe and Dreyer 2015). Similarly, for solids, dedicated texts such as *The mechanics and thermodynamics of continua* by Gurtin, Fried, and Anand provide rigorous frameworks that adhere to thermodynamic principles (Gurtin et al. 2010).

These approaches are complemented by more geometrical methods based on symplectic structures, Lie brackets, and gradient structures, such as the methods for metriplectic, port-Hamiltonian, or GENERIC systems (Morrison 1998, 1986; Gay-Balmaz and Yoshimura 2017; Marsden and Ratiu 2013; Jeltsema et al. 2014; Mielke 2011b). GENERIC, the acronym for *General Equations of Non-Equilibrium Reversible-Irreversible Coupling*, is a thermodynamic formalism that was first introduced by Grmela and Öttinger (Grmela and Öttinger 1997; Öttinger and Grmela 1997), based on the definition of thermodynamic potentials and geometric structures for conservative or reversible and for dissipative and irreversible processes, respectively. Grounded in thermodynamic functions like energy and entropy, the GENERIC formalism uses suitable geometric operators to generate the evolution of the state variables of the system. For over-damped systems, the structure of gradient flows can also ensure Onsager reciprocity by design (Jordan et al. 1998; Onsager 1931a, b; Doi 2011; Peletier 2014). Overall, these different approaches to thermomechanical systems are largely compatible and create a rich landscape of thermodynamic models—one that continues to spark lively debates about their respective advantages and disadvantages, with no final consensus among the thermodynamics community in sight.

In Zafferi et al. (2023a) we previously developed a mechanism to handle transformations between Lagrangian and Eulerian formulations of GENERIC systems and studied their implications on the thermodynamic formulation of systems of reactive multispecies fluids within the framework of GENERIC. In Thomas and Heida (2022) it was shown that the GENERIC formalism extends to thermomechanical systems with state variables interacting between bulk and interfaces in terms of a weak formulation of GENERIC, which naturally encodes coupling conditions along the interfaces. Using such a weak formulation of GENERIC, in Peschka et al. (2022) we studied the transformation behavior suited for a structure-preserving numerical discretization for a fluid-structure interaction problem using the reference map.

The main goal of this work is to provide an Eulerian GENERIC formulation for fluids and solids undergoing large deformations in terms of the reference map and show the feasibility of the weak form for structure-preserving numerical discretization. We establish two equivalent formulations in terms of Lagrangian and Eulerian quantities that are linked via a transformation map where we employ the *reference map*, i.e., the inverse of the deformation or flow map, as the variable to describe deformations. In this spirit, Sect. 3.2 introduces the Lagrangian and Eulerian kinematics and the GENERIC formalism augmented by a brief description of thermodynamics for solid materials. Subsequently, in Sect. 3.3 we present a Lagrangian GENERIC system for thermo-viscoelastic materials and carry out a transformation to the Eulerian setting. We complement it by examples for the energy and entropy functionals and operators, and, by fast equilibration of temperature with a heat reservoir, reduce the GENERIC system to a damped Hamiltonian system. We conclude this work in Sect. 3.4 with a weak formulation of the damped Hamiltonian system at the example of an FSI problem, which is subsequently studied in numerical simulations. This example has two main purposes: Firstly, to show that the GENERIC structures previously introduced can be adapted to thermodynamically closed systems, i.e., systems with no mass exchange with the surroundings (in contrast to isolated systems treated by GENERIC, where no heat exchange with the surroundings is permitted). Secondly, to show that the use of the reference map within this variational framework provides an alternative treatment to multiphase systems.

3.2 Kinematics and Thermodynamics in the Framework of GENERIC

We start by introducing the notations and definitions concerning the Lagrangian (reference) and the Eulerian (actual) configuration. In Sect. 3.2.2, we proceed with a brief discussion on the thermodynamics of solid materials in connection with the previously defined quantities and ultimately we introduce the GENERIC framework.

3.2.1 Lagrangian and Eulerian Kinematics

We consider a material that occupies in the Lagrangian or reference configuration the domain $\bar{\Omega} \subset \mathbb{R}^d$ at time $t = 0$, where typically $d = 1, 2, 3$. At any time $t \in [0, \mathsf{T}]$ the Eulerian or actual domain $\Omega(t)$ is connected to the initial data by the motion, deformation or *flow map* $\boldsymbol{\chi} : [0, \mathsf{T}] \times \bar{\Omega} \to \mathbb{R}^d$ such that $\Omega(t) = \boldsymbol{\chi}(t, \bar{\Omega})$ (Fig. 3.1). Along with the flow map, deformations are measured through the deformation gradient[1] $\bar{\boldsymbol{F}} : [0, \mathsf{T}] \times \bar{\Omega} \to \mathbb{R}^{d\times d}$ defined in referential Cartesian coordinates as

$$\bar{\boldsymbol{F}}_{ij}(t, \bar{x}) := \frac{\partial \boldsymbol{\chi}_i}{\partial \bar{x}_j}(t, \bar{x}) \quad \text{with } 1 \leq i, j \leq d \text{ or simply} \quad \bar{\boldsymbol{F}}(t, \bar{x}) = \bar{\nabla}\boldsymbol{\chi}(t, \bar{x}).$$

Clearly, for the initial datum $\boldsymbol{\chi}(0, \bar{x}) = \bar{x}$ the deformation gradient is the identity matrix $\bar{\boldsymbol{F}}(0, \bar{x}) = \mathbb{I}_d$. An important role in the energy functionals, see Remark 3.3, and in the thermodynamic formalism is played by its Jacobian $\bar{V} := \det \bar{\boldsymbol{F}}$. Lastly,

[1] **Note:** A discussion on the tensorial nature of the deformation gradient can be found in Marsden and Hughes (1994).

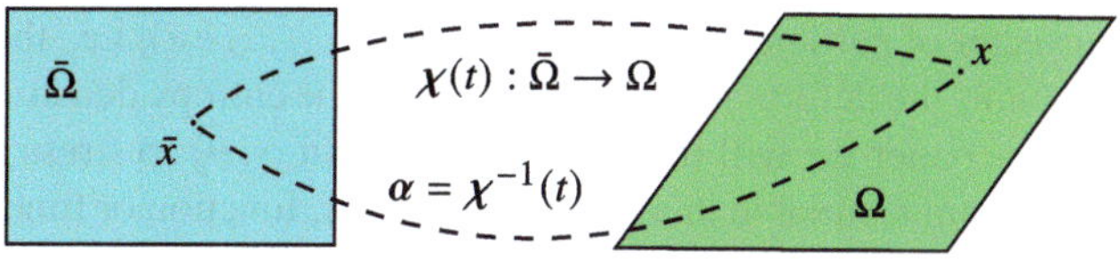

Fig. 3.1 Schematic representation of the reference $\bar{\Omega}$ and actual domain Ω with the flow map $\boldsymbol{\chi}$ and the reference map $\boldsymbol{\alpha}$

the description of a particle motion in Lagrangian coordinates is completed by its velocity field $\bar{\boldsymbol{v}}(t, \bar{x}) = \partial_t \boldsymbol{\chi}(t, \bar{x})$.

Assuming that the deformation $\boldsymbol{\chi}(t, \cdot)$ is invertible for every time $t \in [0, \mathsf{T}]$, we introduce the *reference map* $\boldsymbol{\alpha}(t, \cdot) := \boldsymbol{\chi}^{-1}(t, \cdot)$ that satisfies the identity relation

$$\boldsymbol{\alpha}(t, \boldsymbol{\chi}(t, \bar{x})) = \bar{x}\,, \qquad \text{for any } (t, \bar{x}) \in [0, \mathsf{T}] \times \bar{\Omega}\,. \tag{3.1}$$

From (3.1) one sees that the deformation gradient is the inverse gradient of the reference map, i.e., $\bar{\boldsymbol{F}}(t, \boldsymbol{\alpha}(t, x)) := \boldsymbol{F}(t, x) = \left(\nabla \boldsymbol{\alpha}(t, x)\right)^{-1}$. Throughout this paper we will denote quantities that are "naturally" defined in Lagrangian coordinates with a bar, while we will drop the bar for the Eulerian ones.

3.2.2 *Equilibrium Thermomechanics*

Since the GENERIC formalism relies on thermodynamic functionals and operators we first introduce some basic notions of equilibrium thermodynamics. It originated from the study of gases or fluids, see Truesdell and Toupin (1960) and references therein for a historical review on the subject. One of the main observations of Zafferi et al. (2023a) is to identify the volume $\bar{V}$ being an extensive thermodynamic variable with the determinant of the deformation gradient $\bar{V} := \det \bar{\boldsymbol{F}}$. Observe that in our setting the volume $\bar{V}$ is dimensionless. However, this is consistent with the measurement of extensive quantities in terms of volumetric densities, since the measure of integration $\bar{V}\mathrm{d}\bar{x} = \mathrm{d}x$ carries the unit of volume and upon integration produces an invariant extensive scalar, see Sect. 3.2.3 for further details. To describe fluid rheology only volumetric changes are relevant and thus only $\bar{V}$ enters to the models. Instead, when modeling elastic solids, also deformations beyond pure volumetric changes have to be captured. Therefore, thermomechanical functionals and operators describing solid materials also have to include the dependency on other minors of the deformation gradient $\bar{\boldsymbol{F}}$ besides the determinant. By doing so, we will be able to treat general *hyperelastic* materials.

We follow the treatment of Truesdell and Noll (2004), Noll and Coleman (1974), Gurtin et al. (2010) and consider a state vector

$$\bar{\boldsymbol{q}}_{\bar{s}} := (\boldsymbol{\chi}, \bar{\boldsymbol{c}}, \bar{s}) \tag{3.2}$$

formed by the vector of particle concentrations $\bar{\boldsymbol{c}} = (\bar{c}_1, \ldots, \bar{c}_N)$, i.e., the number of particles per volume, the deformation gradient $\bar{\boldsymbol{F}}$, and the entropy density per volume $\bar{s}$. Throughout this paper we will use a subscript $z \in \{\bar{u}, \bar{s}, \bar{\theta}\}$ to stress the use of a specific thermodynamic variable in a given state vector, function or functional, as in (3.2) or (3.3). We then define the internal energy density (per volume) of the system through an equation of state in natural variables

$$\bar{\mathcal{U}}_{\bar{s}}[\bar{\boldsymbol{q}}_{\bar{s}}] := \bar{\mathcal{U}}_{\bar{s}}(\bar{\boldsymbol{F}}, \bar{\boldsymbol{c}}, \bar{s}) . \tag{3.3}$$

From the first and second law of thermodynamics we obtain the following fundamental relation between the state variables and their conjugate

$$\mathrm{D}\bar{\mathcal{U}}_{\bar{s}} = \bar{\theta}_{\bar{s}} \mathrm{D}\bar{s} + \bar{\boldsymbol{\Sigma}}^{\mathrm{el}}_{\bar{s}} : \mathrm{D}\bar{\boldsymbol{F}} + \sum_{i=1}^{N} (\bar{\boldsymbol{\mu}}_{\bar{s}})_i \mathrm{D}\bar{c}_i \,, \tag{3.4}$$

where D stands for the Frechét derivative. In Sect. 3.2.3, where we introduce the GENERIC formalism, we will be more specific about the functions and the derivatives considered. Similar fundamental relations for solids can be found in De Boer (1977), Baker (2005) and with a different strategy, introducing the concept of thermodynamic tension, in Barron and Munn (1970). From (3.4), we recover the standard definitions of temperature $\bar{\theta}_{\bar{s}}$, chemical potential $\bar{\boldsymbol{\mu}}_{\bar{s}}$ and the elastic stress tensor $\bar{\boldsymbol{\Sigma}}^{\mathrm{el}}_{\bar{s}}$:

$$\bar{\theta}_{\bar{s}} := \frac{\partial \bar{\mathcal{U}}_{\bar{s}}}{\partial \bar{s}}\,, \quad \bar{\boldsymbol{\Sigma}}^{\mathrm{el}}_{\bar{s}} := \frac{\partial \bar{\mathcal{U}}_{\bar{s}}}{\partial \bar{\boldsymbol{F}}}\,, \quad \bar{\boldsymbol{\mu}}_{\bar{s}} := \frac{\partial \bar{\mathcal{U}}_{\bar{s}}}{\partial \bar{\boldsymbol{c}}}\,. \tag{3.5}$$

This definition of elastic stress and its relation to $\bar{\mathcal{U}}_{\bar{s}}$ is usually used in literature as the characterizing condition of hyperelastic materials, see Ciarlet (1988, Chap. 4). Most of the theory of continuum mechanics of elastic bodies was developed independently from thermodynamics, which led to the assumption of constant temperature. For this and other practical reasons, often the Helmholtz free energy is used (or the Massieu potential, if more emphasis on dissipative processes is desired) to design constitutive laws for elastic materials. Let us briefly recall its definition. We define the Helmholtz free energy for the state vector $\bar{\boldsymbol{q}}_{\bar{\theta}} = (\bar{\boldsymbol{F}}, \bar{\boldsymbol{c}}, \bar{\theta})$ in the following way:

$$\bar{\mathscr{F}}_{\bar{\theta}}(\bar{\boldsymbol{q}}_{\bar{\theta}}) = \int_{\bar{\Omega}} \bar{\mathcal{F}}_{\bar{\theta}}[\bar{\boldsymbol{q}}_{\bar{\theta}}]\,\mathrm{d}\bar{x} := \int_{\bar{\Omega}} \left(\bar{\mathcal{U}}_{\bar{\theta}}[\bar{\boldsymbol{q}}_{\bar{\theta}}] - \bar{\theta}\bar{S}_{\bar{\theta}}[\bar{\boldsymbol{q}}_{\bar{\theta}}]\right)\,\mathrm{d}\bar{x}\,, \tag{3.6}$$

which is the Legendre transform of the internal energy $\bar{\mathcal{U}}_{\bar{s}}$ with respect to the entropy $\bar{s}$. By means of the standard relations (3.5) we have:

$$\frac{\partial \bar{\mathcal{F}}_{\bar{\theta}}}{\partial \bar{\theta}} = -\bar{S}_{\bar{\theta}}\,, \qquad \frac{\partial^2 \bar{\mathcal{F}}_{\bar{\theta}}}{\partial \bar{\theta}^2} = -\frac{\bar{C}_{\bar{V}}}{\bar{\theta}}\,, \qquad \frac{\partial \bar{\mathcal{F}}_{\bar{\theta}}}{\partial \bar{\boldsymbol{F}}} = \bar{\boldsymbol{\Sigma}}^{\mathrm{el}}\,,$$

where $\bar{C}_{\bar{V}}$ is the heat capacity at constant volume (isochoric).

While the free energy can and should, in principle, depend arbitrarily on $\bar{\boldsymbol{q}}_{\bar{\theta}}$, we propose a simplified energy model that emphasizes the nonlinear elasticity of the system and adopts idealized assumptions for the dependence on temperature and concentration. This approach follows the strategy of ideal gas relationships for the concentration, assumes a constant (Eulerian) heat capacity, and ensures the distinction between classical extensive and intensive variables in the context of nonlinear elasticity. Accordingly, we propose the following form

$$\bar{\mathcal{F}}_{\bar{\theta}}[\bar{\boldsymbol{q}}_{\bar{\theta}}] := \left[\mathcal{F}_{\text{mech}}(\bar{x}, \bar{\boldsymbol{F}}_{\text{iso}}) + \mathcal{F}_{\text{int}}\left(\bar{x}, \bar{c}/\det(\bar{\boldsymbol{F}}), \bar{\varrho}/\det(\bar{\boldsymbol{F}}), \bar{\theta}\right)\right] \det(\bar{\boldsymbol{F}}), \tag{3.7a}$$

where $\bar{\varrho}$ can be interpreted as the mass density. Its relation to the vector of concentrations $\bar{\boldsymbol{c}}$ will be specified later, cf. (3.16). In (3.7), the isochoric part of the deformation gradient $\bar{\boldsymbol{F}}$ is defined as $\bar{\boldsymbol{F}}_{\text{iso}} = \bar{\boldsymbol{F}}/(\det \bar{\boldsymbol{F}})^{1/d}$ for spatial dimension d so that $\det(\bar{\boldsymbol{F}}_{\text{iso}}) \equiv 1$. Here, elastic contributions (shear forces) are encoded in $\mathcal{F}_{\text{mech}}$, and the remaining contributions due to thermal, chemical, or volumetric changes come from the internal energy $\mathcal{F}_{\text{int}}$. We further distinguish in $\mathcal{F}_{\text{int}}$ contributions due to elasticity and due to the diffusive species by

$$\mathcal{F}_{\text{int}}(\bar{x}, \bar{c}/\det \bar{\boldsymbol{F}}, \bar{\varrho}/\det \bar{\boldsymbol{F}}, \bar{\theta}) := \mathcal{F}_{\text{int}}^{\text{solid}}(\bar{x}, \bar{\varrho}/\det \bar{\boldsymbol{F}}, \bar{\theta}) + \mathcal{F}_{\text{int}}^{\text{diff}}(\bar{x}, \bar{c}/\det \bar{\boldsymbol{F}}, \bar{\theta}). \tag{3.7b}$$

A somewhat idealized choice is

$$\mathcal{F}_{\text{int}}^{\text{solid}}(\bar{x}, \tfrac{\bar{\varrho}}{\det \bar{\boldsymbol{F}}}, \bar{\theta}) := C_{\text{solid}}\bar{\theta}\left[\ln\left(\frac{\bar{\theta}}{\bar{\theta}_{\text{ref}}}\right) - 1\right] + \frac{K\bar{\varrho}^2}{2}\left(\frac{\det \bar{\boldsymbol{F}}}{\bar{\varrho}} - \frac{1}{\bar{\varrho}}\right)^2, \tag{3.7c}$$

$$\mathcal{F}_{\text{int}}^{\text{diff}}(\bar{x}, \tfrac{\bar{c}}{\det \bar{\boldsymbol{F}}}, \bar{\theta}) := \frac{\bar{\boldsymbol{c}}}{\det \bar{\boldsymbol{F}}} k_B \bar{\theta}\left[\ln\left(\frac{\bar{\boldsymbol{c}}}{\det \bar{\boldsymbol{F}}}\lambda_{\bar{\theta}}^3\right) - 1\right], \tag{3.7d}$$

with the (thermal) de Broglie wavelength $\lambda_{\bar{\theta}} \sim \bar{\theta}^{-1/2}$, and further constants C_{solid}, $K, k_B > 0$. Using the above definitions, this gives rise to the simplified additive heat capacity

$$\bar{C}_{\bar{V}} = C_{\text{solid}} \det(\bar{\boldsymbol{F}}) + \tfrac{3}{2} k_B \bar{\boldsymbol{c}}.$$

This implies that $\bar{C}_{\bar{V}}$ is an extensive quantity, i.e., it scales 1-homogeneously with the other extensive quantities $\bar{V} = \det \bar{\boldsymbol{F}}$ and $\bar{\boldsymbol{c}}$. For dilute concentration the heat capacity is usually dominated by $C_{\text{solid}} \det(\bar{\boldsymbol{F}})$, which leads to a constant Eulerian heat capacity $C_V = C_{\text{solid}}$. The above construction relates the Eulerian free energy to its Lagrangian version (3.7) by

$$\mathcal{F}_{\bar{\theta}}[\boldsymbol{q}_{\theta}] = \bar{\mathcal{F}}_{\bar{\theta}}[\bar{\boldsymbol{q}}_{\bar{\theta}}(\boldsymbol{q}_{\theta})]/\det(\bar{\boldsymbol{F}}), \quad \text{where } \boldsymbol{q}_{\theta} := (\boldsymbol{\alpha}, \boldsymbol{c}, \theta),$$

is the Eulerian state vector, where we define $\bar{\boldsymbol{q}}_{\bar{\theta}}(\boldsymbol{q}_{\theta}) := (\boldsymbol{\alpha}, \boldsymbol{c}V, \theta)$. The origin of the relation to the Lagrangian state vector $\bar{\boldsymbol{q}}_{\bar{\theta}}$ via a transformation will be explained in Sect. 3.3.2. With $\varrho = \bar{\varrho} \circ \boldsymbol{\alpha}/V$ we then have

$$\mathscr{F}_\theta(\boldsymbol{q}_\theta) = \int_\Omega \mathcal{F}_\theta[\boldsymbol{q}_\theta]\,\mathrm{d}x, \qquad \mathcal{F}_\theta[\boldsymbol{q}_\theta] = \mathcal{F}_{\mathrm{mech}}(\boldsymbol{\alpha}, \boldsymbol{F}_{\mathrm{iso}}) + \mathcal{F}_{\mathrm{int}}(\boldsymbol{\alpha}, \boldsymbol{c}, \varrho, \theta). \tag{3.8}$$

Another relevant thermodynamic function that we will be using in this paper is the total energy of the system $\bar{\mathscr{E}}$, whose density function is denoted by $\bar{\mathcal{E}}$. In our setting, the total energy density of the system is the sum of two contributions, the internal $\bar{\mathcal{U}}$ and the kinetic energy $\bar{\mathcal{K}} := |\bar{\boldsymbol{m}}|^2/2\bar{\varrho}$, where $\bar{\boldsymbol{m}}$ is the momentum density and relates to the velocity field $\bar{\boldsymbol{v}}$ in a way we shall specify later in Sect. 3.3.1.

3.2.3 Short Introduction to GENERIC

The GENERIC (*General Equation for Non-Equilibrium Reversible-Irreversible Coupling*) formalism was firstly introduced by Grmela and Öttinger in (1997), Öttinger and Grmela (1997). This variational approach aims to describe thermodynamically isolated systems by means of a state space $\bar{Q}$, usually a Banach space with dual $\bar{Q}^*$ and we will denote with $\langle\cdot,\cdot\rangle$ the canonical dual pairing between $\bar{Q}^*$ and $\bar{Q}$. The framework is based on two thermodynamic potentials, the total energy $\bar{\mathscr{E}} : \bar{Q} \to \mathbb{R}$, i.e., the sum of kinetic and internal energy, and the total entropy $\bar{\mathscr{S}} : \bar{Q} \to \mathbb{R}$ of the system, and two geometric structures $\bar{\mathbb{J}}$ and $\bar{\mathbb{K}}$. The geometric structures give rise to two bilinear forms of different natures that characterize the reversible, or Hamiltonian, dynamics and the irreversible, or dissipative, dynamics of the system. The Poisson structure $\bar{\mathbb{J}}(\bar{\boldsymbol{q}}) : \bar{Q}^* \to \bar{Q}$ is antisymmetric, i.e., $\bar{\mathbb{J}}(\bar{\boldsymbol{q}}) = -\bar{\mathbb{J}}^*(\bar{\boldsymbol{q}})$ for all $\bar{\boldsymbol{q}} \in \bar{Q}$, and satisfies Jacobi's identity, whereas the Onsager operator $\bar{\mathbb{K}}(\bar{\boldsymbol{q}}) : \bar{Q}^* \to \bar{Q}$ is symmetric, i.e., $\bar{\mathbb{K}}(\bar{\boldsymbol{q}}) = \bar{\mathbb{K}}^*(\bar{\boldsymbol{q}})$ for all $\bar{\boldsymbol{q}} \in \bar{Q}$, and positively semi-definite, i.e., $\langle \bar{\boldsymbol{\xi}}, \bar{\mathbb{K}}(\bar{\boldsymbol{q}})\bar{\boldsymbol{\xi}} \rangle \geq 0$ for all $\bar{\boldsymbol{\xi}} \in \bar{Q}^*$. The triplet $(\bar{Q}, \bar{\mathscr{E}}, \bar{\mathbb{J}})$ forms a Hamiltonian system for reversible processes, whereas the triplet $(\bar{Q}, \bar{\mathscr{S}}, \bar{\mathbb{K}})$ forms an Onsager system for irreversible dissipative processes. The latter are also called gradient flows and here we follow the nomenclature of Mielke (2011a), since this type of system ensures the Onsager reciprocity relations by design (Jordan et al. 1998; Onsager 1931a, b; Doi 2011; Peletier 2014). A GENERIC quintuple $(\bar{Q}, \bar{\mathscr{E}}, \bar{\mathscr{S}}, \bar{\mathbb{J}}, \bar{\mathbb{K}})$ couples the two via the so-called *Non-Interaction Conditions* (NIC), that can be stated as follows

$$\bar{\mathbb{J}}(\bar{\boldsymbol{q}})\mathrm{D}\bar{\mathscr{S}}(\bar{\boldsymbol{q}}) = 0 = \bar{\mathbb{K}}(\bar{\boldsymbol{q}})\mathrm{D}\bar{\mathscr{E}}(\bar{\boldsymbol{q}}) \tag{3.9a}$$

together with the mass conservation conditions

$$\bar{\mathbb{J}}(\bar{\boldsymbol{q}})\mathrm{D}\bar{\mathscr{M}}(\bar{\boldsymbol{q}}) = 0 = \bar{\mathbb{K}}(\bar{\boldsymbol{q}})\mathrm{D}\bar{\mathscr{M}}(\bar{\boldsymbol{q}}), \tag{3.9b}$$

for the total mass of the system $\bar{\mathscr{M}} : \bar{Q} \to \mathbb{R}$. Altogether, (3.9a) and (3.9b) are called degeneracy conditions.

Above in (3.9) the letter D denotes the Fréchet derivative of the functionals, which is, for a functional $\bar{\mathscr{A}} : \bar{Q} \to \mathbb{R}$ defined by

$$\langle \mathrm{D}\bar{\mathscr{A}}(\bar{\boldsymbol{q}}), \bar{\boldsymbol{\xi}} \rangle := \lim_{0<h\to 0} \tfrac{1}{h} \left(\bar{\mathscr{A}}(\bar{\boldsymbol{q}} + h\bar{\boldsymbol{\xi}}) - \bar{\mathscr{A}}(\bar{\boldsymbol{q}}) \right) , \tag{3.10}$$

where $\mathrm{D}\bar{\mathscr{A}}(\bar{\boldsymbol{q}}) \in \bar{Q}^*$. We similarly introduce the Fréchet derivative $\mathrm{DT}(\bar{\boldsymbol{q}})$ for general maps $\mathrm{T} : \bar{Q} \to Q$. Throughout this paper we work with integral functionals $\bar{\mathscr{A}} : \bar{Q} \to \mathbb{R}$ that act on the state $\bar{\boldsymbol{q}} \in \bar{Q}$. The state is a map $\bar{\boldsymbol{q}} : [0, \mathsf{T}] \times \bar{\Omega} \to \mathbb{R}^k$, $k \in \mathbb{N}$. With a density $\bar{\mathcal{A}} : \bar{\Omega} \times \mathbb{R}^k \times \mathbb{R}^{k\times d} \to \mathbb{R}$, the integral functional $\bar{\mathscr{A}}$ is given by

$$\bar{\mathscr{A}}(\bar{\boldsymbol{q}}) := \int_{\bar{\Omega}} \bar{\mathcal{A}}[\bar{\boldsymbol{q}}] \, \mathrm{d}\bar{x} \, .$$

where we used the abbreviation $\bar{\mathcal{A}}[\bar{\boldsymbol{q}}] := \bar{\mathcal{A}}(\bar{x}, \bar{\boldsymbol{q}}, \nabla\bar{\boldsymbol{q}})$ to simplify notation. Based on this representation, we will use for both the functional derivative (3.10) and for the derivative of the density the symbol D, so that for all $\bar{\boldsymbol{q}}, \bar{\boldsymbol{\xi}} \in \bar{Q}$

$$\begin{aligned} \left\langle \mathrm{D}\bar{\mathscr{A}}(\bar{\boldsymbol{q}}), \bar{\boldsymbol{\xi}} \right\rangle &= \int_{\bar{\Omega}} \mathrm{D}\bar{\mathcal{A}}[\bar{\boldsymbol{q}}] \cdot \bar{\boldsymbol{\xi}} \, \mathrm{d}\bar{x} \, , \quad \text{where} \\ \mathrm{D}\bar{\mathcal{A}}[\bar{\boldsymbol{q}}] \cdot \bar{\boldsymbol{\xi}} &= \left(\partial_{\bar{\boldsymbol{q}}} \bar{\mathcal{A}}(\bar{x}, \bar{\boldsymbol{q}}, \nabla\bar{\boldsymbol{q}}) - \nabla \cdot \partial_{\nabla\bar{\boldsymbol{q}}} \bar{\mathcal{A}}(\bar{x}, \bar{\boldsymbol{q}}, \nabla\bar{\boldsymbol{q}}) \right) \cdot \bar{\boldsymbol{\xi}} \, . \end{aligned} \tag{3.11}$$

Based on above definitions for a GENERIC system $(\bar{Q}, \bar{\mathscr{E}}, \bar{\mathscr{S}}, \bar{\mathbb{J}}, \bar{\mathbb{K}})$ the evolution of a state vector $\bar{\boldsymbol{q}} : [0, \mathsf{T}] \to \bar{Q}$ is governed by the following equation

$$\partial_t \bar{\boldsymbol{q}} = \bar{\mathbb{J}}(\bar{\boldsymbol{q}}) \mathrm{D}\bar{\mathscr{E}}(\bar{\boldsymbol{q}}) + \bar{\mathbb{K}}(\bar{\boldsymbol{q}}) \mathrm{D}\bar{\mathscr{S}}(\bar{\boldsymbol{q}}) \, . \tag{3.12}$$

Conditions (3.9) together with the properties of the operators $\bar{\mathbb{J}}$ and $\bar{\mathbb{K}}$ ensure the compatibility of the system with the laws of thermodynamics. In fact, by assuming that the chain rule holds true, for every solution to (3.12) one has the following inequality for the time derivative of the entropy

$$\begin{aligned} \frac{\mathrm{d}}{\mathrm{d}t} \bar{\mathscr{S}}(\bar{\boldsymbol{q}}(t)) &= \left\langle \mathrm{D}\bar{\mathscr{S}}(\bar{\boldsymbol{q}}), \partial_t \bar{\boldsymbol{q}} \right\rangle = \left\langle \mathrm{D}\bar{\mathscr{S}}(\bar{\boldsymbol{q}}), \bar{\mathbb{J}}(\bar{\boldsymbol{q}}) \mathrm{D}\bar{\mathscr{E}}(\bar{\boldsymbol{q}}) + \bar{\mathbb{K}}(\bar{\boldsymbol{q}}) \mathrm{D}\bar{\mathscr{S}}(\bar{\boldsymbol{q}}) \right\rangle \\ &= \left\langle \mathrm{D}\bar{\mathscr{E}}(\bar{\boldsymbol{q}}), -\bar{\mathbb{J}}(\bar{\boldsymbol{q}}) \mathrm{D}\bar{\mathscr{S}}(\bar{\boldsymbol{q}}) \right\rangle + \left\langle \mathrm{D}\bar{\mathscr{S}}(\bar{\boldsymbol{q}}), \bar{\mathbb{K}}(\bar{\boldsymbol{q}}) \mathrm{D}\bar{\mathscr{S}}(\bar{\boldsymbol{q}}) \right\rangle \\ &= \left\langle \mathrm{D}\bar{\mathscr{S}}(\bar{\boldsymbol{q}}), \bar{\mathbb{K}}(\bar{\boldsymbol{q}}) \mathrm{D}\bar{\mathscr{S}}(\bar{\boldsymbol{q}}) \right\rangle \geq 0 \, . \end{aligned}$$

Analogously, one can prove that the total energy of the system is conserved for a solution of (3.12), i.e.,

$$\frac{\mathrm{d}}{\mathrm{d}t} \bar{\mathscr{E}}(\bar{\boldsymbol{q}}(t)) = 0 \, .$$

3.3 Nonequilibrium Thermomechanical Model with Diffusion

In this section we present a GENERIC model for thermo-viscoelastic materials. Firstly, we provide a Lagrangian formulation $(\bar{Q}, \bar{\mathscr{E}}, \bar{\mathscr{S}}, \bar{\mathbb{J}}, \bar{\mathbb{K}})$. Then we define Eulerian variables and perform a transformation of the functionals and related geometric structures resulting in a corresponding Eulerian GENERIC system $(Q, \mathscr{E}, \mathscr{S}, \mathbb{J}, \mathbb{K})$.

One of the main goals of this article is to construct two equivalent formulations of the physical system, one in terms of the Lagrangian or reference configuration and the other in terms of the Eulerian or actual configuration within the GENERIC framework. Here we collect notions and results about transformations and changes of coordinates for GENERIC systems that we will use later without any further proof. A detailed discussion and its relation to thermodynamics can be found in Zafferi et al. (2023a). The family of relevant transformations consists of Fréchet differentiable maps between two state spaces

$$\mathrm{T}_{\mathrm{LE}} : \bar{Q} \to Q \quad \text{with Fréchet derivative } \mathbb{L}(\bar{\boldsymbol{q}}) := \mathrm{DT}_{\mathrm{LE}}(\bar{\boldsymbol{q}}) . \tag{3.13}$$

Given two GENERIC systems $(\bar{Q}, \bar{\mathscr{E}}, \bar{\mathscr{S}}, \bar{\mathbb{J}}, \bar{\mathbb{K}})$, $(Q, \mathscr{E}, \mathscr{S}, \mathbb{J}, \mathbb{K})$, we say that the *closure conditions* are satisfied if

$$\mathscr{E}(\mathrm{T}_{\mathrm{LE}}(\bar{\boldsymbol{q}})) = \bar{\mathscr{E}}(\bar{\boldsymbol{q}}) , \quad \mathscr{S}(\mathrm{T}_{\mathrm{LE}}(\bar{\boldsymbol{q}})) = \bar{\mathscr{S}}(\bar{\boldsymbol{q}}) , \quad \mathscr{M}(\mathrm{T}_{\mathrm{LE}}(\bar{\boldsymbol{q}})) = \bar{\mathscr{M}}(\bar{\boldsymbol{q}}) \tag{3.14a}$$

$$\mathbb{J}(\mathrm{T}_{\mathrm{LE}}(\bar{\boldsymbol{q}})) = \mathbb{L}(\bar{\boldsymbol{q}})\bar{\mathbb{J}}(\bar{\boldsymbol{q}})\mathbb{L}^*(\bar{\boldsymbol{q}}) , \tag{3.14b}$$

$$\mathbb{K}(\mathrm{T}_{\mathrm{LE}}(\bar{\boldsymbol{q}})) = \mathbb{L}(\bar{\boldsymbol{q}})\bar{\mathbb{K}}(\bar{\boldsymbol{q}})\mathbb{L}^*(\bar{\boldsymbol{q}}) , \tag{3.14c}$$

with $\mathbb{L}^*(\bar{\boldsymbol{q}})$ the adjoint operator of $\mathbb{L}(\bar{\boldsymbol{q}})$. This transformation mechanism is shown in the following diagram 3.2, which illustrates the existence of a transformed structure $(Q, \mathscr{E}, \mathscr{S}, \mathbb{J}, \mathbb{K})$ whose solutions agree with the mapped solutions $\boldsymbol{q}(t) = \mathrm{T}_{\mathrm{LE}}\big(\bar{\boldsymbol{q}}(t)\big)$. In this sense, the *closure conditions* specify the requirements needed in order to obtain a commutative diagram between the two GENERIC systems as in Fig. 3.2. Therefore *closure* here does not refer to the concept of thermodynamically closed systems, but rather the idea of *closure* used in model reduction, where a necessary assumption or a specific pick of the system is required to make the reduction valid.

3.3.1 Lagrangian GENERIC System

States. We consider a time-evolving thermo-viscoelastic solid material in the reference configuration, described by the vector of state variables

$$
\begin{array}{ccc}
\partial_t \bar{\boldsymbol{q}} = \bar{\mathbb{J}}(\bar{\boldsymbol{q}})\mathrm{D}\bar{\mathcal{E}}(\bar{\boldsymbol{q}}) + \bar{\mathbb{K}}(\bar{\boldsymbol{q}})\mathrm{D}\bar{\mathcal{S}}(\bar{\boldsymbol{q}}) & \xrightarrow{\text{solution}} & \bar{\boldsymbol{q}}(t) \in \bar{Q} \\
\Big\downarrow \mathrm{L} & & \Big\downarrow \mathrm{T}_{\mathrm{LE}} \\
\partial_t \boldsymbol{q} = \mathbb{J}(\boldsymbol{q})\mathrm{D}\mathcal{E}(\boldsymbol{q}) + \mathbb{K}(\boldsymbol{q})\mathrm{D}\mathcal{S}(\boldsymbol{q}) & \xrightarrow{\text{solution}} & \boldsymbol{q}(t) \in Q
\end{array}
$$

Fig. 3.2 Diagram showing the mapping relations between two GENERIC systems $(\bar{Q}, \bar{\mathscr{E}}, \bar{\mathscr{S}}, \bar{\mathbb{J}}, \bar{\mathbb{K}})$ and $(Q, \mathscr{E}, \mathscr{S}, \mathbb{J}, \mathbb{K})$ and between their respective solutions $\bar{\boldsymbol{q}}, \boldsymbol{q}$

$$
\bar{\boldsymbol{q}} := \begin{pmatrix} \boldsymbol{\chi} : \bar{\Omega} \to \mathbb{R}^d \\ \bar{\boldsymbol{m}} : \bar{\Omega} \to \mathbb{R}^d \\ \bar{\boldsymbol{c}} : \bar{\Omega} \to \mathbb{R}^N \\ \bar{\theta} : \bar{\Omega} \to \mathbb{R} \end{pmatrix} \in \bar{Q}, \tag{3.15}
$$

with the deformation $\boldsymbol{\chi}$ from Sect. 3.2.1, the linear momentum $\bar{\boldsymbol{m}}$, the vector of concentrations $\bar{\boldsymbol{c}}$ of N species and the temperature $\bar{\theta}$. We assume the existence of a constant N-tuple of molar masses $\boldsymbol{M} \in \mathbb{R}^N$, such that the mass density $\bar{\varrho} : \bar{\Omega} \to \mathbb{R}$ can be expressed as a function of $\bar{\boldsymbol{c}}$ via

$$
\bar{\varrho}(t, \bar{x}) = \bar{\varrho}_0(\bar{x}) + \bar{\boldsymbol{c}}(t, \bar{x}) \cdot \boldsymbol{M}\,. \tag{3.16}
$$

Condition (3.9b) imposes a strict restriction on the evolution of the function $\bar{\boldsymbol{c}}$ and its relation with $\boldsymbol{M}$. In fact, the operators $\bar{\mathbb{J}}$ and $\bar{\mathbb{K}}$ have to be constructed in such a way that (3.9b) is satisfied, which in our barycentric setting enforces the condition $\bar{\boldsymbol{c}}(t, \bar{x}) \cdot \boldsymbol{M} = 0$ for any $(t, \bar{x}) \in [0, \mathrm{T}] \times \bar{\Omega}$.

The linear momentum relates to the velocity field by

$$
\bar{\boldsymbol{m}}(t, \bar{x}) := \bar{\varrho}(t, \bar{x})\bar{\boldsymbol{v}}(t, \bar{x}) \;\; \text{for every } (t, \bar{x}) \in [0, \mathrm{T}] \times \bar{\Omega}\,.
$$

The choice of the temperature as the "thermodynamic" variable is clearly not exclusive but it is made here in order to compare our result with the literature and to be closer to thermodynamic potentials that are used in this scenario, see Eq. (3.7) and Remarks 3.2, 3.3. In the Lagrangian and Eulerian formulation of this system we will not perform a change of the thermodynamic variable to $\bar{u}$ or $\bar{s}$ and thus we drop the $\bar{\theta}$-subscript from the state variable from now on.

Driving functionals. We consider the total energy functional $\bar{\mathscr{E}} : \bar{Q} \to \mathbb{R}$ as the sum of the kinetic energy $\bar{\mathscr{K}} : \bar{Q} \to \mathbb{R}$ and the internal energy $\bar{\mathscr{U}} : \bar{Q} \to \mathbb{R}$, i.e.,

$$\bar{\mathscr{E}}(\bar{\boldsymbol{q}}) := \bar{\mathscr{K}}(\bar{\boldsymbol{q}}) + \bar{\mathscr{U}}(\bar{\boldsymbol{q}}) \qquad \text{with} \tag{3.17a}$$

$$\bar{\mathscr{K}}(\bar{\boldsymbol{q}}) := \int_{\bar{\Omega}} \bar{\mathcal{K}}[\bar{\boldsymbol{q}}]\,\mathrm{d}\bar{x} = \int_{\bar{\Omega}} \frac{|\bar{\boldsymbol{m}}|^2}{2\bar{\varrho}(t,\bar{x})}\,\mathrm{d}\bar{x} \qquad \text{kinetic energy,} \tag{3.17b}$$

$$\bar{\mathscr{U}}(\bar{\boldsymbol{q}}) := \int_{\bar{\Omega}} \bar{\mathcal{U}}[\bar{\boldsymbol{q}}]\,\mathrm{d}\bar{x} = \int_{\bar{\Omega}} \bar{\mathcal{U}}(\bar{x},\bar{\boldsymbol{F}},\bar{\boldsymbol{c}},\bar{\theta})\,\mathrm{d}\bar{x} \qquad \text{internal energy.} \tag{3.17c}$$

We emphasize here the $\bar{\theta}$-dependence of the internal energy density $\bar{\mathcal{U}}$, in view of the computation of the driving force $\mathrm{D}\bar{\mathscr{E}}$ and the verification of the non-interaction conditions (3.9a). Moreover, we highlight the dependence of $\bar{\mathcal{U}}$ on the deformation gradient $\bar{\boldsymbol{F}}$, since this is a feature of solid materials, as discussed previously. Similarly to the total energy, we consider an entropy functional of the following form:

$$\bar{\mathscr{S}}(\bar{\boldsymbol{q}}) := \int_{\bar{\Omega}} \bar{\mathcal{S}}[\bar{\boldsymbol{q}}]\,\mathrm{d}\bar{x} = \int_{\bar{\Omega}} \bar{\mathcal{S}}(\bar{x},\bar{\boldsymbol{F}},\bar{\boldsymbol{c}},\bar{\theta})\,\mathrm{d}\bar{x}\,. \tag{3.18}$$

We now compute the Fréchet derivatives for the densities of the functionals in (3.17)

$$\mathrm{D}\bar{\mathcal{E}}(\bar{\boldsymbol{q}}) = \begin{pmatrix} \mathrm{D}_{\boldsymbol{\chi}}\bar{\mathcal{U}}[\bar{\boldsymbol{q}}] \\ \mathrm{D}_{\bar{\boldsymbol{m}}}\bar{\mathcal{K}}[\bar{\boldsymbol{q}}] \\ \mathrm{D}_{\bar{\boldsymbol{c}}}(\bar{\mathcal{K}}[\bar{\boldsymbol{q}}]+\bar{\mathcal{U}}[\bar{\boldsymbol{q}}]) \\ \mathrm{D}_{\bar{\theta}}\bar{\mathcal{U}}[\bar{\boldsymbol{q}}] \end{pmatrix} = \begin{pmatrix} -\bar{\nabla}\cdot\left(\partial_{\bar{\boldsymbol{F}}}\bar{\mathcal{U}}\right) \\ \bar{\boldsymbol{v}} \\ -\frac{|\bar{\boldsymbol{v}}|^2}{2} + \partial_{\bar{\boldsymbol{c}}}\bar{\mathcal{U}} \\ \bar{C}_{\bar{V}} \end{pmatrix}, \tag{3.19a}$$

where $\bar{C}_{\bar{V}}$ is the heat capacity. For the total entropy in (3.18) we get

$$\mathrm{D}\bar{\mathcal{S}}(\bar{\boldsymbol{q}}) = \begin{pmatrix} \mathrm{D}_{\boldsymbol{\chi}}\bar{\mathcal{S}}[\bar{\boldsymbol{q}}] \\ \mathrm{D}_{\bar{\boldsymbol{m}}}\bar{\mathcal{S}}[\bar{\boldsymbol{q}}] \\ \mathrm{D}_{\bar{\boldsymbol{c}}}\bar{\mathcal{S}}[\bar{\boldsymbol{q}}] \\ \mathrm{D}_{\bar{\theta}}\bar{\mathcal{S}}[\bar{\boldsymbol{q}}] \end{pmatrix} = \begin{pmatrix} -\bar{\nabla}\cdot\left(\partial_{\bar{\boldsymbol{F}}}\bar{\mathcal{S}}\right) \\ 0 \\ \partial_{\bar{\boldsymbol{c}}}\bar{\mathcal{S}} \\ \frac{\bar{C}_{\bar{V}}}{\bar{\theta}} \end{pmatrix}. \tag{3.19b}$$

Geometric structures. For reversible dynamics, we consider the Poisson operator

$$\bar{\mathbb{J}}(\bar{\boldsymbol{q}}) := \begin{pmatrix} 0 & \mathbb{I}_d & 0 & 0 \\ -\mathbb{I}_d & 0 & 0 & \bar{\mathbb{J}}^{\bar{\boldsymbol{m}}\bar{\theta}} \\ 0 & 0 & 0 & 0 \\ 0 & \bar{\mathbb{J}}^{\bar{\theta}\bar{\boldsymbol{m}}} & 0 & 0 \end{pmatrix}, \quad \text{with}$$

$$\bar{\mathbb{J}}^{\bar{\boldsymbol{m}}\bar{\theta}}(\bar{\boldsymbol{q}})[\bar{\eta}_{\bar{\theta}}] := -\bar{\nabla}\cdot\left(\bar{\eta}_{\bar{\theta}}\bar{\theta}\bar{C}_{\bar{V}}^{-1}\partial_{\bar{\boldsymbol{F}}}\bar{\mathcal{S}}\right), \quad \bar{\mathbb{J}}^{\bar{\theta}\bar{\boldsymbol{m}}}(\bar{\boldsymbol{q}})[\bar{\boldsymbol{\eta}}_{\bar{\boldsymbol{m}}}] := -\bar{\theta}\bar{C}_{\bar{V}}^{-1}\bar{\nabla}\bar{\boldsymbol{\eta}}_{\bar{\boldsymbol{m}}} : \partial_{\bar{\boldsymbol{F}}}\bar{\mathcal{S}}\,, \tag{3.20}$$

and where $\bar{\boldsymbol{\eta}} = (\bar{\boldsymbol{\eta}}_{\boldsymbol{\chi}}, \bar{\boldsymbol{\eta}}_{\bar{\boldsymbol{m}}}, \bar{\boldsymbol{\eta}}_{\bar{\boldsymbol{c}}}, \bar{\eta}_{\bar{\theta}})$ is used here as a placeholder for the functional derivatives. This operator conforms with the non-canonical reversible operator for fluids and it can be derived from the canonical Poisson operator, e.g., see Morrison (1998), by means of the transformation and closure conditions (3.14b), when mapping the entropy $\bar{s}$ to the temperature $\bar{\theta}$. A detailed transformation is carried out in Zafferi et al. (2023a). We can readily verify that $\bar{\mathbb{J}}$ is linear and skew-symmetric, and that it satisfies the non-interaction conditions (3.9a), i.e., $\bar{\mathbb{J}}(\bar{\boldsymbol{q}})\mathrm{D}\bar{\mathscr{S}}(\bar{\boldsymbol{q}}) = 0$, where one needs to use the thermodynamic relation $\partial_{\bar{\theta}}\bar{\mathcal{S}} = \bar{C}_{\bar{V}}/\bar{\theta}$.

The irreversible dynamics is characterized in our considered problem by three processes: heat conduction, diffusion of species, and viscous dissipation in the solid. In order to write a suitable operator we exploit the additive property of the Onsager systems. As introduced for reaction diffusion systems (Mielke 2011a), we split the Onsager operator $\bar{\mathbb{K}}$ into contributions and write $\bar{\mathbb{K}}(\bar{\boldsymbol{q}}) = \bar{\mathbb{K}}^{\text{heat}}(\bar{\boldsymbol{q}}) + \bar{\mathbb{K}}^{\text{diff}}(\bar{\boldsymbol{q}}) + \bar{\mathbb{K}}^{\text{vi}}(\bar{\boldsymbol{q}})$. Heat conduction is modeled

$$\bar{\mathbb{K}}^{\text{heat}}(\bar{\boldsymbol{q}}) := \begin{pmatrix} 0\,0\,0 & 0 \\ 0\,0\,0 & 0 \\ 0\,0\,0 & 0 \\ 0\,0\,0 & \bar{\mathbb{K}}^{\text{heat}}_{\bar{\theta}\bar{\theta}}(\bar{\boldsymbol{q}}) \end{pmatrix}, \qquad \text{where} \tag{3.21a}$$

$$\bar{\mathbb{K}}^{\text{heat}}_{\bar{\theta}\bar{\theta}}(\bar{\boldsymbol{q}})[\bar{\eta}_{\bar{\theta}}] := -\frac{1}{\bar{C}_{\bar{V}}(\bar{\boldsymbol{q}})}\bar{\nabla}\cdot\left[\operatorname{Cof}(\bar{\boldsymbol{F}})\bar{\theta}^2\bar{k}_{\text{heat}}(\bar{\boldsymbol{q}})\bar{\boldsymbol{F}}^{-\top}\bar{\nabla}\left(\frac{\bar{\eta}_{\bar{\theta}}}{\bar{C}_{\bar{V}}}\right)\right], \tag{3.21b}$$

with $\bar{k} = \bar{k}_{\text{heat}}(\bar{\boldsymbol{q}}) > 0$ and the cofactor matrix $\operatorname{Cof}(\bar{\boldsymbol{F}}) = \det(\bar{\boldsymbol{F}})\bar{\boldsymbol{F}}^{-\top}$. The dependence on $\bar{\boldsymbol{F}}$ will become clearer later, when mapping to Eulerian coordinates, since it will be used to relate Lagrangian gradient $\bar{\nabla}$ to Eulerian ∇. Symmetry and positive semi-definiteness are readily verified and by setting $\bar{\eta}_{\bar{\theta}} = \mathrm{D}_{\bar{\theta}}\bar{\mathcal{E}} \equiv \partial_{\bar{\theta}}\bar{\mathcal{U}} = \bar{C}_{\bar{V}}$, one sees immediately that the NIC $\bar{\mathbb{K}}^{\text{heat}}\mathrm{D}\bar{\mathscr{E}} \equiv \mathbf{0}$ is satisfied.

The contribution due to diffusion of species, although different in nature, shares a structure similar to (3.21). We set

$$\bar{\mathbb{K}}^{\text{diff}}(\bar{\boldsymbol{q}}) := \begin{pmatrix} 0\,0 & 0 & 0 \\ 0\,0 & 0 & 0 \\ 0\,0 & \bar{\mathbb{K}}^{\text{diff}}_{\bar{c}\bar{c}}(\bar{\boldsymbol{q}}) & 0 \\ 0\,0 & 0 & 0 \end{pmatrix}, \tag{3.22a}$$

$$\bar{\mathbb{K}}^{\text{diff}}_{\bar{c}\bar{c}}(\bar{\boldsymbol{q}}) := -\bar{\nabla}\cdot\left[\bar{\mathbb{M}}(\bar{\boldsymbol{q}})\bar{\nabla}\left(\bar{\eta}_{\bar{c}}\right)\bar{\boldsymbol{F}}^{-1}\operatorname{Cof}(\bar{\boldsymbol{F}})\right]. \tag{3.22b}$$

Here $\bar{\mathbb{M}}(\bar{\boldsymbol{q}}) \in \mathbb{R}^{N\times N}$ is the mobility matrix. In order to ensure mass conservation and (3.9a) we ask the N-tuple of molar masses $\boldsymbol{M}$ to lie in the kernel of $\bar{\mathbb{M}}(\bar{\boldsymbol{q}})$. The construction of diffusion coefficients satisfying this requirement is explored in Zafferi et al. (2023a).

The viscous dissipation operator together with the driving force $\mathrm{D}\bar{\mathscr{S}}$ generates the viscous stress $\bar{\boldsymbol{\Sigma}}^{\text{vi}}$. However, finding physically acceptable viscous stress can be an insidious task, as pointed out in Antman (1998). We tackle this problem similarly to the heat conduction and construct the viscous Onsager operator by "pulling back" the Eulerian Onsager operator for a Stokes fluid, which is defined, e.g., in Pavelka et al. (2018, Ch. 4.6). We recall the notation for the symmetric gradient $\bar{\nabla}_{\text{s}}\bar{\boldsymbol{a}} :=$

$\frac{1}{2}(\bar{\nabla}\bar{\boldsymbol{a}}\bar{\boldsymbol{F}}^{-1} + \bar{\boldsymbol{F}}^{-\top}\bar{\nabla}\bar{\boldsymbol{a}}^\top)$ and introduce a positive semi-definite fourth-order tensor $\bar{\boldsymbol{\Lambda}}(\bar{\boldsymbol{q}})$. We then represent the viscous Onsager operator $\bar{\mathbb{K}}^{\mathrm{vi}}$ in the following block-matrix structure:

$$\bar{\mathbb{K}}^{\mathrm{vi}}(\bar{\boldsymbol{q}}) := \begin{pmatrix} 0 & 0 & 0 & 0 \\ 0 & \bar{\mathbb{K}}^{\mathrm{vi}}_{\bar{\boldsymbol{m}}\bar{\boldsymbol{m}}}(\bar{\boldsymbol{q}}) & 0 & \bar{\mathbb{K}}^{\mathrm{vi}}_{\bar{\boldsymbol{m}}\bar{\theta}}(\bar{\boldsymbol{q}}) \\ 0 & 0 & 0 & 0 \\ 0 & \bar{\mathbb{K}}^{\mathrm{vi}}_{\bar{\theta}\bar{\boldsymbol{m}}}(\bar{\boldsymbol{q}}) & 0 & \bar{\mathbb{K}}^{\mathrm{vi}}_{\bar{\theta}\bar{\theta}}(\bar{\boldsymbol{q}}) \end{pmatrix}, \tag{3.23a}$$

$$\bar{\mathbb{K}}^{\mathrm{vi}}_{\bar{\boldsymbol{m}}\bar{\boldsymbol{m}}}(\bar{\boldsymbol{q}})[\bar{\boldsymbol{\eta}}_{\bar{\boldsymbol{m}}}] := -\bar{\nabla}\cdot\left[\bar{\theta}\bar{\boldsymbol{\Lambda}}(\bar{\boldsymbol{q}}) : \bar{\nabla}_{\mathrm{s}}\bar{\boldsymbol{\eta}}_{\bar{\boldsymbol{m}}}\mathrm{Cof}(\bar{\boldsymbol{F}})\right], \tag{3.23b}$$

$$\bar{\mathbb{K}}^{\mathrm{vi}}_{\bar{\boldsymbol{m}}\bar{\theta}}(\bar{\boldsymbol{q}})[\bar{\eta}_{\bar{\theta}}] := \bar{\nabla}\cdot\left[\bar{\theta}\bar{\boldsymbol{\Lambda}}(\bar{\boldsymbol{q}}) : \bar{\nabla}_{\mathrm{s}}\bar{\boldsymbol{v}}\mathrm{Cof}(\bar{\boldsymbol{F}})\frac{\bar{\eta}_{\bar{\theta}}}{\bar{C}_{\bar{V}}}\right], \tag{3.23c}$$

$$\bar{\mathbb{K}}^{\mathrm{vi}}_{\bar{\theta}\bar{\boldsymbol{m}}}(\bar{\boldsymbol{q}})[\bar{\boldsymbol{\eta}}_{\bar{\boldsymbol{m}}}] := -\frac{1}{\bar{C}_{\bar{V}}}\bar{\nabla}_{\mathrm{s}}\bar{\boldsymbol{v}} : \bar{\theta}\bar{\boldsymbol{\Lambda}}(\bar{\boldsymbol{q}}) : \bar{\nabla}_{\mathrm{s}}\bar{\boldsymbol{\eta}}_{\bar{\boldsymbol{m}}}\bar{V}, \tag{3.23d}$$

$$\bar{\mathbb{K}}^{\mathrm{vi}}_{\bar{\theta}\bar{\theta}}(\bar{\boldsymbol{q}})[\bar{\eta}_{\bar{\theta}}] := \frac{\bar{\eta}_{\bar{\theta}}}{\bar{C}_{\bar{V}}}\bar{\nabla}_{\mathrm{s}}\bar{\boldsymbol{v}} : \bar{\theta}\bar{\boldsymbol{\Lambda}}(\bar{\boldsymbol{q}}) : \bar{\nabla}_{\mathrm{s}}\bar{\boldsymbol{v}}\,\bar{V}. \tag{3.23e}$$

The operator so defined satisfies the symmetry condition, i.e., $\bar{\mathbb{K}} \equiv \bar{\mathbb{K}}^*$ and the positive semi-definiteness, i.e., $\langle\bar{\boldsymbol{\eta}}, \bar{\mathbb{K}}^{\mathrm{vi}}\bar{\boldsymbol{\eta}}\rangle \geq 0$ for all $\bar{\boldsymbol{\eta}} \in \bar{Q}^*$. As before, by taking $\bar{\boldsymbol{\eta}}_{\bar{\boldsymbol{m}}} = \mathrm{D}_{\bar{\boldsymbol{m}}}\bar{\mathscr{E}} \equiv \bar{\boldsymbol{v}}$ and $\bar{\eta}_{\bar{\theta}} = \mathrm{D}_{\bar{\theta}}\bar{\mathscr{E}} \equiv \bar{C}_{\bar{V}}$ we check that the non-interaction condition $\bar{\mathbb{K}}^{\mathrm{vi}}\mathrm{D}\bar{\mathscr{E}} \equiv \boldsymbol{0}$ is satisfied. In order to obtain the Stokes viscosity, we choose $\bar{\boldsymbol{\Lambda}}$ as

$$\bar{\boldsymbol{\Lambda}}(\bar{\boldsymbol{q}}) := 2\bar{\lambda}\mathbb{I}^4_d + \bar{\xi}\left(\mathbb{I}^2_d \otimes \mathbb{I}^2_d\right), \tag{3.23f}$$

where $\bar{\lambda} > 0$ and $\bar{\xi} + \frac{2\bar{\lambda}}{d} > 0$ are the shear and bulk viscosity, respectively, and might be state-dependent. The fourth- and second-order identity tensors are denoted by $\mathbb{I}^4_d$ and $\mathbb{I}^2_d$. With this choice of $\bar{\boldsymbol{\Lambda}}$, the viscous Onsager operator for fluids is retrieved. The product of a fourth-order tensor A with a second-order tensor B gives the second-order tensor with components in the following sense $(A : B)_{ij} = A_{ijkl}B_{kl}$.

By collecting the above definitions we have a GENERIC quintuple $(\bar{Q}, \bar{\mathscr{E}}, \bar{\mathscr{S}}, \bar{\mathbb{J}}, \bar{\mathbb{K}})$ and by applying the operators (3.20)–(3.23) to the driving forces (3.19), the evolution law $\partial_t\bar{\boldsymbol{q}} = \bar{\mathbb{J}}(\bar{\boldsymbol{q}})\mathrm{D}\bar{\mathscr{E}}(\bar{\boldsymbol{q}}) + \bar{\mathbb{K}}(\bar{\boldsymbol{q}})\mathrm{D}\bar{\mathscr{S}}(\bar{\boldsymbol{q}})$ results in the PDE system

$$\partial_t\boldsymbol{\chi} = \frac{\bar{\boldsymbol{m}}}{\bar{\varrho}}, \tag{3.24a}$$

$$\partial_t\bar{\boldsymbol{m}} = \bar{\nabla}\cdot\left(\bar{\boldsymbol{\Sigma}}^{\mathrm{el}} + \bar{\boldsymbol{\Sigma}}^{\mathrm{vi}}\right) - \bar{\boldsymbol{W}}, \tag{3.24b}$$

$$\partial_t\bar{\boldsymbol{c}} = -\bar{\nabla}\cdot\left[\mathrm{Cof}(\bar{\boldsymbol{F}})\bar{\mathbb{M}}\bar{\boldsymbol{F}}^{-\top}\bar{\nabla}(\partial_{\bar{\boldsymbol{c}}}\bar{\mathcal{S}})\right] \tag{3.24c}$$

$$\bar{C}_{\bar{V}}\partial_t\bar{\theta} = \bar{\theta}\bar{\nabla}\bar{\boldsymbol{v}} : \partial_{\bar{\boldsymbol{F}}}\bar{\mathcal{S}} + \bar{\nabla}_{\mathrm{s}}\bar{\boldsymbol{v}} : \bar{\boldsymbol{\Sigma}}^{\mathrm{vi}}\bar{\boldsymbol{F}}^\top + \bar{\nabla}\cdot\left(\mathrm{Cof}(\bar{\boldsymbol{F}})\bar{k}\bar{\boldsymbol{F}}^{-\top}\bar{\nabla}\bar{\theta}\right), \tag{3.24d}$$

where $\bar{\boldsymbol{\Sigma}}^{\mathrm{vi}} := \bar{\boldsymbol{\Lambda}}\bar{\nabla}_{\mathrm{s}}\bar{\boldsymbol{v}}\mathrm{Cof}(\bar{\boldsymbol{F}})$ is the viscous stress and $\bar{\boldsymbol{\Sigma}}^{\mathrm{el}} := \partial_{\bar{\boldsymbol{F}}}(\bar{\mathcal{U}} - \bar{\theta}\bar{\mathcal{S}})$. By including the second gradient of the deformation $\boldsymbol{\chi}$, i.e., $\bar{\nabla}\bar{\boldsymbol{F}}$ in the entropy and energy we would extend our model to the description of the so-called *2nd-grade non-simple materials* and by taking the over-damped limit we recover the equations and the structures presented in Mielke and Roubíček (2020).

Remark 3.1 (*Frame indifference of stresses*) One of the requirements for having a physically consistent model is for the elastic and viscous stresses to be invariant under rigid motions. For the elastic stress tensor $\bar{\boldsymbol{\Sigma}}^{\mathrm{el}}$, this is ensured if, for example, the internal energy $\bar{\mathcal{U}}$ and the entropy $\bar{\mathcal{S}}$ depend on $\bar{\boldsymbol{F}}$ only through the right Cauchy-Green tensor $\bar{\boldsymbol{F}}^{\top}\bar{\boldsymbol{F}}$, see Ciarlet (1988, Theorem 4.2-1) for more details. For the viscous stress tensor being frame indifferent is equivalent to satisfy the following condition for some $\tilde{\boldsymbol{\Sigma}} : \mathbb{R}^{d\times d} \times \mathbb{R}^{d\times d} \to \mathbb{R}^{d\times d}$, cf. Mielke et al. (2014):

$$\bar{\boldsymbol{\Sigma}}^{\mathrm{vi}}(\bar{\boldsymbol{F}}, \partial_t \bar{\boldsymbol{F}}) = \bar{\boldsymbol{F}}\tilde{\boldsymbol{\Sigma}}(\bar{\boldsymbol{F}}^{\top}\bar{\boldsymbol{F}}, \bar{\boldsymbol{F}}^{\top}\partial_t\bar{\boldsymbol{F}} + \partial_t\bar{\boldsymbol{F}}^{\top}\bar{\boldsymbol{F}}).$$

For dissipation structures as in (3.23f) one can see that this condition is satisfied for $\tilde{\boldsymbol{\Sigma}}(\bar{\boldsymbol{F}}, \partial_t\bar{\boldsymbol{F}}) := \bar{\mathbb{A}}(\bar{q})(\bar{\boldsymbol{F}}^{\top}\bar{\boldsymbol{F}})^{-1}\frac{1}{2}(\bar{\boldsymbol{F}}^{\top}\partial_t\bar{\boldsymbol{F}} + \partial_t\bar{\boldsymbol{F}}^{\top}\bar{\boldsymbol{F}})(\bar{\boldsymbol{F}}^{\top}\bar{\boldsymbol{F}})^{-1}\sqrt{\det(\bar{\boldsymbol{F}}^{\top}\bar{\boldsymbol{F}})}$ where we recall that $\bar{\nabla}\bar{\boldsymbol{v}} = \partial_t\bar{\boldsymbol{F}}$. $*$

Remark 3.2 (*Hyperelastic material*) In this setting, the dependence of $\bar{\mathcal{U}}$ on $\bar{\boldsymbol{F}}$ is the main distinguishing trait in modeling elastic responses of the solid material. Following Ciarlet (1988, Chap. 4), the so-called Ogden's materials are described by a Helmholtz free energy, whose mechanical contribution is

$$\bar{\mathcal{F}}(\bar{\boldsymbol{F}}) := \sum_{i=1}^{M}\alpha_i \mathrm{tr}(\bar{\boldsymbol{F}}^{\top}\bar{\boldsymbol{F}})^{\gamma_i/2} + \sum_{j=1}^{N}\beta_j \mathrm{tr}(\mathrm{Cof}(\bar{\boldsymbol{F}}^{\top}\bar{\boldsymbol{F}}))(\bar{\boldsymbol{F}}^{\top})^{\delta_j/2} + \kappa(\det\bar{\boldsymbol{F}}),$$

where $\alpha_i > 0$, $\beta_j > 0$, $\gamma_i \geq 1$, $\delta_j \geq 1$, and $\kappa : (0, +\infty) \to \mathbb{R}$ is a convex function satisfying $\kappa(y) \to +\infty$ as $y \to 0^+$ and subjected to suitable growth as $y \to +\infty$. Other types of hyperelastic materials are the compressible Mooney-Rivlin materials, whose energy can be expressed as follows

$$\bar{\mathcal{F}}(\bar{\boldsymbol{F}}) := \alpha \mathrm{tr}|\bar{\boldsymbol{F}}|^2 + \beta|\mathrm{Cof}(\bar{\boldsymbol{F}})|^2 + \kappa(\det\bar{\boldsymbol{F}}),$$

where $\kappa(y) = cy^2 - d\log(y)$, $c \geq 0$, $d \geq 0$. If $\beta = 0$, then one recovers an equation for a compressible neo-Hookean material. With the function $\kappa(\cdot)$ related to volumetric variations and $c = 0$, $d = 0$ one recovers the incompressible limit.

Remark 3.3 (*Pure volumetric deformation*) Murnaghan's equation of state is widely used in geosciences, see, e.g., Holland and Powell (1998), to model the behavior of a solid under high pressures and it relates the pressure $\bar{\pi}$ to the volume $\bar{V}$, the temperature $\bar{\theta}$ and the bulk modulus $\kappa_{\bar{\theta}}$ through the following equation of state

$$\bar{\pi}(\bar{V}, \bar{\theta}) = \frac{\kappa_{\bar{\theta}}(\bar{\theta})}{\kappa'_{\bar{\theta}}}\left[\left(\frac{\bar{V}_{\mathrm{ref}}(\bar{\theta})}{\bar{V}}\right)^{\partial_{\bar{\pi}}\kappa_{\bar{\theta}}} - 1\right] \tag{3.25}$$

with $\bar{V}_{\rm ref}(\bar{\theta})$ being a reference value of the volume obtained from

$$\bar{V}_{\rm ref}(\bar{\theta}) = \bar{V}(\bar{\pi}_{\rm ref}, \bar{\theta}_{\rm ref}) \left[1 + a(\bar{\theta} - \bar{\theta}_{\rm ref}) - 20a(\sqrt{\bar{\theta}} - \sqrt{\bar{\theta}_{\rm ref}})\right],$$

where $\bar{\pi} = 1\text{bar}$, $\bar{\theta}_{\rm ref} = 298\text{K}$ and $a > 0$. We recall that the bulk modulus is defined through the following standard thermodynamic relation

$$\kappa_{\bar{\theta}} := -\bar{V} \left(\frac{\partial \bar{\pi}}{\partial \bar{V}}\right)_{\bar{\theta}}.$$

If we assume $\partial_{\bar{\pi}} \kappa_{\bar{\theta}}$ to be constant (usually one has $\partial_{\bar{\pi}} \kappa_{\bar{\theta}} \approx 4$) then the equation of state (3.25) is induced by a free energy density of the following type

$$\bar{\mathcal{F}}(\bar{V}, \bar{\theta}) = \frac{\kappa_{\bar{\theta}}(\bar{\theta})}{\partial_{\bar{\pi}} \kappa_{\bar{\theta}}} \bar{V} \left[\frac{1}{\partial_{\bar{\pi}} \kappa_{\bar{\theta}} - 1} \left(\frac{\bar{V}_{\rm ref}}{\bar{V}}\right)^{\partial_{\bar{\pi}} \kappa_{\bar{\theta}}} + 1\right] - \frac{\kappa_{\bar{\theta}}(\bar{\theta}) \bar{V}_{\rm ref}(\bar{\theta})}{\partial_{\bar{\pi}} \kappa_{\bar{\theta}} - 1}. \tag{3.26}$$

This fact can be easily checked by computing the pressure, which is given as the derivative of the free energy density $\bar{\mathcal{F}} = \bar{\mathcal{F}}(\bar{V}, \bar{\theta})$

$$\bar{\pi} = -\frac{\partial \bar{\mathcal{F}}}{\partial \bar{V}}.$$

Let us remark that this energy density involves the deformation gradient only through purely volumetric contributions via $\bar{V} = \det \bar{\boldsymbol{F}}$.

3.3.2 Eulerian GENERIC System $(\boldsymbol{Q}, \mathscr{E}, \mathscr{S}, \mathbb{J}, \mathbb{K})$

In this subsection we consider the GENERIC system $(\bar{\boldsymbol{Q}}, \bar{\mathscr{E}}, \bar{\mathscr{S}}, \bar{\mathbb{J}}, \bar{\mathbb{K}})$ previously defined in Sect. 3.3.1 and perform a change of variables from Lagrangian to Eulerian coordinates using the map $\mathrm{T}_{\rm LE}$ introduced in (3.13). We will use the reference map $\boldsymbol{\alpha}$ defined in (3.1) as one of the state variables. We recall that $\boldsymbol{\alpha}(t, \cdot)$ is the inverse of the deformation $\boldsymbol{\chi}(t)$ at time t, i.e., $\boldsymbol{\alpha}(t, \boldsymbol{\chi}(t, \bar{x})) = \bar{x}$ for every $(t, \bar{x}) \in [0, \mathrm{T}] \times \bar{\Omega}$. Eulerian fields will be denoted without a bar. For intensive variables, the Eulerian fields result by simply composing the Lagrangian field with $\boldsymbol{\alpha}$, as for $\boldsymbol{F} = \bar{\boldsymbol{F}} \circ \boldsymbol{\alpha}$, $V = \bar{V} \circ \boldsymbol{\alpha}$, $\boldsymbol{v} = \bar{\boldsymbol{v}} \circ \boldsymbol{\alpha}$, where $\circ$ denotes the composition of the spatial arguments of two functions, i.e., $\boldsymbol{v}(t, x) = (\bar{\boldsymbol{v}} \circ \boldsymbol{\alpha})(t, x) = \bar{\boldsymbol{v}}(t, \boldsymbol{\alpha}(t, x))$ for all $(t, x) \in [0, \mathrm{T}] \times \Omega$. For extensive variables the transformation additionally involves a division by $\bar{V}$, as can be seen in particular in Eqs. (3.27b) and (3.27c). The new Eulerian state vector $\boldsymbol{q}$ is defined through the map $\mathrm{T}_{\rm LE} : \bar{\boldsymbol{Q}} \to \boldsymbol{Q}$

$$q = \begin{pmatrix} \alpha : \Omega \to \mathbb{R}^d \\ m : \Omega \to \mathbb{R}^d \\ c : \Omega \to \mathbb{R}^N \\ \theta : \Omega \to \mathbb{R} \end{pmatrix} = \mathrm{T}_{\mathrm{LE}}(\bar{q}) \in Q, \text{ where} \tag{3.27a}$$

$$m = \frac{\bar{m}}{\bar{V}} \circ \alpha, \tag{3.27b}$$

$$c = \frac{\bar{c}}{\bar{V}} \circ \alpha, \tag{3.27c}$$

$$\theta = \bar{\theta} \circ \alpha. \tag{3.27d}$$

Here, based on the results of Zafferi et al. (2023a), momentum and concentrations are mapped as extensive variables, while the temperature as an intensive variable is transformed by simply applying the reference map. Since we are not going to perform any further change of variables, we omit the subscript that would mark the dependence on the temperature θ, as previously introduced in (3.2) and below. One can verify that the driving functionals (3.17) and (3.18) indeed satisfy the closure conditions (3.14a) under the mapping (3.27):

$$\mathscr{E}(q) := \mathscr{K}(q) + \mathscr{U}(q) \qquad \text{with} \tag{3.28a}$$

$$\mathscr{K}(q) := \int_\Omega \frac{|m|^2}{2\varrho}\,\mathrm{d}x, \text{ where } \varrho = \frac{\bar{\varrho} \circ \alpha}{V} + c \cdot M, \qquad \text{kinetic energy,} \tag{3.28b}$$

$$\mathscr{U}(q) := \int_\Omega \mathcal{U}(\alpha, F, c, \theta)\,\mathrm{d}x \qquad \text{internal energy,} \tag{3.28c}$$

$$\mathscr{S}(q) := \int_\Omega \mathcal{S}(\alpha, F, c, \theta)\,\mathrm{d}x \qquad \text{entropy,} \tag{3.28d}$$

where $\bar{\mathcal{U}}(\alpha, F, cV, \theta) = \mathcal{U}(\alpha, F, c, \theta)V$ and $\bar{\mathcal{S}}(\alpha, F, cV, \theta) = \mathcal{S}(\alpha, F, c, \theta)V$, cf. with the free energy (3.8). For the densities of the internal energy $\mathcal{U}$ and entropy $\mathcal{S}$ we have highlighted the α-dependency and distinguished it from F. This distinction will simplify the computation of their functional derivative and will clarify that only the derivatives of these functionals with respect to F contribute in the final PDE system. Furthermore, observe that the third component of $\mathcal{U}$ and $\mathcal{S}$ is $\bar{c} = cV$ from (3.27c). The driving forces are given by the Fréchet derivatives of (3.28):

$$\mathrm{D}\mathcal{E}[q] = \left(\mathrm{D}_\alpha\left(\frac{|m|}{2\varrho^2} + \frac{\bar{\mathcal{U}}}{V}\right), v, -\frac{|m|^2}{2\varrho^2}M + \partial_{\bar{c}}\bar{\mathcal{U}}, \frac{\partial_\theta \bar{\mathcal{U}}}{V}\right)^{\mathsf{T}}, \quad \text{where } \bar{c} = cV \tag{3.29a}$$

$$\mathrm{D}\mathcal{S}[q] = \left(\mathrm{D}_\alpha \mathcal{S}, 0, \partial_{\bar{c}}\bar{\mathcal{S}}, \frac{C_V}{\theta}\right)^{\mathsf{T}}. \tag{3.29b}$$

Following (3.13) and below, the Lagrangian rates $\partial_t \bar{q}$ are mapped through the derivative of the Lagrangian-Eulerian transformation:

$$
\mathrm{DT}_{\mathrm{LE}}(\bar{\boldsymbol{q}}) : \bar{\boldsymbol{Q}} \to \boldsymbol{Q}, \quad \mathrm{DT}_{\mathrm{LE}}(\bar{\boldsymbol{q}}) = \begin{pmatrix} -\nabla\boldsymbol{\alpha} & 0 & 0 & 0 \\ -\nabla\cdot(\bar{\boldsymbol{m}}\otimes\square) & \frac{\square}{\bar{V}} & 0 & 0 \\ -\nabla\cdot(\bar{\boldsymbol{c}}\otimes\square) & 0 & \frac{\square}{\bar{V}} & 0 \\ -\square\cdot\nabla\bar{\theta} & 0 & 0 & 1 \end{pmatrix}, \tag{3.30a}
$$

with $\square$ being a placeholder. This mapping induces the following relations between Lagrangian $\partial_t\bar{\boldsymbol{q}}$ and Eulerian velocities $\partial_t\boldsymbol{q}$:

$$
\begin{pmatrix} \partial_t\boldsymbol{\alpha} \\ \partial_t\boldsymbol{m} \\ \partial_t\boldsymbol{c} \\ \partial_t\theta \end{pmatrix} = \mathrm{DT}_{\mathrm{LE}}(\bar{\boldsymbol{q}}) \begin{pmatrix} \partial_t\boldsymbol{\chi} \\ \partial_t\bar{\boldsymbol{m}} \\ \partial_t\bar{\boldsymbol{c}} \\ \partial_t\bar{\theta} \end{pmatrix} = \begin{pmatrix} -(\nabla\boldsymbol{\alpha})\partial_t\boldsymbol{\chi} \\ -\bar{V}^{-1}\partial_t\bar{\boldsymbol{m}} - \nabla\cdot\left(\bar{V}^{-1}\bar{\boldsymbol{m}}\otimes\partial_t\boldsymbol{\chi}\right) \\ -\bar{V}^{-1}\partial_t\bar{\boldsymbol{c}} - \nabla\cdot\left(\bar{V}^{-1}\bar{\boldsymbol{c}}\otimes\partial_t\boldsymbol{\chi}\right) \\ \partial_t\bar{\theta} - \partial_t\boldsymbol{\chi}\cdot\nabla\bar{\theta} \end{pmatrix}, \tag{3.30b}
$$

where the Eulerian gradient is acting on Lagrangian variables at the point $\bar{x} = \boldsymbol{\alpha}(x)$. The adjoint operator $\mathrm{DT}_{\mathrm{LE}}{}^* : \boldsymbol{Q}^* \to \bar{\boldsymbol{Q}}^*$ maps driving forces and can be obtained by integration by parts of the expression $\langle\boldsymbol{\eta}, \mathrm{DT}_{\mathrm{LE}}\bar{\boldsymbol{v}}\rangle_{\boldsymbol{Q}} = \langle\mathrm{DT}_{\mathrm{LE}}{}^*\boldsymbol{\eta}, \bar{\boldsymbol{v}}\rangle_{\bar{\boldsymbol{Q}}}$ valid for any sufficiently smooth $\boldsymbol{\eta} \in \boldsymbol{Q}^*$ and any $\bar{\boldsymbol{v}} \in \bar{\boldsymbol{Q}}$, see (3.11) for the dual pairing. Using $\mathrm{DT}_{\mathrm{LE}}$ from (3.30a) this leads to

$$
\mathrm{DT}_{\mathrm{LE}}{}^*(\bar{\boldsymbol{q}}) = \begin{pmatrix} -\bar{V}\nabla\boldsymbol{\alpha}^\top & (\bar{\nabla}\square)^\top\bar{\boldsymbol{F}}^{-1}\bar{\boldsymbol{m}} & (\bar{\nabla}\square)^\top\bar{\boldsymbol{F}}^{-1}\bar{\boldsymbol{c}} & -\square\bar{V}\bar{\boldsymbol{F}}^{-\top}\bar{\nabla}\bar{\theta} \\ 0 & \square & 0 & 0 \\ 0 & 0 & \square & 0 \\ 0 & 0 & 0 & \bar{V}\square \end{pmatrix}. \tag{3.30c}
$$

The transformed geometric structures $\mathbb{J}(\boldsymbol{q}), \mathbb{K}(\boldsymbol{q}) : \boldsymbol{Q}^* \to \boldsymbol{Q}$ are directly computed using the transformation operators $\mathrm{DT}_{\mathrm{LE}}, \mathrm{DT}_{\mathrm{LE}}{}^*$ as follows

$$
\begin{aligned}
\mathbb{J}(\boldsymbol{q}) &= \mathrm{DT}_{\mathrm{LE}}(\bar{\boldsymbol{q}})\bar{\mathbb{J}}(\bar{\boldsymbol{q}})\mathrm{DT}_{\mathrm{LE}}{}^*(\bar{\boldsymbol{q}}) \\
&= \begin{pmatrix} 0 & -\nabla\boldsymbol{\alpha} & 0 & 0 \\ \nabla\boldsymbol{\alpha}^\top & -(\nabla\square)^\top\boldsymbol{m} - \nabla\cdot(\boldsymbol{m}\otimes\square) & -(\nabla\square)^\top\boldsymbol{c} & \nabla\theta + \frac{1}{V}\bar{\mathbb{J}}^{\bar{\boldsymbol{m}}\bar{\theta}}[\square V] \\ 0 & -\nabla\cdot(\boldsymbol{c}\otimes\square) & 0 & 0 \\ 0 & -\square\cdot\nabla\theta + \bar{\mathbb{J}}^{\bar{\theta}\bar{\boldsymbol{m}}} & 0 & 0 \end{pmatrix},
\end{aligned} \tag{3.31}
$$

where the momentum-temperature-related entries are obtained from (3.20) in terms of

$$
\frac{1}{V}\bar{\mathbb{J}}^{\bar{\boldsymbol{m}}\bar{\theta}}[\eta_\theta V] = -\nabla\cdot\left(\frac{\theta}{\bar{C}_V}\eta_\theta\partial_{\boldsymbol{F}}\boldsymbol{S}\boldsymbol{F}^\top\right), \quad \bar{\mathbb{J}}^{\bar{\theta}\bar{\boldsymbol{m}}}[\boldsymbol{\eta}_{\boldsymbol{m}}] = -\frac{\theta}{C_V}\nabla\boldsymbol{\eta}_{\boldsymbol{m}} : \frac{\partial_{\boldsymbol{F}}S_\theta\boldsymbol{F}^{-\top}}{V}
$$

by exploiting Piola's identity $\bar{\nabla}\cdot(\bar{V}\bar{\boldsymbol{F}}^{-\top}) = \boldsymbol{0}$ as well as the transformation rule $(\bar{\nabla}\cdot\bar{\boldsymbol{A}})\bar{\boldsymbol{F}}^{-\top} = \nabla\cdot(\bar{\boldsymbol{A}}\circ\boldsymbol{\alpha})$ for any $\bar{\boldsymbol{A}} \in \mathbb{R}^{d\times d}$.

Similarly, we obtain or the Onsager operator

$$\mathbb{K}(\boldsymbol{q}) = \mathrm{DT_{LE}}(\bar{\boldsymbol{q}})\bar{\mathbb{K}}(\bar{\boldsymbol{q}})\mathrm{DT_{LE}}^*(\bar{\boldsymbol{q}}) = \begin{pmatrix} 0 & 0 & 0 & 0 \\ 0 & \mathbb{K}_{\boldsymbol{mm}}(\boldsymbol{q}) & 0 & \mathbb{K}_{\boldsymbol{m}\theta}(\boldsymbol{q}) \\ 0 & 0 & \mathbb{K}_{\boldsymbol{cc}}(\boldsymbol{q}) & 0 \\ 0 & \mathbb{K}_{\theta\boldsymbol{m}}(\boldsymbol{q}) & 0 & \mathbb{K}_{\theta\theta}(\boldsymbol{q}) \end{pmatrix}, \tag{3.32a}$$

where the entries have been transformed to

$$\mathbb{K}_{\boldsymbol{mm}}(\boldsymbol{q})\boldsymbol{\eta}_{\boldsymbol{m}} = -\nabla \cdot \left[\theta \boldsymbol{\Lambda}(\boldsymbol{q}) \nabla_{\mathrm{s}} \boldsymbol{\eta}_{\boldsymbol{m}}\right], \tag{3.32b}$$

$$\mathbb{K}_{\boldsymbol{m}\theta}(\boldsymbol{q})\eta_\theta = \nabla \cdot \left[\theta \boldsymbol{\Lambda}(\boldsymbol{q}) \frac{\eta_\theta}{C_V} \nabla_{\mathrm{s}} \boldsymbol{v}\right], \tag{3.32c}$$

$$\mathbb{K}_{\boldsymbol{cc}}(\boldsymbol{q})\boldsymbol{\eta}_{\boldsymbol{c}} = -\nabla \cdot \left(\mathbb{M}(\boldsymbol{q})\boldsymbol{\eta}_{\boldsymbol{c}}\right), \tag{3.32d}$$

$$\mathbb{K}_{\theta\boldsymbol{m}}(\boldsymbol{q})\boldsymbol{\eta}_{\boldsymbol{m}} = -\frac{\theta}{C_V} \nabla_{\mathrm{s}} \boldsymbol{v} : \boldsymbol{\Lambda}(\boldsymbol{q}) : \frac{\eta_\theta}{C_V} \nabla_{\mathrm{s}} \boldsymbol{\eta}_{\boldsymbol{m}}, \tag{3.32e}$$

$$\mathbb{K}_{\theta\theta}(\boldsymbol{q})\eta_\theta = \frac{\theta}{C_V} \nabla_{\mathrm{s}} \boldsymbol{v} : \boldsymbol{\Lambda}(\boldsymbol{q}) : \nabla_{\mathrm{s}} \boldsymbol{v} - \frac{1}{C_V} \nabla \cdot \left(k_{\mathrm{heat}}(\boldsymbol{q})\theta^2 \nabla \frac{\eta_\theta}{C_V}\right). \tag{3.32f}$$

It can be readily checked that the NIC are satisfied and thus the transformation has led to a GENERIC quintuple $(Q, \mathscr{E}, \mathscr{S}, \mathbb{J}, \mathbb{K})$ in the Eulerian frame. We obtain the evolution by replacing $\boldsymbol{\eta}$ in (3.32) with $\boldsymbol{\eta} = \mathrm{D}\mathcal{S}[\boldsymbol{q}]$ from (3.29b) and correspondingly let $\mathrm{D}\mathcal{E}[\boldsymbol{q}]$ from (3.29a) act on the Poisson structure (3.31). This leads to PDE system

$$\partial_t \boldsymbol{\alpha} + (\nabla \boldsymbol{\alpha})\boldsymbol{v} = \boldsymbol{0}, \tag{3.33a}$$

$$\partial_t \boldsymbol{m} + \nabla \cdot (\boldsymbol{m} \otimes \boldsymbol{v}) = \nabla \cdot \left(\boldsymbol{\sigma}^{\mathrm{el}} + \boldsymbol{\sigma}^{\mathrm{vi}}\right), \tag{3.33b}$$

$$\partial_t \boldsymbol{c} + \nabla \cdot (\boldsymbol{c} \otimes \boldsymbol{v}) = -\nabla \cdot \left(\mathbb{M}\partial_{\bar{\boldsymbol{c}}} \bar{\mathcal{S}}\right) \tag{3.33c}$$

$$C_V\left(\partial_t \theta + \boldsymbol{v} \cdot \nabla\theta\right) = -\theta \nabla \boldsymbol{v} : \partial_{\boldsymbol{F}} \bar{\mathcal{S}} \frac{\boldsymbol{F}^{-\top}}{V} + \nabla_{\mathrm{s}} \boldsymbol{v} : \boldsymbol{\sigma}^{\mathrm{vi}} + \nabla \cdot \left(k_{\mathrm{heat}} \nabla\theta\right), \tag{3.33d}$$

$$\text{with } \boldsymbol{\sigma}^{\mathrm{el}} = \partial_{\boldsymbol{F}}(\bar{\mathcal{U}} - \theta\bar{\mathcal{S}})\boldsymbol{F}^\top V^{-1} \text{ and } \boldsymbol{\sigma}^{\mathrm{vi}} = \boldsymbol{\Lambda} : \nabla_{\mathrm{s}} \boldsymbol{v}. \tag{3.33e}$$

where the viscous contribution to the Cauchy stress is generated by $\boldsymbol{\sigma}^{\mathrm{vi}} = \mathbb{K}_{\boldsymbol{m}\theta}(\boldsymbol{q})\eta_\theta$ with $\eta_\theta = C_V/\theta$. In order to obtain (3.33b) some non-trivial computations are involved, which we explain in the following. For this, we look more detailed at the terms that arise by the computation of the second component $(\mathbb{J}(\boldsymbol{q})\mathrm{D}\mathcal{E}[\boldsymbol{q}])_2$ with $\mathrm{D}\mathcal{E}[\boldsymbol{q}]$ from (3.29a) and the Poisson operator $\mathbb{J}(\boldsymbol{q})$ from (3.31). In particular, it is

$$
\begin{aligned}
&(\mathbb{J}(\boldsymbol{q})\mathrm{D}\mathcal{E}[\boldsymbol{q}])_2 \\
&= \left(\nabla\boldsymbol{\alpha}^\top \partial_{\boldsymbol{\alpha}} \left(\frac{|\boldsymbol{m}|^2}{2\varrho}\right) - \left(\nabla\frac{\boldsymbol{m}}{\varrho}\right)^\top \boldsymbol{m} - \nabla\cdot\left(\boldsymbol{m}\otimes\frac{\boldsymbol{m}}{\varrho}\right) + \left(\nabla\frac{|\boldsymbol{m}|^2}{2\varrho^2}\boldsymbol{M}\right)^\top \boldsymbol{c} \right) \\
&\quad + \left(\boldsymbol{F}^{-\top}\mathrm{D}_{\boldsymbol{\alpha}}\frac{\bar{\mathcal{U}}}{V} - \nabla(\partial_{\bar{\boldsymbol{c}}}\bar{\mathcal{U}})^\top \boldsymbol{c} + \frac{\partial_\theta\bar{\mathcal{U}}}{V}\nabla\theta - \nabla\cdot\left(\frac{\theta\,\partial_\theta\bar{\mathcal{U}}}{\bar{C}_V}\frac{\partial_{\boldsymbol{F}}\bar{\mathcal{S}}}{V}\boldsymbol{F}^\top\right)\right),
\end{aligned}
\tag{3.34}
$$

where we have used that $\boldsymbol{v} = \boldsymbol{m}/\varrho$ as well as $\nabla\boldsymbol{\alpha} = \boldsymbol{F}^{-1}$. We will verify now that the first term on the right-hand side of (3.34) gives the transport term $\nabla\cdot(\boldsymbol{m}\otimes\boldsymbol{v})$ in the momentum balance (3.33b) and that the second term on the right-hand side of (3.34) results in the elastic part $\boldsymbol{\sigma}^{\mathrm{el}}$ of the Cauchy stress given in (3.33e).

Indeed, from the first term on the right-hand side of (3.34) we calculate the transport term $\nabla\cdot(\boldsymbol{m}\otimes\boldsymbol{v})$ as follows:

$$
\begin{aligned}
&\nabla\boldsymbol{\alpha}^\top \partial_{\boldsymbol{\alpha}} \left(\frac{|\boldsymbol{m}|^2}{2\varrho}\right) - \left(\nabla\frac{\boldsymbol{m}}{\varrho}\right)^\top \boldsymbol{m} - \nabla\cdot\left(\boldsymbol{m}\otimes\frac{\boldsymbol{m}}{\varrho}\right) + \left(\nabla\frac{|\boldsymbol{m}|^2}{2\varrho^2}\boldsymbol{M}\right)^\top \boldsymbol{c} \\
&= -\frac{\boldsymbol{F}^\top}{V}\frac{|\boldsymbol{m}|^2}{2\varrho^2}\partial_{\boldsymbol{\alpha}}\bar{\varrho}_0 + \nabla\left(\frac{\boldsymbol{m}}{\varrho}\right)^\top \frac{\boldsymbol{m}}{\varrho}\frac{\bar{\varrho}_0}{V} + \frac{\boldsymbol{F}^\top}{V}\frac{|\boldsymbol{m}|^2}{2\varrho^2}\partial_{\boldsymbol{\alpha}}\bar{\varrho}_0 \\
&\quad - \left(\nabla\frac{\boldsymbol{m}}{\varrho}\right)^\top \boldsymbol{m} - \nabla\cdot\left(\boldsymbol{m}\otimes\frac{\boldsymbol{m}}{\varrho}\right) + \left(\nabla\frac{\boldsymbol{m}}{\varrho}\right)^\top\frac{\boldsymbol{m}}{\varrho}(\boldsymbol{M}\cdot\boldsymbol{c}) = -\nabla\cdot(\boldsymbol{m}\otimes\boldsymbol{v}),
\end{aligned}
$$

which then moves with minus to the left-hand side of (3.33b). In the calculation above we have used Piola's identity in Eulerian coordinates $\nabla\cdot(\boldsymbol{F}^\top V^{-1}) = \boldsymbol{0}$, the definition of ϱ and the relation $\boldsymbol{v} = \boldsymbol{m}\varrho^{-1}$.

To see that the elastic part $\boldsymbol{\sigma}^{\mathrm{el}}$ of the Cauchy stress in (3.33b) arises from the second term on the right-hand side of (3.34) we calculate

$$
\begin{aligned}
&\boldsymbol{F}^{-\top}\mathrm{D}_{\boldsymbol{\alpha}}\frac{\bar{\mathcal{U}}}{V} - \nabla(\partial_{\bar{\boldsymbol{c}}}\bar{\mathcal{U}})^\top \boldsymbol{c} + \frac{\partial_\theta\bar{\mathcal{U}}}{V}\nabla\theta - \nabla\cdot\left(\frac{\theta\,\partial_\theta\bar{\mathcal{U}}}{\bar{C}_V}\frac{\partial_{\boldsymbol{F}}\bar{\mathcal{S}}}{V}\boldsymbol{F}^\top\right) \\
&\quad - \sum_k c_k\nabla\partial_{\bar{c}_k}\bar{\mathcal{U}} + \frac{\partial_\theta\bar{\mathcal{U}}}{V}\nabla\theta - \nabla\cdot\left(\theta\frac{\partial_{\boldsymbol{F}}\bar{\mathcal{S}}}{V}\boldsymbol{F}^\top\right) \\
&= \boldsymbol{F}^{-\top}\frac{\partial_{\boldsymbol{\alpha}}\bar{\mathcal{U}}}{V} + \nabla\cdot(\frac{\partial_{\boldsymbol{F}}\bar{\mathcal{U}}}{V}\boldsymbol{F}^\top) + \sum_{i,j}\frac{\partial_{\boldsymbol{F}_{ij}}\bar{\mathcal{U}}}{V}\nabla\boldsymbol{F}_{ij} + \sum_k\frac{\nabla(\partial_{\bar{c}_k}\bar{\mathcal{U}}c_k V)}{V} - \frac{\nabla\bar{\mathcal{U}}}{V} \\
&\quad - \sum_k c_k\nabla\partial_{\bar{c}_k}\bar{\mathcal{U}} + \frac{\partial_\theta\bar{\mathcal{U}}}{V}\nabla\theta - \nabla\cdot\left(\theta\frac{\partial_{\boldsymbol{F}}\bar{\mathcal{S}}}{V}\boldsymbol{F}^\top\right) \\
&= \nabla\cdot\left(\partial_{\boldsymbol{F}}\left(\bar{\mathcal{U}} - \theta\bar{\mathcal{S}}\right)\boldsymbol{F}^\top V^{-1}\right) = \nabla\cdot\boldsymbol{\sigma}^{\mathrm{el}}.
\end{aligned}
$$

Above, we explicitly computed $\mathrm{D}_{\boldsymbol{\alpha}}\mathcal{U}$ and exploited $\partial_\theta \mathcal{U} = \bar{C}_V$. We have also repeatedly used Piola's identity in Eulerian coordinates $\nabla \cdot (\boldsymbol{F}^\top V^{-1}) = \mathbf{0}$, and collect all terms that are generated by $V^{-1}\nabla\mathcal{U}$.

3.3.3 Damped Hamiltonian Limit

The GENERIC formalism was introduced to describe isolated systems, i.e., systems with no exchange of mass or energy with the surroundings. Yet, the geometric structures and thermodynamic potentials presented here can be adapted to the case of closed system, i.e., systems that exchange heat/energy but no mass with the surroundings. Following the work of Mielke (2011b) we show that a damped Hamiltonian system arises by coupling the GENERIC system with a heat bath and in particular that the corresponding evolution equation has a similar structure as the GENERIC evolution Eq. (3.12).

To do so, we consider a GENERIC quintuple $(Q, \mathscr{E}, \mathscr{S}, \mathbb{J}, \mathbb{K})$ and we split the state vector in order to highlight the thermodynamic variable in last position $\boldsymbol{q}_z = (\boldsymbol{w}, z)$, where $z \in \{e, s, \theta\}$ is a placeholder for the total energy e, the total entropy s or the temperature θ while $\boldsymbol{w}$ collects all remaining variables; cf. (3.27) where $\boldsymbol{w} = (\boldsymbol{\alpha}, \boldsymbol{m}, \boldsymbol{c})$ and $z = \theta$. We highlight that for $z = e$, resp. $z = s$, the total energy, resp. entropy, of the system is simply $\mathscr{E}_e(\boldsymbol{q}_e) := \int_\Omega e \,\mathrm{d}x$, resp. $\mathscr{S}_s(\boldsymbol{q}_s) := \int_\Omega s \,\mathrm{d}x$, which produces the driving force

$$\mathrm{D}\mathcal{E}_e(\boldsymbol{q}_e) = (0, 1)^\top, \quad \text{resp. } \mathrm{D}\mathcal{S}_s(\boldsymbol{q}_s) = (0, 1)^\top .$$

This choice of thermodynamic variables simplifies the implementation of the non-interaction conditions (3.9a), $\mathbb{J}_s(\boldsymbol{q}_s)\mathrm{D}\mathscr{S}_s(\boldsymbol{q}_s) \equiv 0$, $\mathbb{K}_e(\boldsymbol{q}_e)\mathrm{D}\mathscr{E}_e(\boldsymbol{q}_e) \equiv 0$. This can be ensured by the two geometric operators $\mathbb{J}_s = \begin{pmatrix} \mathbb{J}_{\boldsymbol{ww}} & \mathbb{J}_{\boldsymbol{w}s} \\ \mathbb{J}_{s\boldsymbol{w}} & 0 \end{pmatrix}$ and $\mathbb{K}_e = \begin{pmatrix} \mathbb{K}_{\boldsymbol{ww}} & \mathbb{K}_{\boldsymbol{w}e} \\ \mathbb{K}_{e\boldsymbol{w}} & \mathbb{K}_{ee} \end{pmatrix}$, where the off-diagonal entries are a combination of first- and second-order differential operators and the entry $\mathbb{K}_{ee}$ relates to heat conduction processes, cf. (3.32f). We refer to Zafferi et al. (2023a) for more examples.

For $z \in \{e, s\}$, we now consider the two transformations

$$\mathrm{T}_z : Q_z \to Q_\theta \,, \quad \boldsymbol{q}_z = (\boldsymbol{w}, z) \mapsto \boldsymbol{q} = \boldsymbol{q}_\theta = (\boldsymbol{w}, \theta) \,,$$

that map the total energy and total entropy to the temperature. Similarly to (3.30a) and (3.30c), their Fréchet derivatives $\mathrm{DT}_z(\boldsymbol{q}) = \mathbb{L}_z(\boldsymbol{q})$ and adjoints $\mathbb{L}_z^*(\boldsymbol{q})$ are given by

$$\mathbb{L}_z(\boldsymbol{q}) = \mathrm{DT}_z(\boldsymbol{q}) := \begin{pmatrix} \mathbb{I}_{\boldsymbol{w}} & 0 \\ \mathrm{D}_{\boldsymbol{w}} z(\boldsymbol{q}) & \partial_\theta z(\boldsymbol{q}) \end{pmatrix}^{-1} = \begin{pmatrix} \mathbb{I}_{\boldsymbol{w}} & 0 \\ -\mathrm{D}_{\boldsymbol{w}} z(\boldsymbol{q})/\partial_\theta z(\boldsymbol{q}) & 1/\partial_\theta z(\boldsymbol{q}) \end{pmatrix}, \tag{3.35}$$

$$\mathbb{L}_z^*(\boldsymbol{q}) = \mathrm{DT}_z^*(\boldsymbol{q}) := \begin{pmatrix} \mathbb{I}_{\boldsymbol{w}} & -\mathrm{D}_{\boldsymbol{w}}^* z(\boldsymbol{q})/\partial_\theta z(\boldsymbol{q}) \\ 0 & 1/\partial_\theta z(\boldsymbol{q}) \end{pmatrix}. \tag{3.36}$$

Here, in contrast to (3.30a) and (3.30c), we have represented these operators in terms of the variables of the image $\boldsymbol{q} = \boldsymbol{q}_\theta$. Given a GENERIC system written in terms of $\boldsymbol{q} = \boldsymbol{q}_\theta$, we consider the closure conditions (3.14) applied to the GENERIC evolution equation

$$\partial_t \boldsymbol{q} = \mathbb{L}_s(\boldsymbol{q})\mathbb{J}_s(\boldsymbol{q})\mathbb{L}_s^*(\boldsymbol{q})\mathrm{D}\mathscr{E}(\boldsymbol{q}) + \mathbb{L}_e(\boldsymbol{q})\mathbb{K}_e(\boldsymbol{q})\mathbb{L}_e^*(\boldsymbol{q})\mathrm{D}\mathscr{S}(\boldsymbol{q}), \tag{3.37}$$

By applying the adjoint DT_z^* to the driving forces, one can see that, for the remaining variables $\boldsymbol{w}$ the Helmholtz free energy and the Massieu potential are the thermodynamic driving potentials:

$$\mathrm{DT}_s^*(\boldsymbol{q})\mathrm{D}\mathcal{E}[\boldsymbol{q}] = \begin{pmatrix} \mathrm{D}_{\boldsymbol{w}}\mathcal{E}[\boldsymbol{q}] - \frac{\partial_\theta \mathcal{E}}{\partial_\theta \mathcal{S}} \mathrm{D}_{\boldsymbol{w}}\mathcal{S}[\boldsymbol{q}] \\ \frac{\partial_\theta \mathcal{E}}{\partial_\theta \mathcal{S}}[\boldsymbol{q}] \end{pmatrix}, \tag{3.38a}$$

$$\mathrm{DT}_e^*(\boldsymbol{q})\mathrm{D}\mathcal{S}[\boldsymbol{q}] = \begin{pmatrix} \mathrm{D}_{\boldsymbol{w}}\mathcal{S}[\boldsymbol{q}] - \frac{\partial_\theta \mathcal{S}}{\partial_\theta \mathcal{E}} \mathrm{D}_{\boldsymbol{w}}\mathcal{E}[\boldsymbol{q}] \\ \frac{\partial_\theta \mathcal{S}}{\partial_\theta \mathcal{E}}[\boldsymbol{q}] \end{pmatrix}. \tag{3.38b}$$

In fact, we recall that the fundamental thermodynamic relation *(Gibbs relation)* implies $\partial_\theta\mathcal{E}/\partial_\theta\mathcal{S} = \theta$ and that the Helmholtz free energy is obtained from a Legendre transformation of the total energy resulting in the densities $\mathcal{F} = \mathcal{E} - \theta\mathcal{S}$. Hence, the first entry of (3.38a) can be interpreted as the derivative of the Helmholtz free energy density $\mathcal{F}$ with respect to $\boldsymbol{w}$, while Eq. (3.38b) presents the derivative of the so-called Massieu potential $\mathcal{M} = -\theta^{-1}\mathcal{F} = \mathcal{S} - \theta^{-1}\mathcal{E}$. Overall, Eq. (3.37) can be rewritten as

$$\begin{aligned} \begin{pmatrix} \partial_t \boldsymbol{w} \\ \partial_t \theta \end{pmatrix} &= \begin{pmatrix} \mathbb{J}_{\boldsymbol{w}\boldsymbol{w}} & \mathbb{J}_{\boldsymbol{w}S} \\ -\frac{\mathrm{D}_{\boldsymbol{w}} S}{\partial_\theta S}\mathbb{J}_{\boldsymbol{w}\boldsymbol{w}} + \frac{\mathbb{J}_{S\boldsymbol{w}}}{\partial_\theta S} & -\frac{\mathrm{D}_{\boldsymbol{w}} S}{\partial_\theta S}\mathbb{J}_{\boldsymbol{w}S} \end{pmatrix} \begin{pmatrix} \mathrm{D}_{\boldsymbol{w}}\mathcal{F} \\ \theta \end{pmatrix} \\ &+ \begin{pmatrix} \mathbb{K}_{\boldsymbol{w}\boldsymbol{w}} & \mathbb{K}_{\boldsymbol{w}e} \\ -\frac{\mathrm{D}_{\boldsymbol{w}}\mathcal{E}}{\partial_\theta \mathcal{E}}\mathbb{K}_{\boldsymbol{w}\boldsymbol{w}} + \frac{\mathbb{K}_{e\boldsymbol{w}}}{\partial_\theta \mathcal{E}} & -\frac{\mathrm{D}_{\boldsymbol{w}}\mathcal{E}}{\partial_\theta \mathcal{E}}\mathbb{K}_{\boldsymbol{w}e} + \frac{\mathbb{K}_{ee}}{\partial_\theta \mathcal{E}} \end{pmatrix} \begin{pmatrix} -\mathrm{D}_{\boldsymbol{w}}\mathcal{F}/\theta \\ 1/\theta \end{pmatrix}. \end{aligned} \tag{3.39}$$

Observe that the operators so derived do not satisfy the properties required for a Poisson structure $\mathbb{J}$ or an Onsager operator $\mathbb{K}$. We further assume that the system is in thermal equilibrium, i.e., $\theta \equiv \theta^* \in (0, \infty)$. This procedure can be done in different ways: by adding a further dissipative contribution to the operators, see Mielke (2011b), by coupling the system with an external heat bath resulting in independent evolution on the boundary, see Thomas and Heida (2022), or by considering a two-phase system with one phase being a thermal source, see Zafferi et al. (2023b). By solely focusing on the evolution of $\boldsymbol{w}$, we obtain the following

$$\partial_t \boldsymbol{w} = \left(\mathbb{J}_{\boldsymbol{ww}}(\boldsymbol{w}, \theta^*) - \frac{1}{\theta^*} \mathbb{K}_{\boldsymbol{ww}}(\boldsymbol{w}, \theta^*) \right) \mathrm{D}_{\boldsymbol{w}} \mathcal{F}(\boldsymbol{w}, \theta^*), \tag{3.40}$$

where θ^* plays the role of an external parameter. Observe that in this case $\mathbb{J}_{\boldsymbol{ww}}$ is again skew-symmetric and satisfies the Jacobi's identity and $\mathbb{K}_{\boldsymbol{ww}}$ is symmetric and positively semi-definite, which would allow us to identify once more $\mathbb{J}_{\boldsymbol{ww}}$ with the conservative and $\mathbb{K}_{\boldsymbol{ww}}$ with the dissipative dynamics. The symmetry and skew-symmetry properties are still valid since $\mathbb{K}_{\boldsymbol{ww}}$ and $\mathbb{J}_{\boldsymbol{ww}}$ are principal minors of $\mathbb{K}$ and $\mathbb{J}$. The Jacobi's identity is satisfied in view of Mielke and Roubíček (2025, Prop. A.1) and the positive semi-definiteness thanks to Sylvester's criterion, see, e.g., Horn and Johnson (2012, Thm. 7.2.5).

3.4 Eulerian Diffusion and Finite Strain in a Multiphase System

We dedicate this section to introduce a weak formulation of (3.40) based on the definition of suitable bilinear forms derived from the operators $\mathbb{J}$ and $\mathbb{K}$ previously discussed. Additionally, we provide a numerical example inspired by the theory of fluid-structure interaction (FSI) problems to highlight the benefits of this variational framework. In the spirit of Betsch and Schiebl (2019) or Egger and Sabouri (2021), for example, this construction has the goal to construct a class of structure-preserving schemes.

Weak form of damped Hamiltonian systems. Based on previous work (Peschka et al. 2022; Schmeller and Peschka 2023), we can reformulate a damped Hamiltonian evolution, i.e., (3.40), for the state variable $\boldsymbol{q} : [0, T] \to Q$ and the free energy $\mathscr{F}$

$$\partial_t \boldsymbol{q} = (\mathbb{J}(\boldsymbol{q}) - \mathbb{K}(\boldsymbol{q}))\, \mathrm{D}\mathscr{F}(\boldsymbol{q}), \tag{3.41}$$

as a saddle-point problem. Therefore, we introduce variables $\boldsymbol{\eta} = (\eta_\alpha, \eta_m, \eta_c) \in \mathcal{V}$ and an operator $\mathbb{B} : Q \to \mathcal{V}^*$ such that for every $\boldsymbol{q} \in Q$ we find $\boldsymbol{\eta} \in \mathcal{V}$ such that $\mathbb{B}^*\eta = \mathrm{D}\mathscr{F}(\boldsymbol{q}) \in Q^*$. Multiplying (3.41) with $\mathbb{B}$ from the left ans inserting $\boldsymbol{\eta}$ then results in the operator equation in saddle-point form

$$\mathbb{A}(\boldsymbol{q})\eta - \mathbb{B}\partial_t \boldsymbol{q} = 0, \tag{3.42a}$$

$$\mathbb{B}^*\eta = \mathrm{D}\mathscr{F}(\boldsymbol{q}), \tag{3.42b}$$

which holds in $\mathcal{V}^* \times Q^*$ and where $\mathbb{A}(\boldsymbol{q}) = \mathbb{B}\big[\mathbb{J}(\boldsymbol{q}) - \mathbb{K}(\boldsymbol{q})\big]\mathbb{B}^* : \mathcal{V} \to \mathcal{V}^*$. This operator equation is equivalent to $\partial_t \boldsymbol{q}$ and $\boldsymbol{\eta}$ being a stationary points of the functional

$$\mathscr{R}(\cdot, \cdot; \boldsymbol{q}) : \mathcal{V}^* \times Q^* \to \mathbb{R}, \quad \mathscr{R}(\boldsymbol{v}, \boldsymbol{\xi}; \boldsymbol{q}) = -\tfrac{1}{2}\langle \mathbb{A}(\boldsymbol{q})\boldsymbol{\xi}, \boldsymbol{\xi}\rangle_{\mathcal{U}} + \langle \mathbb{B}\boldsymbol{v}, \boldsymbol{\xi}\rangle_{\mathcal{U}} + \langle \mathrm{D}\mathscr{F}(\boldsymbol{q}), \boldsymbol{v}\rangle_Q$$

with respect to $\boldsymbol{v}, \boldsymbol{\xi}$ for given $\boldsymbol{q}$. In Peschka et al. (2022) we show how the operators need to be chosen for an Eulerian approach to nonlinear elasticity that is generated from a corresponding Lagrangian model. Furthermore, in Zafferi et al. (2023a) a similar construction is presented for compressible flows with diffusion and reactions. Similar structures have been analyzed in the context of poroelasticity (Egger and Sabouri 2021).

Free energy and bilinear forms. As an example we consider here the dynamics $\boldsymbol{q} : [0, T] \to Q$ of a compressible isothermal multiphase system with state variables

$$\boldsymbol{q} = \begin{pmatrix} \boldsymbol{\alpha} : \Omega \to \mathbb{R}^d \\ \boldsymbol{m} : \Omega \to \mathbb{R}^d \\ \boldsymbol{c} : \Omega \to \mathbb{R}^N \end{pmatrix} \in Q; \tag{3.43}$$

lateron, in the numerical example we set $d = 2$ and $N = 1$. We showcase an Eulerian formulation of nonlinear elasticity in a multiphase setting coupled to diffusion. The multiphase character is introduced by a Lagrangian indicator function

$$\bar{\iota}_0(\bar{x}) = \begin{cases} 1 & \text{if } x \in \Omega_s, \\ 0 & \text{if } x \in \Omega_l, \end{cases} \tag{3.44}$$

where $\bar{\Omega} = \bar{\Omega}_s \cup \bar{\Omega}_l$ is built from the mutually disjoint solid and liquid phase. Similarly, we use $\bar{\iota}_\varepsilon : \bar{\Omega} \to [0, 1]$ a smooth approximation of $\bar{\iota}_0$ as $\varepsilon \to 0$, see Fig. 3.3. Note that if $\bar{\iota}_\varepsilon$ can be expressed explicitly, then ι_ε can also be written explicitly as its composition with $\boldsymbol{\alpha}$. Consequently, derivatives of ι_ε with respect to x, t and also $\boldsymbol{\alpha}$ can be computed explicitly. The Eulerian indicator function $\iota_\varepsilon := \bar{\iota}_\varepsilon \circ \boldsymbol{\alpha}$ satisfies a continuity equation $\partial_t \iota_\varepsilon + \boldsymbol{v} \cdot \nabla \iota_\varepsilon = 0$. One major advantage of the reference map is that we can explicitly express the time and space dependence of material parameters $p_s, p_l \in \mathbb{R}$ of the solid or liquid phase via the smoothed indicator function by

$$p_\varepsilon(t, x) = p_s \bar{\iota}_\varepsilon(\boldsymbol{\alpha}(t, x)) + p_l\big(1 - \bar{\iota}_\varepsilon(\boldsymbol{\alpha}(t, x))\big), \tag{3.45}$$

which can also be simply expressed as $p_\varepsilon = p_s \iota_\varepsilon + p_l(1 - \iota_\varepsilon)$. More generally, any spatial heterogeneity in the Lagrangian frame can be explicitly represented by composition with the reference map $\boldsymbol{\alpha}$. For this example, we use the subscript ε to indicate the interpolation of two material parameters (p_s, p_l) via the smoothed indicator and the return map for general tuples of real-valued parameters.

This allows us to write the Eulerian energy of the isothermal system as

$$\mathscr{F}(\boldsymbol{q}) = \int_\Omega \frac{|\boldsymbol{m}|^2}{2\varrho_\varepsilon} + \mathcal{F}_\varepsilon[\boldsymbol{q}]\,\mathrm{d}x\,, \qquad \mathcal{F}_\varepsilon[\boldsymbol{q}] = \mathcal{F}_\varepsilon^{\mathrm{mech}}(\boldsymbol{F}_{\mathrm{iso}}) + \mathcal{F}_\varepsilon^{\mathrm{int}}(\boldsymbol{c}), \tag{3.46a}$$

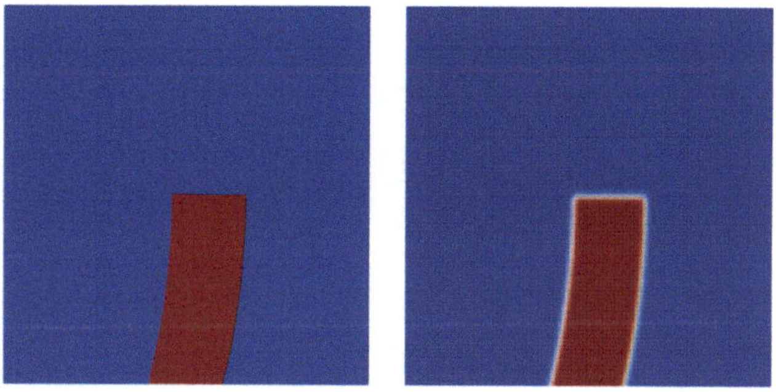

Fig. 3.3 (left) Transformed sharp indicator $\iota_0 = \bar{\iota}_0 \circ \boldsymbol{\alpha}$ on deformed mesh with initial displacement $\bar{\boldsymbol{u}}(t = 0, \bar{x}, \bar{y}) = \bar{x}\bar{y}(1 - \bar{x})(1 - \bar{y})$ and corresponding diffuse indicator ι_ε for $\varepsilon = 10^{-2}$ on a domain $\Omega = \bar{\Omega} = [0, 1]^2$ with solid $\bar{\Omega}_s = [0.4, 0.6] \times [0, 0.5]$ (red) and liquid $\bar{\Omega}_l = \bar{\Omega} \setminus \Omega_s$ (blue)

and, following (3.45), we define the mass density as

$$\varrho_\varepsilon(t, x) = V(t, x)^{-1} \left(\bar{\varrho}_s \bar{\iota}_\varepsilon(\boldsymbol{\alpha}(t, x)) + \bar{\varrho}_l \big(1 - \bar{\iota}_\varepsilon(\boldsymbol{\alpha}(t, x))\big) \right) \tag{3.46b}$$

with referential mass densities $\bar{\varrho}_s, \bar{\varrho}_l \in (0, \infty)$. Similarly, we use an elastic energy

$$\mathcal{F}_\varepsilon^{\text{mech}}(\boldsymbol{F}_{\text{iso}}) = \tfrac{1}{V} \left[\tfrac{G_\varepsilon}{2} \text{tr}(\mathbb{C}_{\text{iso}} - \mathbb{I}) + \tfrac{K_\varepsilon}{2} (V - 1)^2 \right], \tag{3.46c}$$

$$\mathcal{F}_\varepsilon^{\text{int}}(\boldsymbol{c}) = k_\varepsilon \boldsymbol{c} \left(\log \left(\tfrac{c}{c_{\text{ref},\varepsilon}} \right) - 1 \right), \tag{3.46d}$$

with the deformation gradient $\boldsymbol{F} = \nabla \boldsymbol{\alpha}^{-1}$, its Jacobian $V = \det \boldsymbol{F}$ and the isochoric part $\boldsymbol{F}_{\text{iso}} = V^{-1/d} \boldsymbol{F}$. Furthermore, $\mathbb{C}_{\text{iso}} = \boldsymbol{F}_{\text{iso}}^T \boldsymbol{F}_{\text{iso}}$ denotes the isochoric part of the right Cauchy-Green tensor $G_\varepsilon = G_s \iota_\varepsilon + G_l(1 - \iota_\varepsilon)$ captures the energy due to distortion of the solid or liquid phase, $K_\varepsilon = K_s \iota_\varepsilon + K_l(1 - \iota_\varepsilon)$ encodes compressibility and $k_\varepsilon = k_s \iota_\varepsilon + k_l(1 - \iota_\varepsilon)$ encodes the contribution of $\boldsymbol{c}$ to the heat capacity. Note, when calculating $\mathrm{D}\mathcal{F}$, then the derivatives of functions such as ϱ_ε with respect to $\boldsymbol{\alpha}$ need to be taken fully into account to produce a consistent force balance. This form of a saddle-point structure is suitable for numerical treatment, as by testing (3.42) directly translates into the weak form

$$\boldsymbol{a}(\boldsymbol{\eta}, \boldsymbol{\xi}; \boldsymbol{q}) - \boldsymbol{b}(\boldsymbol{\xi}, \partial_t \boldsymbol{q}) = 0, \tag{3.47a}$$

$$\boldsymbol{b}(\boldsymbol{\eta}, \boldsymbol{v}) = \langle \mathrm{D}\mathcal{F}(\boldsymbol{q}), \boldsymbol{v} \rangle, \tag{3.47b}$$

where $\boldsymbol{a}(\boldsymbol{\eta}, \boldsymbol{\xi}; \boldsymbol{q}) := \boldsymbol{j}(\boldsymbol{\eta}, \boldsymbol{\xi}; \boldsymbol{q}) - \boldsymbol{k}(\boldsymbol{\eta}, \boldsymbol{\xi}; \boldsymbol{q})$ with the skew-symmetric bilinear form $\boldsymbol{j}(\boldsymbol{\eta}, \boldsymbol{\xi}; \boldsymbol{q}) = \langle \mathbb{BJ}(\boldsymbol{q})\mathbb{B}^* \boldsymbol{\eta}, \boldsymbol{\xi} \rangle$ and $\boldsymbol{k}(\boldsymbol{\eta}, \boldsymbol{\xi}; \boldsymbol{q}) := \langle \mathbb{BK}(\boldsymbol{q})\mathbb{B}^* \boldsymbol{\eta}, \boldsymbol{\xi} \rangle$ is a symmetric positive bilinear form. We have $\boldsymbol{v} = (v_\alpha, v_m, v_c) \in \boldsymbol{Q}$, $\partial_t \boldsymbol{q} = (\partial_t \boldsymbol{\alpha}, \partial_t \boldsymbol{m}, \partial_t \boldsymbol{c}) \in \boldsymbol{Q}$ and $\boldsymbol{\eta} = (\eta_\alpha, \eta_m, \eta_c) \in \mathcal{V}$ and $\boldsymbol{\xi} = (\xi_\alpha, \xi_m, \xi_c) \in \mathcal{V}$, where

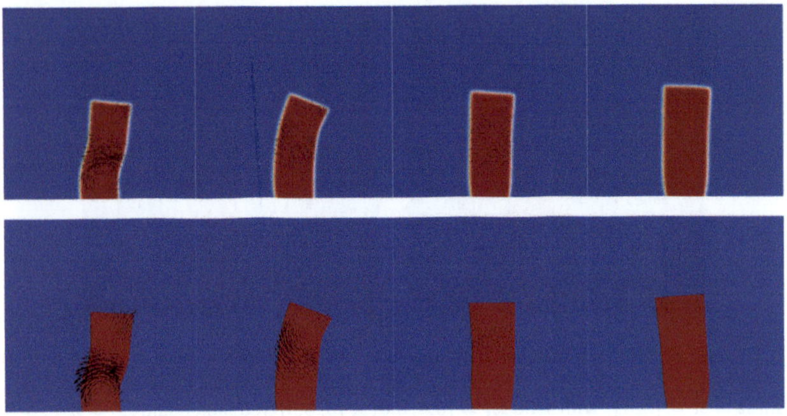

Fig. 3.4 (**top row**) Eulerian solution of Eq. (3.49) with parameters from Table 3.1 showing the indicator function ι_ε (shading blue $\iota_\varepsilon = 0$ and red $\iota_\varepsilon = 1$) and momentum (black arrows) for different times increasing from left to right and (**bottom row**) Lagrangian solution with sharp indicator ι_0 on deformed mesh (shading) and momentum (black arrows) at the same times as in top row

$$j(\eta, \xi; q) = \int_\Omega \left[\nabla \alpha^T \xi_\alpha - \nabla \xi_m^T m\right] \cdot \eta_m - \left[\nabla \alpha^T \eta_\alpha - \nabla \eta_m^T m\right] \cdot \xi_m + c\left[\nabla \xi_c \cdot \eta_m - \nabla \eta_c \cdot \xi_m\right] \mathrm{d}x \tag{3.48a}$$

$$k(\eta, \xi; q) = \int_\Omega \mu_\varepsilon \nabla_s \eta_m \cdot \nabla_s \xi_m + D_\varepsilon(c) \nabla \eta_c \cdot \nabla \xi_c \, \mathrm{d}x \tag{3.48b}$$

$$b(\eta, v) = \int_\Omega \eta_\alpha \cdot v_\alpha + \eta_m \cdot v_m + \eta_c \cdot v_c \, \mathrm{d}x \tag{3.48c}$$

We note that the bilinear forms so introduced are consistent with the previous definition of reversible (3.31) and irreversible (3.32) operators after transformation to the isothermal system (3.40).

Numerical example. This structure-preserving weak form is discretized in space using finite elements and in time with a semi-implicit scheme, where $\mathrm{D}\mathscr{F}$ is computed fully implicit, whereas the dependence of a on η is treated implicitly and the dependence on the state q is treated explicitly. Therefore, for given q^{n-1}, η^{n-1} we seek q^n, η^n such that

$$a(\eta^n, \xi; q^{n-1}) - b\big(\xi^n, \tfrac{1}{\tau}(q^n - q^{n-1})\big) = 0 \tag{3.49a}$$

$$b(\eta^n, v) = \langle \mathrm{D}\mathscr{F}(q^n), v\rangle, \tag{3.49b}$$

for all v, ξ. For this Eulerian formulation scalar and vectorial P_1 finite elements are used throughout.

Numerical solutions for this Eulerian formulation (3.49) based on the material parameters in Table 3.1 are compared with their Lagrangian counterparts in

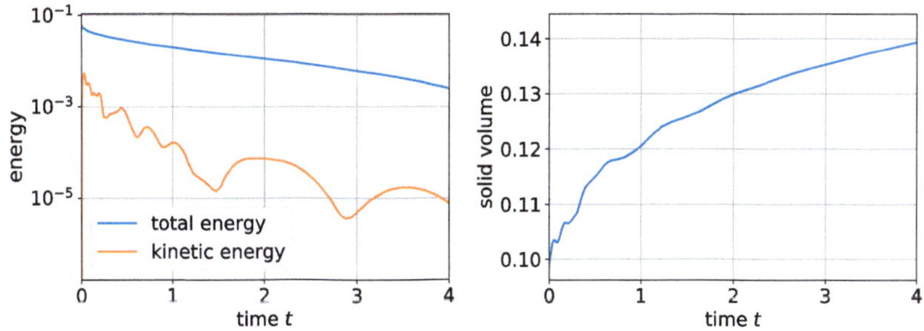

Fig. 3.5 (left panel) Total energy $\mathscr{F}(\boldsymbol{q}(t))$ (shifted by a constant) and kinetic energy $\mathscr{K}(\boldsymbol{q}(t))$ and (right panel) solid volume $\int_\Omega \iota_\varepsilon(t,x)\,\mathrm{d}x$ as functions of time $0 \le t \le 4$. The solid domains $\bar{\Omega}_s = \Omega_s(t=0) = [0.4, 0.6] \times [0, 0.5]$ initial volume is 0.1, which is well approximated by ι_ε

Figs. 3.4 and 3.7. We omit the Lagrangian formulation of the problem and refer instead to Peschka et al. (2022) for details. In the example shown in Fig. 3.4 and the corresponding energies in Fig. 3.5 one can observe an oscillatory motion of the system due to inertia. The aspect ratio of mass densities $\bar{\varrho}_s : \bar{\varrho}_l = 100 : 1$ causes the momentum to be concentrated in the solid phase, i.e., the black arrows are mainly visible in the solid phase. Slight differences between Lagrangian and Eulerian solutions are due to the finite ε and rather large time steps. The oscillation ceases quite quickly due to the large viscosity in solid and liquid phase. Furthermore, Fig. 3.5 clearly shows that during the evolution the (total) energy decreases whereas the solid phase expands significantly by almost 40%. To understand the expansion, one can assume a homogeneous state. In the solid phase, one can then use the stationary state variables as $V(x) = V_s \in (0, \infty)$ and $\boldsymbol{c}(x) = \boldsymbol{c}_s$, while in the liquid phase, the state variables are $V(x) = V_l \in (0, \infty)$ and $\boldsymbol{c}(x) = \boldsymbol{c}_l$. Importantly, the evolution conserves total concentration C and total volume V, which are expressed as

$$C = \boldsymbol{c}_s V_s |\bar{\Omega}_s| + \boldsymbol{c}_l V_l |\bar{\Omega}_l|, \quad V = V_s |\bar{\Omega}_s| + V_l |\bar{\Omega}_l| = 1\,. \tag{3.50}$$

This conservation property allows $\boldsymbol{c}_l$ and V_l to be expressed in terms of $\boldsymbol{c}_s$ and V_s. Consequently, the free energy of the homogeneous system can be written as a function of the two real parameters $\boldsymbol{c}_s$ and V_s. As shown in Fig. 3.6, the free energy exhibits a minimum at $\boldsymbol{c}_s \approx \boldsymbol{c}_{\text{ref,s}} = 2$ and $V_s \approx 2$ and explains the solids expansion. The example above demonstrates that a thermodynamically consistent, unified formulation can be constructed for a multiphase system that allows for large deformations. The usage of the reference map $\boldsymbol{\alpha}$ allows to state of the weak formulation of the nonlinear PDE in spatial (Eulerian) coordinates while retaining information about the reference configuration, which can be useful when Lagrangian material properties need to be accessed, e.g., via material indicator functions. However, for fluid flows with large distortion, it might be non-trivial to ensure the invertibility of $\boldsymbol{\alpha}$, resp. $\boldsymbol{\chi}$, and a more classical discretization of compressible flows in terms of density ϱ and momentum $\boldsymbol{m}$ as the state variables might be more suitable.

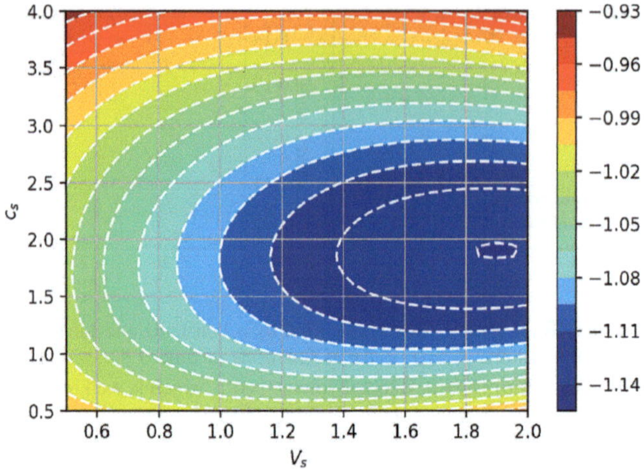

Fig. 3.6 Contour plot of energy $\mathscr{F}(\boldsymbol{q})$ as a function of an assumed homogeneous (constant in space) solid concentration $\boldsymbol{c}_s$ and volume V_s for parameters in Table 3.1. Also the liquid concentration $\boldsymbol{c}_l$ and volume V_l are assumed homogeneous and are determined from conservation laws

Table 3.1 Non-dimensional simulation parameters for fluid and solid phases for $\bar{\Omega} = \Omega = [0, 1]^2$, $\bar{\Omega}_s = [0.4, 0.6] \times [0, 0.5]$ and phase-indicator $\bar{\iota}_\varepsilon$ with $\varepsilon = 0.01$

Parameter	G_i	K_i	μ_i	$\bar{\varrho}_i$	k_i	$c_{\mathrm{ref},i}$	D_i
Fluid ($i = \mathrm{f}$)	10^{-4}	1	10^{-4}	10^{-2}	1	1	10^{-2}
Solid ($i = \mathrm{s}$)	10^{1}	1	10^{-4}	10^{0}	1	2	10^{-2}

By using Eq. (3.47) together with the bilinear forms (3.48), a natural structure-preserving discretization is obtained that ensures thermodynamic consistency, although one needs to be careful in a multiphase setting to compute all derivatives with respect to the indicator functions $\iota_\varepsilon = \bar{\iota}_\varepsilon \circ \boldsymbol{\alpha}$ consistently. This can be an intricate process, which is best handled by discretization frameworks that support automatic differentiation of expressions and functionals, e.g., FEniCS (Logg et al. 2012). Once such a formulation is available, multiphase and multiphysics problems, e.g., the coupling of multiphase flows and diffusion, can be discretized in space with relative ease via finite element methods.

The proposed structure introduces additional terms, which make the equations appear more complicated at first glance, and the usual structure of the momentum balance becomes only apparent after certain terms cancel. While this might imply additional complexity, it remains an open question whether these additional terms, which stem from structure preservation, could be beneficial for numerical stability.

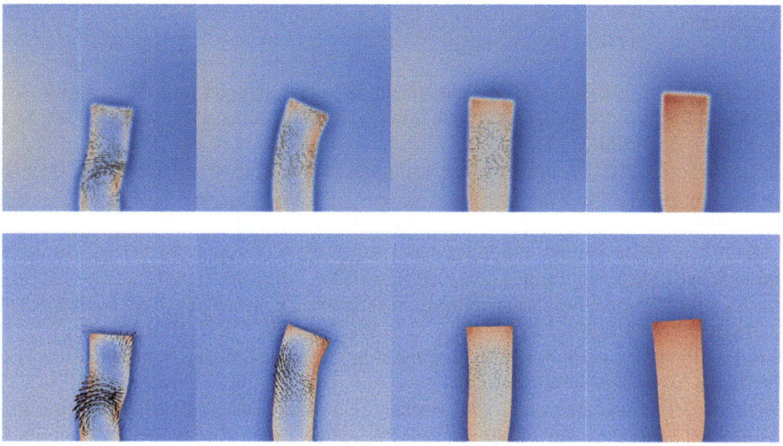

Fig. 3.7 (**top row**) Eulerian solution of Eq. (3.49) with parameters from Table 3.1 showing the concentration $\boldsymbol{c}$ (increasing from blue to red) and momentum (black arrows) for different times increasing from left to right and (**bottom row**) Lagrangian solution with concentration on deformed mesh (shading) and momentum (black arrows) at the same times

3.5 Conclusions and Summary

We presented a GENERIC quintuple $(Q, \mathscr{E}, \mathscr{S}, \mathbb{J}, \mathbb{K})$ for materials at large strain with diffusion processes in Eulerian coordinates, where the *reference* map $\boldsymbol{\alpha}$ (3.1) is used as the main variable governing the deformation. We also introduced the corresponding Lagrangian GENERIC system $(\bar{Q}, \bar{\mathscr{E}}, \bar{\mathscr{S}}, \bar{\mathbb{J}}, \bar{\mathbb{K}})$ and showed its connection to $(Q, \mathscr{E}, \mathscr{S}, \mathbb{J}, \mathbb{K})$ via a transformation mapping Lagrangian to Eulerian state variables $\mathrm{T}_{\mathrm{LE}} : \bar{Q} \to Q$. The thermodynamical functionals and geometric structures are then linked through a linearization of this map $\mathrm{DT}_{\mathrm{LE}}(\bar{\boldsymbol{q}}) = \mathbb{L}(\bar{\boldsymbol{q}})$ and through the *closure conditions* (3.14). In order to systematically structure this approach we introduced kinematics and thermodynamics in the Eulerian and Lagrangian frame. Finally, we constructed a structure-preserving discretization based on a saddle-point structure relying on symmetry-preserving bilinear forms $a = j - k$ with skew-symmetric j and symmetric positive k. We presented an example with fluid-structure interaction and diffusion based on indicator functions, which is similar to other contemporary methods for fluid-structure interaction that explicitly evolve level sets (Rycroft et al. 2020).

Acknowledgements DP thanks for the funding within the DFG Priority Program SPP 2171 *Dynamic Wetting of Flexible, Adaptive, and Switchable Substrates*, project #422792530. MT and AZ acknowledge the funding by the DFG-Collaborative Research Center 1114 *Scaling Cascades in Complex Systems*, project no. 235221301, C09 *Dynamics of rock dehydration on multiple scales*.

References

Abels H, Garcke H, Grün G (2012) Thermodynamically consistent, frame indifferent diffuse interface models for incompressible two-phase flows with different densities. Math Models Methods Appl Sci 22(03):1150013

Aland S, Mokbel D (2021) A unified numerical model for wetting of soft substrates. Int J Numer Methods Eng 122(4):903–918

Alt HW (2009) The entropy principle for interfaces. Fluids and solids. Adv Math Sci Appl 19(2):585

Anand L (2012) A Cahn-Hilliard-type theory for species diffusion coupled with large elastic-plastic deformations. J Mech Phys Solids 60(12):1983–2002

Antman SS (1998) Physically unacceptable viscous stresses. Zeitschrift für angewandte Mathematik und Physik 49(6):980–988

Badal R, Friedrich M, Kružík M (2023) Nonlinear and linearized models in thermoviscoelasticity. Arch Ration Mech Anal 247(1):5

Baker G (2005) Thermodynamics in solid mechanics: a commentary. Philos Trans R Soc A: Math Phys Eng Sci 363(1836):2465–2477

Barron THK, Munn RW (1970) Thermodynamics of solids under stress. Pure Appl Chem 22(3–4):527–534

Beris AN, Edwards BJ (1994) Thermodynamics of flowing systems: with internal microstructure. Number 36. Oxford University Press, USA

Betsch P, Schiebl M (2019) Energy-momentum-entropy consistent numerical methods for large-strain thermoelasticity relying on the generic formalism. Int J Numer Methods Eng 119(12):1216–1244

Biot MA (1984) New variational-Lagrangian irreversible thermodynamics with application to viscous flow, reaction-diffusion, and solid mechanics. Adv Appl Mech 24:1–91

Bothe D, Dreyer W (2015) Continuum thermodynamics of chemically reacting fluid mixtures. Acta Mechanica 226:1757–1805

Ciarlet PG (1988) Mathematical elasticity I: three-dimensional elasticity. North Holland

Constantin P (2001) An Eulerian-Lagrangian approach for incompressible fluids: local theory. J Am Math Soc 14(2):263–278

Cotter CJ, Holm DD, Hydon PE (2007) Multisymplectic formulation of fluid dynamics using the inverse map. Proc R Soc A: Math Phys Eng Sci 463(2086):2671–2687

Datta R, Vilekar SA (2010) The continuum mechanical theory of multicomponent diffusion in fluid mixtures. Chem Eng Sci 65(22):5976–5989

De Boer RB (1977) On the thermodynamics of pressure solution-interaction between chemical and mechanical forces. Geochimica et Cosmochimica acta 41(2):249–256

De Groot SR, Mazur P (2013) Non-equilibrium thermodynamics. Courier Corporation

Ding H, Spelt PDM, Shu C (2007) Diffuse interface model for incompressible two-phase flows with large density ratios. J Comput Phys 226(2):2078–2095

Doi M (2011) Onsager's variational principle in soft matter. J Phys Condens Matter 23(28):284118

Egger H, Sabouri M (2021) On the structure preserving high-order approximation of quasistatic poroelasticity. Math Comput Simul 189:237–252

Gay-Balmaz F, Yoshimura H (2017) A Lagrangian variational formulation for nonequilibrium thermodynamics. Part II: continuum systems. J Geom Phys 111:194–212

Green AE, Naghdi PM (1934) A unified procedure for construction of theories of deformable media. i. classical continuum physics. Proc R Soc London. Series A: Math Phys Sci 448:335–356

Grmela M, Öttinger HC (1997) Dynamics and thermodynamics of complex fluids. i. development of a general formalism. Phys Rev E 56(6):6620

Gurtin ME, Fried E, Anand L (2010) The mechanics and thermodynamics of continua. Cambridge University Press

Holland TJB, Powell R (1998) An internally consistent thermodynamic data set for phases of petrological interest. J Metamorph Geol 16(3):309–343

Horn RA, Johnson CR (2012) Matrix analysis. Cambridge University Press

Ishii M, Hibiki T (2010) Thermo-fluid dynamics of two-phase flow. Springer
Jordan R, Kinderlehrer D, Otto F (1998) The variational formulation of the Fokker-Planck equation. SIAM J Math Anal 29(1):1–17
Kamrin K, Rycroft CH, Nave J-C (2012) Reference map technique for finite-strain elasticity and fluid–solid interaction. J Mech Phys Solids 60(11):1952–1969
Lebon G, Jou D, Casas-Vázquez J (2008) Understanding non-equilibrium thermodynamics, vol 295. Springer
Lions PL (1996a) Mathematical topics in fluid mechanics: volume 1: incompressible models. Mathematical topics in fluid mechanics. Oxford University Press
Lions PL (1996b) Mathematical topics in fluid mechanics: volume 2: compressible models. Mathematical topics in fluid mechanics. Oxford University Press
Logg A, Mardal K-A, Wells G (2012) Automated solution of differential equations by the finite element method: the FEniCS book, vol 84. Springer
Mao Y, Lin S, Zhao X, Anand L (2017) A large deformation viscoelastic model for double-network hydrogels. J Mech Phys Solids 100:103–130
Marsden JE, Hughes TJR (1994) Mathematical foundations of elasticity. Courier Corporation
Marsden JE, Ratiu TS (2013) Introduction to mechanics and symmetry: a basic exposition of classical mechanical systems, vol 17. Springer
Mielke A (2011a) A gradient structure for reaction-diffusion systems and for energy-drift-diffusion systems. Nonlinearity 24(4):1329
Mielke A (2011b) Formulation of thermoelastic dissipative material behavior using GENERIC. Continuum Mech Thermodyn 23(3):233–256
Mielke A, Roubíček T (2020) Thermoviscoelasticity in kelvin-voigt rheology at large strains. Arch Ration Mech Anal 238(1):1–45
Mielke A, Roubíček T (2025) A general thermodynamical model for finitely-strained continuum with inelasticity and diffusion, its GENERIC derivation in Eulerian formulation, and some application. Zeitschrift für angewandte Mathematik und Physik 76(1):1–28
Mielke A, Ortner C, Sengul Y (2014) An approach to nonlinear viscoelasticity via metric gradient flows. SIAM J Math Anal 46(2):1317–1347
Morrison PJ (1986) A paradigm for joined Hamiltonian and dissipative systems. Physica D: Nonlinear Phenomena 18(1-3):410–419
Morrison PJ (1998) Hamiltonian description of the ideal fluid. Rev Modern Phys 70(2):467
Müller I, Müller WH (2009) Fundamentals of thermodynamics and applications: with historical annotations and many citations from Avogadro to Zermelo. Springer
Noll W, Coleman BD (1974) The thermodynamics of elastic materials with heat conduction and viscosity. In: The foundations of mechanics and thermodynamics: selected papers, pp 145–156
Oldroyd JG (1950) On the formulation of rheological equations of state. Proc R Soc London. Series A. Math Phys Sci 200(1063):523–541
Onsager L (1931a) Reciprocal relations in irreversible processes. i. Phys Rev 37(4):405
Onsager L (1931b) Reciprocal relations in irreversible processes. ii. Phys Rev 38(12):2265
Öttinger HC, Grmela M (1997) Dynamics and thermodynamics of complex fluids. ii. illustrations of a general formalism. Phys Rev E 56(6):6633
Pavelka M, Klika V, Grmela M (2018) Multiscale thermo-dynamics: introduction to GENERIC. Walter de Gruyter GmbH & Co KG
Peletier MA (2014) Variational modelling: energies, gradient flows, and large deviations. arXiv preprint arXiv:1402.1990
Peschka D, Zafferi A, Heltai L, Thomas M (2022) Variational approach to fluid-structure interaction via GENERIC. J Non-Equilib Thermodyn 47(2):217–226
Richter T (2013) A fully Eulerian formulation for fluid-structure-interaction problems. J Comput Phys 233:227–240
Roubíček T (2024) Thermodynamics of viscoelastic solids, its Eulerian formulation, and existence of weak solutions. Zeitschrift für angewandte Mathematik und Physik 75(2):51

Roubíček T, Stefanelli U (2023) Viscoelastodynamics of swelling porous solids at large strains by an Eulerian approach. SIAM J Math Anal 55(4):2677–2703

Rycroft CH, Wu C-H, Yu Y, Kamrin K (2020) Reference map technique for incompressible fluid–structure interaction. J Fluid Mech 898:A9

Schmeller L, Peschka D (2023) Gradient flows for coupling order parameters and mechanics. SIAM J Appl Math 83(1):225–253

Thomas M, Heida M (2022) GENERIC for dissipative solids with bulk–interface interaction. Research in mathematics of materials science. Springer, pp 333–364

Truesdell C (1984) Rational thermodynamics. Springer

Truesdell C, Noll W (2004) The non-linear field theories of mechanics. Springer, Berlin, Heidelberg, pp 1–579

Truesdell C, Toupin R (1960) The classical field theories. Springer

Van Der Schaft A, Jeltsema D, et al (2014) Port-Hamiltonian systems theory: an introductory overview. Found Trends® Syst Control 1(2-3):173–378

Wei C, Wu M (2023) An Eulerian nonlinear elastic model for compressible and fluidic tissue with radially symmetric growth. SIAM J Appl Math 83(1):254–275

Zafferi A, Peschka D, Thomas M (2023a) GENERIC framework for reactive fluid flows. Zeitschrift für Angewandte Mathematik und Mechanik 103(7):e202100254

Zafferi A, Huber K, Peschka D, Vrijmoed J, John T, Thomas M (2023b) A porous-media model for reactive fluid–rock interaction in a dehydrating rock. J Math Phys 64(9)

Chapter 4
Pitfalls in the Understanding of Thermomechanics of Continua

Frank Duderstadt

In fondest memory of my two good friends of many years and controversial teachers, Wolfgang Dreyer and Michael Herrmann, who passed away much too soon.

Abstract Language, in general, and physics, in particular, are constantly evolving. When new phenomena are discovered or known phenomena are made accessible to mathematical description, appropriate names are sought to communicate quickly about the phenomenon. Geometric terms such as point, line, or direction are associated with force, time, or motion. Often the first choice of name is not always a happy one, and the names are sometimes replaced only years later by more suitable ones, or they are given the names of people who have distinguished themselves in the field concerned by special achievements in gaining knowledge. Recent examples from continuum mechanics are described below. The text deals with the misleading and changing use of the term deformation, the ambiguity of gradient and divergence, the misleading delimitation of different stress tensors, and definitions of the Gibbs energy of solids.

Keywords Deformation · Transplacement · Stress tensor · Right and left gradients · Gibbs energy

4.1 Motivation

In my many years of collaboration with *Wolfgang Dreyer*, misunderstandings have often arisen in the evaluation of theoretical work by other authors because some authors use the same terms for different phenomena or calculation rules without defining them precisely. The same authors have also changed some of their own

---

F. Duderstadt (✉)
thyssenkrupp Marine Systems (tkMS), Kiel, Germany
e-mail: frank.duderstadt@thyssenkrupp.com

J. Fuhrmann et al. (eds.), *Advances in Continuum Physics*, Advanced Structured Materials 238, https://doi.org/10.1007/978-3-031-93918-1_4

terms in later publications because, after a few years of use, they found the previously chosen terms to be linguistically misleading.

The mechanics volumes of the *Encyclopedia of Physics*, edited by *S. Flügge* from 1960 to 1972, are decisive for many of the terms used today, which were coined under the leadership of *Clifford Truesdell* by authors such as *Toupin, Erickson, Noll, Gurtin, Naghdi*, and *Antman*, who are still pioneers today (Truesdell and Toupin 1960; Truesdell and Noll 1965; Gurtin 1972; Naghdi 1960; Antman 1960). For example, whereas before the literature usually spoke only of stresses, stress vectors and stress tensors, a clear distinction was made between *Cauchy stresses* $\boldsymbol{T}$ and *Piola–Kirchhoff stresses of the first kind* $\boldsymbol{T}_{\mathrm{R}}$ and the *second kind* $\boldsymbol{S}$. Since these works also defined precisely what is meant by the divergence of a tensor of the second order, it is clear from these representations for the stress vectors $\boldsymbol{t} = \boldsymbol{T}\boldsymbol{n}$ and $\boldsymbol{t}_{\mathrm{R}} = \boldsymbol{T}_{\mathrm{R}}\boldsymbol{n}_{\mathrm{R}}$ at a point on the surface of any material partial body with the external surface normals $\boldsymbol{n}$ and $\boldsymbol{n}_{\mathrm{R}}$ that the left-hand side of the stress tensors points in the direction of the acting force and the right-hand side in the direction of the external surface normal. In publications by other authors, however, the reverse order is often used, which can lead to misinterpretations, especially when using the first *Piola–Kirchhoff* tensor, if it is not taken into account that the definitions of divergence used by these other authors usually differ from those in the *Encyclopedia of Physics*. For example, the tensor named *deformation gradient* by *Truesdell* and *Noll* in (1965) in 1965 and abbreviated by $\boldsymbol{F}$ is called *transposed deformation gradient* by other authors because they use a different definition of the gradient, see, e.g., Antman (1995) or H. Altenbach (2012).

Additional linguistic misunderstandings arise from the term *deformation* itself, because in common English and German usage, "deformation" is used synonymously with strain, i.e., a change in shape. In the *Encyclopedia of Physics* (Truesdell and Toupin 1960; Truesdell and Noll 1965), on the other hand, deformation is used synonymously with *change of location*, which is abbreviated as χ in Truesdell and Noll (1965). Even if the new location with respect to a reference location has resulted only in a rigid displacement and rotation, it is still called a deformation, although by definition the form has been preserved.

In the following, an overview of the different representations and the terms used is given in order to make the reader aware of the pitfalls of transferring one and the same fact with different designations and calculation rules. The explanations end with considerations of different definitions of the specific *Gibbs energy* of solids in the literature and their compatibility with the uniformly used definition of the specific *Gibbs* energy of liquids.

4.2 Use of Notation and Mathematical Terms

In the following, the notations used to describe the problems treated are introduced and the corresponding terms are roughly defined. For now, only *real numbers*, *points*, *Euclidean translation vectors*, and *linear mappings* are distinguished and their symbolic notation is defined.

4.2.1 Definitions for Euclidean Point and Vector Spaces

Points, as elements of *topological spaces*, are considered here for the time being only as elements of the *three-dimensional Euclidean point space* $\mathcal{E}$ and appear later again as elements of *material solids*.

In the Euclidean point space $\mathcal{E}$, points are represented by plain (not italicized) Latin letters with serifs (in this text the normal font), e.g., by $\mathrm{X} \in \mathcal{E}$ or $\mathrm{x} \in \mathcal{E}$. A Euclidean point space is always endowed with a *translation space* $\mathcal{V}$, which has an *inner product* and is called *Euclidean vector space* (Noll 1987). The elements of $\mathcal{V}$ are called *Euclidean vectors* (also Euclidean *shifters* or Euclidean *translations*). They map elements in $\mathcal{E}$, i.e., from $\mathcal{E}$ to $\mathcal{E}$, and fulfill the axioms of a *linear space* over the *real number field* $\mathbb{R}$. Euclidean vectors can be identified as special first-order *tensors* and are represented by small Latin letters, bold, italic, with serifs, e.g., by $\boldsymbol{b} \in \mathcal{V}$. The concept of *tensors* will be explained later. Real numbers, also known as real scalars, are represented here as small Latin letters in italics with serifs, e.g., by $a \in \mathbb{R}$, or as small Greek letters, e.g., by $\lambda \in \mathbb{R}$. They can also be identified as zero-order tensors. Real *matrices*, as *lists* or *arrays* of real numbers, are shown in bold italics without serifs, e.g., $\mathsf{X} = [X_1, X_2, X_3] \in \mathbb{R}^3$ or $\mathsf{A} = \begin{bmatrix} A_{11} & A_{12} & A_{13} \\ A_{21} & A_{22} & A_{23} \\ A_{31} & A_{32} & A_{33} \end{bmatrix} \in \mathbb{R}^{3\times 3}$. *Linear mappings* in $\mathcal{V}$, i.e., from $\mathcal{V}$ to $\mathcal{V}$, are special second-order tensors for which Noll proposed the term *lineons* in his 1987 book Noll (1987). The term is derived from the contraction of the word *linear transformations*.

The set of all lineons in $\mathcal{V}$ itself forms a linear space, which is symbolized here by $\mathcal{L}in[\mathcal{V}]$, and whose elements are represented here by large Latin letters in bold italics with serifs, e.g., $\boldsymbol{A} \in \mathcal{L}in[\mathcal{V}]$. Notwithstanding this rule, the letters just described are also used for the *position vectors* $\boldsymbol{X}$, introduced below, which are not lineons and not elements of a linear space.

If $\mathrm{X} \in \mathcal{E}$ is a point in Euclidean space and $\boldsymbol{a}$ is a Euclidean vector, then the shift of the point X by $\boldsymbol{a} \in \mathcal{V}$ into a point $\mathrm{Y} \in \mathcal{E}$ is described by $\mathrm{Y} = \boldsymbol{a}[\mathrm{X}]$ and is represented by the *non-commutative point-vector addition*

$$\boldsymbol{a}[\mathrm{X}] \equiv \mathrm{X} + \boldsymbol{a} = \mathrm{Y}. \tag{4.1}$$

If $\mathrm{X} = \mathrm{X} + \boldsymbol{a}$, then $\boldsymbol{a} = \mathbf{o}$, where $\mathbf{o} \in \mathcal{V}$ is called the *zero vector* in $\mathcal{V}$. If $\boldsymbol{a} + \boldsymbol{b} = \mathbf{o}$, then $\boldsymbol{b} = -\boldsymbol{a}$, where $-\boldsymbol{a}$ is called the *opposite vector* to $\boldsymbol{a}$. Accordingly, $\mathrm{X} = \mathrm{Y} - \boldsymbol{a}$. The vector $\boldsymbol{a}$ can then be represented as a point difference $\boldsymbol{a} = \mathrm{Y} - \mathrm{X}$, so that

$$\mathrm{X} - \mathrm{Y} = -\boldsymbol{a}$$

also holds.

Since $\mathcal{E}$ is by definition a *metric space*, the *Euclidean distance* $\mathrm{d}_{\mathcal{E}}[\mathrm{X},\mathrm{Y}] \geq 0$ can be measured between any X and Y, so that with $\mathbb{P} := \{\lambda \in \mathbb{R} | \lambda \geq 0\}$, the set of all positive real numbers, and (4.1) the relations

$$\mathrm{d}_{\mathcal{E}}[\mathrm{X},\mathrm{Y}]=\mathrm{d}_{\mathcal{E}}[\mathrm{X},\mathrm{X}+\boldsymbol{a}]=\mathrm{d}_{\mathcal{E}}[\mathrm{Y}-\boldsymbol{a},\mathrm{Y}]\in\mathbb{P}$$

hold for any X and Y. For another point $\mathrm{Z}\in\mathcal{E}$, for which $\mathrm{Z}=\mathrm{X}+\boldsymbol{b}$, and therefore also

$$\mathrm{d}_{\mathcal{E}}[\mathrm{X},\mathrm{Z}]=\mathrm{d}_{\mathcal{E}}[\mathrm{X},\mathrm{X}+\boldsymbol{b}]=\mathrm{d}_{\mathcal{E}}[\mathrm{Z}-\boldsymbol{b},\mathrm{Z}],\quad \mathrm{Z}=\mathrm{Y}+\boldsymbol{c}=\mathrm{Y}+\boldsymbol{b}-\boldsymbol{a}$$

and

$$\mathrm{d}_{\mathcal{E}}[\mathrm{Y},\mathrm{Z}]=\mathrm{d}_{\mathcal{E}}[\mathrm{X}+\boldsymbol{a},\mathrm{X}+\boldsymbol{b}]=\mathrm{d}_{\mathcal{E}}[\mathrm{Y},\mathrm{Y}+\boldsymbol{b}-\boldsymbol{a}].$$

The real quantity

$$\boldsymbol{ab}:=\frac{1}{2}\left(\mathrm{d}^2_{\mathcal{E}}[\mathrm{X},\mathrm{X}+\boldsymbol{a}]+\mathrm{d}^2_{\mathcal{E}}[\mathrm{X},\mathrm{X}+\boldsymbol{b}]-\mathrm{d}^2_{\mathcal{E}}[\mathrm{X}+\boldsymbol{a},\mathrm{X}+\boldsymbol{b}]\right)\in\mathbb{R}$$

is then independent of the selected point X and defines the *inner product*, also known as the *scalar product*, in $\mathcal{V}$. The following applies $\boldsymbol{ab}=\boldsymbol{ba}$, $\lambda\boldsymbol{ab}=\lambda(\boldsymbol{ab})=\boldsymbol{a}\lambda\boldsymbol{b}$ for all $\lambda\in\mathbb{R}$, $\boldsymbol{aa}+\boldsymbol{bb}\geq|\boldsymbol{ab}|$, $\boldsymbol{aa}=0\iff\boldsymbol{a}=\boldsymbol{o}$ and $|\boldsymbol{a}|=\sqrt{\boldsymbol{aa}}$, so that the distance between X and Y can be identified with the *magnitude* of $\boldsymbol{a}$

$$\mathrm{d}_{\mathcal{E}}[\mathrm{X},\mathrm{Y}]=|\mathrm{Y}-\mathrm{X}|=|\boldsymbol{a}|.$$

The vector $\frac{\boldsymbol{a}}{|\boldsymbol{a}|}$ normalized to magnitude 1 indicates the *direction* of $\boldsymbol{a}$.

Special vectors $\boldsymbol{e}_i\in\mathcal{V}$ with the property $\boldsymbol{e}_i\boldsymbol{e}_j=\delta_{ij}$ for $i,j\in\{1,2,3\}$ are called *orthonormal*, where δ_{ij} is the *Kronecker* symbol. The set $\{\boldsymbol{e}_1,\boldsymbol{e}_2,\boldsymbol{e}_3\}$ is an *orthonormal basis* in $\mathcal{V}$, and for $a_i:=\boldsymbol{ae}_i$ and $b_i:=\boldsymbol{be}_i$, then using *Einstein's summation convention*, $\boldsymbol{ab}=a_ib_i$ and $|\boldsymbol{a}|=\sqrt{a_ia_i}$.

By selecting an arbitrary point $\mathrm{O}\in\mathcal{E}$ as the reference point, the *position vector* of X with respect to O can be defined by the point difference $\boldsymbol{X}:=\mathrm{X}-\mathrm{O}$. The point O together with a fixed orthonormal base $\{\boldsymbol{e}_1,\boldsymbol{e}_2,\boldsymbol{e}_3\}$ defines a *Cartesian coordinate system*, where

$$X_i:=(\mathrm{X}-\mathrm{O})\boldsymbol{e}_i=\boldsymbol{X}\boldsymbol{e}_i \tag{4.2}$$

are called the *Cartesian coordinates* of X with respect to O.

Remark 4.1 At this point, as in many textbooks, a ***first pitfall*** of a contradictory choice of terms should be pointed out: The set of all position vectors $\mathcal{V}_\mathrm{O}$ from a fixed origin $\mathrm{O}\in\mathcal{E}$ does not have the property of a linear space, since no vector addition is defined for two elements $\boldsymbol{X},\boldsymbol{Y}\in\mathcal{V}_\mathrm{O}$. However, for position vectors $\boldsymbol{X}$ and $\boldsymbol{Y}$ there is always $\boldsymbol{a}\in\mathcal{V}$ for which the non-commutative addition $\boldsymbol{Y}=\boldsymbol{X}+\boldsymbol{a}\in\mathcal{V}_\mathrm{O}$ is defined between a position vector and a Euclidean vector, so that $\boldsymbol{a}=\boldsymbol{Y}-\boldsymbol{X}\in\mathcal{V}$ holds for the position vector difference. If linear spaces $\mathcal{W}$ are always called "vector spaces" and their elements "vectors", this leads to the linguistically questionable statement that "position vectors are not vectors", although they have magnitude and direction. Conversely, there are plenty of examples of elements of linear spaces that have neither magnitude nor direction.

4.2.2 Products in Euclidean Vector Spaces and Linear Transformations

By setting $\boldsymbol{a}\otimes\lambda=\lambda\boldsymbol{a}\in\mathcal{V}$ and $\boldsymbol{a}\otimes\boldsymbol{b}\boldsymbol{c}=(\boldsymbol{b}\boldsymbol{c})\boldsymbol{a}\in\mathcal{V}$, the tensor product

$$\boldsymbol{a}\otimes\boldsymbol{b}\in\mathcal{L}in[\mathcal{V}]$$

between two vectors can be defined, which results in a second-order tensor. Correspondingly, the tensor product $\boldsymbol{a}\otimes\boldsymbol{b}\otimes\boldsymbol{c}$ forms a third-order tensor and the tensor product between $n\in\mathbb{N}$ translation vectors forms a nth-order tensor. With respect to an orthonormal basis, the following applies to a lineon $\boldsymbol{a}\otimes\boldsymbol{b}=a_ib_j\boldsymbol{e}_i\otimes\boldsymbol{e}_j$, where a_ib_j is its component representation of the basis $\boldsymbol{e}_i\otimes\boldsymbol{e}_j$. Its most general form has the representation $\boldsymbol{A}=\boldsymbol{a}_k\otimes\boldsymbol{e}_k$ and the component relation $A_{ij}=a_{ij}$ holds especially with $\boldsymbol{a}_k=(a_l\boldsymbol{e}_l)_k:=a_{lk}\boldsymbol{e}_l$ and $A_{ij}:=(\boldsymbol{A}\boldsymbol{e}_j)\boldsymbol{e}_i=\boldsymbol{e}_i\boldsymbol{A}\boldsymbol{e}_j$. Also, for $\boldsymbol{B}\in\mathcal{L}in[\mathcal{V}]$, result $\boldsymbol{B}(\boldsymbol{A}\boldsymbol{c})=\boldsymbol{B}\boldsymbol{A}\boldsymbol{c}$ with $\boldsymbol{B}\boldsymbol{A}\in\mathcal{L}in[\mathcal{V}]$ and $\boldsymbol{A}^2:=\boldsymbol{A}\boldsymbol{A}\in\mathcal{L}in[\mathcal{V}]$. A *composite* $\boldsymbol{A}\boldsymbol{B}\in\mathcal{L}in[\mathcal{V}]$ of two mappings is the non-commutative tensor multiplication in $\mathcal{L}in[\mathcal{V}]$, i.e., in general $\boldsymbol{A}\boldsymbol{B}\neq\boldsymbol{B}\boldsymbol{A}\in\mathcal{L}in[\mathcal{V}]$ holds.

The (single) *dot product*, also known as the (single) *contraction product*,

$$\boldsymbol{a}\cdot\boldsymbol{b}:=\boldsymbol{a}\boldsymbol{b}\in\mathbb{R},$$

the base pairs enclosing the point are connected to combine the scalar product in $\mathcal{V}$, so that $\boldsymbol{a}\cdot\boldsymbol{A}=\boldsymbol{a}\boldsymbol{A}\in\mathcal{V}$, $\boldsymbol{A}\cdot\boldsymbol{a}=\boldsymbol{A}\boldsymbol{a}\in\mathcal{V}$, and $\boldsymbol{A}\cdot\boldsymbol{B}=\boldsymbol{A}\boldsymbol{B}\in\mathcal{L}in[\mathcal{V}]$ follows. While dot products of two vectors are scalar products, dot products between a vector and a lineon are a linear transformation of the vector. Dot products with a point between two lineons represent a tensor multiplication in $\mathcal{L}in[\mathcal{V}]$. In addition to the single dot product, two different double dot products are defined:

$$\boldsymbol{a}\otimes\boldsymbol{b}\cdot\cdot\,\boldsymbol{c}\otimes\boldsymbol{d}:=(\boldsymbol{b}\cdot\boldsymbol{c})(\boldsymbol{a}\cdot\boldsymbol{d})=(\boldsymbol{b}\boldsymbol{c})(\boldsymbol{a}\boldsymbol{d})\in\mathbb{R}\quad\text{and}\tag{4.3}$$

$$\boldsymbol{a}\otimes\boldsymbol{b}:\boldsymbol{c}\otimes\boldsymbol{d}:=(\boldsymbol{a}\cdot\boldsymbol{c})(\boldsymbol{b}\cdot\boldsymbol{d})=(\boldsymbol{a}\boldsymbol{c})(\boldsymbol{b}\boldsymbol{d})\in\mathbb{R}.\tag{4.4}$$

The scalar quantity $\boldsymbol{A}\cdot\cdot\,\boldsymbol{I}\in\mathbb{R}$ can be identified with the trace of $\boldsymbol{A}$, so that $\mathrm{tr}[\boldsymbol{A}]=\mathrm{tr}[\boldsymbol{A}^{\mathrm{T}}]=\boldsymbol{I}\cdot\cdot\,\boldsymbol{A}$. In contrast, defines

$$\boldsymbol{A}:\boldsymbol{B}=\mathrm{tr}[\boldsymbol{A}\boldsymbol{B}^{\mathrm{T}}]=\mathrm{tr}[\boldsymbol{B}^{\mathrm{T}}\boldsymbol{A}]=\mathrm{tr}[\boldsymbol{B}\boldsymbol{A}^{\mathrm{T}}]$$

the inner product in $\mathcal{L}in[\mathcal{V}]$ and the positive scalar $|\boldsymbol{A}|:=\sqrt{\mathrm{tr}[\boldsymbol{A}^{\mathrm{T}}\boldsymbol{A}]}\in\mathbb{P}$ defines the magnitude of $\boldsymbol{A}$, which is also called the *Frobenius norm* of $\boldsymbol{A}$. Accordingly, $\frac{\boldsymbol{A}}{|\boldsymbol{A}|}$ represents the *direction of the lineon* $\boldsymbol{A}$.

4.2.3 Tensor and Point Fields

Tensor-valued mappings whose values depend on the specific location $\mathrm{X}\in\mathcal{E}$ are called *tensor fields*. $f:\mathcal{E}\to\mathbb{R}$ are called *scalar field's* or tensor field's zeroth order, $\boldsymbol{f}:\mathcal{E}\to\mathcal{V}$ *vector field's* or tensor field's first order, and $\boldsymbol{F}:\mathcal{E}\to\mathcal{L}in[\mathcal{V}]$ *lineon field's* or tensor field's second order. The mappings $\gamma:\mathcal{E}\to\mathcal{E}$ are called *point fields* and are not tensor fields.

Remark 4.2 The values of the tensor fields $f[\mathrm{X}]\in\mathbb{R}$, $\boldsymbol{f}[\mathrm{X}]\in\mathcal{V}$ and $\boldsymbol{F}[\mathrm{X}]\in\mathcal{L}in[\mathcal{V}]$ in $\mathrm{X}\in\mathcal{D}\subset\mathcal{E}$ are independent of whether and which coordinate system $\phi:\mathcal{D}\to\mathbb{R}^3$ was chosen to assign real coordinate values to X, which are also dependent on ϕ. If these ϕ−dependent coordinate values are represented by a real triple $\mathbf{U}_{\mathrm{X}\phi}=\phi[\mathrm{X}]\in\mathbb{R}^3$, the values of the real components are

$$f_i[\mathbf{U}_{\mathrm{X}\phi}],\; f^i[\mathbf{U}_{\mathrm{X}\phi}],\; F_{ij}[\mathbf{U}_{\mathrm{X}\phi}],\; F^i{}_j[\mathbf{U}_{\mathrm{X}\phi}],\; F_i{}^j[\mathbf{U}_{\mathrm{X}\phi}],\; F^{ij}[\mathbf{U}_{\mathrm{X}\phi}] \in \mathbb{R}$$

of $\boldsymbol{f}$ and $\boldsymbol{F}$ at point X, with respect to the two local, to each other *dual, natural, curvilinear coordinate base systems*

$$\{\boldsymbol{g}_{\mathrm{X}\phi 1}, \boldsymbol{g}_{\mathrm{X}\phi 2}, \boldsymbol{g}_{\mathrm{X}\phi 3}\} \quad\text{and}\quad \{\boldsymbol{g}^1_{\mathrm{X}\phi}, \boldsymbol{g}^2_{\mathrm{X}\phi}, \boldsymbol{g}^3_{\mathrm{X}\phi}\},$$

generally depend on the selected coordinate system. By definition,

$$\begin{aligned}
\boldsymbol{f}[\mathrm{X}] &= f_i[\mathbf{U}_{\mathrm{X}\phi}]\,\boldsymbol{g}^i_{\mathrm{X}\phi} = f^i[\mathbf{U}_{\mathrm{X}\phi}]\,\boldsymbol{g}_{\mathrm{X}\phi i},\\
\boldsymbol{F}[\mathrm{X}] &= F_{ij}[\mathbf{U}_{\mathrm{X}\phi}]\,\boldsymbol{g}^i_{\mathrm{X}\phi}\otimes \boldsymbol{g}^j_{\mathrm{X}\phi} = F_i{}^j[\mathbf{U}_{\mathrm{X}\phi}]\,\boldsymbol{g}^i_{\mathrm{X}\phi}\otimes \boldsymbol{g}_{\mathrm{X}\phi j}\\
&= F^{ij}[\mathbf{U}_{\mathrm{X}\phi}]\,\boldsymbol{g}_{\mathrm{X}\phi i}\otimes \boldsymbol{g}_{\mathrm{X}\phi j} = F^i{}_j[\mathbf{U}_{\mathrm{X}\phi}]\,\boldsymbol{g}_{\mathrm{X}\phi}\otimes \boldsymbol{g}^j_{\mathrm{X}\phi}
\end{aligned}$$

and $\boldsymbol{g}_{\mathrm{X}\phi j}\boldsymbol{g}^j_{\mathrm{X}\phi}=\delta^j_i$. In contrast, not all real functions

$$a,\; a_i,\; a^i,\; A_{ij},\; A^i{}_j,\; A_i{}^j,\; A^{ij} : \mathbb{R}^3 \to \mathbb{R}$$

are components of scalar, vector, or lineon fields.

Let $f:\mathcal{E}\to\mathcal{V}$ be a scalar field over $\mathcal{E}$ such that $f[\mathrm{X}]\in\mathbb{R}$ holds for $\mathrm{X}\in\mathcal{E}$. For $\mathrm{Y}=\mathrm{X}+\boldsymbol{a}\in\mathcal{E}$ with $\boldsymbol{a}\in\mathcal{V}$ independent of X, i.e., constant, the function

$$\Delta f[\mathrm{X},\boldsymbol{a}] := f[\mathrm{X}+\boldsymbol{a}] - f[\mathrm{X}] \in \mathbb{R}$$

defines the difference between the function values of f at Y and X. The *limit*

$$\lim_{\lambda\to 0}\frac{\Delta f[\mathrm{X},\lambda\boldsymbol{a}]}{\lambda} := \big(\mathbf{GRAD}\, f[\mathrm{X}]\big)\,\boldsymbol{a} \in \mathbb{R}$$

is linear in $\boldsymbol{a}$, and the vector

$$\mathbf{GRAD}\, f[\mathrm{X}] \in \mathcal{V}$$

defines the *gradient* of f at X if f is smooth enough in the *neighborhood* of X for the limit to exist.

$$\big(\mathbf{GRAD}\, f[\mathrm{X}]\big)\, \boldsymbol{a} = \boldsymbol{a}\ \mathbf{GRAD} f[\mathrm{X}], \tag{4.5}$$

i.e., **GRAD** $f[\mathrm{X}]$ maps $\boldsymbol{a}$ at point $X \in \mathcal{E}$ linearly from the right and also from the left to $\mathbb{R}$.

Remark 4.3 The definition of the gradient does not require the introduction of coordinate systems.

The mapping $\phi^i : \mathcal{D} \to \mathbb{R}$ is the i-th *coordinate* of ϕ for a generally *curvilinear coordinate system* $\phi = [\phi^1, \phi^2, \phi^2] : \mathcal{D} \to \mathbb{R}^3$, with $\mathcal{D} \subset \mathcal{E}$. For its gradient in X, $g^i_{\mathrm{X}\phi} := \mathbf{GRAD}\, \phi^i[\mathrm{X}] \in \mathcal{V}$. $\{g^1_{\mathrm{X}\phi}, g^2_{\mathrm{X}\phi}, g^3_{\mathrm{X}\phi}\}$ provides a *natural basis* in $\mathcal{V}$ for ϕ. It is therefore called a *gradient basis* in $\mathcal{V}$ with respect to ϕ, or, because it is *perpendicular* to the *tangent planes* of the *coordinate surfaces* for $\phi_i = \text{const}$ at each point $\mathrm{X} \in \mathcal{E}$, it is also called a *contravariant basis*. The basis dual to this basis $\{g_{\mathrm{X}\phi 1}, g_{\mathrm{X}\phi 2}, g_{\mathrm{X}\phi 3}\}$ is called the *tangent basis* or *covariant basis*, respectively, and forms the second natural basis in $\mathcal{V}$ with respect to φ.

If (4.2) a *Cartesian* coordinate system is defined and $\mathbf{X} := [X_1, X_2, X_3] \in \mathbb{R}^3$ is the matrix of *Cartesian* coordinates defined in this way, then

$$f[\mathrm{X}] = \bar{f}[\boldsymbol{X}] = \bar{f}[X_i \boldsymbol{e}_i] = \hat{f}[\mathbf{X}]$$

gives the special alternative representations of the gradient of a scalar field, by

$$\mathbf{GRAD}\, f[\mathrm{X}] = \frac{\partial \bar{f}[X_i \boldsymbol{e}_i]}{\partial X_j} \boldsymbol{e}_j = \boldsymbol{e}_j \frac{\partial \bar{f}[X_i \boldsymbol{e}_i]}{\partial X_j} = \frac{\partial \hat{f}[\mathbf{X}]}{\partial X_j} \boldsymbol{e}_j = \boldsymbol{e}_j \frac{\partial \hat{f}[\mathbf{X}]}{\partial X_j} \tag{4.6}$$

using the real *partial derivatives* according to X_i. A component representation of the gradient of a scalar field with respect to the natural bases of a *curvilinear* coordinate system is not necessary for the following explanations.

In this context here we want to look at the Nabla symbol ∇ in more detail. ∇ is used differently in the literature in connection with gradients, and thus ***hides pitfalls***:

- as a general, coordinate system-independent abbreviation for **GRAD**, with

$$\nabla := \mathbf{GRAD}, \text{ so that } \quad \nabla f[\mathrm{X}] = \mathbf{GRAD} f[\mathrm{X}] \quad \text{symbolizes;}$$

- as a special vector-valued, coordinate system-dependent *differential operator* based on the special representation (4.6) with

$$\nabla := \boldsymbol{e}_i \frac{\partial}{\partial X_i}, \text{ so that } \quad \nabla f[\mathrm{X}] = \frac{\partial \hat{f}[\mathbf{X}]}{\partial X_i} \boldsymbol{e}_i \quad \text{symbolizes.}$$

Remark 4.4 Instead of an own commentary on the flexible applicability of these two different uses of ∇, different comments by *Albert Einstein, Walter Noll*, and *Wolfgang Dreyer* on the representation of physical phenomena with and without the introduction of coordinate systems are presented and concluded with a statement by *Gustav Kirchhoff* from 1876 that is still valid today:

Albert Einstein 1949 in Einstein (1949): Autobiographisches

"Warum brauchte es weitere 7 Jahre für die Aufstellung der allgemeinen Relativitätsheorie? Der hauptsächliche Grund liegt darin, dass man sich nicht so leicht von der Auffassung befreit, dass den Koordinaten eine unmittelbare metrische Bedeutung zukommen müsse."

English translation by the editor of Einstein (1949), *Paul Schilpp*: Autobiographical Notes

"Why were another seven years required for the construction of the general theory of relativity? The main reason is the fact that it is not so easy to free oneself from the idea that coordinates must have an immediate metrical meaning."

Walter Noll 1987 in Noll (1987):

"The approach of this treatise is conceptual, geometric, and uncompromisingly coordinate-free. In some of the literature "tensors" are still defined in terms of coordinates and their transformations. To me, this is like looking at shadows dancing on the wall rather than at reality itself."

Wolfgang Dreyer 2014: unpublished script of his lecture "Tensoranalysis" (4 chapters), held at TU Berlin

"Das Wesen von Produkträumen wird in der *multilinearen Algebra* studiert. Diese liefert ein System von Rechenregeln, welche es gestatten, Tensoren nebst zugehörigen Tensorbasen unabhängig von den Koordinatenlinien einzuführen. Ein Tensor wird hier nicht über sein Transformationsverhalten charakterisiert, sondern als lineare Abbildung eingeführt, die beispielsweise Vektoren in Vektoren oder auf Skalare abbildet. Nach meiner subjektiven Meinung bringt dies, weder visuell anschaulich noch rechentechnisch, irgendwelche Vorteile und wird deshalb unterdrückt."

Übung 1.12:	"Besorge ein Buch, welches eine gegenteilige Meinung vertritt, und referiere etwa 15-20 Minuten über die Einführung von Tensoren 2-ter und 3-ter Stufe ohne die Verwendung von Koordinatenlinien. Zeige insbesondere, dass bei Akzeptanz von Koordinatenlinien beide Vorgehensweisen equivalent sind. Beispielsweise könnten als Grundlage die Kapitel ... des Vorlesungsskriptes Festigkeitslehre von A. Bertram dienen, ... http://www.uni-magdeburg.de/ifme/festigkeit.html ... ."

English translation by the author:

"The nature of product spaces is studied in multilinear algebra. This provides a system of computational rules, which allows the introduction of tensors and the corresponding tensor bases, independent of the lines of coordinates. A tensor is not characterized by its transformation behavior, but is introduced as a linear mapping, e.g. mapping vectors into vectors or onto scalars. In my subjective opinion this has no advantages, neither visually nor computationally, and is therefore suppressed."

Exercise 1.12:	"Get a book that argues the opposite and give a 15-20 minute lecture on the introduction of 2nd and 3rd order tensors without the use of coordinate lines. In particular, show that both approaches are equivalent if coordinate lines are accepted. For example, the chapters ... of the lecture notes on strength of materials by A. Bertram could serve as a basis, ... http://www.uni-magdeburg.de/ifme/festigkeit.html ... ."

Gustav Kirchhoff's prescient statement in 1876 in his lectures on mechanics seems appropriate here.

Gustav Kirchhoff 1876 (Kirchhoff 1876): Vorlesungen über Mechanik. Erste Vorlesung, §1.

"Die Mechanik ist die Wissenschaft von der Bewegung; als ihre Aufgabe bezeichnen wir: die in der Natur vor sich gehenden Bewegungen *vollständig* und *auf die einfachste Weise* zu beschreiben.

...

... Nicht so klar ist die Bedeutung der zweiten Forderung, dass die Beschreibung die *einfachste* sei. Es ist von vorn herein sehr wohl denkbar, dass Zweifel darüber bestehen können, ob eine oder eine andere Beschreibung gewisser Erscheinungen die einfachere ist; es ist auch denkbar, dass eine Beschreibung gewisser Erscheinungen, die heute unzweifelhaft die einfachste ist, die man geben kann, später, bei weiterer Entwicklung der Wissenschaft, durch eine noch einfachere ersetzt wird. ..."

English translation by the author: Lectures on Mechanics. First lecture, §1.

"Mechanics is the science of motion; its object is to describe completely and in the simplest manner the motions which occur in nature.

...

... The meaning of the second requirement, that the description be the simplest, is not so clear. It is quite conceivable that there may be doubts from the outset as to whether one or another description of certain phenomena is the simplest; it is also conceivable that a description of certain phenomena, which today is undoubtedly the simplest that can be given, may later, with the further development of science, be replaced by an even simpler one. ..."

Similarly, vector fields $\boldsymbol{f}: \mathcal{E} \to \mathcal{V}$ can be defined on $\mathcal{E}$, so that $\boldsymbol{f}[\mathrm{X}] \in \mathcal{V}$ holds. For $\mathrm{Y} = \mathrm{X} + \boldsymbol{a} \in \mathcal{E}$ and constant $\boldsymbol{a} \in \mathcal{V}$, the difference between the function values for Y and X is

$$\Delta \boldsymbol{f}[\mathrm{X}, \boldsymbol{a}] := \boldsymbol{f}[\mathrm{X} + \boldsymbol{a}] - \boldsymbol{f}[\mathrm{X}] \in \mathcal{V}. \tag{4.7}$$

Its limit

$$\lim_{\lambda \to 0} \frac{\Delta \boldsymbol{f}[\mathrm{X}, \lambda \boldsymbol{a}]}{\lambda} := (\mathbf{GRAD}_{\mathrm{R}} \boldsymbol{f}[\mathrm{X}])\, \boldsymbol{a} = \boldsymbol{a}\ \mathbf{GRAD}_{\mathrm{L}} \boldsymbol{f}[\mathrm{X}] \in \mathcal{V} \tag{4.8}$$

is also linear in $\boldsymbol{a}$. Although its limit is unique, the representation as the result of a linear mapping is not unique, since there is always the possibility of a mapping from the left or from the right. The lineons

$$\mathbf{Grad}_{\mathrm{R}} f[\mathrm{X}], \mathbf{Grad}_{\mathrm{L}} f[\mathrm{X}] \in \mathcal{L}in[\mathcal{V}]$$

define the two gradients of f at X if f is smooth enough near X that the limit exists. $\mathbf{Grad}_{\mathrm{R}} f$ maps $\boldsymbol{a}$ from the right and is called the *right gradient* of f. Correspondingly, $\mathbf{Grad}_{\mathrm{L}} f$ is called the *left gradient* of f and, due to the *uniqueness* of the limit, $\mathbf{Grad}_{\mathrm{L}} f = \mathbf{Grad}_{\mathrm{R}}^{\mathrm{T}} f$ and, in general, $\mathbf{Grad}_{\mathrm{R}} f[\mathrm{X}] \neq \mathbf{Grad}_{\mathrm{L}} f[\mathrm{X}]$.

Let $\gamma\colon \mathcal{E} \to \mathcal{E}$ be a *point field*, i.e., a mapping in $\mathcal{E}$, and $\gamma[X], \gamma[Y] \in \mathcal{E}$ for $\mathrm{Y} = X + \boldsymbol{a}$ and constant $\boldsymbol{a} \in \mathcal{V}$, then the point difference

$$\Delta\gamma[\mathrm{X}, \boldsymbol{a}] := \gamma[\mathrm{X}+\boldsymbol{a}] - \gamma[\mathrm{X}] \in \mathcal{V}$$

is a vector and also its limit

$$\lim_{\lambda \to 0} \frac{\Delta\gamma[\mathrm{X}, \lambda\, \boldsymbol{a}]}{\lambda} := (\mathbf{GRAD}_{\mathrm{R}}\gamma[\mathrm{X}])\, \boldsymbol{a} = \boldsymbol{a}\ \mathbf{GRAD}_{\mathrm{L}}\gamma[\mathrm{X}] \in \mathcal{V}. \qquad (4.9)$$

In the special case

$$\gamma[\mathrm{X}] - \mathrm{X} = \boldsymbol{b} = \mathit{constant} \quad \text{für alle} \quad \mathrm{X} \in \mathcal{E}, \qquad (4.10)$$

γ becomes a translation, i.e., a Euclidean vector $\gamma \equiv \boldsymbol{b} \in \mathcal{V}$, so that the relations $\Delta\gamma[\mathrm{X}, \boldsymbol{a}] = \boldsymbol{a}$ and $\mathbf{GRAD}_{\mathrm{R}}\mathrm{X} = \mathbf{GRAD}_{\mathrm{L}}\mathrm{X} = \boldsymbol{I} \in \mathcal{L}in[\mathcal{V}]$ hold for all $X \in \mathcal{E}$.

Remark 4.5 A new ***pitfall*** that can lead to misinterpretation is the fact that in the literature there is usually no distinction between $\mathbf{GRAD}_{\mathrm{L}}$ and $\mathbf{GRAD}_{\mathrm{R}}$, but only **GRAD**, where **GRAD** stands for $\mathbf{GRAD}_{\mathrm{L}}$ or $\mathbf{GRAD}_{\mathrm{R}}$, depending on the author's preference and tradition.

Examples of $\mathbf{GRAD} \equiv \mathbf{GRAD}_{\mathrm{L}}$

- *A. Duschek* and *A. Hochrainer* 1950, 1961 (Duschek and Hochrainer 1961) in *Tensoranalysis* (*Grundzüge der Tensorrechnung in analytischer Darstellung*, vol. II);
- *H. Tietz* 1955 (Tietz 1955) in *Geometrie* (*Encyclopedia of Physics* vol. II);
- *J. Serrin* 1959 (Serrin 1959) in *Mathematical Principles of Classical Fluid Mechanics* (*Encyclopedia of Physics*, vol. VIII/1);
- *A. I. Lurie* 1970 (1980) (Lurie 1970, 1980) in *(Nonlinear) Theory of Elasticity*;
- *S. Antman* 1995 (Antman 1995) in *Nonlinear Problems of Elasticity*;
- *H. Altenbach* 2012 (Altenbach 2012) in *Kontinuumsmechanik*.

Examples of $\mathbf{GRAD} \equiv \mathbf{GRAD}_\mathrm{R}$

- *C. A. Truesdell* 1952–1991 (Truesdell 1952; Truesdell and Toupin 1960; Truesdell and Noll 1965; Truesdell 1977), *W. Noll* 1955–2010 (Noll 1955; Truesdell and Noll 1965; Noll 1972, 1987) both also with other co-authors in the *Encyclopedia of Physics* as well as before and after publication of the encyclopedia;
- *M. E. Gurtin* 1966–2009 (Gurtin and Williams 1966; Gurtin 1972, 1981; Gurtin et al. 2009) in the *Encyclopedia of Physics* as well as with other co-authors before and after publication of the encyclopedia,
- *J. T. Oden* 1972 (Oden 1972) in *Finite elements of nonlinear continua*;
- *Ph. G. Ciarlet* 1988 (Ciarlet 1988): *Three-dimensional elasticity*;
- *J. E. Marsden* and *T. J. R. Hughes* 1983 (Marsden and Hughes 1983) in *Mathematical foundations of elasticity*;
- *I. Müller* 1985 (Müller 1985) in *Thermodynamics*, *A. Krawietz* 1986 (Krawietz 1986) in *Materialtheorie*;
- *A. Bertram* 2005–2012 (Bertram 2005) in *Elasticity and Plasticity of Large Deformations*;
- *M. Šilhavý* 1997 (Šilhavý 1997) in *The Mechanics and Thermodynamics of Continuous Media;*
- *R. W. Ogden* 1984 (Ogden 1984) in *Non-Linear Elastic Deformations.*

The Nabla symbol ∇ appears in the literature in a variety of ways:

- ∇ as a general, coordinate system-independent abbreviation, $\nabla := \mathbf{GRAD}$:
  - ◇ Without emphasizing,

$$\mathbf{GRAD}\, f \equiv \mathbf{GRAD}_\mathrm{R}\, f = \nabla f$$

is set so that with (4.7)

$$\lim_{\lambda \to 0} \frac{\Delta f[\mathrm{X}, \lambda\, \boldsymbol{a}]}{\lambda} = (\mathbf{GRAD}\, f[\mathrm{X}])\, \boldsymbol{a} = \nabla f[\mathrm{X}]\, \boldsymbol{a} \in \mathcal{V}$$

follows. In the *Encyclopedia of Physics* and beyond, this is used by *Truesdell* and *Noll* in (1965), *Gurtin* in (1972), and also by *Šilhavý* (1997).

  - ◇ Without emphasizing,

$$\mathbf{GRAD}\, f \equiv \mathbf{GRAD}_\mathrm{L}\, f = \nabla f$$

is set so that with (4.7)

$$\lim_{\lambda \to 0} \frac{\Delta f[\mathrm{X}, \lambda \boldsymbol{a}]}{\lambda} = \boldsymbol{a}\, \mathbf{GRAD} \boldsymbol{f}[\mathrm{X}] = \boldsymbol{a}\, \nabla \boldsymbol{f}[\mathrm{X}] \in \mathcal{V}$$

follows, see *Lurie* in (1970; 1980), *Antman* in (1995), and *Altenbach* in (2012).

- As a vector-valued operator with left-hand action

$$\mathbf{GRAD} \boldsymbol{f} \equiv \mathbf{GRAD}_{\mathrm{R}} \boldsymbol{f} = \boldsymbol{f} \otimes \nabla,$$

see *Krawietz* (1986) and *Bertram* (2005).

- As a vector-valued operator with left-hand action and with right-hand action

$$\mathbf{GRAD}_{\mathrm{R}} \boldsymbol{f} = \boldsymbol{f} \otimes \nabla \quad \text{and} \quad \mathbf{GRAD}_{\mathrm{L}} \boldsymbol{f} = \nabla \otimes \boldsymbol{f},$$

see *Schade* (1996).

- ***Pitfall***, due to different notation from all other authors. ∇ is written as a left operator, but evaluated as a right operator:

$$\mathbf{GRAD} \boldsymbol{f} \equiv \mathbf{GRAD}_{\mathrm{R}} \boldsymbol{f} := \nabla \otimes \boldsymbol{f},$$

see *Ogden* (1984) and *Holzapfel* (2000).

The trace of the gradient of a vector field $\boldsymbol{f}$ in X gives the divergence of a vector field in X:

$$\mathrm{DIV}\, \boldsymbol{f}[\mathrm{X}] := \mathrm{tr}\big[\mathbf{GRAD} \boldsymbol{f}[\mathrm{X}]\big] \equiv \mathrm{tr}\big[\mathbf{GRAD}_{\mathrm{L}} \boldsymbol{f}[\mathrm{X}]\big] = \mathrm{tr}\big[\mathbf{GRAD}_{\mathrm{R}} \boldsymbol{f}[\mathrm{X}]\big] \in \mathbb{R}.$$

In particular, for a point field describing a translation according to (4.10),

$$\mathrm{DIV}[\mathrm{X} + \boldsymbol{b}] = \mathrm{DIV}[\mathrm{X}] = 3,$$

since $\boldsymbol{b} \in \mathcal{V}$ is independent of X there.

Remark 4.6 When using the Nabla operator,

$$\mathrm{DIV}\, \boldsymbol{f}[\mathrm{X}] = \nabla \cdot \boldsymbol{f}[\mathrm{X}] = \boldsymbol{f}[\mathrm{X}] \cdot \nabla.$$

If *Cartesian* coordinates are used, $\mathrm{DIV}\, \boldsymbol{f}[\mathrm{X}] = \frac{\partial \hat{f}_i[\mathbf{X}]}{\partial X_i}$.

The gradients of a lineon field are third-order tensors. They are elements of the space of trilinear mappings $\mathcal{L}in_3[\mathcal{V}, \mathbb{R}]$, which can be identified with the space of bilinear mappings $\mathcal{L}in_2[\mathcal{L}in[\mathcal{V}],\mathcal{V}]$ and the linear mappings $\mathcal{L}in[\mathcal{V},\mathcal{L}in[\mathcal{V}]]$. The latter can be used to define three different gradients of a lineon field $\boldsymbol{G}[\mathrm{X}]$. In addition to the right and left gradients $\mathbf{GRAD}_\mathrm{R}\boldsymbol{G}[\mathrm{X}]$ and $\mathbf{GRAD}_\mathrm{L}\boldsymbol{G}[\mathrm{X}]$, there is also the *center gradient* $\mathbf{GRAD}_\mathrm{C}\boldsymbol{G}[\mathrm{X}]$ and it holds for fixed $\boldsymbol{a}, \boldsymbol{b}, \boldsymbol{c} \in \mathcal{V}$

$$\begin{aligned} & =: (\boldsymbol{a}\,\mathbf{GRAD}_\mathrm{L}\boldsymbol{G}[\mathrm{X}]\,\boldsymbol{b})\boldsymbol{c} \\ \lim_{\lambda\to 0}\frac{\boldsymbol{b}\big(\boldsymbol{G}[\mathrm{X}+\lambda\,\boldsymbol{a}]-\boldsymbol{G}[\mathrm{X}]\big)\boldsymbol{c}}{\lambda} & =: (\boldsymbol{b}\,\mathbf{GRAD}_\mathrm{C}\boldsymbol{G}[\mathrm{X}]\,\boldsymbol{c})\boldsymbol{a} \\ & =: (\boldsymbol{b}\,\mathbf{GRAD}_\mathrm{R}\boldsymbol{G}[\mathrm{X}]\,\boldsymbol{a})\boldsymbol{c} \in \mathbb{R}, \end{aligned}$$

so that the vector field equations

$$\boldsymbol{a}\,\mathbf{GRAD}_\mathrm{L}\boldsymbol{G}[\mathrm{X}]\,\boldsymbol{b} = \boldsymbol{b}\,\mathbf{GRAD}_\mathrm{R}\boldsymbol{G}[\mathrm{X}]\,\boldsymbol{a} \qquad \text{and}$$

$$(\mathbf{GRAD}_\mathrm{C}\boldsymbol{G}[\mathrm{X}]\,\boldsymbol{c})\boldsymbol{a} = (\mathbf{GRAD}_\mathrm{R}\boldsymbol{G}[\mathrm{X}]\,\boldsymbol{a})\boldsymbol{c}$$

result. With the three gradients of a lineon field, only two different divergences can be defined: the *right divergence of a lineon field*

$$\mathrm{DIV}_\mathrm{L}\boldsymbol{G}[\mathrm{X}] := \boldsymbol{I}\cdot\cdot\,\mathbf{GRAD}_\mathrm{L}\boldsymbol{G}[\mathrm{X}] = \boldsymbol{I}\cdot\cdot\,\mathbf{GRAD}_\mathrm{C}\boldsymbol{G}[\mathrm{X}] \in \mathcal{V}$$

and the *left divergence of a lineon field*

$$\mathrm{DIV}_\mathrm{R}\boldsymbol{G}[\mathrm{X}] := \mathbf{GRAD}_\mathrm{R}\boldsymbol{G}[\mathrm{X}]\cdot\cdot\,\boldsymbol{I} = \mathbf{GRAD}_\mathrm{C}\boldsymbol{G}[\mathrm{X}]\cdot\cdot\,\boldsymbol{I} \in \mathcal{V}. \tag{4.11}$$

Since the author of these explanations is not aware of the use of $\mathbf{GRAD}_\mathrm{C}\boldsymbol{G}$ in the literature, this gradient will be excluded from further considerations.

Remark 4.7 Another ***pitfall*** that can lead to misinterpretation is the fact that in the literature it is usually written only $\mathrm{DIV}\,\boldsymbol{G}$, which can stand for $\mathrm{DIV}_\mathrm{L}\boldsymbol{G}$ for some authors and $\mathrm{DIV}_\mathrm{R}\boldsymbol{G}$ for others.

- *Truesdell* and *Noll* in (1965), *Gurtin* in (1972), *I. Müller* (1985) and *Šilhavý* in (1997) use

  $$\mathrm{DIV}\,\boldsymbol{G} := \mathrm{DIV}_\mathrm{R}\boldsymbol{G} = (\mathbf{GRAD}_\mathrm{R}\boldsymbol{G})\cdot\cdot\,\boldsymbol{I} := (\mathbf{GRAD}\,\boldsymbol{G})\cdot\cdot\,\boldsymbol{I},$$

  which in *Krawietz* Krawietz (1986) is described by $\boldsymbol{G}\cdot\nabla = (\boldsymbol{G}\otimes\nabla)\cdot\cdot\,\boldsymbol{I} = \mathrm{DIV}_\mathrm{R}\boldsymbol{G}$.

- *Antman* in (1995), *Lurie* in (1970; 1980), and *H. Altenbach* (2012) use

  $$\mathrm{DIV}\,\boldsymbol{G} = \mathrm{DIV}_\mathrm{L}\boldsymbol{G} = \boldsymbol{I}\cdot\cdot\,(\mathbf{GRAD}_\mathrm{L}\boldsymbol{G}) := \boldsymbol{I}\cdot\cdot\,(\mathbf{GRAD}\,\boldsymbol{G})$$

described by $\nabla \cdot \boldsymbol{G} = \boldsymbol{I} \cdot\cdot (\nabla \otimes \boldsymbol{G}) = \mathrm{DIV_L}\,\boldsymbol{G}$.

- *Schade* in (1996) uses

$$\mathrm{DIV_L}\boldsymbol{G} = \boldsymbol{I} \cdot\cdot (\mathbf{GRAD_L}\boldsymbol{G}) \quad \text{and} \quad \mathrm{DIV_R}\boldsymbol{G} = (\mathbf{GRAD_R}\boldsymbol{G}) \cdot\cdot \boldsymbol{I}$$

described by

$$\nabla \cdot \boldsymbol{G} = \boldsymbol{I} \cdot\cdot (\nabla \otimes \boldsymbol{G}) = \mathrm{DIV_L}\boldsymbol{G} \quad \text{and} \quad \boldsymbol{G} \cdot \nabla = (\boldsymbol{G} \otimes \nabla) \cdot\cdot \boldsymbol{I} = \mathrm{DIV_R}\boldsymbol{G}.$$

- *Holzapfel* in (2000) uses

$$\mathrm{DIV}\,\boldsymbol{G} =: \mathrm{DIV_R}\boldsymbol{G} = (\mathbf{GRAD_R}\boldsymbol{G}) \cdot\cdot \boldsymbol{I} := (\mathbf{GRAD}\,\boldsymbol{G}) \cdot\cdot \boldsymbol{I}$$

described by

$$\nabla \cdot \boldsymbol{G} =: (\nabla \otimes \boldsymbol{G}) \cdot\cdot \boldsymbol{I} = \mathrm{DIV_R}\boldsymbol{G},$$

where ∇ is written as a left operator, but evaluated as a right operator. ***Pitfall***, due to different notation from all other authors.

- *Ogden* in (1984) uses the left divergence with the right gradient

$$\mathrm{DIV}\,\boldsymbol{G} =: \mathrm{DIV_L}\boldsymbol{G} = (\mathbf{GRAD_R}\boldsymbol{G}^{\mathrm{T}}) \cdot\cdot \boldsymbol{I} := (\mathbf{GRAD}\,\boldsymbol{G}^{\mathrm{T}}) \cdot\cdot \boldsymbol{I}$$

described by

$$\nabla \cdot \boldsymbol{G} =: (\nabla \otimes \boldsymbol{G}^{\mathrm{T}}) \cdot\cdot \boldsymbol{I} = \mathrm{DIV_L}\boldsymbol{G},$$

where ∇ is written as a left operator, but evaluated as a right operator. ***Two Pitfalls***, due to different notation from all other authors.

If $\boldsymbol{G}$ is *symmetric*, i.e., for $\boldsymbol{G} = \boldsymbol{G}^{\mathrm{T}}$, there is no need to distinguish between $\mathrm{DIV_L}$ and $\mathrm{DIV_R}$, because then $\mathrm{DIV_R}\boldsymbol{G} = \mathrm{DIV_L}\boldsymbol{G}$ holds.

Let $\Omega \subsetneq \mathcal{E}$ be an *open domain* in *Euclidean* space with the *boundary* $\partial\Omega \subsetneq \mathcal{E}$ for which there exists a *volume* $V[\Omega] > 0$ such that

$$V[\Omega] = \int_{\Omega} \mathrm{d}V[\mathrm{X}] = \frac{1}{3}\int_{\partial\Omega} \mathrm{d}a[\mathrm{X}] = \frac{1}{3}\int_{\partial\Omega} \boldsymbol{n}[\mathrm{X}]\mathrm{d}\boldsymbol{a}[\mathrm{X}],$$

where $\mathrm{d}a = |\mathrm{d}\boldsymbol{a}|$ is the absolute value and $\boldsymbol{n} = \frac{\mathrm{d}\boldsymbol{a}}{\mathrm{d}a}$ is the outer normal of the *surface vector* $\mathrm{d}\boldsymbol{a}$. Furthermore, there should exist a quantity $\beta[\Omega] \in \mathcal{V}$ that can be represented as a *volume integral* over a *volume density* $\beta_V \in \mathcal{V}$,

$$\beta[\Omega] = \int_{\Omega} \beta_V[\mathrm{X}]\,\mathrm{d}V[\mathrm{X}],$$

and there exists a lineon field $\boldsymbol{B}[\mathrm{X}] \in \mathcal{L}in[\mathbf{V}]$ for which

$$\mathrm{DIV_R}\,\boldsymbol{B}[\mathrm{X}] = \mathrm{DIV_L}\,\boldsymbol{B}^{\mathrm{T}}[\mathrm{X}] = \beta_V[\mathrm{X}]$$

the following applies

$$\begin{aligned}\beta[\Omega] &= \int_{\Omega} \mathrm{DIV_R}\,\boldsymbol{B}[\mathrm{X}]\,\mathrm{d}V[\mathrm{X}] = \int_{\Omega} \mathrm{DIV_L}\,\boldsymbol{B}^{\mathrm{T}}[\mathrm{X}]\,\mathrm{d}V[\mathrm{X}] \\ &= \int_{\partial\Omega} \boldsymbol{B}[\mathrm{X}]\boldsymbol{n}[\mathrm{X}]\,\mathrm{d}a[\mathrm{X}] = \int_{\partial\Omega} \boldsymbol{n}[\mathrm{X}]\boldsymbol{B}^{\mathrm{T}}[\mathrm{X}]\,\mathrm{d}a[\mathrm{X}] = \int_{\partial\Omega} \boldsymbol{b}[\mathrm{X}]\,\mathrm{d}a[\mathrm{X}].\end{aligned} \tag{4.12}$$

Therefore, $\boldsymbol{B}$ is linear in $\boldsymbol{n}$. If $\boldsymbol{A} := \boldsymbol{B}^{\mathrm{T}}$ and b_i, A_{ij}, $B_{ij} \in \mathbb{R}$ describe the direction of $\boldsymbol{b}$ and the respective other indices describe the direction of the surface normal $\boldsymbol{n}$.

4.3 Material Bodies in Euclidean Space

In the most general concepts of *material bodies* $\mathcal{B}$, each body $\mathcal{B}$, and each *subbody* $\mathcal{B}_{(I)} \subset \mathcal{B}$ and $\mathcal{B}_{(J)} \subset \mathcal{B}$ are clearly associated with strictly positive numbers

$$m[\mathcal{B}] > 0,\, m[\mathcal{B}_{(I)}] > 0 \quad \text{and} \quad m[\mathcal{B}_{(J)}] > 0$$

by definition, which are called the *mass* of $\mathcal{B}$ and the mass of its *subbodies* $\mathcal{B}_{(I)}$ and $\mathcal{B}_{(J)}$, where $I, J \in \mathcal{M}$ are elements of an index set $\mathcal{M}$. It holds

$$m[\mathcal{B}_{(I)} \cup \mathcal{B}_{(J)}] = m[\mathcal{B}_{(I)}] + m[\mathcal{B}_{(J)}], \quad \text{if} \quad \mathcal{B}_{(I)} \cap \mathcal{B}_{(J)} = \emptyset. \tag{4.13}$$

If $\mathcal{B} = \bigcup\limits_{I \in \mathcal{M}} \mathcal{B}_{(I)}$ and $\mathcal{B}_{(I)} \cap \mathcal{B}_{(J)} = \emptyset$ for all $I, J \in \mathcal{M}$, then $m[\mathcal{B}] = \sum\limits_{I \in \mathcal{M}} m[\mathcal{B}_{(I)}]$. The elements $\mathcal{x} \in \mathcal{B}$ of a material body are here called *particles*, as they were also called, physically motivated, in the *Encyclopedia of Physics* (Truesdell and Toupin 1960; Truesdell and Noll 1965) in 1960 and 1965.

Remark 4.8

Noll used the mathematically motivated term of the "*material point*" later in 1965 in Noll (1972), just as *Kirchhoff* had used it 100 years earlier in 1876 in Kirchhoff (1876). Since this term was also used by other authors to denote reference points $X \in \Omega_R \subsetneq \mathcal{E}$, which are used to identify $x \in \mathcal{B}$ in $\mathcal{E}$ by means of the invertible mapping κ_R: $\mathcal{B} \rightarrow \mathcal{E}$, *Truesdell* then proposed to use the term "*substantial point*" to distinguish it from *Euclidean* points (Truesdell 1977) in 1977. *W. Dreyer*, like *I. Müller* in Dreyer et al. (1982), Müller (1985), preferred the term particle for an element of the material body.

4.3.1 *Identification of Particles in Euclidean Space*

Remark 4.9 Understanding the following terms is difficult and full of ***pitfalls*** because they are defined differently in the literature:

1. A term for a phenomenon has been subsequently changed by the introductory authors because it does not match the terms used in everyday language for that phenomenon. It becomes problematic if the publication with the new term has a significantly smaller readership than the old one.
2. The mathematical meaning of the terms is changed in that no clear distinction is made between the elements of the metricless, differentiable manifold $\mathcal{B}$, the Euclidean point space $\mathcal{E}$, the Euclidean vector space $\mathcal{V}$, and the space of real number triples $\mathbb{R}^3$. Instead, $\mathbb{R}^3$ is introduced as both Euclidean point space and vector space, where $\mathcal{B}$ is then understood as a subset of $\mathbb{R}^3$ with volume.
3. A term already introduced by other authors is redefined for a phenomenon which is very similar to the original phenomenon, but not the same. In the best case, the redefinition is mentioned in the published text, but usually not.

A mapping $\kappa_a : \mathcal{B} \rightarrow \mathcal{E}$ is called a *placement* (German: *Platzierung*). It uniquely assigns a place, i.e., a position, in $\mathcal{E}$ to a particle from $\mathcal{B}$ for a fixed index a. The subscript a to the right of κ is an index that is a time index for $a \equiv t$ or $a \equiv \tau$ and is the reference index for $a \equiv \mathrm{R}$.

Remark 4.10

In the 1960s, the value $\kappa_a[x] \in \kappa_a[\mathcal{B}]$ of the mapping κ_a is still called a *configuration* in the *Encyclopedia of Physics* (Truesdell and Toupin 1960; Truesdell and Noll 1965). This term has since been adopted by numerous authors and is still used today. The authors of the Encyclopedia, *Noll* in 1972 (Noll 1972) and *Truesdell* in 1977 (Truesdell 1977), changed the term to *placement*, which they felt was more appropriate.

A placement that is introduced only for reference purposes to identify particles in Euclidean space is called a *reference placement*. If such a placement is completely independent of the present time $t \in \mathbb{R}$ or the past time $\tau \in \mathbb{R}$ with $\tau < t$, it is symbolized by κ_R, so that

$$\mathrm{X} = \kappa_\mathrm{R}[x] \in \kappa_\mathrm{R}[\mathcal{B}] = \Omega_\mathrm{R} \subsetneqq \mathcal{E}.$$

Correspondingly, the mapping

$$\kappa_t \colon \mathcal{B} \times [t_0, t_\mathrm{end}] \to \Omega_{[t_0, t_\mathrm{end}]} \subsetneqq \mathcal{E}$$

is called the *present placement*. For the present time $t \in [t_0, t_\mathrm{end}] \subsetneqq \mathbb{R}$ from a real-time interval bounded by the start time t_0 and the end time t_end, then

$$\mathrm{x}_t = \kappa_t[x] \in \Omega_t = \kappa_t[\mathcal{B}] \subsetneqq \mathcal{E}. \tag{4.14}$$

Remark 4.11 Again, the notation κ_0 can lead to misunderstandings. On the one hand, $\kappa_0 := \kappa_{t_0}$ is sensibly defined as the initial placement; on the other hand, $\kappa_0 := \kappa_\mathrm{R}$ is misleadingly defined as the reference placement, which can be freely chosen and therefore generally does not coincide with the initial placement. Therefore, only in the special case of $\kappa_0 = \kappa_{t_0} = \kappa_\mathrm{R}$ does it make sense to specify the reference placement.

If $\overleftarrow{\kappa_\mathrm{R}} \colon \Omega_\mathrm{R} \to \mathcal{B}$ denotes the inverse mapping of κ_R, then the mapping

$$\chi_t := \kappa_t \overleftarrow{\kappa_\mathrm{R}} : \Omega_\mathrm{R} \to \Omega_t \tag{4.15}$$

called *transplacement* (German: *Umplatzierung*) can be defined for a current time $t \in [t_0, t_\mathrm{end}]$, which represents a *change of placement* (German: *Platzierungswechsel*, see Krawietz (1986)) from the reference placement to the present placement, which in particular can take place without any change of shape. Strictly speaking, it is also not the result of a movement from the reference placement, since the material body does not have to be occupied by its reference placement at any time. For (4.14) and (4.15) therefore applies

$$x_t = \kappa_t\big[x\big] = \kappa_t\big[\overleftarrow{\kappa_\mathrm{R}}[X]\big] = \kappa_t \overleftarrow{\kappa_\mathrm{R}}[\mathrm{X}] = \chi_t[\mathrm{X}].$$

Relative transplacements, which refer to the positions really occupied at the time $\tau \in [t_0, t_{\text{end}}]$ are defined by $\chi_{t\tau} := \kappa_t \kappa_\tau^{\leftarrow}$. They are not discussed further here.

Remark 4.12 *Encyclopedia of Physics* (Truesdell and Toupin 1960; Truesdell and Noll 1965), the change in position is still called "*deformation*". This old term has since been adopted by numerous authors and is still used today. The use of this term for a change of position of a material body, which describes any rotation of a *rigid body*, then leads to the linguistically questionable statement of a "*large or finite deformation*", although by definition a rigid body cannot have any deformation at all in common parlance. Because of this imprecision, *Noll* suggested in later publications, e.g., 1972 in Noll (1972), to use the term "*transplacement*" instead of "*deformation*". This was adopted by *Truesdell* in 1977 in Truesdell (1977).

With the above definitions, time-dependent quantities can be described on material bodies at time t as field quantities f with respect to the selected reference placement or the present placement:

$$f[t, \mathrm{X}] = f\big[t, \chi_{\mathrm{t}}^{\leftarrow}[\mathrm{x}_t]\big] = \tilde{f}[t, \mathrm{x}_t]. \tag{4.16}$$

The description with respect to the reference placement $f[t, \mathrm{X}]$ was first called *Lagrangian description* and *later material* description (Truesdell and Toupin 1960; Truesdell and Noll 1965), because with each point $\mathrm{X} \in \mathcal{E}$ in the Euclidean space $\mathcal{E}$ a particle $x \in \mathcal{B}$ of the material body $\mathcal{B}$ can be identified and with $f[t, \mathrm{X}]$, for a fixed X its temporal change can be described. In Šilhavý (1997) *Šilhavý* speaks of a *referential description*. In contrast, the description of present placement was first called *Eulerian description* and later *spatial description* (Truesdell and Toupin 1960; Truesdell and Noll 1965). In Šilhavý (1997)] *Šilhavý* speaks of *actual description*. Here $\mathrm{x} \equiv \mathrm{x}_t$ and $\tilde{f}[t, \mathrm{x}] \equiv \tilde{f}[t, \mathrm{x}_t]$ for fixed $\mathrm{x} \in \mathcal{E}$ can be used to describe the temporal change of the field quantity $\tilde{f}[t, \mathrm{x}]$ at fixed locations in $\mathcal{E}$. If, for times $\tau, t \in [t_0, t_{\text{end}}]$, tends with $\tau \neq t$, the particle $x \in \mathcal{B}$ with $\kappa_{\mathrm{R}}[x] = \mathrm{X}$ is at a fixed location $\mathrm{x} \in \mathcal{E}$ defined by $\mathrm{x} = \mathrm{x}_t = \kappa_{\mathrm{t}}[x]$ at time t, and a particle $y \in \mathcal{B}$ is at the same location at time τ, for which $y \neq x \in \mathcal{B}$ and thus $\kappa_{\mathrm{R}}[y] \neq \mathrm{X}$ holds in general.

The time-continuous sequence of transplacements $\chi[t, \mathrm{X}] \equiv \chi_t[\mathrm{X}]$ represents the *motion* of the particle corresponding to the point $\mathrm{X} \in \mathcal{E}$ in the selected reference placement for fixed $\mathrm{X} \in \Omega_{\mathrm{R}}$. Thus,

$$\chi : [t_0, t_{\text{end}}] \times \Omega_{\mathrm{R}} \to \Omega_t$$

with $[t_0, t_{\text{end}}] \subsetneq \mathbb{R}$ and $\Omega_{\mathrm{R}}, \Omega_t \subsetneq \mathcal{E}$.

4.3.2 Referential and Spatial Gradients and Time Derivatives of Field Quantities

The partial time derivatives of time-dependent field quantities in the referential description $f[t, \mathrm{X}] \in \mathbb{R}$, $\gamma[t, \mathrm{X}] \in \mathcal{E}$, $\boldsymbol{f}[t, \mathrm{X}] \in \boldsymbol{\mathcal{V}}$ and $\boldsymbol{A}[t, \mathrm{X}] \in \mathcal{L}in[\boldsymbol{\mathcal{V}}]$ are called substantial, material, or referential time derivatives, for which the notation

$$\dot{f}[t, \mathrm{X}] := \frac{\partial f[t, \mathrm{X}]}{\partial t} \in \mathbb{R}, \quad \dot{\gamma}[t, \mathrm{X}] := \frac{\partial \gamma[t, \mathrm{X}]}{\partial t} \in \boldsymbol{\mathcal{V}}, \tag{4.17}$$

$$\dot{\boldsymbol{f}}[t, \mathrm{X}] := \frac{\partial \boldsymbol{f}[t, \mathrm{X}]}{\partial t} \in \boldsymbol{\mathcal{V}} \text{ and } \dot{\boldsymbol{A}}[t, \mathrm{X}] := \frac{\partial \boldsymbol{A}[t, \mathrm{X}]}{\partial t} \in \mathcal{L}in[\boldsymbol{\mathcal{V}}]$$

is common. The right gradient of a vector field $\boldsymbol{f}[t, \mathrm{X}] \equiv \boldsymbol{f}_t[\mathrm{X}] \in \boldsymbol{\mathcal{V}}$ is according to (4.8) represented by

$$\mathbf{Grad}\, \boldsymbol{f}[t, \mathrm{X}] \equiv \mathbf{GRAD}_{\mathrm{R}}\, \boldsymbol{f}_t[\mathrm{X}] \in \mathcal{L}in[\boldsymbol{\mathcal{V}}]$$

and is called the *material* or *referential* (right) *gradient.*

A point field $\gamma[t, \mathrm{X}] \equiv \gamma_t[\mathrm{X}] \in \mathcal{E}$ has according to (4.9) the referential (right) gradient

$$\mathbf{Grad}\, \gamma[t, \mathrm{X}] \equiv \mathbf{GRAD}_{\mathrm{R}} \gamma_t[\mathrm{X}] \in \mathcal{L}in[\boldsymbol{\mathcal{V}}].$$

The right divergence of $\boldsymbol{A}[t, \mathrm{X}] \in \mathcal{L}in[\boldsymbol{\mathcal{V}}]$ is according to (4.11) determined by

$$\mathrm{Div} \boldsymbol{A}[t, \mathrm{X}] \equiv \mathrm{DIV}_{\mathrm{R}} \boldsymbol{A}[t, \mathrm{X}] \in \boldsymbol{\mathcal{V}}.$$

For fixed $\mathrm{x} \in \mathcal{E}$, the partial time derivatives $\frac{\partial \tilde{f}[t,\mathrm{x}]}{\partial t}$, $\frac{\partial \tilde{\gamma}[t,\mathrm{x}]}{\partial t}$, $\frac{\partial \tilde{\boldsymbol{f}}[t,\mathrm{x}]}{\partial t}$, and $\frac{\partial \tilde{\boldsymbol{A}}[t,\mathrm{x}]}{\partial t}$ are called *spatial time derivatives*. According to (4.8), the right gradient of $\tilde{\boldsymbol{f}}[t, \mathrm{x}] \in \boldsymbol{\mathcal{V}}$ is represented by

$$\mathbf{grad}\, \tilde{\boldsymbol{f}}[t, \mathrm{x}] \equiv \mathbf{GRAD}_{\mathrm{R}} \tilde{\boldsymbol{f}}[t, \mathrm{x}] \in \mathcal{L}in[\mathcal{V}]$$

and is called the *spatial* (right) *gradient* of $\tilde{\boldsymbol{f}}[t, \mathrm{x}]$.

In the same way $\mathbf{grad}\, \tilde{\gamma}[t, \mathrm{x}] \in \mathcal{L}in[\mathcal{V}]$ and $\mathrm{div} \tilde{\boldsymbol{A}}[t, \mathrm{x}] \in \mathcal{V}$ are defined with the right gradient and divergence according to (4.9) and (4.11).

4.3.3 Velocity, Deformation, and Configuration: Time Derivatives and Gradients of Transplacements

The material time derivative of the motion point field $\chi[t, \mathrm{X}] \in \mathcal{E}$ results in the vectorial *velocity field*

$$v[t, \mathrm{X}] := \dot{\chi}[t, \mathrm{X}]$$

of the moving particle $x \in \mathcal{B}$, whose reference placement is just $\mathrm{X} \in \mathcal{B}$.

The referential gradient of the transplacement at time t is called *transplacement gradient* (German: *Umplatzierungsgradient*):

$$\boldsymbol{F}[t, \mathrm{X}] := \mathbf{Grad}\chi[t, \mathrm{X}] \equiv \mathbf{Grad}_{\mathrm{R}}\chi[t, \mathrm{X}]. \tag{4.18}$$

The lineon

$$\boldsymbol{C}[t, \mathrm{X}] := \boldsymbol{F}^{\mathrm{T}}\boldsymbol{F}[t, \mathrm{X}] \tag{4.19}$$

is called the *right Cauchy–Green tensor*. If $X, Y \in \mathcal{E}$ and also $\boldsymbol{C}[t, \mathrm{X}]=\boldsymbol{C}[t, \mathrm{Y}]$, then X and Y have the same *configuration*. If $\boldsymbol{C}[t, \mathrm{X}]=\boldsymbol{C}[t, \mathrm{Y}]$ holds for arbitrary $\mathrm{X}, \mathrm{Y} \in \Omega_{\mathrm{R}} = \kappa_{\mathrm{R}}[\mathcal{B}]$, $\mathcal{B}$ and each subbody of $\mathcal{B}$ has the same *shape* at the time t. If $\boldsymbol{C}[t, \mathrm{X}] - \boldsymbol{I} \neq \mathbf{O}$ holds, there is a *deformation* in the neighborhood of X with respect to the reference configuration. If the right *Cauchy–Green* tensor does not change in a certain time interval, its configuration in X does not change. If this is true for all $\mathrm{X} \in \Omega_{\mathrm{R}}$, then the shape of $\mathcal{B}$ in $\mathcal{E}$ does not change in that time interval. The body moves undeformed or rigidly in a certain time interval if it has the same shape at every time t in this time interval and $\boldsymbol{C} = \boldsymbol{I}$ holds.

Remark 4.13 $\boldsymbol{F}$ is usually referred to as "*deformation gradient*" in today's literature, a term first used by *Truesdell*, *Toupin*, and *Noll* in volumes III/1 and III/3 of the *Encyclopedia of Physics* (Truesdell and Toupin 1960; Truesdell and Noll 1965). Five years before the publication of volume III/1 of the encyclopedia, the same authors still used the term "*displacement gradient*" instead of "*deformation gradient*" in their publications, see, e.g., *Truesdell* 1952 (Truesdell 1952) for both $\boldsymbol{F}$ and $\boldsymbol{F}-\boldsymbol{I}$ and *Noll* 1955 (Noll 1955) only for $\boldsymbol{F}$. By referring only to $\boldsymbol{F}-\boldsymbol{I}$ as the "*displacement gradient*" and $\boldsymbol{F}$ as the "*deformation gradient*" in the encyclopedia, it was possible to establish clarity regarding the choice of term, but the linguistic imprecision in the use of the term "*deformation*" remained. This was only removed when *Noll* renamed it to "*transplacement gradient*" in Noll (1972) 1972. *Truesdell*, who adopted *Noll*'s term "*transplacement*" in the first edition of Truesdell (1977) in 1977 but retained "*deformation gradient*", also changed the term to "*transplacement gradient*" in the second edition of Truesdell (1977) in 1991.

Remark 4.14 According to the authors of the encyclopedia, $\boldsymbol{F}$ is a gradient by definition, because these authors imply $\mathbf{Grad} := \mathbf{GRAD}_{\mathrm{R}}$, i.e., gradients are always right gradients. *Antman* in (1995) defines $\mathbf{Grad} := \mathbf{GRAD}_{\mathrm{L}}$, i.e., gradients are left gradients, so for him $\boldsymbol{F}[t, \mathrm{X}] = \left(\mathbf{Grad}\chi[t, \mathrm{X}]\right)^{\mathrm{T}} = \left(\mathbf{GRAD}_{\mathrm{L}}\chi[t, \mathrm{X}]\right)^{\mathrm{T}}$ becomes a transposed deformation gradient, see also *Altenbach* (2012).

Remark 4.15 Replacing the term "*configuration*" by the term "*placement*" freed *Noll* to use the term "*configuration*", which he used from 1972 (Noll 1972) for the *class of placements that differ only in rigid body modifications* (*translations* and *rotations*). *Deformations* are present in a transplacement, e.g., a placement change, only if the placements before and after the change belong to different configurations. This created a ***new pitfall*** in the use of the word "*configuration*", because since then it has to be clear in more recent literature whether "*configuration*" is used in the sense of the encyclopedia or in the sense of *Noll*'s redefinition in 1972 (Noll 1972).

With (4.16), the spatial representation of velocity is

$$\tilde{\boldsymbol{v}}[t, \mathrm{x}_t] = \boldsymbol{v}\big[t, \chi_t^{\leftarrow}[\mathrm{x}_t]\big] = \boldsymbol{v}[t, X] = \dot{\chi}[t, \mathrm{X}],$$

so that (4.17) becomes

$$\dot{f}[t, \mathrm{X}] = \dot{\tilde{f}}[t, \mathrm{x}] = \frac{\partial \tilde{f}[t, \mathrm{x}]}{\partial t} + \mathbf{grad}\,\tilde{f}[t, \mathrm{x}]\,\dot{\chi}[t, \mathrm{X}] = \frac{\partial \tilde{f}[t, \mathrm{x}]}{\partial t} + \mathbf{grad}\,\tilde{f}[t, \mathrm{x}]\,\tilde{\boldsymbol{v}}[t, \mathrm{x}].$$

If $\Omega_t = \chi[t, \Omega_R]$ and $V[\Omega_\mathrm{R}] = \int_{\Omega_\mathrm{R}} \mathrm{d}V_\mathrm{R}[\mathrm{X}]$, then

$$V[\Omega_t] = \int_{\Omega_t} \mathrm{d}V_t[\mathrm{x}] = \int_{\Omega_\mathrm{R}} \det\big[\boldsymbol{F}[t, X]\big]\,\mathrm{d}V_R[\mathrm{X}]$$

and therefore

$$\mathrm{d}V_t[\mathrm{x}] = \det\big[\boldsymbol{F}[t, X]\big]\,\mathrm{d}V_R[\mathrm{X}]. \tag{4.20}$$

4.4 Balance Equations of Material Volumes Without Singularities

In the following, only those continuous material bodies $\mathcal{B}$ are considered, which consist exclusively of volumes with mass. For (4.13), this means that

$$m[\mathcal{B}_{(I)}] = 0 \iff V\big[\Omega_\mathrm{R}[\mathcal{B}_{(I)}]\big] = 0. \tag{4.21}$$

This excludes surfaces, lines, or single points with mass, as used in the structure mechanics of infinitely thin shells and strings, e.g., in Antman (1995). Let $m[\mathcal{B}] > 0$ be the mass of such a body and $\mathrm{x} \in \Omega_t$ be a point of its placement at time t, then for each point $\mathrm{x} \in \Omega_t$ there exists a strictly positive *volume mass density* $\tilde{\varrho}[t, \mathrm{x}] > 0$, such that

$$m[\mathcal{B}] = \int_{\Omega_t} \tilde{\varrho}[t, \mathrm{x}]\mathrm{d}V_t[\mathrm{x}] \tag{4.22}$$

holds. Furthermore, such a reference placement $\Omega_\mathrm{R} = \kappa_\mathrm{R}[\mathcal{B}] \subsetneq \mathcal{E}$ with the volume $V_\mathrm{R} = V[\Omega_\mathrm{R}]$ can always be chosen for which the volume mass density $\varrho_\mathrm{R} > 0$ is constant, so that $m[\mathcal{B}] = \varrho_\mathrm{R}\, V_\mathrm{R} = \int_{\Omega_\mathrm{R}} \varrho_\mathrm{R} \mathrm{d}V_R[\mathrm{X}]$. With (4.22) and (4.20) it follows that

$$\varrho_\mathrm{R} = \varrho[t, \mathrm{X}] \det\big[\boldsymbol{F}[t, \mathrm{X}]\big].$$

Time-dependent scalar quantities $\Phi[t, \mathcal{B}] \in \mathbb{R}$ or vectorial quantities $\boldsymbol{\Phi}[t, \mathcal{B}] \in \mathcal{V}$, associated with a material body $\mathcal{B}$ can be balanced if they are *additive*, i.e., if

$$\begin{aligned} \Phi[t, \mathcal{B}_{(I)} \cup \mathcal{B}_{(J)}] &= \Phi[t, \mathcal{B}_{(I)}] + \Phi[t, \mathcal{B}_{(J)}] \quad \text{and} \\ \boldsymbol{\Phi}[t, \mathcal{B}_{(I)} \cup \mathcal{B}_{(J)}] &= \boldsymbol{\Phi}[t, \mathcal{B}_{(I)}] + \boldsymbol{\Phi}[t, \mathcal{B}_{(J)}] \end{aligned}$$

hold for *separate* partial bodies $\mathcal{B}_{(I)} \cap \mathcal{B}_{(J)} = \emptyset$, just as for the mass in (4.13).

If, for arbitrary quantities $\Phi[t, \mathcal{B}]$ and $\boldsymbol{\Phi}[t, \mathcal{B}]$ with the above properties,

$$\Phi[t, \mathcal{B}_{(I)}] = 0, \; \boldsymbol{\Phi}[t, \mathcal{B}_{(I)}] = 0 \quad \Longleftrightarrow \quad V\big[\Omega_\mathrm{R}[\mathcal{B}_{(I)}]\big] = 0,$$

as in (4.21), material volumes without singularities exist. In this case, there are volume densities $\phi_\mathrm{R}[t, \mathrm{X}], \tilde{\phi}[t, \mathrm{x}] \in \mathbb{R}, \boldsymbol{\phi}_\mathrm{R}[t, \mathrm{X}], \tilde{\boldsymbol{\phi}}[t, \mathrm{x}] \in \mathcal{V}$ as well as mass-specific densities with $\tilde{\phi}_\mathrm{s}[t, \mathrm{x}] = \phi_\mathrm{s}[t, \mathrm{X}] \in \mathbb{R}$ and $\tilde{\boldsymbol{\phi}}_\mathrm{s}[t, \mathrm{x}] = \boldsymbol{\phi}_\mathrm{s}[t, \mathrm{X}] \in \mathcal{V}$ such that

$$\begin{aligned} \Phi[t, \mathcal{B}] &= \int_{\Omega_t} \tilde{\phi}[t, \mathrm{x}] \mathrm{d}V_t[\mathrm{x}] = \int_{\Omega_t} \tilde{\phi}_\mathrm{s}[t, \mathrm{x}] \tilde{\varrho}[t, \mathrm{x}] \mathrm{d}V_t[\mathrm{x}] \\ &= \int_{\Omega_\mathrm{R}} \phi_\mathrm{R}[t, \mathrm{X}] \mathrm{d}V_\mathrm{R}[\mathrm{X}] = \int_{\Omega_\mathrm{R}} \phi_\mathrm{s}[t, \mathrm{X}] \varrho_\mathrm{R} \mathrm{d}V_\mathrm{R}[\mathrm{X}] \end{aligned}$$

and

$$\begin{aligned} \boldsymbol{\Phi}[t, \mathcal{B}] &= \int_{\Omega_t} \tilde{\boldsymbol{\phi}}[t, \mathrm{x}] \mathrm{d}V_t[\mathrm{x}] = \int_{\Omega_t} \tilde{\boldsymbol{\phi}}_\mathrm{s}[t, \mathrm{x}] \tilde{\varrho}[t, \mathrm{x}] \mathrm{d}V_t[\mathrm{x}] \\ &= \int_{\Omega_\mathrm{R}} \boldsymbol{\phi}_\mathrm{R}[t, \mathrm{X}] \mathrm{d}V_\mathrm{R}[\mathrm{X}] = \int_{\Omega_\mathrm{R}} \boldsymbol{\phi}_\mathrm{s}[t, \mathrm{X}] \varrho_\mathrm{R} \mathrm{d}V_\mathrm{R}[\mathrm{X}]. \end{aligned}$$

The temporal changes of an additive quantity of a material sub-volume are recorded in balance equations and are composed of four additive parts:

1. the *flux* into the sub-volume via the common *contact surface* with neighboring sub-volumes;
2. the *production* within the sub-volume;
3. the *supply* to the sub-volume from outside the volume by remote actions; and
4. the *mutual interaction* with the other sub-volumes.

In contrast to the first three contributions, the last one can only be considered non-locally. A general presentation mutual interactions is given in Noll (2005). In particular, the representation of a generalized stress concept with surface contact (flux) and volume interactions is outlined in Noll (2010). In *peridynamics*, non-local volume interactions are assumed and surface contact interactions are completely omitted, so that also a stress concept is completely omitted. If, as in Noll (2005), both surface contact and volume interaction components are assumed to exist, the additional volume interaction components are usually negligible. In the present work, the existence of flux terms, including the associated *contactors*, is assumed in the classical way, and therefore the volume interactions are neglected.

For the inner production and the outer remote effect, volume densities can be specified so that their total amount can be represented as a volume integral. Under certain conditions, the flux through the contact surface can be represented as a surface density, which, if it is linear with respect to the outer surface normal, can also be converted into a volume integral. In this way, balance equations can be represented as local partial differential equations in the absence of mutual interaction components.

When production is excluded as the cause of the change in an additive quantity by definition or postulate, the quantity itself is called a *conservation quantity*, and the corresponding balance equation is called a conservation equation. For example, in classical continuum physics, the mass and *energy* balances are scalar *conservation equations*. The *translational and rotational momentum* balances are vectorial conservation equations. In contrast, the *entropy* balance, for example, is a scalar balance that is not a conservation equation.

If there is a representation of the flux as a surface density in a placement, the flux surface density is called the *proto-contactor* of that placement. It can then be shown that the proto-contactor at a point on the surface is linear to the outer surface normal at the placement of that point, subject to certain assumptions about the continuity of the surface at that point. The linear mapping that maps the surface normal into the proto-contactor is then called the *contactor*.

Remark 4.16 ***Pitfalls in understanding*** these differential equations can occur especially when balancing vectorial quantities. This happens because the representation of the volume integral resulting from the transformation of a surface integral by divergence formulation depends on whether the divergence was used as left or right divergence. A prominent example is the representation of the stress divergence in the momentum balance.

4.5 Stress Tensors

The balance of the change of the *translational momentum* without mutual forces can be understood as the *equilibrium of the forces* of each partial body $\mathcal{P} \subset \mathcal{B}$ with the boundary $\partial\mathcal{P}$:

$$\boldsymbol{f}_{\mathcal{P}}[t] = \boldsymbol{f}_{\partial\mathcal{P}}[t]. \tag{4.23}$$

Instead of translational moment, the term *linear moment* is also used.

The volume forces $\boldsymbol{f}_{\mathcal{P}}$ include the inertial forces, whose specific quantities

$$\boldsymbol{b}_{\text{in}}[t, \mathrm{X}] = \tilde{\boldsymbol{b}}_{\text{in}}[t, \mathrm{x}]$$

in an *inertial frame* are the material accelerations $\ddot{\chi}[t, \mathrm{X}]$, and the remote action forces, whose specific quantities $\boldsymbol{b}_{\mathrm{out}}[t, \mathrm{X}] = \tilde{\boldsymbol{b}}_{\mathrm{out}}[t, \mathrm{x}]$ are, for example, gravitational accelerations. By postulate, the volume forces $\boldsymbol{f}_{\mathcal{P}}$ have no production included.

Under certain surface conditions, such as smoothness and others Noll (2005, 2010), the contact force $\boldsymbol{f}_{\partial\mathcal{P}}$ can be represented as the surface integral of *force proto-contactor*, called *stress vector*:

$$\boldsymbol{f}_{\partial\mathcal{P}}[t] = \int\limits_{\partial\Omega_t[\mathcal{P}]} \tilde{\boldsymbol{t}}\big[t, \mathrm{x}, \boldsymbol{n}[t, \mathrm{x}]\big]\, \mathrm{d}a_t[x] = \int\limits_{\partial\Omega_{\mathrm{R}}[\mathcal{P}]} \boldsymbol{t}_{\mathrm{R}}\big[t, \mathrm{X}, \boldsymbol{n}_{\mathrm{R}}[\mathrm{X}]\big]\, \mathrm{d}a_{\mathrm{R}}[X],$$

where $\mathrm{d}a$ and $\boldsymbol{n}$ are the magnitude and direction of the surface vector $\mathrm{d}\boldsymbol{a}$. $\mathrm{d}a_{\mathrm{R}}$, $\boldsymbol{n}_{\mathrm{R}}$, and $\mathrm{d}\boldsymbol{a}_{\mathrm{R}}$ behave accordingly.

In almost all continuum mechanics textbooks, it is shown that $\tilde{\boldsymbol{t}}\big[t, \mathrm{x}, \boldsymbol{n}[t, \mathrm{x}]\big]$ is linear in $\boldsymbol{n}$. This is usually done with a rectangular tetrahedron using a rectangular coordinate system. More general explanations of this, also in coordinate-free representation, can be found in Truesdell (1977), Noll (2005, 2010), Šilhavý (2009). The *force contactor* that linearly maps an outer surface normal into its force proto-contactor is called a *stress tensor*. However, it is not clearly defined whether the stress tensor maps the normal from the left or from the right into the stress vector. The *Encyclopedia of Physics* (Truesdell and Toupin 1960; Truesdell and Noll 1965; Gurtin 1972) chooses the right one:

$$\boldsymbol{t} = \boldsymbol{T}\boldsymbol{n}. \tag{4.24}$$

While in 1960 in volume III/1 of the *Encyclopedia of Physics* (Truesdell and Toupin 1960) the force contactor of the present placement $\boldsymbol{T}$ was simply called the "stress tensor", in 1960 in volume III/1 of the Encyclopedia (Truesdell and Noll 1965) the contactor is $\boldsymbol{T}$ called the *Cauchy stress tensor* to distinguish it from the 1*stPiola–Kirchhoff stress tensor* $\boldsymbol{T}_{\mathrm{R}}$ defined by $\boldsymbol{t}_{\mathrm{R}} = \boldsymbol{T}_{\mathrm{R}}\boldsymbol{n}_{\mathrm{R}}$ and from the second *Piola–Kirchhoff stress tensor* $\boldsymbol{S}$ defined by $\boldsymbol{T}_{\mathrm{R}} = \boldsymbol{F}\boldsymbol{S}$. If T_{ij} and $T_{\mathrm{R}ij}$ are the components of the natural Cartesian coordinate system, the left index corresponds to the current direction of the force and the right index corresponds to the direction of the surface. For $\boldsymbol{S}$, the left index indicates the direction of the force direction transformed into the reference placement. Since $\mathrm{div}[\mathrm{x}] \equiv \mathrm{DIV}_{\mathrm{R}}[\mathrm{x}]$ and $\mathrm{Div}[\mathrm{X}] \equiv \mathrm{DIV}_{\mathrm{R}}[\mathrm{X}]$ are used in the *Encyclopedia*, the two local versions of (4.23) with (4.12) are as follows:

$$\begin{aligned} \tilde{\varrho}[t, \mathrm{x}]\big(\tilde{\boldsymbol{b}}_{\mathrm{in}}[t, \mathrm{x}] + \tilde{\boldsymbol{b}}_{\mathrm{out}}[t, \mathrm{x}]\big) &- \mathrm{div}\, \tilde{\boldsymbol{T}}[t, \mathrm{x}] \quad \text{and} \\ \varrho_{\mathrm{R}}\big(\boldsymbol{b}_{\mathrm{in}}[t, \mathrm{X}] + \boldsymbol{b}_{\mathrm{out}}[t, \mathrm{X}]\big) &= \mathrm{Div}\, \boldsymbol{T}_{\mathrm{R}}[t, \mathrm{X}]. \end{aligned} \tag{4.25}$$

The same terms for stress tensors, including the use of right gradients and right divergences, are used by *I. Müller* (1985), *M. Šilhavý* (1997), *Ph. Ciarlet* (1988), *T. Oden* (1972), *J. Ball* (1976), and others.

Remark 4.17 Compatible with this representation is *Antman* (Antman 1995), who uses

$$\mathrm{DIV} \equiv \mathrm{DIV}_{\mathrm{L}} \equiv \nabla \cdot \, ,$$

and instead of (4.25) uses the equations $\tilde{\varrho}[t, \mathrm{x}]\big(\tilde{\boldsymbol{b}}_{\mathrm{in}}[t, \mathrm{x}] + \tilde{\boldsymbol{b}}_{\mathrm{out}}[t, \mathrm{x}]\big) = \nabla \cdot \tilde{\boldsymbol{T}}^{\mathrm{T}}[t, \mathrm{x}]$ and $\varrho_{\mathrm{R}}\big(\boldsymbol{b}_{\mathrm{in}}[t, \mathrm{X}] + \boldsymbol{b}_{\mathrm{out}}[t, \mathrm{X}]\big) = \nabla \cdot \boldsymbol{T}_{\mathrm{R}}^{\mathrm{T}}[t, \mathrm{X}]$.

Pitfalls

Malvern (1969) uses both left and right gradients and writes $\boldsymbol{F}$ compatible with the *Encyclopedia of Physics* (Truesdell and Noll 1965) as a right gradient. He uses the mapping of the surface normal from the left to define his stress tensors and uses the representation as a left divergence. He uses the names, Cauchy stress tensor and first Piola–Kirchhoff stress tensor, introduced by *Truesdell* and *Noll* for the stress tensors and simply redefines them to their transpose $\boldsymbol{T}^{\mathrm{T}}$ and $\boldsymbol{T}_{\mathrm{R}}^{\mathrm{T}}$. He makes this redefinition clear in the text with reference to the *Encyclopedia of Physics* (Truesdell and Noll 1965). ***The different stress tensors with the same names are incompatible.***

However, it is incomprehensible why he does not introduce new names for the transposed tensors. Thus, the left side of both tensors indicates the direction of the surface and the right side indicates the direction of the force.

Belytschko et al. also use left divergence in Belytschko et al. (2014) and give the transposed first Piola–Kirchhoff stress tensor the name *nominal stress* tensor, which is compatible with the *Encyclopedia of Physics* Truesdell and Noll (1965). However, he calls the transpose $\boldsymbol{T}^{\mathrm{T}}$ the Cauchy stress tensor, so that the left sides of the nominal stress tensor and its redefined Cauchy stress tensor at a surface are pointing in the outward direction of the surface. Rivlin 1969 redefines the Cauchy stress tensor in the same way (Rivlin 1969). The formal renaming of the Cauchy stress tensor is only irrelevant for non-polar media, where the Cauchy stress tensor is symmetric due to the angular momentum balance. For polar media, whose particles have rotational as well as translational degrees of freedom, ***the now different Cauchy stress tensors are incompatible***.

4.6 Specific Gibbs Free Energy

In the last section, we would like to point out a particular ***pitfall*** that *W. Dreyer* and *F. Duderstadt* already pointed out in 2008 in Dreyer and Duderstadt (2008). There they considered a *thermoelastic mixture* with *a singular surface between the solid and* *liquid phases*. For this purpose, we needed the *Gibbs free energies* for both phases. For the description of this pitfall, the character of the mixture with many constituents is not important and can be explained here using the single-constituent system.

In a thermoelastic fluid, the *specific energy* u, the *specific entropy* s, and the *hydrostatic pressure* $p = -\frac{1}{3}\mathrm{tr}\,\boldsymbol{T}$ depend only on the *temperature field* $\tilde{T}[t, \mathrm{x}]$ and the mass density field $\tilde{\varrho}[t, \mathrm{x}]$, where $\boldsymbol{T}$ is the Cauchy stress tensor according to (4.24). If its reciprocal ϱ^{-1} is used as the state variable instead of ϱ, then $u = u_{\mathrm{L}}[T, \varrho^{-1}]$, $s = s_{\mathrm{L}}[T, \varrho^{-1}]$, and $p = p_{\mathrm{L}}[T, \varrho^{-1}]$. The *Helmholtz free energy* is defined by $\psi = u - Ts$. From the internal energy balance, with the spatial heat flux $\boldsymbol{q}$ and the specific production $-(\varrho^{-1})^{\bullet} p$, and the entropy balance, which is also called the *entropy inequality* because of the requirement of non-negative production, an entropy flow of $\boldsymbol{q}/T$ is obtained and the relationship is

$$\mathrm{d}\psi_{\mathrm{L}}[T, \varrho^{-1}] = -s_{\mathrm{L}}[T, \varrho^{-1}]\,\mathrm{d}T - p_{\mathrm{L}}[T, \varrho^{-1}]\,\mathrm{d}\varrho^{-1}. \tag{4.26}$$

With the introduction of the *Gibbs free energy*

$$g = \psi + \frac{p}{\varrho}, \tag{4.27}$$

a variable exchange from ϱ^{-1} to p is possible by Legendre transformation and (4.26) becomes

$$\mathrm{d}g_{\mathrm{L}}[T, p] = -s_{\mathrm{L}}[T, p]\,\mathrm{d}T + \varrho_{\mathrm{L}}^{-1}[T, p]\,\mathrm{d}p. \tag{4.28}$$

It can be shown that for small elastic strains, i.e., for $0 \leq \varepsilon = |\varrho_{\mathrm{R}}/\varrho - 1| \ll 1$, the leading term of p is linear in ε, while that of ψ is quadratic in ε. This means that g in the leading term is also linear in ε section, we would like to point out a particular ***pitfall*** that *W. Dreyer* and *F. Duderstadt* already pointed out in 2008 in Dreyer and Duderstadt (2008).

In a thermoelastic solid, the specific energy u, the specific entropy s, and the second *Piola–Kirchhoff* stress tensor $\boldsymbol{S}$ depend on the temperature field $T[t, \mathrm{X}]$ and the right *Cauchy–Green* tensor $\boldsymbol{C}[t, \mathrm{X}]$. The balance of the internal energy with the referential heat flux $\boldsymbol{q}_{\mathrm{R}}$ and the specific production $\varrho_{\mathrm{R}}^{-1}\boldsymbol{S}\cdot\cdot\dot{\boldsymbol{C}}/2$ and the entropy equation then gives a referential entropy flux $\boldsymbol{q}_{\mathrm{R}}/T$ and

$$\mathrm{d}\psi_{\mathrm{S}}[T, \boldsymbol{C}] = -s_{\mathrm{S}}[T, \boldsymbol{C}]\,dT + \frac{1}{2\varrho_{\mathrm{R}}}\boldsymbol{S}[T, \boldsymbol{C}] : \mathrm{d}\boldsymbol{C}, \tag{4.29}$$

which, with the *Green strain tensor* $\boldsymbol{E} := \frac{1}{2}(\boldsymbol{C} - \boldsymbol{I})$, can also be written as

$$\mathrm{d}\hat{\psi}_{\mathrm{S}}[T, \boldsymbol{E}] = -\hat{s}_{\mathrm{S}}[T, \boldsymbol{E}]\,dT + \frac{1}{\varrho_{\mathrm{R}}}\hat{\boldsymbol{S}}[T, \boldsymbol{E}] : \mathrm{d}\boldsymbol{E}. \tag{4.30}$$

Instead of $\boldsymbol{F}$ the first *Piola–Kirchhoff* stress tensor $\boldsymbol{T}_{\mathrm{R}}$ can be used as state variable:

$$\mathrm{d}\check{\psi}_S[T, \boldsymbol{F}] = -\check{s}_S[T, \boldsymbol{F}]\,dT + \frac{1}{\varrho_R}\check{\boldsymbol{T}}_R[T, \boldsymbol{F}] : \mathrm{d}\boldsymbol{F}. \tag{4.31}$$

Remark 4.18 ***Pitfall***

The representations (4.29), (4.30), (4.31) lead to the temptation to exchange the variables $\boldsymbol{C}$ and $\boldsymbol{S}$, or between $\boldsymbol{E}$ and $\boldsymbol{S}$, or between $\boldsymbol{T}_R$ and $\boldsymbol{F}$. Landau and Lifschitz in (1986) named the quantity $g_{LL} := \psi - \mathrm{tr}[\boldsymbol{E}\boldsymbol{S}]/\varrho_R$ "*Gibbs free energy of a solid*". The proposed variable exchange is motivated there by a variant of (4.30) modified for the linear theory of small displacement gradients, i.e., for $0 \leq |\boldsymbol{F} - \boldsymbol{I}| \ll 1$. However, a comparison with (4.27) shows that $\psi - \mathrm{tr}[\boldsymbol{E}\boldsymbol{S}]/\varrho_R$ is quadratic in ε for small strains, for $0 \leq \varepsilon = |\boldsymbol{E}| \ll 1$. Therefore, the potential g_{LL} cannot be identified with the *Gibbs free energy* according to (4.27).

Remark 4.19 ***Pitfall***

Truedell and *Toupin* started from (4.31) in Truesdell and Toupin (1960) and called the quantity

$$g_{TT} = \psi - \mathrm{tr}[\boldsymbol{T}_R \boldsymbol{F}^T]/\varrho_R = \psi + 3p/\varrho$$

"*Gibbs free energy*". Although this quantity is linear in ε, it is larger by a factor of 3 than in (4.27). Therefore, the potential g_{TT} is not suitable for identification with the *Gibbs free energy* according to (4.27).

Since $\boldsymbol{C}$ can be represented as a tensor function of the density ϱ and its unimodular part $\boldsymbol{C}_U = (\det \boldsymbol{C})^{-\frac{1}{3}}\boldsymbol{C}$, we have

$$\mathrm{d}\boldsymbol{C} = \varrho_R \frac{2}{3}\left(\frac{\varrho}{\varrho_R}\right)^{\frac{1}{3}} \boldsymbol{C}_U \mathrm{d}\varrho^{-1} + \left(\frac{\varrho_R}{\varrho}\right)^{\frac{2}{3}} \mathrm{d}\boldsymbol{C}_U.$$

Because $\det[\boldsymbol{C}_U] = 1$, holds $\mathrm{tr}[\boldsymbol{C}_U^{-1}\,\mathrm{d}\boldsymbol{C}_U] = 0$ and for all $\boldsymbol{A} \in \mathcal{L}in[\mathcal{V}]$

$$\mathrm{tr}[\boldsymbol{A}\,\mathrm{d}\boldsymbol{C}_U] = \mathrm{tr}\Big[\mathrm{dev}[\boldsymbol{A}\boldsymbol{C}_U]\boldsymbol{C}_U^{-1}\mathrm{d}\boldsymbol{C}_U\Big].$$

Here $\mathrm{dev}[\boldsymbol{A}] = \boldsymbol{A} - \frac{1}{3}\mathrm{tr}[\boldsymbol{A}]\boldsymbol{I}$ is the trace-free component of $\boldsymbol{A}$. The total difference of the specific Helmholtz free energy $\psi = \bar{\psi}_S[T, \varrho, \boldsymbol{C}_U]$ is then

$$\mathrm{d}\psi_S = -s_S \mathrm{d}T - p\mathrm{d}\varrho^{-1} + \boldsymbol{S}_D : \mathrm{d}\boldsymbol{C}_U \tag{4.32}$$

and defines the symmetrical stress tensor

$$\boldsymbol{S}_{\mathrm{D}} = \left(\frac{\varrho_{\mathrm{R}}}{\varrho}\right)^{\frac{2}{3}} \operatorname{dev}[\boldsymbol{S}\boldsymbol{C}_{\mathrm{U}}]\boldsymbol{C}_{\mathrm{U}}^{-1} \quad \text{mit} \quad \operatorname{dev}[\boldsymbol{T}] = \frac{\varrho}{\varrho_{\mathrm{R}}}\boldsymbol{F}_{\mathrm{U}}\boldsymbol{S}_{\mathrm{D}}\boldsymbol{F}_{\mathrm{U}}^{\mathrm{T}}, \tag{4.33}$$

which disappears for thermoelastic fluids. With (4.27) and $g = \bar{g}[T, p, \boldsymbol{C}_{\mathrm{U}}]$ the Gibbs equation

$$\mathrm{d}g = -s\mathrm{d}T + \frac{1}{\varrho}\mathrm{d}p + \frac{1}{2\varrho_{\mathrm{R}}}\boldsymbol{S}_{\mathrm{D}} : \mathrm{d}\boldsymbol{C}_{\mathrm{U}}. \tag{4.34}$$

then follows. Since $\boldsymbol{S}_{\mathrm{D}}{:}\boldsymbol{C}_{\mathrm{U}} = \operatorname{tr}\big[\operatorname{dev}[\boldsymbol{S}]\big] = 0$ holds

$$\mathrm{d}\,\operatorname{tr}[\boldsymbol{S}_{\mathrm{D}}\boldsymbol{C}_{\mathrm{U}}] = \operatorname{tr}[\boldsymbol{S}_{\mathrm{D}}\mathrm{d}\boldsymbol{C}_{\mathrm{U}}] + \operatorname{tr}[\boldsymbol{C}_{\mathrm{U}}\mathrm{d}\boldsymbol{S}_{\mathrm{D}}] = 0$$

and therefore

$$\operatorname{tr}[\boldsymbol{S}_{\mathrm{D}}\mathrm{d}\boldsymbol{C}_{\mathrm{U}}] = -\operatorname{tr}[\boldsymbol{C}_{\mathrm{U}}\mathrm{d}\boldsymbol{S}_{\mathrm{D}}].$$

Instead of (4.34), $g = \bar{\bar{g}}[T, p, \boldsymbol{S}_{\mathrm{D}}]$ now gives the Gibbs equation with temperature and stresses as variables as

$$\mathrm{d}g = -s\mathrm{d}T + \frac{1}{\varrho}\mathrm{d}p - \frac{1}{2\varrho_{\mathrm{R}}}\operatorname{tr}[\boldsymbol{C}_{\mathrm{U}}\mathrm{d}\boldsymbol{S}_{\mathrm{D}}], \tag{4.35}$$

with

$$s = \bar{\bar{s}}[T, p, \boldsymbol{S}_{\mathrm{D}}] = -\frac{\partial\bar{\bar{g}}[T, p, \boldsymbol{S}_{\mathrm{D}}]}{\partial T}, \quad \varrho^{-1} = \bar{\bar{\varrho}}^{-1}[T, p, \boldsymbol{S}_{\mathrm{D}}] = \frac{\partial\bar{\bar{g}}[T, p, \boldsymbol{S}_{\mathrm{D}}]}{\partial p}$$

and

$$\boldsymbol{C}_{\mathrm{U}} = -2\varrho_{\mathrm{R}}\frac{\partial\bar{\bar{g}}[T, p, \boldsymbol{S}_{\mathrm{D}}]}{\partial\boldsymbol{S}_{\mathrm{D}}} = \bar{\bar{\boldsymbol{C}}}_{\mathrm{U}}[T, p, \boldsymbol{S}_{\mathrm{D}}].$$

It has been shown that the Gibbs energies for both liquids and solids have the same structure (4.27) and that at phase equilibrium on the interface, where

$$\begin{aligned} g_{\mathrm{L}} = g_{\mathrm{S}} - \boldsymbol{n}\frac{\operatorname{dev}[\boldsymbol{T}_{\mathrm{S}}]}{\varrho_{\mathrm{S}}}\boldsymbol{n} \quad &= \psi_{\mathrm{S}} - \boldsymbol{n}\frac{\boldsymbol{T}_{\mathrm{S}}}{\varrho_{\mathrm{S}}}\boldsymbol{n} \\ = g_{\mathrm{S}} - \frac{1}{\boldsymbol{n}_{\mathrm{R}}\,\boldsymbol{C}_{\mathrm{U}}\,\boldsymbol{n}_{\mathrm{R}}}\boldsymbol{n}_{\mathrm{R}}\frac{\boldsymbol{S}_{\mathrm{DS}}}{\varrho_{\mathrm{RS}}}\boldsymbol{n}_{\mathrm{R}} &= \psi_{\mathrm{S}} - \frac{1}{\boldsymbol{n}_{\mathrm{R}}\,\boldsymbol{C}\,\boldsymbol{n}_{\mathrm{R}}}\boldsymbol{n}_{\mathrm{R}}\frac{\boldsymbol{S}_{\mathrm{S}}}{\varrho_{\mathrm{RS}}}\boldsymbol{n}_{\mathrm{R}}, \end{aligned}$$

the leading mechanical parts of both phases are linear in the strains for small elastic strains. Only the differential of g gets an additional part, which depends on the distortion change of the solid (Fig. 4.1).

Fig. 4.1 *Wolfgang Dreyer* after a lecture in 2008 in Berlin-Mitte at *WIAS* with a wine glass on a sheet of paper with the note: *Please put nothing on this desk.* In his lectures on phase transitions between liquid and gas phases, the movement of the interface between them as a singular surface under surface tension, he liked to use a wine glass as an illustrative example on which the *Marangoni* effect becomes visible. The visible effect is known to wine or cognac connoisseurs as the tears of wine (or cognac) and is known in German as the Kirchenfenster (church window) effect during wine tasting. The *Marangoni* effect describes the movement of the interface due to surface tension gradients in the direction of the greatest tension. The gradients are caused by differences in temperature and concentration

Acknowledgements The idea to discuss the correct use of the terms placement, deformation, and configuration in continuum mechanics came up together with *Michael Herrmann*, who always considered *Wolfgang Dreyer* as one of his most important teachers and actually wanted to be co-editor of this memorial volume. With this article I also commemorate my friend Michael.

The suggestion to discuss the different use of gradients and divergences in their right and left variants in the literature came from *Kerstin Weinberg*. The starting point was a textbook illustration of a cuboid on whose surfaces the three Cartesian components of the stress vector were drawn as Cartesian components of the stress tensor in such a way that they corresponded to the explanatory text only when the left divergence was used. I owe my awareness of this topic to my tensor analysis teacher, *Heinz Schade*. Like *Michael Herrmann* and *Wolfgang Dreyer*, he died in 2024. In his lectures and exercises in the winter of 1992, *H. Schade* introduced me to the equal use of right and left operators, later published in a textbook (Schade 1996). He ended his lecture on this topic with the memorable statement: In his youth he preferred "left", and now, in his old age, for various reasons, he tends more and more to "right".

The insight that the definition of the Gibbs energy in fluids and solids should be the same comes from Wolfgang Dreyer himself Dreyer and Duderstadt (2008). The above formulations of these consequences, in particular, the definition of the stress tensor $\boldsymbol{S}_\text{D}$ according to (4.33), originate from my company internal *tkMS* (thyssenkrupp Marine Systems) report on rubber elasticity (Duderstadt 2013), which has been available there as an internal schooling document since the first version in 2013. This report has been the basis for numerous discussions with *Wolfgang Dreyer* and *Thomas Böhlke*. The publication of a detailed analysis of this liquid compatible definition of the Gibbs energy of solids is now in preparation together with *Thomas Böhlke* (Böhlke and Duderstadt 2025).

References

Altenbach H (2012) Kontinuumsmechanik. Einführung in die materialunabhängigen und materialabhängigen Gleichungen. Springer Vieweg, Berlin, Heidelberg, 1st edn

Antman SS (1960) Theory of rods, ed. by C. Truesdell, chief ed. by S. Flügge. Mechanics of solids II, Encyclopedia of physics, vol VI/a-2. Springer, Berlin, pp 641–699

Antman SS (1995) Nonlinear problems of elasticity. Springer, New York, 1st edn., 1995, 2nd supplemented ed., 2005

Ball J (1976) Convexity conditions and existence theorems in nonlinear elasticity. Archive Rational Mech Anal 63:337–403

Belytschko T, Liu WK, Moran B, Elkhodary KI (2014) Nonlinear finite elements for continua and structures. Wiley, Chichester, 1st ed. without Elkhodary 2000, 2nd ed. 2014

Bertram A (2005) Elasticity and plasticity of large deformations. Springer, New York, 1st edn., 2005, 2nd revised ed. 2008, 3rd enlarged ed. 2012

Böhlke T, Duderstadt F. A definition of the Gibbs energy that applies to liquids and to solids

Ciarlet PhG (1988) Three-dimensional elasticity, Mathematical elasticity, vol I. Elsevier, Amsterdam

Dreyer W, Duderstadt F (2008) On the modelling of semi-insulating GaAs including surface tension and bulk stresses. Proc R Soc A 464, with data dupplement of chapter 7-10 in http://rspa.royalsocietypublishing.org/content/suppl/2009/02/17/464.2098.2693.DC1.html, pp 2693–2720

Dreyer W, Müller I, Strehlow P (1982) A study of equilibria of interconnected ballons. Q J Mech Appl Math 35(3):419–440

Duderstadt F (2013) Theorie thermo-elastischer Materialien mit Anmerkungen zu entropieclastischen Gummimodellen. Company internal TKMS, unpublished (Kiel)

Duschek A, Hochrainer A (1961) Tensoranalysis, in Grundzüge der Tensorrechnung in analytischer Darstellung II. Springer, Wien, 1st edn. 1950, 2nd revised edn. 1961

Einstein A (1949) Autobigraphisches. In: Albert Einstein: Philosopher-Scientist vol VII, ed. and translated by P. Schilpp. The Library of living Philosophers. (Open Court, Lasalle, 1949, 1951, 1969, 1970), pp 1–94

Gurtin ME (1972) The linear theory of elasticity, ed. by C. Truesdell, chief ed. by S. Flügge. Mechanics of solids II, Encyclopedia of physics, vol VI/a-2. Springer, Berlin, pp 1–295

Gurtin ME (1981) An introduction to continuum mechanics. Academic Press, London

Gurtin ME, Fried E, Anand L (2009) The mechanics and thermodynamics of continua. Cambridge University Press, Cambridge

Gurtin ME, Williams WO (1966) On the Clausius-Duhem inequality. ZAMP 17:626–633

Holzapfel GA (2000) Nonlinear solid mechanics – a continuum approach for engineering. Wiley, Chichester

Kirchhoff G (1876) Vorlesungen über Mechanik, Vorlesungen über mathematische Physik vol I. Teubner, Leipzig, 1st und 2nd edition 1876, 3rd edition 1883, 4th ed. by W. Wien, 1897

Krawietz A (1986) Materialtheorie – Mathematische Beschreibung des phänomenologischen thermomechanischen Verhaltens. Springer, Berlin

Landau LD, Lifschitz EM (Л. Д. Ландау и Е. М. Лифсчиц) (1986) Theory of elasticity (Теория упругости), (Original in Russian, Nauka, 1944, revisited Russ. ed. 1953, 1965 and 1987 englarged by A. M. Kosewitsch und L. P. Pitajewski, English Translation by J. B. Sykes and W. H. Reid, Pergamon Press 1959, 1970, Elsevier 1986)

Lurie A (А. И. Лурье) (1970) Theory of elasticity (Original in Russian, Nauka. Moscow, 1970, translated to English by A. Belyaev, Springer, Berlin, 2005)

Lurie A (А. И. Лурье) (1980) Nonlinear theory of elasticity (Original in Russian, Nauka. Moscow 1980, translated to English by K. A. Lurie, North-Holland, Amsterdam, 1990)

Malvern LE (1969) Introduction to the mechanics of a continuous medium. Prentice-Hall, Englewood Cliffs

Marsden JE, Hughes TJR (1983) Mathematical foundations of elasticity. Prentice-Hall, London

Müller I (1985) Thermodynamics. Pitman, Boston

Naghdi PM (1960) Theory of shells and plates, ed. by C. Truesdell, chief ed. by S. Flügge. Mechanics of solids II, Encyclopedia of physics, vol VI/a-2. Springer, Berlin, pp 425–640

Noll W (1955) On the continuity of the solid and fluid states. J Rational Mech Anal 4:3–81

Noll W (1972) A new mathematical theory simple materials. Archive Rational Mech Anal 52:1–50

Noll W (1987) Finite-dimensional spaces, vol I. (1st edition, Martinus Nijhoff Publishers, Dortrecht, 1987, a corrected reprint, https://www.math.cmu.edu/~wn0g, 2006–2010)

Noll W (2005) Theory of surface interaction, (https://www.math.cmu.edu/~wn0g, 2005), 14 pages

Noll W (2010) Thoughts on the concept of stress. J Elast 100:25–32

Oden JT (1972) Finite elements of nonlinear continua. McGraw-Hill, New York

Ogden RW (1984) Non-linear elastic deformations (first publication, E. Harwood, Chichester, 1984, corrected republication, Dover, 1997)

Rivlin RS (1969) Red herrings and sundry unidentified fish in nonlinear continuum mechanics. Lehigh University, Bethlehem

Schade H (1996) Tensoranalysis. de Gruyter, Berlin

Serrin J (1959) Mathematical principles of classical fluid mechanics, ed. by S. Flügge and C. Truesdell. Fluid dynamics I, Encyclopedia of physics, vol VIII/1. Springer, Berlin, pp 125–263

Šilhavý M (1997) The mechanics and thermodynamics of continuous media. Springer, Berlin

Šilhavý M (2009) Cauchy's stress theorem for stresses represented by measures. Continuum Mech Thermodyn 20:75–96

Tietz H (1955) Geometrie, ed. by S. Flügge. Mathematical methods II, Encyclopedia of physics, vol II. Springer, Berlin, pp 117–197

Truesdell CA (1952) The mechanical foundations of elasticity and fluid dynamics, Part I-III, J Rational Mech (1952, 1953, 1954, 1962)

Truesdell CA (1977) A first course in rational continuum mechanics. Academic Press, Bosten, 1st ed 1977, 2nd corrected, revised, and augmented ed. 1991

Truesdell CA, Noll W (1965) The non-linear field theories of mechanics, ed. by S. Flügge. Encyclopedia of physics, vol III/3. Springer, Berlin

Truesdell CA, Toupin R (1960) Principles of classical mechanics and field theory, ed. by S. Flügge. Encyclopedia of physics, vol III/1. Springer, Berlin, pp 226–793

Chapter 5
Compressibility and Volume Variations Due to Composition in Multicomponent Fluids

Pierre-Étienne Druet

Abstract For single components fluids, vanishing isothermal compressibility implies that the mass density is constant, but the same conclusion is unknown for multicomponent fluids. Here the volume remains affected by changes of the composition. In the present paper we discuss an apparently natural way to conceptualise, based on derivatives of the Gibbs function g, this 'volume change due to composition' as a compression. In this way the phenomenon becomes quantitatively comparable to the usual coefficients of compressibility. This is a first step to investigate the range of validity of the constant density approximation of multicomponent fluids. As an illustration, three different aqueous solutions are discussed.

5.1 Preliminary Words

The occasion of the present paper is bidding Wolfgang Dreyer farewell. I met Wolfgang first in fall 2005 as, fresh graduated from the Humboldt University in Berlin, I started my first position at the Weierstrass Institute, in the group *Thermodynamic Modelling and Analysis of Phase-Transitions* of which he was the leader. We worked six months together on a linearised model problem in magnetohydrodynamics. At that time, we did not find a way to each other. I realise nowadays that with my mainly theoretical background in functional analysis I unfortunately did not understand Wolfgang's way in applied mathematics.

After some years, nevertheless, the project on crystal growth from the melt using magnetic fields had been advanced by many people's contributions, and there was a joined work (Dreyer et al. 2012). The model for fluid flow, heat transfer, and magnetic field distribution which is proposed there was—from Wolfgang's viewpoint, I guess—not especially original. However the fact that he accepted to endorse the Boussinesq approximation of compressible fluids is in itself worth to be noticed. In accordance with the observations of I. Müller, Wolfgang used to reject the possibility of thermal expansion in incompressible fluids as contradiction to the second law of thermodynamics. In his own words, he regarded the incompressible limit as 'a bad

P.-É. Druet (✉)
Berlin, Germany

J. Fuhrmann et al. (eds.), *Advances in Continuum Physics*, Advanced Structured Materials 238, https://doi.org/10.1007/978-3-031-93918-1_5

limit'. That he cosigned the paper (Dreyer et al. 2012) could have been due to his knowledge of a possible rescaling of the Navier-Stokes equations which 'rescues' Boussinesq's approximation, exposed later in Bothe and Dreyer (2015), certainly one of Wolfgang's major works. Anyway, already in our very first collaboration, the compressibility of mixtures was a core question.

From 2014 started a second phase in our relationship. After the initial breakthrough in setting up his electrolyte models (Dreyer et al. 2013) in the context of continuum mechanics, Wolfgang helped me to obtain a position to perform some mathematical analysis on these models. Our joined research on the modelling and analysis of partial differential equations for liquid electrolytes was finalised in Dreyer et al. (2020).

On the basis of this result I was able to obtain a research grant to further investigate the multicomponent *incompressible* case. Here, Wolfgang's definition of incompressibility for a mixture—as given for instance in Bothe and Dreyer (2015), was the starting point. Basically, incompressibility means zero isothermal compressibility. This definition entails that the mass density (the volume) of an incompressible mixture needs not being constant, which in turn leads to interesting mathematical structures (Druet 2021; Bothe and Druet 2021; Druet 2023).

The main point of our last joined paper (Bothe et al. 2023) was to understand better this definition of incompressibility and to investigate the consequences with a rigorous asymptotic for the free energy function.

Thank you Wolfgang for guiding me through this stimulating journey!

5.2 Introduction

According to the thermal equation of state, the density ϱ of a single component fluid (the specific volume $\upsilon = 1/\varrho$) is a function of temperature T and pressure p, and the isothermal compressibility is defined via

$$\beta_T = -\frac{1}{\upsilon(T, p)} \, \partial_p \upsilon(T, p) \, .$$

If $\beta_T = 0$, that is "compressibility equals to zero", then the fluid is *in*compressible. Its density also does not depend on temperature—even if this last conclusion should not be absolutely valid, see Gouin et al. (2012); Gouin and Ruggeri (2012), or in Dreyer's work (Bothe and Dreyer 2015; Bothe et al. 2023). Hence, the density of a fluid with zero compressibility is constant, at least in a certain range of thermodynamic conditions.

In a multicomponent fluid, the volume varies with temperature and pressure but also with the composition. In this paper, the composition shall be expressed by the mass fractions of the components, denoted by $y_1, \ldots, y_N$ if we consider a mixture with $N \geq 2$ constituents $A_1, \ldots, A_N$. For instance, a solution of water A_1 and ethanol A_2 under standard temperature and pressure conditions possesses at $y_1 = 1$ a density of 997 kg/m^3 and at $y_1 = 0$ of 798 kg/m^3. Of course, we would at first not think of this phenomenon as a compression.

Nonetheless, in the multicomponent case, the isothermal compressibility is likewise defined as

$$\beta_T = \beta_{(T,y)} = -\frac{1}{\upsilon(T, p, y_1, \dots, y_N)} \partial_p \upsilon(T, p, y_1, \dots, y_N). \tag{5.1}$$

Here however, $\beta_T = 0$ does not lead to the conclusion that volume or mass density are constant. In the paper Bothe et al. (2023) we show that $\beta_T = 0$ leads, even under maximal assumptions, to a representation $\upsilon(T, p, y_1, \dots, y_N) = \sum_{i=1}^N \bar{\upsilon}_i \, y_i$ with constants $\bar{\upsilon}_1, \dots, \bar{\upsilon}_N$. This means that the volume is independent of temperature and pressure, but continues to depend (linearly) on the composition. This conclusion looks natural, since the measure defined in (5.1) takes only pressure variations into account.

In the present paper, we discuss the possibility to measure the relative importance of the true compression effects, measured by β_T (or, more exactly, $\beta_{(T,y)}$) and the volume effects resulting from composition. In order to quantify the latter we consider the function (see below formula (5.55))

$$\hbar = \hbar(T, p, \mathbf{y}) := \partial_p g(T, p, \mathbf{y}) \left(\mathbf{D}^2_{\mathbf{y},\mathbf{y}} g(T, p, \mathbf{y})\right)^{-1} \left(\frac{1}{\upsilon} \partial_{\mathbf{y}} \upsilon\right) \cdot \left(\frac{1}{\upsilon} \partial_{\mathbf{y}} \upsilon\right), \tag{5.2}$$

in which g is the Gibbs function, and derivatives with respect to $\mathbf{y}$ are taken tangentially to the hyperplane $\sum_{i=1}^N y_i = 1$. We will show that this coefficient occurs naturally as a compressibility if instead of fixing the composition in (5.1) we fix the chemical potentials. In fact, we show that

$$\beta_{(T,y)} + \hbar \stackrel{!}{=} -\frac{\partial_p \upsilon(T, p, \mu_1 - \mu_N, \dots, \mu_{N-1} - \mu_N)}{\upsilon(T, p, \mu_1 - \mu_N, \dots, \mu_{N-1} - \mu_N)}$$

with the vector $\boldsymbol{\mu} = (\mu_1, \dots, \mu_N)$ of chemical potentials. With the help of its projection $\boldsymbol{\mu}^t$ onto the tangent hyperplane of $\sum_{i=1}^N y_i = 1$, the overall compressibility coefficient of the fluid mixture is

$$\beta_{(T,\mu^t)} = -\frac{1}{\upsilon(T, p, \boldsymbol{\mu}^t)} \partial_p \upsilon(T, p, \boldsymbol{\mu}^t) = \beta_{(T,y)} + \hbar. \tag{5.3}$$

We shall moreover illustrate the theory with three examples of aqueous solutions: Water + salt, sucrose, ethanol. In the two first cases, the ratio of $\hbar/\beta_{(T,y)}$ can become large for a dense mixture while, for the third case, the maximal ratio over the whole range of compositions is slightly above one.

Our conclusion is that we should attempt to distinguish multicomponent incompressible fluids from constant density fluids. A quantitative characterisation of the difference could be based on the ratio $\hbar/\beta_{(T,y)}$. Obviously, this will have to be studied further using wider data sets and more performant computational toolboxes in which full Gibbs functions are available.

Our plan is as follows. In Sect. 5.3 we introduce the basic thermodynamic setting and recall some definitions. Then in Sect. 5.4 we derive the computation rules for the derivatives of functions depending on density/volume and chemical potentials. We show how the definition (5.2) arises as a pressure derivative of density, hence a compressibility. Section 5.5 is devoted to the three illustrations. Finally, in an Appendix, we have collected some auxiliary materials and complements.

5.3 The Thermodynamic Setting

5.3.1 A Few Definitions

We shall use the thermodynamic setting of non-equilibrium thermodynamics to describe the mixture. We refer to the work of W. Dreyer, for instance to the Appendix of Bothe et al. (2023) or to the major study (Bothe and Dreyer 2015) on the modelling of reactive fluid mixtures. In the present paper, we shall regard the multicomponent fluid as mechanically and energetically homogeneous—hence being subject to N partial mass balances, but satisfying the principle of momentum and energy balance *as a whole*. This is the lowest complexity level in the description of fluid mixtures and, accordingly, the approach is affected by certain limitations. For the distinction of three complexity levels in the modelling of mixtures (class I, II, and III), Bothe and Dreyer refer to Hutter and Jöhnk (2004). In class II modelling, N partial momentum balances are admitted, allowing for additional phenomena like different imposed velocities for the constituents or like Korteweg stresses. In class III modelling, N different species temperatures are admitted. The class I models nevertheless cover not only all examples of the present paper, but also far reaching applications, see for instance the book Giovangigli (2012).

Variables. The thermodynamic conditions are locally expressed by $N + 1$ variables

$$T \text{ – absolute temperature,} \quad \boldsymbol{\rho} = (\rho_1, \ldots, \rho_N) \text{ – partial mass densities.} \tag{5.4}$$

The partial mass density ρ_i is the amount of mass of A_i available per unit volume of the mixture. The mass density of the fluid, is defined as

$$\varrho = \sum_{i=1}^{N} \rho_i \text{ – total mass density,} \quad y_i = \frac{\rho_i}{\varrho} \text{ –}i^{\text{th}} \text{ mass fractions.} \tag{5.5}$$

Beside the main variables T and $\rho_1, \ldots, \rho_N$, different sets of thermodynamic variables usually occur. The partial mass densities and the internal energy density, denoted by ϱu, are sometimes called *conservative variables* (Giovangigli 2012, Chaps. 8 and 9)

$$(\rho_1, \ldots, \rho_N, \varrho u) \; - \text{conservative variables.} \tag{5.6}$$

Further, with the molar masses (or with the molecular masses) $M_1, \ldots, M_N > 0$ of the constituents, assumed to be constants throughout this investigation, we introduce

$$n_i := \frac{\rho_i}{M_i}, \qquad n = \sum_{i=1}^{N} n_i, \qquad x_i = \frac{n_i}{n}, \tag{5.7}$$

and we call n_i = the partial mole density (the number density), n the total mole density (the number density) and x_i = the mole fraction (the number fraction). Another set of variables uses the temperature, the pressure and the mole or number fractions. These are the variables of the specific Gibbs energy g which are typically the quantities controlled in experiments:

$$(T, \; p, \; x_1, \ldots, x_N)\,. \tag{5.8}$$

The set (5.8) occurs in connection with the discussion of ideal mixtures (5.23) and in the illustration of Sect. 5.5. Owing to the relationships

$$y_i = \frac{M_i\, x_i}{\sum_{j=1}^{N} M_j\, x_j}, \qquad x_i = \frac{y_i}{M_i\, \sum_{j=1}^{N} (y_j / M_j)},$$

a set of variables completely equivalent ot (5.8) is given as

$$(T, \; p, \; y_1, \ldots, y_N)\,. \tag{5.9}$$

Thermodynamic stability and the consequences. In the *stable fluid phase*, the sets of state variables (5.4), (5.6), (5.8) and (5.9) are all equivalent, in the sense that there are smooth bijections transforming these vectors into one another (cp. among others Bothe et al. (2023), Giovangigli and Matuszewski (2012)). We let ϱs be the bulk density of the mixture entropy of the special form

$$\varrho s = -h(\rho_1, \ldots, \rho_N, \varrho u)\,, \tag{5.10}$$

where s is the specific entropy and $h : \mathcal{D} \subseteq \mathbb{R}^N_+ \times \mathbb{R} \to \mathbb{R}$ with $\mathbb{R}^N_+ :=]0, +\infty[^N$ the (material dependent) given constitutive function, assumed strictly convex and sufficiently smooth in its domain $\mathcal{D}$. It is usual to call h the mathematical entropy function—which is convex.

The concavity-postulate is usually derived from the requirement of increasing entropy in all thermodynamic processes, as for instance in the first chapter of the book Müller (1985). In Bothe and Dreyer (2015), Sect. 4, number 7., another motivation of the concavity postulate is given, starting from a stability principle for an homogeneous system with variable volume. As W. Dreyer used to underline, it is wrong to infer that a convex h alone guarantees that the entropy is a stability functional for a thermodynamic process. This is mainly true only for insulated systems. Moreover, in reality, no physical system remains stable under arbitrary thermodynamic conditions, so that the convexity/concavity and smoothness of the entropy function are always local assumptions valid in some neighbourhood of a reference state. For instance, the paper Gouin and Ruggeri (2012) shows that stability is violated if the compressibility attains a certain critical lower threshold.

The following definitions relate the (mass-based) chemical potentials $\mu_1, \ldots, \mu_N$ and the internal energy density ϱu to the state variables

$$-\frac{1}{T} = \partial_{\varrho u} h(\rho_1, \ldots, \rho_N, \varrho u) \text{ and } \frac{\mu_i}{T} = \partial_{\rho_i} h(\rho_1, \ldots, \rho_N, \varrho u) \text{ for } i = 1, \ldots, N\,. \tag{5.11}$$

Further, the thermodynamic pressure p obeys the *Gibbs-Duhem equation*

$$p = -\varrho u + T\, \varrho s + \sum_{i=1}^{N} \rho_i\, \mu_i\,. \tag{5.12}$$

The Helmholtz free energy is defined via $\varrho\psi := \varrho u - T\,\varrho s$. Its constitutive function has the form

$$\varrho\psi = f(T,\, \rho_1, \ldots, \rho_N)\,, \tag{5.13}$$

and then the (mass-based) chemical potentials are given by[1]

$$\mu_i = \partial_{\rho_i} f \quad \text{for } i = 1, 2, \ldots, N\,. \tag{5.14}$$

Other basic quantities can be calculated from f, via

$$p = -f + \sum_{i=1}^{N} \rho_i\, \mu_i\,, \quad \varrho s = -\partial_T f \tag{5.15}$$

$$\varrho u = -T^2\, \partial_T \frac{f}{T} =: \epsilon(T,\, \rho_1, \ldots, \rho_N)\,. \tag{5.16}$$

[1] In Giovangigli (2012), see Sect. 6.2.3, the same object is called *species Gibbs function* denoted by $\mathcal{G}_i$, while the notion of *chemical potential* with symbol μ_i is reserved for the quantity $\mathcal{G}_i/(RT)$.

If the constitutive functions f and h are twice differentiable, the following relationships are valid

$$\mathbf{D}^2_{\rho,\rho} f(T,\, \boldsymbol{\rho}) = T\left(\mathbf{D}^2_{\rho,\rho} h(\boldsymbol{\rho},\, \varrho u) - \frac{\mathbf{D}^2{}_{\rho,\varrho u} h(\boldsymbol{\rho},\, \varrho u) \otimes \mathbf{D}^2{}_{\rho,\varrho u} h(\boldsymbol{\rho},\, \varrho u)}{\partial^2_{\varrho u} h(\boldsymbol{\rho},\, \varrho u)}\right) \tag{5.17}$$

$$\partial^2_T f(T,\, \boldsymbol{\rho}) = -\frac{\partial_T \epsilon(T,\, \boldsymbol{\rho})}{T} = -\frac{1}{T^3\, \partial^2_{\varrho u} h(\boldsymbol{\rho},\, \varrho u)}\,. \tag{5.18}$$

In the main variables (5.4), stability is hence expressed by the two positivity conditions

$$\{\partial^2_{\rho_i,\rho_j} f\} > 0 \quad \text{and} \quad \partial_T \epsilon > 0\,,$$

where >0 applied to a matrix means positive definite. Another important function is the specific Gibbs free enthalpy $g = \psi + p/\varrho$ using the variables (5.9) hence

$$g = g(T,\, p,\, y_1, \ldots, y_N)\,, \tag{5.19}$$

where the derivatives satisfy

$$\partial_T g = -s, \quad \partial_p g = \frac{1}{\varrho}\,. \tag{5.20}$$

The compositional, $y-$derivative is special, because it makes sense only along the hypersurface $\sum_{i=1}^N y_i = 1$. Hence, it is a tangential derivative, which we express with the symbol $\partial^t_{\mathbf{y}}$, and get

$$\partial^t_{\mathbf{y}} g = \boldsymbol{\mu}^t\,. \tag{5.21}$$

Using particular parametrisations of the hypersurface, for instance $y_N = 1 - \sum_{j=1}^{N-1} y_j$, we get special representation of the tangent vector $\boldsymbol{\mu}^t$:

$$\partial^t_{y_i} g = \mu_i - \mu_N \quad \text{for} \quad i = 1, \ldots, N-1\,. \tag{5.22}$$

However, in the general meaning expressed by (5.21), $\boldsymbol{\mu}^t$ is the N dimensional projection of the vector of chemical potentials onto the orthogonal complement of the one vector $\mathbf{1} = (1, \ldots, 1)^\mathsf{T}$. For the Gibbs function the stability conditions then read

$$\{\partial^2_{(T,p)} g\} < 0 \quad \text{and} \quad \{\partial^2_{y_i,y_j} g\}^t > 0\,.$$

Ideal mixture. Following Müller, for instance in Eq. (7.22) of Müller (2013), the concept of an ideal mixture refers to a concrete form of the chemical potentials

$$\mu_i = \mu_i(T,\ p,\ x_i) = g_i(T,\ p) + \frac{R\,T}{M_i}\,\ln x_i\,. \tag{5.23}$$

Here R is the gas constant. In (5.23), $g_i(T,\ p)$ denotes the Gibbs free enthalpy of the constituent A_i. In particular, with $\hat{\rho}_i(T,\ p)$ being the mass density of the constituent A_i as a function of temperature and pressure, we have the identity $\partial_p g_i(T,\ p) = 1/\hat{\rho}_i(T,\ p) =: v_i(T,\ p)$ with the specific volume of A_i. How to construct a full thermodynamic model (particular constitutive models) respecting (5.23) was shown in Druet (2022), Sect. 5.3.

From the Gibbs–Duhem–Euler equation (5.12) and from (5.23) we can derive the identity

$$\sum_{i=1}^{N} \rho_i\, \partial_p g_i(T,\ p) = 1\,, \tag{5.24}$$

which we call the volume additivity of ideal mixtures. It implicitly characterises the pressure as a function of $(T,\ \rho)$, hence we can call the latter relation the equation of state of ideal mixtures.

Compressibility and heat capacity. The isothermal and isocompositional compressibility is defined as $\beta_{(T,y)} = \partial_p \varrho/\varrho$, where the variables (5.9) are considered. Hence the $p-$derivative is built while the temperature T and the composition vector $\mathbf{y}$ are fixed. Using the Gibbs function and in particular $(5.20)_2$, we have

$$\beta_{(T,y)} = -\frac{\partial_p^2 g(T,\ p,\ \mathbf{y})}{\partial_p g(T,\ p,\ \mathbf{y})} > 0 \quad \text{– isothermal, isocompositional compressibility,} \tag{5.25}$$

where the inequality follows from thermodynamic stability. Using the Helmholtz function f, we have equivalently

$$\beta_{(T,\,y)} = \frac{1}{\mathbf{D}^2_{\rho,\rho} f(T,\ \rho)\rho \cdot \rho}\,, \tag{5.26}$$

while (5.17) can be used to express $\beta_{(T,\,y)}$ in terms of the entropy potential.

Another important function is the isothermal and isocompositional heat capacity at constant volume, which is primarily defined as $c_{(T,y)} = \partial_T \varrho u(T,\ \rho)$, hence

$$c_{(v,y)} := \frac{1}{\varrho}\,\partial_T \epsilon(T,\ \rho) > 0 \quad \text{– heat capacity at constant volume and composition.} \tag{5.27}$$

For an ideal mixture, these quantities can be computed from the data $g_1, \ldots, g_N$ of the species, and adopt the following form (see Druet (2022), Lemma 3.1)

$$\beta_{(T,\,y)} = -\sum_{i=1}^{N} \rho_i\, \partial_p^2 g_i(T,\ p)\,, \tag{5.28}$$

$$c_{(\upsilon,\,y)} = -\frac{T}{\varrho}\left(\sum_{i=1}^{N} \partial_T^2 g_i(T,\ p)\,\rho_i - \frac{\left(\sum_{i=1}^{N} \partial_{T,p}^2 g_i(T,\ p)\,\rho_i\right)^2}{\sum_{i=1}^{N} \partial_p^2 g_i(T,\ p)\,\rho_i}\right). \tag{5.29}$$

With the isothermal compressibility of the species given as

$$\beta_T^i(T,\, p) := -\hat{\rho}_i(T,\, p)\, \partial_p^2 g_i(T,\ p)\,,$$

and with the volume fractions $\phi_i = \rho_i / \hat{\rho}_i(T,\, p)$ which sum up to one owing to (5.24), we can write $(5.28)_1$ as

$$\beta_{(T,\,y)} = \sum_{i=1}^{N} \phi_i\, \beta_T^i(T,\ p)\,.$$

The relationship between the heat capacity $c_{(\upsilon,y)}$ and the heat capacities of the species at constant volume is more complex, see Lemma 3.1 in Druet (2022).

5.3.2 *Entropic (Normal Form) Variables*

For I. Müller, the chemical potentials (together with the temperature) are—more fundamentally than the partial densities ρ, the pressure or the composition vector **y**—the continuous quantities expressing the equilibrium state of the mixture, which one can measure. And thus, see Sect. 7.1.2 in Müller (2013),

> (...) we have to get acquainted with the chemical potential, even if that is not pleasant.

The Theory of Irreversible Processes postulates that the mass and heat fluxes in a nonequilibrium multicomponent systems are proportional to gradients of chemical potentials and temperature, hence these are the driving forces thermodynamic systems in non equilibrium. For a recent introduction to multicomponent diffusion, let us refer to Bothe and Dreyer (2015) and Bothe and Druet (2023) where the reader can find many references to the primary literature. In the general theory of diffusion there is the subtle point that not $\mu_1, \ldots, \mu_N$ drive the mass and heat diffusion, but rather $\boldsymbol{\mu}^t$ (see (5.21), (5.22)). Since the mass fractions are subject to the condition $\sum_{i=1}^{N} y_i = 1$, the vector $\boldsymbol{\mu}^t$ has zero average and possesses only $N-1$ independent components. Particular representations can be obtained from bases of the hyperplane $\sum_{i=1}^{N} y_i = 1$, for instance $\boldsymbol{\eta}^k = \mathbf{e}^k - \mathbf{e}^N$ for $k = 1, \ldots, N-1$, with the Euclidian basis vectors. In this case

$$\boldsymbol{\eta}^i \cdot \boldsymbol{\mu}^t = \mu_i - \mu_N \quad \text{for} \quad i = 1, \ldots, N-1\,. \tag{5.30}$$

With the paper Dreyer et al. (2013) we see that one can speak of $\boldsymbol{\mu}^t$ as "relative chemical potentials".

In the context of mathematical treatment for the partial differential equations of multicomponent fluid dynamics (Druet 2022), it has proven convenient to separate the mass density ϱ subject to the continuity equation (hyperbolic conservation law) and $N-1$ variables contributing to entropic dissipation (parabolic system):

$$\left.\begin{array}{ll} \frac{\mu_1-\mu_N}{T}, \ldots, \frac{\mu_{N-1}-\mu_N}{T} & \text{(relative) chemical potentials} \\ -\frac{1}{T} & \text{reciprocal of temperature} \end{array}\right\} \quad \text{the parabolic variables,} \tag{5.31}$$

$$\varrho \quad \text{total mass density} \qquad \text{the hyperbolic variable.} \tag{5.32}$$

Up to the choice of $-1/T$ instead of T, this set of variables belongs to the *intermediate normal form* of multicomponent fluid dynamics: See Giovangigli (2012), Sect. 8.7 for more details.

The transformation rules. Following Druet (2022), we shall in fact introduce the change of variables in a slightly more general way than (5.31). This turns especially useful in the analysis of incompressible models, see Druet (2021, 2023) and Bothe and Druet (2021). First we define the $(N+1)-$vector of extended state variables

$$w_i := \rho_i \quad \text{for } i = 1, \ldots, N, \qquad w_{N+1} := \varrho u, \tag{5.33}$$

to denote the conservative variables (5.6), the ones occurring in the entropy functional. Exploiting the relations (5.11), the combinations

$$w_i^* := \frac{\mu_i}{T} \quad \text{for } i = 1, \ldots, N, \qquad w_{N+1}^* := -\frac{1}{T}, \tag{5.34}$$

are the dual variables, also called the "entropic variables". Suppose that the entropy function of (5.10) is strict convex in its domain. Then, with the help of the conjugate convex function h^* to h, we can invert the relations

$$\frac{\mu_i}{T} = \partial_{\rho_i} h(\boldsymbol{\rho}, \varrho u) \quad \text{for } i = 1, \ldots, N, \qquad -\frac{1}{T} = \partial_{\varrho u} h(\boldsymbol{\rho}, \varrho u),$$

which more compactly now read as

$$\mathbf{w}^* = \nabla_{\mathbf{w}} h(\mathbf{w}). \tag{5.35}$$

In order to guarantee the possibility to switch between conservative and entropic variables, the notion of a function of Legendre-type is essential. We assume that h is such a function of Legendre-type on $\mathcal{D} \subseteq \mathbb{R}_+^N \times \mathbb{R}$ open, convex. We define $\mathcal{D}^*$ to be the image of $\nabla_{\mathbf{w}} h$ on $\mathcal{D}$. If $\mathcal{D}^*$ is open and convex, then the conjugate $h^*(w^*) :=$

$\sup_{w\in\mathcal{D}}\{w^*\cdot w - h(w)\}$ is a function of Legendre-type on $\mathcal{D}^*$. The gradients $\nabla_{\mathbf{w}} h$ on $\mathcal{D}$ and $\nabla_{\mathbf{w}^*} h^*$ on $\mathcal{D}^*$ are inverse to each other, see for instance the Lemma 2.3 of Druet (2022).

By means of a fixed linear transformation, we next separate the parabolic and hyperbolic variables. To this aim, we choose new axes $\boldsymbol{\xi}^1,\dots,\boldsymbol{\xi}^N,\ \boldsymbol{\xi}^{N+1}$ of $\mathbb{R}^{N+1}$ in the following way:

$$\begin{cases} \boldsymbol{\xi}^N := \mathbf{e}^{N+1} = (0,\dots,0,1) \quad \text{and} \quad \boldsymbol{\xi}^{N+1} := \bar{e} := (1,\dots,1,0), \\ \boldsymbol{\xi}^1,\dots,\boldsymbol{\xi}^{N-1} \in \left\{x\in\mathbb{R}^{N+1} \,:\, x_{N+1}=0,\ \sum_{i=1}^N x_i = 0\right\} = \{\boldsymbol{\xi}^N,\ \boldsymbol{\xi}^{N+1}\}^{\perp}. \end{cases} \tag{5.36}$$

As seen, typical is the choice $\boldsymbol{\xi}^k = \mathbf{e}^k - \mathbf{e}^N$ for $k=1,\dots,N-1$. We let $\boldsymbol{\eta}^1,\dots,\boldsymbol{\eta}^{N+1}\in\mathbb{R}^{N+1}$ be the dual basis for $\boldsymbol{\xi}^1,\dots,\boldsymbol{\xi}^{N+1}$. Then

$$\boldsymbol{\eta}^N = \mathbf{e}^{N+1},\quad \boldsymbol{\eta}^{N+1} = \frac{1}{N}\,\boldsymbol{\xi}^{N+1},\quad \eta^k_{N+1} = 0 \text{ and } \sum_{i=1}^N \eta^k_i = 0 \text{ for } k=1,\dots,N-1. \tag{5.37}$$

For $\mathbf{w}^* = (\mu_1/T,\dots,\mu_N/T,\ -1/T)$, we define the projections (see (5.30))

$$q_\ell := \boldsymbol{\eta}^\ell\cdot\mathbf{w}^* := \sum_{i=1}^{N+1} \eta^\ell_i\, w^*_i \quad \text{for} \quad \ell = 1,\dots,N. \tag{5.38}$$

Due to the properties of the chosen basis, in particular to (5.37), we have a relationship

$$\mathbf{w}^* = \sum_{\ell=1}^N q_\ell\,\boldsymbol{\xi}^\ell + (\mathbf{w}^*\cdot\boldsymbol{\eta}^{N+1})\,\boldsymbol{\xi}^{N+1} \quad \text{implying that} \quad w^*_{N+1} = q_N. \tag{5.39}$$

Now, since the coordinate w^*_{N+1} has the physical meaning of $-1/T$, the relevant domain for the new variable q is the half-space

$$\mathcal{H}^N_- := \left\{(q_1,\dots,q_N)\in\mathbb{R}^N \,:\, q_N < 0\right\} = \mathbb{R}^{N-1}\times\mathbb{R}_-.$$

Since $(\mu_1/T,\dots,\mu_N/T,\ -1/T) = \nabla_{\mathbf{w}} h(\rho,\ \varrho u)$, use of the conjugate convex function yields

$$(\rho_1,\ \dots,\rho_N,\varrho u) = \nabla_{\mathbf{w}^*} h^*(\mu_1/T,\dots,\mu_N/T,\ -1/T).$$

In order to isolate the hyperbolic component ϱ (total mass density), we now express

$$\varrho = \sum_{i=1}^{N} \rho_i = \sum_{i=1}^{N+1} \xi_i^{N+1} \, \partial_{w_i^*} h^*(w_1^*, \ldots, w_{N+1}^*)$$
$$= \boldsymbol{\xi}^{N+1} \cdot \nabla_{\mathbf{w}^*} h^*\Big(\sum_{\ell=1}^{N} q_\ell \, \boldsymbol{\xi}^\ell + (\mathbf{w}^* \cdot \boldsymbol{\eta}^{N+1}) \, \boldsymbol{\xi}^{N+1}\Big) .$$

This is an algebraic equation of the form $F(\mathbf{w}^* \cdot \boldsymbol{\eta}^{N+1}, \, q_1, \ldots, q_N, \, \varrho) = 0$. We notice that

$$\partial_{\mathbf{w}^* \cdot \boldsymbol{\eta}^{N+1}} F(\mathbf{w}^* \cdot \boldsymbol{\eta}^{N+1}, \, q_1, \ldots, q_N, \, \varrho) = \mathbf{D}^2 h^*(\mathbf{w}^*) \boldsymbol{\xi}^{N+1} \cdot \boldsymbol{\xi}^{N+1} > 0 \, ,$$

due to the strict convexity of the conjugate function. It can be shown (see Lemma 2.4 in Druet (2022)) that the latter algebraic equation defines the component $\mathbf{w}^* \cdot \boldsymbol{\eta}^{N+1}$ implicitly as a differentiable function of ϱ and $q_1, \ldots, q_N$. We call this function $\mathscr{M}$, satisfying by definition

$$\varrho = \boldsymbol{\xi}^{N+1} \cdot \nabla_{\mathbf{w}^*} h^*\Big(\sum_{\ell=1}^{N} q_\ell \, \boldsymbol{\xi}^\ell + \mathscr{M}(\varrho, \, \mathbf{q}) \, \boldsymbol{\xi}^{N+1}\Big) . \tag{5.40}$$

We obtain the equivalent formulae

$$\mathbf{w}^* = \sum_{\ell=1}^{N} q_\ell \, \boldsymbol{\xi}^\ell + \mathscr{M}(\varrho, \, q_1, \ldots, q_N) \, \boldsymbol{\xi}^{N+1} \, , \tag{5.41}$$

$$\rho_i = \partial_{w_i^*} h^*\Big(\sum_{\ell=1}^{N} q_\ell \, \boldsymbol{\xi}^\ell + \mathscr{M}(\varrho, \, q_1, \ldots, q_N) \, \boldsymbol{\xi}^{N+1}\Big) =: \mathscr{R}_i(\varrho, \, \mathbf{q}), \quad i = 1, \ldots, N \, , \tag{5.42}$$

$$\varrho u = \partial_{w_{N+1}^*} h^*\Big(\sum_{\ell=1}^{N} q_\ell \, \boldsymbol{\xi}^\ell + \mathscr{M}(\varrho, \, q_1, \ldots, q_N) \, \boldsymbol{\xi}^{N+1}\Big) =: \mathscr{U}(\varrho, \, \mathbf{q}) \, , \tag{5.43}$$

with ϱ and $q_1, \ldots, q_N$ as the free variables. Since p obeys (5.12), we have $p = T \, g(\nabla_{\mathbf{w}} h(\mathbf{w}))$, and here g denotes the Legendre transform of h. Since g coincides with h^* for a function h of Legendre-type, we find that

$$p = T \, h^*(\mathbf{w}^*) = -h^*(\mathbf{w}^*)/w_{N+1}^* \, . \tag{5.44}$$

We combine the latter with (5.41) to obtain that

$$p = \mathscr{P}(\varrho,\, \mathbf{q}) := -\frac{1}{q_N}\, h^*\Big(\sum_{\ell=1}^{N} q_\ell\, \boldsymbol{\xi}^\ell + \mathscr{M}(\varrho,\, q_1, \ldots, q_N)\, \boldsymbol{\xi}^{N+1}\Big)\,. \tag{5.45}$$

Not only the pressure, but all thermodynamic quantities can now be introduced as functions of the variables $\varrho,\, \mathbf{q}$. Indeed, considering a function $f = f(T,\, \rho_1, \ldots, \rho_N)$ of the main variables, we use that $T = -1/q_N$ and $\rho = \mathscr{R}(\varrho,\, \mathbf{q})$ (see (5.42)), and we define

$$f(\varrho,\, \mathbf{q}) := f\Big(-\frac{1}{q_N},\, \mathscr{R}_1(\varrho,\, \mathbf{q}), \ldots, \mathscr{R}_N(\varrho,\, \mathbf{q})\Big) \tag{5.46}$$

to obtain the equivalent representation in the entropic variables.

5.4 Transformed Thermodynamic Functions and Their Derivatives

In (5.38), we introduced the variables $q_1, \ldots, q_N$ as linear combinations of the entropic variables $w_1^*, \ldots, w_{N+1}^*$. We next want to compute some derivatives of the transformed coefficient functions: The map $\mathscr{R}(\varrho,\, \mathbf{q})$ of (5.42), the function $\mathscr{U}(\varrho,\, \mathbf{q})$ of (5.43) and $\mathscr{P}(\varrho,\, \mathbf{q})$ of (5.45). Recall that the fundament of the change of variables is the Eq. (5.40).

5.4.1 General Formulas

Differentiation directly gives the following expressions for the derivatives of $\mathscr{M}$:

$$\begin{aligned}
\partial_\varrho \mathscr{M}(\varrho,\, \mathbf{q}) &= \frac{1}{\mathbf{D}^2 h^*(\mathbf{w}^*)\boldsymbol{\xi}^{N+1} \cdot \boldsymbol{\xi}^{N+1}}\,,\\
\partial_{q_k} \mathscr{M}(\varrho,\, \mathbf{q}) &= -\frac{\mathbf{D}^2 h^*(\mathbf{w}^*)\boldsymbol{\xi}^{N+1} \cdot \boldsymbol{\xi}^{k}}{\mathbf{D}^2 h^*(\mathbf{w}^*)\boldsymbol{\xi}^{N+1} \cdot \boldsymbol{\xi}^{N+1}} \quad \text{for} \quad k = 1, \ldots, N-1\,,\\
\partial_{q_N} \mathscr{M}(\varrho,\, \mathbf{q}) &= -\frac{\mathbf{D}^2 h^*(\mathbf{w}^*)\boldsymbol{\xi}^{N+1} \cdot \mathbf{e}^{N+1}}{\mathbf{D}^2 h^*(\mathbf{w}^*)\boldsymbol{\xi}^{N+1} \cdot \boldsymbol{\xi}^{N+1}}\,,
\end{aligned} \tag{5.47}$$

where we recall that $\boldsymbol{\xi}^{N+1} = (1, \ldots, 1,\, 0)^{\mathsf{T}}$ and $\mathbf{w}^* = \mathbf{w}^*(\varrho,\, \mathbf{q})$ is given by (5.41). Note that the expression $\mathbf{D}^2 h^*(\mathbf{w}^*)\boldsymbol{\xi}^{N+1} \cdot \boldsymbol{\xi}^{N+1}$ is strictly positive owing to the strict convexity of h^*. Moreover, we notice that the $\varrho-$derivative and the $q_N := -1/T$

derivative are intrinsic expressions, while the q_k derivative for $k = 1, \ldots, N-1$ depends on the choice of the basis.

Systematic computations are performed in the Appendix-Section A. Here we retain two special expressions. At first, by differentiating the representation (5.45), and writing $\bar{e}$ instead of $\boldsymbol{\xi}^{N+1}$, we obtain the ϱ−derivative of the pressure function $\mathscr{P}$ as

$$\partial_\varrho \mathscr{P}(\varrho, \mathbf{q}) = -\frac{\varrho}{q_N} \frac{1}{\mathbf{D}^2 h^*(\mathbf{w}^*)\bar{e} \cdot \bar{e}} \,. \tag{5.48}$$

At second, use of (5.43) yields

$$\partial_{q_N} \mathscr{U}(\varrho, \mathbf{q}) = \partial^2_{w^*_{N+1}} h^*(\mathbf{w}^*) - \frac{(\mathbf{D}^2 h^*(\mathbf{w}^*)\mathbf{e}^{N+1} \cdot \bar{e})^2}{\mathbf{D}^2 h^*(\mathbf{w}^*)\bar{e} \cdot \bar{e}} \,. \tag{5.49}$$

Recall in these formula that $\mathbf{w}^* = \mathbf{w}^*(\varrho, \mathbf{q})$. These derivatives possess a sign:

$$\partial_\varrho \mathscr{P} > 0 \quad \text{and} \quad \partial_{q_N} \mathscr{U} > 0\,. \tag{5.50}$$

The quotient $1/(\varrho\, \partial_\varrho \mathscr{P})$ possesses the dimension of a compressibility, while $\partial_{q_N} \mathscr{U} / (T^2\, \varrho)$ has the dimension of a heat capacity.

At the present stage, we can already compare these quantities to the compressibility $\beta_{(T,y)}$ and the heat capacity $c_{(\upsilon,y)}$ obtained by fixing the composition variable. Exploiting that $\mathbf{D}^2 h(\mathbf{w})$ and $\mathbf{D}^2 h^*(\mathbf{w}^*)$ are inverse to each other, we can prove for $\mathbf{y_a} = (\boldsymbol{\rho}, a)^\mathsf{T}$ with $a \in \mathbb{R}$ arbitrary that

$$\varrho = (\mathbf{D}^2 h(\mathbf{w}))^{\frac{1}{2}} \mathbf{y_a} \cdot (\mathbf{D}^2 h^*(\mathbf{w}^*))^{\frac{1}{2}} \bar{e} \leq \sqrt{\mathbf{D}^2 h(\mathbf{w}) \mathbf{y_a} \cdot \mathbf{y_a}}\, \sqrt{\mathbf{D}^2 h^*(\mathbf{w}^*)\bar{e} \cdot \bar{e}}\,.$$

Minimising in a and squaring the result yields

$$\frac{\varrho^2}{\mathbf{D}^2 h^*(\mathbf{w}^*)\bar{e} \cdot \bar{e}} \leq \mathbf{D}^2 h(\mathbf{w})(\boldsymbol{\rho}, 0)^\mathsf{T} \cdot (\boldsymbol{\rho}, 0)^\mathsf{T} - \frac{(\mathbf{D}^2 h(\mathbf{w})(\boldsymbol{\rho}, 0)^\mathsf{T} \cdot \mathbf{e}^{N+1})^2}{\partial^2_{\varrho u} h(\mathbf{w})}\,.$$

Now, use of the identities (5.17) implies that

$$\mathbf{D}^2 h(\mathbf{w})(\boldsymbol{\rho}, 0)^\mathsf{T} \cdot (\boldsymbol{\rho}, 0)^\mathsf{T} - \frac{(\mathbf{D}^2 h(\mathbf{w})(\boldsymbol{\rho}, 0)^\mathsf{T} \cdot \mathbf{e}^{N+1})^2}{\partial^2_{\varrho u} h(\mathbf{w})} = \frac{1}{T} \mathbf{D}^2 f(T, \boldsymbol{\rho})\boldsymbol{\rho} \cdot \boldsymbol{\rho}\,.$$

Thus, invoking $(5.48)_1$ with $-1/q_N = T$, we see that

$$\varrho\, \partial_\varrho \mathscr{P} \leq \mathbf{D}^2 f(T, \boldsymbol{\rho})\boldsymbol{\rho} \cdot \boldsymbol{\rho} = \frac{1}{\beta_{(T,y)}}\,.$$

Defining $\beta^{-1}_{(q)} := \varrho\, \partial_{\varrho} \mathscr{P}$ as a compressibility with fixed $\mathbf{q}$, we obtain the relationship

$$\beta_{(q)} \geq \beta_{(T,y)} \,. \tag{5.51}$$

Similarly, we prove in Druet (2022), Eq. (71) for the heat capacities that

$$c_{(q)} := \frac{1}{T^2\, \varrho}\, \partial_{q_N} \mathscr{U} \geq c_{(\upsilon,y)} \,. \tag{5.52}$$

Note that a more intrinsic notation as $\beta_{(q)}$ is the one used in (5.3), with no reference to a particular basis. Clearly $\beta_{(T,\,\mu')}$ is the same object as $\beta_{(q)}$.

5.4.2 *Natural Representation with the Gibbs Function*

We use (5.20), second formula, and (5.22) to obtain that

$$\frac{1}{\varrho} = \partial_p g(T,\, p,\, \mathbf{y}), \quad q_i = \frac{d}{dy_i} g(T,\, p,\, \mathbf{y}') \text{ for } i = 1, \ldots, N-1$$

where we use a particular basis and with $\mathbf{y}' = (y_1, \ldots, y_{N-1}, 1 - \sum_{i=1}^{N-1} y_i)$. For simplicity, we write from now $\frac{d}{dy_i} g(T,\, p,\, \mathbf{y}') = \partial_{y_i} g'$ for $i = 1, \ldots, N-1$, abusing notation. Regarding $(\varrho, \mathbf{q})$ as the main variables, we next build the ϱ−derivatives:

$$\begin{aligned} -\frac{1}{\varrho^2} &= \partial_p^2 g\, \partial_{\varrho} \mathscr{P} + \mathbf{D}^2_{\mathbf{p},\mathbf{y}} g' \cdot \partial_{\varrho} \mathbf{y}(\varrho,\, \mathbf{q}) \,, \\ 0 &= \mathbf{D}^2_{\mathbf{y},\mathbf{p}} g'\, \partial_{\varrho} \mathscr{P} + \mathbf{D}^2_{\mathbf{y},\mathbf{y}} g' \cdot \partial_{\varrho} \mathbf{y}(\varrho,\, \mathbf{q}) \,. \end{aligned}$$

Here the Gibbs function is evaluated at its natural variables. It follows that

$$\frac{1}{\varrho^2} = \Big(- \partial_p^? g' + (\mathbf{D}^2_{\mathbf{y},\mathbf{y}} g')^{-1} \mathbf{D}^2_{\mathbf{y},\mathbf{p}} g' \cdot \mathbf{D}^2_{\mathbf{y},\mathbf{p}} g' \Big)\, \partial_{\varrho} \mathscr{P} \,,$$

which we can rewrite (see also (5.25)) as

$$\begin{aligned} \beta_{(q)} &= \frac{1}{\varrho\, \partial_{\varrho} \mathscr{P}} = \varrho \Big(- \partial_p^2 g + (\mathbf{D}^2_{\mathbf{y},\mathbf{y}} g')^{-1}\, \mathbf{D}^2_{\mathbf{y},\mathbf{p}} g' \cdot \mathbf{D}^2_{\mathbf{y},\mathbf{p}} g' \Big) \\ &= -\frac{\partial_p^2 g}{\partial_p g} + \frac{(\mathbf{D}^2_{\mathbf{y},\mathbf{y}} g')^{-1}\, \mathbf{D}^2_{\mathbf{y},\mathbf{p}} g' \cdot \mathbf{D}^2_{\mathbf{y},\mathbf{p}} g'}{\partial_p g} \\ &= \beta_{(T,y)} + \frac{(\mathbf{D}^2_{\mathbf{y},\mathbf{y}} g')^{-1}\, \mathbf{D}^2_{\mathbf{y},\mathbf{p}} g' \cdot \mathbf{D}^2_{\mathbf{y},\mathbf{p}} g'}{\partial_p g} \,. \end{aligned} \tag{5.53}$$

The latter expression is valid independent of the choice of a basis and we identify

$$\beta_{(T,\mu')} - \beta_{(T,y)} = \frac{(\mathbf{D}^2_{\mathbf{y},\mathbf{y}}g)^{-1}\,\mathbf{D}^2_{\mathbf{y},\mathbf{p}}g \cdot \mathbf{D}^2_{\mathbf{y},\mathbf{p}}g}{\partial_p g}\,, \tag{5.54}$$

as a natural definition of the compressibility of mixing. In fact, due to (5.20), equation 2, in the variables of the Gibbs function (5.9), we have $\partial_{\mathbf{y}}\varrho/\varrho = -\varrho\,\partial^2_{\mathbf{y},\mathbf{p}}g$, and another equivalent representation for (5.54) is

$$\beta_{(T,\mu')} - \beta_{(T,y)} = \partial_p g\,(\mathbf{D}^2_{\mathbf{y},\mathbf{y}}g)^{-1}\left(\frac{\partial_{\mathbf{y}}\varrho}{\varrho}\right)\cdot\left(\frac{\partial_{\mathbf{y}}\varrho}{\varrho}\right), \tag{5.55}$$

as a measure for how volume changes due to composition. Note that $\beta_{(T,\mu')} - \beta_{(T,y)}$ is another name for $\hbar$ defined in (5.2). Here the derivatives with respect to $\mathbf{y}$ are tangential derivatives.

5.4.3 Explicit Formula for Ideal Mixtures

For ideal mixtures, the inverse Hessian of the entropy was computed in the Proposition 3.11 of Druet (2022). (In these computations we employed rescaled molar masses $M_i \to M_i/R$). Recall that $(\mathbf{D}^2 h(\mathbf{w}))^{-1} = \mathbf{D}^2 h^*(\mathbf{w}^*)$, where $\mathbf{w}^* = \nabla_{\mathbf{w}} h(\mathbf{w})$. In particular

$$\sum_{i,k=1}^{N} \partial^2_{w_i^*,w_k^*} h^* = \frac{1}{R}\sum_{i=1}^{N} M_i\,\rho_i\,(1-\partial_p g_i\,\varrho)^2 + T\,\varrho^2 \sum_{i=1}^{N}\rho_i\,|\partial^2_p g_i|\,. \tag{5.56}$$

Combination with the identity (5.48) yields

$$\begin{aligned}\beta_{(T,\mu')} =& \sum_{i=1}^{N}\rho_i\,|\partial^2_p g_i| + \frac{1}{RT}\sum_{i=1}^{N} M_i\,\rho_i\left(\frac{1}{\varrho} - \partial_p g_i\right)^2\\ =& \beta_{(T,y)} + \frac{1}{RT}\sum_{i=1}^{N} M_i\,\rho_i\left(\frac{1}{\varrho} - \partial_p g_i\right)^2.\end{aligned} \tag{5.57}$$

With a bit more computational work, we can obtain a similar formula for the heat capacities, with the result $c_{(q)} \geq c_{(T,y)}$, with equality only for a pure fluid. Since this point is not at the focus, we spare the computation for the sake of brevity.

5.4.4 Incompressibility and Constant Density

In the case of an ideal mixture the relation (5.57) is particularly clear. Interestingly, we can have $\beta_{(q)} = \beta_{(T,y)}$ only in one of the following two cases:

(i) The mixture reduces to a pure fluid, that is, there is an index i_0 such that $\rho_{i_0} = \varrho$ and $\rho_i = 0$ for all $i \neq i_0$;
(ii) All species possess the same specific volume $\partial_p g_1(T, p) = \ldots = \partial_p g_N(T, p)$.

In both cases, the varying composition produces no change in specific volume, and the equation of state (5.24) implies that the pressure is a function of density and temperature, but not of composition.

Using the specific volumes of the species $\partial_p g_i = \upsilon_i$, a closer look at the nonnegative quantity

$$\beta_{(T,\mu')} - \beta_{(T,y)} = \frac{1}{RT} \sum_{i=1}^{N} M_i \, \rho_i \left(\frac{1}{\varrho} - \upsilon_i(T, p) \right)^2$$

reveals that it precisely measures the volume change of mixing. Indeed, passing to the incompressible limit $\beta_{(T,y)} \to 0$, which is studied in Bothe et al. (2023) and Druet (2023), implies that $\upsilon_i(T, \ p)$ must converge to a constant $\bar{\upsilon}_i$ for each i. In the limit, the coefficient $\beta_{(T,\mu')}$ does not vanish, but we have

$$\beta_{(T,\mu')} = \frac{1}{RT} \sum_{i=1}^{N} M_i \, \rho_i \left(\frac{1}{\varrho} - \bar{\upsilon}_i \right)^2, \tag{5.58}$$

as a measure of density variations due to composition. Only in the cases (i) and (ii) we can expect that $\beta_{(T,\mu')}$ is very small too. But these two scenario correspond, in the incompressible limit, precisely to a density ϱ which is nearly constant.

Hence, already in the relatively simple theoretical context of ideal mixtures, we must distinguish between two forms of the incompressible limit:

(1) $\beta_{(T,y)} \to 0$ yielding the incompressibility constraint $\sum_{i=1}^{N} \bar{\upsilon}_i \, \rho_i = 1$ which is studied in Bothe et al. (2023) and in Druet (2021, 2023), Bothe and Druet (2021). Here a volume/density change due to mixing remains possible;
(2) $\beta_{(T,\mu')} \to 0$ yielding the classical incompressibility constraint $\varrho = \text{Const.}$ with its implication $\operatorname{div} \mathbf{v} = 0$ for the velocity field.

Whether (2) is valid for an incompressible multicomponent fluid depends on the concrete constellation under study. In the next section we shall discuss three illustrations for the relative importance of $\beta_{(T,y)}$ and $\beta_{(T,\mu')} - \beta_{(T,y)}$.

The reader interested in yet more thoughts of heuristic nature can consult the Appendix-Section B, where the coefficient $\beta_{(T,\mu')}$ is put in relation to the speed of sound in the fluid.

5.5 Numerical Examples for Three Aqueous Solutions

In this section we consider three binary solutions based on water as the solvent:

- Water + salt, seawater;
- Water + sucrose, sugar water, syrup;
- Water + ethanol.

Seawater or ocean water is in fact not a binary mixture. However, in the main theoretical approach, we regard it as a mixture of pure water with a certain *salinity*. Salinity quantifies the total amount of dissolved substances—where ions of sodium and chloride are dominant.

Also, seawater is not an ideal mixture. At first, it is not possible to rely on data of the pure solute—neither on data of pure ionic species, nor on data of pure 'salt ', which is a granular porous aggregate. However, our strategy is to nevertheless use a binary ideal model, where we fit the data of the solute in order to obtain a good approximation of the true density of seawater *in the range of relevant salinity*—about 3% mass.

For the water + sucrose case, water can dissolve a much larger quantity of sucrose than of salt. Still, for the same reason as for seawater, we cannot use data of the pure sugar to reliably understand volume effects. We adopt the same strategy as in the first example (to fit an ideal model), in this case with a very good agreement for up to 80% of sucrose mass.

In the last case of the water + ethanol solution, the mixture exists over the whole range of compositions, but it is far from being volume-additive and thus, it does not exhibit ideal behaviour. Here we discuss both the binary ideal model and a ternary ideal model with one chemical transformation as proposed in Bothe et al. (2023). The result concerning compressibility numbers seems to make sense.

Considering thus an ideal mixture, we start with a few formula tailored on the binary case. Using (5.24) we have the identity

$$\sum_{i=1}^{N} y_i \, \partial_p g_i(T, p) = \frac{1}{\varrho} \,, \tag{5.59}$$

with the mass fractions $y_1, \ldots, y_N$. This allows to re-express

$$\beta_{(T,y)} = - \frac{\sum_{i=1}^{N} y_i \, \partial_p^2 g_i(T, p)}{\sum_{i=1}^{N} y_i \, \partial_p g_i(T, p)} \,. \tag{5.60}$$

For a mixture of two species $A_1 = S$ (solute) and $A_2 = W$ (water), we denote y the fraction of S. Recall that $\partial_p g_i = 1/\hat{\rho}_i$ for $i = 1, 2$ with the density $\hat{\rho}_i$ of the constituents as function of temperature and pressure. We introduce $\rho^S := \hat{\rho}_1$ and $\rho^W := \hat{\rho}_2$. Then, by means of (5.59)

$$\varrho = \frac{\rho^W}{1 + y\,(\frac{\rho^W}{\rho^S} - 1)}\,. \tag{5.61}$$

Moreover, (5.60) implies that

$$\beta_{(T,y)} = -\frac{\partial_p^2 g^W + y\,(\partial_p^2 g^S - \partial_p^2 g^W)}{\frac{1}{\rho^W} + y\left(\frac{1}{\rho^S} - \frac{1}{\rho^W}\right)}$$

With the isothermal compressibility $\beta_T^{S,W}$ of the substances, we get

$$\beta_{(T,y)} = \frac{\beta_T^W + y\left(\frac{\rho^W}{\rho^S}\,\beta_T^S - \beta_T^W\right)}{1 + y\left(\frac{\rho^W}{\rho^S} - 1\right)}\,, \tag{5.62}$$

and if the solute is assumed incompressible

$$\beta_{(T,y)} = \frac{(1-y)\,\beta_T^W}{1 + y\left(\frac{\rho^W}{\rho^S} - 1\right)}\,. \tag{5.63}$$

Similarly, we compute first that

$$\begin{aligned}
\beta_{(q)} - \beta_{(T,y)} =& \frac{\varrho}{RT}\sum_{i=1}^{N} M_i\,y_i\left(\sum_{k=1}^{N} y_k\,\partial_p g_k - \partial_p g_i\right)^2\\
=& \frac{1}{RT\,\sum_{k=1}^{N} y_k\,\partial_p g_k}\sum_{i=1}^{N} M_i\,y_i\left(\sum_{k=1}^{N} y_k\,\partial_p g_k - \partial_p g_i\right)^2.
\end{aligned}$$

For the binary mixture, we have

$$\sum_{k=1,2} y_k\,\partial_p g_k - \partial_p g_1 = \frac{y-1}{\rho^S} + \frac{1-y}{\rho^W}\,,\quad \sum_{k=1,2} y_k\,\partial_p g_k - \partial_p g_2 = \frac{y}{\rho^S} - \frac{y}{\rho^W}\,.$$

Hence

$$\begin{aligned}
&\sum_{i=1,2} M_i\,y_i\,(\sum_{k=1}^{N} y_k\,\partial_p g_k - \partial_p g_i)^2\\
&= (M^S\,y\,(1-y)^2 + M^W\,(1-y)\,y^2)\left(\frac{1}{\rho^W} - \frac{1}{\rho^S}\right)^2.
\end{aligned}$$

Thus, for the difference of the compressibility coefficients we get

$$\beta_{(q)} - \beta_{(T,y)} = \frac{y\,(1-y)\,(M^S + y\,(M^W - M^S))}{R\,T\,\rho^W\,(1 + y\,(\rho^W/\rho^S - 1))}\left(1 - \frac{\rho^W}{\rho^S}\right)^2. \tag{5.64}$$

5.5.1 Seawater

To obtain data we use the open Python modules *seawater 3.3.5* and *gsw 3.6.19* (Gibbs Seawater Oceanographic Package) from https://pypi.org/project/seawater and https://pypi.org/project/gsw. These provide tools to compute several thermodynamic quantities for ocean water from temperature, pressure and salinity input data. In the documentation of the seawater package, we read that

> The package uses the formulas from Unesco's joint panel on oceanographic tables and standards, UNESCO 1981 and UNESCO 1983 (EOS-80).

The latter are to be found in Fofonoff and Millard (1983). Note that this is not the most recent version of the EOS for seawater:

> The EOS-80 library is considered now obsolete; it is provided here for compatibility with old scripts, and to allow a smooth transition to the new TEOS-10.

We here refer interested readers to IOC (2010) or the website https://www.teos-10.org of TEOS-10. In particular, the newer equation of state incorporates aspects of thermodynamic consistency by using a Gibbs function as general potential. Several computational routines associated with this approach are implemented in the gsw package. However, the older seawater package offers functions with standard temperature input, which are not yet available in the gsw package. This motivates our choice to use both libraries.

In order to construct the binary ideal model, we at first get the density ρ^W of pure water as a function of temperature and pressure, available in both modules. Next, our strategy is to understand the salinity as an incompressible component, hence $\beta_T^S = 0$ (see (5.63)) and ρ^S is a constant, independent on (T, p), in the formulas (5.61) and (5.63). We fit the constant ρ^S in order that the density function (5.61) of our ideal model is exact on a standard vector (T^0, p^0, y^0) of reference conditions. Here T is the temperature in degree Celsius, p the pressure in dbar and y the salinity mass fraction.

To obtain realistic conditions, we use data from *Global Seawater Oxygen-18 Levels* which, among other, contains (T, p, y) measurements of ocean conditions done between 1949 and 2009 at different points all over the earth, see Schmidt et al. (1999). After filtering incomplete data, our database still contains 20295 measurements. Our reference conditions are 3.956 percent of salinity, a temperature of 24.3 °C and a pressure of 20.21 dbar. After fitting ρ^S, we calculate the approximate ideal density using the formula (5.61). We compare the exact density ϱ^{sw} (resp. ϱ^{gsw}) of the seawater library (resp. the Gibbs seawater library) with our calculated density ϱ^{id} according to the ideal model. For the former case, the relative error over the data set is

$$\mathrm{Err}(\varrho) := \frac{\max_{\mathrm{data}} |\varrho^{\mathrm{sw}} - \varrho^{\mathrm{id}}|}{\max_{\mathrm{data}} \varrho^{\mathrm{sw}} - \min_{\mathrm{data}} \varrho^{\mathrm{sw}}} = 0.029\,. \tag{5.65}$$

The plot 5.1 shows different estimations of the density at reference conditions. For comparison, we also plotted some data for a concentrated solution (dead sea) obtained from Krumgalz and Millero (1982). For the ideal model, we hence observe a good fit for the reference ocean waters conditions, and an overestimation of density at larger concentrations. Note that there is no complete agreement between EOS-80 and TEOS-10 for larger concentrations. As a next step we discuss the compressibility. We evaluate isothermal compressibility according to (5.63) and compare it with a computation using the seawater-module. The latter does not provide the isothermal compressibility directly, but with the available functions: density, speed of sound U, heat capacity at constant pressure c_p and thermal expansion coefficient α, we can compute the exact isothermal compressibility via

$$\beta^{\mathrm{sw}}_{(T,y)} = \frac{1}{\varrho\, U^2} + \frac{\alpha^2\, T}{\varrho\, c_p}\,. \tag{5.66}$$

We compare the ideal and the exact isothermal compressibility over the data set with the formula (5.65) and we obtain a relatively large error $\mathrm{Err}(\beta_{(T,y)}) = 0.279$. This is clearly not as good as for the density.

The compressibility of composition $\hbar = \beta_{(q)} - \beta_{(T,y)}$ of the ideal model is next evaluated with the formula (5.64). In order to obtain a comparison, we use the Gibbs seawater library. It allows to compute the full Gibbs potential as function of (T, p, y) using the formula $g = h - T\, s$, where the specific enthalpy h is given by the function *gsw.enthalpy_t_exact* and the specific entropy s by *gsw.entropy_from_t*. Then, we evaluate $\hbar$ by means of the formula (5.54), where the (p, y)-derivatives of g are calculated numerically with a centre-difference scheme.

The plot 5.2 gives the different compressibility numbers relatively to the compressibility of pure water $1.52 \times 10^{-10}\,\mathrm{Pa}^{-1}$ as reference value. We see that for normal ocean conditions, $\beta_{(q)} - \beta_{(T,y)}$ and $\beta_{(T,y)}$ are of the same small order according to all models. For larger salt contents however, the compositional compressibility $\hbar = \beta_{(q)} - \beta_{(T,y)}$ turns to be about six times larger than the isothermal compressibility $\beta_{(T,y)}$ according to the ideal model. However the value obtained from the Gibbs seawater library attains earlier a significantly smaller maximum. Up to which concentration are we allowed to trust the conclusions of the different models? Further investigations will be necessary to better estimate the value of $\hbar$.

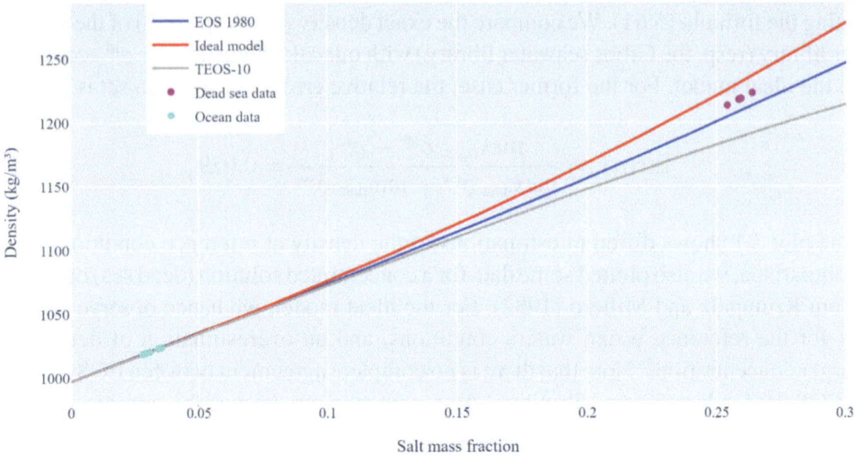

Fig. 5.1 Density of seawater at reference $T = 25\,°\mathrm{C}$ and $p = 10.1325\,\mathrm{dbar}$

5.5.2 *Water and Sucrose*

This type of mixture is very important in food industry but also in biology. Our data source for the density of the water+sucrose are:

- Room temperature, atmospheric pressure in the table on page 5–145 of Haynes (2014), for mass fractions of solute from 0 to 80%;
- Varying temperature, atmospheric pressure in the Table C1 of Barbosa (2003), experimental data;
- Varying temperature, atmospheric pressure in Table 5 of Randall (1932).

For the isothermal compressibility, we rely on the values computed in Tables E1-3 of Barbosa (2003). Overall our database contains 204 entries. We fit the density (5.61) of our ideal model in order to match the measured density at the reference measurement with $T^0 = 20\,°\mathrm{C}$, $p^0 = 10.1325\,\mathrm{dbar}$ and $y^0 = 0.8$. The relative error of this density on the (reduced) data set is $\mathrm{Err}(\varrho) = 0.041$.[2] We refer to Fig. 5.3 for a plot of the density under reference conditions.

For the isothermal compressibility, the relative error is $\mathrm{Err}(\beta_{(T,y)}) = 0.072$. We again plot 5.4 the relative compressibilities with respect to the isothermal compressibility of pure water.

We can observe around a mass fraction of sucrose of 0.5 that the ratio of the compressibility of mixing over the isothermal compressibility attains a maximal value of more than 10.

[2] We here filter from the data set the pressure-values from Barbosa (2003) exceeding atmospheric pressure of more than 1000 bars.

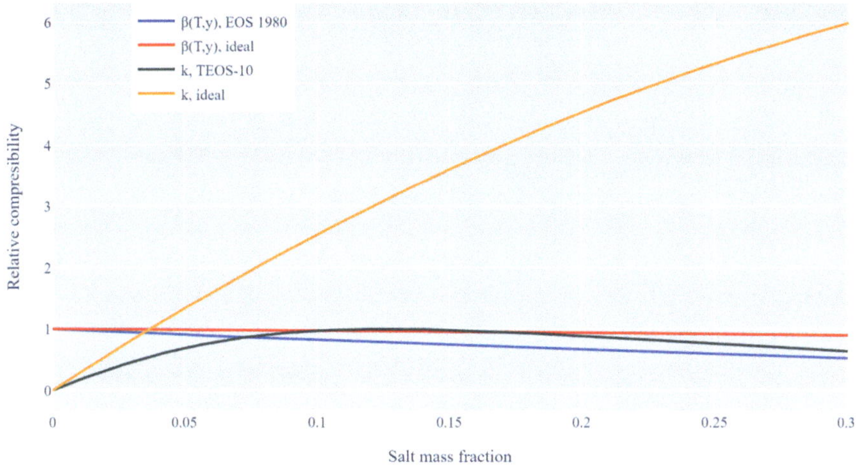

Fig. 5.2 Compressibility of seawater relative to $4.52 \times 10^{-10}\,\text{Pa}^{-1}$ at reference $T = 25\,°\text{C}$ and $p = 10.1325\,\text{dbar}$

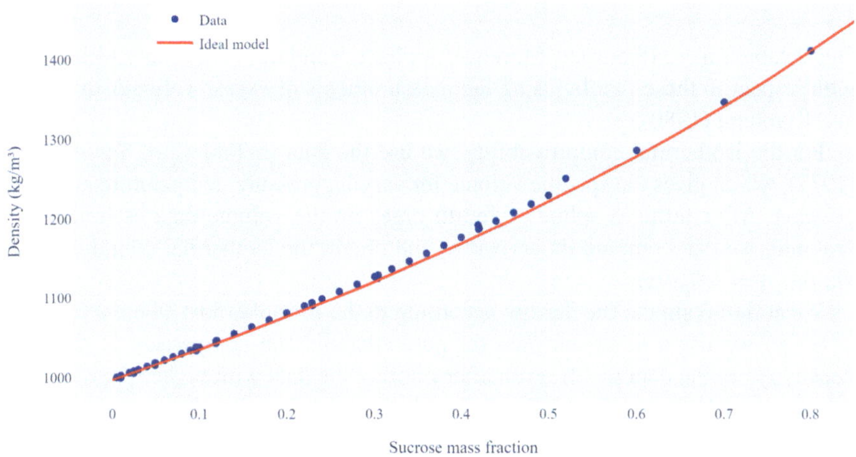

Fig. 5.3 Density of sugarwater at reference $T = 25\,°\text{C}$ and $p = 10.1325\,\text{dbar}$

5.5.3 *Water and Ethanol*

For this third type of solution, we can rely on data for ϱ^S and β_T^S for pure ethanol. Our data for the mixture are:

- Density data at room temperature, atmospheric pressure in the table on page 5–127, 5–128 of Haynes (2014), for mass fractions of solute from 0 to 100%;
- Varying temperature and pressure data in the Table 2 of Tanaka et al. (1977), for mass fractions of solute from 0 to 100%.

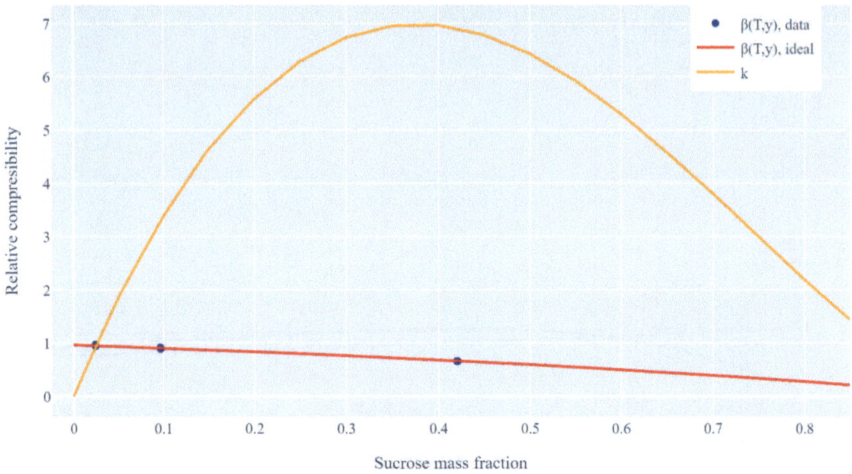

Fig. 5.4 Compressibility of sugarwater relative to $4.52 \times 10^{-10}\,\mathrm{Pa}^{-1}$ at reference $T = 25\,°\mathrm{C}$ and $p = 10.1325\,\mathrm{dbar}$

Our database has 318 entries. Moreover we have 'validated' our binary ideal model with respect to these that by using the measurements of excess volumes in Benson and Kiyohara (1980).

For the isothermal compressibility, we use the data in Table 2 of Tanaka et al. (1977), which gives the specific volume for varying pressure, temperature and composition. After fitting a spline of fourth order on the volume data as function of pressure, we can compute its derivative and obtain the isothermal compressibility via $\beta_{(T,y)} = -\partial_p v/v$.

We at first compute the density according to the formula (5.61), where for $\varrho^S = \varrho^S(T, p)$ we use a fit of our data for pure ethanol. The comparison of the ideal density ϱ^{id} to the data yields, even after filtering the data with higher pressure, to a relative error $\mathrm{Err}(\varrho) = 0.14$. Hence, the ideal model is not really good. Due to the well known effect of volume loss for this mixture, the volume additivity assumption turns critical. For the compressibility, the relative error is even $\mathrm{Err}(\beta_{(T,y)}) = 0.35$. In order to obtain a better estimation of the compressibility coefficients, we switch to a three species ideal model. The procedure is well known for the water+ethanol mixture, see for instance Roux and Desnoyers (1987), Sect. 3, or Sect. 5 in Bothe et al. (2023). We have recalled some details of this model in the Appendix-Section C. Results for the density and different compressibilities under standard conditions of temperature and pressure are plotted in the Figs. 5.5 and 5.6. Although the density of the ternary ideal model does not completely match the initial slope, it does much better, at least optically, than the binary one. As to the compressibility, we observe that for the binary ideal model, the compressibility $\mathscr{k}$ is widely dominated by the isothermal compressibility over the whole range. However, the approximations for the real density and for the isothermal compressibility in this model are rather poor.

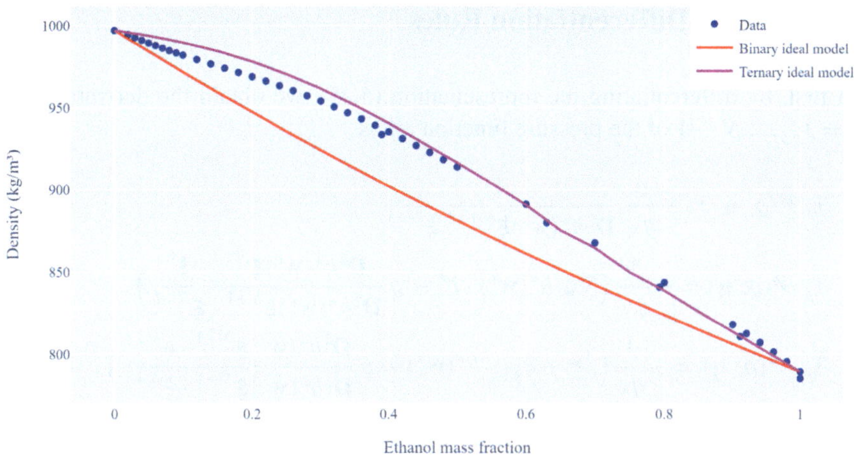

Fig. 5.5 Density of ethanol water near reference $T = 25\,°\mathrm{C}$ and $p = 10.1325\,\mathrm{dbar}$

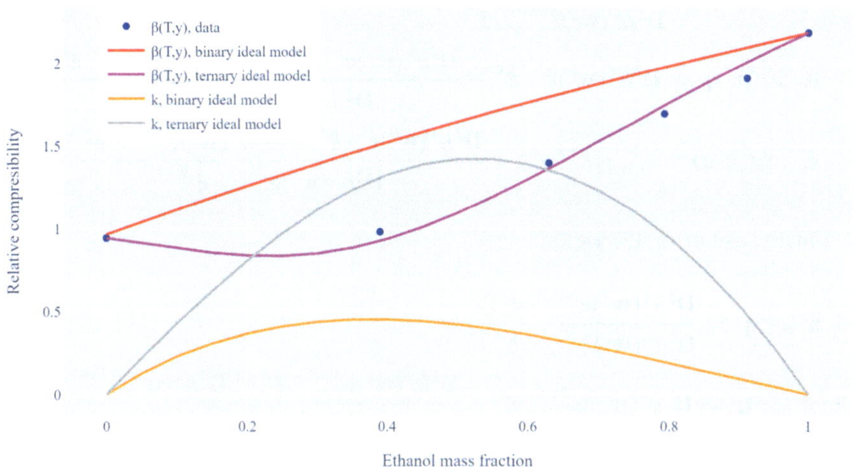

Fig. 5.6 Compressibility of ethanol water relative to $4.52 \times 10^{-10}\,\mathrm{Pa}^{-1}$ at reference $T = 25\,°\mathrm{C}$ and $p = 10.1325\,\mathrm{dbar}$

The ternary model is in the latter respect much better, and it shows that k and $\beta_{(T,y)}$ are approximately of the same order.

Hence for the case of water and ethanol, compressibility effects due on the one hand to pressure and on the other hand to composition look comparable. For low Mach numbers, approximation of the fluid as a constant density fluid could hence do well.

Appendix A. Differentiation Rules

At first, by differentiating the representation (5.45), we obtain the derivatives for $k = 1, \ldots, N-1$ of the pressure function $\mathscr{P}$ as

$$\partial_\varrho \mathscr{P}(\varrho, \mathbf{q}) = -\frac{\varrho}{q_N} \frac{1}{\mathbf{D}^2 h^*(\mathbf{w}^*)\boldsymbol{\xi}^{N+1} \cdot \boldsymbol{\xi}^{N+1}},$$

$$\partial_{q_k} \mathscr{P}(\varrho, \mathbf{q}) = -\frac{1}{q_N} \Big(\nabla_{\mathbf{w}^*} h^*(\mathbf{w}^*) \cdot \boldsymbol{\xi}^k - \varrho \frac{\mathbf{D}^2 h^*(\mathbf{w}^*)\boldsymbol{\xi}^{N+1} \cdot \boldsymbol{\xi}^k}{\mathbf{D}^2 h^*(\mathbf{w}^*)\boldsymbol{\xi}^{N+1} \cdot \boldsymbol{\xi}^{N+1}}\Big),$$

$$\partial_{q_N} \mathscr{P}(\varrho, \mathbf{q}) = -\frac{1}{q_N} \Big(\mathscr{P} + \partial_{w^*_{N+1}} h^*(\mathbf{w}^*) - \varrho \frac{\mathbf{D}^2 h^*(\mathbf{w}^*)\boldsymbol{\xi}^{N+1} \cdot \mathbf{e}^{N+1}}{\mathbf{D}^2 h^*(\mathbf{w}^*)\boldsymbol{\xi}^{N+1} \cdot \boldsymbol{\xi}^{N+1}}\Big). \quad (5.67)$$

Using (5.42), we similarly see for $i = 1, \ldots, N$ and $k = 1, \ldots, N-1$ that

$$\partial_\varrho \mathscr{R}_i(\varrho, \mathbf{q}) = \frac{\mathbf{D}^2 h^*(\mathbf{w}^*)\mathbf{e}^i \cdot \boldsymbol{\xi}^{N+1}}{\mathbf{D}^2 h^*(\mathbf{w}^*)\boldsymbol{\xi}^{N+1} \cdot \boldsymbol{\xi}^{N+1}},$$

$$\partial_{q_k} \mathscr{R}_i(\varrho, \mathbf{q}) = \mathbf{D}^2 h^*(\mathbf{w}^*)\mathbf{e}^i \cdot \boldsymbol{\xi}^k - \frac{\mathbf{D}^2 h^*(\mathbf{w}^*)\mathbf{e}^i \cdot \boldsymbol{\xi}^{N+1}\, \mathbf{D}^2 h^*(\mathbf{w}^*)\boldsymbol{\xi}^{N+1} \cdot \boldsymbol{\xi}^k}{\mathbf{D}^2 h^*(\mathbf{w}^*)\boldsymbol{\xi}^{N+1} \cdot \boldsymbol{\xi}^{N+1}},$$

$$\partial_{q_N} \mathscr{R}_i(\varrho, \mathbf{q}) = \partial^2_{i,N+1} h^*(\mathbf{w}^*) - \frac{\mathbf{D}^2 h^*(\mathbf{w}^*)\mathbf{e}^i \cdot \boldsymbol{\xi}^{N+1}\, \mathbf{D}^2 h^*(\mathbf{w}^*)\boldsymbol{\xi}^{N+1} \cdot \mathbf{e}^{N+1}}{\mathbf{D}^2 h^*(\mathbf{w}^*)\boldsymbol{\xi}^{N+1} \cdot \boldsymbol{\xi}^{N+1}}.$$

Finally, use of (5.43) yields

$$\partial_\varrho \mathscr{U}(\varrho, \mathbf{q}) = \frac{\mathbf{D}^2 h^*(\mathbf{w}^*)\mathbf{e}^{N+1} \cdot \boldsymbol{\xi}^{N+1}}{\mathbf{D}^2 h^*(\mathbf{w}^*)\boldsymbol{\xi}^{N+1} \cdot \boldsymbol{\xi}^{N+1}},$$

$$\partial_{q_k} \mathscr{U}(\varrho, \mathbf{q}) = \mathbf{D}^2 h^*(\mathbf{w}^*)\mathbf{e}^{N+1} \cdot \boldsymbol{\xi}^k - \frac{\mathbf{D}^2 h^*(\mathbf{w}^*)\mathbf{e}^{N+1} \cdot \boldsymbol{\xi}^{N+1}\, \mathbf{D}^2 h^*(\mathbf{w}^*)\boldsymbol{\xi}^{N+1} \cdot \boldsymbol{\xi}^k}{\mathbf{D}^2 h^*(\mathbf{w}^*)\boldsymbol{\xi}^{N+1} \cdot \boldsymbol{\xi}^{N+1}},$$

$$\partial_{q_N} \mathscr{U}(\varrho, \mathbf{q}) = \partial^2_{N+1} h^*(\mathbf{w}^*) - \frac{(\mathbf{D}^2 h^*(\mathbf{w}^*)\mathbf{e}^{N+1} \cdot \boldsymbol{\xi}^{N+1})^2}{\mathbf{D}^2 h^*(\mathbf{w}^*)\boldsymbol{\xi}^{N+1} \cdot \boldsymbol{\xi}^{N+1}}.$$

We can verify the following the relationships

$$\partial_{q_N} \mathscr{M} = -\partial_\varrho \mathscr{U}, \quad \partial_\varrho \mathscr{M} = T\, \varrho\, \partial_\varrho \mathscr{P}, \quad \partial_{q_N} \mathscr{P} = T\, (\mathscr{P} + \mathscr{U} - \varrho\, \partial_\varrho \mathscr{U}).$$

Appendix B. Wave Equation for a Multicomponent Fluid

In a fluid near equilibrium (zero velocity and constant thermodynamic state), compression waves might propagate. As is well known, the speed of compression waves is underestimated by the isothermal compressibility number, as the temperature gradient can align with the density gradient to accelerate the wave. The speed of sound is thus $1/(\varrho\, \beta_s)$ with the isentropic (adiabatic) compressibility β_s.

This regime is described for instance in Sect. 10.4 Müller (2013) for a single component fluid in one space dimension. Here we shall briefly consider in the same spirit the multicomponent case, where in addition mass diffusion gradients can also align with the pressure gradient.

We start from the partial differential equations of multicomponent fluid dynamics where diffusion fluxes, viscosity and external forces are overall neglected. The equations are posed in $\mathbb{R}^3 \times (0, +\infty)$, the position variable is denoted by $x = (x_1, x_2, x_3)$, the time is t. Thus

$$\partial_t \rho_i + \operatorname{div}(\rho_i\, \mathbf{v}) = 0 \tag{5.68}$$

$$\partial_t \varrho u + \operatorname{div}(\varrho u\, \mathbf{v}) = -p \operatorname{div} \mathbf{v} \tag{5.69}$$

$$\partial_t \varrho \mathbf{v} + \operatorname{div}(\varrho\, \mathbf{v} \otimes \mathbf{v}) + \nabla_x p = 0 \tag{5.70}$$

with the velocity field $\mathbf{v} = (v_1, v_2, v_3)$. We look for solutions in the form of small perturbations of a constant state solution: $\rho_i = \rho_i^0 + \bar{\rho}_i, T = T^0 + \bar{T}$, with a velocity field $\mathbf{v} = (\bar{v}_1, \bar{v}_2, \bar{v}_3)$ which is a small perturbation of zero. Notice that not only T and $\boldsymbol{\rho}$ are small perturbation of constant values, but also other thermodynamic variables $\boldsymbol{\mu}$, $\mathbf{y}$, etc.

After dropping from the Eqs. (5.68), (5.69), (5.70) all products of perturbations which are second order small, we deduce in well known manner from (5.68) and (5.70) the equation $\partial_t^2 \varrho - \Delta p = 0$. We use the vectors $\mathbf{w} = (\boldsymbol{\rho},\ \varrho u)$ to rewrite (5.68), (5.69) as

$$\dot{\mathbf{w}} = -(\boldsymbol{\rho},\ \varrho u + p)^\mathsf{T} \operatorname{div} \mathbf{v}\,,$$

and since $\mathbf{w} = \boldsymbol{\partial}_{\mathbf{w}^*} h^*(\mathbf{w}^*)$ also

$$\dot{\mathbf{w}}^* = -\mathbf{D}^2 h(\mathbf{w})\, (\boldsymbol{\rho},\ \varrho u + p)^\mathsf{T} \operatorname{div} \mathbf{v}\,.$$

Due to the continuity equation, we have $-\varrho \operatorname{div} \mathbf{v} = \dot{\varrho}$ and hence

$$\dot{\mathbf{w}}^* = \frac{1}{\varrho}\, \mathbf{D}^2 h(\mathbf{w})\, (\boldsymbol{\rho},\ \varrho u + p)^\mathsf{T}\, \dot{\varrho}\,.$$

Using that the perturbations vanish for large time and integrating the latter yields

$$\bar{\mathbf{w}}^* = \frac{1}{\varrho^0}\, \mathbf{D}^2 h(\mathbf{w}^0)\, (\boldsymbol{\rho}^0,\ \varrho^0 u^0 + p^0)^\mathsf{T}\, \bar{\varrho}\,.$$

Applying the spacial gradient yields

$$\nabla_x \mathbf{w}^* = \frac{1}{\varrho^0} \mathbf{D}^2 h(\mathbf{w}^0)\, (\boldsymbol{\rho}^0,\, \varrho^0 u^0 + p^0)^\mathsf{T}\, \nabla_x \bar{\varrho}\,, \tag{5.71}$$

and we see that the gradients of all thermodynamic variables align to the density gradient. The identity $p = T\, h^*(\mathbf{w}^*) = -h^*(\mathbf{w}^*)/w^*_{N+1}$ is next differentiated, and we get

$$\nabla_x p = \frac{T^0}{\varrho^0} \mathbf{D}^2 h(\mathbf{w}^0)(\boldsymbol{\rho}^0,\, \varrho^0 u^0 + p^0)^\mathsf{T} \cdot (\boldsymbol{\rho}^0,\, \varrho^0 u^0 + p^0)^\mathsf{T}\, \nabla_x \varrho\,.$$

Hence with

$$c^2 = \frac{T}{\varrho} \mathbf{D}^2 h(\mathbf{w})(\boldsymbol{\rho},\, \varrho u + p)^\mathsf{T} \cdot (\boldsymbol{\rho},\, \varrho u + p)^\mathsf{T} \tag{5.72}$$

we have identified the speed of compression waves in the fluid.

"Compression waves " appear in this argument (5.71) to be be accelerated by mass and heat diffusion processes. To show this, we use the function $\mathscr{P}$ to compute that

$$\begin{aligned}\nabla_x p =& \mathscr{P}_\varrho(\varrho,\, \mathbf{q})\, \nabla_x \varrho + \mathscr{P}_\mathbf{q}(\varrho,\, \mathbf{q}) \cdot \nabla_x \mathbf{q}\\ =& \frac{1}{\varrho\, \beta_{(q)}}\, \nabla_x \varrho + \mathscr{P}_\mathbf{q}(\varrho,\, \mathbf{q}) \cdot \nabla_x \mathbf{q}\,.\end{aligned}$$

We can interpret $\nabla_x q_1, \ldots, \nabla_x q_N$ as diffusion driving forces. Mainly, the gradients of the $N-1$ first variables are driving the mass diffusion while $\nabla_x q_N$ is responsible for the heat diffusion—provided that cross effects named after Soret and Dufour are neglected. In this case, the Maxwell-Stefan equations relate the mass diffusion fluxes $\mathbf{j}_1, \ldots, \mathbf{j}_N$ and the driving forces via

$$-\sum_{k=1}^N f_{ik}\, (y_k\, \mathbf{j}_i - y_i\, \mathbf{j}_k) = \rho_i\, \big(\nabla_x \frac{\mu_i}{RT} - \sum_{k=1}^N y_k\, \nabla_x \frac{\mu_k}{RT}\big) \quad \text{for } i = 1, \ldots, N\,, \tag{5.73}$$

in which $\{f_{ik}\}$ is a matrix of phenomenological coefficients describing the friction between the species (reciprocal diffusivities), and R is the gas-constant. In tensorial form

$$\mathcal{B}\mathbf{j} = -\mathcal{R}\mathcal{P}\, \nabla_x \frac{\boldsymbol{\mu}}{T} \quad \text{with} \quad \mathcal{B}_{ik} = \begin{cases} -f_{ik}\, y_i & \text{for } i \neq k\\ \sum_{j\neq i} f_{ij}\, y_j & \text{for } i = k\,, \end{cases} \tag{5.74}$$

with $\mathcal{R} := \operatorname{diag}(\rho_1, \ldots, \rho_N)$ and $\mathcal{P} := \mathbf{Id} - \mathbf{1} \otimes \mathbf{y}$, with $\mathbf{1} = 1^N$. Using the properties (5.37) of the basis $\{\eta^\ell\}$ and the properties of $\mathcal{B}$ we have $-\nabla_x q_\ell = \boldsymbol{\eta}^\ell \cdot \mathcal{R}^{-1}\, \mathcal{B}\mathbf{j}$ for $\ell = 1, \ldots, N-1$. By means of (5.67) we then see that

$$\sum_{\ell=1}^{N-1} \mathscr{P}_{q_\ell}(\varrho,\, \mathbf{q})\, \nabla_x q_\ell = -\frac{T\,\varrho}{\mathbf{D}^2 h^* \bar{e} \cdot \bar{e}}\, \mathcal{B}^\mathsf{T}\, \mathcal{R}^{-1}\, \mathbf{D}^2 h^* \bar{e} \cdot \mathbf{j}\,. \tag{5.75}$$

Similarly, if the heat diffusion flux is given by a Fourier law $\mathbf{j}^h = -\kappa\, \nabla_x T$, then use of (5.67) yields

$$\begin{aligned} \mathscr{P}_{q_N}(\varrho,\, \mathbf{q})\, \nabla_x q_N &= -\frac{\mathbf{j}^h}{\kappa\, T^2}\, \mathscr{P}_{q_N}(\varrho,\, \mathbf{q}) \\ &= -\frac{\varrho u + p - \varrho\, \mathbf{D}^2 h^* \bar{e} \cdot \mathbf{e}^{N+1} / \mathbf{D}^2 h^* \bar{e} \cdot \bar{e}}{\kappa\, T}\, \mathbf{j}^h\,. \end{aligned} \tag{5.76}$$

To resume, a general decomposition of the pressure gradient yields a compressive contribution and a diffusive enhancement:

$$\begin{aligned} \nabla_x p =& \frac{1}{\varrho\, \beta_{(q)}}\, \nabla_x \varrho - \frac{T\, \varrho\, \mathcal{B}^\mathsf{T}\, \mathcal{R}^{-1}\, \mathbf{D}^2 h^* \bar{e}}{\mathbf{D}^2 h^* \bar{e} \cdot \bar{e}} \cdot \mathbf{j} \\ & - \frac{\varrho u + p - \varrho\, \mathbf{D}^2 h^* \bar{e} \cdot \mathbf{e}^{N+1} / \mathbf{D}^2 h^* \bar{e} \cdot \bar{e}}{\kappa\, T}\, \mathbf{j}^h\,. \end{aligned}$$

The compressive part is expressed by the total compressibility coefficient $\beta_{(q)}$ rather than by the more usual isothermal compressibility.

Appendix C. Water+Ethanol Mixture as Ternary Ideal System

We follow closely Sect. 5 of Bothe et al. (2023). See also Roux and Desnoyers (1987), Sect. 3 for a variant of the same model. Through a chemical transformation $\gamma_1\, \mathrm{A}_1 + \gamma_2\, \mathrm{A}_2 \to \gamma_3\, \mathrm{A}_3$ the two species combine to a third one. Here the numbers γ_i are positive integers. The transformation conserve mass, hence with $a_i = \gamma_i/\gamma_3$ for $i = 1, 2$, we must have $M_3 = a_1\, M_1 + a_2\, M_2$. However, since the volume needs not to be conserved, the model can explain deviation from volume additivity of the binary ideal model.

The system is assumed in chemical equilibrium, which is satisfied if $\gamma_1 M_1\, \mu_1 + \gamma_2 M_2\, \mu_2 = \gamma_3 M_3\, \mu_3$. More simply, with $\theta = a_2 M_2/(a_1\, M_1 + a_2\, M_2)$

$$\mu_3 = (1-\theta)\, \mu_1 + \theta\, \mu_2\,.$$

Assuming now the ideal form (5.23) of the chemical potentials we can solve the latter equation for the mole fraction x_3:

$$x_3 = x_1^{a_1}\, x_2^{a_2}\, \exp\Big(\frac{M_3}{RT}\, \big((1-\theta)\, g_1(T,\, p) + \theta\, g_2(T,\, p) - g_3(T,\, p)\big)\Big)\,.$$

The function under the exponential represents the deviation from volume additivity. Since we need to fit both the density and the compressibility, we make a quadratic Ansatz

$$\begin{aligned}&\frac{M_3}{RT}\Big((1-\theta)\,g_1(T,p)+\theta\,g_2(T,p)-g_3(T,p)\Big)\\ =:&\tilde{g}(T,p)\overset{!}{=}(p-p_0)\,(D(T)+D'(T)\,(p-p_0)/2)\,,\end{aligned}$$

with two functions $D(T),\ D'(T)$ to be determined. Hence, we will have

$$\begin{aligned}g_3(T,p)=&(1-\theta)\,g_1(T,p)+\theta\,g_2(T,p)\\ &-\frac{RT}{M_3}\,(p-p_0)\,(D(T)+D'(T)\,(p-p_0)/2)\,,\end{aligned}$$

and have complete data for an ideal ternary model. Using the variables $y =$ mass fraction of A_2, T and p, we can parametrise the complete state. Indeed, we obtain the three mole fractions $x = x(y, T, p)$ by solving for x the three equations

$$x_3 = x_1^{a_1}\,x_2^{a_2}\,e^{\tilde{g}(T,p)},\quad M_2\,x_2 = y\sum_{i=1}^{3}M_i\,x_i,\quad x_3 = 1-x_1-x_2\,. \tag{5.77}$$

Straightforward calculations and (5.24) yield

$$\begin{aligned}\frac{1}{\varrho}=&(1-y)\,\partial_p g_1(T,p)+y\,\partial_p g_2(T,p)\\ &+y_3(y,T,p)\,(\theta\,\partial_p(g_2-g_1)(T,p)-\frac{RT}{M_3}\,(D(T)+D'(T)\,(p-p_0)))\,.\end{aligned} \tag{5.78}$$

Here $y_3(y, T, p)$ is obtained via solution of (5.77). Having measurement $V_0(T)$ for the volume of the binary mixture at some reference point (y_0, p_0), we get for D the equation

$$\begin{aligned}V_0(T)=&(1-y_0)\,\partial_p g_1(T,p_0)+y_0\,\partial_p g_2(T,p_0)\\ &+y_3(y_0,T,p_0)\Big(\theta\,\partial_p(g_2-g_1)(T,p_0)-\frac{RT}{M_3}\,D(T)\Big)\,.\end{aligned} \tag{5.79}$$

Moreover, using differentiation in p of (5.78), we can compute the relation

$$\begin{aligned}\frac{\beta_{(T,y)}}{\rho}=&(1-y)\,\beta_T^1\,\partial_p g_1+y\,\beta_T^2\,\partial_p g_2+y_3\Big(\theta\,(\beta_T^2\partial_p g_2-\beta_T^1\,\partial_p g_1)-\frac{RT}{M_3}\,D'(T)\Big)\\ &+\partial_p y_3\Big(\theta\,\partial_p(g_2-g_1)-\frac{RT}{M_3}\,(D+D'\,(p-p_0))\Big)\,.\end{aligned} \tag{5.80}$$

with the compressibilities β_T^i of the pure species. Restricting to a reference state (y_0, p_0) for which we possess a measurement of the true compressibility of the binary mixture, we determine $D'(T)$. Hence a ternary model is finally fitted.

References

Barbosa RD (2003) High pressure and temperature dependence of thermodynamic properties of model food solutions obtained from in situ ultrasonic measurements. Ph.D. thesis, University of Florida

Benson GC, Kiyohara O (1980) J Solution Chem 9:791. https://doi.org/10.1007/BF00646798

Bothe D, Dreyer W (2015) Acta Mech 226(6):1757. https://doi.org/10.1007/s00707-014-1275-1

Bothe D, Druet PE (2021) J Evol Equ 21(4):4039. https://doi.org/10.1007/s00028-021-00712-3

Bothe D, Druet PÉ (2023) Int J Eng Sci 184:33. https://doi.org/10.1016/j.ijengsci.2023.103818. Id/No 103818

Bothe D, Dreyer W, Druet PE (2023) ZAMM, Z Angew Math Mech 103(7):48. https://doi.org/10.1002/zamm.202100174. Id/No e202100174

Dreyer W, Druet PÉ, Klein O, Sprekels J (2012) Milan J Math 80(2):311. https://doi.org/10.1007/s00032-012-0184-9

Dreyer W, Guhlke C, Müller R (2013) Phys Chem Chem Phys 15:7075. https://doi.org/10.1039/C3CP44390F

Dreyer W, Druet PÉ, Gajewski P, Guhlke C, Angew Z (2020) Math Phys 71(4):68. https://doi.org/10.1007/s00033-020-01341-5. Id/No 119

Druet PE (2023) Nonlinear Anal Real World Appl 72:68. https://doi.org/10.1016/j.nonrwa.2023.103859. Id/No 103859

Druet PE (2021) J Math Anal Appl 499(2):56. https://doi.org/10.1016/j.jmaa.2021.125059. Id/No 125059

Druet PE (2022) Nonlinearity 35(7):3812. https://doi.org/10.1088/1361-6544/ac5679

Fofonoff NP, Millard Jr R (1983) Algorithms for the computation of fundamental properties of seawater. UNESCO Technical Papers in Marine Sciences; 44 (Unesco, Paris, 1983). http://hdl.handle.net/11329/10910.25607/OBP-1450

Giovangigli V (2012) Sci China, Math 55(2):285. https://doi.org/10.1007/s11425-011-4346-y

Giovangigli V, Matuszewski L (2012) Physica D 241(6):649. https://doi.org/10.1016/j.physd.2011.12.002

Gouin H, Ruggeri T (2012) Int J Non-Linear Mech 47(6):688. https://doi.org/10.1016/j.ijnonlinmec.2011.11.005

Gouin H, Muracchini A, Ruggeri T (2012) Continuum Mech Thermodyn 24:505. https://doi.org/10.1007/s00161-011-0201-1

Haynes WE (2014) CRC handbook of chemistry and physics, 95th edn. CRC Press. https://doi.org/10.1201/b17118

Hutter K, Jöhnk K (2004) Continuum methods of physical modeling. Continuum mechanics, dimensional analysis, turbulence. Springer, Berlin

IOC, SCOR, IAPSO, *The international thermodynamic equation of seawater – 2010: Calculation and use of thermodynamic properties*. Intergovernmental Oceanographic Commission, Manuals and Guides No. 56 (UNESCO: United Nations Educational, Scientific and Cultural Organisation, 2010). https://www.teos-10.org

Krumgalz BS, Millero FJ (1982) Marine Chem 11(5):477. https://doi.org/10.1016/0304-4203(82)90012-3

Müller I (1985) Thermodynamics. Interaction of mechanics and mathematics series. Pitman. https://books.google.de/books?id=-Jd6AAAAIAAJ

Müller I (2013) Grundzüge der Thermodynamik: mit historischen Anmerkungen. Springer. https://doi.org/10.1007/978-3-642-56474-1

Randall CR (1932) Bur Stand J Res 8(1):79. https://nvlpubs.nist.gov/nistpubs/jres/8/jresv8n1p79_a2b.pdf

Roux A, Desnoyers J (1987) In: Proceedings of the Indian academy of sciences-chemical sciences, vol 98. Springer, pp 435–451. https://doi.org/10.1007/BF02861539

Schmidt GA, Bigg GR, Rohling EJ (1999). https://data.giss.nasa.gov/o18data

Tanaka Y, Yamamoto T, Satomi Y, Kubota H, Makita T (1977) Rev Phys Chem Jpn 47(1):12. https://repository.kulib.kyoto-u.ac.jp/dspace/bitstream/2433/47042/1/rpcjpnv47p012.pdf

Chapter 6
Incorporating Dissipation in the Lagrange Equations: Part I—Theory

Wolfgang H. Müller, Ekaterina A. Podolskaya, Alexey S. Smirnov, and Boris A. Smolnikov

Dedication
... et malheureusement dans ce monde chacun a son point de vue à soi qui l'empêche de voir le point de vue des autres.
Alexandre Dumas, Le Comte de Monte Cristo
This is for the late Wolfgang Dreyer, a good friend and true scientist whom I owe so much. (WHM)

Abstract In this work we introduce the so-called dissipative function of internal volumetric friction in the Lagrange equations. Unlike the Rayleigh dissipation function, which has been used to describe dissipation due to velocity-dependent friction on surfaces of discrete (rigid) bodies within the Lagrange formalism for a long time, this paper presents a new rheological model—the so-called dissipative gyrostat—which allows to study dissipation within the volume of (rigid) bodies by means of the aforementioned dissipation function. In a series of two papers we will, first, introduce the new concept and outline the theory. To this end, we will initially explain the physical ideas behind the new rheological model in words and formulae as illustrative as possible. Since our intention is also to build a bridge between the world of Lagrange and Truesdellian rational mechanics we will then compare and relate terms

W. H. Müller (✉)
Institut für Mechanik, Fachgebiet Kontinuumsmechanik und Materialtheorie, Technische Universität Berlin, Berlin, Germany
e-mail: wolfgang.h.mueller@tu-berlin.de

E. A. Podolskaya
Polycentrum Solutions SL, Barcelona, Spain
e-mail: kate@polycentrum.com

A. S. Smirnov · B. A. Smolnikov
Peter the Great St. Petersburg Polytechnic University, St. Petersburg, Russia
e-mail: smirnov.alexey.1994@gmail.com

B. A. Smolnikov
e-mail: smoln_ba@spbstu.ru

Institute for Problems in Mechanical Engineering of the Russian Academy of Sciences, St. Petersburg, Russia

J. Fuhrmann et al. (eds.), *Advances in Continuum Physics*, Advanced Structured Materials 238, https://doi.org/10.1007/978-3-031-93918-1_6

in the equations pertinent to both points of view. This will culminate in a rational explanation of the volumetric dissipation function. In the second of this series of papers we shall apply the theory to a study of various beam pendulum problems and also use it for an investigation of the locking phenomenon of planetary bodies.

6.1 Review of Lagrange Mechanics Reversible and Irreversible

While continuum mechanics and continuum thermodynamics based on the pioneering work from the school of Truesdell and Toupin (1960), Truesdell (1969, 1992), Truesdell and Noll (2004) provide a rational basis for describing and quantifying dissipative processes in gaseous, liquid, as well as solid bodies the Lagrange formalism only reluctantly addresses this topic.

Lagrangian mechanics is based on extremum principles and variational analysis, and this is the reason for that. For example we read in the introduction to the famous book by Lanczos (1949) "Forces of a frictional nature, which have no work function, are outside the realm of variational principles, while the Newtonian scheme has no difficulty in including them." In fact, as we shall see, this statement is too pessimistic, because the Lagrange equations can be generalized such that they include certain aspects of dissipation.

We start our discussion by writing them in the usual form known to every student of analytical mechanics[1]:

$$\frac{\mathrm{d}}{\mathrm{d}t}\frac{\partial L}{\partial \dot{q}_i} - \frac{\partial L}{\partial q_i} = 0\,, \quad i = 1, ..., f, \tag{6.1}$$

with the Lagrange function $L = T - U$, T being the kinetic and U the potential energy of the system of rigid bodies, which are connected and related to each other by constraints, such as inextensible (massless) ropes, (massless) rods, supports, bearings, joints, Hookean (elastic) springs, but no dashpots or ratchets, which already belong to the world of dissipative rheological elements.

In the equation $\boldsymbol{q}^{\mathsf{T}} = [q_1, ..., q_f]^{\mathsf{T}}$ denotes the row of generalized (independent) coordinates, where the number f corresponds to the degrees of freedom of the system of interconnected rigid bodies. It is, of course, not a vector in the sense of rational continuum mechanics although Lagrange afficionados (see, e.g., Hahn (2013)) sometimes like to call it that. The generalized coordinates are functions of time, hence, the set of generalized velocities can be obtained by direct differentiation with respect to time t, which we indicate by a dot, $\dot{\boldsymbol{q}}^{\mathsf{T}} = [\dot{q}_1, ..., \dot{q}_f]^{\mathsf{T}}$. The sets can be used to detail the expression for the kinetic and the potential energy in the form of matrix products (see Smirnov and Smolnikov (2017), pg. 9). In general the kinetic energy is a cumbersome function of the generalized coordinates and of the

[1] We use the nomenclature employed in the famous book by Goldstein et al. (2002).

generalized velocities, $T = (\boldsymbol{q}, \dot{\boldsymbol{q}})$. However, for small deviations from equilibrium, the kinetic energy can be written as Smirnov and Smolnikov (2017, 2020)

$$T = \frac{1}{2}\dot{\boldsymbol{q}}^{\mathsf{T}} \cdot \mathbf{A} \cdot \dot{\boldsymbol{q}}, \tag{6.2}$$

where $\mathbf{A}$ is known as the (symmetric) matrix of inertial coefficients. Equation (6.2) looks as if the kinetic energy T is a quadratic form of the generalized velocities alone. This is deceiving. In general, it depends on the generalized coordinates and on the generalized velocities (if so-called scleronomic coordinate relations are considered, which are not explicitly time dependent), and—maybe—explicitly on time.[2]

The potential energy, U, is—by definition—the potential to conservative forces. In other words, it is related to non-dissipative effects. For mechanics, two types of potential energies are important to note. We write

$$U = U^{\mathrm{el}} + U^{\mathrm{g}}\,, \tag{6.3}$$

where indices el and g refer to elastic and gravitational effects.

In general, the elastic part, U^{el}, is a complicated function of the generalized coordinates, $\boldsymbol{q}$. However, for small deviations from equilibrium, it is possible to assume a quadratic form (in the generalized coordinates), analogously to Eq. (6.2):

$$U^{\mathrm{el}} = \frac{1}{2}\boldsymbol{q}^{\mathsf{T}} \cdot \mathbf{C} \cdot \boldsymbol{q}, \tag{6.4}$$

where $\mathbf{C}$ is referred to as the (symmetric) matrix of quasi-elastic coefficients.

The gradient of the elastic potential energy U^{el} yields the typical form of a (restitutive, note the minus sign) Hookean spring force,[3]

$$Q_i^{\mathrm{el}} = -\frac{\partial U^{\mathrm{el}}}{\partial q_i} = -\mathrm{C}_{ij} q_j . \tag{6.5}$$

We may refer to Q_i^{el} as a generalized elastic force, which—*a priori*—is of non-dissipative nature. The gravitational potential U^{g} is also of a non-dissipative nature, but it is not quadratic in the (generalized) coordinates. For an illustration of the whole concept let us consider a free rigid body of mass m and (symmetric) inertia tensor $\boldsymbol{J}$ in the near-field $\boldsymbol{g}$ of the Earth. Its generalized coordinates are then simply the coordinates of its center of mass $q_i \equiv x_i^{\mathrm{c}}$, $i = 1, 2, 3$ (in Cartesian form)[4] and we write:

$$U^{\mathrm{g}} = -m\boldsymbol{g} \cdot \boldsymbol{x}^{\mathrm{s}} \quad \Rightarrow \quad Q_i^{\mathrm{g}} = -\frac{\partial U^{\mathrm{g}}}{\partial x_i^{\mathrm{s}}} = m g_i\,, \quad i = 1, 2, 3. \tag{6.6}$$

---

[2] In Eq. (6.24) below we will also allow explicit time dependence.

[3] Einstein's summation convention for multi-indices applies.

[4] And the three Euler angles ϕ, θ, ψ, which play no role here.

It is often said that a potential U depends only on generalized coordinates and not on generalized velocities (see for example Gross et al. (2011), pg. 224). However, there is (at least) one important exception: just like the gravitational potential depends linearly on position, the potential for the electromagnetic Lorentz force depends linearly on velocity. The Lorentz force is conservative, albeit not in the classical sense, because a scalar potential is insufficient. Nevertheless, a more complex potential function exists. We shall discuss this now, even though we will not explicitly solve electromagnetic problems in this article. Following Goldstein et al. (2002), pp. 22 or Strocchi (2018), Sect. 2.5, we first note that the Lorentz force $\boldsymbol{F}^{\mathrm{L}}$ on a charged point particle (charge Q in Coulomb) can be written in terms of the electric field $\boldsymbol{E}$ and the magnetic field $\boldsymbol{B}$ as follows:

$$\boldsymbol{F}^{\mathrm{L}} = Q\,(\boldsymbol{E} + \dot{\boldsymbol{x}} \times \boldsymbol{B})\,. \tag{6.7}$$

Here we consider an unconstrained case, hence the point charge Q is moving through space with its "ordinary" velocity $\dot{\boldsymbol{x}}$.[5] The two electromagnetic fields are thought as being known and given. They stem from (unknown) external sources, just like the acceleration $\boldsymbol{g}$ was thought as given. They can be obtained from a scalar potential ϕ and a vector potential $\boldsymbol{A}$, which both depend on position $\boldsymbol{x}$ and time t:

$$\boldsymbol{E} = -\left(\nabla\phi + \frac{\partial \boldsymbol{A}}{\partial t}\right)\,, \quad \boldsymbol{B} = \nabla \times \boldsymbol{A}. \tag{6.8}$$

We now define the electromagnetic potential for the Lagrange formalism by:

$$U^{\mathrm{em}} = Q\,(\phi - \dot{\boldsymbol{x}} \cdot \boldsymbol{A})\,. \tag{6.9}$$

Then if the double vector product rule is observed,

$$\begin{aligned} &\dot{\boldsymbol{x}} \times \boldsymbol{B} = \dot{\boldsymbol{x}} \times (\nabla \times \boldsymbol{A}) = (\nabla \boldsymbol{A}) \cdot \dot{\boldsymbol{x}} - \dot{\boldsymbol{x}} \cdot (\nabla \boldsymbol{A}) \quad \Leftrightarrow \\ &\left[\dot{\boldsymbol{x}} \times (\nabla \times \boldsymbol{A})\right]_i = \frac{\partial A_j}{\partial x_i}\dot{x}_j - \dot{x}_j \frac{\partial A_i}{\partial x_j}, \end{aligned} \tag{6.10}$$

and the chain rule for $\boldsymbol{A}$ is taken into account,

$$\frac{\mathrm{d}A_i}{\mathrm{d}t} = \frac{\partial A_i}{\partial x_j}\dot{x}_j + \frac{\partial A_i}{\partial t}. \tag{6.11}$$

Now it is a matter of algebra to show that:

$$Q_i^{\mathrm{em}} = \frac{\mathrm{d}}{\mathrm{d}t}\frac{\partial U^{\mathrm{em}}}{\partial \dot{x}_i} - \frac{\partial U^{\mathrm{em}}}{\partial x_i} \equiv F_i^{\mathrm{L}}. \tag{6.12}$$

[5] Without saying it explicitly all coordinates and velocities that were mentioned so far are with respect to an inertial observer. The question of what changes if a non-inertial observer tries to use the Lagrange formalism will be touched upon later.

Thus we conclude by saying that the potential in the Lagrange function can be split into three parts,

$$U = U^{\mathrm{el}} + U^{\mathrm{g}} + U^{\mathrm{em}} \,, \tag{6.13}$$

if we want to cover mechanical and electromechanical problems. The electromechanical potential U^{em} depends on generalized coordinates and generalized velocities, and the Lorentz force is not a conservative force in the classical sense of mechanics, because more than a scalar potential is needed.

As an example, we attempt to obtain the equations of motion of an unconstrained rigid body by means of the Lagrange equations. Its kinetic energy is given additively by a translational and rotational part,

$$T = \frac{1}{2} m \dot{\boldsymbol{x}}^{\mathrm{s}} \cdot \dot{\boldsymbol{x}}^{\mathrm{s}} + \frac{1}{2} \boldsymbol{\omega} \cdot \boldsymbol{J} \cdot \boldsymbol{\omega}. \tag{6.14}$$

In this particular case we choose as generalized velocities $\dot{\boldsymbol{q}}$ the Cartesian components of the center of gravity velocity vector $\dot{\boldsymbol{x}}^{\mathrm{s}}$ and the Cartesian components of the angular velocity vector $\boldsymbol{\omega}$, hence $\dot{\boldsymbol{q}}^{\mathsf{T}} = \left[\dot{x}_1^{\mathrm{s}}, \dot{x}_2^{\mathrm{s}}, \dot{x}_3^{\mathrm{s}}, \omega_1, \omega_2, \omega_3\right]$. Then, following Eq. (6.2), the inertial matrix $\mathbf{A}$ is given by:

$$\mathbf{A} = \begin{bmatrix} m & 0 & 0 & 0 & 0 & 0 \\ 0 & m & 0 & 0 & 0 & 0 \\ 0 & 0 & m & 0 & 0 & 0 \\ 0 & 0 & 0 & J_{11} & J_{12} & J_{13} \\ 0 & 0 & 0 & J_{21} & J_{22} & J_{23} \\ 0 & 0 & 0 & J_{31} & J_{32} & J_{33} \end{bmatrix} \tag{6.15}$$

with the components J_{ij} of the symmetric inertia tensor with respect to the center of gravity, $\boldsymbol{J} = \boldsymbol{J}^{\mathsf{T}}$.

This in mind the equations of motion can be obtained from (6.1) componentwise. We convert them to vector form:

$$m\ddot{\boldsymbol{x}}^{\mathrm{s}} = m\boldsymbol{g} \,, \quad \frac{\mathrm{d}}{\mathrm{d}t}(\boldsymbol{J} \cdot \boldsymbol{\omega}) = \boldsymbol{0}. \tag{6.16}$$

In conclusion: we obtain Newton's and Euler's laws of motion for a rigid body in the gravitational field, free of external moments, so that the moment of momentum $\boldsymbol{J} \cdot \boldsymbol{\omega}$ is conserved.

A few comments are in order:

- As evident from (6.15) the components of the inertial matrix $\boldsymbol{A}$ do not all need to have the same units. The same holds for the quasi-elastic coefficients of $\boldsymbol{C}$. For example, they can be spring constants of linear springs in N/m or spring constants of torsion springs in N·m. A similar remark holds for other coupling matrices to come.

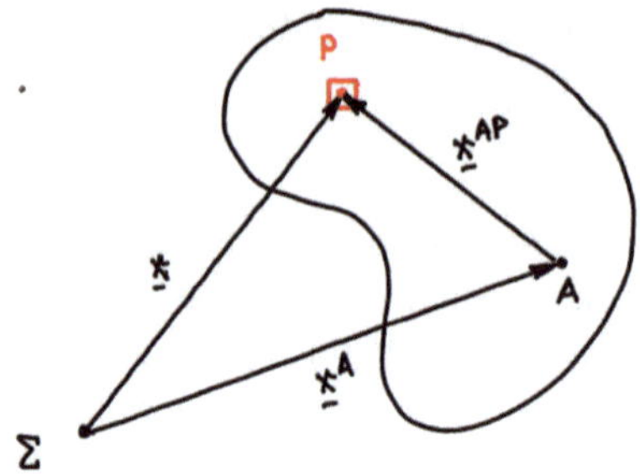

Fig. 6.1 Rigid body kinematics

- The components of the angular velocity vector $\boldsymbol{\omega}$ and the three Euler angles are linked to each other in a complicated manner, see for example Jazar (2011), Chap. 4. This is because the proper way to analyze rigid body kinematics is to use rotation tensors. Then the Euler angles arise rationally during three successive rotations about independent axes. In order to analyze the connection between the angular velocity vector and the rotation tensor and in preparation for the forthcoming analysis of the accelerations experienced by the so-called dissipative gyrostat consider Fig. 6.1. We write from the standpoint of the inertial observer Σ:

$$\boldsymbol{x}(t) = \boldsymbol{x}^{A}(t) + \boldsymbol{x}^{AP}(t). \tag{6.17}$$

The time dependence of the vector $\boldsymbol{x}^{AP}$ within the rigid body is a very peculiar one: this vector can rotate, but it does not change its length, because of the requirement that A and P are two points on the rigid body. We can acknowledge this fact by turning the AP vector from a reference position at time 0 to the current position by means of an orthogonal rotation tensor $\boldsymbol{Q}(t)$ such that:

$$\boldsymbol{x}^{AP}(t) = \boldsymbol{Q}(t) \cdot \boldsymbol{x}^{AP}(0) \ , \quad \boldsymbol{Q} \cdot \boldsymbol{Q}^{\mathsf{T}} = \mathbf{1} = \boldsymbol{Q}^{\mathsf{T}} \cdot \boldsymbol{Q}. \tag{6.18}$$

Then it is easy to show by successive differentiation with respect to time that for the velocities $\boldsymbol{v}$ and accelerations $\boldsymbol{a}$[6] we have

$$\begin{aligned} \boldsymbol{v}(t) &= \boldsymbol{v}^{A}(t) + \boldsymbol{\Omega}(t) \cdot \boldsymbol{x}^{AP}(t) \ , \quad \boldsymbol{\Omega} = \dot{\boldsymbol{Q}}(t) \cdot \boldsymbol{Q}^{\mathsf{T}}(t) \ , \\ \boldsymbol{a}(t) &= \boldsymbol{a}^{A}(t) + \dot{\boldsymbol{\Omega}}(t) \cdot \boldsymbol{x}^{AP}(t) + \boldsymbol{\Omega}(t) \cdot \boldsymbol{\Omega}(t) \cdot \boldsymbol{x}^{AP}(t). \end{aligned} \tag{6.19}$$

Because of $(6.18)_2$ the left spin tensor is antisymmetric, $\boldsymbol{\Omega} = -\boldsymbol{\Omega}^{\mathsf{T}}$. Hence it can be replaced by its vector invariant, the angular velocity vector $\boldsymbol{\omega} = -\frac{1}{2}\boldsymbol{\Omega}_{\times} \Leftrightarrow \boldsymbol{\Omega} = -\boldsymbol{\epsilon} \cdot \boldsymbol{\omega}$, $\boldsymbol{\epsilon}$ being the Levi-Civitá tensor third rank.[7] Therefore (6.19) can alternatively be written using vector products:

---

[6] All with respect to the inertial observer Σ.

[7] For a tensor second rank $\boldsymbol{A} = A_{ij}\boldsymbol{e}_i \otimes \boldsymbol{e}_j$ the vector invariant is given by $\boldsymbol{A}_{\times} = A_{ij}\boldsymbol{e}_i \times \boldsymbol{e}_j$.

$$\begin{aligned} \boldsymbol{v}(t) &= \boldsymbol{v}^A(t) + \boldsymbol{\omega}(t) \times \boldsymbol{x}^{AP}(t) \, , \\ \boldsymbol{a}(t) &= \boldsymbol{a}^A(t) + \dot{\boldsymbol{\omega}}(t) \times \boldsymbol{x}^{AP}(t) + \boldsymbol{\omega}(t) \times \left[\boldsymbol{\omega}(t) \times \boldsymbol{x}^{AP}(t)\right] . \end{aligned} \tag{6.20}$$

We emphasize once more that Eqs. (6.17), (6.19), and (6.20) hold from the standpoint of the inertial observer Σ shown in Fig. 6.1. They contain vectorial and tensorial objects that belong to Σ proper. As we shall see modifications are required if one switches to a non-inertial observer Σ', who is firmly attached to the accelerating and turning rigid body.

In summary: the Lagrange equations in the form (6.1) hold up to this point for reversible processes and we make this explicit by writing them in the following form,

$$\frac{\mathrm{d}}{\mathrm{d}t}\frac{\partial T}{\partial \dot{q}_i} - \frac{\partial T}{\partial q_i} = Q_i^{\mathrm{rev}} \, , \quad i = 1, ..., f, \tag{6.21}$$

where the generalized forces $Q_i^{\mathrm{rev}} = Q_i^{\mathrm{el}} + Q_i^{\mathrm{g}}$ are exclusively of reversible nature and we restrict ourselves to mechanical phenomena. The former we shall change now. One of the first to incorporate dissipation on the right-hand side of the Lagrange equations (6.21) was Rayleigh.[8] His idea was to incorporate surface friction effects, where the frictional force acting on the surface of the rigid body is related to the speed. This led to the introduction of the Rayleigh function R, which, in general, depends on *relative* generalized velocities in a highly nonlinear manner, and in particular for Stokes type of friction quadratically, such that

$$R = \frac{1}{2}\int_a \dot{\boldsymbol{q}}^{\mathsf{T}} \cdot \mathbf{B} \cdot \dot{\boldsymbol{q}} \, \mathrm{d}a \, , \quad Q_i^{\mathrm{f_a}} = -\frac{\partial R}{\partial \dot{q}_i}. \tag{6.22}$$

Following Smolnikov and Belyaev (2008), Smolnikov (2008) **B** is a non-negative (symmetric) matrix of "viscous friction coefficients," and the integration is performed over all rubbing surfaces a. Physically speaking, the function R expresses the power of (generalized) dissipative forces $Q_i^{\mathrm{f_a}}$ of surface friction. It should be pointed out that in other works the integration in (6.22) is not explicitly mentioned, for example in Goldstein et al. (2002), pp. 23, Bersani and Caressa (2021), or Minguzzi (2015). By studying Lurie (2013)[9] or Minguzzi (2015) it becomes clear what the coefficients of **B** can be. For example, they can be the coefficient of dry (Coulomb) friction, which is dimensionless, or it is the dynamic shear viscosity (in Pa·s) in case of Stokes' friction. Generalizing (6.21) the equations of motion now read

$$\frac{\mathrm{d}}{\mathrm{d}t}\frac{\partial T}{\partial \dot{q}_i} - \frac{\partial T}{\partial q_i} = Q_i^{\mathrm{rev}} + Q_i^{\mathrm{f_a}} \, , \quad i = 1, ..., f. \tag{6.23}$$

[8] The original literature by Rayleigh is cited and discussed in Minguzzi (2015).

[9] Initially Lurie also does not mention the integration involved in (6.22), but later he does, cf., Lurie (2013), Sects. 5.11 versus 5.12.

Previous work (Smolnikov 2008; Smolnikov 2008; Smolnikov and Belyaev 2008; Podolskaya and Smolnikov 2010a, b, c; Smirnov and Smolnikov 2017) extends Rayleigh's ideas to viscous dissipative forces within a volume. This new rheological body combines rigid and viscous material aspects and is called *dissipative gyrostat* model. We proceed to explain the concept.

In the cited papers we find an interesting classification of various models of technical mechanics, conservative and dissipative ones, namely:

Type	Rigid body	Deformable body
Conservative/reversible	Euler's model	Cauchy's elastic model
Dissipative/irreversible	Dissipative gyrostat	Viscoelastic model

The true rigid body model of the table relies on the mathematical description of its motion for its center of gravity and its rotation about three independent axes, cf., (6.16).[10] It was described first in the work of Leonard Euler, for example in Euler (1765), Euler (1776). Due to its simplicity (no deformation at all) there is no constitutive equation or material parameter involved. In Euler's work on rigid bodies the question of reversibility or irreversibility of their motion is not addressed: the total energy of a rigid body, which comprises kinetic and (gravitational) potential energy, remains constant during the motion. In fact, this can also be seen in the Lagrange formalism for a conservative system. We consider a general Lagrange function of the form $L = L(\boldsymbol{q}, \dot{\boldsymbol{q}}, t)$. Forming the total differential and incorporating Lagrange's equations for conservative systems (6.1) yields after some algebra:

$$\frac{\mathrm{d}E}{\mathrm{d}t} + \frac{\partial L}{\partial t} = 0\,, \quad E = \dot{q}_j \frac{\partial L}{\partial \dot{q}_j} - L\,. \tag{6.24}$$

In the case of the single unconstrained rigid body that was discussed above, the Lagrange function does not explicitly depend on time and the total energy E is conserved,

$$\frac{\mathrm{d}E}{\mathrm{d}t} = 0\,, \tag{6.25}$$

so that we may write by observing $L = T - U^{\mathrm{g}}$, (6.6), and (6.14):

$$E = T + U^{\mathrm{g}} = \mathrm{const.}_t\,. \tag{6.26}$$

In fact, the result (6.26) is not only valid for the case of a single unconstrained rigid body. This is easy to see, if the following assumptions (Goldstein et al. 2002, pg. 68) are made, first, the potential U of reversible forces in the Lagrange function L depends only on the generalized coordinates and, second, the kinetic energy is a

[10] Where it was restricted to the case of no external moments.

homogeneous function of the generalized velocities of degree 2,[11]

$$\dot{q}_i \frac{\partial T}{\partial \dot{q}_i} = 2T \quad \Rightarrow \quad E = 2T - (T - U) = T + U. \tag{6.27}$$

The next complication addressed in the table from above is the elastic model introduced in the work of Cauchy. Essentially what we have in mind here are the Navier-Lamé equations of linear elasticity, sometimes also referred to as Cauchy-Lamé equations (see Wikipedia, Navier-Cauchy-Gleichungen (2024)), which hold for small reversible (= elastic) deformations, Cauchy (1828). The deformation requires us to introduce a constitutive relation (Hooke's law) and (for isotropic materials) two elastic material constants (which we would call the two Lamé constants today). It should be noted that Cauchy's work was extended later to large elastic deformations (for rubbery materials) and lives on in the deformation measure used there, the left and right Cauchy-Green tensors. However, the elastic deformations addressed in U^{el} of Eq. (6.5) are small ones. Now the arguments leading to (6.24) prevail and we may extend (6.26) on the basis of (6.4) and (6.5) to read:

$$E = T + U^{\mathrm{el}} + U^{\mathrm{g}} = \mathrm{const.}_{,t}, \tag{6.28}$$

if we decide to incorporate the rheological model of a (reversible) Hookean spring as part of Cauchy's world, which rightfully is a continuum mechanics and not a discrete body approach.[12]

The viscoelastic models addressed in the table have several fathers. At least the following two rheological models should be mentioned here, the Maxwell and the Kelvin-Voigt body, which can be viewed as a dashpot and a (Hookean) spring in series or parallel configuration. The dashpot, whose deformation depends linearly on velocity (Newton (2008) in his *Principia*), constitutes the dissipative part. In fact, we find the differential equation connecting time-dependent deformation and force of the Maxwell body in Maxwell (1868). The paper by Thomson (Lord Kelvin) (1865) and the paper of Voigt (1892) present both measurements on damping. They are commonly cited in context with the Kelvin-Voigt model. However, the differential equation normally discussed in a beginner's class on viscoelasticity cannot be found there.

Let us now introduce the *dissipative gyrostat*. To this end we first remind the reader of the meaning of the term *gyrostat*: a technical gyrostat is a rigid system with built-in rotating masses (flywheels) that influence its dynamic behavior. Such systems are commonly used in satellite control, gyroscopic mechanics, and the theory of rotating bodies.

---

[11] This assumption is not as awkward as it may seem on first glance. It is straightforward to prove, for example, for a not explicitly time-dependent relationship between the position vector and the generalized coordinates, $x_i = x_i\left(q_j\right)$, $i = 1, \ldots, 3N$, $j = 1, \ldots, f$ where i characterizes all (original) coordinates for the collective of N rigid bodies involved.

[12] A more proper formulation would make use of a Lagrangian density, see Lagrangian mechanics (2024), Section Classical field theory.

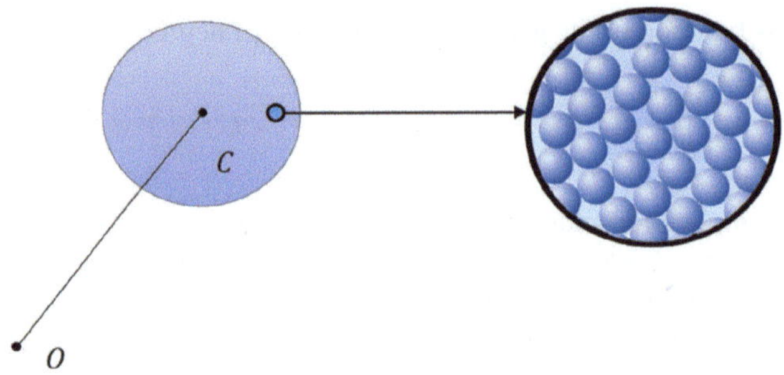

Fig. 6.2 A dissipative gyrostat (adapted from Loseva (2019)): top view of a bucket (center C) filled with sand (left) and the microstructure within, sand grains with pores (right)

Clearly internal friction or energy-consuming processes will affect the motion of such devices. This can occur through various mechanisms, such as: (a) internal viscous friction (e.g., in fluid-filled gyro housings), (b) mechanical losses due to inelastic deformation or contact friction, (c) electromagnetic damping in rotating systems, etc.

So far the real technical system. Of course, the rheological model of a dissipative gyrostat is more abstract, but the phenomena observed in technical systems provide some guidance for explanation what it means. In particular, we want to use this model in context with a rigid body, the motion of which is stopped. Hence imagine a bucket (the rigid body) filled with sand. The microstructure of the sand-filled bucket is illustrated in the cartoon on the right of Fig. 6.2. We can see the sand grains and some pores in between. When the bucket moves non-stationary, this is to say it swings or turns, then relative microdisplacements between the sand grains will lead to (dry) friction, which is associated with internal dissipation that will affect the macroscopic motion. Note that the velocity of a free-falling, sand-filled bucket will not be influenced by its internal microstructure. It will not slow down. To reduce its velocity during free fall *surface friction* is required, which is described by the Rayleigh function. Here we are dealing with *internal friction* and that is triggered by "time-dependent gravitation" so-to-speak, because this is "felt" by the sand medium when the bucket is swinging. The same effect results if the bucket is rotating or turning due to inertia effects (of the d'Alembert type). In short we need the (combined) action of non-stationary inertial-gravitational forces.

Let us now besides a sand filling describe other possible microstructures, where microdisplacements would lead to the same effect. For example we could replace the sand grains and pores with very small beads swimming in a viscous liquid. Moreover, imagine that inside the (rigid) bucket we place a (rigid) 3D grid, such that the grid cells are filled with sand and pores or beads and liquid. This is what was referred to in Smolnikov (2008) as некоторая достаточно жесткая несущая структура(ее «скелета») и внутреннего наполнителя.[13] There the (rigid)

[13] Some sufficiently rigid supporting structure (its "skeleton") and internal filler material.

skeleton with both types of fillings were also introduced as микропористая структура, поры которой заполнены диссипативной средой (сыпучей или вязкой).[14] In continuum mechanics speech such a skeleton cell corresponds to a material particle with certain constitutive properties, which are partly rigid and partly viscous.

Moreover, the mass distribution in the global structure actually remains constant up to the value of microdisplacements within the cells, the scales of which are several orders of magnitude smaller than the scales of motion of the body itself. Therefore, when writing down the equations of motion of bodies, one can neglect the microdisplacements of internally rubbing particles. The influence of microdisplacements on the evolution of the system motion can be evaluated integrally by introducing the dissipative function (6.29) shown below. In this case dissipation is what is called "slow," for example in Smolnikov (2008), and its influence on the behavior of a body is noticeable with a long time interval.

Another quote from the original literature provides additional understanding of the dissipative mechanism of the new rheological model (Smolnikov and Belyaev 2008): "Any other rotation [= a non-stationary one of the rigid body] results in a non-stationary field of internal forces, this field yielding internal microdisplacements and thus the internal dissipation of energy. Clearly, for an adequate description of the process one can make use of the viscoelastic model by introducing internal degrees of freedom. This way is rather broadly presented in the literature, see, e.g., [7] [Chernousko (1965) in this article here]; however, it complicates the mathematical model of the object."

Our suggestion for a mathematical description of the dissipative gyrostat will be a very pragmatic one, Smolnikov and Belyaev (2008), pg. 370: "Let us now proceed to the construction of the dissipative function for the force of internal (volumetric) friction which would reflect the dependence on the general character of the body motion. To this end, we select an elementary volume $\mathrm{d}v$[15] in the body. Let $\mathrm{d}S$ denote the power of the dissipative force in this volume. If the dissipative force $\boldsymbol{N}$ (or the corresponding stresses) is constant in the local reference frame then the relative microdisplacements of volume $\mathrm{d}v$ corresponding to these stresses are also constant and thus no dissipation of the mechanical energy takes place. This is the case for the permanent rotation without external gravitational forces with a constant axipetal acceleration. If the relative acceleration in volume $\mathrm{d}v$ is time dependent then force $\boldsymbol{N}_\mathrm{r}$[16] and, in turn, displacement $\boldsymbol{u}_\mathrm{r}$ are also time dependent. Expressing the elementary power $\mathrm{d}S$ as a quadratic form of the relative velocities $\dot{\boldsymbol{u}}_\mathrm{r}$ and assuming that $\dot{\boldsymbol{u}}_\mathrm{r} = \boldsymbol{D} \cdot \dot{\boldsymbol{N}}_\mathrm{r}$ we arrive at the following result,"

$$S = \frac{1}{2} \int\limits_v \dot{\boldsymbol{u}}_\mathrm{r}^\mathsf{T} \cdot \hat{\mathbf{B}} \cdot \dot{\boldsymbol{u}}_\mathrm{r} \, \mathrm{d}v = \frac{1}{2} \int\limits_v \dot{\boldsymbol{N}}_\mathrm{r}^\mathsf{T} \cdot \mathbf{B} \cdot \dot{\boldsymbol{N}}_\mathrm{r} \, \mathrm{d}v \,, \quad Q_i^{\mathrm{f_v}} = -\frac{\partial S}{\partial \dot{q}_i} . \tag{6.29}$$

---

[14] A microporous structure whose pores are filled with a dissipative medium (loose or viscous).

[15] The symbols of the quote have been adjusted slightly to meet the requirements of our nomenclature.

[16] The index r stands for relative.

In these relations $\hat{\mathbf{B}}$ and $\mathbf{B} = \mathbf{D}^{\mathsf{T}} \cdot \hat{\mathbf{B}} \cdot \mathbf{D}$ are (symmetric) non-negative matrices of dissipation coefficients, and $\mathbf{D}$ is a matrix of compliances.

Generalizing (6.23) the equations of motion now also include the volume-based dissipative force $Q_i^{\mathrm{f_v}}$ and read

$$\frac{\mathrm{d}}{\mathrm{d}t}\frac{\partial T}{\partial \dot{q}_i} - \frac{\partial T}{\partial q_i} = Q_i^{\mathrm{rev}} + Q_i^{\mathrm{f_a}} + Q_i^{\mathrm{f_v}}\,, \quad i = 1, ..., f. \tag{6.30}$$

In order to extend the energy theorem (6.28) to the dissipative case, we note that based on Eqs. $(6.22)_1$ and $(6.29)_1$ the functions R and S are homogeneous functions of degree 2 in the generalized velocities. Hence,

$$\dot{q}_i \frac{\partial\,(R+S)}{\partial \dot{q}_i} = 2\,(R+S) \quad \Rightarrow \quad \frac{\mathrm{d}\,(T+U)}{\mathrm{d}t} = -2\,(R+S)\,. \tag{6.31}$$

The forces $\boldsymbol{N}_{\mathrm{r}}$ included in (6.29) are relatively easy to express through the generalized coordinates and velocities, $\boldsymbol{q}$, $\dot{\boldsymbol{q}}$ (and time derivatives thereof) for a filler experiencing only inertia-gravity loads. Let us consider the special case of a dissipative gyrostat with a constant mass density ρ_0 and the same viscous filler everywhere. Then the matrix $\mathbf{B}$ in Eq. (6.29) is the unit matrix multiplied by one viscous coefficient B. The quantities $\boldsymbol{N}$ turn into force densities which we may simply convert into accelerations $\boldsymbol{w}$ by taking the (constant) mass densities ρ_0 of the dissipative gyrostat out and combining the latter with the viscous coefficient to form a constant $b = \rho_0^2 \mathrm{B}$. Thus (also see Podolskaya and Smolnikov (2010a, b, c), Smirnov and Smolnikov (2024), Smirnov and Smolnikov (2023))

$$S = \frac{b}{2} \int\limits_{v} \dot{\boldsymbol{w}}^2 \,\mathrm{d}v. \tag{6.32}$$

But how to compute the accelerations $\boldsymbol{w}$ and their time derivatives $\dot{\boldsymbol{w}}$? We adjust some symbols in Smolnikov (2008), pg. 42 to the nomenclature of this paper and write down the expression for the force $\boldsymbol{N}$ per unit volume of the filler:

$$\boldsymbol{N} = \rho_0 \Big[-\big(\dot{\boldsymbol{\Omega}}^{\mathrm{r}\prime} + \boldsymbol{\Omega}^{\mathrm{r}\prime} \cdot \boldsymbol{\Omega}^{\mathrm{r}\prime}\big) \cdot \boldsymbol{x}' + \boldsymbol{g}' \Big] = \rho_0 \boldsymbol{w}, \tag{6.33}$$

where ρ_0 is the (constant) mass density of the filler, $\boldsymbol{\Omega}^{\mathrm{r}\prime}$ and $\dot{\boldsymbol{\Omega}}^{\mathrm{r}\prime}$ are the right anti-symmetric tensors of angular velocity and angular acceleration of the dissipative gyrostat, $\boldsymbol{x}'$ is the radius vector to the considered elementary volume, and $\boldsymbol{g}'$ is its gravitational acceleration (if present). The dashes indicate that all quantities belong to a non-inertial observer Σ' attached to the accelerating and rotating body.[17] Taking the time derivative of $\boldsymbol{N}$ we obtain[18]

$$\dot{\boldsymbol{N}} = -\rho_0\Big(\big(\dot{\boldsymbol{\Omega}}^{\mathrm{r}\prime} + \boldsymbol{\Omega}^{\mathrm{r}\prime} \cdot \boldsymbol{\Omega}^{\mathrm{r}\prime}\big)^{\cdot} \cdot \boldsymbol{x} + \boldsymbol{\Omega}^{\mathrm{r}\prime} \cdot \boldsymbol{g}' \Big) = \rho_0 \dot{\boldsymbol{w}}. \tag{6.34}$$

In the next section we will explain why these relations hold by using concepts of continuum mechanics.

6.2 The Continuum Mechanics Perspective

Recall the balance of kinetic energy within a (material) deformable volume $v(t)$[19]:

$$\frac{\mathrm{d}}{\mathrm{dt}} \int\limits_{v(t)} \frac{\rho}{2} \boldsymbol{v}^2 \,\mathrm{dv} = \oint\limits_{\partial v(t)} \boldsymbol{t} \cdot \boldsymbol{v} \,\mathrm{da} - \int\limits_{v(t)} \boldsymbol{\sigma} : (\nabla \boldsymbol{v}) \,\mathrm{dv} + \int\limits_{v(t)} \boldsymbol{k} \cdot \boldsymbol{v} \,\mathrm{dv}. \tag{6.35}$$

All fields are space and time dependent in the current configuration: ρ is the mass density, $\boldsymbol{v}$ is the material velocity, $\boldsymbol{t} = \boldsymbol{n} \cdot \boldsymbol{\sigma}$ is the traction, $\boldsymbol{n}$ is the outward unit normal on the closed surface $\partial v(t)$, $\boldsymbol{\sigma}$ is the Cauchy stress tensor, and $\boldsymbol{k}$ is the body force density.

If we compare this with Eq. $(6.31)_2$ we must of course restrict ourselves to rigid volumes, v, and two terms can immediately be related to symbols from Eq. $(6.31)_2$:

$$\int\limits_{v} \frac{\rho}{2} \boldsymbol{v}^2 \,\mathrm{dv} \quad \to \quad T\,, \quad \oint\limits_{\partial v} \boldsymbol{t} \cdot \boldsymbol{v} \,\mathrm{da} \quad \to \quad R. \tag{6.36}$$

While the first assignment is rather obvious the second deserves some comments:

- Clearly the power of the tractions $\boldsymbol{t}$ is a surface related term, and so is the Rayleigh function R (see Eq. $(6.22)_1$).
- The traction term comprises all actions from the surroundings of the (rigid) body v according to the principle of the free body diagram.

---

[17] To be sure: the rigid body is accelerating and rotating from the standpoint of the inertial observer Σ. For the non-inertial observer Σ' it stands still and experiences fictitious forces. Also see further explanations in Sect. 6.2.

[18] For the readers who study the original reference Smolnikov (2008): observe that the symbols for the Levi-Civitá tensor and the angular velocity vector in the present paper, $\boldsymbol{\epsilon}$ and $\boldsymbol{\omega}$, must not be confused with the notation for the angular acceleration and angular velocity tensors, $\boldsymbol{\varepsilon}$ and ω, used in there. Also note that in the relation before Eq. (5) of that reference the symbol $\boldsymbol{\omega}$ is strictly speaking the vector invariant of the angular velocity tensor, *i.e.*, the angular velocity vector.

[19] $\boldsymbol{\sigma} : (\nabla \boldsymbol{v}) = \sigma_{ij} \frac{\partial v_j}{\partial x_i}$.

- What about the reversibility or irreversibility of these actions? The situation is somewhat alike to the one of the famous rigid body problem of an Atwood wheel with an internal motor (see the section "Masse an Trommel (nicht-konservativ)" in the German wiki on Lagrange mechanics, Lagrange-Formalismus (2024)): this motor creates a rotary moment, which is counted as a non-conservative force on the right-hand side of the Lagrange equations (6.30).
- This is how the assignment $(6.36)_2$ must be understood and it should be kept in mind that the traction term, in principle, can contain more than just frictional forces. However, other types of forces have not been specified in the previous section.

As far as the body forces are concerned we shall discuss only gravitation here and in this context only the near field of the Earth. Then $\boldsymbol{k} = \rho\boldsymbol{g} = -\rho\nabla\varphi$ with $\varphi = -\boldsymbol{g}\cdot\boldsymbol{x}$ and the acceleration $\boldsymbol{g}$ is a vector constant. Moreover, in Eulerian material field description we have $\nabla\varphi\cdot\boldsymbol{v} = \frac{\mathrm{d}\varphi}{\mathrm{d}t}$ and therefore, for a material volume v:

$$\int\limits_{v} \boldsymbol{k}\cdot\boldsymbol{v}\,\mathrm{d}v = -\frac{\mathrm{d}}{\mathrm{d}t}\int\limits_{v} \rho\varphi\,\mathrm{d}v \quad\rightarrow\quad -\frac{\mathrm{d}U^{\mathrm{g}}}{\mathrm{d}t}. \tag{6.37}$$

Moreover, the following sequence of equations holds by definition of the center of mass, $\boldsymbol{x}^{\mathrm{s}} = \frac{1}{m}\int\limits_{v}\rho\boldsymbol{x}\,\mathrm{d}v$, $m = \int\limits_{v}\rho\,\mathrm{d}v$ (compare also Eq. (6.6)):

$$U^{\mathrm{g}} \equiv \int\limits_{v} \rho\varphi\,\mathrm{d}v = -\boldsymbol{g}\cdot\int\limits_{v}\rho\boldsymbol{x}\,\mathrm{d}v = -m\boldsymbol{g}\cdot\boldsymbol{x}^{\mathrm{s}}. \tag{6.38}$$

But this is not the only reversible contribution to the potential U in $(6.31)_2$ from Eq. (6.35). In order to find the second one we decompose the Cauchy stress tensor additively into a reversible (index el(astic)) and irreversible (index f(riction)) bit, $\boldsymbol{\sigma} = \boldsymbol{\sigma}^{\mathrm{el}} + \boldsymbol{\sigma}^{\mathrm{f}}$. For the elastic part we choose Hooke's law for the isotropic case,

$$\boldsymbol{\sigma}^{\mathrm{el}} = \lambda\,\mathrm{tr}\,\boldsymbol{\varepsilon} + 2\mu\boldsymbol{\varepsilon}\ ,\quad \boldsymbol{\varepsilon} = \frac{1}{2}(\nabla\boldsymbol{u} + \boldsymbol{u}\nabla), \tag{6.39}$$

with the two Lamé constants λ, μ and the displacement vector $\boldsymbol{u} = \boldsymbol{x} - \boldsymbol{X}$, relating the position of material particles in the current and reference placement. In the case of small deformations $\boldsymbol{\varepsilon}$[20] we may exchange the substantial time derivative and the gradient operator, hence

$$\frac{1}{2}(\nabla\boldsymbol{v} + \boldsymbol{v}\nabla) = \frac{\mathrm{d}}{\mathrm{d}t}\left[\frac{1}{2}(\nabla\boldsymbol{u} + \boldsymbol{u}\nabla)\right] \equiv \frac{\mathrm{d}\boldsymbol{\varepsilon}}{\mathrm{d}t}. \tag{6.40}$$

[20] The symbol used for the strain tensor of small deformations must not be confused with the Levi-Civitá tensor or the tensor of angular acceleration that were defined before Eq. (6.20) and (6.33), respectively.

Thus by observing Eq. (6.39)$_1$ we obtain as part of the remaining power term on the right-hand side of Eq. (6.35),

$$-\int_v \boldsymbol{\sigma}^{\mathrm{el}} : (\nabla \boldsymbol{v})\, \mathrm{d}v = -\frac{\mathrm{d}}{\mathrm{d}t}\left[\frac{1}{2}\int_v \boldsymbol{\sigma}^{\mathrm{el}} : \boldsymbol{\varepsilon}\, \mathrm{d}v\right] \quad \rightarrow \quad -\frac{\mathrm{d}U^{\mathrm{el}}}{\mathrm{d}t} \tag{6.41}$$

with the elastically stored energy,

$$\frac{1}{2}\int_v \boldsymbol{\sigma}^{\mathrm{el}} : \boldsymbol{\varepsilon}\, \mathrm{d}v \quad \rightarrow \quad U^{\mathrm{el}}. \tag{6.42}$$

Thus the last correlation to make reads:

$$\int_v \boldsymbol{\sigma}^{\mathrm{f}} : \dot{\boldsymbol{\varepsilon}}\, \mathrm{d}v \quad \rightarrow \quad S, \tag{6.43}$$

where we have limited ourselves to small strain rates $\dot{\boldsymbol{\varepsilon}}$. In order to get some intuitive understanding of S in continuum mechanics terms let us assume that the irreversible part of the stress tensor can be described by an incompressible Navier-Stokes fluid, $\boldsymbol{\sigma}^{\mathrm{f}} = 2\mu\dot{\boldsymbol{\varepsilon}}$, where μ denotes the shear viscosity. Then we obtain the following correspondences in deformation space,

$$2\mu \int_v \dot{\boldsymbol{\varepsilon}} : \dot{\boldsymbol{\varepsilon}}\, \mathrm{d}v \quad \rightarrow \quad S = \frac{1}{2}\int_v \dot{\boldsymbol{u}}_{\mathrm{r}}^{\mathsf{T}} \cdot \hat{\mathbf{B}} \cdot \dot{\boldsymbol{u}}_{\mathrm{r}}\, \mathrm{d}v, \tag{6.44}$$

or in stress space,

$$\frac{1}{2\mu}\int_v \boldsymbol{\sigma}^{\mathrm{f}} : \boldsymbol{\sigma}^{\mathrm{f}}\, \mathrm{d}v \quad \rightarrow \quad S = \frac{1}{2}\int_v \dot{\boldsymbol{N}}_{\mathrm{r}}^{\mathsf{T}} \cdot \mathbf{B} \cdot \dot{\boldsymbol{N}}_{\mathrm{r}}. \tag{6.45}$$

We conclude that $\hat{\mathbf{B}}$ and $\mathbf{B}$ have the meaning of a viscosity and an inverse viscosity parameter, respectively. Moreover, the columns of relative microvelocities $\dot{\boldsymbol{u}}_{\mathrm{r}}$ or relative microdisplacements $\boldsymbol{u}_{\mathrm{r}}$ correspond to strain rates and strains between liquid layers gliding on each other. Therefore they are performing a relative movement. Finally, the column of the time derivative of the internal forces, $\dot{\boldsymbol{N}}_{\mathrm{r}}$, are viscous stresses, but not their time derivative.

Some *caveats* are due at this point:

- The intention of our correlations was to bring intuitive meaning to the abstract notion of the internal dissipation function S and to the notion of the dissipative gyrostat. However, such correlations have their limits and should not be overstrained.

- For example, the dissipative gyrostat was thought to consist of a rigid skeleton in which a viscous liquid is acting and dissipating energy. Such a material is certainly not an incompressible Navier-Stokes fluid.
- A homogenization procedure should be used to create a suitable constitutive equation for the dissipative gyrostat which takes its rigid as well as the viscous aspects into account. This is easier said than done, because the concept of a rigid body is alien to rational constitutive theory not to mention to homogenization methods, see Grioli (1983), Fosdick and Royer-Carfagni (2004), Bessonov and Litvinova (2023).

Let us now attempt a rational derivation of Eqs. (6.33) and (6.34). To this end we draw a free body diagram of the red volume element $\mathrm{d}v$ shown in Fig. 6.1 by applying (unknown) surface tractions $\boldsymbol{t}$ on its surface $\partial \mathrm{d}v$. Since Σ is an inertial observer, Newton's second law holds in the following form:

$$\rho \ddot{\boldsymbol{x}} \mathrm{d}v = \lim_{\Delta v \to \mathrm{d}v} \oint_{\partial(\Delta v)} \boldsymbol{t} \, \mathrm{d}a + \rho \boldsymbol{g} \mathrm{d}v, \tag{6.46}$$

where Δv is a (small) material volume element. Note there is no need for inertial or additional fictitional forces on the right-hand side of the equation. The surface term can be transformed into the material response of the matter within the volume element. This material response is expressed in terms of the Cauchy stress tensor, which enters because of Cauchy's law $\boldsymbol{t} = \boldsymbol{n} \cdot \boldsymbol{\sigma}$, $\boldsymbol{n}$ being the outward unit normal. Now the Gauss-Ostrogradski theorem can be applied and thus localization yields:

$$\rho \ddot{\boldsymbol{x}} = \nabla \cdot \boldsymbol{\sigma} + \rho \boldsymbol{g}. \tag{6.47}$$

In fact this is nothing else but the balance of momentum in regular points for an inertial observer (see Müller et al. (2024), pg. 239). For the left-hand side we can now use our knowledge on rigid bodies from $(6.19)_1$ or $(6.20)_1$, so that we obtain[21]:

$$\nabla \cdot \boldsymbol{\sigma} = \rho \big[\boldsymbol{a}^A(t) + \big(\dot{\boldsymbol{\Omega}}(t) + \boldsymbol{\Omega}(t) \cdot \boldsymbol{\Omega}(t) \big) \cdot \boldsymbol{x}^{AP}(t) - \boldsymbol{g} \big]. \tag{6.48}$$

First note that in the examples that will be discussed in the second part of this work, the arbitrarily chosen point A on the rigid dissipative gyrostat will always be fixed and not moving with respect to the inertial frame, for example by a joint or a bearing. Then we may position the inertial observer Σ in A and $\boldsymbol{x}^A = \boldsymbol{0}$, $\boldsymbol{v}^A = \boldsymbol{0}$, $\boldsymbol{a}^A = \boldsymbol{0}$. Moreover, the mass density of the gyrostat is constant, $\rho = \rho_0$.

Now, let the repositioned inertial observer Σ decide (for convenience) to use time-dependent unit vectors, for example, $\boldsymbol{x}^{AP}(t) = r\boldsymbol{e}_r(t)$. Here, in fact, we have

---

[21] As it was mentioned before: introducing stresses in context with rigid bodies might come as a surprise (see for example the discussion in Grioli (1983)). However, one should keep in mind that the rheological model of the dissipative gyrostat consists of a rigid skeleton with a (viscous) filling capable of microdisplacements.

introduced the length r of the radius vector $\boldsymbol{x}'$ from Eq. (6.33), because $\boldsymbol{x}^{AP}(t)$ leads to a material point of the dissipative gyrostat. All time dependencies are explicitly shown. Note in this context that r is *not* time dependent, because of the rigid skeleton of the dissipative gyrostat. Also note that this is a change of coordinates and *not* a change of observer. Hence the previous relation is simplified:

$$\nabla \cdot \boldsymbol{\sigma} = \gamma \left[\left(\dot{\boldsymbol{\Omega}}(t) + \boldsymbol{\Omega}(t) \cdot \boldsymbol{\Omega}(t) \right) \cdot \boldsymbol{r}(t) - \boldsymbol{g} \right]. \tag{6.49}$$

Indeed, this relation bears a certain resemblance to the relation (6.33) proposed in Smolnikov (2008), and we are tempted to identify the bracket as the acceleration $\boldsymbol{w}$. But this would be wrong. We proceed to explain.

First note that the sign in front of the gravitational acceleration, $\boldsymbol{g}$. Second, according to (6.34), $\boldsymbol{g}'$ in Eq. (6.33) must be (relatively) time dependent or otherwise it would not "survive" differentiation with respect to time leading to Eq. (6.34). Looking for an escape we could argue as follows: If the inertial observer Σ so wishes, he may describe $\boldsymbol{g}$ by using cylindrical base vectors, two of which are time dependent, $\boldsymbol{e}_r(t)$ and $\boldsymbol{e}_\varphi(t)$ (see Fig. 6.3), and one is not, $\boldsymbol{e}_z \equiv \boldsymbol{e}_3$. For (planar) rotation problems we obtain the typical form $\boldsymbol{g} = -g \sin[\alpha + \varphi(t)]\boldsymbol{e}_r(t) - g \cos[\alpha + \varphi(t)]\boldsymbol{e}_\varphi(t)$ (also see Fig. 6.3). Thus $\boldsymbol{g}$ becomes implicitly time dependent. However, recall the time-independent base of the inertial observer $\boldsymbol{e}_{1,2}$. Now it is a matter of trigonometry to show that $\boldsymbol{e}_r(t) = \cos[\alpha + \varphi(t)]\boldsymbol{e}_1 + \sin[\alpha + \varphi(t)]\boldsymbol{e}_2$, $\boldsymbol{e}_\varphi(t) = -\sin[\alpha + \varphi(t)]\boldsymbol{e}_1 + \cos[\alpha + \varphi(t)]\boldsymbol{e}_2$ and by insertion it follows (again) that $\boldsymbol{g} = -g\boldsymbol{e}_2$. So we may hand-wave as much as we like, the $\boldsymbol{g}$ in (6.49) is not time dependent *and* it has a different sign when compared to $\boldsymbol{g}'$ from Eq. (6.33).

To find answers to what this is all about we now turn our attention to the rotation tensor $\boldsymbol{Q}(t)$ from Eq. (6.18) for the situation shown in Fig. 6.3. Note that $\boldsymbol{Q}(t)$ is an object of the inertial observer Σ. Hence we should represent it in his proper non-moving base $\boldsymbol{e}_i$, $i = 1, 2, 3$. By looking at the figure it becomes clear that the final position of the rigid rectangle results from two rotations, first by the angle α to

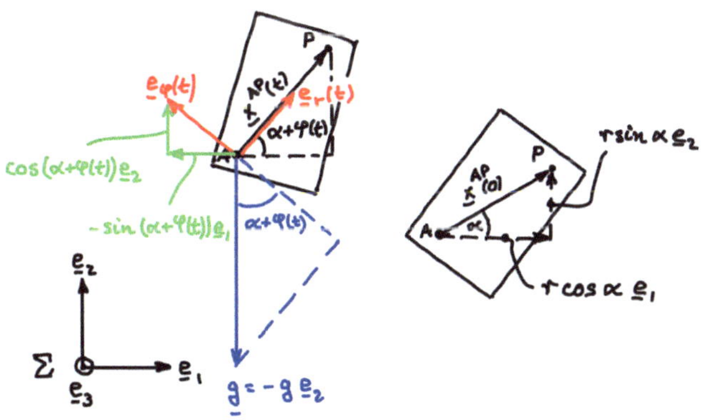

Fig. 6.3 Turning a rigid body from a reference position to the left

bring $\boldsymbol{x}^{AP}$ into its reference position $\boldsymbol{x}^{AP}(0)$ and, second, by the angle $\varphi(t)$ to turn it into the current position $\boldsymbol{x}^{AP}(t)$. Hence the rotation tensor $\boldsymbol{Q}(t)$ can be obtained by $\boldsymbol{Q}(t) = \boldsymbol{Q}(\varphi(t)) \cdot \boldsymbol{Q}(\alpha)$. We evaluate this as follows:

$$\begin{bmatrix} \cos\big(\alpha+\varphi(t)\big) & -\sin\big(\alpha+\varphi(t)\big) & 0 \\ \sin\big(\alpha+\varphi(t)\big) & \cos\big(\alpha+\varphi(t)\big) & 0 \\ 0 & 0 & 1 \end{bmatrix} = \begin{bmatrix} \cos\varphi(t) & -\sin\varphi(t) & 0 \\ \sin\varphi(t) & \cos\varphi(t) & 0 \\ 0 & 0 & 1 \end{bmatrix} \cdot \begin{bmatrix} \cos\alpha & -\sin\alpha & 0 \\ \sin\alpha & \cos\alpha & 0 \\ 0 & 0 & 1 \end{bmatrix}. \tag{6.50}$$

In this matrix notation we have compiled the components $Q_{ij}(t)$ of the rotation tensor $\boldsymbol{Q}(t) = Q_{ij}(t)\boldsymbol{e}_i \otimes \boldsymbol{e}_j$. It is a representation with respect to the non-moving base $\boldsymbol{e}_i, i = 1, 2, 3$ of the inertial observer Σ.[22] And therefore, $\boldsymbol{\Omega}$ in (6.49) is also an object of the inertial observer Σ. And this is exactly what we have to realize: The quantities $\dot{\boldsymbol{\Omega}}^{\mathrm{r}'}$ and $\boldsymbol{\Omega}^{\mathrm{r}'}$ in Eq. (6.33) are *not*.

To find out what exactly they are, we consider the Euclidean transformation depicted in Fig. 6.4. A first remark: the blue arrow $\boldsymbol{x}$ represents the radius vector of the inertial observer Σ to a material point P. In contrast to that the red arrows are objects of the non-inertial observer Σ': $\boldsymbol{x}'$ points from its origin to the material point P and $\boldsymbol{b}$ to the origin of Σ. Now note that from the standpoint of the non-inertial observer accelerations are related according to[23]:

$$\Sigma': \;\; \boldsymbol{a}' = \tilde{\boldsymbol{a}} - \dot{\boldsymbol{\Omega}}^{\mathrm{r}'} \cdot \big[\boldsymbol{x}' - \boldsymbol{b}\big] - 2\boldsymbol{\Omega}^{\mathrm{r}'} \cdot \big[\boldsymbol{v}' - \boldsymbol{V}\big] - \boldsymbol{\Omega}^{\mathrm{r}'} \cdot \boldsymbol{\Omega}^{\mathrm{r}'} \cdot \big[\boldsymbol{x}' - \boldsymbol{b}\big] + \boldsymbol{A}. \tag{6.51}$$

The symbols in this equation deserve some explanations. To begin with, $\boldsymbol{\Omega}^{\mathrm{r}'}(t)$ stands for the right spin tensor of the observer $\boldsymbol{\Sigma}'$, which is defined by $\boldsymbol{\Omega}^{\mathrm{r}'}(t) =$

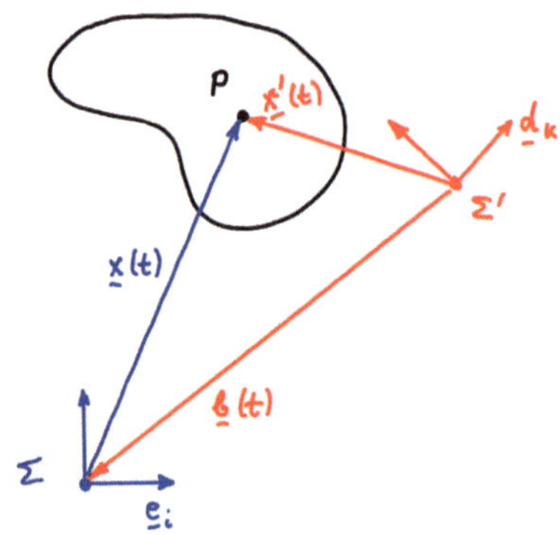

Fig. 6.4 Euclidean transformation between two observers Σ and Σ'

[22] Rigid body specialists take such details often for granted but this can be treacherous for outsiders and also for people coming from continuum mechanics.

[23] See Müller et al. (2024) for details of the proof.

$\boldsymbol{Q}'^{\mathsf{T}}(t) \cdot \dot{\boldsymbol{Q}}'(t)$. In there $\boldsymbol{Q}'(t) = \boldsymbol{d}_i \otimes \tilde{\boldsymbol{e}}_i(t)$ is the rotation tensor of the observer Σ'. Note the tilde on the unit base $\boldsymbol{e}_i$. This is Zhilin's idea of the image vector (see Zhilin (2001), Sect. 4.8, Zhilin (2015), Sect. 1.1.4, Zhilin (2012), Sects. 4.1–4.5, Zhilin (2001), Sect. 1.4, Ivanova et al. (2017)): the base $\boldsymbol{e}_i$ (see Fig. 6.4) belongs to the inertial observer Σ and is time-independent. But for the non-inertial observer Σ' this base is turning and time dependent. To make it his own Σ' must create time-dependent images $\tilde{\boldsymbol{e}}_i(t)$. We conclude that $\boldsymbol{Q}'(t)$ turns $\tilde{\boldsymbol{e}}_i(t)$ onto the (time-independent) base $\boldsymbol{d}_i$ proper to Σ' and write $\boldsymbol{d}_i = \boldsymbol{Q}'(t) \cdot \tilde{\boldsymbol{e}}_i$. Indeed, *all* quantities in Eq. (6.51) are objects of Σ' as we have clearly stated at the beginning of Eq. (6.51). The time derivatives of $\boldsymbol{b}(t)$ were denoted by $\boldsymbol{V}(t) = \dot{\boldsymbol{b}}(t)$ and $\boldsymbol{A}(t) = \ddot{\boldsymbol{b}}(t)$. Moreover, $\boldsymbol{a}'(t) = \ddot{\boldsymbol{x}}'(t)$.

In fact, Eq. (6.51) can also be written such that it holds for the inertial observer. To this end, the image operation $\widetilde{(\cdot)}$ must be applied to each of the terms in Eq. (6.51). The result reads:

$$\Sigma: \ \tilde{\boldsymbol{a}}' = \boldsymbol{a} + \dot{\boldsymbol{\Omega}}^{\mathrm{r}} \cdot \left(\tilde{\boldsymbol{x}}' - \tilde{\boldsymbol{b}}\right) + 2\boldsymbol{\Omega}^{\mathrm{r}} \cdot \left(\tilde{\boldsymbol{v}}' - \tilde{\boldsymbol{V}}\right) - \boldsymbol{\Omega}^{\mathrm{r}} \cdot \boldsymbol{\Omega}^{\mathrm{r}} \cdot \left(\tilde{\boldsymbol{x}}' - \tilde{\boldsymbol{b}}\right) + \tilde{\boldsymbol{A}}, \tag{6.52}$$

because $\widetilde{\boldsymbol{\Omega}}^{\mathrm{r}'} = -\boldsymbol{\Omega}^{\mathrm{r}}$. At this point it is necessary to introduce a few concepts. To begin with, let us rewrite the last equation:

$$\Sigma: \ -\boldsymbol{a} = -\tilde{\boldsymbol{a}}' + \dot{\boldsymbol{\Omega}}^{\mathrm{r}} \cdot \left(\tilde{\boldsymbol{x}}' - \tilde{\boldsymbol{b}}\right) + 2\boldsymbol{\Omega}^{\mathrm{r}} \cdot \left(\tilde{\boldsymbol{v}}' - \tilde{\boldsymbol{V}}\right) - \boldsymbol{\Omega}^{\mathrm{r}} \cdot \boldsymbol{\Omega}^{\mathrm{r}} \cdot \left(\tilde{\boldsymbol{x}}' - \tilde{\boldsymbol{b}}\right) + \tilde{\boldsymbol{A}}. \tag{6.53}$$

It is customary to refer to the term $-\boldsymbol{a}(t)$ as *inertial acceleration* (see Seeler (2014), Sect. 3.2.3.1) and this includes, most importantly, the minus sign (see, e.g., Taylor (2005)). It was d'Alembert's idea[24] to reduce dynamic problems to static ones, in the sense that the sum of *all* forces including the inertial one is equal to zero. This is referred to as *dynamic equilibrium*. In continuum mechanics speech this means we must rewrite the local balance of momentum shown in Eq. (6.47), which is valid in the inertial frame, as follows:

$$\nabla \cdot \boldsymbol{\sigma} + \rho \boldsymbol{g} - \rho \boldsymbol{a} = \boldsymbol{0}. \tag{6.54}$$

Moreover, it is customary to refer to the terms on the right-hand side of Eq. (6.53) as d'Alembert accelerations. Also, $\dot{\boldsymbol{\Omega}}^{\mathrm{r}} \cdot \left(\tilde{\boldsymbol{x}}' - \tilde{\boldsymbol{b}}\right)$ is called Euler acceleration, $2\boldsymbol{\Omega}^{\mathrm{r}} \cdot \left(\tilde{\boldsymbol{v}}' - \tilde{\boldsymbol{V}}\right)$ is known as Coriolis acceleration, and $-\boldsymbol{\Omega}^{\mathrm{r}} \cdot \boldsymbol{\Omega}^{\mathrm{r}} \cdot \left(\tilde{\boldsymbol{x}}' - \tilde{\boldsymbol{b}}\right)$ is the centrifugal acceleration. Note that there is nothing "fictitious" in the d'Alembert accelerations if we look at them from the standpoint of the inertial observer. They appear naturally in Eq. (6.53) as a breakdown of the inertial acceleration $-\boldsymbol{a}$, which has no fictitious character either, because, if multiplied by the mass density, it is nothing else but the negative temporal change of momentum: the inertial observer does not "feel" it/them.

---

[24] It is said that d'Alembert introduced it in his famous book d'Alembert (1743). However, when studying the book such a crystal clear statement cannot be found.

In the case of the non-inertial observer the situation is different: if we look at the images of the d'Alembert accelerations in Eq. (6.51) we may rightfully call them *fictitious accelerations*, as it is often done (e.g., Clapham and Nicholson (2014), pg. 312), especially if the non-inertial observer Σ' is "attached" to the rigid body. He—and in particular the dissipative gyrostat material—can feel them, and they have no clue where they come from, because they believe to be at rest.

In this spirit recall now the balance of momentum for a non-inertial observer Σ',

$$\Sigma' : \quad \rho' \, \boldsymbol{a}' = \nabla' \cdot \boldsymbol{\sigma}' + \rho' \left(\boldsymbol{g}' - \boldsymbol{i} \right), \tag{6.55}$$

with the definition of the so-called inertial or fictitious accelerations,

$$\begin{aligned} \boldsymbol{i} =& \dot{\boldsymbol{\Omega}}^{\mathrm{r}'} \cdot \left[\boldsymbol{x}' - \boldsymbol{b} \right] + 2\boldsymbol{\Omega}^{\mathrm{r}'} \cdot \left[\boldsymbol{v}' - \boldsymbol{V} \right] + \boldsymbol{\Omega}^{\mathrm{r}'} \cdot \boldsymbol{\Omega}^{\mathrm{r}'} \cdot \left[\boldsymbol{x}' - \boldsymbol{b} \right] - \boldsymbol{A} = \\ & \dot{\boldsymbol{\omega}}^{\mathrm{r}'} \times \left[\boldsymbol{x}' - \boldsymbol{b} \right] + 2\boldsymbol{\omega}^{\mathrm{r}'} \times \left[\boldsymbol{v}' - \boldsymbol{V} \right] + \boldsymbol{\omega}^{\mathrm{r}'} \times \left(\boldsymbol{\omega}^{\mathrm{r}'} \times \left[\boldsymbol{x}' - \boldsymbol{b} \right] \right) - \boldsymbol{A}. \end{aligned} \tag{6.56}$$

Now let us specialize this general scenario, which is valid for deformable bodies, to the needs of the dissipative gyrostat, which is essentially rigid. To this end we, first, attach the unit base $\boldsymbol{d}_i$ firmly in a point A of the (now rigid) body. Second, we put a joint in A and make it unmovable in the sense that it stays put in space but the gyrostat can rotate about it freely. Third, we relocate the inertial observer to point A. Also note that the density is now a constant ρ_0'.[25] Thus Eqs. (6.55), (6.56) turn into:

$$\boldsymbol{0} = \nabla' \cdot \boldsymbol{\sigma}' + \rho_0' \left(\boldsymbol{g}' - \boldsymbol{i} \right) \quad \Rightarrow \quad \nabla' \cdot \boldsymbol{\sigma}' = \rho_0' \left(\boldsymbol{i} - \boldsymbol{g}' \right), \tag{6.57}$$

with

$$\boldsymbol{i} = \left(\dot{\boldsymbol{\Omega}}^{\mathrm{r}'} + \boldsymbol{\Omega}^{\mathrm{r}'} \cdot \boldsymbol{\Omega}^{\mathrm{r}'} \right) \cdot \boldsymbol{x}', \tag{6.58}$$

and therefore:

$$\nabla' \cdot \boldsymbol{\sigma}' = \rho_0' \left[\left(\dot{\boldsymbol{\Omega}}^{\mathrm{r}'} + \boldsymbol{\Omega}^{\mathrm{r}'} \cdot \boldsymbol{\Omega}^{\mathrm{r}'} \right) \cdot \boldsymbol{x}' - \boldsymbol{g}' \right]. \tag{6.59}$$

Note that $\boldsymbol{x}'$—an object of the non-inertial observer Σ', who is now rotating with the dissipative gyrostat—is *not* time dependent for Σ' because of the rigid body assumption. In fact, this is finally the radius vector of Eq. (6.33). Moreover, $\boldsymbol{g}'$ is explicitly time dependent: the gravity vector is "turning" for Σ', at least so the non-inertial observer thinks. $\nabla' \cdot \boldsymbol{\sigma}'$ is the reaction of the matter constituting the dissipative gyrostat to all accelerations. The matter of the body is concerned and it is only proper to investigate this from the standpoint of the co-moving observer Σ'. So, at last, we introduce the images of the relevant d'Alembert accelerations from Eq. (6.53), namely, $\dot{\boldsymbol{\Omega}}^{\mathrm{r}} \cdot \left(\tilde{\boldsymbol{x}}' - \tilde{\boldsymbol{b}} \right)$ and $-\boldsymbol{\Omega}^{\mathrm{r}} \cdot \boldsymbol{\Omega}^{\mathrm{r}} \cdot \left(\tilde{\boldsymbol{x}}' - \tilde{\boldsymbol{b}} \right)$, observe that $\widetilde{\dot{\boldsymbol{\Omega}}}^{\mathrm{r}'} = -\boldsymbol{\Omega}^{\mathrm{r}}$, and arrive at their counterparts with negative sign, $-\dot{\boldsymbol{\Omega}}^{\mathrm{r}'} \cdot \boldsymbol{x}'$ and $-\boldsymbol{\Omega}^{\mathrm{r}'} \cdot \boldsymbol{x}'$, which

[25] The density is a Euclidean scalar and therefore $\rho' = \rho$. We just kept the dash to point out that everything is written with respect to Σ'.

when multiplied by the mass density ρ_0' result in the fictitious forces.[26] Then by introducing all required minus signs in Eq. (6.59) we conclude:

$$-\nabla' \cdot \boldsymbol{\sigma}' = \rho_0' \left[-\left(\dot{\boldsymbol{\Omega}}^{\mathrm{r}\prime} + \boldsymbol{\Omega}^{\mathrm{r}\prime} \cdot \boldsymbol{\Omega}^{\mathrm{r}\prime} \right) \cdot \boldsymbol{x}' + \boldsymbol{g}' \right] \equiv \rho_0 \boldsymbol{w}. \tag{6.60}$$

Let us also investigate the time derivative of $\boldsymbol{g}'$ with the concept of images. Note that $\dot{\tilde{\boldsymbol{e}}}_i(t) = -\boldsymbol{\Omega}^{\mathrm{r}}(t)' \cdot \tilde{\boldsymbol{e}}_i(t)$ and we must write:

$$\boldsymbol{g}'(t) = \tilde{\boldsymbol{g}}(t) = g_i \tilde{\boldsymbol{e}}_i(t) \quad \Rightarrow \quad \dot{\boldsymbol{g}}'(t) = -\boldsymbol{\Omega}^{\mathrm{r}}(t)' \cdot \boldsymbol{g}'(t), \tag{6.61}$$

which explains the expression for $\dot{\boldsymbol{w}}$ from Eq. (6.34) to be inserted in Eq. (6.32) in order to compute the internal dissipation function S.

In this context recall that stresses can, in principle, be decomposed additively into reversible (elastic) and irreversible (friction) parts, $\boldsymbol{\sigma} = \boldsymbol{\sigma}^{\mathrm{el}} + \boldsymbol{\sigma}^{\mathrm{f}}$. Also recall that such a decomposition led to the assignments (6.41) and (6.43). Consequently, the symbol $\boldsymbol{\sigma}'$ in Eq. (6.60) refers exclusively to the irreversible part, because, by definition, a dissipative gyrostat does not allow elastic deformation due to its rigid skeleton and the filling is purely viscous. In other words, Eq. (6.60) is a conservative estimate according to which all inertial-gravitational accelerations cause irreversible action. In what follows we do not emphasize this any more and do not add an index f to $\boldsymbol{\sigma}'$.

Moreover, note that the minus sign issue is important for principal reasons, but in the end not in practice, because the dissipation function S is based on squares of $\dot{\boldsymbol{w}}$, see Eq. (6.32). It should also be pointed out that in Podolskaya and Smolnikov (2010a, b, c) it is written that $\boldsymbol{w} = -\tilde{\boldsymbol{w}} + \boldsymbol{g}$. There it is pointed out that $\tilde{\boldsymbol{w}}$ "is (the) absolute acceleration of a volume element (the minus sign corresponds to the force of inertia)." This confirms the statement above.

Also note that strictly speaking that the integration in Eq. (6.32) should be carried out with respect to the volume v' of the non-inertial observer Σ'. Then the function S of internal volumetric dissipation is an object of the non-inertial observer. Introducing it into the Lagrange equations (which were originally formulated with respect to an inertial system) on the right-hand side of Eq. (6.30), introduces a mix of quantities from both worlds. However, this might be acceptable from the standpoint of rational mechanics, because S is essentially a scalar quantity, see Eqs. (6.32). The proof that it is an Euclidean scalar is left to future work. This concerns especially the time derivative that is used. Note that it was originally used on a vector and not on a scalar.

[26] More on the d'Alembert concept and fictitious forces can be found in Müller et al. (2024). We will not repeat it here.

6.3 Conclusions

The purpose of this first part was mostly pedagogical:

- The concept of Lagrange equations with and without irreversible terms has been reviewed.
- The concept of the rheological element known as the dissipative gyrostat was introduced. It describes volume-related dissipation within the framework of Lagrange's equations similarly as the Rayleigh function addresses surface friction in them. Mathematically it is described by the internal dissipation function S.
- The viewpoints of Lagrange and continuum mechanics have been compared and an intuitive interpretation of the dissipative gyrostat and the internal dissipation function S has been given.
- The way of how to calculate S as it was introduced heuristically in previous papers has been interpreted in terms of change of observers. It was pointed out that S is essentially an object of a non-inertial observer connected to the (rigid) dissipative gyrostat.

Acknowledgements The authors want to thank Messrs. Erik Brauer, B.Sc. and Sanjay Mahamagamoorthy, B.Sc. for helpful discussions on the topic of this paper. Three reviewers also provided valuable comments.

References

Lagrange-Formalismus. Wikipedia. https://de.wikipedia.org/wiki/Lagrange-Formalismus. Accessed 15 Dec 2024

Lagrangian mechanics. Wikipedia. https://en.wikipedia.org/wiki/Lagrangian_mechanics. Accessed 15 Dec 2024

Navier-Cauchy-Gleichungen. Wikipedia. https://de.wikipedia.org/wiki/Navier-Cauchy-Gleichungen. Accessed 15 Dec 2024

Bersani AM, Caressa P (2021) Lagrangian descriptions of dissipative systems: a review. Math Mech Solids 26(6):785–803

Bessonov N, Litvinova Y (2023) On the issue of the stress in incompressible and rigid media: numerical modeling. In: Altenbach H, Eremeyev V (eds) Advances in linear and nonlinear continuum and structural mechanics. Springer, pp 69–85

Cauchy A (1828) Sur les équations qui expriment les conditions d'équilibre ou les lois du mouvement intérieur d'un corps solide, élastique ou non élastique. ex. de math 3, 160–187 (1828). Comptes Rendus de l'Académie des Sciences, Paris 36:454–459

Chernousko FL (1965) Движение твердого тела с полостями, заполненными вязкой жидкостью, при малых числах Рейнольдса. Журнал вычислительной математики и математической физики 5(6):1049–1070

Clapham C, Nicholson JR (2014) The concise Oxford dictionary of mathematics. Oxford University Press, USA

d'Alembert JLR (1743) Traité de dynamique. David l'aîné

Euler L (1765) Du mouvement de rotation des corps solides autour d'un axe variable. Mémoires de l'académie des sciences de Berlin, pp 154–193

Euler L (1776) Nova methodus motum corporum rigidorum degerminandi. Novi commentarii academiae scientiarum Petropolitanae, pp 208–238

Fosdick R, Royer-Carfagni G (2004) Stress as a constraint reaction in rigid bodies. J Elast 74:265–276

Goldstein H, Poole C, Safko J (2002) Classical mechanics, 3rd ed. Addison Wesley, San Francisco

Grioli G (1983) On the stress in rigid bodies. Meccanica 18:3–7

Gross D, Hauger W, Schröder J, Wall WA, Govindjee S (2011) Engineering mechanics 3: dynamics. Springer

Hahn H (2013) Rigid body dynamics of mechanisms 1 theoretical basis. Springer Science & Business Media

Ivanova E, Vilchevskaya E, Müller WH (2017) A study of objective time derivatives in material and spatial description. Mechanics for Materials and Technologies, pp 195–229

Jazar RN (2011) Advanced dynamics: rigid body, multibody, and aerospace applications. Wiley

Lanczos C (1949) The variational principles of mechanics. Univ. Toronto Press, Toronto

Loseva TA (2019) Влияние внутренней диссипации на вращение диска в поле силы тяжести. Bachelor thesis, Peter the Great Polytechnical University

Lurie AI (2013) Analytical mechanics. Springer Science & Business Media

Maxwell JC (1868) XV. On the dynamical theory of gases. Lond Edinb Dublin Philos Mag J Sci 35(235):129–145

Minguzzi E (2015) Rayleigh's dissipation function at work. Eur J Phys 36(3):035014

Müller WH, Ivanova EA, Vilchevskaya EN (2024) Momentum and spin balance in an inertial frame obtained by using the concept of images. In: Müller W, Noe A, Ferber F (eds) New achievements in mechanics: a tribute to Klaus peter Herrmann. Springer, pp 217–261

Newton I (2008) Principia mathematica (1687). Trans. Andrew Motte

Podolskaya E, Smolnikov BA (2010a) Evolutionary dynamics of two linked bodies' plane motion. In: Evtim Zahariev MC (ed) Euromech colloquium 515 advanced applications and perspectives of multibody dynamics. Curran Associates, Inc., Red Hook, NY, pp 72–73

Podolskaya EA, Smolnikov BA (2010b) Плоское эволюционное движение связки двух тел. In: Материалы лучших докладов XXXVIII Недели науки СПБГПУ. СПБ.: Изд-во Политехн. ун-та, pp 124–126b

Podolskaya EA, Smolnikov BA (2010c) Эволюционная динамика диссипативного диска, вращающегося в гравитационном поле. In: XXXIX Неделя науки СПБГПУ: материалы международной научно-практической конференции. Часть B. СПБ.: Изд-во Политехн. ун-та, pp 113–115

Seeler KA (2014) Introduction to system dynamics. Springer

Smirnov AS, Smolnikov BA (2017) О соответствии эволюционной и классической моделей внутренней диссипации. In: Неделя науки СПБПУ, pp 142–144

Smirnov AS, Smolnikov BA (2017) Оптимальное гашение свободных колебаний в линейных механических системах. Машиностроение и инженерное образование 3:8–15

Smirnov AS, Smolnikov BA (2022) Колебания двойного математического маятника с внутренним трением. In: Современное машиностроение: наука и образование: материалы 11-й Международной научной конференции / Под ред. А.Н. Евграфова и А.А. Поповича. СПБ.: ПОЛИТЕХ-ПРЕСС, pp 113–125

Smirnov AS, Smolnikov BA (2023) Oscillations of double mathematical pendulum with internal friction. In: International conference modern engineering: science and education. Springer, Cham, pp 153–163

Smirnov AS, Smolnikov BA (2024) Движение сферического маятника со слабой нелинейностью при наличии внешнего или внутреннего трения. In: Современное машиностроение: Наука и образование 2024: материалы 13-й Международной научной конференции, 20 июня 2024 года / Под ред. А.Н. Евграфова и А.А. Поповича. СПБ.: ПОЛИТЕХ-ПРЕСС, pp 129–148

Smolnikov B, Belyaev AK (2008) Evolutional dynamics and stability of dissipative solids. Acta Mech 195(1):365–377

Smolnikov BA (2008a) Фрактальное твердое тело в задачах аналитической механики. In: A.L. Smirnova, E.F. Zhigalko (eds.) Труды семинара «Компьютерные методы в механике сплошной среды». СПБ: изд-во СПБГУ, pp 17–32

Smolnikov BA (2008b) Эволюционная динамика маятниковых систем. Теория механизмов и машин 6(1):41–47

Strocchi F (2018) A primer of analytical mechanics. Springer

Taylor JR (2005) Classical mechanics. Sausalito

Thomson W (1865) Iv. on the elasticity and viscosity of metals. Proc R Soc Lond 14:289–297

Truesdell C (1969) Rational thermodynamics a course of lectures on selected topics. McGraw-Hill New York

Truesdell C, Noll W (2004) The non-linear field theories of mechanics, edited by Stuart S. Antman, 3rd ed. Springer, Berlin Heidelberg New York

Truesdell C, Toupin RA (1960) The classical field theories. Springer, Heidelberg

Truesdell CA (1992) A first course in rational continuum mechanics, vol 1. Academic Press

Voigt W (1892) Ueber innere Reibung fester Körper, insbesondere der Metalle. Ann Phys 283(12):671–693

Zhilin PA (2001) Векторы и тензоры второго ранга в трехмерном пространстве (Vectors and second-rank tensors in three-dimensional space, in Russian). Nestor, St. Petersburg

Zhilin PA (2001) Теоретическая механика: Учеб. пособие (Theoretical mechanics: A textbook, in Russian). Санкт–Петербург Издательство СПБГТУ, St. Petersburg

Zhilin PA (2012) Рациональная механика сплошных сред (Rational Continuum Mechanics, in Russian). Санкт-Петербург Издательство Политехнического университета, St. Petersburg

Zhilin PA (2012) Рациональная механика сплошных сред (Rational Continuum Mechanics, in Russian). Санкт-Петербург ИздательствоПолитехнического университета, St. Petersburg

Chapter 7
Incorporating Dissipation in the Lagrange Equations: Part II—Examples

Wolfgang H. Müller, Ekaterina A. Podolskaya, Alexey S. Smirnov, and Boris A. Smolnikov

Abstract The volumetric or internal dissipation function S is computed for different scenarios. By insertion in the dissipative Lagrange equations the equations of motion are obtained. Various (dissipative) pendulum problems will be studied: the mathematical pendulum, the rod pendulum, the hammer pendulum, and the sector pendulum. Each of these problems allows to get new inside into how to set up the internal dissipation function. Finally we will use this approach for a study of the tidal locking problem of celestial objects, such as the Moon.

7.1 A Reminder of the Irreversible Lagrange Equations

The Lagrange equations including dissipative forces read:

$$\frac{\mathrm{d}}{\mathrm{d}t}\frac{\partial T}{\partial \dot{q}_i} - \frac{\partial T}{\partial q_i} = Q_i^{\mathrm{rev}} + Q_i^{\mathrm{f}_a} + Q_i^{\mathrm{f}_v}\ , \quad i = 1, ..., f, \tag{7.1}$$

W. H. Müller (✉)
Institut für Mechanik, Fachgebiet Kontinuumsmechanik und Materialtheorie, Technische Universität Berlin, Berlin, Germany
e-mail: wolfgang.h.mueller@tu-berlin.de

E. A. Podolskaya
Polycentrum Solutions SL, Barcelona, Spain
e-mail: kate@polycentrum.com

A. S. Smirnov · B. A. Smolnikov
Peter the Great St. Petersburg Polytechnic University, St. Petersburg, Russia
e-mail: smirnov.alexey.1994@gmail.com

B. A. Smolnikov
e-mail: smoln_ba@spbstu.ru

Institute for Problems in Mechanical Engineering of the Russian Academy of Sciences, St. Petersburg, Russia

J. Fuhrmann et al. (eds.), *Advances in Continuum Physics*, Advanced Structured Materials 238, https://doi.org/10.1007/978-3-031-93918-1_7

where $\boldsymbol{q}^\mathsf{T} = [q_1, ..., q_f]^\mathsf{T}$ denotes the row of generalized (independent) coordinates with the number f corresponding to the degrees of freedom of the system of interconnected rigid bodies, and $\dot{\boldsymbol{q}}^\mathsf{T} = [\dot{q}_1, ..., \dot{q}_f]^\mathsf{T}$ being the generalized velocities. The kinetic energy T for a system of rigid bodies (for small deviations from equilibrium) or its continuous form for a single rigid body of volume v is given by

$$T = \frac{1}{2}\dot{\boldsymbol{q}}^\mathsf{T} \cdot \mathbf{A} \cdot \dot{\boldsymbol{q}} \ , \quad T = \frac{\rho_0}{2} \int\limits_v \boldsymbol{v} \cdot \boldsymbol{v} \, \mathrm{d}v, \tag{7.2}$$

where $\boldsymbol{v}$ is the material velocity of an element of the continuous rigid body of volume v and (constant) mass density ρ_0. $\mathbf{A}$ is the (symmetric) matrix of inertial coefficients.

Three reversible generalized forces $\boldsymbol{Q}^{\mathrm{rev}}$ are distinguished, gravitational, elastic, and electromagnetic ones. They follow by differentiation from the corresponding (reversible) potential energies with respect to the generalized coordinates and velocities. For gravitation we have for the case of the near-field $\boldsymbol{g}$ of the Earth:

$$U^{\mathrm{g}} = -m\boldsymbol{g} \cdot \boldsymbol{x}^{\mathrm{s}} \quad \Rightarrow \quad Q_i^{\mathrm{g}} = -\frac{\partial U^{\mathrm{g}}}{\partial x_i^{\mathrm{s}}} = mg_i \ , \quad i = 1, 2, 3, \tag{7.3}$$

or in continuous form for a single homogeneous rigid body of mass m:

$$U^{\mathrm{g}} = -\rho_0 \boldsymbol{g} \cdot \int\limits_v \boldsymbol{x} \mathrm{d}v \quad \Rightarrow \quad \boldsymbol{Q}^{\mathrm{g}} = -\nabla U^{\mathrm{g}} = m\boldsymbol{g}. \tag{7.4}$$

For the reversible elastic force we write (in case of small deviations from equilibrium)

$$U^{\mathrm{el}} = \frac{1}{2}\boldsymbol{q}^\mathsf{T} \cdot \mathbf{C} \cdot \boldsymbol{q} \quad \Rightarrow \quad Q_i^{\mathrm{el}} = -\frac{\partial U^{\mathrm{el}}}{\partial q_i} = -\mathrm{C}_{ij} q_j, \tag{7.5}$$

where $\mathbf{C}$ is referred to as the (symmetric) matrix of quasi-elastic coefficients.

And finally for the electromagnetic potential for a point mass charged by the amount Q:

$$U^{\mathrm{em}} = Q\left(\phi - \dot{\boldsymbol{x}} \cdot \boldsymbol{A}\right) \quad \Rightarrow \quad Q_i^{\mathrm{em}} = \frac{\mathrm{d}}{\mathrm{d}t}\frac{\partial U^{\mathrm{em}}}{\partial \dot{x}_i} - \frac{\partial U^{\mathrm{em}}}{\partial x_i} \equiv F_i^{\mathrm{L}}, \tag{7.6}$$

with the (reversible) Lorentz force $\boldsymbol{F}^{\mathrm{L}}$ involving the electric and magnetic fields $\boldsymbol{E}$ and $\boldsymbol{B}$, respectively:

$$\boldsymbol{F}^{\mathrm{L}} = Q\left(\boldsymbol{E} + \dot{\boldsymbol{x}} \times \boldsymbol{B}\right) \ , \quad \boldsymbol{E} = -\nabla\phi - \frac{\partial A}{\partial t} \ , \quad \boldsymbol{B} = \nabla \times \boldsymbol{A}, \tag{7.7}$$

where ϕ and $\boldsymbol{A}$ denote the electric and the magnetic potential, respectively. In the following examples only gravity will be important, the two other expressions are left to further studies.

Moreover we distinguish two irreversible forces, $\boldsymbol{Q}^{\mathrm{f}_a}$ and $\boldsymbol{Q}^{\mathrm{f}_v}$, which are related to friction along the surface a:

$$R = \frac{1}{2}\int_a \dot{\boldsymbol{q}}^{\mathsf{T}} \cdot \mathbf{B} \cdot \dot{\boldsymbol{q}} \, \mathrm{d}a \ , \quad Q_i^{\mathrm{f}_a} = -\frac{\partial R}{\partial \dot{q}_i}, \tag{7.8}$$

R being the so-called Rayleigh function, which is quadratic in the generalized velocities, and to dissipation within the volume v, namely,

$$S = \frac{1}{2}\int_v \dot{\boldsymbol{u}}_{\mathrm{r}}^{\mathsf{T}} \cdot \hat{\boldsymbol{B}} \cdot \dot{\boldsymbol{u}}_{\mathrm{r}} \, \mathrm{d}v = \frac{1}{2}\int_v \dot{\boldsymbol{N}}_{\mathrm{r}}^{\mathsf{T}} \cdot \mathbf{B} \cdot \dot{\boldsymbol{N}}_{\mathrm{r}} \, \mathrm{d}v \ , \quad Q_i^{\mathrm{f}_v} = -\frac{\partial S}{\partial \dot{q}_i}, \tag{7.9}$$

where S is known as the internal dissipation function for the rheological model known as the dissipative gyrostat. Suggestively speaking the microstructure of this body can be imagined as a microporous skeleton (the rigid body) filled with a dissipative medium showing dry or viscous friction. This medium can move under the action of a non-stationary field of internal inertial and gravitational forces.[1]

In these relations $\hat{\boldsymbol{B}}$ and $\mathbf{B} = \mathbf{D}^{\mathsf{T}} \cdot \hat{\boldsymbol{B}} \cdot \mathbf{D}$ are (symmetric) non-negative matrices of dissipation coefficients, and $\mathbf{D}$ is a matrix of compliances. Moreover, the column of relative microvelocities $\dot{\boldsymbol{u}}_{\mathrm{r}}$ or column of relative microdisplacements $\boldsymbol{u}_{\mathrm{r}}$ can be interpreted as strain rates and strains between liquid layers of the filling gliding on each other. Finally, the column of the time-derivative of the internal forces, $\dot{\boldsymbol{N}}_{\mathrm{r}}$, are viscous stresses, but not their time derivative.

An important alternative formula for calculating the volumetric dissipation function S is[2]

$$S = \frac{b}{2}\int_{v'} \dot{\boldsymbol{w}}^2 \, \mathrm{d}v', \tag{7.10}$$

where the acceleration $\boldsymbol{w}$ is obtained from:

$$-\nabla' \cdot \boldsymbol{\sigma}' = \rho'_0\left[-\left(\dot{\boldsymbol{\Omega}}^{\mathrm{r}'} + \boldsymbol{\Omega}^{\mathrm{r}'} \cdot \boldsymbol{\Omega}^{\mathrm{r}'}\right) \cdot \boldsymbol{x}' + \boldsymbol{g}'\right] \equiv \rho_0 \boldsymbol{w}, \tag{7.11}$$

and b is a viscosity related dissipation coefficient (in SI-units of $\frac{\mathrm{kg\,s^3}}{\mathrm{m^3}}$), dashes refer to quantities related to a non-inertial observer Σ' firmly attached to the rigid body,

---

[1] Also see further explanations in the first part of the article.

[2] Rightfully the volume element was called $\mathrm{d}v'$ and not $\mathrm{d}v$, because it belongs to the non-inertial observer Σ', but being a polar Euclidean scalar it does not change value. The same remark holds for the volume integrals in Eq. (7.9) where in hindsight v should be replaced by v'.

hence $\boldsymbol{\sigma}'$ is its Cauchy stress tensor, and $\boldsymbol{g}'$ is its (time-dependent) gravitational acceleration. The right spin tensor $\dot{\boldsymbol{\Omega}}^{\mathrm{r}}$ is given by:

$$\boldsymbol{\Omega}^{\mathrm{r}'}(t) = \boldsymbol{Q}'^{\mathsf{T}}(t) \cdot \dot{\boldsymbol{Q}}'(t), \tag{7.12}$$

where $\boldsymbol{Q}'(t) = \boldsymbol{d}_i \otimes \tilde{\boldsymbol{e}}_i(t)$ is the rotation tensor proper to Σ'. It maps the image of the orthogonal base $\boldsymbol{e}_i$ of the inertial observer Σ onto the orthogonal unit base $\boldsymbol{d}_i$ of the non-inertial one. Of course Σ' conceives $\boldsymbol{d}_i$ as time-independent. This image base is referred to as $\tilde{\boldsymbol{e}}_i(t)$. It is time-dependent, because for Σ' the base of Σ is turning. The negative term $-\left(\dot{\boldsymbol{\Omega}}^{\mathrm{r}'} + \boldsymbol{\Omega}^{\mathrm{r}'} \cdot \boldsymbol{\Omega}^{\mathrm{r}'}\right) \cdot \boldsymbol{x}'$ in (7.11) can be interpreted as "fictitious" accelerations in the sense of d'Alembert. It is for these reasons that $\boldsymbol{w}$ is called inertial-gravitational acceleration. It is an object of Σ'.[3]

Equation (7.11) implies that inertial-gravitational accelerations are transformed into (dissipative) stresses within a volume element of the dissipative gyrostat. For future purpose it is useful to represent the negative divergence of stresses $\boldsymbol{\sigma}'$ (or rather the inertial-gravitational accelerations $\boldsymbol{w}$) by tractions $\boldsymbol{t}'$ on the surface $\partial(\mathrm{d}v')$ of the volume element $\mathrm{d}v'$, all with respect to the non-inertial observer Σ'[4]:

$$\rho_0 \boldsymbol{w} \mathrm{d}v' = -\nabla' \cdot \boldsymbol{\sigma}' \mathrm{d}v' = - \lim_{\Delta v \to \mathrm{d}v} \oint_{\partial(\Delta v')} \boldsymbol{t}' \, \mathrm{d}a'. \tag{7.13}$$

This is a consequence of the Cauchy theorem $\boldsymbol{t}' = \boldsymbol{n}' \cdot \boldsymbol{\sigma}'$ and the Gauss-Ostrogradski theorem when applied so-to-speak locally to a volume element $\mathrm{d}v'$. Moreover, recall that stresses can, in principle, be decomposed additively into reversible (elastic) and irreversible (friction) parts, $\boldsymbol{\sigma}' = \boldsymbol{\sigma}^{\mathrm{el}} + \boldsymbol{\sigma}^{\mathrm{f}}$. However, the symbol $\boldsymbol{\sigma}'$ in Eq. (7.13) refers exclusively to the irreversible part, because, by definition, a dissipative gyrostat does not allow elastic stresses due to its rigid skeleton and viscous filling. In other words, Eq. (7.13) is a conservative estimate according to which all inertial- gravitational accelerations cause irreversible action. In what follows we do not emphasize this any more and do not add an index f to $\boldsymbol{\sigma}'$.

7.2 The Mathematical Pendulum

Consider the situation shown in Fig. 7.1 (left). In the case of a mathematical pendulum the mass of the thread of length l is neglected when compared to the point mass m attached at its end. The problem was studied before in Smolnikov and Belyaev (2008),

[3] For historical reasons it carries no dash. It appeared first in earlier work, e.g., Smolnikov (2008b), where a change of observer was no issue of discussion.

[4] Alternatively the surface integral could be replaced by a sum over the (small) surface elements that enclose the volume Δv. This formula will prove useful later when evaluation normal and shear force diagrams for beam structures.

Smolnikov (2008a, b). In what follows we will point out a few things that were not mentioned in there. Moreover, the mathematical pendulum is an excellent starting point in order to understand the effects of the volumetric dissipation function S.

The kinetic and the gravitational potential energy are determined for the inertial observer Σ. Hence the images of the unit vectors $\boldsymbol{d}_1$ and $\boldsymbol{d}_2$ are used, which are time-dependent, $\tilde{\boldsymbol{d}}_1(t), \tilde{\boldsymbol{d}}_2(t)$. Moreover, the mass density of the pendulum system is given by a delta function, $\rho = m\delta\left(\boldsymbol{x} - l\tilde{\boldsymbol{d}}_1\right)$. Therefore:

$$\begin{aligned} T &= \frac{1}{2}\int\limits_v m\delta\left(\boldsymbol{x} - l\tilde{\boldsymbol{d}}_1\right) l^2\dot{\varphi}^2 \mathrm{d}^3\boldsymbol{x} = \frac{1}{2}ml^2\dot{\varphi}^2, \\ U^{\mathrm{g}} &= -\boldsymbol{g}\cdot\int\limits_v m\delta\left(\boldsymbol{x} - l\tilde{\boldsymbol{d}}_1\right)\boldsymbol{x}\,\mathrm{d}^3\boldsymbol{x} = -m\boldsymbol{g}\cdot\left(l\tilde{\boldsymbol{d}}_1\right) = -mgl\cos\varphi. \end{aligned} \tag{7.14}$$

For the inertial-gravitational acceleration $\boldsymbol{w}$ and its time derivative the time-independent vectors $\boldsymbol{d}_1$ and $\boldsymbol{d}_2$ of the non-inertial observer Σ' are used in context with the d'Alembert accelerations as shown in Fig. 7.1:

$$\boldsymbol{w} = \left(l\dot{\varphi}^2 + g\cos\varphi\right)\boldsymbol{d}_1 \quad \Rightarrow \quad \dot{\boldsymbol{w}} = \dot{\varphi}\left(2l\ddot{\varphi} - g\sin\varphi\right)\boldsymbol{d}_1. \tag{7.15}$$

Note that in all pendulum problems that are studied here we assume that the pendulum is in the process of swinging to the right. This is important for drawing the direction of the d'Alembert forces in Fig. 7.1 correctly. Hence the volumetric dissipation function and the corresponding dissipative force are given by:

$$S = \frac{bAl}{2}\dot{\varphi}^2\left(2l\ddot{\varphi} - g\sin\varphi\right)^2 \Rightarrow Q^{\mathrm{f}_v} = -\frac{\partial S}{\partial\dot{\varphi}} = -bAl^3\dot{\varphi}\left(2\ddot{\varphi} - k^2\sin\varphi\right)^2 \tag{7.16}$$

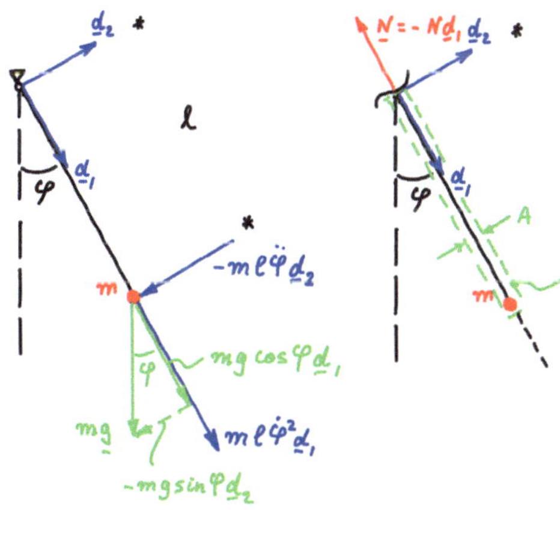

Fig. 7.1 Mathematical pendulum

with the cross-section of the rod, A, and the eigenfrequency $k = \sqrt{\frac{g}{l}}$. The latter can be used to introduce a dimensionless time $\tau = kt$, a derivative with respect to which will be denoted by circles. Note that only the radial accelerations $\left(l\dot{\varphi}^2 + g\cos\varphi\right)\boldsymbol{d}_1$ contribute to S, because only these are "transmitted" into the rod. The circumferential ones $-(l\ddot{\varphi} + g\sin\varphi)\,\boldsymbol{d}_2$ have no effect; they belong to m and do not enter the rod so-to-speak (see Fig. 7.1).

Hence the dimensionless equation of motion reads:

$$\overset{\circ\circ}{\varphi} + \sin\varphi = -b\overset{\circ}{\varphi}\left(2\overset{\circ\circ}{\varphi} - \sin\varphi\right)^2 \ , \quad b \rightarrow \frac{bAgk}{m}, \tag{7.17}$$

where the now dimensionless frictional parameter b was reassigned.

Some numerical solutions of the differential equation are shown in Fig. 7.2 for initial conditions $\varphi(0) = 0.1$ and $\overset{\circ}{\varphi}(0) = 0.1$. The plots on the left show the dissipation-free case. The plots on the right are for $b = 1$. The amplitude goes down initially. However, a limit cycle is obtained. This is because the dissipation function converges to zero at large times (see Fig. 7.3).

We now explore different ways of setting up the dissipation function. First we turn to Eq. (7.13) in order to obtain the inertial-gravitational accelerations. For evaluation we consider (in Fig. 7.1) an infinitesimal piece of rod with cross-section A and radial length $\mathrm{d}r$. Then $\mathrm{d}v = A\mathrm{d}r$. Its hull $\partial(\mathrm{d}v)$ consists of two front surfaces of size A

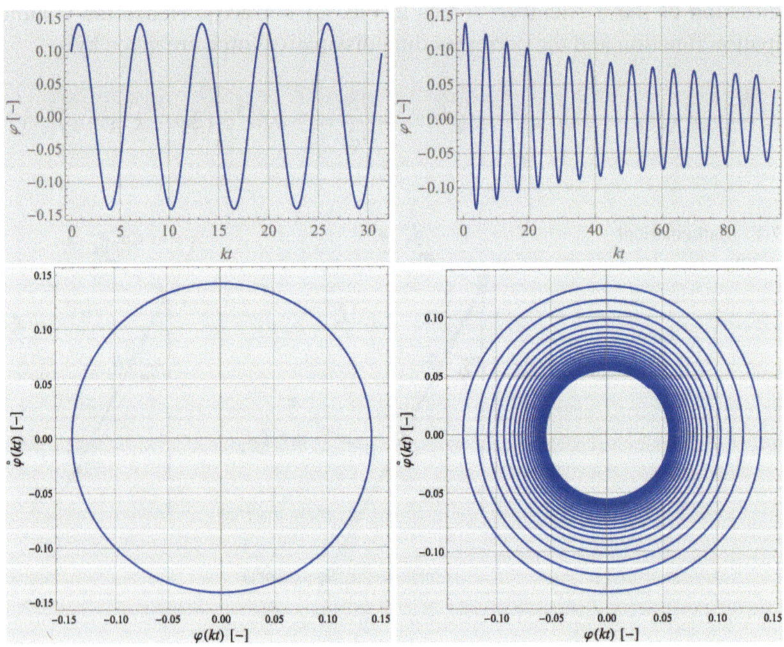

Fig. 7.2 Angle versus time and phase plots for the mathematical pendulum where $k = \sqrt{g/l}$

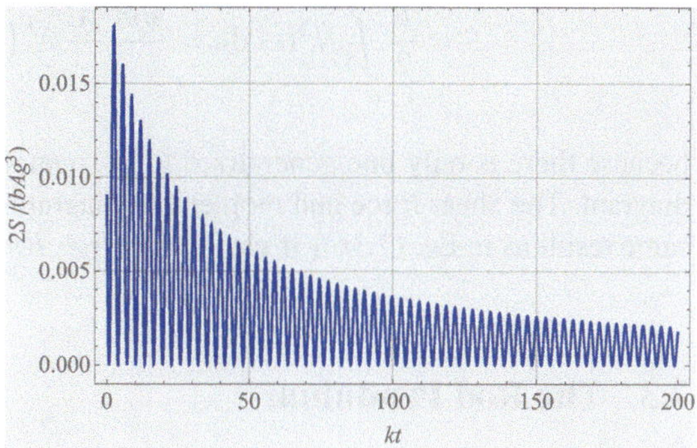

Fig. 7.3 Dissipation function versus time for the mathematical pendulum

and the mantle of size $2\sqrt{\pi A}\mathrm{d}r$ if we assume the cross-section to be circular. The mantle is traction-free. Hence it does not contribute to the surface integral in (7.13). The front surfaces are subject to tensile forces $\pm m\left(l\dot{\varphi}^2 + g\cos\varphi\right)\boldsymbol{d}_1$. The sum of these forces vanishes and we conclude from Eq. (7.13),

$$\rho_0 \boldsymbol{w}\mathrm{d}v' = -\lim_{\Delta v' \to \mathrm{d}v'} \oint\limits_{\partial(\Delta v')} \boldsymbol{t}'\,\mathrm{d}a' = \boldsymbol{0}, \tag{7.18}$$

which, on first glance, is in contradiction to our previous result $(7.15)_1$. However, this is deceiving, because we neglect the mass of the rod and, therefore, we must put $\rho_0 = 0$ in the last equation and the trivial identity $\boldsymbol{0} = \boldsymbol{0}$ results.

Now if we wish to use Eq. (7.13) to compute the dissipation function S we must choose the volume v' shown in Fig. 7.1_2. Note that it wraps around the point mass m. The mantle is now $2\sqrt{\pi A}l$. It is still traction-free and does not contribute to the surface integral. Because of the wrapping the bottom of the rod is also traction free there and it does not contribute either. However, after drawing the free body diagram, there is a tensile force contribution $\boldsymbol{N} = -m\left(l\dot{\varphi}^2 + g\cos\varphi\right)\boldsymbol{d}_1$ at the supported end—a traction that is not zero. We now adjust Eq. (7.13) to the new scenario and find:

$$\int\limits_{v'} m\delta\left(\boldsymbol{x} - l\boldsymbol{d}_1\right)\boldsymbol{w}\,\mathrm{d}v = -\oint\limits_{\partial v'} \boldsymbol{t}'\,\mathrm{d}a' \quad \Rightarrow \quad m\boldsymbol{w} - +m\left(l\dot{\varphi}^2 + g\cos\varphi\right)\boldsymbol{d}_1, \tag{7.19}$$

which agrees with the previous result $(7.15)_1$. The problem of the $\boldsymbol{0} = \boldsymbol{0}$ result in Eq. (7.18) arises only because of the point mass. As we shall see in the case of the (non-mathematical) rod pendulum, the mass of which is homogeneously distributed, no such artifacts occur.

There is yet another way of looking at the problem: obviously the normal force diagram along the rod is constant and given by $N(r) = m\left(l\dot{\varphi}^2 + g\cos\varphi\right)$. It is positive because of the tensile force. We now turn to Eq. (7.9), which for the case of the rod reduces to

$$S = \frac{B}{2}\int\limits_{v} \dot{N}^2(r)\,\mathrm{d}v = \frac{Bm^2Al}{2}\dot{\varphi}^2\left(2\ddot{\varphi} - k^2\sin\varphi\right)^2, \tag{7.20}$$

because there is only one generalized force, namely the one from the normal force diagram. The shear force and momentum diagrams are identically zero. This is the same result as in Eq. $(7.16)_1$ if we identify $b = Bm^2$.

7.3 The Rod Pendulum

Consider the situation shown in Fig. 7.4 (left). The homogeneous rod of cross-section A and length l starts swinging because of its own mass. There are no point masses attached. Moreover, we will assume that the rod is thin.

The kinetic and the gravitational potential energy are determined with respect to the inertial observer Σ. Hence the images of the unit vectors $\boldsymbol{d}_1$ and $\boldsymbol{d}_2$ are used, which are time-dependent, $\tilde{\boldsymbol{d}}_1(t), \tilde{\boldsymbol{d}}_2(t)$. Moreover, the mass density ρ_0 of the rod pendulum is constant. Consider the mass element $\mathrm{d}m = \rho_0 A\mathrm{d}r$ at the position $\boldsymbol{x}(t) = r\tilde{\boldsymbol{d}}_1(t)$ in the figure. Because of $\dot{\tilde{\boldsymbol{d}}}_1(t) = \dot{\varphi}\tilde{\boldsymbol{d}}_2(t)$ we have $\boldsymbol{v} \equiv \dot{\boldsymbol{x}}(t) = r\dot{\varphi}\tilde{\boldsymbol{d}}_2(t)$ and therefore:

$$\begin{aligned} T &= \frac{\rho_0 A\dot{\varphi}^2}{2}\int\limits_{r=0}^{l} r^2\mathrm{d}r = \frac{ml^2}{6}\dot{\varphi}^2 = \frac{J}{2}\dot{\varphi}^2, \\ U^{\mathrm{g}} &= -\rho_0 A\boldsymbol{g}\cdot\int\limits_{r=0}^{l} r\tilde{\boldsymbol{d}}_1\mathrm{d}r = -g\cos\varphi\rho_0 A\frac{l^2}{2} = -mg\frac{l}{2}\cos\varphi, \end{aligned} \tag{7.21}$$

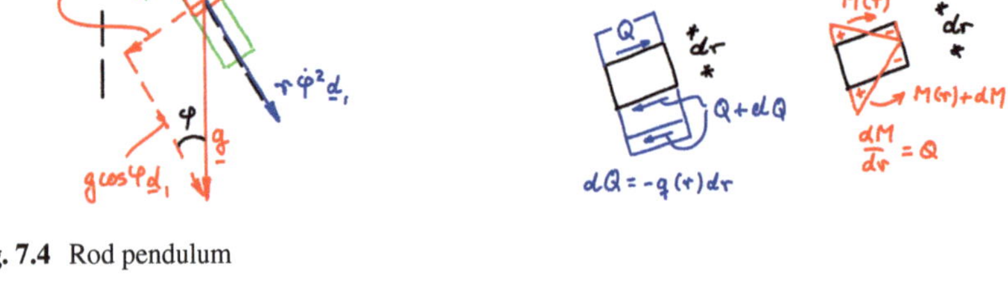

Fig. 7.4 Rod pendulum

where the total mass $m = \rho_0 Al$ and the moment of inertia $J = \frac{ml^2}{3}$ with respect to the location of the pivot were introduced.[5]

For the inertial-gravitational acceleration $\boldsymbol{w}$ and its time derivative the time-independent vectors $\boldsymbol{d}_1$ and $\boldsymbol{d}_2$ of the non-inertial observer Σ' are used in context with the d'Alembert accelerations as shown in the figure:

$$\begin{aligned} \boldsymbol{w} &= \left(r\dot{\varphi}^2 + g\cos\varphi\right)\boldsymbol{d}_1 - \left(r\ddot{\varphi} + g\sin\varphi\right)\boldsymbol{d}_2 \quad \Rightarrow \\ \dot{\boldsymbol{w}} &= \dot{\varphi}\left(2r\ddot{\varphi} - g\sin\varphi\right)\boldsymbol{d}_1 - \left(r\dddot{\varphi} + g\dot{\varphi}\cos\varphi\right)\boldsymbol{d}_2. \end{aligned} \tag{7.22}$$

Hence the volumetric dissipation function and the corresponding dissipative force are given by:

$$\begin{aligned} S &= \frac{bAl}{2}\left[\frac{l^2}{3}\left(\dddot{\varphi}^2 + 4\dot{\varphi}^2\ddot{\varphi}^2\right) + gl\dot{\varphi}\left(\dddot{\varphi}\cos\varphi - 2\dot{\varphi}\ddot{\varphi}\sin\varphi\right) + g^2\dot{\varphi}^2\right] \quad \Rightarrow \\ Q^{\mathrm{f}_v} &= -\frac{\partial S}{\partial\dot{\varphi}} = -\frac{bAl}{2}\left[\frac{8}{3}l^2\dot{\varphi}\ddot{\varphi}^2 + gl\left(\dddot{\varphi}\cos\varphi - 4\dot{\varphi}\ddot{\varphi}\sin\varphi\right) + 2g^2\dot{\varphi}\right]. \end{aligned} \tag{7.23}$$

Recall the eigenfrequency for this problem, $k = \sqrt{\frac{mgl}{2J}}$. As in the previous example we use it to introduce a dimensionless time $\tau = kt$, and derivatives with respect to this quantity will be denoted by circles. Then the equation of motion reads:

$$\overset{\circ\circ}{\varphi} + \sin\varphi = -b\left[2\overset{\circ}{\varphi}\overset{\circ\circ}{\varphi}^2 + \frac{3}{2}\left(\overset{\circ\circ\circ}{\varphi}\cos\varphi - 4\overset{\circ}{\varphi}\overset{\circ\circ}{\varphi}\sin\varphi\right) + 2\overset{\circ}{\varphi}\right], \quad b \to \frac{bAlg^2}{2Jk}, \tag{7.24}$$

where a now dimensionless frictional parameter b was reassigned, which differs from the one shown in Eq. (7.17).

Now, before we discuss the numerical solution of the differential equation, we shall try to rederive it by using, first, the surface integral approach from Eq. (7.13) and, second, an elementary beam approach and compute normal, shear and moment diagrams in order to exploit Eq. (7.9). In preparation we first note that the (dynamic) normal and shear force densities (both in $\frac{\mathrm{N}}{\mathrm{m}}$) are given by:

$$n(r) = A\rho_0\left(r\dot{\varphi}^2 + g\cos\varphi\right), \quad q(r) = A\rho_0\left(r\ddot{\varphi} + g\sin\varphi\right), \quad 0 \le r \le l. \tag{7.25}$$

The differential equations for the (dynamic) normal, shear force and moment diagrams are just like the ones in statics, if we neglect inertial effects as we should in this "slow" dissipation approach,[6] because these quantities will be used to compute S:

$$\frac{\mathrm{d}N}{\mathrm{d}r} = -n(r), \quad \frac{\mathrm{d}Q}{\mathrm{d}r} = -q(r), \quad \frac{\mathrm{d}M}{\mathrm{d}r} = Q(r). \tag{7.26}$$

[5] All of this and what follows is correct under the thin rod assumption. Also see the remarks regarding the in-plane extensions of the hammer in the next example.

[6] See the discussion in Part I of this work.

With the boundary conditions $N(l) = 0$, $Q(l) = 0$, and $M(0) = 0$ we obtain by integration when using Eq. (7.25):

$$
\begin{aligned}
N(r) &= -A\rho_0 \left[\frac{1}{2}(r^2 - l^2)\dot{\varphi}^2 + g\,(r - l)\cos\varphi \right], \\
Q(r) &= -\frac{A\rho_0}{2} \left[(r^2 - l^2)\ddot{\varphi} + 2g\,(r - l)\sin\varphi \right], \\
M(r) &= -\frac{A\rho_0}{2} \left[\frac{r}{3}(r^2 - 3l^2)\ddot{\varphi} + gr\,(r - 2l)\sin\varphi \right].
\end{aligned}
\tag{7.27}
$$

Now consider the right insets of Fig. 7.4 where free body diagrams of a small piece of the rod are shown. Its mantle is traction free as in the case of the mathematical pendulum. On the front surfaces the normal, shear and moment contributions are shown separately (based on the standard positive sign convention). In the case of the normal forces only $\mathrm{d}N\boldsymbol{d}_1$ contributes to the traction $\boldsymbol{t}$. In the case of shear forces it is $-\mathrm{d}Q\boldsymbol{d}_2$, and in the case of the moments it is nothing at all, because of positive and negative bending stresses. Hence Eq. (7.13) yields:

$$
\begin{aligned}
\rho_0 \boldsymbol{w} A \mathrm{d}r &= - \lim_{\Delta v' \to \mathrm{d}v'} \oint_{\partial(\Delta v')} \boldsymbol{t}'\,\mathrm{d}a' = -\mathrm{d}N\boldsymbol{d}_1 + \mathrm{d}Q\boldsymbol{d}_2 = \\
&\rho_0 A \left[\left(r\dot{\varphi}^2 + g\cos\varphi \right) \boldsymbol{d}_1 - \left(r\ddot{\varphi} + g\sin\varphi \right) \boldsymbol{d}_2 \right] \mathrm{d}r,
\end{aligned}
\tag{7.28}
$$

where Eqs. $(7.26)_{1,2}$ and (7.25) were used in the last step. The result agrees completely with the previous *ansatz* from Eq. $(7.22)_1$. Hence the result for S is exactly the same.

Now on to the alternative relation (7.9). There are three candidates for generalized forces $\boldsymbol{N}_\mathrm{r}$, namely the surfaces $N(r)$, $Q(r)$, and $M(r)$, each of which need to be differentiated in time, squared and integrated with respect to the volume of the beam:

$$
S = \frac{1}{2} \int_{v'} \dot{\boldsymbol{N}}_\mathrm{r}^\mathsf{T} \cdot \mathbf{B} \cdot \dot{\boldsymbol{N}}_\mathrm{r}\,\mathrm{d}v' = \frac{A}{2} \int_{r=0}^{l} \left(\mathrm{B}^N \dot{N}^2 + \mathrm{B}^Q \dot{Q}^2 + \mathrm{B}^M \dot{M}^2 \right) \mathrm{d}r,
\tag{7.29}
$$

where three different dissipation coefficients were used, B^N, B^Q, and B^M, respectively. Hence after integration the resulting expression is obviously longer than (7.23). However, it does contain the same dependencies on φ and its time derivatives *but* with different prefactors. Hence we conclude that the structure of the ODE does not change crucially.

Now we start exploring numerical solutions of Eq. (7.24). The fundamental difference in comparison with Eq. (7.17) for the mathematical pendulum is that the dissipative term now adds third order time derivatives in φ. What do we need to do in order to obtain a numerical solution? Formally speaking we need to prescribe also a value for the initial acceleration, $\ddot{\varphi}(0)$, but how is this achieved in practice? Indeed, we are so used to solving second order differential equations for oscillators that we hardly ponder how to realize the corresponding ordinary initial conditions.

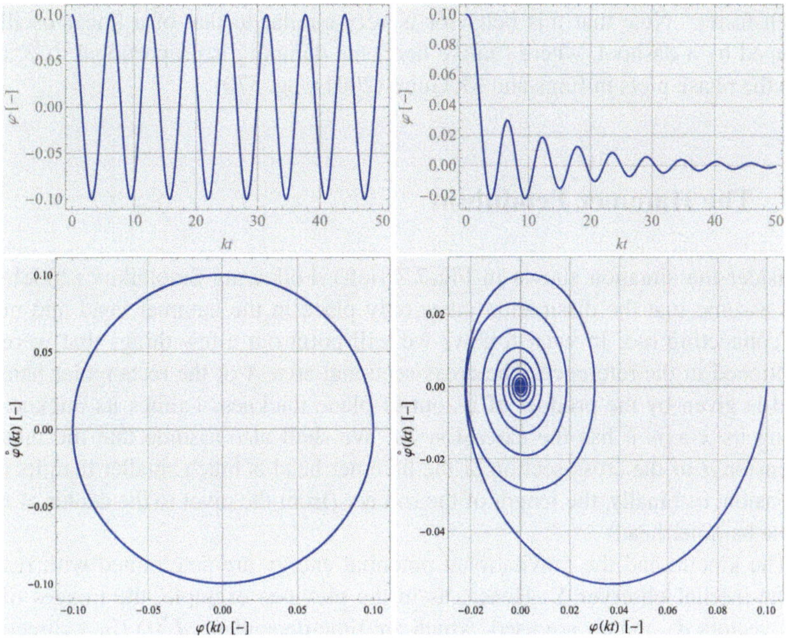

Fig. 7.5 Angle versus time and phase plots for the rod pendulum

Whilst it is obvious how to impose an initial displacement $\varphi(0)$ on a rod pendulum, the prescription of a precise initial (angular) velocity $\dot{\varphi}(0) \neq 0$ is easily done on the computer but much less so in reality. The only initial impulse that is easy to realize is $\dot{\varphi}(0) = 0$, and so is $\ddot{\varphi}(0) = 0$, in simple words, we just do nothing. Therefore let us see what happens if we choose $\varphi(0) = 0.1$, $\dot{\varphi}(0) = 0$, and $\ddot{\varphi}(0) = 0$. The results are shown in Fig. 7.5 with $b = 0$ on the left and $b = 1$ on the right.

This time no limit cycle results: The dissipation is so strong that the oscillation comes to a standstill. This is also visible by looking at the behavior of S over time: Fig. 7.6. In comparison with Fig. 7.3 the volumetric dissipation function S decreases

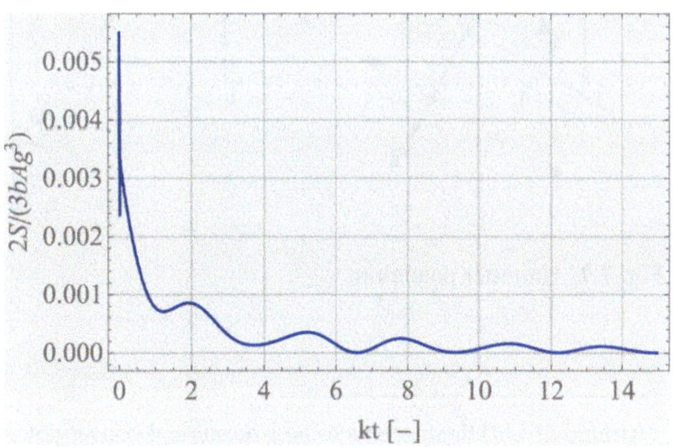

Fig. 7.6 Dissipation function versus time for the rod pendulum obtained from Eq. (7.23)$_1$

much faster.[7] Note that this behavior is very similar to that of a linear oscillator damped by a dashpot, where (unlike here) the damping is proportional to $\dot{\varphi}$ alone (see the phase plots in Enns and McGuire (2001), pg. 570).

7.4 The Hammer Pendulum

Consider the situation shown in Fig. 7.7 (left). Following Smolnikov (2008b) we will assume that the dissipation takes only place in the hammer head and not in the connecting rod. In what follows we will point out a few things that were not mentioned in the reference. The cross-sectional area A of the rectangular hammer head is given by the product of its out-of-plane thickness t times its thickness d. Along its x-axis it has the extension $2a$. We shall also assume that the in-plane extension d of the cross-section of the hammer head is much smaller than its axial extension, a. Finally, the length of the rod is l (from the pivot to the center of mass of the hammer head).

The kinetic and the gravitational potential energy are determined with respect to the inertial observer Σ. Hence, as in the previous example, the images of the unit vectors $\boldsymbol{d}_1$ and $\boldsymbol{d}_2$ are used, which are time-dependent, $\tilde{\boldsymbol{d}}_1(t)$ (in y-direction), $\tilde{\boldsymbol{d}}_2(t)$ (in x-direction). Moreover, the mass density ρ_0 of the hammer pendulum is constant. Consider the point P at position $\boldsymbol{x}(t) = (l + y)\,\tilde{\boldsymbol{d}}_1(t) + x\tilde{\boldsymbol{d}}_2(t)$ in the mass element $\mathrm{d}m = \rho_0 t \mathrm{d}x\mathrm{d}y$ of the figure. Strictly speaking, because of $\dot{\tilde{\boldsymbol{d}}}_1(t) = \dot{\varphi}\tilde{\boldsymbol{d}}_2(t)$ and $\dot{\tilde{\boldsymbol{d}}}_2(t) = -\dot{\varphi}\tilde{\boldsymbol{d}}_1(t)$ we then have $\boldsymbol{v} \equiv \dot{\boldsymbol{x}}(t) = \dot{\varphi}\left[(l + y)\,\tilde{\boldsymbol{d}}_2(t) - x\tilde{\boldsymbol{d}}_1(t)\right]$

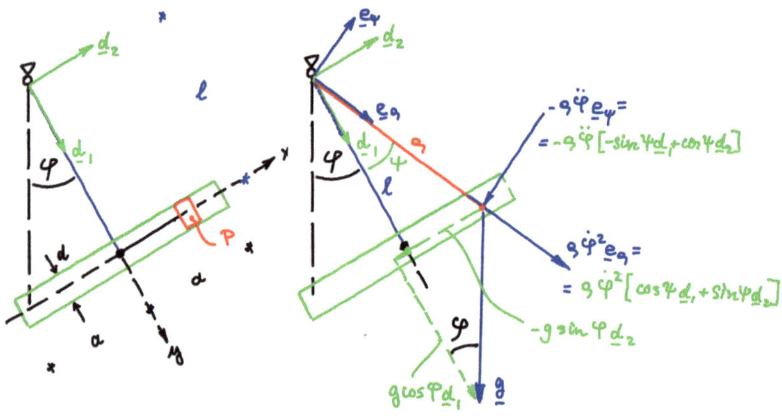

Fig. 7.7 Hammer pendulum

[7] Around $kt = 0$ there seems to be a numerical convergence problem.

and therefore[8]:

$$T = \frac{\rho_0 A \dot{\varphi}^2}{2} \int\limits_{x=-a}^{+a} \int\limits_{y=-d/2}^{+d/2} [(l+y)^2 + x^2] \mathrm{d}y \mathrm{d}x \approx \frac{m}{2}\left(l^2 + \frac{a^2}{3}\right)\dot{\varphi}^2 = \frac{J}{2}\dot{\varphi}^2,$$

$$U^{\mathrm{g}} = -\rho_0 \boldsymbol{g} \cdot \int\limits_{v} \boldsymbol{x} \mathrm{d}v = -\rho_0 t \left(g \cos\varphi \tilde{\boldsymbol{d}}_1 - g \sin\varphi \tilde{\boldsymbol{d}}_2\right) \cdot$$

$$\int\limits_{x=-a}^{+a} \int\limits_{y=-d/2}^{+d/2} \left[(l+y)\tilde{\boldsymbol{d}}_1 + x\tilde{\boldsymbol{d}}_2\right] \mathrm{d}y \mathrm{d}x = -mgl\cos\varphi, \tag{7.30}$$

where the total mass $m = \rho_0 A l$ and the moment of inertia $J \approx m\left(l^2 + \frac{a^2}{3}\right)$ with respect to the location of the pivot were introduced. The y-contribution in the integral for T is neglected. In the integral for U^{g} this assumption is unnecessary.

For the inertial-gravitational acceleration $\boldsymbol{w}$ the time-independent vectors $\boldsymbol{e}_\rho$ and $\boldsymbol{e}_\psi$ of the non-inertial observer Σ' are used first for the two d'Alembert accelerations which is then converted into $\boldsymbol{d}_1$ and $\boldsymbol{d}_2$. From Fig. 7.7 (right) we obtain:

$$\begin{aligned} \boldsymbol{w} &= \rho\dot{\varphi}^2 \boldsymbol{e}_\rho + g\cos\varphi \boldsymbol{d}_1 - \rho\ddot{\varphi}\boldsymbol{e}_\psi - g\sin\varphi \boldsymbol{d}_2 = \\ &\left(l\dot{\varphi}^2 + x\ddot{\varphi} + g\cos\varphi\right)\boldsymbol{d}_1 + \left(x\dot{\varphi}^2 - l\ddot{\varphi} - g\sin\varphi\right)\boldsymbol{d}_2. \end{aligned} \tag{7.31}$$

Here we put $l = \rho\cos\psi$ and $x = \rho\sin\psi$. Note that this is not quite true because, strictly speaking, the vector $\rho\boldsymbol{e}_\rho$ does not lead to a point on the $x-x$-axis. Rather it is given by $(l+y)\boldsymbol{d}_1 + x\boldsymbol{d}_2$, and hence, $\rho = \sqrt{(l+y)^2 + x^2}$. In what follows the integration with respect to y will be simplified such that, in effect, we will neglect terms of d that are small in comparison to a.

The time derivative of $\boldsymbol{w}$ is given by:

$$\dot{\boldsymbol{w}} = \left(2l\dot{\varphi}\ddot{\varphi} + x\dddot{\varphi} - g\dot{\varphi}\sin\varphi\right)\boldsymbol{d}_1 + \left(2x\dot{\varphi}\ddot{\varphi} - l\dddot{\varphi} - g\dot{\varphi}\cos\varphi\right)\boldsymbol{d}_2. \tag{7.32}$$

Hence the volumetric dissipation function and the corresponding dissipative force are (approximately) given by:

$$\begin{aligned} S = \frac{bAa}{3}&\left[\dot{\varphi}^2\left(3g^2 - 12gl\ddot{\varphi}\sin\varphi + 4\left(a^2+3l^2\right)\ddot{\varphi}^2\right) + \right. \\ &\left. 6gl\dot{\varphi}\dddot{\varphi}\cos\varphi + \left(a^2+3l^2\right)\dddot{\varphi}^2\right] \quad \Rightarrow \end{aligned} \tag{7.33}$$

$$Q^{\mathrm{f}_v} = -\frac{\partial S}{\partial\dot{\varphi}} = -\frac{2bAa}{3}\left[\dot{\varphi}\left(3g^2 - 12gl\ddot{\varphi}\sin\varphi + 4\left(a^2+3l^2\right)\ddot{\varphi}^2\right) + 3gl\dddot{\varphi}\cos\varphi\right].$$

[8] It is instructive to remember where the approximation of small lateral extension $d \ll a$ of the hammer enters, because we shall use it again below.

The eigenfrequency for this problem is $k = \sqrt{\frac{3gl}{a^2+3l^2}}$. As before we use it to introduce a dimensionless time $\tau = kt$. Derivatives with respect to this quantity will be denoted by circles. The equation of motion reads:

$$\overset{\circ\circ}{\varphi} + \sin\varphi = -b\left[\overset{\circ}{\varphi}\left(3 + \frac{a^2}{l^2} - 12\overset{\circ\circ}{\varphi}\sin\varphi + 12\overset{\circ\circ}{\varphi}^2\right) + 3\overset{\circ\circ\circ}{\varphi}\cos\varphi\right], \tag{7.34}$$

where a dimensionless frictional parameter $b \to \frac{6bAg^2l^3}{m\left(a^2+3l^2\right)^2 k}$ was reassigned.

This ODE was solved for $a/l = 2/3, b = 1$ and $\varphi(0) = 0.1, \dot{\varphi}(0) = 0$, and $\ddot{\varphi}(0) = 0$. The result is shown in Fig. 7.8. As in the case of the rod pendulum there is some similarity to the phase plot for a positive dashpot damper. However, the oscillation does not seem to die out completely. The dissipation function is shown in Fig. 7.9.

Now, we shall try to rederive the result for the inner dissipation function by using, first, the surface integral approach from Eq. (7.13) and, second, an elementary beam approach where we compute the normal, shear and moment diagrams in order to exploit Eq. (7.9). In preparation we conclude from Eq. (7.31) that the (dynamic) normal and shear force densities (both in $\frac{\text{N}}{\text{m}}$) are given by:

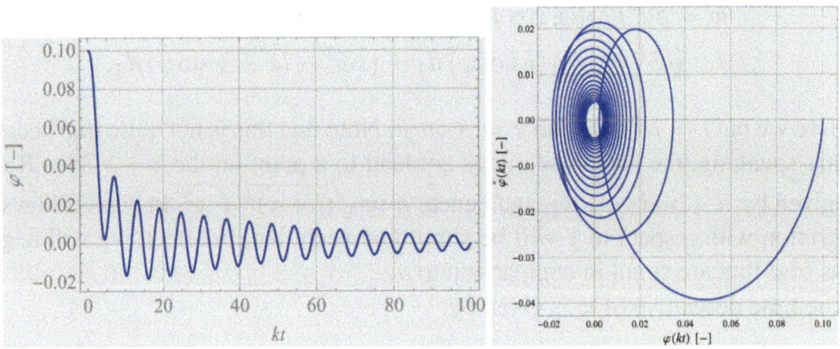

Fig. 7.8 Angle versus time and phase plots for the hammer pendulum

Fig. 7.9 Dissipation function S versus time for the hammer pendulum

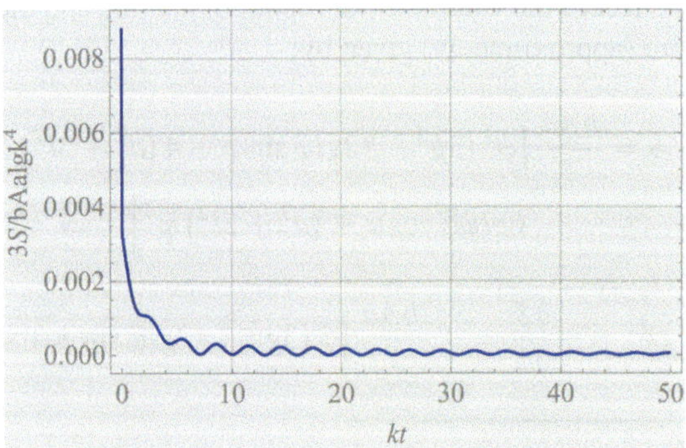

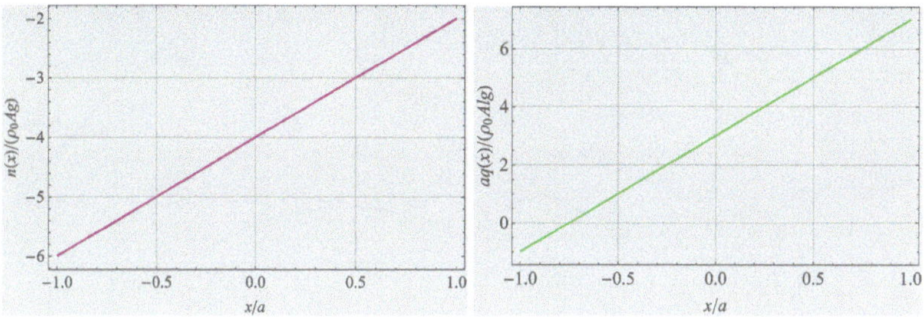

Fig. 7.10 Normal and shear force densities over position along the hammer

$$n(x) = \rho_0 A \left(x\dot{\varphi}^2 - (l\ddot{\varphi} + g \sin\varphi) \right) ,$$
$$q(x) = \rho_0 A \left(l\dot{\varphi}^2 + g \cos\varphi + x\ddot{\varphi} \right) , \quad -a \le x \le a. \tag{7.35}$$

These functions are shown in normalized form in Fig. 7.10. For this purpose the dimensionless factors $\alpha = \frac{a\dot{\varphi}^2}{g} + \frac{a}{l}\cos\varphi$, $\beta = \frac{a^2\ddot{\varphi}}{lg}$, $\gamma = \sin\varphi + \frac{l\ddot{\varphi}}{g}$, and $\delta = \frac{a\dot{\varphi}^2}{g}$ were introduced. For principal illustration the plots were drawn for the choice $\alpha = 3$, $\beta = 4$, $\gamma = 4$, $\delta = 2$.

The differential equations for the normal, shear force and moment diagrams read[9]:

$$\frac{\mathrm{d}N}{\mathrm{d}x} = -n(x) , \quad \frac{\mathrm{d}Q}{\mathrm{d}x} = -q(x) , \quad \frac{\mathrm{d}M}{\mathrm{d}x} = Q(x). \tag{7.36}$$

They are just like the ones in statics, if we neglect inertial effects as in the example of the rod pendulum. However, we must compute them section-wise now, and there will be jumps at $x = 0$ because of the supporting rod l, which is stiffly attached to the hammer. The boundary conditions are $N(\pm a) = 0$, $Q(\pm a) = 0$, and $M(\pm a) = 0$. We obtain by integration of Eq. (7.35):

$$\begin{aligned}
&0 \le x \le +a : \\
&N(x) = -\rho_0 A \left[l\ddot{\varphi} + g \sin\varphi - \frac{\dot{\varphi}^2}{2}(a + x) \right] (a - x) , \\
&Q(x) = \rho_0 A \left[l\dot{\varphi}^2 + g \cos\varphi + \frac{\ddot{\varphi}}{2}(a + x) \right] (a - x) , \\
&M(x) = -\rho_0 A \left[l\dot{\varphi}^2 + g \cos\varphi + \frac{\ddot{\varphi}}{3}(2a + x) \right] \frac{(a - x)^2}{2} , \\
&-a \le x \le 0 : \\
&N(x) = \rho_0 A \left[l\ddot{\varphi} + g \sin\varphi + \frac{\dot{\varphi}^2}{2}(a - x) \right] (a + x) ,
\end{aligned} \tag{7.37}$$

[9] Note that these diagrams are dynamic unlike to what is usually taught in a traditional strength of materials course. On the other hand, here we made explicit use of d'Alembert's idea to treat dynamics problems as if they were static.

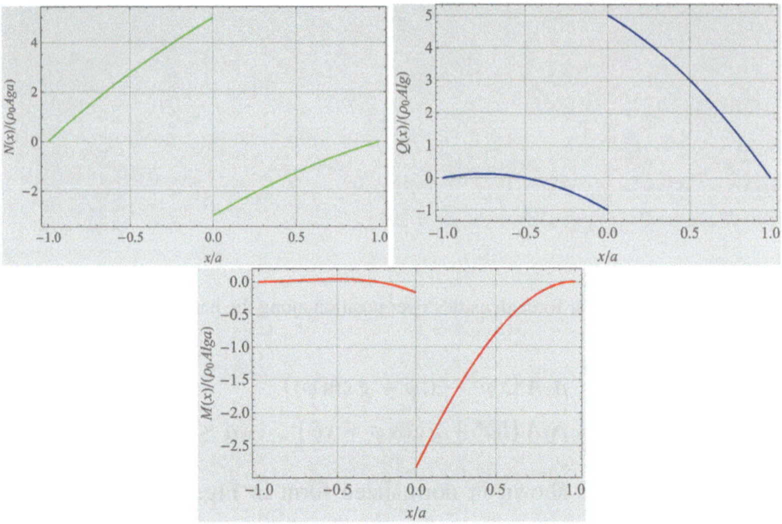

Fig. 7.11 Normal, shear force, and momentum diagrams over position along the hammer

$$Q(x) = -\rho_0 A \left[l\dot{\varphi}^2 + g\cos\varphi - \frac{\ddot{\varphi}}{2}(a-x) \right](a+x) \ ,$$

$$M(x) = -\rho_0 A \left[l\dot{\varphi}^2 + g\cos\varphi - \frac{\ddot{\varphi}}{3}(2a-x) \right]\frac{(a+x)^2}{2} \ .$$

These functions are shown in Fig. 7.11 in normalized form for the choice of α, β, γ, δ, from above. For the jumps at $x = 0$ we note that[10]:

$$\frac{N^+ - N^-}{\rho_0 Aga} = -2\gamma \ , \quad \frac{Q^+ - Q^-}{\rho_0 Ag} = 2\alpha \ , \quad \frac{M^+ - M^-}{\rho_0 Alga} = -\frac{2}{3}\beta. \tag{7.38}$$

There is an intuitive explanation for this behavior. Consider Fig. 7.12. On the left we can see the fixed lever l which due to its support of the total mass $m = \rho_0 A2a$ carries a normal force $-m\,(g\sin\varphi + l\ddot{\varphi})$ and a shear force $m\left(l\dot{\varphi}^2 + g\cos\varphi\right)$. The moment is due to the distribution $q(x)$. As shown in the inset on the right, there is a clockwise (negative) moment contribution: the shear load $q(x)$ can be decomposed into a rectangle and a triangle. The rectangle and the gravitational loads (from the two halves of the hammer head) are both symmetric about $x = 0$ and give no moment. Only the triangle does. The triangle corresponds to an equivalent load $2\rho_0 Aa^2\ddot{\varphi}$, which is applied at the position $x = \frac{a}{3}$. Hence the strength of the resulting (clockwise) moment is $\frac{2}{3}\rho_0 Aa^3\ddot{\varphi}$.

Now consider Fig. 7.13 where a free body diagram of a small piece of the hammer head is shown. The outer surfaces are all traction free. On the left and right inner

[10] + and − refer to right and left-sided limit values.

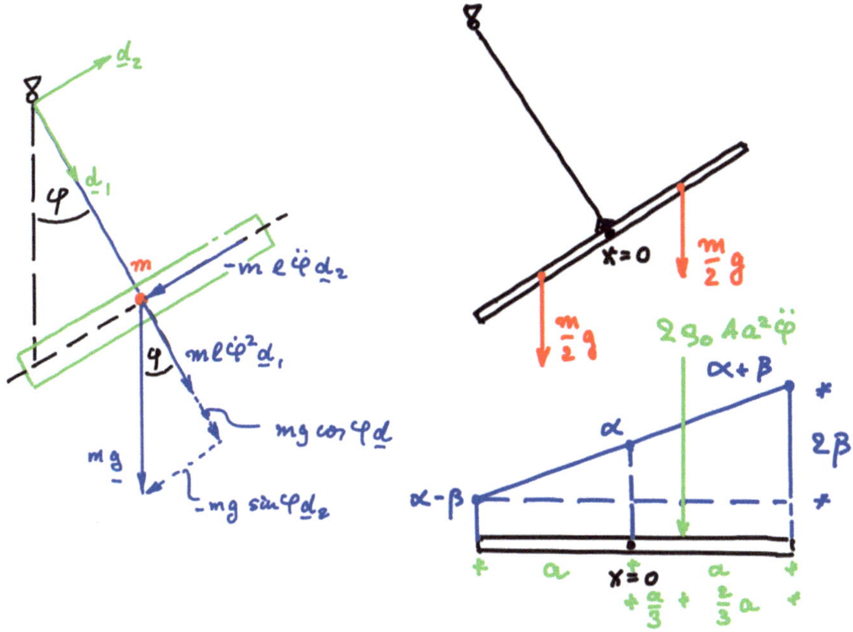

Fig. 7.12 Explanation for jumps of forces and moment

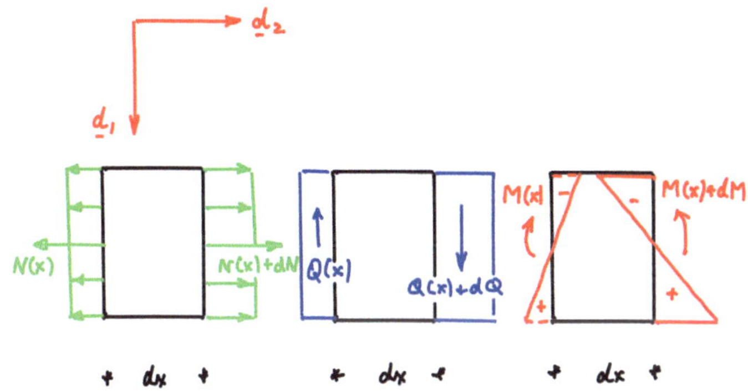

Fig. 7.13 Evaluation of the surface integral

surfaces the normal, shear and moment contributions are shown separately (based on the standard positive sign convention). In the case of the normal forces only $\mathrm{d}N\boldsymbol{d}_2$ contributes to the traction $\boldsymbol{t}$. In the case of shear forces it is $\mathrm{d}Q\boldsymbol{d}_1$, and in the case of the moments there is no contribution at all, because of positive and negative bending stresses.

Hence Eq. (7.13) yields:

$$\rho_0 \boldsymbol{w} A \mathrm{d}x = -\oint_{\partial(\mathrm{d}v')} \boldsymbol{t}' \,\mathrm{d}a' = -\mathrm{d}N\boldsymbol{d}_2 - \mathrm{d}Q\boldsymbol{d}_1 = \\ \rho_0 A \left[\left(x\dot{\varphi}^2 - (l\ddot{\varphi} + g\sin\varphi)\right)\boldsymbol{d}_2 + \left(l\dot{\varphi}^2 + g\cos\varphi + x\ddot{\varphi}\right)\boldsymbol{d}_1\right]\mathrm{d}x, \tag{7.39}$$

where Eqs. $(7.36)_{1,2}$ and (7.35) were used in the last step. The result agrees completely with Eq. (7.31). Hence the result for S is exactly the same.

Now on to the alternative relation (7.9). There are three candidates for generalized forces $\boldsymbol{N}_\mathrm{r}$, namely the surfaces $N(r)$, $Q(r)$, and $M(r)$, each of which need to be differentiated in time, squared and integrated with respect to the volume of the beam:

$$S = \frac{1}{2}\int_{v'} \dot{\boldsymbol{N}}_\mathrm{r}^\mathsf{T} \cdot \boldsymbol{B} \cdot \dot{\boldsymbol{N}}_\mathrm{r} \,\mathrm{d}v' = \frac{A}{2}\int_{x=-a}^{x=+a} \left(B^N \dot{N}^2 + B^Q \dot{Q}^2 + B^M \dot{M}^2\right)\mathrm{d}x, \tag{7.40}$$

where as in the example of the rod pendulum three different dissipation coefficients were used, B^N, B^Q, and B^M, respectively.

It should be noted that for the calculation of S in Smolnikov (2008b) only the influence of the moments was taken into account. Whilst the expressions for the left and right moment contributions in this reference agree with Eqs. $(7.37)_{3,6}$ (up to the signs) the integration result corresponding to (7.40) is faulty, also because *both* parts were integrated from $-a$ to $+a$. Using Mathematica now reveals:

$$S^M = \frac{B^M A^3 \rho_0^2 a^5}{20}\left[\dot{\varphi}^2\left(2l\ddot{\varphi} - g\sin\varphi\right)^2 + \frac{11}{21}a^2\dddot{\varphi}^2\right] \tag{7.41}$$

Also, in Smolnikov (2008b) it is pointed out that in order to get to this result only the shear load is taken into account. Hence had we used in the expression for the accelerations in Eq. (7.31) only the $\boldsymbol{d}_1$-term for computing S in Eq. (7.10), we would have obtained,

$$S = bAa\left[\dot{\varphi}^2\left(2l\ddot{\varphi} - g\sin\varphi\right)^2 + \frac{1}{3}a^2\dddot{\varphi}^2\right]. \tag{7.42}$$

This contains the same dependencies on φ and its time derivatives as Eq. (7.41), albeit with different prefactors. Hence the structure of the ODE of motion does not crucially change. A similar remark holds if we take the $\dot{N}$ and $\dot{Q}$-terms in the integral (7.40) into account. Then the dependencies on φ and time derivatives which we encountered in Eq. (7.33) also appear but with different prefactors.

7.5 The Sector Pendulum

Consider the pendulum shown in Fig. 7.14, which consists of a circular sector of radius R, angular width 2α, and out-of-plane thickness t.

Fig. 7.14 Sector pendulum

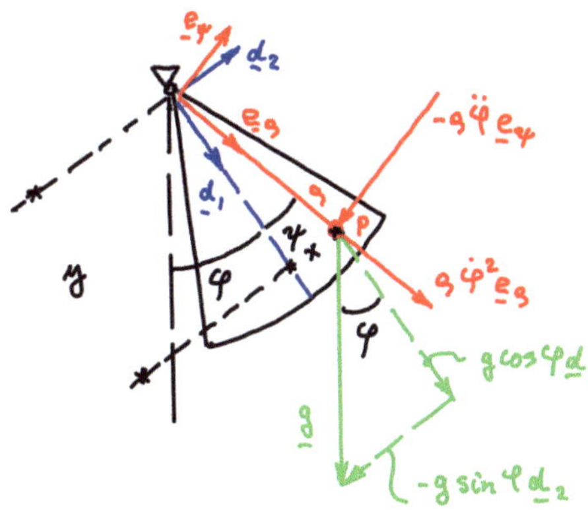

As always the kinetic and the gravitational potential energy are determined with respect to the inertial observer Σ. We consider a mass element $\mathrm{d}m = \rho_0 t \rho \, \mathrm{d}\rho \mathrm{d}\psi$ at position $\boldsymbol{x}(t) = \rho \tilde{\boldsymbol{e}}_\rho(t)$ with constant mass density ρ_0. Hence, as in the previous examples, the images of the unit vectors $\boldsymbol{e}_\rho$ and $\boldsymbol{e}_\psi$ are used, which are time-dependent, $\tilde{\boldsymbol{e}}_\rho(t)$, $\tilde{\boldsymbol{e}}_\psi(t)$. Similarly, the images of the vectors $\boldsymbol{d}_1$, $\boldsymbol{d}_2$ can be used to describe the position, $\boldsymbol{x}(t) = y\tilde{\boldsymbol{d}}_1(t) + x\tilde{\boldsymbol{d}}_2(t)$. Note that because we consider a point P on a rigid body, x, y, ρ do not depend on time. Also, $\dot{\tilde{\boldsymbol{d}}}_1(t) = \dot{\varphi}(t)\tilde{\boldsymbol{d}}_2(t)$ and $\dot{\tilde{\boldsymbol{d}}}_2(t) = -\dot{\varphi}(t)\tilde{\boldsymbol{d}}_1(t)$. Therefore, $\boldsymbol{v}(t) \equiv \dot{\boldsymbol{x}}(t) = \dot{\varphi}(t)\left(y\tilde{\boldsymbol{d}}_2(t) - x\tilde{\boldsymbol{d}}_1(t)\right)$. Hence, because of $\rho^2 = x^2 + y^2$, we have $\boldsymbol{v}^2 = \dot{\varphi}^2\rho^2$ and the kinetic and potential energies are

$$
\begin{aligned}
T &= \frac{\rho_0 t \dot{\varphi}^2}{2} \int_{\rho=0}^{R} \int_{\psi=-\alpha}^{+\alpha} \rho^3 \mathrm{d}\psi \mathrm{d}\rho = \frac{\rho_0 t \dot{\varphi}^2 R^4 \alpha}{4} = \frac{J}{2}\dot{\varphi}^2, \\
U^{\mathrm{g}} &= -\rho_0 \int_v \boldsymbol{g} \cdot \boldsymbol{x} \mathrm{d}v = \\
&- \rho_0 t g \int_{\rho=0}^{R} \int_{\psi=-\alpha}^{+\alpha} \rho^2 \left(\cos(\varphi+\psi)\,\tilde{\boldsymbol{e}}_\rho - \sin(\varphi+\psi)\,\tilde{\boldsymbol{e}}_\psi\right) \cdot \tilde{\boldsymbol{e}}_\psi \mathrm{d}\psi \mathrm{d}\rho = \\
&- \frac{2mgR\sin\alpha}{3\alpha}\cos\varphi,
\end{aligned}
\tag{7.43}
$$

where the total mass $m = \rho_0 \alpha R^2 t$ and the moment of inertia

$$
J = \rho_0 t \int_{\rho=0}^{R} \int_{\psi=-\alpha}^{+\alpha} \rho^3 \mathrm{d}\psi \mathrm{d}\rho = \frac{\rho_0 t R^4 \alpha}{2} = \frac{1}{2} m R^2, \tag{7.44}
$$

with respect to the location of the pivot were introduced.

For the inertial-gravitational acceleration $\boldsymbol{w}$ the time-independent vectors $\boldsymbol{e}_\rho$ and $\boldsymbol{e}_\psi$ of the non-inertial observer Σ' are used first in order to express the two d'Alembert accelerations efficiently. This is then converted into expression using $\boldsymbol{d}_1$ and $\boldsymbol{d}_2$. From Fig. 7.14 we obtain:

$$\boldsymbol{w} = \rho\dot{\varphi}^2\boldsymbol{e}_\rho - \rho\ddot{\varphi}\boldsymbol{e}_\psi + g\cos\varphi\boldsymbol{d}_1 - g\sin\varphi\boldsymbol{d}_2 = \\ \left(\rho\cos\psi\dot{\varphi}^2 + \rho\sin\psi\ddot{\varphi} + g\cos\varphi\right)\boldsymbol{d}_1 + \left(\rho\sin\psi\dot{\varphi}^2 - \rho\cos\psi\ddot{\varphi} - g\sin\varphi\right)\boldsymbol{d}_2. \tag{7.45}$$

These expression bear a certain resemblance to the Eqs. (7.31) of the hammer pendulum. However, this time no approximations were made.

The square of the time derivative of $\boldsymbol{w}$ is given by:

$$\dot{\boldsymbol{w}}^2 = 4\rho^2\dot{\varphi}^2\ddot{\varphi}^2 + \rho^2\dddot{\varphi}^2 + g^2\dot{\varphi}^2 - 4\rho g\dot{\varphi}^2\ddot{\varphi}\sin(\varphi+\psi) + 2\rho g\dot{\varphi}\dddot{\varphi}\cos(\varphi+\psi). \tag{7.46}$$

Integration yields the volumetric dissipation function and the corresponding dissipative force:

$$S = \frac{bt}{2}\int\limits_{\rho=0}^{R}\int\limits_{\psi=-\alpha}^{+\alpha}\rho\dot{\boldsymbol{w}}^2\mathrm{d}\psi\mathrm{d}\rho = \\ btR^2\left[\alpha\left(\dot{\varphi}^2\ddot{\varphi}^2R^2 + \frac{\dot{\varphi}^2}{2}g^2 + \frac{\dddot{\varphi}^2}{4}R^2\right) + \frac{2gR}{3}\dot{\varphi}\sin\alpha\left(\dddot{\varphi}\cos\varphi - 2\dot{\varphi}\ddot{\varphi}\sin\varphi\right)\right], \\ Q^{\mathrm{f}_v} = -\frac{\partial S}{\partial\dot{\varphi}} = \\ -btR^2\left[\alpha\dot{\varphi}\left(2\ddot{\varphi}^2R^2 + g^2\right) + \frac{2gR}{3}\sin\alpha\left(\dddot{\varphi}\cos\varphi - 4\dot{\varphi}\ddot{\varphi}\sin\varphi\right)\right]. \tag{7.47}$$

The eigenfrequency for this problem is $k = \sqrt{\frac{4g\sin\alpha}{3R\alpha}}$. As before we use it to introduce a dimensionless time $\tau = kt$. Derivatives with respect to this quantity will be denoted by circles. The equation of motion reads:

$$\overset{\circ\circ}{\varphi} + \sin\varphi = -b\left[\alpha\overset{\circ}{\varphi}\left(4\frac{\sin^2\alpha}{\alpha^2}\overset{\circ\circ}{\varphi}{}^2 + \frac{9}{8}\right) + \frac{\sin^2\alpha}{\alpha}\left(\overset{\circ\circ\circ}{\varphi}\cos\varphi - 4\overset{\circ}{\varphi}\overset{\circ\circ}{\varphi}\sin\varphi\right)\right], \tag{7.48}$$

where a dimensionless frictional parameter $b \to \frac{16btg^2}{9mk}$ was reassigned.

This ODE was solved for $a/l = 2/3, b = 1$ and $\varphi(0) = 0.1, \dot{\varphi}(0) = 0$, and $\ddot{\varphi}(0) = 0$. The result is shown in Fig. 7.15. As in the case of the rod pendulum there is some similarity to the phase plot for a positive dashpot damper. However, the oscillation does not seem to die out completely. The dissipation function is shown in Fig. 7.16.

In summary: the pendulum problems were only rudimentarily evaluated in this work. The large number of parameters deserves a more detailed study. A comparison

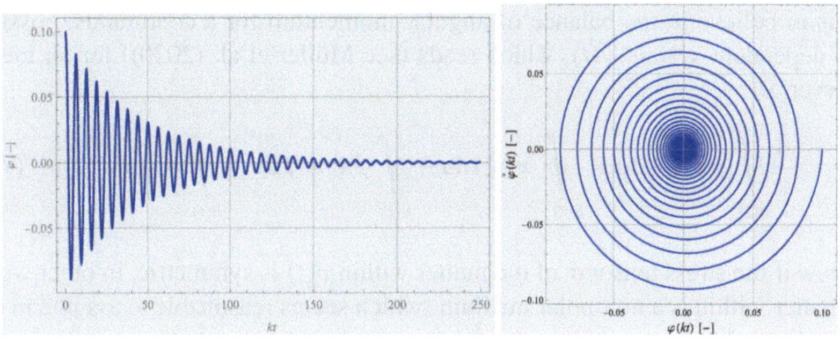

Fig. 7.15 Angle versus time and phase plots for the sector pendulum

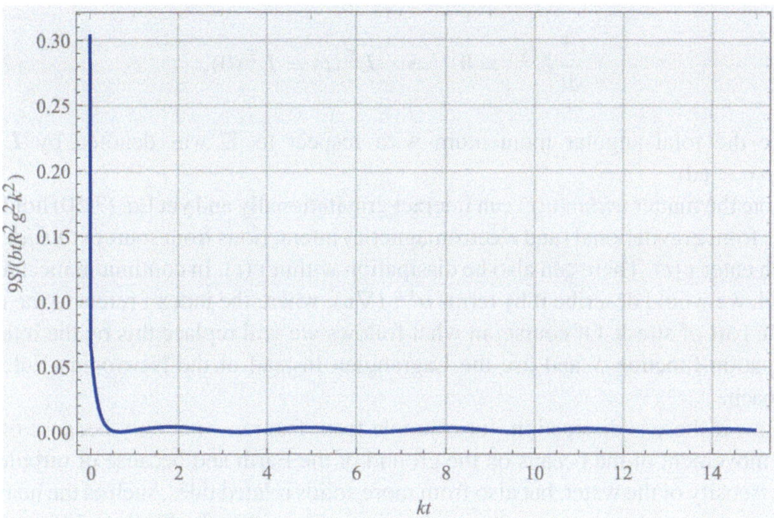

Fig. 7.16 Dissipation function versus time for the sector pendulum

with the results of the oscillation behavior of the traditional pendulum subjected to friction proportional to velocity is also pending. All of this is reserved for future studies.

7.6 Tidal Effects in the Earth-Moon System

7.6.1 Phenomenology

From the perspective of an inertial observer the moment of momentum vector of a system free of external moments remains constant in time. This is (surprisingly) independent of dissipative effects within the system. The clearest way to realize this is to believe in the balance of angular momentum for a continuous, possibly-time dependent system $v(t)$, which reads (see Müller et al. (2020)) for an inertial observer Σ:

$$\frac{\mathrm{d}}{\mathrm{d}t}\int\limits_{v(t)} \rho \boldsymbol{x} \times \boldsymbol{v} \mathrm{d}v = \oint\limits_{\partial v(t)} \boldsymbol{x} \times \boldsymbol{t} \mathrm{d}a + \int\limits_{v(t)} \rho \boldsymbol{x} \times \boldsymbol{f} \mathrm{d}v - \int\limits_{v(t)} \boldsymbol{\sigma}_{\times} \mathrm{d}v. \tag{7.49}$$

Now if the stress tensor $\boldsymbol{\sigma}$ of the matter within $v(t)$ is symmetric, in other words, the matter within is a non-polar medium (which seems reasonable to assume in case of the Earth-Moon system) its vector invariant vanishes, $\boldsymbol{\sigma}_{\times} = \sigma_{ij}\boldsymbol{e}_i \times \boldsymbol{e}_j \equiv \boldsymbol{0}$.[11] Moreover, we pretend the influence of other celestial bodies acting from outside is negligible, such that $\boldsymbol{t} = \boldsymbol{0}$ on $\partial v(t)$ and $\boldsymbol{f} = \boldsymbol{0}$. Then, indeed,

$$\frac{\mathrm{d}}{\mathrm{d}t}\boldsymbol{L}^{(\Sigma)} = \boldsymbol{0} \quad \Rightarrow \quad \boldsymbol{L}^{\Sigma}(t) = \boldsymbol{L}^{\Sigma}(0), \tag{7.50}$$

where the total angular momentum with respect to Σ was denoted by $\boldsymbol{L}^{\Sigma} = \int_{v(t)} \rho \boldsymbol{x} \times \boldsymbol{v} \mathrm{d}v$.

Note the matter *within* $v(t)$ can interact gravitationally and yet Eq. (7.50) holds: $\boldsymbol{f}$ stems from gravitational (and electromagnetic) interactions from sources *outside* and which enter $v(t)$. There can also be dissipation within $v(t)$. In continuum mechanics speech we would describe it by terms $\boldsymbol{\sigma}^{\mathrm{f}} : (\nabla \boldsymbol{v})$, where the index f refers to the non-elastic part of stress. Of course, in what follows we will replace this by the internal dissipation function S and use the Lagrangian instead of the Newtonian-Eulerian approach.

Now, if there is dissipation, for example from friction—not only because of the tidal movement of the oceans on the ground of the Earth and because of turbulence and viscosity of the water, but also from more solids related tides, such as the heaving of the continental crust or the kneading of magma within the Earth and (formerly) within the Moon—then kinetic energy will be lost. And if kinetic energy is lost, then the material velocity field $\boldsymbol{v}$ within the system is affected. This in turn means that the global movement of both Earth and the Moon is at stake: the rotational movement about their own axis (let us call it *spin*) will change, in fact, slow down. But that spin is part of the moment of momentum of the whole system, which is required to stay constant. Hence in order to do so the orbital moment of momentum has to increase, which means the Moon has to recede from Earth. In Stacey and Davis (2008), pg. 102 we read: "The inner planets and the Moon are solid and their tides dissipate rotational energy. This has the effect of slowing and eventually stopping relative

[11] There are two ways to look at the balance of moment of momentum. First, there is the fraction that "derives" is by vector multiplication from the balance of momentum (see Müller et al. (2020)). By doing so the production term $\boldsymbol{\sigma}_{\times}$ is obtained and it is concluded that there is no production of moment of momentum, if the stress tensor is symmetric. The other fraction postulates it as Euler's second law and rewrites the left hand side for the temporal change of angular momentum by using the balance of momentum (for example Zhilin (2012)). The conclusion is that the stress tensor *must* be symmetric. In any case in the end we have $\boldsymbol{\sigma}_{\times} = \boldsymbol{0}$ (for classical continua).

rotations. The large tide raised in the Moon by the Earth has completely stopped its rotation relative to the Earth. The tide raised by the Moon in the Earth is slowing the Earth's rotation with a consequent transfer of angular momentum to the lunar orbit, causing the Moon's orbit to expand at a rate that is currently about 3.8 cm/year."

Indeed, all of this are just words or hand-waving arguments, so one might think. But this first impression is deceiving. There is an immense literature on dissipative effects of the tides. Chapter 8 of Stacey and Davis (2008) and the literature cited therein may serve as a starting point for further reading on this subject. However, in this paper it is *not* our intention to repeat the arguments of the pertinent literature. We will also *not* include all the complications that are involved, such as the Earth being a spheroid, the orbital plane of the moon being inclined, the Earth's axis of rotation being tilted, the shifted barycenter of the Earth-Moon system about which both move along a (different) Kepler ellipse, etc. We would like to suggest a new approach to the dissipative tide problem. To appreciate this one should know that the use of viscoelastic models is very popular for use in simulations of the current literature, for example the Maxwell rheological body in the work of the group of Efroimsky (e.g., Efroimsky (2018)) or (quite recently) Kelvin-Voigt rheology (see Wang et al. (2022)).

Our proposition is to use the rheological model of the dissipative gyrostat and we will test it on modeling the tidal locking of the celestial satellites. We proceed to describe our approach.

7.6.2 Modeling Tidal Locking

The following deliberations are based on former work, namely (Podolskaya and Smolnikov 2010b, a). However, here we shall provide many more details.

The planar tidal locking model is illustrated in Fig. 7.17. This figure focuses on the object with internal dissipation, called m (the Moon), a circular disk of radius r_m and out-of-plane thickness t. It rotates counterclockwise at a fixed distance R_m about the common center of gravity with the object M (the Earth) called c. The extensions of M are not shown, but later we will assume that it is also a disk of radius r_M circling about c at a constant distance R_M.

This circling motion needs to be explained in more detail. To this end see Fig. 7.18. Think of Moon and Earth as a dumbbell of two different masses connected by a rod. The length of the rod is fixed and its mass is ignored. In reality the rod is provided by the gravitational action between Moon and Earth and its distance changes since we know that the Moon recedes from Earth by roughly 3 cm per year. This small change is ignored in our model and the corresponding degree of freedom is suppressed. Moreover the dumbbell turns by the angle $\theta(t)$ about the center of gravity c, where an inertial system is placed. Both Moon and Earth follow this rigid body motion. Thus their centers m and M describe a circular path. The sizes of these circles are R_m and R_M, respectively. In reality the movement is not circular but along a Kepler ellipse with very small eccentricity. For example in the case of the Moon it is roughly

0.06. This effect is also ignored in the model. Moreover, the path of the real Moon around the Earth is within a plane that is tilted by about 5 °C with respect to the ecliptic plane of the solar system. Also the equatorial plane of the Earth is inclined by 23.4 °C. Both complications are ignored in our model, which has only one plane.

Hence the time it takes for the angle $\theta(t)$ of our model to complete a full circle 2π corresponds in reality to 27.3 days. This time is known in astronomy as the synodic period of the Moon. In reality the Moon also rotates once about its own axis in that time—and therefore always turns the same side towards the Earth. This is the so-called *bound rotation* or *tidal locking*. And this is where our model departs: We do *not* assume that the locking is already in place. Rather we wish to see it develop because of internal dissipation. Therefore we assign to the Moon not only the angular velocity $\dot{\theta}(t)$ but on top of that a second one, $\dot{\varphi}(t)$, which gives rise to the additional angle $\varphi(t)$ shown in Fig. 7.17.

As it was mentioned before the dissipation will develop only in the Moon disk. Therefore the self-rotation of the Earth (our "day") does not enter the model at all. The angular velocity of the Earth with respect to the inertial system in c is $\dot{\theta}$ and this is it.

For future use let us consider the relation between the various distances that occur in Fig. 7.17. First of all we shall determine the location of the so-called barycenter c, which was characterized by $R_m = \overline{cm}$ and $R_M = \overline{cM}$ in Figs. 7.17 and 7.18. We want to relate its location to the center-to-center distance d of both objects and argue as follows: Let us pretend for the moment that the bodies m and M shown in both figures are spherical. Then we evoke Newton's second law with respect to the inertial system located in c (see Fig. 7.17) as follows:

Fig. 7.17 Moon m and Earth M circling around their common center of gravity c

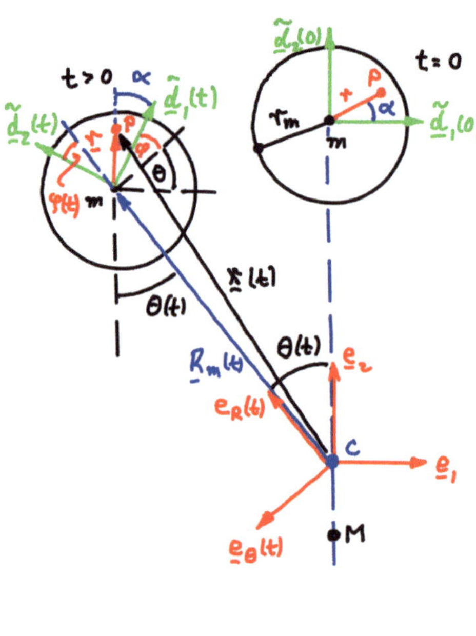

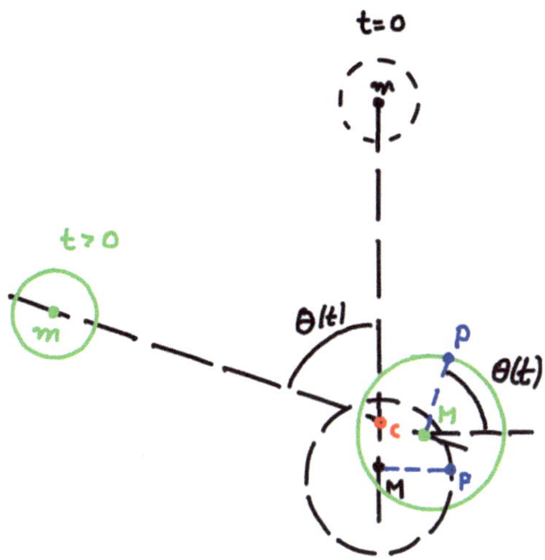

Fig. 7.18 Rigid body motion of m and M about c

$$m\ddot{\boldsymbol{R}}_m = -G\frac{mM}{d^2}\boldsymbol{e}_R\ ,\quad M\ddot{\boldsymbol{R}}_M = G\frac{mM}{d^2}\boldsymbol{e}_R, \tag{7.51}$$

where we have assumed that the gravitational force (G is Newton's constant) can be computed as if the distributed masses m and M are point masses. This is true if the mass distribution within each (now spherical) object is just a function of radial distance from the corresponding center, which is a fair assumption for celestial bodies. Moreover,

$$\dot{\boldsymbol{e}}_R = \dot{\theta}\boldsymbol{e}_\theta\ ,\quad \dot{\boldsymbol{e}}_\theta = -\dot{\theta}\boldsymbol{e}_R, \tag{7.52}$$

and therefore,

$$R_M = \frac{m}{m+M}d. \tag{7.53}$$

Note that this results holds independently of whether $\dot{\theta}$ is constant or not. If we use $m : M = 1 : 81$ and (the current value of) $d \approx 384000\,\text{km}$ we have $R_M \approx 4700\,\text{km}$, which is the famous result that the barycenter of the Moon-Earth system is located within the Earth. Let us continue to work with spherical objects for a while. The potential outside of a gravitating sphere with a total mass M distributed such that the mass density depends on radial distance from the center only enjoys full radial symmetry and is given by

$$\phi(\xi) = -\frac{GM}{\xi}\ ,\quad r_m \le \xi < \infty, \tag{7.54}$$

where r_M is the outer radius of the gravitating sphere.

Knowing this we now wish to compute the potential energy U^g of a second sphere m the mass of which shall also be radially distributed, $\rho(r)$ (see Fig. 7.19). To this

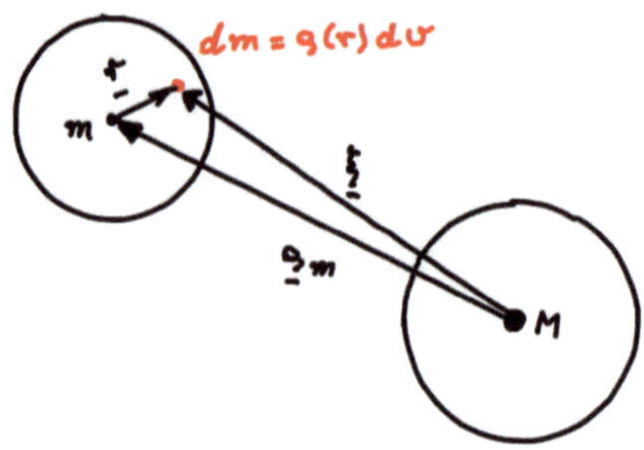

Fig. 7.19 Sphere M attracting sphere m

end we write:

$$U^{g} = \int_{v_m} \rho(r)\phi(\xi)\mathrm{d}v_m \ , \tag{7.55}$$

where v_m is the volume of the second sphere, namely the Moon. Surprisingly the integration can be performed in closed form involving elementary integrals only and by using vector relations, such as $\boldsymbol{\xi} = \boldsymbol{\rho}_m + \boldsymbol{r}$. The result reads:

$$U^{g} = \frac{GMm}{\rho_m} \ , \quad m = 4\pi \int_0^{r_m} \rho(r) r^2 \mathrm{d}r. \tag{7.56}$$

In words: the potential energy U^{g} of the sphere m is the same as the one of an equivalent point mass put in its center of gravity. And what is more: if m circles about the center of M it does not change. Furthermore note, the result is independent of how close both spheres are to each other.[12]

But what if m were a cylinder of in-plane thickness t attracted by a sphere M?[13] Then things get complicated. The integration process leads to elliptic integrals (see for a start Singh (1977) and explore the literature cited therein) and only if ρ_m is much greater then the diameters of both spheres the simple result (7.56) prevails. Things get even worse when two cylinders interact gravitationally with each other. But this was assumed in the original model shown in Figs. 7.17 and 7.18. What to do? In this case we may argue that because m is so far away from M the gravitational acceleration $\boldsymbol{g}$ that each of its material points experiences is a vector of the same magnitude and the direction of that vector is given by $-\boldsymbol{e}_R$. It is actually unnecessary to characterize $\boldsymbol{g}$ any further in terms of a gravitational law, because we will use it for nondimensionalization in our final equations. But if we wish to specify it then its magnitude will essentially be given by $\frac{GmM}{d^2} \approx \frac{GmM}{R_m^2}$. Hence within this approximation the potential energy of our model moon is given by:

[12] Of course penetration is not allowed.

[13] Assume in Fig. 7.19 that the circular cylinder m extends by $\pm t/2$ perpendicularly to the drawing plane.

$$U^{\mathrm{g}} = \frac{GMm}{R_m}, \tag{7.57}$$

which is a constant if it circles about c as we have assumed. It will therefore not contribute in the Lagrange equations. Note that if we wish to study the receding of the Moon from Earth this holds no longer true.

In hindsight this result is quite obvious. The situation is the same as in the case of the near-Earth gravitational field: if we move the center of gravity of a body parallel to the ground then its potential energy will not change. The parallel movement corresponds to a circular movement in the case of a far-away Moon. On the other hand, our study will be helpful for future work when we start with a fast spinning Moon close to Earth that is slowly receding and advancing toward tidal locking due to internal dissipation. Then there will be a contribution of its potential energy in the Lagrange equations that, most likely, must be evaluated numerically unless we go fully 3D and operate with gravitating spheres.

Let us now specify the kinetic energy T of the system shown in Fig. 7.17. There is an easy way for its computation and a more tedious one, which will be helpful when we compute the dissipation function S.

According to the parallel axis theorem we have:

$$\begin{aligned} T &= \frac{J_m}{2}\left(\dot{\theta}+\dot{\varphi}\right)^2 + \frac{m}{2}\left(R_m\dot{\theta}\right)^2 + \frac{J_M}{2}\dot{\theta}^2 + \frac{M}{2}\left(R_M\dot{\theta}\right)^2 , \\ J_m &= \frac{m}{2}r_m^2 , \quad m = \pi r_m^2 t\rho_0 , \quad J_M = \frac{M}{2}r_M^2 , \quad M = \pi r_M^2 t\rho_0 , \end{aligned} \tag{7.58}$$

where J_m and J_M are the moments of inertia of two (homogeneous) cylinders of outer radius r_m and r_M (and constant mass density ρ_0 which could differ between m and M) with respect to the center of gravity, respectively. They are responsible for the spin contributions of the Moon and of the Earth. Note that in the case of the Earth we take only $\dot{\theta}$ and not an additional (daily) spin into account—as it was mentioned before—but in the case of the Moon we do, $\dot{\varphi}$. All rotations are assumed to occur counterclockwise, which is important for the direction of the d'Alembert accelerations.

A more elaborate calculation of this result makes use of the various unit vectors shown in Fig. 7.17. For future use we note the following vector relations:

$$\begin{aligned} \boldsymbol{x}(t) &= \left[R_m \sin\varphi(t) + r\cos\alpha\right]\tilde{\boldsymbol{d}}_1(t) + \left[R_m\cos\varphi(t) + r\sin\alpha\right]\tilde{\boldsymbol{d}}_2(t) , \\ \dot{\tilde{\boldsymbol{d}}}_1(t) &= \left(\dot{\theta}+\dot{\varphi}\right)\tilde{\boldsymbol{d}}_2(t) , \quad \dot{\tilde{\boldsymbol{d}}}_2(t) = -\left(\dot{\theta}+\dot{\varphi}\right)\tilde{\boldsymbol{d}}_1(t) . \end{aligned} \tag{7.59}$$

It should be noted that $\tilde{\boldsymbol{d}}_{1,2}(t)$ are the images of the unit vectors $\boldsymbol{d}_{1,2}$ that are firmly attached to the Moon disk. From the perspective of the Moon the latter are time-independent. They will become important later in context with the dissipation function S. But here we compute the kinetic energy T from the perspective of the inertial system in c and, therefore, we must use the time-dependent vectors $\tilde{\boldsymbol{d}}_{1,2}(t)$. It is now a matter of performing the integrations correctly to show that the contribution of the Moon to the total kinetic energy T is

$$T_m = \frac{t}{2} \int_{r=0}^{r_m} \int_{\alpha=0}^{2\pi} \rho_0 \, \dot{\boldsymbol{x}} \cdot \dot{\boldsymbol{x}} \; r \mathrm{d}r \mathrm{d}\alpha = \frac{J_m}{2} \left(\dot{\theta} + \dot{\varphi}\right)^2 + \frac{m}{2} \left(R_m \dot{\theta}\right)^2 . \tag{7.60}$$

The contribution of the Earth can be calculated analogously. However, there is only the rigid body contribution as explained above.

Similarly we calculate the angular momentum with respect to the inertial observer in c for the Moon:

$$\begin{aligned} \boldsymbol{L}_m^c = \int_{v_m} \boldsymbol{x} \times (\rho_0 \dot{\boldsymbol{x}}) \, \mathrm{d}v = \\ \rho_0 t \int_{r=0}^{r_m} \int_{\alpha=0}^{2\pi} \boldsymbol{x} \times \dot{\boldsymbol{x}} \; r \mathrm{d}r \mathrm{d}\alpha \; \boldsymbol{e}_z = \left[m R_m^2 \dot{\theta}(t) + J_m \left(\dot{\theta}(t) + \dot{\varphi}(t) \right) \right] \boldsymbol{e}_z , \end{aligned} \tag{7.61}$$

where $\boldsymbol{e}_z$ is the unit vector perpendicular to the drawing plane. There are two contributions to angular momentum, one from the orbital movement and one due to the spin of the Moon, which involves two angular velocities. The situation is analogous for the Earth, with the exception that there is no $\dot{\varphi}(t)$ contribution. The total angular momentum is therefore:

$$\boldsymbol{L}^c(t) = \left[\left(m R_m^2 + M R_M^2 + J_m + J_M \right) \dot{\theta}(t) + J_m \dot{\varphi}(t) \right] \boldsymbol{e}_z . \tag{7.62}$$

There are no external moments acting on the system. The angular momentum is therefore conserved, and we conclude that we can express $\dot{\theta}(t)$ by $\dot{\varphi}(t)$ and initial data ω_0:

$$\dot{\theta}(t) = \omega_0 - \varepsilon \dot{\varphi}(t) \; , \quad \varepsilon = \frac{J_m}{m R_m^2 + M R_M^2 + J_m + J_M} \; , \quad \omega_0 = \dot{\theta}(0) + \varepsilon \dot{\varphi}(0). \tag{7.63}$$

In other words, we have just one variable, $\varphi(t)$. ε is a small dimensionless parameter. If we assume the current distances R_m =379300 km, R_M =4700 km, r_m =1737 km, r_M =6378 km and take the mass ratio $M/m = 81$ then $\varepsilon = 0.00001$. If we believe the giant impact theory then it was 24000 km away from the Earth at the time of its creation (Lunar 2025). Then $\varepsilon = 0.0004$, which is still a small number. This is now used to rewrite the expression for the kinetic energy T from (7.58) and in preparation for the Lagrange equations we compute[14]

[14] The second term on the left does not contribute, because T does not depend on φ.

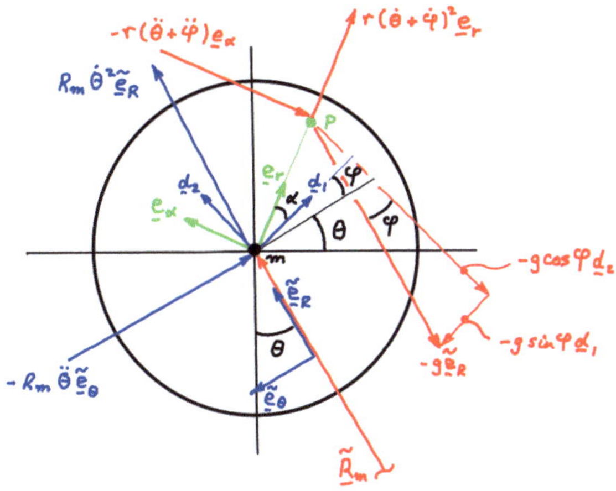

Fig. 7.20 D'Alembert accelerations and coordinates for the Moon m

$$\frac{\mathrm{d}}{\mathrm{d}t}\left(\frac{\partial T}{\partial \dot{\varphi}}\right) - \frac{\partial T}{\partial \varphi} = (1-\varepsilon)\, J_m \ddot{\varphi}, \tag{7.64}$$

a relatively simple result. This is not so for the right hand side of the Lagrange equations. The term $= -\frac{\partial S}{\partial \dot{\varphi}}$ is complicated. First the dissipation function. We note from Fig. 7.20 for the accelerations:

$$\begin{aligned}
&\boldsymbol{w} = \\
&\left[R_m \sin\varphi \dot{\theta}^2 + r\left(\ddot{\theta}+\ddot{\varphi}\right)\sin\alpha + R_m\cos\varphi\ddot{\theta} + r\left(\dot{\theta}+\dot{\varphi}\right)^2\cos\alpha - g\sin\varphi\right]\boldsymbol{d}_1 + \\
&\left[R_m \cos\varphi \dot{\theta}^2 - r\left(\ddot{\theta}+\ddot{\varphi}\right)\cos\alpha - R_m\sin\varphi\ddot{\theta} + r\left(\dot{\theta}+\dot{\varphi}\right)^2\sin\alpha - g\cos\varphi\right]\boldsymbol{d}_2.
\end{aligned} \tag{7.65}$$

It is noteworthy that the R_m-force terms hold for all material points within m. For convenience they were assigned to the center of m in Fig. 7.20 (blue arrows): in order to reach the point P from there we have to add $r\cos\theta\tilde{\boldsymbol{e}}_R$ and $r\sin\theta\tilde{\boldsymbol{e}}_\theta$, $r \in [0, r_m]$. But these contributions are much smaller than those from R_m and therefore neglected. Differentiation with respect to t and observing Eq. (7.63) allows us to calculate the dissipation function S:

$$
\begin{aligned}
S = \frac{b\pi r_m^2 t}{4}\Bigg[& 2\dot{\varphi}^2\Bigg(g^2 - 2gR_m\left(\omega_0 - \varepsilon\dot{\varphi}\right)^2 + 2\left(1-\varepsilon\right)^2 r_m^2\ddot{\varphi}^2 + \\
& R_m^2\left(\left(\omega_0 - \varepsilon\dot{\varphi}\right)^4 + \varepsilon^2\ddot{\varphi}^2\right)\Bigg) + \\
& \left(\left(1-\varepsilon\right)^2 r_m^2 + 2\varepsilon^2 R_m^2\right)\left(4\left(\omega_0 - \varepsilon\dot{\varphi}\right)^2\ddot{\varphi}^2 + \dddot{\varphi}^2\right) + \\
& 4\dot{\varphi}\left(2\left(1-\varepsilon\right)^2 r_m^2\left(\omega_0 - \varepsilon\dot{\varphi}\right)\ddot{\varphi}^2 + \right. \\
& \quad \left. R_m\left(g\varepsilon\dddot{\varphi} - \varepsilon R_m\left(\omega_0 - \varepsilon\dot{\varphi}\right)\left(2\varepsilon\ddot{\varphi}^2 + \left(\omega_0 - \varepsilon\dot{\varphi}\right)\dddot{\varphi}\right)\right)\right)\Bigg]
\end{aligned}
\tag{7.66}
$$

Differentiation yields:

$$
\begin{aligned}
\frac{\partial S}{\partial\dot{\varphi}} = \frac{bm}{\rho_0}\Bigg[& g^2\dot{\varphi} + 2\left(1-\varepsilon\right)^3 r_m^2\left(\dot{\varphi}\left(1-\varepsilon\right) + \omega_0\right)\ddot{\varphi}^2 + \\
& gR_m\Bigg(-2\omega_0^2\dot{\varphi} + 6\varepsilon\omega_0\dot{\varphi}^2 + \varepsilon\left(-4\varepsilon\dot{\varphi}^3 + \dddot{\varphi}\right)\Bigg) + \\
& R_m^2\Bigg(\omega_0^4\dot{\varphi} - 6\varepsilon\omega_0^3\dot{\varphi}^2 + \varepsilon\omega_0^2\left(12\varepsilon\dot{\varphi}^3 - \dddot{\varphi}\right) - \\
& \quad 2\varepsilon^2\omega_0\left(5\varepsilon\dot{\varphi}^4 + \left(1+2\varepsilon\right)\ddot{\varphi}^2 - 2\dot{\varphi}\dddot{\varphi}\right) + \\
& \quad \varepsilon^2\dot{\varphi}\left(3\varepsilon^2\dot{\varphi}^4 + \left(1+2\varepsilon\right)^2\ddot{\varphi}^2 - 3\varepsilon\dot{\varphi}\dddot{\varphi}\right)\Bigg)\Bigg]
\end{aligned}
\tag{7.67}
$$

We now follow Eqs. (7.1) and $(7.9)_2$ and combine Eqs. (7.64) and (7.67). Then we normalize time t by ω_0, hence $\tilde{t} = \omega_0 t$. Differentiations with respect to the dimensionless time are denoted by circles. The result reads:

$$
\begin{aligned}
\overset{\circ\circ}{\varphi} = -\frac{2bg^2}{\rho_0 r_m^2\left(1-\varepsilon\right)\omega_0}\Bigg[& \overset{\circ}{\varphi} + 2\left(1-\varepsilon\right)^3 C_1\left(\overset{\circ}{\varphi}\left(1-\varepsilon\right) + 1\right)\overset{\circ\circ}{\varphi}{}^2 + \\
& C_2\Bigg(-2\overset{\circ}{\varphi} + 6\varepsilon\overset{\circ}{\varphi}{}^2 + \varepsilon\left(-4\varepsilon\overset{\circ}{\varphi}{}^3 + \overset{\circ\circ\circ}{\varphi}\right)\Bigg) + \\
& C_3\Bigg(\overset{\circ}{\varphi} - 6\varepsilon\overset{\circ}{\varphi}{}^2 + \varepsilon\left(12\varepsilon\overset{\circ}{\varphi}{}^3 - \overset{\circ\circ\circ}{\varphi}\right) - \\
& \quad 2\varepsilon^2\left(5\varepsilon\overset{\circ}{\varphi}{}^4 + \left(1+2\varepsilon\right)\overset{\circ\circ}{\varphi}{}^2 - 2\overset{\circ}{\varphi}\overset{\circ\circ\circ}{\varphi}\right) + \\
& \quad \varepsilon^2\overset{\circ}{\varphi}\left(3\varepsilon^2\overset{\circ}{\varphi}{}^4 + \left(1+2\varepsilon\right)^2\overset{\circ\circ}{\varphi}{}^2 - 3\varepsilon\overset{\circ}{\varphi}\overset{\circ\circ\circ}{\varphi}\right)\Bigg)\Bigg].
\end{aligned}
\tag{7.68}
$$

Three dimensionless constants were defined:

$$
C_1 = \frac{r_m^2\omega_0^4}{g^2}\,,\quad C_2 = \frac{R_m\omega_0^2}{g}\,,\quad C_3 = \frac{R_m^2\omega_0^4}{g^2}. \tag{7.69}
$$

If we take today's distances and angular velocities for the Moon, we have $g = \frac{GM}{R_m^2} \approx 0.0028 \frac{\mathrm{m}}{\mathrm{s}^2}$, $\omega_0 \approx \frac{2\pi}{27.3\mathrm{d}} = 2.7 \cdot 10^{-6} \frac{1}{\mathrm{s}}$ and therefore $C_1 \approx 2 \cdot 10^{-11}$, $C_2 \approx 0.001$, $C_3 \approx 10^{-6}$. Therefore (7.68) can be easily integrated in closed form:

$$\overset{\circ\circ}{\varphi} = -\beta \overset{\circ}{\varphi} \quad \Rightarrow \quad \overset{\circ}{\varphi}(\tilde{t}) = \overset{\circ}{\varphi}(0) \exp\left(-\beta \tilde{t}\right) \ , \ \ \beta = \frac{2bg^2}{\rho_0 r_m^2 \left(1 - \varepsilon\right) \omega_0} > 0. \tag{7.70}$$

We reach the conclusion that the additional spin of the Moon will exponentially decay in time. This is the tidal locking effect.

7.7 Conclusions

The purpose of this second part of our work was mostly pedagogical:

- The concept of Lagrange equations with internal dissipation was given an intuitive meaning by studying various examples of pendulums and the locking phenomenon of celestial bodies. They were modeled by the rheological model of the dissipative gyrostat.
- The internal dissipation function S, which is always positive, was explicitly calculated for the aforementioned cases.
- In the case of the rod and the hammer pendulum the dissipation function S was also obtained by using concepts of elementary beam theory. Dynamic (!) normal, shear force and moment diagrams were constructed.
- The solution to all problems contains various (dimensionless) parameters. However, only selected case studies were performed. A more systematic analysis is still pending.
- The tidal locking effect of the Moon was explained by using the dissipative gyrostat model where the Moon stays at a constant distance. The full problem where the Moon is also receding is still waiting for a thorough treatment. A first attempt in this direction was made in Podolskaya and Smolnikov (2010c).

Acknowledgements The authors want to thank Messrs. Erik Brauer, B.Sc. and Sanjay Mahamagamoorthy, B.Sc. for helpful discussions on the topic of this paper. Three reviewers provided additional helpful comments.

References

Lunar distance. Wikipedia. https://en.wikipedia.org/wiki/Lunar_distance. Accessed 15 Jan 2025
Efroimsky M (2018) Tidal viscosity of Enceladus. Icarus 300:223–226
Enns RH, McGuire GC (2001) Nonlinear physics with Mathematica for scientists and engineers. Birkhauser, Boston

Müller WH, Rickert W, Vilchevskaya EN (2020) Thence the moment of momentum. ZAMM-J Appl Math Mech/Zeitschrift für Angewandte Mathematik und Mechanik 100(5):e202000117

Podolskaya E, Smolnikov BA (2010a) Evolutionary dynamics of two linked bodies' plane motion. In: Evtim Zahariev MC (ed) Euromech colloquium 515 advanced applications and perspectives of multibody dynamics. Curran Associates, Inc., Red Hook, NY, pp 72–73

Podolskaya EA, Smolnikov BA (2010b) Плоское эволюционное движение связки двух тел. In: Материалы лучших докладов XXXVIII Недели науки СПБГПУ. СПБ.: Изд-во Политехн. ун-та, pp 124–126b

Podolskaya EA, Smolnikov BA (2010c) Эволюционная динамика диссипативного диска, вращающегося в гравитационном поле. In: XXXIX Неделя науки СПБГПУ: материалы международной научно-практической конференции. Часть B. СПБ.: Изд-во Политехн. ун-та, pp 113–115

Singh SK (1977) Gravitational attraction of a vertical right circular cylinder. Geophys J Int 50(1):243–246

Smolnikov B, Belyaev AK (2008) Evolutional dynamics and stability of dissipative solids. Acta Mech 195(1):365–377

Smolnikov BA (2008a) Фрактальное твердое тело в задачах аналитической механики. In: A.L. Smirnova, E.F. Zhigalko (eds.) Труды семинара «Компьютерные методы в механике сплошной среды». СПБ: изд-во СПБГУ, pp 17–32

Smolnikov BA (2008b) Эволюционная динамика маятниковых систем. Теория механизмов и машин 6(1):41–47

Stacey FD, Davis PM (2008) Physics of the earth. Cambridge University Press

Wang Y, Mora P, Liang Y (2022) A direct numerical verification of tidal locking mechanism using the discrete element method. Celest Mech Dyn Astron 134(5):42

Zhilin PA (2012) Рациональная механика сплошных сред (Rational Continuum Mechanics, in Russian). Санкт-Петербург ИздательствоПолитехнического университета, St. Petersburg

Chapter 8
On Jump Conditions at Phase Boundaries for Ordered and Disordered Phases

Wolfgang Dreyer

Abstract This is a study on jump conditions across the interface between two adjacent phases. The interface behaves as a free boundary, and in sharp interface models jump conditions are used to determine the values of thermodynamic fields at the free boundaries. In this study the jump conditions are derived from balance equations for singular surfaces that do not have singular lines, i.e., triple junctions are not considered here. At first we present the most general form of jump conditions to give a general framework, from where we consider various special cases with a focus on the influence of mechanical fields on the interfacial processes. The special cases include the Hoffmann/Cahn capillarity vector theory and jump conditions for interfaces where order/disorder transitions are involved. Furthermore we discuss interfacial chemical reaction laws, and in particular the creation and annihilation of vacancies at a liquid/solid interface.

8.1 Introduction

An example for the need of interfacial jump conditions is given by stress-assisted diffusion in single-crystal Gallium Arsenide (GaAs) that leads to the formation and growth of unwanted liquid droplets in a solid surrounding, Dreyer and Duderstadt (2003a), Wenzl et al. (1990, and 1993). The modeling of this thermodynamic process relies in its essential parts on jump conditions at the liquid/solid interface. There is already an extensive literature on jump conditions, see in particular Alexander and Johnson (1985), Müller (1985), Moeckel (1975), Grauel (1980), Johnson (1987), Alt (2001), and Gurtin and Voorhees (1996) for a general thermodynamic treatment, Liu (1992), where pure mechanical jump conditions are derived and Meirmanov (1992) where exclusively the Stefan jump condition is explored. However, the jump conditions that appropriately represent the phenomena of the described example

W. Dreyer (deceased)
Formerly with Weierstrass-Institut für Angewandte Analysis und Stochastik, Mohrenstr. 39, D-10117, Berlin, Germany
e-mail: wolfgang.h.mueller@tu-berlin.de (Wolfgang H. Müeller, serving as curator and corresponding editor)

J. Fuhrmann et al. (eds.), *Advances in Continuum Physics*, Advanced Structured Materials 238, https://doi.org/10.1007/978-3-031-93918-1_8

are not among the results from the literature, because these ignore order/disorder transitions and the creation and annihilation of vacancies are also not considered.

Here the general objective is a study on the interface between two adjacent bodies. Different kinds of interfaces may appear: (i) Phase boundaries: The interface separates two phases of a single substance, say liquid and gas. A more complex example is as follows. The interface separates two phases consisting of a mixture of several constituents. Some of them may cross the interface, while it behaves impermeable for other constituents. Interfacial chemical reactions may also happen, so that incoming constituents are created or annihilated on the interface. (ii) Contact areas: Two adjacent bodies are made from different materials, say copper in contact with steel. In this case, often only mechanical forces are transmitted across the interface. (iii) Membranes and shells: It may happen, that the interface between two adjacent bodies is in fact a thin body. In the zero thickness limit, the thin body can usually be represented by a surface, which is called *membrane* or *shell*, respectively. A rubber balloon which separates two gases provides a well-known example for a membrane.

In this study, we are especially interested in phase boundaries, however, many of the calculated relations are also valid for contact areas and thin bodies. In the following we will use the general notion *interface*. Sometimes, the interface is called a *singular surface*, because in general the interfacial limit values of the thermodynamic fields of the adjacent bodies are discontinuous, when they are formed from different sides of the interface.

There are different strategies to derive jump conditions. They can be established by minimization of the Gibbs free energy, see Alexander and Johnson (1985), for a quite general treatment in case that fluids and only disordered solids are involved, and Dreyer and Duderstadt (2003a), where a plane interface between a liquid and the ordered solid, viz., semi-insulating GaAs, is considered. In the current study, the interfacial jump conditions rely on global equations of balance for a test volume $V = V_+ \cup V_-$, where the partial volumes V_+ and V_- are occupied by two phases of a material body consisting of $a_c^{+/-}$ chemically different constituents. For disordered solids, the same strategy is explored by Gurtin and Voorhees (1996, 1998); however, these authors additionally introduce the so-called configurational forces. These quantities are avoided in the following.

The thermodynamic constitutive theory of chemically reacting mixtures is not among the objectives of this study. However, even so we need some thermodynamic relations for the bulk materials: the Gibbs equation, the Gibbs/Duhem equation, and the relation between energy and entropy flux. The underlying thermodynamic mixture theory is carefully and extensively developed and discussed in Müller (1985 and 2001) by Ingo Müller. The few necessary generalizations to mixtures of solids are explicitly given in the current study.

In the past, the anisotropy of interfaces is often described by the capillarity vector ξ according to Hoffmann and Cahn (1972, 1974). In Sect. 13.2, we relate ξ to the normal component of the surface stress vector and we show that its appearance induce a nonconserved interfacial angular momentum. The Cahn/Hoffmann jump conditions ignore interfacial elasticity and deviatoric stresses of the bulk materials. Both phenomena are included in the corresponding jump conditions of the current study.

In contrast to disordered solids, there are privileged lattice sites in ordered solids, and this phenomenon is described by a partition of the crystal lattice into various sublattices. Interfacial jump conditions must determine the distribution of the constituents on the sublattices at a, say, liquid/ordered solid interface. They are thus different from jump conditions that describe a liquid/disordered solid interface. Both cases are discussed in detail in Sects. 15.3 and 15.4.

8.2 The Content of the Classical Gibbs/Thomson Law

The most well-known jump conditions concern the interface between a liquid droplet and vapor, and determine (i) the pressure jump at the interface and (ii) the vapor pressure by the curvature of the droplet. The equations are known as the Gibbs/Thomson law, Thomson (1870), Gibbs (1940). In order to illustrate the content of the Gibbs/Thomson law we consider the system from Fig. 8.1. A pressure vessel is closed by a movable piston and contains a spherical liquid droplet with radius r consisting of a single substance in contact with its vapor at uniform temperature T_0 and constant outer pressure p_0.

The conditions of (unstable) equilibrium are, see Müller (1985, 2001), Dreyer and Duderstadt (2003b) for details,

$$p_+ = p_0, \quad p_- = p_+ + \frac{2\sigma}{r}, \quad g_+(T_0, p_+) = g_-(T_0, p_-). \tag{8.1}$$

Here σ is the surface tension and the functions $g_{+/-}$ denote the specific Gibbs free energies of vapor and liquid, respectively. In Sects. 13.1 and 15.3.2, the laws $(8.1)_2$ and $(8.1)_3$ will be rederived.

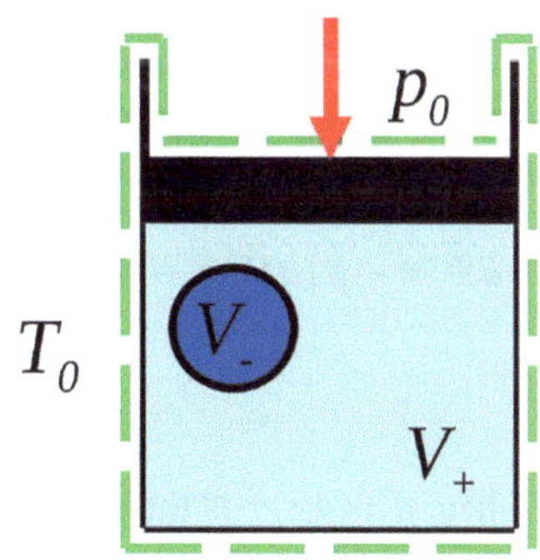

Fig. 8.1 Device for demonstration of the Gibbs/Thomson law

The conditions $(8.1)_{1,2}$ describe mechanical equilibrium, while $(8.1)_3$ determines thermodynamic equilibrium. The condition $(8.1)_2$ relates the pressure difference to the curvature of the droplet and was first formulated by Laplace, Laplace (1806). In $(8.1)_3$, we may eliminate the liquid pressure by means of $(8.1)_2$. If we consider an incompressible liquid, i.e., $\partial v/\partial p = 0$, where v is the specific volume of the liquid, we obtain due to $(\partial g/\partial p) = v$

$$g_+(T_0, p_+) - g_-(T_0, p_+) = v_- \frac{2\sigma}{r}. \tag{8.2}$$

This is the classical Gibbs Thomson law, which can be used to calculate the vapor pressure p_+ as a function of temperature and curvature of the droplet.

It is important to compare $(8.1)_3$ with (8.2). Both laws have the same content. However, the pressure arguments in $(8.1)_3$ correspond to the vapor phase and liquid phase, respectively, while in (8.2) the vapor pressure appears in g_+ as well as in g_-. The pressure difference between liquid and vapor is explicitly given on the right-hand side of (8.2).

A simple kinetic consideration allows even the calculation of a growth law for the droplet, Dreyer and Duderstadt (2003b). It reads

$$\begin{aligned} \frac{dr}{dt} = v_- \sqrt{\frac{kT_0}{2\pi m}} \frac{p_V(T_0)}{kT_0} \exp\left(\frac{m(g_+(T_0, p_+) - g_-(T_0, p_+))}{kT_0}\right) \times \\ \times \left(1 - \exp\left(\frac{m(g_+(T_0, p_+) - g_-(T_0, p_-))}{kT_0}\right)\right) \end{aligned} \tag{8.3}$$

The newly introduced quantities are the atomic mass, m, the Boltzmann constant, k, and the vapor pressure, p_V, that corresponds to a flat interface. The vapor pressure p_V solves (8.2) for $r \to \infty$.

For a given initial radius $r(0)$, the growth law determines the evolution of the droplet. For given T_0, p_0, we can calculate from $g_+(T_0, p_0) - g_-(T_0, p_0) = v_- 2\sigma/r_c$ a critical radius r_c, and for $r(0) < r_c$, the droplet will shrink while it will grow for $r(0) > r_c$.

Finally, we give a simple calculation that relates the surface tension to the free energy density of the interface. To this end, we calculate the necessary work to change the total volume of the system in an isothermal process. This work is equal to a change of the total free energy Ψ. We write

$$\begin{aligned} d\Psi = -p_+ dV = -p_+ d(V_+ + V_-) = -p_+ dV_+ - p_- dV_- + (p_- - p_+) dV_- = \\ = d\Psi_+ + d\Psi_- + \sigma da, \end{aligned} \tag{8.4}$$

where $a = 4\pi r^2$ is the surface of the droplet. We conclude that the total change of the free energy consists of three parts corresponding to the two bulk phases and to the interface. In particular, we may identify $d\Psi_s = \sigma da$, where $d\Psi_s$ gives the isothermal change of the interfacial free energy Ψ_s.

A slight generalization of this result can be obtained as follows: Let T_s and S_s denote surface temperature and surface entropy, respectively, and let us assume that a change of T_s gives rise to a change of the surface free energy in the usual manner, so that there results an interfacial Gibbs equation, viz.

$$d\Psi_s = -S_s dT_s + \sigma da. \tag{8.5}$$

There follows

$$S_s = -\frac{\partial \Psi_s}{\partial T_s}, \quad \sigma = \frac{\partial \Psi_s}{\partial a} \quad \text{and the integrability condition} \quad \frac{\partial S_s}{\partial a} = -\frac{\partial \sigma}{\partial T_s}. \tag{8.6}$$

Furthermore, let us consider a small surface element a with a uniform distribution of temperature T_s, and let us assume

$$\Psi_s = \psi_s(T_s)a, \quad S_s = s_s(T_s)a. \tag{8.7}$$

We conclude

$$\sigma = \psi_s, \quad s_s = -\frac{\partial \sigma}{\partial T_s}, \tag{8.8}$$

and this form of the interfacial contribution to the free energy is the historical reason to call the introduced *surface tension* σ also *interfacial free energy density*.

The described model is obviously too simple in many circumstances. For example, rubber membranes do not behave in this simple way, see Dreyer (1981), Dreyer et al. (1982) for details. Furthermore, interfaces that represent phase boundaries are also not adequately described by this model if anisotropic solids are involved. In Sect. 8.12, we will generalize this result to anisotropic and elastic deformable surfaces.

Atomistic models for the calculation of the surface tension are found to be in Dreyer (1981) for elastic membranes, in Ono and Kondo (1960) for fluid/fluid interfaces and in Theil (2002), where the calculation starts from a lattice model.

8.3 A Piece of Thermodynamics for Liquids and Solids

8.3.1 Euler and Lagrange Description of the Variables

We consider a body B whose volume $V(t)$ may depend on time t. We may use a fixed Cartesian coordinate system to indicate a space point, which at some time t is occupied by a *material point* P of B, by its Cartesian coordinates $x = (x^i) = (x^1, x^2, x^3)$. The notion *material point* means the smallest units of B that cannot further be resolved on the chosen space scale.

If the body B is a mixture of a_c constituents, the variables to describe the thermodynamic state of B are the fields

$$\begin{aligned}
\rho_a &= \rho_a(t,x), \quad a \in \{1,2,...,a_c\}, && \text{partial mass densities,}\\
\upsilon^i &= \upsilon^i(t,x), && \text{barycentric velocity of the mixture,}\\
u &= u(t,x), && \text{specific internal energy of the mixture.}
\end{aligned} \tag{8.9}$$

The total mass density of the mixture, ρ, is given by the sum over the partial mass densities, and for some purposes, it is useful to calculate the mixture velocity, υ^i, from the velocities of the constituents, υ^i_a. Thus, we write

$$\rho = \sum_{a=1}^{a_c} \rho_a \quad \text{and} \quad \upsilon^i = \sum_{a=1}^{a_c} \frac{\rho_a}{\rho} \upsilon^i_a. \tag{8.10}$$

The given description of B is called spatial or Euler description, and we call the pair (t, x) Euler coordinates.

In order to describe the deformation of the considered body, it is useful to indicate any material point P of B at a fixed reference time t_0 by its coordinates $X = (X^i) = (X^1, X^2, X^3)$ with respect to a rectangular Cartesian coordinate system.

The motion of the point P, is then described by the field $\chi(t, X) = \chi^i(t, X) = (\chi^1(t, X), \chi^2(t, X), \chi^3(t, X))$, which gives the coordinates of P at time $t > t_0$:

$$x^i = \chi^i(t, X). \tag{8.11}$$

The field χ is called *motion*, and its first derivatives

$$\hat{\upsilon}^i = \frac{\partial \chi^i}{\partial t} \quad \text{and} \quad F^{ij} = \frac{\partial \chi^i}{\partial X^j} \quad \text{with} \quad J = \det(F^{ij}) \neq 0 \tag{8.12}$$

are the *velocity* and the *deformation gradient*, respectively. The condition on the Jacobian J guarantees that (8.11) is invertible.

Later on, we will need an identity, that reads

$$\frac{\partial}{\partial X^k}(J \overset{-1}{F^{ki}}) = 0. \tag{8.13}$$

This identity follows from the definitions (8.12) by a straightforward calculation.

If the body B is a mixture consisting of a_c constituents, the variables are the fields of motion, partial mass densities and internal energy, respectively,

$$x^i = \chi^i(t, X), \quad \rho_a = \hat{\rho}_a(t, X), \quad u = \hat{u}(t, X). \tag{8.14}$$

The description of the body B by fields that map the coordinates X into some function space, is called material or Lagrange description, and we call the pair (t, X) Lagrange coordinates.

We may use the field of motion to relate the Euler description to the Lagrange description by

$$\hat{\rho}_a(t,X) = J\rho_a(t,\chi(t,X)), \quad \hat{v}^i(t,X) = v^i(t,\chi(t,X)), \qquad \hat{u}(t,X) = Ju(t,\chi(t,X)). \tag{8.15}$$

The Lagrange description is quite natural for solids, whereas, in liquids and gases, the Euler description is more convenient. In this study on interfacial boundary conditions, both descriptions are needed, because it often happens that the adjacent substances at an interface are liquid and solid.

8.3.2 Gibbs Equation for Fluids and Solids as Bulk Materials

There are further important quantities to describe properties of mixtures. These are: temperature, T, specific entropy, s, free energy, $\psi = u - Ts$, and the chemical potentials of the constituents, μ_a, Müller (1985, 2001).

Furthermore, there are three measures of stress, viz. the Cauchy stress, σ^{ik}, and the first and second Piola Kirchhoff stresses, respectively, $\hat{\sigma}^{ik}$ and t^{ik}:

$$\sigma^{ik} = \sigma^{ki}, \quad \hat{\sigma}^{ik} = J\overset{-1}{F}{}^{ij}\sigma^{jk}, \quad t^{ik} = J\overset{-1}{F}{}^{ij}\overset{-1}{F}{}^{kl}\sigma^{jl}. \tag{8.16}$$

The Cauchy stress appears naturally in the Euler description, the first Piola Kirchhoff stress is a measurable quantity, and appears naturally in the Lagrange description. Finally, the second Piola Kirchhoff stress is needed to formulate constitutive laws, which relate the stress to the variables, Truesdell and Toupin (1960). Let us decompose the Cauchy stress into its trace and the traceless part, the latter one is indicated by angle brackets around the indices. We write

$$\sigma^{ik} = -p\delta^{ik} + \sigma^{<ik>} \quad \text{with} \quad p = -\frac{1}{3}\sigma^{nn}, \tag{8.17}$$

and we call the quantity p the pressure.

There also are various measures of the deformation of a body. An important one is the right Cauchy Green tensor, C^{ik}, and its unimodular restriction, which we denote by c^{ik}:

$$C^{ik} = F^{mi}F^{mk} \quad \text{and} \quad c^{ik} = J^{-3/2}C^{ik} \quad \text{so that} \quad \det(c^{ik}) = 1. \tag{8.18}$$

All these quantities are related to each other by the Gibbs equation, which read in fluids as well as in thermoelastic solids

$$d\rho\psi = -\rho s dT + \frac{1}{2} J^{-1/3} t^{ik} dc^{ik} + \sum_{a=1} \mu_a d\rho_a. \tag{8.19}$$

The Gibbs equation implies

$$\rho s = -\frac{\partial \rho\psi}{\partial T}, \quad t^{ik} = 2J^{1/3} \frac{\partial \rho\psi}{\partial c^{ik}}, \quad \text{and} \quad \mu_a = \frac{\partial \rho\psi}{\partial \rho_a}. \tag{8.20}$$

Furthermore, there holds the Gibbs/Duhem equation

$$p = -\rho\psi + \sum_{a=1} \mu_a \rho_a. \tag{8.21}$$

We conclude that the knowledge of the free energy density $\rho\psi(T, \rho_1, ..., \rho_{a_c}, c^{ik})$ is sufficient to calculate all the introduced quantities. In this study, we assume the free energy density to be given.

In a mixture of isotropic fluids, the free energy does not depend on c^{ik}, and the second Piola/Kirchhoff stress drops out of the Gibbs equation. Thus, for a fluid phase, the knowledge of the free energy density $\rho\psi(T, \rho_1, ..., \rho_{a_c})$ is sufficient to relate the introduced quantities to the variables.

The proof of the Gibbs equation and of the Gibbs/Duhem equation for fluid mixtures and for those solids that only consist of a single substance can be found in Müller (1985). The generalization to a solid mixture, viz. Eqs. (8.19) and (8.21), has not been published elsewhere, but it can be derived along the strategies that are outlined by Müller (1985).

8.3.3 *Entropy Flux/Heat Flux Relation*

Later on, we need a further result of thermodynamics of mixtures, which gives the relation between the entropy flux, ϕ^k, and the heat flux, q^k. In a chemically reacting mixture of fluids and even in solids, which may be exposed to the influence of mechanical stress fields, the entropy flux/heat flux relation is given by

$$\phi^k = \frac{1}{T}\left(q^k - \sum_{a=1}^{a_c} \mu_a J_a^k\right). \tag{8.22}$$

Here J_a^k denote the components of the diffusion flux of the constituents, see Müller (1985) and Dreyer and Duderstadt (2003a) for details.

8.4 Singular Surfaces

The common boundary between two adjacent phases is described by a surface $\mathfrak{s}$. We call $\mathfrak{s}$ a singular surface, because the thermodynamic fields may suffer discontinuities if the phase boundary is crossed. The conditions that relate the limiting values of the fields at a singular surface are called jump conditions.

8.4.1 *Various Descriptions of Surfaces*

In Euler coordinates, a surface $\mathfrak{s}$ may be described by the equation

$$\mathfrak{s}(t,x)=0. \tag{8.23}$$

Under some assumptions, we may also describe $\mathfrak{s}$ in Lagrange coordinates.

Throughout this study, we consider exclusively so called *coherent interface boundaries*. These are defined by a field of motion which is continuous across the surface $\mathfrak{s}$. Thus, we are allowed to use $x^i=\chi^i(t,X)$ to describe $\mathfrak{s}$ in the reference configuration according to

$$\hat{\mathfrak{s}}(t,X):=\mathfrak{s}(t,\chi(t,X))=0. \tag{8.24}$$

There are two more possibilities to describe the geometry of a surface. To this end, we introduce Gaussian parameters U^Γ, $\Gamma\in\{1,2\}$, and describe the surface by

$$x^i_s=\hat{x}^i_s(t,U^\Gamma) \quad\text{and}\quad X^i_s=\hat{X}^i_s(t,U^\Gamma) \tag{8.25}$$

in the actual and in the reference configuration, respectively.

8.4.2 *Euler and Lagrange Description of Surfaces*

We consider the surface in Euler coordinates and assign to each point of the surface a normal vector, ν^i, and its normal speed, w_ν, respectively. In Lagrange coordinates, we write N^i and W_N. The normal vectors can be calculated from (8.23) and (8.24) by

$$\nu^i=\frac{\mathfrak{s}_{x_i}}{\sqrt{\mathfrak{s}_{x_k}\mathfrak{s}_{x_k}}} \quad\text{and}\quad N^i=\frac{\hat{\mathfrak{s}}_{X_i}}{\sqrt{\hat{\mathfrak{s}}_{X_k}\hat{\mathfrak{s}}_{X_k}}}, \tag{8.26}$$

while the normal speeds follow according to

$$w_\nu = -\frac{\mathfrak{s}_t}{\sqrt{\mathfrak{s}_{x_k}\mathfrak{s}_{x_k}}} \quad \text{and} \quad W_N = -\frac{\hat{\mathfrak{s}}_t}{\sqrt{\hat{\mathfrak{s}}_{X_k}\hat{\mathfrak{s}}_{X_k}}}. \tag{8.27}$$

Here partial derivatives are indicated by indices.

The side of $\mathfrak{s}$ into which the normal points is called the "+" region, the other side is the "–" region. The brackets

$$[[\psi]] = \psi_+ - \psi_- \quad \text{and} \quad \{\psi\} = \frac{1}{2}(\psi_+ + \psi_-), \tag{8.28}$$

denote the jump and the mean value of any generic quantity ψ, where $\psi_{+/-}$ represent the limiting values of ψ formed in the "+" region and "–" region, respectively:

$$\psi_{+/-} = \lim_{x \in \text{"}+/-\text{"},\, x \to x^s} \psi(t, x). \tag{8.29}$$

The continuity condition, $[[\chi_i]] = 0$, implies that the temporal and spatial derivatives of χ_i, giving the velocity υ_i and the deformation gradient F_{ij}, obeys the Hadamard conditions

$$[[\upsilon_i]] = -W_N[[F_{ij}]]N_j, \quad [[F_{ij}]] = a_i N_j. \tag{8.30}$$

A proof of (8.30) is found in Truesdell and Toupin (1960).

8.4.3 Gaussian Description of Surfaces

Another possible description of the surface $\mathfrak{s}$ results by the introduction of Gaussian surface parameters U^Γ, $\Gamma \in \{1, 2\}$, so that any point P on $\mathfrak{s}$ is indicated by the function $x_s^i = \hat{x}_s^i(t, U^\Gamma)$. In P, we define tangential vectors, the metric tensor and the normal vector, respectively, according to

$$\tau_\Gamma^i = \frac{\partial \hat{x}_s^i}{\partial U^\Gamma}, \quad g_{\Delta\Gamma} = \tau_\Delta^i \tau_\Gamma^i, \quad \nu^i = \frac{\varepsilon^{ijk}\tau_1^j \tau_2^k}{\sqrt{g}}, \quad \text{with} \quad g = \det(g_{\Delta\Gamma}). \tag{8.31}$$

The spatial derivative of τ_Γ^i may be decomposed with respect to tangential and normal components as

$$\frac{\partial \tau_\Gamma^i}{\partial U^\Delta} = \Gamma_{\Gamma\Delta}^\Lambda \tau_\Lambda^i + b_{\Gamma\Delta}\nu^i \quad \text{with} \quad \Gamma_{\Gamma\Delta}^\Lambda = g^{\Lambda\Sigma}\frac{\partial \tau_\Gamma^i}{\partial U^\Delta}\tau_\Sigma^i \quad \text{and} \quad b_{\Gamma\Delta} = \frac{\partial \tau_\Gamma^i}{\partial U^\Delta}\nu^i. \tag{8.32}$$

The quantities $\Gamma^{\Lambda}_{\Gamma\Delta}$ and $b_{\Gamma\Delta}$ are the Christoffel symbols and the curvature tensor, respectively. The spatial derivative of the normal vector ν^i can be calculated to

$$\frac{\partial \nu^i}{\partial U^\Delta} = -g^{\Sigma\Gamma} b_{\Delta\Gamma} \tau^i_\Sigma. \tag{8.33}$$

The mean curvature k_M is defined by

$$k_M = \frac{1}{2} b_{\Gamma\Delta} g^{\Gamma\Delta}. \tag{8.34}$$

Next, we define the speed w^i of $\mathfrak{s}$ and its decomposition into tangential and normal parts by

$$w^i = \frac{\partial \hat{x}^s_i}{\partial t} = w^\Gamma_\tau \tau^i_\Gamma + w_\nu \nu^i. \tag{8.35}$$

The calculation of the spatial derivative of w^i yields

$$\frac{\partial w^i}{\partial U^\Delta} = (w^\Gamma_{\tau;\Delta} - w_\nu g^{\Gamma\Sigma} b_{\Sigma\Delta}) \tau^i_\Gamma + \left(\frac{\partial w_\nu}{\partial U^\Delta} + w^\Gamma_\tau b_{\Gamma\Delta} \right) \nu^i, \tag{8.36}$$

where the semicolon denotes covariant derivatives.

For later use, we need the time derivatives of the metric tensor and the normal vector, which can also easily be calculated. They read

$$\frac{\partial g_{\Delta\Gamma}}{\partial t} = w^\Sigma_{\tau;\Delta} g_{\Gamma\Sigma} + w^\Sigma_{\tau;\Gamma} g_{\Delta\Sigma} - 4 w_\nu k_M \quad \text{and} \quad \frac{\partial \nu^i}{\partial t} = -\left(\frac{\partial w_\nu}{\partial U^\Gamma} + w^\Sigma_\tau b_{\Sigma\Gamma} \right) g^{\Gamma\Delta} \tau^i_\Delta. \tag{8.37}$$

Obviously, we likewise may start from the function $\hat{X}^i_s(t, U^\Gamma)$ in order to describe the surface $\mathfrak{s}$ with respect to Lagrange coordinates. In this case, we end up with formulas that have the same structure as above, but in order to indicate the Lagrange description, all quantities are now written with capital letters. For example, we write

$$W^i = \frac{\partial \hat{X}^s_i}{\partial t}, \quad T^i_\Gamma = \frac{\partial \hat{X}^s_i}{\partial U^\Gamma}, \quad G_{\Delta\Gamma} = T^i_\Delta T^i_\Gamma, \quad G = \det(G_{\Delta\Gamma}), \tag{8.38}$$

and the other quantities are defined accordingly.

There are simple relations involving normal vectors and interfacial speeds of both descriptions. These read

$$N^k = \sqrt{\frac{g}{G}} \frac{1}{J^{+/-}} F^{ik}_{+/-} \nu^i, \quad w^i = \upsilon^i_{+/-} + F^{ik}_{+/-} W^k, \quad W_N = \sqrt{\frac{g}{G}} \frac{1}{J^{+/-}} (w_\nu - \upsilon^k_{+/-} \nu^k). \tag{8.39}$$

We may conclude from $(8.39)_1$ that the expression $J\overset{-1}{F^{ki}}N^k$ is continuous across the surface:

$$[[J\overset{-1}{F^{ki}}]]N^k = 0. \tag{8.40}$$

Later on, this identity will be used to transform the jump conditions between the Euler and the Lagrange description.

8.5 General Equations of Balance

The general equations of balance will be formulated for a material volume, which has the property that no mass flux through its boundary ∂V is allowed. In other words: The boundary of a material volume moves with the velocity of the particles of the body. A material volume is useful to describe the global equations of balance.

Figure 8.2 shows a material volume which is decomposed into two regions with volumes V_+ and V_- by a singular surface $\mathfrak{s}$. The normal vector of the material surfaces ∂V_+, ∂V_- and of the singular surface are denoted by n^k and ν^k, respectively. The unit vector $\tilde{e}^k = e^\Delta \tau^k_\Delta$ indicates a vector which is tangential to $\mathfrak{s}$ and normal to $\partial\mathfrak{s}$.

Note that ∂V_+, ∂V_- are material surfaces, so that no mass flux is allowed through ∂V_+ , ∂V_-, while the singular surface $\mathfrak{s}$ is generally not material, so that the exchange of mass between the volumes V_+ and V_- is possible.

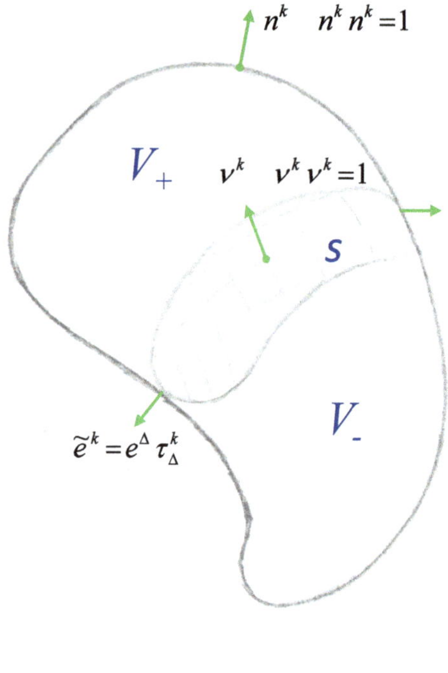

Fig. 8.2 A material volume is divided into two parts by a singular surface

8.5.1 General Global Equation of Balance

We consider the volume, which is indicated in Fig. 8.2 and formulate the general equation of balance for a generic additive quantity Ψ:

$$\frac{d\Psi}{dt} = \text{Flux}(\Phi) + \text{Source}(\Xi). \tag{8.41}$$

The quantity Ψ may have a volume density ψ_υ and a surface density ψ_s, and we decompose Ψ as

$$\Psi = \int\limits_{V_+ + V_-} \psi_\upsilon d^3x + \int\limits_{\mathfrak{s}} \psi_s da. \tag{8.42}$$

Accordingly, we introduce a volume flux density φ_υ^k and surface flux density $\tilde{\varphi}_s^k = \varphi_s^\Delta \tau_\Delta^k$ in order to decompose the flux Φ of Ψ through $\partial V_+ \backslash \mathfrak{s}$, $\partial V_- \backslash \mathfrak{s}$ and through $\partial \mathfrak{s}$:

$$\Phi = - \int\limits_{\partial V_+ \backslash \mathfrak{s}} \varphi_\upsilon^k \nu_k da - \int\limits_{\partial V_- \backslash \mathfrak{s}} \varphi_\upsilon^k \nu_k da - \int\limits_{\partial \mathfrak{s}} \varphi_s^\Delta e_\Delta dl. \tag{8.43}$$

The first two contributions to the flux describe nonconvective fluxes across the material surfaces $\partial V_+ \backslash \mathfrak{s}$, $\partial V_- \backslash \mathfrak{s}$, while the third contribution describes a flux, which is tangential to $\mathfrak{s}$ and normal to its boundary $\partial \mathfrak{s}$. The sign is chosen, so that incoming fluxes induce positive contributions to Φ.

Let Ξ be the source of Ψ with volume density ξ_υ and surface density ξ_s:

$$\Xi = \int\limits_{V_+ + V_-} \xi_\upsilon d^3x + \int\limits_{\mathfrak{s}} \xi_s da. \tag{8.44}$$

The general equation of balance (8.41) thus read

$$\frac{d}{dt}\Big(\int\limits_{V_+ + V_-} \psi_\upsilon d^3x + \int\limits_{s} \psi_s da\Big) = - \int\limits_{\partial V_+ \backslash \mathfrak{s}} \varphi_\upsilon^k n_k da - \int\limits_{\partial V_- \backslash \mathfrak{s}} \varphi_\upsilon^k n_k da - \int\limits_{\partial \mathfrak{s}} \varphi_s^\Delta e_\Delta dl +$$
$$\int\limits_{V_+ + V_-} \xi_\upsilon d^3x + \int\limits_{\mathfrak{s}} \xi_s da. \tag{8.45}$$

Next, we will exploit (8.45) separately for points lying in $V_\pm$ and $\mathfrak{s}$, respectively. To this end, we need so-called *transport theorems*, which are rules to interchange the time differentiation in (8.45) with the appearing time dependent volume and surface integrals.

8.5.2 Transport Theorems for Volume and Surface Integrals, Surface Divergence Theorem

The identity

$$\frac{d}{dt}\int\limits_{V_+ + V_-} \psi_\upsilon d^3x = \int\limits_{V_+ + V_-} \frac{\partial \psi_\upsilon}{\partial t} d^3x + \int\limits_{\partial V_+ \backslash \mathfrak{s} \cup \partial V_- \backslash \mathfrak{s}} \psi_\upsilon \upsilon^k n^k da \mp \int\limits_{\mathfrak{s}} [[\psi_\upsilon]] w_\nu da, \tag{8.46}$$

is called transport theorem for volume integrals, and its proof is an immediate consequence of Reynolds transport theorem for volume integrals for a region with volume Ω, whose surface $\partial\Omega$ moves with the velocity w_k, see Müller (1985), Aris (1996), and Truesdell and Toupin (1960) for details:

$$\frac{d}{dt}\int\limits_{\Omega} \psi_\upsilon d^3x = \int\limits_{\Omega} \frac{\partial \psi_\upsilon}{\partial t} d^3x + \int\limits_{\partial\Omega} \psi_\upsilon w^k n^k da. \tag{8.47}$$

In order to formulate the transport theorem for surface integrals, we assume at first that the surface fields are given as functions of Gaussian surface parameters U^Γ, $\Gamma \in \{1, 2\}$. The transport theorem for surface integrals is the identity

$$\frac{d}{dt}\int\limits_{\mathfrak{s}} \psi_s da = \int\limits_{\mathfrak{s}} \left(\frac{\partial \psi_s(t, U^\Gamma)}{\partial t} + \psi_s (w^\Delta_{\tau;\Delta} - 2k_M w_\nu) \right) da, \tag{8.48}$$

and its proof is given in Aris (1996) and Müller (1985).

Finally, there is a divergence theorem for surfaces, which reads

$$\int\limits_{\mathfrak{s}} \varphi^\Delta_{;\Delta} da = \oint\limits_{\partial\mathfrak{s}} \varphi^\Delta e_\Delta dl. \tag{8.49}$$

The proof of the identity (8.49) relies on the three-dimensional Stokes theorem, and its proof can also be found in Aris (1996) and in Müller (1985).

8.5.3 General Local Equation of Balance in Euler Coordinates

The following local versions of the global balance (8.45) will be given here in Euler coordinates. The transition to Lagrange coordinates will be carried out in Sect. 5.4.

In regular points, i.e., $x \in V_\pm$, the local version of (8.45) reads

$$\frac{\partial \psi_\upsilon}{\partial t} + \frac{\partial}{\partial x_k}(\psi_\upsilon \upsilon^k + \varphi^k_\upsilon) = \xi_\upsilon. \tag{8.50}$$

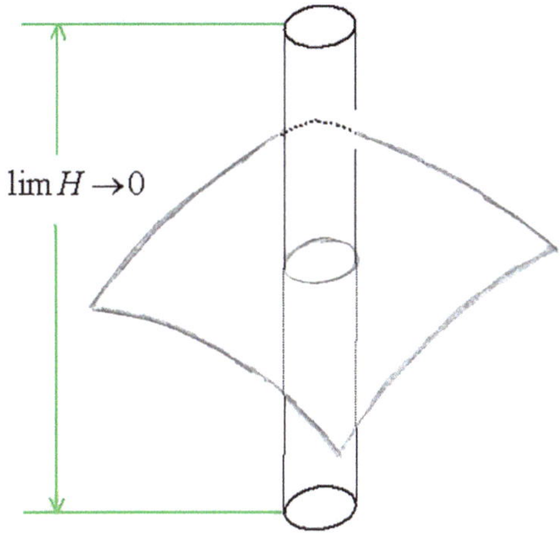

Fig. 8.3 Cylindrical volume to be used to derive jump conditions across a singular surface

The proof relies on (8.46), which will be applied to an arbitrary material volume that is contained in, V_+ and V_-, respectively.

Regarding points, lying on the singular surface $\mathfrak{s}$, the local version of (8.45) reads

$$-w_\nu[[\psi_\upsilon]] + [[\psi_\upsilon \upsilon^k + \varphi_\upsilon^k]]\nu_k = -\frac{\partial \psi_s(t, U^\Gamma)}{\partial t} - \psi_s(w^\Delta_{\tau;\Delta} - 2k_M w_\nu) - \varphi^\Delta_{s;\Delta} + \xi_s. \tag{8.51}$$

The proof of Eq. (8.51) relies on the so-called *pillbox argument*: The global equation of balance (8.45) is applied to the cylindrical volume of Fig. 8.3. The limit $H \to 0$ implies immediately Eq. (8.51).

The generic Eqs. (8.50) and (8.51) set the general frame of equations of balance. The conservation laws for mass, momentum, energy, and further special equations of balance follow by special choices for the generic densities, fluxes, and productions.

8.5.4 *General Local Equation of Balance in Lagrange Coordinates*

The transition of the local equations of balance from Euler to Lagrange coordinates rely on the definitions

$$\hat{\psi}_\upsilon = J\psi_\upsilon, \quad \hat{\varphi}_\upsilon^k = J\overset{-1}{F^{kj}}\varphi_\upsilon^j, \quad \hat{\xi}_\upsilon = J\xi_\upsilon. \tag{8.52}$$

In regular points, i.e., $X \in V_\pm$, the local version of (8.45) reads

$$\frac{\partial \hat{\psi}_\upsilon}{\partial t} + \frac{\partial \hat{\varphi}_\upsilon^k}{\partial X_k} = \hat{\xi}_\upsilon. \tag{8.53}$$

In order to proof Eq. (8.53), we insert in (8.50) the definitions $(8.52)_{1,3}$, and apply the chain rule to the divergence. Due to the identity (8.13) there appears the quantity $\hat{\varphi}_{\upsilon}^{k}$ under the divergence.

In order to derive the local analogue of (8.53) in points of the singular surface $\mathfrak{s}$, we start from (8.51). Here, we insert the definitions $(8.52)_{1,2}$, then we take care for the identities (8.39) and (8.40) and obtain

$$-W_N[[\hat{\psi}_{\upsilon}]] + [[\hat{\varphi}_{\upsilon}^{k}]]N^k = \sqrt{\frac{g}{G}}\left(-\frac{\partial\psi_s(t,U^{\Gamma})}{\partial t} - \psi_s(w_{\tau;\Delta}^{\Delta} - 2k_M w_{\nu}) - \varphi_{s;\Delta}^{\Delta} + \xi_s\right). \tag{8.54}$$

This is the final form of the generic equation of balance in singular points in Lagrange coordinates. Note that only its left-hand side has been transformed from Euler to Lagrange coordinates. In this study, it is not necessary to describe also the surface terms on the right-hand side of (8.54) with respect to the reference configuration of the considered body.

8.5.5 A Remark on the Equivalence of Balance Equations in Euler And Lagrange Description

The last two sections have revealed a complete equivalence of the generic representation of balance equations in Euler and Lagrange description. Thus, at a first glance, one might think that it is sufficient to consider only one of the descriptions. However, both are needed, because the special cases, which are related to special physical situations, destroy the equivalence.

For example, if we restrict the velocity of a substance, which crosses the interface, to normal motion, we must decide whether we are interested in normal motion with respect to the actual surface, i.e., $(\upsilon^i)^{+/-} = (\upsilon_{\nu})^{+/-}\nu^i$, or to define normal motion with respect to the reference configuration, i.e., $(\upsilon^i)^{+/-} = (\upsilon_N)^{+/-}N^i$.

8.6 The Conservation Laws for Mass, Momentum, and Energy

We are interested exclusively in interfacial boundary condition, i.e., special equations of balance at singular surfaces. For this reason, we do not consider the well-known special forms of the equations of balance in regular points in detail.

In this chapter, we identify the generic densities, fluxes, and sources with physical quantities and their properties. In turn the generic equations of balance (8.51) become physical laws. In particular, the equations of balance become conservation laws if the production densities ξ_{υ} and ξ_s are set equal to zero.

The conservation laws for mass, momentum, and energy are the most important ones. They result with special choices for the generic densities, fluxes, and productions appearing in (8.51).

The balance of mass relies on the physical experience that there are neither nonconvective mass fluxes nor mass sources, and we write

$$\begin{aligned} \psi_\upsilon &= \rho && \text{bulk mass density,} \\ \psi_s &= \rho_s && \text{surface mass density,} \\ \varphi_\upsilon^k &= 0, && \varphi_s^\Delta = 0, \quad \xi_\upsilon = 0, \quad \xi_s = 0. \end{aligned} \tag{8.55}$$

With these settings, the generic jump condition (8.51) becomes the balance of mass:

$$[[\dot{m}]] = -\frac{\partial \rho_s}{\partial t} - \rho_s(w^\Delta_{\tau;\Delta} - 2k_M w_\nu), \quad \text{with} \quad \dot{m} = \rho(\upsilon^k \nu^k - w_\nu). \tag{8.56}$$

In order to derive the balance of momentum, we ignore external gravitational and Lorentz forces, and in this case physical experience states that there is no source of momentum. We identify

$$\psi_\upsilon = \rho \upsilon^i \quad \text{-bulk momentum density, } \upsilon^i \text{ velocity of particles,}$$

$$\psi_s = \rho_s w^i \quad \text{-surface momentum density, } w^i \text{ surface velocity,} \tag{8.57}$$

$$\varphi_\upsilon^k = -\sigma^{ik} \quad \text{-(Cauchy) stress, force per area,} \tag{8.58}$$

$$\varphi_s^\Delta = -t_s^{i\Delta} \quad \text{-surface stress vector} \equiv \text{ force per line,} \tag{8.59}$$

$$\xi_\upsilon = 0, \qquad \xi_s = 0. \tag{8.60}$$

With these identifications, we obtain the momentum balance. Note that we have eliminated two parts on its right-hand side by means of the mass balance:

$$[[\dot{m}(\upsilon^i - w^i)]] - [[\sigma^{ik}]]\nu^k = -\rho_s \frac{\partial w^i}{\partial t} + t^{i\Delta}_{s;\Delta}. \tag{8.61}$$

For later use, we decompose the surface stress vector into tangential and normal components. We write

$$t_s^{i\Delta} = S^{\Delta\Gamma} \tau^i_\Gamma + S^\Delta \nu^i. \tag{8.62}$$

The newly introduced quantities $S^{\Delta\Gamma}$ and S^Δ are called surface stress tensor and normal stress, respectively.

Finally, we obtain the balance of energy if we identify

$$
\begin{aligned}
\psi_\upsilon &= \rho(u + \frac{1}{2}\upsilon^2) && \text{bulk energy density,} \\
\psi_s &= u_s + \frac{\rho_s}{2}w^2 && \text{surface energy density,} \\
\varphi_\upsilon^k &= q^k - \sigma^{ik}\upsilon^i && \text{energy flux of the bulk,} \\
\varphi_s^\Delta &= q_s^\Delta - t_s^{i\Delta}w^i && \text{surface energy flux,} \\
\xi_\upsilon &= 0, \quad \xi_s = 0.
\end{aligned}
\tag{8.63}
$$

We ignore again external gravitational and Lorentz forces and additionally we ignore energy sources resulting from radiation. In this case, physical experience states that there is no further source of energy. The following special form (8.64) of the energy balance results because again we have eliminated two parts on its right-hand side by means of the mass balance:

$$
\left[\left[\dot{m}\left(u + \frac{1}{2}\upsilon^2 - \frac{1}{2}w^2\right)\right]\right] + [[q^k - \sigma^{ik}\upsilon^i]]\nu^k = -\rho_s\frac{\partial \frac{1}{2}w^2}{\partial t} - \frac{\partial u_s}{\partial t} - u_s(w^\Delta_{\tau;\Delta} - 2k_M w_\nu) - (q_s^\Delta - t_s^{i\Delta}w^i)_{;\Delta}.
\tag{8.64}
$$

The decompositions of the energy into internal energy, $(\rho u,\ u_s)$, and kinetic energy, $(\rho/2\upsilon^2,\ \rho_s/2w^2)$ and of the energy flux into heat flux $(q^k,\ q_s^\Delta)$ and power of stress $(-\sigma^{ik}\upsilon^i,\ -t_s^{i\Delta}w^i)$ is induced by the requirement that equations of balance must be invariant with respect to Galilean transformation.

Note that we have not assumed that the internal surface energy is proportional to the surface mass density. It is most important that we do not make this assumption, because it will turn out that even in the case, where inertia of the singular surface is ignored, there still may be a nonzero contribution to the internal surface energy density.

8.7 Conservation of Moment of Momentum

Forming the cross-product of the momentum balance with the surface position vector $\hat{x}_s^i$ leads to the balance of the moment of momentum, $\rho_s\varepsilon^{ijk}\hat{x}_s^j w^k$, which is in general a nonconserved quantity. The corresponding production ξ_s can be identified with $\varepsilon^{ijk}t_s^{j\Delta}\tau_\Delta^k$, see Müller (1985) for more details.

Only if we assume that the surface does not carry any internal moment of momentum, the moment of momentum is a conserved quantity and we have

$$
\varepsilon^{ijk}t_s^{j\Delta}\tau_\Delta^k = 0.
\tag{8.65}
$$

We introduce here the decomposition (8.62) of the surface stress vector and conclude

$$
S^{\Delta\Gamma} = S^{\Gamma\Delta} \quad \text{and} \quad S^\Delta = 0.
\tag{8.66}
$$

This case is well known in the literature, e.g., see Naghdi (1960), and a model, where (8.66) is satisfied is called *membrane model* in contrast to the *shell model*, where (8.66) is not satisfied. In physical terms: If a surface is exposed to mechanical loads and behaves approximately rigid, then $(8.66)_2$ cannot be valid. Alternatively, if the external load is exclusively balanced by tangential surface stresses, the surface changes its shape, and the restrictions are valid. A rubber membrane provides a prominent example for this case, Dreyer (1981) and Dreyer et al. (1982).

8.8 Equations of Balance for Kinetic and Internal Energy

Multiplying the momentum balance with the surface velocity w^i yields the balance of kinetic energy

$$[[\dot{m}(w^i v^i - w^2)]] - [[\sigma^{ik} w^i]]\nu^k = -\rho_s \frac{\partial w^2/2}{\partial t} + (w^i t_s^{i\Delta})_{;\Delta} - t_s^{i\Delta} \frac{\partial w^i}{\partial U^\Delta}. \quad (8.67)$$

We subtract now the balance of kinetic energy from the balance of (total) energy (8.64) and obtain the balance of internal energy

$$\begin{aligned} &\left[\left[\dot{m}\left(u + \frac{1}{2}(v-w)^2\right)\right]\right] + [[q^k]]\nu^k - [[\sigma^{ik}(v^i - w^i)]]\nu^k = \\ &= -\frac{\partial u_s}{\partial t} - u_s(w^\Delta_{\tau;\Delta} - 2k_M w_\nu) - q^\Delta_{s;\Delta} + t_s^{i\Delta} \frac{\partial w^i}{\partial U^\Delta}. \end{aligned} \quad (8.68)$$

The Stefan condition and its generalizations will turn out as special cases of the balance of internal energy.

8.9 Equations of Balance for the Partial Masses

Next, we form balance equations for the partial masses. These read

$$[[\dot{m}(c_a - c_{sa})]] + [[J_a^k]]\nu^k = -\rho_s \frac{\partial c_{sa}}{\partial t} - J^\Delta_{s;\Delta} + \zeta_{sa}. \quad (8.69)$$

The balance of partial mass (8.69) follows from the general equation of balance (8.51) if we identify

$\psi_\upsilon = \rho c_a$ partial bulk mass density, c_a concentration of constituent a,

$\psi_s = \rho_s c_{sa}$ partial surface mass density, c_{sa} surface concentration of constituent a,

$\varphi^k_\upsilon = J^k_a = \rho c_a(\upsilon^k_a - \upsilon^k)$, diffusion flux, υ^k_a velocity of constituent a (8.70)

with $\sum_{a=1}^{a_c} J^k_a = 0,$

$\varphi^\Delta_s = J^\Delta_{sa}$, surface diffusion flux with $\sum_{a=1}^{a_c} J^\Delta_{sa} = 0,$ (8.71)

$\xi_\upsilon = \zeta_{\upsilon a}$, production rate of constituent a with $\sum_{a=1}^{a_c} \zeta_a = 0,$

$\xi_s = \zeta_{sa}$, surface production rate of constituent a with $\sum_{a=1}^{a_c} \zeta_{sa} = 0.$

In order to obtain the balance equations (8.69), we have again eliminated two parts on the right-hand side of (8.51) by means of the mass balance.

The partial masses are nonconserved quantities, neither in the bulk nor on a singular surface, because chemical reactions might take place. However, the total mass is conserved, and this fact implies the given restrictions for the diffusion fluxes and the productions rates, respectively.

8.10 Equations of Balance for the Entropy

Another important quantity is the entropy, which is nonconserved in nonequilibrium processes. The equation of balance for the entropy at singular surfaces reads

$$[[\dot{m}s]] + [[\varphi^k]]\nu^k = -\frac{\partial s_s}{\partial t} - s_s(w^\Delta_{\tau;\Delta} - 2k_M w_\nu) - (\varphi^\Delta_s)_{;\Delta} + \varsigma_s. \tag{8.72}$$

The balance of the entropy (8.72) follows from the general equation of balance (8.51) if we identify

$$\begin{aligned}
\psi_\upsilon &= \rho s && \text{bulk entropy density,} \quad s \text{ specific entropy ,}\\
\psi_s &= s && \text{surface entropy density,}\\
\varphi^k_\upsilon &= \phi^k && \text{(nonconvective) entropy flux in the bulk,}\\
\varphi^\Delta_s &= \phi^\Delta_s && \text{surface entropy flux,}\\
\xi_\upsilon &= \varsigma_\upsilon && \text{local bulk entropy production with} \quad \varsigma_\upsilon \geq 0,\\
\xi_s &= \varsigma_s && \text{local surface entropy production with} \quad \varsigma_s \geq 0.
\end{aligned} \tag{8.73}$$

We ignore entropy sources due to radiation. Physical experience states that the local entropy production rates are nonnegative. We call this statement the *strong second law of thermodynamics*, because this law goes beyond the famous statement that was proved by Clausius (1850), Müller (1985): The temporal change of the total entropy, S, of a body, whose outer surface is at uniform temperature T cannot be smaller but the ratio of heat supply $\dot{Q}$ divided by T:

$$\frac{dS}{dt} \geq \frac{\dot{Q}}{T}. \tag{8.74}$$

Note that, similarly to the surface energy density, we have not assumed that the surface entropy density is proportional to the surface mass density. It is most important that we do not make this assumption, because it will turn out that even in the case, where inertia of the singular surface is ignored, there is a nonzero contribution to the surface entropy density.

8.11 A Simplified Interfacial Model

If we ignore singular lines on $\mathfrak{s}$, the given singular equations of balance for mass, momentum, internal energy, partial masses, and entropy represent the most general case. Radiation, external gravitational forces, and Lorentz forces are represented by volume densities, and thus exclusively enter the regular equations of balance but not their singular counterparts.

Now we will introduce some simplifications which bring the singular equations of balance in a simpler form. The following choices of simplifications are made with the objective to present illustrative examples which yield (i) well known jump conditions at interfacial boundaries and the underlying assumptions and (ii) generalizations regarding mechanical effects and jump conditions where ordered solids and vacancies are involved.

The most general case that we will study from now on relies on the restrictions

$$\begin{aligned}
\rho_s &= 0, \quad \text{and} \quad c_{sa} = 0, \quad \text{no interfacial inertia,} \\
J_{sa}^{\Delta} &= 0, \quad \text{no tangential diffusion flux at the interface,} \\
q_{sa}^{\Delta} &= 0, \quad \text{no tangential heat flux at the interface,} \\
\phi_{sa}^{\Delta} &= 0, \quad \text{no tangential entropy flux at the interface,} \\
[[T]] &= 0, \quad \text{the temperature is continuous across the interface.}
\end{aligned} \tag{8.75}$$

Note that chemical production and entropy production within the interface are still allowed.

Interfacial chemical reactions are described by ζ_{sa}, which gives the interfacial mass production rate of constituent a. We assume that N_R different chemical reactions might happen within the interface, and these are indexed by $r \in \{1, 2, ..., N_R\}$. The stoichiometric coefficient of constituent a in the reaction r is denoted by γ_a^r. Let Γ^r be the net rate of reaction r, so that the mass production rate can be written

$$\zeta_{sa} = \sum_{r=1}^{N_R} \gamma_a^r M_a \Gamma^r, \tag{8.76}$$

where M_a demotes the molecular weight of constituent a.

Finally, we summarize the general singular balance equations according to the given restrictions. There follows the balance of mass, balance of momentum, balance of internal energy, balance of partial masses, and the balance of entropy, respectively:

$$\begin{aligned}
&[[\dot{m}]] = 0, \quad \text{with} \quad \dot{m} = \rho(\upsilon^k \nu^k - w_\nu) \\
&\dot{m}[[\upsilon^i]] - [[\sigma^{ik}]]\nu^k = (S^{\Delta\Gamma}_{;\Delta} - S^\Delta b^\Gamma_\Delta)\tau^i_\Gamma + (S^{\Delta\Gamma} b_{\Delta\Gamma} + S^\Delta_{;\Delta})\nu^i \\
&\dot{m}\left[\left[\left(u + \frac{1}{2}(\upsilon - w)^2\right)\right]\right] + [[q^k]]\nu^k - [[\sigma^{ik}(\upsilon^i - w^i)]]\nu^k = \\
&-\frac{\partial u_s}{\partial t} - u_s(w^\Delta_{\tau;\Delta} - 2k_M w_\nu) + S^{\Delta\Gamma}(g_{\Gamma\Sigma} w^\Sigma_{\tau;\Delta} - b_{\Delta\Gamma} w_\nu) + S^\Delta\left(\frac{\partial w_\upsilon}{\partial U^\Delta} + w^\Sigma_\tau b_{\Sigma\Delta}\right) \\
&\dot{m}[[c_a]] + [[J^k_a]]\nu^k = \sum_{r=1}^{N_R} \gamma^r_a M_a \Gamma^r \\
&\dot{m}[[s]] + \frac{1}{T}\left[\left[\left(q^k - \sum_{a=1}^{a_c} \mu_a J^k_a\right)\right]\right]\nu^k = -\frac{\partial s_s}{\partial t} - s_s(w^\Delta_{\tau;\Delta} - 2k_M w_\nu) + \varsigma_s \quad \text{with} \quad \varsigma_s \geq 0.
\end{aligned} \tag{8.77}$$

The mass balance now implies the continuity of the mass flux $\dot{m}$ through the interface. Thus, we may remove $\dot{m}$ out of the jump brackets in the other balance equations.

8.12 The Second Law of Thermodynamics for Singular Surfaces

In this section, we derive the Gibbs equation for surfaces. In general, its derivation is cumbersome, but follows the strategy that is carefully described by I. Müller in the monograph Müller (1985). For that reason, we start here from the simplified surface model of the last section.

We multiply the entropy balance $(8.77)_5$ by the temperature T, which is continuous in the simplified case, and subtract the result from the internal energy balance $(8.77)_4$. There follows

$$\begin{aligned}
-T\varsigma_s = &-\frac{\partial \psi_s}{\partial t} - s_s\frac{\partial T}{\partial t} + (S^{\Delta\Gamma} - \psi_s g^{\Delta\Gamma})(g_{\Gamma\Sigma} w^\Sigma_{\tau;\Delta} - b_{\Delta\Gamma} w_\nu) + S^\Delta\left(\frac{\partial w_\upsilon}{\partial U^\Delta} + w^\Sigma_\tau b_{\Sigma\Delta}\right) \\
&+\dot{m}\left[\left[\left(\psi + \frac{1}{2}(\upsilon - w)^2\right)\right]\right] - [[\sigma^{ik}(\upsilon^i - w^i)]]\nu^k + \left[\left[\sum_{a=1}^{a_c} \mu_a J^k_a\right]\right]\nu^k \leq 0.
\end{aligned} \tag{8.78}$$

Herein, we have introduced the specific free energy of the bulk $\psi = u - Ts$ and the surface free energy density $\psi_s = u_s - Ts_s$ with the physical units $Joule/kg$ and $Joule/m^2$, respectively. We recall from Sect. 8.2 that by definition $\psi_s \equiv \sigma$, where σ is called surface tension, which is sometimes a misleading designation.

We assume now that the surface free energy density, the surface entropy density, and the surface stress depend at most on the set

$$T, \frac{\partial T}{\partial U^\Delta}, g_{\Delta\Gamma}, g^0_{\Delta\Gamma}, b_{\Delta\Gamma}, b^0_{\Delta\Gamma}, \nu^i, \nu^{0i}. \tag{8.79}$$

The quantities $g^0_{\Delta\Gamma}, b^0_{\Delta\Gamma}, \nu^{0i}$ are metric tensor, curvature tensor, and normal vector, respectively, at the reference time $t = 0$, so that the pairs $(g_{\Delta\Gamma}, g^0_{\Delta\Gamma})$, $(b_{\Delta\Gamma}, b^0_{\Delta\Gamma})$, (ν^i, ν^{0i}) are measures of the deformation of the surface. See Dreyer (1981) and Dreyer et al. (1982), where the deformation of a rubber balloon may serve as an illustrative example.

After carrying out the necessary differentiations in (8.78), we observe that the surface entropy production depends linearly on

$$\frac{\partial T}{\partial t}, \frac{\partial^2 T}{\partial U^\Delta \partial t}, w^\Sigma_{\tau;\Delta}, \frac{\partial w_\upsilon}{\partial U^\Delta}, \frac{\partial b_{\Delta\Gamma}}{\partial t}. \tag{8.80}$$

However, the inequality must hold for arbitrary values of these fields, and thus we conclude that the factors of the derivatives (8.80) must vanish, because otherwise we could violate the inequality. This statement is discussed at length for various circumstances and materials in Müller (1985). The consequences are: (i) surface free energy density, surface entropy density, surface stress, and normal stress do not depend on

$$\frac{\partial T}{\partial U^\Delta} \quad \text{and} \quad b_{\Delta\Gamma}, \tag{8.81}$$

(ii) there holds the Gibbs equation for surfaces, which may be written as

$$s_s = -\frac{\partial\sigma}{\partial T}, \quad S^{\Delta\Gamma} = \sigma g^{\Delta\Gamma} + \frac{1}{2}\frac{\partial\sigma}{\partial g_{\Delta\Gamma}}, \quad S^\Delta = -g^{\Delta\Gamma}\tau^k_\Gamma\frac{\partial\sigma}{\partial\nu^k}, \tag{8.82}$$

(iii) there remains the inequality

$$-T\varsigma_s = \dot{m}\left[\left[(\psi + \frac{1}{2}(\upsilon - w)^2)\right]\right] - [[\sigma^{ik}(\upsilon^i - w^i)]]\nu^k + \left[\left[\sum_{a=1}^{a_c}\mu_a J^k_a\right]\right]\nu^k \leq 0, \tag{8.83}$$

which will be further exploited in Sect. 8.15.

8.13 Mechanical Equilibrium

8.13.1 Equilibrium Conditions

Mechanical equilibrium is established if the barycentric velocity υ^i is zero. On the other hand, impermeable interfaces are defined by $(\upsilon^k)^{+/-}\nu^k = w_\nu$ so that there is no (total) mass flux across the interface. Note that impermeable interfaces may still allow a flux of partial masses.

Both cases lead to the same mechanical equilibrium conditions, which follow from $(8.77)_2$ and can

$$-[[\sigma^{ik}]]\nu^i\nu^k = S^{\Delta\Gamma}b_{\Delta\Gamma} + S^{\Delta}_{;\Delta} \quad \text{and} \quad -[[\sigma^{ik}]]\tau^i_\Sigma\nu^k = g_{\Gamma\Sigma}(S^{\Delta\Gamma}_{;\Delta} - S^{\Delta}b^{\Gamma}_{\Delta}). \tag{8.84}$$

We conclude from the constitutive laws $(8.83)_{1,2}$ and the conditions (8.84) for mechanical equilibrium that the jump $[[\sigma^{ik}]]\nu^i\nu^k$ of the Cauchy stress can be calculated if the surface free energy density is known. The determination of the jump $-[[\sigma^{ik}]]\tau^i_\Sigma\nu^k$ needs additional information on the antisymmetric part $S^{[\Delta\Gamma]}$ of the surface stress $S^{\Delta\Gamma}$. However, it turns out that this jump is of minor importance, in particular it is not needed to determine the jumps of the chemical potentials, see Sects. 15.3 and 15.4 for more details.

We consider two interesting special cases. (i) If the bulk materials are mixtures of nonviscous fluids, their stress is represented by a pressure, $\sigma^{ik} = -p\delta^{ik}$. If furthermore the surface free energy is a function of temperature only, $\sigma(T)$, the equilibrium conditions reduce to

$$[[p]] = 2\sigma k_M \quad \text{and} \quad \frac{\partial\sigma}{\partial T}\frac{\partial T}{\partial U^\Delta} = 0. \tag{8.85}$$

We conclude that the pressure jump is proportional to the mean curvature of the interface, see Sect. 8.12. This law was first derived by Laplace for spheres, Laplace (1806). In addition, the temperature must be uniform on the interface.
(ii) Now we assume that the surface tension depends exclusively on the normal vector, $\sigma(\nu^i)$, which itself is a function of the surface parameters U^Δ. Let us introduce the function $\hat{\sigma}(t, U^\Delta) = \sigma(\nu^i(t, U^\Delta))$, and by means of the identity

$$\frac{\partial\hat{\sigma}}{\partial U^\Delta} = -b^{\Gamma}_{\Delta}\tau^k_\Gamma\frac{\partial\sigma}{\partial\nu^k}, \tag{8.86}$$

we may rewrite the conditions (8.85) according to

$$-[[\sigma^{ik}]]\nu^i\nu^k = 2\hat{\sigma}k_M + (\overset{-1}{b})^{\Delta\Gamma}\hat{\sigma}_{;\Delta\Gamma} + (\overset{-1}{b})^{\Delta\Gamma}_{;\Delta}\hat{\sigma}_{;\Gamma} \quad \text{and} \quad -[[\sigma^{ik}]]\tau^i_\Sigma\nu^k = g_{\Sigma\Gamma}S^{[\Delta\Gamma]}_{;\Delta}. \tag{8.87}$$

8.13.2 Discussion and Comparison with the Literature

In 1950, C. Herring introduced a surface free energy density that depends on the interfacial normal vector in order to study sintering processes where small particles decrease their free energy by growing together, Herring (1951). In the same article, Herring also considered interfacial elasticity and he showed that the surface stress tensor should be calculated according to the relation $(8.82)_2$, which follows here as a consequence of the second law of thermodynamics.

In 1972 and 1974, J. W. Cahn & D. W. Hoffmann put forward the studies on anisotropy of interfaces, Hoffmann and Cahn (1972), Hoffmann and Cahn (1974). They ignored interfacial elasticity and considered the case where the interfacial free energy density σ exclusively depends on the interfacial normal vector ν^i. Cahn and Hoffmann substituted σ by the capillarity vector ξ^i, which nowadays is called Hoffmann/Cahn vector, and they equipped ξ^i with the properties

$$\sigma = \xi^i \nu^i, \quad d\sigma = \xi^i d\nu^i, \quad \nu^i d\xi^i = 0. \tag{8.88}$$

In Hoffmann and Cahn (1974), Cahn and Hoffmann established the equilibrium law

$$[[p]] = div_s \cdot \xi. \tag{8.89}$$

In our notation, the operator $div_s \cdot$ reads

$$div_s \cdot \xi = g^{\Gamma\Delta} \tau^i_\Delta \frac{\partial \xi^i}{\partial \nu^i}, \tag{8.90}$$

While the explicit meaning of $div_s \cdot$ is not generally given in Hoffmann and Cahn (1974), it appears explicitly in Garcke et al. (1998), where a sharp interface limit of a phase field model is carried out.

We now compare our results with the Cahn/Hoffmann law (8.89). To this end, we start from (8.84) and assume according to Cahn/Hoffmann that σ exclusively depends on ν^i. There follows with (8.88)

$$-[[\sigma^{ik}]]\nu^i \nu^k = 2\sigma k_M - g^{\Gamma\Delta} \tau^l_\Delta \left(\frac{\partial \xi^k \nu^k}{\partial \nu^l} \right)_{;\Delta} = -div_s \cdot \xi, \tag{8.91}$$

which gives the Cahn/Hoffmann law (8.89) if the bulk stress fields are isotropic.

The tangential components of $[[\sigma^{ik}]]\nu^k$ are related to the antisymmetric part of the surface stress tensor, $S^{[\Gamma\Sigma]}$, because there results

$$[[\sigma^{ik}]]\tau^i_\Delta \nu^k = g_{\Delta\Gamma} S^{[\Gamma\Sigma]}_{;\Sigma}. \tag{8.92}$$

Next, we calculate the surface stress vector $t^{i\Delta}$, which gives the force per line element of the interface. We introduce the surface vector

$$X^i_\Gamma = \epsilon^{ijk}\tau^j_\Gamma \nu^k, \tag{8.93}$$

which lies in the tangential plane and is perpendicular to the direction Γ. We obtain

$$t^{i\Delta} = \epsilon^{ijk}\xi^j X^k_\Gamma g^{\Gamma\Delta}. \tag{8.94}$$

This is the Hoffmann/Cahn representation of the surface stress vector, which is best suited to study triple junctions.

Note that both the Cahn/Hoffmann relations (8.89) and (8.94) only hold if interfacial elasticity is ignored and if temperature is uniform on the interface.

In the literature, the Hoffmann/Cahn framework is used to compare the sharp interface limit of phase field models with pure sharp interface models and to relate the higher gradient coefficients of phase field models with the surface free energy density σ. Various examples are found to be in Wheeler and McFadden (1997), Nestler and Wheeler (1998), Garcke et al. (1998), and Dreyer and Wagner (2003).

Interfacial elastic effects have also already been considered by other authors. Alexander and Johnson calculated jump conditions for disordered solid/liquid interfaces including interfacial elasticity via the minimization of the Gibbs free energy Alexander and Johnson (1985). Gurtin and Voorhees propose balance equations to derive jump conditions with elastic effects, and they give a systematic evaluation of the second law of thermodynamics, Gurtin and Voorhees (1996 and 1998). However, their reasonings rely on the so-called configurational forces, which are not needed in the current study, where we exclusively deal with the classical forces of mechanics.

8.14 The Stefan Condition and its Generalizations

In this section, we discuss the assumptions that lead to the well-known Stefan condition

$$-w_\nu\lambda + [[q^k]]\nu^k = 0. \tag{8.95}$$

The quantity λ may depend on temperature and is called latent heat in the literature, Meirmanov (1992).

The Stefan condition must obviously result as a special case of the internal energy balance $(8.77)_3$, and we will now analyze the necessary assumptions. To this end, let us at first assume (i) $\sigma^{ik} = -p\delta^{ik}$, (ii) $\upsilon^i = 0$, (iii) $\psi_s = \sigma(T)$, and (iv) $w^i = w_\nu\nu^i$ so that the energy balance $(8.77)_3$ can be reduced to

$$-w_\nu\rho\left[\left[(u + \frac{p}{\rho} + \frac{1}{2}w^2\right]\right] + [[q^k]]\nu^k = T\frac{\partial^2\sigma}{\partial T^2}\frac{\partial T}{\partial t} - T\frac{\partial\sigma}{\partial T}2k_M w_\nu. \tag{8.96}$$

In case that we neglect the kinetic energy of the interface, we may identify the latent heat as $\lambda = \rho[[(u + \frac{p}{\rho}]]$, which is in accordance to classical thermodynamics. However, even now, the Stefan condition does not result. The first contribution on the right-hand side of (8.96) only vanishes either in processes with constant temperature, or if the surface entropy density does not depend on T. However, this is an exceptional case. But even then, there is a second contribution which is proportional to the mean curvature and to the interfacial normal speed. Finally, it is important to note that the surface parameters must be chosen so that (iv) is satisfied. See Dreyer and Wagner (2003), where an example is given that a different choice is often prefered due to numerical aspects.

8.15 Interfacial Equilibrium Conditions and Growth Laws

8.15.1 On the General Structure of the Interfacial Entropy Production

The physical law whereupon the interfacial entropy production is positive in irreversible processes and zero in equilibrium gives rise to further important jump conditions involving the chemical potentials at the interface. The derivation of these conditions relies on the interfacial entropy inequality (8.83). In the next section, we will show that the interfacial entropy inequality can be brought into a sum of binary products:

$$-T\varsigma_s = \sum_{A=1}^{N} F_A D_A \leq 0. \tag{8.97}$$

The quantities D_A and F_A are called driving forces and fluxes, respectively, see also Abeyaratne and Knowles (1990) for a very interesting example.

Driving forces that result from sharp interface limits of phase field models a derived and discussed in Caginalp (1991), Cahn et al. (1996), and Dreyer and Wagner (2003).

In equilibrium, the interfacial entropy production is zero and always positive in nonequilibrium processes. If the interfacial processes do produce interfacial entropy, we exploit the inequality (8.97) as follows, see also Gurtin and Voorhees (1996) and Liu (1992): We assume that there are constitutive laws that relate the driving forces $D_1, D_2, \ldots D_N$ to the interfacial fluxes F_A as $F_A = \tilde{F}_A(D_B, W)$, where W abbreviates other variables that are not among the driving forces. We conclude from (8.97) that ς_s assumes a minimum for $D_A = 0$, and furthermore, we have

$$\left(\frac{\partial}{\partial D_B}\left(\sum_{A=1}^{N} \tilde{F}_A D_A\right)\right)_{|equ.} = \tilde{F}_{B|equ.} = 0, \quad B \in \{1, 2, .., N\}. \tag{8.98}$$

In other words,

$$D_B = 0 \Leftrightarrow F_B = 0 \quad \text{in equilibrium.} \tag{8.99}$$

Thus, it follows that the fluxes can be represented by

$$F_A = -\sum_{B=1}^{N} M_{AB} D_B \quad \text{with} \quad M_{AB} \quad \text{– positive definite.} \tag{8.100}$$

The matrix M_{AB} contains the surface mobilities of the interface, which must be determined experimentally. However, they also can be calculated from an underlying kinetic theory. Section 8.2 provides an example for the latter case. The constitutive law (8.100) is called *growth law*.

There exist an interesting limiting case. We recall that $D_A = 0$ implies $F_A = 0$. However, if the surface mobilities are much larger than the bulk mobilities, we may consider the limit $M_{AB} \to \infty$, and then we must have $D_B = 0$, so that the fluxes F_A remain finite. In this case, the equilibrium conditions $D_A = 0$ still holds, but the fluxes F_A will be calculated from interfacial balance equations, and the process under consideration is called *diffusion controlled.*

On the other hand, if the interface mobilities are comparable or smaller as the bulk mobilities, the considered process relies on the law (8.100) and is called *interface controlled.*

Next, we identify the driving forces and fluxes explicitly.

8.15.2 Alternative Representation of the Interfacial Entropy Inequality

The exploitation of the interfacial entropy production (8.83) yields simple results only if we consider the two special cases: (i) either the bulk materials consist of isotropic fluids or (ii) the bulk materials move at the interface exclusively in normal direction, i.e.,

$$\text{in case (i):} \quad \sigma^{ik} = -p\delta^{ik} \quad \text{or in case (ii):} \quad \upsilon^i = \upsilon_\nu \nu^i. \tag{8.101}$$

In both cases, the inequality (8.83) reduces to

$$-T\varsigma_s = \dot{m}\left[\!\left[\left(\psi + \frac{1}{2}(\upsilon - w)^2\right)\delta^{ik} - \frac{1}{\rho}\sigma^{ik}\right]\!\right]\nu^i\nu^k + \left[\!\left[\sum_{a=1}^{a_c}\mu_a J_a^k\right]\!\right]\nu^k \le 0. \tag{8.102}$$

From now on, we consider exclusively the special case (ii).

For a further simplification of the inequality (8.102), we define the quantity

$$\dot{m}_a^{+/-} = (\dot{m}c_a + J_a^k \nu^k)^{+/-}, \tag{8.103}$$

which gives the total one sided flux of constituent a across the interface. The singular mass balance $(8.77)_5$ can now be written

$$[[\dot{m}_a]] = \sum_{r=1}^{N_R} M_a \gamma_a^r \Gamma^r, \tag{8.104}$$

and there holds due to $(8.71)_3$

$$\sum_{a=1}^{a_c^{+/-}} \dot{m}_a^{+/-} = \dot{m}. \tag{8.105}$$

Next, we eliminate the diffusion flux in the inequality (8.102) and take care for the Gibbs/Duhem equation (8.21) to obtain

$$-T\varsigma_s = \dot{m}\left[\left[\frac{1}{2}(\upsilon - w)^2 - \frac{1}{\rho}\sigma^{<ik>}\nu^i\nu^k\right]\right] + \left[\left[\sum_{a=1}^{a_c}\dot{m}_a\mu_a\right]\right] \leq 0. \tag{8.106}$$

For a further exploitation of this form of the entropy inequality, we need to know the behavior of the various constituents at the interface. In general, there are four different kinds of constituents, and we put them into four groups as follows. The first group contains constituents that may cross the interface without being subjected to chemical reactions, i.e.,

$$\Lambda_1 = \{a \,|\, \dot{m}_a^+ = \dot{m}_a^-, \dot{m}_a^\pm \neq 0\}. \tag{8.107}$$

The members of the second group of constituents cannot cross the interface, because they experience the interface as an impermeable wall, i.e.,

$$\Lambda_2 = \{a \,|\, \dot{m}_a^+ = 0,\ \dot{m}_a^- = 0\}. \tag{8.108}$$

The third group contains constituents appearing on both sides at the interface, and they are involved in interfacial chemical reactions, i.e.,

$$\Lambda_3 = \{a \,|\, \dot{m}_a^+ - \dot{m}_a^- = \sum_{r=1}^{N_R} M_a \gamma_a^r \Gamma^r\}. \tag{8.109}$$

Finally, there is a fourth group $\Lambda_4 = \Lambda_4^+ \cup \Lambda_4^-$ of constituents that are also involved in interfacial chemical reactions but they appear either on the "+" side or on the "–" side of the interface, i.e.,

$$\Lambda_4^{+/-} = \{a \,|\, \dot{m}_a^{+/-} = (+/-)\sum_{r=1}^{N_R} M_a \gamma_a^r \Gamma^r\}. \tag{8.110}$$

We proceed to rewrite the inequality (8.106) and in particular the second jump bracket, which now can be written

$$\left[\left[\sum_{a=1}^{a_c}\dot{m}_a\mu_a\right]\right] = \sum_{a\in\Lambda_1}\dot{m}_a[[\mu_a]] + \sum_{a\in\Lambda_3} < \dot{m}_a > [[\mu_a]] + \sum_{r=1}^{N_R}\Gamma^r\sum_{a\in\Lambda_3}M_a\gamma_a^r < \mu_a > +$$
$$\sum_{r=1}^{N_R}\Gamma^r\left(\sum_{a\in\Lambda_4^+}M_a\gamma_a^r\mu_a^+ + \sum_{a\in\Lambda_4^-}M_a\gamma_a^r\mu_a^-\right). \tag{8.111}$$

Next, we assume that the constituent $a = 1$ belongs to Λ_1, and we eliminate $\dot{m}_1$ by means of the side condition (8.106). The final result reads

$$-T\varsigma_s = \dot{m}\left[\left[\left(\mu_1 + \frac{1}{2}(\upsilon - w)^2\right) - \frac{1}{\rho}\sigma^{<ik>}\nu^i\nu^k\right]\right] +$$
$$\sum_{a\in\Lambda_1}\dot{m}_a[[\mu_a - \mu_1]] + \sum_{a\in\Lambda_3} < \dot{m}_a > [[\mu_a - \mu_1]] + \tag{8.112}$$
$$\sum_{r=1}^{N_R}\Gamma^r\left(\sum_{a\in\Lambda_3}M_a\gamma_a^r < \mu_a > + \sum_{a\in\Lambda_4^+}M_a\gamma_a^r(\mu_a^+ - \frac{1}{2}[[\mu_1]]) + \sum_{a\in\Lambda_4^-}M_a\gamma_a^r(\mu_a^- + \frac{1}{2}[[\mu_1]]\right).$$

Note that the representations (8.106) and (8.112) of the interfacial entropy production are equivalent. However, the representation (8.112) will be only applied to identify driving forces and fluxes for the special case that (i) the solid phase is a *disordered solid*, i.e., there are no privileged lattice sites and (ii) no vacancies appear.

The other interesting case is (i) the solid has a sublattice structure, so that privileged lattice sites can be described, and (ii) vacancies are taken into account. This case will be treated on the basis of the representation (8.106).

8.15.3 Identification of Fluxes and Driving Forces for Disordered Solids Without Vacancies

In this section, we consider interfaces between two fluids, between fluids and disordered solids without vacancies, and between two disordered solid phases without vacancies, respectively. In these cases and unlike when ordered solids are involved, the only side condition is given by (8.105). Thus, we may directly identify the fluxes and driving forces in the inequality (8.112) by its comparison with the general inequality (8.97). We read off the fluxes

$$F_A = \left\{\dot{m}, \dot{m}_{a\in\Lambda_1}, < \dot{m}_{a\in\Lambda_3} >, \dot{m}^+_{a\in\Lambda_4^+}, \dot{m}^-_{a\in\Lambda_4^-}, \Gamma^{r\in\{1,2...,N_R\}}\right\}, \tag{8.113}$$

and the corresponding driving forces

$$D_A = \{ [[(\mu_1 + \frac{1}{2}(\upsilon - w)^2) - \frac{1}{\rho}\sigma^{<ik>}\nu^i\nu^k]], [[\mu_a - \mu_1]]_{a\in\Lambda_1}, [[\mu_a - \mu_1]]_{a\in\Lambda_3},$$
$$\left(\sum_{a\in\Lambda_3} M_a\gamma_a^r < \mu_a > + \sum_{a\in\Lambda_4^+} M_a\gamma_a^r(\mu_a^+ - \frac{1}{2}[[\mu_1]]) + \sum_{a\in\Lambda_4^-} M_a\gamma_a^r(\mu_a^- + \frac{1}{2}[[\mu_1]])\right)_{r\in\{1,2...,N_R\}} \}. \tag{8.114}$$

8.15.3.1 Evolution to Equilibrium, Growth Laws

If all the interfacial processes happen without entropy productions, and if N constituents are able to cross the interface, there follow $N + N_R$ conditions for the interfacial values of concentrations and stresses:

$$\left[\left[\mu_1 + \frac{1}{2}(\upsilon - w)^2)\right]\right] = \left[\left[\frac{1}{\rho}\sigma^{<ik>}\right]\right]\nu^i\nu^k$$
$$[[\mu_a - \mu_1]] = 0 \quad \text{for} \quad a \in \Lambda_1 \cup \Lambda_3,\ a = 1 \in \Lambda_1,$$
$$\left(\sum_{a\in\Lambda_3} M_a\gamma_a^r < \mu_a > + \sum_{a\in\Lambda_4^+} M_a\gamma_a^r(\mu_a^+ - \frac{1}{2}[[\mu_1]]\right) + \tag{8.115}$$
$$\sum_{a\in\Lambda_4^-} M_a\gamma_a^r\left(\mu_a^- + \frac{1}{2}[[\mu_1]]\right)_{r\in\{1,2...,N_R\}} = 0 \quad \text{for} \quad r \in \{1, 2, ...N_R\}$$

Recall that these interfacial equilibrium conditions hold if either no interfacial entropy production appears or if all the surface mobilities are large, so that we may consider the limit $M_{AB} \to \infty$. In this case, we must set $D_B = 0$ to obtain finite fluxes F_A, and the fluxes will be calculated from the interfacial balance equations. Recall that we have called such process *diffusion controlled.*

Obviously, we also may meet situations where the various surface mobilities are of different order. For example, it may happen that chemical reactions establish equilibrium much faster as the two other processes. In this case, we have

$$\left(\sum_{a\in\Lambda_3} M_a\gamma_a^r < \mu_a > + \sum_{a\in\Lambda_4^+} M_a\gamma_a^r(\mu_a^+ - \frac{1}{2}[[\mu_1]]\right) +$$
$$\sum_{a\in\Lambda_4^-} M_a\gamma_a^r\left(\mu_a^- + \frac{1}{2}[[\mu_1]]\right)_{r\in\{1,2...,N_R\}} = 0 \quad \text{for} \quad r \in \{1, 2, ...N_R\} \quad \text{and} \tag{8.116}$$

$$F_A = -\sum_{B=1}^{N} M_{AB} D_B \quad \text{for} \quad A, B \in \{1, 2, ...N\}.$$

8.15.3.2 Examples for Interfacial Equilibrium Conditions

In this section, we assume that all considered processes are diffusion controlled so that the driving forces are zero. The following examples serve for an illustration of the resulting conditions.

Single Substance

The interface separates two different phases, which both consist of a single substance. In this case we have $c_1 = 1$ and the chemical potential μ_1 becomes the specific Gibbs free energy, g. Correspondingly, there follows only a single condition from (8.115), which read

$$\left[\left[\mu_1 + \frac{1}{2}(\upsilon - w)^2)\right]\right] = \left[\left[\frac{1}{\rho}\sigma^{<ik>}\right]\right]\nu^i\nu^k \quad \text{with} \quad \mu_1 = g = \psi - \frac{1}{3\rho}\sigma^{kk}. \tag{8.117}$$

If we neglect the kinetic energy and if the substance is mechanically described by an isotropic stress, i.e., $\sigma^{ik} = -p\delta^{ik}$, there follows the classical result

$$[[g]] = 0 \quad \text{with} \quad g = \psi + \frac{p}{\rho}. \tag{8.118}$$

In Sect. 8.12, we have already recognized the jump condition (8.118) as the base for the classical Gibbs/Thomson law. We conclude now that the Gibbs/Thomson law only holds in fluids, because, in general, we have $\sigma^{<ik>} \neq 0$ in solids.

Binary mixture

Now let us consider an interface that separates two phases consisting of binary mixtures. We assume additionally that both constituents may cross the interface without chemical reactions. From (8.115), we read off the two conditions

$$\left[\left[\mu_1 + \frac{1}{2}(\upsilon - w)^2\right]\right] = \left[\left[\frac{1}{\rho}\sigma^{<ik>}\right]\right]\nu^i\nu^k \quad \text{and} \quad -[[\mu_1]] + [[\mu_2]] = 0. \tag{8.119}$$

Thus, only if we neglect the kinetic energy and assume again an isotropic stress, we obtain the known result

$$[[\mu_1]] = [[\mu_2]] = 0. \tag{8.120}$$

Solid/liquid interface with five constituents

In this example, we consider a solid/liquid interface, $(+/-)$. There are three constituents, A, B, C, in the solid and four constituents, A, B, D, E, in the liquid. The constituents A, B are found to be in both phases, while C, D, E cannot cross the interface. Furthermore, we assume that within the interface there is a chemical reaction between the constituents C and D, i.e.,

$$\gamma_C C_+ + \gamma_D D_- = 0. \tag{8.121}$$

For example, if oxygen appears in the liquid as the molecule O_2, it dissociates into $2O$ within the interface before it can occupy a lattice site of the solid.

Thus, we read off from Eqn. (8.115) three conditions, viz.

$$\begin{aligned} &\left[\left[\mu_A + \frac{1}{2}(\upsilon - w)^2\right]\right] = \left[\left[\frac{1}{\rho}\sigma^{<ik>}\right]\right] \nu^i \nu^k, \\ &[[\mu_A]] - [[\mu_B]] = 0, \\ &\gamma_C M_C \mu_C^- + \gamma_D M_D \mu_D^+ = 0. \end{aligned} \tag{8.122}$$

Furthermore, there are three special flux conditions, viz.

$$(J_C^k)^- = -\dot{m} c_C^-, \quad (J_D^k)^- = -\dot{m} c_D^-, \quad (J_E^k)^- = -\dot{m} c_E^-. \tag{8.123}$$

This example may also serve to illustrate that a sufficient number of conditions are available to determine all unknowns at the interface. To this end, we simplify the considered process, by assuming (i) mechanical and interfacial chemical equilibrium is established, (ii) the temperature is uniform and constant, (iii) the barycentric velocity, υ^i, vanishes, and (iv) the bulk fluxes are given by Fick's law

$$J_a^k = -\sum_{b=1}^{a_c} m_{ab} \frac{\partial(\mu_b - \mu_1)}{\partial x^k}. \tag{8.124}$$

Let us further assume that the mechanical problem, that we discussed above have already been solved. Then there remain eight unknowns at the interface, viz. seven concentrations, c_A^+, c_A^-, c_B^+, c_B^-, c_C^-, c_D^-, c_E^-, and the normal speed, w_ν, of the interface. For its determination, we thus need eight equations. The first six equations are already be given by the conditions (8.122) and (8.123), and there are two further jump conditions for the constituents A and B that may cross the interface, viz.

$$-\rho w_\nu (c_A^+ - c_A^-) + ((J_A^k)^+ - (J_A^k)^-)\nu^k = 0, \quad -\rho w_\nu (c_B^+ - c_B^-) + ((J_B^k)^+ - (J_B^k)^-)\nu^k = 0. \tag{8.125}$$

8.15.4 Identification of Driving Forces and Fluxes for Ordered Solids and with Vacancies

8.15.4.1 Statement of the Problem: Solid/Liquid Interfaces in Semi-insulating Gallium Arsenide

In this section, we consider the interface between a liquid and an ordered solid with sublattice structure and vacancies. Interfaces in semi-insulating Gallium Arsenide (GaAs) will serve as an important example, where the distinctions to the previous case with disordered solids can at best be illustrated.

Single-crystal GaAs contains the major elements gallium and arsenic, and additionally various trace elements, which are of most importance in order to fabricate semi-conducting or semi-insulating GaAs, respectively. All constituents occupy three sublattices of face-centered cubic (fcc) symmetry. The sublattices are indicated by three Greek letters: α, β, γ. The lattice α is dominantly occupied by gallium, while the arsenic is the major substance on the lattice sites of sublattice β. The sublattice γ indicates an interstitial lattice, where the dominant elements are vacancies. For a proper application of single-crystal semi-conducting or semi-insulating GaAs as a wafer material, it is crucial that the trace elements are distributed homogeneously on the lattice sites of the three sublattices. In order to remove dislocations, which appear during the process of crystal growth, a special heat treatment of the wafer becomes necessary. However, the heat treatment might start a diffusion process, where the trace elements move preferably toward the vicinity of the dislocations and find here thermodynamic conditions so that the formation of liquid droplets, dominantly formed by the arsenic, sets in. The complete thermodynamic model that takes care for 16 constituents is described in Wenzl et al. (1990, 1993), Dreyer and Duderstadt (2003a, b).

Here, we consider a simplified case with a solid/liquid interface. The three sublattices of the solid are occupied by gallium (Ga), arsenic (As), and vacancies (V). We assume that Ga is only found on sublattice α, while the other constituents can be found on all three sublattices. Thus, there are seven constituents in the solid, viz.

$$Ga_\alpha,\ As_\alpha,\ As_\beta,\ As_\gamma,\ V_\alpha,\ V_\beta,\ V_\gamma. \tag{8.126}$$

and Ga_L, As_L in the liquid, so that Ga and As can cross the interface, while the vacancies are created or annihilated here.

We are confronted here with two new problems: (i) There are several side conditions due to the fact that all three sublattices have the same number densities, n_{SL}, of sublattice sites. (ii) The creation or annihilation of vacancies at the interface is obviously not accompanied by creation or annihilation of mass.

8.15.4.2 Interfacial Entropy Inequality for Ordered Solid/Liquid Interfaces

The two mentioned problems require a change of the thermodynamic description by a transition from the one sided mass fluxes to one sided mole fluxes. There holds

$$\dot{m}_a^{+/-} = (\dot{m}c_a + J_a^k \nu^k)^{+/-} = M_a \dot{n}_a^{+/-} = M_a(\dot{n}y_a + j_a^k \nu^k)^{+/-} \quad \text{with} \quad \dot{m} = \sum_{a=1}^{a_c^{+/-}} M_a \dot{n}_a^{+/-}. \tag{8.127}$$

The newly introduced quantities are M_a—molecular weight, $\dot{n}$—total mole flux, $\dot{n}_a^{+/-}$—one sided mole flux, y_a—mole fraction and j_a^k—mole diffusion flux.

Next, we re-scale the chemical potentials according to $\tilde{\mu}_a = M_a \mu_a$. The unit of the chemical potential has thus been changed from *Joule/kg* to *Joule/mole*. We consider vacancies as massless particles, i.e., we set $M_{V_\alpha} = M_{V_\beta} = M_{V_\gamma} = 0$, however, nevertheless they are carrier of energy and entropy. For this reason, we introduce the following rule: The rescaled chemical potentials $\tilde{\mu}_{V_\alpha}$, $\tilde{\mu}_{V_\beta}$, $\tilde{\mu}_{V_\gamma}$ of the vacancies are nonzero.

The general interfacial entropy inequality (8.102) can thus be written

$$-T\varsigma_s = \sum_{a=1}^{a_c^{+/-}} M_a \dot{n}_a^{+/-} \left[\left[\frac{1}{2}(\upsilon - w)^2 - \frac{1}{\rho}\sigma^{<ik>} \nu^i \nu^k \right]\right] + \left[\left[\sum_{a=1}^{a_c} \dot{n}_a \tilde{\mu}_a \right]\right] \leq 0. \tag{8.128}$$

We assume that the $+/-$ refers to solid and liquid, respectively, and we write in detail

$$\begin{aligned} -T\varsigma_s = {} & (M_{Ga}\dot{n}_{Ga_\alpha} + M_{As}(\dot{n}_{As_\alpha} + \dot{n}_{As_\beta} + \dot{n}_{As_\gamma})) \left[\left[\frac{1}{2}(\upsilon - w)^2 - \frac{1}{\rho}\sigma^{<ik>} \nu^i \nu^k \right]\right] + \\ & \dot{n}_{Ga_\alpha}\tilde{\mu}_{Ga_\alpha} - \dot{n}_{Ga_L}\tilde{\mu}_{Ga_L} + \dot{n}_{As_\alpha}\tilde{\mu}_{As_\alpha} + \dot{n}_{As_\beta}\tilde{\mu}_{As_\beta} + \dot{n}_{As_\gamma}\tilde{\mu}_{As_\gamma} - \dot{n}_{As_L}\tilde{\mu}_{As_L} + \\ & \dot{n}_{V_\alpha}\tilde{\mu}_{V_\alpha} + \dot{n}_{V_\beta}\tilde{\mu}_{V_\beta} + \dot{n}_{V_\gamma}\tilde{\mu}_{V_\gamma} \leq 0. \end{aligned} \tag{8.129}$$

There are four additional side conditions:

(i) Conservation of mass

$$\dot{n}_{Ga_\alpha} = \dot{n}_{Ga_L}, \quad \dot{n}_{As_\alpha} + \dot{n}_{As_\beta} + \dot{n}_{As_\gamma} = \dot{n}_{As_L}. \tag{8.130}$$

(ii) Equal number densities of lattice sites

$$\dot{n}_{Ga_\alpha} + \dot{n}_{As_\alpha} + \dot{n}_{V_\alpha} = \dot{n}_{As_\beta} + \dot{n}_{V_\beta} = \dot{n}_{As_\gamma} + \dot{n}_{V_\gamma}. \tag{8.131}$$

These conditions serve to eliminate four one sided fluxes in the interfacial entropy inequality. We chose $\dot{n}_{Ga_L}$, $\dot{n}_{As_L}$, $\dot{n}_{V_\alpha}$, $\dot{n}_{V_\gamma}$, and obtain

$$
\begin{aligned}
-T\varsigma_s = {} & \dot{n}_{Ga_\alpha}\left(\tilde{\mu}_{Ga_\alpha} - \tilde{\mu}_{Ga_L} - \tilde{\mu}_{V_\alpha} + \left[\left[\frac{M_{Ga}}{2}(\upsilon - w)^2 - \frac{M_{Ga}}{\rho}\sigma^{<ik>}\nu^i\nu^k\right]\right]\right) + \\
& \dot{n}_{As_\alpha}\left(\tilde{\mu}_{As_\alpha} - \tilde{\mu}_{As_L} - \tilde{\mu}_{V_\alpha} + \left[\left[\frac{M_{As}}{2}(\upsilon - w)^2 - \frac{M_{As}}{\rho}\sigma^{<ik>}\nu^i\nu^k\right]\right]\right) + \\
& \dot{n}_{As_\beta}(\tilde{\mu}_{As_\beta} - \tilde{\mu}_{As_L} + \tilde{\mu}_{V_\alpha} + \tilde{\mu}_{V_\gamma} + \left[\left[\frac{M_{As}}{2}(\upsilon - w)^2 - \frac{M_{As}}{\rho}\sigma^{<ik>}\nu^i\nu^k\right]\right]) + \\
& \dot{n}_{As_\gamma}\left(\tilde{\mu}_{As_\gamma} - \dot{n}_{As_L}\tilde{\mu}_{As_L} - \tilde{\mu}_{V_\gamma} + \left[\left[\frac{M_{As}}{2}(\upsilon - w)^2 - \frac{M_{As}}{\rho}\sigma^{<ik>}\nu^i\nu^k\right]\right]\right) + \\
& \dot{n}_{V_\beta}(\tilde{\mu}_{V_\alpha} + \tilde{\mu}_{V_\beta} + \tilde{\mu}_{V_\gamma}) \leq 0.
\end{aligned}
\tag{8.132}
$$

8.15.4.3 Identification of Fluxes and Driving Forces

We compare the interfacial entropy inequality ((8.132)) with the general form (8.97) to identify five fluxes

$$
F_A = \{\dot{n}_{Ga_\alpha}, \dot{n}_{As_\alpha}, \dot{n}_{As_\beta}, \dot{n}_{As_\gamma}, \dot{n}_{V_\beta}\} \tag{8.133}
$$

and five driving forces

$$
\begin{aligned}
D_A = {} & \{\tilde{\mu}_{Ga_\alpha} - \tilde{\mu}_{Ga_L} - \tilde{\mu}_{V_\alpha} + \left[\left[\frac{M_{Ga}}{2}(\upsilon - w)^2 - \frac{M_{Ga}}{\rho}\sigma^{<ik>}\nu^i\nu^k\right]\right], \\
& \dot{n}_{As_\alpha}(\tilde{\mu}_{As_\alpha} - \tilde{\mu}_{As_L} - \tilde{\mu}_{V_\alpha} + \left[\left[\frac{M_{As}}{2}(\upsilon - w)^2 - \frac{M_{As}}{\rho}\sigma^{<ik>}\nu^i\nu^k\right]\right], \\
& \dot{n}_{As_\beta}(\tilde{\mu}_{As_\beta} - \tilde{\mu}_{As_L} + \tilde{\mu}_{V_\alpha} + \tilde{\mu}_{V_\gamma} + \left[\left[\frac{M_{As}}{2}(\upsilon - w)^2 - \frac{M_{As}}{\rho}\sigma^{<ik>}\nu^i\nu^k\right]\right], \\
& \tilde{\mu}_{As_\gamma} - \dot{n}_{As_L}\tilde{\mu}_{As_L} - \tilde{\mu}_{V_\gamma} + \left[\left[\frac{M_{As}}{2}(\upsilon - w)^2 - \frac{M_{As}}{\rho}\sigma^{<ik>}\nu^i\nu^k\right]\right], \\
& \tilde{\mu}_{V_\alpha} + \tilde{\mu}_{V_\beta} + \tilde{\mu}_{V_\gamma}\}.
\end{aligned}
\tag{8.134}
$$

8.15.4.4 Interfacial Equilibrium Conditions for Solid/Liquid Interfaces in GaAs

In this section, we assume that all considered processes are diffusion controlled, so that the driving forces are zero. We neglect as usual the kinetic energy, and in this case, the resulting conditions read

$$
\begin{aligned}
&\tilde{\mu}_{Ga_\alpha} - \tilde{\mu}_{Ga_L} - \tilde{\mu}_{V_\alpha} = \left[\left[\frac{M_{Ga}}{\rho}\sigma^{<ik>}\nu^i\nu^k\right]\right],\\
&\tilde{\mu}_{As_\alpha} - \tilde{\mu}_{As_L} - \tilde{\mu}_{V_\alpha} = \left[\left[\frac{M_{As}}{\rho}\sigma^{<ik>}\nu^i\nu^k\right]\right],\\
&\tilde{\mu}_{As_\beta} - \tilde{\mu}_{As_L} + \tilde{\mu}_{V_\alpha} + \tilde{\mu}_{V_\gamma} = \left[\left[\frac{M_{As}}{\rho}\sigma^{<ik>}\nu^i\nu^k\right]\right],\\
&\tilde{\mu}_{As_\gamma} - \tilde{\mu}_{As_L} - \tilde{\mu}_{V_\gamma} = \left[\left[\frac{M_{As}}{\rho}\sigma^{<ik>}\nu^i\nu^k\right]\right],\\
&\tilde{\mu}_{V_\alpha} + \tilde{\mu}_{V_\beta} + \tilde{\mu}_{V_\gamma} = 0.
\end{aligned}
\tag{8.135}
$$

These five conditions together with the four side conditions (8.130) and (8.131) yield nine algebraic equations, that may be used to determine the nine interfacial number densities of the constituents if the mechanical part of the problem has been solved.

In case that the bulk phases are homogeneous and in equilibrium, there exist an alternative method to derive the jump conditions (8.135). This method relies on the exploitation of the statement, that for given and fixed values of temperature and external load, equilibrium in a two phase mixture is established by the minimum of the available free energy. This approach is carried out for the *GaAs* problem without mechanical phenomena by Wenzl et al. (1990), and Wenzl et al. (1990), and with mechanical/chemical coupling by Dreyer and Duderstadt (2003a, b).

In either case, the jump conditions (8.135) that apply when ordered solids are involved, cannot be obtained as a special case from the jump conditions (8.115), which give the classical textbook results for interfacial conditions. We have shown here that the latter only hold in disordered mixtures, a fact which is not mentioned in many textbooks.

8.16 Local Equations of Balance in Lagrange Coordinates

8.16.1 Equations of Balance for Partial Masses, Momentum, Energy, and Entropy

The special jump conditions (8.77), which are formulated by means of Euler coordinates, will now form the base for a derivation of the corresponding jump conditions in Lagrange coordinates.

We start with the relations $(8.77)_1$, (8.39), (8.40) involving normal vectors and interfacial normal speeds in Euler and Lagrange coordinates:

$$
\dot{m} = -\rho J\sqrt{\frac{G}{g}}W_N, \quad \text{with} \quad [[\dot{m}]] = 0, \quad \nu^i = \sqrt{\frac{G}{g}}J\overset{-1}{F}{}^{ki}N^k, \quad w^i = v^i_{+/-} + F^{ij}_{+/-}W^j. \tag{8.136}
$$

We conclude from $(8.136)_1$ and $(8.136)_3$ that the products ρJ and $J\overset{-1}{F^{ki}}N^k$ are continuous across the interface:

$$[[\rho J]] = 0, \quad [[J\overset{-1}{F^{ki}}]]N^k = 0. \tag{8.137}$$

Recall that we consider as before exclusively the case, where the barycentric velocity is restricted to normal motion, i.e., $\upsilon^i = \upsilon_\nu \nu^i$ at the interface. Furthermore, we restrict the surface parameters so that a surface point moves exclusively in normal direction, i.e., we set $w^i = w_\nu \nu^i$. In this case, the eqn. $(8.136)_4$ implies $\tau^i_\Delta F^{ij} W^j = 0$.

Finally, we recall the definitions (8.52), and define accordingly

$$\rho_0 = \rho J, \quad \rho_{0a} = \rho_a J, \quad \hat{J}^k_a = J\overset{-1}{F^{ki}} J^i_a, \quad \hat{q}^k = J\overset{-1}{F^{ki}} q^i. \tag{8.138}$$

We insert these relations and definitions into the jump conditions (8.77). There follow the corresponding jump conditions in Lagrange coordinates:

$$\begin{aligned}
&-\rho_0 W_N[[\upsilon^i]] - [[\hat{\sigma}_{ik}]]N^k = \sqrt{\frac{g}{G}}((S^{\Delta\Gamma}_{;\Delta} - S^\Delta b^\Gamma_\Delta)\tau^i_\Gamma + (S^{\Delta\Gamma} b_{\Delta\Gamma} + S^\Delta_{;\Delta})\nu^i)\\
&-\rho_0 W_N \left[\left[\left(u + \frac{1}{2}W_N^2 F^{ij}F^{ij}\right)\delta^{ks} - \frac{1}{\rho_0}\hat{\sigma}^{ik}F^{is}\right]\right] N^k N^s + [[\hat{q}^k]]\nu^k =\\
&\sqrt{\frac{g}{G}}(-\frac{\partial u_s}{\partial t} - u_s(w^\Delta_{\tau;\Delta} - 2k_M w_\nu) + S^{\Delta\Gamma}(g_{\Gamma\Sigma} w^\Sigma_{\tau;\Delta} - b_{\Delta\Gamma} w_\nu) + S^\Delta(\frac{\partial w_\upsilon}{\partial U^\Delta} + w^\Sigma_\tau b_{\Sigma\Delta}))\\
&-\rho_0 W_N[[c_a]] + [[\hat{J}^k_a]]N^k = \sqrt{\frac{g}{G}}\sum_{r=1}^{N_R} \gamma^r_a M_a \Gamma^r\\
&-\rho_0 W_N[[s]] + \left[\left[\frac{1}{T}(\hat{q}^k - \sum_{a=1}^{as} \mu_a \hat{J}^k_a)\right]\right] N^k = \sqrt{\frac{g}{G}}(-\frac{\partial s_s}{\partial t} - s_s(w^\Delta_{\tau;\Delta} - 2k_M w_\nu) + \varsigma_s)\\
&\qquad \text{with} \quad \varsigma_s \geq 0.
\end{aligned} \tag{8.139}$$

Note that the right-hand sides of the eqn. (8.139) are not transformed into Lagrange coordinates.

The form of the kinetic energy in the internal energy balance $(8.139\,)_3$ results as follows:

$$\begin{aligned}
\left[\left[\frac{1}{2}(\upsilon - W)^2\right]\right] &= \left[\left[\frac{1}{2}F^{ij}F^{ik}\right]\right] W^j W^k = \left[\left[\frac{1}{2}F^{ij}F^{ik}\right]\right] N^j N^k W_N^2 =\\
&= \langle F^{ij} \rangle [[F^{ik}]]N^j N^k W_N^2 = \frac{1}{2}[[W_N^2 F^{ij} F^{ij}]],
\end{aligned} \tag{8.140}$$

where the last equality relies on the Hadamard condition $(8.30)_2$, which implies

$$< F^{ij} > [[F^{ik}]]N^j N^k =< F^{ij} > [[F^{ij}]]. \tag{8.141}$$

8.16.2 *Interfacial Entropy Production*

In this section, we derive the interfacial entropy production corresponding to its counterpart in Euler coordinates. As before we consider exclusively, the special case where the temperature is continuous across the interface, $[[T]] = 0$. We multiply the interfacial entropy balance $(8.139)_4$ by T, and subtract the resulting equation from the balance of internal energy $(8.139)_4$. There follows the interfacial entropy production in Lagrange coordinates

$$-\rho_0 W_N \left[\left[(\psi + \frac{1}{2} W_N^2 F^{ij} F^{ij})\delta^{ks} - \frac{1}{\rho_0}\hat{\sigma}^{ik} F^{is}\right]\right] N^k N^s + \left[\left[\sum_{a=1}^{a_S} \mu_a \hat{J}_a^k)\right]\right] N^k = \sqrt{\frac{g}{G}}(-T\varsigma_s) \leq 0. \tag{8.142}$$

The quantity $\psi\delta^{ks} - \frac{1}{\rho_0}\hat{\sigma}^{ik} F^{is}$ is known as the Eshelby tensor in the mechanical literature. Eshelby has introduced this tensor in Eshelby (1975), see also Liu (1992) and Mura (1982) for further examples that illustrate the role of the Eshelby tensor.

In order to obtain from (8.142) useful equilibrium conditions and growth laws, we have at first to specialize the second bracket involving chemical potentials and diffusion fluxes. However, this procedure can be done according to the same strategy that we have developed in Sect. 8.15, and will not carried out here.

Acknowledgements The author gratefully acknowledges most fruitful discussions with Ingo Müller and Dietmar Hömberg, who suggested to study the topic of this paper.

References

Abeyaratne R, Knowles JK (1990) On the driving traction acting on a surface of strain discontinuity in a continuum. J Mech Phys Solids 38(3):345–360

Alexander JID, Johnson WC (1985) Thermomechanical equilibrium in solid-fluid systems with curved interfaces. J Appl Phys 58(2):816–824

Alt HW (29 Aug 2001) On the entropy principle for surface diffusion. Lecture given in the workshop multiscale problems and phase transitions at WIAS Berlin

Aris R (1996) Vectors, tensors and the basic equations of fluid mechanics. Prentice Hall, Eaglewood Cliffs, N.J.

Caginalp G (1991) Phase field models and sharp interfaces: some differences in subtle situations. Rocky Mt J Math 21(2):617–629

Cahn JW, Elliott CM, Novick-Cohen A (1996) The Cahn-Hilliard equation with a concentration dependent mobility: motion by minus the Laplacian of the mean curvature. Euro J Appl Math 7

Clausius R (1850) Über eine veränderte Form des zweiten Hauptsatzes der mechanischen Wärmetheorie. Poggendorf's Annalen der Physik 93
Dreyer W (1981) On thermodynamics and kinetic theory of ideal rubber membranes. J Appl Mech 48(2)
Dreyer W, Duderstadt F (2003) Modeling of the appearance of arsenic droplets in galliumarsenide. Preprint, WIAS
Dreyer W, Duderstadt F (2003) On the Becker/Döring theory of nucleation of liquid droplets in solids. Preprint, WIAS
Dreyer W, Müller I, Strehlow P (1982) A study of equilibria of interconnected rubber balloons. Q J Mech Appl Math 25(3)
Dreyer W, Wagner B (2003) Sharp interface model for eutectic alloys. Part I: concentration dependent surface tension, Preprint, WIAS
Eshelby JD (1975) The elastic energy momentum tensor. J Elast 5:321–335
Garcke H, Nestler B, Stoth B (1998) On anisotropic order parameter models for multi-phase systems and their sharp interface limits. Phys D 115:87–108
Gibbs JW (1940) On the equilibrium of heterogeneous substances. In: Collected works, vol 1. Yale University Press, New Haven
Grauel A (1980) Thermodynamics of an interfacial fluid membrane. Phys A 103:468–520
Gurtin ME, Voorhees PW (1996) The thermodynamics of evolving interfaces far from equilibrium. Acta Mater 44(1)
Gurtin ME, Voorhees PW (1998) On the effects of elastic stress on the motion of fully faceted interfaces. Acta Mater 46(6)
Herring C (1951) Surface tension as a motivation for sintering. In: Kingston WE (ed) The physics of powder metallurgy, Chap. 8. McGraw-Hill, New York
Hoffmann DW, Cahn JW (1972) A vector thermodynamics for anisotropic surfaces I: fundamentals and applications to plane surface junctions. Surf Sci 31
Hoffmann DW, Cahn JW (1974) A vector thermodynamics for anisotropic surfaces II: Curved and faceted surfaces. Acta Mater 22
Johnson WC (1987) Precipitate shape evolution under applied stress-thermodynamics and kinetics. Met Trans 18A
Laplace PS (1806) Traité de méchanique céleste, vol 4. Courcier, Paris
Liu IS (1992) On interface equilibrium and inclusion problems. Contin Mech Thermodyn 4:177–186
Meirmanov AM (1992) The Stefan problem. de Gruyter, Berlin, New York
Moeckel GP (1975) Thermodynamics of an interface. Arch Rat Mech Anal 57:255–280
Müller I (1985) Thermodynamics. Pitman advanced publishing program, Boston. Interaction of mechanics and mathematics series
Müller I (2001) Grundzüge der Thermodynamik, 3. Auflage. Springer, Berlin
Mura T (1982) Micromechanics of defects in solids. Martinus Nijhoff Publishers, The Hague, Boston, New York
Naghdi PM (1960) The theory of shells and plates. Handbuch der Physik, vol VIa/2. Springer, Berlin, Göttingen, Heidelberg
Nestler B, Wheeler AA (1998) Anisotropic multi-phase-field model: interfaces and junctions. Phys Rev E 57(3):2602–2609
Ono S, Kondo S (1960) Molecular theory of surface tension, Handbuch der Physik, vol X. Springer, Berlin, Göttingen, Heidelberg
Theil F (20 Sep 2002) Surface energies in lattice models. Lecture given in the workshop multiscale problems in quantum and classical mechanics, averaging techniques and young measures WIAS Berlin
Thomson W (1870) On the equilibrium of Vapour at a curved surface of liquid. Proc Roy Soc Edinb 7(63)
Truesdell C, Toupin R (1960) The classical field theories. Handbuch der Physik, vol III/1. Springer, Berlin, Göttingen, Heidelberg

Wenzl H, Mika K, Henkel D (1990) Thermochemical analysis of native point defects and substitutional carbon and baron in GaAs crystal growth. Cryst Res Technol

Wenzl H, Oates WA, Mika K (1993) Defect thermodynamics and phase diagrams in compound crystal growth process. Handbook of crystal growth, vol 1. Elsevier

Wheeler AA, McFadden GB (1997) On the notion of a ξ vector and a stress tensor for a general class of anisotropic diffuse interface models. Proc Roy Soc Lond A 453(1963):1611–1630

Chapter 9
Balance Laws and Transport Theorems for Flows with Singular Interfaces

Ferdinand Thein and Gerald Warnecke

Dedication: This paper is dedicated to the memory of Wolfgang Dreyer. For both authors he was a very important source of knowledge as well as a partner for fruitful discussions on the topics of continuum mechanics and thermodynamics, especially concerning phase transitions.

Abstract This paper gives a concise but rigorous mathematical description of a material control volume that is separated into two parts by a singular surface at which physical states are discontinuous. The geometrical background material is summarized in a unified manner. Transport theorems for use in generic balance laws are given with proofs since they provide some insight into the results. Also the step from integral balances to differential equations is given in some detail.

9.1 Introduction

In this work, we present a general mathematical derivation of balance laws involving curved interfaces. These may be shock fronts, phase boundaries, reaction fronts, or surfaces containing surfactants. This includes the possibility of modeling physical processes, such as phase transitions or chemical reactions, and flow of substances within the interface. We want to provide a compact, self-contained reference for mathematical modeling with hyperbolic balance laws, mathematical analysis, and numerical computations of engineering applications using such balance laws.

F. Thein
Institut für Mathematik, Johannes-Gutenberg-Universität Mainz, Mainz, Germany
e-mail: fthein@uni-mainz.de

G. Warnecke (✉)
Institute for Analysis and Numerics, Otto-von-Guericke University Magdeburg, Magdeburg, Germany
e-mail: gerald.warnecke@ovgu.de

J. Fuhrmann et al. (eds.), *Advances in Continuum Physics*, Advanced Structured Materials 238, https://doi.org/10.1007/978-3-031-93918-1_9

We consider transport theorems, jump conditions as well as the influence of curvature of the surfaces.

The main aim of this paper is a rigorous mathematical derivation of some general surface differential equations in Sect. 9.4.5 that combine differences in bulk quantities with dynamics on the surface. We give a unified treatment of volume and surface balances from a mathematical point of view that draws on multiple sources in the literature. They only provide partial aspects of our exposition. Also these sources are not easily compared due to differences in assumptions, notation, and presentation. We will show how they are related. An important point is the clarification of various derivatives, that are in use, of quantities defined on surfaces. Our proofs make heavy use of parametrizations of the surfaces. But the results that we obtain are independent of the choice of parametrization.

Our work was motivated by a preprint of Dreyer (2003) that has been included in this book. This again was based on Müller (1985). We focus on a detailed mathematical description of moving surfaces and the derivation of generic balance laws in the volume as well as on the surface. This includes some proofs of fundamental theorems which we include because they give insight on how certain terms in the resulting differential equations are generated. For example, in the derivation of Müller (1985, (3.7)) a flux correction is postulated. In the proof of the generalized Reynolds Transport Theorem 9.4.4, one can see how this correction arises from the description of a moving surface by parametrization and flow maps. Further sources of background material are Aris (1989), Cermelli et al. (2005), Dziuk and Elliott (2007), Grinfeld (2013), Gurtin (1993), Petryk and Mróz (1986) as well as Slattery et al. (2007).

The dynamics of the fluid motion can be described by certain physical laws. These laws are in general referred to as *balance laws*. They are formulated by applying basic physical principles to moving material volumes and surfaces of arbitrary shape and size. Transport theorems and mathematical arguments are needed to derive differential equations from the integral balances. Our derivation of differential equations on volumes and surfaces is based on Dreyer (2003) and Müller (1985). We also recommend Aris (1989) as well as Truesdell and Toupin (1960). The latter is also very interesting from a historical perspective. Truesdell and Toupin (1960) present the results in the context of original works. Thus the influences of mathematicians like *Christoffel, Euler, Hadamard, Hugoniot, Reynolds* on this topic become visible. A treatment of this topic in two dimensions can be found in Gurtin (1993). Truesdell and Toupin (1960) did not treat dynamics on surfaces. Dziuk and Elliott (2007) only considered moving hypersurfaces in $\mathbb{R}^{n+1}$ and no volume dynamics. Another very useful reference on interfacial transport is Slattery et al. (2007).

We consider here the transport of control volumes and moving surfaces that remain inside some domain $\Omega \subseteq \mathbb{R}^3$. An important aspect that we do not take into account is the additional complication that arises when these interact with the boundary $\partial\Omega$ of a bounded domain. The interaction of the surface with the boundary of a domain was studied by Dell'Isola and Romano (1986), Gurtin et al. (1989) as well as Estrada and Kanwal (1991).

From a mathematical point of view, balance laws are often conservation laws in the sense that the differential equations are given in divergence form. The field of conservation laws, even when restricted to particular equations, is far too extensive to be completely discussed here. For a brief overview of the mathematical results for balance laws, in particular *hyperbolic conservation laws*, we refer to the following literature. Evans (1998) contains a compact description of the topic, especially of systems of conservation laws. Further we recommend Dafermos (2016), LeFloch (2002), and Smoller (2012). Especially Dafermos (2016) starts with a very general approach, introducing the balance laws in terms of suitable measures and further gives a nice historical review of the topic. Smoller (2012) does not focus on hyperbolic conservation laws alone, reaction–diffusion equations are also treated. An introduction and discussion of certain analytical methods for conservation laws can be found in Warnecke (1999). Some more recent analytical results were presented in Feireisl et al. (2021). For additional literature, see the references in the literature just mentioned.

In this paper, we do not provide specific balances for physical quantities. The latter can be found in Bedeaux (1986), Dreyer (2003), Müller (1985) as well as Truesdell and Toupin (1960). Mavrovouniotis and Brenner (1993) discussed a micro–macro approach to deriving surface balances as surface excess properties, a concept introduced by Gibbs. Further derivations of balance laws in the presence of interfaces may be found in Anderson et al. (2007), Bedeaux et al. (1976), Dell'Isola and Romano (1987) as well as Slattery et al. (2007).

The key contribution from physics to this topic is provided by the theory of irreversible processes and non-equilibrium thermodynamics. Some theory of non-equilibrium thermodynamics is given in Bedeaux (1986), Mauri (2013) as well as De Groot and Mazur (2013). An introduction to balance laws from a thermodynamic point of view can be found in Landau and Lifshitz (1987), Müller (1985) as well as Müller and Müller (2009).

We begin by introducing in Sect. 9.2 various concepts for stationary surfaces. This is then followed by the extension to moving surfaces. In the next section, we formulate generic integral balance laws that can be used for physical states such as mass, momentum, or energy. Section 9.4 contains various integral theorems for the volume and for surfaces that are needed in order to derive differential equations from the integral balances. These include balances for singular moving surfaces in the volume domain at which physical states may be discontinuous. A final section contains some comments on the lower dimensional cases.

9.2 Calculus of Stationary and Moving Surfaces

The aim of the present section is to briefly summarize the most important results for moving surfaces. We refer to the works by Aris (1989), Dziuk and Elliott (2007), Grinfeld (2013) and the references therein. But, first we have to introduce various concepts for stationary surfaces.

We will make extensive use of tensors. For a detailed introduction refer to the literature, e.g., Aris (1989, Chap. 7) or Grinfeld (2013, Chap. 6). A tensor is a quantity which has certain properties under the transformation of the coordinate system. The order of a tensor is given by the number of indices. A tensor with upper indices is called *contravariant* and with lower indices *covariant*. A tensor may also be of mixed type with upper and lower indices. The indices in a term consisting of indexed quantities that appear only once are called *free indices*. They should appear in each term on each side of an equation exactly once.

In the following, we will use the notational conventions:

(i) Lowercase Greek indices refer to the surface and may take the values $\{1, 2\}$. In contrast, lowercase Latin indices refer to the ambient space, i.e. the standard $\mathbb{R}^3$ in our case.

(ii) Bold lowercase letters refer to vectors or tensors of order 1 in $\mathbb{R}^3$, bold uppercase letters to matrices or tensors of order 2.

(iii) We use the Einstein summation convention, i.e., summation applies to every index which appears twice. Once as superscript and once as subscript. Such an index is often called *dummy index* and may be renamed. See the following two examples of the scalar product in two and three space dimensions

$$\mathbf{a} \cdot \mathbf{b} = a^1 b_1 + a^2 b_2 = a^\alpha b_\alpha = a^\beta b_\beta,$$

$$\mathbf{x} \cdot \mathbf{y} = \sum_{i=1}^{3} \mathrm{x}_i \mathrm{y}_i = \mathrm{x}^i \mathrm{y}_i = \mathrm{x}^j \mathrm{y}_j.$$

We will later also apply the convention to some summations from 1 to n using lowercase Latin letters. We call the summation of a dummy index a *contraction*. Tensors may change in a certain way when transforming the coordinate system. A contraction is invariant under the transformation and thus it is one of the fundamental operations in tensor calculus, see Grinfeld (2013, Chap. 6). Also the contraction of tensors produces again a tensor.

9.2.1 Stationary Surfaces

A stationary two-dimensional *surface* Σ embedded in the ambient space $\mathbb{R}^3$ may be described through a parametrization depending on a two-dimensional coordinate system. We want to specify this in the following. Let $\mathcal{U} \subseteq \mathbb{R}^2$ as well as $\Omega \subseteq \mathbb{R}^3$ be open subsets, $\mathbf{u} \in \mathcal{U}$ and $\boldsymbol{\Phi} : \mathcal{U} \to \Omega$ a bijective smooth parametrization such that

$$\Sigma \equiv \boldsymbol{\Phi}(\mathcal{U}) = \left\{ \mathbf{x} \in \Omega \mid \mathbf{x} = \boldsymbol{\Phi}(\mathbf{u}) = (\Phi^1(\mathbf{u}), \Phi^2(\mathbf{u}), \Phi^3(\mathbf{u}))^T \text{ for } \mathbf{u} \in \mathcal{U} \right\}. \quad (9.1)$$

By a *smooth* function on an open set we will always mean that it is continuously differentiable on the set as often as is necessary for formulas to be meaningful. It may also imply continuous bounded extensions to the boundary of the set when values on the boundary are needed.

Alternatively, such a surface can be defined implicitly as the level set of a smooth function $f : \Omega \to \mathbb{R}$. We will introduce such a description more generally for moving surfaces in Sect. 9.2.3. This is useful in integration theory, see, e.g. Thomas et al. (2004, Sect. 16.5), and in some applications, especially when dealing with a changing topology of moving surfaces. A drawback is that for such a description of a surface one can only determine a normal velocity vector. In contrast, an explicit description of a moving surface determines via (9.34) a velocity vector of the surface. This explicit description will be a time-dependent extension of the one given above.

On the surface we must consider three kinds of quantities, such as tensors. There are those that are defined on the surface only and are described by the surface coordinates $\mathbf{u}$ and depend only on them. Then there are those, such as tangent and normal vectors, that have coordinates in the ambient space but depend on the surface coordinates. Further, there are those that have extensions in a neighborhood around the surface. Their nature determines what kind of operations are possible with or on them.

The *tangent vector fields* $\boldsymbol{\tau}_\alpha \in \mathbb{R}^3$ and the *unit normal vector field* $\boldsymbol{\nu} \in \mathbb{R}^3$ to the surface Σ embedded in $\mathbb{R}^3$ are given by

$$\boldsymbol{\tau}_\alpha = \frac{\partial \boldsymbol{\Phi}}{\partial u_\alpha}, \ \alpha = 1, 2 \quad \text{and} \quad \boldsymbol{\nu} = \frac{\boldsymbol{\tau}_1 \times \boldsymbol{\tau}_2}{|\boldsymbol{\tau}_1 \times \boldsymbol{\tau}_2|}. \tag{9.2}$$

Since we want the tangents to be linearly independent the Jacobian of $\boldsymbol{\Phi}$ should be of rank two. The components of the *surface metric tensor*[1] $\underline{G}$ are given as

$$g_{\alpha\beta} = \tau_{\alpha;i}\tau_{\beta;i} = \boldsymbol{\tau}_\alpha \cdot \boldsymbol{\tau}_\beta, \tag{9.3}$$

cp. Aris (1989, (9.13.3)) or Grinfeld (2013, (5.7)). Thus the metric tensor $g_{\alpha\beta}$ is positive definite and symmetric. The inverse of the metric tensor is denoted by $g^{\alpha\beta}$. With the *Kronecker delta tensor* satisfying $\delta^\alpha_\gamma = 1$ for $\alpha = \gamma$ and $= 0$ for $\alpha \neq \gamma$ we have

$$g^{\alpha\beta} g_{\beta\gamma} = \delta^\alpha_\gamma \quad \text{and} \quad g = \det(g_{\alpha\beta}) = g_{11}g_{22} - g_{12}^2. \tag{9.4}$$

We consider the ϵ_{ijk} or *permutation tensor* for the cross product and determinant, see Aris (1989, (2.32.4)) or Grinfeld (2013, (9.8)). Its components are $\epsilon_{123} = \epsilon_{231} = \epsilon_{312} = 1$, $\epsilon_{132} = \epsilon_{321} = \epsilon_{213} = -1$ and all other components equal 0. So we have 1 for the even cyclic permutations of 123, -1 for the odd permutations, and 0 for all repeated indices like ϵ_{112}. This gives $(\mathbf{a} \times \mathbf{b})_i = \epsilon_{ijk} a_j b_k$ and $\det \mathbf{A} = \epsilon_{ijk} a_{jk}$ for a matrix $\mathbf{A} = \left(a_{ij}\right)_{1 \le i,j \le 3}$. With the ϵ-δ-identity

[1] Aris (1989, Sect. 9.13) denoted this tensor by $a_{\alpha\beta}$ and introduced an additional metric tensor g_{ij} of $\mathbb{R}^3$ in an exercise. For simplicity, we assume the usual Euclidean scalar product on $\mathbb{R}^3$ where this tensor corresponds to the identity matrix. In Aris (1989, Chap. 10), he used such an extra metric tensor in an intrinsic formulation of the surface in $\mathbb{R}^2$.

$$\epsilon_{ijk}\epsilon_{irs} = \delta_{jr}\delta_{ks} - \delta_{js}\delta_{kr}, \tag{9.5}$$

see, e.g., Aris (1989, Exercise 2.32.1), we have

$$\begin{aligned}|\boldsymbol{\tau}_1 \times \boldsymbol{\tau}_2|^2 =& \epsilon_{ijk}\tau_{\alpha;j}\tau_{\beta;k}\epsilon_{irs}\tau_{\alpha;r}\tau_{\beta;s} = (\delta_{jr}\delta_{ks} - \delta_{js}\delta_{kr})\tau_{\alpha;j}\tau_{\beta;k}\tau_{\alpha;r}\tau_{\beta;s}\\ =& \tau_{\alpha;j}\tau_{\beta;k}\tau_{\alpha;j}\tau_{\beta;k} - \tau_{\alpha;j}\tau_{\beta;k}\tau_{\alpha;k}\tau_{\beta;j} = g_{\alpha\alpha}g_{\beta\beta} - (g_{\alpha\beta})^2 = g.\end{aligned}$$

This gives

$$|\boldsymbol{\tau}_1 \times \boldsymbol{\tau}_2| = \sqrt{g}. \tag{9.6}$$

Let $\psi : \Sigma \to \mathbb{R}$ be a suitable function such that $\psi \circ \boldsymbol{\Phi}$ is integrable. We define the surface integral over Σ as

$$\int_{\Sigma} \psi(\mathbf{x})\,\mathrm{d}S = \int_{\mathcal{U}} (\psi \circ \boldsymbol{\Phi})(\mathbf{u})\,|\boldsymbol{\tau}_1 \times \boldsymbol{\tau}_2|\,\mathrm{d}u^1\mathrm{d}u^2 = \int_{\mathcal{U}} (\psi \circ \boldsymbol{\Phi})(\mathbf{u})\,\sqrt{g}\,\mathrm{d}u^1\mathrm{d}u^2, \tag{9.7}$$

cp. Thomas et al. (2004, Sect. 16.6) or Bourne and Kendall (1992, Sect. 5.6).

Another important observation is that a vector may be decomposed into its tangential and normal components. This is due to the fact that the surface tangents and the surface normal form a basis of $\mathbb{R}^3$. For a vector field $\mathbf{a} : \Omega \to \mathbb{R}^3$, we thus may write

$$\mathbf{a} = a_{\tau}^{\alpha}\boldsymbol{\tau}_{\alpha} + a_{\nu}\boldsymbol{\nu} \tag{9.8}$$

with the *tangential components* a_{τ}^{α} and *normal component* $a_{\nu} = \mathbf{a} \cdot \boldsymbol{\nu}$.

When we move along the surface the tangents and the normal will change according to the coordinates. First we want to discuss the tangents. From geometric intuition, it is clear that in general the tangent will change in tangential and normal direction. Thus we define according to (9.8) a representation of the derivatives of $\boldsymbol{\tau}_{\alpha}$ as

$$\frac{\partial \boldsymbol{\tau}_{\alpha}}{\partial u_{\beta}} = \Gamma_{\alpha\beta}^{\gamma}\boldsymbol{\tau}_{\gamma} + b_{\alpha\beta}\boldsymbol{\nu}. \tag{9.9}$$

The normal components $b_{\alpha\beta}$ of the tangent vectors are called the *curvature tensor* and are given as

$$b_{\alpha\beta} = \frac{\partial \boldsymbol{\tau}_{\alpha}}{\partial u_{\beta}} \cdot \boldsymbol{\nu} = \frac{\partial^2 \boldsymbol{\Phi}}{\partial u_{\alpha}\partial u_{\beta}}\boldsymbol{\nu}. \tag{9.10}$$

Roughly speaking, they give the change of the tangent vectors in normal direction with respect to the coordinates. Again the geometric view is quite clear since we expect a zero normal component of the derivative of a tangent for a flat surface, which has curvature zero.

Along with the curvature tensor one can define the *mean curvature*[2] κ_M given by

$$\kappa_M = \frac{1}{2} g^{\alpha\beta} b_{\alpha\beta}, \tag{9.11}$$

see Aris (1989, (9.37.1) and (9.38.7)).

The tangential components in (9.9) are called the *Christoffel symbols*. They can be calculated by multiplying (9.9) with $g^{\delta\varepsilon}$ from the left and $\boldsymbol{\tau}_\delta$ from the right as well as using (9.3) to give

$$g^{\delta\varepsilon} \frac{\partial \boldsymbol{\tau}_\alpha}{\partial u_\beta} \cdot \boldsymbol{\tau}_\delta = \Gamma^\gamma_{\alpha\beta} g_{\gamma\delta} g^{\delta\varepsilon} = \Gamma^\varepsilon_{\alpha\beta}. \tag{9.12}$$

We expand the left-hand side and use the fact that $\boldsymbol{\tau}_\alpha$ was obtained as a derivative to obtain

$$\begin{aligned} g^{\delta\varepsilon} \frac{\partial \boldsymbol{\tau}_\alpha}{\partial u_\beta} \cdot \boldsymbol{\tau}_\delta &= \frac{1}{2} g^{\delta\varepsilon} \left(\frac{\partial \boldsymbol{\tau}_\alpha}{\partial u_\beta} \cdot \boldsymbol{\tau}_\delta + \frac{\partial \boldsymbol{\tau}_\alpha}{\partial u_\beta} \cdot \boldsymbol{\tau}_\delta \right) = \frac{1}{2} g^{\delta\varepsilon} \left(\frac{\partial \boldsymbol{\tau}_\alpha}{\partial u_\beta} \cdot \boldsymbol{\tau}_\delta + \frac{\partial \boldsymbol{\tau}_\beta}{\partial u_\alpha} \cdot \boldsymbol{\tau}_\delta \right) \\ &= \frac{1}{2} g^{\delta\varepsilon} \left(\frac{\partial (\boldsymbol{\tau}_\alpha \cdot \boldsymbol{\tau}_\delta)}{\partial u_\beta} - \frac{\partial \boldsymbol{\tau}_\delta}{\partial u_\beta} \cdot \boldsymbol{\tau}_\alpha + \frac{\partial (\boldsymbol{\tau}_\beta \cdot \boldsymbol{\tau}_\delta)}{\partial u_\alpha} - \frac{\partial \boldsymbol{\tau}_\delta}{\partial u_\alpha} \cdot \boldsymbol{\tau}_\beta \right) \\ &= \frac{1}{2} g^{\delta\varepsilon} \left(\frac{\partial g_{\alpha\delta}}{\partial u_\beta} + \frac{\partial g_{\beta\delta}}{\partial u_\alpha} - \frac{\partial (\boldsymbol{\tau}_\alpha \cdot \boldsymbol{\tau}_\beta)}{\partial u_\delta} \right). \end{aligned}$$

With (9.12) we finally have, cp. Aris (1989, (9.91.2)),

$$\Gamma^\gamma_{\alpha\beta} = \frac{1}{2} g^{\gamma\delta} \left(\frac{\partial g_{\alpha\delta}}{\partial u_\beta} + \frac{\partial g_{\beta\delta}}{\partial u_\alpha} - \frac{\partial g_{\alpha\beta}}{\partial u_\delta} \right). \tag{9.13}$$

The curvature tensor and the Christoffel symbols are quantities which are intrinsic to the surface, see Aris (1989, Sects. 9.13 and 9.21) or Grinfeld (2013, Sects. 5.9, 5.12, and 10.9). From (9.13) or the fact that $\frac{\partial \boldsymbol{\tau}_\alpha}{\partial u_\beta} = \frac{\partial \boldsymbol{\tau}_\beta}{\partial u_\alpha} = \frac{\partial^2 \boldsymbol{\Phi}}{\partial u_\alpha \partial u_\beta}$ in (9.9), it is clear that the Christoffel symbol is symmetric in the lower indices.

Now we discuss how the change of the unit normal can be calculated. Intuitively there should be no change in their normal component. Indeed this can be seen differentiating $\boldsymbol{\nu} \cdot \boldsymbol{\nu} = 1$. We therefore may write

$$\frac{\partial \boldsymbol{\nu}}{\partial u_\alpha} = T^\beta_\alpha \boldsymbol{\tau}_\beta. \tag{9.14}$$

In the following, we want to determine the tangential components T^β_α. Therefore, we differentiate $\boldsymbol{\nu} \cdot \boldsymbol{\tau}_\gamma = 0$ with respect to u^α

[2] The factor $1/2$ coming from the arithmetic mean of the principal curvatures is the usual convention. It is also possible to define the mean curvature without this factor, cf. Bothe (2020, 2022), Cermelli et al. (2005, Sect. 2), Grinfeld (2013, Sect. 12.4), or Dziuk and Elliott (2007).

$$0 = \frac{\partial \boldsymbol{\nu}}{\partial u_\alpha} \cdot \boldsymbol{\tau}_\gamma + \frac{\partial \boldsymbol{\tau}_\gamma}{\partial u_\alpha} \cdot \boldsymbol{\nu} \quad \text{or equivalently} \quad \frac{\partial \boldsymbol{\nu}}{\partial u_\alpha} \cdot \boldsymbol{\tau}_\gamma = -\frac{\partial \boldsymbol{\tau}_\gamma}{\partial u_\alpha} \cdot \boldsymbol{\nu}.$$

If now (9.14) is multiplied by $\boldsymbol{\tau}_\gamma$ and compared to the previous formula, we obtain

$$\frac{\partial \boldsymbol{\nu}}{\partial u_\alpha} \cdot \boldsymbol{\tau}_\gamma = T_\alpha^\beta \boldsymbol{\tau}_\beta \cdot \boldsymbol{\tau}_\gamma = T_\alpha^\beta g_{\beta\gamma} = -\frac{\partial \boldsymbol{\tau}_\gamma}{\partial u_\alpha} \cdot \boldsymbol{\nu}$$

or equivalently

$$T_\alpha^\beta = -g^{\beta\gamma} \frac{\partial \boldsymbol{\tau}_\gamma}{\partial u_\alpha} \cdot \boldsymbol{\nu}.$$

Inserting this into (9.14) and using (9.9) finally gives

$$\frac{\partial \boldsymbol{\nu}}{\partial u_\alpha} = -g^{\beta\gamma} b_{\gamma\alpha} \boldsymbol{\tau}_\beta. \tag{9.15}$$

Covariant Derivatives

So far we only used partial derivatives with respect to the surface coordinates. This however might lead to problems since the result may depend on the coordinates chosen for some quantities. Therefore, one needs a new derivative which is independent of the chosen parametrization. This is called the *covariant surface derivative* ∇_α. We will omit the details of the derivation and interpretation. We again refer to Aris (1989, Sect. 9.25) and Grinfeld (2013, Chap. 11). In short, one can think of the covariant derivative as respecting the change in the coordinates as well as the change with respect to the tangents.

In some generality the covariant derivative for a quantity with mixed indices having coordinates in the ambient space $\mathbb{R}^3$ and on the surface we have, see Grinfeld (2013, (11.27)),

$$\nabla_\gamma T_{j\beta}^{i\alpha} = \frac{\partial T_{j\beta}^{i\alpha}}{\partial u^\gamma} + \tau_{\gamma;k} \Gamma_{mk}^i T_{j\beta}^{m\alpha} - \tau_{\gamma;k} \Gamma_{jk}^m T_{m\beta}^{i\alpha} + \Gamma_{\delta\gamma}^\alpha T_{j\beta}^{i\delta} - \Gamma_{\gamma\beta}^\delta T_{j\delta}^{i\alpha}. \tag{9.16}$$

See also Aris (1989, Sects. 9.25 and 9.34). The basic rules are that for every upper index one has a + Christoffel term and a − for every lower index. The term for every Latin index involves the tangent vector. If an index does not appear the whole Christoffel term with this index vanishes. For example, the covariant derivative of a contravariant surface vector **a** or covariant spatial vector **b** is

$$\nabla_\gamma a^\alpha = \frac{\partial a^\alpha}{\partial u^\gamma} + \Gamma_{\delta\gamma}^\alpha a^\delta \quad \text{and} \quad \nabla_\alpha b_j = \frac{\partial b_j}{\partial u_\alpha} - \tau_{\alpha;k} \Gamma_{jk}^m b_m.$$

The covariant derivative of a scalar function ψ is $\nabla_\alpha \psi = \frac{\partial \psi}{\partial u_\alpha}$.

The covariant derivative has the following properties, more may be found in the given literature:

(i) The covariant derivative coincides with the partial derivative when applied to invariants, e.g. the normal $\boldsymbol{\nu}$ and its components.
(ii) The covariant derivative of a tensor is again a tensor. The order of the tensor is increased by one.
(iii) The covariant derivative satisfies sum and product rules, but different derivatives do not commute.
(iv) The covariant derivative commutes with contraction.
(v) The covariant derivatives $\nabla_\gamma g_{\alpha\beta}$ and $\nabla_\gamma g^{\alpha\beta}$ vanish. This is also called *metrinilic property*,[3] see Grinfeld (2013, Sect. 8.6.7).

For the tangent vectors $\boldsymbol{\tau}_\alpha$ we obtain from (9.16) using (9.9)

$$\nabla_\beta \boldsymbol{\tau}_\alpha = \frac{\partial \boldsymbol{\tau}_\alpha}{\partial u_\beta} - \Gamma^\gamma_{\alpha\beta} \boldsymbol{\tau}_\gamma = b_{\alpha\beta} \boldsymbol{\nu}, \tag{9.17}$$

cp. Grinfeld (2013, (11.43)). In this result, the main property of the covariant derivative becomes apparent, i.e., the result is independent from the chosen parametrization of the surface. Further we have that $\nabla_\beta \boldsymbol{\tau}_\alpha = \nabla_\alpha \boldsymbol{\tau}_\beta$.

Differentiating $\boldsymbol{\nu} \cdot \boldsymbol{\nu} = 1$ gives $\nabla_\alpha \boldsymbol{\nu} \cdot \boldsymbol{\nu} = 0$. So $\nabla_\alpha \boldsymbol{\nu}$ is a tangent vector, i.e. $\nabla_\alpha \boldsymbol{\nu} = \nu^\beta_\alpha \boldsymbol{\tau}_\beta$. Further we have from $\boldsymbol{\tau}_\gamma \cdot \boldsymbol{\nu} = 0$ that

$$0 = \nabla_\alpha \left(\boldsymbol{\tau}_\gamma \cdot \boldsymbol{\nu} \right) = \nabla_\alpha \boldsymbol{\tau}_\gamma \cdot \boldsymbol{\nu} + \boldsymbol{\tau}_\gamma \cdot \nabla_\alpha \boldsymbol{\nu}$$

or using (9.17)

$$g_{\beta\gamma} \nu^\beta_\alpha = \nu^\beta_\alpha \boldsymbol{\tau}_\beta \cdot \boldsymbol{\tau}_\gamma = \nabla_\alpha \boldsymbol{\nu} \cdot \boldsymbol{\tau}_\gamma = -\nabla_\alpha \boldsymbol{\tau}_\gamma \cdot \boldsymbol{\nu} = -b_{\alpha\gamma} \boldsymbol{\nu} \cdot \boldsymbol{\nu} = -b_{\alpha\gamma}.$$

This gives

$$\nu^\beta_\alpha = -g^{\beta\gamma} b_{\gamma\alpha} \quad \text{or} \quad \nabla_\alpha \boldsymbol{\nu} = -g^{\beta\gamma} b_{\gamma\alpha} \boldsymbol{\tau}_\beta, \tag{9.18}$$

cp. Aris (1989, (9.36.2) and (9.36.3)).

9.2.2 *Surface Differential Operators*

We now want to study surface differential operators. For this we assume that all mathematical pathologies are excluded, in particular, that the surface is sufficiently smooth. Let $\nabla_{\mathbf{x}} = (\frac{\partial}{\partial x_1}, \frac{\partial}{\partial x_2}, \frac{\partial}{\partial x_3}) = (\partial^1, \partial^2, \partial^3)$ be the spatial gradient of a function. For a smooth function $\psi : \Omega \to \mathbb{R}$, with $\Omega \subseteq \mathbb{R}^3$ open, we define the *surface*

[3] Sometimes also called metrilinic property.

gradient as the projection[4] in tangential direction

$$\nabla_\Sigma \psi = \nabla_{\mathbf{x}} \psi - (\nabla_{\mathbf{x}} \psi \cdot \boldsymbol{\nu})\boldsymbol{\nu} \tag{9.19}$$

with components $\partial^i_\Sigma \psi = \partial^i \psi - (\nabla_{\mathbf{x}} \psi \cdot \boldsymbol{\nu})\nu^i$. For a surface quantity we have $\nabla_\Sigma \psi \cdot \boldsymbol{\nu} = 0$. Setting $\psi = a_i$ in the ith component, we can define the *surface divergence* for covariant vector fields $\mathbf{a} : \Omega \to \mathbb{R}^3$ as

$$\nabla_\Sigma \cdot \mathbf{a} = \partial^i_\Sigma a_i = \left[\partial^i a_i - (\nabla_{\mathbf{x}} a_i \cdot \boldsymbol{\nu})\nu^i\right] = \nabla_{\mathbf{x}} \cdot \mathbf{a} - (\nabla_{\mathbf{x}} a_i \cdot \boldsymbol{\nu})\nu^i. \tag{9.20}$$

From both definitions we obtain, using the notations $\psi_\nu = \nabla_{\mathbf{x}} \psi \cdot \boldsymbol{\nu}$ and $a_\nu = \mathbf{a} \cdot \boldsymbol{\nu}$, the product rule

$$\begin{aligned} \nabla_\Sigma \cdot (\psi \mathbf{a}) &= \nabla_{\mathbf{x}} \cdot (\psi \mathbf{a}) - (\nabla_{\mathbf{x}}(\psi a_i) \cdot \boldsymbol{\nu})\nu^i \\ &= \mathbf{a} \cdot \nabla_{\mathbf{x}} \psi + \psi \nabla_{\mathbf{x}} \cdot \mathbf{a} - (\nabla_{\mathbf{x}} \psi \cdot \boldsymbol{\nu}) a_i \nu^i - \psi (\nabla_{\mathbf{x}} a_i) \cdot \boldsymbol{\nu})\nu^i \\ &= \mathbf{a} \cdot \nabla_\Sigma \psi + \psi_\nu a_\nu + \psi \nabla_\Sigma \cdot \mathbf{a} - \psi_\nu a_\nu = \mathbf{a} \cdot \nabla_\Sigma \psi + \psi \nabla_\Sigma \cdot \mathbf{a}. \end{aligned} \tag{9.21}$$

Now we want to deduce the equivalent coordinate formulation of the differential operators. Therefore, we calculate the tangential derivative using (9.19) and (9.3)

$$\begin{aligned} \nabla_\Sigma \psi \cdot \boldsymbol{\tau}_\gamma &= (\nabla_{\mathbf{x}} \psi - (\nabla_{\mathbf{x}} \psi \cdot \boldsymbol{\nu})\boldsymbol{\nu}) \cdot \boldsymbol{\tau}_\gamma = \nabla_{\mathbf{x}} \psi \cdot \boldsymbol{\tau}_\gamma = \nabla_{\mathbf{x}} \psi \cdot \frac{\partial \boldsymbol{\Phi}}{\partial u^\gamma} \\ &= \frac{\partial \psi}{\partial u^\gamma} = \frac{\partial \psi}{\partial u_\alpha} \delta^\alpha_\gamma = g^{\alpha\beta} \frac{\partial \psi}{\partial u_\alpha} g_{\beta\gamma} = g^{\alpha\beta} \frac{\partial \psi}{\partial u_\alpha} \boldsymbol{\tau}_\beta \cdot \boldsymbol{\tau}_\gamma. \end{aligned} \tag{9.22}$$

Note that we found the natural relations

$$\nabla_\Sigma \psi \cdot \boldsymbol{\tau}_\alpha = \frac{\partial \psi}{\partial u_\alpha} = \nabla_\alpha \psi \quad \text{and} \quad \nabla_\Sigma \psi \cdot \boldsymbol{\tau}_\alpha = \nabla_{\mathbf{x}} \psi \cdot \boldsymbol{\tau}_\alpha. \tag{9.23}$$

We also have $\nabla_\Sigma \psi \cdot \boldsymbol{\nu} = 0$ and all three vectors $\boldsymbol{\tau}_1, \boldsymbol{\tau}_2, \boldsymbol{\nu}$ form a basis of $\mathbb{R}^3$. Therefore, (9.22) implies that

$$\nabla_\Sigma \psi = g^{\alpha\beta} \frac{\partial \psi}{\partial u_\alpha} \boldsymbol{\tau}_\beta = g^{\alpha\beta} \frac{\partial \psi}{\partial u_\alpha} \frac{\partial \boldsymbol{\Phi}}{\partial u_\beta} \tag{9.24}$$

with components $\partial^i_\Sigma \psi = g^{\alpha\beta} \frac{\partial \psi}{\partial u_\alpha} \frac{\partial \Phi^i}{\partial u_\beta} = g^{\alpha\beta} \frac{\partial \psi}{\partial u_\alpha} \tau_{\beta;i}$. Thus we can again take $\psi = a_i$ and write the surface divergence of a covariant vector field $\mathbf{a}$ as

---

[4] One can define the order 2, rank 1 tensor $\mathbf{N} = \left(\nu^i \nu_j\right)_{1 \le i,j \le 3}$. It satisfies $\mathbf{N}^2 = \mathbf{N}$ since $\boldsymbol{\nu} \cdot \boldsymbol{\nu} = 1$. Then for any covariant tensor $\mathbf{a}$ of order 1 we have $\mathbf{Na} = (\mathbf{a} \cdot \boldsymbol{\nu})\boldsymbol{\nu} = a_\nu \boldsymbol{\nu}$, i.e., $\mathbf{N}$ is the normal projection. With the Kronecker delta tensor $\mathbf{I}$ for $\mathbb{R}^3$ the complementary projection in the tangential direction is given by $\mathbf{T} = \mathbf{I} - \mathbf{N}$. This gives $\mathbf{a}^{\shortparallel} = \mathbf{Ta}$ and $\nabla_\Sigma = \mathbf{T}\nabla$. Mavrovouniotis and Brenner (1993, Sect. 2) denoted $\mathbf{T}$ as the surface idemfactor.

$$\nabla_\Sigma \cdot \mathbf{a} = \partial^i_\Sigma a_i(\mathbf{x}) = g^{\alpha\beta}\frac{\partial a_i}{\partial u_\alpha}\frac{\partial \Phi^i}{\partial u_\beta} = g^{\alpha\beta}\frac{\partial a_i}{\partial u_\alpha}\tau_{\beta;i}. \tag{9.25}$$

This can be compared to the surface divergence defined in Aris (1989, (9.41.6)). The latter can be written as

$$\nabla_\Sigma \cdot \mathbf{a} = g^{\alpha\beta}\boldsymbol{\tau}_\beta \nabla_\alpha \mathbf{a} = g^{\alpha\beta}\frac{\partial \Phi^i}{\partial u_\beta}\nabla_\alpha a_i. \tag{9.26}$$

Since the covariant derivative of a scalar function is $\nabla_\alpha a_i = \frac{\partial a_i}{\partial u_\alpha}$ both (9.25) and (9.26) are equivalent.

So far we considered derivatives of functions ψ and vector fields $\mathbf{v}$ defined in the volume surrounding the surface. The formulas (9.19) and (9.20) are only defined for such functions and vector fields, respectively. But we can also consider functions and vector fields that are only defined on the surface Σ, i.e., $\psi : \Sigma \to \mathbb{R}$. We then say that such a function is smooth iff $\psi(\boldsymbol{\Phi}(\cdot)) : \mathcal{U} \to \mathbb{R}$ is a smooth function. For such functions, the surface gradient (9.24) is well defined. So thereby the surface gradient (9.19) may be extended to such functions. We can proceed analogously for vector fields and define their divergence via (9.25).

For further calculations, it is important to obtain the surface divergence of the normal $\boldsymbol{\nu}$. We start with (9.26), use (9.18), (9.3), and (9.11) to obtain

$$\begin{aligned}\nabla_\Sigma \cdot \boldsymbol{\nu} &= g^{\alpha\beta}\boldsymbol{\tau}_\alpha \cdot \nabla_\beta \boldsymbol{\nu} = -g^{\alpha\beta}\boldsymbol{\tau}_\alpha \cdot g^{\varepsilon\gamma} b_{\gamma\beta}\boldsymbol{\tau}_\varepsilon = -g^{\alpha\beta}(\boldsymbol{\tau}_\alpha \cdot \boldsymbol{\tau}_\varepsilon) g^{\varepsilon\gamma} b_{\gamma\beta} \\ &= -g^{\alpha\beta} g_{\alpha\varepsilon} g^{\varepsilon\gamma} b_{\gamma\beta} = -g^{\alpha\beta}\delta^\gamma_\alpha b_{\gamma\beta} = -g^{\alpha\beta} b_{\alpha\beta} = -2\kappa_M.\end{aligned} \tag{9.27}$$

Let us decompose the vector field $\mathbf{a} : \mathbb{R}^3 \to \mathbb{R}^3$ into a *tangential part* $\mathbf{a}^{\shortparallel}$ and a *normal part* $\mathbf{a}^\perp$ as $\mathbf{a} = \mathbf{a}^{\shortparallel} + \mathbf{a}^\perp$, cp. (9.8), with

$$\mathbf{a}^\perp = (\mathbf{a} \cdot \boldsymbol{\nu})\,\boldsymbol{\nu} = a_\nu\,\boldsymbol{\nu} \qquad \text{and} \qquad \mathbf{a}^{\shortparallel} = \mathbf{a} - \mathbf{a}^\perp = a^\alpha_\tau \boldsymbol{\tau}_\alpha. \tag{9.28}$$

This implies $\mathbf{a}^{\shortparallel} \cdot \boldsymbol{\nu} = 0$. We can reformulate the surface divergence as follows. We calculate the divergence of the normal part using (9.20)

$$\nabla_\Sigma \cdot \mathbf{a}^\perp = \nabla_\Sigma \cdot [(\mathbf{a} \cdot \boldsymbol{\nu})\,\boldsymbol{\nu}] = \nabla_\Sigma(\mathbf{a} \cdot \boldsymbol{\nu}) \cdot \boldsymbol{\nu} + (\mathbf{a} \cdot \boldsymbol{\nu})\nabla_\Sigma \cdot \boldsymbol{\nu}. \tag{9.29}$$

With (9.24) we determine the first term to be

$$\nabla_\Sigma(\mathbf{a} \cdot \boldsymbol{\nu}) \cdot \boldsymbol{\nu} = g^{\alpha\beta}\frac{\partial(a_j \nu_j)}{\partial u_\alpha}\boldsymbol{\tau}_\beta \cdot \boldsymbol{\nu} = 0. \tag{9.30}$$

So now (9.29) gives $\nabla_\Sigma \cdot \mathbf{a}^\perp = (\mathbf{a} \cdot \boldsymbol{\nu})\nabla_\Sigma \cdot \boldsymbol{\nu}$. With this and (9.27), we obtain

$$\nabla_\Sigma \cdot \mathbf{a} = \nabla_\Sigma \cdot \left(\mathbf{a}^{\shortparallel} + \mathbf{a}^\perp\right) = \nabla_\Sigma \cdot \mathbf{a}^{\shortparallel} + (\mathbf{a} \cdot \boldsymbol{\nu})\nabla_\Sigma \cdot \boldsymbol{\nu} = \nabla_\Sigma \cdot \mathbf{a}^{\shortparallel} - 2\kappa_M a_\nu, \tag{9.31}$$

cp. Dziuk and Elliott (2007, (2.6)).

In the following, we want to derive a further useful identity for the surface divergence of the tangential part of a vector field. Let us first consider the product $\boldsymbol{\nu} \cdot (\nabla_{\mathbf{x}} \times (\boldsymbol{\nu} \times \mathbf{a}))$. We apply the $\epsilon - \delta$-identity (9.5), $\boldsymbol{\nu} \cdot \boldsymbol{\nu} = 1$, $\nabla_{\mathbf{x}}(\boldsymbol{\nu} \cdot \boldsymbol{\nu}) = 0$ and (9.20) to obtain

$$\begin{aligned}
\boldsymbol{\nu} \cdot (\nabla_{\mathbf{x}} \times (\boldsymbol{\nu} \times \mathbf{a})) &= \nu^i \epsilon_{ijk} \partial^j (\epsilon_{krs} \nu_r a_s) = \nu^i \epsilon_{kij} \epsilon_{krs} \partial^j (\nu_r a_s) \\
&= \nu^i (\delta_{ir}\delta_{js} - \delta_{is}\delta_{jr}) \partial^j (\nu_r a_s) = \nu^i \partial^j (\nu^i a_j) - \nu_i \partial^j (\nu_j a_i) \\
&= \nu^i a_j \partial^j \nu_i + \nu^i \nu_i \partial^j a_j - \nu^i a_i \partial^j \nu_j - \nu^i \nu_j \partial^j a_i \\
&= \mathbf{a} \cdot \frac{1}{2} \nabla_{\mathbf{x}} (\boldsymbol{\nu} \cdot \boldsymbol{\nu}) + \nabla_{\mathbf{x}} \cdot \boldsymbol{a} - (\mathbf{a} \cdot \boldsymbol{\nu}) \nabla_{\mathbf{x}} \cdot \boldsymbol{\nu} - (\nabla_{\mathbf{x}} a_i \cdot \boldsymbol{\nu}) \nu^i \\
&= \nabla_{\Sigma} \cdot \mathbf{a} - (\mathbf{a} \cdot \boldsymbol{\nu}) \nabla_{\mathbf{x}} \cdot \boldsymbol{\nu}.
\end{aligned}$$

Now we compute the surface divergence of the tangential part of the vector field **a**. We use the definition (9.28) and the identity (9.30) to get

$$\begin{aligned}
\nabla_{\Sigma} \cdot \mathbf{a}^{\shortparallel} &= \nabla_{\Sigma} \cdot (\mathbf{a} - (\mathbf{a} \cdot \boldsymbol{\nu}) \boldsymbol{\nu}) = \nabla_{\Sigma} \cdot \mathbf{a} - \nabla_{\Sigma} (\mathbf{a} \cdot \boldsymbol{\nu}) \cdot \boldsymbol{\nu} - (\mathbf{a} \cdot \boldsymbol{\nu}) \nabla_{\Sigma} \cdot \boldsymbol{\nu} \\
&= \nabla_{\Sigma} \cdot \mathbf{a} - (\mathbf{a} \cdot \boldsymbol{\nu}) \nabla_{\mathbf{x}} \cdot \boldsymbol{\nu}.
\end{aligned}$$

Combining the two calculations, we obtain the formula

$$\nabla_{\Sigma} \cdot \mathbf{a}^{\shortparallel} = (\nabla_{\mathbf{x}} \times (\boldsymbol{\nu} \times \mathbf{a})) \cdot \boldsymbol{\nu}. \tag{9.32}$$

9.2.3 *Moving Surfaces*

So far we only considered a fixed surface $\Sigma \subset \Omega$ with $\Omega \subseteq \mathbb{R}^3$ open. Now we want to extend the results to moving surfaces. A moving surface Σ can be considered as a family of surfaces $\Sigma(\mathrm{t})$ with a parameter $\mathrm{t} \in [0, T] = I$ for some $T > 0$. The stationary *initial reference surface* $\Sigma_0 = \Sigma(0)$ may be described as before via a smooth bijective parametrization $\boldsymbol{\Phi}_{\Sigma_0} : \mathcal{U} \to \Sigma_0$ with $\mathcal{U} \subseteq \mathbb{R}^2$ an open subset. We further assume that we have a smooth mapping $\boldsymbol{\chi}_{\Sigma} : I \times \Sigma_0 \to \mathbb{R}^3$ with $\boldsymbol{\chi}_{\Sigma}(\mathrm{t}, \mathbf{y}) \in \Sigma(\mathrm{t})$. We define the *flow map* $\boldsymbol{\chi}_{\Sigma}^{\mathrm{t}}(\mathbf{y}) = \boldsymbol{\chi}_{\Sigma}(\mathrm{t}, \mathbf{y})$ of the surface and assume that $\boldsymbol{\chi}_{\Sigma}^{\mathrm{t}}$ is a diffeomorphism, i.e. a bijective smooth map, from the initial surface Σ_0 to $\Sigma(\mathrm{t})$. So it parametrizes the moving surface $\Sigma(\mathrm{t})$ over the stationary initial surface Σ_0. Note that $\boldsymbol{\chi}_{\Sigma}^0 = \boldsymbol{\chi}_{\Sigma}(0, \cdot)$ is the identity map on Σ_0 and that $\boldsymbol{\chi}_{\Sigma}$ is not directly related to the map $\boldsymbol{\chi}$ considered in Sect. 9.2.4. The latter represents infinitesimal particles that may move through the surface $\Sigma(\mathrm{t})$. But $\boldsymbol{\chi}_{\Sigma}$ describes the movement of points on the surface.

Now we may also describe the surface $\Sigma(\mathrm{t})$ in terms of the surface parameters $\mathbf{u} = (u^1, u^2)^T \in \mathcal{U}$ using the mapping

$$\boldsymbol{\Phi} = \chi_\Sigma(\cdot, \boldsymbol{\Phi}_{\Sigma_0}(\cdot)) : I \times \mathcal{U} \to \Sigma(\mathrm{t}).$$

The results from the previous subsections remain valid and we assume that the dependence on the parameter t is smooth. We introduce the *velocity* of a *surface point* $\mathbf{x} \in \Sigma(\mathrm{t})$ that is given by

$$\mathbf{w}(\mathrm{t}, \mathbf{x}) = \frac{\partial \chi_\Sigma}{\partial \mathrm{t}}(\mathrm{t}, \mathbf{y}) \quad \text{for} \quad \mathbf{x} = \chi_\Sigma(\mathrm{t}, \mathbf{y}) \in \Sigma(\mathrm{t}) \tag{9.33}$$

or

$$\mathbf{w}(\mathrm{t}, \boldsymbol{\Phi}(\mathrm{t}, \mathbf{u})) = \frac{\partial \chi_\Sigma}{\partial \mathrm{t}}(\mathrm{t}, \boldsymbol{\Phi}_{\Sigma_0}(\mathbf{u})) = \frac{\partial \boldsymbol{\Phi}}{\partial \mathrm{t}}(\mathrm{t}, \mathbf{u}). \tag{9.34}$$

It describes the movement of the surface points and can be independent of the surrounding particle flow velocity $\mathbf{v}$ that will be introduced in Sect. 9.2.4. We have for the trajectory in $\mathbb{R}^3$ of such a point on the surface

$$\frac{\mathrm{d}\mathbf{x}}{\mathrm{dt}}(\mathrm{t}) = \mathbf{w}(\mathrm{t}, \mathbf{x}(\mathrm{t})) = \mathbf{w}(\mathrm{t}, \boldsymbol{\Phi}(\mathrm{t}, \mathbf{u})).$$

The parametrization and the flow map are mathematical tools used in order to have a better understanding of the surface movement and to prove theorems. We only need to have a parametrization of the initial surface and the flow map. In practical applications, the velocity or the normal velocity of the surface will be determined from physical relations. The moving surface geometry or the flow map would have to be determined by integrating such vector fields. Mathematical theory for this integration was given by Bothe (2020).

As in (9.28) we can decompose the velocity as $\mathbf{w} = \mathbf{w}^{\shortparallel} + \mathbf{w}^{\perp}$. The vector $\mathbf{w}^{\shortparallel}$ is sometimes called *tangential coordinate velocity*. The term $w_\nu = \mathbf{w} \cdot \boldsymbol{\nu}$ can be called *velocity of the surface* since the normal direction is implied. Then $w_\nu \boldsymbol{\nu}$ is the *vector normal velocity*, see Grinfeld (2013, (15.32)), and w_ν the normal speed. It is important to note that w_ν is an invariant of the surface. A nice description and geometric motivation is given in Grinfeld (2013, Sects. 15.4 and 15.5).

Implicit Description and an Explicit Parametrization of Surfaces

Let us assume that a surface $\Sigma(\mathrm{t}) \subset \Omega$ is given as an implicit function $f(\mathrm{t}, \mathbf{x}) = 0$ with $f : I \times \mathbb{R}^3 \to \mathbb{R}$ and $|\nabla_\mathbf{x} f| > 0$, which implics $|\nabla_{\mathrm{t},\mathbf{x}} f| > 0$. Thc timc and spacc unit normal vector field is given as $\mathbf{n} = \frac{\nabla_{\mathrm{t},\mathbf{x}} f}{|\nabla_{\mathrm{t},\mathbf{x}} f|}$ since the gradient is orthogonal to level sets. We have from (9.33)

$$\begin{aligned} 0 &= \frac{\mathrm{d}}{\mathrm{dt}} f(\mathrm{t}, \chi_\Sigma(\mathrm{t}, \mathbf{x})) = \left(f_\mathrm{t} + \nabla_\mathbf{x} f \cdot \frac{\partial \chi_\Sigma}{\partial \mathrm{t}} \right)(\mathrm{t}, \chi_\Sigma(\mathrm{t}, \mathbf{x})) \\ &= (f_\mathrm{t} + \nabla_\mathbf{x} f \cdot \mathbf{w})\,(\mathrm{t}, \chi_\Sigma(\mathrm{t}, \mathbf{x})), \end{aligned}$$

and for $i = 1, 2, 3$

$$0 = \frac{\partial}{\partial u_\alpha} f(\mathrm{t}, \boldsymbol{\Phi}(\mathrm{t}, \mathbf{u})) = \left(\nabla_{\mathbf{x}} f \cdot \frac{\partial \boldsymbol{\Phi}}{\partial u_\alpha}\right)(\mathrm{t}, \boldsymbol{\Phi}(\mathrm{t}, \mathbf{u})) = (\nabla_{\mathbf{x}} f \cdot \boldsymbol{\tau}_\alpha)(\mathrm{t}, \boldsymbol{\Phi}(\mathrm{t}, \mathbf{u})).$$

The latter equation proves that $\boldsymbol{\nu} = \frac{\nabla_{\mathbf{x}} f}{|\nabla_{\mathbf{x}} f|}$ is the spatial unit normal vector field that is orthogonal to the tangent vectors $\boldsymbol{\tau}_\alpha$ in $\mathbb{R}^3$ for any $\mathrm{t} \in I$. The former gives with the notation $f_t = \frac{\partial}{\partial \mathrm{t}} f$

$$w_\nu = \mathbf{w} \cdot \boldsymbol{\nu} = \mathbf{w} \cdot \frac{\nabla_{\mathbf{x}} f}{|\nabla_{\mathbf{x}} f|} = \frac{-f_t}{|\nabla_{\mathbf{x}} f|}. \tag{9.35}$$

Suppose that we can describe the points $\mathbf{x} \in \Sigma(\mathrm{t})$ on the surface by a function $\sigma : I \times \Omega^2 \to \mathbb{R}$ with a suitable set $\Omega^2 \subseteq \mathbb{R}^2$ as $\mathbf{x} = (\mathrm{x}_1, \mathrm{x}_2, \sigma(\mathrm{t}, \mathrm{x}_1, \mathrm{x}_2))$. Then it is given implicitly by $0 = f(\mathrm{t}, \mathbf{x}) = \mathrm{x}_3 - \sigma(\mathrm{t}, \mathrm{x}_1, \mathrm{x}_2)$ for $(\mathrm{x}_1, \mathrm{x}_2) \in \Omega^2$. We set $\mathbf{x}_\sigma = (\mathrm{x}_1, \mathrm{x}_2, \sigma(\mathrm{t}, \mathrm{x}_1, \mathrm{x}_2)) \in \mathbb{R}^3$. For the normal velocity w_ν of the surface Σ we have from (9.35)

$$w_\nu = \frac{-f_\mathrm{t}}{|\nabla_{\mathbf{x}} f|} = \frac{\sigma_\mathrm{t}}{\sqrt{(\sigma_{\mathrm{x}_1})^2 + (\sigma_{\mathrm{x}_2})^2 + 1}}.$$

This implies that

$$\sigma_\mathrm{t}(\mathrm{t}, \mathrm{x}_1, \mathrm{x}_2) = w_\nu(\mathrm{t}, \mathbf{x}_\sigma)\sqrt{(\sigma_{\mathrm{x}_1})^2 + (\sigma_{\mathrm{x}_2})^2 + 1}. \tag{9.36}$$

Intrinsic Time Derivatives

A major issue when working with moving surfaces is how to define an intrinsic time derivative that is invariant under a change of parametrization. We cannot keep $\mathbf{x}$ fixed, since $\mathbf{x}$ has moved when we consider a time $\mathrm{t} + \Delta\mathrm{t}$. For this purpose Thomas (1957, Sect. 4), Thomas (1961, Sect. II,3) introduced a time derivative for functions $\psi : I \times \Sigma(\mathrm{t}) \to \mathbb{R}$ defined at $\mathbf{x} \in \Sigma(\mathrm{t})$ as follows. Consider the unit normal vector $\boldsymbol{\nu}(\mathbf{x})$ and the straight line

$$\mathcal{L} = \{\mathbf{z} \in \mathbb{R}^3 | \mathbf{z} = \lambda \boldsymbol{\nu}_\Sigma(\mathbf{x}) + \mathbf{x} \text{ for } \lambda \in \mathbb{R}\}.$$

For a small time $\Delta\mathrm{t}$ we take the moved surface $\Sigma(\mathrm{t} + \Delta\mathrm{t})$ and let $\mathbf{z}^\mathbf{x}$ be the point where $\mathcal{L}$ intersects $\Sigma(\mathrm{t} + \Delta\mathrm{t})$. Then we define the *Thomas time derivative*, assuming that the limit exists,

$$\frac{\delta}{\delta \mathrm{t}} \psi(\mathrm{t}, \mathbf{x}) = \lim_{\Delta\mathrm{t} \to 0} \frac{\psi(\mathrm{t} + \Delta\mathrm{t}, \mathbf{z}^\mathbf{x}) - \psi(\mathrm{t}, \mathbf{x})}{\Delta\mathrm{t}}. \tag{9.37}$$

Truesdell and Toupin (1960, Sect. 179), Bowen and Wang (1971), Petryk and Mróz (1986) as well as Bothe (2022) defined this time derivative by taking the curve that is obtained by integrating the normal velocity field $w_\nu \boldsymbol{\nu}_\Sigma$ instead of the line $\mathcal{L}$. Both curves have the velocity $w_\nu(\mathrm{t}, \mathbf{x})\boldsymbol{\nu}_\Sigma(\mathrm{t}, \mathbf{x})$ at $\mathbf{x}$. So both approaches give the same result. The definition is independent of a parametrization of the surface. Truesdell and Toupin (1960, Sect. 179) as well as Bowen and Wang (1971) discussed the extension of this derivative to tensors.

We can interpret this derivative by assuming that $\widehat{\psi} : I \times \widehat{\Omega} \to \mathbb{R}$ is an extended smooth function with $\Sigma(t) \subset \widehat{\Omega} \subseteq \Omega \subseteq \mathbb{R}^3$ an open neighborhood of $\Sigma(t)$. A simple extension of ψ to a neighborhood of the surface may be obtained by a function that is constant in the directions normal to the surface on both sides of the surface, see e.g. Cermelli et al. (2005, Sect. 3.4). This would lead to a vanishing normal derivative. Thus, we then have $\mathbf{w} \cdot \nabla_{\mathbf{x}} \widehat{\psi}_\Sigma = w^{\shortparallel} \cdot \nabla_\Sigma \widehat{\psi}_\Sigma$ and $\widehat{\psi}_\nu = 0$. We do not assume such a specific extension, so we may have $\widehat{\psi}_\nu \neq 0$. Let $\mathbf{u}_{\Delta\mathrm{t}}$ be chosen so that $\mathbf{z}^{\mathbf{x}} = \boldsymbol{\Phi}(\mathrm{t} + \Delta\mathrm{t}, \mathbf{u}_{\Delta\mathrm{t}})$. Inserting the parametrization $\mathbf{x} = \boldsymbol{\Phi}(\mathrm{t}, \mathbf{u})$ into (9.37) implies that

$$
\begin{aligned}
\frac{\delta}{\delta \mathrm{t}} \widehat{\psi}(\mathrm{t}, \boldsymbol{\Phi}(\mathrm{t}, \mathbf{u})) &= \lim_{\Delta\mathrm{t} \to 0} \frac{\widehat{\psi}(\mathrm{t} + \Delta\mathrm{t}, \boldsymbol{\Phi}(\mathrm{t} + \Delta\mathrm{t}, \mathbf{u}_{\Delta\mathrm{t}})) - \widehat{\psi}(\mathrm{t}, \boldsymbol{\Phi}(\mathrm{t}, \mathbf{u}))}{\Delta\mathrm{t}} \\
&= \lim_{\Delta\mathrm{t} \to 0} \frac{\widehat{\psi}(\mathrm{t} + \Delta\mathrm{t}, \boldsymbol{\Phi}(\mathrm{t} + \Delta\mathrm{t}, \mathbf{u}_{\Delta\mathrm{t}})) - \widehat{\psi}(\mathrm{t} + \Delta\mathrm{t}, \boldsymbol{\Phi}(\mathrm{t}, \mathbf{u}))}{\Delta\mathrm{t}} \\
&\quad + \lim_{\Delta\mathrm{t} \to 0} \frac{\widehat{\psi}(\mathrm{t} + \Delta\mathrm{t}, \boldsymbol{\Phi}(\mathrm{t}, \mathbf{u})) - \widehat{\psi}(\mathrm{t}, \boldsymbol{\Phi}(\mathrm{t}, \mathbf{u}))}{\Delta\mathrm{t}} \\
&= \lim_{\Delta\mathrm{t} \to 0} \left(\sum_{i=1}^{3} \frac{\widehat{\psi}(\mathrm{t} + \Delta\mathrm{t}, \boldsymbol{\Phi}(\mathrm{t} + \Delta\mathrm{t}, \mathbf{u}_{\Delta\mathrm{t}})) - \widehat{\psi}(\mathrm{t} + \Delta\mathrm{t}, \boldsymbol{\Phi}(\mathrm{t}, \mathbf{u}))}{\Phi^i(\mathrm{t} + \Delta\mathrm{t}, \mathbf{u}_{\Delta\mathrm{t}}) - \Phi^i(\mathrm{t}, \mathbf{u})} \right. \\
&\qquad \left. \cdot \frac{\Phi^i(\mathrm{t} + \Delta\mathrm{t}, \mathbf{u}_{\Delta\mathrm{t}}) - \Phi^i(\mathrm{t}, \mathbf{u})}{\Delta\mathrm{t}} \right) + \frac{\partial}{\partial \mathrm{t}} \widehat{\psi}(\mathrm{t}, \boldsymbol{\Phi}(\mathrm{t}, \mathbf{u})) \\
&= \left[\nabla_{\mathbf{x}} \widehat{\psi} \cdot w_\nu \boldsymbol{\nu}_\Sigma + \frac{\partial}{\partial \mathrm{t}} \widehat{\psi} \right] (\mathrm{t}, \boldsymbol{\Phi}(\mathrm{t}, \mathbf{u})) = \left[\frac{\partial}{\partial \mathrm{t}} \widehat{\psi} + \widehat{\psi}_\nu w_\nu \right] (\mathrm{t}, \boldsymbol{\Phi}(\mathrm{t}, \mathbf{u})).
\end{aligned}
\tag{9.38}
$$

We see that for functions that are only defined on a surface the Thomas derivative is a generalization of the partial derivative with respect to time for such extended functions, as the term $\widehat{\psi}_\nu$ will vanish if the extension by constants along normal directions is used.

For $\psi : I \times \Sigma \to \mathbb{R}$ we can proceed analogously to the Thomas time derivative when we replace the normal velocity $w_\nu(\mathrm{t}, \mathbf{x})\boldsymbol{\nu}_\Sigma(\mathrm{t}, \mathbf{x})$ by the velocity $\mathbf{w}(\mathrm{t}, \mathbf{x})$ of the surface. We can use the flow map and set $\widehat{\mathbf{z}}^{\mathbf{x}} = \chi_\Sigma^{\Delta\mathrm{t}}(\mathbf{x})$. Then we define the *Lagrangian time derivative* for surfaces, assuming that the limit exists,

$$
\frac{D}{D\mathrm{t}} \psi(\mathrm{t}, \mathbf{x}) = \lim_{\Delta\mathrm{t} \to 0} \frac{\psi(\mathrm{t} + \Delta\mathrm{t}, \widehat{\mathbf{z}}^{\mathbf{x}}) - \psi(\mathrm{t}, \mathbf{x})}{\Delta\mathrm{t}}.
\tag{9.39}
$$

With $\Delta \mathbf{z}^{\mathbf{x}} = \widehat{\mathbf{z}}^{\mathbf{x}} - \mathbf{z}^{\mathbf{x}}$ we then obtain

$$\begin{aligned}\frac{D}{Dt}\psi(t,\mathbf{x}) &= \lim_{\Delta t\to 0}\frac{\psi(t+\Delta t,\widehat{\mathbf{z}}^{\mathbf{x}}) - \psi(t+\Delta t,\mathbf{z}^{\mathbf{x}})}{\Delta t} + \lim_{\Delta t\to 0}\frac{\psi(t+\Delta t,\mathbf{z}^{\mathbf{x}}) - \psi(t,\mathbf{x})}{\Delta t}\\ &= \lim_{\Delta t\to 0}\sum_{i=1}^{3}\frac{\psi(t+\Delta t,\widehat{\mathbf{z}}^{\mathbf{x}}) - \psi(t+\Delta t,\mathbf{z}^{\mathbf{x}})}{\Delta z_i^{\mathbf{x}}}\frac{\Delta z_i^{\mathbf{x}}}{\Delta t} + \frac{\delta}{\delta t}\psi(t,\mathbf{x})\\ &= \frac{\delta}{\delta t}\psi(t,\mathbf{x}) + \mathbf{w}^{\shortparallel}(t,\mathbf{x})\cdot\nabla_{\Sigma}\psi(t,\mathbf{x}).\end{aligned} \tag{9.40}$$

So the Lagrangian derivative $\frac{D}{Dt}$ is equal to the Thomas derivative $\frac{\delta}{\delta t}$ plus an advective term.[5]

Intrinsic time derivatives for general tensors can be found in Truesdell and Toupin (1960, Sect. 179). They are called displacement derivatives there. For a more recent discussion of these derivatives, see Bowen and Wang (1971).

Let us set $\widetilde{\psi}(t,\mathbf{u}) = \psi(t,\boldsymbol{\Phi}(t,\mathbf{u}))$ for a function $\psi : I\times\Sigma\to\mathbb{R}$. We can define, assuming that the limit exists, the time derivative

$$\overset{\circ}{\psi}(t,\mathbf{u}) = \frac{\partial}{\partial t}\widetilde{\psi}(t,\mathbf{u}) = \lim_{\Delta t\to 0}\frac{\widetilde{\psi}(t+\Delta t,\mathbf{u}) - \widetilde{\psi}(t,\mathbf{u})}{\Delta t}. \tag{9.41}$$

This derivative was used by Dreyer (2003, (48)) and Müller (1985, (3.9)) in transport theorems for surfaces. Using $\widehat{\mathbf{z}}^{\mathbf{x}} = \chi_{\Sigma}^{\Delta t}(\boldsymbol{\Phi}(t,\mathbf{u})) = \boldsymbol{\Phi}(t+\Delta t,\mathbf{u})$ and the definition of $\widetilde{\psi}$ it satisfies

$$\begin{aligned}\overset{\circ}{\psi}(t,\mathbf{u}) &= \frac{\partial}{\partial t}\widetilde{\psi}(t,\mathbf{u}) = \lim_{\Delta t\to 0}\frac{\psi(t+\Delta t,\boldsymbol{\Phi}(t+\Delta t,\mathbf{u})) - \psi(t,\boldsymbol{\Phi}(t,\mathbf{u}))}{\Delta t}\\ &= \lim_{\Delta t\to 0}\frac{\psi(t+\Delta t,\widehat{\mathbf{z}}^{\mathbf{x}}) - \psi(t,\mathbf{x})}{\Delta t} = \frac{D}{Dt}\psi(t,\mathbf{x}).\end{aligned} \tag{9.42}$$

So the derivative (9.41) is equivalent to the Lagrangian time derivative (9.39). This justifies the notation $\overset{\circ}{\psi}(t,\mathbf{x}) = \frac{D}{Dt}\psi(t,\mathbf{x})$ for the Lagrangian time derivative.

The Time Derivative of the Metric

Since the metric is time dependent now we also want to calculate the time derivative of the determinant g of the metric. We apply (9.78) to the symmetric matrix $\mathbf{G}(t) = g_{\alpha\beta}(t)$ with $\mathbf{G}(t)^{-1} = g^{\alpha\beta}(t)$ and $\det(\mathbf{G}) = g$. This gives, using the symmetry of $\mathbf{G}(t)$,

---

[5] In Grinfeld (2013, (15.34)), the Lagrangian time derivative is written as a partial time derivative. The Thomas time derivative is introduced as the invariant time derivative $\dot{\nabla}$. Therefore, in the definition of $\dot{\nabla}$ via Grinfeld (2013, (15.37)) the advective term has a minus sign.

$$\frac{\mathrm{d}g}{\mathrm{d}t} = g^{\alpha\beta}\frac{\mathrm{d}g_{\alpha\beta}}{\mathrm{d}t}g. \tag{9.43}$$

By using (9.3) and then (9.2), we have

$$\frac{\mathrm{d}g_{\alpha\beta}}{\mathrm{d}t} = \frac{\partial^2\boldsymbol{\Phi}}{\partial u_\alpha\partial \mathrm{t}}\frac{\partial\boldsymbol{\Phi}}{\partial u_\beta} + \frac{\partial\boldsymbol{\Phi}}{\partial u_\alpha}\frac{\partial^2\boldsymbol{\Phi}}{\partial u_\beta\partial \mathrm{t}} = \frac{\partial w_k}{\partial u_\alpha}\frac{\partial\Phi^k}{\partial u_\beta} + \frac{\partial\Phi^k}{\partial u_\alpha}\frac{\partial w_k}{\partial u_\beta}.$$

With (9.43), (9.25) and the symmetry of the inverse metric tensor $g^{\alpha\beta}$, we have the following formula

$$\begin{aligned}\frac{\mathrm{d}}{\mathrm{d}t}\sqrt{g} =& \frac{1}{2\sqrt{g}}\frac{\mathrm{d}}{\mathrm{d}t}g = \frac{1}{2\sqrt{g}}g^{\alpha\beta}\frac{\mathrm{d}g_{\alpha\beta}}{\mathrm{d}t}g\\ =&\frac{1}{2}\left(g^{\alpha\beta}\frac{\partial w_k}{\partial u_\alpha}\frac{\partial\Phi^k}{\partial u_\beta} + g^{\beta\alpha}\frac{\partial w_k}{\partial u_\beta}\frac{\partial\Phi^k}{\partial u_\alpha}\right)\sqrt{g} = (\nabla_\Sigma\cdot\mathbf{w})\sqrt{g}.\end{aligned} \tag{9.44}$$

This is the time derivative of the surface element needed in the next section when the integrals (9.7) are taken over moving surfaces and differentiated with respect to time. It is quite natural that the evolution of this element is determined by the surface divergence of the velocity field of the surface.

Bothe (2020, 2022) considered moving interfaces in a stationary domain. In Bothe (2020), he gave various results for the trajectories $\mathbf{x}(\mathrm{t}) = \chi^{\mathbf{x}}(\mathrm{t}) = \chi(\mathrm{t}, \mathbf{y})$ for $\mathrm{t} \in I$ of points on the surface given as solutions to systems of ordinary differential equations.

9.2.4 *Moving Control Volumes with Singular Moving Surfaces*

Motions and Control Volumes

For the description of particle flows in $\mathbb{R}^n$, it is very common to introduce a smooth map $\chi : I \times \mathbb{R}^n \to \mathbb{R}^n$ with $\chi(0, \mathbf{x}) = \mathbf{x} \in \mathbb{R}^n$. By smooth we mean that the map is continuously differentiable as often as needed. We assume it to be at least a diffeomorphism. For any fixed $\mathbf{y} \in \mathbb{R}^n$, the map defines a trajectory $\chi^{\mathbf{x}} : I \to \mathbb{R}^n$, via $\mathbf{x}(\mathrm{t}) = \chi^{\mathbf{x}}(\mathrm{t}) = \chi(\mathrm{t}, \mathbf{y})$. These are the trajectories of infinitesimal particles that are initially at $\mathbf{y}$ and then at $\mathbf{x}(\mathrm{t})$ for $\mathrm{t} \in [0, T]$. No particles should appear or disappear. So we assume that the map $\chi^{\mathrm{t}} : \mathbb{R}^n \to \mathbb{R}^n$, defined as $\chi^{\mathrm{t}}(\mathbf{y}) = \chi(\mathrm{t}, \mathbf{y})$ for any fixed $\mathrm{t} \in I$ must be a smooth bijection. We have $\chi^0(\mathbf{y}) = \mathbf{y}$, i.e., this map is the identity map in $\mathbb{R}^n$. We define

$$\mathbf{v}(\mathrm{t}, \mathbf{x}) = \frac{\partial}{\partial \mathrm{t}}\chi(\mathrm{t}, \mathbf{x})$$

to be the *velocity field* of the particles. This description of a particle flow is known as the formulation in *Lagrangian* coordinates $\boldsymbol{\chi}$. In this section, we assume that it is continuous. The mappings $\boldsymbol{\chi}$ or the family of mappings $\boldsymbol{\chi}^{\mathrm{t}}$ are called *motions* or *flow maps*. The invertible Jacobian matrices $\mathbf{D}_{\mathbf{x}}\boldsymbol{\chi}^{\mathrm{t}} = \left(\frac{\partial \chi_j^{\mathrm{t}}}{\partial \mathrm{x}_k}\right)_{1\le j,k\le 3}$ are the *deformation gradients*.

Let $\mathcal{V}(0) \subset \mathbb{R}^n$ be any open subset. We call it the *reference volume*. Then the moving volume $\mathcal{V}(\mathrm{t}) = \boldsymbol{\chi}^{\mathrm{t}}[\mathcal{V}(0)] \subset \mathbb{R}^n$ is a *material control volume*. This means that we assume that there is no mass flux through the boundary $\mathcal{S}(\mathrm{t}) = \partial\mathcal{V}(\mathrm{t})$ of $\mathcal{V}(\mathrm{t})$. Thus the boundary $\mathcal{S}(\mathrm{t})$ moves with the velocity of the particles sitting on it. With particle one does not actually refer to the molecules or atoms of the material which occupy a positive volume. In a continuum at any time t we interpret any spatial point as the location of an *infinitesimal particle*. The dynamics of the control volume and its boundary are given by the space time Lagrangian mapping $\boldsymbol{\chi}$. For simplicity, we will not always emphasize the time dependence of such a Lagrangian volume $\mathcal{V}$ or surface $\mathcal{S}$. A stationary Eulerian volume will later be denoted by $\Omega \subseteq \mathbb{R}^n$.

Moving Surfaces in Control Volumes

Let us consider such a control volume $\mathcal{V} \subset \mathbb{R}^3$ that contains a singular moving surface Σ separating it into two open volumes denoted by $\mathcal{V}_1$ and $\mathcal{V}_2$. By singular we mean that physical states may be discontinuous across the surface. We have $\mathcal{V} = \mathcal{V}_1 \cup \Sigma \cup \mathcal{V}_2$. Such a situation is sketched in Fig. 9.1. Note that whereas the boundary surface $\mathcal{S}$ is assumed to be a material surface, the internal surface Σ may allow particle transfer between the two regions. We assume that we have a smooth mapping $\boldsymbol{\chi}_\Sigma : I \times \Sigma_0 \to \mathbb{R}^3$ from the initial reference surface $\Sigma_0 = \Sigma(0)$ to the moving surface $\Sigma = \Sigma(\mathrm{t})$, as introduced in Sect. 9.2.3. The velocity field $\mathbf{w}$ on Σ is independent of the velocity field of particles defined above.

We take $\boldsymbol{\nu}$ to be outward pointing unit normal vector field on $\mathcal{S} = \partial\mathcal{V}$. The unit normal field on the surface Σ pointing from $\mathcal{V}_1$ to $\mathcal{V}_2$ is denoted by $\boldsymbol{\nu}_\Sigma$. Let $\boldsymbol{\tau}$ be

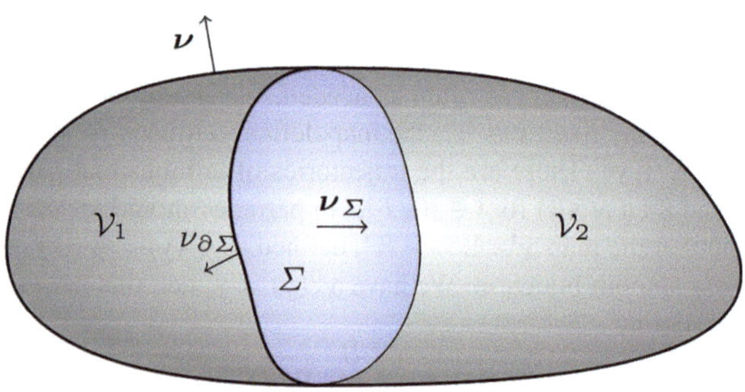

Fig. 9.1 An arbitrary material volume $\mathcal{V}$ separated by an immaterial internal surface Σ into the sub-volumes $\mathcal{V}_1$ and $\mathcal{V}_2$

a tangent vector field to the boundary curve $\partial\Sigma$. We assume that the unit normal vector field $\boldsymbol{\nu}_\Sigma$ to the surface and $\boldsymbol{\tau}$ satisfy the usual right-hand rule. Looking at the surface from the side to which the normal points, the curve is oriented in the counter-clockwise direction. We set $\boldsymbol{\nu}_{\partial\Sigma} = \boldsymbol{\tau} \times \boldsymbol{\nu}_\Sigma$ to be the outward pointing boundary normal that is tangential to the surface.

Without loss of generality we assume that all components of the normal field $\boldsymbol{\nu}_\Sigma$ are positive and that $w_\nu > \mathbf{v} \cdot \boldsymbol{\nu}$. So the volume $\mathcal{V}_1$ becomes larger at the expense of $\mathcal{V}_2$ since the surface Σ is moving faster in the normal direction than the local material velocity of particles and into $\mathcal{V}_2(\mathrm{t})$. We denote by $\widehat{\mathcal{V}}_i(\mathrm{t}) = \boldsymbol{\chi}^{\mathrm{t}}[\mathcal{V}_i(0)]$ for $i = 1, 2$ the material volumes of the particles that are initially in $\mathcal{V}_i(0)$ and introduce the difference volume $\mathcal{V}_D(\mathrm{t}) = \mathcal{V}_1(\mathrm{t})\setminus\widehat{\mathcal{V}}_1(\mathrm{t})$. So the volume $\mathcal{V}_1(\mathrm{t})$ is divided into two parts. The initial set $\mathcal{V}_2(0)$ is also divided into two parts, $\mathcal{V}^{\mathrm{t}}_{21} \subset \mathcal{V}_2(0)$ and $\mathcal{V}^{\mathrm{t}}_{22} \subset \mathcal{V}_2(0)$. Note that they contain infinitesimal particles at time 0. But their composition changes with time. The set $\mathcal{V}^{\mathrm{t}}_{21}$ contains the particles that move from $V_2(0)$ to $\mathcal{V}_1 \cup \Sigma$ in the time interval $[0, \mathrm{t}]$ i.e., $\mathcal{V}_D(\mathrm{t}) = \boldsymbol{\chi}^{\mathrm{t}}[\mathcal{V}^{\mathrm{t}}_{21}]$. In $\mathcal{V}^{\mathrm{t}}_{22}$ are the particles that make up $\mathcal{V}_2(\mathrm{t}) = \boldsymbol{\chi}^{\mathrm{t}}[\mathcal{V}^{\mathrm{t}}_{22}]$.

Now we want to parametrize the moving boundary $\Sigma_0(\mathrm{t})$ between $\mathcal{V}_{21}$ and $\mathcal{V}_{22}$. We have $\Sigma(\mathrm{t}) = \boldsymbol{\chi}^{\mathrm{t}}[\Sigma_0(\mathrm{t})]$ and assume that its unit normal vector field $\boldsymbol{\nu}_{\Sigma_0}$ points toward $\mathcal{V}^{\mathrm{t}}_{22}$. We denote by $\boldsymbol{\chi}^{-\mathrm{t}}$ the inverse of the bijective map $\boldsymbol{\chi}^{\mathrm{t}}$. For $\mathrm{t} \in I$ let $\boldsymbol{\chi}_{\Sigma_0}(\mathrm{t}, \cdot) = \boldsymbol{\chi}^{-\mathrm{t}}(\mathrm{t}, \boldsymbol{\chi}_\Sigma(\mathrm{t}, \cdot)) : \Sigma_0 \to \mathbb{R}^3$. Then for $\mathbf{y} \in \Sigma_0$ we have $\boldsymbol{\chi}_{\Sigma_0}(\mathrm{t}, \mathbf{y}) \in \Sigma_0(\mathrm{t})$ and $\boldsymbol{\chi}_\Sigma(\mathrm{t}, \mathbf{y}) = \boldsymbol{\chi}(\mathrm{t}, \boldsymbol{\chi}_{\Sigma_0}(\mathrm{t}, \mathbf{y}))$. The mapping $\boldsymbol{\chi}_{\Sigma_0}$ parametrizes the moving boundary $\Sigma_0(\mathrm{t})$ between $\mathcal{V}^{\mathrm{t}}_{21}$ and $\mathcal{V}^{\mathrm{t}}_{22}$ over the stationary surface Σ_0.

The Jacobian matrix or deformation gradient $\mathbf{D_x}\boldsymbol{\chi}^{\mathrm{t}} = \left(\frac{\partial \chi^{\mathrm{t}}_j}{\partial x_k}\right)_{1\le j,k\le n}$ is assumed to be invertible. Then we have

$$
\begin{aligned}
\frac{\partial}{\partial \mathrm{t}}\boldsymbol{\chi}_\Sigma(\mathrm{t}, \mathbf{y}) &= \frac{\mathrm{d}}{\mathrm{dt}}\boldsymbol{\chi}(\mathrm{t}, \boldsymbol{\chi}_{\Sigma_0}(\mathrm{t}, \mathbf{y})) \\
&= \frac{\partial \boldsymbol{\chi}}{\partial \mathrm{t}}(\mathrm{t}, \boldsymbol{\chi}_{\Sigma_0}(\mathrm{t}, \mathbf{y})) + \mathbf{D_x}\boldsymbol{\chi}^{\mathrm{t}}(\mathrm{t}, \boldsymbol{\chi}_{\Sigma_0}(\mathrm{t}, \mathbf{y}))\frac{\partial \boldsymbol{\chi}_{\Sigma_0}}{\partial \mathrm{t}}(\mathrm{t}, \mathbf{y}).
\end{aligned}
$$

We set $\mathbf{y} = \boldsymbol{\chi}_{\Sigma_0}(\mathrm{t}, \mathbf{y}_0)$, $\mathbf{x} = \boldsymbol{\chi}^{\mathrm{t}}(\mathbf{y})$, and $\mathbf{w}_0(\mathrm{t}, \mathbf{y}) = \frac{\partial \boldsymbol{\chi}_{\Sigma_0}}{\partial \mathrm{t}}(\mathrm{t}, \mathbf{y}_0)$. Further, we have $\mathbf{w}(\mathbf{x}) = \frac{\partial}{\partial \mathrm{t}}\boldsymbol{\chi}_\Sigma(\mathrm{t}, \mathbf{y}_0)$ and $\mathbf{v}(\mathbf{x}) = \frac{\partial \boldsymbol{\chi}}{\partial \mathrm{t}}(\mathrm{t}, \boldsymbol{\chi}_{\Sigma_0}(\mathrm{t}, \mathbf{y}_0))$. Then we obtain

$$
\mathbf{w}(\mathrm{t}, \mathbf{x}) = \mathbf{v}(\mathrm{t}, \mathbf{x}) + \mathbf{D_x}\boldsymbol{\chi}^{\mathrm{t}}(\mathrm{t}, \mathbf{y})\mathbf{w}_0(\mathrm{t}, \mathbf{y}).
$$

Solving for $\mathbf{w}_0(\mathrm{t}, \mathbf{y})$, we have

$$
\mathbf{w}_0(\mathrm{t}, \mathbf{y}) = \left(\mathbf{D_x}\boldsymbol{\chi}^{\mathrm{t}}\right)^{-1}(\mathrm{t}, \mathbf{x})\left[\mathbf{w}(\mathrm{t}, \mathbf{x}) - \mathbf{v}(\mathrm{t}, \mathbf{x})\right]. \tag{9.45}
$$

The velocity of the surface $\Sigma_0(\mathrm{t})$ is a transformation of the relative velocity $\mathbf{w} - \mathbf{v}$ of the surface $\Sigma(\mathrm{t})$.

Integration Formulas

We want to assume that the boundary $\mathcal{S}_2(0)$ of the initial control volume $\mathcal{V}_2(0)$ can be parametrized over the closure $\overline{\Omega} \subseteq \mathbb{R}^2$ of a fixed open shadow domain Ω in the following way. There are suitable functions $\alpha, \beta : \Omega \to \mathbb{R}$ that satisfy $\alpha(\mathrm{y}_1, \mathrm{y}_2) = \beta(\mathrm{y}_1, \mathrm{y}_2)$ for $(\mathrm{y}_1, \mathrm{y}_2) \in \partial\Omega$, $\alpha(\mathrm{y}_1, \mathrm{y}_2) < \beta(\mathrm{y}_1, \mathrm{y}_2)$ for $(\mathrm{y}_1, \mathrm{y}_2) \in \Omega$ and

$$\mathcal{V}_2(0) = \{\, \mathbf{y} \in \mathbb{R}^3 \mid (\mathrm{y}_1, \mathrm{y}_2) \in \Omega \text{ and } \alpha(\mathrm{y}_1, \mathrm{y}_2) < \mathrm{y}_3 < \beta(\mathrm{y}_1, \mathrm{y}_2) \,\}. \tag{9.46}$$

The functions α, β parametrize the boundary $\mathcal{S}_2(0)$. Under these circumstances we have for any integrable function $h : \mathcal{V}_2(0) \to \mathbb{R}$ the formula

$$\int_{\mathcal{V}_2(0)} h(\mathbf{y})\, \mathrm{d}V_0 = \int_{\Omega} \int_{\alpha(\mathrm{y}_1,\mathrm{y}_2)}^{\beta(\mathrm{y}_1,\mathrm{y}_2)} h(\mathbf{y}) \mathrm{d}\mathrm{y}_3 \mathrm{d}A. \tag{9.47}$$

For many arguments with integrals over control volumes, one can make such an assumption. One can possibly use another selection of the two variables or divide the control volume into pieces that satisfy such an assumption.

We already assume that $\nu_\Sigma^3 > 0$ and we additionally assume that $\nu_{\Sigma_0}^3 > 0$. This implies that the moving surface $\Sigma_0(\mathrm{t})$ can be described by a function $\sigma : I \times \Omega \to \mathbb{R}$. Since we define it over Ω there may be parts of $\Sigma_0(\mathrm{t})$ that lie outside of $\mathcal{V}_2(0)$. But this does not matter. The variables $\mathrm{y}_1, \mathrm{y}_2$ correspond to the parameters u^1, u^2 in Sect. 9.2.1. With t as an extra parameter we have the parametrization $\boldsymbol{\Phi}_{\Sigma_0}(\mathrm{t}, \mathrm{y}_1, \mathrm{y}_2) = (\mathrm{y}_1, \mathrm{y}_2, \sigma(\mathrm{t}, \mathrm{y}_1, \mathrm{y}_2))$. The tangent vectors are $\boldsymbol{\tau}_1^0 = (1, 0, \sigma_{\mathrm{y}_1})$ and $\boldsymbol{\tau}_2^0 = (0, 1, \sigma_{\mathrm{y}_2})$ giving

$$\boldsymbol{\tau}_1^0 \times \boldsymbol{\tau}_2^0 = (-\sigma_{\mathrm{y}_1}, -\sigma_{\mathrm{y}_2}, 1)$$

and

$$\boldsymbol{\nu}_{\Sigma_0} = \frac{\boldsymbol{\tau}_1^0 \times \boldsymbol{\tau}_2^0}{|\boldsymbol{\tau}_1^0 \times \boldsymbol{\tau}_2^0|} = \frac{(-\sigma_{\mathrm{y}_1}, -\sigma_{\mathrm{y}_2}, 1)}{\sqrt{(\sigma_{\mathrm{y}_1})^2 + (\sigma_{\mathrm{y}_2})^2 + 1}}.$$

The surface integrals (9.7) become, using the plane surface element $\mathrm{d}A = \mathrm{d}\mathrm{y}_1 \mathrm{d}\mathrm{y}_2$,

$$\begin{aligned} \int_{\Sigma_0} \psi(\mathbf{x}) \mathrm{d}S &= \int_{\Omega} \psi(\boldsymbol{\Phi}_{\Sigma_0}(\mathbf{y})) |\boldsymbol{\tau}_1^0 \times \boldsymbol{\tau}_2^0|\, \mathrm{d}A \\ &= \int_{\Omega} \psi(\boldsymbol{\Phi}_{\Sigma_0}(\mathbf{y})) \sqrt{(\sigma_{\mathrm{y}_1})^2 + (\sigma_{\mathrm{y}_2})^2 + 1}\, \mathrm{d}A. \end{aligned} \tag{9.48}$$

We obtain a parametrization of the singular surface $\Sigma(\mathrm{t})$ by the mapping $\boldsymbol{\Phi}_\Sigma = \boldsymbol{\chi}^{\mathrm{t}} \circ \boldsymbol{\Phi}_{\Sigma_0}$. It gives the tangent vectors as $\boldsymbol{\tau}_\alpha = \mathbf{D}_{\mathbf{y}} \boldsymbol{\chi}^{\mathrm{t}} \boldsymbol{\tau}_\alpha^0$ for $\alpha = 1, 2$. We use the determinant $\mathcal{J}^{\mathrm{t}}$ of the Jacobian matrix of $\boldsymbol{\chi}^{\mathrm{t}}$ as given by (9.76). From Eq. (9.80), we have

$$\boldsymbol{\tau}_1 \times \boldsymbol{\tau}_2 = \mathcal{J}^{\mathrm{t}} (\mathbf{D}_{\mathbf{x}} \boldsymbol{\chi}^{-\mathrm{t}})^T \left(\boldsymbol{\tau}_1^0 \times \boldsymbol{\tau}_2^0\right)$$

and using (9.2) this gives

$$\boldsymbol{\nu}_\Sigma |\boldsymbol{\tau}_1 \times \boldsymbol{\tau}_2| = \boldsymbol{\tau}_1 \times \boldsymbol{\tau}_2 = \left((\mathbf{D}_\mathbf{x}\chi^{-t})^T \boldsymbol{\nu}_{\Sigma_0} \right) \mathcal{J}^t \, |\boldsymbol{\tau}_1^0 \times \boldsymbol{\tau}_2^0|.$$

For any vector field $\mathbf{a}$ with $\mathbf{a} \cdot \boldsymbol{\nu}_\Sigma$ integrable on $\Sigma(t)$, we obtain the integral transformation

$$\int_\Sigma \mathbf{a} \cdot \boldsymbol{\nu}_\Sigma \, dS = \int_\Omega \mathbf{a} \cdot \boldsymbol{\nu}_\Sigma |\boldsymbol{\tau}_1 \times \boldsymbol{\tau}_2| \, dA = \int_\Omega \left((\mathbf{D}_\mathbf{x}\chi^{-t})\mathbf{a} \right) \cdot \boldsymbol{\nu}_{\Sigma_0} \mathcal{J}^t |\boldsymbol{\tau}_1^0 \times \boldsymbol{\tau}_2^0| \, dA. \tag{9.49}$$

9.3 Generic Balance Laws

In this short section, we want to consider generic balance laws that may be used to describe the dynamics of a fluid in the presence of interfaces. A general balance equation for an additive, i.e., extensive, physical quantity Ψ states that the total change in time is equal to the flux across the boundary and the sources

$$\frac{d\Psi}{dt} = \underbrace{\Phi}_{\text{Flux}} + \underbrace{\Xi}_{\text{Source}}. \tag{9.50}$$

We take a control volume $\mathcal{V} \subset \mathbb{R}^3$ that is separated into $\mathcal{V}_1$ and $\mathcal{V}_2$ by a surface Σ, as in Fig. 9.1. On $\mathcal{V}$ we have a volume density ψ and we assume an additional surface density ψ_Σ on Σ. Then Ψ may be written as

$$\Psi(t) = \int_{\mathcal{V}_1 \cup \mathcal{V}_2} \psi \, dV + \int_\Sigma \psi_\Sigma \, dS. \tag{9.51}$$

We use the color green to mark volume quantities and blue for surface quantities. The terms marked with mixed colors are volume quantities that will later contribute to the surface balance.

We introduce a *volume flux density* $\mathbf{j}$ and a *surface flux density* $\mathbf{j}_\Sigma$. But, we write them with boundary flux integrals as

$$\Phi(t) = -\int_S \mathbf{j} \cdot \boldsymbol{\nu} \, dS - \int_{\partial\Sigma} \mathbf{j}_\Sigma \cdot \boldsymbol{\nu}_{\partial\Sigma} \, dl, \tag{9.52}$$

as in Dreyer (2003, Sect. 5.1) and Müller (1985, Sect. 3.1). The fluxes may include differential terms such as Fick's or Fourier's laws. The resulting differential equations can therefore be diffusive, i.e., parabolic.

The first integral accounts for the non-convective flux across the closed surface of any arbitrarily small material volume. Whereas the second integral describes the flux which is tangential to the surface Σ and normal to its closed boundary $\partial\Sigma$. The unit normal vector field $\boldsymbol{\nu}_{\partial\Sigma}$ on the curve $\partial\Sigma$ is chosen to be tangential to the surface Σ and outward with respect to the surface Σ. Since the unit normal vector fields $\boldsymbol{\nu}$ and $\boldsymbol{\nu}_{\partial\Sigma}$ are pointing outward, the sign takes care that incoming fluxes contribute positively to Φ. Using the Gauss–Green theorem both boundary integrals can later be converted to appropriate domain integrals of the divergence of the fluxes.

Finally, the source Ξ can be decomposed using a *volume density* ξ and a *surface density* ξ_Σ. Thus it is given as

$$\Xi(\mathrm{t}) = \int\limits_{\mathcal{V}_1\cup\mathcal{V}_2} \xi \,\mathrm{d}V + \int\limits_{\Sigma} \xi_\Sigma \,\mathrm{d}S. \tag{9.53}$$

In the case of a zero right-hand side, Eq. (9.50) becomes a physical conservation law in a strict sense. This is the case for conservation of mass leading to the continuity equation. For other generic quantities Ψ we have the additional terms on the right-hand side making them balance laws. The boundary flux density integrals with $\mathbf{j}$ and $\mathbf{j}_\Sigma$ can be transformed into divergence integrals over the volume that together with the terms from the left-hand side of (9.50) lead to partial differential equations in divergence form, then it has become a custom that the resulting differential equations are called conservation laws. Such differential equations are called balance laws if they additionally contain non-differential source terms or differential terms that cannot be brought into divergence form.

Inserting Eqs. (9.51), (9.52), and (9.53) into the general balance law (9.50) gives

$$\begin{aligned}\frac{\mathrm{d}}{\mathrm{dt}}\left(\int\limits_{\mathcal{V}_1\cup\mathcal{V}_2} \psi \,\mathrm{d}V + \int\limits_{\Sigma} \psi_\Sigma \,\mathrm{d}S\right) \\ = -\int\limits_{\mathcal{S}} \mathbf{j}\cdot\boldsymbol{\nu}\,\mathrm{d}S - \int\limits_{\partial\Sigma} \mathbf{j}_\Sigma\cdot\boldsymbol{\nu}_{\partial\Sigma}\,\mathrm{d}l + \int\limits_{\mathcal{V}_1\cup\mathcal{V}_2} \xi \,\mathrm{d}V + \int\limits_{\Sigma} \xi_\Sigma \,\mathrm{d}S.\end{aligned} \tag{9.54}$$

Examples for physical quantities and their corresponding densities may be found in Bedeaux (1986), Dreyer (2003, Sect. 6) as well as Truesdell and Toupin (1960). Some are nicely summarized in a table in Müller (1985, Sect. 3.2).

The basic assumption that surface quantities contribute to the total quantity Ψ is a major point in our considerations. In fact, this assumption will be responsible for additional terms in the equations on singular surfaces with discontinuous weak solutions. This includes differential terms, see, e.g. (9.71), that are absent in the standard case of shock or contact discontinuities where the jump conditions are purely algebraic.

9.4 Transport Theorems, Volume, and Surface Balance Equations

In the following, the time derivative on the left-hand side of Eq. (9.54) is evaluated using transport theorems since the coordinates of volume and surface depend on time. The resulting relations then lead to differential equations in the volume and on the surface.

9.4.1 Volume Transport

We start with the well-known Reynolds Transport Theorem for volume integrals, see also Aris 1989, Sect. 4.22, Truesdell and Toupin 1960, Sect. 81 as well as Warnecke 1999, Satz 1.3. The proof is given since it allows some insight into the result and its later extensions. We state the theorem for any dimension $n = 1, 2, 3, \ldots$ since it does not depend on dimension. For dimension $n = 1$ (9.56) gives versions of Leibniz's rule for integrals, see, e.g., Thomas et al. 2004, Chap. 5 for a version that is missing the integral term arising due to a parameter-dependent integrand.

Theorem *(The Reynolds Transport Theorem in $\mathbb{R}^n$.) Let $\psi : I \times \mathbb{R}^n \to \mathbb{R}$ be a continuously differentiable density function and $\mathcal{V}(\mathrm{t}) \in \mathbb{R}^n$ be a closed volume moving with the local particle velocity $\mathbf{v}(\mathrm{t}, \mathbf{x})$. We assume that the boundary $\mathcal{S}(\mathrm{t}) = \partial\mathcal{V}(\mathrm{t})$ consists of finitely many smooth pieces. Let $\boldsymbol{\nu}$ denote the outer unit normal field of the hypersurface $\mathcal{S}(\mathrm{t})$. Then the following equations hold*

$$\frac{\mathrm{d}}{\mathrm{dt}}\Psi(\mathrm{t}) = \frac{\mathrm{d}}{\mathrm{dt}}\int\limits_{\mathcal{V}(\mathrm{t})} \psi(\mathrm{t}, \mathbf{x})\, \mathrm{d}V = \int\limits_{\mathcal{V}(\mathrm{t})} \left[\frac{\partial}{\partial \mathrm{t}}\psi + \nabla_{\mathbf{x}} \cdot (\psi \mathbf{v})\right](\mathrm{t}, \mathbf{x})\, \mathrm{d}V \tag{9.55}$$

and

$$\frac{\mathrm{d}}{\mathrm{dt}}\Psi(\mathrm{t}) = \int\limits_{\mathcal{V}(\mathrm{t})} \frac{\partial}{\partial \mathrm{t}}\psi(\mathrm{t}, \mathbf{x})\, \mathrm{d}V + \int\limits_{\mathcal{S}(\mathrm{t})} [\psi\, \mathbf{v} \cdot \boldsymbol{\nu}](\mathrm{t}, \mathbf{x})\, \mathrm{d}S. \tag{9.56}$$

Proof The key idea is that at any time the volume $\mathcal{V}(\mathrm{t})$ may be described applying a smooth transformation to the initial volume $\mathcal{V}(0)$. Let the initial volume $\mathcal{V}(0)$ be described using the coordinates $\mathbf{y} = (\mathrm{y}_1, \ldots, \mathrm{y}_n)^T$. Since we consider a closed material volume no particles are removed from or added to the volume. Thus we assume the existence of a diffeomorphism $\chi : I \times \mathbb{R}^n$ such that at every time $\mathrm{t} \in I$ we have $\mathbf{x}(\mathrm{t}) = \chi(\mathrm{t}, \mathbf{y}_0) = \chi^{\mathrm{t}}(\mathbf{y}_0)$, see Appendix 9.6. This is the Lagrangian description of the flow or motion.

The Jacobian matrix or deformation gradient of χ^{t} for fixed t is given by $\mathbf{D}\chi^{\mathrm{t}}(\mathbf{y}) = (\partial\chi_j^{\mathrm{t}}/\partial \mathrm{y}_k)_{1\le j,k\le n}$. Further, let $\mathcal{J}^{\mathrm{t}} = \det(\mathbf{D}\chi)$ denote the corresponding determinant. The determinant $\mathcal{J}^{\mathrm{t}}$ is needed to relate the volume element $\mathrm{d}V$ at time t to the volume element of the initial configuration $\mathrm{d}V_0$, i.e., $\mathrm{d}V = \mathcal{J}^{\mathrm{t}}\mathrm{d}V_0$. The derivative with respect to time of $\mathcal{J}^{\mathrm{t}}$ is given by (9.79) as $\frac{\mathrm{d}\mathcal{J}^{\mathrm{t}}}{\mathrm{dt}} = (\nabla_{\mathbf{x}} \cdot \mathbf{v})\,\mathcal{J}^{\mathrm{t}}$. Now we obtain

$$
\begin{aligned}
\frac{\mathrm{d}}{\mathrm{dt}}\Psi(\mathrm{t}) &= \frac{\mathrm{d}}{\mathrm{dt}}\int\limits_{\mathcal{V}(\mathrm{t})} \psi(\mathrm{t}, \mathbf{y})\,\mathrm{d}V = \frac{\mathrm{d}}{\mathrm{dt}}\int\limits_{\mathcal{V}(0)} \psi(\mathrm{t}, \chi^{\mathrm{t}}(\mathbf{y}))\mathcal{J}^{\mathrm{t}}(\mathrm{t}, \mathbf{y})\,\mathrm{d}V_0 \\
&= \int\limits_{\mathcal{V}(0)} \frac{\mathrm{d}}{\mathrm{dt}}\left[\psi(\mathrm{t}, \chi^{\mathrm{t}}(\mathbf{y}))\mathcal{J}^{\mathrm{t}}(\mathrm{t}, \mathbf{y})\right]\mathrm{d}V_0 \\
&= \int\limits_{\mathcal{V}(0)} \left[\frac{\mathrm{d}}{\mathrm{dt}}\psi(\mathrm{t}, \chi^{\mathrm{t}}(\mathbf{y})) + \psi(\mathrm{t}, \chi^{\mathrm{t}}(\mathbf{y}))[\nabla_{\mathbf{x}} \cdot \mathbf{v}(\mathrm{t}, \chi^{\mathrm{t}}(\mathbf{y}))]\right]\mathcal{J}^{\mathrm{t}}(\mathrm{t}, \mathbf{y})\,\mathrm{d}V_0 \\
&= \int\limits_{\mathcal{V}(\mathrm{t})} \left[\frac{\mathrm{d}}{\mathrm{dt}}\psi(\mathrm{t}, \mathbf{x}) + \psi(\mathrm{t}, \mathbf{x})(\nabla_{\mathbf{x}} \cdot \mathbf{v}(\mathrm{t}, \mathbf{x}))\right]\mathrm{d}V.
\end{aligned}
\tag{9.57}
$$

Using the chain rule and $\mathbf{v}(\mathrm{t}, \mathbf{y}) = \mathrm{d}\mathbf{y}/\mathrm{dt} = \partial\chi^{\mathrm{t}}(\mathbf{y})/\partial\mathrm{t}$ leads to

$$
\frac{\mathrm{d}}{\mathrm{dt}}\psi(\mathrm{t}, \mathbf{x}) = \frac{\partial}{\partial \mathrm{t}}\psi(\mathrm{t}, \mathbf{x}) + \nabla_{\mathbf{x}}\psi(\mathrm{t}, \mathbf{x}) \cdot \mathbf{v}(\mathrm{t}, \mathbf{x}).
$$

This inserted into the last integral gives (9.55). The second Eq. (9.56) follows by applying the Gauss–Green theorem to the integral with the divergence for any arbitrary but fixed t. In this derivation t is just a parameter. □

Note that the divergence term in (9.55) is a combination of two contributions, of the advective or Lagrangian derivative resulting from the total time derivative of ψ and of the divergence of the velocity field coming from the time derivative of the volume element. The first part is the advection of the physical quantity ψ along stream lines of the flow. The second is from the local expansion or compression of the control volume due to the flow. Both combine nicely to give the divergence term which is mathematically very appreciated when weak solutions are considered in conjunction with jump discontinuities of physical quantities in flows, such as shock waves, contact discontinuities, reaction fronts, or phase transitions.

9.4.2 Divergence Theorem and Transport Theorem for Embedded Moving Surfaces

The case $n = 2$ of the Reynolds Transport Theorem 9.4.1 would only cover moving surfaces in the plane $\mathbb{R}^2$. In order to treat moving surfaces inside of moving control volumes in $\mathbb{R}^3$ we need an additional transport theorem. First, we give a divergence or Gauss–Green theorem for curved surfaces in $\mathbb{R}^3$. For this we consider any open subset $\Omega \subseteq \mathbb{R}^3$ and a bounded smooth surface $\Sigma \subset \mathbb{R}^3$ that is bounded by a smooth curve $\partial\Sigma$. Let $\boldsymbol{\tau}$ be a tangent vector field to the boundary curve. We assume that the vector field $\boldsymbol{\nu}_\Sigma$, which is normal to the surface, and the tangent vector field $\boldsymbol{\tau}$ satisfy the usual right-hand rule. We take $\boldsymbol{\nu}_{\partial\Sigma} = \boldsymbol{\tau} \times \boldsymbol{\nu}_\Sigma$ to be the outward pointing boundary normal that is tangential to the surface. Now we want to show the divergence theorem for surfaces in Ω, cp. Lee (2013, Theorem 16.32) for a more general version of (9.58).

Theorem (Divergence Theorem or Gauss–Green Theorem for Surfaces in $\mathbb{R}^3$) *Let $\Sigma \subset \Omega \subseteq \mathbb{R}^3$ be a bounded smooth surface. Further, $\mathbf{a} : \Sigma \to \mathbb{R}^3$ is a continuously differentiable vector field that is either defined on the boundary $\partial\Sigma$ or has a bounded continuous extension to this boundary. Like in* (9.28) *it may be decomposed into tangential and normal components as follows $\mathbf{a} = \mathbf{a}^{\shortparallel} + a_\nu \boldsymbol{\nu}_\Sigma$. By $\mathrm{d}l$ we denote the line element on the curve $\partial\Sigma$. We assume that the curve is continuous and consists of finitely many smooth pieces. Then the following divergence formula for surface integrals holds*

$$\int_{\Sigma} \left[\nabla_\Sigma \cdot \mathbf{a}^{\shortparallel}\right](\mathbf{x})\,\mathrm{d}S = \int_{\partial\Sigma} [\mathbf{a} \cdot \boldsymbol{\nu}_{\partial\Sigma}]\,(\mathbf{x})\,\mathrm{d}l. \tag{9.58}$$

From this we obtain the formula

$$\int_{\Sigma} [\nabla_\Sigma \cdot \mathbf{a}]\,(\mathbf{x})\,\mathrm{d}S = \int_{\partial\Sigma} [\mathbf{a} \cdot \boldsymbol{\nu}_{\partial\Sigma}]\,(\mathbf{x})\,\mathrm{d}l - \int_{\Sigma} [2\kappa_M a_\nu]\,(\mathbf{x})\,\mathrm{d}S. \tag{9.59}$$

Proof The proof is basically an application of the Kelvin–Stokes theorem, cp. Aris (1989, (3.32.2)), Bourne and Kendall (1992, Sects. 6.5 and 6.6), Grinfeld (2013, (14.53)), or Thomas et al. (2004, Sect. 16.7), to a particular vector field $\mathbf{b} : \Sigma \to \mathbb{R}^3$ and suitable surface Σ with boundary curve $\partial\Sigma$. We use it in the form

$$\int_{\Sigma} [(\nabla_{\mathbf{x}} \times \mathbf{b}) \cdot \boldsymbol{\nu}_\Sigma]\,(\mathbf{x})\,\mathrm{d}S = \int_{\partial\Sigma} [\mathbf{b} \cdot \boldsymbol{\tau}](\mathbf{x})\,\mathrm{d}l. \tag{9.60}$$

Here the normals $\boldsymbol{\nu}$ on the surface and the tangents $\boldsymbol{\tau}$ of the curve are oriented by the right-hand rule. If you look at the surface from the side to which the normals point, the curve is traversed counter-clockwise.

Now we take the particular vector field $\mathbf{b} = \boldsymbol{\nu}_\Sigma \times \mathbf{a}$. Using (9.32), we obtain for the left-hand side of (9.60)

$$\int_\Sigma [(\nabla_\mathbf{x} \times (\boldsymbol{\nu}_\Sigma \times \boldsymbol{a})) \cdot \boldsymbol{\nu}_\Sigma](\mathbf{x})\,\mathrm{d}S = \int_\Sigma \left[\nabla_\Sigma \cdot \boldsymbol{a}^{\shortparallel}\right](\mathbf{x})\,\mathrm{d}S.$$

Using the circular symmetry of the triple product, we have for the right-hand side of (9.60)

$$\int_{\partial\Sigma} [(\boldsymbol{\nu}_\Sigma \times \mathbf{a}) \cdot \boldsymbol{\tau}](\mathbf{x})\,\mathrm{d}l = \int_{\partial\Sigma} [(\boldsymbol{\tau} \times \boldsymbol{\nu}_\Sigma) \cdot \mathbf{a}](\mathbf{x})\,\mathrm{d}l = \int_{\partial\Sigma} [\mathbf{a} \cdot \boldsymbol{\nu}_{\partial\Sigma}](\mathbf{x})\,\mathrm{d}l.$$

Combining them gives formula (9.58). Equation (9.59) follows by using (9.31). □

Next we obtain an analog of the Reynolds Transport Theorem 9.4.1 for moving surfaces.

Theorem (Transport Theorem for Moving Surfaces in $\mathbb{R}^3$) *Let $\Sigma(\mathrm{t})$, $\mathrm{t} \in [0, T]$ be a family of smooth surfaces in $\Omega \subseteq \mathbb{R}^3$ with the velocity field $\mathbf{w}$ of surface points and the surface area element $\mathrm{d}S = \sqrt{g}\mathrm{d}u^1\mathrm{d}u^2$ for the surface coordinates $\mathbf{u} = (u^1, u^2) \in \mathcal{U}$, see Sect. 9.2.1. Here $g(\mathrm{t}, \mathbf{u})$ denotes the determinant of the metric tensor (9.4) of the surface at time* t.

Further let $\psi : I \times \Sigma(\mathrm{t}) \to \mathbb{R}$ be a smooth function in the sense that the Lagrangian time derivative $\mathring{\psi}(\mathrm{t}, \mathbf{x}) = \frac{D}{D\mathrm{t}}\psi(\mathrm{t}, \mathbf{x})$ exists, see (9.39) and (9.42). Recall that the surface divergence $\nabla_\Sigma \cdot \mathbf{w}$ is given by (9.20). We set $w_\nu = \mathbf{w} \cdot \boldsymbol{\nu}_\Sigma$. We obtain the following transport equations

$$\frac{\mathrm{d}}{\mathrm{dt}} \int_{\Sigma(\mathrm{t})} \psi(\mathrm{t}, \mathbf{x})\,\mathrm{d}S = \int_{\Sigma(\mathrm{t})} \left[\mathring{\psi} + \psi\,\nabla_\Sigma \cdot \mathbf{w}\right](\mathrm{t}, \mathbf{x})\,\mathrm{d}S \tag{9.61}$$

$$= \int_{\Sigma(\mathrm{t})} \left[\mathring{\psi} + \psi\,\left(\nabla_\Sigma \cdot \mathbf{w}^{\shortparallel} - 2\kappa_M w_\nu\right)\right](\mathrm{t}, \mathbf{x})\,\mathrm{d}S \tag{9.62}$$

$$= \int_{\Sigma(\mathrm{t})} \left[\frac{\delta}{\delta \mathrm{t}}\psi + \nabla_\Sigma \cdot \left(\psi\,\mathbf{w}^{\shortparallel}\right) - 2\kappa_M w_\nu\right](\mathrm{t}, \mathbf{x})\,\mathrm{d}S. \tag{9.63}$$

Now we consider $\psi : I \times \Omega \to \mathbb{R}$ is a smooth function. We set $\psi_\nu = \boldsymbol{\nu}_\Sigma \cdot \nabla_\mathbf{x}\psi$. Then we can write the transport equations as

$$\frac{\mathrm{d}}{\mathrm{dt}} \int_{\Sigma(\mathrm{t})} \psi(\mathrm{t}, \mathbf{x})\,\mathrm{d}S = \int_{\Sigma(\mathrm{t})} \left[\frac{\partial\psi}{\partial \mathrm{t}} + \nabla_\Sigma \cdot (\psi\,\mathbf{w}) + \psi_\nu \mathbf{w}_\nu\right](\mathrm{t}, \mathbf{x})\,\mathrm{d}S \tag{9.64}$$

$$= \int_{\Sigma(\mathrm{t})} \left[\frac{\partial\psi}{\partial \mathrm{t}} + \nabla_\Sigma \cdot (\psi\mathbf{w}^{\shortparallel}) + (\psi_\nu - 2\kappa_M\psi)w_\nu\right](\mathrm{t}, \mathbf{x})\,\mathrm{d}S. \tag{9.65}$$

Proof The proof of the transport theorem for the surface is quite analogous to the proof of Theorem 9.4.1. Important preparations were made in Sect. 9.2.3 by defining the parametrization of the initial reference surface $\Sigma_0 \subset \Omega$, the flow map χ_Σ^t as well as the appropriate time derivatives for moving surfaces and clarifying the latter. First, we go via the mapping $\boldsymbol{\Phi}_{\Sigma_0}$ from the surface parameters in $\mathcal{U}$ to the initial reference configuration Σ_0 and then via the flow map $\chi_\Sigma^t = \chi_\Sigma(\mathrm{t}, \cdot)$ to the current surface $\Sigma(\mathrm{t})$. This mapping is given by $\boldsymbol{\Phi}_\Sigma = \chi_\Sigma^t \circ \boldsymbol{\Phi}_{\Sigma_0} : \mathcal{U} \to \Sigma(\mathrm{t})$, see Sect. 9.2.3. We recall the surface integral given by (9.7), the function $\widetilde{\psi}(\mathrm{t}, \mathbf{u}) = \psi(\mathrm{t}, \boldsymbol{\Phi}_\Sigma(\mathrm{t}, \mathbf{u}))$, the time derivative of the metric g satisfying (9.44) and (9.42) for the Lagrangian derivative. Thus, we obtain

$$
\begin{aligned}
\frac{\mathrm{d}}{\mathrm{dt}} \int_{\Sigma(\mathrm{t})} \psi(\mathrm{t}, \mathbf{x})\, \mathrm{d}S &= \frac{\mathrm{d}}{\mathrm{dt}} \int_{\mathcal{U}} \psi(\mathrm{t}, \boldsymbol{\Phi}_\Sigma(\mathrm{t}, \mathbf{u}))\, \sqrt{g(\mathrm{t}, \mathbf{u})}\, \mathrm{d}u^1 \mathrm{d}u^2 \\
&= \frac{\mathrm{d}}{\mathrm{dt}} \int_{\mathcal{U}} \widetilde{\psi}(\mathrm{t}, \mathbf{u})\, \sqrt{g(\mathrm{t}, \mathbf{u})}\, \mathrm{d}u^1 \mathrm{d}u^2 \\
&= \int_{\mathcal{U}} \left[\frac{D}{D\mathrm{t}} \psi + \psi \nabla_\Sigma \cdot \mathbf{w} \right] (\mathrm{t}, \boldsymbol{\Phi}_\Sigma(\mathrm{t}, \mathbf{u})) \sqrt{g(\mathrm{t}, \mathbf{u})}\, \mathrm{d}u^1 \mathrm{d}u^2 \\
&= \int_{\Sigma(\mathrm{t})} \left(\mathring{\psi}(\mathrm{t}, \mathbf{x}) + \psi(\mathrm{t}, \mathbf{x}) \nabla_\Sigma \cdot \mathbf{w}(\mathrm{t}, \mathbf{x}) \right) \mathrm{d}S.
\end{aligned}
$$

This is the first Eq. (9.61). The second Eq. (9.62) follows using (9.31). For the third Eq. (9.63) we use (9.40). The fourth Eq. (9.64) is an application of (9.40), (9.38), and (9.21) to (9.61). Finally, we use (9.31) in (9.64) to obtain (9.65). □

The formula (9.61) can be found in Dziuk and Elliott (2007, Lemma 2.2, Appendix A). But, for the proof they assume that $\psi : I \times \Omega \to \mathbb{R}$. They called this transport equation a Leibniz formula. The formula (9.65) is Dziuk and Elliott (2007, (2.10)). They used a level set description of the surface. The resulting transport equation is the same.

The formula (9.65) is also given in Cermelli et al. (2005, (3.14)) for the case $\psi_\nu = 0$ and using (9.58) of the Divergence Theorem 9.4.2. They used an extension of ψ to a neighborhood of $\Sigma(\mathrm{t})$. See also the literature cited there. The formula (9.62) was given in Petryk and Mróz (1986, (2.34)).

In particular note that the time derivative of the metric corresponds to the surface divergence of the velocity field $\mathbf{w}$ in (9.61). This highlights that the surface transport theorem is an analog to the Reynolds Transport Theorem 9.4.1. The time derivative of the integrated quantity is the integral of the Lagrangian time derivative of a quantity plus the quantity times the divergence of the velocity field. The formulas (9.62) and (9.65) show how a non-flat geometry comes into play through the mean curvature term.

Note that Dreyer (2003) and Müller (1985) considered only the two maps $\boldsymbol{\Phi}_{\Sigma_0}$ and $\boldsymbol{\Phi}_{\Sigma}$, see Sect. 9.2.3. Then $\boldsymbol{\chi}_{\Sigma} = \boldsymbol{\Phi}_{\Sigma}(\cdot, \boldsymbol{\Phi}_{\Sigma}^{-1}(\cdot))$ is only implied. This leads to slightly more complicated formulas. They introduced an extra notation for the metric on Σ_0 which is actually only $\mathbf{G}(0)$ in our notation. For the Transport Theorem 9.4.2 they refer back to the stationary surface Σ_0 whereas we refer to its stationary parameter set $\mathcal{U}$. This is mathematically equivalent. It seems that they wanted to stay closer to the proof of the Reynolds Transport Theorem 9.4.1 for volumes.

We find our approach somewhat simpler, as it does not involve the reciprocal of $g(0)$ in the integrals. Also the introduction of $\boldsymbol{\chi}_{\Sigma}$ seems to add clarity to the concepts. Also the formulas Dreyer (2003, (48)) and Müller (1985, (3.9)) were given with respect to the parameter variables. The time derivative of ψ is then the Lagrangian time derivative $\mathring{\psi}$ given as (9.42).

Aris (1989, Sects. 10.12 and 10.31) has a different formulation and proof. He considered a dynamic function from $\mathbb{R}^2$ to $\mathbb{R}^2$. Both are equipped with Riemannian geometries. The densities ψ are not functions of $\mathbf{x} \in \mathbb{R}^3$ but of $\mathbf{u}(t) \in \mathbb{R}^2$. Instead of using (9.44) for the differentiated area element he used $\frac{\mathrm{d}}{\mathrm{dt}}\sqrt{g} = \frac{1}{2g}\frac{\mathrm{d}g}{\mathrm{dt}}\sqrt{g}$ instead. Thereby some insight into the transport is lost. His formulas are completely intrinsic to the surface, which is mathematically nice, but not so useful for practical applications to the embedded moving surfaces we are interested in. In some sections, he did discuss the embedding into $\mathbb{R}^3$, but only in a somewhat sketchy manner. An intrinsic formulation does not have normal transport of the surface but only the transport within the surface.

If we have only intrinsic information on the surface, then there are no normal velocities at all. So we would also have $w_\nu = 0$. Then Eq. (9.65) becomes a two-dimensional analog of Reynolds Transport Theorem 9.4.1. It has a non-Cartesian divergence of the intrinsic curved geometry of the surface. This case is of interest here only if the surface has a purely tangential motion.

9.4.3 Separate Differential Balance Equations

We consider an open subset $\Omega \subseteq \mathbb{R}^3$ and any family of open control volumes $\mathcal{V}(\mathrm{t})$ with $\mathcal{V}(\mathrm{t}) \subset \Omega$ for $\mathrm{t} \in [0, T]$ with $T > 0$. We take the generic balance law (9.54) on a time-dependent control volume $\mathcal{V}(\mathrm{t})$. We use formula (9.55) of the Reynolds Transport Theorem 9.4.1 and (9.65) of the Surface Transport Theorem 9.4.2 to determine the time derivatives. One could take (9.63) instead with a different time derivative and one term less. The latter makes it more general. We assume that the boundary flux $\mathbf{j}$ is defined everywhere in the volume and $\mathbf{j}_{\Sigma}$ everywhere on the surface. To the volume boundary flux term, we apply the usual Gauss–Green theorem and to the surface boundary flux term the surface version (9.58) in order to obtain

$$0 = \int\limits_{\mathcal{V}(t)} \left[\frac{\partial}{\partial t}\psi + \nabla_{\mathbf{x}} \cdot [\psi \mathbf{v} + \mathbf{j}] - \xi \right] (t, \mathbf{x}) \, dV$$
$$+ \int\limits_{\Sigma(t)} \left[\frac{\partial \psi}{\partial t} + \nabla_{\Sigma} \cdot (\psi \mathbf{w}^{\shortparallel}) + (\psi_\nu - 2\kappa_M \psi) w_\nu + \nabla_\Sigma \cdot \mathbf{j}^{\shortparallel}_\Sigma - \xi_\Sigma \right] (t, \mathbf{x}) \, dS. \tag{9.66}$$

This equation is satisfied for any $t \in [0, T]$ and any control volume $\mathcal{V}(t) \subset \Omega$ of any shape and appropriate size. If a control volume does not contain a surface $\Sigma(t)$ then we take the integral over $\Sigma(t)$ to be 0.

We will make use of the Lebesgue differentiation theorem of real analysis. For this we consider for some fixed $t \in]0, T[$ a family of open balls

$$B^n_{\mathbf{x},r} = \{\mathbf{y} \in \mathbb{R}^n \mid |\mathbf{y} - \mathbf{x}| < r\} \subset \mathcal{V}(t)$$

with radius small $r > 0$ enough. Their volumes are denoted as $\mathrm{vol}(B^n_{\mathbf{x},r})$. Any other shapes with a scaling factor like the radius $r > 0$ can be used for such arguments.

Take $(t, \mathbf{x}) \in]0, T[\times\Omega$. Then we can find an open control volume with $\mathbf{x} \in \mathcal{V}(t) \subset \Omega$ and there are infinitely many balls $B^n_{\mathbf{x},r} \subset \mathcal{V}(t)$ with $r > 0$ sufficiently small. Then the Lebesgue differentiation theorem states that for any integrable function $f : \Omega \to \mathbb{R}$ the right-hand side of the inequalities

$$\left| f(\mathbf{x}) - \frac{1}{\mathrm{vol}(B^n_{\mathbf{x},r})} \int\limits_{B^n_{\mathbf{x},r}} f(\mathbf{y}) \, d\mathbf{y} \right| = \left| \frac{1}{\mathrm{vol}(B^n_{\mathbf{x},r})} \int\limits_{B^n_{\mathbf{x},r}} [f(\mathbf{x}) - f(\mathbf{y})] \, d\mathbf{y} \right|$$
$$\leq \frac{1}{\mathrm{vol}(B^n_{\mathbf{x},r})} \int\limits_{B^n_{\mathbf{x},r}} |f(\mathbf{x}) - f(\mathbf{y})| \, d\mathbf{y}$$

goes to 0 for $r \to 0$ for almost any $\mathbf{x} \in \Omega$, see, e.g., Evans (1998, Appendix E.4) or Rudin (1987, Theorem 7.7). If f is a continuous function this holds for all $\mathbf{x} \in \Omega$. This implies that

$$\frac{1}{\mathrm{vol}(B^n_{\mathbf{x},r})} \int\limits_{B^n_{\mathbf{x},r}} f(\mathbf{y}) \, d\mathbf{y} \;\to\; f(\mathbf{x}) \qquad \text{for} \quad r \to 0.$$

In our application, all integrals equal 0 implying $f(\mathbf{x}) = 0$. The variable t is just a parameter in this argument. We assume here that the integrands are continuous. Then the results hold for all points $(t, \mathbf{x}) \in]0, T] \times \Omega$.

We consider any $\mathbf{x}_0 \in \Sigma(t)$ with $\mathbf{u}_0 \in \mathcal{U}$ be such that $\boldsymbol{\Phi}(\mathbf{u}_0) = \mathbf{x}_0$. We take a family of balls $B^2_{\mathbf{u}_0,r} \subset \mathcal{U}$ for $r > 0$ sufficiently small and set $B^2_{\Sigma,\mathbf{x}_0,r} = \boldsymbol{\Phi}[B^2_{\mathbf{u}_0,r}]$ to give curved disks on the surface over which we will integrate. We write the surface integrals over the open parameter set $\mathcal{U}$ using (9.7). We define the curved *pillbox set* over the disk $B^2_{\Sigma,\mathbf{x}_0,r}$ as

$$\mathcal{V}_{\Sigma,r} = \{\mathbf{x} \in \mathbb{R}^3 \mid \mathbf{x} = \lambda \boldsymbol{\nu}_\Sigma + \mathbf{y} \text{ with } \mathbf{y} \in B^2_{\Sigma,\mathbf{x}_0,r},\ \lambda \in]-r, r[\} \subset \Omega.$$

We take Eq. (9.66) over $\mathcal{V}_{\Sigma,r}$, then $\Sigma(\mathrm{t}) = B^2_{\Sigma,\mathbf{x}_0,r}$ and divide it by $\mathrm{vol}(B^2_{\mathbf{u}_0,r})$. Further, we write the volume integral as an average multiplied by the factor $\frac{\mathrm{vol}(V_{\Sigma,r})}{\mathrm{vol}(B^2_{\mathbf{u}_0,r})}$. The average will be bounded and the factor will converge to 0 for $r \to 0$. This gives the differential equation

$$\frac{\partial \psi}{\partial \mathrm{t}} + \nabla_\Sigma \cdot (\psi \mathbf{w}^{\shortparallel}) + (\psi_\nu - 2\kappa_M \psi) w_\nu + \nabla_\Sigma \cdot \mathbf{j}^{\shortparallel}_\Sigma = \xi_\Sigma \tag{9.67}$$

satisfied for any $(\mathrm{t}, \mathbf{x}) \in]0, T] \times \Omega$ with $\mathbf{x} \in \Sigma(\mathrm{t})$ on a surface. Using the integrand of (9.61) it can also be written as

$$\mathring{\psi}_\Sigma + \psi_\Sigma \nabla_\Sigma \cdot \mathbf{w} + \nabla_\Sigma \cdot \mathbf{j}^{\shortparallel}_\Sigma = \xi_\Sigma$$

in case $\psi_\nu = 0$ or ψ_ν does not exist. Now that we know that the integrand of the surface integral vanishes. We can do the analogous argument with averages of the volume integral. This gives us the differential equation for the volume

$$\frac{\partial}{\partial \mathrm{t}} \psi + \nabla_{\mathbf{x}} \cdot (\psi \mathbf{v} + \mathbf{j}) = \xi. \tag{9.68}$$

for any $(\mathrm{t}, \mathbf{x}) \in]0, T] \times \Omega$. We obtained two separate equations.

9.4.4 A Generalized Transport Theorem

As in Sect. 9.2.4, we want to consider an open control volume $\mathcal{V}(\mathrm{t}) \subset \Omega \subseteq \mathbb{R}^3$ for $\mathrm{t} \in I = [0, T]$. We assume that it is divided into two open subsets $\mathcal{V}_1(\mathrm{t})$ and $\mathcal{V}_2(\mathrm{t})$ by an internal open surface $\Sigma(\mathrm{t})$ with $\mathcal{V}(\mathrm{t}) = \mathcal{V}_1(\mathrm{t}) \cup \Sigma(\mathrm{t}) \cup \mathcal{V}_2(\mathrm{t})$. This is depicted in Fig. 9.1. We also take into account functions $\psi : I \times \Omega \to \mathbb{R}$ which are continuous in $\mathcal{V}_1(\mathrm{t}) \cup \mathcal{V}_2(\mathrm{t})$ with limiting values from each side at the surface $\Sigma(\mathrm{t})$. In order to keep the notation compact, we introduce the jump brackets $[\![\cdot]\!]$. For points $\mathbf{x}_\Sigma \in \Sigma$, we write

$$\psi_i(\mathrm{t}, \mathbf{x}_\Sigma) = \lim_{\mathcal{V}_i \ni \mathbf{x} \to \mathbf{x}_\Sigma} \psi(\mathrm{t}, \mathbf{x}),\ i = 1, 2 \text{ and } [\![\psi]\!](\mathrm{t}, \mathbf{x}_\Sigma) = \psi_2(\mathrm{t}, \mathbf{x}_\Sigma) - \psi_1(\mathrm{t}, \mathbf{x}_\Sigma).$$

The plus sign in the jump bracket is always on the side to which the unit normal vector field $\boldsymbol{\nu}_\Sigma$ on the surface points. Usually we will not write out the arguments since either the equations containing jump brackets should hold in any point of the surface or it is clear from the context which point is meant.

Further, we assume that the velocity field $\mathbf{v}$ is also discontinuous at Σ with bounded limiting values from each side. In shock waves, the normal velocity field $\mathbf{v}^\perp = \mathbf{v} \cdot \boldsymbol{\nu}_\Sigma$ is discontinuous, whereas the tangential velocity field $\mathbf{v}^{\shortparallel}$ is continuous, see, e.g., Anderson (2003, Sect. 4.3) or Toro (2009, Sect. 3.2). The same is true for contact discontinuities, see Anderson (2003, Sect. 7.1) or Toro (2009, p. 97). There may also be shear waves, see Toro (2009, pp. 107,111), in which the tangential velocity jumps. The velocity field has bounded one-sided limits at $\Sigma(\mathrm{t})$.

The fact that the velocity field is discontinuous at $\Sigma(\mathrm{t})$ means that the trajectories $\mathbf{x}(\mathrm{t}) = \chi^{\mathrm{t}}(\mathbf{x}_0)$ are only continuous there and abruptly change direction at an angle. We always assume that the motions χ are at least continuous with respect to t and $\mathbf{x}$. Let $\mathbf{v}_i$ for $i = 1, 2$ distinguish the limiting values at $\Sigma(\mathrm{t})$ of the velocity field on $\mathcal{V}_i(\mathrm{t})$. The surface $\Sigma(\mathrm{t})$ moves with the smooth velocity field $\mathbf{w} : I \times \Sigma(\mathrm{t}) \to \mathbb{R}^3$ that may be independent of $\mathbf{v}$. Under these assumptions we want to give a generalization of Theorem 9.4.1 and obtain a coupled balance law on the surface.

The proof of Theorem 9.4.1 revealed that the total change in time of a physical quantity or state is balanced by the local change of the density function and two additional effects. First, there is the convective part, and second there is a contribution due to the deformation of the material volume. We now want to extend it to the case of a singular surface along which the volume functions may be discontinuous.

Theorem (Generalized Reynolds Transport Theorem in $\mathbb{R}^3$) *Let $\mathcal{V}(\mathrm{t}) \subset \Omega \subseteq \mathbb{R}^3$ for $\mathrm{t} \in [0, T]$ be an open material volume as in Fig. 9.1 with boundary $\mathcal{S}(\mathrm{t}) = \partial\mathcal{V}(\mathrm{t})$. We assume that the boundary $\mathcal{S}(\mathrm{t})$ consists of finitely many smooth pieces. The moving internal two-dimensional surface $\Sigma(\mathrm{t})$ separates $\mathcal{V}(\mathrm{t})$ into two volumes $\mathcal{V}_1(\mathrm{t})$ and $\mathcal{V}_2(\mathrm{t})$.*

The function $\psi : I \times \Omega \to \mathbb{R}$ and the velocity field $\mathbf{v} : I \times \Omega \to \mathbb{R}^3$ are continuously differentiable on $\mathcal{V}_1(\mathrm{t})$ and $\mathcal{V}_2(\mathrm{t})$ but discontinuous with bounded one-sided limits at the moving surface $\Sigma(\mathrm{t})$ in the manner introduced above. Then the following generalization of (9.55) *holds*

$$
\begin{aligned}
\frac{\mathrm{d}}{\mathrm{dt}} \int\limits_{\mathcal{V}(\mathrm{t})} \psi(\mathrm{t}, \mathbf{x})\, \mathrm{d}V = & \int\limits_{\mathcal{V}_1(\mathrm{t})} \left[\frac{\partial}{\partial \mathrm{t}} \psi + \nabla_{\mathbf{x}} \cdot (\psi \mathbf{v}) \right] (\mathrm{t}, \mathbf{x})\, \mathrm{d}V \\
& + \int\limits_{\mathcal{V}_2(\mathrm{t})} \left[\frac{\partial}{\partial \mathrm{t}} \psi + \nabla_{\mathbf{x}} \cdot (\psi \mathbf{v}) \right] (\mathrm{t}, \mathbf{x})\, \mathrm{d}V \\
& - \int_{\Sigma(\mathrm{t})} \Big[[\![\psi(\mathbf{w} - \mathbf{v})]\!] \cdot \boldsymbol{\nu}_\Sigma \Big] (\mathrm{t}, \mathbf{x})\, \mathrm{d}S. \qquad (9.69)
\end{aligned}
$$

We can also generalize (9.56) *as*

$$\frac{\mathrm{d}}{\mathrm{dt}}\int\limits_{\mathcal{V}(\mathrm{t})}\psi(\mathrm{t},\mathbf{x})\,\mathrm{d}V=\int\limits_{\mathcal{V}_1(\mathrm{t})\cup\mathcal{V}_2(\mathrm{t})}\frac{\partial}{\partial\mathrm{t}}\psi(\mathrm{t},\mathbf{x})\,\mathrm{d}V+\int\limits_{\mathcal{S}(\mathrm{t})}\Big[\psi(\mathbf{v}\cdot\boldsymbol{\nu})\Big](\mathrm{t},\mathbf{x})\,\mathrm{d}S$$
$$-\int\limits_{\Sigma(\mathrm{t})}\Big[[\![\psi]\!](\mathbf{w}\cdot\boldsymbol{\nu}_\Sigma)\Big](\mathrm{t},\mathbf{x})\,\mathrm{d}S. \tag{9.70}$$

Proof The integrand on the right-hand side of (9.55) is singular on $\Sigma(\mathrm{t})$. The same holds for Σ_0. So in the argument (9.57) we have to split the integrals over $\mathcal{V}(\mathrm{t})$ and $\mathcal{V}(0)$ into the three parts $\widehat{\mathcal{V}}_1(\mathrm{t})$, $\mathcal{V}_D(\mathrm{t})$, $\mathcal{V}_2(\mathrm{t})$ as well as $\mathcal{V}_1(0)$, $\mathcal{V}_{21}^{\mathrm{t}}$, $\mathcal{V}_{22}^{\mathrm{t}}$. They are defined in Sect. 9.2.4.

We set $h(\mathrm{t},\mathbf{y})=\psi(\mathrm{t},\boldsymbol{\chi}^{\mathrm{t}}(\mathbf{y}))\mathcal{J}^{\mathrm{t}}(\mathrm{t},\mathbf{y})$ for $\mathbf{y}\in\mathcal{V}(0)$ and obtain

$$\frac{\mathrm{d}}{\mathrm{dt}}\left(\int\limits_{\mathcal{V}_1(\mathrm{t})}+\int\limits_{\mathcal{V}_2(\mathrm{t})}\right)\psi(\mathrm{t},\mathbf{x})\,\mathrm{d}V=\frac{\mathrm{d}}{\mathrm{dt}}\left(\int\limits_{\widehat{\mathcal{V}}_1(\mathrm{t})}+\int\limits_{\mathcal{V}_D(\mathrm{t})}+\int\limits_{\mathcal{V}_2(\mathrm{t})}\right)\psi(\mathrm{t},\mathbf{x})\,\mathrm{d}V$$
$$=\int\limits_{\mathcal{V}_1(0)}\frac{\partial}{\partial\mathrm{t}}h(\mathrm{t},\mathbf{y})\,\mathrm{d}V_0+\frac{\mathrm{d}}{\mathrm{dt}}\left(\int\limits_{\mathcal{V}_{21}(t)}+\int\limits_{\mathcal{V}_{22}(t)}\right)h(\mathrm{t},\mathbf{y})\,\mathrm{d}V_0.$$

The first integral gives the $\widehat{\mathcal{V}}_1(\mathrm{t})$ part of the integral over $\mathcal{V}_1(\mathrm{t})$ in (9.69).

We assume for the moment that the function ψ and the velocity field $\mathbf{v}$ are continuous at Σ. We recall the notations and assumptions from Sect. 9.2.4. Specifically we use the description of the initial volume $\mathcal{V}_2(0)$ by (9.46) over the open shadow domain $D\subseteq\mathbb{R}^2$. The moving surface Σ_0 in $\mathcal{V}(0)$ is parametrized using a function $\sigma: I\times D\to\mathbb{R}$. Since we are interested in using this theorem locally, this is not a serious restriction. The assumption can always be achieved when the surface is small enough. But the proof can be extended to surfaces that do not satisfy the assumption via a piecewise approach by cutting up the surface into suitable parts that do. We omit this technical step.

We divide D into two domains, namely, D_1 where $\sigma\le\alpha$ or $\beta\le\sigma$ and D_2 where $\alpha<\sigma<\beta$. The first may be empty. We also introduce the short-hand notation $\mathbf{y}_{\sigma^i}=(\mathrm{y}_1,\mathrm{y}_2,\sigma^i(\mathrm{t},\mathrm{y}_1,\mathrm{y}_2))$ with $i=1,2$ in order to indicate the one-sided limiting values of h on Σ. For the moment we assume that they are identical, but we keep track of them. We can now use the Leibniz integral rule with respect to y_3 to obtain

$$\frac{\mathrm{d}}{\mathrm{dt}}\Bigg(\int\limits_{\mathcal{V}_{21}(t)} + \int\limits_{\mathcal{V}_{22}(t)}\Bigg) h(\mathrm{t}, \mathbf{y})\, \mathrm{d}V_0 = \int\limits_{D_1}\Bigg(\int\limits_{\alpha}^{\beta} \frac{\mathrm{d}}{\mathrm{dt}} h(\mathrm{t}, \mathbf{y})\, \mathrm{dy}_3\Bigg)\, \mathrm{d}A$$

$$+ \int\limits_{D_2} \frac{\mathrm{d}}{\mathrm{dt}}\Bigg(\int\limits_{\alpha}^{\sigma} h(\mathrm{t}, \mathbf{y})\, \mathrm{dy}_3\Bigg)\, \mathrm{d}A + \int\limits_{D_2} \frac{\mathrm{d}}{\mathrm{dt}}\Bigg(\int\limits_{\sigma}^{\beta} h(\mathrm{t}, \mathbf{y})\, \mathrm{dy}_3\Bigg)\, \mathrm{d}A$$

$$= \int\limits_{D}\int\limits_{\alpha}^{\beta} \frac{\partial}{\partial \mathrm{t}} h(\mathrm{t}, \mathbf{y})\, \mathrm{dy}_3\, \mathrm{d}A + \int\limits_{D_2} \left[h(\mathbf{y}_{\sigma^1})\sigma_{\mathrm{t}}^1 - h(\mathbf{y}_{\sigma^2})\sigma_{\mathrm{t}}^2\right] \mathrm{d}A$$

$$= \Bigg(\int\limits_{\mathcal{V}_{21}(t)} + \int\limits_{\mathcal{V}_{22}(t)}\Bigg) \frac{\partial}{\partial \mathrm{t}} h(\mathrm{t}, \mathbf{y})\, \mathrm{d}V_0 + \int\limits_{D_2} \left[h(\mathbf{y}_{\sigma^1})\sigma_{\mathrm{t}}^1 - h(\mathbf{y}_{\sigma^2})\sigma_{\mathrm{t}}^2\right] \mathrm{d}A.$$

The volume integrals give the remaining contributions to volume integrals in (9.69). For the surface integral we have using (9.36), (9.7), and (9.48)

$$\int\limits_{D_2} h(\mathbf{y}_{\sigma^1})\sigma_{\mathrm{t}}^1\, \mathrm{d}A = \int\limits_{D_2} \psi(\mathrm{t}, \boldsymbol{\chi}^{\mathrm{t}}(\mathbf{y}_{\sigma^1}))\mathcal{J}^{\mathrm{t}}(\mathrm{t}, \mathbf{y}_{\sigma^i})\, u_\nu(\mathrm{t}, \mathbf{y}_{\sigma^1})|\boldsymbol{\tau}_1^0 \times \boldsymbol{\tau}_2^0|(\mathrm{t}, \mathbf{y}_{\sigma^1})\, \mathrm{d}A.$$

Now we apply (9.45) and (9.49) to obtain

$$\int\limits_{D_2} h(\mathbf{y}_{\sigma^1})\sigma_{\mathrm{t}}^1\, \mathrm{d}A$$

$$= \int\limits_{D_2} \psi(\mathrm{t}, \boldsymbol{\chi}^{\mathrm{t}}(\mathbf{y}_{\sigma^1}))\mathcal{J}^{\mathrm{t}}(\mathrm{t}, \mathbf{y}_{\sigma^i})\left(\left(\mathbf{D}_{\mathbf{x}}\boldsymbol{\chi}^{-\mathrm{t}}(\mathrm{t}, \mathbf{x})\right)(\mathbf{w}(\mathrm{t}, \mathbf{x}) - \mathbf{v}^i(\mathrm{t}, \mathbf{x}))\right)$$

$$\cdot\, \boldsymbol{\nu}_{\Sigma_0}(\mathrm{t}, \mathbf{y}_{\sigma^1})|\boldsymbol{\tau}_1^0 \times \boldsymbol{\tau}_2^0|(\mathrm{t}, \mathbf{y}_{\sigma^1})\, \mathrm{d}A$$

$$= \int\limits_{D_2} \left[\psi(\mathbf{w} - \mathbf{v}^1) \cdot \boldsymbol{\nu}_\Sigma |\boldsymbol{\tau}_1 \times \boldsymbol{\tau}_2|\right](\mathrm{t}, \boldsymbol{\chi}^{\mathrm{t}}(\mathbf{y}_{\sigma^1}))\, \mathrm{d}A$$

$$= \int\limits_{\Sigma} \left[\psi(\mathbf{w} - \mathbf{v}^1) \cdot \boldsymbol{\nu}_\Sigma\right](\mathrm{t}, \mathbf{x})\, \mathrm{d}S.$$

This leads to (9.69) since we obtain

$$\int\limits_{D_2} \left[h(\mathbf{y}_{\sigma^1})\sigma_{\mathrm{t}}^1 - h(\mathbf{y}_{\sigma^2})\sigma_{\mathrm{t}}^2\right] \mathrm{d}A = \int\limits_{\Sigma} \left[\,[\![\psi(\mathbf{w} - \mathbf{v})]\!] \cdot \boldsymbol{\nu}_\Sigma\right](\mathrm{t}, \mathbf{x})\, \mathrm{d}S.$$

Since we assumed that the function ψ and the velocity field $\mathbf{v}$ are continuous at Σ the latter integral vanishes. But the formula (9.69) remains true in the discontinuous case since we never used derivatives of ψ and $\mathbf{v}$ on Σ in its derivation, but only their one-sided limits.

We can apply the Gauss–Green theorem to the integrals over $\mathcal{V}_1(t)$ and $\mathcal{V}_2(t)$. On Σ this gives an integrand $[\![\psi\mathbf{v}]\!]$ that cancels out leaving the velocity field $\mathbf{w}$ to give (9.70). □

This extension of the Reynolds Transport Theorem 9.4.1 is found in Thomas (1949), Truesdell and Toupin (1960, Sect. 192), and Müller (1985, Sect. 3.1) with rather heuristic proofs. There it is stated in the form of formula (9.70) which differs from (9.69) only by an application of the Gauss–Green theorem.

9.4.5 Derivation of Surface Equation with Jump Conditions

To obtain the balance equation for the singular points on a surface, there is a so-called *pillbox argument*. It was used by Truesdell and Toupin (1960, Sect. 193), Müller (1985, Sect. 3.1) and for the two-dimensional case by Gurtin (1993, Sect. 3.2). The result generalizes the derivation of jump conditions on surfaces of discontinuity that goes back to the seminal paper of Riemann (1990). Zemplén (1905) contributed a more formal and general derivation of jump conditions for quasi-one-dimensional flows in cylindrical ducts. This was then extended by Kotchine (1926) to more general surfaces. He used pillbox-type arguments and took jumps in the tangential velocity into account.

We consider an open subset $\Omega \subseteq \mathbb{R}^3$ and any surface of singular points $\Sigma(t) \subset \Omega$ for $t \in [0, T]$. The idea is to choose a family of cylindrical type sets $\mathcal{V}_{\Sigma,\varepsilon} \subset \Omega$ as neighborhoods around the surface as control volume. We define analogous to Sect. 9.4.3 a new curved *pillbox set*, for any $\varepsilon > 0$ small enough, over the curved disk $B^2_{\Sigma,\mathbf{x}_0,r}$ for any $\mathbf{x}_0 \in \Sigma(t)$ and some suitable $r > 0$ as

$$\mathcal{V}_{\Sigma,\varepsilon} = \{\mathbf{x} \in \mathbb{R}^3 \mid \mathbf{x} = \lambda\boldsymbol{\nu}_\Sigma + \mathbf{y} \text{ with } \mathbf{y} \in B^2_{\Sigma,\mathbf{x}_0,r},\ \lambda \in]-\varepsilon,\varepsilon[\,\} \subset \Omega$$

with the bottom surface $\mathcal{S}_1(\varepsilon)$ and top surfaces $\mathcal{S}_2(\varepsilon)$ given as

$$\mathcal{S}_i(\varepsilon) = \{\mathbf{x} \in \mathbb{R}^3 \mid \mathbf{x} = (-1)^i\varepsilon\boldsymbol{\nu}_\Sigma + \mathbf{y} \text{ with } \mathbf{y} \in B^2_{\Sigma,\mathbf{x}_0,r}\}$$

as well as side surfaces

$$\mathcal{S}_{i,\varepsilon} = \{\mathbf{x} \in \mathbb{R}^3 \mid \mathbf{x} = (-1)^i\lambda\boldsymbol{\nu}_\Sigma + \mathbf{y} \text{ with } \mathbf{y} \in \partial B^2_{\Sigma,\mathbf{x}_0,r},\ \lambda \in]0,\varepsilon[\,\}.$$

Here we do not want to shrink the curved disk on the surface, as we did in Sect. 9.4.3, but only the normal direction. We divide $\mathcal{V}_{\Sigma,\varepsilon}$ into $\mathcal{V}_{1,\varepsilon}$, defined as the subset with $\lambda \in]-\varepsilon, 0[$ and analogously $\mathcal{V}_{2,\varepsilon}$ with $\lambda \in]0,\varepsilon[$. The unit normal vector field $\boldsymbol{\nu}_\Sigma$ on the surface Σ points into $\mathcal{V}_{2,\varepsilon}$. A sketch for a flat surface is given in Fig. 9.2.

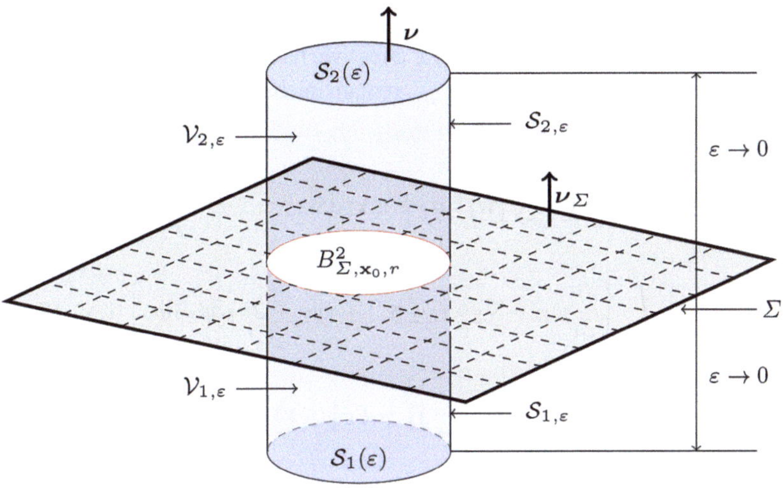

Fig. 9.2 Sketch of the pillbox argument

We consider the general balance law (9.54) on such an arbitrary material control volume $\mathcal{V}_{\Sigma,\varepsilon}$. Additionally, we assume that the flux $\mathbf{j}$ is discontinuous on Σ with bounded one-sided limits. For the time derivative of the volume terms, we use (9.70) and for the surface term (9.65) to obtain

$$\begin{aligned}
&\int\limits_{\mathcal{V}_{1,\varepsilon}\cup\mathcal{V}_{2,\varepsilon}} \frac{\partial}{\partial t}\psi \,\mathrm{d}V + \int\limits_{\mathcal{S}_{1,\varepsilon}\cup\mathcal{S}_{2,\varepsilon}} \psi(\mathbf{v}\cdot\boldsymbol{\nu})\,\mathrm{d}S + \int\limits_{\mathcal{S}_1(\varepsilon)\cup\mathcal{S}_2(\varepsilon)} \psi(\mathbf{v}\cdot\boldsymbol{\nu})\,\mathrm{d}S \\
&\quad - \int\limits_{B^2_{\Sigma,\mathbf{x}_0,r}} [\![\psi]\!]\,(\mathbf{w}\cdot\boldsymbol{\nu}_\Sigma)\,\mathrm{d}S \\
&\quad + \int\limits_{B^2_{\Sigma,\mathbf{x}_0,r}} \left[\frac{\partial}{\partial t}\psi_\Sigma + \nabla_\Sigma\cdot(\psi_\Sigma \mathbf{w}^{\shortparallel}) + (\psi_\nu - 2\kappa_M\psi_\Sigma)w_\nu\right]\mathrm{d}S \\
&= - \int\limits_{\mathcal{S}_{1,\varepsilon}\cup\mathcal{S}_{2,\varepsilon}\cup\mathcal{S}_1(\varepsilon)\cup\mathcal{S}_2(\varepsilon)} \mathbf{j}\cdot\boldsymbol{\nu}\,\mathrm{d}S - \int\limits_{\partial B^2_{\Sigma,\mathbf{x}_0,r}} \mathbf{j}_\Sigma\cdot\boldsymbol{\nu}_{\partial\Sigma}\,\mathrm{d}l + \int\limits_{\mathcal{V}_{1,\varepsilon}\cup\mathcal{V}_{2,\varepsilon}} \xi\,\mathrm{d}V \\
&\quad + \int\limits_{B^2_{\Sigma,\mathbf{x}_0,r}} \xi_\Sigma\,\mathrm{d}S.
\end{aligned}$$

Note that here $\psi_\nu = \nabla_{\mathbf{x}}\psi_\Sigma\cdot\boldsymbol{\nu}_\Sigma$ and $w_\nu = \mathbf{w}\cdot\boldsymbol{\nu}_\Sigma$. In case $\psi_\nu = 0$ or ψ_ν does not exist, we would replace the use of (9.65) by (9.63). Now as ε goes to zero, the volume integrals over $\mathcal{V}_{1,\varepsilon}$ and $\mathcal{V}_{2,\varepsilon}$ as well as the surface integrals over $\mathcal{S}_{1,\varepsilon}$ and $\mathcal{S}_{2,\varepsilon}$ go to

0. These are the green and white terms. The surface integrals over $\mathcal{S}_1(\varepsilon)$ and $\mathcal{S}_2(\varepsilon)$ become integrals over $B^2_{\Sigma,\mathbf{x}_0,r}$ where for $\mathcal{S}_1(\varepsilon)$ we have $\boldsymbol{\nu} \to -\boldsymbol{\nu}_\Sigma$ and $\boldsymbol{\nu} \to \boldsymbol{\nu}_\Sigma$ for $\mathcal{S}_2(\varepsilon)$. The integrals connected to $B^2_{\Sigma,\mathbf{x}_0,r}$ remain unchanged. But, to the integral over $\partial B^2_{\Sigma,\mathbf{x}_0,r}$ we apply (9.58). This gives

$$
\begin{aligned}
0 = &\int\limits_{B^2_{\Sigma,\mathbf{x}_0,r}} \left[(\llbracket \psi \mathbf{v} \rrbracket - \llbracket \psi \rrbracket \mathbf{w}) \cdot \boldsymbol{\nu}_\Sigma \right] (\mathrm{t}, \mathbf{x}) \, \mathrm{d}S \\
&+ \int\limits_{B^2_{\Sigma,\mathbf{x}_0,r}} \left[\frac{\partial}{\partial \mathrm{t}} \psi_\Sigma + \nabla_\Sigma \cdot (\psi_\Sigma \mathbf{w}^{\shortparallel}) + (\psi_\nu - 2\kappa_M \psi_\Sigma) w_\nu \right] (\mathrm{t}, \mathbf{x}) \, \mathrm{d}S \\
&+ \int\limits_{B^2_{\Sigma,\mathbf{x}_0,r}} \left[\llbracket \mathbf{j} \rrbracket \cdot \boldsymbol{\nu}_\Sigma + \nabla_\Sigma \cdot \mathbf{j}^{\shortparallel}_\Sigma - \xi_\Sigma \right] (\mathrm{t}, \mathbf{x}) \, \mathrm{d}S.
\end{aligned}
$$

Now we use the argument from Sect. 9.4.3, since $\mathbf{x}_0$ and $r > 0$ can be arbitrarily chosen, that the equality holds point-wise at $(\mathrm{t}, \mathbf{x})$ for the integrands. Thus we obtain the point-wise balance law on the surface

$$
\frac{\partial}{\partial \mathrm{t}} \psi_\Sigma + \nabla_\Sigma \cdot (\psi_\Sigma \mathbf{w}^{\shortparallel}) + (\psi_\nu - 2\kappa_M \psi_\Sigma) w_\nu + \nabla_\Sigma \cdot \mathbf{j}^{\shortparallel}_\Sigma - \xi_\Sigma = \llbracket \psi \rrbracket \, w_\nu - \llbracket \psi \mathbf{v} + \mathbf{j} \rrbracket \cdot \boldsymbol{\nu}_\Sigma. \tag{9.71}
$$

On the right-hand side is the usual jump condition for conservation laws. The left-hand side consists of the extra surface terms that we included.

In case we have ψ_Σ only defined on the surfaces Σ we can use (9.61) in the balance law (9.54) and follow the same steps. We can replace (9.71) by the rather concise form

$$
\mathring{\psi}_\Sigma + \psi_\Sigma \nabla_\Sigma \cdot \mathbf{w} + \nabla_\Sigma \cdot \mathbf{j}^{\shortparallel}_\Sigma - \xi_\Sigma = \llbracket \psi \rrbracket \, w_\nu - \llbracket \psi \mathbf{v} + \mathbf{j} \rrbracket \cdot \boldsymbol{\nu}_\Sigma. \tag{9.72}
$$

We may substitute $\nabla_\Sigma \cdot \mathbf{w} = \nabla_\Sigma \cdot \mathbf{w}^{\shortparallel} - 2\kappa_M w_\nu$ from (9.31). Further we use (9.26), (9.28), and (9.3) to have $\nabla_\Sigma \cdot \mathbf{w}^{\shortparallel} = \nabla_\alpha w^\alpha_\tau$ and $\nabla_\Sigma \cdot \mathbf{j}^{\shortparallel}_\Sigma = \nabla_\alpha j^\alpha_\tau$. This gives these equations in the form

$$
\mathring{\psi}_\Sigma + \psi_\Sigma (\nabla_\alpha w^\alpha_\tau - 2\kappa_M w_\nu) + \nabla_\alpha j^\alpha_\tau - \xi_\Sigma = \llbracket \psi \rrbracket \, w_\nu - \llbracket \psi \mathbf{v} + \mathbf{j} \rrbracket \cdot \boldsymbol{\nu}_\Sigma. \tag{9.73}
$$

This is the formula in Müller (1985, (3.16)) in the variables $(\mathrm{t}, \mathbf{u})$. There the source term ξ_Σ is split into a production and a supply term, see also Mavrovouniotis and Brenner (1993, (5.1)). Dreyer (2003, (51)) used $\frac{\partial}{\partial \mathrm{t}} \widetilde{\psi}_\Sigma(\mathrm{t}, \mathbf{u})$ instead of the identical $\mathring{\psi}_\Sigma(\mathrm{t}, \mathbf{u})$, see (9.42).

9.5 Moving Curves in Two Dimensions, Points in One Dimension

So far we derived the equations in $\mathbb{R}^3$. But the lower dimensional cases may also be derived from these equations. In particular, we focus on (9.71). We briefly want to *sketch* how this may be done in $\mathbb{R}^2$. Details of the theory of moving piecewise smooth curves in the plane together with physical properties are presented in Gurtin (1993).

We assume a symmetry such that the surface density is constant in one surface direction and that the normal vector fields have no component in this direction. For example, we may choose our coordinate system such that the density is constant in x_3-direction and the normal vector fields have the x_3-component zero. Then the problem reduces to a two-dimensional problem, where the moving singular surface $\Sigma(t)$ reduces to a moving curve $c(t)$ in the x_1-x_2-plane. The remaining surface parameter may be taken to be the *arc-length* s of the curve. Then the tangent vector is a unit vector. Further the mean curvature, as the arithmetic mean of the principle curvatures, reduces to the curvature κ of the curves. Finally, the divergence of the tangential component of the surface velocity reduces to the derivative with respect to the arc-length s. The Lagrangian derivative $\overset{\circ}{\psi}$ is an advective tangential derivative with respect to arc-length. Thus, we obtain as a modification of (9.71) a formula, for the case in which ψ_c is also defined in the plane, as

$$\left[\overset{\circ}{\psi}_c + \psi_c\left(\frac{\partial}{\partial s}w_\tau - \kappa w_\nu\right) + \psi_{c,\nu}w_\nu - \xi_c + \frac{\partial}{\partial s}\mathbf{j}_c^{\parallel}\right](t, s)$$
$$= \left[w_\nu[\![\psi]\!] - [\![\psi\mathbf{v} + \mathbf{j}]\!]\boldsymbol{\nu}\right](t, s). \qquad (9.74)$$

If ψ_c is defined only on the moving curves we can use the formula without the normal derivative of ψ_c, i.e. $\psi_{c,\nu} = 0$.

Now for the one-dimensional case we assume a planar interface, which stays planar in time, with a spatially constant surface density. We have a moving point of singularity $p(t) \in \mathbb{R}$. Since the metric is constant in time, the derivative $\partial_t g$ vanishes and there is no curvature. The velocity is the scalar function $w(t, p(t)) = \frac{d}{dt}p(t)$. Thus, we obtain

$$\frac{d}{dt}\psi_p(t, p(t)) - \xi_p(t, p(t)) = \left[w[\![\psi]\!] - [\![\psi v + j]\!]\right](t, p(t)). \qquad (9.75)$$

This type of jump condition is also known as *generalized Rankine–Hugoniot jump conditions*, cf. Yang (1999, (3.7)).

9.6 Time-Dependent Determinants

We want to calculate the time derivative of the determinant of the Jacobian matrix or deformation gradient of the motion χ^t from Sect. 9.2.4. It is denoted by $\mathcal{J}^t$ and given as

$$\mathcal{J}^t = \det \mathbf{D}_\mathbf{x} \chi^t = \det \left(\frac{\partial \chi_j^t}{\partial x_k} \right)_{1 \le j,\, k \le n} . \tag{9.76}$$

For this let $\mathbf{A} = (a_{jk})_{1 \le j,\, k \le n} \in \mathbb{R}^{n \times n}$ a matrix or tensor of order 2 with t dependent coefficients. For any pair of indices j, k we define the matrix $\mathbf{A}_{jk}$ as the matrix obtained from $\mathbf{A}$ by replacing row j and column k zero entries except for $a_{jk} = 1$, i.e.

$$\mathbf{A}_{jk} = \begin{pmatrix} a_{11} & \cdots 0 \cdots & a_{1n} \\ \vdots & & \vdots \\ 0 & \cdots 1 \cdots & 0 \\ \vdots & & \vdots \\ a_{n1} & \cdots 0 \cdots & a_{nn} \end{pmatrix} .$$

With the *co-factors* $\det \mathbf{A}_{jk}$ of $\mathbf{A}$ the inverse matrix $\mathbf{A}^{-1}$ is given as, see, for instance Meyer (2001, Sect. 6.2) or Strang (1988, Sect. 4.4),[6]

$$\mathbf{A}^{-1} = \left(a_{jk}^{-1} \right)_{1 \le j,\, k \le n} = \left(a^{jk} \right)_{1 \le j,\, k \le n} = \left(\frac{\det \mathbf{A}_{kj}}{\det \mathbf{A}} \right)_{1 \le j,\, k \le n} ,$$

i.e., we have

$$\det \mathbf{A}_{jk} = a_{kj}^{-1} \det \mathbf{A} = a^{kj} \det \mathbf{A} \tag{9.77}$$

and $a_{jl} a^{lk} = \delta_j^k$. We use the summation convention that indices that appear twice are summed from 1 to n and $\delta_{jk} = \delta_j^k = 0$ for $j \ne k$ and $= 1$ for $j = k$ are the *Kronecker delta* tensors. We introduce the nth-order *permutation tensor* $\epsilon_{i_1 \cdots i_n}$ using indexed indices. It is equal to 1 when $i_1 \cdots i_n$ is an even permutation of the numbers $123 \cdots n$, equal to -1 for an odd permutation and otherwise equal to 0. The determinant of $\mathbf{A}$ is

$$\det \mathbf{A} = \epsilon_{i_1 \cdots i_n} a_{1 i_1} \cdots a_{n i_n}$$

and that of $\mathbf{A}_{jk}$

$$\det \mathbf{A}_{jk} = \epsilon_{i_1 \cdots i_n} a_{1 i_1} \cdots a_{j-1\, i_{j-1}} \cdot \delta_{i_j k} \cdot a_{j+1\, i_{j+1}} \cdots a_{n i_n} .$$

[6] These references do not use the matrix $\mathbf{A}_{jk}$ but the $(n-1) \times (n-1)$ matrix where the j-th row and k-th column are deleted. The determinant then involves an extra factor of $(-1)^{j+k}$ that we can avoid.

This implies that $\frac{\partial \det \mathbf{A}}{\partial a_{jk}} = \det \mathbf{A}_{jk}$. Now with (9.77) we have

$$\frac{d \det \mathbf{A}}{dt} = \frac{\partial \det \mathbf{A}}{\partial a_{jk}} \frac{\partial a_{jk}}{dt} = \det \mathbf{A}_{jk} \frac{da_{jk}}{dt} = a^{kj} \frac{da_{jk}}{dt} \det \mathbf{A}. \tag{9.78}$$

This we apply to $\mathcal{J}^{\mathrm{t}}$ and note that in this case $a_{kj}^{-1} = a^{kj} = \frac{\partial x_k}{\partial \chi_j^{\mathrm{t}}}$. We obtain

$$\frac{d\mathcal{J}^{\mathrm{t}}}{d\mathrm{t}} = \frac{\partial x_k}{\partial \chi_j^{\mathrm{t}}} \frac{d}{d\mathrm{t}} \left(\frac{\partial \chi_j^{\mathrm{t}}}{\partial x_k} \right) \mathcal{J}^{\mathrm{t}} = \frac{\partial x_k}{\partial \chi_j^{\mathrm{t}}} \frac{\partial v^j}{\partial x_k} \mathcal{J}^{\mathrm{t}} = \frac{\partial v^j}{\partial \chi_j^{\mathrm{t}}} \mathcal{J}^{\mathrm{t}} = \frac{\partial v^j}{\partial x_j} \mathcal{J}^{\mathrm{t}}$$

or

$$\frac{d\mathcal{J}^{\mathrm{t}}}{dt} = (\nabla_{\mathbf{x}} \cdot \mathbf{v})\, \mathcal{J}^{\mathrm{t}}. \tag{9.79}$$

The two-dimensional analog is given in (9.44).

9.7 Cross-Product Transformation

Let $\mathbf{A} \in \mathbb{R}^{3\times 3}$ be an invertible matrix and $\boldsymbol{\tau}_1, \boldsymbol{\tau}_2 \in \mathbb{R}^3$ two vectors. We denote by $\mathbf{C} = \left(\det \mathbf{A}_{jk}\right)_{1\le j,k\le 3}$ the matrix of co-factors. Then we have the following formula for the cross product

$$\mathbf{A}\boldsymbol{\tau}_1 \times \mathbf{A}\boldsymbol{\tau}_2 = \mathbf{C}(\boldsymbol{\tau}_1 \times \boldsymbol{\tau}_2) = \det \mathbf{A}(\mathbf{A}^{-1})^T (\boldsymbol{\tau}_1 \times \boldsymbol{\tau}_2). \tag{9.80}$$

From (9.77) we have $\mathbf{A}^{-1} = \frac{1}{\det \mathbf{A}} \mathbf{C}^T$. This gives the second equality. The first equality is a slightly tedious calculation for each of the three components. Starting from the left-hand side some terms cancel to give the right-hand side.

Acknowledgements F. Thein gratefully acknowledges the Deutsche Forschungsgemeinschaft (DFG, German Research Foundation) for the financial support through 525939417/SPP2410.

References

Anderson DM, Cermelli P, Fried E, Gurtin ME, McFadden GB (2007) General dynamical sharp-interface conditions for phase transformations in viscous heat-conducting fluids. J Fluid Mech 581:323–370

Anderson J (2003) Modern compressible flow: with historical perspective. Aeronautical and aerospace engineering series, McGraw-Hill Education

Aris R (1989) Vectors, tensors, and the basic equations of fluid mechanics. Dover Publications, New York

Bedeaux D (1986) Nonequilibrium thermodynamics and statistical physics of surfaces. Adv Chem Phys 64:47–109
Bedeaux D, Albano A, Mazur P (1976) Boundary conditions and non-equilibrium thermodynamics. Phys A: Stat Mech Its Appl 82(3):438–462
Bothe D (2020) On moving hypersurfaces and the discontinuous ODE-system associated with two-phase flows. Nonlinearity 33(10):5425–5456
Bothe D (2022) Sharp-interface continuum thermodynamics of multicomponent fluid systems with interfacial mass. Internat J Eng Sci, 179. Paper No. 103731, 31
Bourne D, Kendall P (1992) Vector analysis and cartesian tensors, 3rd edn. Taylor & Francis
Bowen RM, Wang CC (1971) On displacement derivatives. Quart Appl Math 29:29–39
Cermelli P, Fried E, Gurtin ME (2005) Transport relations for surface integrals arising in the formulation of balance laws for evolving fluid interfaces. J Fluid Mech 544:339–351
Dafermos CM (2016) Hyperbolic conservation laws in continuum physics, vol 325. Grundlehren der mathematischen Wissenschaften. Springer, Berlin Heidelberg
De Groot S, Mazur P (2013) Non-equilibrium thermodynamics. Dover books on physics. Dover Publications
Dell'Isola F, Romano A (1986) On a general balance law for continua with an interface. Ricerche Mat 35(2):325–337
Dell'Isola F, Romano A (1987) On the derivation of thermomechanical balance equations for continuous systems with a nonmaterial interface. Internat J Eng Sci 25(11–12):1459–1468
Dreyer W (2003) On jump conditions at phase boundaries for ordered and disordered phases. Preprint 869, Weierstrass institute for applied analysis and stochastics (WIAS), Berlin. Reprinted in this volume
Dziuk G, Elliott CM (2007) Finite elements on evolving surfaces. IMA J Numer Anal 27(2):262
Estrada R, Kanwal RP (1991) Non-classical derivation of the transport theorems for wave fronts. J Math Anal Appl 159(1):290–297
Evans LC (1998) Partial differential equations. American Mathematical Society, Providence, RI
Feireisl E, Lukáčová-Medvid'ová M, Mizerová H, She B (2021) Numerical analysis of compressible fluid flows. Modeling, simulation and applications, vol 20. Springer, Cham
Grinfeld P (2013) Introduction to tensor analysis and the calculus of moving surfaces. Springer, New York
Gurtin ME (1993) Thermomechanics of evolving phase boundaries in the plane. Oxford University Press
Gurtin ME, Struthers A, Williams WO (1989) A transport theorem for moving interfaces. Quart Appl Math 47(4):773–777
N. Kotchine (1926) Sur la théorie des ondes de choc dans un fluide. *Rendiconti del Circolo Matematico di Palermo Series 2* 50(2):305–344
Landau L, Lifshitz E (1987) Fluid mechanics—volume 6 of course of theoretical physics. Elsevier Science
Lee JM (2013) Introduction to smooth manifolds, vol 218. Graduate texts in mathematics. Springer, New York
LeFloch P (2002) Hyperbolic systems of conservation laws: the theory of classical and nonclassical shock waves. Lectures in mathematics. Birkhäuser Verlag
Mauri R (2013) Non-equilibrium thermodynamics in multiphase flows. Soft and biological matter. Springer, Berlin Heidelberg
Mavrovouniotis GM, Brenner H (1993) A micromechanical investigation of interfacial transport processes. i. interfacial conservation equations. Philos Trans R Soc London Ser A: Phys Eng Sci 345(1675):165–207
Meyer C (2001) Matrix analysis and applied linear algebra. SIAM, Philadelphia
Müller I (1985) Thermodynamics. Interaction of Mechanics and Mathematics Series. Pitman
Müller I, Müller WH (2009) Fundamentals of thermodynamics and applications. Springer, Berlin Heidelberg

Petryk H, Mróz Z (1986) Time derivatives of integrals and functionals defined on varying volume and surface domains. Arch Mech (Arch Mech Stos) 38(5–6):697–724

Riemann B (1990) Über die Fortpflanzung ebener Luftwellen von endlicher Schwingingsweite. In: Collected papers. Springer, pp 188–207

Rudin W (1987) Real and complex analysis, 3rd edn. McGraw-Hill Book Co., New York

Slattery JC, Sagis L, Oh E-S (2007) Interfacial transport phenomena. Springer Science & Business Media

Smoller J (2012) Shock waves and reaction-diffusion equations, vol 258. Springer Science & Business Media

Strang G (1988) Linear Algebra and its Applications. Harcourt Brace Jovanovich Coll. Publ., Fort Worth—Philadelphia—London

Thomas G, Weir M, Hass J, Giordano F (2004) Thomas' calculus. Pearson Addison Wesley

Thomas TY (1949) The fundamental hydrodynamical equations and shock conditions for gases. Math Mag 22:169–189

Thomas TY (1957) Extended compatibility conditions for the study of surfaces of discontinuity in continuum mechanics. J Math Mech 6:311–322

Thomas TY (1961) Plastic flow and fracture in solids by Tracy Y Thomas. Elsevier

Toro EF (2009) Riemann solvers and numerical methods for fluid dynamics. Springer, Berlin Heidelberg

Truesdell C, Toupin R (1960) The classical field theories. In: Flügge S (ed) Principles of classical mechanics and field theory/Prinzipien der Klassischen Mechanik und Feldtheorie. Springer, Berlin Heidelberg, pp 226–858

Warnecke G (1999) Analytische Methoden in der Theorie der Erhaltungsgleichung. B.G. Teubner, Stuttgart-Leipzig

Yang H (1999) Riemann problems for a class of coupled hyperbolic systems of conservation laws. J Differ Equ 159(2):447–484

Zemplén G (1905) Kriterien für die physikalische Bedeutung der unstetigen Lösungen der hydrodynamischen Bewegungsgleichungen. Math Ann 61(3):437–449

Chapter 10
Multi-velocity Sharp-Interface Continuum Thermodynamics of Fluid Systems with Adsorption

Dieter Bothe

In memoriam Wolfgang Dreyer

Abstract We revisit the sharp-interface continuum thermodynamics of two-phase multicomponent fluid systems, accounting for partial mass and momentum balances both in the bulk phases and on the interface. This allows to describe the transfer of species between the bulk phases and the interface, i.e. ad- and desorption processes. In fact, the transfer of any constituent between the two bulk-phases is considered as a series of ad- and desorption processes. In this framework, all species transfer processes are coupled via the interfacial thermodynamics. As a consequence, the influence of surface active species on the transfer of other constituents can be captured in detail. The derivation of this model class relies on an axiomatic form of the entropy principle which, at the same time, allows for an efficient closure process. This form of the entropy principle has been introduced for one-phase fluid systems in Bothe and Dreyer (2015) as the result of intense joint work of the late Wolfgang Dreyer and the present author.

10.1 Introduction

Multicomponent two-phase fluid systems are ubiquitous in nature, science and engineering. A prototypical case is that of an air bubble in water. For this system, water, nitrogen, oxygen and carbon dioxide, among many more, are the involved constituents. A generic two-phase fluid system thus is composed of two multicomponent mixtures which form the two bulk phases, being in contact at their common interface. We assume that the chemical species are homogeneously mixed on the molecular scale, where the composition of the mixture depends on the position **x** and

D. Bothe (✉)
Department of Mathematics and Profile Topic Thermofluids and Interfacial Phenomena, Technical University of Darmstadt, Darmstadt, Germany
e-mail: bothe@mma.tu-darmstadt.de

J. Fuhrmann et al. (eds.), *Advances in Continuum Physics*, Advanced Structured Materials 238, https://doi.org/10.1007/978-3-031-93918-1_10

on the time instant t. In a continuum physical description, balance equations for the partial masses, i.e. individual balance equations for the species mass densities ρ_i, are required, where the species are indexed with $i = 1, \ldots, N$ in case of N different chemical components.

We consider two-phase fluid systems inside a fixed domain $\Omega \subset \mathbf{R}^3$ in which two bulk phases, occupying the subdomains $\Omega^+(t)$ and $\Omega^-(t)$, are separated by an interface $\Sigma(t)$. This interface represents an extremely thin transition zone between the two bulk phases, in which the composition will change abruptly. As time evolves, the interface may move and deform. We employ the sharp-interface modeling framework, in which $\Sigma(t)$ is represented by a family of hypersurfaces, i.e. this transition layer is modeled as having zero thickness. This is a very sensible approximation under conditions (temperature, pressure) away from the so-called critical point at which this transition zone becomes more and more thicker before it finally disappears as the fluid system changes into a single-phase supercritical state. For water, the critical point is at about 647 K and 22 MPa ($\approx$ 220 bar); see Wagner and Pruß (2002). Under so-called normal conditions (i.e., at a temperature of 20 °C = 293.15 K and a pressure of $p = 1$ bar $= 101\,325$ Pa), the water-air interface, for instance, has a thickness of about 0.6–0.8 nm, defined as the thickness of the zone of anisotropic structure. While this surprisingly small thickness was already known from molecular dynamics simulations for some time, it has only been measured very recently in Fellows et al. (2024). This underlines the validity of the sharp-interface approximation. In this idealized scenario, the partial mass densities as well as all other intensive quantities, display a jump discontinuity at Σ. For example, at the (planar) air-water interface and at normal conditions, the oxygen mass density jumps from $\rho^L_{O_2} = 278.6\ \mathrm{g\,m^{-3}}$ in the liquid phase to $\rho^G_{O_2} = 9.1\ \mathrm{g\,m^{-3}}$ in the gas phase. Hence the oxygen density jumps down to only 3.3%. Therefore, an integral part of any sharp-interface model is formed by so-called jump conditions, relating the two one-sided limits of, e.g., the species mass densities.

In many real-world fluid systems, surface active substances (so-called *surfactants*) are present, which adsorb at the fluid interface, changing the interfacial free energy. If A_i denotes such a surfactant species, the interface will locally carry a certain mass of A_i per interfacial area, hence an area-specific mass density ρ_i^Σ is to be used to describe the presence of A_i in the adsorbed state. This is mandatory in order to add a proper interfacial thermodynamical description. The process of transfer of a species from one of the bulk phases to the interface is called *adsorption*. In order to distinguish between a species either as present in the bulk phases or in the adsorbed state on the interface, we will use the superscripts $+$, $-$ and Σ, such that A_i^Σ denotes the i-th constituent in its adsorbed form. If the adsorbed form A_i^Σ is present, this species will typically also be transferred into one or both of the adjacent bulk phases, a process called *desorption*.

Two-phase fluid systems, which are out of equilibrium, exchange in particular, mass, momentum and energy. In multicomponent two-phase fluid systems away from 'chemical equilibrium' (to be made precise below), matter will be exchanged across the interface, i.e. a transfer of chemical constituents takes place. This process will be called *mass transfer* throughout. For instance, if air is brought into contact with pure water, then oxygen and nitrogen (as well as all other gaseous species inside the air) will be transferred into the water, i.e. the bubble will start to dissolve. In a more realistic scenario, the water will already carry some dissolved gaseous components. Once the gas and the liquid phase come into contact, these dissolved components will be transferred from the surrounding liquid into the bubble unless they are already present there at sufficiently high concentrations. Hence different species are transferred simultaneously and, possibly, in different directions—so-called conjugate mass transfer. This has been quantitatively demonstrated in Hosoda et al. (2014), where the dissolution of a CO_2-bubble in water has been investigated. The experiment showed that the bubble will not disappear completely and by gas chromatography it was shown that the residual bubble was composed of air, coming out of the liquid phase, where it was dissolved in the water.

Adsorption of surfactant strongly influences the transfer rates of, e.g., gaseous components into a liquid phase. It is commonly accepted that there are two different ways, how this influence is mediated. First, since surfactant is inhomogeneously distributed along the interface, the interfacial tension displays non-zero surface gradients. This induces so-called Marangoni stresses which act as tangential forces at the interface, thus changing the flow field locally, resulting for instance in a reduced terminal rise velocity of buoyant bubbles. Second, even in the absence of fluid flow, the partial coverage of the interface with surfactant molecules constitutes a barrier against the passage of other molecules across the interface. This leads to an additional *hindrance effect*, also referred to as mass transfer resistance, sometimes also denoted a steric effect. Experimental results on this hindrance effect can be found, e.g., in Sardeing et al. (2006), Hébrard et al. (2009), Aoki et al. (2015), Hori et al. (2019), Hori et al. (2020). Let us note that the analogous phenomenon on the rate of evaporation, being strongly influenced by the presence of surfactants, has been studied much earlier by Langmuir; see Langmuir and Langmuir (1927), where a so-called energy barrier model was introduced to explain this phenomenon, at least qualitatively. A survey on the surfactant-induced hindrance effect onto evaporation and mass transfer can be found in Barnes (1986).

The influence of surfactants on local mass transfer rates due to Marangoni effects has been studied recently by means of numerical simulations on different levels of detail, see Kentheswaran et al. (2023), Garbina et al. (2025) and the references given there. In the underlying mathematical models, the presence of surfactant changes the local surface tension. The resulting Marangoni stresses strongly impact the local hydrodynamics and, hence, the mass transfer rates mainly due to a modified interplay of diffusive and convective characteristic time scales. However, the direct hindrance effect of surfactant on mass transfer processes is not included in these models. If it is accounted for at all, then via global mass transfer correlations; see, e.g., Sardeing et al. (2006). A possible reason behind is a fundamental difficulty to include such hindrance effects as a *local* mechanism into continuum physical models: in the standard models, the mass transfer of, for example, a gaseous species is driven by the difference of the one-sided limits of the chemical potential at the interface. These limit values depend on bulk quantities such as, with a common choice of primitive variables, the local temperature, pressure and composition at the respective position. If surfactants are included in this standard model, their effect is a change of the local surface tension. While this affects the local hydrodynamics via the momentum jump condition, a model without *interfacial* (molar) mass for all relevant species does not provide a channel for a direct interaction between the local coverage by surfactant and the local rate of mass transfer of another constituent.

To implement such a channel of interaction between adsorbed surfactant and a transfer species, the model needs to be enriched. Note that, if a species is transferred from one side of the interface to the other one, it has to pass the transition layer in between, hence will be present within this layer for some positive residence time. See Fig. 10.1 for an illustration of this conceptual consideration. This implies that, within the framework of a sharp-interface model, the average number of particles (atoms or molecules) of this transfer species within a certain surface area within Σ is positive. In other words, if A_k is this transfer species, than $\rho_k^\Sigma > 0$. The point here is not about possible inertia effects of this contribution to the mass on the interface, but on the implications for the thermodynamical state of the interface as a multicomponent interfacial mixture. Consequently, to allow for proper coupling of mass transfer and sorption phenomena, *all constituents* should be included in the model with partial mass densities in the bulk phases and also on the interface. Of course, this still allows for a specific constituent which is only present in one bulk phase or even just on the interface. This idea has been introduced and investigated in Bothe (2022) by the present author. Let us note in passing that surface concentrations in this model class are necessarily non-negative, while so-called excess mass densities for the dividing interface due to Gibbs can be negative; see Guggenheim and Adam (1933). This non-negativity is a qualitative property of solutions of our model, which implies certain structural properties of the transfer coefficients; cf. Bothe and Druet (2023) for the implications of positivity requirements on (cross-)diffusion coefficients. The price to pay for having this useful property is that the spatial region which is lumped into the sharp interface is still considered as bulk volume with the bulk densities.

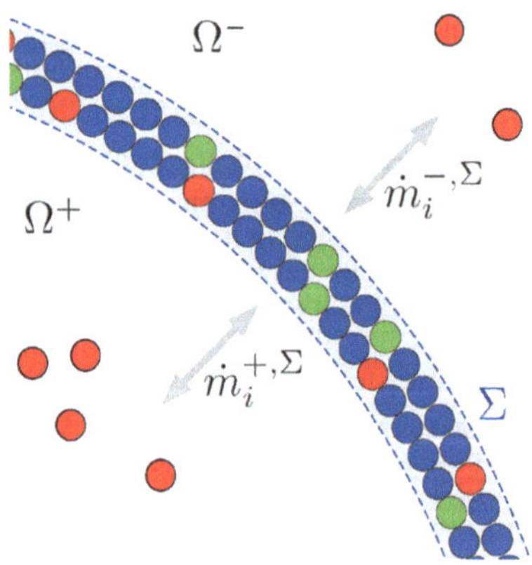

Fig. 10.1 Mass transfer between two bulk phases as a series of two one-sided bulk-interface transfer processes. Reproduced (in adapted form) from Bothe (2022) with permission

This introduces a negligible error, unless the volume of the transition zone becomes a significant fraction of the total volume.

This conceptual picture, as illustrated by Fig. 10.1, also implies that mass transfer must be considered as the succession of two bulk-interface transfer processes, taking place one after the other. Furthermore, both of these one-sided mass transfer processes are bi-directional, which opens a more flexible and realistic way of modeling. In this picture, the interfacial thermodynamical state automatically influences both one-sided bi-directional mass transfer rates, hence also the resulting total rate of the series of transfer steps.

Mathematical models for multicomponent two-phase fluid systems that account for the interfacial mass densities of all constituents have only been rarely investigated in the literature. Building on prior sharp-interface modeling of irreversible processes given in Waldmann (1967) and Bedeaux et al. (1976), in the seminal paper Bedeaux (1986), Dick Bedeaux derived such a sharp-interface model with entropy inequality in the spirit of T.I.P. (Theory of Irreversible Processes, cf. De Groot and Mazur (1984)). A few years later, the monograph (Slattery 1990) by Slattery appeared, which contains a similar sharp-interface model with interfacial mass densities, deriving and evaluating bulk and interfacial entropy production rates in the spirit of Truesdell and Noll; cf. Truesdell and Noll (2003). In these works, and later in Sagis (2014), Dreyer et al. (2018), Bothe (2022), an interfacial entropy production rate is given. Most of the given entropy production rates are not equivalent and those which are correct are often arranged in a disadvantageous way such that a resulting closure, if given at all, would not result in a physically sound model.

In the present paper, we derive a general sharp-interface model for multicomponent two-phase fluid systems in which not only partial mass densities in the bulk phases and on the interface are included for all constituents, but also *partial* momenta of all species in the bulk phases and on the interface are individually balanced. Hence, we develop the *multi-velocity* sharp-interface model class for chemically reacting fluid systems composed of N chemical constituents. In the notion of Hutter, see Hutter and Jöhnk (2004), this yields a *class-II* model for fluid systems with interfacial mass, where every constituent has its individual velocity, with independent bulk and interface fields. Our main motivation to include partial momentum balances with their individual, species-specific velocities is the quest for a model class with a minimum number of necessary cross-couplings in order to obtain a practically useful PDE model. This requires to properly relate all terms appearing in the entropy production with corresponding dissipative mechanisms, which in turn relies on a clarification of the structure of the entropy production. Here it turns out that the class-II, i.e. the multi-velocity model reveals such relations that are hidden if a more standard class-I model (with the common, barycentric velocity) is employed. The latter case was investigated in Bothe (2022), which treated the multicomponent two-phase fluid systems with interfacial mass, putting main emphasis on the surfactant hindrance effect on mass transfer. There, in order to clarify the structure of the entropy production, the class-II balance equations and the entropy production rates for bulk phases and the sharp interface has been collected in an appendix. Here, we resume the investigations from Bothe (2022), presenting a self-contained derivation of the model and adding in particular a full set of constitutive equations. The multi-velocity approach also allows for a refined and more detailed modeling of interfacial transport processes. In principle, a further extension of this model to account for individual temperatures is also possible; see Ruggeri (2009) concerning such multi-temperature modeling in case of single-phase fluid mixtures.

Many more details about the general strategy of deriving the class-II continuum thermodynamics can be found in the joint work (Bothe and Dreyer 2015) of Wolfgang Dreyer and the present author, where many new ideas on the continuum thermodynamics of single-phase multicomponent chemically reacting fluid mixtures have been developed. The present work also profited from Wolfgang's 'jump paper' (Dreyer 2003).

10.2 Sharp-Interface Mathematical Modeling

We assume the two-phase multicomponent fluid system to fill the considered domain $\Omega \subset \mathbf{R}^3$ in such a way that two bulk phases $\Omega^{\pm}(t)$ are separated by a sharp interface, i.e. by a surface $\Sigma(t)$ in $\mathbf{R}^3$:

$$\Omega = \Omega^+(t) \cup \Omega^-(t) \cup \Sigma(t),$$

a disjoint decomposition of Ω. For technical simplicity, we assume the $\Sigma(t)$ to be closed surfaces without contact to the domain boundary $\partial\Omega$, thus avoiding the appearance of contact lines. Let us note in passing that this setup assumes the total domain to be fixed, hence the total volume is conserved if Ω is a bounded set. This is a standard assumption in essentially all mathematical models, as it allows to work with a set of PDEs inside a fixed volume. A more realistic approach would allow (some part of) the boundary to have some flexibility and, e.g., to introduce a pressure control; cf. Bothe and Soga (2016).

The family of moving hypersurfaces $\{\Sigma(t)\}_{t\in I}$, with $I = (a, b) \subset \mathbf{R}$ an open time interval, is assumed to form a C^1-hypersurface in space-time $\mathbf{R}^4$ such that the instantaneous surfaces $\Sigma(t)$ are C^2-surfaces in $\mathbf{R}^3$ so that curvature is well-defined on $\Sigma(t)$. In particular, the total curvature (twice the mean curvature) $\kappa_\Sigma = \mathrm{div}_\Sigma\,(-\mathbf{n}_\Sigma)$ is a well-defined continuous quantity. In addition, we assume that the normal field $\mathbf{n}_\Sigma(t, \cdot)$ of $\Sigma(t) \subset \mathbf{R}^3$ is continuously differentiable, jointly in both variables (t, x). Under these assumptions, we call $\{\Sigma(t)\}_{t\in I}$ a $C^{1,2}$-family of moving hypersurfaces. We also denote by $\mathcal{M} = \mathrm{gr}\,(\Sigma)$ the graph of the family $\{\Sigma(t)\}_{t\in I}$, defined as

$$\mathcal{M} := \{(t, \mathbf{x}) \in \mathbf{R}^4 : t \in I,\ \mathbf{x} \in \Sigma(t)\} = \bigcup_{t\in I} \big(\{t\} \times \Sigma(t)\big). \tag{10.1}$$

These definitions are also employed in Kimura (2008), Prüss and Simonett (2016), Bothe (2022) and in a similar form in Giga (2006).

For a $C^{1,2}$-family $\{\Sigma(t)\}_{t\in I}$ of moving hypersurfaces, V_Σ denotes the *speed of normal displacement* of $\Sigma(\cdot)$ and is defined via the relation

$$V_\Sigma(t,\mathbf{x}) = \langle \gamma'(t), \mathbf{n}_\Sigma(t,\gamma(t))\rangle \tag{10.2}$$

for any C^1-curve γ with $\gamma(t) = \mathbf{x}$ and $\mathrm{gr}\,(\gamma) \subset \mathcal{M}$, where $\langle\cdot,\cdot\rangle$ denotes the inner product in $\mathbf{R}^3$. It is not difficult to show that the definition via (10.2) is equivalent to

$$V_\Sigma(t,\mathbf{x}) = -\mathbf{n}^t_{\mathcal{M}}(t,\mathbf{x})/||\mathbf{n}^{\mathbf{x}}_{\mathcal{M}}(t,\mathbf{x})||, \tag{10.3}$$

where $\mathbf{n}_{\mathcal{M}} = (\mathbf{n}^t_{\mathcal{M}}, \mathbf{n}^{\mathbf{x}}_{\mathcal{M}})$ is the normal field of $\mathcal{M}$, oriented in such a way that $\mathbf{n}_\Sigma(t,\mathbf{x}) = \mathbf{n}^{\mathbf{x}}_{\mathcal{M}}(t,\mathbf{x})/||\mathbf{n}^{\mathbf{x}}_{\mathcal{M}}(t,\mathbf{x})||$. The characterization via (10.3) shows that V_Σ is a purely kinematic quantity, determined only by the family $\{\Sigma(t)\}_{t\in I}$ of moving interfaces. Hence, the value of V_Σ from (10.2) does not depend on the choice of the specific curve; cf. Chap. 2.5 in Prüss and Simonett (2016). The computation of V_Σ is especially simple if $\{\Sigma(t)\}_{t\in I}$ is given by a level set description, i.e.

$$\Sigma(t) = \{\mathbf{x} \in \mathbf{R}^3 : \phi(t,\mathbf{x}) = 0\} \tag{10.4}$$

with $\phi \in C^{1,2}(\mathcal{N})$ for some open neighborhood $\mathcal{N} \subset \mathbf{R} \times \mathbf{R}^3$ of $\mathcal{M}$ such that $\nabla_{\mathbf{x}}\phi \neq 0$ on $\mathcal{M}$. In this case, $\mathbf{n}_{\mathcal{M}} = \nabla_{(t,\mathbf{x})}\phi/||\nabla_{(t,\mathbf{x})}\phi|| = (\partial_t\phi, \nabla_{\mathbf{x}}\phi)/||\nabla_{(t,\mathbf{x})}\phi||$, hence

$$V_\Sigma(t,\mathbf{x}) = -\frac{\partial_t\phi(t,\mathbf{x})}{\|\nabla_{\mathbf{x}}\phi(t,\mathbf{x})\|} \quad \text{for } t \in J,\ \mathbf{x} \in \Sigma(t). \tag{10.5}$$

In the literature, V_Σ is often called normal velocity of $\Sigma(\cdot)$, but we prefer to call it the speed of normal displacement since V_Σ is a scalar quantity. In fact, the speed of normal displacement gives rise to a natural interface velocity field, given as $\mathbf{w}_\Sigma := V_\Sigma\mathbf{n}_\Sigma$, which therefore is intrinsic to any $C^{1,2}$-family $\{\Sigma(t)\}_{t\in I}$ of moving hypersurfaces. According to Corollary 1 together with Theorem 1 in Bothe (2020), the associated initial value problem

$$\dot{\mathbf{x}}(t) = \mathbf{w}_\Sigma(t,\mathbf{x}(t)) \quad \text{for } t \in I, \quad \mathbf{x}(t_0) = \mathbf{x}_0 \tag{10.6}$$

is uniquely solvable (at least locally in time) for every $t_0 \in I$ and $\mathbf{x}_0 \in \Sigma(t_0)$. The unique (local) solution $\mathbf{x}(\cdot\,; t_0, \mathbf{x}_0)$ of (10.6), starting in $(t_0, \mathbf{x}_0)$, follows the moving surface along a path which always runs normal to $\Sigma(t)$. Consequently, the intrinsic interface velocity $\mathbf{w}_\Sigma$ (hence V_Σ) induces a Lagrangian-type derivative, the so-called Thomas derivative ∂_t^Σ (cf. Thomas (1957)), defined as

$$\partial_t^\Sigma\phi^\Sigma(t_0,\mathbf{x}_0) := \frac{d}{dt}\phi^\Sigma(t,\mathbf{x}(t;t_0,\mathbf{x}_0))_{|t=t_0}, \tag{10.7}$$

where ϕ^Σ denotes a quantity which is defined on $\mathcal{M}$. According to Lemma 2 together with Theorem 1 in Bothe (2020), unique (local) solvability of the associated initial value problems also holds true for any velocity field of the type[1]

$$\mathbf{v}^\Sigma = V_\Sigma\, \mathbf{n}_\Sigma + \mathbf{v}_{||}^\Sigma, \tag{10.8}$$

where the additional tangential part $\mathbf{v}_{||}^\Sigma$ is continuous on $\mathcal{M}$ and the maps $\mathbf{v}_{||}^\Sigma(t,\cdot)$ are locally Lipschitz continuous on $\Sigma(t)$ for every $t \in I$. This gives rise to a Lagrangian derivative $\frac{D^\Sigma}{Dt}$ associated with $\mathbf{v}^\Sigma$, defined in analogy to the Thomas derivative above, for which one can show that

$$\frac{D^\Sigma \phi^\Sigma}{Dt} = \partial_t^\Sigma \phi^\Sigma + \mathbf{v}_{||}^\Sigma \cdot \nabla_\Sigma \phi^\Sigma \tag{10.9}$$

with ∇_Σ denoting the surface gradient (w.r.t. to $\Sigma(t)$, where t is understood); cf. Bothe (2020). Note also that $\mathbf{v}_{||}^\Sigma \cdot \nabla_\Sigma \phi^\Sigma = \mathbf{v}^\Sigma \cdot \nabla_\Sigma \phi^\Sigma$, since $\nabla_\Sigma \phi^\Sigma$ is tangential to Σ.

The Thomas derivative allows to formulate a general surface transport theorem. For this purpose, let $V \subset \Omega$ be a fixed control volume with outer normal field $\mathbf{n}$ and let $\Sigma_V(t) := \Sigma(t) \cap V$. Then the identity

$$\frac{d}{dt} \int_{\Sigma_V(t)} \phi^\Sigma\, do = \int_{\Sigma_V(t)} \left(\partial_t^\Sigma \phi^\Sigma - \phi^\Sigma \kappa_\Sigma V_\Sigma\right) do - \int_{\partial\Sigma_V(t)} \frac{\phi^\Sigma V_\Sigma\, \mathbf{n}\cdot\mathbf{n}_\Sigma}{\sqrt{1-(\mathbf{n}\cdot\mathbf{n}_\Sigma)^2}}\, ds \tag{10.10}$$

holds, see Dell'Isola and Romano (1986), Gurtin et al. (1989), Alke and Bothe (2009). For a particular control volume V such that $\mathbf{n}$ is tangential to $\Sigma_V(t)$ on $\partial\Sigma_V(t)$, i.e. $\mathbf{n}\cdot\mathbf{n}_\Sigma = 0$ on $\partial\Sigma_V(t)$, the transport identity (10.10) simplifies to

$$\frac{d}{dt} \int_{\Sigma_V(t)} \phi^\Sigma\, do = \int_{\Sigma_V(t)} \left(\partial_t^\Sigma \phi^\Sigma - \phi^\Sigma \kappa_\Sigma V_\Sigma\right) do. \tag{10.11}$$

Below, any control volume which intersects $\Sigma(t)$ is assumed to be of this type, which is sufficient for deriving the local balance equations. From here on, the time dependence of $\Sigma_V(t)$ and $\partial\Sigma_V(t)$ in integration domains is suppressed for better readability.

To obtain a generic balance equation, we start with *integral balances*, since they are valid across the interface, while a direct formulation of interfacial transmission and jump conditions is not obvious. Let Φ denote an extensive, i.e. a mass additive quantity such as mass, momentum, energy, with volumetric density ϕ (i.e. ϕ^+ and ϕ^- in the phase Ω^+ and Ω^-, respectively) and interfacial, area specific density ϕ^Σ. Given a fixed control volume V, let $\mathbf{N}$ denote the outer unit normal to $\partial\Sigma_V$ which is at the same time tangential to Σ; cf. Fig. 10.2. With these notations, the generic

[1] We use Σ as a subscript for intrinsic, geometrically defined quantities, while we use Σ as superscript for all other interface quantities.

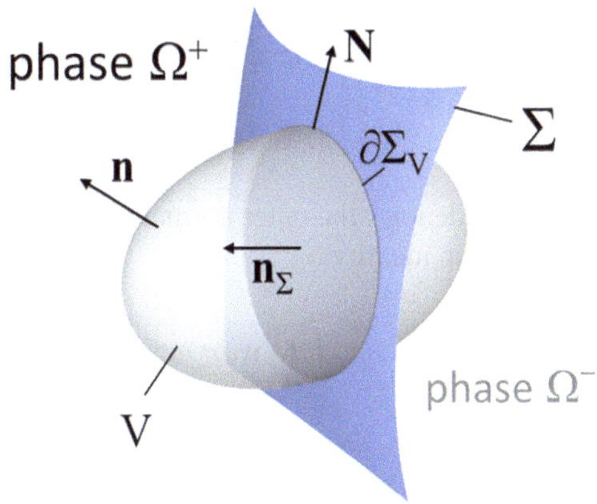

Fig. 10.2 Configuration of the phases and basic notations for a control volume, intersecting the interface

integral balance for the amount of an extensive quantity Φ inside a *fixed* control volume V is given by

$$\frac{d}{dt}\Big[\int_V \phi \, dx + \int_{\Sigma_V} \phi^\Sigma \, do\Big] = -\int_{\partial V} \mathbf{j}_{\text{tot}} \cdot \mathbf{n} \, do + \int_V f \, dx \\ - \int_{\partial \Sigma_V} \mathbf{j}^\Sigma_{\text{tot}} \cdot \mathbf{N} \, ds + \int_{\Sigma_V} f^\Sigma \, do, \tag{10.12}$$

where $\mathbf{j}_{\text{tot}}$ is the flux of Φ inside the bulk phases, $\mathbf{j}^\Sigma_{\text{tot}}$ the corresponding flux inside the interface, f the source terms acting in the bulk and, finally, f^Σ the interfacial source term. It is important to note that both $\mathbf{j}_{\text{tot}}$ and $\mathbf{j}^\Sigma_{\text{tot}}$ are total fluxes with respect to the reference frame in which the fixed control volume is given. Below, they will be split into a convective plus a diffusive contribution. We require the interfacial fluxes to be tangential to the interface. This is no restriction, since only $\mathbf{j}^\Sigma_{\text{tot}} \cdot \mathbf{N}$ enters the balance equation. In order to obtain the local form of the generic balance equation, some mathematical tools are required. We briefly recall the relevant identities, while more details can be found in Slattery et al. (2007); for proofs see Kimura (2008); Prüss and Simonett (2016) or Barrett et al. (2020). The two-phase transport theorem in the version for fixed control volumes V states that

$$\frac{d}{dt}\int_V \phi \, dx = \int_{V \setminus \Sigma} \partial_t \phi \, dx - \int_{\Sigma_V} [\![\phi]\!] V_\Sigma \, do, \tag{10.13}$$

where

$$[\![\phi]\!](t, \mathbf{x}) := \lim_{h \to 0+} \big(\phi(t, \mathbf{x} + h\mathbf{n}_\Sigma) - \phi(t, \mathbf{x} - h\mathbf{n}_\Sigma)\big) \tag{10.14}$$

denotes the jump of a quantity ϕ across the interface in a form which makes the jump conditions orientation-invariant. For later use, note that

$$[\![\phi]\!]\,\mathbf{n}_\Sigma = -\left(\phi^+\mathbf{n}^+ + \phi^+\mathbf{n}^+\right), \tag{10.15}$$

where $\mathbf{n}^\pm$ denotes the outer normal to $\Omega^\pm$ at Σ. The two-phase divergence theorem provides the identity

$$\int_{\partial V} \mathbf{f}\cdot\mathbf{n}\,do = \int_{V\setminus\Sigma} \operatorname{div}\mathbf{f}\,dx + \int_{\Sigma_V} [\![\mathbf{f}\cdot\mathbf{n}_\Sigma]\!]\,do. \tag{10.16}$$

We intentionally write $V\setminus\Sigma$ as the domain of integration to avoid confusion with distributional derivatives which would include a Dirac delta contribution supported on the interface; the latter would in fact correspond to the jump bracket term in (10.16). Finally, the surface divergence theorem states that

$$\int_{\partial\Sigma_V} \mathbf{f}^\Sigma\cdot\mathbf{N}\,ds = \int_{\Sigma_V} \operatorname{div}_\Sigma \mathbf{f}^\Sigma\,do \tag{10.17}$$

for vector fields $\mathbf{f}^\Sigma$ being tangential to Σ.

Applying the two-phase transport theorem to the first term in (10.12), the extended surface transport theorem in the form (10.11) to the second term, the two-phase divergence theorem to the third term and the surface divergence theorem to the fifth term yields

$$\int_{V\setminus\Sigma} (\partial_t\phi + \operatorname{div}\mathbf{j}_{\text{tot}})\,dx + \int_{\Sigma_V} (-[\![\phi]\!]V_\Sigma + [\![\mathbf{j}_{\text{tot}}\cdot\mathbf{n}_\Sigma]\!])\,do$$

$$+\int_{\Sigma_V} \left(\partial_t^\Sigma\phi^\Sigma + \operatorname{div}_\Sigma(\phi^\Sigma V_\Sigma\mathbf{n}_\Sigma + \mathbf{j}^\Sigma_{\text{tot}})\right)do = \int_V f\,dx + \int_{\Sigma_V} f^\Sigma\,do.$$

Collecting bulk and interface terms, we obtain the differential form of the generic balance equation in an Eulerian form by means of the usual localization procedure. The result reads as

$$\partial_t\phi + \operatorname{div}\mathbf{j}_{\text{tot}} = f \quad \text{in } \Omega\setminus\Sigma, \tag{10.18}$$

$$\partial_t^\Sigma\phi^\Sigma + \operatorname{div}_\Sigma(\phi^\Sigma V_\Sigma\mathbf{n}_\Sigma + \mathbf{j}^\Sigma_{\text{tot}}) - [\![\phi]\!]V_\Sigma + [\![\mathbf{j}_{\text{tot}}\cdot\mathbf{n}_\Sigma]\!] = f^\Sigma \quad \text{on } \Sigma. \tag{10.19}$$

Recall that both $\mathbf{j}_{\text{tot}}$ and $\mathbf{j}^\Sigma_{\text{tot}}$ are total fluxes with respect to the reference frame in which the chosen control volumes are fixed, and $\mathbf{j}^\Sigma_{\text{tot}}$ is tangential to $\Sigma(t)$. If these are decomposed into a convective and a diffusive (also called molecular) flux according to

$$\mathbf{j}_{\text{tot}} = \phi\,\mathbf{v} + \mathbf{j}, \qquad \mathbf{j}^{\Sigma}_{\text{tot}} = \phi^{\Sigma}\,\mathbf{v}^{\Sigma}_{||} + \mathbf{j}^{\Sigma}, \tag{10.20}$$

the generic local balance equations for an extensive quantity Φ with bulk density ϕ and interfacial density ϕ^{Σ} read as

$$\partial_t \phi + \operatorname{div}\left(\phi\,\mathbf{v} + \mathbf{j}\right) = f \quad \text{in } \Omega \setminus \Sigma, \tag{10.21}$$

$$\partial_t^{\Sigma} \phi^{\Sigma} + \operatorname{div}_{\Sigma}(\phi^{\Sigma}\,\mathbf{v}^{\Sigma} + \mathbf{j}^{\Sigma}) + [\![\phi(\mathbf{v} - \mathbf{v}^{\Sigma})]\!] \cdot \mathbf{n}_{\Sigma} + [\![\mathbf{j} \cdot \mathbf{n}_{\Sigma}]\!] = f^{\Sigma} \quad \text{on } \Sigma. \tag{10.22}$$

Concerning the jump bracket notation, we treat interfacial quantities as having one-sided bulk limits of the same value, e.g. $[\![\mathbf{j} \cdot \mathbf{n}_{\Sigma}]\!]$ above is the same as $[\![\mathbf{j}]\!] \cdot \mathbf{n}_{\Sigma}$. Hence, the generic interfacial balance can be written somewhat more compactly as

$$\partial_t^{\Sigma} \phi^{\Sigma} + \operatorname{div}_{\Sigma}(\phi^{\Sigma}\,\mathbf{v}^{\Sigma} + \mathbf{j}^{\Sigma}) + [\![\phi(\mathbf{v} - \mathbf{v}^{\Sigma}) + \mathbf{j}]\!] \cdot \mathbf{n}_{\Sigma} = f^{\Sigma} \quad \text{on } \Sigma. \tag{10.23}$$

Note that the velocities $\mathbf{v}$, $\mathbf{v}^{\Sigma}$ appearing in the convective flux contributions are not yet specified and can also refer to individual velocities $\mathbf{v}_i$ of constituent A_i.

10.3 Sharp-Interface Balance Equations

We suppose that the fluid phases are composed of N different constituents (chemical species) $A_1, \ldots, A_N$. Typically, all species will be present at least in one bulk phase and on the interface. We do not separately index the constituents in the different phases; if a species is not present, its mass density simply vanishes. We account for bulk chemical reactions between the A_i according to

$$\alpha_1^a\,A_1 + \ldots + \alpha_N^a\,A_N \rightleftharpoons \beta_1^a\,A_1 + \ldots + \beta_N^a\,A_N \quad \text{for } a = 1, \ldots, N_R \tag{10.24}$$

with stoichiometric coefficients $\alpha_i^a, \beta_i^a \in \mathbb{N}_0$. Note that we allow for separate chemical reaction networks in the different bulk phases, i.e. in full detail the stoichiometric coefficients read $\alpha_i^{+,a}, \beta_i^{+,a} \in \mathbb{N}_0$ for $a - 1, \ldots, N_R^+$ to describe the chemistry in $\Omega^+(t)$ and $\alpha_i^{-,a}, \beta_i^{-,a} \in \mathbb{N}_0$ for $a = 1, \ldots, N_R^-$ for the chemistry in $\Omega^-(t)$. Since the aspect of chemical reactions is not thc main point, here, we suppress the phase index for better readability whenever this is possible without loss of understanding. All reactions are considered with their forward and backward direction. The molar reaction rate, i.e. the number of chemical conversions in multiples of the Avogadro number per time and volume, of the ath reaction is denoted R_a^f for the forward (from left to right in (10.24)) and R_a^b for the backward direction. The total molar reaction rate of the ath reaction hence is $R_a = R_a^f - R_a^b$ and the mass production rate of constituent A_i due to the ath reaction is

$$r_i = \sum_{a=1}^{N_R} M_i \nu_i^a R_a \quad \text{with } \nu_i^a := \beta_i^a - \alpha_i^a, \tag{10.25}$$

where M_i is the molar mass of A_i. Because mass is conserved in every single reaction, it holds that

$$\sum_{i=1}^{N} M_i \nu_i^a = 0 \quad \text{for all } a = 1, \ldots, N_R. \tag{10.26}$$

We write A_i^Σ for the adsorbed form of constituent A_i in order to distinguish it from the latter, i.e. from $A_i^\pm$ in the adjacent bulk phases. We then account for interfacial chemical reactions between the A_i^Σ according to

$$\alpha_1^{\Sigma,a}\, A_1^\Sigma + \ldots + \alpha_N^{\Sigma,a}\, A_N^\Sigma \rightleftharpoons \beta_1^{\Sigma,a}\, A_1^\Sigma + \ldots + \beta_N^{\Sigma,a}\, A_N^\Sigma \quad \text{for } a = 1, \ldots, N_R^\Sigma. \tag{10.27}$$

In full analogy to the bulk case, we let

$$r_i^\Sigma = \sum_{a=1}^{N_R^\Sigma} M_i \nu_i^{\Sigma,a} R_a^\Sigma \quad \text{with } \nu_i^{\Sigma,a} := \beta_i^{\Sigma,a} - \alpha_i^{\Sigma,a} \tag{10.28}$$

to obtain r_i^Σ for the interfacial production rate of partial mass of A_i^Σ and have

$$\sum_{i=1}^{N} M_i \nu_i^{\Sigma,a} = 0 \quad \text{for all } a = 1, \ldots, N_R^\Sigma. \tag{10.29}$$

Building on Bothe and Dreyer (2015) and on the appendix in Bothe (2022), we derive a full multi-velocity model, in which the constituents' partial mass as well as their partial momentum are balanced individually, both for the bulk phases and for the interface. Starting point for this class-II continuum thermodynamics of such systems therefore are the balance equations for partial and total mass, partial and total momentum as well as total internal energy, both in the bulk phases and on the interface. We also include the balance equations for partial internal energy since it determines certain structural properties.

We start with the balance of partial mass, where we insert

$$\begin{aligned} &\phi = \rho_i, \quad &&\mathbf{v} = \mathbf{v}_i, \quad &&\mathbf{j} = 0, \quad &&f = r_i, \\ &\phi^\Sigma = \rho_i^\Sigma, \quad &&\mathbf{v}^\Sigma = \mathbf{v}_i^\Sigma, \quad &&\mathbf{j}^\Sigma = 0, \quad &&f^\Sigma = r_i^\Sigma \end{aligned}$$

into the local form of the master balance (10.21) and (10.23), respectively. This yields the

Partial mass balance

$$\partial_t \rho_i + \operatorname{div}(\rho_i \mathbf{v}_i) = r_i \qquad \text{in } \Omega \setminus \Sigma, \tag{10.30}$$
$$\partial_t^\Sigma \rho_i^\Sigma + \operatorname{div}_\Sigma(\rho_i^\Sigma \mathbf{v}_i^\Sigma) + [\![\rho_i(\mathbf{v}_i - \mathbf{v}_i^\Sigma) \cdot \mathbf{n}_\Sigma]\!] = r_i^\Sigma \qquad \text{on } \Sigma. \tag{10.31}$$

Below, we occasionally skip the domains for the balance equations to save space, since the time derivative determines whether the balance is a bulk or an interface balance. In order to maintain a single common interface between the two bulk phases, the individual interface velocities $\mathbf{v}_i^\Sigma$ need to fulfil the *consistency requirement* (cf. Bothe (2020))

$$\mathbf{v}_i^\Sigma \cdot \mathbf{n}_\Sigma = V_\Sigma \quad \text{for all } i = 1, \ldots, N. \tag{10.32}$$

In other words, the surfaces of discontinuity of the individual mass densities move with the same normal speed as Σ. But note that the individual interfacial velocities will in general have different tangential components.

The jump term in (10.31) represents two one-sided bulk-interface mass exchange terms. In the physico-chemical sciences, these bulk-interface mass (species) exchange processes are named sorption, with *adsorption* being the bulk-to-interface mass exchange and *desorption* for the reverse direction. We therefore let

$$\dot{m}_i^{\pm,\Sigma} := \rho_i^\pm(\mathbf{v}_i^\pm - \mathbf{v}_i^\Sigma) \cdot \mathbf{n}^\pm \tag{10.33}$$

with $\mathbf{n}^\pm$ the outer normals to $\Omega^\pm$ denote these one-sided bulk-to-interface exchange rates of partial mass. Then the *sorption formulation* of the partial mass balances reads as

$$\partial_t \rho_i + \operatorname{div}(\rho_i \mathbf{v}_i) = r_i \qquad \text{in } \Omega \setminus \Sigma, \tag{10.34}$$
$$\partial_t^\Sigma \rho_i^\Sigma + \operatorname{div}_\Sigma(\rho_i^\Sigma \mathbf{v}_i^\Sigma) = \dot{m}_i^{+,\Sigma} + \dot{m}_i^{-,\Sigma} + r_i^\Sigma \qquad \text{on } \Sigma. \tag{10.35}$$

As a consequence of (10.26), (10.29), *conservation of total mass* holds true, i.e.

$$\sum_{i=1}^N r_i = 0, \qquad \sum_{i=1}^N r_i^\Sigma = 0. \tag{10.36}$$

Summation of the partial mass balances (10.30) and (10.31), together with (10.36), yields the

Balance of total mass

$$\partial_t \rho + \operatorname{div}(\rho \mathbf{v}) = 0 \qquad \text{in } \Omega \setminus \Sigma, \tag{10.37}$$
$$\partial_t^\Sigma \rho^\Sigma + \operatorname{div}_\Sigma(\rho^\Sigma \mathbf{v}^\Sigma) - [\![\rho(\mathbf{v} - \mathbf{v}^\Sigma) \cdot \mathbf{n}_\Sigma]\!] = 0 \qquad \text{on } \Sigma, \tag{10.38}$$

where

$$\rho := \sum_{i=1}^{N} \rho_i, \quad \rho \mathbf{v} := \sum_{i=1}^{N} \rho_i \mathbf{v}_i, \quad \rho^\Sigma := \sum_{i=1}^{N} \rho_i^\Sigma, \quad \rho^\Sigma \mathbf{v}^\Sigma := \sum_{i=1}^{N} \rho_i^\Sigma \mathbf{v}_i^\Sigma. \tag{10.39}$$

We continue with the balance of partial momenta, where we insert

$$\begin{aligned} \phi &= \rho_i \mathbf{v}_i, & \mathbf{v} &= \mathbf{v}_i, & \mathbf{j} &= -\mathbf{S}_i, & f &= \rho_i \mathbf{b}_i + \mathbf{f}_i, \\ \phi^\Sigma &= \rho_i^\Sigma \mathbf{v}_i^\Sigma, & \mathbf{v}^\Sigma &= \mathbf{v}_i^\Sigma, & \mathbf{j}^\Sigma &= -\mathbf{S}_i^\Sigma, & f^\Sigma &= \rho_i^\Sigma \mathbf{b}_i^\Sigma + \mathbf{f}_i^\Sigma \end{aligned}$$

into the local form of the master balance (10.21) and (10.23), respectively. This yields the

Partial momentum balance

$$\partial_t(\rho_i \mathbf{v}_i) + \operatorname{div}(\rho_i \mathbf{v}_i \otimes \mathbf{v}_i - \mathbf{S}_i) = \rho_i \mathbf{b}_i + \mathbf{f}_i \qquad \text{in } \Omega \setminus \Sigma, \tag{10.40}$$

$$\partial_t^\Sigma(\rho_i^\Sigma \mathbf{v}_i^\Sigma) + \operatorname{div}_\Sigma(\rho_i^\Sigma \mathbf{v}_i^\Sigma \otimes \mathbf{v}_i^\Sigma - \mathbf{S}_i^\Sigma) + [\![\dot{m}_i \mathbf{v}_i - \mathbf{S}_i \mathbf{n}_\Sigma]\!] = \rho_i^\Sigma \mathbf{b}_i^\Sigma + \mathbf{f}_i^\Sigma \qquad \text{on } \Sigma \tag{10.41}$$

with $\mathbf{f}_i$ and $\mathbf{f}_i^\Sigma$ the rate of momentum exchange between the species and the abbreviation

$$\dot{m}_i := \rho_i(\mathbf{v}_i - \mathbf{v}_i^\Sigma) \cdot \mathbf{n}_\Sigma = \rho_i(\mathbf{v}_i - \mathbf{v}^\Sigma) \cdot \mathbf{n}_\Sigma, \tag{10.42}$$

where the second equality comes from (10.32). We assume *conservation of total momentum* to be valid, hence

$$\sum_{i=1}^{N} \mathbf{f}_i = 0, \qquad \sum_{i=1}^{N} \mathbf{f}_i^\Sigma = 0. \tag{10.43}$$

Employing the abbreviation

$$\dot{m} := \sum_{i=1}^{N} \dot{m}_i = \rho(\mathbf{v} - \mathbf{v}^\Sigma) \cdot \mathbf{n}_\Sigma, \tag{10.44}$$

this is applied to the sum of the partial momentum balances (10.40), (10.41) to obtain the

Balance of total momentum

$$\partial_t(\rho\mathbf{v}) + \operatorname{div}(\rho\mathbf{v}\otimes\mathbf{v} - \mathbf{S}) = \rho\mathbf{b} \qquad \text{in } \Omega\setminus\Sigma, \tag{10.45}$$

$$\partial_t^\Sigma(\rho^\Sigma\mathbf{v}^\Sigma) + \operatorname{div}_\Sigma(\rho^\Sigma\mathbf{v}^\Sigma\otimes\mathbf{v}^\Sigma - \mathbf{S}^\Sigma) + [\![\dot{m}\mathbf{v} - \mathbf{S}\mathbf{n}_\Sigma]\!] = \rho^\Sigma\mathbf{b}^\Sigma \qquad \text{on } \Sigma, \tag{10.46}$$

with the total bulk and interface stresses

$$\mathbf{S} := \sum_{i=1}^N(\mathbf{S}_i - \rho_i\mathbf{u}_i\otimes\mathbf{u}_i), \qquad \mathbf{S}^\Sigma := \sum_{i=1}^N(\mathbf{S}_i^\Sigma - \rho_i^\Sigma\mathbf{u}_i^\Sigma\otimes\mathbf{u}_i^\Sigma), \tag{10.47}$$

containing the bulk and interface diffusion velocities

$$\mathbf{u}_i := \mathbf{v}_i - \mathbf{v}, \qquad \mathbf{u}_i^\Sigma := \mathbf{v}_i^\Sigma - \mathbf{v}^\Sigma, \tag{10.48}$$

and the bulk and interface body force densities

$$\rho\mathbf{b} := \sum_{i=1}^N \rho_i\mathbf{b}_i, \qquad \rho^\Sigma\mathbf{b}^\Sigma := \sum_{i=1}^N \rho_i^\Sigma\mathbf{b}_i^\Sigma. \tag{10.49}$$

Below, we employ the stress decompositions

$$\mathbf{S}_i = -P_i\mathbf{I} + \mathbf{S}_i^\circ, \qquad \mathbf{S}_i^\Sigma = -P_i^\Sigma\mathbf{I}_\Sigma + \mathbf{S}_i^{\Sigma,\circ} \tag{10.50}$$

with the *mechanical partial pressures* P_i and P_i^Σ defined as

$$P_i := -\frac{1}{3}\operatorname{trace}(\mathbf{S}_i), \qquad P_i^\Sigma := -\frac{1}{2}\operatorname{trace}(\mathbf{S}_i^\Sigma), \tag{10.51}$$

where the interfacial 'identity tensor' $\mathbf{I}_\Sigma$ denotes the projection onto the local tangential plane, i.e.

$$\mathbf{I}_\Sigma = \mathbf{I} - \mathbf{n}_\Sigma\otimes\mathbf{n}_\Sigma. \tag{10.52}$$

In equilibrium, the mechanical partial pressures reduce to the *thermodynamical partial pressures* p_i and p_i^Σ, respectively, i.e.

$$P_{i|\text{equ}} = p_i, \qquad P_{i|\text{equ}}^\Sigma = p_i^\Sigma, \tag{10.53}$$

as the partial thermodynamical pressures are functions solely of the respective thermodynamical state variables. Let

$$\pi_i := P_i - p_i, \qquad \pi_i^\Sigma := P_i^\Sigma - p_i^\Sigma \tag{10.54}$$

denote the *irreversible part of the bulk and interface partial pressures*, respectively. Then

$$\mathbf{S}_i^{\mathrm{irr}} = -\pi_i \mathbf{I} + \mathbf{S}_i^{\circ}, \qquad \mathbf{S}_i^{\Sigma,\mathrm{irr}} = -\pi_i^{\Sigma} \mathbf{I}_{\Sigma} + \mathbf{S}_i^{\Sigma,\circ} \tag{10.55}$$

are the irreversible stress contributions.

We continue with the balance of energy, where we include partial energy balances since they provide structural information even if they are not part of the final multi-velocity model. For simplicity we do not include external energy supply by radiation. Insertion of

$$\phi = \rho_i\big(e_i + \frac{\mathbf{v}_i^2}{2}\big), \qquad \mathbf{v} = \mathbf{v}_i, \quad \mathbf{j} = \mathbf{q}_i - \mathbf{v}_i \cdot \mathbf{S}_i, \quad f = \rho_i \mathbf{v}_i \cdot \mathbf{b}_i + k_i,$$

$$\phi^{\Sigma} = \rho_i^{\Sigma}\big(e_i^{\Sigma} + \frac{(\mathbf{v}_i^{\Sigma})^2}{2}\big), \;\; \mathbf{v}^{\Sigma} = \mathbf{v}_i^{\Sigma}, \;\; \mathbf{j}^{\Sigma} = \mathbf{q}_i^{\Sigma} - \mathbf{v}_i^{\Sigma} \cdot \mathbf{S}_i^{\Sigma}, \;\; f^{\Sigma} = \rho_i^{\Sigma} \mathbf{v}_i^{\Sigma} \cdot \mathbf{b}_i^{\Sigma} + k_i^{\Sigma}$$

into the local form of the master balance (10.21) and (10.23), respectively, yields the

Partial energy balance

$$\partial_t\big(\rho_i\big(e_i + \frac{\mathbf{v}_i^2}{2}\big)\big) + \operatorname{div}\big(\rho_i\big(e_i + \frac{\mathbf{v}_i^2}{2}\big)\mathbf{v}_i + \mathbf{q}_i - \mathbf{v}_i \cdot \mathbf{S}_i\big) = \rho_i \mathbf{v}_i \cdot \mathbf{b}_i + k_i, \tag{10.56}$$

$$\partial_t^{\Sigma}\big(\rho_i^{\Sigma}\big(e_i^{\Sigma} + \frac{(\mathbf{v}_i^{\Sigma})^2}{2}\big)\big) + \operatorname{div}_{\Sigma}\big(\rho_i^{\Sigma}\big(e_i^{\Sigma} + \frac{(\mathbf{v}_i^{\Sigma})^2}{2}\big)\mathbf{v}_i^{\Sigma} + \mathbf{q}_i^{\Sigma} - \mathbf{v}_i^{\Sigma} \cdot \mathbf{S}_i^{\Sigma}\big) \tag{10.57}$$
$$+ [\![\rho_i\big(e_i + \frac{\mathbf{v}_i^2}{2}\big)(\mathbf{v}_i - \mathbf{v}_i^{\Sigma}) \cdot \mathbf{n}_{\Sigma}]\!] + [\![\mathbf{q}_i \cdot \mathbf{n}_{\Sigma}]\!] - [\![\mathbf{v}_i \cdot \mathbf{S}_i \mathbf{n}_{\Sigma}]\!] = \rho_i^{\Sigma} \mathbf{v}_i^{\Sigma} \cdot \mathbf{b}_i^{\Sigma} + k_i^{\Sigma},$$

where k_i and k_i^{Σ} denote the rate of energy exchange between the different constituents. We assume *conservation of total energy*, hence

$$\sum_{i=1}^{N} k_i = 0, \qquad \sum_{i=1}^{N} k_i^{\Sigma} = 0. \tag{10.58}$$

Subtraction of the local balance of partial kinetic energy (as obtained from the local balance of partial mass and momentum) yields the

Partial internal energy balance

$$\partial_t(\rho_i e_i) + \operatorname{div}(\rho_i e_i \mathbf{v}_i + \mathbf{q}_i) = \nabla \mathbf{v}_i : \mathbf{S}_i - \mathbf{v}_i \cdot \mathbf{f}_i + \frac{1}{2} r_i \mathbf{v}_i^2 + k_i \quad \text{in } \Omega \setminus \Sigma, \tag{10.59}$$

$$\partial_t^{\Sigma}(\rho_i^{\Sigma} e_i^{\Sigma}) + \operatorname{div}_{\Sigma}(\rho_i^{\Sigma} e_i^{\Sigma} \mathbf{v}_i^{\Sigma} + \mathbf{q}_i^{\Sigma}) + [\![\dot{m}_i (e_i + \frac{(\mathbf{v}_i - \mathbf{v}_i^{\Sigma})^2}{2})]\!] + [\![\mathbf{q}_i \cdot \mathbf{n}_{\Sigma}]\!] \tag{10.60}$$
$$- [\![(\mathbf{v}_i - \mathbf{v}_i^{\Sigma}) \cdot \mathbf{S}_i \mathbf{n}_{\Sigma}]\!] = \nabla_{\Sigma} \mathbf{v}_i^{\Sigma} : \mathbf{S}_i^{\Sigma} - \mathbf{v}_i^{\Sigma} \cdot \mathbf{f}_i^{\Sigma} + \frac{1}{2} r_i^{\Sigma} (\mathbf{v}_i^{\Sigma})^2 + k_i^{\Sigma} \quad \text{on } \Sigma.$$

We next consider the full mixture internal energy, where the latter has to be carefully defined because there are different options. In general, the total internal energy is defined to be the remainder if the total kinetic energy is subtracted from the total energy. In a class-I context, this means to subtract $\rho\mathbf{v}^2/2$ from the total energy density, while all partial kinetic energy densities $\rho_i\mathbf{v}_i^2/2$ should be subtracted in the class-II context; cf. Bothe and Dreyer (2015). We congruously employ the latter concept, i.e. we define

$$\rho e = \sum_{i=1}^{N} \rho_i e_i, \qquad \rho^\Sigma e^\Sigma = \sum_{i=1}^{N} \rho_i^\Sigma e_i^\Sigma \tag{10.61}$$

to be the densities of the total internal energy in the bulk and on the interface. This is applied to the sum of equations (10.59) and (10.60) to obtain the

Balance of total internal energy

$$\partial_t(\rho e) + \operatorname{div}(\rho e\mathbf{v} + \tilde{\mathbf{q}}) = \sum_{i=1}^{N} \mathbf{S}_i : \nabla\mathbf{v}_i - \sum_{i=1}^{N} \mathbf{u}_i \cdot \big(\mathbf{f}_i - r_i\mathbf{v}_i + \frac{r_i}{2}\mathbf{u}_i\big) \ \text{ in } \Omega\setminus\Sigma, \tag{10.62}$$

$$\begin{aligned} &\partial_t^\Sigma(\rho^\Sigma e^\Sigma) + \operatorname{div}_\Sigma(\rho^\Sigma e^\Sigma \mathbf{v}^\Sigma + \tilde{\mathbf{q}}^\Sigma) + [\![\dot{m}\, e]\!] + [\![\tilde{\mathbf{q}}\cdot\mathbf{n}_\Sigma]\!] \\ &+ [\![\sum_{i=1}^{N} \dot{m}_i\big(\frac{(\mathbf{v}_i - \mathbf{v}_i^\Sigma)^2}{2} - \mathbf{n}_\Sigma \cdot \frac{\mathbf{S}_i}{\rho_i}\cdot\mathbf{n}_\Sigma\big)]\!] - [\![\sum_{i=1}^{N} (\mathbf{v}_i - \mathbf{v}_i^\Sigma)_{||}\cdot(\mathbf{S}_i\mathbf{n}_\Sigma)_{||}]\!] \\ &= \sum_{i=1}^{N} \mathbf{S}_i^\Sigma : \nabla_\Sigma \mathbf{v}_i^\Sigma - \sum_{i=1}^{N} \mathbf{u}_i^\Sigma \cdot \big(\mathbf{f}_i^\Sigma - r_i^\Sigma(\mathbf{v}_i^\Sigma - \frac{\mathbf{u}_i^\Sigma}{2})\big) \qquad \text{on } \Sigma \end{aligned} \tag{10.63}$$

with

$$\tilde{\mathbf{q}} := \sum_{i=1}^{N} (\mathbf{q}_i + \rho_i e_i \mathbf{u}_i), \quad \tilde{\mathbf{q}}^\Sigma = \sum_{i=1}^{N} \big(\mathbf{q}_i^\Sigma + \rho_i^\Sigma e_i^\Sigma \mathbf{u}_i^\Sigma\big). \tag{10.64}$$

We will later rewrite the latter balances, employing the final form of the heat fluxes which will only become clear after the entropy production rates have been formulated in an appropriate form.

All constitutive quantities need to be related to the independent variables by material-dependent functions, i.e. by closure relations. These functions cannot be arbitrary, but need to satisfy certain principles, namely the principle of material frame indifference and the second law of thermodynamics, i.e. the entropy principle including the entropy inequality. We focus on the latter and only refer to Müller (1985), Bothe and Dreyer (2015) and the references given there concerning the principle of material frame indifference and other principles which might be imposed.

10.4 Entropy Principle – The Second Law of Thermodynamics

We employ the entropy principle in a sharpened version, which has been introduced in Bothe and Dreyer (2015), here extended from single-phase to two-phase multicomponent fluid systems. In order to state the entropy principle in an axiomatic form, we assign to every physical quantity an associated *parity*, called *positive* (+1) or *negative* (–1). The original concept of parity for a time-dependent quantity characterizes its behavior under time reversal (see Meixner (1973)), but in the continuum setting, the symmetry of the solutions under time reversal is lost, because the partial differential equations (PDEs) describe irreversible processes. However, for studying the structure of the (unclosed) balance equations, the concept of reversibility does not play a role. This observation allows to adapt and extend the notion of parity to the balance equations of continuum mechanics. It turns out (see Bothe and Dreyer (2015) for details) that the parity of a physical quantity is obtained by assigning the factor -1 if the unit of time, 'second' (s), appears with an uneven power and +1 if it appears with even power. Here we rely on the SI base units m, kg, s, K, mol, A, cd. The unit 'ampere' (A) for the electric current does not occur in the present mixture theory[2]. If the latter unit is involved, a further factor -1 is assigned if the unit ampere appears with uneven power. For instance, the combination A s leads to even parity. For example, the parity of the densities of mass, momentum and internal energy are

$$[\rho] = \frac{\text{kg}}{\text{m}^3} \to +1, \qquad [\rho \mathbf{v}] = \frac{\text{kg}}{\text{m}^2\text{s}} \to -1, \qquad [\rho e] = \frac{\text{kg}}{\text{m s}^2} \to +1. \tag{10.65}$$

Evidently, time derivatives alter the parity, while spatial derivatives keep the parity unchanged.

Any solution of the above system of partial differential equations, composed of the balance Eqs. (10.30), (10.31), (10.40), (10.41), (10.59) and (10.60), is called a *thermodynamic process*. Here, by a solution we mean functions which satisfy the equations in a local sense, i.e. they do not form the solution of a complete initial boundary value problem. In particular, the value of a quantity and of its spatial derivative can thus be chosen independently by adjusting the time derivative appropriately. With this notion, the *second law of thermodynamics* consists of the following axioms.

(I) There are entropy/entropy-flux pairs $(\rho s, \Phi)$ and $(\rho^\Sigma s^\Sigma, \Phi^\Sigma)$ as material-dependent quantities, where ρs, $\rho^\Sigma s^\Sigma$ are objective scalars and Φ, Φ^Σ are objective vectors.

(II) The pairs $(\rho s, \Phi)$, $(\rho^\Sigma s^\Sigma, \Phi^\Sigma)$ satisfy the balance equations

$$\partial_t(\rho s) + \nabla \cdot (\rho s \mathbf{v} + \Phi) = \zeta \quad \text{in } \Omega \setminus \Sigma, \tag{10.66}$$

$$\partial_t^\Sigma(\rho^\Sigma s^\Sigma) + \nabla_\Sigma \cdot (\rho^\Sigma s^\Sigma \mathbf{v}^\Sigma + \Phi^\Sigma) + [\![\rho s(\mathbf{v} - \mathbf{v}^\Sigma) + \Phi]\!] \cdot \mathbf{n}_\Sigma = \zeta^\Sigma \quad \text{on } \Sigma, \tag{10.67}$$

[2] cf. Dreyer et al. (2018) for an extension to electrothermodynamics.

where ζ and ζ^Σ are the entropy production rates in the bulk and on the interface, respectively. The specific entropies s, s^Σ have the physical dimension $\mathrm{J\,kg^{-1}\,K^{-1}} = \mathrm{m^2 s^{-2} K^{-1}}$, hence are of positive parity. The entropy fluxes and the entropy productions thus have negative parity.

(III) Any admissible pair of entropy fluxes Φ, Φ^Σ is such that

(i) the resulting entropy production rates ζ and ζ^Σ both consist of sums of binary products according to

$$\zeta = \sum_m \mathcal{N}_m \mathcal{P}_m, \qquad \zeta^\Sigma = \sum_{m^\Sigma} \mathcal{N}^\Sigma_{m^\Sigma} \mathcal{P}^\Sigma_{m^\Sigma}, \tag{10.68}$$

where the $\mathcal{N}_m$, $\mathcal{N}^\Sigma_{m^\Sigma}$ denote objective quantities of negative parity, while $\mathcal{P}_m$, $\mathcal{P}^\Sigma_{m^\Sigma}$ refer to objective quantities of positive parity.

(ii) Each binary product corresponds to a dissipative mechanism, and the set of these must be fixed in advance. Then,

$$\mathcal{N}_m \mathcal{P}_m \geq 0 \text{ and } \mathcal{N}^\Sigma_{m^\Sigma} \mathcal{P}^\Sigma_{m^\Sigma} \geq 0 \tag{10.69}$$

for all m, m^Σ and for every thermodynamic process.

In addition to these universal axioms, we impose two more specific ones which refer to the most general constitutive models we are interested in. These are:

(IV) For the class of fluid mixtures under consideration, we restrict the dependence of the entropy according to

$$\rho s = \rho s(\rho e, \rho_1, \ldots, \rho_N) \quad \text{and} \quad \rho^\Sigma s^\Sigma = \rho^\Sigma s^\Sigma(\rho^\Sigma e^\Sigma, \rho_1^\Sigma, \ldots, \rho_N^\Sigma), \tag{10.70}$$

where ρs and $\rho^\Sigma s^\Sigma$ are *strictly concave functions* which are strictly increasing in the first argument and satisfy the principle of material frame indifference. By means of these functions, we define the *(absolute) temperature* T, T^Σ and *chemical potentials* μ_i, μ_i^Σ according to

$$\frac{1}{T} := \frac{\partial \rho s}{\partial \rho e}, \quad -\frac{\mu_i}{T} := \frac{\partial \rho s}{\partial \rho_i} \quad \text{and} \quad \frac{1}{T^\Sigma} := \frac{\partial \rho^\Sigma s^\Sigma}{\partial \rho^\Sigma e^\Sigma}, \quad -\frac{\mu_i^\Sigma}{T^\Sigma} := \frac{\partial \rho^\Sigma s^\Sigma}{\partial \rho_i^\Sigma}. \tag{10.71}$$

(V) The following *dissipative mechanisms* occur in the fluid mixtures under consideration: *mass diffusion*, *chemical reaction*, *viscous flow* (including bulk and shear viscosity) and *heat conduction* both in the bulk phases and in the interface as well as *sliding friction*, *energy transfer* and *ad- and desorption* between the two bulk phases and the interface.

Axiom (IV) implicitly contains an assumption of local thermodynamical equilibrium in order to formulate relations of the type (10.70). But note that this does not include a statement of local thermodynamic equilibrium between the interface and the adjacent bulk locations.

A comprehensive discussion of this form of the entropy principle can be found in Bothe and Dreyer (2015), Bothe (2022) and the references given there.

In order to utilize the entropy principle, we need to calculate the structure of the entropy production rates by means of the left-hand sides of the bulk and interface entropy balance equations. For this purpose, we insert the entropy representations from (10.70) into the entropy balances (10.66) and (10.67). Then, using the chain rule and (10.71), the balance equations for mass, momentum and internal energy are employed to eliminate all time derivatives. At this point, we make a general simplification by assuming the fluids to be non-polar in the sense that the stresses are symmetric, i.e.

$$\mathbf{S}_i^{\mathsf{T}} = \mathbf{S}_i \quad \text{and} \quad \left(\mathbf{S}_i^{\Sigma}\right)^{\mathsf{T}} = \mathbf{S}_i^{\Sigma} \quad \text{for } i = 1, \ldots, N. \tag{10.72}$$

Recall that we split the stresses into isotropic and traceless parts according to (10.50).

We first consider the bulk entropy production. We insert $\rho s = \rho s(\rho e, \rho_1, \ldots, \rho_N)$ into the first two terms and carry out the differentiations. Then we eliminate the resulting time derivatives by means of the balance equations (10.30) and (10.63), introducing temperature and chemical potentials (defined as in (10.71)) via the chain rule. After straightforward computations, a first representation of the bulk entropy production is

$$\begin{aligned}\zeta = \operatorname{div}\Big(\Phi - \frac{\tilde{\mathbf{q}}}{T} + \sum_{i=1}^{N} \frac{\rho_i \mu_i \mathbf{u}_i}{T}\Big) - \frac{1}{T}\Big(\rho e - T\rho s - \sum_{i=1}^{N} \rho_i \mu_i\Big) \operatorname{div} \mathbf{v} \\ -\frac{1}{T}\sum_{i=1}^{N} P_i \operatorname{div} \mathbf{v}_i + \frac{1}{T}\sum_{i=1}^{N} \mathbf{S}_i^{\circ} : \mathbf{D}_i^{\circ} - \frac{1}{T}\sum_{a=1}^{N_R} R_a \mathcal{A}_a + \tilde{\mathbf{q}} \cdot \nabla \frac{1}{T} \\ - \sum_{i=1}^{N} \mathbf{u}_i \cdot \Big(\rho_i \nabla \frac{\mu_i}{T} + \frac{1}{T}\big(\mathbf{f}_i - r_i \mathbf{v}_i + \frac{1}{2} r_i \mathbf{u}_i\big)\Big).\end{aligned} \tag{10.73}$$

In (10.73), the term $\sum_{i=1}^{N} r_i \mu_i$ has been rewritten as $\sum_{a=1}^{N_R} R_a \mathcal{A}_a$, introducing the so-called *affinities*

$$\mathcal{A}_a = \sum_{i=1}^{N} \nu_i^a M_i \mu_i. \tag{10.74}$$

This reflects the fact that each chemical reaction relates to a dissipative mechanism.

Now, for the evaluation of the entropy production in the class-II model one has to account for the fact that, while for instance $\operatorname{div} \mathbf{u}_i$ can be chosen independently of $\mathbf{u}_i$ in exploiting the entropy principle, the defining relations $\mathbf{u}_i = \mathbf{v}_i - \mathbf{v}$ lead to the constraints $\operatorname{div} \mathbf{u}_i = \operatorname{div} \mathbf{v}_i - \operatorname{div} \mathbf{v}$. These constraints are incorporated into the exploitation using Lagrange multipliers $\Lambda_i = \Lambda_i(T, \rho_1, \ldots, \rho_N)$ and adding the terms

$$\sum_{i=1}^{N} \frac{\Lambda_i}{T}(\operatorname{div} \mathbf{v}_i - \operatorname{div} \mathbf{v} - \operatorname{div} \mathbf{u}_i)$$

to the entropy production. Indeed, by Lemma 7.3.1 in Liu (2002), the entropy inequality with this addition and without constraints is equivalent to the original inequality, but evaluated under the constraints. Then, after rearrangement of terms, the entropy production reads as

$$\begin{aligned}\zeta = \operatorname{div}\Big(\Phi - \frac{\tilde{\mathbf{q}}}{T} + \sum_{i=1}^{N} \frac{(\rho_i\mu_i - \Lambda_i)\mathbf{u}_i}{T}\Big) - \frac{1}{T}\sum_{a=1}^{N_R} R_a \mathcal{A}_a + \frac{1}{T}\sum_{i=1}^{N} \mathbf{S}_i^\circ : \mathbf{D}_i^\circ \\ -\frac{1}{T}\sum_{i=1}^{N}(P_i - \Lambda_i)\operatorname{div} \mathbf{v}_i - \frac{1}{T}\Big(\rho\psi - \sum_{i=1}^{N}(\rho_i\mu_i - \Lambda_i)\Big)\operatorname{div} \mathbf{v} \\ + \tilde{\mathbf{q}} \cdot \nabla\frac{1}{T} - \sum_{i=1}^{N} \mathbf{u}_i \cdot \Big(\rho_i \nabla\frac{\mu_i}{T} - \nabla\frac{\Lambda_i}{T} + \frac{1}{T}\big(\mathbf{f}_i - r_i\mathbf{v}_i + \frac{1}{2}r_i\mathbf{u}_i\big)\Big),\end{aligned} \tag{10.75}$$

where we introduced the *free energy* $\rho\psi = \rho e - \rho s T$. In order to satisfy axiom (III-i) in the simplest possible manner, we choose the entropy flux as

$$\Phi = \frac{\tilde{\mathbf{q}}}{T} - \sum_{i=1}^{N} \frac{(\rho_i\mu_i - \Lambda_i)\mathbf{u}_i}{T}, \tag{10.76}$$

where the Λ_i will be determined next. For this purpose, we make use of axiom (III-ii), which requires $\mathcal{N}_m \mathcal{P}_m \geq 0$ for each dissipative mechanism, individually. Therefore, we can first restrict to thermodynamic processes that run with

$$R_a = 0, \quad \nabla T = 0, \quad \mathbf{u}_i = 0, \quad \mathbf{D}_i^\circ = 0$$

and still obtain information about the remaining binary products. With $\operatorname{tr}(\mathbf{D}_i) = \operatorname{div} \mathbf{v}_i$, the entropy inequality implies

$$T\zeta = -\sum_{i=1}^{N}(P_i - \Lambda_i)\operatorname{div} \mathbf{v}_i - \Big(\rho\psi + \sum_{i=1}^{N}(\Lambda_i - \rho_i\mu_i)\Big)\operatorname{div} \mathbf{v} \geq 0$$

for this restricted class of processes, where $\operatorname{div} \mathbf{v}$ and all $\operatorname{div} \mathbf{v}_i$ can be varied independently in the exploitation. For $\operatorname{div} \mathbf{v} = \operatorname{div} \mathbf{v}_i = 0$ the processes are in equilibrium, where the entropy production assumes a minimum. Independent of any closure relation to be chosen later, this implies

$$\rho\psi + \sum_{i=1}^{N}(\Lambda_i - \rho_i\mu_i) = 0, \tag{10.77}$$

as the left-hand side is a function solely of the thermodynamic state, i.e. a function of $(T, \rho_1, \ldots, \rho_N)$. Hence

$$-\sum_{i=1}^{N}(P_i - \Lambda_i)\,\mathrm{div}\,\mathbf{v}_i \geq 0, \tag{10.78}$$

where the binary product represents the rate of entropy production due to local, species-specific volume variations within the mixture. This shows that Λ_i necessarily equals the partial thermodynamical pressure of constituent A_i introduced in (10.53), i.e.

$$\Lambda_i = P_{i|\mathrm{equ}} = p_i. \tag{10.79}$$

Consequently, $\sum_{i=1}^{N} \Lambda_i = \sum_{i=1}^{N} p_i = p$ is the thermodynamical pressure of the mixture, and (10.77) becomes the *Gibbs-Duhem equation*, viz.

$$\rho\psi + p - \sum_{i=1}^{N} \rho_i\mu_i = 0. \tag{10.80}$$

Equation (10.80) also holds for general thermodynamic processes, since it only involves equilibrium quantities. It is sometimes also called *Euler* relation. Note that only a *single* Gibbs-Duhem equation results, although individual stress contributions are accounted for.

For deriving a closed PDE model with simple closure relations, we shall be guided by the fundamental relevance of the entropy derivatives, i.e. the terms $1/T$, $1/T^{\Sigma}$ and $-\mu_i/T$, $-\mu_i^{\Sigma}/T^{\Sigma}$, as the amplification factors, which connect changes in the distribution of the conserved quantities with the rate of entropy production. Knowing $\Lambda_i = p_i$, we therefore split the term $\nabla p_i/T$ in the entropy production and distribute it accordingly. This yields the final form of the reduced[3] entropy production rate, which reads as

$$\begin{aligned}\zeta = \frac{1}{T}\sum_{i=1}^{N}\mathbf{S}_i^{\circ} : \mathbf{D}_i^{\circ} - \frac{1}{T}\sum_{i=1}^{N}\pi_i\,\mathrm{div}\,\mathbf{v}_i - \frac{1}{T}\sum_{a=1}^{N_R} R_a\mathcal{A}_a + \sum_{i=1}^{N}\mathbf{q}_i\cdot\nabla\frac{1}{T} \\ - \sum_{i=1}^{N}\mathbf{u}_i\cdot\Big(\rho_i\nabla\frac{\mu_i}{T} + \frac{1}{T}(\mathbf{f}_i - r_i\mathbf{v}_i + \frac{1}{2}r_i\mathbf{u}_i - \nabla p_i) - (\rho_i e_i + p_i)\nabla\frac{1}{T}\Big).\end{aligned} \tag{10.81}$$

Inserting $\Lambda_i = p_i$ into (10.76) yields

[3] Reduced entropy production refers to the outcome after the entropy flux has been chosen.

$$\Phi = \sum_{i=1}^{N} \Big(\frac{\mathbf{q}_i}{T} + \frac{(\rho_i e_i + p_i - \rho_i \mu_i)\mathbf{u}_i}{T} \Big). \tag{10.82}$$

As a side remark, observe that even with a single entropy balance, one might now define a partial entropy via

$$\rho_i s_i := \frac{\rho_i e_i + p_i - \rho_i \mu_i}{T}. \tag{10.83}$$

Then $\rho s = \sum_{i=1}^{N} \rho_i s_i$ by (10.80), hence the total entropy flux attains the form

$$\rho s \mathbf{v} + \Phi = \sum_{i=1}^{N} \Big(\frac{\mathbf{q}_i}{T} + \rho_i s_i \mathbf{v}_i \Big). \tag{10.84}$$

One might then introduce individual entropy balances according to

$$\partial_t (\rho_i s_i) + \nabla \cdot (\rho_i s_i \mathbf{v}_i + \frac{\mathbf{q}_i}{T}) = \zeta_i, \tag{10.85}$$

in which case the second law of thermodynamics (10.66) attains the form

$$\sum_{i=1}^{N} \zeta_i \geq 0 \quad \text{for any thermodynamic process.} \tag{10.86}$$

Let us note in passing that we can achieve consistency with the classical class-I entropy flux, given as

$$\Phi = \frac{\mathbf{q}}{T} - \sum_{i=1}^{N} \frac{\rho_i \mu_i \mathbf{u}_i}{T}, \tag{10.87}$$

if we define the *heat flux* for the multi-temperature class-II model as

$$\mathbf{q} := \sum_{i=1}^{N} \big(\mathbf{q}_i + (\rho_i e_i + p_i)\mathbf{u}_i \big). \tag{10.88}$$

Defining the partial enthalpy densities according to

$$\rho_i h_i := \rho_i e_i + p_i, \tag{10.89}$$

this heat flux contains the diffusive transport of the partial enthalpies. This allows for an intuitive interpretation as a fluid element carries energy composed of a kinetic part and an enthalpic part. Employing the heat flux from (10.88) leads to an alternative form of the total internal energy balance, which for the bulk phases reads as

$$\partial_t(\rho e) + \operatorname{div}(\rho e \mathbf{v} + \mathbf{q}) = \sum_{i=1}^{N} \mathbf{S}_i^{\mathrm{irr}} : \nabla \mathbf{v}_i - p \operatorname{div} \mathbf{v}$$
$$- \sum_{i=1}^{N} \mathbf{u}_i \cdot (\mathbf{f}_i - r_i \mathbf{v}_i + \frac{r_i}{2} \mathbf{u}_i - \nabla p_i). \tag{10.90}$$

Similarly, a lengthy calculation along the same line of arguments yields the reduced interfacial entropy production rate as

$$\begin{aligned}
\zeta^\Sigma = & \frac{1}{T^\Sigma} \sum_{i=1}^{N} \mathbf{S}_i^{\Sigma,\circ} : \mathbf{D}_i^{\Sigma,\circ} - \frac{1}{T^\Sigma} \sum_{i=1}^{N} \pi_i^\Sigma \nabla_\Sigma \cdot \mathbf{v}_i^\Sigma - \frac{1}{T^\Sigma} \sum_a R_a^\Sigma \mathcal{A}_a^\Sigma + \mathbf{q}^\Sigma \cdot \nabla_\Sigma \frac{1}{T^\Sigma} \\
& - \sum_{i=1}^{N} \mathbf{u}_i^\Sigma \cdot \Big(\rho_i^\Sigma \nabla_\Sigma \frac{\mu_i^\Sigma}{T^\Sigma} + \frac{1}{T^\Sigma} (\mathbf{f}_i^\Sigma - r_i^\Sigma (\mathbf{v}_i^\Sigma - \frac{\mathbf{u}_i^\Sigma}{2}) - \nabla_\Sigma p_i^\Sigma) \Big) \\
& - \frac{1}{T^\Sigma} \sum_{i=1}^{N} (\mathbf{v}_i^+ - \mathbf{v}_i^\Sigma)_{||} \cdot (\mathbf{S}_i^{+,\mathrm{irr}} \mathbf{n}^+)_{||} - \frac{1}{T^\Sigma} \sum_{i=1}^{N} (\mathbf{v}_i^- - \mathbf{v}_i^\Sigma)_{||} \cdot (\mathbf{S}_i^{-,\mathrm{irr}} \mathbf{n}^-)_{||} \\
& - \Big(\frac{1}{T^+} - \frac{1}{T^\Sigma} \Big) (\dot{m}^{+,\Sigma} h^+ + \mathbf{q}^+ \cdot \mathbf{n}^+) - \Big(\frac{1}{T^-} - \frac{1}{T^\Sigma} \Big) (\dot{m}^{-,\Sigma} h^- + \mathbf{q}^- \cdot \mathbf{n}^-) \\
& + \sum_{i=1}^{N} \dot{m}_i^{+,\Sigma} \Big(\frac{\mu_i^+}{T^+} - \frac{\mu_i^\Sigma}{T^\Sigma} + \frac{1}{T^\Sigma} \Big(\frac{(\mathbf{v}_i^+ - \mathbf{v}_i^\Sigma)^2}{2} - \mathbf{n}^+ \cdot \frac{\mathbf{S}_i^{+,\mathrm{irr}}}{\rho_i^+} \cdot \mathbf{n}^+ \Big) \Big) \\
& + \sum_{i=1}^{N} \dot{m}_i^{-,\Sigma} \Big(\frac{\mu_i^-}{T^-} - \frac{\mu_i^\Sigma}{T^\Sigma} + \frac{1}{T^\Sigma} \Big(\frac{(\mathbf{v}_i^- - \mathbf{v}_i^\Sigma)^2}{2} - \mathbf{n}^- \cdot \frac{\mathbf{S}_i^{-,\mathrm{irr}}}{\rho_i^-} \cdot \mathbf{n}^- \Big) \Big),
\end{aligned} \tag{10.91}$$

where we have employed the constitutive relation

$$\Phi^\Sigma := \frac{\mathbf{q}^\Sigma}{T^\Sigma} - \sum_{i=1}^{N} \frac{\rho_i^\Sigma \mu_i^\Sigma \mathbf{u}_i^\Sigma}{T^\Sigma} \tag{10.92}$$

for the entropy flux. Moreover, the *interfacial heat flux* $\mathbf{q}^\Sigma$ is defined as

$$\mathbf{q}^\Sigma = \sum_{i=1}^{N} \big(\mathbf{q}_i^\Sigma + (\rho_i^\Sigma e_i^\Sigma + p_i^\Sigma) \mathbf{u}_i^\Sigma \big). \tag{10.93}$$

The first five terms of the interfacial entropy production are in full analogy to the bulk case, while the remaining once represent dissipative mechanisms due to bulk-interface transmission fluxes. For the latter, we expanded the jump brackets into the one-sided terms according to (10.15). This needs to be done in order to decompose the jump terms into two binary products, thus arriving at the postulated structure of the entropy production rate. The derivation of the reduced interfacial entropy production rate also yields the interfacial Gibbs-Duhem (or, Euler) relation, i.e.

$$\rho^\Sigma e^\Sigma + p^\Sigma - \rho^\Sigma s^\Sigma T^\Sigma = \sum_{i=1}^N \rho_i^\Sigma \mu_i^\Sigma. \tag{10.94}$$

We finally record an alternative form of the total internal energy balance for the interface, which employs the heat fluxes $\mathbf{q}^\Sigma$ from (10.93). It reads as

$$\begin{aligned} &\partial_t^\Sigma(\rho^\Sigma e^\Sigma) + \operatorname{div}_\Sigma(\rho^\Sigma e^\Sigma \mathbf{v}^\Sigma + \mathbf{q}^\Sigma) + [\![\dot m\, h + \mathbf{q}\cdot\mathbf{n}_\Sigma]\!] \\ &+ [\![\sum_{i=1}^N \dot m_i \big(\frac{(\mathbf{v}_i - \mathbf{v}_i^\Sigma)^2}{2} - \mathbf{n}_\Sigma\cdot\mathbf{S}_i^{\rm irr}\cdot\mathbf{n}_\Sigma\big)]\!] - [\![\sum_{i=1}^N (\mathbf{v}_i - \mathbf{v}_i^\Sigma)_{||}\cdot(\mathbf{S}^{\rm irr}\mathbf{n}_\Sigma)_{||}]\!] \\ &= \sum_{i=1}^N \mathbf{S}_i^{\Sigma,\rm irr} : \nabla_\Sigma \mathbf{v}_i^\Sigma - p^\Sigma \operatorname{div}_\Sigma \mathbf{v}^\Sigma - \sum_{i=1}^N \mathbf{u}_i^\Sigma\cdot\Big(\mathbf{f}_i^\Sigma - r_i^\Sigma\big(\mathbf{v}_i^\Sigma - \frac{\mathbf{u}_i^\Sigma}{2}\big) - \nabla_\Sigma p_i^\Sigma\Big). \end{aligned} \tag{10.95}$$

The remaining step is the modeling of the constitutive quantities via material-depend functions, where the entropy principle gives restrictions to the constitutive relations for these dissipative mechanisms. Recall that the axiom (III-ii) requires all individual contributions to be non-negative.

10.5 Closure Relations

We group the set of constitutive quantities into ‘bulk constitutive quantities’, ‘interface constitutive quantities’ and ‘transmission constitutive quantities’. We start with

Closure relations for bulk constitutive quantities.

The first two terms in the bulk entropy production rate (10.81) refer to *viscous stresses*, where we employ a linear (in the co-factors) closure. The traceless (deviatoric) part of the partial stress is thus modeled as

$$\mathbf{S}_i^\circ = 2\sum_{k=1}^N \eta_{ik}\mathbf{D}_k^\circ, \tag{10.96}$$

while the irreversible contribution to the partial pressure is modeled as

$$\pi_i = -\sum_{k=1}^N \lambda_{ik} \operatorname{div} \mathbf{v}_k. \tag{10.97}$$

The matrices $[\eta_{ik}]$ and $[\lambda_{ik}]$ are symmetric and positive (semi-)definite. All phenomenological coefficients depend on the basic thermodynamic variables which are, after a Legendre transformation, $T, \rho_1, \ldots, \rho_N$ inside the bulk phases. The deviatoric part $\mathbf{S}_i^\circ$ models the partial stress due to shear, where the η_{ik} are the partial shear viscosities. The π_i model viscous pressure contributions due to volume variations, where the material-dependent parameters λ_{ik} are the partial bulk (or, volume) viscosities.

Next, we recall the closure of the mass production rates according to Bothe and Dreyer (2015), i.e. the modeling of the *chemical reaction rates*. This needs to guarantee non-negativity of the corresponding contribution in the entropy production rate, i.e. it must ensure that

$$\zeta_R := -\sum_{a=1}^{N_R} R_a \frac{\mathcal{A}_a}{RT} =: -\langle \mathbf{R}, \mathbf{A} \rangle \geq 0 \quad \text{with } R_a = R_a^f - R_a^b, \tag{10.98}$$

where it is convenient to divide the affinities by RT (with R the universal gas constant) in order to obtain dimensionless co-factors associated to the (molar-based) chemical potentials. Since chemical reactions are activated processes which often run far away from equilibrium, a linear (in the chemical affinities) closure for R_a is not appropriate. Instead, we exploit the structure $R_a = R_a^f - R_a^b$, where the unidirectional rates satisfy $R_a^f, R_a^b > 0$. Note that $R_a^f, R_a^b \geq 0$ holds by definition, but also the strict inequalities are valid. The latter follows from the fact that ρ_i cannot vanish locally, unless the constituent A_i becomes extinct. But as we include the forward and backward direction for each chemical reaction, and also account for non-degenerate mass diffusion, this cannot happen. Therefore, we can employ the nonlinear closure[4]

$$\ln \frac{R_a^f}{R_a^b} = -\gamma_a \frac{\mathcal{A}_a}{RT} \quad \text{with } \gamma_a > 0. \tag{10.99}$$

This closure implies a non-negative contribution to the entropy production by chemical reactions in the bulk phases, since the logarithm is monotone increasing. Because of the *strict* monotonicity of the logarithm, the reactive contribution to the entropy production only vanishes if *all reaction are separately in equilibrium*, i.e. all forward and corresponding backward rates coincide. This is an instance of the *principle of detailed balance*, called Wegscheider's condition in the context of chemical reaction kinetics. Note that one of the rates—either for the forward or the backward path—still needs to be modeled, while the form of the other one then follows from (10.99).

The standard closure for chemical reaction rates, used in so-called *elementary reactions* that do not involve intermediate steps, is referred to as *mass action kinetics*. There, based on the idea of 'reactive collisions', the rate for the forward reaction, say, is modeled as $R_a^f = k_a^f \prod_i x_i^{\alpha_i^a}$ with rate constants $k_a^f > 0$; recall that $x_i = c_i/c$

[4] This form of the closure can be traced back to work by Marcelin (1910), Kohnstamm and, later, by De Donder; cf. Lengyel (1989). I am grateful to V. Giovangigli for pointing out the reference Marcelin (1910).

denotes the molar fractions. To obtain the backward rate via (10.99), the chemical potentials need to be modeled by material functions since they appear in the affinities. For the important class of so-called *ideal mixtures*, the chemical potentials obey the relations

$$\mu_i(T, p, x_k) = g_i(T, p) + \frac{RT}{M_i} \log x_i \quad \text{for } i = 1, \ldots, N, \tag{10.100}$$

where $g_i(T, p)$ is the Gibbs free energy of the pure component A_i under the temperature and pressure of the mixture. The variable x_k inside μ_i in (10.100) stands for the full composition $(x_1, \ldots, x_N)$, but constrained to satisfy $\sum_k x_k = 1$. As only the single molar fraction x_i appears on the right-hand side, this constraint is not effective. This class of mixtures defined by (10.100) is motivated by (ideal) mixtures of ideal gases, where g_i is explicitly given as a function of T, p. For ideal mixtures, it follows that

$$R_a^b = R_a^f \exp\Big(\frac{1}{RT} \sum_{i=1}^N g_i(T, p)\Big) \prod_i x_i^{\nu_i^a}. \tag{10.101}$$

Following the above mentioned idea of 'reactive collisions', the forward rate is

$$R_a^f = k_a^f \prod_i x_i^{\alpha_i^a} \quad \text{with } k_a^f > 0. \tag{10.102}$$

This leads to a net reaction rate according to

$$R_a = k_a^f \prod_i x_i^{\alpha_i^a} - k_a^b \prod_i x_i^{\beta_i^a} \quad \text{with } \frac{k_a^b}{k_a^f} = \exp(\frac{1}{RT} \sum_{i=1}^N g_i(T, p)), \tag{10.103}$$

showing also the consistency of the present framework in that 'forward' and 'backward' directions can be exchanged. Note also that the 'rate constants' $k_a^f, k_a^b > 0$ will depend on the primitive variables, in fact they depend strongly on the temperature. It is important to observe that only one of k_a^f or k_a^b can be modeled in this manner, while the other one is then determined by (10.103).

The result (10.103) is restricted to the class of ideal mixtures. In the case of non-ideal mixtures, the literature usually starts from (10.103) as well, but with activities instead of molar fractions; for a rather exhaustive review see Pekar (2005); cf. also Lengyel (1989). Note that the chemical potentials are then rewritten in the form

$$\mu_i(T, p, x_k) = g_i(T, p) + \frac{RT}{M_i} \log a_i \quad \text{for } i = 1, \ldots, N. \tag{10.104}$$

As the so-called activities a_i are allowed to depend on all primitive variables, the introduction of activities just replaces one unknown function by another one. It is nevertheless widely used in chemical engineering. With this notation, the general (non-ideal) case leads to a reaction rate of the form

$$R_a = k_a^f \prod_i a_i^{\alpha_i^a} - k_a^b \prod_i a_i^{\beta_i^a} \quad \text{with } \frac{k_a^b}{k_a^f} = \exp(\frac{1}{RT} \sum_{i=1}^N g_i(T,p)). \tag{10.105}$$

Thus the same structure as in (10.103) results—one motivation to work with activities.

Cross-effects between different chemical reactions can also be introduced on the basis of the nonlinear closure for the rate functions from (10.99). A detailed derivation within the framework of the entropy principle in the form as given above was provided in Bothe and Dreyer (2015). We recall the final result, viz.

$$R_a = \sum_{b=1}^{N_R} \alpha_b L_{ab} \Big(1 - \exp\big(\sum_{c=1}^{N_R} L_{cb} \frac{\mathcal{A}_c}{RT}\big)\Big) \tag{10.106}$$

with $\alpha_b > 0$ and non-negative L_{ab} such that the matrix $L := [L_{ab}]_{a,b=1}^{N_R}$ is symmetric and positive (semi-)definite, where all coefficients are functions of $(T, \rho_1, \ldots, \rho_N)$. Note that

$$\zeta_R = -\langle L^{-1}\mathbf{R}, L^{\mathsf{T}}\mathbf{A}\rangle = \sum_{b=1}^{N_R} \alpha_b \big(e^{(L^{\mathsf{T}}\mathbf{A})_b} - 1\big)(L^{\mathsf{T}}\mathbf{A})_b \geq 0 \;\Leftrightarrow\; \sum_{b=1}^{N_R} (L^{\mathsf{T}}\mathbf{A})_b^2 \geq 0. \tag{10.107}$$

Concerning *mass diffusion and heat conduction*, we first specialize to the case without chemical reactions, i.e. we first restrict to processes such that $R_a = 0$ for all a. The relevant contribution in the entropy production rate then simplifies to read

$$\zeta_{\text{TD}} = \sum_{i=1}^N \mathbf{q}_i \cdot \nabla\frac{1}{T} - \sum_{i=1}^N \mathbf{u}_i \cdot \Big(\rho_i \nabla\frac{\mu_i}{T} + \frac{1}{T}(\mathbf{f}_i - \nabla p_i) - \rho_i h_i \nabla\frac{1}{T}\Big). \tag{10.108}$$

Observe that the term $\rho_i h_i \mathbf{u}_i \nabla\frac{1}{T}$ naturally fits to both binary products. It hence induces a natural cross-coupling between mass diffusion and heat conduction. Therefore, both phenomena should be treated as a single 'super-mechanism'—the so-called thermo-diffusion. Consequently, we will account for general cross-coupling between the two effects, which can be introduced by entropy-neutral mixing, a concept that has been developed in Bothe and Dreyer (2015). For this purpose, we introduce a zero-addition according to

$$\zeta_{\text{TD}} = \sum_{i=1}^N (\mathbf{q}_i - d_i^T \mathbf{u}_i) \cdot \nabla\frac{1}{T} - \sum_{i=1}^N \mathbf{u}_i \cdot \Big(\rho_i \nabla\frac{\mu_i}{T} + \frac{\mathbf{f}_i - \nabla p_i}{T} - (\rho_i h_i + d_i^T)\nabla\frac{1}{T}\Big). \tag{10.109}$$

In order to apply a standard closure strategy for the second term, which has to account for the fact that the co-factors are not all independent, we add $\lambda \sum_i y_i \mathbf{u}_i \nabla\frac{1}{T} = 0$ with a Lagrange multiplier $\lambda \in \mathbf{R}$. This yields

$$\zeta_{\rm TD} = \sum_{i=1}^{N} (\mathbf{q}_i - d_i^T \mathbf{u}_i) \cdot \nabla \frac{1}{T} - \sum_{i=1}^{N} \mathbf{u}_i \cdot \Big(\rho_i \nabla \frac{\mu_i}{T} + \frac{\mathbf{f}_i - \nabla p_i}{T} - (\rho_i h_i + d_i^T - y_i \lambda) \nabla \frac{1}{T} \Big). \tag{10.110}$$

We then choose λ such that the co-factors of the $\mathbf{u}_i$ add up to zero, which leads to $\lambda = \sum_{i=1}^N d_i^T$, where the Gibbs-Duhem relation (10.80) has been used. Now, replacing the d_i^T by $d_i^T - y_i \lambda$, we see that the introduction of entropy-neutral mixing between the two binary products in $\zeta_{\rm TD}$ leads to (10.109), but with the additional constraint

$$\sum_{i=1}^{N} d_i^T = 0. \tag{10.111}$$

Now, we are in position to apply the closure in the strict form according to axiom (III), in which every dissipative mechanism is treated separately. Linear closure yields

$$\sum_{i=1}^{N} (\mathbf{q}_i - d_i^T \mathbf{u}_i) = \alpha \, \nabla \frac{1}{T}$$

with a coefficient $\alpha(T, \rho_1, \ldots, \rho_N) \geq 0$, hence the heat flux from (10.88) ís modeled as

$$\mathbf{q} = \alpha \, \nabla \frac{1}{T} + \sum_{i=1}^{N} (\rho_i h_i + d_i^T) \mathbf{u}_i. \tag{10.112}$$

The closure of the momentum exchange terms $\mathbf{f}_i$, intimately related to the diffusion velocities $\mathbf{u}_i$ and, hence, to mass diffusion, further assumes binary interactions between the species, cf. Bothe and Dreyer (2015) for a detailed derivation. Here, let us only record the result, viz.

$$\frac{1}{T} \mathbf{f}_i = -\rho_i \nabla \frac{\mu_i}{T} + \frac{1}{T} \nabla p_i + (\rho_i h_i + d_i^T) \nabla \frac{1}{T} - \sum_j f_{ij} \rho_i \rho_j (\mathbf{v}_i - \mathbf{v}_j), \tag{10.113}$$

where the $f_{ij}(T, \rho_1, \ldots, \rho_N)$ for $i \neq j$ are strictly positive with $f_{ij} = f_{ji}$.

If chemical reactions occur, the $\mathbf{f}_i$ are no longer objective vectors, but the $\mathbf{f}_i - r_i \mathbf{v}_i$ are the constitutive quantities. For this case, an elaborate derivation has been introduced in Bothe and Dreyer (2015), which for the first time gave a consistent treatment of chemical reactions in the class-II framework for single-phase multicomponent fluid systems. The resulting closure law for the interaction forces is

$$\frac{1}{T} (\mathbf{f}_i - r_i \mathbf{v}_i) = -\rho_i \nabla \frac{\mu_i}{T} + \frac{1}{T} \nabla p_i + (\rho_i h_i + d_i^T) \nabla \frac{1}{T} - \sum_{j=1}^{N} (f_{ij} \rho_i \rho_j + C_{ij})(\mathbf{v}_i - \mathbf{v}_j) \tag{10.114}$$

with symmetric $f_{ij}(T, \rho_1, \ldots, \rho_N) > 0$, where the *chemical interaction matrix* has the entries

$$C_{ij} = \frac{1}{T} \sum_{a=1}^{N_R} \frac{M_i M_j}{\sum_k \alpha_k^a M_k} \big(R_a^f \beta_i^a \alpha_j^a + R_a^b \alpha_i^a \beta_j^a \big). \tag{10.115}$$

Note that the matrix $[C_{ij}]$ is, in general, *not* symmetric. But it attains a symmetric form if the system is in *detailed chemical reaction equilibrium*, a notion meaning that $R_a^f = R_a^b$ for *all* reactions $a = 1, \ldots, N_R$. In Bothe and Dreyer (2015), besides a motivation of the structure of the C_{ij}, it is also shown that this choice is consistent with the entropy principle from above, especially with axiom (III-ii).

Let us note in passing, that the cross-effects modeled by inclusion of the coupling terms $d_i^T \mathbf{u}_i \nabla \frac{1}{T}$ account for the Soret effect (also called Ludwig-Soret effect), i.e. the relative motion of particles due to a temperature gradient, and the Dufour effect, i.e. a diffusive contribution to the heat flux. For a historical review see Rahman and Saghir (2014).

Closure relations for interfacial constitutive quantities.

In full analogy to the bulk closure relations we obtain the following constitutive equations. For the traceless part of the partial interface stresses:

$$\mathbf{S}_i^{\Sigma,\circ} = 2 \sum_{k=1}^{N} \eta_{ik}^{\Sigma} \mathbf{D}_k^{\Sigma,\circ}. \tag{10.116}$$

For the irreversible contribution to the partial interface pressures:

$$\pi_i^{\Sigma} = - \sum_{k=1}^{N} \lambda_{ik}^{\Sigma} \nabla_{\Sigma} \cdot \mathbf{v}_k^{\Sigma}. \tag{10.117}$$

The matrices $[\eta_{ik}^{\Sigma}]$ and $[\lambda_{ik}^{\Sigma}]$ are symmetric and positive (semi-)definite. All phenomenological coefficients depend on the basic thermodynamic variables which are, after a Legendre transformation, $T^{\Sigma}, \rho_1^{\Sigma}, \ldots, \rho_N^{\Sigma}$. The $\mathbf{S}_i^{\Sigma,\circ}$ are the partial stresses due to shear inside the interface, where the η_{ik}^{Σ} are the partial interface shear viscosities. The π_i^{Σ} are the viscous pressure contributions due to area variations inside the interface, where the material-dependent parameters λ_{ik}^{Σ} are the partial dilatational viscosities. The interface shear and dilatational viscosities are also referred to as *intrinsic surface viscosities* in order to distinguish them from apparent viscosities that can be induced by Marangoni-type effects due to inhomogeneous interfacial tension. At this point, it should be noted that the interfacial tension, usually denoted as γ in the physico-chemical interface sciences and as σ in the hydrodynamical theory of capillary flows, is related to the interface pressure via

$$\gamma = \sigma = -p^{\Sigma}.$$

Let us also note in passing that in the physico-chemical interface sciences, the notion of interface (or surface) pressure often refers to the difference $\gamma^{\mathrm{clean}} - \gamma^{\mathrm{contam}}$ between the interfacial tension of the clean interface and that of the interface with, e.g., surface active agents adsorbed to it.

Partial interfacial viscosities have not been studied so far. In the class-I interface fluid model with single (barycentric) velocity, a closure according to (10.116) and (10.117) gives rise to the so-called *Boussinesq surface fluid*, having linear 'Newtonian' rheology. The resulting Boussinesq surface stress (Scriven (1960), Slattery et al. (2007), cf. also Bothe and Prüss (2010)) reads as

$$\mathbf{S}^{\Sigma} = (-p^{\Sigma} + (\lambda^{\Sigma} - \eta^{\Sigma})\nabla_{\Sigma} \cdot \mathbf{v}^{\Sigma})\,\mathbf{I}_{\Sigma} + 2\eta^{\Sigma}\,\mathbf{D}^{\Sigma} \tag{10.118}$$

with the interface dilatational and shear viscosities $\lambda^{\Sigma}, \eta^{\Sigma} \geq 0$.

Closure relations for the chemical reactions rates with cross effects between the different interface reactions:

$$R_a^{\Sigma} = \sum_{b=1}^{N_R} \alpha_b^{\Sigma} L_{ab}^{\Sigma}\Big(1 - \exp\big(\sum_{c=1}^{N_R} L_{cb}^{\Sigma} \frac{\mathcal{A}_c^{\Sigma}}{RT}\big)\Big) \tag{10.119}$$

with $\alpha_b^{\Sigma} > 0$ and non-negative L_{ab}^{Σ} such that the matrix $L^{\Sigma} := [L_{ab}^{\Sigma}]_{a,b=1}^{N_R^{\Sigma}}$ is symmetric and positive (semi-)definite, where all coefficients are functions of $(T^{\Sigma}, \rho_1^{\Sigma}, \ldots, \rho_N^{\Sigma})$.

Closure relation for the interfacial heat flux:

$$\mathbf{q}^{\Sigma} = \alpha^{\Sigma}\,\nabla_{\Sigma} \frac{1}{T^{\Sigma}} + \sum_{i=1}^{N} (\rho_i^{\Sigma} h_i^{\Sigma} + d_i^{T,\Sigma})\mathbf{u}_i^{\Sigma}, \tag{10.120}$$

with a coefficient $\alpha^{\Sigma}(T^{\Sigma}, \rho_1^{\Sigma}, \ldots, \rho_N^{\Sigma}) \geq 0$, where the interfacial thermo-diffusion coefficients $d_i^{T,\Sigma}$ satisfy

$$\sum_{i=1}^{N} d_i^{T,\Sigma} = 0. \tag{10.121}$$

In (10.120), $h_i^{\Sigma} = e_i^{\Sigma} + p_i^{\Sigma}/\rho_i^{\Sigma}$ denotes the specific partial interfacial enthalpy.

Closure relations for the interaction forces between the different constituents, where we focus on the case without chemical reactions for simplicity of presentation:

$$\frac{1}{T^{\Sigma}}\mathbf{f}_i^{\Sigma} = -\rho_i^{\Sigma}\nabla_{\Sigma}\frac{\mu_i^{\Sigma}}{T^{\Sigma}} + \frac{\nabla_{\Sigma} p_i^{\Sigma}}{T^{\Sigma}} + (\rho_i^{\Sigma} h_i^{\Sigma} + d_i^{T,\Sigma})\nabla_{\Sigma}\frac{1}{T^{\Sigma}} - \sum_j f_{ij}^{\Sigma} \rho_i^{\Sigma} \rho_j^{\Sigma} (\mathbf{v}_i^{\Sigma} - \mathbf{v}_j^{\Sigma}), \tag{10.122}$$

where $f_{ij}^{\Sigma}(T^{\Sigma}, \rho_1^{\Sigma}, \ldots, \rho_N^{\Sigma}) > 0$ (for $i \neq j$) is symmetric. The case with interfacial chemistry can be treated in full analogy to (10.114) and (10.115).

Closure relations for bulk-interface transmission constitutive quantities.

The remaining binary products in the interfacial entropy production rate from (10.91) are associated with the transfer of partial mass, partial momentum and (internal) energy between one of the bulk phases and the interface.

The 6th and 7th binary products in (10.91) correspond to entropy production due to momentum transfer between the bulk phases and the interface. This is related to dissipation by friction in case of slippage of (at least one of) the bulk phases against the interface. Linear (in the co-factors) closure yields individual one-sided Navier slip conditions according to

$$\alpha_i^{\pm}(\mathbf{v}_i^{\pm} - \mathbf{v}_i^{\Sigma})_{||} + (\mathbf{S}_i^{\pm} \cdot \mathbf{n}^{\pm})_{||} = 0 \quad \text{with} \quad \alpha_i^{\pm} \geq 0. \tag{10.123}$$

There are two common special choices: the first one is $\alpha_i^{\pm} = 0$, which leads to a free (or, perfect) slip condition, i.e. $(\mathbf{S}_i^{\pm} \cdot \mathbf{n}^{\pm})_{||} = 0$ at Σ. The second one corresponds to the limit $\alpha_i^{\pm} \to \infty$, which yields continuity of the tangential velocity at Σ, i.e. $[\![\mathbf{v}_{i,||}]\!] = 0$, and $\mathbf{v}_{i,||}^{\Sigma} = \mathbf{v}_{i,||}$. In the general case of $\alpha_i^{\pm} \in (0, \infty)$, the relations (10.123) link the one-sided limits of the bulk velocity fields to the tangential interface velocities, where the latter are determined from solving a PDE system, containing a closed version of (10.41).

Let us note in passing that a more general, coupled closure is or course possible. If cross-effects inside the momentum transmission are accounted for, a corresponding generalisation reads as

$$\sum_{k=1}^{N} \alpha_{ik}^{\pm}(\mathbf{v}_k^{\pm} - \mathbf{v}_k^{\Sigma})_{||} + (\mathbf{S}_i^{\pm} \cdot \mathbf{n}^{\pm})_{||} = 0 \qquad (i = 1, \ldots, N), \tag{10.124}$$

where the matrix $[\alpha_{ik}]$ or slip coefficients is symmetric and positive (semi-)definite.

The 8th and 9th binary products in (10.91) refer to entropy production due to energy transmission from one of the bulk phases to the interface. This is accompanied with a jump in the temperature, i.e. the interface temperature T^{Σ} is, in general, different from the one-sided limits of the bulk temperature; see the review paper Persad and Ward (2016) and the references given there. Linear closure yields the one-sided closure relations

$$\frac{1}{T^{\Sigma}} - \frac{1}{T^{\pm}} = \beta^{\pm}\big(\dot{m}^{\pm,\Sigma}h^{\pm} + \mathbf{q}^{\pm} \cdot \mathbf{n}^{\pm}\big) \quad \text{with} \quad \beta^{\pm} \geq 0, \tag{10.125}$$

where $h^{\pm} = e^{\pm} + p^{\pm}/\rho^{\pm}$ are the specific bulk enthalpies. A common special choice is $\beta^{\pm} = 0$, implying continuity of the temperature at Σ, while $\beta^{\pm} \to \infty$ leads to an energetically isolating (adiabatic) interface.

The 10th and 11th binary products in (10.91) refer to entropy production due to one-sided mass transfer between the bulk phases and the interface. If cross-effects between the different species transfer processes are ignored, a linear closure gives the relations

$$\dot{m}_i^{\pm,\Sigma} = \gamma_i^{\pm}\Big(\frac{\mu_i^{\pm}}{T^{\pm}} - \frac{\mu_i^{\Sigma}}{T^{\Sigma}} + \frac{1}{T^{\Sigma}}\big(\frac{(\mathbf{v}_i^{\pm} - \mathbf{v}_i^{\Sigma})^2}{2} - \mathbf{n}^{\pm} \cdot \frac{\mathbf{S}_i^{\pm,\text{visc}}}{\rho_i^{\pm}} \cdot \mathbf{n}^{\pm}\big)\Big) \tag{10.126}$$

with $\gamma_i^{\pm} \geq 0$. As already noted above in the context of chemical reactions, one of the rates—either the ad- or the desorption rate—has to be modeled based on a micro-theory or experimental knowledge. Then the other rate follows from (10.126), up to a state-dependent coefficient. In the limiting cases, as $\gamma_i^{\pm} \to \infty$, one obtains

$$\frac{\mu_i^{\pm}}{T^{\pm}} = \frac{\mu_i^{\Sigma}}{T^{\Sigma}} - \frac{1}{T^{\Sigma}}\big(\frac{(\mathbf{v}_i^{\pm} - \mathbf{v}_i^{\Sigma})^2}{2} - \mathbf{n}^{\pm} \cdot \frac{\mathbf{S}_i^{\pm,\text{visc}}}{\rho_i^{\pm}} \cdot \mathbf{n}^{\pm}\big). \tag{10.127}$$

These limiting cases correspond to vanishing (one-sided) interfacial resistance against mass transfer. If the temperature is assumed to be continuous, an often imposed assumption, and if the kinetic and viscous terms can be neglected compared to the chemical potentials, then (10.127) yields

$$\mu_i^{+} = \mu_i^{-} = \mu_i^{\Sigma}; \tag{10.128}$$

cf. Bothe and Fleckenstein (2013) for an estimation of the strength of the different contributions in (10.127). Let us note that the first identity in (10.128), which is usually employed to describe the concentration jump at the interface, is often associated with 'chemical equilibrium' between the bulk phases. This identity has to be evaluated with care, since not only the respective one-sided limits of bulk compositions enter the chemical potentials, but also the different pressures, so that, in detail, it reads as

$$\mu_i^{+}(T, p^{+}, x_1^{+}, \ldots, x_N^{+}) = \mu_i^{-}(T, p^{-}, x_1^{-}, \ldots, x_N^{-}). \tag{10.129}$$

Since the pressure also has a jump at Σ and the height of this jump depends on the curvature, the pressure dependence of the chemical potentials introduces a curvature influence into (10.129), implying in particular that smaller bubbles display an increased solubility. This is the Gibbs-Thomson effect, while in the context of thermally driven phase transfer, the analogous effect is described by the Kelvin equation; see Buff (1956), Powles (1985).

Equation (10.126) shows that mass transfer is driven by chemical potential differences but, in addition, a kinetic term and viscous forces contribute to the driving force. In mass transfer applications, the systems are usually far from equilibrium in particular concerning the mass transfer processes. Then, in analogy to the closure of chemical reaction rates, a non-linear closure turns out to be more appropriate.

Actually, together with the conception of mass transfer as a series of two one-sided bulk-interface transfer processes, an enriched model for mass transfer results, which includes the local effects of adsorbed surface active agents. This model was first introduced by the present author in Bothe (2022) and goes far beyond existing mass transfer models. In the current framework of a multi-velocity model, this will be elaborated on in more detail in the subsequent section. We refrain from considering cross-couplings between mass transfer of different constituents, the formulation of which would follow along the same lines as already explained above.

Let us finally check that all constitutive quantities as listed above have been treated. This is clear for all intrinsic bulk and interface quantities, hence we only need to check the bulk-interface transfer terms. In this regard, we first obtain the $\dot{m}_i^{\pm}$ from (10.126), say, or a more refined closure for this term as given below. Summing up the $\dot{m}_i^{\pm}$, this also yields $\dot{m}^{\pm}$. Next, the tangential components of $\mathbf{S}^{\pm}\mathbf{n}_\Sigma$ follow from (10.123), unless it is replaced by the condition $\mathbf{v}_{||}^{\pm} = \mathbf{v}_{||}^{\Sigma}$ in the limiting case of (one-sided) no-slip between bulk and interface. Finally, in the general case of interfacial temperature jumps, $\mathbf{q}^{\pm} \cdot \mathbf{n}_\Sigma$ then follows from (10.125), since all other quantities therein are either thermodynamic state variables or belong to the primitive variables, i.e. the balanced quantities.

10.6 Mass Transfer Influenced by Adsorbed Species

In order to understand the influence of surface coverage by surfactant onto the mass transfer of another chemical species, we have to employ a general closure like (10.126). However, in mass transfer applications, the system is not in local equilibrium at the fluid interface: while the interface is assumed to be in local equilibrium in itself as an interfacial phase, i.e. in the tangential directions, the thermodynamic state in adjacent locations in the bulk phases can differ significantly. Hence, in normal direction across the interface, the system can be far away from local equilibrium. Consequently, a linear closure of mass transfer processes is inappropriate. We therefore employ a non-linear, logarithmic closure in analogy to chemical reactions. This leads to constitutive relations for the bulk-interface mass transfer that are consistent to experimentally confirmed closures. A first version of the derivation below has been given in Bothe (2015).

We split the one-sided mass transfer terms into an adsorption (superscript 'ad') and a desorption ('de') term, according to

$$\dot{m}_i^{+,\Sigma} = s_i^{ad,+} - s_i^{de,+}, \quad \dot{m}_i^{-,\Sigma} = s_i^{ad,-} - s_i^{de,-}. \tag{10.130}$$

The crucial point is that the transfer of A_i from bulk phase $+$ to bulk phase $-$, say, is modeled as a series of two sorption processes, one from phase $+$ onto the interface, followed by another sorption process from Σ to bulk phase $-$. This is consistent with the assumption that the interface, representing a thin transition layer, will also carry the transfer component A_i with positive interfacial mass density ρ_i^{Σ}, cf. Fig. 10.1.

Hence we need to model the one-sided sorption rates, using (non-linear) constitutive relations.

For technical simplicity, we neglect the kinetic and viscous terms inside the mass transfer entropy production contribution, i.e. we exploit the reduced binary products

$$\zeta^{\pm}_{\text{TRANS}} = \sum_{i=1}^{N} \left(s_i^{ad,\pm} - s_i^{de,\pm}\right)\left(\frac{\mu_i^{\pm}}{T^{\pm}} - \frac{\mu_i^{\Sigma}}{T^{\Sigma}}\right) \tag{10.131}$$

to derive the closure relations, ignoring also direct cross-effects between different species. We hence let

$$\ln \frac{s_i^{ad,\pm}}{s_i^{de,\pm}} = a_i^{\pm}\left(\frac{\mu_i^{\pm}}{RT^{\pm}} - \frac{\mu_i^{\Sigma}}{RT^{\Sigma}}\right) \quad \text{with} \quad a_i^{\pm} \geq 0. \tag{10.132}$$

As mentioned above in the context of chemical reactions, one of the rates—either the ad- or the desorption rate—has to be modeled based on a micro-theory or experimental knowledge. Then the other rate has to obey (10.132). Below, we let $a_i^{\pm} = 1$ for simplicity as this already suffices to obtain an enriched mass transfer model.

We first consider the case of a soluble surfactant which, for simplicity, is only present in Ω^+, where we suppress the bulk phase index, and on Σ. Desorption is often more easy to model, as there is more space available in the adjacent bulk phase then on the interface. The simplest realistic rate function is

$$s_i^{de} = k_i^{de} x_i^{\Sigma} \tag{10.133}$$

with the interfacial molar fraction $x_i^{\Sigma} = c_i^{\Sigma}/c^{\Sigma}$, where $c_i^{\Sigma} = \rho_i^{\Sigma}/M_i$ and $c^{\Sigma} = \sum_{i=1}^{N} c_i^{\Sigma}$. According to (10.132) with $a_i^{\pm} = 1$, the associated adsorption rate is

$$s_i^{ad} = k_i^{de} x_i^{\Sigma} \exp\left(\frac{\mu_i^{\pm}}{RT^{\pm}} - \frac{\mu_i^{\Sigma}}{RT^{\Sigma}}\right). \tag{10.134}$$

To obtain a concrete model, we assume ideal mixtures both in the bulk and on the interface, i.e. (we include all x_k as independent variables, where $\sum_{k=1}^{N} x_k = 1$ is implicitly assumed)

$$\mu_i^{\pm}(T, p, x_k) = g_i^{\pm}(T, p) + RT \ln x_i^{\pm} \tag{10.135}$$

with $g_i^{\pm}(T, p)$ denoting the bulk Gibbs free energy of component A_i under the temperature and pressure of the mixture. Analogously, we employ

$$\mu_i^{\Sigma}(T^{\Sigma}, p^{\Sigma}, x_k^{\Sigma}) = g_i^{\Sigma}(T^{\Sigma}, p^{\Sigma}) + RT^{\Sigma} \ln x_i^{\Sigma} \tag{10.136}$$

with $g_i^{\Sigma}(T^{\Sigma}, p^{\Sigma})$ denoting the interface Gibbs free energy of adsorbed component A_i^{Σ} under the interface temperature and pressure of the mixture. Insertion of the chemical potentials into (10.134) yields

$$s_i^{ad} = k_i^{de} \exp\big(\frac{g_i^+}{RT} - \frac{g_i^\Sigma}{RT^\Sigma}\big) x_i^+ =: k_i^{ad} x_i^+, \tag{10.137}$$

where k_i^{ad} depends in particular on the surface pressure. Together, (10.133) and (10.137) yield the simplest ad- and desorption rates, leading to the so-called Henry isotherm; see Kralchevsky and Danov (2015) for more details on sorption isotherms of surfactants.

Next, we consider a transfer component A_i, like a dissolving gas, which does not accumulate at the interface as a surfactant does, but nevertheless needs to pass through the transmission zone between the bulk phases, mathematically represented by the sharp interface. Rewriting the interfacial mass balance (10.31) for A_i^Σ in non-dimensional form, we obtain

$$\frac{l_{\mathrm{n}}^{\mathrm{ref}}}{l_{\mathrm{tan}}^{\mathrm{ref}}} \big(\partial_{t^*}^\Sigma \rho_i^{\Sigma,*} + \nabla_\Sigma^* \cdot (\rho_i^{\Sigma,*} \mathbf{v}_i^{\Sigma,*})\big) + [\![\dot{m}_i^*]\!] = 0 \text{ on } \Sigma, \tag{10.138}$$

where $l_{\mathrm{tan}}^{\mathrm{ref}}$ denotes a characteristic length scale along which ρ_i^Σ is expected to display significant changes. The other length scale is $l_{\mathrm{n}}^{\mathrm{ref}} := \rho_{\mathrm{ref}}^\Sigma / \rho_{\mathrm{ref}}$. For non-accumulating species it will characterize the typical thickness of a layer in the bulk, being adjacent to the interface, such that this layer contains a similar amount of A_i than is adsorbed on the interface (i.e. inside the thin transition zone represented by the interface). So, while $l_{\mathrm{tan}}^{\mathrm{ref}}$ will represent a mesoscopic length scale, say a bubble or droplet diameter, $l_{\mathrm{n}}^{\mathrm{ref}}$ will typically be on a sub-micrometer scale. In this situation, equation (10.138) and, hence, (10.31) can be approximated very accurately by

$$[\![\dot{m}_i]\!] = 0 \quad \Leftrightarrow \quad \dot{m}_i^{+,\Sigma} + \dot{m}_i^{-,\Sigma} = 0. \tag{10.139}$$

Thus, while the interfacial mixture thermodynamics is accounted for, the capacity of the interface to store species mass is neglected.

Consequently, employing (10.130) and (10.132) with $a_i^\pm = 1$, we obtain

$$s_i^{de,+}\Big(\exp\big(\frac{\mu_i^+}{RT^+} - \frac{\mu_i^\Sigma}{RT^\Sigma}\big) - 1\Big) + s_i^{de,-}\Big(\exp\big(\frac{\mu_i^-}{RT^-} - \frac{\mu_i^\Sigma}{RT^\Sigma}\big) - 1\Big) = 0. \tag{10.140}$$

With the same simplifying assumption of ideal mixtures and (10.133), this implies

$$k_i^{de,+}\Big(\exp\big(\frac{g_i^+}{RT^+} - \frac{g_i^\Sigma}{RT^\Sigma}\big)x_i^+ - x_i^\Sigma\Big) + k_i^{de,-}\Big(\exp\big(\frac{g_i^-}{RT^-} - \frac{g_i^\Sigma}{RT^\Sigma}\big)x_i^- - x_i^\Sigma\Big) = 0, \tag{10.141}$$

hence

$$x_i^\Sigma = \frac{k_i^{de,+} \exp\big(\frac{g_i^+}{RT^+} - \frac{g_i^\Sigma}{RT^\Sigma}\big)x_i^+ + k_i^{de,-} \exp\big(\frac{g_i^-}{RT^-} - \frac{g_i^\Sigma}{RT^\Sigma}\big)x_i^-}{k_i^{de,+} + k_i^{de,-}}. \tag{10.142}$$

Inserting this value into the first expression in (10.141), Eq. (10.130) yields

$$\dot{m}_i^{+,\Sigma} = \frac{k_i^{de,+} k_i^{de,-}}{k_i^{de,+} + k_i^{de,-}} \exp\big(-\frac{g_i^\Sigma}{RT^\Sigma}\big)\Big(\exp\big(\frac{g_i^+}{RT^+}\big)x_i^+ - \exp\big(\frac{g_i^-}{RT^-}\big)x_i^-\Big). \tag{10.143}$$

Note that $\dot{m}_i^{+,\Sigma} = -\dot{m}_i^{-,\Sigma}$ due to (10.139). The key point about this closure—as compared to a closure without interfacial mass densities for transfer components—is that here, the mass transfer rate is influenced by the interface pressure via the interface Gibbs free energy. This allows to account for the effect of surfactants on the mass transfer of the considered transfer component according to the following causal chain: The presence of surfactant adsorbed at the interface changes the interface pressure. Since the interface chemical potentials of *all* species depend (significantly) on the interface pressure, those change accordingly. As the mass transfer rates are not only determined by the one-sided limits of the bulk chemical potentials but by their mutual difference to the interface chemical potential, the transfer rates become dependent on surfactant concentration. This is independent of the hydrodynamics inside the fluid system.

Concrete forms of the mass transfer relation (10.143) of course depend on the employed model for the free energies. For example, in order to account for the different area demands of the adsorbed species, a possible surface equation of state is

$$p^\Sigma = -K^\Sigma + \frac{RT}{1-\theta}\sum_{i=1}^N \delta_i c_i^\Sigma \quad \text{with } \theta = \sum_{i=1}^N c_i^\Sigma / c_i^{\Sigma,\infty}, \tag{10.144}$$

where θ is the total coverage of the interface and $\delta_i > 0$ model the different area demand of A_i^Σ; cf. Bothe (2016). Then the interface analog of the construction of a consistent free energy as explained in §15 in Bothe and Dreyer (2015) yields the corresponding interface free energy as

$$\rho^\Sigma \psi^\Sigma = -(1-\theta)p^\Sigma + RT^\Sigma \sum_{i=1}^N \delta_i c_i^\Sigma \ln\big(1 + \frac{p^\Sigma}{K^\Sigma}\big) + RT^\Sigma \sum_{i=1}^N c_i^\Sigma \ln x_i^\Sigma. \tag{10.145}$$

The corresponding (molar based) interfacial chemical potentials then are

$$\mu_i^\Sigma = g_i^\Sigma(T^\Sigma, p^\Sigma) + RT^\Sigma \ln x_i^\Sigma \quad \text{for } i = 1, \dots, N \tag{10.146}$$

with

$$g_i^\Sigma(T^\Sigma, p^\Sigma) = \frac{p^\Sigma}{c_i^{\Sigma,\infty}} + RT^\Sigma \delta_i \ln\big(1 + \frac{p^\Sigma}{K^\Sigma}\big) \quad \text{for } i = 1, \dots, N. \tag{10.147}$$

Insertion of (10.147) into (10.143) implies the relation

$$\dot{m}_i^{+,\Sigma} = \frac{k_i}{(1 + p^\Sigma / K^\Sigma)^{\delta_i}} \exp\big(\frac{-p^\Sigma}{c_i^{\Sigma,\infty} R T^\Sigma}\big)\Big(\exp\big(\frac{g_i^+}{RT^+}\big)x_i^+ - \exp\big(\frac{g_i^-}{RT^-}\big)x_i^-\Big). \tag{10.148}$$

Notice that the above derivation treats the adsorbed form A_i^Σ of a non-accumulating ('surface inactive') constituent A_i as being an intermediate species in a transition state, when passing through the interface. It is thus related to transition state theory, often used to derive reaction rates of chemical reactions involving intermediates; cf. Laidler and King (1983). Let us also note in passing that in mass transfer applications, a continuous temperature field is usually assumed, i.e. $T^+_{|\Sigma} = T^-_{|\Sigma} = T^\Sigma$.

Taking the clean interface as the reference state, and denoting by $\gamma = -p^\Sigma$ the interfacial tension, (10.148) implies

$$\dot{m}_i^{\text{contam}} = \Big(\frac{K^\Sigma - \gamma^{\text{clean}}}{K^\Sigma - \gamma^{\text{contam}}}\Big)^{\delta_i} \exp\Big(-\frac{\gamma^{\text{clean}} - \gamma^{\text{contam}}}{c_i^{\Sigma,\infty} R T^\Sigma}\Big)\, \dot{m}_i^{\text{clean}}. \tag{10.149}$$

The mass transfer reduction in (10.149) contains as the main damping factor the exponential term. This corresponds to an exponential damping factor of Boltzmann type, i.e. a factor of the form $\exp(-a\,\Delta p^\Sigma / RT^\Sigma)$, in accordance with the energy barrier model due to Langmuir; see Langmuir and Langmuir (1927), Ciani et al. (2004) and the references given there. An intuitive explanation of the energy barrier model assumes that, in order to cross the interface, a molecule first has to 'open a hole' in the contaminated interface against the increased surface pressure. Only a fraction of the molecules carries sufficient energy in order to perform this work, hence the frequency of transfer of molecules across the interface is reduced by a corresponding Boltzmann factor.

Acknowledgements I got to know Wolfgang Dreyer in the afternoon of February 5 in 2009 at the 4th Workshop 'Micro-Macro Modelling and Simulation of Liquid-Vapour Flows', where he chaired my talk on 'Modelling and simulation of two-phase flows with mass transfer'. His insightful and good-natured critical comments immediately proved to me that he has the deep knowledge of continuum thermodynamics of (fluid) interfaces that I was eagerly looking for. In the following years, I had the privilege to intensely work with Wolfgang and to learn from him about the science and history of continuum thermodynamics. I cannot overestimate how much I benefited from our friendship.
The author gratefully acknowledges financial support provided by the German Research Foundation (DFG) within the Collaborative Research Center SFB 1194 (project ID 265191195).

References

Alke A, Bothe D (2009) 3D numerical modeling of soluble surfactant at fluidic interfaces based on the Volume-of-Fluid method. Fluid Dyn Mater Process 5:345–372

Aoki J, Hayashi K, Hosokawa S, Tomiyama A (2015) Effects of surfactants on mass transfer from single carbon dioxide bubbles in vertical pipes. Chem Eng Technol 38:1955–1964

Aoki J, Hori Y, Hayashi K, Hosokawa S, Tomiyama A (2017) Mass transfer from single carbon dioxide bubbles in alcohol aqueous solutions in vertical pipes. Int J Heat Mass Tran 108:1991–2001

Barnes GT (1986) The effects of monolayers on the evaporation of liquids. Adv Colloid Interfac 25:89–200

Barrett JW, Garcke H, Nürnberg R (2020) Parametric finite element approximations of curvature-driven interface evolutions, Chapter 4. In: Bonito A, Nochetto RH (eds) Handbook of numerical analysis, vol **21**. Elsevier, pp 275–423

Bedeaux D (1986) Nonequilibrium thermodynamics and statistical physics of surfaces. In: Prigogine I, Rice SA (eds) Advance in chemical physics, vol 64. Wiley, pp 47–109

Bedeaux D, Albano AM, Mazur P (1976) Boundary conditions and non-equilibrium thermodynamics. Phys A 82:438–462

Bothe D (2015) On the multi-physics of mass transfer across fluid interfaces. In: Schindler F-P, Kraume M (eds) 7th international workshop—IBW7 on transport phenomena with moving boundaries. Fortschr.-Ber. VDI Reihe 3, Nr. 947, VDI-Verlag Düsseldorf, pp 1–23

Bothe D (2016) Mass transfer across fluid interfaces—modeling the influence of adsorbed surfactant. In: Hishida T (ed) Proceedings of the RIMS symposium on mathematical analysis of viscous incompressible fluids. Kyoto University, pp 12–17

Bothe D (2020) On moving hypersurfaces and the discontinuous ODE-system associated with two-phase flows. Nonlinearity 33:5425–5456

Bothe D (2022) Sharp-interface continuum thermodynamics of multicomponent fluid systems with interfacial mass. Int J Eng Sci 179:103731

Bothe D, Dreyer W (2015) Continuum thermodynamics for chemically reacting fluid mixtures. Acta Mech 226:1757–1805

Bothe D, Druet P-E (2023) On the structure of continuum thermodynamical diffusion fluxes—a novel closure scheme and its relation to the Maxwell-Stefan and the Fick-Onsager approach. Int J Eng Sci 184:103818

Bothe D, Fleckenstein S (2013) Modeling and VOF-based numerical simulation of mass transfer processes at fluidic particles. Chem Eng Sci 101:283–302

Bothe D, Prüss J (2010) On the two-phase Navier-Stokes equations with Boussinesq-Scriven surface fluid. J Math Fluid Mech 12:133–150

Bothe D, Soga K (2016) Thermodynamically consistent modeling for dissolution/growth of bubbles in an incompressible solvent. In: Amann H, Giga Y, Kozono H, Okamoto H, Yamazaki M (eds) Recent developments of mathematical fluid mechanics. Springer Basel, pp 111–134

Buff FP (1956) Curved fluid interfaces. I. The Generalized Gibbs-Kelvin equation. J Chem Phys 25:146–153

Ciani I, Burt DP, Daniele S, Unwin PR (2004) Effect of surface pressure on oxygen transfer across molecular monolayers at the air/water interface: Scanning electrochemical microscopy investigations using a mercury hemispherical microelectrode probe. J Phys Chem B 108:3801–3809

Dell'Isola F, Romano A (1986) On a general balance law for continua with an interface. Ricerche di Matematica 35(2):325–337

Dreyer W (2003) On jump conditions at phase boundaries for ordered and disordered phases. WIAS Preprint 869. Reprinted in this volume

Dreyer W, Guhlke C, Müller R (2018) Bulk-surface electrothermodynamics and applications to electrochemistry. Entropy 20(939):1–44

Fellows AP, Duque AD, Balos V, Lehmann L, Netz RR, Wolf M, Thämer M (2024) How thick is the air-water interface?—A direct experimental measurement of the decay length of the interfacial structural anisotropy. Langmuir 40:18760–18772

Garbina V, Bothe D, Brenn G, Casciola CM, Colin C, Marengo M, Risso F, Tryggvason G, Lohse D (2025) Bubbles and bubbly flows. Int J Multiphas Flow, 105240

Giga Y (2006) Surface evolution equations: a level set approach. Monogr Math 99. Birkhäuser, Basel

De Groot SR, Mazur P (1984) Non-equilibrium thermodynamics. Dover Publications

Guggenheim EA, Adam NK (1933) The thermodynamics of adsorption at the surface of solutions. Proc Roy Soc A 139:218–236

Gurtin M, Struthers A, Williams W (1989) A transport theorem for moving interfaces. Q Appl Math 47(4):773–777

Hébrard G, Zeng J, Loubiere K (2009) Effect of surfactants on liquid side mass transfer coefficients: a new insight. Chem Eng J 148:132–138

Hori Y, Hirota Y, Hayashi K, Hosokawa S, Tomiyama A (2019) Combined effects of alcohol and electrolyte on mass transfer from single carbon-dioxide bubbles in vertical pipes. Int J Heat Mass Tran 136:521–530

Hori Y, Bothe D, Hayashi K, Hosokawa S, Tomiyama A (2020) Mass transfer from single carbon-dioxide bubbles in surfactant-electrolyte mixed aqueous solutions in vertical pipes. Int J Multiphas Flow 124:103207

Hosoda S, Abe S, Hosokawa S, Tomiyama A (2014) Mass transfer from a bubble in a vertical pipe. Int J Heat Mass Tran 69:215–222

Hutter K, Jöhnk K (2004) Continuum methods of physical modeling. Springer, Heidelberg

Jimenez M, Dietrich N, Grace JR, Hébrard G (2014) Oxygen mass transfer and hydrodynamic behaviour in wastewater: determination of local impact of surfactants by visualization techniques. Water Res 58:111–121

Kentheswaran K, Antier W, Dietrich N, Lalanne B (2023) Impact of surfactants on the rise of deformable bubbles and interfacial gas-liquid mass transfer. J Fluid Mech 970:A5

Kimura M (2008) Geometry of hypersurfaces and moving hypersurfaces in $\mathbf{R}^m$—for the study of moving boundary problems. In: Benes M, Feireisl E (eds) Lecture notes volume IV, topics in mathematical modeling, Jindřich Nečas center for mathematical modeling. Matfyzpress , pp 39–93

Kralchevsky PA, Danov KD (2015) Chemical physics of colloid systems and interfaces. In: Birdi KS (ed) Handbook of surface and colloid chemistry 4th ed. CRC Press, pp 247–414

Laidler KJ, King MC (1983) The development of transition-state theory. J Phys Chem 87(15):2657–2664

Langmuir I, Langmuir DB (1927) The effect of monomolecular films on the evaporation of ether solutions. J Phys Chem 31:1719–1731

Lengyel S (1989) Chemical kinetics and thermodynamics—a history of their relationship. Symmetry 2:443–455

Liu I-S (2002) Continuum mechanics. Springer

Marcelin R (1910) Sur la mécanique des phénomenes irréversibles. C R Hebd Seances Acad Sci Paris 5:1052–1055

Meixner J (1973) Consistency of the Onsager-Casimir reciprocal relations. Adv Mol Relax Process 5:319–331

Müller I (1985) Thermodynamics. Pitman

Pekar M (2005) Thermodynamics and foundations of mass-action kinetics. Prog React Kinet Mech 30:3–113

Persad AH, Ward CA (2016) Expressions for the Evaporation and Condensation Coefficients in the Hertz-Knudsen relation. Chem Rev 116(14):7727–7767

Powles JG (1985) On the validity of the Kelvin equation. J Phys A: Math Gen 18:1551–1560

Prüss J, Simonett G (2016) Moving interfaces and quasilinear parabolic problems. Monographs in mathematics, Birkhäuser

Rahman MA, Saghir MZ (2014) Thermodiffusion or Soret effect: historical review. Int J Heat Mass Tran 73:693–705
Ruggeri T (2009) Multi-temperature mixture of fluids. Theor Appl Mech 36(3):207–238
Sagis LMC (2014) Dynamic behavior of interfaces: modeling with nonequilibrium thermodynamics. Adv Colloid Interf Sci 206:328–343
Sardeing R, Painmanakul P, Hébrard G (2006) Effect of surfactants on liquid-side mass transfer coefficients in gas-liquid systems: a first step to modeling. Chem Eng Sci 61:6249–6260
Scriven LE (1960) Dynamics of a fluid interface. Chem Eng Sci 12:98–108
Slattery JC (1990) Interfacial transport phenomena. Springer, New York
Slattery JC, Sagis L, Oh E-S (2007) Interfacial transport phenomena, 2nd ed. Springer, New York
Thomas TY (1957) Extended compatibility conditions for the study of surfaces of discontinuity in continuum mechanics. J Math Mech 6:311–322
Truesdell C, Noll W (2003) The non-linear field theories of mechanics, 3rd ed. Springer
Wagner W, Pruß A (2002) The IAPWS formulation 1995 for the thermodynamic properties of ordinary water substance for general and scientific use. J Phys Chem Ref Data 31:387–535
Waldmann L (1967) Non-equilibrium thermodynamics of boundary conditions. Zeitschrift für Naturforschung A 22(8):1269–1280

Chapter 11
Damage Modeling in Batteries by Coupling Lithiation and Mechanics by Means of a Finite Element Method

Bilen Emek Abali and Per Isaksson

Abstract Li-ion batteries work owing to migration, diffusion, and lithiation of ions within the electrode and electrolyte. This process is described by governing equations given by a mixture theory and Maxwell's equations at the macroscale. Coupling these two scales poses theoretical difficulties and also significant computational challenges. We aim to address these issues by discussing challenges in a formulation and use a continuum mechanics approach capable of showing a so-called pitting phenomenon observed experimentally.

11.1 Damage Modeling

The complex chemistry and structure of Li-ion batteries lead to challenging phenomena within a cell during operation. During charging and discharging of a battery, from a broad perspective, ions are carrying charges and shuffle back and forth between the cathode and the anode. This cycling process is irreversible and a hysteresis causes a dissipation in the system (Dreyer et al. 2010). Modeling such a system at the length scale of ions (Dreyer et al. 2013) has a great potential but it may be computationally demanding and one often uses some sort of homogenization by mixture theory (Bothe and Dreyer 2015) that increases the complexity in constitutive equations yet reducing the degrees of freedom in a numerical environment, thus, reducing the computation time.

The motion of ions is coupled with electromagnetism, in general, even the governing equations are under discussion (Muller 2023) since there are several possibilities to combine mechanics-related and electromagnetism-related forces (Reich et al. 2018). Furthermore, there are challenges in choosing the configuration in the formulation since a material frame is suitable for mass and momentum balance, but

B. E. Abali (✉) · P. Isaksson
Division of Applied Mechanics, Department of Materials Science and Engineering, Uppsala University, Uppsala, Sweden
e-mail: b.emek.abali@angstrom.uu.se

P. Isaksson
e-mail: per.isaksson@angstrom.uu.se

J. Fuhrmann et al. (eds.), *Advances in Continuum Physics*, Advanced Structured Materials 238, https://doi.org/10.1007/978-3-031-93918-1_11

an open domain like a spatial frame is needed by electromagnetic fields (Abali and Queiruga 2019). By reducing the complexity under assumptions like no magnetic flux or chemical reaction, the system is simplified to be seen as one species electro-migrating and causing mechanical stresses. Such induced stresses are small in one charging and discharging cycle, yet they become influential in fatigue life due to material strength losses. One possible approach is to model the damage within an electrode particle at micrometer length scale by using an additional field denoting microcracks at a lower length scale. In such an approach, we understand damage in a bulk of contiguous molecules or atomic lattices as a fracture, or discontinuity of connection, between them. Continuum modeling is not at the molecular length scale but the description in this manner visualizes the basic mechanism. Yet a damage formulation is challenging to model with a constitutive theory. Microcracks are often modeled as a scalar field (order parameter) (Murakami 2012; Altenbach and Sadowski 2015; Ochsner 2016). This damage variable increases monotonously, in other words, there is no healing allowed, based on the original idea in Rabotnov (1959). Either we begin with an initial 1 increasing to infinity (Kachanov 1986; Lemaitre and Desmorat 2005); or equivalently, we start from 1 to 0 (inverse of the damage variable); or even one minus the variable makes starting from 0 until 1 that denotes fracture that is a loss of electrical contact leading to trapping or a capacity loss in the battery. In this way, two length scales are connected by a scalar damage variable (Provatas and Elder 2011). This modeling approach is known as the phase-field approach that is used herein as well for a diffusive fracture modeling by means of a staggered (Miehe et al. 2010; Kiendl et al. 2016; Kastner 2016; Teichtmeister et al. 2017) or monolithic approach (Clayton and Knap 2016; Levitas 2018; Babaei and Levitas 2018; Amirian et al. 2022).

An evolution equation is often used for modeling the damage variable with a phase field as an order parameter based on a potential (Mumford and Shah 1989) developed in Francfort and Marigo (1998) as a minimization problem or by a Γ-convergence analysis (Ambrosio and Tortorelli 1990). A bump function is 1 within a finite domain (compact support) and 0 otherwise; this function has sufficient regularity such that a differential equation is formally proposed in the form of an evolution equation. Hence, a sharp crack is regularized by a crack surface density functional, $\gamma = \gamma(\omega, \omega_{,i})$, where, and henceforth, a comma notation denotes a (partial) spatial derivative. The crack density functional γ, typically chosen as either first order or second order in ω, can be chosen with a great freedom from the family of elliptic functionals approximating the Mumford–Shah functional (Braides 1998).

Using such approximation, we obtain in a continuum body, $\mathcal{B}$, an integral relation with an accumulated (dissipative) energy density, ${}^{\mathrm{d}}\psi$, and the critical energy release rate, G_{c}, as follows:

$$\int_{\mathcal{B}} {}^{\mathrm{d}}\psi \, \mathrm{d}V \approx \int_{\mathcal{B}} G_{\mathrm{c}} \gamma \, \mathrm{d}V, \tag{11.1}$$

where

$$\gamma = \frac{1}{c_\alpha \ell} \Big(\alpha + \ell^2 \omega_{,i} \omega_{,i} \Big). \tag{11.2}$$

Here, the constant ℓ is a regularization parameter, a kind of characteristic length, determining the width of the regularized crack while $c_\alpha = 4\int_0^1 \alpha^{0.5}\,\mathrm{d}\omega$ is a normalization constant, which depends on the choice of crack geometry function $\alpha = \alpha(\omega) \in [0, 1]$. We use index notation in Cartesian coordinates; and understand EINSTEIN's summation convention over repeated indices. A variational formulation is used

$$\begin{aligned} &\int_{\mathcal{B}} {}^{\mathrm{d}}\psi\,\mathrm{d}V = R\,, \\ &\delta \int_{\mathcal{B}} {}^{\mathrm{d}}\psi\,\mathrm{d}V = \int_{\mathcal{B}} R\delta\omega\,\mathrm{d}V\,,\quad R = \frac{\partial \mathcal{L}}{\partial \omega}, \end{aligned} \tag{11.3}$$

where the right-hand side is connected to a LAGRANGE function

$$\mathcal{L} = -f + \rho f_i u_i\,,\quad f = g\,{}^{\mathrm{e}}\psi. \tag{11.4}$$

A potential energy is caused by gravitation, where $|f_i| = 9.81\,\mathrm{N/kg}$ is the specific body force. Free energy density, f, is given by a degradation function, $g = g(\omega)$, and stored energy density, ${}^{\mathrm{e}}\psi$, which is chosen according to the bulk material.

Now, let us consider the case of linear elasticity, ${}^{\mathrm{e}}\psi = {}^{\mathrm{e}}\psi(\boldsymbol{\varepsilon})$. To prevent crack growth under compression, the strain energy density ${}^{\mathrm{e}}\psi$ is decomposed into a tensile/positive part and a compressive/negative part, where only the positive part affects damage, cf. Ambati et al. (2015), i.e.,

$$\int_{\mathcal{B}} {}^{\mathrm{e}}\psi\,\mathrm{d}V = \int_{\mathcal{B}} \left(g\,{}^{\mathrm{e}}\psi^{+} + {}^{\mathrm{e}}\psi^{-} \right) \mathrm{d}V. \tag{11.5}$$

The consistent elastic stiffness tensor C_{ijkl} is given by

$$C_{ijkl} = g \frac{\partial^2\,{}^{\mathrm{e}}\psi^{+}}{\partial \varepsilon_{ij} \partial \varepsilon_{kl}} + \frac{\partial^2\,{}^{\mathrm{e}}\psi^{-}}{\partial \varepsilon_{ij} \partial \varepsilon_{kl}}, \tag{11.6}$$

and the Cauchy stress tensor reads $\sigma_{ij} = C_{ijkl}\varepsilon_{kl}$. In the literature many decompositions have been suggested, cf. Miehe et al. (2010), Amor et al. (2009), van Dijk et al. (2020), Hesammokri and Isaksson (2023, 2024). Here, we apply a hydrostatic-deviatoric decomposition, introduced by Amor et al. (2009), who argue that crack growth is driven by shear and tensile bulk deformation, which is well justified from a theoretical point of view. Assuming a linear elastic isotropic material and using the bulk modulus, K, and shear modulus, G, the strain energy density is written ${}^{\mathrm{e}}\psi = \frac{1}{2}K\varepsilon_{kk}^2 + G\varepsilon_{ij}^{\mathrm{dev}}\varepsilon_{ji}^{\mathrm{dev}}$, where the deviatoric strain tensor $\varepsilon_{ij}^{\mathrm{dev}} = \varepsilon_{ij} - \frac{1}{\delta_{ll}}\varepsilon_{kk}\delta_{ij}$. Following Amor et al. (2009), we decompose ${}^{\mathrm{e}}\psi$ into a negative part

$${}^{\mathrm{e}}\psi^{-} = \frac{1}{2}K\,\langle \varepsilon_{kk} \rangle_{-}^{2}\,, \tag{11.7}$$

and a crack-driving positive part

$$ {}^{\mathrm{e}}\psi^{+} = \frac{1}{2} K \left\langle \varepsilon_{kk} \right\rangle_{+}^{2} + G \varepsilon_{ij}^{\mathrm{dev}} \varepsilon_{ji}^{\mathrm{dev}} . \tag{11.8} $$

Here, $\langle \cdot \rangle_{+}$ and $\langle \cdot \rangle_{-}$ selects the positive and negative parts of their argument, i.e.,

$$ \langle x \rangle_{+} = \begin{cases} x, & \text{if } x \geq 0 \\ 0, & \text{else} \end{cases} , \quad \langle x \rangle_{-} = \begin{cases} x, & \text{if } x \leq 0 \\ 0, & \text{else} \end{cases} . \tag{11.9} $$

Now, using a decomposed strain energy density, we obtain the damage evolution by

$$ \begin{aligned} \delta \int_{\mathcal{B}} {}^{\mathrm{d}}\psi \, \mathrm{d}V &= \int_{\mathcal{B}} \frac{\partial \mathcal{L}}{\partial \omega} \delta\omega \, \mathrm{d}V , \\ \int_{\mathcal{B}} \left(\frac{\partial\, {}^{\mathrm{d}}\psi}{\partial \omega} \delta\omega + \frac{\partial\, {}^{\mathrm{d}}\psi}{\partial \omega_{,i}} \delta\omega_{,i} \right) \mathrm{d}V &= - \int_{\mathcal{B}} \frac{\partial g}{\partial \omega} {}^{\mathrm{e}}\psi^{+} \delta\omega \, \mathrm{d}V . \end{aligned} \tag{11.10} $$

We will discuss two different formalism possibilities. It is assumed that $\omega = 1$ represents broken material and $\omega = 0$ represents intact material. One usual choice for the degradation function is a quadratic one, $g = (1 - \omega)^2$, and a linear geometric relation $\alpha = \omega$ as in Pham et al. (2011) leading to $c_\alpha = 8/3$ such that we obtain

$$ {}^{\mathrm{d}}\psi = \frac{3 G_c}{8 \ell} \Big(\omega + \ell^2 \omega_{,i} \omega_{,i} \Big) . \tag{11.11} $$

After inserting this choice, called Ambrosio–Tortorelli 1 (AT1), we obtain a weak form necessary for the numerical implementation

$$ \begin{aligned} \int_{\mathcal{B}} \left(\frac{3 G_c}{8 \ell} \delta\omega + \frac{3 G_c \ell}{4} \omega_{,i} \delta\omega_{,i} \right) \mathrm{d}V &= - \int_{\mathcal{B}} \frac{\partial g}{\partial \omega} {}^{\mathrm{e}}\psi^{+} \delta\omega \, \mathrm{d}V , \\ \int_{\mathcal{B}} \left(\frac{3 G_c}{8 \ell} \delta\omega + \frac{3 G_c \ell}{4} \omega_{,i} \delta\omega_{,i} - 2(1 - \omega)\, {}^{\mathrm{e}}\psi^{*} \delta\omega \right) \mathrm{d}V &= 0 , \end{aligned} \tag{11.12} $$

where ${}^{\mathrm{e}}\psi^{*} = \max({}^{\mathrm{e}}\psi^{+})$ is used algorithmically in order to ensure a monotonously increasing damage variable.

Another common choice is to use a quadratic degradation function, $g = (1 - \omega)^2$, and a quadratic geometric relation, $\alpha = \omega^2$, resulting in $c_\alpha = 2$ with

$$ {}^{\mathrm{d}}\psi = \frac{G_c}{2 \ell} \Big(\omega^2 + \ell^2 \omega_{,i} \omega_{,i} \Big) . \tag{11.13} $$

In this case, called Ambrosio–Tortorelli 2 (AT2), we acquire the weak form

$$ \int_{\mathcal{B}} \left(\frac{G_c}{\ell} \omega \delta\omega + G_c \ell \omega_{,i} \delta\omega_{,i} - 2(1 - \omega)\, {}^{\mathrm{e}}\psi^{*} \delta\omega \right) \mathrm{d}V = 0 . \tag{11.14} $$

We implement both the AT1 and AT2 crack density functionals in order to ignite a discussion about their differences in the case of microcracks and fractures in electrodes during charging and discharging.

Moreover, mechanical equilibrium is assumed to be established instantly such that it is modeled as instantaneous regarding the ion motion; this simplification is indeed suitable with the chosen LAGRANGEan in Eq. (11.4), $\mathcal{L} = \mathcal{L}(\omega, \boldsymbol{u}, \nabla \boldsymbol{u})$, and we extend the problem to be solved

$$\begin{gathered}\delta \int_{\mathcal{B}} \mathcal{L} \mathrm{d}V = \delta \int_{\mathcal{B}} {}^{\mathrm{d}}\psi \, \mathrm{d}V , \\ \int_{\mathcal{B}} \left(\frac{\partial \mathcal{L}}{\partial \omega} \delta\omega + \frac{\partial \mathcal{L}}{\partial u_i} \delta u_i + \frac{\partial \mathcal{L}}{\partial u_{i,j}} \delta u_{i,j} \right) \mathrm{d}V = \int_{\mathcal{B}} \left(\frac{\partial\, {}^{\mathrm{d}}\psi}{\partial \omega} \delta\omega + \frac{\partial\, {}^{\mathrm{d}}\psi}{\partial \omega_{,i}} \delta\omega_{,i} \right) \mathrm{d}V ,\end{gathered} \tag{11.15}$$

leading to

$$\begin{gathered}\int_{\mathcal{B}} \left(\frac{\partial \mathcal{L}}{\partial u_i} \delta u_i + \frac{\partial \mathcal{L}}{\partial u_{i,j}} \delta u_{i,j} \right) \mathrm{d}V = \int_{\mathcal{B}} \left(\frac{\partial\, {}^{\mathrm{d}}\psi}{\partial \omega} \delta\omega + \frac{\partial\, {}^{\mathrm{d}}\psi}{\partial \omega_{,i}} \delta\omega_{,i} - \frac{\partial \mathcal{L}}{\partial \omega} \delta\omega \right) \mathrm{d}V , \\ \int_{\mathcal{B}} \left(\rho f_i \delta u_i - g \frac{\partial\, {}^{\mathrm{e}}\psi^{+}}{\partial u_{i,j}} \delta u_{i,j} \right) \mathrm{d}V = \int_{\mathcal{B}} \left(\frac{\partial\, {}^{\mathrm{d}}\psi}{\partial \omega} \delta\omega + \frac{\partial\, {}^{\mathrm{d}}\psi}{\partial \omega_{,i}} \delta\omega_{,i} - 2(1-\omega)\, {}^{\mathrm{e}}\psi^{*} \delta\omega \right) \mathrm{d}V .\end{gathered} \tag{11.16}$$

It is emphasized that such a simplification is of importance to distinguish between fast propagation of flux in momentum and slow propagation of flux of ions in order to introduce the coupling between mechanical stress and concentration in a phenomenological manner, which will be evident in the following. Without going into the detail, we refer to Dreyer et al. (2016) for a more general formulation than used herein.

11.2 Lithiation Modeling

Lithium insertion into the electrode is governed by a mass balance and mixture theory,

$$\frac{\partial c}{\partial t} + (v_i c)_{,i} + J_{i,i} = r , \tag{11.17}$$

where the number density, c in mol/m^3, is called concentration, we refer to Abali (2020) for a discussion about different terminologies in the literature and a simple derivation. The velocity of the mass (barycentric motion of the mixture), here the host particle (electrode), $\boldsymbol{v}$, causes transport of ions via advection, this term is often called the lattice velocity or rate of displacement of the bulk (mixture). No chemical reaction is imposed, i.e., $r = 0$, within the electrode such that the motion of ions is due to the flux, $\boldsymbol{J}$. This simplification is effected by the understanding that the reaction is at the interface between electrode and electrolyte. There are cathode and anode, composed of different materials, for example, a widely used example is a graphite anode, such as C_6, and a transition metal oxide cathode, such as lithium cobalt oxide $LiCoO_2$, where during charging Li^+ ions leave one electrode

$$LiCoO_2 \rightleftarrows CoO_2 + Li^+ + e^- \tag{11.18}$$

and move to the other electrode

$$C_6 + Li^+ + e^- \rightleftarrows LiC_6 \tag{11.19}$$

where the separation/merging of positive ion and electrode is a reaction at the interface between electrode and electrolyte. This process is based on the intercalation/insertion of lithium ions for operation. There are other mechanisms such as conversion reaction and alloying as well. Among alloy-type electrode materials, silicon (Si) is a promising candidate studied because its reaction with lithium occurs at low discharge potentials and it possesses a large theoretical specific capacity (Taiwo et al. 2017). In such alloying Si-electrode, in the lithiation process, Li^+-ions diffuse into the Si forming a low-density Li_xSi_y phase. Sequential phase transformations have been observed, e.g., the phases $Li_{12}Si_7$, Li_7Si_3, $Li_{13}Si_4$, and $Li_{22}Si_5$, Kim et al. (2021).

However, independent of the lithiation process, we model only the electrode, thus, at the boundary this reaction is modeled by a given concentration or flux. Then, within the electrode, the host material occupied with lithium, $LiCoO_2$, LiC_6 or Li_xSi_y, which we call shortly LiS for lithium occupied host site, and Li^+ + e^-, which we call Li for lithium-ion, are modeled at each location by their concentration, c_{LiS} and c_{S}, respectively. The reaction of separation of an electron from Li is ignored since it happens at the boundary, although critical, this assumption performs adequately for isothermal processes. In general, there are two partial mass balances for LiS and Li, but as we know that their sum, $c_{\mathrm{tot}} = c_{\mathrm{LiS}} + c_{\mathrm{S}}$, is a conserved quantity (even simpler herein since a reaction term is suppressed, if not, then their reaction rate is the same but opposite signs), modeling one is enough to get the other one once we know the relation c_{tot}. For this relation, we follow Christensen and Newman (2006) and use a mole fraction, $x_{\mathrm{LiS}} = c_{\mathrm{LiS}}/c_{\mathrm{tot}}$, which is called the State Of Charge (SOC). The motion of ions is related to the electrochemical potential, composed of a chemical and electrical term,

$$\bar{\mu}_{\mathrm{LiS}} = \mu_{\mathrm{LiS}} + z_{\mathrm{LiS}} F \phi , \tag{11.20}$$

according to IUPAC definition (McNaught and Wilkinson 1669), it is the partial molar Gibbs energy, $\mathcal{G}$. In a mixture, we use the energy per volume (of the mixture), g, and therefore relate

$$\mu_{\mathrm{LiS}} = \frac{\partial g}{\partial c_{\mathrm{LiS}}} , \tag{11.21}$$

for constant stress (isothermal processes). In the case of (uncharged) lithium migrating within the electrode, $z = 0$, leading to chemical and electrochemical potentials identical. For modeling this term, an ideal solution would use

$$\mu_{\text{LiS}} = \mu^0_{\text{LiS}}(T, \boldsymbol{\sigma}) + RT \ln(x_{\text{LiS}}), \tag{11.22}$$

with the chemical potential of LiS depending only on the mole fraction of LiS but neglecting intermolecular interaction between LiS and S. There is a term called activity, γ_{LiS}, usually added for correction,

$$\mu_{\text{LiS}} = \mu^0_{\text{LiS}}(T, \boldsymbol{\sigma}) + RT \ln(x_{\text{LiS}}) + RT \ln(\gamma_{\text{LiS}}) = \mu^0_{\text{LiS}}(T, \boldsymbol{\sigma}) + RT \ln(x_{\text{LiS}}\gamma_{\text{LiS}}). \tag{11.23}$$

Dependence on stress is the coupling between concentration and fracture. One possible, phenomenological approach is to use a linear relation (Yang 2005) between the lithium concentration causing a volumetric expansion

$$\varepsilon_{\text{vol}} = \frac{1}{\delta_{ii} K}\Big[\sigma_{\text{vol}} + \Omega_{\text{LiS}}(c_{\text{LiS}} - c^{\text{ref}})\Big], \tag{11.24}$$

due to its molar volume, Ω_{LiS}, for a 2D, $\delta_{ii} = 2$, or a 3D, $\delta_{ii} = 3$, geometry. Analogously, for a small displacement assumption, strain reads

$$\varepsilon_{ij} = \frac{1}{2}\Big(u_{i,j} + u_{j,i}\Big) - \frac{\Omega_{\text{LiS}}}{\delta_{kk}}(c_{\text{LiS}} - c^{\text{ref}})\delta_{ij}. \tag{11.25}$$

The volumetric part of the strain is affected by the concentration. The aforementioned stress definition from this strain may be seen also as decomposed in volumetric and deviatoric parts

$$\sigma_{ij} = \sigma_{\text{vol}}\delta_{ij} + \sigma^{\text{dev}}_{ij}. \tag{11.26}$$

The linear relation in Eq. (11.24) is then utilized to justify

$$\mu^0_{\text{LiS}}(T, \boldsymbol{\sigma}) = \mu^{\text{ref}}_{\text{LiS}}(T) - \Omega_{\text{LiS}}\sigma_{\text{vol}}. \tag{11.27}$$

In order to bring the (electro)chemical potential in relation with the motion of ions, we relate the velocity of ions to the gradient of potential,

$$v^{\text{LiS}}_i = -M_{\text{LiS}}\mu^{\text{LiS}}_{,i}, \tag{11.28}$$

where M_{LiS} is the mobility. The latter is indeed a fundamental approach leading to an isotropic form. Obviously, tracking ions may be challenging, so we homogenize and use flux,

$$
\begin{aligned}
J_i^{\text{LiS}} &= c_{\text{LiS}} v_i^{\text{LiS}} = -c_{\text{LiS}} M_{\text{LiS}} \mu_{,i}^{\text{LiS}} \\
&= -c_{\text{LiS}} M_{\text{LiS}} \bigg[\big(\mu_{\text{LiS}}^{\text{ref}} - \Omega \sigma_{\text{vol}}\big)_{,i} + \frac{RT}{x_{\text{LiS}} \gamma_{\text{LiS}}} \big(x_{\text{LiS}} \gamma_{\text{LiS}}\big)_{,i} \bigg].
\end{aligned}
\tag{11.29}
$$

By following Zhang et al. (2007), in an isothermal process, $\nabla T = 0$, the first term vanishes, $\nabla \mu_{\text{LiS}}^{\text{ref}} = \partial \mu_{\text{LiS}}^{\text{ref}} / \partial T \nabla T = 0$, and we obtain

$$
J_i^{\text{LiS}} = -c_{\text{LiS}} M_{\text{LiS}} \bigg[-\Omega_{\text{LiS}} \sigma_{,i}^{\text{vol}} + \frac{RT}{x_{\text{LiS}} \gamma_{\text{LiS}}} \big(x_{\text{LiS}} \gamma_{\text{LiS}}\big)_{,i} \bigg].
\tag{11.30}
$$

In general, total concentration depends on the composition and stress. We simplify by using a constant total concentration to rewrite, $x_{\text{LiS}} = c_{\text{LiS}} / c_{\text{tot}}$, which results in

$$
J_i^{\text{LiS}} = -c_{\text{LiS}} M_{\text{LiS}} \bigg[-\Omega_{\text{LiS}} \sigma_{,i}^{\text{vol}} + \frac{RT}{c_{\text{LiS}} \gamma_{\text{LiS}}} \big(c_{,i}^{\text{LiS}} \gamma_{\text{LiS}} + c_{\text{LiS}} \gamma_{,i}^{\text{LiS}}\big) \bigg].
\tag{11.31}
$$

A further simplification is to assume that the activity variation within the electrode is negligible regarding concentration variation. By introducing a so-called diffusion parameter, $D_{\text{LiS}} = M_{\text{LiS}} RT$, we obtain

$$
J_i^{\text{LiS}} = D_{\text{LiS}} \bigg[\frac{c_{\text{LiS}} \Omega_{\text{LiS}}}{RT} \sigma_{,i}^{\text{vol}} - c_{,i}^{\text{LiS}} \bigg].
\tag{11.32}
$$

We use a model assumption that the lattice velocity is zero, in other words, only diffusive Li transport is modeled such that from Eq. (11.17), we obtain the weak form after integration by parts

$$
\int_{\mathcal{B}} \Big(\frac{\partial c}{\partial t} \delta c - J_i \delta c_{,i} \Big) \, \mathrm{d}V + \int_{\partial \mathcal{B}} J_{\text{BV}} \delta c \, \mathrm{d}A = 0.
\tag{11.33}
$$

On the boundaries, the mass flux is given by a so-called extended BUTLER–VOLMER equation divided by the FARADAY constant, resulting in J_{BV}. The extended BUTLER–VOLMER equation models the electric current for a simple, unimolecular redox reaction. In a battery cell with a cathode and anode, two reactions occur and BUTLER–VOLMER considers that both a cathodic and anodic reaction occur at the same electrode. This simplification is useful for a measurement of the whole battery cell but less realistic for a local simulation model. Therefore, in this work, we use it as a given quantity and model negative flux demonstrating ions enter the electrode that has initially no lithium ions. Such a lithiation process represents the first cycle after manufacturing. Especially for Si-based electrodes, this first cycle is known as problematic leading to cracks (or even pulverization). One possible model assumption is to use the flux term multiplied by the degradation function in order to inhibit diffusion across cracks. This assumption is indeed difficult to justify by phenomenological means since in-operando experiments are seldom to track damage and diffusion simultaneously. We use herein degradation only on the mechanical part where the stiffness of the structure is decreased that increases locally strain and thus a hydrostatic stress gradient causing ion flux in a coupled manner.

11.3 Numerical Implementation

We solve the introduced equations by discretizing them in space by the finite element method and in time by the finite difference method. We use equidistant time steps, Δt, for a simpler implementation. In each position, unknowns are

- displacement, $\boldsymbol{u}$, after neglecting own weight's deformation, $\boldsymbol{f}=0$, from Eq. (11.16), with the degradation function, $g=(1-\omega)^2$, as follows:

$$F_u = \int_{\mathcal{B}} \Big(g\sigma_{ij}\delta u_{i,j} \Big) \mathrm{d}V, \tag{11.34}$$

- lithium in the host site concentration, $c=c_{\mathrm{LiS}}$, with c^0 indicating the concentration one time step before,

$$F_c = \int_{\mathcal{B}} \Big(\frac{c-c^0}{\Delta t}\delta c - gJ_i\delta c_{,i} \Big) \mathrm{d}V + \int_{\partial\mathcal{B}} J_{\mathrm{BV}}\delta c \,\mathrm{d}A, \tag{11.35}$$

- and damage field, ω, with AT1 formulation:

$$F_\omega^{\mathrm{AT1}} = \int_{\mathcal{B}} \Big(\frac{3G_c}{8\ell}\delta\omega + \frac{3G_c\ell}{4}\omega_{,i}\delta\omega_{,i} - 2(1-\omega)^{\mathrm{e}}\psi^*\delta\omega \Big) \mathrm{d}V, \tag{11.36}$$

  or AT2 formulation:

$$F_\omega^{\mathrm{AT2}} = \int_{\mathcal{B}} \Big(\frac{G_c}{\ell}\omega\delta\omega + G_c\ell\omega_{,i}\delta\omega_{,i} - 2(1-\omega)^{\mathrm{e}}\psi^*\delta\omega \Big) \mathrm{d}V, \tag{11.37}$$

We use a staggered scheme for implementing all equations in order to select differing solvers for them. By choosing a charging rate, C_r, that is a full charge is aimed for in $1/C_\mathrm{r}$ hour, and then selecting either AT1 or AT2 as the damage model, we use a standard formalism as described in Algorithm 1. We emphasize that the length scale micrometer is chosen as the main unit and all units are consistent as given, the conversion of parameters is done by simple multiplication. The staggered method has the order of solving for damage within a time step and then an additional loop for displacement and concentration. Damage is only affected by $H_{\max}$ that is constant within a time step. Point interpolation is used for conditional operators that limit concentration and damage variables between given bounds.

Algorithm 1: Implementing a staggered computational algorithm

Data: Units:μm, ms, mol, mg, nA, K
Data: Derived units:μN≙mg μm/ms^2, MPa≙μN/μm^2, pJ≙μN μm, nW≙pJ/ms, V=nW/nA, GOhm=V/nA, nS=1/GOhm
Data: $t = 0$: $c = c_{\text{ref}}$, $\boldsymbol{u} = 0$, $\omega = 0$
$V = \int \mathrm{d}V \ , \ A = \int \mathrm{d}A \ , \ J_{\text{BV}} = -c_{\max} C_{\text{r}} \frac{V}{A}$
while $t \le t_{\text{end}}$ **do**
  if AT1 is selected **then**
    $F_\omega^{\text{AT1}} = 0$
  if AT2 is selected **then**
    $F_\omega^{\text{AT2}} = 0$
  if $\omega \geq 0$ **then**
    $\omega \leftarrow \omega$
  else
    $\omega \leftarrow 0$
  if $\omega \leq 1$ **then**
    $\omega \leftarrow \omega$
  else
    $\omega \leftarrow 1$
  $c_{\text{last}} \leftarrow c$
  while $c_{\text{error}} > 0.001$ **do**
    $F_u = 0$
    $F_c = 0$
    if $c \geq c_{\min}$ **then**
      $c \leftarrow c$
    else
      $c \leftarrow c_{\min}$
    if $c \leq c_{\max}$ **then**
      $c \leftarrow c$
    else
      $c \leftarrow c_{\max}$
    $c_{\text{error}} = \int \left(\frac{c - c_{\text{last}}}{c_{\max}} \right)^2 \mathrm{d}V$
  if $H_{\max} \geq \psi^+$ **then**
    $H_{\max} \leftarrow H_{\max}$
  else
    $H_{\max} \leftarrow \psi^+$
  Result: c, $\boldsymbol{u}$, ω write out for postprocessing
  $c^0 \leftarrow c$, $\boldsymbol{u}^0 \leftarrow \boldsymbol{u}$, $\omega^0 \leftarrow \omega$
  $t \leftarrow t + \Delta t$

There are some key issues to allow a convergence by means of the finite element method (FEM), the solution space is HILBERTian–SOBOLEV, $\mathcal{H}^1$, for details we refer to Zohdi (2018). We discretize the domain by standard finite elements. Concentration and damage variables are discretized by $\mathcal{H}^1$-conforming $\mathcal{P}_1$ elements, linear within and continuous across. The displacement variable is discretized by $\mathcal{H}^1$-conforming $\mathcal{P}_2$ elements, quadratic within and continuous across. As known as the Galerkin approach (Ottosen and Petersson 1992), the same space is applied for all test functions, δc, $\delta \boldsymbol{u}$, $\delta \omega$.

For the implementation, i.e., linearizing, assembling, and solving, we exploit Firedrake open-source packages (Ham et al. 2023) with PetSc solvers (Balay et al. 1997, 2024a, b). The algorithm is based on Python scripting that is then compiled in C++ and solved efficiently. We use a docker environment and mpiexec library for parallelization. The geometry and mesh are generated by Salome and Netgen algorithms implemented in Salome, respectively.

The geometry is shown in Fig. 11.1 that is an estimation of a realistic particle in a slurry with ideally spherical particles. We emphasize that spherical particles are never possible for such small dimensions. Tomography studies reveal 3D particle shapes; however, even in particle accelerators, voxel sizes are a limiting factor, and usually 0.5 μm is the best possible precision in particles of around 50 μm in size. We craft a particle by simplifying in 2D and use a fine mesh with element size between

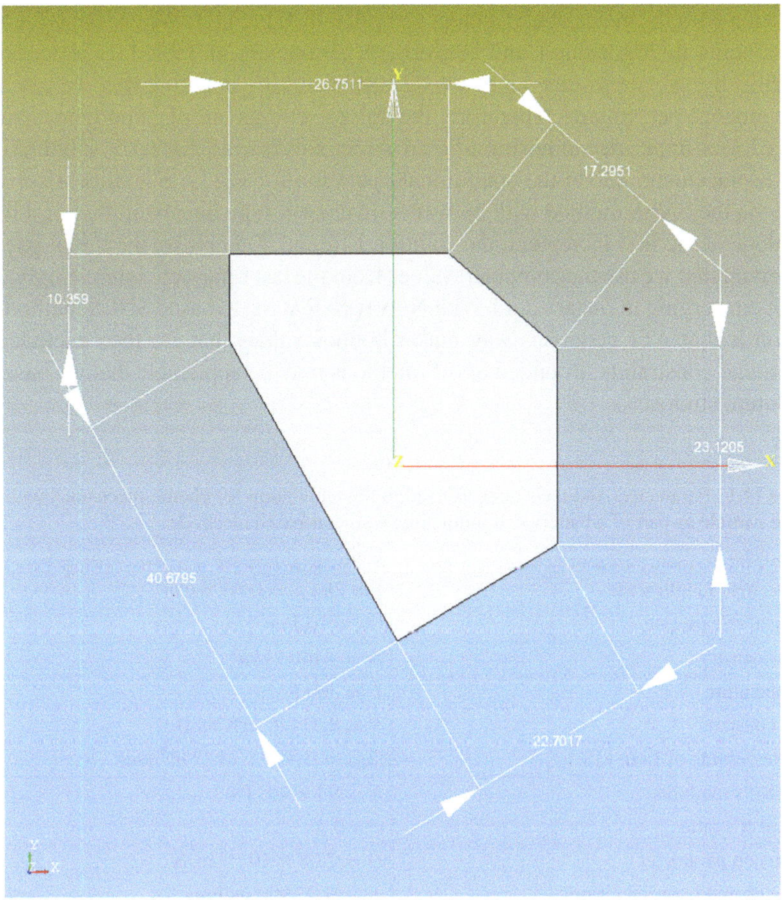

Fig. 11.1 Geometry is a 2D plane with a shape mimicking a particle inspired by tomography study in Taiwo et al. (2017), dimensions are given in μm

0.1 and 0.2 μm. In this way, by using 37 096 nodes, we generate 73 485 triangular elements with a high quality. The aspect ratio (longest edge by shortest edge of each element) is less than 1.1 for 95% of elements. Small size variations and high-quality elements are needed for phase-field simulations that we use for damage modeling. In the computations, the regularization characteristic length ℓ is set equal to two times the maximum element edge length $h_{\max}$ in the mesh, i.e., $\ell = 2h_{\max}$, see Hesammokri and Isaksson (2023).

11.4 Results and Discussion

Simulations are conducted for a continuous lithiation process from zero initial concentration, c_{ref}, for a silicon-based electrode that is known to pulverize even in one full lithiation with the material parameters compiled in Table 11.1. We use the aforementioned units in Algorithm 1 and convert the parameters in Table 11.1 accordingly. Specifically, we stress that 1 mol $\hat{=}$ 6.02214076 × 10^{23} denoting that we calculate with amount per volume. Therefore, the solver settings are of importance and we use a lower-upper decomposition-based direct solver with KRYLOV subiterations. For displacement, due to the conditional operation for $\langle \varepsilon_{kk} \rangle_+$, a nonlinear solver is used via the SNES method with NEWTON–RAPHSON type linearization. For a linear problem, at the first increment, the solution is found. Otherwise, the initial guess is important that we use the computed values from the last time step. Interestingly, AT1 is not converging in linearization with NEWTON–RAPHSON based SNES method, the reason is that AT1 generates way out-of-bounds values that are then excluded by additional constraints. Bounded SNES methods may be applicable that we have left for future studies.

Table 11.1 Parameters and their units as used in the simulation for chemo-mechanics within an active particle as part of a battery cell mimicking a silicon-based electrode

Parameter	Unit
Phase field length scale	$\ell = 2h_{\max} = 0.6676\,\mu\mathrm{m}$
Critical release rate	$G_c = 10$ J/m^2
Mass density	$\rho_0 = 4140$ kg/m^3
Temperature	$T = 300$ K
Gas constant	$R = 8.31446$ J/(K mol)
Molar volume of Li in LiS	$\Omega_{\mathrm{LiS}} = 3.497 \times 10^{-6}$ m^3/mol
YOUNG's modulus	$E = 93 \times 10^9$ Pa
POISSON's ratio	$\nu = 0.3$
Diffusion parameter	$D = 7.08 \times 10^{-15}$ m^2/s
Fully charged concentration	$c_{\max} = 22\,900$ mol/m^3
Reference and initial concentration	$c_{\mathrm{ref}} = 0$ mol/m^3

For stability, we use the degradation function as $g = (1 - \omega)^2 + 10^{-3}$ such that a zero value on the diagonal of the stiffness matrix is prevented. For the weak form for displacement, if we use pure NEUMANN boundary conditions, the operator is singular. Hence, we define a minimal area to hold in x and y directions, around the middle, less than 0.001 of the volume is fixed to prevent a rigid body rotation.

The simulation begins with zero lithiated particles and tries to insert more lithium by a mass flux on the boundary. In this way, we approximate a situation that enough reactants and products are available and the only limiting factor is the possible speed of the migrating particles. By using a moderate charging rate, $C_\mathrm{r} = 2$, we expect that charging and discharging are electrochemically stable but still quick enough to initiate fracture in this electrode. Indeed the concentration distribution shows no significant irregularities for the first lithium insertion shown in Fig. 11.2. This result denotes a heterogeneous (also called polarized) lithiation process mainly due to the geometry irregularities. A perfectly round particle would result in a radially equal lithification preventing concentration gradients. Apart from the irregularities around the vicinity of edges, the solution has some disturbances due to the stress-assisted ion flux. This perturbation is large enough to initiate irregular damage even after 2 min of lithiation. The crack initiation and propagation are qualitatively the same for AT1 and AT2, we present in Fig. 11.3 the damage field, ω, in the current configuration (by adding the displacement on the nodal positions).

Obviously, particle expansion while lithiation creates a detachment from the electrode/electrolyte interface with an introduced damage field along the interface, which is the boundary of the computational domain. In general, the geometry effect is evident that a concentration gradient is causing a (volumetric) strain heterogeneity and thus damage leading to a detachment from the interface. We emphasize that the simulation starts without damage and the crack is initiated with an instability caused by

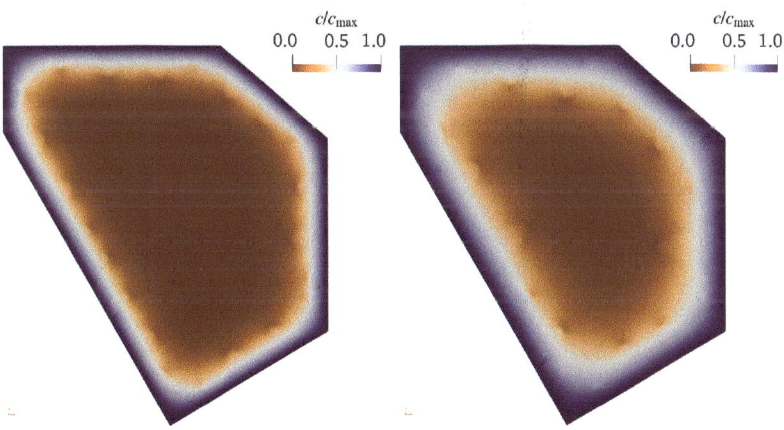

Fig. 11.2 Concentration as a percentage to the maximum possible value, left: after 12 min and right: after 30 min, both with AT1 where AT2 has qualitatively the same solution

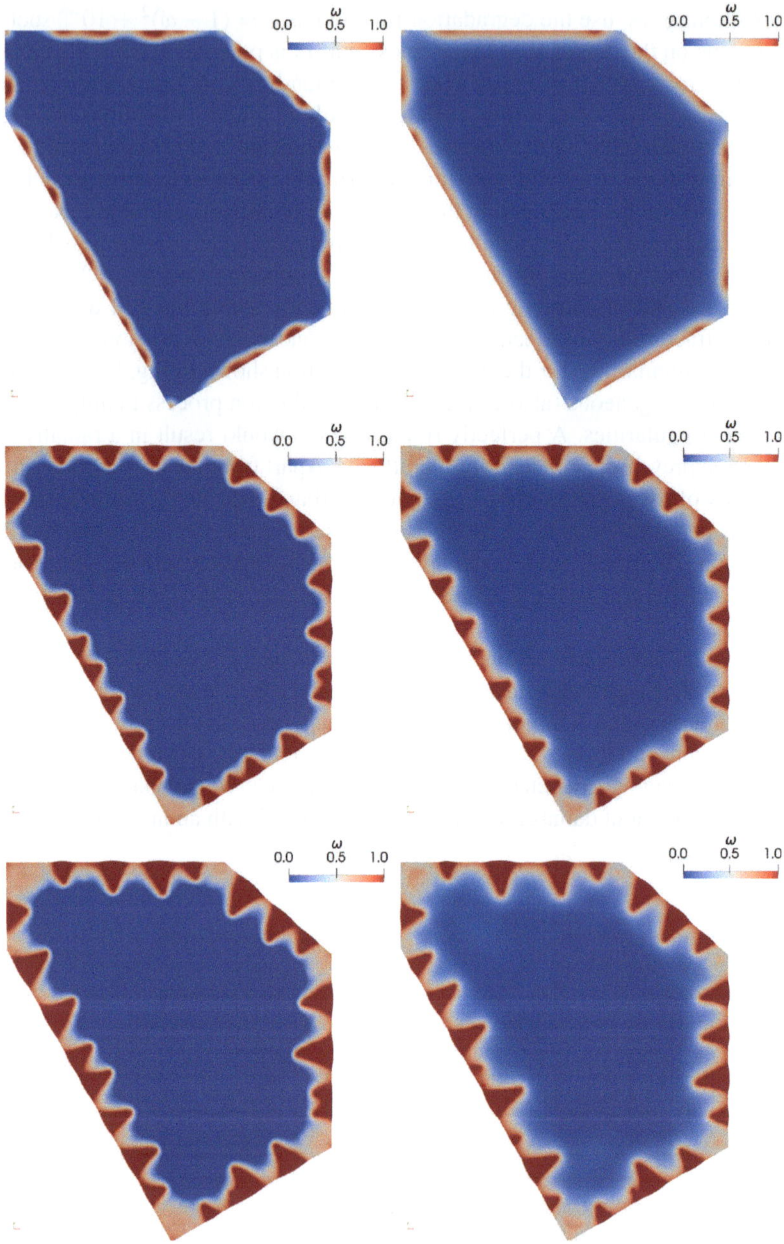

Fig. 11.3 Damage variable, after 2 min (top), 6 min (middle), 12 min (bottom) of lithiation, left: AT1 and right: AT2, in all cases, deformation is scaled by 1 such that the current configuration is presented

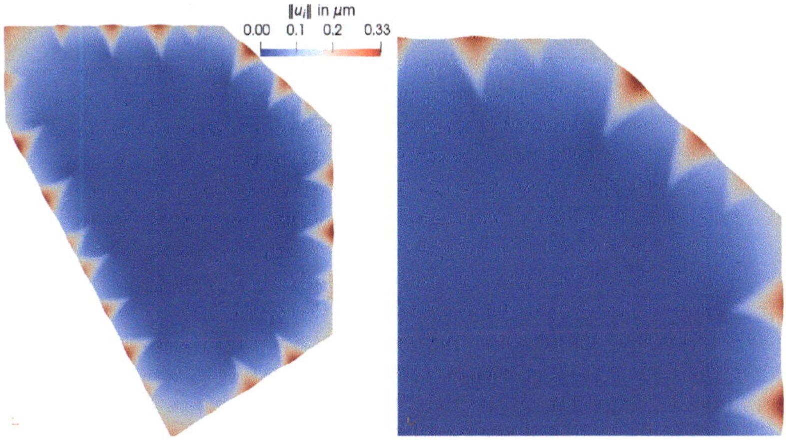

Fig. 11.4 Displacement at 12 min, with AT1, magnitude of displacement is shown in colors on the current configuration, zoom on the right presents pitting phenomenon

the geometry itself leading to a heterogeneous hydrostatic stress gradient. In the literature, specifically for battery research, crack simulations begin with a given damage initially. For example, an analogous implementation in FEniCS has been done in Noii et al. (2024), and a large displacement formulation is in Miehe et al. (2016). We stress that the concentration flux involves stress gradient meaning that the second derivative of displacement is a key driving force for the shown inhomogeneity.

The interface detachment is interpreted as an electrical isolation such that this phenomenon leads to a capacity loss. We have used an assumption that the lithiation is not limited by the degradation directly. Yet mechanical degradation inhibits lithiation indirectly since fracture increases strain and stress that steers flux. AT1 and AT2 demonstrate a similar pattern in damage evolution. AT2 begins a crack initiation immediately, thus, causing a more spread-out solution with respect to AT1. However, no significant difference has been observed in general.

A key outcome is modeling the so-called pitting phenomenon that is visible in Fig. 11.4. Experimental evidence is demonstrated of this phenomenon (Taiwo et al. 2017) yet the computational modeling via continuum mechanics seems to be missing in the literature. The localization of the damage field creates regions on the surface detached from each other owing to the hydrostatic-deviatoric decomposition and geometry irregularities. In this manner, the expansion caused by the lithiation generates larger deformation around these relatively softer regions due to the mechanical degradation related to the damage field. These "bumps" are the first stage of the irregularity leading to a crack formation from the surface toward the particle interior. We stress that this phenomenon is related to the geometry and materials response under tension and compression. It is not because of material heterogeneity or anisotropy or trapping mechanisms. Multiphysics simulations are beneficial to isolate different mechanisms and analyze experimentally observed phenomena to increase the understanding of complex systems.

All developed codes are publicly available in Abali (2025) to be used under the GNU Public license (GNU 2007).

11.5 Conclusion

Lithiation in a battery particle is a multiphysics problem with stress-assisted lithium insertion leading to a volume change. Depending on the composition, these volume changes may be large and destructive. For crack evolution in battery electrodes, geometry plays a significant role. A formulation and a numerical implementation have different challenges. In a formulation, several material parameters are difficult to determine such that simplifications are needed. Yet their significance is not well understood. In this work, we shed light upon a possible choice of damage modeling, namely AT1 and AT2, both are extensively used in phase-field models for crack initiation and propagation. With a 2D simplification, we have observed interface detachment known well in solid-state batteries where the electrolyte is a solid. In gel-type electrolytes, this detachment may be compensated by the electrolyte deformation. Solely electrode simulation needs to be extended to electrode and electrolyte with an interface model accurate enough to tackle problems in the electrification of the transport system and beyond.

Acknowledgements Both authors acknowledge the support by the battery initiative Compel (COMPetitiveness for the ELectrification of the transport system) to the Uppsala University.

References

Abali BE (2020) Modeling mechanochemistry in Li-ion batteries. In: Nicosia G, Romano V (eds) Scientific computing in electrical engineering, mathematics in industry, vol 32. Springer Nature, Cham, Chap. 8, pp 79–91

Abali BE (2025) Supply code for computations. http://bilenemek.abali.org/

Abali BE, Queiruga AF (2019) Theory and computation of electromagnetic fields and thermomechanical structure interaction for systems undergoing large deformations. J Comput Phys Ser 394:200–231

Altenbach H, Sadowski T (eds) (2015) Failure and damage analysis of advanced materials, CISM international centre for mechanical sciences, vol 560. Springer, Vienna

Ambati M, Gerasimov T, De Lorenzis L (2015) A review on phase-field models of brittle fracture and a new fast hybrid formulation. Comput Mech series 55(2):383–405

Ambrosio L, Tortorelli VM (1990) Approximation of functional depending on jumps by elliptic functional via t-convergence. Commun Pure Appl Math Ser 43(8):999–1036

Amirian B, Jafarzadeh H, Abali BE, Reali A, Hogan JD (2022) Phase-field approach to evolution and interaction of twins in single crystal magnesium. Comput Mech Ser 70(4):803–818

Amor H, Marigo JJ, Maurini C (2009) Regularized formulation of the variational brittle fracture with unilateral contact: numerical experiments. J Mech Phys Solid Ser 57(8):1209–1229
Babaei H, Levitas VI (2018) Phase-field approach for stress-and temperature-induced phase transformations that satisfies lattice instability conditions. Part 2. Simulations of phase transformations Si I ↔ Si II. Int J Plast Ser 107:223–245
Balay S, Abhyankar S, Adams MF, Benson S, Brown J, Brune P, Buschelman K, Constantinescu EM, Dalcin L, Dener A, Eijkhout V, Faibussowitsch J, Gropp WD, Hapla V, Isaac T, Jolivet P, Karpeev D, Kaushik D, Knepley MG, Kong F, Kruger S, May DA, McInnes LC, Mills RT, Mitchell L, Munson T, Roman JE, Rupp K, Sanan P, Sarich J, Smith BF, Zampini S, Zhang H, Zhang H, Zhang J (2024a) PETSc web page. https://petsc.org/
Balay S, Abhyankar S, Adams MF, Benson S, Brown J, Brune P, Buschelman K, Constantinescu E, Dalcin L, Dener A, Eijkhout V, Faibussowitsch J, Gropp WD, Hapla V, Isaac T, Jolivet P, Karpeev D, Kaushik D, Knepley MG, Kong F, Kruger S, May DA, McInnes LC, Mills RT, Mitchell L, Munson T, Roman JE, Rupp K, Sanan P, Sarich J, Smith BF, Suh H, Zampini S, Zhang H, Zhang H, Zhang J (2024b) PETSc/TAO users manual. Technical report ANL-21/39—Revision 3.22, Argonne National Laboratory
Balay S, Gropp WD, McInnes LC, Smith BF (1997) Efficient management of parallelism in object oriented numerical software libraries. In: Arge E, Bruaset AM, Langtangen HP (eds) Modern software tools in scientific computing. Birkhäuser Press, pp 163–202
Bothe D, Dreyer W (2015) Continuum thermodynamics of chemically reacting fluid mixtures. Acta Mech Ser 226:1757–1805
Braides A (1998) Approximation of free-discontinuity problems. 1694. Springer
Christensen J, Newman J (2006) Stress generation and fracture in lithium insertion materials. J Solid State Electrochem Ser 10:293–319
Clayton J, Knap J (2016) Phase field modeling and simulation of coupled fracture and twinning in single crystals and polycrystals. Comput Methods Appl Mech Eng Ser 312:447–467
Dreyer W, Jamnik J, Guhlke C, Huth R, Moškon J, Gaberšček M (2010) The thermodynamic origin of hysteresis in insertion batteries. Nat Mater Ser 9(5):448–453
Dreyer W, Guhlke C, Müller R (2013) Overcoming the shortcomings of the Nernst-Planck model. Phys Chem Chem Phys Ser 15(19):7075–7086
Dreyer W, Guhlke C, Müller R (2016) A new perspective on the electron transfer: recovering the butler–volmer equation in non-equilibrium thermodynamics. Phys Chem Chem Phys 18(36):24,966–24,983
Francfort GA, Marigo JJ (1998) Revisiting brittle fracture as an energy minimization problem. J Mech Phys Solids Ser 46(8):1319–1342
GNU Operating System (2007) GNU general public license. http://www.gnu.org/copyleft/gpl.html
Ham DA, Kelly PHJ, Mitchell L, Cotter CJ, Kirby RC, Sagiyama K, Bouziani N, Vorderwuelbecke S, Gregory TJ, Betteridge J, Shapero DR, Nixon-Hill RW, Ward CJ, Farrell PE, Brubeck PD, Marsden I, Gibson TH, Homolya M, Sun T, McRae ATT, Luporini F, Gregory A, Lange M, Funke SW, Rathgeber F, Bercea GT, Markall GR (2023) Firedrake user manual, 1st ed. Imperial College London and University of Oxford and Baylor University and University of Washington
Hesammokri P, Isaksson P (2024) Numerical analyses of brittle crack growth experiments in compression using a modified phase-field theory. Int J Solids Struct Ser 296(112):815
Hesammokri P, Yu H, Isaksson P (2023) An extended hydrostatic-deviatoric strain energy density decomposition for phase-field fracture theories. Int J Solids Struct Ser 262(112):080
Kachanov L (1986) Introduction to continuum damage mechanics, mechanics of elastic stability, vol 10. Springer Science & Business Media
Kästner M, Hennig P, Linse T, Ulbricht V (2016) Phase-field modelling of damage and fracture-convergence and local mesh refinement. In: Naumenko K, Aßmus M (eds) Advanced methods of continuum mechanics for materials and structures, advanced structured materials, vol 60. Springer, pp 307–324
Kiendl J, Ambati M, De Lorenzis L, Gomez H, Reali A (2016) Phase-field description of brittle fracture in plates and shells. Comput Methods Appl Mech Eng Ser 312:374–394

Kim M, Yang Z, Bloom I (2021) Review-the lithiation/delithiation behavior of SI-based electrodes: a connection between electrochemistry and mechanics. J Electrochem Soc 168(1):010,523
Lemaitre J, Desmorat R (2005) Engineering damage mechanics: ductile, creep, fatigue and brittle failures. Springer Science & Business Media
Levitas VI (2018) Phase field approach for stress-and temperature-induced phase transformations that satisfies lattice instability conditions. Part I. General theory. Int J Plast Ser 106:164–185
McNaught AD, Wilkinson A et al (1997) Compendium of chemical terminology, vol 1669. Blackwell Science Oxford
Miehe C, Welschinger F, Hofacker M (2010) Thermodynamically consistent phase-field models of fracture: variational principles and multi-field FE implementations. Int J Numer Methods Eng Ser 83(10):1273–1311
Miehe C, Hofacker M, Welschinger F (2010) A phase field model for rate-independent crack propagation: robust algorithmic implementation based on operator splits. Comput Methods Appl Mech Eng Ser 199(45–48):2765–2778
Miehe C, Dal H, Schänzel LM, Raina A (2016) A phase-field model for chemo-mechanical induced fracture in lithium-ion battery electrode particles. Int J Numer Methods Eng Ser 106(9):683–711
Müller I, Müller WH (2023) Electrodynamics and rational thermodynamics. ZAMM-J Appl Math Mech/Zeitschrift für Angewandte Mathematik und Mechanik 103(4):e202300,209
Mumford DB, Shah J (1989) Optimal approximations by piecewise smooth functions and associated variational problems. Commun Pure Appl Math Ser 42(5):577–685
Murakami S (2012) Continuum damage mechanics: a continuum mechanics approach to the analysis of damage and fracture, vol 185. Springer Science & Business Media
Noii N, Milijasevic D, Waisman H, Khodadadian A (2024) Fatigue failure theory for lithium diffusion induced fracture in lithium-ion battery electrode particles. Comput Methods Appl Mech Eng Ser 428(117):068
Öchsner A (2016) Continuum damage mechanics. In: Continuum damage and fracture mechanics. Springer, pp 65–84
Ottosen NS, Petersson H (1992) Introduction to the finite element method. Prentice Hall
Pham K, Amor H, Marigo JJ, Maurini C (2011) Gradient damage models and their use to approximate brittle fracture. Int J Damage Mech Ser 20(4):618–652
Provatas N, Elder K (2011) Phase-field methods in materials science and engineering. Wiley
Rabotnov YN (1959) Mekhanizm dlitelnogo razrusheniya. Sbornik Voprosy prochnosti materialov I konstruktsii, pp 5–7
Reich FA, Rickert W, Müller WH (2018) An investigation into electromagnetic force models: differences in global and local effects demonstrated by selected problems. Contin Mech Thermodyn Ser 30:233–266
Taiwo OO, Paz-García JM, Hall SA, Heenan TM, Finegan DP, Mokso R, Villanueva-Pérez P, Patera A, Brett DJ, Shearing PR (2017) Microstructural degradation of silicon electrodes during lithiation observed via operando x-ray tomographic imaging. J Power Sources Ser 342:904–912
Taiwo OO, Paz-García JM, Hall SA, Heenan TM, Finegan DP, Mokso R, Villanueva-Pérez P, Patera A, Brett DJ, Shearing PR (2017) Microstructural degradation of silicon electrodes during lithiation observed via operando x-ray tomographic imaging. J Power Sources Ser 342:904–912
Teichtmeister S, Kienle D, Aldakheel F, Keip MA (2017) Phase field modeling of fracture in anisotropic brittle solids. Int J Non-Linear Mech Ser 97:1–21
van Dijk NP, Espadas-Escalante JJ, Isaksson P (2020) Strain energy density decompositions in phase-field fracture theories for orthotropy and anisotropy. Int J Solids Struct Ser 196:140–153
Yang F (2005) Interaction between diffusion and chemical stresses. Mater Sci Eng: Ser 409(1–2):153–159
Zhang X, Shyy W, Sastry AM (2007) Numerical simulation of intercalation-induced stress in li-ion battery electrode particles. J Electrochem Soc Ser 154(10):A910
Zohdi TI (2018) Finite element primer for beginners. Springer

Chapter 12
On the Non-constant Dielectric Susceptibility in Continuum Models for Electrolytes

Rüdiger Müller

Dedicated to the memory of Wolfgang Dreyer, whom I was fortunate enough to experience as an outstanding scientist, passionate teacher and great personality.

Abstract A recently developed Debye–Hückel-type ion–ion interaction free energy model based on concentration and temperature-dependent dielectric susceptibility is applied to homogeneous electrolyte bulk regions where it is known that non-constant susceptibility is necessary for a correct description of the heat capacity and the apparent molar volume. The model imposes a thermodynamic compatibility condition that prescribes the temperature dependence of the susceptibility. Here, this condition is successfully tested against experimental data. Contrary to the electrolyte bulk, where in equilibrium the electric field vanishes, the field strength can rise extremely high in charged boundary layers. Therefore, we consider in the layer an explicit polarization free energy contribution instead of the ion–ion interaction energy and a dependence of the susceptibility on the electric field to the model. The strong influence of combined field and concentration dependence of the susceptibility on the double-layer structure with its spatial distribution of ions and the solvent is illustrated. The impact of the field-dependent dielectric saturation on the boundary layer capacitance is studied numerically.

12.1 Introduction

In continuum electro-thermodynamics, cf. de Groot and Mazur (1984); Müller (1985); Dreyer et al. (2018), the dielectric susceptibility, along with the magnetic susceptibility, is the crucial link that establishes the coupling of the classical balance

---

R. Müller (✉)
Formerly with WIAS Berlin, Berlin, Germany
e-mail: rue.m@gmx.de; Ruediger.Mueller@thws.de

Fakultät Angewandte Natur- und Geisteswissenschaften, Technische Hochschule Würzburg-Schweinfurt, Münzstraße 12, 97070 Würzburg, Germany

J. Fuhrmann et al. (eds.), *Advances in Continuum Physics*, Advanced Structured Materials 238, https://doi.org/10.1007/978-3-031-93918-1_12

equations of matter with the Maxwell equations of electrodynamics. Notwithstanding this, in applications of electrochemical models, the dielectric susceptibility is usually treated as just a constant parameter, most likely for the sake of simplicity, see, e.g., the standard textbook (Newman and Thomas-Alyea 2004). There are therefore good reasons to study in detail here various important effects that arise in electrochemical applications when the dielectric susceptibility is not simply a constant but depends on the temperature, the mixture composition and the electric field. Before we state the aims and results of this contribution, let us first lay out the conceptual background relevant for the modeling.

Polarization and susceptibility in general

Susceptibility relates the *electric field* $\boldsymbol{E}$ to the *vector of polarization* $\boldsymbol{P}$ which represents effects originating from a smaller scale than the scales that are resolved in the macroscopic continuum description of matter, cf. e.g., Kovetz (2000). Similarly, magnetic susceptibility relates the *magnetic flux density* $\boldsymbol{B}$ to the *Lorentz magnetization* $\boldsymbol{M}$, but due to our focus on application to liquid electrolytes we will not consider magnetic effects further. While on the macroscopic scale e.g., water is modeled as an electrically neutral continuum, the unequal distribution of positive and negative charge within the molecule gives rise to a fixed electric dipole. In the absence of an electric field, these dipoles are randomly oriented due to thermal noise and they do not contribute an effect on the macroscopic scale. In the presence of an outer electric field the thermal noise is superimposed by a force that tends to align the microscopic dipoles with the field. The resulting effect on the macroscopic scale is described by the vector of polarization $\boldsymbol{P}$. Within a certain range, one can assume that the degree of alignment grows linearly with the electric field strength and therefore write $\boldsymbol{P} = \chi\, \varepsilon_0 \boldsymbol{E}$. As soon as almost full alignment of the dipoles is achieved however, polarization can no longer increase linearly with the field strength and the proportionality factor χ must depend on $\boldsymbol{E}$. The basic theoretical derivation of this *dielectric saturation* was developed in Debye (1929); Onsager (1936); Kirkwood (1939); Booth (1951) and has undergone numerous extensions since then, see, e.g., Fulton (2009). In addition to the fixed dipoles, the electric field induces further dipoles in molecules by shifting the charge center of the electrons. This induced polarization also undergoes saturation at high field strengths. However, the necessary field strength to observe dielectric saturation is in the order of $10^8\ \mathrm{V\,m^{-1}}$ and is thus too large to be realized on the macroscopic scale. It can, however, be achieved in the immediate vicinity of ions or near charged surfaces in the double layer.

Since the alignment of the dipoles competes with the thermal motion, the dielectric susceptibility decreases with increasing temperature. In addition, thermal expansion leads to a lower density of dipoles and thereby also reduces the susceptibility. The dynamic process of (re-)orientation of dipoles in a time depending and possibly alternating electric field is governed by the Debye equation for dielectric relaxation, cf. de Groot and Mazur (1984). While at very high frequencies only the induced dipoles can follow the rapidly alternating field orientation, the resonance behavior of the dipoles at low frequencies is used to extrapolate to an equilibrium configuration. Within the framework of non-equilibrium thermodynamics, the susceptibility χ, which is also called static dielectric permittivity, is an equilibrium quantity that is determined by the free energy and restricted by the second law of thermodynamics, whereas non-equilibrium quantities are governed by kinetic coefficients. This raises the question of the legitimacy of determining the susceptibility by extrapolation from high-frequency measurements. At least for aqueous electrolytes however, it is shown in Chandra (2000) that any remaining dynamic contribution to the static permittivity after extrapolation must be negligibly small.

Ion solvation and susceptibility in mixtures

The strong electric field near an ion attracts the oppositely charged pole of surrounding molecules with a dipole. In so-called polar solvents such as water, the ions therefore form a solvation shell of oriented dipolar solvent molecules around themselves. The solvent molecules in the solvation shell are not free to move independently but can only move along with the center ion. This ion solvation has profound implications for thermodynamic modeling because only the solvated ion as a whole participates in the entropy of the mixture and not the individual solvent molecules in the solvation shell. Furthermore, diffusion is to be understood as the movement of solvated ions relative to the barycentric velocity of the mixture. The concept of electrolytes as mixtures containing solvated ions is described in Landstorfer et al. (2016); Dreyer et al. (2018, 2019).

In addition, there is another aspect of solvation, namely the bound solvent molecules cannot align their dipoles according to an external electric field, as is the case of free solvent molecules. The presence of solvated ions therefore reduces the susceptibility of the mixture below the value of the pure solvent. This effect is known as *dielectric decrement*, cf. Hasted et al. (1948); Buchner et al. (1999); Marcus (2013). Within a certain range, this decrement is linearly proportional to the ion concentration and thus also to the volume of the ions in the mixture. The interaction between ions and solvent and the formation of complexes such as solvation shells significantly influence the volume of electrolytes, see Redlich and Meyer (1964); Millero (1971); Marcus (2011). For a volume V_S in m^3 of the pure solvent and addition of N_E^0 molecules of a salt, the *apparent molar volume* Φ_V with the dimensions $\mathrm{L\,mol^{-1}}$ is used to express the actual volume of the mixture as $V = V_S + \frac{1}{1000}\Phi_V\,\frac{N_E^0}{1000 N_A}$.

Here N_A denotes the Avogadro number as specified in Table 12.1. Modeling of Φ_V is based on non-constant susceptibility in the Debye–Hückel theory that is discussed next.

Table 12.1 Physical constants

Boltzmann constant	$k_B = 1.380649 \times 10^{-23}\ \mathrm{J\,K^{-1}}$
Elementary charge	$e_0 = 1.602176634 \times 10^{-19}\ \mathrm{C}$
Avogadro number	$N_A = 6.02214076 \times 10^{23}\ \mathrm{mol^{-1}}$
Vacuum dielectric constant	$\varepsilon_0 = 8.8541878188 \times 10^{-12}\ \mathrm{C\,V^{-1}\,m^{-1}}$

Susceptibility in the Debye–Hückel theory

The Debye–Hückel theory (Debye and Hückel 1923) established the electric ion–ion interaction as a cause of non-ideal behavior of dilute electrolytes, without having to rely on incomplete dissociation as an explanation. The theory is often considered too simplified, but nevertheless useful at least in the limit of strong dilution. It is commonly applied in models in the form of limiting laws for infinite dilution, cf. e.g., Newman and Thomas-Alyea (2004). The original derivation in Debye and Hückel (1923) first considers a mesoscopic problem to calculate the electric potential of a selected ion surrounded by an electrolyte that is characterized only by its susceptibility χ. Then the energy contributions of all ions are averaged to obtain the inner energy density from which the free energy density is subsequently derived by integrating a thermodynamic relation with respect to temperature. However, it was soon recognized that a non-constant and, in particular, temperature-dependent susceptibility is required, since for the specific heat capacity or the apparent molar volume the limit law otherwise even predicted the wrong sign of the slope (Redlich and Meyer 1964) and the correct sign could only be obtained if the susceptibility depends on the pressure (Redlich and Rosenfeld 1931). Because a susceptibility dependent on pressure or temperature is not compatible with the derivation (Debye and Hückel 1923), the thought experiment of a hypothetical charging process was introduced in Debye (1924). Since this charging process leads to the same result for the free energy as previously in Debye and Hückel (1923), it is now generally accepted, although the justification of the charging process remains somewhat controversial, cf. e.g., Groß and Halpern (1925); Kontogeorgis et al. (2018). Various modifications and extensions with regard to non-constant susceptibility and the charging process are derived and reviewed, e.g., in Vincze et al. (2010); Shilov and Lyashchenko (2015); Kontogeorgis et al. (2018); Silva et al. (2023); Valiskó and Boda (2023). The use of solvated ion radii is discussed and rejected in Vincze et al. (2010); Valiskó and Boda (2014). In Dreyer and Müller (2025), an alternative derivation of an electric ion–ion interaction energy is given based on the thermodynamic framework of Dreyer et al. (2018). This derivation does not involve the integration of a thermodynamic relation and does not depend on any charging process in order to allow the susceptibility to depend on temperature and concentration. However, this does not allow the temperature dependence to be freely chosen, but instead it is predetermined by a thermodynamic consistency constraint.

Susceptibility in boundary layers and double-layer capacitance

The early electrolyte theory from before (Debye and Hückel 1923) based on the analogy of dilute electrolytes to ideal gases, leads to a double layer according to a Poisson–Boltzmann (PB) model. The PB model leads to the Gouy–Chapman (GC) double-layer capacitance, which contains the susceptibility as the only model parameter. Its reliance on the dilute electrolyte assumption makes the GC model only useful near the zero charge potential E^{PZC} of an interface as the model provides no bounds for the capacitance at large applied potentials. Therefore two ways of extending the GC capacitance model have been applied. The first way is to introduce in addition to an "outer" or "diffuse" layer according to the GC model an "inner" or "compact" Stern layer with a capacitance independent of the electrolyte concentration. The second way consists of so-called steric Poisson–Boltzmann models (sPB), that introduce size parameters into the model and handle volume exclusion effects of finite size ions by a lattice gas model cf. e.g., Kralj-Iglič and Iglič (1996); Borukhov et al. (1997); Kilic et al. (2007). Extensions of lattice gas models to ions of different sizes have also been developed, cf. e.g., Chu et al. (2007); Li (2009); Liu and Eisenberg (2014) and also the general thermodynamic model of Dreyer et al. (2013); Landstorfer et al. (2016); Dreyer et al. (2018) reduces in the incompressible limit to this class of models. Both ways of extension can also be combined, see, e.g., Bazant et al. (2009, Fig. 8). In all these mentioned sPB approaches the susceptibility is constant. When tuning their size parameters to experimental data, like (Valette 1981, 1982, 1989), the models tend to imply solvated ion radii that appear unrealistic large without adding an additional Stern-layer capacitance, unless some "effective" constant χ^{eff} considerably lower than the susceptibility of the solvent is used, cf. Landstorfer et al. (2016); Landstorfer and Müller (2022).

In Bazant et al. (2009), also a different way is provided to obtain a Stern like thin layer from a sPB model by integrating dielectric saturation into the model. The idea of dielectric saturation is also included to a PB model in Gongadze et al. (2011) and further adapted to sPB models in Gongadze et al. (2011); Gongadze and Iglič (2012); Berg and Benjaminsen (2014). The relation of these models to concentration-dependent dielectric decrement is demonstrated in Dubtsov et al. (2018). Steric modified Poisson–Boltzmann models accounting for the concentration dependence with linear dielectric decrement are developed in Ben-Yaakov et al. (2011); Hatlo et al. (2012); Nakayama and Andelman (2015); Figliuzzi et al. (2016) and extended to different ion sizes in Gupta and Stone (2018). The model of Dreyer et al. (2013); Landstorfer et al. (2016); Dreyer et al. (2018) is extended in Landstorfer and Müller (2022) to account for both concentration- and field-dependent susceptibility.

Objectives and results

The electrode|electrolyte interface and its differential capacitance are still not fully understood due to their complexity, cf. e.g., Schmickler (2020). By considering only salts dissociating into non-adsorbing ions, such as KPF_6, or to lesser extend NaF or $NaClO_4$, specific electrolyte–surface interactions can be excluded from the modeling and the resulting model for the boundary layer capacitance only depends on the electrolyte properties. Most of the parameters of a steric electrolyte model for the boundary layer capacitance are also present in a bulk electrolyte model in equilibrium. *In this paper*, we propose to identify whenever possible the electrolyte model parameters from bulk electrolyte experiments. We apply the novel Debye–Hückel-like ion interaction model from Dreyer and Müller (2025) which builds on concentration and temperature-dependent susceptibility to model the apparent molar volume, ion solvation and the dissociation degree. From this, solvated ion radii, solvation number and Gibbs energies of the dissociation reaction are determined. In particular, it is demonstrated that the model can handle the negative apparent molar volume of NaF and associated negative radii of unsolvated ions. Where available, independent experimental data for the nonlinear concentration dependence of the susceptibility is used to confirm the determined model parameters. We separately study the thermodynamic consistency condition on the temperature dependence of the susceptibility that is imposed by Dreyer and Müller (2025). By adjusting only the temperature dependence of the Gibbs energy, remarkably good reproduction of experimental data of NaCl solutions is reached for concentrations up to $5\,\mathrm{mol\,L^{-1}}$ and temperatures between 5°C and 35°C. Then, we couple the fully parameterized bulk model to the boundary layer. Here, dielectric saturation in locally extremely strong electric fields has to be added to the model. We study numerically the impact of dielectric decrement combined with dielectrics saturation on the double layer and its capacitance. In comparison to Landstorfer and Müller (2022), new features of the capacitance curves are observed, due to the different parameter range related to the rather small solvation number $\kappa_\alpha = 6$ that is implied by the bulk model. Moreover, we benefit from the in comparison considerably smaller parameter space, since only parameters of the dielectric saturation remain to be determined from experimental capacitance data.

Before that, we first state in Sect. 12.2 the model equations and the underlying model assumptions. Then, the applied constitutive models are summarized in Sect. 12.3, before they are applied to the bulk electrolyte in Sect. 12.4 and to the layers in Sect. 12.5.

12.2 General Model Equations

We follow the bulk-surface electro-thermodynamic modeling approach of Guhlke (2015); Dreyer et al. (2018); Müller and Landstorfer (2023). It is based on the particular form of an the entropy principle as developed in Bothe and Dreyer (2015), where also differences to similar earlier variants, e.g., de Groot and Mazur (1984); Müller (1985) are reviewed. In Guhlke (2015); Dreyer et al. (2018), the classical non-relativistic balance equations of matter (Bothe and Dreyer 2015) are coupled to the general Maxwell equations with universal Maxwell–Lorentz aether relations from Truesdell and Toupin (1960). Since this coupling comes with several problems for the construction of constitutive equations, it is proposed in Müller and Landstorfer (2023) to restrict from the beginning the electromagnetic field to one of the two non-overlapping Galileo limits of the Maxwell equations, cf. Le Bellac and Lévy-Leblond (1973); Rousseaux (2013). For electrochemical applications, the quasi-electrostatic Galileo limit of Maxwell's equations is applied such that finally the same electrochemical model is reached in Müller and Landstorfer (2023) as in Dreyer et al. (2018), but all the conceptual difficulties can be circumvented.

Maxwell's equations are valid across all scales. Thus, conservation of charge is expressed in an universal way as

$$\operatorname{div}(\varepsilon_0 \boldsymbol{E}) = n^{\mathrm{e}}, \tag{12.1}$$

where $\boldsymbol{E}$ is the *electric field*, $\boldsymbol{D} = \varepsilon_0 \boldsymbol{E}$ is the *charge potential* and n^{e} denotes the *(total) electric charge*. Going to the macroscopic scale of the balance equations of matter, some internal electric structure might have been neglected and therefore the macroscopic *free charge density* n^{F} defined below in (12.5) does not necessarily coincide with the total charge density n^{e}. For the coupling of field and matter, (Guhlke 2015; Dreyer et al. 2018) emphasizes the purely formal introduction of the *vector of polarization* $\boldsymbol{P}$ from this difference of the charge densities as

$$\mathrm{div}(\boldsymbol{P}) = n^{\mathrm{F}} - n^{\mathrm{e}} \; = -n^{\mathrm{P}}, \qquad (12.2)$$

and leaves the derivation of a macroscopic constitutive material model related to $\boldsymbol{P}$ to a later step after the application of the entropy principle. The entropy principle in Guhlke (2015); Dreyer et al. (2018); Müller and Landstorfer (2023) is based on the choice of $\boldsymbol{P}$ as an independent variable for the constitutive function of the entropy. However, by the assumptions formulated below, it is possible to apply a change of variables from $\boldsymbol{P}$ to $\boldsymbol{E}$, cf. Landstorfer and Müller (2022); Müller and Landstorfer (2023), which eases the constitutive modeling but requires some care in relation to the free energy density, the definition of pressure and formulation of the balances for momentum and energy.

12.2.1 Assumptions

- All relevant features in the considered electrochemical applications can be well described in the *quasi-electrostatic* limit of Maxwell equations. Thus, the electromotoric force equals the electric field $\boldsymbol{E}$ and there exists an *electrostatic potential* φ, such that

$$\boldsymbol{E} = -\nabla\varphi, \qquad [\![\varphi]\!] = \text{const. on intersecting surfaces}, \qquad (12.3)$$

  and the constant can be chosen as zero.
- The timescale of polarization relaxation in liquid electrolytes is typically in the range of 10^{-8} s (cf. Buchner et al. (1999)), and therefore orders of magnitude smaller than the relevant observed time scales of the considered applications. Thus, we assume quasi-equilibrium of polarization relaxation.
- We assume that the dielectric properties of the liquid electrolyte are reasonably well represented by a scalar susceptibility χ which may be a function of temperature, electrolyte composition and the electric field strength $|\boldsymbol{E}|$. The vector of polarization $\boldsymbol{P}$ thus satisfies (also see Table 12.1)

$$\boldsymbol{P} = \chi(T, (n_\alpha)_\alpha, |\boldsymbol{E}|) \cdot \varepsilon_0 \boldsymbol{E}. \qquad (12.4)$$

Table 12.2 Reference values and material parameters for water

$T^{\text{ref}} = 25°\text{C}$	$n^{\text{ref}} = 1000\, N_A$
$m_S = 18.0152833\,\text{g}\,\text{mol}^{-1}$	$z_S = 0$
$\upsilon_S^{\text{ref}} = (55.345\, n^{\text{ref}})^{-1}$	$K = 2.2 \times 10^{+9}\,\text{J}\,\text{m}^{-3}$
$\chi_S^{\text{ref}} = 78.38$	

12.2.2 Balance Equations of Charged Mixtures

We summarize here the model equations adapted to the above simplifying assumptions, i.e., fast polarization relaxation. We consider an electrolyte that occupies at time $t > 0$ the spatial domain $\Omega(t) \subseteq \mathbb{R}^3$. The electrolyte is modeled as a mixture of different constituents and we use an index set I to refer to the different constituents of the mixture in Ω. We denote by n_α for $\alpha \in I$ the particle number densities per m^3. Each constituent has the (atomic) mass m_α and the net charge $z_\alpha e_0$, where z_α is the charge number of the constituent and e_0 is the elementary charge (also see Table 12.1).

The *total mass density* and the free charge density are defined as

$$\rho = \sum_{\alpha \in I} m_\alpha n_\alpha, \qquad n^{\text{F}} = e_0 \sum_{\alpha \in I} z_\alpha n_\alpha. \tag{12.5}$$

Neglecting gravitational forces, the balance equations within the volume domain Ω for time $t > 0$ are

$$-\operatorname{div}((1+\chi)\varepsilon_0 \nabla\varphi) = n^{\text{F}}, \tag{12.6a}$$

$$\partial_t n_\alpha + \operatorname{div}(n_\alpha \boldsymbol{v} + \tfrac{1}{m_\alpha}\boldsymbol{J}_\alpha) = \sum_k \nu_\alpha^k R^k, \tag{12.6b}$$

$$\partial_t(\rho\boldsymbol{v}) + \operatorname{div}\left(\rho\boldsymbol{v}\otimes\boldsymbol{v} - \boldsymbol{\Sigma}\right) = 0, \tag{12.6c}$$

$$\partial_t(\rho u) + \operatorname{div}\left(\rho u\, \boldsymbol{v} + \boldsymbol{Q}\right) = \boldsymbol{J}^{\text{F}}\cdot\boldsymbol{E} + \left(\boldsymbol{\Sigma} - \boldsymbol{\sigma}^{\text{Mw}} - \boldsymbol{E}\otimes\boldsymbol{P} + (\boldsymbol{E}\cdot\boldsymbol{P})\mathbf{1}\right) : \nabla\boldsymbol{v}$$

Table 12.3 Salt specific parameters for aqueous solutions at the reference temperature $T^{\text{ref}} = 25°\text{C}$

	NaF	KF	NaCl	$NaClO_4$	$KClO_4$	KPF_6
Φ_V^0 / [$\text{L}\,\text{mol}^{-1}$]	−3.04	7.18	16.29	42.53	52.75	65.79
$\Delta\upsilon$ / υ_S^{ref}	−0.5	−0.6	−0.4	−0.35	−0.2	−0.7
d	12	11	12	14	13	13
Δg / [eV]	−0.105	−0.085	−0.78	−0.085	−0.14	−0.15

$$+ \Big(\partial_t \boldsymbol{P} + (\boldsymbol{v} \cdot \nabla)\boldsymbol{P}\Big) \cdot \boldsymbol{E}. \tag{12.6d}$$

Here $\boldsymbol{v}$, $\boldsymbol{J}_\alpha$ and R^k denote the *barycentric velocity*, the non-convective mass fluxes and reaction rates, respectively. Not all of the diffusive mass fluxes $\boldsymbol{J}_\alpha$ are linearly independent and for each of the reactions, the reaction rate R^k has to conserve mass and charge. We thus have for the stoichiometric coefficients ν_α^k and the fluxes the constraints

$$\sum_{\alpha\in I} m_\alpha \nu_\alpha^k = 0, \qquad \sum_{\alpha\in I} z_\alpha \nu_\alpha^k = 0, \qquad \sum_{\alpha\in I} \boldsymbol{J}_\alpha = 0. \tag{12.7}$$

We denote the *total stress tensor* by $\boldsymbol{\Sigma}$ to write the momentum balance in the generic way (12.6d), $\boldsymbol{\Sigma}$ in particular contains the *Maxwell stress tensor* $\boldsymbol{\sigma}^{\mathrm{Mw}}$

$$\boldsymbol{\sigma}^{\mathrm{Mw}} = \varepsilon_0 \boldsymbol{E} \otimes \boldsymbol{E} - \tfrac{\varepsilon_0}{2}|\boldsymbol{E}|^2 \cdot \mathbf{1} \tag{12.8}$$

to account for the Lorentz force. We denote the unit tensor by $\mathbf{1}$ and ρu is the *inner energy density*. For balance equations that provide boundary conditions, we refer to Dreyer et al. (2018); Müller and Landstorfer (2023).

12.2.3 Thermodynamic Consistent Constitutive Framework

The entropy principle is applied in Dreyer et al. (2018); Müller and Landstorfer (2023) with an *entropy density* ρs which is a function of the independent variables ρu, n_α and $\boldsymbol{P}$. Then, the further constitutive modeling is greatly eased by the introduction of the *temperature* T and the *free energy density* $\rho\psi$ and the already mentioned change of variables from $\boldsymbol{P}$ to $\boldsymbol{E}$. We introduce

$$\rho\psi = \rho u - T\,\rho s, \qquad \text{with} \qquad \frac{1}{T} = \frac{\partial}{\partial \rho u}\rho s \tag{12.9}$$

and obtain the consistency conditions

$$\rho s = -\frac{\partial}{\partial T}\rho\psi, \qquad \rho u = -T^2 \frac{\partial}{\partial T}\left(\frac{\rho\psi}{T}\right). \tag{12.10}$$

Due to the change of variables, the assumed polarization equilibrium implies

$$\boldsymbol{P} = -\frac{\partial \rho\psi}{\partial \boldsymbol{E}}. \tag{12.11}$$

The *chemical potentials* and *pressure* are defined as

$$\mu_\alpha = \frac{\partial \rho\psi}{\partial n_\alpha}, \qquad p = -\rho\psi + \sum_{\alpha\in\mathcal{I}} n_\alpha \mu_\alpha. \tag{12.12}$$

We note that in general μ_α and p contain electric field-dependent contributions and the definition of the pressure, that has the form a Gibbs–Duhem relation in the absence of electric field, is related to the change of variables from $\boldsymbol{P}$ to $\boldsymbol{E}$, cf. Landstorfer and Müller (2022); Müller and Landstorfer (2023).

We let $0 \in \mathcal{I}$ refer to a selected constituent. Typically the solvent is chosen as a reference constituent. The constitutive equations following Dreyer et al. (2018); Müller and Landstorfer (2023) are for each counting index k and $\alpha \in \mathcal{I} \setminus \{0\}$

$$R^k = R_0^k \cdot \left(1 - \exp\left(\frac{A^k}{k_B T} \sum_{\alpha\in\mathcal{I}^\pm} \nu_\alpha^k \mu_\alpha\right)\right), \tag{12.13a}$$

$$\boldsymbol{J}_\alpha = -\sum_{\gamma\in\mathcal{I}\setminus\{0\}} M_{\alpha\gamma} \cdot \left(\nabla\left(\frac{\mu_\gamma}{T} - \frac{\mu_0}{T}\frac{m_\gamma}{m_0}\right) + \frac{e_0}{T}\left(z_\gamma - z_0 \frac{m_\gamma}{m_0}\right)\nabla\varphi\right), \tag{12.13b}$$

$$\boldsymbol{\Sigma} = p\cdot\mathbf{1} + \boldsymbol{\sigma}^{\mathrm{visc}} + \boldsymbol{\sigma}^{\mathrm{Mw}} + \boldsymbol{E}\otimes\boldsymbol{P}, \tag{12.13c}$$

$$\boldsymbol{\sigma}^{\mathrm{visc}} = \eta_b \operatorname{div}(\boldsymbol{v})\,\mathbf{1} + \eta_s \cdot \left(\nabla\boldsymbol{v} + (\nabla\boldsymbol{v})^T\right), \tag{12.13d}$$

$$\boldsymbol{Q} = -\frac{\kappa}{T^2}\nabla T. \tag{12.13e}$$

The phenomenological coefficients satisfy $A^k > 0$, $\kappa > 0$, $(\eta_b + \frac{3}{2}\eta_s) > 0$, $\eta_s > 0$ and $M_{\alpha\gamma}$ form a positive definite matrix.

12.2.4 Equilibrium

In thermodynamic equilibrium, i.e., $0 = R^k$, $0 = \boldsymbol{J}_\alpha$, and $0 = \boldsymbol{\sigma}^{\mathrm{visc}}$, (12.13a), (12.13b) and balance equations (12.6a) and (12.6c) imply

$$0 = \sum_{\alpha\in\mathcal{I}} \nu_\alpha^k \mu_\alpha \qquad \text{for all } k, \tag{12.14a}$$

$$0 = \nabla\left(\mu_\alpha - \tfrac{m_\alpha}{m_0}\mu_0\right) + z_\alpha e_0 \boldsymbol{E} \qquad \text{for } \alpha \in \mathcal{I}\setminus 0, \tag{12.14b}$$

$$\nabla p = n^{\mathrm{F}}\boldsymbol{E} + (\nabla\boldsymbol{E})\boldsymbol{P}. \tag{12.14c}$$

While taking the gradient of (12.12) and applying (12.11) yields

$$\nabla p = \sum_{\alpha\in\mathcal{I}} n_\alpha \nabla\mu_\alpha + (\nabla\boldsymbol{E})\boldsymbol{P}, \tag{12.15}$$

we get from summing the equilibrium conditions (12.14b)

$$\sum_{\alpha \in I} n_\alpha \nabla \mu_\alpha = n^{\mathrm{F}} \boldsymbol{E} + \frac{\rho}{m_0} \nabla \mu_0. \tag{12.16}$$

Thus, combining (12.16) and (12.15) with (12.14c), we can first conclude that $\nabla \mu_0 = 0$. Then we get in a next step, for the electrochemical potential $\mu_\alpha + e_0 z_\alpha \varphi$ of all species, where $\boldsymbol{E} = -\nabla\varphi$, the well-known equilibrium conditions

$$\nabla(\mu_\alpha + e_0 z_\alpha \varphi) = 0 \quad \text{for all } \alpha \in I. \tag{12.17}$$

Since this condition follows as a consequence from the consistent coupling of the diffusional equilibrium with the momentum balance and the mechanical and polarization equilibrium, (12.17) is called *extended diffusional equilibrium*, cf. Landstorfer and Müller (2022).

Electric field-pressure relation and boundary layer charge

In the case of a planar one-dimensional setting, the $x-$component of the total stress tensor $\boldsymbol{\Sigma}$ reduces to

$$\Sigma = -p + (1+\chi)\,\varepsilon_0 |\partial_x \varphi|^2 - \tfrac{\varepsilon_0}{2} |\partial_x \varphi|^2. \tag{12.18}$$

The momentum balance (12.6c) implies the constancy of the total stress in equilibrium. Moreover, as shown asymptotically in Dreyer et al. (2015), the electric field vanishes in the electrolyte bulk in equilibrium. Thus, in each point in the electrolyte domain it holds

$$p - p^{\text{bulk}} = \frac{1+2\chi}{2}\,\varepsilon_0 |\partial_x \varphi|^2, \tag{12.19}$$

where p^{bulk} denotes the pressure in the electrolyte bulk. These relations can be used to determine the boundary layer charge

$$Q^{\text{BL}} = -\int_0^{x^{\text{bulk}}} n^{\mathrm{F}} \, \mathrm{d}x = \Big((1+\chi)\cdot \varepsilon_0 \partial_x \varphi\Big)\Big|_{x=0} \tag{12.20}$$

as function of the potential difference $U = \varphi|_{x=0} - \varphi^{\text{bulk}}$, viz.

$$Q^{\text{BL}}(U) = -\operatorname{sign}(U) \cdot \sqrt{2\varepsilon_0 \left(\frac{(1+\chi)^2}{1+2\chi} \cdot (p - p^{\text{bulk}})\right)\Big|_{x=0}}. \tag{12.21}$$

12.3 The Constitutive Model

The electrolyte model of Dreyer et al. (2013); Landstorfer et al. (2016); Dreyer et al. (2018); Landstorfer and Müller (2022) employes a simple mixture model with a free energy $\rho\psi(T, (n_\alpha)_{\alpha \in I})$ including entropy of mixing, elastic interactions between the constituents and solvation of ions. A general construction of a thermodynamically consistent simple mixture models from admissible data is provided in Bothe et al. (2023), along with an asymptotic analysis of the incompressible limit. For a reference pressure small compared to the elastic bulk modulus, i.e., $p^{\text{ref}} \ll K$, a constraint $\sum_{\alpha\in\mathcal{I}} \upsilon_\alpha^{\text{ref}} \beta_\alpha(T)\, n_\alpha = 1$ arises and p takes the role of an independent variable. The action of the electric field on the electrolyte is modeled by an additional free energy contribution that vanishes in the absence of an electric field. However, as the electric field vanishes in bulk domains in equilibrium, otherwise neglected higher order effects like the electric interaction of ions can become relevant. To account for the bulk electrolyte as a mixture of charged particles also in this case, a Debye–Hückel-type electric ion interaction energy is introduced and thermodynamically consistent coupled to the simple mixture in Dreyer and Müller (2025). Since it is not obvious, if and how this new interaction energy contribution could be transferred from the electro-neutral bulk to charged boundary layers, a splitting into two different constitutive models in the different regions is proposed in Dreyer and Müller (2025).

Boundary layers

At domain boundaries or intersecting surfaces, there are typically thin layers where locally there is a macroscopic charge imbalance, i.e., $n^{\text{F}} \neq 0$. The according strong electric field, that reaches the order of $10^9\,\text{V}\,\text{m}^{-1}$ by far dominates the electric interaction of individual ions. Within boundary layers, we propose a constitutive model based on the simple mixture model and an explicit energy contribution due to polarization by the electric field, viz.

$$(\text{BL}) \begin{cases} \rho\psi = \rho\psi^{\text{sim}}(T, (n_\alpha)_{\alpha\in\mathcal{I}}) \;+ \rho\psi^{\text{pol}}(T, (n_\alpha)_{\alpha\in\mathcal{I}}, |\boldsymbol{E}|), \\ \mu_\alpha = \;\mu_\alpha^{\text{sim}}(T, (n_\alpha)_{\alpha\in\mathcal{I}}) \;+\; \mu_\alpha^{\text{pol}}(T, (n_\alpha)_{\alpha\in\mathcal{I}}, |\boldsymbol{E}|), \\ p = p^{\text{sim}} + p^{\text{pol}}, \end{cases} \tag{12.22}$$

Homogeneous bulk domains

An external electric field is screened from the bulk domains by boundary layers such that $\boldsymbol{E} \approx 0$ and an electrolyte thus appears locally electro-neutral on a macroscopic scale, i.e., $n^{\text{F}} = 0$. In homogeneous bulk domains, where on the macroscopic scale the electric field vanishes, we propose

$$(\text{HB}) \begin{cases} \rho\psi = \rho\psi^{\text{sim}}(T, (n_\alpha)_{\alpha\in\mathcal{I}}) + \rho\psi^{\text{D}}(T, (n_\alpha)_{\alpha\in\mathcal{I}}), \\ \mu_\alpha = \;\mu_\alpha^{\text{sim}}(T, (n_\alpha)_{\alpha\in\mathcal{I}}) \;+\; \mu_\alpha^{\text{D}}(T, (n_\alpha)_{\alpha\in\mathcal{I}}), \\ p = p^{\text{sim}} + p^{\text{D}}, \end{cases} \tag{12.23}$$

where $\rho\psi^{\text{sim}}$ and $\rho\psi^{\text{D}}$ denote the contributions of the simple mixture model and of the Debye–Hückel-like model, respectively. Since both contributions p^{sim} and p^{D} in general need to balance each other, the consideration of a (quasi-)incompressible limit is not suitable in this context.

12.3.1 The Simple Mixture Model

The derivation of a free energy is greatly eased by a change of variables. We denote the total number density and its inverse by

$$n = \sum_{\alpha \in I} n_\alpha, \qquad\qquad \nu = \frac{1}{n}, \tag{12.24}$$

respectively, where ν has the dimensions of a volume. Then, we introduce as new variables the mole fractions

$$y_\alpha = \frac{1}{n}\, n_\alpha = \nu\, n_\alpha \qquad \text{implying the constraint} \quad \sum_{\alpha \in I} y_\alpha = 1. \tag{12.25}$$

For abbreviation, we set $\boldsymbol{y} = (y_\alpha)_{\alpha \in I}$ and introduce $M(\boldsymbol{y}) = \sum_{\alpha \in I} m_\alpha y_\alpha$.

Due to the coupling with the electric ion interaction energy, Dreyer and Müller (2025) employs the state space $(T, \nu, \boldsymbol{y})$, instead of $(T, p, \boldsymbol{y})$ that is used in Bothe et al. (2023). As a consequence, the derivation of the free energy needs to build on the specific heat at constant volume c_υ here, instead of the specific heat at constant pressure c_p which often can be measured more easily in experiments and is therefore used in Bothe et al. (2023). According to Bothe et al. (2023) the important difference between the specific heats satisfies

$$c_p - c_\upsilon = \frac{T}{M(\boldsymbol{y})} \frac{\partial \nu}{\partial T} \frac{\partial p}{\partial T} \geq 0. \tag{12.26}$$

For the simple mixture model, we make the following assumptions **[SM1]** - **[SM3]**, cf. Bothe et al. (2023); Dreyer and Müller (2025):

[SM1] Thermal equation of state. To describe changes of the molar volume due to thermal expansion, elastic compression and changes of the composition, we assume the following simple constitutive law for the pressure

$$p^{\text{sim}}(T, \nu, \boldsymbol{y}) = p^{\text{ref}} + K \cdot \left(\frac{\nu^{\text{sim}}(T, \boldsymbol{y})}{\nu} - 1 \right), \tag{12.27a}$$

$$\nu^{\text{sim}}(T, \boldsymbol{y}) = \sum_{\alpha \in I} \upsilon_\alpha^{\text{ref}} \beta_\alpha(T) \cdot y_\alpha. \tag{12.27b}$$

Here, $K > 0$ denotes the isotropic compression modulus, and $\upsilon_\alpha^{\text{ref}}$ for $\alpha \in I$ are the molar volumes in a reference state with $T = T^{\text{ref}}$ and $p^{\text{sim}} = p^{\text{ref}}$. The functions $\beta_\alpha(T)$ describing the thermal expansion of constituent $\alpha \in I$ have to satisfy $\beta_\alpha(T^{\text{ref}}) = 1$.

[SM2] Specific heat at constant volume. We choose a simple linear approach with respect to y_α, that does not depend on temperature, i.e.,

$$M(\boldsymbol{y})\, c_v(T, \nu^{\text{sim}}, \boldsymbol{y}) = \sum_{\alpha\in I} m_\alpha c_\alpha^{\text{ref}}\, y_\alpha, \tag{12.28}$$

where the constant $c_\alpha^{\text{ref}} = c_\alpha^{\text{ref}}(\nu^{\text{sim}})$ are the specific heat constants of the pure constituents.

[SM3] Specific entropy and specific internal energy at constant temperature and constant volume. The essential ingredient of a simple mixture is a Boltzmann-typ logarithmic mixing contribution in addition to the purely linear approach, viz.

$$M(\boldsymbol{y})\, u(T^{\text{ref}}, \nu^{\text{sim}}, \boldsymbol{y}) = \sum_{\alpha\in I} m_\alpha u_\alpha^{\text{ref}}\, y_\alpha, \tag{12.29a}$$

$$M(\boldsymbol{y})\, s(T^{\text{ref}}, \nu^{\text{sim}}, \boldsymbol{y}) = \sum_{\alpha\in I} m_\alpha s_\alpha^{\text{ref}}\, y_\alpha - \sum_{\alpha\in I} k_B \ln(y_\alpha)\, y_\alpha, \tag{12.29b}$$

where the constants $u_\alpha^{\text{ref}} = u_\alpha(T^{\text{ref}}, \nu^{\text{sim}})$ and $s_\alpha^{\text{ref}} = s_\alpha(T^{\text{ref}}, \nu^{\text{sim}})$, for $\alpha \in I$, are the energy constants and the entropy constants for the pure constituents, respectively.

Free energy density and chemical potentials

The above the simple mixture assumptions (12.27)–(12.29) yield for the free energy density written as a function of the variable $(T, (n_\alpha)_{\alpha\in I})$

$$\begin{aligned}\rho\psi^{\text{sim}} &= \sum_{\alpha\in I} n_\alpha\, m_\alpha \psi_\alpha^{\text{ref}} + \sum_{\alpha\in I} n_\alpha\, m_\alpha c_\alpha^{\text{ref}} \cdot \left(T - T^{\text{ref}} - T\ln\left(\tfrac{T}{T^{\text{ref}}}\right)\right)\\ &+ k_B T \sum_{\alpha\in I} n_\alpha \ln\left(\tfrac{n_\alpha}{n}\right) - \sum_{\alpha\in I} n_\alpha\, m_\alpha s_\alpha^{\text{ref}} \cdot (T - T^{\text{ref}})\\ &+ (K - p^{\text{ref}})\Big(1 - \sum_{\alpha\in I} n_\alpha\, \upsilon_\alpha^{\text{ref}} \beta_\gamma(T)\Big) + K \sum_{\alpha\in I} n_\alpha\, \upsilon_\alpha^{\text{ref}} \beta_\alpha(T) \cdot \ln\Big(\sum_{\alpha\in I} n_\alpha \upsilon_\alpha^{\text{ref}} \beta_\gamma(T)\Big)\end{aligned} \tag{12.30}$$

where we set the reference energies

$$\psi_\alpha^{\text{ref}} = u_\alpha^{\text{ref}} - T^{\text{ref}} s_\alpha^{\text{ref}}. \tag{12.31}$$

With the definition of the reference Gibbs energies $g_\alpha^{\text{ref}} = m_\alpha\, \psi_\alpha^{\text{ref}} + \upsilon_\alpha^{\text{ref}}\, p^{\text{ref}}$, the chemical potentials then are

$$\begin{aligned}\mu_\alpha^{\text{sim}} &= g_\alpha^{\text{ref}} + p^{\text{ref}} \upsilon_\alpha^{\text{ref}} \cdot (\beta_\alpha(T) - 1) + m_\alpha c_\alpha^{\text{ref}} \cdot \left(T - T^{\text{ref}} - T\ln\left(\tfrac{T}{T^{\text{ref}}}\right)\right)\\ &+ k_B T \ln\left(\tfrac{n_\alpha}{n}\right) - m_\alpha s_\alpha^{\text{ref}} \cdot (T - T^{\text{ref}})\\ &+ K\, \upsilon_\alpha^{\text{ref}} \beta_\alpha(T) \cdot \ln\left(\sum_{\gamma\in I} n_\gamma\, \upsilon_\gamma^{\text{ref}} \beta_\gamma(T)\right).\end{aligned} \tag{12.32}$$

12.3.2 Ion Interaction Energy Contribution

For given reference temperature T^{ref} and reference number density n^{ref}, we introduce the Debye screening length ℓ as a reference length and define the inverse length Λ by

$$\ell = \sqrt{\frac{\varepsilon_0 k_B T^{\text{ref}}}{e_0^2 n^{\text{ref}}}}, \qquad (\Lambda\ell)^2 = \frac{T^{\text{ref}}}{T}\frac{1}{1+\chi}\sum_{\gamma\in I} z_\gamma^2 \frac{n_\gamma^\infty}{n^{\text{ref}}}. \tag{12.33}$$

In Dreyer and Müller (2025) a linearized Poisson–Boltzmann problem is solved for a single selected ion and its contribution to the inner energy and to the entropy are obtained from integrating over the space. Collecting the contributions of all ions then yields the densities of internal energy and entropy

$$\rho u^{\text{D}} = -\frac{a}{2}\frac{\chi + T\frac{\partial}{\partial T}\chi}{(1+\chi)^2}\sum_{\gamma\in I} z_\gamma^2 \frac{n_\gamma}{n^{\text{ref}}} F(R_\gamma, \lambda), \tag{12.34a}$$

$$\rho s^{\text{D}} = -\frac{a}{2}\frac{\frac{\partial}{\partial T}\chi}{(1+\chi)^2}\sum_{\gamma\in I} z_\gamma^2 \frac{n_\gamma}{n^{\text{ref}}} F(R_\gamma, \lambda), \tag{12.34b}$$

where R_γ denotes the radius of a sphere with the volume $\upsilon_\gamma^{\text{ref}}\beta_\gamma(T)$ and we defined for abbreviation

$$F(R_\gamma, \Lambda\ell) := \Lambda\ell\frac{3+2\Lambda R_\gamma}{(1+\Lambda R_\gamma)^2}, \qquad a := \frac{e_0^2}{8\pi\varepsilon_0}\frac{n^{\text{ref}}}{\ell}, \qquad \lambda = \Lambda\ell. \tag{12.35}$$

Setting $\rho\psi^{\text{D}} = \rho u^{\text{D}} - T\rho s^{\text{D}}$, the resulting free energy density is

$$\rho\psi^{\text{D}} = -\frac{a}{2}\frac{\chi}{(1+\chi)^2}\sum_{\gamma\in I} z_\gamma^2 \frac{n_\gamma}{n^{\text{ref}}} F(R_\gamma, \lambda). \tag{12.36}$$

To guarantee thermodynamic consistency, the constitutive functions (12.34) and (12.36) need to satisfy the thermodynamic conditions $\rho s^{\text{D}} = -\frac{\partial}{\partial T}\rho\psi^{\text{D}}$, thereby implying for the temperature dependence of the susceptibility the constraint

$$0 = \frac{\partial}{\partial T}\left(\left(\frac{\chi}{1+\chi}\right)^2 \sum_{\gamma\in I} z_\gamma^2 \frac{n_\gamma}{n^{\text{ref}}} F(R_\gamma, \lambda)\right). \tag{12.37}$$

Contributions to chemical potentials and pressure

We introduce the linear operator $\mathcal{L}$, such that

$$\mathcal{L}_\chi = \sum_{\alpha\in I} n_\alpha \frac{\partial}{\partial n_\alpha}\chi, \tag{12.38}$$

and note that whenever the above operator $\mathcal{L}$ acts on a homogeneous function h of degree one with respect to the number densities, then $\mathcal{L}_h = h$. For abbreviation, we set

$$\mathcal{F} = \sum_{\gamma\in I} z_\gamma^2 \frac{n_\gamma}{n^{\mathrm{ref}}} F(R_\gamma, \lambda), \qquad d\mathcal{F} = \sum_{\gamma\in I} z_\gamma^2 \frac{n_\gamma}{n^{\mathrm{ref}}} \frac{\partial}{\partial\lambda} F(R_\gamma, \lambda). \tag{12.39}$$

Then, the contributions to the chemical potentials and pressure, respectively, are

$$\mu_\alpha^{\mathrm{D}} = -\frac{a}{2}\frac{\chi}{(1+\chi)^2}\left[\frac{z_\alpha^2}{n^{\mathrm{ref}}}\left(F(R_\alpha,\lambda) + \frac{1}{\tau(1+\chi)\lambda^2}\frac{\lambda}{2}d\mathcal{F}\right) - \left(\frac{\chi-1}{\chi}\mathcal{F} + \frac{\lambda}{2}d\mathcal{F}\right)\frac{\frac{\partial}{\partial n_\alpha}\chi}{(1+\chi)}\right], \tag{12.40a}$$

$$p^{\mathrm{D}} = -\frac{a}{2}\frac{\chi}{(1+\chi)^2}\left(\frac{\lambda}{2}d\mathcal{F} - \left(\frac{\chi-1}{\chi}\mathcal{F} + \frac{\lambda}{2}d\mathcal{F}\right)\frac{\mathcal{L}_\chi}{(1+\chi)}\right). \tag{12.40b}$$

12.3.3 Polarization Contribution

Different polarization contributions to the free energy are analyzed in Landstorfer and Müller (2022), related to the way the susceptibility depends on the variables n_α and $|\boldsymbol{E}|$. According to the constraint (12.37), we are only free to choose the dependence of susceptibility on the composition of the electrolyte at the chosen reference temperature T^{ref}. Here, we set

$$\chi(T^{\mathrm{ref}}, (n_\alpha)_{\alpha\in I}, |\boldsymbol{E}|) = \sum_{\alpha\in I} \frac{\chi_\alpha^{\mathrm{ref}}\, \upsilon_\alpha^{\mathrm{ref}} n_\alpha}{\sqrt{1 + a_\alpha|\boldsymbol{E}|^2}}, \tag{12.41}$$

where a_α are parameters that control the onset of the dielectric saturation. The field strength dependence in (12.41) corresponds to the case $m = 1/2$ in Grahame (1950). The contribution of each species to the susceptibility is proportional to the specific volume $\upsilon_\alpha^{\mathrm{ref}}$. This choice is motivated by Debye (1929) and also fits well to an explanation of linear dielectric decrement by excluded volume effects, cf. Dubtsov et al. (2018). The assumption (12.4) and (12.11) imply

$$\chi = -\frac{\partial}{\partial|\boldsymbol{E}|}\rho\psi^{\mathrm{pol}}(T^{\mathrm{ref}}, (n_\alpha)_{\alpha\in I}, |\boldsymbol{E}|) \cdot \frac{1}{\varepsilon_0|\boldsymbol{E}|}, \tag{12.42}$$

such that we get by integration the contributions to the free energy density, to the chemical potential and to the pressure as

$$\rho\psi^{\text{pol}}(T^{\text{ref}}, (n_\alpha)_{\alpha\in I}, |\boldsymbol{E}|) = -\varepsilon_0 \sum_{\alpha\in I} \chi_\alpha^{\text{ref}}\ \upsilon_\alpha^{\text{ref}} n_\alpha \frac{\sqrt{1+a_\alpha|\boldsymbol{E}|^2}-1}{a_\alpha}, \tag{12.43a}$$

$$\mu_\alpha^{\text{pol}}(T^{\text{ref}}, (n_\alpha)_{\alpha\in I}, |\boldsymbol{E}|) = -\varepsilon_0\ \chi_\alpha^{\text{ref}}\ \upsilon_\alpha^{\text{ref}} \frac{\sqrt{1+a_\alpha|\boldsymbol{E}|^2}-1}{a_\alpha}, \tag{12.43b}$$

$$p^{\text{pol}}(T^{\text{ref}}, (n_\alpha)_{\alpha\in I}, |\boldsymbol{E}|) = -\rho\psi^{\text{pol}}(T^{\text{ref}}, (n_\alpha)_{\alpha\in I}, |\boldsymbol{E}|) + \mathcal{L}_{\rho\psi^{\text{pol}}} \quad = 0. \tag{12.43c}$$

12.4 Homogeneous Bulk

We consider an electrolyte that is prepared by dissolving N_E^0 molecules of the electrically neutral salt in initially N_S^0 molecules of a neutral solvent S. To characterize the electrolyte mixture, we introduce the dimensionless *total input ratio* σ and the *molality* m in $\text{mol}\,\text{kg}^{-1}$ as

$$\sigma = \frac{N_E^0}{N_S^0}, \qquad m = \frac{1}{m_S N_A}\,\sigma, \tag{12.44}$$

respectively. A molality $m = 1\,\text{mol}\,\text{kg}^{-1}$ of a salt in water corresponds to an input ratio $\sigma \approx 0.018$.

The salt dissociates into negatively charged anions and positively charged cations which then become solvated by a shell of solvent molecules. We refer to the solvated anions and cations as A and C, respectively. Neglecting any other species, the index set of the electrolyte is then given as $I = \{\text{A, C, E, S}\}$. The net reaction of dissociation and subsequent solvation reaction is

$$\text{E} + (\nu_A\kappa_A + \nu_C\kappa_C)\text{S} <\!\!\rightleftharpoons\!\!> \nu_A\text{A} + \nu_C\text{C}, \tag{12.45}$$

where κ_A and κ_C are the numbers of solvent molecules in the solvation shell of A and C, respectively. The stoichiometric coefficients of the reaction are given by ν_A and ν_C on the right-hand side of are (12.45) and

$$\nu_E = -1, \qquad \nu_S = -(\nu_A\kappa_A + \nu_C\kappa_C) \tag{12.46}$$

taken from the left-hand side are (12.45). For homogeneous systems, the *dissociation degree* δ is formally introduced such that the actual number N_E of undissociated salt in the mixture is

$$N_E = (1-\delta)\,N_E^0 \quad \text{and} \quad N_\alpha = \delta\,\nu_\alpha\,N_E^0 \ \text{ for } \alpha \in \{A, C\}. \tag{12.47}$$

The equilibrium condition for the reaction (12.45) is according to (12.13a) given as a mass action law

$$\sum_{\alpha\in I} \nu_\alpha(\mu_\alpha - g_\alpha^{\text{ref}}) = \Delta g \qquad \text{with} \quad \Delta g = -\sum_{\alpha\in I} \nu_\alpha\, g_\alpha^{\text{ref}}. \tag{12.48}$$

If the Gibbs energy Δg is in the order of $+1$ eV, then the electrolyte will be effectively completely dissociated up to a certain concentration, depending on the solvation numbers κ_α, where due to a lack of remaining free solvent molecules the solvation of further center ions is not possible any more.

The local number densities n_α for $\alpha \in I$ are given in a homogeneous bulk domain by dividing the actual particle number N_α by the a-priori unknown volume V of the domain. The initial volume V_S^0 of the pure solvent and the actual volume V of the mixture are

$$V_S^0 = N_S^0 \cdot \upsilon_S^{\text{ref}}\, \beta_S(T), \qquad V = V_S^0 + \frac{1}{1000}\Phi_V \frac{N_E^0}{1000 N_A}, \tag{12.49a}$$

where Φ_V is the *apparent molar volume*. Then, the initial mass density ρ_S^0 of the pure solvent and the mass density ρ of the homogeneous mixture are

$$\rho_S^0 = \frac{m_S N_S}{V_S^0}, \qquad \rho = \frac{1 + m_E N_A\, m}{\frac{V_S^0}{m_S N_S^0} + \frac{1}{1000}\Phi_V \frac{1}{1000} m}. \tag{12.50a}$$

Without applying any model assumption, it is possible to determine the apparent molar volume form measurements of the mass density of the solution as

$$\Phi_V = 1000 \cdot \left(\frac{m_E\, N_A}{\rho} + \left(\frac{1000}{\rho} - \frac{1000}{\rho_S^0} \right) \frac{1}{m} \right). \tag{12.51}$$

12.4.1 Application of the Constitutive Model

We apply the constitutive model (12.23), i.e., $\boldsymbol{E} \approx 0$, $\mu_\alpha = \mu_\alpha^{\text{sim}} + \mu_\alpha^{\text{D}}$ and $p = p^{\text{sim}} + p^{\text{D}}$.

We use reference values for temperature and number densities and material parameters for water according to Table 12.2. Then, the Debye screening length in the ion interaction energy model and the radius of the solvent species are

$$\ell \approx 0.04856\,\text{nm}, \qquad R_S(T^{\text{ref}}) = \sqrt[3]{\frac{3}{4\pi}\upsilon_S^{\text{ref}}} \approx 0.1928\,\text{nm} \approx 4\ell. \tag{12.52}$$

At the reference temperature $T = T^{\text{ref}}$, we can prescribe the concentration dependence of the susceptibility in the absence of an electric field. For any other temperature $T \neq T^{\text{ref}}$, then χ has to be determined from (12.37). As in Landstorfer and Müller (2022); Dreyer and Müller (2025), we set

Table 12.4 Ion specific parameters at $T^{\text{ref}} = 25°\text{C}$ based on the choice $\kappa_A = \kappa_C = 6$

	Na^+	K^+	F^-	Cl^-	ClO4$^-$	PF_6^-
$\upsilon_\alpha^{\text{ref}}\, 1000 N_A$ / [L mol^{-1}]	0.1224	0.13262	0.09138	0.11071	0.13695	0.15
R_α / [nm]	0.3647	0.3746	0.3309	0.3527	0.3787	0.3903
$\widetilde{r}_\alpha$ / [pm]	177	212.5	−189	97	224.5	245.5
d_α	7	6	5	5	7	7
χ_α^{ref}	21.19	33.14	23.66	33.22	27.27	31.71

$$\underline{\text{at } T = T^{\text{ref}}:} \qquad \chi = \sum_{\alpha \in I} \chi_\alpha^{\text{ref}} \upsilon_\alpha^{\text{ref}} n_\alpha, \tag{12.53a}$$

$$\text{with } \chi_\alpha^{\text{ref}} = \chi_S^{\text{ref}} - \frac{d_\alpha}{\upsilon_\alpha^{\text{ref}} n^{\text{ref}}} \text{ for } \alpha \in \{A, C\}. \tag{12.53b}$$

As noted in Dreyer and Müller (2025), this approach is consistent with experimental evidence, as it approximates for dilute solutions a linear dielectric decrement of the form $\chi = \chi_S^{\text{ref}} - d \cdot c$, where c is the electrolyte concentration in mol L^{-1} and $d = d_A + d_C$, cf. Hasted et al. (1948); Buchner et al. (1999); Marcus (2013). We introduce the volume change due to the dissociation reaction as

$$\Delta\upsilon(T) := \sum_{\alpha \in I} \nu_\alpha\, \upsilon_\alpha^{\text{ref}} \beta_\alpha(T) \tag{12.54}$$

and choose the external pressure as the reference pressure, i.e., $p^{\text{ref}} = p^{\text{ext}}$. The thermal equation of state (12.27) can then be expressed as

$$p^{\text{sim}} = p^{\text{ext}} + K \cdot \left(\frac{V_S^0}{V} \cdot \left(1 + \frac{\upsilon_E^{\text{ref}} \beta_E(T) + \Delta\upsilon(T)\,\delta}{\upsilon_S^{\text{ref}} \beta_S(T)}\, \sigma \right) - 1 \right). \tag{12.55}$$

In equilibrium, the dissociation reaction satisfies the mass action law (12.48), and the pressure in the bulk equals the external pressure, see, e.g., Dreyer et al. (2013). For a given salt input ratio σ and the external pressure p^{ext}, the apparent molar volume Φ_V of the mixture is determined from

$$p^{\text{ext}} = p^{\text{sim}} + p^{\text{D}}, \tag{12.56a}$$

$$\Delta g = \sum_{\alpha \in I} \nu_\alpha (\mu_\alpha^{\text{sim}} - g_\alpha^{\text{ref}}) + \sum_{\alpha \in I} \nu_\alpha \mu_\alpha^{\text{D}}. \tag{12.56b}$$

12.4.2 Solvated Ion Size

The model parameters can be identified from comparison with experimental data. It can be shown, cf. Dreyer and Müller (2025), that in the strong dilution limit $m \to 0$ the electrolyte is completely dissociated, i.e., $\delta \to 1$, and $\Phi_V \to \Phi_V^0$ with

$$\frac{1}{1000\mathcal{N}_A}\frac{\Phi_V^0}{1000} = \upsilon_E^{\text{ref}}\beta_E + \Delta\upsilon \quad = (\upsilon_A^{\text{ref}}\beta_A - \kappa_A\upsilon_S^{\text{ref}}\beta_S) + (\upsilon_C^{\text{ref}}\beta_C - \kappa_C\upsilon_S^{\text{ref}}\beta_S). \tag{12.57}$$

Values of the extrapolated Φ_V^0 are listed in Table 12.3 together with values of $\Delta\upsilon$ and Δg deduced from fit to data from mass density measurements. We note that Φ_V^0 values deduced here are slightly lower than in the literature, see, e.g., Millero (1971), because the constitutive model based on μ_α^{D} here implies steeper limit slopes of Φ_V^0 compared to standard Debye–Hückel theory, cf. Redlich and Meyer (1964); Dreyer and Müller (2025). Measurements of the dielectric decrement d in the linear regime below 1 mol L^{-1} and the additive splitting into $d = d_A + d_C$ have been reviewed in Marcus (2013), where the accuracy of the literature data is found to be rather low, i.e., in the order unity.

The right-hand side in (12.57) can be interpreted as the sum of specific volumes of the unsolvated ions. We introduce the radius $\widetilde{r}_\alpha$ of an unsolvated center ion together with the related radius R_α of the solvated ion as

$$\widetilde{r}_\alpha = \sqrt[3]{\frac{3}{4\pi}\cdot(\upsilon_\alpha^{\text{ref}}\beta_\alpha - (\kappa_\alpha\upsilon_S^{\text{ref}}\beta_S)}, \qquad R_\alpha = \sqrt[3]{\frac{3}{4\pi}\cdot\upsilon_\alpha^{\text{ref}}\beta_\alpha}. \tag{12.58}$$

For a group of several related salts sharing either the same anion or the same cation, a set of unsolvated ion radii $\widetilde{r}_\alpha$ approximating the Φ_V^0 data according to (12.57) and (12.58) can be found without having to specify the solvation numbers κ_α. After this, κ_α has to be determined from fit to data for Φ_V and χ for salt input $m > 0$. Here, $\kappa_\alpha = 6$ is chosen for all ionic species and the resulting values of R_α, $\upsilon_\alpha^{\text{ref}}$ and χ_α^{ref} are listed in Table 12.4. All results presented in this subsection are related to the temperature $T = T^{\text{ref}} = 25°$C.

Sodium chloride and Perchlorate salts

The constitutive model has been applied and extensively numerically tested in Dreyer and Müller (2025) over a wider range of salt concentrations and temperatures for aqueous solutions of NaCl, KCl, NaClO4 and KClO4. The model parameters for NaCl listed in Tables 12.3 and 12.4 were obtained from fit to mass density data from Chen et al. (1980); Allred and Woolley (1981); Gates and Wood (1985); Apelblat and Manzurola (1999) and measurements of concentration-dependent susceptibility in Buchner et al. (1999). For the perchlorate salts, mass density measurements of Hnedkovsky and Hefter (2021) were used, but no direct measurement data for the dielectric decrement was found. The value $d_{\text{ClO}_4^-} = 7$ chosen here is smaller than the additive decrement 10 that is suggested in Marcus (2013) mainly based on data for $LiClO_4$, but it is still within the estimated tolerance for the data listed there for $NaClO_4$. No susceptibility data for $KClO_4$ was found in the literature.

Fluoride salts

For the fluoride salts, we tested the model against mass density data of Pedersen et al. (1984); Buchner et al. (1994) and the data for concentration-dependent χ in Buchner et al. (1994); Loginova et al. (2006). The value $d_{F^-} = 5$ applied here is larger than the suggested value 4 in Marcus (2013), but still within the tolerance estimated there. The apparent molar volume computed with the model for aqueous solutions of NaF and of KF agree well with the experimental data as shown in Fig. 12.1. We observe that the apparent molar volume Φ_V, which is derived from mass density without the application of any model, is negative for dilute NaF solutions, as it is also in the case of LiF, see Pedersen et al. (1984). Thus the derived unsolvated ion radius $\widetilde{r}_{F^-} = -189\,\text{pm}$ in Table 12.4 is negative. This should be interpreted in the way that the bound water molecules in the solvation shell around the F^- ion are more densely packed, compared to the free solvent molecules. For the chosen solvation number $\kappa_\alpha = 6$, the specific volume $\upsilon_\alpha^{\text{ref}}$ and the radius R_α that we use in the model, are positive in the case of F^-, see Table 12.4.

The computed susceptibility according to (12.53) is shown in Fig. 12.1 for KF solutions. The nonlinearity is related to incomplete dissociation which is controlled by the Gibbs energy Δg. We observe that for more negative Δg, the computed susceptibility follows the linear decrement over a wider range of concentrations. As the impact of the parameter χ_E^{ref} is not very significant below $4\,\text{mol}\,\text{L}^{-1}$ of KF, no values of χ_E^{ref} are listed in Table 12.4 and we used $2\chi_E^{\text{ref}} = \chi_A^{\text{ref}} + \chi_C^{\text{ref}}$. The impact of rather extreme choices $\chi_E^{\text{ref}} = 0$ or $\chi_E^{\text{ref}} = 80$ are also shown in Fig. 12.1. For NaF, the available data of Buchner et al. (1994) only spans a small concentration range where the susceptibility remains in the linear regime.

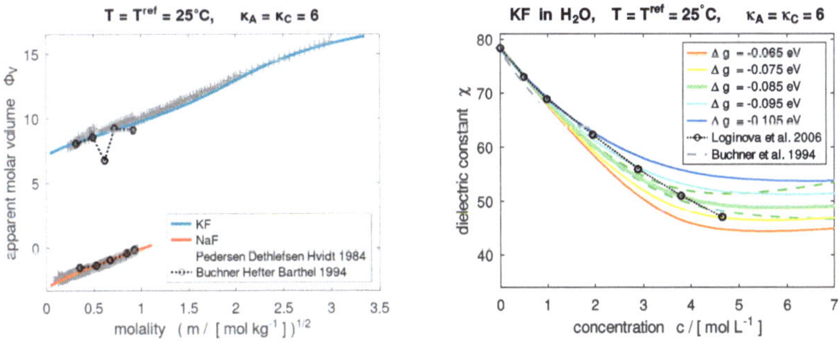

Fig. 12.1 Left: Apparent molar volume for aqueous solutions of NaF and KF, and comparison with data from Pedersen et al. (1984); Buchner et al. (1994). Right: computed susceptibility based on (12.53) for different choices of Δg and comparison with data from Buchner et al. (1994); Loginova et al. (2006). Dashed green lines are related to alternatively setting $\chi_E^{\text{ref}} = 80$ for the upper dashed line and $\chi_E^{\text{ref}} = 0$ for the lower one

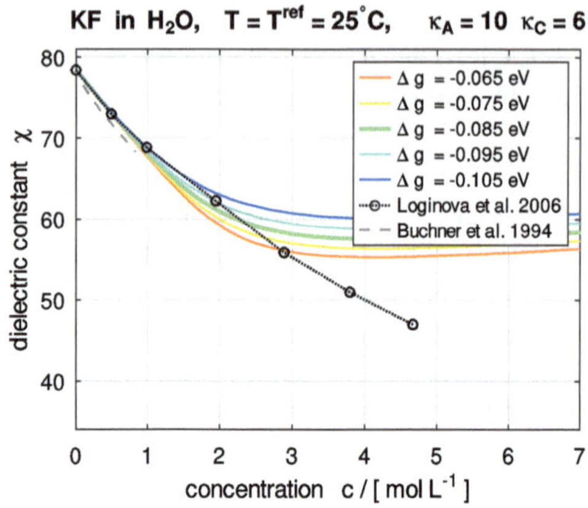

Fig. 12.2 Computed susceptibility based on (12.53), but in difference to Fig. 12.1 here with $\kappa_A = 10$, $\kappa_C = 6$ shows a more abrupt transition from linear decrement to a plateau of χ

In addition to the Gibbs energy Δg also the solvation numbers κ_α have an influence on the dissociation degree δ. There is no direct experimental data for δ, but based on the linear approach for χ we can use the concentration-dependent susceptibility for the identification of the model parameters κ_α. While we here propose $\kappa_\alpha = 6$ for all univalent ions, Fig. 12.2 shows the numerical results for χ when for the fluoride anion $\kappa_A = 10$ is chosen instead. We observe in this case that χ turns into a plateau like regime more early and more abrupt than in the case $\kappa_A = 6$ and the susceptibility does not decrease below $\chi \approx 55$.

Potassium hexafluorophosphate

The limited available experimental data for solutions of KPF_6 makes parameter identification more difficult. Mass density data is given in Robinson et al. (1961). No appropriate susceptibility data for aqueous KPF_6 solutions was found to additionally adjust the parameters Δg and κ_A. Figure 12.3 shows the computed Φ_V curves with the parameters from Table 12.4. In addition there are also curves shown for stronger dissociation related to a larger $\Delta g = -0.08\,\text{eV}$. Due to the then more flat slope, the extrapolation of $\Phi_V(m)$ for $m \to 0$ leads to a larger unsolvated ion radius $\widetilde{r}_A$ in this case. To the contrary, a stronger formation of undissociated ion pairs due to a more negative $\Delta g = -0.18\,\text{eV}$ leads to a steeper slope of the Φ_V curve, and thus smaller $\widetilde{r}_A$. Moreover, the impact of the solvation number κ_A can be tested. When $\kappa_A = 0$ is chosen, $d_A \leq 3$ is necessary in order to ensure $\chi_A^{\text{ref}} \geq 0$ according to (12.53). The resulting curves of Φ_V for this alternative parameter choice are also show in Fig. 12.3 and only minor differences to the case $\kappa_A = 6$ can be observed. Due to the rather low solubility of KPF_6, only up to around $0.5\,\text{mol}\,\text{kg}^{-1}$ in water, the computed concentration-dependent susceptibility for KPF_6 remains in the regime of linear dielectric decrement.

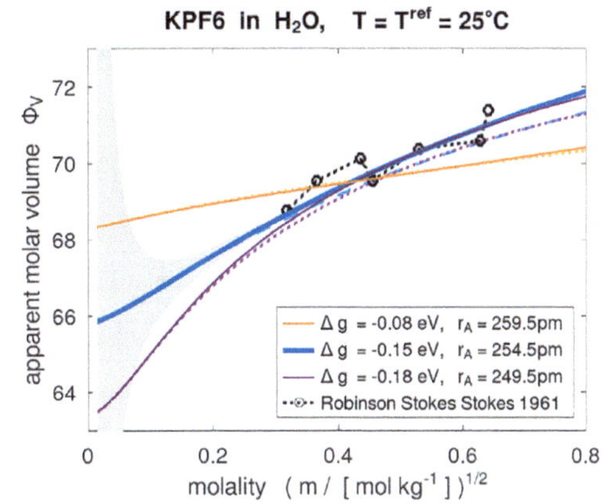

Fig. 12.3 Apparent molar volume for aqueous solutions of KPF_6 for different Δg and comparison with data of Robinson et al. (1961). Dashed and dotted lines correspond to computations with alternative setting of $\kappa_A = 0$ and $d_A = 3$. The shaded area indicates the confidence interval for a uncertainty in the mass density data of $10^{-5}\,\text{kg}\,\text{L}^{-1}$. The shaded grey area represents the experimental uncertainty

12.4.3 Temperature Dependence of the Susceptibility

A distinct feature of the electric ion interaction energy model applied here is the imposed constraint (12.37) on the temperature dependence of the susceptibility. It requires that

$$\left(\frac{\chi}{\chi+1}\right)^2 \sum_{\gamma\in I} z_\gamma^2 \frac{n_\gamma}{n^{\text{ref}}} F(R_\gamma(\tau), \lambda) = \left(\left(\frac{\chi}{\chi+1}\right)^2 \sum_{\gamma\in I} z_\gamma^2 \frac{n_\gamma}{n^{\text{ref}}} F(R_\gamma, \lambda)\right)\Bigg|_{T=T^{\text{ref}}} . \tag{12.59}$$

The implications of this constraint are studied in Dreyer and Müller (2025) under the assumption of complete dissociation in the dilute solution regime. It is found there that i) the model is compatible with the measurements of Buchner et al. (1999) for NaCl solutions of less than $1\,\text{mol}\,\text{L}^{-1}$ concentration, ii) the deviation from linearity is small enough to allow a linear approach such as (12.53) for any chosen reference temperature, and iii) extrapolation of the relation (12.59) to infinite dilution $c \to 0$ ensures that the temperature dependence of the susceptibility χ_S of the pure solvent is independent of the dissolved salt, but only depends on the thermal expansion $\beta_S(T)$ of the solvent.

For the fluoride salts studied in the previous subsection, there is only susceptibility data available at $T = 25°\text{C}$ and no data was found for the perchlorate salts and for KPF_6. Thus, we test the thermodynamic constraint (12.59) only for NaCl solutions with the data of Buchner et al. (1999), as the data for the related salt KCl provided in Chen et al. (2003) is also restricted to $T = 25°\text{C}$.

For the thermal expansion $\beta_S(T)$ of the solvent, we use the precise measurements of the mass density of pure water between 4°C and 70°C provided in Apelblat and Manzurola (1999). For all other species, we apply a linearization of the thermal

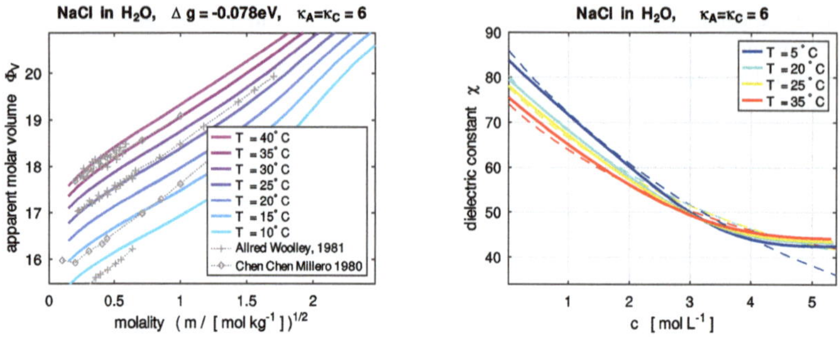

Fig. 12.4 Left: apparent molar volume of aqueous NaCl solution for various temperatures and comparison with experimental data of Allred and Woolley (1981) for $T = 10°\text{C}$, $T = 25°\text{C}$ and $T = 40°\text{C}$ and Chen et al. (1980) for $T = 15°\text{C}$ and $T = 35°\text{C}$. Right: computed susceptibility based on (12.59) and dashed lines of the same color according to Buchner et al. (1999) for comparison

expansion around T^{ref}. Based on data from Apelblat and Manzurola (1999); Hnedkovsky and Hefter (2021), we set for NaCl the coefficients to $\beta_A^{\text{ref}} = 319.7 \cdot 10^{-6}$ and $\beta_C^{\text{ref}} = 754.5 \cdot 10^{-6}$. For the undissociated salt we set $\beta_E^{\text{ref}} = 255.75 \cdot 10^{-6}$ corresponding to the thermal expansion of water at $T = 25°\text{C}$.

In contrast to Dreyer and Müller (2025), we now let the Gibbs energy Δg depend linearly on T, viz.

$$\Delta g(T) = \Delta g(T^{\text{ref}}) + 10^{-3}\,\frac{\text{eV}}{\text{K}} \cdot (T - T^{\text{ref}}). \tag{12.60}$$

Figure 12.4 shows the resulting curves of Φ_V and χ at different temperatures, also in the less dilute concentration range $c > 1\,\text{mol}\,\text{L}^{-1}$. The computed curves agree well with the experimental data, although the linearization of the thermal expansion becomes inappropriate for larger deviations from the reference temperature, see Dreyer and Müller (2025). We observe that the susceptibility shows a linear dielectric decrement for low concentrations $c < 2\,\text{mol}\,\text{L}^{-1}$, where in contrast to Dreyer and Müller (2025) the slope here increases for lower temperatures, due to the temperature dependence of Δg.

12.5 Double Layer at $T = T^{\text{ref}}$

We consider the double layer at the interface between an aqueous bulk electrolyte as described in the previous section and a well-defined surface of a metal single-crystal electrode. The constitutive model can be restricted to the electrolyte by the assumptions of (i) an idealized electrode material model such that the charged boundary layer in the metal is independent of the applied voltage between the electrode and the electrolyte and (ii) a non-adsorbing electrolyte such that except for the metal electrons no constitutive model is needed for the surface. NaF and $NaClO_4$ are considered only weakly adsorbing on silver surfaces and KPF_6 is considered non-adsorbing, cf. Valette (1981). The only remaining material property of the metal electrode than is the *potential of zero charge* E^{PZC} such that the potential difference between the electrode surface and the electrolyte bulk is $U = E - E^{\text{PZC}}$, where E is the measurable electrode potential.

Typically, the boundary layer in aqueous electrolytes is in the order of nanometers and thus thin compared to the other dimensions in an experimental setup. We thus can assume an infinite planar electrode surface located at the coordinate plane $x = 0$ and describe the layer in the electrolyte in a semi infinite one-dimensional setup. Then, the idealized 1D boundary layer can be coupled to electrolyte bulk by matched formal asymptotics, cf. Dreyer et al. (2015).

12.5.1 Constitutive Model in the Layer

We apply inside the layer the constitutive model (12.22), i.e., $\mu_\alpha = \mu_\alpha^{\text{sim}} + \mu_\alpha^{\text{pol}}$ and $p^{\text{pol}} = 0$ such that $p = p^{\text{sim}}$. By restricting the considerations to the reference temperature $T = T^{\text{ref}}$, the thermal expansion coefficients are all equal to $\beta_\alpha(T^{\text{ref}}) = 1$ and the model simplifies to

$$\mu_\alpha = g_\alpha^{\text{ref,BL}} + k_B T^{\text{ref}} \ln(y_\alpha) + K\, \upsilon_\alpha^{\text{ref}} \cdot \ln\left(1 + \frac{p^{\text{sim}} - p^{\text{ref,BL}}}{K}\right) + \mu_\alpha^{\text{pol}}, \tag{12.61a}$$

$$\mu_\alpha^{\text{pol}} = -\upsilon_\alpha^{\text{ref}}\, \chi_\alpha^{\text{ref}}\, \varepsilon_0 \frac{\sqrt{1 + a_\alpha |\boldsymbol{E}|^2} - 1}{a_\alpha}, \tag{12.61b}$$

$$n \sum_{\alpha \in I} \upsilon_\alpha^{\text{ref}}\, y_\alpha = 1 + \frac{p^{\text{sim}} - p^{\text{ref,BL}}}{K}, \tag{12.61c}$$

$$\chi = \sum_{\alpha \in I} \frac{\chi_\alpha^{\text{ref}}\, \upsilon_\alpha^{\text{ref}}\, n_\alpha}{\sqrt{1 + a_\alpha |\boldsymbol{E}|^2}}. \tag{12.61d}$$

This boundary layer model is coupled with model (12.23) for the homogeneous bulk, where the total pressure $p = p^{\text{sim}} + p^{\text{D}}$ is assumed by condition (12.56a) to coincide with the external pressure, i.e. $p = p^{\text{ref}}$. At $T = T^{\text{ref}}$, the bulk chemical potentials then simplify to

$$\mu_\alpha^{\text{bulk}} = g_\alpha^{\text{ref,bulk}} + k_B T^{\text{ref}} \ln\left(y_\alpha^{\text{bulk}}\right) + K\, \upsilon_\alpha^{\text{ref}} \cdot \ln\left(1 - \frac{p^{\text{D}}}{K}\right) + \mu_\alpha^{\text{D}}. \tag{12.62}$$

The model parameters $\upsilon_\alpha^{\text{ref}}$, χ_α^{ref} coincide in both models (12.61) and (12.62) and are given by Table 12.4 from the previous section, but the reference Gibbs energies need to be distinguished. For the coupling of both models, we require that $\varphi \to \varphi^{\text{bulk}}$, $\boldsymbol{E} \to 0$ and thus $\mu_\alpha^{\text{pol}} \to 0$ if $x \to x^{\text{bulk}}$. Moreover, the extended diffusional equilibrium (12.17) then requires $\mu_\alpha \to \mu_\alpha^{\text{bulk}}$ and $y_\alpha \to y_\alpha^{\text{bulk}}$ for $x \to x^{\text{bulk}}$. We thus set

$$g_\alpha^{\text{ref,BL}} = g_\alpha^{\text{ref,bulk}} + K\, \upsilon_\alpha^{\text{ref}} \cdot \ln\left(1 - \tfrac{p^{\text{D}}}{K}\right) + \mu_\alpha^{\text{D}} \quad \text{and} \quad p^{\text{ref,BL}} = p^{\text{ref}} - p^{\text{D}}. \tag{12.63}$$

More general, (12.17) implies that the electrochemical potentials coincide in each point $x \in \Omega^{\text{BL}}$ with the boundary value from the bulk model, i.e.,

$$\begin{aligned} k_B T^{\text{ref}} \ln(y_\alpha) = k_B T^{\text{ref}} \ln\left(y_\alpha^{\text{bulk}}\right) - z_\alpha e_0 \cdot \left(\varphi(x) - \varphi^{\text{bulk}}\right) \\ - K\, \upsilon_\alpha^{\text{ref}} \cdot \ln\left(1 + \tfrac{p^{\text{sim}} + p^{\text{D}} - p^{\text{ref}}}{K}\right) - \mu_\alpha^{\text{pol}}. \end{aligned} \tag{12.64}$$

We conclude that $p^{\text{sim}} \to p^{\text{ref}} - p^{\text{D}}$ for $x \to x^{\text{bulk}}$ and by using the one-dimensional momentum balance (12.18), we infer

$$|\partial_x \varphi|^2 = \frac{2}{1 + 2\chi} \frac{p^{\text{sim}} + p^{\text{D}} - p^{\text{ref}}}{\varepsilon_0}, \tag{12.65}$$

and we can eliminate the electric field $\partial_x \varphi$ from the definitions of (12.61b) and (12.61d) for μ_α^{pol} and χ, respectively. Given $(\varphi - \varphi^{\text{bulk}},\ p^{\text{ref}},\ p^{\text{D}})$, we can determine the variables $(p^{\text{sim}}, n, (y_\alpha)_\alpha, (\mu_\alpha^{\text{pol}})_\alpha, \chi)$ from the coupled system

$$1 = \sum_{\alpha \in \mathcal{I}} y_\alpha, \tag{12.66a}$$

$$1 = n \sum_{\alpha \in \mathcal{I}} \upsilon_\alpha^{\text{ref}}\, y_\alpha - \frac{p^{\text{sim}} + p^{\text{D}} - p^{\text{ref}}}{K}, \tag{12.66b}$$

$$\begin{aligned} y_\alpha = y_\alpha^{\text{bulk}} \cdot \exp\Bigg(- z_\alpha \tfrac{e_0}{k_B T^{\text{ref}}} \left(\varphi - \varphi^{\text{bulk}}\right) \\ - K \tfrac{\upsilon_\alpha^{\text{ref}}}{k_B T^{\text{ref}}} \ln\left(1 + \tfrac{p^{\text{sim}} + p^{\text{D}} - p^{\text{ref}}}{K}\right) - \tfrac{1}{k_B T^{\text{ref}}} \mu_\alpha^{\text{pol}} \Bigg), \end{aligned} \tag{12.66c}$$

$$\mu_\alpha^{\text{pol}} = -\upsilon_\alpha^{\text{ref}}\, \chi_\alpha^{\text{ref}} \varepsilon_0 \frac{\sqrt{1 + \frac{a_\alpha}{\varepsilon_0} \frac{2}{1+2\chi} (p^{\text{sim}} + p^{\text{D}} - p^{\text{ref}})} - 1}{a_\alpha}, \tag{12.66d}$$

$$\chi = n \sum_{\alpha \in \mathcal{I}} \frac{\chi_\alpha^{\text{ref}}\, \upsilon_\alpha^{\text{ref}}\, y_\alpha}{\sqrt{1 + \frac{a_\alpha}{\varepsilon_0} \frac{2}{1+2\chi} (p^{\text{sim}} + p^{\text{D}} - p^{\text{ref}})}}. \tag{12.66e}$$

Note that in contrast to Dreyer et al. (2013); Landstorfer et al. (2016); Landstorfer and Müller (2022), we do not apply the incompressible limit for large bulk modulus K. For water, the atmospheric pressure satisfies $p^{\text{ref}} \ll K$ and, as shown in Dreyer and Müller (2025), typically also $p^{\mathrm{D}} \ll K$. Thus, as long as $p = p^{\mathrm{sim}}$ remains moderate in the layer, it is possible to neglect the pressure terms in (12.66b) and to linearize the logarithmic term in (12.66c) as $-\frac{\upsilon_\alpha^{\mathrm{ref}}}{k_B T^{\mathrm{ref}}} \cdot (p^{\mathrm{sim}} + p^{\mathrm{D}} - p^{\mathrm{ref}})$. However, the incompressible limit can not be applied here because i) in the bulk it is necessary to balance the elastic versus the electric interaction contribution to the pressure and ii) in boundary layers p^{sim} can grow extremely large, as can already be seen from Dreyer et al. (2013); Landstorfer and Müller (2022).

To compute the spatial structure of the layer for a given potential difference $U = \varphi(0) - \varphi^{\mathrm{bulk}}$, we choose a sufficiently large domain $\Omega^{\mathrm{BL}} = (0, x^{\mathrm{bulk}})$ that approximates the asymptotically infinite layer and complement the extended diffusional equilibrium system (12.66) by the Poisson boundary value problem

$$-\partial_x(\chi\, \varepsilon_0 \partial_x \varphi) = n^{\mathrm{F}} \qquad \text{in } \Omega^{\mathrm{BL}}, \tag{12.67a}$$

$$\varphi(0) = \varphi^{\mathrm{bulk}} + U, \qquad \varphi(x^{\mathrm{bulk}}) = \varphi^{\mathrm{bulk}}. \tag{12.67b}$$

12.5.2 Boundary Layers

To test the model, we solve the system (12.66), (12.67) by a piecewise linear finite element method with mass lumping and use material parameters for aqueous NaClO4 solutions determined from the bulk model in the Sect. 12.4 an listed in Tables 12.3 and 12.4. For a fixed set of parameters, i.e., $a_\alpha = 0.01 \left(\frac{e_0\, L^{\mathrm{ref}}}{k_B T^{\mathrm{ref}}}\right)^2$ for all $\alpha \in \mathcal{I}$, the computed spatial profiles $\chi(x)$ of the susceptibility and of the number densities $n_A(x)$ and $n_S(x)$ of the anions and the solvent, respectively, are shown in Fig. 12.5. To some extend, we can repeat the observations made in Landstorfer and Müller (2022) for the combination of concentration-dependent dielectric decrement and field-dependent dielectric saturation. While the spatial profiles of $\varphi(x)$ and $p^{\mathrm{sim}}(x)$ always remain monotone, as it is required for (12.66), the number densities show non-monotone spatial profiles of $n_A(x)$ and $n_S(x)$. We note that the local maximum height of $n_A(x)$ is below the limit $(1000 N_A\, \upsilon_A^{\mathrm{ref}})^{-1} \approx 7.3\,\mathrm{mol\,L^{-1}}$ related to the incompressible model. If the applied potential difference is larger than $U \approx 0.4\,\mathrm{V}$, then $n_A(x)$ passes through a local minimum and increases monotonously thereafter. Also the solvent density undergoes a local minimum and than increases again towards the electrode surface at $x = 0$. The pressure p^{sim} at the surfaces reaches into the order of 10^9 Pa, rendering the incompressible limit inadequate. Alternative computations with the incompressible limit model of Landstorfer and Müller (2022) show in comparison lower local minimum of $n_A(x)$ which also is in a larger distance from the local maximum.

Dependence on a_α

The only model parameters that are not already determined by the bulk model are a_α for $\alpha \in I$ controlling the strength of the dielectric saturation. We normalize

$$a_\alpha = \left(\frac{e_0\, L^{\mathrm{ref}}}{k_B T^{\mathrm{ref}}} \right)^2 A_\alpha \tag{12.68}$$

and study the impact of the dielectric saturation when all a_α coincide for all species. The computed spatial profiles are shown in Fig. 12.6. For small $A_\alpha \le 10^{-4}$, the susceptibility is mainly determined by the concentration-dependent dielectric decrement and we observe monotonous profiles of the number densities n_A and n_S. The susceptibility $\chi(x)$ remains larger than 40, and in agreement with Landstorfer and Müller (2022) the solvent density n_S at the surface at $x = 0$ is bounded from below by

$$n_S|_{x=0} \to \tfrac{1}{2} \max \left(0, \frac{\chi_S^{\mathrm{ref}} - 2\chi_A^{\mathrm{ref}} - 1}{(\chi_S^{\mathrm{ref}} - \chi_A^{\mathrm{ref}})\, v_S^{\mathrm{ref}}} \right) \approx 12.368\,\mathrm{mol\,L^{-1}} \quad \text{for } U \to \infty. \tag{12.69}$$

For large $A_\alpha \ge 0.2$ on the other hand, the field-dependent dielectric saturation dominates and we again observe monotonous profiles of n_A and n_S, but in this case the solvent number density n_S vanishes at the surface. In the intermediate range of A_α, where concentration and field-dependent effects of χ compete, the solvent density n_S undergoes a local minimum and then increases again to almost the bulk density. In difference to Landstorfer and Müller (2022), there is no saturation of $n_A(x)$ and densities larger than the limit density of the incompressible case can be attained.

Dependence on a_A

We fix the parameter $A_S = 0.01$ for the solvent and vary A_A. The impact of the parameter A_C can also be studied in analogous way by applying negative potential differences between the electrode and the electrolyte bulk. For small $A_A \le 0.002$ we see in Fig. 12.7 that the solvent gets pushed away from the surface and the density

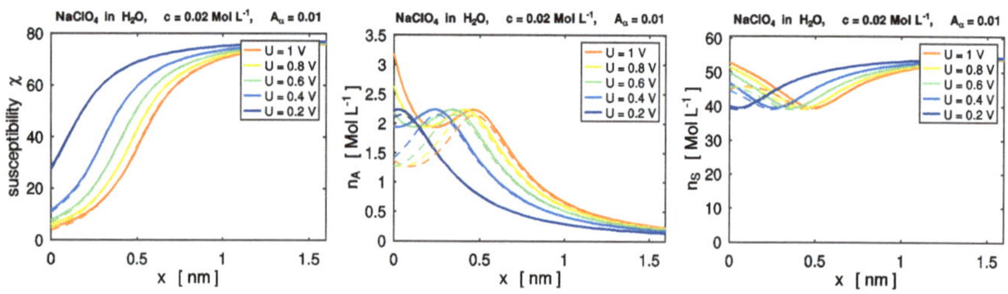

Fig. 12.5 Computed spatial profiles for increasing potential difference U: susceptibility $\chi(x)$ (left) and non-monotone number densities $n_A(x)$ of anions (middle) and $n_S(x)$ of the solvent (right). The electrode surface is located at $x = 0$. Parameters of the electrolyte were chosen to represent an aqueous solution of $NaClO_4$ with bulk concentration $c = 0.02$ and $a_\alpha = 0.01 \left(\frac{e_0\, L^{\mathrm{ref}}}{k_B T^{\mathrm{ref}}} \right)^2$. For comparison, the dashed lines show the corresponding curves for the incompressible model

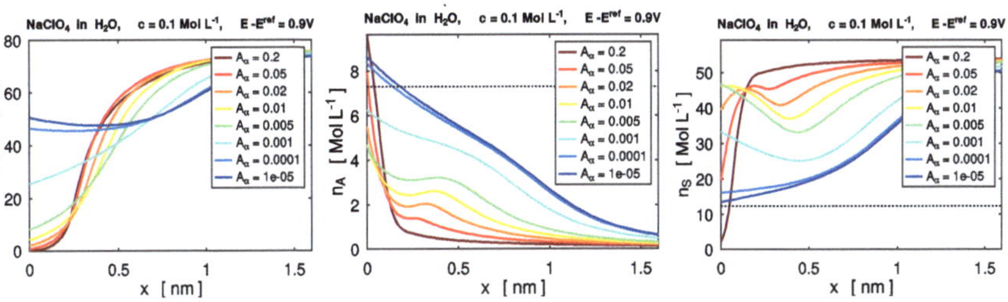

Fig. 12.6 Dependence on a_α: spatial boundary layer profiles of $\chi(x)$ (left), $n_A(x)$ (middle) and $n_S(x)$ (right). Parameters for a $NaClO_4$ solution of bulk concentration $c = 0.1$ are used and a potential difference of $U = 0.9$ V is applied. The dotted lines indicate bounds related to the incompressible model

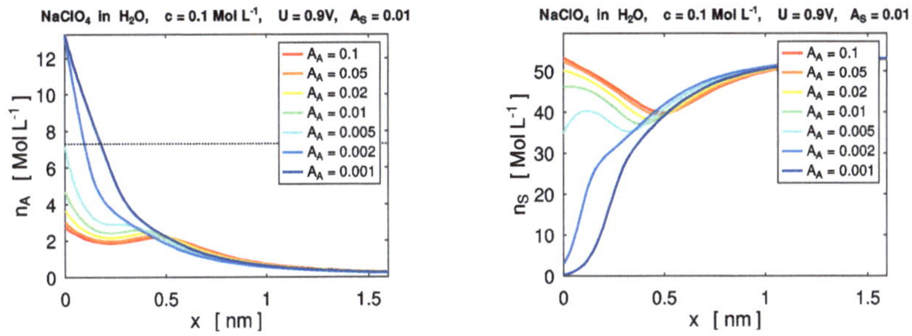

Fig. 12.7 Dependence on a_A for fixed $a_S = 0.01\left(\frac{e_0 L^{\mathrm{ref}}}{k_B T^{\mathrm{ref}}}\right)^2$: spatial profiles of $n_A(x)$ (left) and $n_S(x)$ (right). Parameters for a $NaClO_4$ solution of bulk concentration $c = 0.1$ are used and a potential difference of $U = 0.9$ V is applied. The dotted line indicates the bound related to the incompressible model

n_S drops to zero at $x = 0$. The local maximum of the anion density n_A disappears and n_A grows far beyond the saturation level of the incompressible limit. For large $A_A \geq A_S$ on the other hand, the solvent density n_S grow stronger after n_S has passed through the local minimum and the anion density n_A remains rather small near the surface.

12.5.3 Double-Layer Capacitance

The computed densities in the layer can be integrated in space to obtain the boundary layer charge. As well, the boundary layer charge can directly obtained from (12.66) alone by applying (12.21). The capacitance is then obtained from taking the derivative $C(U) = \frac{d}{dU} Q^{\mathrm{BL}}(U)$.

Dependence on a_α

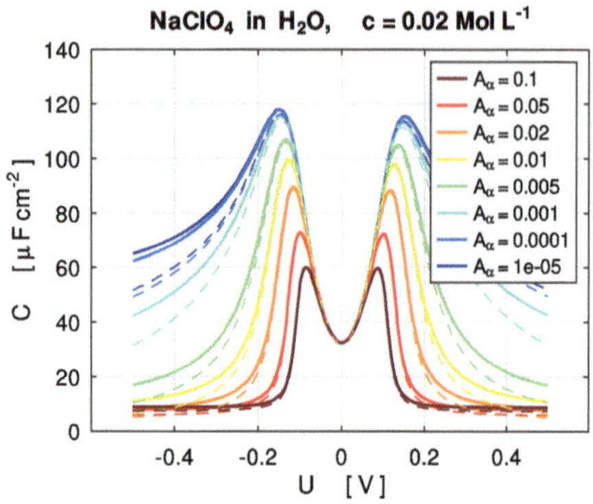

Fig. 12.8 Computed camel shaped differential capacitance for different parameter values a_α controlling the strength of the dielectric saturation. Parameters for a solution of NaClO4 with bulk concentration $c = 0.02\,\mathrm{mol\,L^{-1}}$ have been used. Dashed lines show alternative computations for the incompressible limit model

Figure 12.8 shows the computed differential capacitance curves for different values assigned uniformly to all parameters a_α. All curves agree qualitatively with the camel shape of the measured capacitance in Valette (1981). Also, all curves coincide close to $U = 0$. When increasing a_α, the capacitance branches of more early from the curve related to $A_\alpha = 10^{-5}$, thereby attaining a lower local maximum at a smaller distance from $U = 0$. For the height of the local maximum to agree with the measurements of Valette (1981) a parameter $A_\alpha \geq 0.1$ for the dielectric saturation is needed. Then however, the local maxima are located at $U \approx \pm 0.1\,\mathrm{V}$ and the capacitance very sharply drops to a rather low plateau of $C \approx 10\,\mu\mathrm{F\,cm^{-1}}$. By comparing the results with the incompressible model, we observe that compressibility increases the capacitance for large applied potential difference. This is to be expected, since the boundary layer profiles of the ions increase beyond the bounds of the incompressible model. However this effect is not very pronounced for small $A_\alpha \leq 10^{-3}$ and diminishes as the dielectric saturation gets stronger.

Dependence on a_A and a_C

We first set $A_S = 0.01$, allowing the solvent to remain at the surface even for large applied potential differences. Then varying A_A, and letting $A_C = A_A$, we see little effect on the capacitance curves, cf. Fig. 12.9 (left). There is only a slight increase of the maxima height and cf. the capacitance beyond the local maxima, if $A_A = A_C \leq 10^{-3}$. In the case of strong dielectric saturation of the solvent however, i.e., $A_S = 1$, we observe the occurrence of secondary local maxima. The height of these maxima increases with decreasing parameter size $A_A = A_C$, see Fig. 12.9 (right). A similar observation is reported in Bazant et al. (2009), and it is attributed there to a transition from small χ to large χ. But, while in Bazant et al. (2009) a single parameter controls the secondary maxima, the effect is here related to the difference in strength of the saturation effect between the solvent and the ionic species.

Capacitance for different salts

In Valette (1981, 1982, 1989), the double-layer capacitance is compared between the salts NaF, $NaClO_4$ and KPF_6. While for KPF_6 the capacitance shows remarkable symmetry with respect to the potential of zero charge at $U = 0$, the local maxima are higher on the anionic branch for $U > 0$ for NaF, $NaClO_4$ and the local minima depend on the bulk concentration, indicating some remaining ainion adsorption to the electrode surface. Figure 12.10 shows the computed capacitance curves for solution of the mentioned salts of various concentration. We observe that the differences between the different salts are rather small, and only in the case of NaF the local maximum for $U > 0$ is higher compared to other salts. Close to $U = 0$ the curves coincide with the Guy-Chapman capacitance and only for the case of the most concentrated solution of $0.1\,\mathrm{mol\,L^{-1}}$ KPF_6, we observe some reduction of the local minimum height due to incomplete dissociation in the bulk.

12.6 Discussion, Conclusions, and Outlook

The electrolyte model used in this work covers a wide range of different physical phenomena. This generality is achieved at the cost of a considerable number of free parameters. Not all of these parameters are of relevance for the present study and the number of parameters is at least somewhat limited by the predominantly made assumption of linear material laws. The linear laws are only a viable modeling option for systems that show nonlinear behavior because of the inclusion of incomplete dissociation and solvated ions in the model. The main parameters controlling the dissociation degree δ are the Gibbs energy Δg and the solvation numbers κ_α. These parameters have negligibly influence on the limiting apparent molar volume Φ_V^0 and the linear dielectric decrement d in the strong dilution limit $m \to 0$. In the range of larger molality m of the solution, the influences of incomplete dissociation and ion–ion interaction on the apparent molar volume largely overlap. Thus, combination of mass density data with independent data of concentration-dependent susceptibility can help to separate these overlapping effects and their related parameters. The nonlinearity in the observed concentration-dependent susceptibility for larger salt

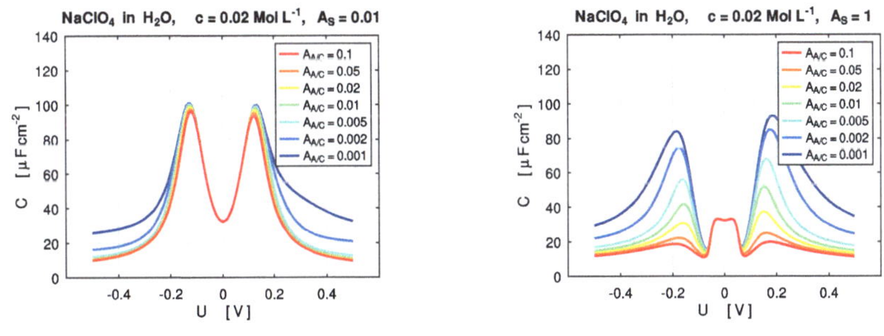

Fig. 12.9 Dependence on a_A for fixed a_S: Capacitance for $A_S = 0.01$ (left) and $A_S = 1$ (right). Model parameters for a solution of NaClO4 with bulk concentration $c = 0.02\,\mathrm{mol\,L^{-1}}$ were used

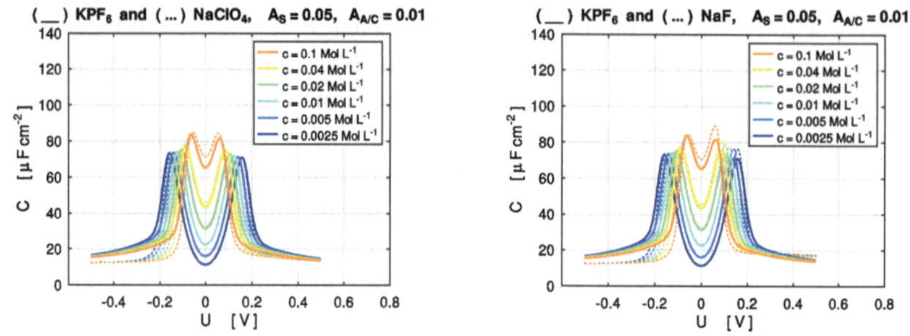

Fig. 12.10 Left: computed differential capacitance for solutions of KPF_6 (solid lines) and of $NaClO_4$ (dotted lines) for different bulk concentrations c. Each individual curve shows a high degree of symmetry with respect to the potential of zero charge $U = 0$ and both sets show a good agreement throughout the concentration range. Right: same plot as left, but this time KPF_6 (solid lines) is compared to NaF (dotted lines)

concentrations can be used to determine dissociation degree δ and the controlling parameters Δg and κ_α.

A remarkable consequence of the model used here is the fact that the temperature dependence of the susceptibility cannot be modeled freely, but is dictated by a thermodynamic consistency condition. We tested the model for solutions of NaCl, as this is the only salt for which sufficient data were found. For different temperatures, the dependence of the slope of χ as a function of concentration can be captured by the model if the Gibbs energy depends on temperature. To gain more insight into the temperature dependence of Δg, it would be desirable to investigate additional independent experiments, particularly related to heat capacity and heat of dilution. This is left for further work.

We studied the molar volume and the dielectric decrement for solutions of $NaClO_4$, NaF, KF, KPF_6, which are prototypes of non-adsorbing salts. From mass density data for these and several related salts, specific volume and susceptibility parameters of the ions have been extracted and, in the case of KF, also confirmed by data of concentration-dependent susceptibility. For KPF_6, more data is needed. The negative apparent molar volume associated with the fluoride ion is a notable aspect that supports the concept of solvated ions as the constituents of electrolytic mixtures. Instead of having a mixture of constituents with possibly negative volume, the solvated ions always have positive specific volumes, although these can also be smaller than the sum of the volumes of as many free solvent molecules as there are bound solvent molecules in the solvation shell.

In the boundary layer, dielectric saturation is added as another phenomenon to the mathematical model. In combination with the dielectric decrement, it is possible to realize very different structures of the boundary layer. In particular, the combination of both effects can cause the solvent in a boundary layer to remain directly at the surface and not to be displaced by the counter-ions. The strong reduction in susceptibility at the surface is reminiscent of the idea of a Stern-layer made of oriented solvent dipoles. The pressures that occur at large potential differences across the boundary layer can become extremely high, so that the assumption of incompressibility cannot be maintained. This allows additional charge accumulation at the interface. The resulting effect on the capacitance curves is, however, comparatively weak and diminishes at high dielectric saturation. Strong dielectric saturation not only leads to an early, strong suppression of the boundary layer capacitance C^{BL}, which reduces the height of the local maxima, but also moves these maxima closer together. A very striking effect occurs if the dielectric saturation has a significantly weaker effect on the solvated ions than on the solvent. In this case, secondary local maxima appear, the height of which is determined by the dielectric saturation.

Experiments on the capacitance at the interface between silver single-crystal surfaces and aqueous solutions of $NaClO_4$, NaF and KPF_6, Valette (1981, 1982, 1989) show agreement between the different salts at potential differences U more negative than the location of the local minimum, but on the other hand, differences between them for U more positive than the location of the minimum. Due to the dependence of the local minimum position on the bulk electrolyte concentration, the larger capacitance of $NaClO_4$, NaF on the more positive side can be attributed to the adsorption of anions on the surface. In contrast, in the case of KPF_6, the minimum position is independent of the bulk concentration, indicating no residual adsorption of anions, and the capacitance is highly symmetric with respect to the PZC at $U = 0$. Assuming that the height of the local maximum depends only on the specific volumes $\upsilon_\alpha^{\text{ref}}$, this striking symmetry motivates the assumption of a unique specific volume of the ions, which depends only on the valence of the ions. We propose, on the other hand, that, without adding any new features to the electrolyte model in the layer other than dielectric saturation, the size parameters in the layer should coincide with those from the bulk, where the determination of the specific volumes from mass density measurements is the most natural way. For the salts considered and with $\kappa_\alpha = 6$, this leads to a specific volume $\upsilon_{\text{F}^-}^{\text{ref}}$ for the solvated fluoride ion being about 40% smaller than $\upsilon_{\text{KPF}_6^-}^{\text{ref}}$. Due to the field and concentration-dependent susceptibility of the model, the resulting difference in the height of the local maxima is small enough to maintain the interpretation of the differences between the experimental curves of the different salts as mainly caused by adsorption.

However, the results also show limitations of the model. The drop in capacity beyond the local maxima is too steep compared to experiments and the remaining capacity is very small. This behavior, known from Landstorfer and Müller (2022), could not be avoided here by taking compressibility into account in the model. Since it is not obvious that the ions in highly concentrated layers still retain their complete solvation shell that they have in the bulk, the model could be extended to include partially solvated ions as additional species. However, taking partial solvation into account has the disadvantage that it may not be possible to reliably determine their size parameters and dielectric properties from bulk experiments.

It can be seen that the model used here still corresponds to the GC model near the PZC, i.e., the electrical interaction energy $\rho\psi^{\text{D}}$ of the ions and the incomplete dissociation are not sufficient to significantly reduce the capacity at the rather low volume concentrations of $c \leq 0.1\,\mathrm{mol\,L^{-1}}$ used in the experiment. While incomplete dissociation shows some but insufficient, effect for the capacity for solutions of $0.1\,\mathrm{mol\,L^{-1}}$ of KPF_6^-, it is not generally accepted that incomplete dissociation exists at all for solutions of alkali halides such as NaCl or NaF. This suggest that modeling of additional effects might be required, e.g., of specific solvent–surface interaction that account for differences between solid and liquid electrodes or between different solvents, cf. e.g., Emets and Damaskin (2000); Shatla et al. (2021). If these effects help to improve the model and reduce the capacitance near the PZC, then the experimentally observed capacitance maxima could possibly be related to the secondary maxima observed here in the case of a stronger dielectric saturation of the solvent compared to the ionic species.

References

Apelblat A, Manzurola E (1999) Volumetric properties of water, and solutions of sodium chloride and potassium chloride at temperatures from $T = 277.15K$ to $T = 343.15K$ at molalities of (0.1, 0.5, and 1.0) $mol \cdot kg^{-1}$. J Chem Thermodyn 31(7):869–893

Allred GC, Woolley EM (1981) Heat capacities of aqueous HCl, NaOH, and NaCl at 283.15, 298.15 and 313.15 K: ΔC_p° for ionization of water. J Chem Thermodyn 13(2):147–154

Borukhov I, Andelman D, Orland H (1997) Steric effects in electrolytes: a modified Poisson-Boltzmann equation. Phys Rev Lett 79:435–438

Berg P, Benjaminsen BE (2014) Effects of finite-size ions and relative permittivity in a nanopore model of a polymer electrolyte membrane. Electrochim Acta 120:429–438

Bothe D, Dreyer W (2015) Continuum thermodynamics of chemically reacting fluid mixtures. Acta Mech 226(6):1757–1805

Bothe D, Dreyer W, Druet P-E (2023) Multicomponent incompressible fluids-An asymptotic study. J Appl Math Mech 103(7):e202100174

Buchner R, Hefter GT, Barthel J (1994) Dielectric relaxation of aqueous NaF and KF solutions. J Chem Soc, Faraday Trans 90:2475–2479

Buchner R, Hefter GT, May PM (1999) Dielectric relaxation of aqueous NaCl solutions. J Phys Chem A 103(1):1–9

Bazant MZ, Kilic MS, Storey BD, Ajdari A (2009) Towards an understanding of induced-charge electrokinetics at large applied voltages in concentrated solutions. Adv Coll Interface Sci 152(1–2):48–88

Booth F (1951) The dielectric constant of water and the saturation effect. J Chem Phys 19(4):391–394

Ben-Yaakov D, Andelman D, Podgornik R (2011) Dielectric decrement as a source of ion-specific effects. J Chem Phys 134(7):074705

Chu VB, Bai Y, Lipfert J, Herschlag D, Doniach S (2007) Evaluation of ion binding to DNA duplexes using a size-modified Poisson-Boltzmann theory. Biophys J 93(9):3202–3209

Chen C-TA, Chen JH, Millero FJ (1980) Densities of NaCl, $MaCl_2$, Na_2SO_4, and $MgSO_4$ aqueous solutions at 1 atm from 0 to 50°C and from 0.001 to 1.5m. J Chem Eng Data 25(4):307–310

Chandra A (2000) Static dielectric constant of aqueous electrolyte solutions: is there any dynamic contribution? J Chem Phys 113(3):903–905

Chen T, Hefter G, Buchner R (2003) Dielectric spectroscopy of aqueous solutions of KCl and CsCl. J Phys Chem A 107(20):4025–4031

Debye P (1924) Osmotische Zustandsgleichung und Aktivität verdünnter starker Elektrolyte. Physik Z 25:97–107

Debye P (1929) Polar molecules. Chemical Catalogue Company, New York

de Groot SR, Mazur P (1984) Non-equilibrium thermodynamics. Dover Publications, New York

Dreyer W, Guhlke C, Müller R (2013) Overcoming the shortcomings of the Nernst-Planck model. Phys Chem Chem Phys 15:7075–7086

Dreyer W, Guhlke C, Müller R (2015) Modeling of electrochemical double layers in thermodynamic non-equilibrium. Phys Chem Chem Phys 17:27176–27194

Dreyer W, Guhlke C, Müller R (2018) Bulk-surface electro-thermodynamics and applications to electrochemistry. Entropy 20(12):939

Dreyer W, Guhlke C, Müller R (2019) The impact of solvation and dissociation on the transport parameters of liquid electrolytes: continuum modeling and numerical study. Eur Phys J-Spec Top 227(18):2515–2538

Debye P, Hückel E (1923) Zur Theorie der Elektrolyte I. Gefrierpunktserniedrigung und verwandte Erscheinungen. Physik Z 24(9):185–206

Dreyer W, Müller R (2025) Derivation and thermodynamically consistent coupling of a Debye-Hückel type energy to a steric electrolyte model and application to the apparent molar volume and phase boundaries. arXiv:2502.04344

Dubtsov AV, Pasechnik SV, Shmeliova DV, Saidgaziev AS, Gongadze E, Iglič A, Kralj S (2018) Liquid crystalline droplets in aqueous environments: electrostatic effects. Soft Matter 14:9619–9630

Emets VV, Damaskin BB (2000) Electrical double layer on the Cd-Ga liquid alloy in solutions of surface-inactive electrolytes in water, methanol and propylene carbonate. J Electroanal Chem 491(1):30–38

Figliuzzi B, Chan WHR, Buie CR, Moran JL (2016) Nonlinear electrophoresis in the presence of dielectric decrement. Phys Rev E 94:023115

Fulton RL (2009) The nonlinear dielectric behavior of water: comparisons of various approaches to the nonlinear dielectric increment. J Chem Phys 130(20):204503

Groß P, Halpern O (1925) Über die temperaturabhängigen Parameter in der Statistik und die Debyesche Elektrolyttheorie. Physik Z 26:403–407

Gongadze E, Iglič A (2012) Decrease of permittivity of an electrolyte solution near a charged surface due to saturation and excluded volume effects. Bioelectrochemistry 87:199–203

Grahame DC (1950) Effects of dielectric saturation upon the diffuse double layer and the free energy of hydration of ions. J Chem Phys 18(7):903–909

Gupta A, Stone HA (2018) Electrical double layers: effects of asymmetry in electrolyte valence on steric effects, dielectric decrement, and ion-ion correlations. Langmuir 34(40):11971–11985

Guhlke C (2015) Theorie der elektrochemischen Grenzfläche. PhD thesis, TU-Berlin

Gongadze E, van Rienen U, Iglič A (2011) Generalized Stern models of the electric double layer considering the spatial variation of permittvity and finite size of ions in saturation regime. Gen Physiol Biophys 16(4):576–594

Gongadze E, van Rienen U, Kralj-Iglič V, Iglič A (2011) Langevin Poisson-Boltzmann equation: point-like ions and water dipoles near a charged surface. Gen Physiol Biophys 30(2):130–137

Gates JA, Wood RH (1985) Densities of aqueous solutions of sodium chloride, magnesium chloride, potassium chloride, sodium bromide, lithium chloride, and calcium chloride from 0.05 to 5.0 $mol \cdot kg^{-1}$ and 0.1013 to 40MPa at 298.15K. J Chem Eng Data 30(1):44–49

Hnedkovsky L, Hefter G (2021) Densities and apparent molar volumes of aqueous solutions of $NaClO_4$, $KClO_4$, and KCl at temperatures from 293 to 343 K. J Chem Eng Data 66(9):3645–3658

Hasted JB, Ritson DM, Collie CH (1948) Dielectric properties of aqueous ionic solutions. Parts I and II. J Chem Phys 16(1):1–21

Hatlo MM, van Roij R, Lue L (2012) The electric double layer at high surface potentials: the influence of excess ion polarizability. EPL 97(2):28010

Kilic MS, Bazant MZ, Ajdari A (2007) Steric effects in the dynamics of electrolytes at large applied voltages. I. Double-layer charging. Phys Rev E 75:021502

Kralj-Iglič V, Iglič A (1996) A simple statistical mechanical approach to the free energy of the electric double layer including the excluded volume effect. J Phys II 6(4):477–491

Kirkwood JG (1939) The dielectric polarization of polar liquids. J Chem Phys 7(10):911–919
Kontogeorgis GM, Maribo-Mogensen B, Thomsen K (2018) The Debye-Hückel theory and its importance in modeling electrolyte solutions. Fluid Ph Equilib 462:130–152
Kovetz A (2000) Electromagnetic theory. Oxford University Press, Oxford
Le Bellac M, Lévy-Leblond JM (1973) Galilean electromagnetism. Nuovo Cim B 14(2):217–234
Liu J-L, Eisenberg B (2014) Poisson-Nernst-Planck-Fermi theory for modeling biological ion channels. J Chem Phys 141(22):22D532
Landstorfer M, Guhlke C, Dreyer W (2016) Theory and structure of the metal-electrolyte interface incorporating adsorption and solvation effects. Electrochim Acta 201:187–219
Li B (2009) Continuum electrostatics for ionic solutions with non-uniform ionic sizes. Nonlinearity 22(4):811–833
Loginova DV, Lileev AS, Lyashchenko AK (2006) Microwave dielectric properties of aqueous solutions of potassium and cesium fluorides. Russ J Phys Chem 80:1626–1633
Landstorfer M, Müller R (2022) Thermodynamic models for a concentration and electric field dependent susceptibility in liquid electrolytes. Electrochim Acta 428:140368
Marcus Y (2011) Electrostriction in electrolyte solutions. Chem Rev 111(4):2761–2783
Marcus Y (2013) Evaluation of the static permittivity of aqueous electrolytes. J Solution Chem 42(12):2354–2363
Millero FJ (1971) Molal volumes of electrolytes. Chem Rev 71(2):147–176
Müller R, Landstorfer M (2023) Galilean bulk-surface electrothermodynamics and applications to electrochemistry. Entropy 25(3):416
Müller I (1985) Thermodynamics. Pitman Publishing, London
Nakayama Y, Andelman D (2015) Differential capacitance of the electric double layer: the interplay between ion finite size and dielectric decrement. J Chem Phys 142(4):044706
Newman J, Thomas-Alyea KE (2004) Electrochemical systems. Wiley, Hoboken, NJ
Onsager L (1936) Electric moments of molecules in liquids. J Am Chem Soc 58(8):1486–1493
Pedersen TG, Dethlefsen C, Hvidt A (1984) Volumetric properties of aqueous solutions of alkali halides. Carlsberg Res Commun 49(3):445–455
Redlich O, Meyer DM (1964) The Molal volumes of electrolytes. Chem Rev 64(3):221–227
Rousseaux G (2013) Forty years of Galilean electromagnetism (1973–2013). Eur Phys J Plus 128(8):81–128
Redlich O, Rosenfeld P (1931) Das partielle molare Volumen von gelösten Elektrolyten. I Z Phys Chem 155A(1):65–74
Robinson RA, Stokes JM, Stokes RH (1961) Potassium hexafluorophosphate-An associated electrolyte. J Phys Chem 65(3):542–546
Schmickler W (2020) Double layer theory. J Solid State Electrochem 24(9):2175–2176
Shilov IY, Lyashchenko AK (2015) The role of concentration dependent static permittivity of electrolyte solutions in the Debye-Hückel theory. J Phys Chem B 119(31):10087–10095
Shatla AS, Landstorfer M, Baltruschat H (2021) On the differential capacitance and potential of zero charge of Au(111) in some aprotic solvents. ChemElectroChem 8(10):1817–1835
Silva GM, Liang X, Kontogeorgis GM (2023) How to account for the concentration dependency of relative permittivity in the Debye-Hückel and Born equations. Fluid Ph Equilib 566:113671
Truesdell C, Toupin R (1960) The classical field theories, vol III/1 of Handbuch der Physik. Springer, Berlin, Göttingen, Heidelberg
Valette G (1981) Double layer on silver single-crystal electrodes in contact with electrolytes having anions which present a slight specific adsorption: Part I. the (110) face. J Electroanal Chem 122:285–297
Valette G (1982) Double layer on silver single crystal electrodes in contact with electrolytes having anions which are slightly specifically adsorbed: Part II. The (100) face. J Electroanal Chem 138(1):37–54
Valette G (1989) Double layer on silver single crystal electrodes in contact with electrolytes having anions which are slightly specifically adsorbed: Part III. The (111) face. J Electroanal Chem 269(1):191–203

Valiskó M, Boda D (2014) The effect of concentration- and temperature-dependent dielectric constant on the activity coefficient of NaCl electrolyte solutions. J Chem Phys 140(23):234508
Valiskó M, Boda D (2023) Resurrection of Hückel's idea: decoupling ion-ion and ion-water terms in activity coefficients via the state-dependent dielectric constant. Fluid Ph Equilib 572:113826
Vincze J, Valiskó M, Boda D (2010) The nonmonotonic concentration dependence of the mean activity coefficient of electrolytes is a result of a balance between solvation and ion-ion correlations. J Chem Phys 133(15):154507

Chapter 13
Elastic–Plastic–Creep Deformation and Damage in Sn–Ag–Cu Solder Alloys

Torsten Hauck, Aleksandr Morozov, and Wolfgang H. Müller

Abstract One of the most critical failure mechanisms of solder joints is material fatigue. Repeated inelastic deformations cause the accumulation of microdefects and, after a certain number of load cycles, the final failure. This paper presents a modeling approach for the prediction of damage evolution in Sn–Ag–Cu alloys. Constitutive equations are explicitly derived for combined elasticity, plasticity, creep, and uniform isotropic damage. All required material parameters are identified and possible material test procedures are discussed. The modeling approach is demonstrated for a single solder ball subjected to cyclic mechanical loading.

Keywords Sn-Ag-Cu model · Coupled plasticity–creep-damage · Solder fatigue

13.1 Introduction

Sn–Ag–Cu alloys are leading solder materials in electronics assemblies. These alloys provide an optimal balance of strength and fatigue resistance. However, enabling reliable solder interconnects can still become challenging. Assemblies of components with material mismatch and varying environment temperature often lead to fatigue and degradation of the solder alloy. The degradation can be measured as a softening of the material. Figure 13.1 shows stress–strain curves measured at Sn–Ag–Cu specimen that was subjected to different numbers of tension–compression load cycles. It clearly shows a reduction of yield strength and ultimate strength with increasing cycle numbers. The softening is attributed to repeated inelastic deformation causing the evolution of microstructural damage. Figure 13.2 shows microstructure images at different stages of damage evolution.

This paper presents a material model for solder fatigue analysis of Sn–Ag–Cu alloys. It combines elastic–viscoplastic deformation behavior and continuum damage. The constitutive equations are explicitly derived. The equations are then

T. Hauck (✉) · A. Morozov · W. H. Müller
Fachgebiet Kontinuumsmechanik und Materialtheorie, Technische Universität Berlin, Institut für Mechanik, Berlin, Germany
e-mail: Torsten.Hauck@gmx.de

J. Fuhrmann et al. (eds.), *Advances in Continuum Physics*, Advanced Structured Materials 238, https://doi.org/10.1007/978-3-031-93918-1_13

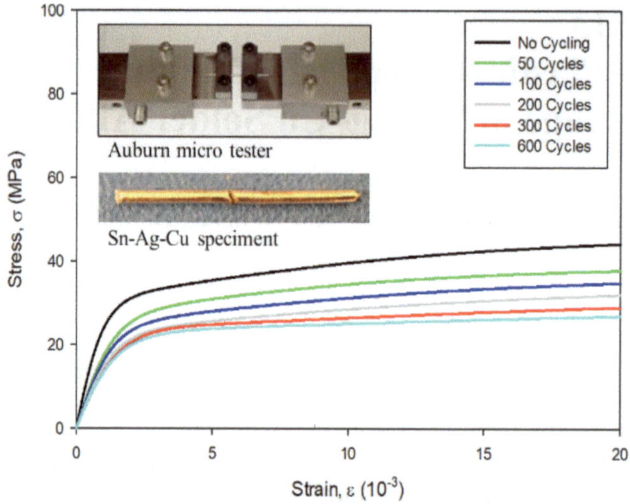

Fig. 13.1 Stress–strain curves at the pristine state and after mechanical cycling (Chen et al. 2022)

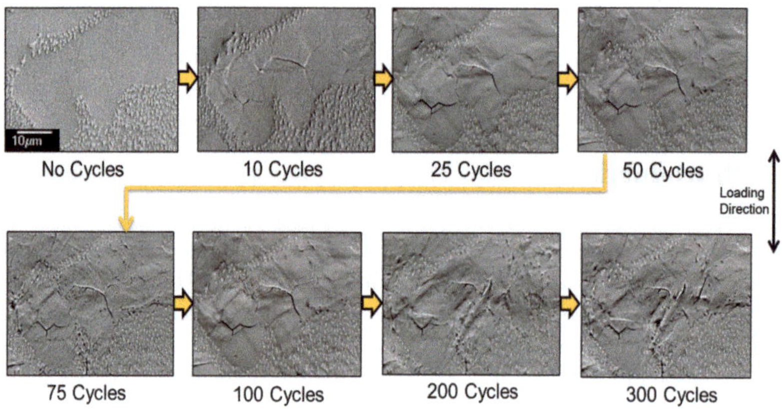

Fig. 13.2 Evolution of microstructure damage in Sn–Ag–Cu alloy, Courtesy of Jeffrey C. Suhling, CAVE3, Auburn University, Auburn, AL USA

specialized for uniaxial tension test conditions. A MATHEMATICA script is developed for the integration of the 1-D rate equations. Various test conditions are simulated, such as monotonic tension, and cyclic tension–compression. Potential material characterization procedures are discussed. One complete set of material properties is applied for the fatigue simulation of a single solder ball subjected to cyclic mechanical loading.

13.2 Material Model

We consider a viscoplastic material model with damage, where the viscoplastic part is decomposed into rate-independent J2-plasticity and rate-dependent creep. The plasticity model is derived from nonlinear kinematic hardening behavior. Creep is introduced in the form of a hyperbolic sine function with an Arrhenius term for the temperature dependency. Damage is included through a scalar damage parameter representing a uniform and isotropic distribution of microdefects.

13.2.1 Constitutive Equations

In order to construct the constitutive equations, the total strain rates are decomposed into elastic, plastic, creep, and thermal components as follows,

$$\dot{\varepsilon}_{ij} = \dot{\varepsilon}_{ij}^{\text{el}} + \dot{\varepsilon}_{ij}^{\text{pl}} + \dot{\varepsilon}_{ij}^{\text{cr}} + \dot{\varepsilon}_{ij}^{\text{th}}, \tag{13.1}$$

and further decomposed into its hydrostatic and deviatoric parts:

$$\dot{\varepsilon}_{ij} = \frac{1}{3}\dot{\varepsilon}_{kk}\delta_{ij} + \dot{e}_{ij}, \tag{13.2}$$

where Einstein's summation rule holds, and δ is the Kronecker delta. Note that the elastic strain rates consist of hydrostatic and deviatoric parts, whereas plastic and creep strain rates only contain deviators, and the thermal strain rate only contains a hydrostatic part. Thus, the hydrostatic and deviatoric parts of the total strain rates read,

$$\dot{\varepsilon}_{kk} = \dot{\varepsilon}_{kk}^{\text{el}} + \dot{\varepsilon}_{kk}^{\text{th}}, \dot{e}_{ij} = \dot{e}_{ij}^{\text{el}} + \dot{\varepsilon}_{ij}^{\text{pl}} + \dot{\varepsilon}_{ij}^{\text{cr}}. \tag{13.3}$$

Stress rates are related to the elastic strain rates by the linear relation,

$$\dot{\sigma}_{ij} = K\dot{\varepsilon}_{kk}^{\text{el}}\delta_{ij} + 2G\dot{e}_{ij}^{\text{el}}, \tag{13.4}$$

where K and G are bulk and shear moduli, respectively. By substituting elastic strain rates with (13.3), stress rates read

$$\dot{\sigma}_{ij} = K\left(\dot{\varepsilon}_{kk} - \dot{\varepsilon}_{kk}^{\text{th}}\right)\delta_{ij} + 2G\left(\dot{e}_{ij} - \dot{\varepsilon}_{ij}^{\text{pl}} - \dot{\varepsilon}_{ij}^{\text{cr}}\right). \tag{13.5}$$

We will now identify each strain rate component in (13.5). We start with the thermal hydrostatic strain,

$$\dot{\varepsilon}_{kk}^{\text{th}} = 3\beta\dot{T}, \tag{13.6}$$

with the linear coefficient of thermal expansion β and temperature rate $\dot{T}$.

Next is the plastic strain rate. Here, we assume J2 incremental plasticity. The constitutive model is developed as described by Zienkiewicz and Taylor in Zienkiewicz and Taylor (2000). For the Sn–Ag–Cu alloy, the Huber-von Mises yield function is chosen including the effect of kinematic hardening as

$$F = \sqrt{(s_{kl} - \alpha_{kl})(s_{kl} - \alpha_{kl})} - \sqrt{\frac{2}{3}}\sigma_{\mathrm{y}} = 0, \tag{13.7}$$

where $\mathbf{s}$, $\boldsymbol{\alpha}$, and σ_{y} are deviatoric stresses, back stresses, and the yield strength, respectively.

The back stresses are defined as a superposition of a number of evolving kinematic back stress tensors, see Chaboche (1989, 1991),

$$\alpha_{ij} = \sum\nolimits_k \alpha_{ij}^{(k)}, \tag{13.8}$$

where the evolution of each back stress tensor $\boldsymbol{\alpha}^{(k)}$ is

$$\dot{\alpha}_{ij}^{(k)} = \frac{2}{3} h^{(k)} \dot{\varepsilon}_{ij}^{\mathrm{pl}} - \gamma^{(k)} \alpha_{ij}^{(k)} \dot{\varepsilon}_{\mathrm{eq}}^{\mathrm{pl}}. \tag{13.9}$$

The first and second term of (13.9) represent strain hardening and dynamic recovery of the back-stresses, respectively. $h^{(k)}$ and $\gamma^{(k)}$ are material parameters, $\dot{\varepsilon}_{\mathrm{eq}}^{\mathrm{pl}}$ is the equivalent plastic strain rate.

The plastic strain rates are derived from the yield function F as

$$\dot{\varepsilon}_{ij}^{\mathrm{pl}} = \dot{\lambda}\frac{\partial F}{\partial s_{ij}} = \dot{\lambda} n_{ij},\ n_{ij} = \frac{s_{ij} - \alpha_{ij}}{\sqrt{(s_{kl} - \alpha_{kl})(s_{kl} - \alpha_{kl})}}, \tag{13.10}$$

where the multiplier $\dot{\lambda}$ and $\mathbf{n}$ represent the norm and the direction of the plastic strain rates, respectively. The equivalent plastic strain rate is derived from

$$\dot{\varepsilon}_{\mathrm{eq}}^{\mathrm{pl}} = \sqrt{\frac{2}{3}\dot{\varepsilon}_{kl}^{\mathrm{pl}}\dot{\varepsilon}_{kl}^{\mathrm{pl}}}. \tag{13.11}$$

With (13.10) and $\mathbf{n}\cdot\mathbf{n} = 1$, Eq. (13.11) becomes

$$\dot{\varepsilon}_{\mathrm{eq}}^{\mathrm{pl}} = \sqrt{\frac{2}{3}}\dot{\lambda}. \tag{13.12}$$

By combining Eqs. (13.8), (13.9) and (13.12) the evolution of back stresses reads

$$\dot{\alpha}_{ij} = \dot{\lambda}\sum\nolimits_k \left(\frac{2}{3}h^{(k)} n_{ij} - \sqrt{\frac{2}{3}}\gamma^{(k)}\alpha_{ij}^{(k)}\right). \tag{13.13}$$

Besides the yield criterion (13.7), the plastic consistency condition must be satisfied as follows

$$\dot{F} = \frac{\partial F}{\partial s_{ij}} \dot{s}_{ij} + \frac{\partial F}{\partial \alpha_{ij}} \dot{\alpha}_{ij} = 0. \tag{13.14}$$

With the back stress rates (13.13) and the deviatoric stress rates,

$$\dot{s}_{ij} = 2G\left(\dot{e}_{ij} - \dot{\varepsilon}^{\mathrm{pl}}_{ij} - \dot{\varepsilon}^{\mathrm{cr}}_{ij}\right), \tag{13.15}$$

the consistency condition (13.14) is rewritten to

$$\dot{F} = 2Gn_{ij}\left(\dot{e}_{ij} - \dot{\varepsilon}^{\mathrm{cr}}_{ij}\right) - \dot{\lambda}\left(2G - \sum\nolimits_k \left(\frac{2}{3}h^{(k)} - \sqrt{\frac{2}{3}}\gamma^{(k)} n_{ij}\alpha^{(k)}_{ij}\right)\right) = 0, \tag{13.16}$$

from which the multiplier $\dot{\lambda}$ is derived as

$$\dot{\lambda} = \frac{G}{G + \frac{1}{3}\sum_{(k)}\left(h^{(k)} - \sqrt{\frac{3}{2}}\gamma^{(k)} n_{ij}\alpha^{(k)}_{ij}\right)} n_{kl}\left(\dot{e}_{kl} - \dot{\varepsilon}^{\mathrm{cr}}_{kl}\right). \tag{13.17}$$

By inserting Eq. (13.17) into Eq. (13.10), and with

$$H = \sum\nolimits_{(k)}\left(h^{(k)} - \sqrt{\frac{3}{2}}\gamma^{(k)} n_{ij}\alpha^{(k)}_{ij}\right) \tag{13.18}$$

the plastic strain rates read

$$\dot{\varepsilon}^{\mathrm{pl}}_{ij} = \frac{G}{G + \frac{1}{3}H}\left(\dot{e}_{kl} - \dot{\varepsilon}^{\mathrm{cr}}_{kl}\right) n_{kl} n_{ij}. \tag{13.19}$$

By inserting Eq. (13.19) into Eq. (13.5) the stress rates follow as

$$\dot{\sigma}_{ij} = K\left(\dot{\varepsilon}_{kk} - \dot{\varepsilon}^{\mathrm{th}}_{kk}\right)\delta_{ij} + 2G\left(\delta_{ik}\delta_{jl} - \frac{G}{G + \frac{1}{3}H} n_{ij} n_{kl}\right)\left(\dot{e}_{kl} - \dot{\varepsilon}^{\mathrm{cr}}_{kl}\right) \tag{13.20}$$

which is known as the Prandtl–Reuss relation.

Still to be defined are the creep strain rates. Here, we choose the Garofalo creep model, (Garofalo 1965). The creep strain rates read

$$\dot{\varepsilon}^{\mathrm{cr}}_{ij} = \dot{\varepsilon}^{\mathrm{cr}}_{\mathrm{eq}} m_{ij} \tag{13.21}$$

with the equivalent strain rate $\dot{\varepsilon}^{\mathrm{cr}}_{\mathrm{eq}}$ and $\mathbf{m}$ representing norm and direction of $\dot{\boldsymbol{\varepsilon}}^{\mathrm{cr}}$, and

$$\dot{\varepsilon}_{\rm eq}^{\rm cr} = A\left(\sinh\left(\alpha\sigma_{\rm eq}\right)\right)^n \exp\left(-\frac{Q}{RT}\right),\ m_{ij} = \frac{s_{ij}}{\sqrt{s_{kl}s_{kl}}}, \tag{13.22}$$

where $\sigma_{\rm eq}$ is the von Mises equivalent stress.

13.2.2 Continuum Damage Model

Damage is introduced as material softening caused by a uniform and isotropic distribution of microdefects. A damage parameter D is introduced that relates the effective stresses acting on the material between the microdefects to the resulting reduced stresses acting on the homogenized material including the microdefects, see Kachanov–Rabotnov (Rabotnov 1968, 1969; Kachanov 1986),

$$\sigma_{ij\,\rm hom} = (1 - D)\sigma_{ij}. \tag{13.23}$$

In this paper, we directly relate damage evolution to the energy dissipation during inelastic deformations as follows:

$$\dot{D} = Z(1 - D)\left(\sigma_{ij}\dot{\varepsilon}_{ij}^{\rm pl} + \sigma_{ij}\dot{\varepsilon}_{ij}^{\rm cr}\right). \tag{13.24}$$

Z is the damage rate coefficient, another material parameter. After time integration of (13.24), the damage parameter follows as

$$D = 1 - \exp\left[-Z\left(\int \sigma_{ij}\,{\rm d}\varepsilon_{ij}^{\rm pl} + \int \sigma_{ij}{\rm d}\varepsilon_{ij}^{\rm cr}\right)\right]. \tag{13.25}$$

13.2.3 Uniaxial Stress State

A uniaxial stress is considered for the comparison with material test results and the identification of material parameters. The stress rate tensor reads,

$$\dot{\sigma}_{ij} = \sigma_{11}\delta_{1j}\delta_{i1} \tag{13.26}$$

with the normal stress component σ_{11}. We skip the modifications of the three-dimensional material model and jump directly to its representation for the uniaxial stress state. By omitting all indices, the stress rate reads

$$\dot{\sigma} = \begin{cases} \frac{EH}{E+H}\left(\dot{\varepsilon} - \dot{\varepsilon}^{\rm cr} - \dot{\varepsilon}^{\rm th}\right), & \left|\sigma - \frac{3}{2}\alpha\right| \geq \sigma_{\rm y} \\ E\left(\dot{\varepsilon} - \dot{\varepsilon}^{\rm cr} - \dot{\varepsilon}^{\rm th}\right), & \left|\sigma - \frac{3}{2}\alpha\right| < \sigma_{\rm y} \end{cases}. \tag{13.27}$$

E, H, α, and σ_Y are the elastic modulus, the kinematic hardening modulus, the back stress, and the yield stress, respectively. Using the Chaboche model, H and α read,

$$H = \sum\nolimits_{(k)} \left(h^{(k)} - \frac{3}{2}\gamma\alpha^{(k)} \text{sign}\left(\sigma - \frac{3}{2}\alpha \right) \right), \tag{13.28}$$

$$\alpha = \sum\nolimits_{(k)} \alpha^{(k)}, \dot{\alpha}^{(k)} = \frac{2}{3} h^{(k)} \dot{\varepsilon}^{\text{pl}} - \gamma^{(k)} \alpha^{(k)} \left| \dot{\varepsilon}^{\text{pl}} \right|, \tag{13.29}$$

where $h^{(k)}$ and $\gamma^{(k)}$ are material constants describing kinematic hardening and dynamic recovery mechanisms during cyclic loading and unloading.

The creep strain rate reads:

$$\dot{\varepsilon}^{\text{cr}} = \text{sign}(\sigma) C_1 (\sinh(C_2|\sigma|))^{C_3} \exp\left(-\frac{C_4}{T} \right). \tag{13.30}$$

13.3 Parameter Identification

Table 13.1 shows the required set of model parameters, which all have to be determined by material characterization. Standard test conditions can be applied from which most of the listed parameters can directly be identified. Some parameters, however, require a nonlinear parameter fit, which includes the numerical integration of the stress rate Eq. (13.27).

13.3.1 Iterative Time Integration

An explicit iterative time integration algorithm is developed for strain-controlled tension–compression loads at constant temperature. The stress $\sigma_{(n+1)}$ at the end of a time step Δt explicitly follows from its value $\sigma_{(n)}$ and the increment $\Delta\sigma_{(n)}$ as

$$\sigma_{(n+1)} = \sigma_{(n)} + \Delta\sigma_{(n)}, \quad \Delta\sigma_{(n)} = \begin{cases} \frac{EH_{(n)}}{E+H_{(n)}} \left(\dot{\varepsilon}_{(n)} - \dot{\varepsilon}^{\text{cr}}_{(n)} \right) \Delta t, & \left| \sigma_{(n)} - \frac{3}{2}\alpha_{(n)} \right| \geq \sigma_{\text{y}} \\ E \left(\dot{\varepsilon}_{(n)} - \dot{\varepsilon}^{\text{cr}}_{(n)} \right) \Delta t, & \left| \sigma_{(n)} - \frac{3}{2}\alpha_{(n)} \right| < \sigma_{\text{y}} \end{cases}. \tag{13.31}$$

The algorithm is programmed in MATHEMATICA. A screenshot of the script is shown in Fig. 13.3. The algorithm requires the input of the total strain-time function and all material parameters as listed in Table 13.1. The output contains all strain components, back stress, effective stress, and nominal stress as functions of time.

Table 13.1 Identified material parameters

Elasticity	
G [MPa]	Shear modulus
K [MPa]	Bulk modulus
E [MPa]	Elastic modulus $E = \frac{9KG}{3K+G}$
$\nu[-]$	Poisson's ratio $\nu = \frac{3K-2G}{2(3K+G)}$
$\beta[1/°\text{C}]$	Coefficient of thermal expansion
Plasticity (Chaboche model)	
σ_y[MPa]	Yield strength
n [−]	Number of kinematic hardening models
h_1 [MPa]	1st strain hardening coefficient
γ_1 [−]	1st recovery coefficient
h_2 [MPa]	2nd strain hardening coefficient
γ_2 [−]	2nd recovery coefficient
Creep (Garofalo model)	
C_1 [1/s]	Creep parameter 1
C_2 [1/MPa]	Creep parameter 2
C_3 [−]	Creep parameter 3
C_4 [1/K]	Creep parameter 4
Damage (Kachanov model)	
Z	Damage rate coefficient

The algorithm is used to investigate load cases, assess parameter significance, and identify material properties by comparison with material characterization data. What follows are three examples. The material properties used in these examples are not validated, some are hypothetical values, and others are intentionally modified.

13.3.2 Tension Test

The tension test is reproduced for a constant strain rate of 0.002·1/s and material properties as shown in Fig. 13.4. Figure 13.5 shows the total strain-time function input and the stress response. Elastic regions and plastic regions can clearly be distinguished. The test can be used for the extraction of elastic modulus and yield stress. The parameters of the nonlinear kinematic hardening model can be identified by expressing stress as a function of plastic strain as follows:

$$\sigma = \sigma_y + \sum\nolimits_{(k)} \frac{h^{(k)}}{\gamma^{(k)}} \left(1 - \exp\left(-\gamma^{(k)} \varepsilon^{\text{pl}}\right)\right). \tag{13.32}$$

```
(* initial conditions *)
σ = 0; α = 0; α1 = 0; α2 = 0; σY = σY0; εp = 0; εc = 0; εe = 0; d = 0; Win = 0;
n = 5; dt = 0.01; n = IntegerPart[n p / dt]; r = ConstantArray[0, {8, n}];
For[i = 1, i ≤ n, i++,
 dε = ε[i dt] - ε[(i - 1) dt]; (* total strain increment *)
 dεc = Sign[σ - α] C1 Sinh[C2 Abs[σ - α]] ^C3 dt; (* creep strain increment *)
 εc = εc + dεc; (* creep strain *)
 If[Abs[σ - 3 / 2 α] ≥ σY && (dε - dεc) (σ - 3 / 2 α) > 0,   (* yield condition *)
  (*plasticity branch *)
  Hk = a1 + a2 - 3 / 2 γ1 α1 Sign[dεp] - 3 / 2 γ2 α2 Sign[dεp]; (* kinmeatic hardening modulus *)
  ET = EM Hk / (EM + Hk); (* tangential modulus *)
  dσ = ET (dε - dεc); (* stress increment *)
  dεp = dσ / Hk; (* plastic strain increment *)
  εp = εp + dεp; (* plastic strain *)
  α1 = α1 + 2 / 3 a1 dεp  - γ1 α1 Abs[dεp]; (* 1st backstress *)
  α2 = α2 + 2 / 3 a2 dεp  - γ2 α2 Abs[dεp]; (* 2nd backstress *)
  α = α1 + α2; (* total back stress *)
  dεe = dε - dεp - dεc; (* elastic strain increment *)
  dWin = σ dεp + σ dεc; (* inelastic stress work density increment *)
  Win = Win + dWin; (* inelastic stress work density *)
  d = 1 - Exp[-Z Win]; (* damage parameter *)
  σ = σ + dσ; (* stress *)
  εe = εe + dεe, (* elastic strain *)
  (* elastic branch *)
  dσ = EM (dε - dεc); (* stress increment *)
  dεe = dσ / EM; (* elastic strain increment *)
  σ = σ + dσ; (* stress *)
  εe = εe + dεe (* elastic strain *)] ;
 (* results *)
 r⟦1, i⟧ = ε[i dt]; (* total strain *)
 r⟦2, i⟧ = σ * (1 - d) ;]
(* result plots *)
plot1 = ListLinePlot[Transpose[{r⟦1, All⟧, r⟦2, All⟧}], GridLines → Automatic]
```

Fig. 13.3 MATHEMATICA script for explicit time integration of the 1-D stress rate Eq. (13.31)

```
Elastic modulus = 26000 MPa

Poisson ratio = 0.3

bulk modulus = 21666.7 MPa

Shear modulus = 10000. MPa

Yield stress = 20 MPa

Hardening constants {h1, γ1, h2, γ2} = {5903.28, 339.011, 1002.05, 11.954}

Creep constants {C1, C2, C3, C4} = {1.482 × 10^-6, 0.01975, 1.548, 0}

Damage rate coefficient Z = 0.1
```

Fig. 13.4 Hypothetical properties for Sn–Ag–Cu alloy, refer (Hauck et al. 2025)

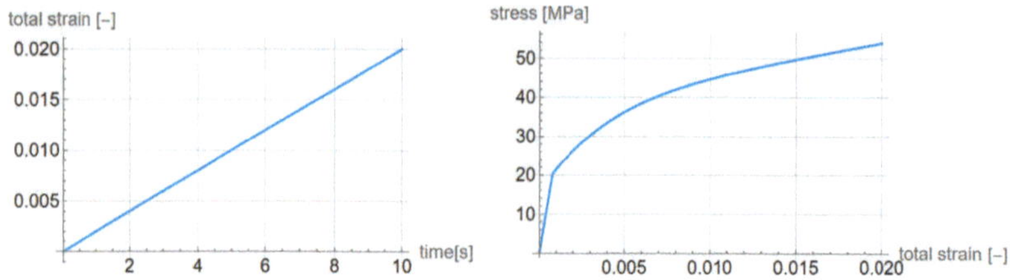

Fig. 13.5 Total strain versus time (left), stress versus total strain (right) for the tension test

and matching with measured yield curves.

13.3.3 Tension–Compression Cycle Test

Repeated strain reversals cause cyclic plasticity that is associated with a stress–strain hysteresis. The area enclosed by a hysteresis loop represents the amount of energy dissipation and is directly related to material damage. In order to achieve good visibility of the impact of damage on the shape of the stress–strain hysteresis, we artificially increase the damage rate coefficient Z of the material. Figure 13.6 shows the applied strain cycle input and the stress–strain hysteresis response for 5 cycle repetitions. It is obvious that the stresses reduce from cycle to cycle. This is attributed to the evolving damage in the material. The damage evolution, however, will run much slower in real materials. It will take hundreds or even thousands of load cycles to failure. The MATHEMATICA script can be used to identify the real damage rate coefficient of the solder alloy from empirical fatigue test results.

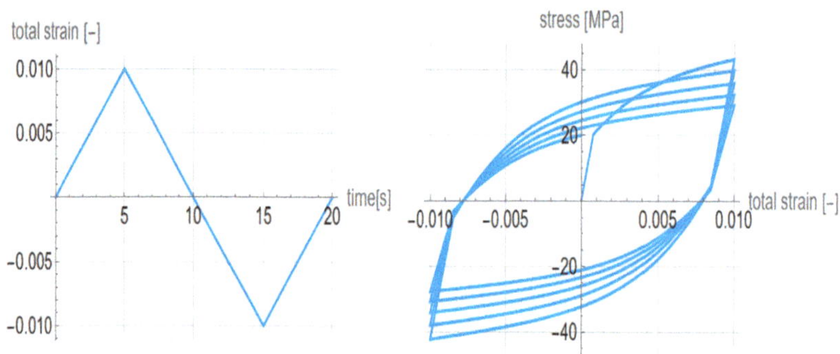

Fig. 13.6 Strain cycle (left), stress–strain hysteresis for 5 cycles (right)

13.3.4 Tension–Compression Cycle Test with Dwell Time

Here we want to visualize the influence of creep deformation on the stress–strain hysteresis. To this end, the creep constant C_1 is artificially magnified. It represents a mechanical cycling test at elevated temperature. In addition, we insert dwell sections of constant strain into the strain cycle. That is giving the material more time for creep and stress relaxation. Figure 13.7 shows strain cycle input and stress–strain hysteresis response again for 5 cycles. These hysteresis loops show additional vertical sections of stress relaxation at constant strain. These loops explicitly show all characteristics of our material model. It starts at the origin with linear elastic deformation until the yield stress is reached, and continues with nonlinear elastic–plastic deformation until the first strain dwell starts. During the dwell, at constant strain, stresses relax. At the end of the dwell, the strain rate reverses direction and we see a linear elastic deformation again. Elastic–plastic deformation starts much earlier due to the translation of the yield function during the nonlinear kinematic hardening. The hysteresis loops are all identical in shape but with a reduction of stress amplitudes. This reduction is caused by damage evolution. The presented algorithm can help design appropriate test procedures for the identification of all required material parameters.

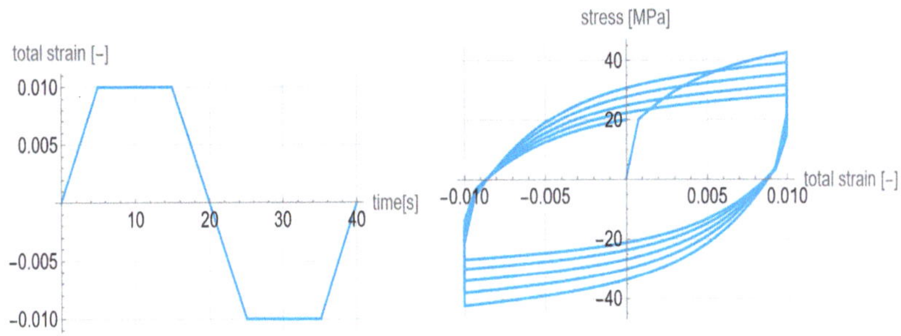

Fig. 13.7 Strain cycle with dwell sections (left), stress–strain hysteresis for 5 cycles (right)

13.4 Solder Ball Subjected to Mechanical Cycling

The material model is applied for a qualitative fatigue analysis of a single Sn–Ag–Cu solder joint. The solder joint has the shape of a truncated sphere. Cyclic horizontal and vertical displacements, u_x and u_y, are applied at the top, while the bottom is fixed in space. The resulting displacement constraints are typical for thermal cycling, where combined shear and tension loads are caused by the thermal mismatch of parts of the microelectronic assemblies. Material interfaces between the solder ball and the connected parts are critical areas for fatigue. ANSYS finite element analysis is used to run a fatigue analysis with our material model. Figure 13.8 shows the finite element mesh of the solder joint with the applied displacement constraints. The red-marked elements are closest to expected stress concentrations in the solder joint and are considered critical locations for damage initiation. Figure 13.9 shows hysteresis loops for normal and shear components in these critical locations. A reduction of stress amplitudes with ongoing cycling is observed. And as discussed before, this is a result of damage evolution along with material softening. Figure 13.10 shows the damage evolution in the cross section of the solder joint. Damage indeed initiates from the stress concentration and then propagates along a diagonal path toward the sphere center. This failure mechanism correlates well with experimental observations.

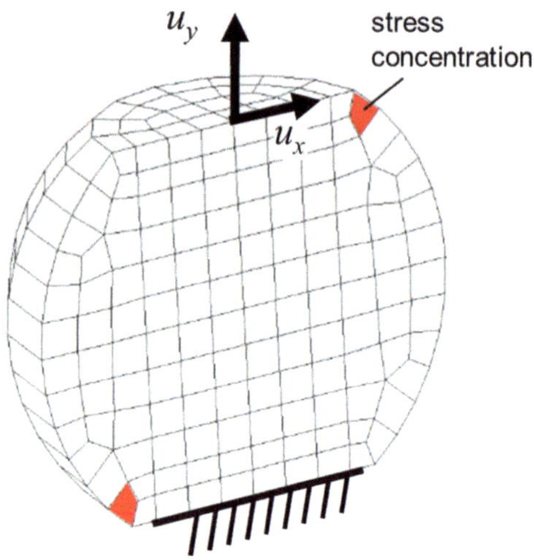

Fig. 13.8 FE-model of a solder ball with prescribed boundary conditions at top and bottom faces. Critical areas of the solder are marked with red color

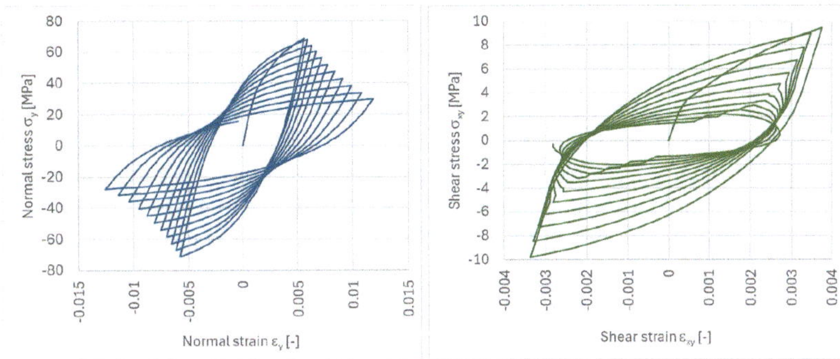

Fig. 13.9 Tension stress–strain-hysteresis (left) and shear-stress–strain hysteresis (right) in the critical area of the solder ball

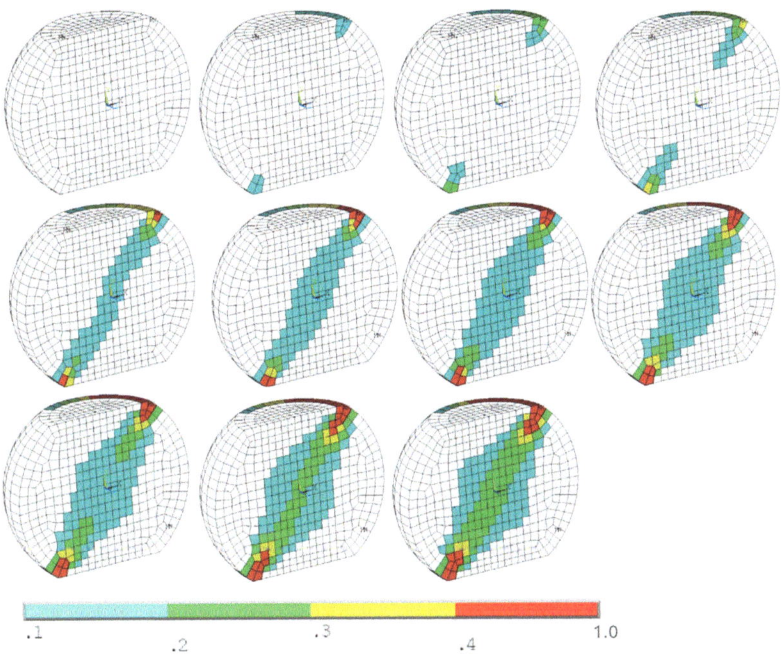

Fig. 13.10 Damage distribution in the solder ball, 11 subsequent load cycles from top left to bottom right, damage rate is highly accelerated, the color bar represents the damage parameter D

13.5 Summary

A combined elastic–plastic–creep-damage model is proposed for solder materials. It is based on von Mises plasticity, Chaboche linear kinematic hardening, Garofalo creep, and Kachanov–Rabotnov damage models. The stress rate equation is specialized for the uniaxial stress state for parameter identification. An iterative integration algorithm is programmed in MATHEMATICA. The algorithm is used to predict the stress response for various strain-controlled monotonic and cycling test conditions. Stress–strain hysteresis loops are investigated and material parameter identification is discussed. The material model is used for the finite element fatigue analysis of a solder ball in ANSYS. The ANSYS model predicts damage initiation points and damage propagation. The simulation results qualitatively match experimental observations. Quantitative prediction of fatigue life, however, requires confirmed material properties from comprehensive material testing.

Acknowledgements The authors acknowledge the experimental work on SAC solder alloys conducted by Prof. Suhling and his students at Auburn University. The authors dedicate this paper to the late Prof. Wolfgang Dreyer, who once made a remarkable statement regarding the damage parameter D, which occurs in this paper as well: "D wie doof!." *Doof* means stupid in German, and he wanted to express his disdain for phenomenological laws and quantities.

References

Chaboche JL (1989) Constitutive equations for cyclic plasticity and cyclic viscoplasticity. Int J Plast 5(3):247–302

Chaboche JL (1991) On some modifications of kinematic hardening to improve the description of ratchetting effects. Int J Plast 7(7):661–678

Chen Y, Hauck T, Fahim A, Haq MA, Hoque MA, Suhling JC (2022) Damage evolution in SAC solder alloys. In: 2022 23rd international conference on thermal, mechanical and multi-physics simulation and experiments in microelectronics and microsystems, EuroSimE 2022. Institute of Electrical and Electronics Engineers Inc., pp 1–9

Garofalo F (1965) Fundamentals of creep and creep-rupture in metals. Macmillan Co., Collier-Macmillan Ltd., New York

Hauck T, Morozov A, Chen Y, Renzi L, Müller WH (2025) Elastic-plastic-creep model with damage for SAC solder alloys. In: 2025 26th international conference on thermal, mechanical and multi-physics simulation and experiments in microelectronics and microsystems (EuroSimE). IEEE, pp 1–6

Kachanov L (1986) Introduction to continuum damage mechanics, vol. 10. Springer Science & Business Media

Rabotnov YN (1968) Creep rupture. In: Hetenyi M, Vincenti M (eds) Proceedings of applied mechanics conference. Stanford University, Springer, Berlin, pp 342–349

Rabotnov YN (1969) Creep problems in structural members. North-Holland, Amsterdam

Zienkiewicz OC, Taylor RL (2000) Finite element method: volume 2, 5th edn. Butterworth-Heinemann Ltd, pp 64–66

Chapter 14
Constitutive Theory and Simulation of Entropy and Enthalpy Relaxation in the Glass Transition Regime

Johannes Keursten, Sabine Enders, and Thomas Böhlke

Abstract The glass transition is a central physical phenomenon in polymeric materials. It is related with significant changes of the thermomechanical properties and pronounced process-dependent behavior. This work deals with process-dependent caloric behavior in the glass transition regime which is known as enthalpy relaxation. Based on Coleman-Noll procedure and the scheme of linear irreversible thermodynamics, a model of enthalpy relaxation is derived and analyzed. Therefore, an internal temperature is introduced as internal variable and the analogy with viscoelasticity is utilized. The non-equilibrium part of the entropy is identified as thermodynamic driving force. For a purely temperature-dependent relaxation time of the internal temperature, the qualitative behavior of the model is discussed. Furthermore, the model parameters are examplarily adapted for polystyrene. In a first step, rate-independent behavior of the heat capacity in glassy and equilibrium is approximated by linear functions. Subsequently, the parameters for the evolution equation of the internal temperature are identified based on the reduced heat capacity. By comparing two different temperature rates, a rate-dependency of the relaxation time is observed. Therefore, the model is extended by a rate-dependency function for the relaxation time where temperature- and rate-dependency are multiplicatively splitted.

Keywords Glass transition · Enthalpy relaxation · Fictive temperature · Polystyrene · Heat capacity

---

In fond memory of Wolfgang Dreyer.

J. Keursten · T. Böhlke (✉)
Institute of Engineering Mechanics, Karlsruhe Institute of Technology (KIT), Karlsruhe, Germany
e-mail: thomas.boehlke@kit.edu

S. Enders
Institute of Technical Thermodynamics and Refrigeration Technology, Karlsruhe Institute of Technology (KIT), Karlsruhe, Germany

J. Fuhrmann et al. (eds.), *Advances in Continuum Physics*, Advanced Structured Materials 238, https://doi.org/10.1007/978-3-031-93918-1_14

14.1 Introduction

In a variety of industrial applications, thermoplastic polymers are processed. Automotive industry or construction industry can be mentioned as examples (Ehrenstein 2011). The advantages of thermoplastics include good processability, low material costs and high mass-related stiffness and strength (Menges et al. 2011). With that, thermoplastics can, e.g., be used as matrix material in fiber-reinforced composites (Böhlke et al. 2019). A central phenomenon in the physical behavior of thermoplastics is the glass transition. In the glass transition regime, a significant change of thermomechanical properties and a pronounced process-dependent behavior can be observed (Ehrenstein 2011). Thermoplastic polymers are processed above the glass transition and applied in the glass transition regime or below. Thus, for a precise prediction of the thermomechanical material behavior, the glass transition must be taken into account in material modeling (Stommel et al. 2018). This work focusses on the caloric behavior. Process-dependent caloric behavior in the glass transition regime is known as enthalpy relaxation (Hodge 1994). For the process simulation and optimization of fiber-reinforced components (Görthofer et al. 2019; Meyer et al. 2023), material models that take into account the above mentioned relaxation phenomena will be an important approach in the future to describe the material behavior beyond the mechanical behavior.

State of the art To characterize caloric behavior in the glass transition regime, differential scanning calorimetry (DSC) is a standard method (Höhne et al. 2003). There is a large number of experimental studies on enthalpy relaxation in thermoplastic polymers. Regarding polystyrene, studies on the influence of temperature rate (Badrinarayanan et al. 2007; Tropin et al. 2015) and temperature history (Hodge and Huvard 1983; Sakatsuji et al. 2013), cyclic temperature loadings (Weyer et al. 1997, 2001), or physical aging (Simon et al. 2001; Hadač et al. 2007) are exemplarily mentioned here.

The Tool-Narayanaswamy-Moynihan (TNM) model (Tool 1945; Narayanaswamy 1971; Moynihan et al. 1976) and the Kovacs-Aklonis-Hutchinson-Ramos (KAHR) model (Kovacs et al. 1979) are commonly used in order to model enthalpy relaxaton or volume recovery (Simon et al. 2001; Badrinarayanan et al. 2007). The TNM model is based on the fictive temperature introduced by Tool (1945; 1946) to account for relaxation processes. In Tool's model, the evolution of the fictive temperature depends on temperature and fictive temperature itself. Thus, nonlinear structural effects are taken into account. Narayanaswamy (1971) further developed the model by splitting the fictive temperature into multiple parts with varying relaxation times. Moynihan et al. (1976) modified the model by introducing a mixing parameter x to weight the temperature- and structural dependency of the relaxation time. In general, the evolution of the fictive temperature θ_{fi} according to all three versions of the TNM model can be summarized as

$$\dot{\theta}_{\mathrm{fi}} = \frac{\theta - \theta_{\mathrm{fi}}}{\tau_0 a_\theta(\theta) a_{\theta\mathrm{fi}}(\theta_{\mathrm{fi}})}, \tag{14.1}$$

where the expressions for the relaxation time τ in the denominator differ. To consider a non-exponential evolution of the fictive temperature, a Kohlrausch-Williams-Watts (KWW) function (Kohlrausch 1854; Williams and Watts 1970) is often used as response function (Moynihan et al. 1976; Simon et al. 2001; Tropin et al. 2015). This function is defined by a stretched exponential factor $\beta \in [0, 1]$ (Weyer et al. 2001). For $\beta = 1$, Eq. (14.1) holds and the response function shows exponential behavior. For the calculation of the evolution of the fictive temperature, a numerical integration scheme must be used (Hempel et al. 1999; Weyer et al. 2001). The heat capacity according to the TNM model is given by Eq. (14.2) (cf. Badrinarayanan et al. 2007; Svoboda and Málek 2013)

$$c_p = c_p^{\mathrm{gl}} + \Delta c_p \frac{\mathrm{d}\theta_{\mathrm{fi}}}{\mathrm{d}\theta}. \tag{14.2}$$

In the studies by Simon (1997) and Liu et al. (2013), the TNM model is coupled with Fourier's law of heat conduction. Svoboda and Malek (2013) discuss parameter identification methods regarding the TNM model.

The KAHR model (Kovacs et al. 1979) defines relaxation equations for fractional departures from equilibrium. This departures can, e.g., be enthalpy quantities. For a single enthalpy departure $\delta = h - h_{\mathrm{eq}}$, the relaxation equation of the KAHR model can be given by Eq. (14.3) (cf. Kovacs et al. 1979; Argatov and Kocherbitov 2021)

$$\dot{\delta} + \frac{\delta}{\tau_0 a_\theta(\theta) a_\delta(\delta)} = -\Delta c_p \dot{\theta}. \tag{14.3}$$

Like the TNM model, the KAHR model accounts for temperature and nonlinear structural dependencies of the relaxation times. Again, instead of Eq. (14.3), a non-exponential response function can be used (Kovacs et al. 1979). A simplified version of the KAHR model that neglects structural dependencies of the relaxation time was discussed by Aragatov et al. (2021). Fictive temperature of the TNM model and enthalpy departure of the KAHR model are related via (Simon et al. 2001)

$$\theta_{\mathrm{fi}} = \theta + \frac{\delta}{\Delta c_p}. \tag{14.4}$$

Three further models of enthalpy relaxation can be found in the study of Tropin et al. (2015). In the extended TNM model, Moynihan's expression for the relaxation time is replaced by the Vogel-Fulcher-Tammann-Hesse expression (VFTH) (Vogel 1921; Fulcher 1925; Tammann and Hesse 1926). This approach was also applied, i.e., by Weyer et al. (2001; 2005). Based on the molecular-kinetic Adam-Gibbs theory (Adam and Gibbs 1965), alternative expressions for the relaxation time of the TNM model can be derived, see, e.g. Scherer (1984); Hodge (1991); Ngai (1999). An alternative to TNM-like formulations is given by the Gutzow-Schmelzer method (Gutzow et al. 2008; Tropin et al. 2011a, b) that is based on a lattice-hole theory.

Dudowicz et al. (2008) developed the so-called generalized entropy theory, where the basic idea consists in the replacement of the entropy by the entropy density within the Adam-Gibbs theory (Adam and Gibbs 1965).

In the studies by Lion et al. (2016; 2017), enthalpy relaxation is considered in a continuum mechanical framework. A dimensionless thermal internal variable is introduced to model enthalpy relaxation. The evolution of this internal variable depends on temperature modeled by the Williams-Landel-Ferry (WLF) function (Williams et al. 1955). Nonlinear structural dependencies of the evolution equation are neglected. The evolution equation of the internal variable γ and the resulting heat capacity are given by Eq. (14.5) (cf. Lion et al. 2017)

$$\dot{\gamma} = \frac{\beta(\theta - \theta_0) - \gamma}{\tau_0 a_\theta(\theta)}, \quad c_p = c_p^{\mathrm{gl}} + \frac{\widetilde{C}(\beta\theta_0 + \gamma)\dot{\gamma}}{\varrho\dot{\theta}}. \tag{14.5}$$

Lion's model serves as basis for the model considered in this work.

Outline This work is structured as follows. First, the constitutive theory of linear thermoviscoelasticity is summarized in Sect. 14.2.1. The governing equations of a specific rate-dependent caloric model are then specified in Sects. 14.2.2 and 14.2.3. In Sect. 14.3, we discuss the qualitative behavior of the model for a test case. In addition, the model parameters are adapted in Sect. 14.4 in order to reproduce experimental DSC curves of polystyrene.

Notation	
a, A, α	Scalar quantity
$\boldsymbol{q}, \boldsymbol{g}$	First order tensor
$\boldsymbol{\sigma}, \boldsymbol{\varepsilon}$	Second order tensor
$\boldsymbol{I}$	Second order unity tensor
$\underline{y}, \underline{z}$	Tuple
$\underline{\underline{A}}$	Linear operator
$\boldsymbol{a} \cdot \boldsymbol{a}$	Scalar product of first or second order tensors
$\dot{(\cdot)}$	Material time derivative
$\mathrm{grad}\,(\cdot)$	Gradient
$\mathrm{div}\,(\cdot)$	Divergence
$(\cdot)_0$	Property with respect to a reference state

14.2 Methods and Fundamentals

14.2.1 Constitutive Theory of Linear Thermoviscoelasticity

Balance equations For the following consideration, the balance of internal energy and the balance of entropy are of interest. In a small strain setup, the local formulation of these balance equations is given by Eqs. (14.6) and (14.7) (cf., e.g., Gurtin et al. 2010)

$$\varrho \dot{e} = \varrho w - \operatorname{div}(\boldsymbol{q}) + \boldsymbol{\sigma} \cdot \dot{\boldsymbol{\varepsilon}}, \tag{14.6}$$

$$\varrho \dot{s} = \varrho w/\theta - \operatorname{div}(\boldsymbol{q}/\theta) + \varrho \phi. \tag{14.7}$$

Mass specific internal energy and mass specific entropy are denoted by e and s, mass density by ϱ, and absolute temperature by θ. Note that entropy flux and entropy supply are assumed to be given by $\boldsymbol{q}/\theta$ and w/θ respectively where $\boldsymbol{q}$ is the heat flux vector and w is a heat source (Coleman and Noll 1963). The entropy production is denoted by ϕ. As we consider isobaric processes, the balance of linear momentum, $\operatorname{div}(\boldsymbol{\sigma}) = \mathbf{0}$, is of minor interest. Furthermore, for simplicity, we restrict to purely volumetric deformations which means that Cauchy stress tensor $\boldsymbol{\sigma}$ and infinitesimal strain tensor $\boldsymbol{\varepsilon}$ are given by Eq. (14.8) (cf., e.g., Liu 2002)

$$\boldsymbol{\sigma} = -p\boldsymbol{I}, \quad \boldsymbol{\varepsilon} = \frac{1}{3}\varepsilon^{\mathrm{vol}}\boldsymbol{I}, \quad \varepsilon^{\mathrm{vol}} = \frac{v - v_0}{v_0} = \varrho_0 v - 1, \tag{14.8}$$

where p is the pressure and $v = 1/\varrho$ the specific volume. The volumetric strain $\varepsilon^{\mathrm{vol}}$ is assumed to be small. Thus, the mass density is approximately constant, $\varrho \approx \varrho_0$. By the second law of thermodynamics, the entropy production is non-negative. With $\phi \geq 0$ and a Legendre transformation, $e = \psi + s\theta$, the Clausius-Duhem inequality can be derived (Liu 2002)

$$-\frac{p}{\varrho_0}\dot{\varepsilon}^{\mathrm{vol}} - \dot{\psi} - s\dot{\theta} - \frac{\boldsymbol{q} \cdot \boldsymbol{g}}{\varrho_0 \theta} \geq 0, \quad \boldsymbol{g} = \operatorname{grad}(\theta). \tag{14.9}$$

The Clausius-Duhem inequality is derived based on the assumptions for the non-convective entropy flux and the entropy supply mentioned above. These are constitutive assumptions with limitations in many cases. Approaches for the exploitation of the second law of thermodynamics without this special assumption can be found, e.g., in Liu (2002); Šilhavý (1997); Papenfuß (2020).

Exploitation of dissipation inequality In addition to the natural state variables, volumetric strain and temperature, a set of internal variables $\underline{z}$ is introduced to account for relaxation processes. Mass specific free energy ψ, pressure, entropy and heat flux are constitutive properties that must be determined constitutively in order to receive a closed set of equations. Furthermore, the evolution of the internal variables is defined by a constitutive relation for the rate of the internal variables, $\dot{\underline{z}} = \underline{f}$. Formally, the constitutive properties are given by

$$\mathcal{K} = \{p, \boldsymbol{q}, \psi, s, \underline{f}\}, \quad \mathcal{K} = \mathcal{K}(\mathcal{E}), \tag{14.10}$$

where $\mathcal{E}$ is the set of input parameters which are assumed to be

$$\mathcal{E} = \{\varepsilon^{\text{vol}}, \dot{\varepsilon}^{\text{vol}}, \theta, \dot{\theta}, \underline{z}, \boldsymbol{g}, \dot{\boldsymbol{g}}\}. \tag{14.11}$$

Purely dissipative parts of stress and entropy that do not enter the free energy are neglected here, e.g., purely viscous stresses. Applying Coleman-Noll procedure (Coleman and Noll 1963; Coleman and Gurtin 1967) to Eq. 14.9, gives the potential relations

$$p = -\varrho_0 \frac{\partial \psi}{\partial \varepsilon^{\text{vol}}}, \quad s = -\frac{\partial \psi}{\partial \theta} \tag{14.12}$$

and

$$\frac{\partial \psi}{\partial \dot{\varepsilon}^{\text{vol}}} = 0, \quad \frac{\partial \psi}{\partial \dot{\theta}} = 0, \quad \frac{\partial \psi}{\partial \boldsymbol{g}} = \boldsymbol{0}, \quad \frac{\partial \psi}{\partial \dot{\boldsymbol{g}}} = \boldsymbol{0}. \tag{14.13}$$

Thus, the free energy does only depend on the natural state variables and the internal variables: $\psi = \psi(\varepsilon^{\text{vol}}, \theta, \underline{z})$. By introducing the thermodynamic driving force with respect to the internal variables, $\underline{y} = -\partial\psi/\partial\underline{z}$, the reduced dissipation inequality reads

$$\underline{y} \cdot \dot{\underline{z}} - \frac{\boldsymbol{q} \cdot \boldsymbol{g}}{\varrho\theta} \geq 0. \tag{14.14}$$

The dissipation inequality is satisfied, e.g., when formulating the constitutive equations following the scheme of linear irreversible thermodynamics. This means that the evolution of the internal variables is given by $\dot{\underline{z}} = \underline{\underline{A}}(\mathcal{E})\underline{y}$ with a semi-positive definite $\underline{\underline{A}}$ for all $\mathcal{E}$. The heat flux vector can be modeled, e.g., by Fourier's law, $\boldsymbol{q} = -\kappa \boldsymbol{g}$, with a positive thermal conductivity coefficient κ. Heat conduction will not be addressed in this work. As a side note, with the approach used here, the Clausius-Duhem inequality is fulfilled by demanding that both terms are non-negative. This is a restriction for the corresponding constitutive models. For more general cases, one could demand that only the sum of both terms is non-negative. In this case, the constitutive laws for the heat flux and the evolution of the internal variables would be coupled.

Modeling of relaxation Next, mechanical and caloric behavior is modeled. These models should account for viscoelastic behavior and rate-dependent caloric behavior. The latter is known as enthalpy relaxation. Thermal expansion will not be considered. We restrict the approach to linear first-order models. This means that only a single relaxation time is considered (Serra-Aguila et al. 2019). For both, mechanical and caloric behavior, an internal variable is introduced. These are the viscous strain $\varepsilon_{\text{v}}^{\text{vol}}$ and the internal temperature θ_{v}. A coupling between mechanical and caloric behavior is neglected, yielding an additive split of the free energy

$$\psi(\varepsilon^{\text{vol}}, \varepsilon_{\text{v}}^{\text{vol}}, \theta, \theta_{\text{v}}) = \psi_{\text{m}}(\varepsilon^{\text{vol}}, \varepsilon_{\text{v}}^{\text{vol}}) + \psi_{\text{c}}(\theta, \theta_{\text{v}}). \tag{14.15}$$

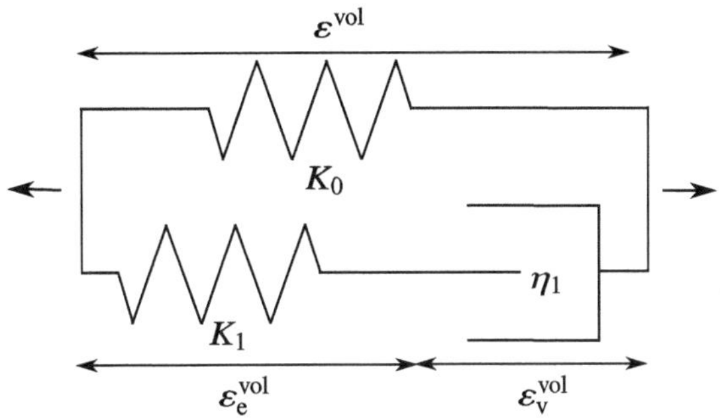

Fig. 14.1 First-order generalized Maxwell model for volumetric deformations. Bulk moduli are denoted by K_0 and K_1, bulk viscosity by η_1. The strain within the Maxwell element is splitted additively into an elastic part ε_e^{vol} and a viscous part ε_v^{vol}

A first-order generalized Maxwell model is used to model viscoelasticity, see Fig. 14.1. This model is a parallel connection of a linear elastic spring and a linear viscoelastic Maxwell model. Using the additive split of the strain within the Maxwell element, the total pressure is

$$p = -K_0\varepsilon^{vol} - K_1(\varepsilon^{vol} - \varepsilon_v^{vol}) = p_e + p_{ve}. \tag{14.16}$$

Using the potential relation in Eq. (14.12), the mechanical part of the free energy can be calculated by integration with respect to the volumetric strain

$$\begin{aligned} \psi_m(\varepsilon^{vol}, \varepsilon_v^{vol}) &= \psi_e(\varepsilon^{vol}) + \psi_{ve}(\varepsilon^{vol} - \varepsilon_v^{vol}) \\ &= \frac{K_0}{2\varrho_0}\varepsilon^{vol2} + \frac{K_1}{2\varrho_0}(\varepsilon^{vol} - \varepsilon_v^{vol})^2. \end{aligned} \tag{14.17}$$

The viscoelastic part ψ_{ve} depends on the difference between total strain and viscous strain. This difference quantifies a departure from equilibrium meaning that ψ_{ve} vanishes in the equilibrium state. The driving force of viscoelasticity equals the negative viscoelastic pressure divided by the mass density,

$$-\frac{\partial\psi}{\partial\varepsilon_v^{vol}} = \frac{K_1}{\varrho_0}\left(\varepsilon^{vol} - \varepsilon_v^{vol}\right) = -\frac{p_{ve}}{\varrho_0}. \tag{14.18}$$

Applying the scheme of linear irreversible thermodynamics, the evolution equation of the viscous strain is

$$\dot{\varepsilon}_v^{vol} = -\frac{p_{ve}}{\eta_1(\theta,\cdot)} = \frac{\varepsilon^{vol} - \varepsilon_v^{vol}}{\lambda(\theta,\cdot)}, \quad \lambda = \frac{\eta_1}{K_1}. \tag{14.19}$$

To satisfy the dissipation inequality, see Eq. (14.14), the volume viscosity η_1 must be a positive quantity. Consequently, the relaxation time λ is also a positive quantity as

K_1 is positive due to convexity of the free energy. As viscosities do strongly depend on temperature, the temperature dependency is explicitely given in the formula. Further dependencies are conceivable. Alternatively, a relaxation equation for the viscoelastic pressure p_{ve} can be derived. Using the time derivative of the viscoelastic pressure from Eq. (14.16) and replacing the rate of the viscous strain via Eq. (14.19), this relaxation equation reads

$$\lambda(\theta, \cdot)\dot{p}_{\mathrm{ve}} + p_{\mathrm{ve}} = -K_1 \lambda(\theta, \cdot)\dot{\varepsilon}^{\mathrm{vol}}. \tag{14.20}$$

When approaching the equilibrium state, the viscous strain tends to the total strain and the viscoelastic part of the pressure tends to zero. It should be noted that the viscoelastic model can be generalized to account for a distribution of relaxation times (Brinson and Brinson 2015). In this case, further Maxwell elements are added to the parallel connection and a viscous strain is introduced for each Maxwell element.

As both enthalpy relaxation and viscoelasticity are relaxation processes, the caloric part of the free energy is formulated analogously

$$\begin{aligned}\psi_{\mathrm{c}}(\theta, \theta_{\mathrm{v}}) &= \psi_\theta(\theta) + \psi_{\mathrm{v}\theta}(\theta - \theta_{\mathrm{v}}) \\ &= \psi_\theta(\theta) + \frac{C}{2}(\theta - \theta_{\mathrm{v}})^2.\end{aligned} \tag{14.21}$$

The temperature-dependent equilibrium part $\psi_\theta(\theta)$ can be determined via

$$-\theta \frac{\partial^2 \psi_\theta}{\partial \theta^2} = c_p^{\mathrm{eq}}, \tag{14.22}$$

where c_p^{eq} is the equilibrium heat capacity that will be specified later. Lion and coworkers (2017) introduced a similar expression where the internal variable is a dimensionless quantity related to a reference temperature, see Eq. (14.5). In this work, the internal temperature is an absolute quantity. Thus, the works differ in the way the internal variables and the material parameters C and $\widetilde{C}$ are defined. However, derived properties like entropy or heat capacity are equivalent. Due to analogy with viscoelasticity, $\psi_{\mathrm{v}\theta}$ is called *viscocaloric* part of the free energy in the following. Evaluating the potential relation for the entropy, see Eq. (14.12), gives

$$s = -\psi_\theta'(\theta) - C(\theta - \theta_{\mathrm{v}}). \tag{14.23}$$

The thermodynamical driving force is the negative of the viscocaloric entropy,

$$-\frac{\partial \psi}{\partial \theta_{\mathrm{v}}} = C(\theta - \theta_{\mathrm{v}}) = -s_{\mathrm{v}\theta}. \tag{14.24}$$

Thus, enthalpy relaxation can also be seen as *entropy relaxation*. By the scheme of linear irreversible thermodynamics, the evolution equation of the internal temperature is

Table 14.1 Analogy of volumetric viscoelasticity and entropy relaxation

Viscoelasticity		Entropy relaxation
$\varepsilon^{\mathrm{vol}}$	Independent state variable	θ
$\varepsilon_{\mathrm{v}}^{\mathrm{vol}}$	Internal variable	θ_{v}
$\psi_{\mathrm{m}} = \psi_{\mathrm{e}}(\varepsilon^{\mathrm{vol}}) + K_1(\varepsilon^{\mathrm{vol}} - \varepsilon_{\mathrm{v}}^{\mathrm{vol}})^2/(2\varrho_0)$	Free energy	$\psi_{\mathrm{c}} = \psi_\theta(\theta) + C(\theta - \theta_{\mathrm{v}})^2/2$
$p = -\varrho_0\psi'(\varepsilon^{\mathrm{vol}}) - K_1(\varepsilon^{\mathrm{vol}} - \varepsilon_{\mathrm{v}}^{\mathrm{vol}})$	Dependent state variable	$s = -\psi_\theta'(\theta) - C(\theta - \theta_{\mathrm{v}})$
$-p_{\mathrm{ve}}/\varrho_0 = K_1(\varepsilon^{\mathrm{vol}} - \varepsilon_{\mathrm{v}}^{\mathrm{vol}})/\varrho_0$	Thermodynamic driving force	$-s_{\mathrm{v}\theta} = C(\theta - \theta_{\mathrm{v}})$
$\dot{\varepsilon}_{\mathrm{v}}^{\mathrm{vol}} = (\varepsilon^{\mathrm{vol}} - \varepsilon_{\mathrm{v}}^{\mathrm{vol}})/\lambda$	Evolution equation	$\dot{\theta}_{\mathrm{v}} = (\theta - \theta_{\mathrm{v}})/\tau$
$\lambda\dot{p}_{\mathrm{ve}} + p_{\mathrm{ve}} = -K_1\lambda\dot{\varepsilon}^{\mathrm{vol}}$	Relaxation equation	$\tau\dot{s}_{\mathrm{v}\theta} + s_{\mathrm{v}\theta} = -C\tau\dot{\theta}$
$\Phi_{\mathrm{m}} = p_{\mathrm{ve}}^2/(2\varrho_0\eta)$	Dissipation potential	$\Phi_{\mathrm{c}} = s_{\mathrm{v}\theta}^2/(2\xi)$

$$\dot{\theta}_{\mathrm{v}} = -\frac{s_{\mathrm{v}\theta}}{\xi(\theta,\cdot)} = \frac{\theta - \theta_{\mathrm{v}}}{\tau(\theta,\cdot)}, \quad \tau = \frac{\xi}{C}. \tag{14.25}$$

The thermal viscosity ξ and, thus, the relaxation time τ are positive quantities. Note that the relaxation time of the viscoelastic model (λ) and the one of the caloric model (τ) may differ from each other. Using the analogy with viscoelasticity, see Eq. (14.20), the relaxation equation for the viscocaloric entropy is given by

$$\tau(\theta,\cdot)\dot{s}_{\mathrm{v}\theta} + s_{\mathrm{v}\theta} = -C\tau(\theta,\cdot)\dot{\theta}. \tag{14.26}$$

Like the viscoelastic model, the viscocaloric model can be generalized to account for a distribution of relaxation times. Then, multiple internal temperatures are introduced (Lion et al. 2017).

Due to the linearity of the evolution Eqs. (14.19) and (14.25) with respect to the driving forces, a dissipation potential Φ can be determined via $\dot{\underline{z}} = \partial\Phi/\partial\underline{y}$ (Šilhavý 1997), giving

$$\Phi = \frac{p_{\mathrm{ve}}^2}{2\varrho_0\eta(\theta,\cdot)} + \frac{s_{\mathrm{v}\theta}^2}{2\xi(\theta,\cdot)}. \tag{14.27}$$

The thermoviscoelastic model follows the theory of generalized standard materials (Halphen and Nguyen 1975) with a pressure and an entropy as thermodynamic driving forces. In Table 14.1, the governing equations of viscoelasticity and entropy relaxation are summarized and compared. The viscocaloric model is discussed in more detail in the following.

14.2.2 Specification of the Viscocaloric Model

Enthalpy In the following, we restrict to purely caloric behavior. Thus, the mechanical part of the free energy ψ_{m} is neglected and the term pv is constant. First, we have to determine the enthalpy. Via Legendre transformation, the mass specific enthalpy is given by

$$h = \psi + s\theta + pv = \psi_\theta(\theta) - \theta\psi_\theta'(\theta) + \frac{C}{2}(\theta - \theta_v)^2 - C\theta(\theta - \theta_v) + pv. \tag{14.28}$$

Like the free energy, the enthalpy is also additively splitted into an equilibrium and a non-equilibrium part.

Equilibrium part As $\theta_v = \theta$ holds in equilibrium, the equilibrium part of the enthalpy is defined by

$$\psi_\theta(\theta) - \theta\psi_\theta'(\theta) + pv := h_{eq}(\theta) = \int c_p^{eq}(\theta)\, d\theta. \tag{14.29}$$

The following function is used for the equilibrium heat capacity

$$c_p^{eq}(\theta) = \frac{m_1 + m_2}{2}\theta - \frac{m_1 - m_2}{2\zeta}\ln\left(\frac{\cosh(\zeta(\theta - \theta_T))}{\cosh(\zeta\theta_T)}\right) + c_0. \tag{14.30}$$

This function is taken from Lion et al. (2017) and extended by the constant c_0. The function has the following properties:

$$\theta \ll \theta_T: \quad c_p^{eq} = m_1\theta + c_0, \tag{14.31}$$

$$\theta \gg \theta_T: \quad c_p^{eq} = m_2\theta + (m_1 - m_2)\theta_T + c_0. \tag{14.32}$$

At the transition temperature θ_T, the slope of the function changes from m_1 to m_2. This allows to model the change in slope of the heat capacity during the glass transition by a continuous function. The parameter ζ defines how broad the transition regime is.

Evolution equation The rate of the internal temperature is

$$\dot{\theta}_v = \frac{\theta - \theta_v}{\tau(\theta, \dot{\theta})}, \tag{14.33}$$

where the relaxation time is assumed to depend on temperature and temperature rate via

$$\tau(\theta, \dot{\theta}) = a_\theta(\theta) b(\dot{\theta}) \tau_0, \quad \log(a_\theta(\theta)) = -\frac{c_1(\theta - \theta_0)}{c_2 + \theta - \theta_0}. \tag{14.34}$$

As simplification, temperature- and rate-dependency are splitted multiplicatively. For the temperature-dependency, the WLF equation (Williams et al. 1955) is used (Lion et al. 2017). The relaxation time is very small for $\theta \gg \theta_0$ and very large for $\theta \ll \theta_0$. The reference temperature θ_0 is often set to the glass transition temperature θ_G. In Fig. 14.2, a plot of the WLF function $a_\theta(\theta)$ for two different values of c_1 is given. The plot compares the WLF function using the parameters suggested by Williams

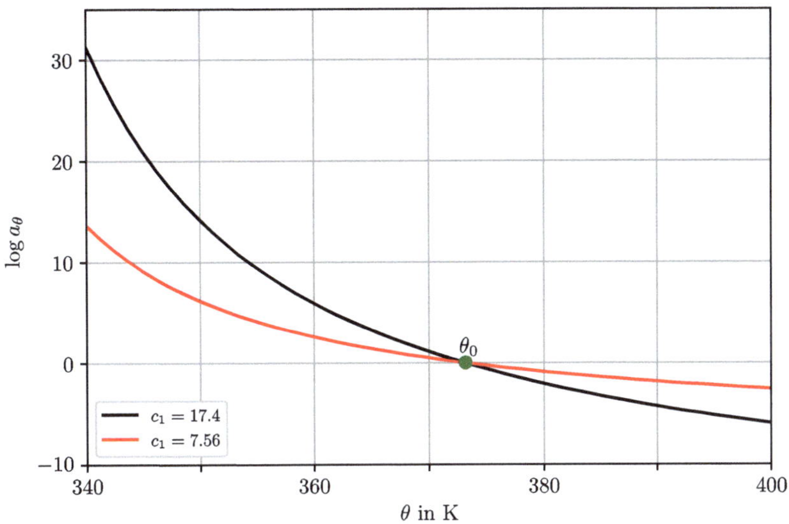

Fig. 14.2 Plot of the WLF function for two values of c_1. At the reference temperature θ_0, the function equals 1. A decrease of c_1 leads to a flattening of the curve. The parameter c_2 is set to 51.6 K in both curves

et al. (1955) (black curve) and the WLF function as used by Lion et al. (2017) (red curve).

Apparent heat capacity By assuming isobaric conditions and neglecting heat sources, the balance of enthalpy is derived via Eq. (14.6) and the Legendre transformation $h = e + pv$. This gives

$$\varrho \dot{h} = -\mathrm{div}\,(\boldsymbol{q})\,. \tag{14.35}$$

By assuming a constant heat power $\dot{Q}$ and a homogeneous field, integrating over a control volume V gives (Lion and Johlitz 2016; Lion et al. 2017)

$$m\dot{h} = \int_V \varrho \dot{h}\,\mathrm{d}V = -\int_V \mathrm{div}\,(\boldsymbol{q}) = \dot{Q} \approx \frac{\Delta Q}{\Delta t}. \tag{14.36}$$

By definition, the heat ΔQ is

$$\Delta Q = c_p m \Delta\theta. \tag{14.37}$$

The apparent heat capacity c_p can then be calculated via

$$c_p = \left.\frac{\dot{h}}{\dot{\theta}}\right|_{p=\mathrm{const}}, \quad \dot{\theta} \approx \frac{\Delta\theta}{\Delta t}. \tag{14.38}$$

This heat capacity is called *apparent* as the standard heat capacity is seen as an equilibrium property for $\dot{\theta} \to 0$. Thus, the apparent heat capacity is the heat capacity that is observed for a specific temperature history. It includes equilibrium and non-equilibrium behavior. With the time-derivative of Eq. (14.28), the heat capacity of the viscocaloric model is

$$c_p = c_p^{\mathrm{eq}} - C\theta + C\frac{\theta_{\mathrm{v}}\dot{\theta}_{\mathrm{v}}}{\dot{\theta}}, \tag{14.39}$$

consisting of a rate-dependent (viscocaloric) and a rate-independent part. Another useful property is the reduced heat capacity (Tropin et al. 2015)

$$c_p^{\mathrm{red}} = \frac{c_p - c_p^{\mathrm{gl}}}{c_p^{\mathrm{eq}} - c_p^{\mathrm{gl}}} = \frac{\theta_{\mathrm{v}}\dot{\theta}_{\mathrm{v}}}{\theta\dot{\theta}}, \quad c_p^{\mathrm{gl}} = c_p^{\mathrm{eq}} - C\theta. \tag{14.40}$$

14.2.3 *Initial Value Problem*

For a given temperature load $\theta(t)$, the evolution Eq. (14.33) must be solved. Therefore, an initial value of the internal temperature is needed, $\theta_{\mathrm{v}}(t = 0) = \theta_{\mathrm{v}0}$. If a temperature load starts above the glass transition, $\theta_{\mathrm{v}0}$ can be set to the equilibrium value of the internal temperature. This is due to the small relaxation time in this temperature regime. The equilibrium value is given by the absolute temperature $\theta(t = 0)$. In contrast, if a process starts below the glass transition temperature in a non-equilibrium state, $\theta_{\mathrm{v}0}$ depends on the temperature history of the material. In the following, we consider the case that a measurement of $c_p(\theta(t))$ is given as input. A heating process from $\theta_{\mathrm{min}} \ll \theta_{\mathrm{G}}$ to $\theta_{\mathrm{max}} \gg \theta_{\mathrm{G}}$ with constant temperature rate $\dot{\theta}$ is the considered load. The temperature history of the material, $\theta(t < 0)$, is unknown. We assume that the rate-independent glassy and equilibrium heat capacity are already known. With that, the reduced heat capacity can be calculated from the experimental input via Eq. (14.40). Reformulating Eq. (14.40) by chain rule gives

$$\frac{1}{2}\gamma := c_p^{\mathrm{red}}\theta\dot{\theta} = \theta_{\mathrm{v}}\dot{\theta}_{\mathrm{v}} = \frac{1}{2}\frac{\mathrm{d}\theta_{\mathrm{v}}^2}{\mathrm{d}t}. \tag{14.41}$$

Integrating over the considered time interval then yields

$$\Gamma := \int_0^T \gamma \,\mathrm{d}t = \theta_{\mathrm{v}}^2(t = T) - \theta_{\mathrm{v}}^2(t = 0). \tag{14.42}$$

From the experimental input, the function $\gamma(t)$ is known at discrete time steps. Thus, the integral can then be solved numerically. As the process ends above the glass transition in the equilibrium state, the internal temperature at $t = T$ can be set to $\theta_{\mathrm{v}}(t = T) = \theta(t = T) = \theta_{\mathrm{max}}$. With that, the initial value for a simulation is given although the temperature history of the material is unknown,

$$\theta_{v0} = \sqrt{\theta_{max}^2 - \Gamma}. \tag{14.43}$$

Due to the nonlinearity of the WLF equation, the initial value problem must be solved numerically. In this work, an implicit Runge-Kutta scheme from the `Python` package `scipy` is used for all simulations. In post-processing, the heat capacity and reduced heat capacity, see Eqs. (14.39) and (14.40), are evaluated based on a finite difference scheme for $\dot{\theta}_v$.

14.2.4 DSC Measurements

As input for this work, DSC tests at a temperature rate of ±0.2 K/min are performed. For the measurement of the apparent heat capacity, a Calvet HT DSC from Setaram KEP Technologies SA is used. The Calvet sensor was calibrated using the Joule effect calibration method. The cells were temperature-calibrated via standard substances, i.e., indium, zinc, bismuth, and tin. The used polystyrene (CAS 9003-53-6) was purchased from Sigma–Aldrich USA (Product Number: 441147, Batch-Number: MKCL2307). According to the data sheet from Sigma-Aldrich, the number average molecular weight is 170 kg/mol and the mass-average molecular weight is 350 kg/mol.

14.3 Qualitative Behavior

14.3.1 Glassy and Equilibrium State

In this section, we want to discuss the qualitative behavior of the viscocaloric model. Glassy and equilibrium state are the limiting cases of the model (Lion et al. 2017). Within these two regimes, the behavior is rate-independent. For the equilibrium state, the rate of the internal temperature in Eq. (14.33) is set to zero. As already mentioned, this means that the equilibrium value of the internal temperature equals the absolute temperature. Due to the small relaxation time at high temperatures, the internal temperature immediately follows the absolute temperature, $\theta_v = \theta$. The reduced heat capacity equals one in equilibrium state, see Eq. (14.40). In contrast, for large relaxation times at low temperatures, the evolution of the internal temperature is frozen, $\dot{\theta}_v = 0$. This means that the equilibrium state cannot be approached within finite time. In literature, this state is known as glassy state. The value of the reduced heat capacity equals zero in glassy state. The difference between equilibrium and glassy heat capacity is given by the parameter C via $C\theta$, see Eq. (14.39).

14.3.2 Glass Transition Regime

The transition between glassy and equilibrium state is known as glass transition. Within the glass transition regime, pronounced process-dependent behavior is observed. In Fig. 14.3, this behavior is illustrated by simulations with a test set of parameters. Figure 14.3a shows a cooling process with constant cooling rate. By the WLF equation, the relaxation time increases monotonically and, thus, the mobility decreases monotonically. Therefore, the frozen state at low temperature is approached by monotonic processes of internal temperature and heat capacity. In contrast, Fig. 14.3b shows non-monotonic behavior for a subsequent heating process. The process begins in the non-equilibrium glassy state, the relaxation time decreases continuously. As soon as the relaxation time is small enough, the internal temperature begins to approach its equilibrium state. However, the internal temperature falls below the equilibrium state because the mobility is still low. In this range, the rate of the internal temperature is negative causing an exothermic undershoot in the heat capacity. As the temperature increases and the relaxation time further decreases, the internal temperature approaches the equilibrium state from below. In this range, the rate of the internal temperature is larger than the heating rate leading to an endothermic overshoot in the heat capacity. Figure 14.4 shows the influence of different temperature rates on the temperature interval of the glass transition. By the blue cooling curves, it can be seen that a higher cooling rate shifts the glass transition to higher temperatures. The faster the cooling rate, the larger is the final value of the internal temperature as there is less time to approach equilibrium, see Fig. 14.4b. The subsequent heating curves differ only in the initial values of the internal temperature θ_{v0}. It can be seen that the local extrema depend on θ_{v0} and, therefore, on the previous temperature history. The exothermic undershoot is more pronounced for larger initial values, the endothermic overshoot for smaller initial values. In extreme cases, one of the two local extrema cannot be observed. Note that an increasing heating rate shifts the glass transition to lower temperatures. Furthermore, the heating rate also influences the local extrema. As mentioned before, the rate-dependent caloric behavior during glass transition is known as enthalpy relaxation. In this work, we call this effect entropy relaxation as the viscocaloric part of the entropy is the thermodynamic driving force. An illustration of entropy relaxation is given in Fig. 14.5. A temperature load consisting of a cooling step and a subsequent constant temperature step is applied. After cooling, the internal temperature deviates from its equilibrium and then approaches its equilibrium value again. Note that the kink in the prescribed temperature also leads to a kink in the internal temperature at the same time. The difference between internal and absolute temperature relaxes to zero and it equals the viscocaloric entropy divided by the parameter C. In literature, this behavior is also known as physical aging.

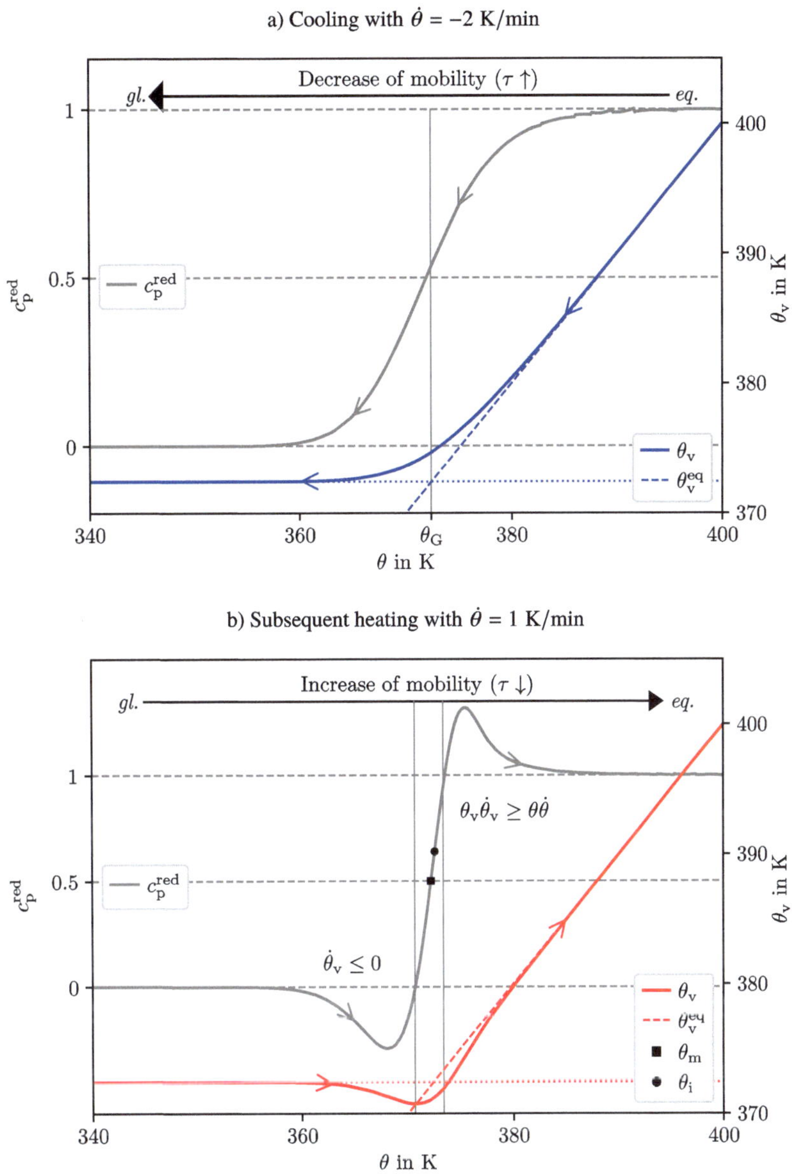

Fig. 14.3 Behavior of internal temperature and reduced heat capacity for a test case with WLF parameters of $c_1 = 7.56$ and $c_2 = 51.6$ K and a relaxation time of $\tau_0 = 100$ s. In the glass transition regime, the model shows monotonic behavior during cooling and non-monotonic behavior, related with endothermic and exothermic effects, during heating. At low temperatures, the model is in glassy (gl.) state and at high temperatures, it is in equilibrium (eq.)

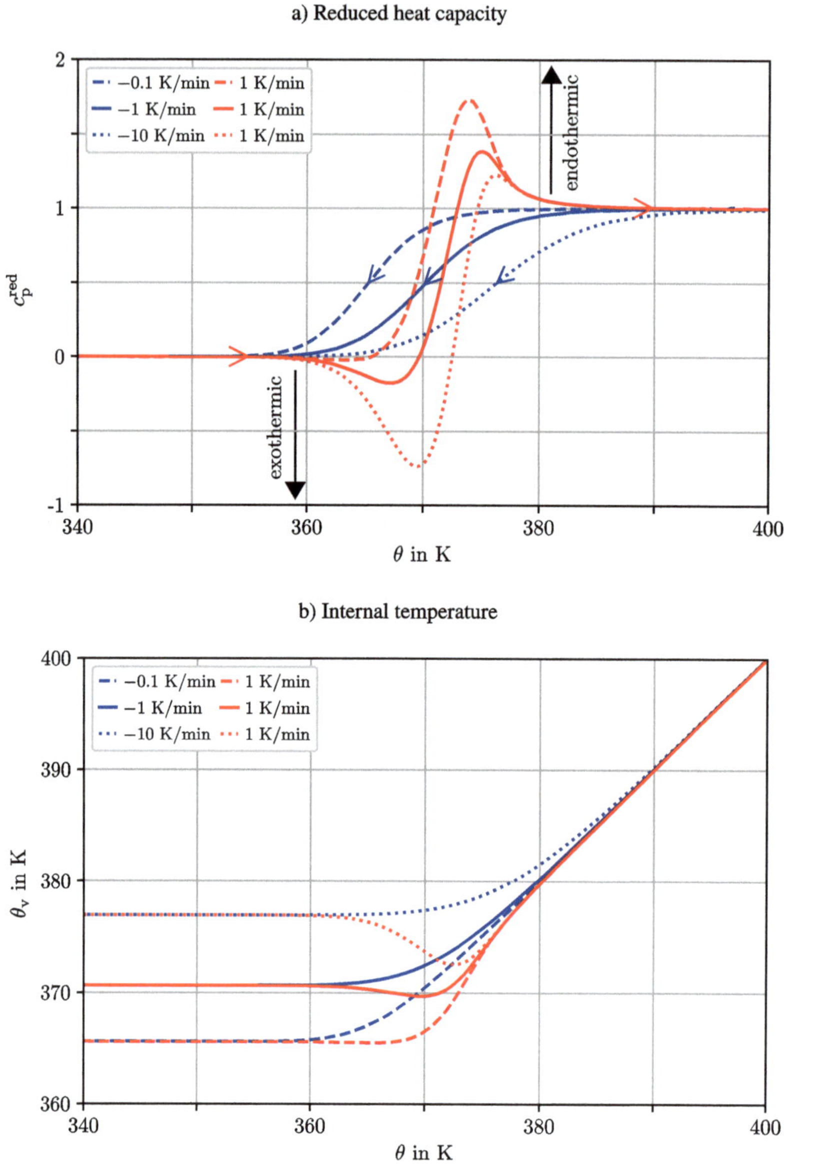

Fig. 14.4 Simulation of cooling with different temperature rates (blue curves) and subsequent heating with a fixed temperature rate (red curves). The parameters are the same as in Fig. 14.3

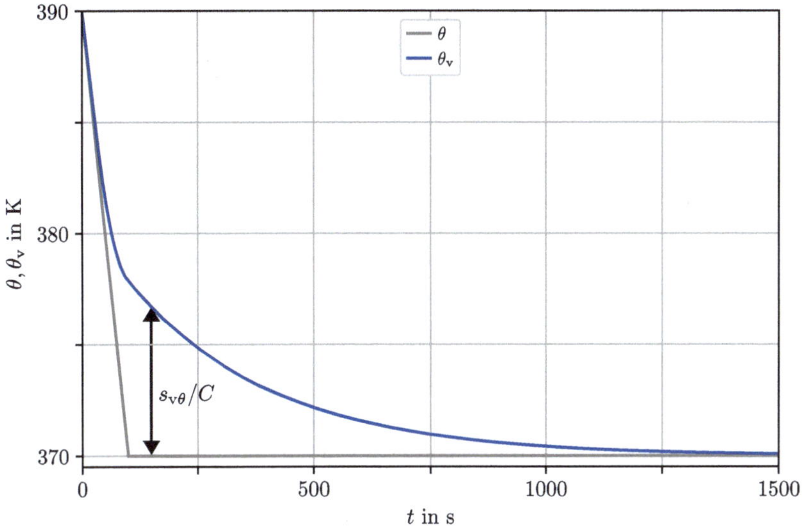

Fig. 14.5 Behavior of the internal temperature for the temperature load depicted in gray. The difference between internal temperature and absolute temperature is proportional to the viscocaloric entropy $s_{v\theta}$ which is relaxing to zero if the temperature is held constant. The parameters are the same as in Fig. 14.3

14.3.3 Definition of the Glass Transition Temperature

The glass transition is observed in a temperature interval that depends on the applied temperature history. Nevertheless, a single glass transition temperature is often specified in literature. Therefore, we will briefly discuss how a glass transition temperature can be defined from the curves of the internal temperature and the reduced heat capacity. At this point, it should be noted that the behavior of the internal temperature is comparable to the behavior of the thermal strain in dilatometric experiments, see, e.g., Badrinarayanan et al. (2007). In cooling processes, see Fig. 14.3a, the intersection of the tangents of the internal temperature regarding equilibrium and glassy state can be defined as the glass transition temperature θ_G. Thus, the final value of the internal temperature equals the glass transition temperature. In the plot of the reduced heat capacity, the glass transition temperature corresponds to the inflection point of the curve. For heating processes, the definition of a glass transition temperature is not unambigous due to the non-monotonic behavior that is caused by the non-equilibrium state at the beginning (ASTM E1356-23 2023). The intersection of the tangents does not necessarily correspond to the inflection point θ_i of the reduced heat capacity. In addition, the points referred to may lie outside of the interval of $c_p^{red} \in [0, 1]$. An alternative option, which is frequently used and easy to evaluate, is to use the midpoint temperature θ_m with $c_p^{red}(\theta_m) = 0.5$. In all cases, the glass transition temperature is a process-dependent parameter.

14.4 Parameter Identification

14.4.1 Experimental Input

In the following, we will determine the model parameters in order to reproduce experimental data. DSC measurements with atactic polystyrene are used as experimental input. The measured apparent heat capacities are depicted in Fig. 14.6a. The curves at 10 K/min are taken from a study of Tropin and coworkers (Tropin et al. 2015). Monotonic behavior during cooling and entropy relaxation during heating can be observed. The local endothermic maximum is strongly pronounced, a local exothermic minimum cannot be seen. The cooling curve at 0.2 K/min shows a kink at approximately 387 K. However, in the equilibrium state, this curve could be continued monotonically if the associated heating curve is considered. Thus, the kink is assumed to be a measuring error. Theoretically, a polymer degradation can be possible leading to a small change in the heat capacity. The measurement was repeated replacing the air by nitrogen or by argon in order to avoid polymer degradation. These repetitions also led to the unphysical kink and, hence, polymer degradation can be excluded as reason for the kink. The possible explanation is the limiting heat transfer at very low heating rates. In the heating curve at 0.2 K/min, both a local minimum and maximum can be observed. Furthermore, both cooling curves show that the glass transition temperature is slightly higher at the faster cooling rate.

14.4.2 Glassy and Equilibrium Behavior

As first step of a parameter identification, glassy and equilibrium heat capacity, c_p^{gl} and c_p^{eq} are to be determined. These functions are approximated by the functions in Eqs. (14.30) and (14.40). The parameters $\theta_{\mathrm{T}}, \zeta, C, m_1, m_2$ and c_0 are to be determined. As an example, the cooling curve at 10 K/min is considered here, see Fig. 14.7. We set the transition temperature θ_{T} to the approximated glass transition temperature of 100 °C and, based on Lion et al. (2017), $\zeta = 0.1$. Glassy and equilibrium behavior is approximated by linear functions, depicted by the dashed tangents in Fig. 14.7,

$$c_p^{\mathrm{gl}} = m_{\mathrm{gl}}\theta + c_0^{\mathrm{gl}}, \quad c_p^{\mathrm{eq}} = m_{\mathrm{eq}}\theta + c_0^{\mathrm{eq}}. \tag{14.44}$$

With the relations from Eqs. (14.31) and (14.32), an equation system can be derived

$$c_p^{\mathrm{gl}} = (m_1 - C)\theta + c_0 \overset{!}{=} m_{\mathrm{gl}}\theta + c_0^{\mathrm{gl}} \tag{14.45}$$

$$c_p^{\mathrm{eq}} = m_2\theta + (m_1 - m_2)\theta_{\mathrm{T}} + c_0 \overset{!}{=} m_{\mathrm{eq}}\theta + c_0^{\mathrm{eq}}. \tag{14.46}$$

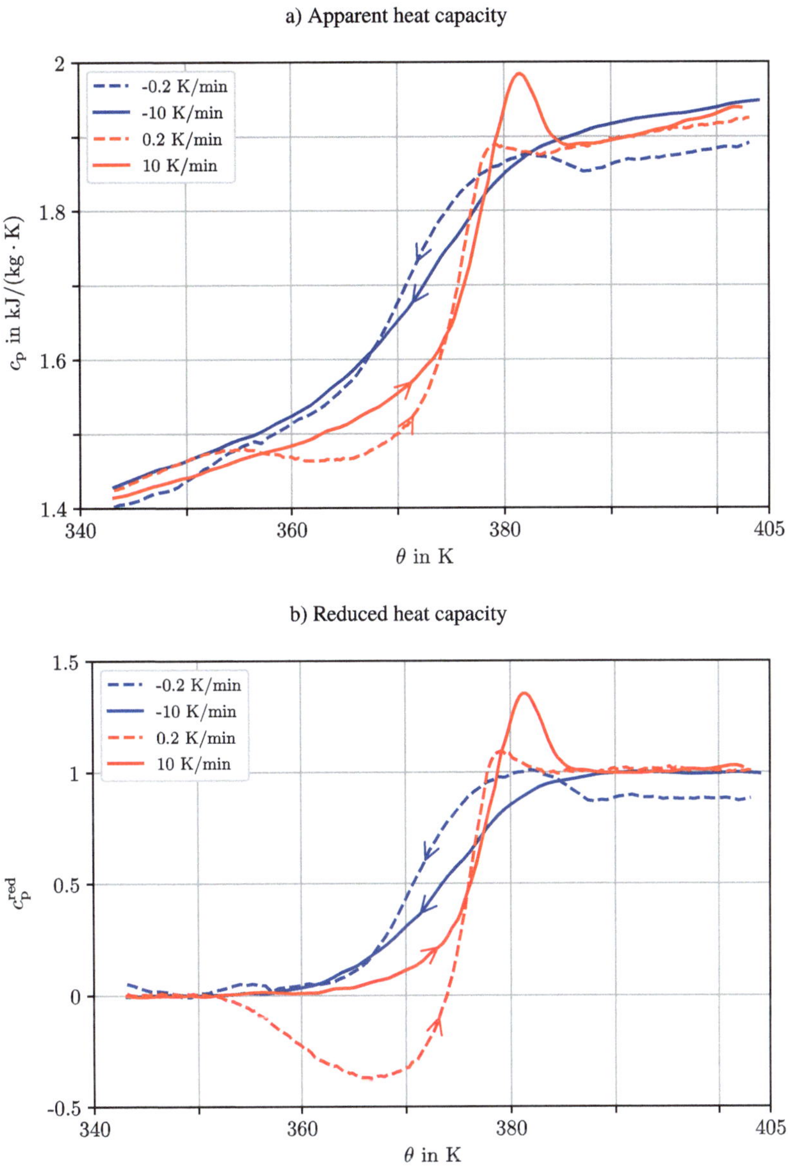

Fig. 14.6 Plot of DSC measurements that are used as input for a parameter identification. The data at 10 K/min (dashed lines) is taken from Tropin et al. (2015)

Table 14.2 Identified parameters to describe equilibrium and glassy heat capacity of the curves in Fig. 14.6a. The transition temperature θ_T and the parameter ζ are predefined

$\dot{\theta}$ in K/min	C in J/(kgK^2)	c_0 in J/(kgK)	m_1 in J/(kgK^2)	m_2 in J/(kgK^2)
0.2	0.67	−696	6.84	2.17
−0.2	0.69	−1012	7.69	2.17
10	0.84	0	4.96	2.67
−10	0.79	−372	6.04	2.17
$\theta_T = 373.15$ K, $\zeta = 0.1$				

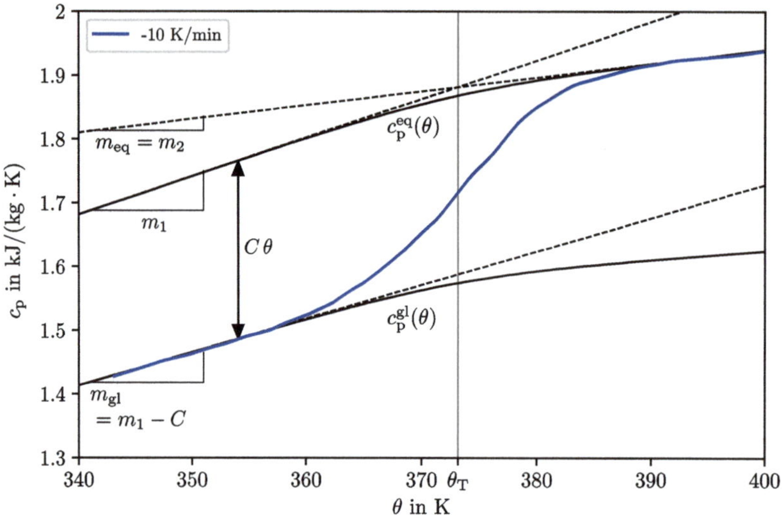

Fig. 14.7 Determination of the parameters $\{m_1, m_2, C\}$ that enter the rate-independent caloric part of the enthalpy. Glassy and equilibrium behavior is approximated by linear functions respectively. The transition temperature θ_T is set to the approximated glass transition temperature

By comparison of coefficients, the remaining parameters are determined,

$$c_0 = c_0^{gl}, \quad m_1 = \frac{c_0^{eq} - c_0^{gl}}{\theta_T} + m_{eq}, \quad m_2 = m_{eq}, \quad C = m_1 - m_{gl}. \tag{14.47}$$

For each experimental curve in Fig. 14.6a, a set of parameters is determined. Note that the equilibrium behavior of the cooling curve at 0.2 K/min is approximated by the corresponding heating curve. The determined parameters are given in Table 14.2. With that, the reduced heat capacity can be computed for each curve, see Fig. 14.6b.

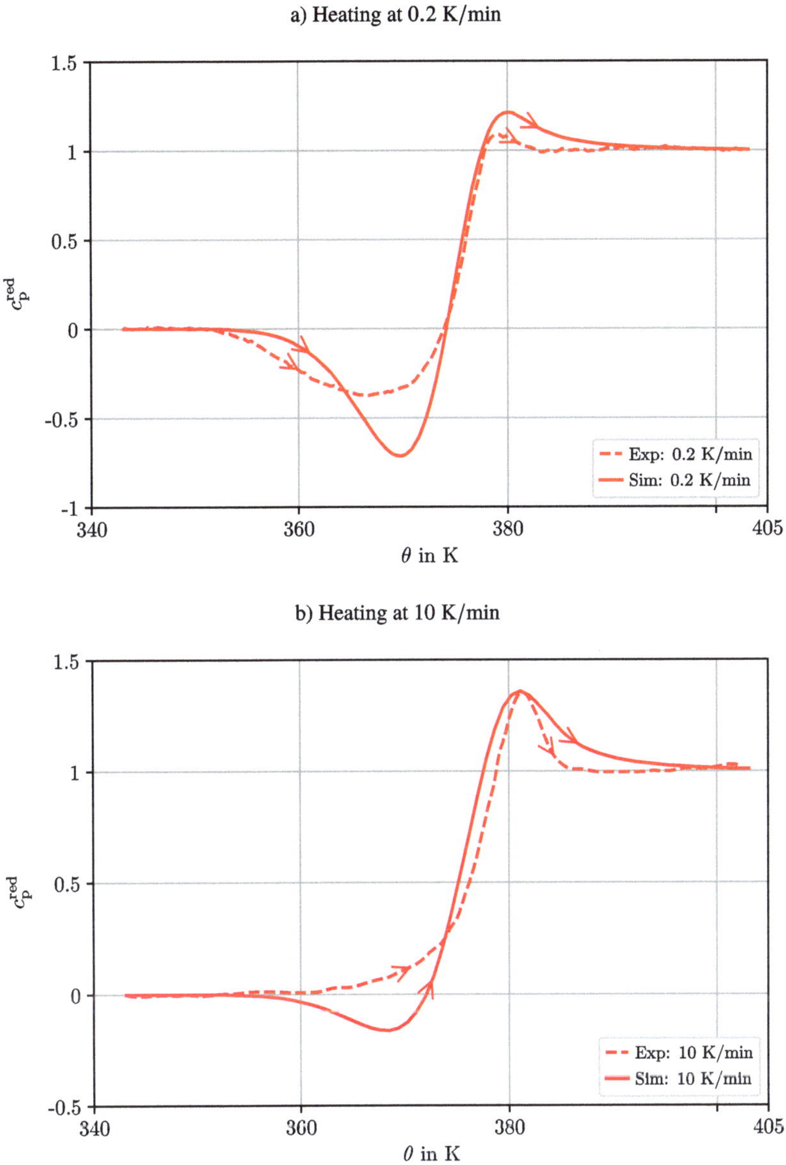

Fig. 14.8 Comparison of experimental data and simulation. WLF parameters and rate-dependent relaxation time are fitted based on the experimental curves

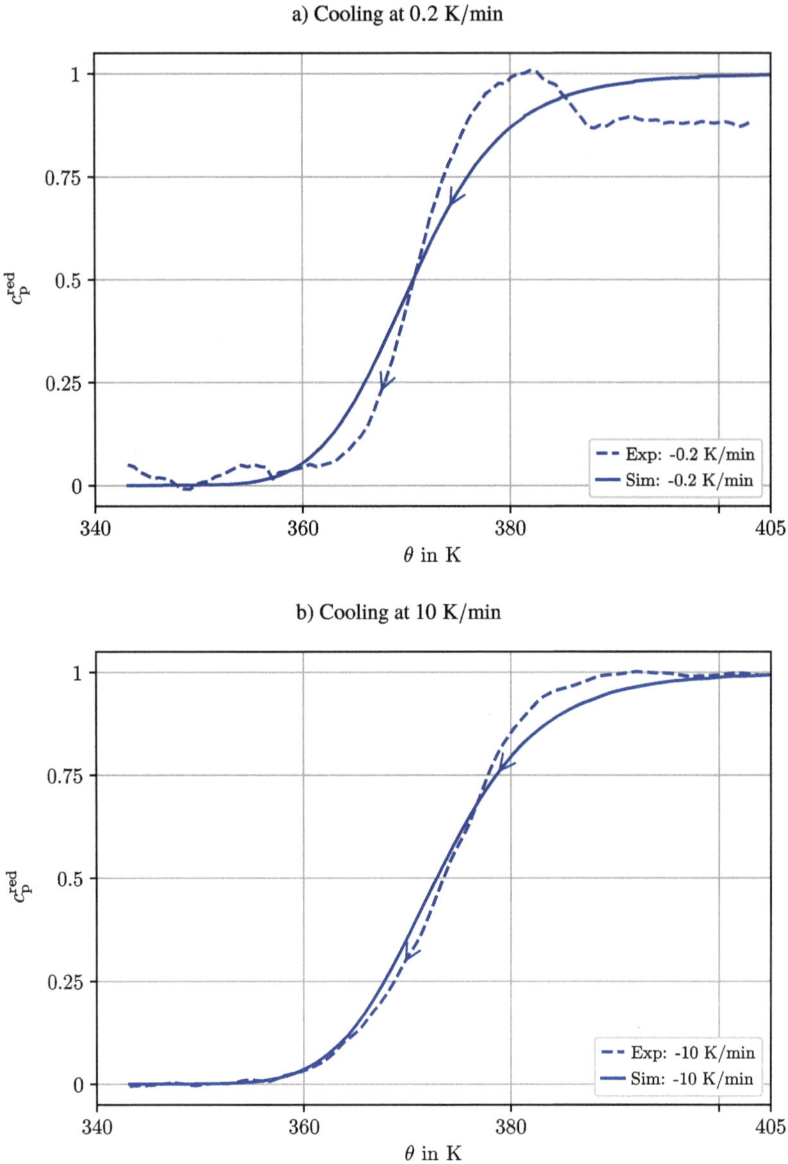

Fig. 14.9 Comparison of experimental data and simulation. WLF parameters and rate-dependent relaxation time are the same as in Fig. 14.8

Table 14.3 Fitted parameters of the evolution equation. The WLF parameter c_2 and the reference temperature θ_0 are predefined

$\lvert\dot{\theta}\rvert$ in K/min	c_1 –	c_2 in K	θ_0 in K	τ_{ref} in s
0.2	5.08	51.6	373.15	1200
10	5.08	51.6	373.15	45

14.4.3 Glass Transition Regime

The WLF parameters $\{c_1, c_2\}$ and the relaxation time τ can now be determined based on the curves of the reduced heat capacity. For the qualitative investigations in Sect. 14.3, the parameters were set to $c_1 = 7.56$ and $c_2 = 51.6$ K. The reference temperature θ_{ref} was set to the approximated glass transition temperature of 100 °C. A comparison of Figs. 14.4a and 14.6b shows that the temperature interval of the glass transition is wider in the experimental curves. Reducing c_1 or increasing c_2 leads to a flattening of the WLF function and, thus, to a wider glass transition interval, see Fig. 14.2. We limit the investigation to an adjustment of the parameter c_1. With constant WLF parameters, a change in the relaxation time leads to a shift of the glass transition temperature. An increase of the relaxation time causes an increase of the glass transition temperature. The parameters are adapted based on the heating curves. Therefore, the initial values of the simulations are calculated via Eqs. (14.42) and (14.43). First, the heating curve at 0.2 K/min is considered. A global least squares optimization gives the result plotted in Fig. 14.8a. It can be seen that the local extrema are more pronounced in the simulative curve. The temperature interval of the glass transition is fitted well. As we introduced a multiplicative split of temperature-dependency and rate-dependency in Eq. (14.34), the parameter c_1 is fixed for the consideration of the faster heating process and only the relaxation time is adapted. In Fig. 14.8b, the result is depicted. The simulation shows a local minimum that cannot be seen in the experimental curve. This corresponds with the overestimation of the undershoot at 0.2 K/min. The glass transition interval is again fitted well. All determined and defined parameters are given in Table 14.3. It can be seen that the relaxation times differ significantly. This motivates to introduce the relaxation time as a rate- and temperature-dependent property. With the determined parameters, the corresponding cooling processes are simulated. Simulative and experimental results are in good agreement if we account for the measuring error at 0.2 K/min mentioned above.

Table 14.4 Identified parameters of the proposed rate-dependency function $b(\dot{\theta})$

b_1	b_2	k
0.02	0.003	1.01

14.4.4 Rate-Dependent Shift

This work is restricted to the consideration of two different temperature rates. Thus, at last, we want to discuss how the rate-dependency of the relaxation time can be interpolated. First, a linear interpolation is applied. The simulation of cooling at two further cooling rates is shown in Fig. 14.10a. With increasing rate, the glass transition temperature first increases and then decreases again. In reality, a monotonic increase in the glass transition temperature is to be expected. Thus, a nonlinear interpolation is necessary. For this purpose, it is assumed that the glass transition temperature increases linearly between the considered cooling rates. The values of the relaxation times are determined for a logarithmic distribution of cooling rates in order to meet this condition, see Fig. 14.11. For the iterative determination of the relaxation times, we utilize the fact that the glass transition temperature during cooling corresponds to the final value of the internal temperature. The resulting relaxation times are plotted in Fig. 14.11. In Fig. 14.10b, a linear increase of the glass transition temperature can now be observed. Based on Fig. 14.11, the rate-dependency function $b(\dot{\theta})$ in Eq. (14.34) could be modeled by a power law function,

$$b(\dot{\theta}) = b_1 + b_2 \left(\frac{|\dot{\theta}|}{\dot{\theta}_0} \right)^{-k}, \quad \dot{\theta}_0 = 1 \text{ K/s}. \tag{14.48}$$

We assume that the relaxation time is the same for both heating and cooling. Therefore, the function depends on the absolute value of the temperature rate. The fitted coefficients are given in Table 14.4. Note that these parameters are related to a reference relaxation time at 0.2 K/min.

14.5 Discussion

Glassy and equilibrium state The apparent heat capacity is given by superposition of a rate-independent equilibrium part and a rate-dependent non-equilibrium part, see Eq. (14.39). In glassy and equilibrium state, the heat capacity is rate-independent and does only depend on temperature. For the given experimental data, this can be adequately approximated by linear functions as shown in Sect. 14.4.2. A continuous function was used to model the equilibrium part of the heat capacity, see Eq. (14.30). This allows for a smooth transition from a larger slope in the glassy state to a smaller slope in the equilibrium state. Furthermore, the linear approximations define the

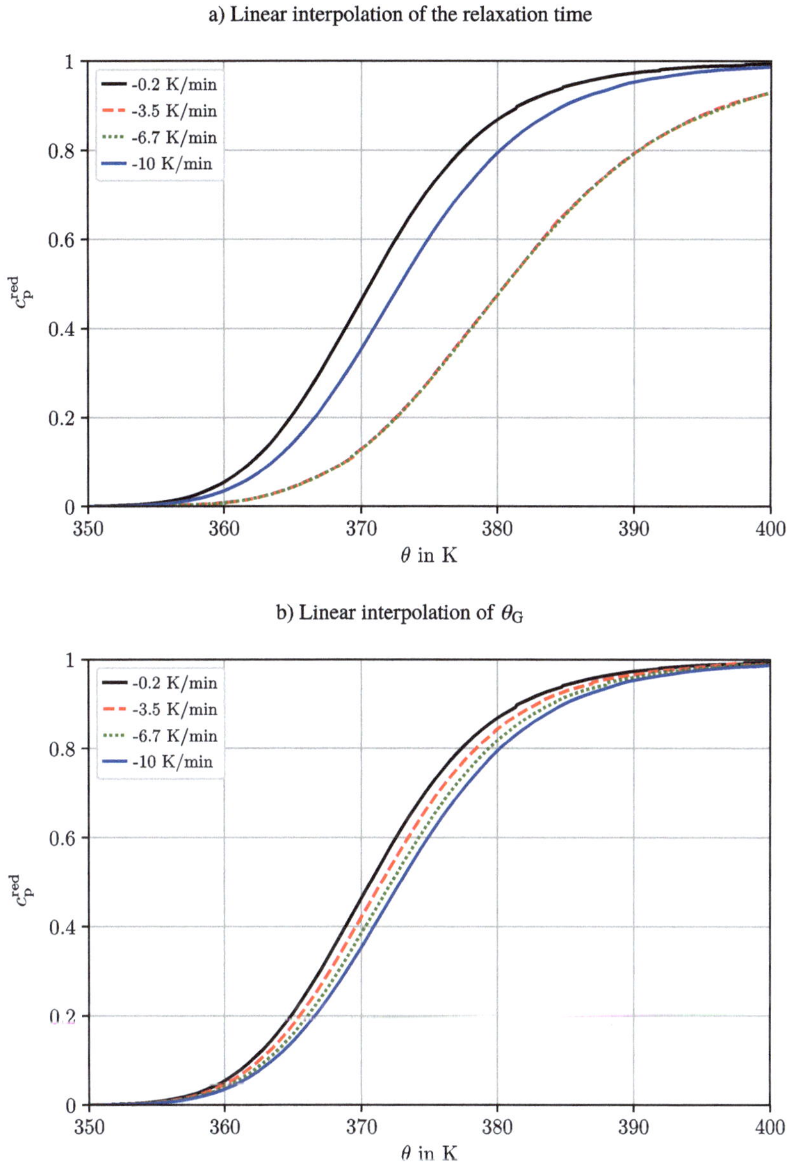

Fig. 14.10 Plot of reduced heat capacity for varying cooling rates and two different approaches to interpolate the rate-dependency of the relaxation time. The black and blue curves result from the previous parameter identification

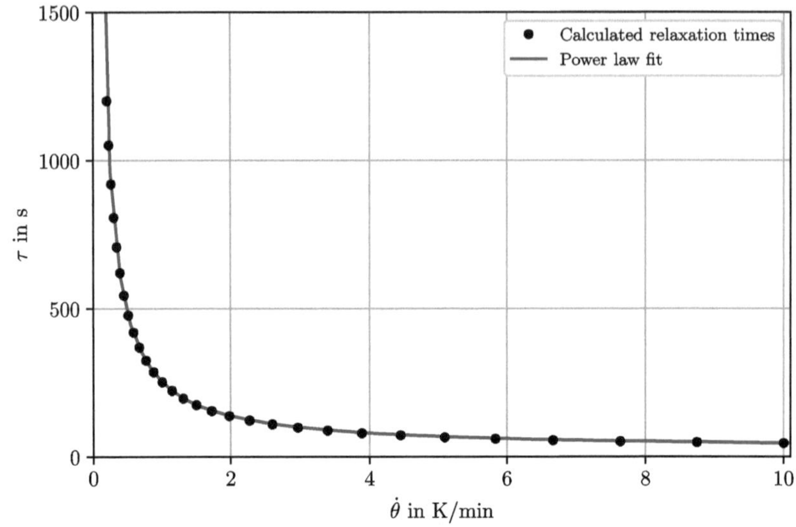

Fig. 14.11 Rate-dependency of the relaxation time for a linear interpolation of the glass transition temperature

material parameter C that is related with the change of heat capacity in the glass transition. Then, the reduced heat capacity, see Eq. (14.40), could be calculated for the experimental curves. The advantage of the reduced heat capacity is that it filters out the temperature-dependencies in glassy and equilibrium state and allows to focus on the viscocaloric behavior within the glass transition.

Glass transition regime As already stated, the considered viscocaloric model is based on the model by Lion et al. (2016; 2017). The model follows the scheme of linear irreversible thermodynamics. In contrast to Lion's model, the thermal internal variable in this work is introduced as an absolute quantity, cf. Eqs. (14.5) and (14.25). The reason is the clearer physical meaning of the internal temperature. Without conversion, it is the absolute temperature at which the current material state would be an equilibrium state. Furthermore, in cooling processes, the final value of the internal temperature equals the glass transition temperature, see Sect. 14.3.3. In addition to the study by Lion et al. (2017), the similar model structure of viscoelasticity and entropy relaxation was demonstrated in our work. A comparison with the TNM model, cf. Eqs. (14.1) and (14.25), shows that internal temperature and fictive temperature are similar concepts. Within a classical continuum thermomechanical framework applied in Sect. 14.2, it becomes clear that the fictive temperature is an internal variable. Therefore, we stick to the term *internal* temperature. The internal temperature is non-observable, accounts for glass forming processes on molecular scale and allows to model relaxation processes. Alternatively, the evolution equation for the internal temperature can be reformulated into a relaxation equation that directly operates on the viscocaloric part of the entropy, see Eq. (14.26). Note that the negative of the viscocaloric entropy was identified as the thermodynamic driving force of viscocaloric behavior. This relaxation equation is comparable with the KAHR relaxation equation in Eq. (14.3). The enthalpy is a linear function of the departure from equilibrium in TNM and KAHR model. In contrast, in the viscocaloric model in this work, the enthalpy is a quadratic function of the departure from equilibrium. The evolution equations of all mentioned models, see Eqs. (14.1), (14.3), (14.5) and (14.25), follow the scheme of linear irreversible thermodynamics and result into exponential response functions. If, instead, a non-exponential response function is used, this is not the case. The main difference of the models is the way the relaxation time is defined. In Lion's model, the relaxation time does only depend on temperature modeled by a WLF function. This means that time-temperature superposition holds and the material model is thermorheological simple (Tschoegl et al. 2002). As shown in Sect. 14.3, the qualitative behavior of enthalpy relaxation observed in experimental studies can be modeled with that. In TNM and KAHR model, besides the temperature-dependency, the relaxation time depends on fictive temperature or departure. With that, a nonlinear structural dependency of the relaxation time is given and the model is thermorheological complex. In Sect. 14.4, the parameters of the viscocaloric model were fitted to experimental data of polystyrene at two different temperature rates. By adapting the WLF parameters, the glass transition temperature interval could be extended and good agreement with experimental results was obtained. However, by comparing the two temperature rates, a rate-dependency of the relaxation time was observed. Therefore, the model was extended by a rate-dependency function, see Eq. (14.34), implying thermorheological complex behavior. Temperature- and rate-dependency of the relaxation time are splitted multiplicatively. This is comparable with TNM and KAHR model where temperature- and structural dependency are splitted multiplicatively. Therefore, the rate-dependency function used in this work can be seen as an alternative to the structural dependency in TNM and KAHR model. In Sect. 14.4.4, it was discussed how to interpolate the relaxation time between the two considered temperature rates. To obtain a reasonable rate-dependency of the glass transition temperature, we assumed that the glass transition temperature increases linearily with increasing temperature rate. With that, the rate-dependency function of the relaxation time must be nonlinear, monotonically decreasing and convex, e.g., a power law function with negative exponent.

14.6 Conclusion and Outlook

In this work, a caloric model to account for enthalpy relaxation in the glass transition regime was considered. The model is an adaption of Lion's model (Lion et al. 2017) and uses an internal temperature as internal variable to account for relaxation processes. Based on Coleman-Noll procedure and linear irreversible thermodynamics, the model was derived in a continuum mechanical framework. In addition, the analogy to viscoelasticity was used in order to formulate similar approaches for the free energy and the evolution equation of the internal temperature. With that, we showed that the negative of the non-equilibrium (viscocaloric) entropy is the thermodynamic driving force of the model. This motivates to use the term entropy relaxation instead of enthalpy relaxation. The qualitative behavior of the model with a pure temperature-dependency of the relaxation time, modeled by a WLF equation, was discussed. Furthermore, the model was applied in order to reproduce experimental DSC measurements of polystyrene at constant heating and cooling rates. Rate-independent behavior of the heat capacity in glassy and equilibrium state was modeled by linear temperature-dependent functions with a smooth transition between both states. The parameters for the rate-dependent part of the model were then adapted based on heating curves of the reduced heat capacity at two different temperature rates. Therefore, we showed that the initial value of the internal temperature for heating simulations can be calculated from experimental data although the temperature history of the material is unknown. In general, the simulations are in good agreement with the experimental results. The glass transition temperature interval was fitted well for heating and cooling, exothermic undershoots during heating were overestimated. By comparing the two considered temperature rates, a rate-dependent relaxation time was observed. Therefore, the model was extended by a rate-dependency function for the relaxation time giving a relaxation time that depends on temperature and temperature rate. By assuming a linearily increasing glass transition temperature with increasing temperature rate, we showed that the rate-dependency function can, e.g., be modeled by a power law function with negative exponent.

In future investigations of the model, experimental data at further temperature rates and temperature histories should be considered. Furthermore, several extensions of the model are conceivable. To account for a distribution of relaxation times, the model could be generalized by introducing multiple internal temperatures. Additionally, a coupling of mechanical and caloric behavior could be of interest. This would include a pressure-dependency of the heat capacity.

Acknowledgements JK and TB gratefully acknowledge the funding by the Deutsche Forschungsgemeinschaft (DFG, German Research Foundation), project number 255730231, within the International Research Training Group *"Integrated engineering of continuous-discontinuous long fiber reinforced polymer structures" (GRK 2078/2).*

Credit author statement *Johannes Keursten*: Investigation, Conceptualization, Methodology, Simulation, Visualization, Writing - original draft. *Sabine Enders*: Writing - review editing. *Thomas Böhlke*: Supervision, Conceptualization, Resources, Funding acquisition, Writing - review editing.

Competing Interests The authors declare that they have no known competing financial interests or personal relationships that could have appeared to influence the work reported in this paper.

References

Adam G, Gibbs J (1965) J Chem Phys 43(1):139. https://doi.org/10.1063/1.1696442
Argatov I, Kocherbitov V (2021) Continuum Mech Thermodyn 33(1):107. https://doi.org/10.1007/s00161-020-00891-3
ASTM E1356-23, Standard test method for assignment of the glass transition temperatures by differential scanning calorimetry. ASTM International, West Conshohocken, PA, USA
Badrinarayanan P, Zheng W, Li Q, Simon S (2007) J Non-Cryst Solids 353(26):2603. https://doi.org/10.1016/j.jnoncrysol.2007.04.025
Böhlke T, Henning F, Hrymak A, Kärger L, Weidenmann K, Wood J (2019) Continous - discontinous fiber-reinforced polymers: an integrated engineering approach. Carl Hanser Verlag, Munich. https://doi.org/10.3139/9781569906934
Brinson H, Brinson L (2015) Polymer engineering science and viscoelasticity, 2nd edn. Springer Science+Business Media, New York. https://doi.org/10.1007/978-1-4899-7485-3
Coleman B, Gurtin M (1967) J Chem Phys 47(2):597. https://doi.org/10.1063/1.1711937
Coleman B, Noll W (1963) Arch Ration Mech Anal 13(1):167. https://doi.org/10.1007/BF01262690
Dudowicz J, Freed K, Douglas J (2008) Adv Chem Phys 137:125. https://doi.org/10.1002/9780470238080.ch3
Ehrenstein G (2011) Polymer werkstoffe, 3rd edn. Carl Hanser Verlag, Munich
Fulcher G (1925) J Am Ceram Soc 8(6):339. https://doi.org/10.1111/j.1151-2916.1925.tb16731.x
Görthofer J, Meyer N, Pallicity T, Schöttl L, Trauth A, Schemmann M, Hohberg M, Pinter P, Elsner P, Henning F, Hrymak A, Seelig T, Weidenmann K, Kärger L, Böhlke T (2019) Compos B Eng 169:133. https://doi.org/10.1016/j.compositesb.2019.04.001
Gurtin M, Fried E, Anand L (2010) The mechanics and thermodynamics of continua. Cambridge University Press, New York. https://doi.org/10.1017/CBO9780511762956
Gutzow I, Schmelzer J, Petroff B (2008) J Non-Cryst Solids 354(2–9):311. https://doi.org/10.1016/j.jnoncrysol.2007.07.042
Hadač J, Slobodian P, Říha P, Sáha P, Rychwalski R, Emri I, Kubát J (2007) J Non-Cryst Solids 353:2681. https://doi.org/10.1016/j.jnoncrysol.2007.05.017
Halphen B, Nguyen Q (1975) J de Mécanique 14(1):39
Hempel E, Kahle S, Unger R, Donth E (1999) Thermochim Acta 329(2):97. https://doi.org/10.1016/S0040-6031(99)00011-8
Hodge I (1991) J Non-Cryst Solids 131–133:435. https://doi.org/10.1016/0022-3093(91)90336-5
Hodge I (1994) J Non-Cryst Solids 169(3):211. https://doi.org/10.1016/0022-3093(94)90321-2
Hodge I, Huvard G (1983) Macromolecules 16:371. https://doi.org/10.1021/ma00237a006
Höhne G, Hemminger W, Flammersheim H (2003) Differential scanning calorimetry, 2nd edn. Springer, Berlin. https://doi.org/10.1007/978-3-662-06710-9
Kohlrausch R (1854) Annalen der Physik und Chemie 91:179
Kovacs A, Aklonis J, Hutchinson J, Ramos A (1979) J Polym Sci Polym Phys Ed 17(7):1097. https://doi.org/10.1002/pol.1979.180170701
Lion A, Johlitz M (2016) Continuum Mech Thermodyn 28(3):799. https://doi.org/10.1007/s00161-015-0415-8
Lion A, Mittermeier C, Johlitz M (2017) Continuum Mech Thermodyn 29(5):1061. https://doi.org/10.1007/s00161-016-0551-9
Liu I (2002) Continuum mechanics. Advanced texts in physics. Springer, Berlin. https://doi.org/10.1007/978-3-662-05056-9
Liu G, Li L, Zheng Y, Zuo Y (2013) J Non-Cryst Solids 365:13. https://doi.org/10.1016/j.jnoncrysol.2013.01.017
Menges G, Haberstroh E, Michaeli W, Schmachtenberg E (2011) Menges werkstoffkunde kunststoffe, 6th edn. Carl Hanser Verlag, Munich
Meyer N, Gajek S, Görthofer J, Hrymak A, Kärger L, Henning F, Schneider M, Böhlke T (2023) Compos B Eng 249:110380. https://doi.org/10.1016/j.compositesb.2022.110380

Moynihan C, Macedo P, Montrose C, Gupta P, DeBolt M, Dill J, Dom B, Drake P, Easteal A, Elterman P, Moeller R, Sasabe H, Wilder J (1976) Ann N Y Acad Sci 279(1):15. https://doi.org/10.1111/j.1749-6632.1976.tb39688.x
Narayanaswamy O (1971) J Am Ceram Soc 54(10):491. https://doi.org/10.1111/j.1151-2916.1971.tb12186.x
Ngai K (1999) J Phys Chem B 103(28):5895. https://doi.org/10.1021/jp990594w
Papenfuß C (2020) Continuum thermodynamics and constitutive theory. Springer Nature Switzerland, Cham. https://doi.org/10.1007/978-3-030-43989-7
Sakatsuji W, Konishi T, Miyamoto Y (2013) Phys Rev E 88(1):012605. https://doi.org/10.1103/PhysRevE.88.012605
Scherer G (1984) J Am Ceram Soc 67(7):504. https://doi.org/10.1111/j.1151-2916.1984.tb19643.x
Serra-Aguila A, Puigoriol-Forcada J, Reyes G, Menacho J (2019) Acta Mech Sin 35(6):1191. https://doi.org/10.1007/s10409-019-00895-6
Šilhavý M (1997) The mechanics and thermodynamics of continuous media. Texts and monographs in physics. Springer, Berlin. https://doi.org/10.1007/978-3-662-03389-0
Simon S (1997) Macromolecules 30(14):4056. https://doi.org/10.1021/ma9614508
Simon S, Sobieski J, Plazek D (2001) Polymer 42(6):2555. https://doi.org/10.1016/S0032-3861(00)00623-6
Stommel M, Stojek M, Korte W (2018) FEM zur Berechnung von Kunststoff- und Elastomerbauteilen, 2nd edn. Carl Hanser Verlag, Munich
Svoboda R, Málek J (2013) J Non-Cryst Solids 378:186. https://doi.org/10.1016/j.jnoncrysol.2013.07.008
Tammann G, Hesse W (1926) Z Anorg Allg Chem 156(1):245
Tool A (1945) J Res Natl Bur Stand 34:199
Tool A (1946) J Am Ceram Soc 29(9):240. https://doi.org/10.1111/j.1151-2916.1946.tb11592.x
Tropin T, Schmelzer J, Schick C (2011) J Non-Cryst Solids 357(4):1291. https://doi.org/10.1016/j.jnoncrysol.2010.11.111
Tropin T, Schmelzer J, Schick C (2011) J Non-Cryst Solids 357(4):1303. https://doi.org/10.1016/j.jnoncrysol.2010.12.005
Tropin T, Schulz G, Schmelzer J, Schick C (2015) J Non-Cryst Solids 409:63. https://doi.org/10.1016/j.jnoncrysol.2014.11.001
Tschoegl N, Knauss W, Emri I (2002) Mech Time-Depend Mater 6(1):53. https://doi.org/10.1023/A:1014421519100
Vogel H (1921) Physikalische Zeitschrift 22:645
Weyer S, Hensel A, Korus J, Donth E, Schick C (1997) Thermochim Acta 304–305:251. https://doi.org/10.1016/S0040-6031(97)00286-4
Weyer S, Merzlyakov M, Schick C (2001) Thermochim Acta 377(1–2):85. https://doi.org/10.1016/S0040-6031(01)00543-3
Weyer S, Merzlyakov M, Schick C (2005) Polymer 46(26):12240. https://doi.org/10.1016/j.polymer.2005.10.097
Williams G, Watts D (1970) Trans Faraday Soc 66:80. https://doi.org/10.1039/TF9706600080
Williams M, Landel R, Ferry J (1955) J Am Chem Soc 77(14):3701. https://doi.org/10.1021/ja01619a008

Chapter 15
Constitutive Equations for Non-linear Isotropic Elastic Solids of Limited Compressibility

Evren Belenlioğlu and Thomas Böhme

Abstract Based on internal constraints depending on the deformations, an initially compressible solid can be mathematically prescribed to become less compressible during the motion and tends to become incompressible for larger deformations. This behavior is defined within this work as *limited compressibility*. Physically and geometrically non-linear constitutive equations for limited compressible elastic solids are derived, both for Cauchy and Green elasticity, which leads to an extension of the material equations for incompressible elastic solids such as Mooney-Rivlin or Neo-Hookean materials. The material response to the uniaxial extension state of a limited compressible elastic solid is analytically calculated as an example. By using the test results of an uniaxial extension for rubber, it is shown that the derived constitutive equations are able to simulate the measured values and deliver acceptable results.

15.1 Introduction

For the detailed mathematical formulation of kinematics, kinetics and the material behavior of continuous media, various textbooks and publications about continuum mechanics or solid mechanics are available, such as Truesdell and Noll (2004); Eringen (1980); Haupt (2002); Holzapfel (2000). The general theory of internal constraints is explored in works e.g. Green et al. (1970); Gurtin and Guidugli (1973); Casey (1995); Antman (2007). For specific theories related to internal constraints, readers may refer to publications such as Nagehban (2007); Klisch (1999); Bertram and Glüge (2016); Zdunek (2020).

In the literature there are different approaches to describe material behavior under internal constraints. One method involves a purely kinematic approach, suggesting

T. Böhme
Department for Wood Engineering, University for Sustainable Development Eberswalde (HNEE), Eberswalde, Germany
e-mail: thomas.boehme@hnee.de

E. Belenlioğlu (✉)
Bursa, Turkey
e-mail: evrenbelenlioglu@gmail.com

J. Fuhrmann et al. (eds.), *Advances in Continuum Physics*, Advanced Structured Materials 238, https://doi.org/10.1007/978-3-031-93918-1_15

relationships between the components of the deformation gradient, as introduced in Sect. 15.4.2. This approach is also the focus of this work to describe limited compressible behavior of materials.

As shown in Sect. 15.4.4, the compressibility, or the volume change of an infinitesimal material element, is described by $J := \det(\mathbf{F})$. Probably the most common and one of the simplest internal constraints is isochoric motion, which describes incompressible behavior: $J = \lambda_1\lambda_2\lambda_3 = 1$, where λ_i are the principal stretches, which describe how much the material is stretched or compressed along the principal directions.

Uniaxial tensile tests are the most common method for identifying the mechanical properties of a solid. For isotropic materials under uniaxial tension, if λ_1 is the stretch in the tensile direction, the deformations in the transverse directions be equal: $\lambda_2 = \lambda_3$. Accordingly, the determinant given by $J = \lambda_1\lambda_2^2$ and for incompressible materials with $J = \lambda_1\lambda_2^2 = 1$ we note:

$$\lambda_2 = \lambda_1^{-1/2}. \tag{15.1}$$

To describe compressible behavior, various kinematic relations in the form $\lambda_2 = f(\lambda_1)$ and $J = g(\lambda_1)$ are proposed as modifications of Eq. (15.1). The following kinematic relations are given for homogeneous isotropic compressible elastic materials in the uniaxial extension state.

Here, a power-law relation is suggested by Blatz and Ko (1962); viz.

$$\lambda_2 = \lambda_1^{-\nu_0} \quad \Rightarrow \quad J = \lambda_1^{1-2\nu_0}. \tag{15.2}$$

Within Blatz and Ko (1962), the constant ν_0 is interpreted as the Poisson's ratio for finite strains. For $\nu_0 = 1/2$, Eq. (15.2) yields to $J = 1$ and $\lambda_2 = \lambda_1^{-1/2}$ of incompressible materials.

Another power-law relation is suggested by Beatty and Stalnaker (1986):

$$\lambda_2 = \lambda_1^{-\frac{1}{2}+\varepsilon_0} \quad \Rightarrow \quad J = \lambda_1^{2\varepsilon_0}. \tag{15.3}$$

The constant ε_0 describes the deviation from the incompressible behavior. Since it is a small number in the range of $0 < \varepsilon_0 \ll 1$, it is suitable for nearly incompressible materials. A formulation of the kinematic relation in the form

$$\lambda_2 = \lambda_1^{-\frac{1}{2}+\frac{n}{2}} \quad \Rightarrow \quad J = \lambda_1^n \tag{15.4}$$

with $n < 1$ is also possible as discussed at Beatty (1987).

Another example for internal constraints for compressible materials is the Bell constraint for so-called Bell materials (Bell 1985; Beatty and Hayes 1992):

$$\mathrm{tr}(\mathbf{V} - \mathbf{I}) = 0 \quad \Leftrightarrow \quad \lambda_1 + \lambda_2 + \lambda_3 = 3. \tag{15.5}$$

As shown in Sect. 15.2, $\mathbf{V}$ is the stretch tensor and $\mathbf{I}$ is the identity tensor.

In this work, as discussed in Sect. 15.4.4, compressibility is described by ansatz functions that depend on the deformations, inspired by the internal constraint given in Gummert (1990), namely

$$J = 1 + \phi(\lambda_1). \tag{15.6}$$

Another approach for describing compressibility is used exclusively in hyperelasticity theory. It models nearly incompressible behavior through an additional term in the strain energy density function, mostly serving as a penalty term to avoid locking phenomena in numerical calculations. For more details and references, the reader is referred to Carlson and Tortorelli (1996); Horgan and Murphy (2009a, b); Doll and Schweizerhof (2000).

In the following sections, the kinematics of continuous media, the constitutive equations and the internal constraints for volume change are briefly introduced. The terms, symbols and their physical meanings are briefly explained to help readers who do not yet have an in-depth understanding of continuum mechanics.

15.2 Kinematics of Deformations

The motion of a material body $\mathcal{B}$ is described as a mapping from its reference configuration $\kappa_0 \equiv \kappa(\mathcal{B}_0, t_0)$ to its current configuration $\kappa \equiv \kappa(\mathcal{B}, t)$ with the time t as parameter ($t > t_0$).

The displacement of a material point $\mathcal{P} \in \mathcal{B}$ is given as the difference between the position vectors of $\mathcal{P}$ in the reference and current configurations

$$\mathbf{u}(\mathcal{P}) := \mathbf{x}(\mathcal{P}) - \mathbf{x}_0(\mathcal{P}). \tag{15.7}$$

Choosing the same Cartesian basis $\mathbf{e}_i$ with an origin O fixed in space as an absolute frame of reference $\{O, \mathbf{e}_i\}$ for both configurations, using the Einstein summation convention, Eq. (15.7) can be written in components

$$u_i\, \mathbf{e}_i := x_i\, \mathbf{e}_i - x_i^0\, \mathbf{e}_i = (x_i - x_i^0)\, \mathbf{e}_i \tag{15.8}$$

or in coordinates

$$u_i := x_i - x_i^0. \tag{15.9}$$

The deformation gradient tensor $\mathbf{F}$ is the essential quantity for the transformations of the elements of the material body $\mathcal{B}$ from its reference configuration κ_0 to its current configuration κ.

For example, the linear transformation of the material line element $\mathrm{d}\mathbf{x}_0 \in \mathcal{B}_0$ to $\mathrm{d}\mathbf{x} \in \mathcal{B}$ is given by

$$\mathrm{d}\mathbf{x} = \mathbf{F} \cdot \mathrm{d}\mathbf{x}_0 \quad \Leftrightarrow \quad \mathrm{d}x_i = F_{ij}\, \mathrm{d}x_j^0 \tag{15.10}$$

with

$$\mathbf{F} = \frac{\partial\,\mathbf{x}}{\partial\,\mathbf{x}_0} \equiv \nabla_0\,\mathbf{x} \qquad \Leftrightarrow \qquad F_{ij} = \frac{\partial\,x_i}{\partial\,x_j^0}\,. \tag{15.11}$$

Another example is the transformation of the material volume element by

$$\mathrm{d}V = \det(\mathbf{F})\,\mathrm{d}V_0. \tag{15.12}$$

Equation (15.12) is important for the formulation of compressibility. $J := \det(\mathbf{F})$ is the value for the volume change of the infinitesimal material volume $\mathrm{d}V_0$ during the motion. Since the volume cannot have negative values, J must always be positive: $J > 0$. The special case $J = 1$ for any t indicates the isochoric motion $\mathrm{d}V = \mathrm{d}V_0$ which describes the incompressible behavior of rubber-like solids.

We note, when the undeformed state of the body is chosen as the reference configuration, we obtain the initial conditions:

$$\mathbf{F}_0 = \mathbf{I} \quad \text{and} \quad \det(\mathbf{F}_0) \equiv J_0 = 1. \tag{15.13}$$

$\mathbf{I} = \mathbf{e}_i\mathbf{e}_i = \mathbf{e}_1\mathbf{e}_1 + \mathbf{e}_2\mathbf{e}_2 + \mathbf{e}_3\mathbf{e}_3$ is the identity tensor.

A multiplicative decomposition of the deformation gradient $\mathbf{F}$ in an isotropic part $\mathbf{F}^{\odot} := J^{1/3}\mathbf{I}$, which describes the dilatation with $\det(\mathbf{F}^{\odot}) = J$, and a remaining part $\overline{\mathbf{F}}$ describing the distortion with the property $\det(\overline{\mathbf{F}}) = 1$, is given as:

$$\mathbf{F} = \mathbf{F}^{\odot} \cdot \overline{\mathbf{F}} = J^{1/3}\mathbf{I} \cdot \overline{\mathbf{F}} = J^{1/3}\overline{\mathbf{F}} \quad \Rightarrow \quad \overline{\mathbf{F}} \equiv J^{(-1/3)}\mathbf{F}. \tag{15.14}$$

With the polar decomposition of $\mathbf{F} = \mathbf{R} \cdot \mathbf{U} = \mathbf{V} \cdot \mathbf{R}$, the mapping can be separated into a purely rotational part $\mathbf{R}$ and a purely stretch part $\mathbf{U}$ or $\mathbf{V}$. The stretch tensors are symmetric: $\mathbf{U} = \mathbf{U}^T$ and $\mathbf{V} = \mathbf{V}^T$.

The rotation tensor $\mathbf{R}$ is a proper orthogonal tensor with the properties $\mathbf{R}^{-1} = \mathbf{R}^T$ and $\det(\mathbf{R}) = +1$. Since $\det(\mathbf{F}) = \det(\mathbf{V} \cdot \mathbf{R}) = \det(\mathbf{V}) \cdot \det(\mathbf{R}) = \det(\mathbf{V})$, we can write for the determinants of the tensors $\det(\mathbf{U}) = \det(\mathbf{V}) = \det(\mathbf{F})$. Further we note the relation $\mathbf{V} = \mathbf{R} \cdot \mathbf{U} \cdot \mathbf{R}^T$ and for the special case $\mathbf{R} = \mathbf{I}$, it follows $\mathbf{F} = \mathbf{V} = \mathbf{U}$.

Any deformation, especially large deformations, can be described by the right Cauchy-Green deformation tensor $\mathbf{C} := \mathbf{F}^T \cdot \mathbf{F} = \mathbf{C}^T$ or alternatively by the left Cauchy-Green deformation tensor $\mathbf{B} := \mathbf{F} \cdot \mathbf{F}^T = \mathbf{B}^T$. By definitions, $\mathbf{B}$ and $\mathbf{C}$ are symmetric and do not contain rotations:

$$\mathbf{B} := \mathbf{F} \cdot \mathbf{F}^T = \mathbf{V} \cdot \mathbf{R} \cdot \mathbf{R}^T \cdot \mathbf{V}^T = \mathbf{V}^2, \tag{15.15}$$

$$\mathbf{C} := \mathbf{F}^T \cdot \mathbf{F} = \mathbf{U}^T \cdot \mathbf{R}^T \cdot \mathbf{R} \cdot \mathbf{U} = \mathbf{U}^2. \tag{15.16}$$

To calculate the stretches $\mathbf{V}$ from the deformations, Eq. (15.15) is rewritten as $\mathbf{V} = \sqrt{\mathbf{B}}$. In order to exploit the root, the components of $\mathbf{B}$ must first be transformed to spectral form by the principal axis transformation. After determining the $\mathbf{V}$, the rotation tensor can be calculated: $\mathbf{R} = \mathbf{V}^{-1} \cdot \mathbf{F}$.

Subtracting the undeformed state $\mathbf{C}_0 = \mathbf{F}_0 \cdot \mathbf{F}_0^T = \mathbf{I}$ from the deformation tensors $\mathbf{C}$ (or $\mathbf{B}$ in analogy), we define the finite strain tensors, right (r) or left (ℓ) Green strain tensors

$$\mathbf{G}^r := \frac{1}{2}\,[\mathbf{C} - \mathbf{I}] \quad \text{and} \quad \mathbf{G}^\ell := \frac{1}{2}\,[\mathbf{B} - \mathbf{I}]. \tag{15.17}$$

The displacement gradient tensor $\mathbf{H}$ is defined by the derivation of the displacement vector from Eq. (15.7), namely $\mathbf{u} := \mathbf{x} - \mathbf{x}_0$:

$$\mathbf{H} := \frac{\partial\,\mathbf{u}}{\partial\,\mathbf{x}_0} \equiv \nabla_0\mathbf{u} = \nabla_0(\mathbf{x} - \mathbf{x}_0) = \nabla_0\mathbf{x} - \nabla_0\mathbf{x}_0 = \mathbf{F} - \mathbf{I}. \tag{15.18}$$

After replacing $\mathbf{F} = \mathbf{I} + \mathbf{H} = \mathbf{I} + \nabla_0\mathbf{u}$ in Eq. (15.17) we obtain the relations:

$$\mathbf{G}^r = \frac{1}{2}\,\left[\mathbf{H} + \mathbf{H}^T + \mathbf{H}^T \cdot \mathbf{H}\right], \tag{15.19}$$

$$\mathbf{G}^\ell = \frac{1}{2}\,\left[\mathbf{H} + \mathbf{H}^T + \mathbf{H} \cdot \mathbf{H}^T\right]. \tag{15.20}$$

It is worth to note these equations using the displacements, known as the strain-displacement relations:

$$\mathbf{G}^r = \frac{1}{2}\,\left[\nabla_0\mathbf{u} + (\nabla_0\mathbf{u})^T + (\nabla_0\mathbf{u})^T \cdot \nabla_0\mathbf{u}\right], \tag{15.21}$$

$$\mathbf{G}^\ell = \frac{1}{2}\,\left[\nabla_0\mathbf{u} + (\nabla_0\mathbf{u})^T + \nabla_0\mathbf{u} \cdot (\nabla_0\mathbf{u})^T\right]. \tag{15.22}$$

The infinitesimal strains (related with so-called small deformations) are obtained by the linearisation of the finite strains in Eqs. (15.19) or (15.20). The second order terms $o(\|\mathbf{H}^2\|)$ are neglected, which yield the strain-displacement relation for small deformations:

$$\mathrm{lin}(\mathbf{G}^\ell) = \mathrm{lin}(\mathbf{G}^r) = \frac{1}{2}\,\left[\mathbf{H} + \mathbf{H}^T\right] \tag{15.23}$$

or with $\mathrm{lin}(\mathbf{G}^\ell) = \mathrm{lin}(\mathbf{G}^r) =: \mathbf{E} = \varepsilon_{ij}\,\mathbf{e}_i\mathbf{e}_j$

$$\mathbf{E} = \frac{1}{2}\,\left[\nabla\mathbf{u} + (\nabla\mathbf{u})^T\right]. \tag{15.24}$$

$\mathbf{E}$ is the infinitesimal strain tensor - widely used in engineering disciplines—with the eigenvalues ε_i; the principal strains for small deformations. We note that ∇_0 is replaced by ∇, since for small deformations it is appropriate to assume that $\mathbf{u}(\mathbf{x}_0, t) \approx \mathbf{u}(\mathbf{x}, t)$, which implies $\mathrm{d}\mathbf{x}_0 \approx \mathrm{d}\mathbf{x}$ (see, e.g., Haupt (2002), Sect. 1.9).

15.2.1 Two Simple Deformations

Among many others, we note two deformation states which are used in further calculations, namely the pure stretch state and the uniaxial extension/compression state. For other deformation states, see, for example, Carroll and Horgan (1990).

Pure Stretch:
In this state of deformation, the specimen is stretched or compressed in three mutually perpendicular directions, which can be described with a Cartesian coordinate system $\mathbf{e}_i$. The stretch in the $\mathbf{e}_i$ direction is denoted by λ_i. The deformation gradient for pure stretch is given as

$$\mathbf{F} = \lambda_1\,\mathbf{e}_1\mathbf{e}_1 + \lambda_2\,\mathbf{e}_2\mathbf{e}_2 + \lambda_3\,\mathbf{e}_3\mathbf{e}_3 = \mathbf{F}^T. \tag{15.25}$$

Since there is no mixed components such $\mathbf{e}_i\mathbf{e}_j$ $(i \neq j)$, $\mathbf{e}_i$ are the principle directions and λ_i are the eigenvalues of $\mathbf{F}$.

According to Eq. (15.18), the displacement gradient tensor can be written as

$$\mathbf{H} = (\lambda_1 - 1)\,\mathbf{e}_1\mathbf{e}_1 + (\lambda_2 - 1)\,\mathbf{e}_2\mathbf{e}_2 + (\lambda_3 - 1)\,\mathbf{e}_3\mathbf{e}_3 = \mathbf{H}^T \tag{15.26}$$

and since $\mathbf{H}$ is symmetric in Eq. (15.26), the infinitesimal strains read according to Eq. (15.23) to

$$\mathbf{E} = \mathbf{H} \quad \Leftrightarrow \quad \varepsilon_i = \lambda_i - 1. \tag{15.27}$$

For large deformations, the left Cauchy-Green deformation tensor results in

$$\mathbf{B} = \mathbf{F}\cdot\mathbf{F}^T = \mathbf{F}^2 = \lambda_1^2\,\mathbf{e}_1\mathbf{e}_1 + \lambda_2^2\,\mathbf{e}_2\mathbf{e}_2 + \lambda_3^2\,\mathbf{e}_3\mathbf{e}_3. \tag{15.28}$$

Since there is no component with a mixed product $\mathbf{e}_i\mathbf{e}_j$ $(i \neq j)$ in Eq. (15.28), the $\mathbf{e}_i$ are the principal directions (eigenvectors) and the eigenvalues λ_i^2 are the principle Cauchy-Green deformations. The left stretch tensor $\mathbf{V}$ can be calculated as

$$\mathbf{V} = \sqrt{\mathbf{B}} = \sqrt{\mathbf{F}^2} = \mathbf{F} = \lambda_1\,\mathbf{e}_1\mathbf{e}_1 + \lambda_2\,\mathbf{e}_2\mathbf{e}_2 + \lambda_3\,\mathbf{e}_3\mathbf{e}_3. \tag{15.29}$$

The rotation tensor results to: $\mathbf{R} = \mathbf{V}^{-1}\cdot\mathbf{F} = \mathbf{F}^{-1}\cdot\mathbf{F} = \mathbf{I}$.

Uniaxial Extension/Compression:
A specimen is stretched (or compressed) only in one direction, namely with λ_1 in the principle direction $\mathbf{e}_1$. Uniaxial extension test results show that the cross-sectional area of the specimen becomes smaller, and if the material is isotropic, the deformations in transverse directions are the same: $\lambda_2 = \lambda_3$ (see for example Batra (1976)). Replacing this relation in Eqs. (15.25) and (15.29) we obtain

$$\mathbf{F} = \lambda_1\,\mathbf{e}_1\mathbf{e}_1 + \lambda_2\,(\mathbf{e}_2\mathbf{e}_2 + \mathbf{e}_3\mathbf{e}_3) = \mathbf{V}. \tag{15.30}$$

For small deformations, Eq. (15.26) simplifies to

$$\mathbf{H} = (\lambda_1 - 1)\,\mathbf{e}_1\mathbf{e}_1 + (\lambda_2 - 1)\;(\mathbf{e}_2\mathbf{e}_2 + \mathbf{e}_3\mathbf{e}_3)\,. \tag{15.31}$$

Introducing the Poisson's ratio $\nu := -\varepsilon_2/\varepsilon_1$ and using Eq. (15.27), the infinitesimal strain tensor $\mathbf{E}$ in spectral form results to

$$\mathbf{E} = \varepsilon_1\,\mathbf{e}_1\mathbf{e}_1 - \nu\,\varepsilon_1\,(\mathbf{e}_2\mathbf{e}_2 + \mathbf{e}_3\mathbf{e}_3). \tag{15.32}$$

For large deformations, the Cauchy-Green deformation tensor from Eq. (15.28) simplifies to

$$\mathbf{B} = \lambda_1^2\,\mathbf{e}_1\mathbf{e}_1 + \lambda_2^2\;(\mathbf{e}_2\mathbf{e}_2 + \mathbf{e}_3\mathbf{e}_3) = \mathbf{V}^2. \tag{15.33}$$

15.2.2 Change of Frame of Reference

Quantities such as the position vector, mass density, or stress tensor can be described in different frames of reference. For example, a position vector $\mathbf{x}_P(t)$ describes the position of a point P with respect to the absolute frame of reference $\{O, \mathbf{e}_i\}$. The same point P can be represented by $\mathbf{x}_P^*(t^*)$ with respect to another frame of reference $\{O^*, \mathbf{e}_i^*\}$, which has a different origin and orientation in space. The relation between $\mathbf{x}_P^*$ and $\mathbf{x}_P$ is given by the Euclidean transformation (see, e.g., Müller and Ruggeri (1998), Sect. 4.1)

$$\mathbf{x}_P^*(t^*) = \mathbf{Q}(t)\cdot\mathbf{x}_P(t) + \mathbf{c}(t)\,, \qquad t^* = t\,, \tag{15.34}$$

where $\mathbf{c}(t)$ is the position vector of the origin O with respect to O^*, and $\mathbf{Q}$ is an orthogonal tensor.

In analogy to Eq. (15.34), the motion of a material point $\mathcal{P}$ can be described in different frames of reference. The transformation in current configuration can be written as

$$\mathbf{x}^* = \mathbf{Q}\cdot\mathbf{x} + \mathbf{c}. \tag{15.35}$$

Applying ∇_0 to Eq. (15.35) from left, we obtain the transformation rule for the deformation gradient:

$$\mathbf{F}^* = \mathbf{Q}\cdot\mathbf{F}. \tag{15.36}$$

Equation (15.36) allow us to derive the transformation rules for various kinematical quantities:

$$\begin{aligned}
\mathbf{R}^* &= \mathbf{Q}\cdot\mathbf{R},\\
\mathbf{B}^* &= \mathbf{F}^*\cdot\left(\mathbf{F}^*\right)^T = \mathbf{Q}\cdot\mathbf{F}\cdot\mathbf{F}^T\cdot\mathbf{Q}^T = \mathbf{Q}\cdot\mathbf{B}\cdot\mathbf{Q}^T,\\
\mathbf{V}^* &= \mathbf{F}^*\cdot\left(\mathbf{R}^*\right)^T = \mathbf{Q}\cdot\mathbf{F}\cdot\mathbf{R}^T\cdot\mathbf{Q}^T = \mathbf{Q}\cdot\mathbf{V}\cdot\mathbf{Q}^T,
\end{aligned} \tag{15.37}$$

$$\mathbf{C}^* = \left(\mathbf{F}^*\right)^T \cdot \mathbf{F}^* = \mathbf{F}^T \cdot \mathbf{Q}^T \cdot \mathbf{Q} \cdot \mathbf{F} = \mathbf{F}^T \cdot \mathbf{F} = \mathbf{C},$$
$$\mathbf{U}^* = \left(\mathbf{R}^*\right)^T \cdot \mathbf{F}^* = \mathbf{R}^T \cdot \mathbf{Q}^T \cdot \mathbf{Q} \cdot \mathbf{F} = \mathbf{R}^T \cdot \mathbf{F} = \mathbf{U}.$$

Any second-order tensor $\mathbf{A}$ that obeys the transformation rule $\mathbf{A}^* = \mathbf{Q} \cdot \mathbf{A} \cdot \mathbf{Q}^T$ is called a *frame-indifferent* or *objective* tensor. The tensors $\mathbf{B}$ and $\mathbf{V}$ are objective tensors. On the other hand, if the tensor remains unchanged by a change of frame of reference, i.e., $\mathbf{A}^* = \mathbf{A}$, such as $\mathbf{C}$ and $\mathbf{U}$, it is called an *invariant* tensor.

An objective scalar s is defined as $s^* = s$, which is also invariant. An objective vector $\mathbf{v}$ is defined as $\mathbf{v}^* = \mathbf{Q} \cdot \mathbf{v}$.

For a detailed explanation of the topic, the reader is referred to Haupt (2002), Sect. 4.

15.2.3 Isotropic Tensor Functions, Principal Invariants

Let $\mathbf{A}$ be a second-order symmetric tensor. A scalar-valued tensor function $f(\mathbf{A})$ is a function that maps the tensor $\mathbf{A}$ to a scalar value. Similarly, a tensor-valued tensor function $\mathbf{f}(\mathbf{A})$ maps the tensor $\mathbf{A}$ to a tensor (see Truesdell and Noll (2004), Sect. 7).

A scalar-valued tensor function $f(\mathbf{A})$ is said to be *isotropic*, if the relation

$$f(\mathbf{A}) = f(\mathbf{Q} \cdot \mathbf{A} \cdot \mathbf{Q}^T) \tag{15.38}$$

holds for all orthogonal tensors $\mathbf{Q}$ (see Truesdell and Noll (2004), Sect. 8).

On the other hand, a scalar-valued tensor function $f(\mathbf{A})$ is an *invariant* if and only if it can be expressed as a function of its principal invariants (see Truesdell and Noll (2004), Sect. 10):

$$f(\mathbf{A}) = f(\mathrm{I_A}, \mathrm{II_A}, \mathrm{III_A}) \tag{15.39}$$

The principal invariants of a second-order tensor—for example, the invariants of $\mathbf{V}$ from Eq. (15.29)—are defined as

$$\begin{aligned} \mathrm{I_V} &:= \mathrm{tr}(\mathbf{V}) = \lambda_1 + \lambda_2 + \lambda_3, \\ \mathrm{II_V} &:= \frac{1}{2}\left[(\mathrm{tr}\,\mathbf{V})^2 - \mathrm{tr}(\mathbf{V}^2)\right] = \lambda_1\lambda_2 + \lambda_2\lambda_3 + \lambda_3\lambda_1, \\ \mathrm{III_V} &:= \det(\mathbf{V}) = \lambda_1\lambda_2\lambda_3 \equiv J \end{aligned} \tag{15.40}$$

where λ_i are the eigenvalues of the tensor $\mathbf{V}$.

Similarly, we note the invariants of $\mathbf{B} = \mathbf{V}^2$ for pure stretch using Eq. (15.28)

$$\begin{aligned} \mathrm{I_B} &:= \mathrm{tr}(\mathbf{B}) = \lambda_1^2 + \lambda_2^2 + \lambda_3^2, \\ \mathrm{II_B} &:= \frac{1}{2}\left[(\mathrm{tr}\,\mathbf{B})^2 - \mathrm{tr}(\mathbf{B}^2)\right] = \lambda_1^2\lambda_2^2 + \lambda_2^2\lambda_3^2 + \lambda_3^2\lambda_1^2, \\ \mathrm{III_B} &:= \det(\mathbf{B}) = \lambda_1^2\lambda_2^2\lambda_3^2 = J^2. \end{aligned} \tag{15.41}$$

It is important to note that the invariants of the stretch tensors **U** and **V** are identically: $\mathrm{I_U} = \mathrm{I_V}$, $\mathrm{II_U} = \mathrm{II_V}$ and $\mathrm{III_U} = \mathrm{III_V}$. The invariants of the deformation tensors **B** and **C** are also identically: $\mathrm{I_C} = \mathrm{I_B}$, $\mathrm{II_C} = \mathrm{II_B}$ and $\mathrm{III_C} = \mathrm{III_B}$.

All these invariants are objective scalars. For example:

$$\mathrm{I}^*_{\mathrm{B}} = \mathrm{I_B}\,, \qquad \mathrm{II}^*_{\mathrm{B}} = \mathrm{II_B}\,, \qquad \mathrm{III}^*_{\mathrm{B}} = \mathrm{III_B}\,. \tag{15.42}$$

For further calculations, we note the following derivatives:

$$\frac{\partial \mathrm{I_B}}{\partial \mathbf{B}} = \mathbf{I}\,, \qquad \frac{\partial \mathrm{II_B}}{\partial \mathbf{B}} = \mathrm{I_B}\,\mathbf{I} - \mathbf{B}\,, \qquad \frac{\partial \mathrm{III_B}}{\partial \mathbf{B}} = \mathrm{III_B}\,\mathbf{B}^{-1}. \tag{15.43}$$

15.3 Kinetics

The stress state of an infinitesimal volume element $\mathrm{d}V$ of a material body $\mathcal{B}$ in the current configuration $\kappa(\mathcal{B}, t)$ is described by the Cauchy stress tensor $\mathbf{S} = \sigma_{ij}\,\mathbf{e}_i\mathbf{e}_j$.

The dynamics of the motion of a material body $\mathcal{B}$ in the current configuration is described by the Euler's first and second balance equations: For solids in *static* or *quasi-static* states, the balance equation for linear momentum (so-called equation of mechanical equilibrium)

$$\nabla \cdot \mathbf{S} = \mathbf{0} \qquad \Leftrightarrow \qquad \frac{\partial \sigma_{ij}}{\partial x_j} = 0\,, \tag{15.44}$$

and the symmetry of **S** resulting from the balance equation for angular momentum

$$\mathbf{S} = \mathbf{S}^T \qquad \Leftrightarrow \qquad \sigma_{ij} = \sigma_{ji}\,, \tag{15.45}$$

both so-called Cauchy axioms describe the motion of the body.

Since different configurations lead to different stress definitions, we can introduce the Piola-Kirchhoff stress tensors at this point.

The 1st Piola-Kirchhoff stress tensor **P** is related to both configurations, where forces in the current configuration act on the geometry of the reference configuration. These stresses are known as engineering stresses. The relation between the stresses **P** and **S** is given as

$$\mathbf{P} = J\,\mathbf{S}\cdot\mathbf{F}^{-T} \qquad \Leftrightarrow \qquad \mathbf{S} = \frac{1}{J}\,\mathbf{P}\cdot\mathbf{F}^T, \tag{15.46}$$

where **F** and $J = \det(\mathbf{F})$ act as transformation operators of the stresses in different configurations.

The stress state of an infinitesimal volume element of a material body in the reference configuration $\mathcal{B}_0$ is described by the 2nd Piola-Kirchhoff stress tensor $\mathbf{T}$, given by the relation

$$\mathbf{T} = J\,\mathbf{F}^{-1}\cdot\mathbf{S}\cdot\mathbf{F}^{-T} \qquad \Leftrightarrow \qquad \mathbf{S} = \frac{1}{J}\,\mathbf{F}\cdot\mathbf{T}\cdot\mathbf{F}^{T}\,. \tag{15.47}$$

We note that $\mathbf{T}$ obeys the Cauchy axioms $\nabla_0\cdot\mathbf{T} = \mathbf{0}$ and $\mathbf{T} = \mathbf{T}^T$ in the reference configuration κ_0.

15.4 Constitutive Equations

For a volume element $\mathrm{d}V$ in motion, we have on one hand the kinematic quantities related with geometry, namely the displacement vector $\mathbf{u}$ (with 3 unknown components) and the symmetric deformation tensor $\mathbf{B}$ or $\mathbf{G}$ (with 6 unknown components), on the other the kinetic quantity related with forces, namely the symmetric stress tensor $\mathbf{S}$ (with 6 unknown components). For these $3 + 6 + 6 = 15$ unknown quantities, we have the strain-displacement relations, for example Eq. (15.22) which implies 6 scalar equations and the equilibrium condition Eq. (15.44) with 3 scalar equations, together $6 + 3 = 9$ equations. To close this equation system, 6 additional scalar equations containing the deformations and the stresses are required, for example in the form

$$\mathbf{S} = \mathcal{F}[\mathbf{B}]\,, \tag{15.48}$$

where $\mathcal{F}$ is tensor-valued tensor functional.

The modelling of material behavior can be realised on three levels: the formulation of constitutive equations, the formulation of material symmetry, and the formulation of internal constraints (see Haupt (2002), Sect. 7.1).

- Constitutive Equations: The kinematics and kinetics of a material body do not depend on the characteristics of the material. However, the relations between kinetic and kinematic quantities, given in the form of Eq. (15.48), generally vary for different groups of materials and are known as constitutive equations. These equations are derived by considering several principles, as described, e.g., in Haupt (2002). Some of these principles are:

  - The Principle of Determinism: The stress state at a material point $\mathcal{P}$ depends only on its past deformation and stress history, not on any future events.
  - The Principle of Local Action: The stress state at a material point $\mathcal{P}$ is influenced only by its neighboring material points at an infinitesimal distance. Other material points in the body do not influence the stress state at $\mathcal{P}$. This principle enables us to define the *simple material*, where only $\mathbf{F}$ is a variable of the functional, excluding its gradients.

– The Principle of Material Objectivity (or Frame-Indifference): The form of the constitutive functional is the same for different observers in different frames of reference:

$$\mathcal{F}^* = \mathcal{F}. \tag{15.49}$$

This means that the constitutive functional is unchanged while their values and the values of their variables do change (see Müller and Ruggeri (1998), Sect. 4.2). This rule also applies to scalar-valued tensor functions: $f^* = f$. Considering Eqs. (15.38), (15.39) and Eq. (15.42), the principle of material objectivity is ensured by isotropic tensor functions, such as:

$$f(\mathbf{B}) = f(\mathrm{I_B}, \mathrm{II_B}, \mathrm{III_B}) \tag{15.50}$$

- Material symmetry: Since we are considering isotropic solids with identical mechanical properties in all directions, the material is symmetric under any rotation.
- Internal constraints: By focusing on the compressibility of solids, specifically the volume change described by the kinematic quantity $\det(\mathbf{F})$, internal constraints and related constitutive equations become the central topic of this work. It is discussed in Sect. 15.4.2.

15.4.1 Cauchy Elasticity

Elastic solids return to their original shapes after loading and unloading without any permanent deformation. The constitutive equations describe the relationship between stresses and corresponding strains in a material throughout the loading history or simply during motion, for example as given in Eq. (15.51).

The constitutive equation of a simple isotropic elastic material without memory is given by a tensor valued tensor function $\mathcal{F}$ (see, e.g., Truesdell and Noll (2004), Sect. 47):

$$\mathbf{S} = \mathcal{F}(\mathbf{B}) = \psi_0\,\mathbf{I} + \psi_1\,\mathbf{B} + \psi_2\,\mathbf{B}^2 \tag{15.51}$$

with the material functions ψ_i, which depend only on the invariants of $\mathbf{B}$, according to the principle of material frame indifference: $\psi_i(\mathrm{I_B}, \mathrm{II_B}, \mathrm{III_B})$.

Materials underlying Eq. (15.48) or in particular Eq. (15.51) are often noted as the Cauchy-elastic materials or simple elastic materials.

15.4.2 Internal Constraints

This work focuses on the application of internal constraints, by using the basic formulation provided in many textbooks (e.g., Truesdell and Noll (2004), Sect. 30). Accordingly, a simple internal constraint is defined as

$$\gamma\,(\mathbf{F}) = 0 \tag{15.52}$$

where γ is a scalar-valued tensor function (or functional) which formulates mathematical restrictions of the motion given as kinematic relations between the elements of the deformation gradient tensor $\mathbf{F}$.

In analogy to the Lagrangian analytical mechanics of constrained rigid systems or mass point systems, internal constraints γ are associated with reaction forces or reaction stresses $\mathbf{P}_R$ (also called reactive stresses or constraint stresses):

$$\mathbf{P}_R = \alpha\,\frac{\partial \gamma}{\partial \mathbf{F}}. \tag{15.53}$$

The scalar α is the Lagrange multiplier. $\mathbf{P}_R$ is the 1st Piola-Kirchhoff reaction stress tensor. The Cauchy reaction stresses can be calculated using Eq. (15.46):

$$\mathbf{S}_R = \frac{1}{J}\,\mathbf{P}_R \cdot \mathbf{F}^T = \frac{\alpha}{J}\frac{\partial \gamma}{\partial \mathbf{F}} \cdot \mathbf{F}^T. \tag{15.54}$$

The internal constraint $\gamma\,(\mathbf{F}) = 0$ is subjected to the principles of the constitutive theory of materials (see, Truesdell and Noll (2004), Sect. 30). To ensure the principle of the material frame-indifference, obeying

$$\gamma^* = \gamma\,, \tag{15.55}$$

γ must be an isotropic tensor function as represented in Eq. (15.50), depending on the invariants of the deformation tensor. For example in the form

$$\gamma\,(\mathbf{F}) = \tilde{\gamma}\,(\mathbf{B}) = \tilde{\gamma}\,(\mathrm{I_B}, \mathrm{II_B}, \mathrm{III_B}) = 0, \tag{15.56}$$

which also obeys the principle of the material frame indifference $\tilde{\gamma}^* = \tilde{\gamma}$.

Considering the total differential of Eq. (15.56) and using the symmetries, we obtain

$$\begin{aligned} \mathrm{d}\gamma(\mathbf{F}) &= \mathrm{d}\tilde{\gamma}\,(\mathbf{B}) \\ \frac{\partial \gamma\,(\mathbf{F})}{\partial \mathbf{F}} : \mathrm{d}\mathbf{F}^T &= \frac{\partial \tilde{\gamma}\,(\mathbf{B})}{\partial \mathbf{B}} : \mathrm{d}\,\mathbf{B} = \frac{\partial \tilde{\gamma}\,(\mathbf{B})}{\partial \mathbf{B}} : \mathrm{d}(\mathbf{F} \cdot \mathbf{F}^T) \\ &= \frac{\partial \tilde{\gamma}\,(\mathbf{B})}{\partial \mathbf{B}} : (\mathrm{d}\mathbf{F} \cdot \mathbf{F}^T + \mathbf{F} \cdot \mathrm{d}\mathbf{F}^T) \\ &= \frac{\partial \tilde{\gamma}\,(\mathbf{B})}{\partial \mathbf{B}} : 2\,(\mathbf{F} \cdot \mathrm{d}\mathbf{F}^T) \\ &= 2\,\frac{\partial \tilde{\gamma}\,(\mathbf{B})}{\partial \mathbf{B}} \cdot \mathbf{F} : \mathrm{d}\mathbf{F}^T\,. \end{aligned} \tag{15.57}$$

Comparing the left and right sides results in

$$\frac{\partial \gamma\,(\mathbf{F})}{\partial \mathbf{F}} = 2\,\frac{\partial \tilde{\gamma}\,(\mathbf{B})}{\partial \mathbf{B}} \cdot \mathbf{F}. \tag{15.58}$$

According to Eqs. (15.53) and (15.54), the reaction stresses can be written as

$$\mathbf{P}_R = 2\alpha\,\frac{\partial \tilde{\gamma}}{\partial \mathbf{B}} \cdot \mathbf{F} \quad \text{and} \quad \mathbf{S}_R = \frac{2\alpha}{J}\,\frac{\partial \tilde{\gamma}}{\partial \mathbf{B}} \cdot \mathbf{B}. \tag{15.59}$$

In order to incorporate the internal constraint of Eq. (15.52), the reaction stress of Eq. (15.59) can be added to the stresses of the unconstrained motion, for example to Eq. (15.51):

$$\mathbf{S} = \mathcal{F}(\mathbf{B}) + \mathbf{S}_R = \mathcal{F}(\mathbf{B}) + \frac{2\alpha}{J}\,\frac{\partial \tilde{\gamma}}{\partial \mathbf{B}} \cdot \mathbf{B}. \tag{15.60}$$

Reaction stresses do not work in any motion that satisfies the constraint (see, for example Casey (1995)). We also note that the stresses $\mathbf{S} - \mathbf{S}_R = \mathcal{F}(\mathbf{B}) =: \mathbf{S}_E$ called active or extra stresses. The sum $\mathbf{S}_E + \mathbf{S}_R = \mathbf{S}$ called total stresses.

For simple materials, the number of the internal constraints are limited with the number of independent components of $\mathbf{B}$. If the number of constraints n at its maximum ($n = n_{\text{max}}$), then $\mathcal{B}$ is a rigid body. In case of more than one constraint, the Cauchy stresses can be calculated with

$$\mathbf{S} = \mathcal{F}(\mathbf{B}) + \frac{2}{J}\sum_{i=1}^{n} \alpha_i\,\frac{\partial \tilde{\gamma}_i}{\partial \mathbf{B}} \cdot \mathbf{B}. \tag{15.61}$$

15.4.3 *Incompressible Materials*

The internal constraint for incompressible materials, $J = 1$, can be written in the form of Eq. (15.52):

$$\gamma\,(\mathbf{F}) = \gamma\,(\mathrm{III}_{\mathrm{F}}) = \mathrm{III}_{\mathrm{F}} - 1 = J - 1 = 0. \tag{15.62}$$

Alternatively, it can be written by using $\mathbf{B}$ as

$$\tilde{\gamma}\,(\mathbf{B}) = \tilde{\gamma}\,(\mathrm{III}_{\mathrm{B}}) = \mathrm{III}_{\mathrm{B}} - 1 = J^2 - 1 = 0. \tag{15.63}$$

Both constraints can be expressed as

$$\mathrm{III}_{\mathrm{F}} = J = 1 \quad \Leftrightarrow \quad \mathrm{III}_{\mathrm{B}} = J^2 = 1\,, \tag{15.64}$$

or by using the stretches

$$J = \lambda_1 \lambda_2 \lambda_3 = 1 \quad \Leftrightarrow \quad J^2 = \lambda_1^2 \lambda_2^2 \lambda_3^2 = 1. \tag{15.65}$$

Considering $\lambda_2 = \lambda_3$ in Eq. (15.65), for the uniaxial extension or compression state we obtain the kinematic relation for the incompressible stretch in transverse direction λ_2^{in} (in: incompressible)

$$\lambda_2^{\text{in}} = \sqrt{\frac{1}{\lambda_1}} = \lambda_1^{-1/2}. \tag{15.66}$$

For the uniaxial extension/compression, according to Eq. (15.33), the Cauchy-Green deformation tensor $\mathbf{B}$ results to

$$\mathbf{B}^{\text{in}} = \lambda_1^2 \, \mathbf{e}_1\mathbf{e}_1 + \frac{1}{\lambda_1} \left(\mathbf{e}_2\mathbf{e}_2 + \mathbf{e}_3\mathbf{e}_3\right). \tag{15.67}$$

According to Eq. $(15.59)_1$, the incompressible 1st Piola-Kirchhoff reaction stress tensor $\mathbf{P}_R^{\text{in}}$ for $\tilde{\gamma} = \text{III}_\text{B} - 1$ from Eq. (15.63), by using Eq. $(15.43)_3$, yields

$$\begin{aligned} \mathbf{P}_R^{\text{in}} &= 2\alpha \left.\frac{\partial \tilde{\gamma}}{\partial \mathbf{B}}\right|_{\text{III}_\text{B}=1} \cdot \mathbf{F} = 2\alpha \left.\frac{\partial \text{III}_\text{B}}{\partial \mathbf{B}}\right|_{\text{III}_\text{B}=1} \cdot \mathbf{F} = 2\alpha \left.\text{III}_\text{B}\, \mathbf{B}^{-1}\right|_{\text{III}_\text{B}=1} \cdot \mathbf{F} \\ &= 2\alpha \, \mathbf{B}^{-1} \cdot \mathbf{F}, \end{aligned} \tag{15.68}$$

and the incompressible Cauchy reaction stress tensor can be written as

$$\mathbf{S}_R^{\text{in}} = J^{-1} \mathbf{P}_R^{\text{in}} \cdot \mathbf{F}^T \Big|_{J=1} = 2\alpha \, \mathbf{B}^{-1} \cdot \mathbf{F} \cdot \mathbf{F}^T = 2\alpha \, \mathbf{I}. \tag{15.69}$$

Adding this relation to Eq. (15.51), we obtain the Cauchy stresses of the incompressible materials

$$\mathbf{S}^{\text{in}} = \mathcal{F}(\mathbf{B}^{\text{in}}) + \mathbf{S}_R^{\text{in}} = (2\alpha + \psi_0) \, \mathbf{I} + \psi_1 \, \mathbf{B} + \psi_2 \, \mathbf{B}^2 \tag{15.70}$$

with $\psi_i = \psi_i(\text{I}_\text{B}, \text{II}_\text{B})$.

15.4.4 Compressible Materials

The material volume change or the compressibility is described by the third invariant J. Therefore, any restrictions or limitations to the compressibility can be described by restrictions on J.

We note the relation between J and the invariants of $\mathbf{H}$:

$$J = \det(\mathbf{F}) = \det(\mathbf{I} + \mathbf{H}) = 1 + \mathrm{I_H} + \mathrm{II_H} + \mathrm{III_H}. \tag{15.71}$$

Since $J = 1$ describes the incompressible behavior, the compressibility of the material is described by the remaining part $\mathrm{I_H} + \mathrm{II_H} + \mathrm{III_H}$ of the Eq. (15.71). Based on this, an approximation for J for constrained compressible materials can be written as given in Gummert (1990) (renaming Δ with ϕ)

$$J = 1 + \phi \qquad \text{or} \qquad \mathrm{III_B} = J^2 = 1 + \varphi. \tag{15.72}$$

The functions ϕ and φ describe the deviation from the isochoric motion, allowing for compressible behavior. For $\phi \to 0$ we obtain the internal constraint $J = 1$ or alternatively $\varphi \to 0$ implies $\mathrm{III_B} = 1$ for incompressible materials.

Rewriting Eq. $(15.72)_2$, the internal constraint for compressible materials can be written as

$$\tilde{\gamma} = \mathrm{III_B} - 1 - \varphi = 0. \tag{15.73}$$

In the following subsections, the internal constraints in the form of Eq. (15.73) will be the primary focus.

15.4.5 Compressibility at Infinitesimal Strains

To describe the material volume change within a consistent linear theory, the same linearisation of Eq. (15.23) is applied to the determinant J as represented in Eq. (15.71). Neglecting second order ($\mathrm{II_H}$) and third order ($\mathrm{III_H}$) terms with respect to the first order term ($\mathrm{I_H}$) and considering $\mathrm{lin}(\mathrm{I_H}) \equiv \mathrm{I_E}$, we obtain the linearised determinant

$$\mathrm{lin}(J) = 1 + \mathrm{I_E}. \tag{15.74}$$

By calculating $\mathrm{I_E} = (1 - 2\nu)\,\varepsilon_1$ from Eq. (15.32) and substituting ε_1 with $\lambda_1 - 1$ as given by Eq. (15.27), we derive the following expression for small deformations:

$$\mathrm{lin}(J) = 1 + (1 - 2\nu)\,(\lambda_1 - 1). \tag{15.75}$$

As expected from a consistent linear theory, J becomes a linear function of λ_1 with a constant slope $1 - 2\nu$. This allows for physically valid values of the Poisson's ratio to be in the range of $0 \leq \nu \leq 1/2$.

This means that a constant value ν alone is the measure for the compressibility of the material for small deformations; having values from fully compressible material with $\nu = 0$ to incompressible material with $\nu = 1/2$.

On the other hand, the allowed values of ν also restrict the slope of the linear function: $0 \leq 1 - 2\nu \leq 1$.

Since ν can not be a function of λ_1 because of the lineraity of J in Eq. (15.75), the formulation of a variable compressibility (a nonlinear J) for infinitesimal strains is not possible.

For incompressible materials, considering $\text{lin}(J) = 1$ in Eq. (15.74), we obtain the internal constraint for infinitesimal strains:

$$\text{I}_\text{E} = \text{tr}(\mathbf{E}) = 0 \quad \Leftrightarrow \quad \varepsilon_1 + \varepsilon_2 + \varepsilon_3 = 0. \tag{15.76}$$

15.5 Limited Compressible Materials

With the term *limited compressibility* we note the limitation of volume change J, that for a certain range of initial deformations the material behaves compressible (J is variable) and for larger deformations it approximates to incompressible behavior (J is constant or nearly constant).

Nearly incompressible or slightly compressible materials can be considered as a special case (or subgroup) of limited compressible materials, whose deviation from the incompressible behavior is very small. This behavior is discussed in Horgan and Murphy (2009b), also using the term limited compressibility.

In addition, for the limited compressibility, it is required that the formulation for finite strains must be compatible with the infinitesimal strains.

15.5.1 *Formulation of the Constitutive Equations*

For the mathematical formulation, we assume that the internal constraints for limited compressible materials has the form of Eq. (15.73). We suggest the general formulation

$$\tilde{\gamma}\,(\text{I}_\text{B}, \text{II}_\text{B}, \text{III}_\text{B}) = \text{III}_\text{B} - 1 - \varphi(\text{I}_\text{B}, \text{II}_\text{B}) = 0, \tag{15.77}$$

or with $\text{III}_\text{B} \equiv J^2$

$$J^2 = \text{III}_\text{B} = 1 + \varphi(\text{I}_\text{B}, \text{II}_\text{B}), \tag{15.78}$$

or J as a function of φ:

$$J(\varphi) = \sqrt{1 + \varphi(\text{I}_\text{B}, \text{II}_\text{B})}\,. \tag{15.79}$$

According to Eq. (15.60), the limited compressible (lc) Cauchy reaction stress tensor can be written as

$$\mathbf{S}_R^{\text{lc}} = \frac{2\alpha}{\sqrt{1+\varphi}} \left[\frac{\partial \text{III}_\text{B}}{\partial \mathbf{B}} - \frac{\partial \varphi(\text{I}_\text{B}, \text{II}_\text{B})}{\partial \mathbf{B}} \right]_{J=\sqrt{1+\varphi}} \cdot \mathbf{B}. \tag{15.80}$$

By using the chain rule the equation yields

$$\mathbf{S}_R^{\text{lc}} = \frac{2\alpha}{\sqrt{1+\varphi}} \left[\frac{\partial \text{III}_\text{B}}{\partial \mathbf{B}} - \frac{\partial \varphi}{\partial \text{I}_\text{B}} \frac{\partial \text{I}_\text{B}}{\partial \mathbf{B}} - \frac{\partial \varphi}{\partial \text{II}_\text{B}} \frac{\partial \text{II}_\text{B}}{\partial \mathbf{B}} \right]_{J=\sqrt{1+\varphi}} \cdot \mathbf{B}. \tag{15.81}$$

By using the identities from Eq. (15.43) we obtain

$$\mathbf{S}_R^{\text{lc}} = \frac{2\alpha}{\sqrt{1+\varphi}} \left[\text{III}_\text{B}\, \mathbf{B}^{-1} - \frac{\partial \varphi}{\partial \text{I}_\text{B}} \mathbf{I} - \frac{\partial \varphi}{\partial \text{II}_\text{B}} (\text{I}_\text{B}\, \mathbf{I} - \mathbf{B}) \right]_{J=\sqrt{1+\varphi}} \cdot \mathbf{B} \tag{15.82}$$

$$= \frac{2\alpha}{\sqrt{1+\varphi}} \left[(1+\varphi)\, \mathbf{B}^{-1} \cdot \mathbf{B} - \frac{\partial \varphi}{\partial \text{I}_\text{B}} \mathbf{B} - \frac{\partial \varphi}{\partial \text{II}_\text{B}} (\text{I}_\text{B}\, \mathbf{I} \cdot \mathbf{B} - \mathbf{B} \cdot \mathbf{B}) \right] \tag{15.83}$$

Rewriting the equation, we obtain further

$$\mathbf{S}_R^{\text{lc}} = \left(2\alpha \sqrt{1+\varphi}\right) \mathbf{I} - \frac{2\alpha}{\sqrt{1+\varphi}} \left(\frac{\partial \varphi}{\partial \text{I}_\text{B}} + \frac{\partial \varphi}{\partial \text{II}_\text{B}} \text{I}_\text{B} \right) \mathbf{B} + \frac{2\alpha}{\sqrt{1+\varphi}} \frac{\partial \varphi}{\partial \text{II}_\text{B}} \mathbf{B}^2 . \tag{15.84}$$

Considering the general form of the constitutive law in Eq. (15.51) and the reaction stresses in Eq. (15.84), the Cauchy stresses can be expressed as

$$\begin{aligned} \mathbf{S}^{\text{lc}} &= \mathcal{F}(\mathbf{B}) + \mathbf{S}_R^{\text{lc}} \\ &= \left[\psi_0 + 2\alpha \sqrt{1+\varphi} \right] \mathbf{I} + \left[\psi_1 - \frac{2\alpha}{\sqrt{1+\varphi}} \left(\frac{\partial \varphi}{\partial \text{I}_\text{B}} + \frac{\partial \varphi}{\partial \text{II}_\text{B}} \text{I}_\text{B} \right) \right] \mathbf{B} + \\ &\quad + \left[\psi_2 + \frac{2\alpha}{\sqrt{1+\varphi}} \left(\frac{\partial \varphi}{\partial \text{II}_\text{B}} \right) \right] \mathbf{B}^2 . \end{aligned} \tag{15.85}$$

One can easily see in Eq. (15.85) that for $\varphi = 0$ we obtain the incompressible material equation, namely Eq. (15.70). The terms including the derivatives of φ describe the extension of the incompressible material equation to limited compressible behavior.

Considering the special case, in which the constraint $\tilde{\gamma}$ does not depend on II_B, but on I_B and III_B, yields:

$$\tilde{\gamma}\,(\text{I}_\text{B}, \text{III}_\text{B}) - \text{III}_\text{B} - 1 - \varphi(\text{I}_\text{B}) - 0 . \tag{15.86}$$

Equation (15.84) simplifies to

$$\mathbf{S}_R^{\text{lc}} = \left[2\alpha \sqrt{1+\varphi} \right] \mathbf{I} - 2\alpha \frac{\varphi'(\text{I}_\text{B})}{\sqrt{1+\varphi}} \mathbf{B} \tag{15.87}$$

with the abbreviation

$$\varphi' = \varphi'(\text{I}_\text{B}) := \frac{\mathrm{d}\varphi(\text{I}_\text{B})}{\mathrm{d}\text{I}_\text{B}} , \tag{15.88}$$

and the Cauchy stresses of Eq. (15.85) results to

$$\mathbf{S}^{\mathrm{lc}} = \left[\psi_0 + 2\alpha\sqrt{1+\varphi}\right]\mathbf{I} + \left[\psi_1 - 2\alpha\frac{\varphi'(\mathrm{I_B})}{\sqrt{1+\varphi}}\right]\mathbf{B} + \psi_2\mathbf{B}^2 . \tag{15.89}$$

As expected we receive a simpler equation and in comparison with the incompressible material equation, only one additional stress term appears in Eq. (15.89), depending on the function φ and its derivative φ'.

In analogy, assuming if the constraint $\tilde{\gamma}$ is not depending on $\mathrm{I_B}$, but on $\mathrm{II_B}$ and $\mathrm{III_B}$, namely

$$\tilde{\gamma}\,(\mathrm{II_B}, \mathrm{III_B}) = \mathrm{III_B} - 1 - \varphi(\mathrm{II_B}) = 0, \tag{15.90}$$

the derivative $\partial\varphi/\partial\mathrm{I_B}$ disappears in Eqs. (15.84) and (15.85) and we obtain

$$\begin{aligned}\mathbf{S}^{\mathrm{lc}} = \left[\psi_0 + 2\alpha\sqrt{1+\varphi}\right]\mathbf{I} + \left[\psi_1 - \frac{2\alpha}{\sqrt{1+\varphi}}\left(\frac{\mathrm{d}\varphi}{\mathrm{dII_B}}\,\mathrm{I_B}\right)\right]\mathbf{B} +\\ + \left[\psi_2 + \frac{2\alpha}{\sqrt{1+\varphi}}\left(\frac{\mathrm{d}\varphi}{\mathrm{dII_B}}\right)\right]\mathbf{B}^2 .\end{aligned} \tag{15.91}$$

15.5.2 Formulation of the Ansatz Function φ

After having the constitutive law, Eq. (15.85) with the internal constraint in the form of Eq. (15.77), we focus on the function φ, which describes compressibility. For this reason, we will seek functions that reflect the expected physical behavior. Accordingly, the ansatz functions must meet the following conditions:

- The motion is initially compressible and the determinant J (or $\mathrm{III_B} = J^2$) varies for initial values of deformations. But for larger (or finite) deformations, the determinant J approximates a constant value, which is the limit of compressible behavior. The functions shown in Fig. 15.1 serve as examples of these types of functions.
- The derivatives of the ansatz functions must exist, as they are used in Eq. (15.85).
- The ansatz functions should be consistent with the linear theory. Consequently, as discussed in Sect. 15.4.5, the slope of the functions in the undeformed state should remain within the range [0, 1]:

$$0 \leq \left.\frac{\mathrm{d}\varphi(\mathrm{I_B})}{\mathrm{dI_B}}\right|_{\lambda_i=1} \leq 1 \quad \text{or} \quad 0 \leq \left.\frac{\mathrm{d}\varphi(\lambda_1)}{\mathrm{d}\lambda_1}\right|_{\lambda_1=1} \leq 1. \tag{15.92}$$

- As a consequence of the requirement in Eq. (15.92), considering J (or J^2) as a function of λ_1 as shown in Fig. 15.1, the limited compressible values of J in the tensile state ($\lambda_1 > 1$) should exceed the incompressible value $J = 1$, while in the compression state ($\lambda_1 < 1$), they should be smaller. This condition is essential to ensure physically meaningful results.
- As another consequence of the requirement in Eq. (15.92), considering λ_2 as a function of λ_1, as shown in Fig. 15.2, the limited compressible λ_2 values in the

tensile state must exceed the incompressible λ_2 values, while in the compression state, they must be smaller. This condition is essential to ensure the compatibility with the linear theory and to produce physically meaningful results.

To obtain the relevant equations, we prefer an inductive method; first using a one dimensional scalar equation for a simple plausible formulation, than generalisation to three dimensional equation using tensor notation. The following steps are required:

- Choosing the variable list of φ; $\varphi(\mathrm{I_B})$, $\varphi(\mathrm{II_B})$ or $\varphi(\mathrm{I_B}, \mathrm{II_B})$, which leads to Eqs. (15.85), (15.89) or (15.91) for further calculations. Then we note one dimensional equivalents of the chosen constraint and the material equation.
- Since $\mathbf{B}$ is quadratic in λ_i, we suggest a uniaxial (one dimensional) kinematic relation $\gamma(\lambda_i^2) = 0$ including an ansatz function for $\varphi(\lambda_i^2)$, which is compatible with the linear theory and can meet other physical requirements.
- Modification of the uniaxial kinematic relation including the ansatz function to a multiaxial (three dimensional) form of internal constraint using tensor variables, namely the deformation tensor $\mathbf{B}$ itself, instead of its components λ_i, in the form of Eq. (15.77).

Choosing the variable list of φ
We choose Eqs. (15.86) and the (15.89). By replacing $\mathrm{I_B}$ with λ_1^2, we receive the following one dimensional relation for limited compressibility for uniaxial extension

$$\tilde{\gamma}\,(\lambda_1^2, \mathrm{III_B}) = \mathrm{III_B} - 1 - \varphi(\lambda_1^2 - 1) = 0 \tag{15.93}$$

with $\mathrm{III_B} = \lambda_1^2\lambda_2^4$. For the undeformed state where $t = 0$, we note $\lambda_i^2(t = 0) = 1$, $\mathrm{III_B}(t = 0) = 1$ and $\varphi(t = 0) = 0$.

Suggesting Ansatz Functions for φ
The simplest ansatz, a constant $\varphi = \varphi_0 = 0$ substituted in Eq. (15.93) results as expected the restriction for the isochoric motion $\mathrm{III_B} - 1 = 0$.

We suggest three different ansatz functions that reflect our expectations of physical behavior. φ_1: a hyperbolic tangent function. φ_2: an arc tangent function. φ_3: a sigmoid function.

$$\begin{aligned}
\varphi_1(\lambda_1^2) &:= \delta_1\ \tanh\left[\beta_1\,(\lambda_1^2 - 1)\right], \\
\varphi_2(\lambda_1^2) &:= \delta_2\ \arctan\left[\beta_2\,(\lambda_1^2 - 1)\right], \\
\varphi_3(\lambda_1^2) &:= \frac{\delta_3}{1 + \zeta_3\ \exp[\beta_3\,(1 - \lambda_1^2)]} - \frac{\delta_3}{1 + \zeta_3},
\end{aligned} \tag{15.94}$$

where δ_i, β_i and ζ_3 are constants. Figure 15.1 shows the plots of $\mathrm{III_B} = 1 + \varphi_i$ over λ_1 for the chosen constants:

$$\{\delta_1 = 0.5\,, \beta_1 = 0.5\}\,; \{\delta_2 = 0.3\,, \beta_2 = 0.5\}\,; \{\delta_3 = 0.4\,, \beta_3 = 0.4\,, \zeta_3 = 1.5\}. \tag{15.95}$$

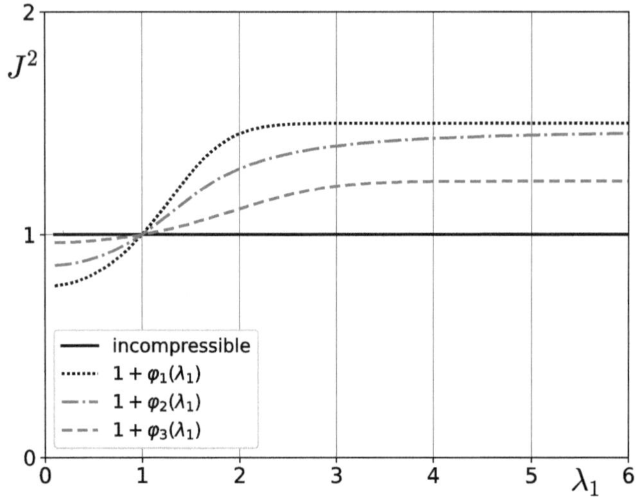

Fig. 15.1 Suggested ansatz functions $\mathrm{III_B} \equiv J^2 = 1 + \varphi_i(\lambda_1)$ and comparison with the incompressible behavior $\mathrm{III_B} \equiv J^2 = 1$

These constants all have the same unit, namely 1.

As seen in the plots, the functions φ_i fulfill the requirement, that the motion is initially compressible, but for larger deformations the determinant $\mathrm{III_B} = J^2$ approximates to a constant value, to the limit of the compressible behavior. The functions φ_i have limited deviations from the incompressible value $\mathrm{III_B} = 1$. For bigger values of λ_1, the functions approximate to constant values which are adjustable by varying δ_i. The constants β_i are related with the slope of the functions φ_i at the undeformed state ($\lambda_1 = 1$, $J^2 = 1$) of Fig. 15.1.

To compare the functions φ_i with Eq. (15.66) of isochoric behavior, we use the relation for uniaxial extension $\mathrm{III_B} = \lambda_1^2 \lambda_2^4$, rearrange Eq. (15.93) and obtain for limited compressible (lc) materials

$$\lambda_2^{\mathrm{lc}} = \sqrt{\lambda_1^{-1} \sqrt{1 + \varphi_i(\lambda_1^2)}}. \tag{15.96}$$

As one can see in Fig. 15.2, for the same values of δ_i, β_i and ζ_3, as expected, we obtain a slight deviation from the incompressible response λ_2^{in}. The curves also reflect the condition that the incompressible stretch in transverse direction λ_2^{in} has smaller values than the limited compressible stretch λ_2^{lc} in the tensile state ($\lambda_1 > 1$) and vice versa, a bigger values in the compression state ($\lambda_1 < 1$).

An additional requirement for the functions $\varphi_i(\lambda_1^2)$ from the perspective of Eq. (15.84) is that the derivatives of φ_i respectively its argument λ_1^2 exist.

To verify whether the requirement from Eq. (15.92) is satisfied, we consider the derivatives of φ_i:

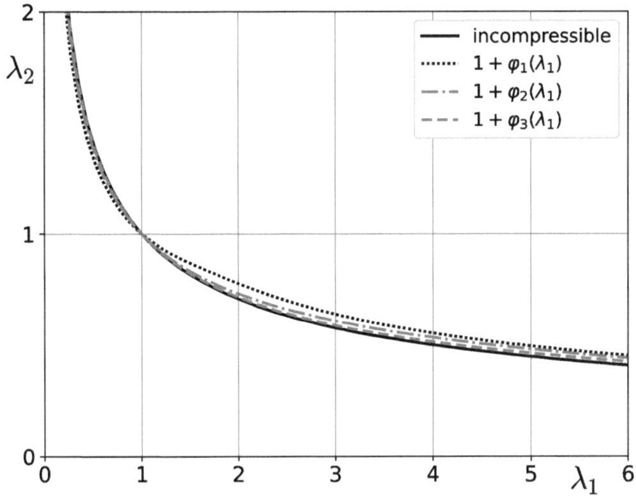

Fig. 15.2 Comparison of the functions $\lambda_2^{\mathrm{lc}} = \sqrt{\lambda_1^{-1}\sqrt{1+\varphi_i\,(\lambda_1^2)}}$ with the incompressible behavior $\lambda_2^{\mathrm{in}} = \sqrt{\lambda_1^{-1}}$

$$\begin{aligned}
\varphi_1'(\lambda_1^2) &\equiv \frac{\mathrm{d}\varphi_1(\lambda_1^2)}{\mathrm{d}\lambda_1^2} = \frac{\delta_1\,\beta_1}{\cosh^2\left[\beta_1\,(\lambda_1^2-1)\right]}\,,\\
\varphi_2'(\lambda_1^2) &\equiv \frac{\mathrm{d}\varphi_2(\lambda_1^2)}{\mathrm{d}\lambda_1^2} = \frac{\delta_2\,\beta_2}{1+\left[\beta_2\,(\lambda_1^2-1)\right]^2}\,,\\
\varphi_3'(\lambda_1^2) &\equiv \frac{\mathrm{d}\varphi_1(\lambda_1^2)}{\mathrm{d}\lambda_1^2} = \frac{\delta_3\,\beta_3\,\zeta_3\,\exp[\beta_3\,(1-\lambda_1^2)]}{\{1+\zeta_3\,\exp[\beta_3\,(1-\lambda_1^2)]\}^2}\,.
\end{aligned} \tag{15.97}$$

The slopes of the functions φ_i at the undeformed state ($t = 0$) are the derivatives φ_1' at $\lambda_i = 1$:

$$\varphi_1'(1) = \delta_1\,\beta_1\,, \qquad \varphi_2'(1) - \delta_2\,\beta_2\,, \qquad \varphi_3'(1) = \frac{\delta_3\,\beta_3\,\zeta_3}{\{1+\zeta_3\}^2}. \tag{15.98}$$

Applying the constants from Eq. (15.95) we obtain

$$\varphi_1'(1) = 0.25\,, \qquad \varphi_2'(1) = 0.15\,, \qquad \varphi_3'(1) = 0.0384. \tag{15.99}$$

The values are within the range [0, 1] and satisfy the condition in Eq. (15.92).

Uniaxial Response

For limited compressible (lc) materials we use Eq. (15.87) and replace $\mathrm{I_B}$ with λ_1^2, the principle Cauchy reaction stress in the tensile direction results to

$$\sigma_{R,1}^{\text{lc}} = 2\alpha\sqrt{1+\varphi(\lambda_1^2)} - 2\alpha\,\frac{\varphi'(\lambda_1^2)}{\sqrt{1+\varphi(\lambda_1^2)}}\,\lambda_1^2. \tag{15.100}$$

With the derivatives of the functions from Eq. (15.97) we obtain the Cauchy principle reaction stress σ_1^{lc} for limited compressible materials (no sum over $i = 1, 2, 3$):

$$\sigma_{R,1}^{\text{lc}} = 2\alpha\sqrt{1+\varphi(\lambda_1^2)} - \frac{2\alpha}{\sqrt{1+\varphi(\lambda_1^2)}}\,\frac{\delta_i\,\beta_i}{\xi_i(\lambda_1^2)}\,\lambda_1^2 \tag{15.101}$$

with the functions $\xi_i(\lambda_1^2)$

$$\begin{aligned}
\xi_1(\lambda_1^2) &:= \cosh^2\left[\beta_1\,(\lambda_1^2-1)\right],\\
\xi_2(\lambda_1^2) &:= 1+\left[\beta_2\,(\lambda_1^2-1)\right]^2,\\
\xi_3(\lambda_1^2) &:= \frac{\{1+\zeta_3\,\exp[\beta_3\,(1-\lambda_1^2)]\}^2}{\zeta_3\,\exp[\beta_3\,(1-\lambda_1^2)]}.
\end{aligned} \tag{15.102}$$

As expected, by considering $\delta_i \to 0$ or $\varphi_i \to 0$ in Eq. (15.101), resulting the Cauchy principle reaction stress $\sigma_{R,1}^{\text{lc}}(\varphi_i) \to \sigma_{R,1}^{\text{in}} = 2\alpha$, the incompressible behavior in Eq. (15.69) is obtained.

Generalisation to Tensor Variables

In order to describe a complete deformation state in Eq. (15.93), on the one hand the argument of $\varphi_i(\lambda_1^2)$ must be extended with the terms λ_2^2 and λ_3^2, considering the tensor $\mathbf{B}$; on the other hand the principle of material objectivity must be ensured in the form $\varphi_i^* = \varphi_i$ or as in Eq. (15.50). Since φ_i must be isotropic tensor functions, the first invariant $\mathrm{I_B} := \mathrm{tr}(\mathbf{B}) = \lambda_1^2 + \lambda_2^2 + \lambda_3^2$ will be used as the argument, instead of λ_1^2.

With $\mathbf{B}_0 = \mathbf{I}$, $\det(\mathbf{B}_0) = 1$, $\mathrm{tr}(\mathbf{B}_0) = \mathrm{tr}(\mathbf{I}) = 3$ and $\tilde{\gamma}_0 = \tilde{\gamma}\,(\mathbf{B}_0) = 0$ at the undeformed configuration, Eq. (15.93) can be written as:

$$\tilde{\gamma}\,(\mathrm{I_B}, \mathrm{III_B}) = \mathrm{III_B} - 1 - \varphi(\mathrm{I_B} - 3) = 0 \tag{15.103}$$

with $\varphi(0) = 0$. Obviously, Eq. (15.103) is a simpler form of Eq. (15.77). This equation can be rearranged to the alternative form:

$$\mathrm{III_B} = 1 + \varphi(\mathrm{I_B} - 3). \tag{15.104}$$

In analogy to Eq. (15.94), the extended ansatz functions are chosen as:

$$\begin{aligned}
\varphi_1(\mathrm{I_B}) &= \delta_1\,\tanh\left[\beta_1\,(\mathrm{I_B}-3)\right],\\
\varphi_2(\mathrm{I_B}) &= \delta_2\,\arctan\left[\beta_2\,(\mathrm{I_B}-3)\right],
\end{aligned} \tag{15.105}$$

$$\varphi_3(\mathrm{I_B}) = \frac{\delta_3}{1+\zeta_3\,\exp[\beta_3\,(3-\mathrm{I_B})]} - \frac{\delta_3}{1+\zeta_3}.$$

According to Eq. (15.88), the derivatives φ_i' are

$$\begin{aligned}\varphi_i'(\mathrm{I_B}) &= \frac{\delta_i\,\beta_i}{\xi_i(\mathrm{I_B})} \quad \text{(no sum over } i\text{)} \quad \text{with} \\ \xi_1(\mathrm{I_B}) &:= \cosh^2[\beta_1\,(\mathrm{I_B}-3)], \\ \xi_2(\mathrm{I_B}) &:= 1+[\beta_2\,(\mathrm{I_B}-3)]^2, \\ \xi_3(\mathrm{I_B}) &:= \frac{\{1+\zeta_3\,\exp[\beta_3\,(3-\mathrm{I_B})]\}^2}{\zeta_3\,\exp[\beta_3\,(3-\mathrm{I_B})]}.\end{aligned} \tag{15.106}$$

The related limited compressible constitutive equation is given in Eq. (15.89). The ansatz functions φ_i and the derivatives φ_i' are given by Eqs. (15.105) and (15.106).

15.6 Parameter Identification

The constants of the ansatz functions φ_i can be determined by mechanical tests, where the complete strain state **B** is measured during the motion. The resulting plots J or $\mathrm{III_B}$ over λ_i help us to analyse the level of the compressibility and also give us the possibility to determine the constants using curve fitting methods.

Another approach is to assume the level of compressibility, such as nearly incompressible or limited compressible behavior and choosing suitable ansatz functions with suitable constants, which show physically plausible results in the plots J or $\mathrm{III_B}$ over λ_i.

On the other hand, Eq. (15.85) has also material functions ψ_i with different constants depending on the choice of these functions. These constants are determined from the results of mechanical tests, where the stresses and strains are measured.

Within the scope of this work a set of suitable constants of φ_i are suggested and the constants of constitutive equation ψ_i are calculated by using curve fitting methods. The chosen constants of φ_i are manually varied for better results of the curve fitting.

For reference, the experimental values of the uniaxial extension test from Treloar's widely used work (Treloar 1944) is preferred. The material is 8% S rubber at 20°C. The tensile Cauchy stress σ_1 is given in MPa, the corresponding plot is given in Fig. 15.3 by the thick dots.

In the examples below, the uniaxial extension state is applied, which requires $\lambda_2 = \lambda_3$ and $\sigma_2 = \sigma_3 = 0$ for isotropic materials with

$$\mathrm{III_B} = \lambda_1^2\lambda_2^4 = 1+\varphi_i(\mathrm{I_B}) \quad \text{and} \quad \mathrm{I_B} = \lambda_1^2 + 2\,\lambda_2^2. \tag{15.107}$$

Since we use exponential or trigonometric functions φ_i with λ_2 in the arguments, it is not possible to rearrange the resulting transcendental equations to λ_2 as in

Eq. (15.96). This is possible with the replacement

$$\varphi(\mathrm{I_B}) \Leftrightarrow \varphi(\lambda_1^2) \quad \text{and} \quad \lambda_2(\lambda_1) = \sqrt{\lambda_1^{-1}\sqrt{1+\varphi_i(\lambda_1^2)}}, \tag{15.108}$$

which can be interpreted as an assumption. Since the concept of the formulation of the limited compressibility is an approximation and based on various assumptions, an additional assumption will be this "switch" between $\mathrm{I_B}$ and λ_1^2 in the arguments of the transcendental equations.

15.6.1 Example 1

From the mathematical aspect, the simplest assumption for material functions ψ_i are constant values: $\psi_i = c_i$. Thus Eq. (15.89), by using Eq. (15.105), can be written as:

$$\mathbf{S}^{\mathrm{lc}} = \left[c_0 + 2\alpha\sqrt{1+\varphi}\right]\mathbf{I} + \left[c_1 - 2\alpha\frac{\varphi'(\mathrm{I_B})}{\sqrt{1+\varphi}}\right]\mathbf{B} + c_2\,\mathbf{B}^2. \tag{15.109}$$

The stress state of the uniaxial extension of an isotropic solid is given with $\sigma_1 \neq 0$, $\sigma_2 = \sigma_3 = 0$. The related deformation state is already noted at Eq. (15.33). The components of Eq. (15.109) can be written as:

$$\sigma_1^{\mathrm{lc}} = c_0 + 2\alpha\sqrt{1+\varphi} + \left[c_1 - \frac{2\alpha\varphi'}{\sqrt{1+\varphi}}\right]\lambda_1^2 + c_2\,\lambda_1^4, \tag{15.110}$$

$$\sigma_2^{\mathrm{lc}} = c_0 + 2\alpha\sqrt{1+\varphi} + \left[c_1 - \frac{2\alpha\varphi'}{\sqrt{1+\varphi}}\right]\lambda_2^2 + c_2\,\lambda_2^4. \tag{15.111}$$

Since $\sigma_2^{\mathrm{lc}} \equiv 0$, a short calculation, Eq. (15.111) delivers the Lagrange parameter:

$$2\alpha(\lambda_1, \lambda_2) = -\frac{c_0 + c_1\,\lambda_2^2 + c_2\,\lambda_2^4}{\sqrt{1+\varphi} - \frac{\varphi'}{\sqrt{1+\varphi}}\,\lambda_2^2}. \tag{15.112}$$

Using the assumption of Eq. (15.108) and eliminating λ_2, we obtain

$$2\alpha(\lambda_1) = -\frac{c_0 + c_1\,\lambda_1^{-1}\sqrt{1+\varphi} + c_2\,\lambda_1^{-2}\,(1+\varphi)}{\sqrt{1+\varphi} - \lambda_1^{-1}\,\varphi'}. \tag{15.113}$$

Replacing 2α in Eq. (15.110), we obtain the principal Cauchy stress in the tensile direction.

With a curve-fitting method, the constants c_i can be determined from an uniaxial extension test with measured values σ_1 and λ_1.

Table 15.1 Constants of the plots belonging to Eq. (15.110). δ_i, β_i, ζ_3 have the same unit 1 (dimensionless), c_0, c_1, c_2 have the same unit (MPa) as the stress σ

	δ_i	β_i	ζ_3	c_0	c_1	c_2
φ_1	0.30	0.10		5.52	0.018	0.0015
φ_2	0.30	0.07		8.80	−0.030	0.0022
φ_3	0.30	0.10	0.3	34.67	−0.013	0.0020

The constants c_i are calculated by using Python's *curve_fit()* function (based on the non-linear least squares method) with the manually adjusted constants δ_i, β_i and ζ_3. All are shown at Table 15.1 and the results can be seen in Fig. 15.3.

As shown in Fig. 15.3, all three functions provide similar and acceptable results. It is evident that the hyperbolic tangent function φ_1 and the sigmoid function φ_3 deliver almost the same curve which has a better fit compared to the curve of the arc tangent function φ_2. The incompressible equation with $\varphi_i = 0$, $\varphi_i' = 0$ and constants c_i, which is also shown in the figure, can not provide satisfactory results because of its parabolic form.

To verify the requirement from Eq. (15.92), we calculate $\varphi_i'(1)$ using Eq. (15.98) and observe that the values lie within the range [0, 1], satisfying the condition in Eq. (15.92). As a consequence, we see at Figs. 15.4 and 15.5 that all three functions φ_i are within the expected values.

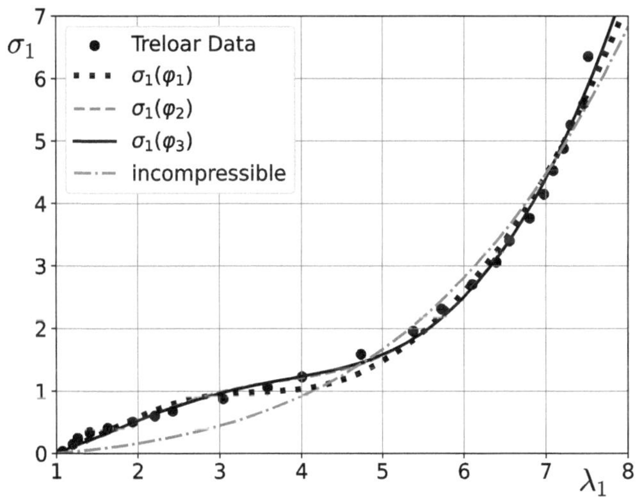

Fig. 15.3 Plots of Eq. (15.110) and the comparison with the reference data measured by Treloar (1944). σ_1 is given in MPa

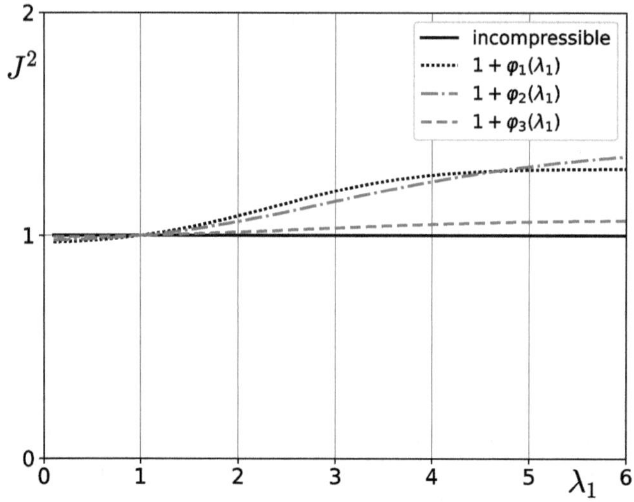

Fig. 15.4 Plots of the ansatz functions from Eq. (15.94) with the constants from Table 15.1 and comparison with the incompressible behavior $J^2 = 1$

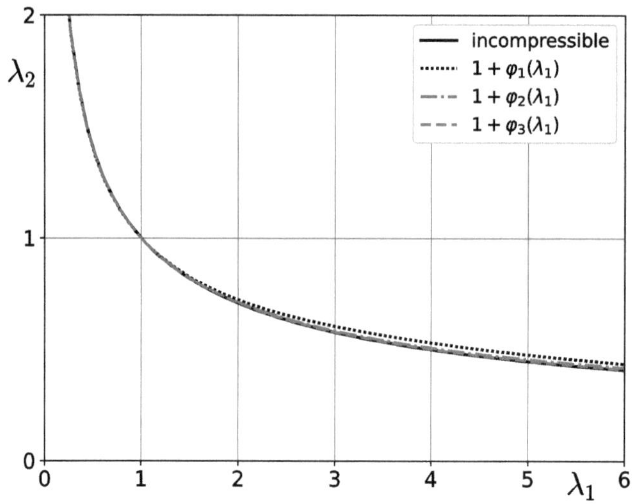

Fig. 15.5 Plots of the relation $\lambda_2^{\text{lc}} = \sqrt{\lambda_1^{-1}\sqrt{1+\varphi_i(\lambda_1^2)}}$ with the constants from Table 15.1 and comparison with the incompressible behavior $\lambda_2^{\text{in}} = \sqrt{1/\lambda_1}$

15.7 Green Elasticity (Hyperelasticity)

A hyperelastic material (or Green-elastic material) is a material whose stress state can be obtained by the derivation of a potential function $W(\mathbf{F})$, a scalar-valued isotropic tensor function, the so-called strain energy density function or elastic potential with the relation

$$\mathbf{P} = \frac{\partial W}{\partial \mathbf{F}}. \tag{15.114}$$

$\mathbf{P}$ denotes the 1st Piola-Kirchhoff stress tensor from Eq. (15.46). The Cauchy stress tensor $\mathbf{S}$ can be calculated with

$$\mathbf{S} = \frac{1}{J}\,\mathbf{P}\cdot\mathbf{F}^T = \frac{1}{J}\frac{\partial W}{\partial \mathbf{F}}\cdot\mathbf{F}^T. \tag{15.115}$$

Alternatively, the hyperelastic constitutive equation can be formulated by using $\mathbf{B}$ instead of $\mathbf{F}$, i.e. $\tilde{W}(\mathbf{B})$ instead of $W(\mathbf{F})$:

$$\mathbf{P} = 2\,\frac{\partial \tilde{W}}{\partial \mathbf{B}}\cdot\mathbf{F} \quad \text{and} \quad \mathbf{S} = \frac{2}{J}\frac{\partial \tilde{W}}{\partial \mathbf{B}}\cdot\mathbf{B}. \tag{15.116}$$

To ensure the principle of material frame indifference, the invariants of $\mathbf{B} := \mathbf{F}\cdot\mathbf{F}^T$ will be used as variables:

$$\tilde{W}(\mathbf{B}) = \tilde{W}(\mathrm{I_B}, \mathrm{II_B}, \mathrm{III_B}). \tag{15.117}$$

The derivative of $\tilde{W}$ respectively $\mathbf{B}$ results to

$$\frac{\partial \tilde{W}}{\partial \mathbf{B}} = \frac{\partial \tilde{W}}{\partial \mathrm{I_B}}\frac{\partial \mathrm{I_B}}{\partial \mathbf{B}} + \frac{\partial \tilde{W}}{\partial \mathrm{II_B}}\frac{\partial \mathrm{II_B}}{\partial \mathbf{B}} + \frac{\partial \tilde{W}}{\partial \mathrm{III_B}}\frac{\partial \mathrm{III_B}}{\partial \mathbf{B}}. \tag{15.118}$$

Considering the identities of Eq. (15.43) and inserting Eqs. (15.118) in $(15.116)_2$, we obtain after a short calculation for the Cauchy stresses:

$$\mathbf{S} = 2\,J\,\frac{\partial \tilde{W}}{\partial \mathrm{III_B}}\,\mathbf{I} + \frac{2}{J}\left(\frac{\partial \tilde{W}}{\partial \mathrm{I_B}} + \mathrm{I_B}\,\frac{\partial \tilde{W}}{\partial \mathrm{II_B}}\right)\mathbf{B} - \frac{2}{J}\frac{\partial \tilde{W}}{\partial \mathrm{II_B}}\,\mathbf{B}^2. \tag{15.119}$$

Comparing this equation for Green-elastic materials with Eq. (15.51) for Cauchy-elastic materials, we obtain the relations

$$\psi_0 = 2\,J\,\frac{\partial \tilde{W}}{\partial \mathrm{III_B}}, \quad \psi_1 = \frac{2}{J}\left(\frac{\partial \tilde{W}}{\partial \mathrm{I_B}} + \mathrm{I_B}\,\frac{\partial \tilde{W}}{\partial \mathrm{II_B}}\right), \quad \psi_2 = -\frac{2}{J}\frac{\partial \tilde{W}}{\partial \mathrm{II_B}}. \tag{15.120}$$

Thus the mathematical focus shifts from the search of suitable material functions ψ_i to the search of a suitable strain energy function $\tilde{W}$. Therefore the main task in the practice of hyperelasticity is to choose the suitable potential function.

A list of elastic potentials for compressible and incompressible solids is given for example in Bertram (2012), Sect. 7.4. A comparative study of 85 different elastic potentials are discussed in He et al. (2022).

In the following sections, rather than discussing application examples of Eq. (15.119), we aim to derive a limited compressible extension of the widely used Mooney-Rivlin material law. First, we will formulate a constitutive equation for compressible solids. Then, by enforcing incompressibility, we will derive the Mooney-Rivlin material equation. Finally, by introducing hyperelasticity with internal constraints and applying limited compressible internal constraints, we will arrive at the targeted constitutive equation.

A widely established theory is the additive decomposition of $\tilde{W}$ into a distorsional (or isochoric) part $\overline{W}$ and a dilatational (or volumetric) part $W^{\odot}$ by using the decomposition given in Eq. (15.14) (see, e.g., Holzapfel (2000), Eq. 6.98):

$$\tilde{W}(\mathbf{B}) = \overline{W}(\overline{\mathbf{B}}) + W^{\odot}(J) \tag{15.121}$$

with

$$\mathbf{B} = \mathbf{F} \cdot \mathbf{F}^T = \overline{\mathbf{F}}\, J^{1/3} \cdot \overline{\mathbf{F}}^T\, J^{1/3} = \overline{\mathbf{F}} \cdot \overline{\mathbf{F}}^T\, J^{2/3} = \overline{\mathbf{B}}\, J^{2/3} \tag{15.122}$$

or $\overline{\mathbf{B}} := J^{-2/3}\,\mathbf{B}$ with its invariants

$$\bar{\mathrm{I}}_{\mathrm{B}} = J^{-2/3}\,\mathrm{I}_{\mathrm{B}}\,, \quad \overline{\mathrm{II}}_{\mathrm{B}} = J^{-4/3}\,\mathrm{II}_{\mathrm{B}}\,, \quad \overline{\mathrm{III}}_{\mathrm{B}} = 1. \tag{15.123}$$

Obviously, for incompressible materials with $J = 1$, we obtain

$$\bar{\mathrm{I}}_{\mathrm{B}} = \mathrm{I}_{\mathrm{B}}\,, \quad \overline{\mathrm{II}}_{\mathrm{B}} = \mathrm{II}_{\mathrm{B}}\,, \quad \overline{\mathrm{III}}_{\mathrm{B}} = \mathrm{III}_{\mathrm{B}} = 1 \quad \text{and} \quad \tilde{W}(\mathbf{B}^{\mathrm{in}}) = \overline{W}(\overline{\mathbf{B}}). \tag{15.124}$$

The Cauchy stresses according to Eq. (15.116) results to

$$\mathbf{S} = \frac{2}{J}\frac{\partial \tilde{W}}{\partial \mathbf{B}} \cdot \mathbf{B} = \frac{2}{J}\left[\frac{\partial \overline{W}}{\partial \bar{\mathrm{I}}_{\mathrm{B}}}\frac{\partial \bar{\mathrm{I}}_{\mathrm{B}}}{\partial \mathbf{B}} + \frac{\partial \overline{W}}{\partial \overline{\mathrm{II}}_{\mathrm{B}}}\frac{\partial \overline{\mathrm{II}}_{\mathrm{B}}}{\partial \mathbf{B}} + \frac{\partial W^{\odot}}{\partial J}\frac{\partial J}{\partial \mathbf{B}}\right] \cdot \mathbf{B}. \tag{15.125}$$

For the derivatives we have:

$$\begin{aligned}
\frac{\partial \bar{\mathrm{I}}_{\mathrm{B}}}{\partial \mathbf{B}} &= \frac{\partial}{\partial \mathbf{B}}\left[J^{-2/3}\,\mathrm{I}_{\mathrm{B}}\right] = J^{-2/3}\left[\mathbf{I} - \frac{1}{3}\,\mathrm{I}_{\mathrm{B}}\,\mathbf{B}^{-1}\right], \\
\frac{\partial \overline{\mathrm{II}}_{\mathrm{B}}}{\partial \mathbf{B}} &= \frac{\partial}{\partial \mathbf{B}}\left[J^{-4/3}\,\mathrm{II}_{\mathrm{B}}\right] = J^{-4/3}\left[\mathrm{I}_{\mathrm{B}}\mathbf{I} - \mathbf{B} - \frac{2}{3}\,\mathrm{II}_{\mathrm{B}}\,\mathbf{B}^{-1}\right], \\
\frac{\partial J}{\partial \mathbf{B}} &= \frac{1}{2}\,J\,\mathbf{B}^{-1}.
\end{aligned} \tag{15.126}$$

By substituting these identities into Eq. (15.125) and rearranging, we obtain a general hyperelastic constitutive equation for compressible materials:

$$\mathbf{S} = \left[-\frac{2}{3} J^{-5/3}\, \mathrm{I_B}\, \frac{\partial \overline{W}}{\partial \bar{\mathrm{I}}_\mathrm{B}} - \frac{4}{3} J^{-7/3}\, \mathrm{II_B}\, \frac{\partial \overline{W}}{\partial \overline{\mathrm{II}}_\mathrm{B}} + \frac{\partial W^{\odot}}{\partial J} \right] \mathbf{I} + \\ + 2\, J^{-5/3} \left[\frac{\partial \overline{W}}{\partial \bar{\mathrm{I}}_\mathrm{B}} + J^{-2/3}\, \mathrm{I_B}\, \frac{\partial \overline{W}}{\partial \overline{\mathrm{II}}_\mathrm{B}} \right] \mathbf{B} - 2\, J^{-7/3}\, \frac{\partial \overline{W}}{\partial \overline{\mathrm{II}}_\mathrm{B}}\, \mathbf{B}^2 \,, \quad (15.127)$$

which is comparable with Eq. (15.51).

To derive a more specific equation while aiming for the lowest level of nonlinearity in each term, we can require

$$\frac{\partial \overline{W}}{\partial \bar{\mathrm{I}}_\mathrm{B}} = \text{const.} \;, \quad \frac{\partial \overline{W}}{\partial \overline{\mathrm{II}}_\mathrm{B}} = \text{const.} \quad \text{and} \quad \frac{\partial W^{\odot}}{\partial J} \ \text{is linear in } J. \quad (15.128)$$

Referring to Horgan and Murphy (2009a), we adopt the following elastic potential which satisfies the requirements of Eq. (15.128):

$$\tilde{W} = \overline{W}(\bar{\mathrm{I}}_\mathrm{B}, \overline{\mathrm{II}}_\mathrm{B}) + W^{\odot}(J) = \frac{c_1}{2}(\bar{\mathrm{I}}_\mathrm{B} - 3) + \frac{c_2}{2}(\overline{\mathrm{II}}_\mathrm{B} - 3) + \frac{c_3}{6}\,(J-1)^2. \quad (15.129)$$

By using this ansatz in Eq. (15.127) we obtain

$$\mathbf{S} = -\frac{1}{3}\left[c_1\, J^{-5/3}\, \mathrm{I_B} - 2\, c_2\, J^{-7/3}\, \mathrm{II_B} - c_3(J-1)\right]\mathbf{I} + J^{-5/3}\left[c_1 + c_2\, J^{-2/3}\, \mathrm{I_B}\right]\mathbf{B} - \\ - c_2\, J^{-7/3} \mathbf{B}^2. \quad (15.130)$$

Eliminating $\mathbf{B}^2$ by means of the Cayley-Hamilton formula $\mathbf{B}^2 = \mathrm{I_B}\,\mathbf{B} - \mathrm{II_B}\,\mathbf{I} + \mathrm{III_B}\,\mathbf{B}^{-1}$ the equation yields to

$$\mathbf{S} = -\frac{1}{3}\left[c_1 J^{-5/3}\mathrm{I_B} \quad c_2 J^{-7/3}\mathrm{II_B} - c_3(J-1)\right]\mathbf{I} + c_1 J^{-5/3}\mathbf{B} - c_2 J^{-1/3}\mathbf{B}^{-1}. \quad (15.131)$$

Equation (15.131) is a possible constitutive equation for compressible elastic solids.

Mooney-Rivlin Material

For incompressible elastic materials with $J = 1$, considering Eqs. (15.124) in (15.129), we obtain the Mooney-Rivlin elastic potential:

$$\tilde{W} = \tilde{W}(\mathrm{I_B}, \mathrm{II_B}) = \frac{c_1}{2}(\mathrm{I_B} - 3) + \frac{c_2}{2}(\mathrm{II_B} - 3) \quad (15.132)$$

with the constants c_1 and c_2. Substituting $c_3 = 0$ and $J = 1$ in Eq. (15.131) we obtain

$$\mathbf{S}^{\mathrm{in}} = -\frac{1}{3}\,[c_1\, \mathrm{I_B} - c_2\, \mathrm{II_B}]\; \mathbf{I} + c_1\, \mathbf{B} - c_2\, \mathbf{B}^{-1}. \quad (15.133)$$

In the literature, the term $\frac{1}{3}(c_1\,\mathrm{I_B} - c_2\,\mathrm{II_B}) \equiv p$ in Eq. (15.133) is defined as the hydrostatic pressure. We can note the equation in the form

$$\mathbf{S}^{\text{in}} = -p\,\mathbf{I} + c_1\,\mathbf{B} - c_2\,\mathbf{B}^{-1}. \tag{15.134}$$

Equation (15.134) is known as the Mooney-Rivlin material model, which is widely used for isotropic incompressible elastic solids such as rubber-like materials.

Neo-Hookean Material

Assuming that the elastic potential depends only on the first invariant; $\tilde{W} = \tilde{W}(\mathrm{I_B})$, we obtain a simpler form, namely the Neo-Hookean elastic potential. By setting $c_2 = 0$ in Eq. (15.132) we note

$$\tilde{W} = \tilde{W}(\mathrm{I_B}) = c_1(\mathrm{I_B} - 3). \tag{15.135}$$

By using $c_2 = 0$ also in Eq. (15.133), we obtain the Cauchy stresses for Neo-Hookean materials

$$\mathbf{S}^{\text{in}} = -\frac{1}{3}\,c_1\,\mathrm{I_B}\,\mathbf{I} + c_1\,\mathbf{B} \tag{15.136}$$

which is linear in $\mathbf{B}$, which means quadratic in stretches λ_i, and still a geometrically non-linear material equation. Again, replacing $\frac{1}{3}\,c_1\,\mathrm{I_B} \equiv \tilde{p}$ as the hydrostatic pressure, we obtain

$$\mathbf{S}^{\text{in}} = -\tilde{p}\,\mathbf{I} + c_1\,\mathbf{B}. \tag{15.137}$$

15.7.1 Hyperelasticity with Internal Constraints

In case of hyperelasticity, in analogy to Eq. (15.61), the reaction stresses are added to the extra stresses. The constitutive equation is given as

$$\mathbf{P} = \mathbf{P}_E + \mathbf{P}_R = \frac{\partial W}{\partial \mathbf{F}} + \sum_{i=1}^{n} \alpha_i\,\frac{\partial \gamma_i}{\partial \mathbf{F}}. \tag{15.138}$$

According to Eq. (15.115) the Cauchy stresses results to

$$\mathbf{S} = \mathbf{S}_E + \mathbf{S}_R = \frac{1}{J}\,\frac{\partial W}{\partial \mathbf{F}} \cdot \mathbf{F}^T + \frac{1}{J}\left(\sum_{i=1}^{n} \alpha_i\,\frac{\partial \gamma_i}{\partial \mathbf{F}}\right) \cdot \mathbf{F}^T. \tag{15.139}$$

Alternatively, considering Eqs. $(15.116)_2$ and (15.61), we obtain

$$\mathbf{S} = \mathbf{S}_E + \mathbf{S}_R = \frac{2}{J}\,\frac{\partial \tilde{W}}{\partial \mathbf{B}} \cdot \mathbf{B} + \frac{2}{J}\left(\sum_{i=1}^{n} \alpha_i\,\frac{\partial \tilde{\gamma}_i}{\partial \mathbf{B}}\right) \cdot \mathbf{B}. \tag{15.140}$$

We note that the internal constraint is not in the strain energy function $\tilde{W}$ directly, but in the term $\mathbf{S}_R$. Indirectly, since the components of the stretch tensor $\mathbf{B}$ are in relation via $\tilde{\gamma}(\mathbf{B}) = 0$, the constraint is considered in all terms depending on $\mathbf{B}$.

For the incompressible hyperelastic materials with $J = 1$ and $J^2 = \mathrm{III_B} = 1$, according to Eqs. (15.140) and (15.69) we can note

$$\mathbf{S}^{\mathrm{in}} = 2\left.\frac{\partial \tilde{W}}{\partial \mathbf{B}}\right|_{J=1} \cdot \mathbf{B} + 2\alpha\,\mathbf{I}. \tag{15.141}$$

15.7.2 Limited Compressible Hyperelastic Materials

First, we note the hyperelastic constitutive equation for $\tilde{W}(\mathbf{B}) = \tilde{W}(\mathrm{I_B}, \mathrm{II_B}, \mathrm{III_B})$. Considering the internal constraint $J^2 = \mathrm{III_B} = 1 + \varphi(\mathrm{I_B}, \mathrm{II_B})$ and substituting $\mathbf{S}_R$ with the $\mathbf{S}_R^{\mathrm{lc}}$ from Eq. (15.84), and a coefficient comparison with Eq. (15.51), we again obtain Eq. (15.85) with the relations

$$\begin{aligned}
\psi_0 &= 2\sqrt{1+\varphi}\left.\frac{\partial \tilde{W}}{\partial \mathrm{III_B}}\right|_{1+\varphi},\\
\psi_1 &= \frac{2}{\sqrt{1+\varphi}}\left.\left(\frac{\partial \tilde{W}}{\partial \mathrm{I_B}} + \mathrm{I_B}\frac{\partial \tilde{W}}{\partial \mathrm{II_B}}\right)\right|_{J=\sqrt{1+\varphi}},\\
\psi_2 &= -\frac{2}{\sqrt{1+\varphi}}\left.\frac{\partial \tilde{W}}{\partial \mathrm{II_B}}\right|_{J=\sqrt{1+\varphi}}.
\end{aligned} \tag{15.142}$$

A simpler form of the constitutive equation we obtain by considering the constraint in the form $\mathrm{III_B} = 1 + \varphi(\mathrm{I_B})$. In analogy to Eq. (15.89), we can write

$$\begin{aligned}
\mathbf{S}^{\mathrm{lc}} &= 2\sqrt{1+\varphi}\left[\left.\frac{\partial \tilde{W}}{\partial \mathrm{III_B}}\right|_{1+\varphi} + \alpha\right]\mathbf{I} + \frac{2}{\sqrt{1+\varphi}}\left[\left.\left(\frac{\partial \tilde{W}}{\partial \mathrm{I_B}} + \mathrm{I_B}\frac{\partial \tilde{W}}{\partial \mathrm{II_B}}\right)\right|_{J=\sqrt{1+\varphi}} - \alpha\varphi'\right]\mathbf{B} -\\
&\quad - \frac{2}{\sqrt{1+\varphi}}\left[\left.\frac{\partial \tilde{W}}{\partial \mathrm{II_B}}\right|_{J=\sqrt{1+\varphi}}\right]\mathbf{B}^2.
\end{aligned} \tag{15.143}$$

Alternatively, assuming the form $\tilde{W}(\mathbf{B}) = \overline{W}(\overline{\mathbf{B}}) + W^{\odot}(J)$ and adding the reaction stresses from Eqs. (15.87) to the (15.127), substituting $J = \sqrt{1+\varphi}$, we note

$$\mathbf{S}^{\mathrm{lc}} = \left[-\frac{2}{3}J^{-5/3}\,\mathrm{I_B}\frac{\partial \overline{W}}{\partial \overline{\mathrm{I}}_{\mathrm{B}}} - \frac{4}{3}J^{-7/3}\,\mathrm{II_B}\frac{\partial \overline{W}}{\partial \overline{\mathrm{II}}_{\mathrm{B}}} + \frac{\partial W^{\odot}}{\partial J} + 2\alpha\sqrt{1+\varphi}\right]_{J=\sqrt{1+\varphi}}\mathbf{I} +$$

$$+\left[2\,J^{-5/3}\,\frac{\partial\overline{W}}{\partial\bar{\mathrm{I}}_{\mathrm{B}}}+2\,J^{-7/3}\,\mathrm{I}_{\mathrm{B}}\,\frac{\partial\overline{W}}{\partial\overline{\mathrm{II}}_{\mathrm{B}}}-2\alpha\,\frac{\varphi'(\mathrm{I}_{\mathrm{B}})}{\sqrt{1+\varphi}}\right]_{J=\sqrt{1+\varphi}}\mathbf{B}\;-$$
$$-\left[2\,J^{-7/3}\,\frac{\partial\overline{W}}{\partial\overline{\mathrm{II}}_{\mathrm{B}}}\right]_{J=\sqrt{1+\varphi}}\mathbf{B}^{2}. \qquad (15.144)$$

15.8 Application Examples

For the following examples, in analogy to the Sect. 15.6.1, we will consider the uniaxial extension with the stress state $\sigma_1 \neq 0$, $\sigma_2 = \sigma_3 = 0$ and the stretches from Eq. (15.30) with $\lambda_2 = \lambda_3$ for isotropic materials.

15.8.1 Limited Compressible Extension of the Mooney-Rivlin Model

Requiring the internal constraint in the form $\mathrm{III}_{\mathrm{B}} = J^2 = 1 + \varphi(\mathrm{I}_{\mathrm{B}})$, choosing the elastic potential given in Eq. (15.129), and substituting it in Eq. (15.144), we derive

$$\mathbf{S}^{\mathrm{lc}} = \left[-\frac{c_1}{3}\,J^{-5/3}\,\mathrm{I}_{\mathrm{B}}-\frac{2c_2}{3}\,J^{-7/3}\,\mathrm{II}_{\mathrm{B}}+\frac{c_3}{3}(J-1)+2\alpha\,\sqrt{1+\varphi}\right]_{J=\sqrt{1+\varphi}}\mathbf{I}\;+$$
$$+\left[c_1\,J^{-5/3}+c_2\,J^{-7/3}\,\mathrm{I}_{\mathrm{B}}-2\alpha\,\frac{\varphi'(\mathrm{I}_{\mathrm{B}})}{\sqrt{1+\varphi}}\right]_{J=\sqrt{1+\varphi}}\mathbf{B}\;-$$
$$-\left[c_2\,J^{-7/3}\right]_{J=\sqrt{1+\varphi}}\mathbf{B}^{2}. \qquad (15.145)$$

Defining $1+\varphi(\lambda_1^2) =: f(\lambda_1^2) = f$, we obtain

$$\mathbf{S}^{\mathrm{lc}} = \left[-\frac{c_1}{3}\,f^{-5/6}\,\mathrm{I}_{\mathrm{B}}-\frac{2c_2}{3}\,f^{-7/6}\,\mathrm{II}_{\mathrm{B}}+\frac{c_3}{3}(f^{1/2}-1)+2\alpha\,f^{1/2}\right]\mathbf{I}\;+$$
$$+\left[c_1\,f^{-5/6}+c_2\,f^{-7/6}\,\mathrm{I}_{\mathrm{B}}-2\alpha\,\varphi'\,f^{-1/2}\right]\mathbf{B}-\left[c_2\,f^{-7/6}\right]\mathbf{B}^2. \qquad (15.146)$$

Alternatively, by using the Cayley-Hamilton formula $\mathbf{B}^2 = \mathrm{I}_{\mathrm{B}}\,\mathbf{B} - \mathrm{II}_{\mathrm{B}}\,\mathbf{I} + \mathrm{III}_{\mathrm{B}}\,\mathbf{B}^{-1}$ the equation yields

$$\mathbf{S}^{\mathrm{lc}} = \left[-\frac{c_1}{3}f^{-5/6}\,\mathrm{I}_{\mathrm{B}}+\frac{c_2}{3}f^{-7/6}\,\mathrm{II}_{\mathrm{B}}+\frac{c_3}{3}(f^{1/2}-1)+2\alpha f^{1/2}\right]\mathbf{I}\;+$$
$$+\left[c_1 f^{-5/6}-2\alpha\,\varphi' f^{-1/2}\right]\mathbf{B}-c_2 f^{-1/6}\,\mathbf{B}^{-1}. \qquad (15.147)$$

For $J = 1 \Rightarrow f = 1, \varphi = 0$ and $\varphi' = 0$ in Eq. (15.147), we obtain the Mooney-Rivlin model in Eq. (15.133). Therefore, Eq. (15.147) or its alternative form Eq. (15.146) can be interpreted as the limited compressible extension of the Mooney-Rivlin model.

We consider again the assumptions of Eq. (15.108) with the "switch" from $\varphi(\mathrm{I_B})$ to $\varphi(\lambda_1^2)$ and use the functions $\varphi_i(\lambda_1^2)$ from Eq. (15.94) and $\varphi_i'(\lambda_1^2)$ from Eq. (15.97). By using $1 + \varphi_i(\lambda_1^2) =: f_i(\lambda_1^2) = f_i$ to simplify the notation, rewrite Eq. (15.96) and the invariants $\mathrm{I_B}, \mathrm{II_B}$:

$$\lambda_2^2 = \lambda_1^{-1} f_i^{1/2}, \tag{15.148}$$

$$\mathrm{I_B} = \lambda_1^2 + 2\lambda_2^2 = \lambda_1^2 + 2\lambda_1^{-1} f_i^{1/2}, \tag{15.149}$$

$$\mathrm{II_B} = \lambda_2^4 + 2\lambda_1^2\lambda_2^2 = \lambda_1^{-2} f_i + 2\lambda_1 f_i^{1/2}. \tag{15.150}$$

The components of Eq. (15.147) can be written as (no sum over i):

$$\begin{aligned}\sigma_1^{\mathrm{lc}} = & \frac{2c_1}{3}\left(\lambda_1^2 f_i^{-5/6} - \lambda_1^{-1} f_i^{-1/3}\right) - \frac{2c_2}{3}\left(\lambda_1^{-2} f_i^{-1/6} - \lambda_1 f_i^{-2/3}\right) + \\ & + \frac{c_3}{3}\left(f_i^{1/2} - 1\right) + 2\alpha\left(f_i^{1/2} - \varphi_i'\lambda_1^2 f_i^{-1/2}\right),\end{aligned} \tag{15.151}$$

$$\begin{aligned}\sigma_2^{\mathrm{lc}} = & -\frac{c_1}{3}\left(\lambda_1^2 f_i^{-5/6} - \lambda_1^{-1} f_i^{-1/3}\right) + \frac{c_2}{3}\left(\lambda_1^{-2} f_i^{-1/6} - \lambda_1 f_i^{-2/3}\right) + \\ & + \frac{c_3}{3}\left(f_i^{1/2} - 1\right) + 2\alpha\left(f_i^{1/2} - \varphi_i'\lambda_1^{-1}\right).\end{aligned} \tag{15.152}$$

Since the stresses in the transverse directions are zero for the uniaxial extension, $\sigma_2^{\mathrm{lc}} = 0$, we obtain from Eq. (15.152)

$$2\alpha(\lambda_1) = \frac{c_1\left(\lambda_1^2 f_i^{-5/6} - \lambda_1^{-1} f_i^{-1/3}\right) - c_2\left(\lambda_1^{-2} f_i^{-1/6} - \lambda_1 f_i^{-2/3}\right) - c_3\left(f_i^{1/2} - 1\right)}{3\left(f_i^{1/2} - \varphi_i'\lambda_1^{-1}\right)}. \tag{15.153}$$

Replacing the Lagrange parameter $2\alpha(\lambda_1)$ in Eq. (15.151), we obtain the limited compressible principle Cauchy stress in the tensile direction.

With a curve-fitting method, the constants c_i can be determined from an uniaxial extension test with the measured/set values σ_1 and λ_1^2.

Analogous to the example in Sect. 15.6.1, Python's *curve_fit()* function (non-linear least squares method) was applied by using the chosen internal constraint constants $\delta_i, \beta_i, \zeta_3$, as presented in Table 15.2. The calculated constants of the constitutive equation c_i are also shown at Table 15.2. The results can be seen at Fig. 15.6.

As shown in Fig. 15.6, all three φ_i provide acceptable results, making them suitable as ansatz functions. Among them, the curve of the sigmoid function, φ_3, appears to fit the test data best. The hyperbolic tangent function, φ_1, and the arc tangent function, φ_2, produce almost identical curves. Furthermore, as shown in Fig. 15.6, the incompressible Mooney-Rivlin law fails to accurately represent the test data.

Table 15.2 Constants of the plots belonging to the limited compressible Mooney-Rivlin Eq. (15.151). $\delta_i, \beta_i, \zeta_3$ have the same unit 1. c_1, c_2, c_3 have the same unit (MPa) as the stress σ_1

	δ_i	β_i	ζ_3	c_1	c_2	c_3
φ_1	0.3	0.03		0.141	0.094	−177.0
φ_2	0.4	0.03		0.254	−0.117	−161.2
φ_3	0.3	0.02	30	−0.013	0.336	17246.1

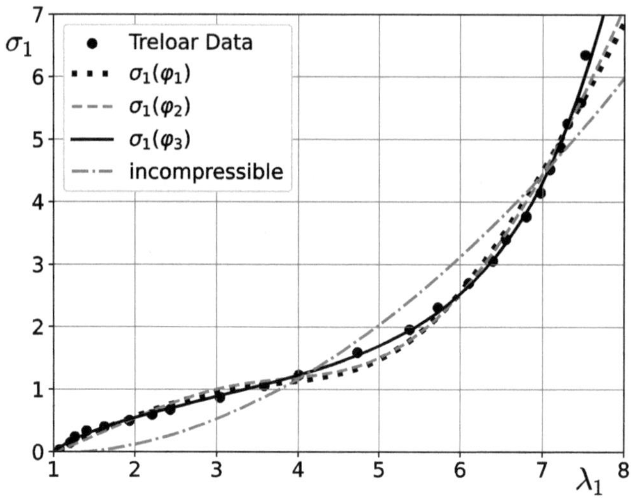

Fig. 15.6 Plots of the limited compressible Mooney-Rivlin Eq. (15.151), incompressible Mooney-Rivlin model and the reference data measured by Treloar (1944). σ_1 is given in MPa

We calculate $\varphi_i'(1)$ using Eq. (15.98) and observe that the values lie within the range [0, 1], satisfying the condition in Eq. (15.92). As a consequence, Figs. 15.7 and 15.8 demonstrate that the proposed constraints remain within physically acceptable limits for limited compressibility. Due to their minor deviations from incompressible behavior, the three functions φ_i, with the constants given in Table 15.2, can be considered as nearly incompressible internal constraints.

Remark: It is a valid question as to whether it is possible to derive limited compressible extension of Mooney-Rivlin model without including the dilatation term in the Mooney-Rivlin elastic potential Eq. (15.132). The distinction to the incompressible case lies in the constitutive equation, specifically in the terms related to the volume change J, which is replaced by $J = \sqrt{1+\varphi}$ to account for limited compressibility. By setting $c_3 = 0$ in Eqs. (15.151) and (15.152), the desired material law is easily obtained, however it was not possible to find a suitable set of constants of φ_i to achieve satisfactory results.

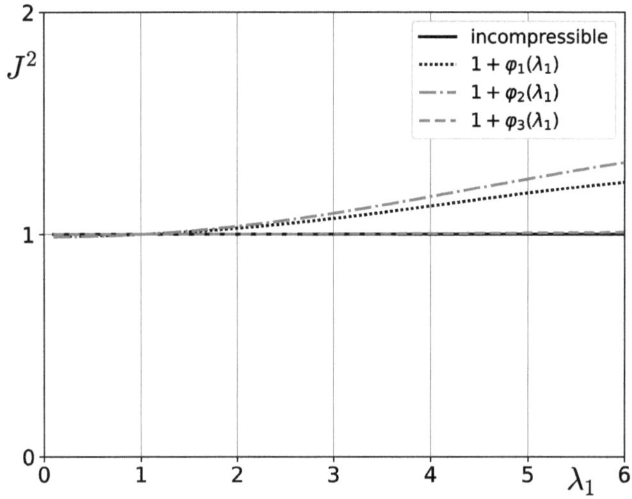

Fig. 15.7 Plots of the ansatz functions from Eq. (15.94) with the constants from Table 15.2 and comparison with incompressible behavior $J^2 = 1$

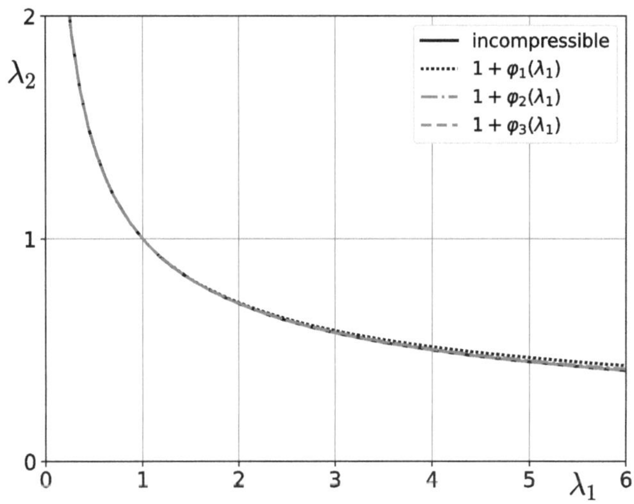

Fig. 15.8 Plots of the relation $\lambda_2^{\mathrm{lc}} = \sqrt{\lambda_1^{-1}\sqrt{1+\varphi_i(\lambda_1)}}$ with the constants from Table 15.2 and comparison with the incompressible behavior $\lambda_2^{\mathrm{in}} = \sqrt{1/\lambda_1}$

15.8.2 Limited Compressible Extension of the Neo-Hookean Model

In analogy to Eq. (15.136), replacing $c_2 = 0$ in Eq. (15.147), and again using the abbreviation $f := 1 + \varphi(\lambda_1^2)$, we obtain

Table 15.3 Constants of the plots belonging to the limited compressible Neo-Hookean Eq. (15.155). δ_i, β_i, ζ_3 have the same unit 1. $\tilde{c}_1$, $\tilde{c}_3$ have the same unit (MPa) as the stress σ_1

	δ_i	β_i	ζ_3	$\tilde{c}_1$	$\tilde{c}_3$
φ_1	0.45	0.026		0.187	−110.7
φ_2	0.39	0.035		0.213	−131.9
φ_3	0.50	0.090	4	0.145	−60.3

$$\mathbf{S}^{\mathrm{lc}} = \left[-\frac{\tilde{c}_1}{3} f^{-5/6} \mathrm{I}_{\mathrm{B}} + \frac{\tilde{c}_3}{3}(\sqrt{f} - 1) + 2\alpha\sqrt{f}\right]\mathbf{I} + \left[\tilde{c}_1 f^{-5/6} - \frac{2\alpha\varphi'}{\sqrt{f}}\right]\mathbf{B}. \quad (15.154)$$

which can be interpreted as the limited compressible extension of the Neo-Hookean material in Eq. (15.136).

The components of Eq. (15.154) can be written substituting $c_2 = 0$ in Eqs. (15.151) and (15.152) and using again φ_i from Eq. (15.105) and φ_i' from Eq. (15.106) (no sum over i):

$$\sigma_1^{\mathrm{lc}} = \frac{2\tilde{c}_1}{3}\left[\lambda_1^2 - \lambda_1^{-1} f_i^{1/2}\right] f_i^{-5/6} + \frac{\tilde{c}_3}{3} + 2\alpha\left(f_i^{1/2} - \varphi_i'\lambda_1^2 f_i^{-1/2}\right), \quad (15.155)$$

$$\sigma_2^{\mathrm{lc}} = -\frac{\tilde{c}_1}{3}\left[\lambda_1^2 - \lambda_1^{-1} f_i^{1/2}\right] f_i^{-5/6} + \frac{\tilde{c}_3}{3} + 2\alpha\left(f_i^{1/2} - \varphi_i'\lambda_1^{-1}\right). \quad (15.156)$$

Since the stresses in transverse directions are zero, $\sigma_2^{\mathrm{lc}} = 0$, we obtain from Eq. (15.156):

$$2\alpha(\lambda_1) = \frac{\tilde{c}_1\left[\lambda_1^2 - \lambda_1^{-1} f_i^{1/2}\right] f_i^{-5/6} - \tilde{c}_3}{3\left(f_i^{1/2} - \varphi_i'\lambda_1^{-1}\right)}. \quad (15.157)$$

Replacing the Lagrange parameter $2\alpha(\lambda_1)$ in Eq. (15.155), we obtain the limited compressible principle Cauchy stress in the tensile direction.

Analogous to the examples in previous sections, Python's *curve_fit()* function was applied to determine the constants. The constants of the internal constraints δ_i, β_i, ζ_3 and the constants of the constitutive equation $\tilde{c}_1$, $\tilde{c}_3$ are given in Table 15.3.

Figure 15.9 illustrates that all three functions capture the main characteristics of the test data. However, the deviations, especially in the range $2 < \lambda_1 < 6$, are larger than desired. However, compared to the incompressible Neo-Hookean law, the limited compressible equations provide improved results.

Considering the initial requirements for an ansatz function, we calculate $\varphi_i'(1)$ using Eq. (15.98) and observe that the values lie within the range [0, 1], satisfying the condition in Eq. (15.92). As a consequence, as demonstrated in Figs. 15.10 and 15.11, the proposed internal constraints remain within physically acceptable limits for limited compressibility. Due to their minor deviations from incompressible behavior, the three functions φ_i, with the constants given in Table 15.3, can be considered as nearly incompressible internal constraints.

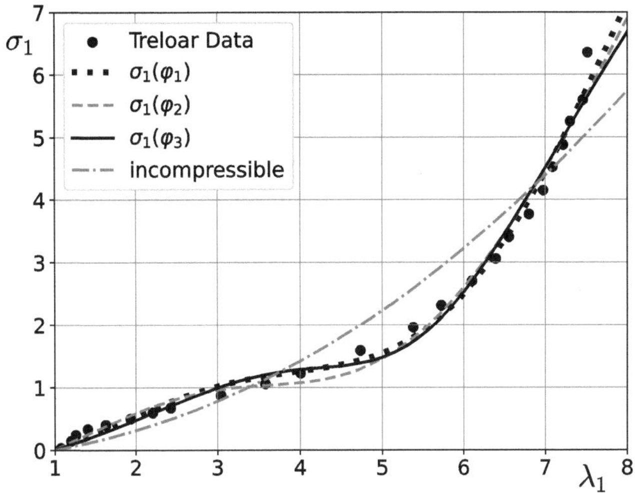

Fig. 15.9 Plots of the limited compressible Neo-Hookean Eq. (15.155), incompressible Neo-Hookean model and the reference data measured by Treloar (1944). σ_1 is given in MPa

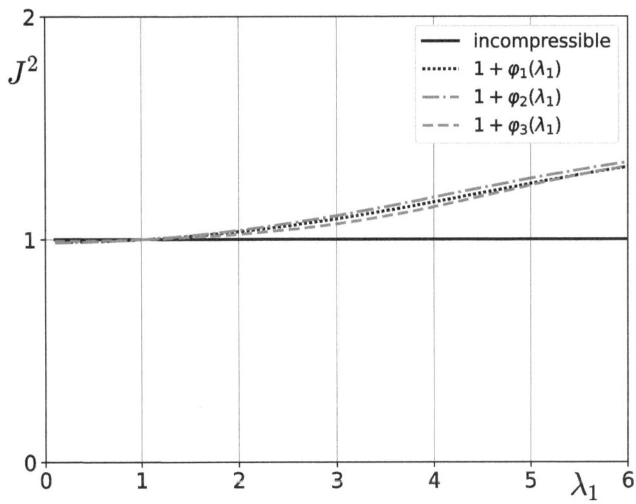

Fig. 15.10 Plots of the ansatz functions from Eq. (15.94) with the constants from Table 15.3 and comparison with $J = 1$

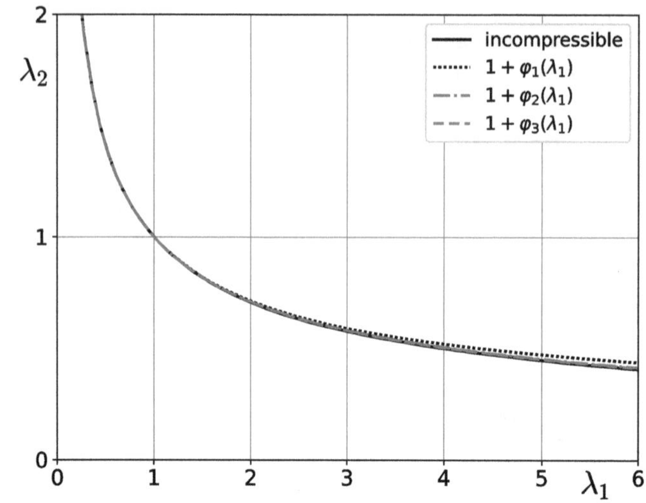

Fig. 15.11 Plots of the relation $\lambda_2^{\mathrm{lc}} = \sqrt{1 + \phi_i(\lambda_1)/\lambda_1}$ with the constants from Table 15.3 and comparison with the incompressible behavior $\lambda_2^{\mathrm{in}} = \sqrt{1/\lambda_1}$

15.9 Experimental Studies with Natural Rubber Compounds

In order to prove the practicability of the derived equations, we finally present experimental one-directional tensile tests with two different, natural rubber compounds (NR), provided by Trelleborg Antivibration Solutions Germany GmbH.

Both compounds are widely used e.g. for rubber springs within mobile machineries to isolate and damp vibrations (e.g. cabin or engine mounts). Here the rubber parts act as elastic decoupling elements to interrupt—or, at least, fault—the transfer path for structure borne noise. On the other hand side, rubber springs are also used as elastic connections to allow for a desired vibration transfer, cf., Fig. 15.12.

In what follows we will focus on natural rubber with two different hardnesses, viz. 50 and 70 Shore-A (ShA). Both compounds have similar recipes but differ by the amount of carbon blacks, which are varied to adjust the hardness—and consequently the stiffness—level. Higher amounts of carbon blacks yields higher hardness values but, on the other hand side, increase rubber damping by internal friction and decrease durability and elastic elongation range.

In order to predict the internal stresses within the rubber of antivibration machine elements, engineers typically perform mechanical Finite Elements (FE) calculations, in which pre-known external loads will be applied to the machine parts. However, correct results require the definition of realistic constitutive equations, i.e., stress-strain relations, during the preprocessing of the FE-simulations. Especially the ability of the stress-strain relation to reproduce the so-called S-flap (change of the initially regressive slope to the final progressive slope) is important to reproduce the correct

Fig. 15.12 Rubber springs at a vibratory plate compactor (elastic connection between compactor steel plate and engine/imbalance mass)

stresses within rubber under large deformations. For this reason, experiments were performed to record the stress-strain response and to use these data to adjust the material parameters δ_i, β_i, ζ_3 and c_1, c_2, c_3 within the limited compressible Mooney-Rivlin Eq. (15.151).

15.9.1 Test Set-Up

The one-directional tensile tests were performed on a universal Zwick/Roell (2.5 KN) tensile testing machine, located at the Institute of Mechanics of the Technical University Berlin, Fig. 15.13 (left panel). Further, so-called S2-shouldered test bars according to DIN 53504, DIN (2017), were used, which were provided by Trelleborg Antivibration Solutions Germany GmbH. These bone-like test specimens, Fig. 15.13 (right panel), consist of NR rubber and were produced in two different hardness versions, namely 50 and 70 ShA.

The tensile test bars were vertically fixed between the (roughened) clamps of the testing machine and tightened with a tensile pre-load of $F_0 = 0.25$ N. After fixing the specimen, three so-called conditioning cycles were executed, during which the bars will be loaded/unloaded three times with $\Delta L = 20$ mm. Note that no external setup for tracking the initial length L_0 was available. Therefore, the initial distance between the upper and lower clamps denotes the reference length L_0. After conditioning, the ultimate test was carried out by shifting the upper clamps upward with 20 mm/s and measuring the corresponding length change ΔL, together with the associated force F.

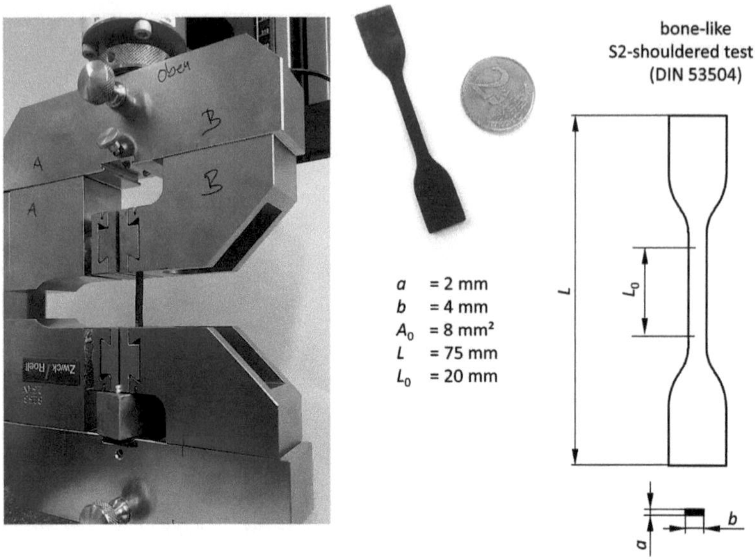

Fig. 15.13 Snapshot of tensile testing and S2-shouldered test bar according to DIN 53504

Six separate tests were performed in total; in particular, three tests for each hardness version. For each hardness the most representative curve were selected and the stretches λ_1 as well as the Cauchy stresses σ_1 were calculated using the following formulae[1]:

$$\lambda_1 = \frac{L}{L_0} = \frac{L_0 + \Delta L}{L_0}, \tag{15.158}$$

$$\sigma_1 = \frac{F}{A} = \frac{F}{A_0 L_0 / L} = \frac{F}{A_0}\lambda_1 . \tag{15.159}$$

Finally, it is worth mentioning that, for increasing deformation, the specimen tends to slide within the clamps. Thus, the experiment was stopped when this phenomenon was observed by an unrealistic decrease of the slope of the measured F-ΔL-curves.

Figure 15.14 illustrates the obtained stress-stretch curves for both investigated NR-compounds. Obviously, the aforementioned S-flap is clearly pronounced for the 70ShA-specimens. However, the 70ShA-specimens start to slide within the clamps already for $\lambda_1 > 1.8$. Therefore the tests were stopped to avoid unrealistic results and, consequently, no data were exploited for larger deformation regimes. Also, the test data of 50ShA-specimens for $\lambda_1 < 2.0$ shows a slightly S-flap behavior, similar to the 70ShA data.

[1] To calculate σ_1 from F, the current cross-sectional area A is required, which is difficult to measure. By assuming (only in this case) incompressibility, one can derive the following relation: $V = V_0 \Rightarrow A \cdot L = A_0 \cdot L_0 \Rightarrow A = A_0 \cdot L_0/L \Rightarrow A = A_0/\lambda_1$.

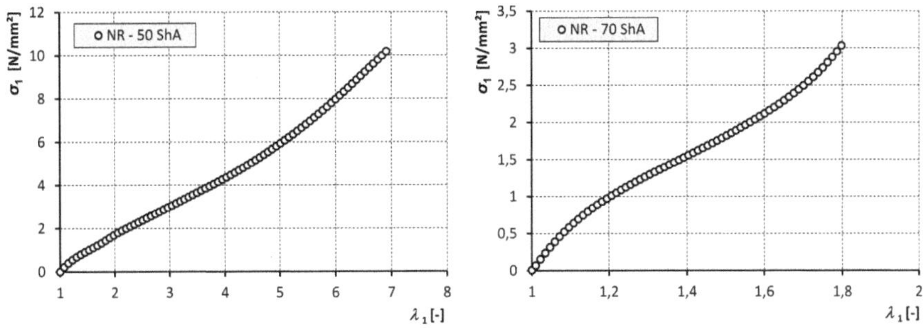

Fig. 15.14 Experimental curves for Cauchy stresses $\sigma_1(\lambda_1)$ of NR rubber in 50 ShA (right) und 70 ShA (left)

Table 15.4 Constants of the plots belonging to the limited compressible Mooney-Rivlin Eq. (15.151) respectively test data NR 70 ShA. δ_i, β_i, ζ_3 have the same unit 1. c_1, c_2, c_3 have the same unit (MPa) as the stress σ_1

	δ_i	β_i	ζ_3	c_1	c_2	c_3
φ_1	0.2	0.4		5.317	−3.099	−427.5
φ_2	0.25	0.45		6.076	−3.716	−267.3
φ_3	0.3	0.3	2.5	−2.990	5.611	3837.6

15.9.2 Curve Fitting and Materials Parameter

With the same strategy as already applied, e.g., in Sect. 15.6.1, we use Python's *curve_fit()* function to adjust the materials specific parameter δ_i, β_i, ζ_3 and c_1, c_2, c_3 within Eq. (15.151). Table 15.4 shows the obtained parameter and Fig. 15.15 compares the experimental curves with the resulting limited compressible Mooney-Rivlin ansatz of Eq. (15.151) for the test data NR 70 ShA.

Fortunately Fig. 15.15 confirms the observations, already done during the fitting procedure of Treloar's data in the previous chapter. All three ansatz-functions φ_i deliver almost the same curves with good fitting results which allow to reproduce the S-flap. As one can see, the incompressible Mooney-Rivlin model significantly deviates from the test data and cannot reproduce the S-pattern behavior due to its parabolic nature.

Figure 15.16 shows that the sigmoid function φ_3 accurately reproduces the test data for NR 50 ShA. In comparison, the limited compressible model using the ansatz functions φ_1 and φ_2, together with the incompressible Mooney-Rivlin model, exhibits considerable deviations in the range $1 < \lambda_1 < 3$. The corresponding constants are provided in Table 15.5.

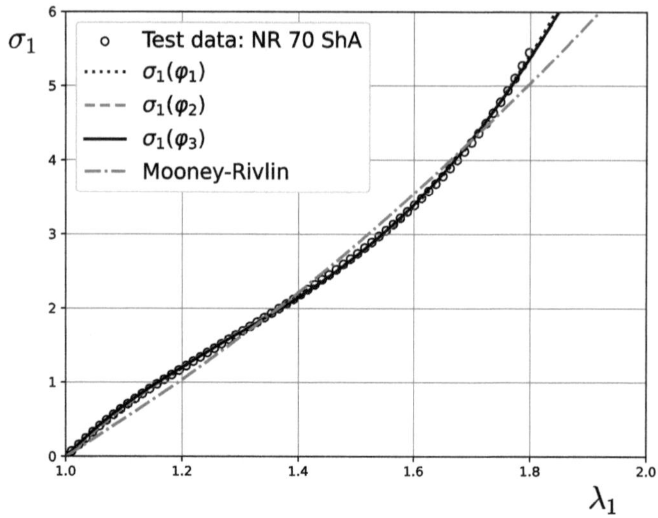

Fig. 15.15 Plots of the limited compressible Mooney-Rivlin Eq. (15.151), incompressible Mooney-Rivlin model and the test data NR 70 ShA. Cauchy stress σ_1 is given in MPa

Table 15.5 Constants of the plots belonging to the limited compressible Mooney-Rivlin Eq. (15.151) respectively test data NR 50 ShA. $\delta_i, \beta_i, \zeta_3$ have the same unit 1. c_1, c_2, c_3 have the same unit (MPa) as the stress σ_1

	δ_i	β_i	ζ_3	c_1	c_2	c_3
φ_1	0.4	0.5		2.414	−3.558	105.9
φ_2	0.2	0.7		2.469	−4.546	423.2
φ_3	0.1	0.1	1.5	1.754	−1.362	−9196.7

15.10 Summary and Outlook

The following outcomes can be noted as the result of this work:

- Assuming that the internal constraint for any compressible solid has the form $J = 1 + \phi$ or $J^2 = \mathrm{III}_\mathrm{B} = 1 + \varphi$, where ϕ or φ describes the compressible behavior, the constitutive Eqs. (15.85), (15.89) and (15.91) were derived.
- It was demonstrated how to propose a suitable ansatz function φ for a one-dimensional stress state, and how to extend it to a three-dimensional stress state.
- Using Treloar's measurement data, the applicability of the derived constitutive equation with the chosen ansatz functions φ_i was shown. There were suitable ansatz functions, and among them, the sigmoid function yielded the best results.
- Hyperelastic constitutive equations for limited compressible solids were introduced in Eq. (15.143) and alternatively in Eq. (15.144). The applicability was shown with two examples: Limited compressible extensions of the Mooney-Rivlin

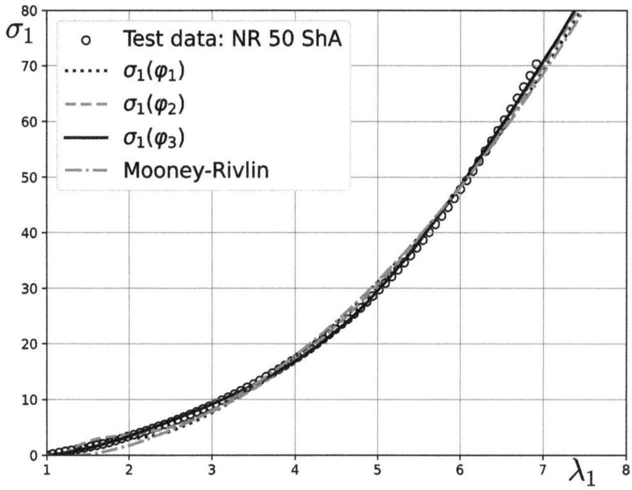

Fig. 15.16 Plots of the limited compressible Mooney-Rivlin Eq. (15.151), incompressible Mooney-Rivlin model and the test data NR 50 ShA. Cauchy stress σ_1 is given in MPa

and Neo-Hookean models. We successfully determined suitable ansatz functions, which have delivered acceptable results. It turned out that the appropriate ansatz functions could vary depending on the strain energy density functions.

- One-dimensional tensile tests with two commercial NR-rubber compounds were performed to illustrate the applicability of the presented formalism, in particular the limited compressible Mooney-Rivlin model.

The primary goal of this work is to use simple functions to model the internal constraints for volume changes in compressible solids. By adjusting the functions to align with expected volume change behavior, we defined a limited compressibility that also allows for describing the nearly incompressible volume changes in solids.

As demonstrated in the examples, it can be successfully applied to both Cauchy and Green elasticity. Incompressible material equations such as Mooney-Rivlin or Neo-Hookean models can be extended to describe nearly incompressible or limited compressible behavior.

As shown in this work, the chosen ansatz functions, result in constitutive equations which can be used to describe both geometrically and physically nonlinearities.

This work represents an initial step toward future research, with potential tasks including:

- Searching for other ansatz functions φ_i, which can better describe the limited compressible behavior and ideally have physical meanings.
- Investigating the ansatz functions with respect to consistency with the second law of thermodynamics.
- Verifying the validity of the constitutive equations considering multi-axial tensile tests and modifying all relevant equations if needed.

- Focusing on the ansatz functions with second invariant: $\varphi(\mathrm{II_B})$ or $\varphi(\mathrm{I_B}, \mathrm{II_B})$ and examining the related constitutive equations.
- Verifying the applicability of the equations for simple and complex structures, both for analytical and numerical solutions.
- Investigating the effect of the limited compressibility on viscoelastic behavior of materials.
- Applying the formalism of limited compressibility to anisotropic solids.

Finally it is worth mentioning that the method introduced in this work is focused on the internal constrains for volume change. Further questions e.g. whether a similar approach is applicable for other internal constraints, such for limited extensibility of solids, are still open.

Acknowledgements Trelleborg Antivibration Solutions Germany GmbH is gratefully acknowledged for providing the bone-like test specimen for the experimental investigations. Furthermore our sincere thanks are given to Arion Juritza, who perform the tensile tests in the laboratory of the Institute of Mechanics at the Technical University of Berlin (TUB).

References

Antman SS (2007) Material constraints in continuum mechanics. Atti della Accademia Nazionale dei Lincei. Classe di Scienze Fisiche, Matematiche e Naturali. Rendiconti Lincei. Matematica e Applicazioni, vol 70. no 6. pp 623–664

Batra RC (1976) Deformation produced by a simple tensile load in an isotropic elastic body. J Elast 6(1):109–111

Beatty MF, Hayes MA (1992) Deformations of an elastic, internally constrained material. Part 1-Homogeneous deformations. J Elast 29:1–84

Beatty MF (1987) Topics in finite elasticity: hyperelasticity of rubber, elastomers, and biological tissues—with examples. Appl Mech Rev 40(12):1699–1734

Beatty MF, Stalnaker DO (1986) The Poisson function of finite elasticity. J Appl Mech 53(4):807–813

Bell JF (1985) Contemporary perspectives in finite strain plasticity. Int J Plast 1:3–27

Bertram A (2012) Elasticity and plasticity of large deformations, 3rd edn. An introduction. Springer, Heidelberg Dordrecht London New York

Bertram A, Glüge R (2016) Gradient materials with internal constraints. Math Mech Complex Syst 4(1):1–15

Blatz PJ, Ko WL (1962) Application of finite elastic theory to the deformation of rubbery materials. Trans Soc Rheol 6:223–251

Carlson DE, Tortorelli DA (1996) On hyperelasticity with internal constraints. J Elast 42(1):91–98

Carroll MM, Horgan CO (1990) Finite strain solutions for a compressible elastic solid. Q Appl Math 48(4):767–780

Casey J (1995) A treatment of internally constrained materials. J Appl Mech 62:542–544

DIN 53504:2017-03: Prüfung von Kautschuk und Elastomeren - Bestimmung von Reißfestigkeit, Zugfestigkeit, Reißdehnung und Spannungswerten im Zugversuch, Beuth Verlag (2017)

Doll S, Schweizerhof K (2000) On the development of volumetric strain-energy functions. J Appl Mech 67:17–21

Eringen AC (1980) Mechanics of continua, 2nd edn. Krieger Publishing Company, Robert E

Green AE, Naghdi PM, Trapp JA (1970) The thermodynamics of continuum with internal constraints. Int J Eng Sci 8:891–908
Gummert P (1990) A constitutive equation for non-linear, compressible viscoelastic materials with fading memory. In: Creep in structures. 4th IUTAM symposium, Cracow, Poland. Springer, Berlin, pp 83–90
Gurtin ME, Guidugli PP (1973) The thermodynamics of constrained materials. Arch Ration Mech Anal 51(3):165–191
Haupt P (2002) Continuum mechanics and theory of materials, 2nd edn. Springer, Berlin, Heildelberg
He H, Zhang Q, Zhang Y, Chen J, Zhang L, Li F (2022) A comparative study of 85 hyperelastic constitutive models for both unfilled rubber and highly filled rubber nanocomposite material. Nano Mater Sci 4(2):64–82
Holzapfel GA (2000) Nonlinear solid mechanics: a continuum approach for engineering. Wiley, Chichester
Horgan CO, Murphy JG (2009a) Constitutive modeling for moderate deformations of slightly compressible rubber. J Rheol 53(6):153–168
Horgan CO, Murphy JG (2009b) On the volumetric part of strain-energy functions used in the constitutive modeling of slightly compressible solid rubbers. Int J Solids Struct 46:3078–3085
Klisch SM (1999) Internally constrained mixtures of elastic continua. Math Mech Solids 4(4):481–498
Müller I, Ruggeri T (1998) Rational extended thermodynamics, 2nd edn. Springer tracts of natural philosophy. Springer, New York
Nagehban M (2007) Single and multiple material constraints in thermoelasticity. Math Mech Solids 12(6):623–664
Treloar LRG (1944) Stress-strain data for vulcanised rubber under various types of deformation. Trans Faraday Soc 40:59–70
Truesdell C, Noll W (2004) The Non-linear field theories of mechanics. In: Antmann SS (ed) 3rd edn. Springer, Berlin, Heilderberg
Zdunek A (2020) On purely mechanical simple kinematic internal constraints. J Elast 139:123–152

Chapter 16
Phase-Field Cosserat Modelling of Grain Boundary-Pore Interaction

Flavien Ghiglione and Samuel Forest

The second author dedicates this work, with admiration and sadness, to the memory of Dr. Wolfgang Dreyer who introduced to him the existence and power of phase field methods thirty years ago.

Abstract The development of additive manufacturing has made it possible to produce geometrically complex parts. However, these processes appear to inevitably introduce voids into the material, which may be responsible for premature failure. Upon thermomechanical treatment, microstructure evolutions such as grain boundary migration and recrystallization are promoted. Therefore, complex interactions between grain boundaries and voids emerge for heat-treated additively manufactured parts. In the literature, these interactions are typically studied using phase-field models that neglect the effect of stress and strain. In this work, a fully coupled phase-field and Cosserat crystal plasticity model reveals that grain boundary migration through a void contained in a bicrystal subjected to compression reduces the stress concentration in the region swept by the interface.

16.1 Introduction

Additive manufacturing allows for fast building of complex metallic parts through a process of layer-by-layer addition of material. However, these techniques introduce detrimental defects into the microstructure, such as porosity. In particular, intragranular voids can lead to stress concentration, which is known to promote the fracture of crystalline materials (Bourcier et al. 1986). Unfortunately, pores seem to appear for any additive manufacturing process employed and any material. As such they cannot be avoided.

F. Ghiglione · S. Forest (✉)
Mines Paris, PSL University, Evry, France
e-mail: samuel.forest@minesparis.psl.eu

J. Fuhrmann et al. (eds.), *Advances in Continuum Physics*, Advanced Structured Materials 238, https://doi.org/10.1007/978-3-031-93918-1_16

Thermomechanical treatment of additively manufactured parts will therefore induce complex interactions between these voids and migrating grain boundaries. For instance, much like the Zener pinning effect, voids can impact grain boundary behaviour during the process of recrystallization (Zhang et al. 2022). As another example, migration of the grain boundary across a pore can reduce stress concentration and prevent crack formation and propagation (Wang et al. 2023). From the industrial viewpoint, it is therefore essential to investigate the interaction between voids and migrating grain boundaries to design dependable safety components through additive manufacturing techniques.

Such complex interactions can be modelled using full-field simulations including individual defects and neighbouring grains. In the field of computational metallurgy, several paradigms have been proposed to tackle grain boundary migration, among which Monte-Carlo Potts (Anderson et al. 1984), cellular automata (Hesselbarth and Göbel 1991), level sets (Bernacki et al. 2008) and phase field models. The latter approach is particularly elegant due to its thermodynamic consistency. It regularizes interfaces by describing them as continuous and diffuse fields varying across a narrow region, and sharp interface properties can be recovered as a limit case (Elder et al. 2001). In the classical approach to phase-field modelling of microstructure evolution of crystalline materials, each grain is assigned one phase field variable (Chen and Yang 1994; Steinbach and Pezzolla 1999). This is referred to as multi-phase-field (MPF) models. Alternatively, the whole polycrystalline microstructure can be described by only two phase-fields in the model proposed by Kobayashi et al. (2000); Warren et al. (2003), referred to as the Kobayashi-Warren-Carter (KWC) model. However, in the two latter approaches to phase-field modelling of microstructure evolution, mechanical effects were not incorporated and thus could not account for microstructural changes induced by viscoplastic deformation. This can be remedied by including other physical contributions, such as elastic strain or stored energy by adding them to the free energy functional (Steinbach and Apel 2006). In this regard, the KWC-Cosserat model (Ask et al. 2018b), that combines the KWC model and Cosserat crystal plasticity, is a strong candidate to simulate microstructure evolution in deformed polycrystals.

The interaction of grain boundaries and pores has been studied within a MPF framework by Hoetzer et al. (2016); Rehn et al. (2018) in the context of sintering. Focus is brought on morphological changes and pore attachment and (partial) detachment in both idealized and realistic grain geometries. Similarly, MPF simulations of void-pinning effects in irradiated zirconium for coarse and fine-grained microstructures were performed by Han (2023). Finally, Zhang et al. (2022) showed, again with MPF simulations, the influence of voids on the kinetics of migration and shape of the grain boundaries during recrystallization in additively manufactured samples. Even though complex geometrical changes are represented in the aforementioned studies, they never consider mechanical effects. Therefore, we propose to investigate pore-grain boundaries interactions with considerations of strain and stresses. To do so, we employ the KWC and KWC-Cosserat models to simulate the stored energy-driven migration of a single grain boundary through a void.

The outline of this chapter is as follows. First, the main equations of the KWC model, Cosserat crystal plasticity and KWC-Cosserat model are recalled in Sect. 16.2. Subsequently, stored energy-driven migration of a grain boundary through a void is first studied in Sect. 16.3 with the KWC model without mechanical effects. In this section, the simulation is carried out once again with the KWC-Cosserat model by adding compression followed by relaxation to account for stress redistribution.

16.2 Orientation Phase-Field Models for Microstructure Evolution

16.2.1 Kobayashi-Warren-Carter Phase-Field Model

The Kobayashi-Warren-Carter phase-field model (Kobayashi et al. 2000; Warren et al. 2003) is two-dimensional and allows full-field simulation of microstructure evolution. Contrary to multi-phase field approaches, it is based on only two scalar order parameters, which are the crystal lattice orientation θ and the crystal order parameter ϕ. The latter varies between zero for a fully disordered phase and one for a perfect crystal. It is worth noting that the presence of grain boundaries and lattice defects such as dislocations locally contribute to lower the value of ϕ. The original model was extended to account for the energy stored by dislocations in Abrivard et al. (2012), which is a driving force for grain boundary migration. The dimensionless free energy $\bar{\mathcal{F}}_{\text{KWC}}$ of the extended KWC model is given by

$$\bar{\mathcal{F}}_{\text{KWC}} = \int_{\bar{\Omega}} \left[\bar{\psi}(\bar{\nabla}\theta, \phi, \bar{\nabla}\phi) + \phi \bar{E} \right] d\bar{V} \tag{16.1}$$

where $\bar{\nabla}$ is a dimensionless gradient and $\bar{E}$ is the dimensionless dislocation-based stored energy (Abrivard et al. 2012). The dimensionless Helmholtz free energy density $\bar{\psi}(\bar{\nabla}\theta, \phi, \bar{\nabla}\phi)$ is frame-invariant and has the following form:

$$\bar{\psi}(\bar{\nabla}\theta, \phi, \bar{\nabla}\phi) = f(\phi) + \frac{\bar{\alpha}^2}{2} \left|\bar{\nabla}\phi\right|^2 + \bar{s} g(\phi) \left|\bar{\nabla}\theta\right| + \frac{\bar{\varepsilon}^2}{2} h(\phi) \left|\bar{\nabla}\theta\right|^2 \tag{16.2}$$

where the model parameters $\bar{\alpha}, \bar{s}, \bar{\varepsilon}$ are dimensionless. It should be noted that the singular term $|\bar{\nabla}\theta|$, not differentiable at 0, is required to localize grain boundaries of finite width whereas the term $|\bar{\nabla}\theta|^2$ is needed for grain boundary motion (Warren et al. 2003). The functions f, g, h are dimensionless by construction and have the following form:

$$f(\phi) = \frac{1}{2}(1-\phi)^2 \tag{16.3}$$

$$g(\phi) = -2\left(\log(1-\phi) + \phi\right) \tag{16.4}$$

$$h(\phi) = \phi^2 \tag{16.5}$$

In this work the logarithmic form of $g(\phi)$ is adopted to recover the Read-Shockley part of the grain boundary energy (Kobayashi et al. 2000). This requires a regularisation for the numerical implementation of the model, which is achieved by adding a small positive constant $\gamma_g \ll 1$ to the argument of the logarithm.

Assuming relaxational dynamics and associated variational derivatives for both phase fields (Finel et al. 2010), the evolution equations are

$$\bar{\eta}_\phi \dot{\phi} = \left(\frac{\partial \bar{\psi}}{\partial \bar{\nabla}\phi}\right) \cdot \bar{\nabla} - \frac{\partial \bar{\psi}}{\partial \phi} = \bar{\alpha}^2 \bar{\nabla}^2 \phi - f_{,\phi} - \bar{s} g_{,\phi} |\bar{\nabla}\theta| - \frac{\bar{\varepsilon}^2}{2} h_{,\phi} |\bar{\nabla}\theta|^2 - \bar{E} \tag{16.6}$$

$$P \bar{\eta}_\theta \phi^2 \dot{\theta} = \left(\frac{\partial \bar{\psi}}{\partial \bar{\nabla}\theta}\right) \cdot \bar{\nabla} - \frac{\partial \bar{\psi}}{\partial \theta} = \bar{\nabla} \cdot \left(\bar{\varepsilon}^2 h \bar{\nabla}\theta + \bar{s} g \frac{\bar{\nabla}\theta}{|\bar{\nabla}\theta|}\right) \tag{16.7}$$

with $\dot{y} = \partial y / \partial \bar{t}$ where $\bar{t}$ is a dimensionless time scale. P is a dimensionless inverse mobility function assumed to depend on ϕ and/or $\bar{\nabla}\theta$. The parameters $\bar{\eta}_\phi$, $\bar{\eta}_\theta$ are constant kinetic factors. The choice $P = 1$ results in a spatially uniform mobility. Alternatively, the function P can be made different in the bulk of the grain and in the diffuse grain boundary region by a suitable dependence of $\bar{\nabla}\theta$, in order to avoid simultaneous shrinkage and rotation during curvature driven migration:

$$P(\bar{\nabla}\theta) = 1 + \left(\frac{\bar{\mu}_p}{\bar{\varepsilon}} - 1\right) \exp\left(-\beta_p \bar{\varepsilon} |\bar{\nabla}\theta|\right) \tag{16.8}$$

The singular term $1/|\bar{\nabla}\theta|$ in Eq. (16.7) requires regularization in the numerical treatment. In accordance with Warren et al. (2003) this is achieved by replacing the term $|\bar{\nabla}\theta|$ in the free energy with a quadratic potential for orientation gradients below a numerical threshold $1/\bar{\gamma}$.

An evolution equation for the stored energy must be specified to account for recovery in the wake of the moving grain boundary. Following Abrivard et al. (2012) the time evolution of the stored energy is given by:

$$\dot{\bar{E}} = \begin{cases} -C_D \bar{E} \tanh(\bar{C}_A |\bar{\nabla}\theta|^2) \dot{\phi}, & \dot{\phi} > 0 \\ 0, & \dot{\phi} \leq 0 \end{cases} \tag{16.9}$$

with C_D and $\bar{C}_A$ some constants settling the recovery behavior.

Note that mass transport is not considered in the present work but can be introduced as usual in phase field/diffusion models (Finel et al. 2010).

16.2.2 Cosserat Crystal Plasticity

The Cosserat medium is endowed with the displacement vector $\underline{\boldsymbol{u}}$ and the microrotation pseudo-vector $\underline{\boldsymbol{\theta}}$ as independent degrees of freedom (Eringen 1976). At small deformation, rotation and curvature the objective deformation measures are the generally non-symmetric Cosserat deformation and curvature (or wryness) tensors $\underset{\sim}{e}$ and $\underset{\sim}{\kappa}$, respectively:

$$\underset{\sim}{e} = \underline{u} \otimes \nabla + \underset{\approx}{\epsilon} \cdot \underline{\theta} \tag{16.10}$$

$$\underset{\sim}{\kappa} = \underline{\theta} \otimes \nabla \tag{16.11}$$

Their work-conjugate counterparts are the non-symmetric force-stress tensor $\underset{\sim}{\sigma}$ and couple-stress tensor $\underset{\sim}{m}$. The balance of momentum and balance of moment of momentum are recalled here in the static case (slow processes) and in the absence of volume forces and moments:

$$\underset{\sim}{\sigma} \cdot \nabla = \underline{0} \tag{16.12}$$

$$\underset{\sim}{m} \cdot \nabla - \underset{\approx}{\epsilon} : \underset{\sim}{\sigma} = \underline{0} \tag{16.13}$$

The total strain tensor is additively decomposed into elastic and plastic contributions:

$$\underset{\sim}{e} = \underset{\sim}{e}^e + \underset{\sim}{e}^p \tag{16.14}$$

The curvature tensor is here assumed to be purely elastic for simplicity.

In the isotropic case the elasticity laws are

$$\underset{\sim}{\sigma} = \lambda \mathrm{Tr}\,(\underset{\sim}{e}^e)\underset{\sim}{I} + 2\mu \mathrm{sym}(\underset{\sim}{e}^e) + 2\mu_c \mathrm{skew}(\underset{\sim}{e}^e) \tag{16.15}$$

$$\underset{\sim}{m} = \alpha \mathrm{Tr}\,(\kappa)\underset{\sim}{I} + 2\beta \mathrm{sym}(\underset{\sim}{\kappa}) + 2\gamma \mathrm{skew}(\underset{\sim}{\kappa}) \tag{16.16}$$

where μ, λ are the Lamé coefficients and the parameters α, β, γ are the torsion-bending stiffnesses. The sym and skew operatores respectively denote the symmetric and skew-symmetric parts of a second rank tensor. The coupling modulus μ_c penalizes the difference between the Cosserat micro-rotation and the lattice rotation $\underset{\sim}{\omega}^e$:

$$\underset{\sim}{\omega}^e = \mathrm{skew}(\underset{\sim}{e}^e) - \underset{\approx}{\epsilon} \cdot \underline{\theta} \tag{16.17}$$

For a sufficiently high μ_c, the Cosserat micro-rotation equates the lattice rotation. Its gradient, the lattice curvature, can be related to the dislocation density tensor $\underset{\sim}{\alpha}$ (Nye 1953):

$$\underset{\sim}{\alpha} \sim \underset{\sim}{\kappa}^T + \mathrm{Tr}\,(\underset{\sim}{\kappa})\underset{\sim}{I} \tag{16.18}$$

The Cosserat curvature can therefore be interpreted as an approximate measure of the dislocation density tensor.

16.2.3 KWC-Cosserat Model

In this section we briefly recall the coupled KWC-Cosserat model proposed by Ask et al. (2018a, b, 2020). For a more in-depth presentation of the model, the reader is referred to the aforementioned works as well as Ghiglione et al. (2024). The model is formulated within the framework of small strains, rotations and curvatures. As it only considers the Read-Shockley part of the grain boundary energy-misorientation curve, it is limited to low angle grain boundaries. The coupling between the KWC phase field model and Cosserat mechanics via the following free energy potential:

$$\bar{\psi}(\phi, \bar{\nabla}\phi, \underline{\theta} \otimes \bar{\nabla}, \underset{\sim}{e}^e, \rho^\alpha) = \underbrace{f(\phi) + \frac{\bar{\alpha}^2}{2}\left|\bar{\nabla}\phi\right|^2 + \bar{s}g(\phi)\|\underline{\theta} \otimes \bar{\nabla}\| + \frac{\bar{\varepsilon}^2}{2}h(\phi)\|\underline{\theta} \otimes \bar{\nabla}\|^2}_{\text{KWC GB energy}}$$
$$+ \underbrace{\frac{1}{2}\mathrm{sym}(\underset{\sim}{e}^e) : \underset{\approx}{\bar{E}}^s : \mathrm{sym}(\underset{\sim}{e}^e) + 2\bar{\mu}_c \mathrm{skew}(\underset{\sim}{e}^e) : \mathrm{skew}(\underset{\sim}{e}^e)}_{\text{Cosserat strain energy}}$$
$$+ \underbrace{\bar{\psi}_\rho(\phi, \rho^\alpha)}_{\text{SSD accumulation}} \tag{16.19}$$

where the total elastic strain tensor has been split into symmetric and non-symmetric parts. The fourth-order tensor $\underset{\approx}{E}^s$ is the classical tensor of the elastic moduli acting on the symmetric elastic strain tensor. The stored energy potential accounting for the accumulation of statistically stored dislocations (SSD) is given by

$$\bar{\psi}_\rho(\phi, \rho^\alpha) = \frac{1}{2}\phi\bar{\mu}\bar{b}^2 \sum_{\alpha=1}^{N}\sum_{\beta=1}^{N} h^{\alpha\beta}\bar{\rho}^\beta \tag{16.20}$$

where $\mu, b, \rho^\alpha, h^{\alpha\beta}$ are respectively the shear modulus, the magnitude of the Burgers vector, the dislocation density of the slip system α and the slip system interaction matrix.

In this coupled formulation, the balance equations in the absence of external forces are then given by

$$\bar{\nabla} \cdot \bar{\underline{\xi}}_\phi + \bar{\pi}_\phi^{sto} + \bar{\pi}_\phi^{dis} = 0 \quad \text{in } \bar{\Omega} \tag{16.21}$$
$$\underset{\sim}{\bar{\sigma}} \cdot \bar{\nabla} = 0 \quad \text{in } \bar{\Omega} \tag{16.22}$$
$$\underset{\sim}{\bar{m}} \cdot \bar{\nabla} + 2\overset{\times}{\underline{\bar{\sigma}}} = 0 \quad \text{in } \bar{\Omega} \tag{16.23}$$

Equation (16.21) is nothing but the evolution equation for ϕ (Eq. (16.6)) in the variational framework of Gurtin (1996), in which $\underline{\xi}_\phi$ and $\pi_\phi = \pi_\phi^{sto} + \pi_\phi^{dis}$ are respectively the generalized forces associated to $\nabla\phi$ and ϕ. The field equations are complemented by Neumann boundary conditions setting the values of the generalized fluxes $\underline{\bar{\xi}}_{\phi} \cdot \underline{n}, \underset{\sim}{\bar{\sigma}} \cdot \underline{n}, \underset{\sim}{\bar{m}} \cdot \underline{n}$ on the boundary of the domain $\partial\bar{\Omega}$.

The various stress measures satisfy the following constitutive equations, after derivation of the free energy potential (16.19):

$$\mathrm{sym}(\underset{\sim}{\bar{\sigma}}) = \underset{\approx}{\bar{E}^{s}} : \underset{\sim}{\varepsilon}^{e} \tag{16.24}$$

$$\underline{\overset{\times}{\bar{\sigma}}} = 2\bar{\mu}_c \underline{\overset{\times}{e}}^{e} \tag{16.25}$$

$$\underset{\sim}{\bar{m}} = \left[\bar{s}g(\phi)\frac{1}{\|\underline{\theta} \otimes \bar{\nabla}\|} + \bar{\varepsilon}^2 h(\phi)\right]\underline{\theta} \otimes \bar{\nabla} \tag{16.26}$$

$$\underline{\bar{\xi}}_{\phi} = \bar{\alpha}^2 \bar{\nabla}\phi \tag{16.27}$$

The inelastic strain rate given by the following Eq. (16.28) is assumed to have contributions from crystal plasticity in the bulk of the grain and viscoelastic/viscoplastic relaxation of the skew-symmetric stresses at the grain boundary in response to lattice reorientation:

$$\underset{\sim}{\dot{e}}^{p} = \underbrace{\sum_{\alpha=1}^{N} \dot{\gamma}\, \underline{\ell}^{\alpha} \otimes \underline{n}^{\alpha}}_{\text{Crystal plasticity in the bulk of the grain}} - \underbrace{\underset{\approx}{\epsilon} \cdot \underline{\dot{\theta}}^{\star}}_{\text{Relaxation of skew-symmetric stress at the GB}} \tag{16.28}$$

where $\underline{\ell}^{\alpha}$ and $\underline{n}^{\alpha}$ are the slip direction and normal to the slip plane for the slip system α. The first part of the additively decomposed plastic strain rate corresponds to elastoplastic deformation in the bulk of the grain. The second one takes place in the diffuse grain boundary, where an orientation change is only induced by the evolution of $\underline{\theta}^{\star}$. The latter can be thought of as the orientation reference with zero skew-symmetric stress. This irreversible mechanism is associated with a relaxation equation whose driving force is the skew-symmetric stress:

$$\underline{\dot{\theta}}^{\star} = \frac{1}{\bar{\eta}^{\star} P(\bar{\nabla}\theta)} \underline{\overset{\times}{\bar{\sigma}}} \tag{16.29}$$

The reader is referred to Ghiglione et al. (2024) for a more detailed explanation and alternative relaxation equations. It is worth noting that due to the introduction of the reference orientation $\underline{\theta}^{\star}$ in Eq. (16.28), the skew-symmetric part of the elastic strain tensor is modified:

$$\underline{\overset{\times}{e}}^{e} = \underline{\overset{\times}{\omega}}^{e} - \underline{\theta} - \underline{\theta}^{\star} \sim \underline{0} \Rightarrow \underline{\overset{\times}{\omega}}^{e} \sim \underline{\theta} + \underline{\theta}^{\star} \tag{16.30}$$

This leads us to adopt the following initial condition to accommodate the initial orientation distribution from a stress-free state:

$$\underline{\theta}^{\star}(t=0) = -\underline{\theta}\,(t=0) \tag{16.31}$$

The slip rate $\dot{\gamma}$ in Eq. (16.28) is given by the power law function:

$$\dot{\gamma} = \sum_{\alpha=1}^{N} \left\langle \frac{|\bar{\tau}^{\alpha}| - \bar{R}^{\alpha}}{\bar{K}} \right\rangle^{n} \operatorname{sign}(\bar{\tau}^{\alpha}),$$

$$\text{with} \quad \bar{\tau}^{\alpha} = \bar{\underset{\sim}{\sigma}} : \underline{\ell}^{\alpha} \otimes \underline{n}^{\alpha} \quad \text{and} \quad \bar{R}^{\alpha} = \bar{\tau}_0 + \chi \bar{\mu} \bar{b} \sqrt{\sum_{\beta=1}^{N} h^{\alpha\beta} \bar{\rho}^{\beta}} \tag{16.32}$$

where $\langle \bullet \rangle = \max(0, \bullet)$, $\bar{K}$, n are viscoplasticity parameters, and $\chi \sim 0.3$ is a coefficient. The normalised resolved shear stress is denoted by $\bar{\tau}^{\alpha}$ and the critical resolved shear stress by $\bar{R}^{\alpha}$. The evolution of the SSD density follows a Kocks-Mecking law modified to account for static recovery in the wake of a moving grain boundary (Ask et al. 2020):

$$\dot{\bar{\rho}}^{\alpha} = \begin{cases} \dfrac{1}{\bar{b}} \left(\dfrac{1}{K_r} \sqrt{\sum_{\beta} \bar{\rho}^{\beta}} - 2\bar{d}\bar{\rho}^{\alpha} \right) |\dot{\gamma}^{\alpha}| \bar{\rho}^{\alpha} C_D \tanh\left(\bar{C}_A |\theta \otimes \bar{\nabla}|\right) \dot{\phi}, & \dot{\phi} > 0 \\ \dfrac{1}{\bar{b}} \left(\dfrac{1}{K_r} \sqrt{\sum_{\beta} \bar{\rho}^{\beta}} - 2\bar{d}\bar{\rho}^{\alpha} \right) |\dot{\gamma}^{\alpha}|, & \dot{\phi} \leq 0 \end{cases} \tag{16.33}$$

16.3 Simulations of Grain Boundary Migration Through a Void

16.3.1 Problem Set-up

Consider a 15×10 rectangular domain with a hole of radius $\bar{R} = 1.25$. The mesh, made of 4800 quadratic elements with reduced integration and 14800 nodes, is visible in Fig. 16.1.

A grain boundary is initialized at the left side of the structure so that the hole is fully contained in the grain on the right side of the boundary. To this end, a homogeneous ϕ field is set to 1 while an orientation difference of 0.3 rad is prescribed across the grain boundary location. The field ϕ is then automatically equilibrated according to the evolution equations. Subsequently, a stored energy difference is applied to provide a driving force for grain boundary migration across the void. Zero flux boundary conditions are enforced for both ϕ and θ at all the boundaries (including the hole surface) except for the top and bottom edges where periodicity is prescribed.

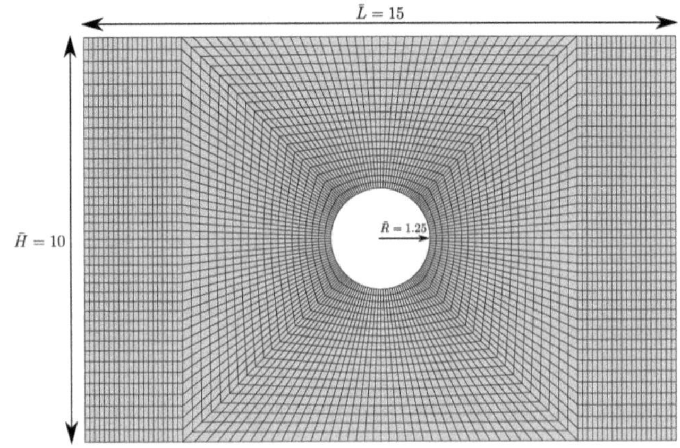

Fig. 16.1 Finite element mesh used in the simulation of stored energy driven migration of a grain boundary in a perforated plate

16.3.2 KWC Simulation

The simulation is first performed with the KWC model, i.e. in the absence of mechanical effects. The parameters given in Table 16.1 correspond to pure copper at about 220°C. Their identification requires consideration of misorientation-dependent grain boundary energy and grain boundary mobility. The corresponding physical properties were taken from experimental and molecular dynamics results available in the literature (Ghiglione et al. 2024).

Snapshots of the orientation and crystal order fields can be seen in Fig. 16.2. The curvature of the grain boundary becomes more noticeable as it approaches the pore, due to the interaction of ϕ with the void's free surface. As it passes through the pore, the zero flux boundary conditions ($\bar{\nabla}\phi \cdot \underline{\boldsymbol{n}} = 0$ with $\underline{\boldsymbol{n}}$ the outwards normal) on the void require the grain boundary to remain orthogonal to the void surface, resulting in the curvature of the grain boundary. A reorientation takes place around the void area, which leads to the reduction of the value of ϕ in that region. Once both the top and bottom portions of the grain boundary have passed through the void, they merge into a curved then straight grain boundary again. At the final state, a single crystal remains due to the grain boundary sweeping across the entire domain. Note that a similar behaviour has been modelled by Zhang et al. (2022) via a modified multi phase-field model of Chen-Yang type. The degrees of freedom in their research represent the recrystallized grains and inhomogeneous microstructure, rather than the grain orientation and crystal order. Another key difference is related to the numerical resolution method. In the present chapter, the finite element method is used so that the hole is represented by a meshless region. In contrast, Zhang et al. (2022) use a finite difference resolution scheme so that the grid points inside the hole are characterised by an additional parameter in the free energy density that modifies the grain boundary and stored energy contributions. Although the approaches used differ fundamentally,

Table 16.1 Model parameters used in the simulation of stored energy driven migration of a grain boundary in a perforated plate

Parameter	$\bar{\alpha}$	$\bar{s}$	$\bar{\varepsilon}$	$\bar{\eta}_\phi$	$\bar{\eta}_\theta$	$\bar{\mu}_P$	β_P	$\bar{\gamma}$	γ_g
Value	0.31	0.75	1	10	1	10^6	10^3	10^4	10^{-4}

there is significant similarity in the qualitative behaviour exhibited by the grain boundary.

16.3.3 *Consideration of Mechanical Effects: KWC-Cosserat Simulation*

The study is now extended by using the KWC-Cosserat model. The initial state, shown in Fig. 16.3, still consists in a grain boundary with a stored energy difference on each side of the interface, the initial dislocation density being higher on the right grain. A face centred cubic crystal symmetry is chosen involving 12 octahedral slip systems. The crystal orientation on the left side is 0 rad and that on the right side is 0.3 rad with respect to the [001] axis. It is obtained from a previous KWC simulation to ensure that ϕ is equilibrated with respect to both the orientation and stored energy difference. In order to roughly simulate a metal forming process, the bicrystal is subjected to compression and subsequent relaxation by imposing a vertical displacement at the top surface that linearly increases from $\bar{u}_y(\bar{t} = 0) = 0$ to $\bar{u}_y(\bar{t} = 20) = -0.1$, corresponding to -0.01 overall compression strain, and is then held constant. The simulation is in "2.5D" such that the structure can uniformly expand axially along the z (out of plane) direction, i.e. ε_{zz} is homogeneous and $\varepsilon_{xz} = \varepsilon_{yz} = 0$. These conditions are also called generalised plane strain conditions. The parameters used are indicative of pure copper at about 200°C and are given in Table 16.2. All 12 FCC slip systems are considered and the interaction matrix $[h]$ is given by

$$h_{ij} = \begin{cases} \omega_1 \text{ if } i \neq j \\ 1 + \omega_1 - \omega_2 \text{ else} \end{cases} \tag{16.34}$$

The deformed state before relaxation is shown in Fig. 16.4. An overall (vertical) compression stress is observed. Stress concentration occurs around the hole, with higher compressive stresses at the equator points and vanishing axial stress at the North and South poles. A tensile loading region is observed above the North pole and below the South pole. During relaxation, migration of the grain boundary is seen in Fig. 16.5 to occur as per the KWC model studied before. The KWC-Cosserat model provides further information. Notably, grain boundary motion through the void is shown to significantly reduce the stress concentration in the region swept by the

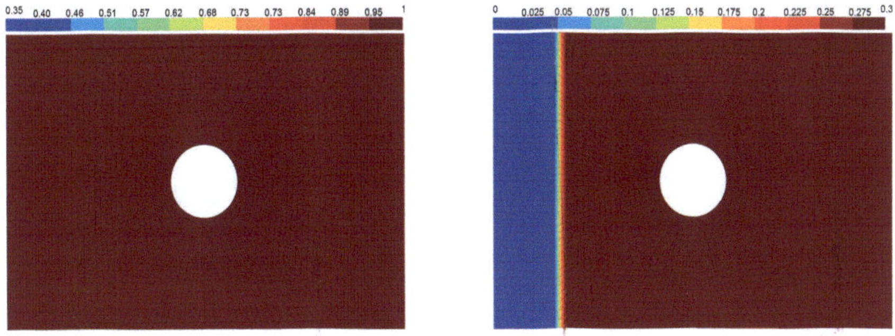

(a) ϕ and θ at $\bar{t} = 0$.

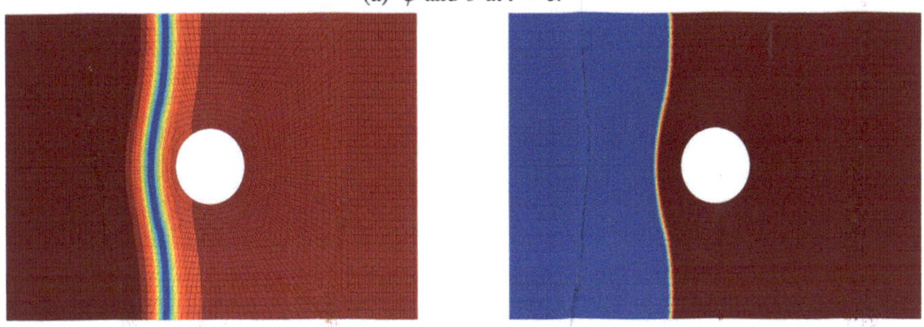

(b) ϕ and θ at $\bar{t} = 7500$.

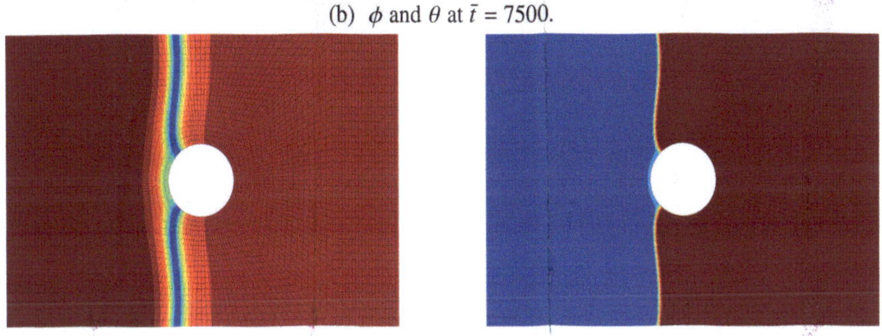

(c) ϕ and θ at $\bar{t} = 7700$.

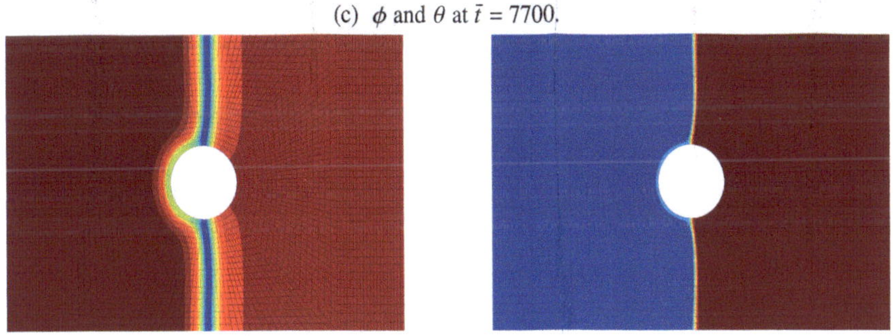

(d) ϕ and θ at $\bar{t} = 8100$.

Fig. 16.2 Snapshots of the crystal order (ϕ, on the left) and orientation (θ, on the right) fields during simulation of grain boundary migration through a void with the KWC model

(e) ϕ and θ at $\bar{t} = 9100$.

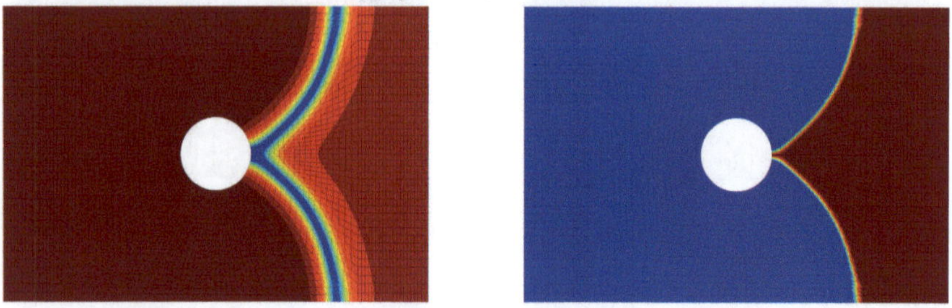

(f) ϕ and θ at $\bar{t} = 10550$.

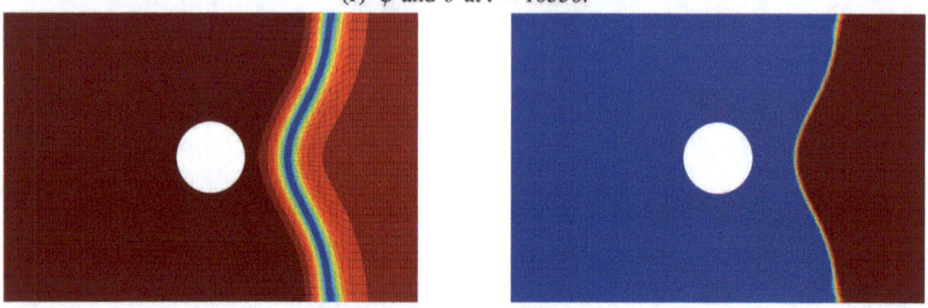

(g) ϕ and θ at $\bar{t} = 10600$.

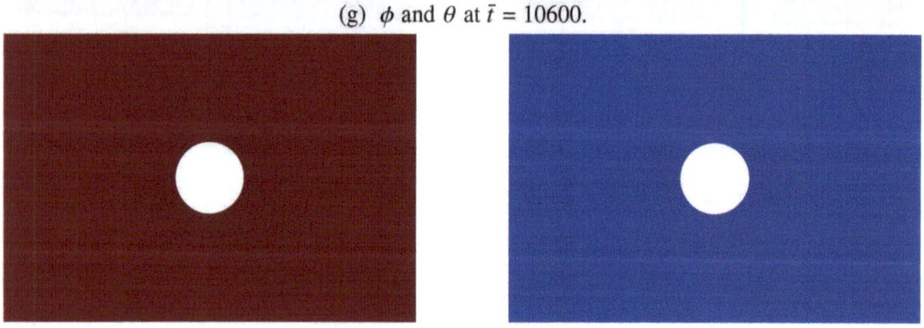

(h) ϕ and θ at $\bar{t} = 16100$.

Fig. 16.2 (continued)

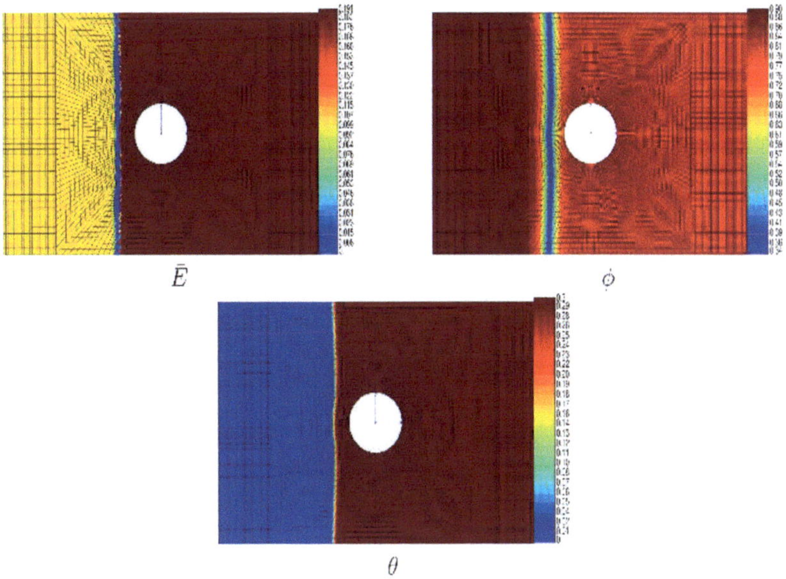

Fig. 16.3 Initial microstructure to study grain boundary migration through a void subjected to compression along the vertical direction and subsequent relaxation

interface. The compressive state can be shown to change to a much lower stress state with a magnitude about five times lower than in the compressive stress concentration regions in the left grain. Conversely, away from this zone, the stress does not exhibit a drastic decrease in magnitude. Once the grain boundary has swept the whole right grain domain very small stress levels remain, in contrast to significant compressive stress on the left side. This stress state emerges from the evolution of the reference orientation accompanied by relaxation of skew-symmetric stresses, according to Eq. (16.29), and subsequent stress components redistribution through the equilibrium equation. The motion of the grain boundary is accompanied by relaxation of elastic strains and related stress components. In the recent work of Wang et al. (2023), experimental investigations showed that grain boundary migration through voids prevented the nucleation and propagation of micro-cracks through the voids. They explained this phenomenon by means of crystal plasticity finite element (CPFE) simulations, by alleviation of stress concentration in the vicinity of the voids, i.e. exactly the case studied in this section. However, in their work grain boundary motion was not explicitly accounted for since their approach was based on a 'hard-soft grain pair'. The simulation with the KWC-Cosserat model therefore provides qualitative agreement with their conclusions, but also allows to go beyond the limitations of CPFE through the incorporation of the dynamics of the grain boundary.

The evolution of the plastic strains during the grain boundary migration is shown in Fig. 16.6. The plastic strain fields arise from the activation of slip systems. It is found that four main slip systems are activated in both grains. The presence of the hole

Table 16.2 Material parameters used for the simulation of a bicrystal containing a void and subjected to compression

Parameter	$\bar{C}_{11}$	$\bar{C}_{12}$	$\bar{C}_{44}$	$\bar{\mu}_c$	$\bar{K}$	n	$\bar{\tau}_c$	χ	$\bar{\mu}$	K_r	$\bar{d}$	$\bar{b}$	ω_1	ω_2
Value	69600	47800	32600	10000	4.35	10	4.35	0.3	20000	71.4	10^{-3}	0.256	1.5	1.2

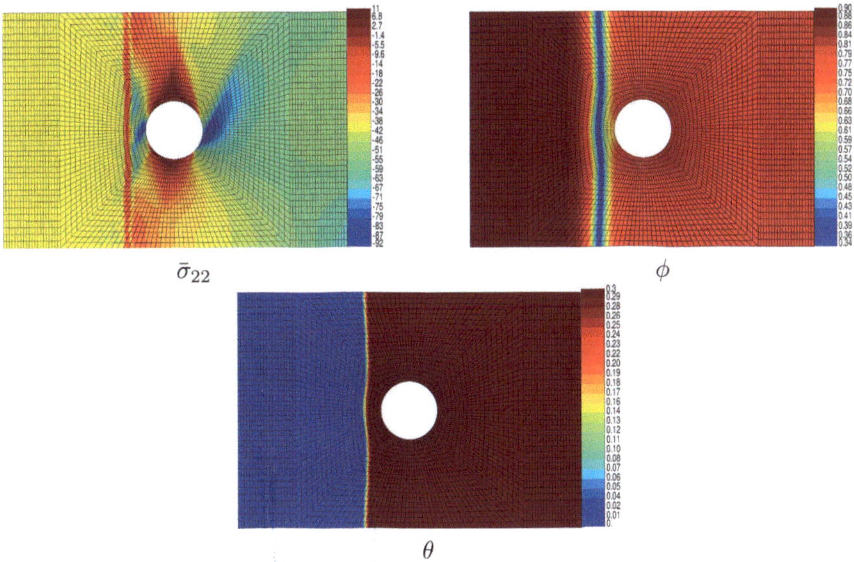

Fig. 16.4 Bicrystal with intragranular void after compression ($\bar{t} = 20$): axial stress field (**a**), phase field (**b**) and orientation field (**c**)

induces the activation of two additional slip systems in the form of strain localization bands close to the equator points, visible in Fig. 16.6 for all plastic strain components. In addition, the sweeping grain boundary region experiences enhanced activation of slip systems. Only slight evolution of plastic strain components is observed during relaxation. This is due to the fact that their variation is directly connected to the variation of elastic strains which are one order of magnitude smaller than the plastic strain induced by the initial compression. The strain redistribution is visible for the plastic strain component 33 because its field is much more homogeneous. The total strain component 33 is uniform due to the generalized plane strain condition. The elastic and plastic strain components 33 are not homogeneous due to the different crystal orientations. The magnitude of plastic strain 33 is seen to decrease and homogenize in the wake of the moving grain boundary. Once the grain boundary has swept the whole domain, heterogeneous patterns of plastic deformation are still visible.

16.4 Conclusion and Prospects

The problem of grain boundary migration through a void was first addressed using the two-variable KWC phase-field model. The morphological changes observed were similar to the ones published in the supplementary material of Zhang et al. (2022), that used a modified Chen-Yang multi-phase-field model. Then, the problem was revisited using the KWC-Cosserat model (Ask et al. 2018b) that accounts for the interplay

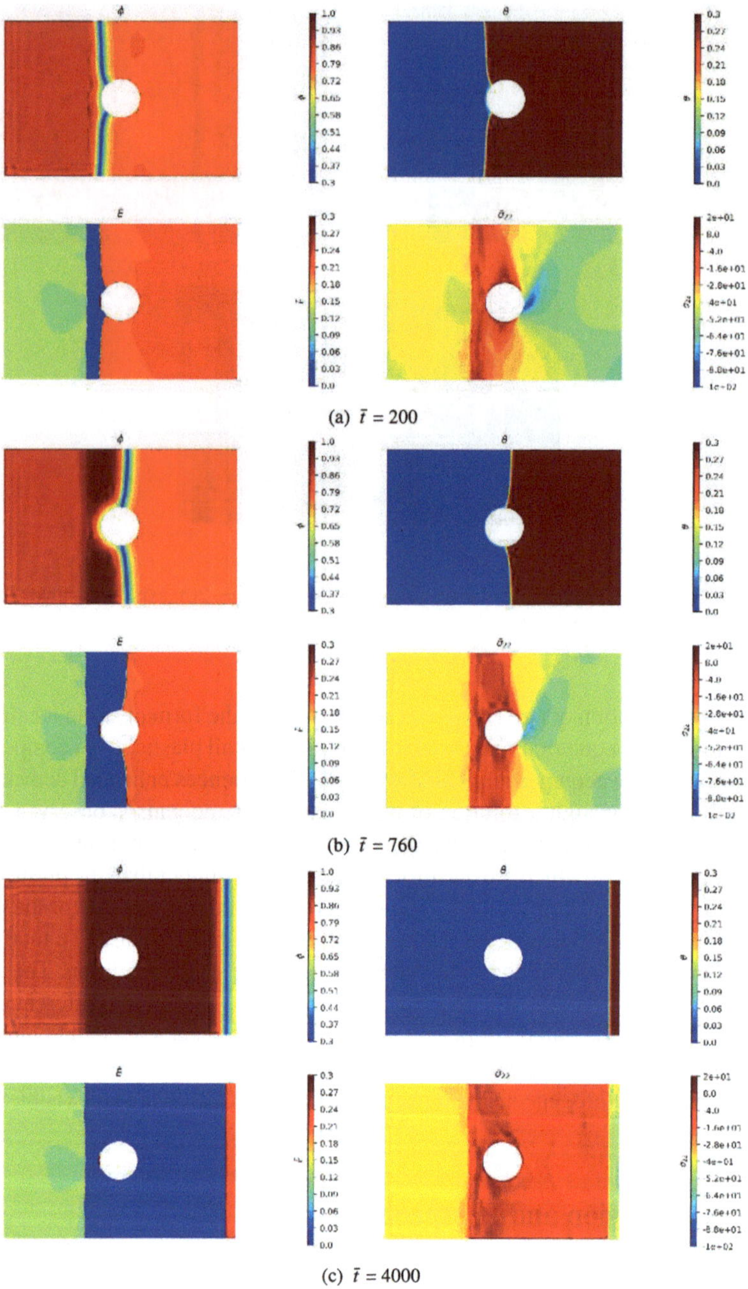

Fig. 16.5 Bicrystal with intragranular void after compression and subsequent relaxation: Maps of crystallinity order parameter, lattice orientation, stored energy and axial stress component at three time steps

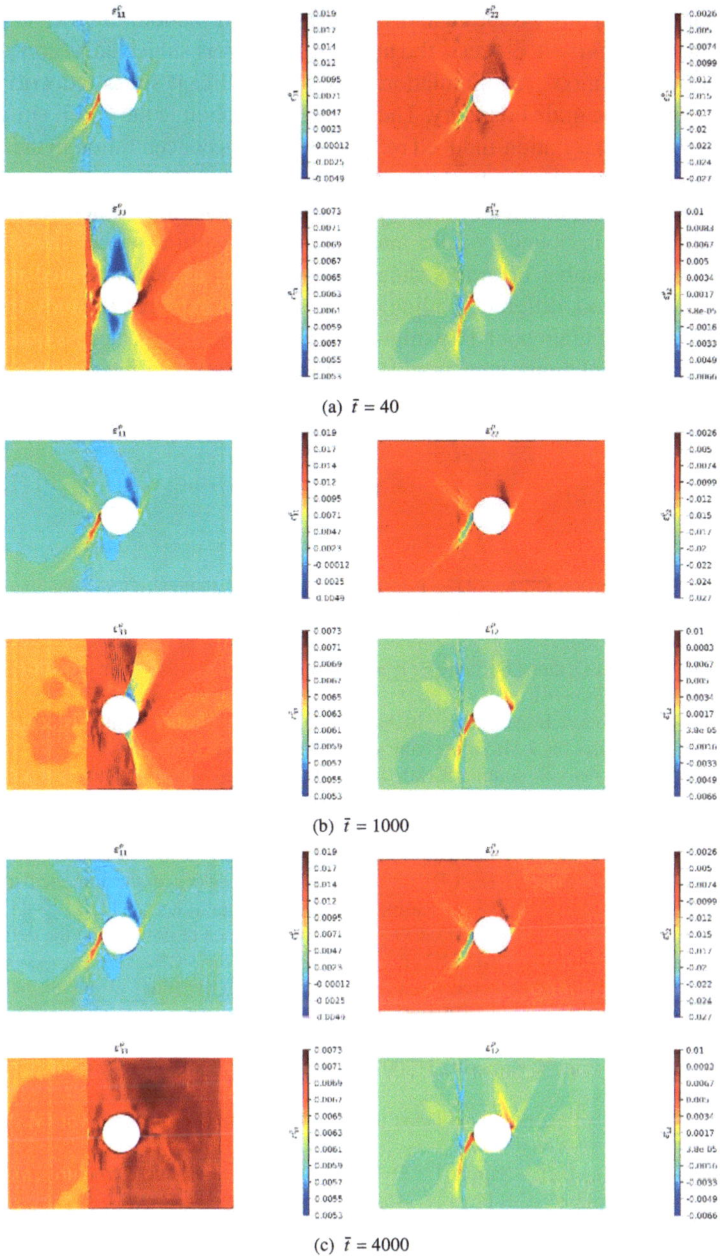

Fig. 16.6 Bicrystal with intragranular void after compression and subsequent relaxation: Maps of four plastic strain components at three time steps

between microstructure evolution and viscoplastic deformation. In order to roughly simulate a metal forming process, the microstructure was subjected to compression followed by relaxation. The simulation shows a significant stress reduction in the region swept by the grain boundary. These results corroborate the CPFE simulations of grain boundary migration-induced reduction of stress concentration in the vicinity of the void performed by Wang et al. (2023).

The simulations presented in this chapter can be regarded as a first step towards understanding grain boundary-pore interactions and require further investigation. Firstly, a larger number of grains should be included, as demonstrated in Zhang et al. (2022). Second, different loading cases should be explored to study the stress relaxation effect of grain boundary migration. Finally, in order to access quantitative and predictive simulations, an experiment-based calibration of the parameters must be performed.

References

Abrivard G, Busso E, Forest S, Appolaire B (2012) Phase field modelling of grain boundary motion driven by curvature and stored energy gradients. Part I: theory and numerical implementation. Phil Mag 92:3618–3642

Anderson M, Srolovitz D, Grest G, Sahni P (1984) Computer simulation of grain growth-I. Kinetics. Acta Metall 32(5):783–791

Ask A, Forest S, Appolaire B, Ammar K (2018a) Cosserat crystal plasticity with dislocation-driven grain boundary migration. J Micromechanics Mol Phys 3(03n04):1840009

Ask A, Forest S, Appolaire B, Ammar K (2020) Microstructure evolution in deformed polycrystals predicted by a diffuse interface Cosserat approach. Adv Model Simul Eng Sci 7(1):1–28

Ask A, Forest S, Appolaire B, Ammar K, Salman OU (2018) A Cosserat crystal plasticity and phase field theory for grain boundary migration. J Mech Phys Solids 115:167–194

Bernacki M, Chastel Y, Coupez T, Logé RE (2008) Level set framework for the numerical modelling of primary recrystallization in polycrystalline materials. Scripta Mater 58(12):1129–1132

Bourcier R, Koss D, Smelser R, Richmond O (1986) The influence of porosity on the deformation and fracture of alloys. Acta Metall 34(12):2443–2453

Chen L-Q, Yang W (1994) Computer simulation of the domain dynamics of a quenched system with a large number of nonconserved order parameters: The grain-growth kinetics. Phys Rev B 50(21):15752

Elder K, Grant M, Provatas N, Kosterlitz J (2001) Sharp interface limits of phase-field models. Phys Rev E 64(2):021604

Eringen A (1976) Polar and non local fields theories. In: Continuum physics, vol IV. Academic Press

Finel A, Le Bouar Y, Gaubert A, Salman U (2010) Phase field methods: microstructures, mechanical properties and complexity. C R Phys 11:245–256

Ghiglione F, Ask A, Ammar K, Appolaire B, Forest S (2024) Cosserat-phase-field modeling of grain nucleation in plastically deformed single crystals. J Mech Phys Solids 187:105628

Gurtin M (1996) Generalized Ginzburg-Landau and Cahn-Hilliard equations based on a microforce balance. Physica D 92:178–192

Han G (2023) Phase-field modeling of interaction between void and grain boundaries and its swelling effect. AIP Adv 13(8)

Hesselbarth HW, Göbel I (1991) Simulation of recrystallization by cellular automata. Acta Metall Mater 39(9):2135–2143

Hoetzer J, Rehn V, Rheinheimer W, Hoffmann MJ, Nestler B (2016) Phase-field study of pore-grain boundary interaction. J Ceram Soc Jpn 124(4):329–339
Kobayashi R, Warren JA, Carter WC (2000) A continuum model of grain boundaries. Physica D 140(1–2):141–150
Nye JF (1953) Some geometrical relations in dislocated crystals. Acta Metall 1(2):153–162
Rehn V, Hötzer J, Kellner M, Seiz M, Serr C, Rheinheimer W, Hoffmann M, Nestler B (2018) The impact of pores on microstructure evolution: a phase-field study of pore-grain boundary interaction. In: High performance computing in science and engineering' 17: transactions of the high performance computing center, Stuttgart (HLRS) 2017. Springer, Berlin, pp 485–502
Steinbach I, Apel M (2006) Multi phase field model for solid state transformation with elastic strain. Physica D 217:153–160
Steinbach I, Pezzolla F (1999) A generalized field method for multiphase transformations using interface fields. Physica D 134(4):385–393
Wang W, Balint DS, Shirzadi AA, Wang Y, Lee J, Aucott L, Jiang J (2023) Imparted benefits on mechanical properties by achieving grain boundary migration across voids. Acta Mater 119103
Warren JA, Kobayashi R, Lobkovsky AE, Carter WC (2003) Extending phase field models of solidification to polycrystalline materials. Acta Mater 51(20):6035–6058
Zhang C, Yadav V, Moelans N, Jensen DJ, Yu T (2022) The effect of voids on boundary migration during recrystallization in additive manufactured samples-A phase field study. Scripta Mater 214:114675

Chapter 17
Phase Field Modeling and Sharp Interface Limits of Multi-component Gas-Liquid Flows with Electrochemical Reactions and Phase Transition Effects

Wolfgang Dreyer, Jan Giesselmann, and Christiane Kraus

Abstract We derive a novel thermodynamically consistent diffuse-interface model for compressible electrolytes undergoing phase transitions. The fluid mixtures may contain N constituents in liquid and vapor phases, with the possibility of phase coexistence. Furthermore, all constituents are assumed to exhibit polarizable and magnetizable properties. Our proposed thermodynamically consistent model serves as a generalized framework of the Allen–Cahn/Navier–Stokes/Poisson type, tailored for multi-component flows involving phase transitions and electrochemical reactions. Using matched asymptotic techniques, we analyze the physically admissible sharp interface limits of the model. Two scaling regimes are considered: a non-coupled regime and a coupled regime, where coupling occurs between the smallness parameter in the Poisson equation and the interface width. In the sharp interface limit, we derive a generalized Allen–Cahn/Euler/Poisson system that governs mixtures with electrochemical reactions in the bulk phases, complemented by admissible interfacial conditions. These conditions include, for example, a generalized Gibbs–Thomson law and a dynamic Young–Laplace law.

17.1 Introduction

This study presents a model for chemically reacting viscous fluid mixtures capable of undergoing phase transitions between liquid and vapor states. The mixture consists of

† W. Dreyer—deceased.

W. Dreyer
Weierstrass Institute, Berlin, Germany

J. Giesselmann
TU Darmstadt, Darmstadt, Germany
e-mail: giesselmann@mathematik.tu-darmstadt.de

C. Kraus (✉)
TH Würzburg-Schweinfurt, Würzburg, Germany
e-mail: christiane.kraus@thws.de

J. Fuhrmann et al. (eds.), *Advances in Continuum Physics*, Advanced Structured Materials 238, https://doi.org/10.1007/978-3-031-93918-1_17

N constituents, which may include polarizable and magnetizable matter. The system is governed by N partial mass balance equations, a single equation for barycentric momentum balance, and a Poisson-type equation. To model phase transitions, we introduce an artificial phase field that distinguishes the liquid and vapor phases by assigning the values 1 and -1 to the liquid and the vapor phases, respectively. Within the transition layer separating adjacent phases, the phase field changes smoothly between 1 and -1. Typically, these layers are very thin, resulting in steep phase field gradients.

The proposed model belongs to the class of diffuse-interface models, which represent phase transitions through smooth changes in the phase field. In contrast, sharp interface models describe these transitions with discontinuities at the interface. While sharp interface models have simpler physical foundations, an essential and non-trivial question is whether the sharp interface limits of diffuse models yield physically valid sharp interface models.

Diffuse interface models solve partial differential equations (PDEs) across the transition region, while sharp interface models impose jump conditions at the phase boundary, sometimes combined with geometric PDEs.

In the isothermal and quasi-static electrodynamic setting, our diffuse-interface model is described by a system of PDEs for the partial mass densities ρ_α, the barycentric velocity $\boldsymbol{v}$, the phase field parameter χ, and the electrical potential φ. The evolution equation for the total mass density $\rho = \sum_{\alpha=1}^{N} \rho_\alpha$ replaces the equation for ρ_N by keeping the same structure for all partial mass densities ρ_α, $\alpha = 1, \ldots N$:

$$
\begin{aligned}
0 &= \partial_t \rho_\alpha + \operatorname{div}(\rho_\alpha \boldsymbol{v}) - \operatorname{div}\Bigg(\sum_{\beta=1}^{N-1} M_{\alpha\beta} \Big(\nabla \frac{\mu_\beta - \mu_N}{T} + \frac{1}{T} \Big(\frac{z_\beta e_0}{m_\beta} - \frac{z_N e_0}{m_N} \Big) \nabla \varphi \Big) \Bigg) \\
&\quad - \sum_{i=1}^{N_R} m_\alpha \gamma_\alpha^i M_{\mathrm{r}}^i \Big(1 - \exp\Big(\frac{1}{kT} \sum_{\beta=1}^{N} m_\beta \gamma_\beta^i \mu_\beta \Big) \Big), \qquad \alpha = 1, \ldots, N-1, \\
0 &= \partial_t \rho + \operatorname{div}(\rho \boldsymbol{v}), \\
0 &= \partial_t (\rho \boldsymbol{v}) + \operatorname{div}(\rho \boldsymbol{v} \otimes \boldsymbol{v}) + \nabla \Big(\sum_{\alpha=1}^{N} \rho_\alpha \mu_\alpha - \rho f - W - \frac{\gamma}{2} |\nabla \chi|^2 \Big) + \gamma \operatorname{div}(\nabla \chi \otimes \nabla \chi) \\
&\quad - \operatorname{div}(\boldsymbol{\sigma}^{\mathrm{NS}}) + \varepsilon_0 \operatorname{div} \Big((1 + s(\chi)) \Big(\frac{1}{2} |\nabla \varphi|^2 \mathbf{1} - \nabla \varphi \otimes \nabla \varphi \Big) \Big), \\
0 &= \partial_t \chi + \boldsymbol{v} \cdot \nabla \chi + \frac{\tau}{\rho} \Big(W' - \gamma \Delta \chi + \frac{\partial(\rho f)}{\partial \chi} - \frac{\varepsilon_0}{2} s'(\chi) |\nabla \varphi|^2 \Big), \\
0 &= \varepsilon_0 \operatorname{div}((1 + s(\chi)) \nabla \varphi) + n^{\mathrm{F}},
\end{aligned}
$$

where $M_{\alpha\beta}$, M_{r}^i are the mobilities; μ_α denotes the chemical potentials; T is the temperature; m_α is the atomic mass; e_0 is the elementary charge; ε_0 is the vacuum permittivity; z_α are integer charge numbers; and k is Boltzmann's constant. The model incorporates the following free energy:

$$\rho\psi := W(\chi) + \frac{\gamma}{2}|\nabla\chi|^2 + h(\chi)\rho\psi_{\mathrm{L}}(\rho_1, \ldots, \rho_N) + (1 - h(\chi))\rho\psi_{\mathrm{V}}(\rho_1, \ldots, \rho_N) - \frac{\varepsilon_0}{2}s(\chi)|\boldsymbol{E}|^2,$$

where $W(\chi) := (\chi - 1)^2(\chi + 1)^2$, $\rho\psi_{\mathrm{L}}$, $\rho\psi_{\mathrm{V}}$ are the free energy functions for the pure phases and $\boldsymbol{E}$ is the electric field. The smooth interpolation function $h : \mathbb{R} \to [0, 1]$ satisfies

$$h(z) = \begin{cases} 1 \text{ for } & z \geq 1, \\ 0 \text{ for } & z \leq -1, \end{cases}$$

such that $h'(z) = 0$ for all $|z| \geq 1$. Similarly, we assume

$$s(\chi) = h(\chi)s_{\mathrm{L}} + (1 - h(\chi))s_{\mathrm{V}},$$

where $s_{\mathrm{L/V}}$ are the susceptibilities of the pure phases. For brevity, we define

$$\begin{aligned}(\rho f)((\rho_\alpha)_\alpha, \chi) &:= (\rho f)(\rho_1, \ldots, \rho_N, \chi) \\ &:= h(\chi)\rho\psi_{\mathrm{L}}(\rho_1, \ldots, \rho_N) + (1 - h(\chi))\rho\psi_{\mathrm{V}}(\rho_1, \ldots, \rho_N).\end{aligned}$$

By definition, the chemical potentials are given by

$$\mu_\alpha := \frac{\partial(\rho\psi)}{\partial\rho_\alpha} = \frac{\partial(\rho f)}{\partial\rho_\alpha}.$$

In addition, $n^{\mathrm{F}} := e_0 \sum_{\alpha=1}^{N} \frac{z_\alpha}{m_\alpha}\rho_\alpha$ is the free charge density, $\boldsymbol{\sigma}^{\mathrm{NS}}$ denotes the Navier–Stokes stress, and τ, γ are (positive) constants.

If electrical effects are neglected, this compressible multi-phase flow model reduces to an Allen–Cahn/Navier–Stokes-type model, which has been studied in Dreyer et al. (2014) and numerically investigated by Hantke et al. (2023). For single-component systems undergoing liquid–vapor transitions, it aligns with the model proposed by Blesgen (1999). Analytical studies of Blesgen's model are available in Kotschote (2012); Feireisl et al. (2010), where existence of strong local-in-time solutions and weak solutions has been shown. A modified version of Blesgen's model is investigated in Witterstein (2011).

Building on the approach outlined in Dreyer et al. (2014), the impingement of a droplet onto a solid surface at high-impact speeds is investigated in Ostrowski et al. (2020), where thermodynamically consistent boundary conditions are developed to account for fluid–wall interactions, including dynamic contact angles.

In the absence of chemical reactions, our work is related to diffuse-interface models for incompressible and quasi-incompressible fluids. A diffuse-interface model of Navier–Stokes–Cahn–Hilliard type for two incompressible, viscous Newtonian fluids, having the same densities, has been introduced by Hohenberg and Halperin in Hohenberg and Halperin (1977). That model has been modified in several thermodynamically consistent ways such that different densities are allowed, see, e.g., Gurtin et al. (1996); Lowengrub and Truskinovsky (1998); Abels et al. (2012). For existence results of strong local-in-time solutions and weak solutions, we refer to Abels

(2009, 2012); Abels et al. (2012). A diffuse-interface model for two incompressible constituents which permits the transfer of mass between the phases due to diffusion and phase transitions has been proposed in Aki et al. (2014, 2012). The densities of the fluids may be different, which leads to quasi-incompressibility of the mixture.

In Ghosh et al. (2024), a phase field model for simulating evaporation processes in porous media, explicitly incorporating a vapor component with the liquid and gas phases, is introduced and sharp interface limits are considered. The model framework integrates conservation laws for mass (including phases and vapor components), momentum, and energy, while the Allen–Cahn equation governs the evolution of the phase field.

In Ten Eikelder et al. (2023), various diffuse-interface Navier–Stokes Cahn–Hilliard models with non-matching densities are unified under a single framework. The approach is based on three principles: (1) a single system of balance laws from continuum mixture theory, (2) a unified energy-dissipation law for quasi-incompressible models, and (3) variations arising only from constitutive choices. This framework consolidates existing models using a single momentum equation for mixture velocity while allowing for further refinement in energy functionals and constitutive dependencies.

Modeling and simulation of compressible multi-component gas-liquid flows with chemical reaction and phase transition effects have been also considered in Xu et al. (2024). Therefore, all volume fraction transport equations with non-conservative characteristics of gaseous components are substituted with a thermal equilibrium assumption inside the gaseous mixture for closure.

A strong thermodynamic foundation for viscous, heat-conducting fluid mixtures undergoing chemical reactions is considered in Bothe and Dreyer (2015). By leveraging the entropy principle, the authors derive a closed system of partial mass and momentum balances along with an internal energy balance.

This work is structured as follows: In the next section, we derive the thermodynamically consistent model for multi-component flows with phase transitions and electrochemical reactions. Section 17.3 addresses the non-dimensionalization and introduces two scaling regimes for asymptotic analysis. Finally, Sects. 17.5 and 17.6 examine the sharp interface limits for the identified scaling regimes.

17.2 The Electrolyte Model

Constituents and phases. We consider a fluid mixture consisting of N constituents $A_1, A_2, \ldots, A_N$ indexed by $\alpha \in \{1, 2, \ldots, N\}$. Each constituent has a (atomic) mass $(m_\alpha)_{\alpha=1,2,\ldots,N}$ and may carry a charge $(z_\alpha e_0)_{\alpha=1,2,\ldots,N}$, where e_0 is the elementary charge and z_α are integers (positive, negative, or zero). All constituents may consist of polarizable and magnetizable matter.

The fluid mixture may exist in two phases: liquid (L) and vapor (V), which may coexist. In this work, we describe these phases using a diffuse-interface approach, where the interface between the liquid and vapor phases is modeled as a thin layer.

Within this layer, thermodynamic quantities transition smoothly between their values in each phase, though typically with steep gradients.

The constituent with index N represents the neutral solvent, for which $z_N = 0$. Among the $N-1$ dissolved substances, there are neutral molecules, as well as positive and negative ions, which are the products of dissociation reactions. There are N_R reactions, indexed by $i \in \{1, 2, \ldots, N_R\}$, of the general type:

$$a_1^i A_1 + a_2^i A_2 + \cdots + a_N^i A_N \rightleftarrows b_1^i A_1 + b_2^i A_2 + \cdots + b_N^i A_N. \tag{17.1}$$

The constants $(a_\alpha^i)_{\alpha=1,2,\ldots,N}$ and $(b_\alpha^i)_{\alpha=1,2,\ldots,N}$ are positive integers and $\gamma_\alpha^i = b_\alpha^i - a_\alpha^i$ denote the stoichiometric coefficients of the reaction i.

Basic quantities and basic variables. Two phase mixtures can be modeled within three different model classes, denoted by Class I–Class III, cf. Hutter and Jöhnk (2004) for the theory of mixtures.

Class I The basic variables are the number densities $(n_\alpha)_{\alpha=1,2,\ldots,N}$ of the constituents, the barycentric velocity $\boldsymbol{v}$, the temperature T of the mixture, the electromagnetic field $(\boldsymbol{E}, \boldsymbol{B})$, and the phase field χ.
Class II The basic variables are the number densities $(n_\alpha)_{\alpha=1,2,\ldots,N}$, the velocities $(\boldsymbol{v}_\alpha)_{\alpha=1,2,\ldots,N}$ of the constituents, the temperature T, the electromagnetic field $(\boldsymbol{E}, \boldsymbol{B})$, and the phase field χ.
Class III Finally, in Class III, we have the number densities $(n_\alpha)_{\alpha=1,2,\ldots,N}$, the velocities $(\boldsymbol{v}_\alpha)_{\alpha=1,2,\ldots,N}$, the temperatures $(T_\alpha)_{\alpha=1,2,\ldots,N}$ of the constituents, the electromagnetic field $(\boldsymbol{E}, \boldsymbol{B})$, and the phase field χ.

In this study, we choose a description within Class I.

The mixture occupies a region $\Omega \subset \mathbb{R}^3$. At any time $t \geq 0$, the thermodynamic state of the mixture is described by N partial mass densities $(\rho_\alpha)_{\alpha=1,2,\ldots,N}$, the barycentric velocity $\boldsymbol{v}$, the temperature T of the mixture, and the electromagnetic field $(\boldsymbol{E}, \boldsymbol{B})$. These quantities may depend on time $t \geq 0$ and space $\boldsymbol{x} = (x_i)_{i=1,\ldots,3} = (x_1, x_2, x_3)$. However, the magnetic field and the temperature appear explicitly only in the modeling part (Sect. 17.2). We focus on isothermal processes and neglect magnetic fields, omitting $\boldsymbol{B}$, while treating T as a constant parameter in the equations.

To indicate the phase at a given point $(t, \boldsymbol{x})$, we introduce the phase field χ as an additional basic variable. The phase field takes values in the interval $[-1, 1]$, where $\chi = 1$ represents the liquid phase and $\chi = -1$ corresponds to the vapor phase.

Multiplying the number densities by m_α and $e_0 z_\alpha$, respectively, we obtain the partial mass densities and the partial free charge densities:

$$\rho_\alpha = m_\alpha n_\alpha, \qquad n_\alpha^{\mathrm{F}} = e_0 z_\alpha n_\alpha. \tag{17.2}$$

Multiplication of the velocities by $n_\alpha m_\alpha$ and $n_\alpha z_\alpha e_0$, respectively, gives the mass fluxes and the free currents:

$$\boldsymbol{j}_\alpha = \rho_\alpha \boldsymbol{v}_\alpha, \qquad \boldsymbol{j}_\alpha^F = e_0 z_\alpha \boldsymbol{v}_\alpha n_\alpha. \tag{17.3}$$

The total mass density of the mixture and the barycentric velocity are defined as

$$\rho = \sum_{\alpha=1}^{N} \rho_\alpha, \qquad \boldsymbol{v} = \frac{1}{\rho}\sum_{\alpha=1}^{N} \rho_\alpha \boldsymbol{v}_\alpha. \tag{17.4}$$

Non-convective mass fluxes and non-convective currents are given by

$$\boldsymbol{J}_\alpha = \rho_\alpha \boldsymbol{u}_\alpha, \qquad \boldsymbol{J}_\alpha^{\mathrm{F}} = \frac{z_\alpha e_0}{m_\alpha}\boldsymbol{J}_\alpha, \quad \text{where} \quad \boldsymbol{u}_\alpha = \boldsymbol{v}_\alpha - \boldsymbol{v} \tag{17.5}$$

denotes the diffusion velocity. From definitions $(17.5)_{1,3}$, it follows that

$$\sum_{\alpha=1}^{N} \boldsymbol{J}_\alpha = 0. \tag{17.6}$$

The total free charge density and total free current are expressed as

$$n^{\mathrm{F}} = \sum_{\alpha=1}^{N} n_\alpha^{\mathrm{F}}, \qquad \boldsymbol{j}^{\mathrm{F}} = n^{\mathrm{F}}\boldsymbol{v} + \sum_{\alpha=1}^{N} \boldsymbol{J}_\alpha^{\mathrm{F}}. \tag{17.7}$$

Applying Maxwell's theory to continuous matter shows that the total electric charge density n^e and the total electric current $\boldsymbol{j}^e$ consist of two contributions:

$$n^{\mathrm{e}} = n^{\mathrm{F}} + n^{\mathrm{P}}, \qquad \boldsymbol{j}^{\mathrm{e}} = \boldsymbol{j}^{\mathrm{F}} + \boldsymbol{j}^{\mathrm{P}}. \tag{17.8}$$

In addition to free charge densities and free currents, there are charge densities and currents due to polarization and magnetization (Müller 1985).

$$n^{\mathrm{P}} = -\mathrm{div}(\boldsymbol{P}), \qquad \boldsymbol{j}^{\mathrm{P}} = \frac{\partial \boldsymbol{P}}{\partial t} + \mathrm{curl}(\boldsymbol{P} \times \boldsymbol{v} + \boldsymbol{M}). \tag{17.9}$$

Here, $\boldsymbol{P}$ and $\boldsymbol{M}$ represent the polarization and magnetization vectors, respectively. Polarization accounts for effects caused by microscopic charges, such as atomic dipoles in atoms and molecules. Microscopic currents are macroscopically represented by the magnetization vector.

Finally, we introduce the total number density of the mixture and the atomic fractions of the constituents.

$$n = \sum_{\alpha=1}^{N} n_\alpha, \qquad y_\alpha = \frac{n_\alpha}{n} \quad \text{with} \quad \sum_{\alpha=1}^{N} y_\alpha = 1. \tag{17.10}$$

Equations of balance for matter. The basic variables of the system are governed by a coupled set of partial differential equations, which are based on the quasi-static

Maxwell equations and the balance equations for matter. We begin by introducing the balance equations for matter within the Class I model, which necessitates equations for the mass balance of the constituents, as well as for the momentum and energy balance of the mixture. These equations are given as follows:

$$\partial_t \rho_\alpha + \operatorname{div}(\rho_\alpha \boldsymbol{v} + \boldsymbol{J}_\alpha) = r_\alpha, \qquad \alpha = 1, 2, \ldots, N, \tag{17.11}$$

$$\partial_t(\rho \boldsymbol{v}) + \operatorname{div}(\rho \boldsymbol{v} \otimes \boldsymbol{v} - \boldsymbol{\sigma}) = \rho \boldsymbol{b} + \boldsymbol{k}, \tag{17.12}$$

$$\partial_t\Big(\rho e + \frac{\rho}{2}\boldsymbol{v}^2\Big) + \operatorname{div}\Big((\rho e + \frac{\rho}{2}\boldsymbol{v}^2)\boldsymbol{v} + \boldsymbol{q} - \boldsymbol{v}\cdot\boldsymbol{\sigma}\Big) = \rho \boldsymbol{b}\cdot\boldsymbol{v} + \pi. \tag{17.13}$$

Additionally, we introduce a balance equation for the phase field:

$$\partial_t(\rho\chi) + \operatorname{div}(\rho\chi\boldsymbol{v} + \boldsymbol{J}_\chi) = \xi_\chi. \tag{17.14}$$

In the above equations, there appear several physical quantities:

- r_α: mass production of constituent A_α;
- $\boldsymbol{\sigma}$: stress;
- ρe: internal energy density;
- $\boldsymbol{q}$: heat flux;
- $\boldsymbol{J}_\chi$: non-convective flux of the phase field;
- ξ_χ: phase field production.

Furthermore, the force density consists of two distinct components:

- $\rho \boldsymbol{b}$: the force density due to gravity and inertia;
- $\boldsymbol{k}$: the Lorentz force density arising from electromagnetic fields.

Similarly, the power of forces is decomposed as

- $\rho \boldsymbol{b}\cdot\boldsymbol{v}$: power due to gravity and inertia;
- π: power associated with Joule heating.

For simplicity, in the following derivations, we neglect the force density $\boldsymbol{b}$ by setting $\boldsymbol{b} = 0$.

The mass production rate of constituent A_α is influenced by forward and backward reactions. The rcaction rates R_{f}^i and R_{b}^i represent the number of forward and backward reactions per unit volume per unit time. Thus,

$$r_\alpha = \sum_{i=1}^{N_{\mathrm{R}}} m_\alpha \gamma_\alpha^i (R_{\mathrm{f}}^i - R_{\mathrm{b}}^i). \tag{17.15}$$

Charge and mass conservation for each reaction $i \in \{1, 2, \ldots, N_{\mathrm{R}}\}$ require

$$\sum_{\alpha=1}^{N} z_\alpha \gamma_\alpha^i = 0 \quad \text{and} \quad \sum_{\alpha=1}^{N} m_\alpha \gamma_\alpha^i = 0, \quad \text{implying} \quad \sum_{\alpha=1}^{N} r_\alpha = 0. \tag{17.16}$$

The last condition in (17.16) ensures the conservation of total mass. Summing the partial mass balance Eq. (17.11) leads to the total mass balance for the mixture:

$$\partial_t \rho + \operatorname{div}(\rho \boldsymbol{v}) = 0. \tag{17.17}$$

Here, we have used definitions from (17.4) alongside conditions $(17.16)_3$ and (17.6).

A brief reminder on Maxwell's equations. The electromagnetic field $(\boldsymbol{E}, \boldsymbol{B})$ is determined by Maxwell's equations, which take the form Müller (1985):

$$\partial_t \boldsymbol{B} + \operatorname{curl}(\boldsymbol{E}) = 0, \qquad \operatorname{div}(\boldsymbol{B}) = 0, \tag{17.18}$$

$$-\frac{1}{c^2}\partial_t \boldsymbol{E} + \operatorname{curl}(\boldsymbol{B}) = \mu_0 \boldsymbol{j}^{\mathrm{e}}, \qquad \operatorname{div}(\boldsymbol{E}) = \frac{1}{\varepsilon_0} n^{\mathrm{e}}. \tag{17.19}$$

Here, the speed of light is related to the vacuum permittivity and permeability via $c^2 = 1/(\varepsilon_0 \mu_0)$. The total electric charge density n^{e} and the total electric current density $\boldsymbol{j}^{\mathrm{e}}$ are provided by expressions (17.7)–(17.10).

Multiplying Maxwell's equations by $\boldsymbol{E}$ and $\boldsymbol{B}$, respectively, leads to the balance equations:

$$\partial_t \boldsymbol{m}^{\mathrm{e}} + \operatorname{div}(-\boldsymbol{\sigma}^{\mathrm{e}}) = -n^{\mathrm{e}} \boldsymbol{E} - \boldsymbol{j}^{\mathrm{e}} \times \boldsymbol{B}, \qquad \partial_t e^{\mathrm{e}} + \operatorname{div}(\boldsymbol{q}^{\mathrm{e}}) = -\boldsymbol{j}^{\mathrm{e}} \cdot \boldsymbol{E}. \tag{17.20}$$

These equations are interpreted as the equations of balance for electromagnetic momentum and electromagnetic energy. The corresponding densities and fluxes have the unique representations:

$$\boldsymbol{m}^{\mathrm{e}} = \varepsilon_0 \boldsymbol{E} \times \boldsymbol{B}, \qquad \boldsymbol{\sigma}^{\mathrm{e}} = \varepsilon_0 \boldsymbol{E} \otimes \boldsymbol{E} + \frac{1}{\mu_0} \boldsymbol{B} \otimes \boldsymbol{B} - \frac{1}{2}\left(\varepsilon_0 \boldsymbol{E}^2 + \frac{1}{\mu_0} \boldsymbol{B}^2\right) \mathbf{1}, \tag{17.21}$$

$$e^{\mathrm{e}} = \frac{\varepsilon_0}{2} \boldsymbol{E}^2 + \frac{1}{2\mu_0} \boldsymbol{B}^2, \quad \boldsymbol{q}^{\mathrm{e}} = \frac{1}{\mu_0} \boldsymbol{E} \times \boldsymbol{B}. \tag{17.22}$$

By adding the electromagnetic momentum and energy balance equations to their respective matter balance equations, we derive the total momentum and total energy equations. The postulate of conservation of total momentum and total energy leads to the identification of the Lorentz force and its associated power:

$$\boldsymbol{k} = n^{\mathrm{e}} \boldsymbol{E} + \boldsymbol{j}^{\mathrm{e}} \times \boldsymbol{B}, \qquad \pi = \boldsymbol{j}^{\mathrm{e}} \cdot \boldsymbol{E}. \tag{17.23}$$

On the quasi-static setting of electrodynamics. In electrochemical literature, there is significant confusion regarding the quasi-static approximation of Maxwell's equations. To clarify this topic, a brief discussion is warranted.

We begin by introducing a rescaling of time, space, the electric and magnetic fields, and the conductivity: σ:

$$t = t_0 \tilde{t}, \quad \boldsymbol{x} = x_0 \tilde{\boldsymbol{x}}, \quad \boldsymbol{E} = E_0 \tilde{\boldsymbol{E}}, \quad \boldsymbol{B} = \frac{E_0}{c} \tilde{\boldsymbol{B}}, \quad \sigma = \sigma_0 \tilde{\sigma}, \quad n^{\mathrm{F}} = n_0^{\mathrm{F}} \tilde{n}^{\mathrm{F}}. \tag{17.24}$$

From Ohm's law, we express the current density as

$$\boldsymbol{j}^{\mathrm{e}} = \sigma_0 \tilde{\sigma} E_0 \tilde{\boldsymbol{E}} =: \sigma_0 E_0 \tilde{\boldsymbol{j}}^{\mathrm{e}}.$$

The rescaled magnetic field has the same dimension as the electric field so that both fields can be compared. Additionally, we define $x_0/t_0 = v_0$, where v_0 represents a characteristic diffusion velocity of matter, i.e., we set

$$n_0^{\mathrm{F}} v_0 = \sigma_0 E_0.$$

Suppressing tildes for notational simplicity, Maxwell's equations in the rescaled form become

$$\tfrac{v_0}{c}\partial_t \boldsymbol{B} + \operatorname{curl}(\boldsymbol{E}) = 0, \qquad \operatorname{div}(\boldsymbol{B}) = 0, \tag{17.25}$$

$$-\tfrac{v_0}{c}\partial_t \boldsymbol{E} + \operatorname{curl}(\boldsymbol{B}) = c\mu_0 x_0 \sigma_0 \boldsymbol{j}^{\mathrm{e}}, \qquad \operatorname{div}(\boldsymbol{E}) = \frac{\sigma_0 t_0}{\varepsilon_0} n^{\mathrm{e}}. \tag{17.26}$$

Since the dimensionless parameter $c\mu_0\sigma_0 x_0$ is of order 1, the time derivatives in Eqs. $(17.18)_1$, $(17.19)_1$ are multiplied by the small factor v_0/c. In the leading-order approximation, returning to dimensional quantities, Maxwell's equations reduce to

$$\operatorname{curl}(\boldsymbol{E}) = 0, \qquad \operatorname{div}(\boldsymbol{B}) = 0, \tag{17.27}$$

$$\operatorname{curl}(\boldsymbol{B}) = \mu_0 \boldsymbol{j}^{\mathrm{e}}, \qquad \operatorname{div}(\boldsymbol{E}) = \frac{1}{\varepsilon_0} n^{\mathrm{e}}, \tag{17.28}$$

which we call the quasi-static version of the Maxwell equations. A similar argument shows that the Lorentz force simplifies to

$$\boldsymbol{k} = n^e \boldsymbol{E}, \tag{17.29}$$

and its power is as before. Since the electric field in the quasi-static setting can be derived from an electrostatic potential φ, we obtain by (17.19)

$$\boldsymbol{E} = \nabla\varphi, \qquad \varepsilon_0 \Delta\varphi = -n^{\mathrm{e}}, \tag{17.30}$$

and the magnetic field follows from $(17.28)_1$.

It is important to note that while the modeling part of this work relies on the full set of Maxwell's equations, in applications, we employ the quasi-static approximation.

The balance equation for the internal energy. Taking the scalar product of the momentum balance Eq. (17.12) with the velocity $\boldsymbol{v}$, we derive the kinetic energy balance. Subtracting this from the energy balance (17.13) yields the internal energy equation:

$$\partial_t(\rho e) + \mathrm{div}(\rho e \boldsymbol{v} + \boldsymbol{q}) = \boldsymbol{\sigma} : D(\boldsymbol{v}) + \left(\sum_{\alpha=1}^{N} \frac{z_\alpha e_0}{m_\alpha} \boldsymbol{J}_\alpha + \dot{\boldsymbol{P}} + \boldsymbol{P}\mathrm{div}(\boldsymbol{v}) - \boldsymbol{P} \cdot \nabla \boldsymbol{v} + \mathrm{curl}(\boldsymbol{M}) \right) \cdot \boldsymbol{\mathcal{E}}. \tag{17.31}$$

Here, $\dot{\boldsymbol{P}} = \partial_t \boldsymbol{P} + \boldsymbol{v} \cdot \nabla \boldsymbol{P}$ represents the material time derivative of the polarization, and the electromotive intensity is given by $\boldsymbol{\mathcal{E}} = \boldsymbol{E} + \boldsymbol{v} \times \boldsymbol{B}$. The right-hand side of (17.31) describes internal energy production due to mechanical stresses, charge diffusion, polarization, and magnetization effects.

Constitutive model: General strategy The model variables are chosen as

$$(\rho_\alpha)_{\alpha=1,2,\ldots,N}, \quad \boldsymbol{v}, \quad T, \quad \chi, \quad \boldsymbol{E} \ \text{ and } \ \boldsymbol{B}.$$

Their evolution is governed by

(i) the balance equations for the partial masses (17.11), the (barycentric momentum) (17.12), and for the phase field (17.14);
(ii) the internal energy balance (17.31);
(iii) Maxwell's equations (17.18) and (17.19).

These equations introduce additional quantities: In the mass balances, we have the reaction rates $R^i_{\mathrm{f,b}}$ and the diffusion fluxes $\boldsymbol{J}_\alpha$. The constitutive quantities of the momentum balance are the stress $\boldsymbol{\sigma}$, the charge density n^{e}, and the electric current $\boldsymbol{j}^{\mathrm{e}}$. The latter quantities also occur in Maxwell's equations. The phase field balance contains the phase field flux $\boldsymbol{j}_\chi$ and the production rate ξ_χ. Finally, the constitutive quantities of the internal energy balance are the internal energy ρe, the heat flux $\boldsymbol{q}$, the stress $\boldsymbol{\sigma}$, and the polarization $\boldsymbol{P}$. These quantities must be defined through constitutive relations, which must be formulated in a thermodynamically consistent manner. This requires satisfying two fundamental principles:

(i) *Material Frame Indifference:* Constitutive functions must remain invariant under Euclidean transformations.
(ii) *Entropy Principle:* The second law of thermodynamics must hold.

Material Frame Indifference states that the constitutive laws of objective tensors must be invariant under transformations of the form:

$$t^* = t + a, \quad x_i^* = O_{ij}(t)x_j + b_i(t), \tag{17.32}$$

where $\mathbf{O}(t)$ is an orthogonal transformation ($\mathbf{O}(t)\mathbf{O}(t)^\mathsf{T} = \mathbf{1}$).

Next we define the notion of *objective scalars, vectors, and tensors (of rank two)* if their components transform according to

$$s^* = \det(\mathbf{O})^p s \qquad \text{for scalars,} \tag{17.33}$$

$$v_i^* = \det(\mathbf{O})^p O_{ij} v_j \qquad \text{for vectors,} \tag{17.34}$$

$$T_{ij}^* = \det(\mathbf{O})^p O_{ik} O_{jl} T_{kl} \qquad \text{for rank two tensors.} \tag{17.35}$$

Scalars and vectors are also called tensor of ranks 0 and 1, respectively. For $p = 0$ the objective tensor is called absolute objective tensor and for $p = 1$ we have an axial objective tensor.

As an example, we consider an objective absolute tensor $\boldsymbol{T}$, i.e., we have $T^*_{ij} = O_{ik}O_{jl}T_{kl}$. Let us further assume that $\mathbf{T}$ is a function of $\nabla\mathbf{v}$, so that in general we have

$$T_{ij} = f_{ij}\Big(\frac{\partial v_k}{\partial x_l}\Big), \quad \text{respectively } T^*_{ij} = f^*_{ij}\Big(\frac{\partial v^*_k}{\partial x^*_l}\Big). \tag{17.36}$$

Then, with $v^*_i = O_{ij}(v_j - \dot{O}_{jl}O_{kl}(x^*_k - b_k) - \dot{b}_j)$, objectivity amounts to

$$f^*(\mathbf{O}(t)\nabla\mathbf{v}\,\mathbf{O}(t)^\mathsf{T} + \dot{\mathbf{O}}(t)\mathbf{O}(t)^\mathsf{T}) = \mathbf{O}(t)f(\nabla\mathbf{v})\mathbf{O}(t)^\mathsf{T}. \tag{17.37}$$

In this case, the principle of material frame indifference states that $f^* = f$. In other words, it implies that f is an *isotropic function*. Moreover, f can only depend on the symmetric part $\mathbf{D}$ of $\nabla\mathbf{v}$ which follows from (17.37) by choosing $\mathbf{O} = \mathbf{1}$ and $\dot{\mathbf{O}} = -\mathbf{R}$, where $\mathbf{R}$ is the antisymmetric part of $\nabla\mathbf{v}$.

Classification 1: Transformation properties of important quantities. In this section, we outline the transformation properties of several fundamental quantities. These can be categorized into kinematic and non-kinematic quantities. The transformation behavior of kinematic quantities follows from their definitions, whereas the transformation properties of non-kinematic quantities must be postulated. For a more detailed discussion and motivation, we refer to Müller (1985) and Truesdell and Toupin (1960).

Kinematic quantities include, for example, the barycentric velocity and diffusion velocities. The diffusion velocities, as well as the symmetric part of the velocity gradient, are absolute objective tensors. In contrast, the barycentric velocity and the antisymmetric part of the velocity gradient are non-objective quantities.

Below is a classification of various important quantities based on their transformation properties:

- Absolute objective scalars: mass densities, internal energy density, phase field, reaction rates, and phase field production.
- Absolute objective vectors: diffusion fluxes, phase field flux, charge potential, Lorentz force, polarization, and electromotive force.
- Axial objective vectors: magnetization, magnetic flux density.
- Absolute objective tensor: (Cauchy) stress tensor.

It is important to note that the internal energy density is an objective scalar only if the stress tensor is symmetric. In that case, the antisymmetric part of the velocity gradient in (17.31) vanishes, and the mechanical power term $\boldsymbol{\sigma} : D(\boldsymbol{v})$ is formulated using the symmetric part $D(\boldsymbol{v})$ of the velocity gradient.

Additionally, the electric field $\boldsymbol{E}$ and the magnetic current potential $\boldsymbol{H}$ are not objective quantities. However, the sum of the electromagnetic terms in (17.31) forms an objective scalar.

Classification 2: Parity of key quantities. The formulation of the 2nd law of thermodynamics necessitates an additional classification of the involved quantities based on their time dependence. This classification is determined by the dimensional analysis of time units:

- If a given quantity q involves the time unit *"second"* with an *odd* exponent, it is assigned a *negative parity*:
$$Pq = -1.$$
- If the time unit appears with an *even* exponent, it is assigned a *positive parity*:
$$Pq = +1.$$

We adhere to SI units, with one exception: the SI unit of charge, $A \cdot s$, is replaced by Coulombs (CC), ensuring that the charge density has a parity of $+1$.
For example, the parity values of several important physical quantities are

$$[\rho] = \frac{\text{kg}}{\text{m}^3} \to +1, \qquad [\rho \mathbf{v}] = \frac{\text{kg}}{\text{m}^2\, s} \to -1, \qquad [\rho e] = \frac{\text{kg}}{\text{m}\, s^2} \to +1 \qquad [\mathbf{B}] = \frac{\text{Vs}}{\text{m}^2} \to -1. \tag{17.38}$$

Evidently, time derivatives of a quantity reverse its parity, while spatial derivatives leave the parity unchanged.

Formulation of the entropy principle. A thermodynamic process is defined as any solution of the previously introduced system of partial differential equations—including (17.11), (17.12), (17.14), (17.31), (17.18), and (17.19). Here, by solution, we refer to functions that locally satisfy the balance equations. Notably, the values of a quantity and its spatial derivatives can be chosen independently.

With this framework in place, the 2nd law of thermodynamics is formulated in terms of four universal axioms and two material-dependent axioms. For further details and motivation, see Bothe and Dreyer (2015).

(I) There exists an entropy/entropy flux pair $(\rho s, \boldsymbol{\Phi})$ as a material-dependent quantity, where ρs is an absolute objective scalar and $\boldsymbol{\Phi}$ is an absolute objective vector. The entropy has the physical dimension
$$\text{J} \cdot \text{kg}^{-1} \cdot \text{K}^{-1} = \text{m}^2\text{s}^{-2}\text{K}^{-1},$$
implying that it has positive parity. Consequently, the entropy flux and entropy production ζ have negative parity.

(II) The entropy pair $(\rho s, \boldsymbol{\Phi})$ satisfies the balance equation:
$$\frac{\partial}{\partial t}(\rho s) + \nabla \cdot (\rho s \boldsymbol{v} + \boldsymbol{\Phi}) = \zeta. \tag{17.39}$$

(III) Any admissible entropy flux satisfies the conditions:

(i) The entropy production ζ can be written as a sum of binary products:

$$\zeta = \sum_m \mathcal{N}_m \mathcal{P}_m, \tag{17.40}$$

where $\mathcal{N}_m$ are quantities of negative parity and $\mathcal{P}_m$ are quantities of positive parity.

(ii) The product $\mathcal{N}_m \mathcal{P}_m \geq 0$ for all m and for every thermodynamic process.

(IV) A thermodynamic process is said to be in *thermodynamic equilibrium* if $\zeta = 0$. This is understood in a pointwise sense, meaning equilibrium need not hold everywhere. A process is *reversible* if $\zeta = 0$ everywhere.

Additionally, we impose two further ones which refer to the most general constitutive models we are interested in.

(V) The following *dissipative mechanisms* are considered for fluid mixtures:

- Diffusion of mass and charge.
- Chemical reaction.
- Viscous flow.
- Heat conduction.
- Phase transition due to diffusion.
- Phase transition due to phase production.

In equilibrium, these mechanisms satisfy the conditions:

$$\boldsymbol{u}_\alpha = 0, \quad R_{\mathrm{f}}^i = R_{\mathrm{b}}^i, \quad D(\boldsymbol{v}) = 0, \quad \boldsymbol{q} = 0, \quad \boldsymbol{J}_\chi = 0, \quad \xi_\chi = 0. \tag{17.41}$$

(VI) For the class of fluid mixtures under consideration, the entropy dependence is restricted to the form:

$$\rho s = \rho\tilde{s}(\rho e - \boldsymbol{\mathcal{E}} \cdot \boldsymbol{P}, \rho_1, \ldots, \rho_N, \boldsymbol{\mathcal{E}}, \boldsymbol{B}, \chi, \nabla\chi), \tag{17.42}$$

where $\rho\tilde{s}$ is a *concave function* that satisfies material frame indifference. Using this function, we define the *(absolute) temperature* T, the *chemical potentials* $(\mu_i)_{i=1,2,\ldots,N}$ of the constituents, and the chemical potential of the phases μ_χ as

$$\frac{1}{T} := \frac{\partial \rho\tilde{s}}{\partial(\rho e - \boldsymbol{\mathcal{E}} \cdot \boldsymbol{P})}, \qquad \frac{\mu_i}{T} := -\frac{\partial \rho\tilde{s}}{\partial \rho_i}, \qquad \frac{\mu_\chi}{T} := -\left(\frac{\partial \rho\tilde{s}}{\partial \chi} - \nabla \cdot \frac{\partial \rho\tilde{s}}{\partial \nabla\chi}\right). \tag{17.43}$$

Identification of the entropy production. To determine the entropy production, we introduce the material time derivative into Eq. (17.39) and incorporate the entropy function given by Eq. (17.42). This yields the intermediate expression:

$$\zeta = \frac{1}{T}((\rho e)^{\cdot} - \dot{\boldsymbol{P}} \cdot \boldsymbol{\mathcal{E}} - \boldsymbol{P} \cdot \dot{\boldsymbol{\mathcal{E}}}) - \sum_{\alpha=1}^{N} \frac{\mu_\alpha}{T} \dot{\rho}_\alpha + \frac{\partial \rho \tilde{s}}{\partial \boldsymbol{\mathcal{E}}} \cdot \dot{\boldsymbol{\mathcal{E}}} + \frac{\partial \rho \tilde{s}}{\partial \boldsymbol{B}} \cdot \dot{\boldsymbol{B}} + \frac{\partial \rho \tilde{s}}{\partial \chi} \dot{\chi} + \frac{\partial \rho \tilde{s}}{\partial \nabla \chi} \cdot (\nabla \chi)^{\cdot}$$
$$+ \rho s \operatorname{div}(\boldsymbol{v}) + \operatorname{div}(\boldsymbol{\Phi}). \tag{17.44}$$

Next, the time derivatives of internal energy, partial mass densities, and the phase field are eliminated using the corresponding balance equations. Additionally, the term $(\nabla\chi)^{\cdot}$ is rewritten using the identity

$$(\nabla \chi)^{\cdot} = \nabla \dot{\chi} - \nabla \boldsymbol{v} \cdot \nabla \chi. \tag{17.45}$$

Furthermore, we use the identity

$$\boldsymbol{\mathcal{E}} \cdot \operatorname{curl}(\boldsymbol{M}) = -\operatorname{div}(\boldsymbol{\mathcal{E}} \times \boldsymbol{M}) + \boldsymbol{M} \cdot \operatorname{curl}(\boldsymbol{\mathcal{E}}) \tag{17.46}$$

and substitute $\operatorname{curl}(\boldsymbol{\mathcal{E}})$ by a variant of $(17.18)_1$, viz.,

$$\operatorname{curl}(\boldsymbol{\mathcal{E}}) = -\dot{\boldsymbol{B}} - \boldsymbol{B} \operatorname{div}(\boldsymbol{v}) + \boldsymbol{B} \cdot \nabla \boldsymbol{v}. \tag{17.47}$$

After appropriate rearrangements, the entropy production takes the form:

$$\zeta = \operatorname{div}\left(\boldsymbol{\Phi} - \frac{1}{T}(\boldsymbol{q} + \boldsymbol{\mathcal{E}} \times \boldsymbol{M}) + \frac{1}{T}\sum_{\alpha=1}^{N} \mu_\alpha \boldsymbol{J}_\alpha + \frac{\partial \rho \tilde{s}}{\partial \nabla \chi} \dot{\chi} - \frac{1}{\rho}\left(\frac{\partial \rho \tilde{s}}{\partial \chi} - \nabla \cdot \frac{\partial \rho \tilde{s}}{\partial \nabla \chi}\right) \boldsymbol{J}_\chi\right)$$
$$+ \left(\frac{\partial \rho \tilde{s}}{\partial \boldsymbol{\mathcal{E}}} - \frac{\boldsymbol{P}}{T}\right) \cdot (\dot{\boldsymbol{\mathcal{E}}} + \nabla \boldsymbol{v} \cdot \boldsymbol{\mathcal{E}}) + \left(\frac{\partial \rho \tilde{s}}{\partial \boldsymbol{B}} - \frac{\boldsymbol{M}}{T}\right) \cdot (\dot{\boldsymbol{B}} - \boldsymbol{B} \cdot \nabla \boldsymbol{v})$$
$$+ \frac{1}{T}\left(\boldsymbol{\sigma} - T \frac{\partial \rho \tilde{s}}{\partial \nabla \chi} \otimes \nabla \chi - T \boldsymbol{\mathcal{E}} \otimes \frac{\partial \rho \tilde{s}}{\partial \boldsymbol{\mathcal{E}}} + T \frac{\partial \rho \tilde{s}}{\partial \boldsymbol{B}} \otimes \boldsymbol{B}\right.$$
$$\left. - \left(\rho e - \boldsymbol{\mathcal{E}} \cdot \boldsymbol{P} - T \rho \tilde{s} - \sum_{\alpha=1}^{N} \rho_\alpha \mu_\alpha + \boldsymbol{M} \cdot \boldsymbol{B}\right) \mathbf{1}\right) : \nabla \boldsymbol{v}$$
$$+ \boldsymbol{q} \cdot \nabla \frac{1}{T} - \sum_{\alpha=1}^{N} \boldsymbol{J}_\alpha \cdot \left(\nabla \frac{\mu_\alpha}{T} - \frac{z_\alpha}{m_\alpha T} \boldsymbol{\mathcal{E}}\right) - \frac{1}{T} \sum_{i=1}^{N_\mathrm{R}} (R_\mathrm{f}^i - R_\mathrm{b}^i) \left(\sum_{\alpha=1}^{N} m_\alpha \gamma_\alpha \mu_\alpha\right)$$
$$- \boldsymbol{J}_\chi \cdot \nabla \frac{\mu_\chi}{T} - \frac{1}{T} \xi_\chi \mu_\chi. \tag{17.48}$$

Remarks on the composition and classification of terms in (17.48):

1. The expressions

$$\dot{\boldsymbol{\mathcal{E}}} + \nabla(\boldsymbol{v}) \cdot \boldsymbol{\mathcal{E}} \quad \text{and} \quad \dot{\boldsymbol{B}} - \boldsymbol{B} \cdot \nabla \boldsymbol{v}$$

represent objective vectors due to the transformation properties:

$$\dot{\mathcal{E}}_i^* + \nabla_i^*(v_k^*)\mathcal{E}_k^* = O_{ij}(\dot{\mathcal{E}}_j + \nabla_j(v_k)\mathcal{E}_k), \quad \dot{B}_i^* + \nabla_k^*(v_i^*)B_k^* = \det(O) O_{ij}(\dot{B}_j + \nabla_k(v_j)B_k). \tag{17.49}$$

2. The entropy function is constrained by the principle of material frame indifference, leading to the representation:

$$\rho\tilde{s}(\rho e - \boldsymbol{E}\cdot\boldsymbol{P}, \rho_1, \ldots, \rho_N, \boldsymbol{E}, \boldsymbol{B}, \chi, \nabla\chi) = \rho\bar{s}(\rho e - \boldsymbol{E}\cdot\boldsymbol{P}, \rho_1, \ldots, \rho_N, |\boldsymbol{E}|^2, |\boldsymbol{B}|^2, (\boldsymbol{E}\cdot\boldsymbol{B})^2, \chi, |\nabla\chi|^2), \quad (17.50)$$

which implies that the terms

$$\frac{\partial\rho\tilde{s}}{\partial\nabla\chi}\otimes\nabla\chi, \qquad -\boldsymbol{\mathcal{E}}\otimes\frac{\partial\rho\tilde{s}}{\partial\boldsymbol{\mathcal{E}}}+\frac{\partial\rho\tilde{s}}{\partial\boldsymbol{B}}\otimes\boldsymbol{B} \quad (17.51)$$

are symmetric objective tensors.
3. For this reason, the factor of $\nabla\boldsymbol{v}$ is symmetric, ensuring that only the symmetric part of the velocity gradient, $D(\boldsymbol{v})$, appears in (17.44).
4. Thus the representation (17.44) consists of a divergence and a sum of binary products with objective factors of negative, respectively, positive parity.

To satisfy Axiom (III-i), the entropy flux is chosen as

$$\boldsymbol{\Phi} = \frac{1}{T}(\boldsymbol{q}+\boldsymbol{\mathcal{E}}\times\boldsymbol{M}) - \frac{1}{T}\left(\sum_{\alpha=1}^{N}\mu_\alpha\boldsymbol{J}_\alpha + \frac{1}{\rho}\mu_\chi\boldsymbol{J}_\chi\right) - \frac{\partial\rho\tilde{s}}{\partial\nabla\chi}\dot{\chi}. \quad (17.52)$$

Then, the remaining part of (17.48) is identified as the entropy production according to Axiom (IV):

$$\begin{aligned}
\zeta = &\left(\frac{\partial\rho\tilde{s}}{\partial\boldsymbol{E}}-\frac{\boldsymbol{P}}{T}\right)\cdot(\dot{\boldsymbol{E}}+\nabla(\boldsymbol{v})\cdot\boldsymbol{\mathcal{E}}) + \left(\frac{\partial\rho\tilde{s}}{\partial\boldsymbol{B}}-\frac{\boldsymbol{M}}{T}\right)\cdot(\dot{\boldsymbol{B}}-\boldsymbol{B}\cdot\nabla\boldsymbol{v})\\
&+\frac{1}{T}\Bigg(\boldsymbol{\sigma} - T\frac{\partial\rho\tilde{s}}{\partial\nabla\chi}\otimes\nabla\chi - T\boldsymbol{\mathcal{E}}\otimes\frac{\partial\rho\tilde{s}}{\partial\boldsymbol{\mathcal{E}}} + T\frac{\partial\rho\tilde{s}}{\partial\boldsymbol{B}}\otimes\boldsymbol{B}\\
&\qquad - \left(\rho e - \boldsymbol{\mathcal{E}}\cdot\boldsymbol{P} - T\rho\tilde{s} - \sum_{\alpha=1}^{N}\rho_\alpha\mu_\alpha + \boldsymbol{M}\cdot\boldsymbol{B}\right)\boldsymbol{1}\Bigg) : D(\boldsymbol{v})\\
&+\boldsymbol{q}\cdot\nabla\frac{1}{T} - \sum_{\alpha=1}^{N}\boldsymbol{J}_\alpha\cdot\left(\nabla\frac{\mu_\alpha}{T} - \frac{z_\alpha e_0}{m_\alpha T}\boldsymbol{\mathcal{E}}\right) - \frac{1}{T}\sum_{i=1}^{N_{\mathrm{R}}}(R_{\mathrm{f}}^i - R_{\mathrm{b}}^i)\left(\sum_{\alpha=1}^{N} m_\alpha\gamma_\alpha\mu_\alpha\right)\\
&-\boldsymbol{J}_\chi\cdot\nabla\frac{\mu_\chi}{T} - \frac{1}{T}\xi_\chi\mu_\chi.
\end{aligned} \quad (17.53)$$

Each product describes a dissipative mechanism and couples a quantity of negative parity with a quantity of positive parity.

This formulation of entropy production facilitates the derivation of constitutive equations for various physical quantities, including polarization, magnetization, stress, heat flux, diffusion fluxes, reaction rates, phase fluxes, and phase production rates. Cross effects between different dissipative mechanisms may be included.

When these effects are introduced by mixing within the same parity class, ensuring the conservation of entropy production, the Onsager symmetry naturally emerges. This symmetry is rigorously established and described in the literature (e.g., Bothe and Dreyer (2015)). For illustrative purposes, we couple heat conduction and diffusion later.

Polarization and magnetization. We now turn to the constitutive equations for the polarization $\boldsymbol{P}$ and magnetization $\boldsymbol{M}$. To satisfy Axiom III-ii, we select the following relationships:

$$\boldsymbol{P} = T\frac{\partial \rho\tilde{s}}{\partial \boldsymbol{E}} - \tau_{\mathcal{E}}(\dot{\boldsymbol{\mathcal{E}}} + \nabla(\boldsymbol{v}) \cdot \boldsymbol{\mathcal{E}}) \quad \text{and} \quad \boldsymbol{M} = T\frac{\partial \rho\tilde{s}}{\partial \boldsymbol{B}} - \tau_B(\dot{\boldsymbol{B}} - \boldsymbol{B} \cdot \nabla(\boldsymbol{v})), \tag{17.54}$$

where $\tau_{\mathcal{E}} \geq 0$ and $\tau_B \geq 0$ are phenomenological coefficients. It is noteworthy that the constitutive quantities $\boldsymbol{P}$ and $\boldsymbol{M}$ depend on the variables of the entropy function, along with $\nabla \boldsymbol{v}$ and the time derivatives of $\boldsymbol{\mathcal{E}}$ and $\boldsymbol{B}$. These equations encapsulate various complex phenomena, such as hysteresis and inertia in free charges, which lead to frequency-dependent refractive indices. A special case arises for $\tau_{\mathcal{E}} = 0$ and $\tau_B = 0$ resulting in the following constitutive equations:

$$\boldsymbol{P} = T\frac{\partial \rho\tilde{s}}{\partial \boldsymbol{E}} \quad \text{and} \quad \boldsymbol{M} = T\frac{\partial \rho\tilde{s}}{\partial \boldsymbol{B}} \tag{17.55}$$

which still account for phenomena such as piezoelectricity and paramagnetism.

Stress. The constitutive equation for stress, which identically satisfies Axiom III-ii, can be directly obtained from the second line of Eq. (17.48). By decomposing the velocity gradient tensor $D(\boldsymbol{v})$ into its trace component and its traceless part, denoted as $D^\circ(\boldsymbol{v})$, and abbreviating the factor of $D(\boldsymbol{v})$ by $\boldsymbol{A}$, the second line of (17.48) reads

$$1/3\ \mathrm{Tr}(\boldsymbol{A})\mathrm{div}(\boldsymbol{v}) + \boldsymbol{A}^\circ : D^\circ(\boldsymbol{v}).$$

Then Axiom III-ii is satisfied for the constitutive equations:

$$\frac{1}{3}\mathrm{Tr}(\boldsymbol{A}) = (\lambda + \frac{2}{3}\eta)\mathrm{div}(\boldsymbol{v}) \quad \text{and} \quad \boldsymbol{A}^\circ = 2\eta D^\circ(\boldsymbol{v}). \tag{17.56}$$

The phenomenological coefficients $\lambda + \frac{2}{3}\eta \geq 0$ and $\eta \geq 0$ are called bulk and shear modulus, respectively. Thus the constitutive equation for the traceless part of the stress reads

$$\begin{aligned}\boldsymbol{\sigma}^\circ = T\Big(\Big(&\frac{\partial \rho\tilde{s}}{\partial \nabla\chi} \otimes \nabla\chi - \frac{1}{3}\frac{\partial \rho\tilde{s}}{\partial \nabla\chi} \cdot \nabla\chi \boldsymbol{1}\Big) \\ &+ \Big(\boldsymbol{\mathcal{E}} \otimes \frac{\partial \rho\tilde{s}}{\partial \boldsymbol{\mathcal{E}}} - \frac{1}{3}\frac{\partial \rho\tilde{s}}{\partial \boldsymbol{\mathcal{E}}} \cdot \boldsymbol{\mathcal{E}}\boldsymbol{1}\Big) - \Big(\frac{\partial \rho\tilde{s}}{\partial \boldsymbol{B}} \otimes \boldsymbol{B} - \frac{1}{3}\frac{\partial \rho\tilde{s}}{\partial \boldsymbol{B}} \cdot \boldsymbol{B}\boldsymbol{1}\Big)\Big) + 2\eta D^\circ(\boldsymbol{v}),\end{aligned} \tag{17.57}$$

and for the trace of the stress we obtain

$$\mathrm{Tr}(\boldsymbol{\sigma}) = \frac{T}{3}\Big(\frac{\partial \rho\tilde{s}}{\partial \nabla\chi} \cdot \nabla\chi + \frac{\partial \rho\tilde{s}}{\partial \boldsymbol{\mathcal{E}}} \cdot \boldsymbol{\mathcal{E}} - \frac{\partial \rho\tilde{s}}{\partial \boldsymbol{B}} \cdot \boldsymbol{B}\Big) - \Big(\rho e - \boldsymbol{\mathcal{E}} \cdot \boldsymbol{P} - T\rho\tilde{s} - \sum_{\alpha=1}^{N} \rho_\alpha\mu_\alpha + \boldsymbol{M} \cdot \boldsymbol{B}\Big)$$
$$+ (3\lambda + 2\eta)\mathrm{div}(\boldsymbol{v}). \qquad (17.58)$$

From the above expressions, it follows that both the trace and the deviatoric part of the stress tensor contain terms that vanish in equilibrium, viz., the components proportional to velocity gradients. Therefore, we further decompose the stress tensor into a viscous and a non-viscous contribution. We define the viscous component as $\boldsymbol{\sigma}^{\mathrm{NS}}$, referring to the Navier–Stokes system, while the non-viscous contribution is denoted as $\boldsymbol{\sigma}^{\mathrm{nv}}$. The Navier–Stokes stress component is then given by

$$\boldsymbol{\sigma}^{\mathrm{NS}} = \lambda \mathrm{div}(\boldsymbol{v}) + 2\eta D(\boldsymbol{v}), \qquad (17.59)$$

while the non-viscous stress component takes the form:

$$\boldsymbol{\sigma}^{\mathrm{nv}} = T\Big(\frac{\partial \rho\tilde{s}}{\partial \nabla\chi} \otimes \nabla\chi + \boldsymbol{\mathcal{E}} \otimes \frac{\partial \rho\tilde{s}}{\partial \boldsymbol{\mathcal{E}}} - \frac{\partial \rho\tilde{s}}{\partial \boldsymbol{B}} \otimes \boldsymbol{B}\Big) - \Big(\rho e - \boldsymbol{\mathcal{E}} \cdot \boldsymbol{P} - T\rho\tilde{s} - \sum_{\alpha=1}^{N} \rho_\alpha\mu_\alpha + \boldsymbol{M} \cdot \boldsymbol{B}\Big)\mathbf{1}. \qquad (17.60)$$

Thermo-diffusion. Bothe and Dreyer (2015) have introduced a novel methodology for incorporating cross effects in thermodynamic systems. In Bothe and Dreyer (2015), it is also shown how (17.5) and (17.61) couples by the different class models, e.g., see Eqs. (152)–(158). To illustrate the approach for cross effects, we examine the interplay between heat and diffusion fluxes. Initially, we consider only the dissipation due to diffusion. The corresponding entropy production is

$$\zeta^{\mathrm{D}} = -\sum_{\alpha=1}^{N} \boldsymbol{J}_\alpha \cdot \Big(\nabla\frac{\mu_\alpha}{T} - \frac{z_\alpha e_0}{m_\alpha T}\boldsymbol{\mathcal{E}}\Big) = -\sum_{\alpha=1}^{N-1} \boldsymbol{J}_\alpha \cdot \Big(\nabla\frac{\mu_\alpha - \mu_N}{T} - \frac{1}{T}\Big(\frac{z_\alpha e_0}{m_\alpha} - \frac{z_N e_0}{m_N}\Big)\boldsymbol{\mathcal{E}}\Big). \qquad (17.61)$$

Due to the constraint (17.6), constitutive equations are required only for $(N-1)$ independent fluxes. The simplest constitutive relation, assuming no coupling, is given by

$$\boldsymbol{J}_\alpha = -M_\alpha\Big(\nabla\frac{\mu_\alpha - \mu_N}{T} - \frac{1}{T}\Big(\frac{z_\alpha e_0}{m_\alpha} - \frac{z_N e_0}{m_N}\Big)\boldsymbol{\mathcal{E}}\Big), \qquad (17.62)$$

where the mobilities $M_\alpha \geq 0$ denote non-negative phenomenological coefficients. To incorporate coupling effects among the constituents of the mixture, we proceed by revisiting (17.61) and defining the prefactors of the diffusion fluxes as

$$\mathcal{P}_\alpha = \nabla\frac{\mu_\alpha - \mu_N}{T} - \frac{1}{T}\Big(\frac{z_\alpha e_0}{m_\alpha} - \frac{z_N e_0}{m_N}\Big)\boldsymbol{\mathcal{E}}.$$

Thus, entropy production can be rewritten as

$$\zeta^{\mathrm{D}} = -\sum_{\alpha=1}^{N-1} \boldsymbol{J}_{\alpha} \cdot \boldsymbol{\mathcal{P}}_{\alpha}.$$

Then we introduce two matrices $\boldsymbol{A}$ and $\boldsymbol{B}$ of dimension $(N-1)^2$. We choose $\boldsymbol{A}^{\mathrm{T}} = \boldsymbol{B}^{-1}$ and obtain

$$\zeta^{\mathrm{D}} = -\sum_{\alpha=1}^{N-1} \boldsymbol{J}_{\alpha} \cdot \boldsymbol{\mathcal{P}}_{\alpha} = -\sum_{\alpha=1}^{N-1} \Big(\sum_{\gamma=1}^{N-1} A_{\alpha\gamma} \boldsymbol{J}_{\gamma}\Big) \cdot \Big(\sum_{\delta=1}^{N-1} B_{\alpha\delta} \boldsymbol{\mathcal{P}}_{\delta}\Big). \tag{17.63}$$

Now we formulate constitutive equations as before, viz.,

$$\sum_{\gamma=1}^{N-1} A_{\alpha\gamma} \boldsymbol{J}_{\gamma} = -\tilde{M}_{\alpha} \sum_{\delta=1}^{N-1} B_{\alpha\delta} \boldsymbol{\mathcal{P}}_{\delta} \quad \text{with} \quad \tilde{M}_{\alpha} \geq 0. \tag{17.64}$$

Solution for the diffusion fluxes yields the constitutive equation:

$$\boldsymbol{J}_{\alpha} = -\sum_{\beta=1}^{N-1} M_{\alpha\beta} \boldsymbol{\mathcal{P}}_{\beta} \quad \text{with} \quad M_{\alpha\beta} = \sum_{\gamma=1}^{N-1} B_{\gamma\alpha} \tilde{M}_{\gamma} B_{\gamma\beta}. \tag{17.65}$$

The newly derived mobility matrix $\boldsymbol{M}$ is both symmetric and positive definite. Consequently, if cross effects do not lead to additional entropy production, Onsager symmetry naturally emerges as a direct implication of the framework.

As an additional example, we now examine the entropy production ζ^{HD} associated with the coupled dissipative mechanisms of heat conduction and diffusion. The entropy production in this case is given by

$$\zeta^{\mathrm{HD}} = \boldsymbol{q} \cdot \nabla \frac{1}{T} - \sum_{\alpha=1}^{N-1} \boldsymbol{J}_{\alpha} \cdot \boldsymbol{\mathcal{P}}_{\alpha}. \tag{17.66}$$

The conventional choice of constitutive functions for $(N-1)$ diffusion fluxes $(\boldsymbol{J}_{\alpha})_{\alpha=1,2,\ldots,N-1}$ and the heat flux $\boldsymbol{q}$ is

$$\boldsymbol{J}_{\alpha} = -\sum_{\beta=1}^{N-1} M_{\alpha\beta} \left(\nabla \frac{\mu_{\beta} - \mu_N}{T} - \frac{1}{T}\Big(\frac{z_{\beta} e_0}{m_{\beta}} - \frac{z_N e_0}{m_N}\Big)\boldsymbol{\mathcal{E}}\right) + L^{\mathrm{J}}_{\alpha} \nabla \frac{1}{T}, \tag{17.67}$$

$$\boldsymbol{q} = -\sum_{\alpha=1}^{N-1} L^{\mathrm{q}}_{\alpha} \left(\nabla \frac{\mu_{\alpha} - \mu_N}{T} - \frac{1}{T}\Big(\frac{z_{\alpha} e_0}{m_{\alpha}} - \frac{z_N e_0}{m_N}\Big)\boldsymbol{\mathcal{E}}\right) + a \nabla \frac{1}{T}, \tag{17.68}$$

where the kinetic coefficients must satisfy the condition that the matrix

$$\begin{pmatrix} M_{\alpha\beta} & L^{\mathrm{J}}_{\alpha} \\ L^{\mathrm{q}}_{\alpha} & a \end{pmatrix} \quad \text{is positive definite.} \tag{17.69}$$

Classically, the cross effects in (17.67) and (17.68) are related to each other by postulating the Onsager symmetry relations:

$$M_{\alpha\beta} = M_{\beta\alpha} \quad \text{and} \quad L^{\mathrm{J}}_{\alpha} = L^{\mathrm{q}}_{\alpha}. \tag{17.70}$$

According to Bothe and Dreyer (2015), the Onsager symmetry is achieved as follows: The symmetry of the diffusion matrix is taken to be granted from the last example. To establish the symmetry $L^{\mathrm{J}}_{\alpha} = L^{\mathrm{q}}_{\alpha}$, we reconsider the entropy production expression (17.63) and introduce the thermo-diffusion coefficients D_α by adding two terms to ξ^{HD} that conserve ξ^{HD}:

$$\xi^{\mathrm{HD}} = \left(\boldsymbol{q} + \sum_{\alpha=1}^{N-1}(D_\alpha - D_N)\boldsymbol{J}_\alpha\right)\cdot\nabla\frac{1}{T} - \sum_{\alpha=1}^{N-1}\boldsymbol{J}_\alpha\cdot\left(\boldsymbol{\mathcal{P}}_\alpha - (D_\alpha - D_N)\nabla\frac{1}{T}\right). \tag{17.71}$$

Then we propose two constitutive equations for $(N-1)$ diffusion fluxes $(\boldsymbol{J}_\alpha)_{\alpha=1,2,\ldots,N-1}$ and the heat flux $\boldsymbol{q}$ by the simple relations:

$$\boldsymbol{J}_\alpha = -\sum_{\beta=1}^{N-1} M_{\alpha\beta}\left(\nabla\left(\frac{\mu_\beta}{T} - \frac{\mu_N}{T}\right) - \frac{1}{T}\left(\frac{z_\beta e_0}{m_\beta} - \frac{z_N e_0}{m_N}\right)\boldsymbol{\mathcal{E}}\right) + \sum_{\beta=1}^{N-1} M_{\alpha\beta}(D_\beta - D_N)\nabla\frac{1}{T}, \tag{17.72}$$

$$\boldsymbol{q} = -\sum_{\alpha-1}^{N-1}(D_\alpha - D_N)\boldsymbol{J}_\alpha + a\nabla\left(\frac{1}{T}\right). \tag{17.73}$$

By comparing these expressions with (17.67) and (17.68), we identify the relationships

$$L^{\mathrm{J}}_{\alpha} = \sum_{\beta=1}^{N-1} M_{\alpha\beta}(D_\beta \quad D_N) \quad \text{and} \quad L^{\mathrm{q}}_{\alpha} - \sum_{\beta=1}^{N-1} M_{\beta\alpha}(D_\beta - D_N).$$

Thus the symmetry $M_{\alpha\beta} = M_{\beta\alpha}$ implies the symmetry $L^{\mathrm{J}}_{\alpha} = L^{\mathrm{q}}_{\alpha}$. This is a further example that the Onsager symmetry is a consequence of cross effects that conserve the entropy production.

A detailed discussion of the creation of cross effects by mixing within the parity classes is found in Bothe and Dreyer (2015). For very special cases, there are various proofs of the Onsager symmetries within the phenomenological setting. For instance,

Truesdell (1969) and Müller (1985) have derived these symmetries under particular assumptions within a Class II model.

Constitutive equations for special cases. In this study, we are not interested in the most general constitutive model possible. Instead, we focus on a specific case characterized by the following assumptions:

1. $\tau_{\mathcal{E}} = 0, \tau_{\boldsymbol{B}} = 0$;
2. no magnetization, i.e., $\boldsymbol{M} = 0$;
3. isothermal processes;
4. no coupling between dissipative mechanisms;
5. the Allen–Cahn equation for the phase field, i.e., $\boldsymbol{J}_\chi = 0$.

Under these conditions, the inequality (17.44) may be identically satisfied by the simple constitutive laws:

$$\boldsymbol{P} = T\frac{\partial \rho\tilde{s}}{\partial \boldsymbol{\mathcal{E}}}, \quad \frac{\partial \rho\tilde{s}}{\partial \boldsymbol{B}} = 0, \tag{17.74}$$

$$\boldsymbol{\sigma}^{\text{NS}} = \lambda \text{div}(\boldsymbol{v}) + 2\eta D(\boldsymbol{v}), \tag{17.75}$$

$$\boldsymbol{\sigma}^{\text{nv}} = T\frac{\partial \rho\tilde{s}}{\partial \nabla\chi} \otimes \nabla\chi + T\boldsymbol{\mathcal{E}} \otimes \frac{\partial \rho\tilde{s}}{\partial \boldsymbol{\mathcal{E}}} - \left(\rho e - \boldsymbol{\mathcal{E}}\cdot\boldsymbol{P} - T\rho\tilde{s} - \sum_{\alpha=1}^{N} \rho_\alpha\mu_\alpha\right)\boldsymbol{1}, \tag{17.76}$$

$$\boldsymbol{J}_\alpha = -\sum_{\beta=1}^{N-1} M_{\alpha\beta}\nabla\left(\frac{\mu_\beta - \mu_N}{T} - \frac{1}{T}\left(\frac{z_\beta e_0}{m_\beta} - \frac{z_N e_0}{m_N}\right)\boldsymbol{\mathcal{E}}\right), \tag{17.77}$$

$$\ln\left(\frac{R_{\text{b}}^i}{R_{\text{f}}^i}\right) = \frac{1}{kT}\sum_{\alpha=1}^{N} m_\alpha\gamma_\alpha^i\mu_\alpha \quad \text{for} \quad i = 1, 2, \ldots, N_{\text{R}}, \tag{17.78}$$

$$\xi_\chi = -\tau\mu_\chi. \tag{17.79}$$

Equation (17.78) serves as an ansatz for chemical reactions far from equilibrium. The associated entropy production remains non-negative, satisfying $-\ln\left(R_{\text{b}}^i/R_{\text{f}}^i\right)(R_{\text{f}}^i - R_{\text{b}}^i) \geq 0$.

Introduction of the (Helmholtz) free energy density. It is often preferable to express thermodynamic relationships in terms of temperature rather than internal energy density. To facilitate this, we introduce the Helmholtz free energy density:

$$\rho\psi = \rho e - \boldsymbol{P}\cdot\boldsymbol{\mathcal{E}} - T\rho s.$$

In the entropy function we change the variable $\rho e - \boldsymbol{P}\cdot\boldsymbol{\mathcal{E}}$ to T with $\rho e = \rho\hat{e}(T, \rho_1, \ldots, \rho_N, \boldsymbol{\mathcal{E}}, \chi, \nabla\chi)$. Then, for $\psi = \hat{\psi}(T, \rho_1, \ldots, \rho_N, \boldsymbol{\mathcal{E}}, \chi)$ we obtain the following relations from (17.43) and (17.74)

$$\rho s = -\frac{\partial \rho\hat{\psi}}{\partial T}, \quad \mu_\alpha = \frac{\partial \rho\hat{\psi}}{\partial \rho_\alpha}, \quad \rho e = -T^2\frac{\partial}{\partial T}\left(\frac{\rho\hat{\psi}}{T}\right), \quad \boldsymbol{P} = -\frac{\partial \rho\hat{\psi}}{\partial \boldsymbol{\mathcal{E}}}, \quad \mu_\chi = \frac{\partial \rho\hat{\psi}}{\partial \chi} - T\nabla\cdot\frac{1}{T}\frac{\partial \rho\hat{\psi}}{\partial \nabla\chi}. \tag{17.80}$$

Using the free energy density formulation, the non-viscous part of the stress tensor (17.76) can be rewritten

$$\boldsymbol{\sigma}^{\text{nv}} = -\frac{\partial \rho\hat{\psi}}{\partial \nabla\chi} \otimes \nabla\chi + \frac{1}{2}(\boldsymbol{\mathcal{E}} \otimes \boldsymbol{P} + \boldsymbol{P} \otimes \boldsymbol{\mathcal{E}}) + \Big(\rho\psi - \sum_{\alpha=1}^{N} \rho_\alpha \mu_\alpha\Big)\mathbf{1}. \tag{17.81}$$

It is convenient to introduce the pressure p by

$$p = -\rho\psi + \sum_{\alpha=1}^{N} \rho_\alpha \mu_\alpha. \tag{17.82}$$

Then, this relation is called *Gibbs–Duhem equation*. A detailed description of the pressure, i.e., the partial pressure and its traceless part can be found in Bothe and Dreyer (2015), Eqs. (11)–(13).

The free energy density plays a central role in characterizing the thermodynamic behavior of mixtures containing both charged and neutral constituents. Its explicit definition is given in the following section.

17.3 Choice of Energy Density and Non-linear Stability

We consider the quasi-static regime of electrodynamics in which $\boldsymbol{B}$ drops out and $\mathcal{E} = \boldsymbol{E} = -\nabla\varphi$. Furthermore, we choose the following free energy density:

$$\rho\psi := W(\chi) + \frac{\gamma}{2}|\nabla\chi|^2 + h(\chi)\rho\psi_{\text{L}}(\rho_1, \ldots, \rho_N) + (1 - h(\chi))\rho\psi_{\text{V}}(\rho_1, \ldots, \rho_N) - \frac{\varepsilon_0}{2}s(\chi)|\boldsymbol{E}|^2, \tag{17.83}$$

where $W(\chi) := (\chi - 1)^2(\chi + 1)^2$, $h : \mathbb{R} \to [0, 1]$ is a smooth interpolation function satisfying

$$h(z) = \begin{cases} 1 \text{ for } & z \geq 1, \\ 0 \text{ for } & z \leq -1, \end{cases} \tag{17.84}$$

such that $h'(z) = 0$ for all $|z| \geq 1$. In (17.83), $\rho\psi_{\text{L}}, \rho\psi_{\text{V}} : (0, \infty)^N \longrightarrow [0, \infty)$ represents the free energy functions of the pure phases which we assume to be given by a combination of isotropic elastic response and entropy of mixing, i.e.:

$$\rho\psi_{\text{L/V}} = \sum_\alpha \rho_\alpha \psi_\alpha^{\text{R}} + (K_{L/V} - p^{\text{R}})\Big(1 - \frac{n}{n^{\text{R}}}\Big) + K_{L/V}\frac{n}{n^{\text{R}}} \ln\Big(\frac{n}{n^{\text{R}}}\Big) + kT\sum_\alpha n_\alpha \ln\Big(\frac{n_\alpha}{n}\Big),$$

where $K_{L/V}$ are the bulk moduli and n^{R}, p^{R}, ψ_α^{R} are reference number density, reference pressure, and reference energies, respectively. Note that (17.83) and (17.84) imply that $\chi = +1(-1)$ corresponds to liquid (vapor). Similarly, we assume

$$s(\chi) = h(\chi)s_{\mathrm{L}} + (1 - h(\chi))s_{\mathrm{V}}, \tag{17.85}$$

where $s_{\mathrm{L/V}}$ are the susceptibilities of the pure phases. For brevity, we define

$$\begin{aligned}(\rho f)((\rho_\alpha)_\alpha, \chi) &:= (\rho f)(\rho_1, \ldots, \rho_N, \chi) \\ &:= h(\chi)\rho\psi_{\mathrm{L}}(\rho_1, \ldots, \rho_N) + (1 - h(\chi))\rho\psi_{\mathrm{V}}(\rho_1, \ldots, \rho_N).\end{aligned} \tag{17.86}$$

By definition, the chemical potentials μ_α are given by

$$\mu_\alpha := \frac{\partial(\rho\psi)}{\partial\rho_\alpha} = \frac{\partial(\rho f)}{\partial\rho_\alpha}. \tag{17.87}$$

Thus, the following system of partial differential equations (PDEs) governs the evolution of the partial mass densities ρ_α, the barycentric velocity $\boldsymbol{v}$, the phase field parameter χ, and the electrical potential φ, with the equation for ρ_N replaced by the evolution equation for the total mass density $\rho = \sum_{\alpha=1}^{N} \rho_\alpha$:

$$\begin{aligned}
0 &= \partial_t \rho_\alpha + \mathrm{div}(\rho_\alpha \boldsymbol{v}) - \mathrm{div}\Big(\sum_{\beta=1}^{N-1} M_{\alpha\beta}\Big(\nabla\frac{\mu_\beta - \mu_N}{T} + \frac{1}{T}\Big(\frac{z_\beta e_0}{m_\beta} - \frac{z_N e_0}{m_N}\Big)\nabla\varphi\Big)\Big) \\
&\quad - \sum_{i=1}^{N_R} m_\alpha \gamma_\alpha^i M_{\mathrm{r}}^i \Big(1 - \exp\Big(\frac{1}{kT}\sum_{\beta=1}^{N} m_\beta \gamma_\beta^i \mu_\beta\Big)\Big), \qquad \alpha = 1, \ldots, N-1, \\
0 &= \partial_t \rho + \mathrm{div}(\rho \boldsymbol{v}), \\
0 &= \partial_t(\rho\boldsymbol{v}) + \mathrm{div}(\rho\boldsymbol{v} \otimes \boldsymbol{v}) + \nabla\Big(\sum_{\alpha=1}^{N} \rho_\alpha \mu_\alpha - \rho f - W - \frac{\gamma}{2}|\nabla\chi|^2\Big) + \gamma\,\mathrm{div}(\nabla\chi \otimes \nabla\chi) \\
&\quad - \mathrm{div}(\boldsymbol{\sigma}^{\mathrm{NS}}) + \varepsilon_0 \mathrm{div}\Big((1 + s(\chi))\Big(\frac{1}{2}|\nabla\varphi|^2\mathbf{1} - \nabla\varphi \otimes \nabla\varphi\Big)\Big), \\
0 &= \partial_t \chi + \boldsymbol{v} \cdot \nabla\chi + \frac{\tau}{\rho}\Big(W' - \gamma\Delta\chi + \frac{\partial(\rho f)}{\partial\chi} - \frac{\varepsilon_0}{2}s'(\chi)|\nabla\varphi|^2\Big), \\
0 &= \varepsilon_0\,\mathrm{div}((1 + s(\chi))\nabla\varphi) + n^{\mathrm{F}},
\end{aligned} \tag{17.88}$$

where $n^{\mathrm{F}} := e_0 \sum_{\alpha=1}^{N} \frac{z_\alpha}{m_\alpha}\rho_\alpha$ represents the free charge density and $M_{\mathrm{r}}^i := R_{\mathrm{f}}^i$.

17.3.1 Energy Inequality

Regarding the stability of the system (17.88), we can prove an energy inequality. Let $\Omega(t) \subset \mathbb{R}^3$ be a moving open and bounded material domain with C^1-boundary, and $T_f > 0$ be a time up to which we assume that classical solutions of the system (17.88) exist.

Lemma 17.1 (Energy inequality) *Let $((\rho_\alpha)_\alpha, \boldsymbol{v}, \chi, \varphi)$ be a classical solution of* (17.88) *in the domain $\Omega_{T_f} := \bigcup_{t\in(0,T_f)} \Omega(t) \times \{t\}$ and let φ satisfy the Laplace equation in $\mathbb{R}^3 \setminus \Omega(t)$ for all $t \in (0, T_f)$. In addition, let the following boundary conditions be satisfied on $\partial\Omega(t)$ for all $t \in (0, T_f)$:*

$$\boldsymbol{J}_\alpha = 0, \nabla\chi \cdot \boldsymbol{n} = 0,$$

where $\boldsymbol{n}$ denotes the normal vector to $\partial\Omega(t)$. Then the following relation holds:

$$\begin{aligned}&\frac{\mathrm{d}}{\mathrm{d}t}\Big(\int_\Omega W(\chi) + \frac{\gamma}{2}|\nabla\chi|^2 + (\rho f)((\rho_\alpha)_\alpha, \chi) + \frac{\varepsilon_0}{2}(1+s(\chi))|\nabla\varphi|^2 + \frac{\rho}{2}|\boldsymbol{v}|^2 \,\mathrm{d}\boldsymbol{x} + \int_{\mathbb{R}^3\setminus\Omega} \frac{\varepsilon_0}{2}|\nabla\varphi|^2 \,\mathrm{d}\boldsymbol{x}\Big)\\ &= \int_{\partial\Omega} \varepsilon_0 \boldsymbol{v} \cdot \Big(\nabla\varphi_+ \otimes \nabla\varphi_+ - \nabla\varphi_- \otimes \nabla\varphi_- + (|\nabla\varphi_-|^2 - |\nabla\varphi_+|^2)\mathbf{1}\Big) \cdot \boldsymbol{n} \,\mathrm{d}\boldsymbol{a} - T\int_\Omega \zeta \,\mathrm{d}\boldsymbol{x}.\end{aligned} \tag{17.89}$$

Proof By definition of $\rho\psi$, the left-hand side of Eq. (17.89) can be expressed as

$$\frac{\mathrm{d}}{\mathrm{d}t}\Big(\int_\Omega \rho e - T\rho s + e^{\mathrm{e}} + \frac{\rho}{2}|\boldsymbol{v}|^2 \,\mathrm{d}\boldsymbol{x} + \int_{\mathbb{R}^3\setminus\Omega} e^{\mathrm{e}} \,\mathrm{d}\boldsymbol{x}\Big) =: \frac{\mathrm{d}}{\mathrm{d}t}A, \tag{17.90}$$

where e^{e} was defined in (17.22). Using transport theorems, the time derivative $\frac{\mathrm{d}}{\mathrm{d}t}A$ can be rewritten as

$$\begin{aligned}\frac{\mathrm{d}}{\mathrm{d}t}A = &\int_\Omega \Big(\Big(\rho e + \frac{\rho}{2}|\boldsymbol{v}|^2 + e^{\mathrm{e}}\Big)_t + \mathrm{div}\Big((\rho e + \frac{\rho}{2}|\boldsymbol{v}|^2 + e^{\mathrm{e}})\boldsymbol{v}\Big)\Big) \,\mathrm{d}\boldsymbol{x}\\ &+ \int_{\partial\Omega} (e^{\mathrm{e}}_- - e^{\mathrm{e}}_+)\, \boldsymbol{v} \cdot \boldsymbol{n} \,\mathrm{d}\boldsymbol{a} - T\int_\Omega \Big((\rho s)_t + \mathrm{div}(\rho s \boldsymbol{v})\Big) \,\mathrm{d}\boldsymbol{x},\end{aligned} \tag{17.91}$$

where e^{e}_- and e^{e}_+ denote the trace of e^{e} on $\partial\Omega$ taken from the interior and exterior of Ω, respectively. By employing the local balance Eqs. (17.13) and $(17.20)_2$, it follows that

$$\frac{\mathrm{d}}{\mathrm{d}t}A = \int_{\partial\Omega} \Big(-\boldsymbol{q} \cdot \boldsymbol{n} + \boldsymbol{v} \cdot \boldsymbol{\sigma}\boldsymbol{n} + (e^{\mathrm{e}}_- - e^{\mathrm{e}}_+)\boldsymbol{v} \cdot \boldsymbol{n} + T\boldsymbol{\Phi} \cdot \boldsymbol{n}\Big) \,\mathrm{d}\boldsymbol{a} - T\int_\Omega \zeta \,\mathrm{d}\boldsymbol{x}.$$

Inserting the definition of $\boldsymbol{\Phi}$ in (17.52) and noting that $\boldsymbol{J}_\alpha = 0$ on the boundary and $\boldsymbol{J}_\chi = 0$ due to the chosen Allen–Cahn-type model, we obtain

$$\frac{\mathrm{d}}{\mathrm{d}t}A = \int_{\partial\Omega} \boldsymbol{v} \cdot \Big(\boldsymbol{\sigma} + (e^{\mathrm{e}}_- - e^{\mathrm{e}}_+)\mathbf{1}\Big) \cdot \boldsymbol{n} \,\mathrm{d}\boldsymbol{a} - T\int_\Omega \zeta \,\mathrm{d}\boldsymbol{x}.$$

From the momentum balance, we infer

$$\boldsymbol{\sigma}^{\mathrm{e}}_+ = \boldsymbol{\sigma} + \boldsymbol{\sigma}^{\mathrm{e}}_-, \tag{17.92}$$

where $\boldsymbol{\sigma}^{\mathrm{e}}_{-}$, $\boldsymbol{\sigma}^{\mathrm{e}}_{+}$ are the corresponding traces. Note that in the quasi-static regime, the stress tensor takes the form:

$$\boldsymbol{\sigma}^{\mathrm{e}} = \varepsilon_0 \nabla\varphi \otimes \nabla\varphi - e^{\mathrm{e}}\mathbf{1}. \tag{17.93}$$

Therefore, $\frac{\mathrm{d}}{\mathrm{d}t}A$ simplifies to

$$\frac{\mathrm{d}}{\mathrm{d}t}A = \int_{\partial\Omega} \varepsilon_0 \boldsymbol{v} \cdot \Big(\nabla\varphi_+ \otimes \nabla\varphi_+ - \nabla\varphi_- \otimes \nabla\varphi_- + (|\nabla\varphi_-|^2 - |\nabla\varphi_+|^2)\mathbf{1}\Big) \cdot \boldsymbol{n}\, \mathrm{d}\boldsymbol{a} - T\int_{\Omega} \zeta\, \mathrm{d}\boldsymbol{x}.$$

Remark 17.1 The available free energy A serves as a Lyapunov function only in the case where $\boldsymbol{v} = 0$ on $\partial\Omega$. In the present context, the admissible boundary conditions for Maxwell's equations are given by

$$\boldsymbol{n} \times \nabla\varphi_+ = \boldsymbol{n} \times \nabla\varphi_-, \quad \text{and} \quad \varepsilon_0(1 + s(\chi_-))\nabla\varphi_- \cdot \boldsymbol{n} = \varepsilon_0\nabla\varphi_+ \cdot \boldsymbol{n} + n^F_{\partial\Omega},$$

where $n^F_{\partial\Omega}$ represents the free charges on $\partial\Omega$. Thus, even if $n^F_{\partial\Omega} = 0$ the normal component of the electric field is generally not continuous.

17.4 Non-dimensionalization

To ensure physically meaningful scalings, we non-dimensionalize problem (17.88). For this purpose, we introduce characteristic reference quantities, denoted by superscript ^{c} and non-dimensional quantities denoted by *, i.e.:

$$\boldsymbol{x} = x^c\boldsymbol{x}^*,\ t = t^c t^*,\ \rho_\alpha = \rho^c\rho^*_\alpha,\ \boldsymbol{v} = v^c\boldsymbol{v}^*,\ \lambda_{1,2} = \lambda^c\lambda^*_{1,2},\ \tau = \tau^c\tau^*,\ M_{\alpha\beta} = M^c M^*_{\alpha\beta},$$
$$M^i_{\mathrm{r}} = M^c_{\mathrm{r}}(M^i_{\mathrm{r}})^*,\ \gamma_\beta = \gamma^c_{\mathrm{r}}(\gamma^i_\beta)^*,\ m_\beta = m^c m^*_\beta,\ W = W^c W^*,\ \rho f = (\rho f)^c(\rho f)^*,\ \gamma = \gamma^c\gamma^*,$$
$$\frac{\mu_\beta}{T} = \mu^c\mu^*_\beta,\ \varphi = \varphi^c\varphi^*,\ s = s^c s^*,\ \varepsilon_0 = \varepsilon^c_0\varepsilon^*_0,\ e_0 z_\alpha = z^c z^*_\alpha.$$

Note that χ and $h(\chi)$ do not need to be non-dimensionalized. The density is expressed as $\rho = \rho^c\rho^*$, where ρ^* is given by $\rho^* = \sum_\alpha \rho^*_\alpha$.

As we are interested in hyperbolic scalings we set $x^c = v^c t^c$ and $(\rho f)^c = \rho^c\mu^c$ and define the following Mach and Reynolds numbers:

$$\mathrm{M}_W := v^c\sqrt{\frac{\rho^c}{W^c}},\quad \mathrm{M}_{\rho f} := v^c\sqrt{\frac{\rho^c}{(\rho f)^c}},\quad \mathrm{Re} := \frac{\rho^c v^c x^c}{\lambda^c}. \tag{17.94}$$

Additionally, we introduce non-dimensional quantities related to the reaction and diffusion rates

$$\bar{M}_{\mathrm{d}} := \frac{M^c\mu^c}{v^c x^c\rho^c},\quad \bar{M}_{\mathrm{r}} := \frac{M^c_{\mathrm{r}}\gamma^c_{\mathrm{r}} m^c t^c}{\rho^c},\quad \bar{A} := \frac{m^c\gamma^c_{\mathrm{r}}\mu^c}{kT},\quad \bar{\tau} := \frac{\tau^c t^c W^c}{\rho^c}. \tag{17.95}$$

For the electrical effects, we define the following non-dimensional quantities:

$$\bar{M}_{\mathrm{e}} := \frac{M^c z^c \varphi^c}{\rho^c m^c v^c x^c}, \quad \bar{\varepsilon} := \frac{\varepsilon_0^c (\varphi^c)^2}{\rho^c (v^c)^2 (x^c)^2}, \quad \underline{\varepsilon} := \frac{\varepsilon_0^c m^c \varphi^c}{(x^c)^2 z^c \rho^c}. \tag{17.96}$$

We assume that the small parameter

$$\delta := \sqrt{\frac{\gamma^c}{(x^c)^2\, W^c}} \tag{17.97}$$

is proportional to the width of the interfacial layer. This assumption can be justified by Γ-limit techniques, cf. Sternberg (1988); Owen et al. (1990); Dreyer and Kraus (2010); Luckhaus and Modica (1989). Then, suppressing * in the notation, the non-dimensionalized version of (17.88) reads

$$\begin{aligned}
0 &= \partial_t \rho_\alpha + \operatorname{div}(\rho_\alpha \boldsymbol{v}) - \operatorname{div}\left(\sum_{\beta=1}^{N-1} M_{\alpha\beta}\left(\bar{M}_{\mathrm{d}} \nabla(\mu_\beta - \mu_N) + \bar{M}_{\mathrm{e}}\left(\frac{z_\beta}{m_\beta} - \frac{z_N}{m_N}\right)\nabla\varphi\right)\right) \\
&\quad - \bar{M}_{\mathrm{r}} \sum_{i=1}^{N_R} m_\alpha \gamma_\alpha^i \left(1 - \exp\left(\bar{A} \sum_{\beta=1}^{N} m_\beta \gamma_\beta^i \mu_\beta\right)\right), \\
0 &= \partial_t \rho + \operatorname{div}(\rho \boldsymbol{v}), \\
0 &= \partial_t(\rho \boldsymbol{v}) + \operatorname{div}(\rho \boldsymbol{v} \otimes \boldsymbol{v}) + \frac{1}{\mathrm{M}_{\rho f}^2} \nabla\left(\sum_{\alpha=1}^{N} \rho_\alpha \mu_\alpha - \rho f\right) - \frac{1}{\mathrm{M}_W^2} \nabla\left(W + \frac{\gamma}{2}\delta^2 |\nabla\chi|^2\right) \\
&\quad + \frac{\gamma\delta^2}{\mathrm{M}_W^2} \operatorname{div}(\nabla\chi \otimes \nabla\chi) - \frac{1}{\mathrm{Re}} \operatorname{div}(\boldsymbol{\sigma}^{\mathrm{NS}}) + \bar{\varepsilon}\varepsilon_0 \operatorname{div}\left((1 + s^c s(\chi))\left(\frac{|\nabla\varphi|^2}{2}\mathbf{1} - \nabla\varphi \otimes \nabla\varphi\right)\right), \\
0 &= \partial_t \chi + \boldsymbol{v} \cdot \nabla\chi + \bar{\tau}\frac{\tau}{\rho}\left(W' - \gamma\delta^2 \Delta\chi + \frac{\mathrm{M}_W^2}{\mathrm{M}_{\rho f}^2}\frac{\partial \rho f}{\partial \chi} - \bar{\varepsilon}\,\mathrm{M}_W^2\, s^c \frac{\varepsilon_0}{2} s'(\chi) |\nabla\varphi|^2\right), \\
0 &= \underline{\varepsilon}\varepsilon_0 \operatorname{div}((1 + s^c s(\chi))\nabla\varphi) + n^F.
\end{aligned} \tag{17.98}$$

In the sequel, we will consider two scaling regimes. In both cases, we set

$$\bar{A} = 1,\ s^C = 1,\ \bar{M}_{\mathrm{d}} = 1,\ \bar{M}_{\mathrm{r}} = 1,\ \bar{M}_{\mathrm{e}} = 1,\ \mathrm{M}_W = \sqrt{\delta},\ \mathrm{Re} = \frac{1}{\delta^2},\ \mathrm{M}_{\rho f} = 1,\ \bar{\tau} = \frac{1}{\delta^2}.$$

In the *uncoupled regime*, we consider

$$\bar{\varepsilon} = \underline{\varepsilon} = 1, \tag{17.99}$$

while we consider

$$\bar{\varepsilon} = \underline{\varepsilon} = \delta \tag{17.100}$$

in the *coupled regime*.

17.4.1 Assumptions and Definitions for Formal Asymptotics

We provide the necessary assumptions and definitions related to formal asymptotic expansions to ensure this paper is self-contained.

17.4.1.1 Outer Setting

We define the two bulk phases for $t \in [0, T_f)$ by

$$\Omega^-(t;\varepsilon) := \{\boldsymbol{x} \in \Omega : \chi_\varepsilon(t,\boldsymbol{x}) < 0\} \quad \text{and} \quad \Omega^+(t;\varepsilon) := \{\boldsymbol{x} \in \Omega : \chi_\varepsilon(t,\boldsymbol{x}) > 0\}.$$

We assume that the solutions $((\rho_\alpha)_\alpha, \boldsymbol{v}, \chi, \varphi)$ of the considered scalings of (17.88) have expansions in ε in the outer regions $\Omega^\pm(t;\varepsilon)$, (in fact, we only need expansions up to the first two summands of each series):

$$\rho_{\alpha,\varepsilon}(t,\boldsymbol{x}) = \sum_{i=0}^{\infty} \varepsilon^i \rho_{\alpha,i}(t,\boldsymbol{x}), \quad \boldsymbol{v}_\varepsilon(t,\boldsymbol{x}) = \sum_{i=0}^{\infty} \varepsilon^i \boldsymbol{v}_i(t,\boldsymbol{x}) \tag{17.101}$$

$$\chi_\varepsilon(t,\boldsymbol{x}) = \sum_{i=0}^{\infty} \varepsilon^i \chi_i(t,\boldsymbol{x}) \text{ and } \varphi(t,\boldsymbol{x}) = \sum_{i=0}^{\infty} \varepsilon^i \varphi_i(t,\boldsymbol{x}). \tag{17.102}$$

We call a family $(\hat{\Gamma}(t))_{t\in[0,T_f)}$ an oriented $C^{1,2}$-family of hypersurfaces if for each point $(t_0, \boldsymbol{x}_0) \in (0, T_f) \times \mathbb{R}^d$ with $\boldsymbol{x}_0 \in \hat{\Gamma}(t_0)$ the following properties are satisfied:

(i) There exists an open subset $O \subset \mathbb{R}^d$ containing $\boldsymbol{x}_0$, $\delta > 0$ and a function $u \in C^{1,2}((t_0 - \delta, t_0 + \delta) \times O)$ such that

$$O \cap \hat{\Gamma}(t) = \{\boldsymbol{x} \in O \mid u(t,\boldsymbol{x}) = 0\} \quad \text{and}$$
$$\nabla u(t,\boldsymbol{x}) \neq 0 \quad \text{for } \boldsymbol{x} \in O \cap \hat{\Gamma}(t), t \in (t_0 - \delta, t_0 + \delta).$$

(ii) There exists a unit normal field $\boldsymbol{\nu}$ for $\hat{\Gamma}$ such that $\boldsymbol{\nu} \in C^0\left(\bigcup_{0<t<T_f}(\{t\} \times \hat{\Gamma}(t)), \mathbb{R}^d\right)$ and $\boldsymbol{\nu}(t,\cdot) \in C^1(\hat{\Gamma}(t), \mathbb{R}^d)$.

We assume that for $\varepsilon > 0$ small enough,

$$\Gamma_\varepsilon := \left\{(t,\boldsymbol{x}) \in [0,T_f) \times \Omega \;:\; \chi_\varepsilon(t,\boldsymbol{x}) = 0\right\}$$

is a set of smoothly evolving oriented $C^{1,2}$-hypersurfaces in $[0, T_f) \times \mathbb{R}^{\mathrm{d}}$. In addition, we assume the existence of a limiting $C^{1,2}$-family of oriented hypersurfaces Γ for ε going to zero. The curve Γ is the zeroth order of the interface. We denote the limiting bulk regions by $\Omega^+(t)$ and $\Omega^-(t)$. Further orders of Γ_ε are not required here. They would be needed if we considered higher order jump conditions, see Dreyer et al. (2012).

17.4.1.2 Inner Setting

In a neighborhood of Γ, we introduce a new coordinate system. To this end, let ϱ be a local parameterization of Γ:

$$\varrho : [0, T_f) \times U \to \mathbb{R}^{\mathrm{d}},$$

where $[0, T_f) \subset \mathbb{R}$ and $U \subset \mathbb{R}^{d-1}$ are the time interval and the spatial parameter domain, respectively.

Next, we parameterize a neighborhood of $\varrho(t, U)$ in $\mathbb{R}^{\mathrm{d}}$ as follows:

$$(t, \boldsymbol{x}) = (t, \varrho(t, \boldsymbol{s}) + \varepsilon z \boldsymbol{\nu}(t, \boldsymbol{s})) \tag{17.103}$$

with $0 < \varepsilon \leq \varepsilon_0$ for some $\varepsilon_0 > 0$ and $z \in \mathbb{R}$. The normal and tangential velocities of the interface Γ are related to the parameterization via

$$\boldsymbol{w}_\nu := w_\nu \boldsymbol{\nu} := (\partial_t \varrho \cdot \boldsymbol{\nu})\boldsymbol{\nu} \quad \text{and} \quad \boldsymbol{w}_\tau := \partial_t \varrho - (\partial_t \varrho \cdot \boldsymbol{\nu})\boldsymbol{\nu}. \tag{17.104}$$

For a generic function, depending on outer variables f we denote the corresponding function in inner variables by capital F, i.e.:

$$F(t, \boldsymbol{s}, z) = f(t, \boldsymbol{x}).$$

The partial derivatives of these functions transform as follows:

$$\begin{pmatrix} \nabla f \\ \partial_t f \end{pmatrix} = \begin{pmatrix} \frac{1}{|\boldsymbol{T}|^2}\boldsymbol{T} & \varepsilon^{-1}\boldsymbol{\nu} & 0 \\ -\boldsymbol{w}_\tau & -\varepsilon^{-1} w_\nu & 1 \end{pmatrix} \begin{pmatrix} \nabla_\Gamma F \\ \partial_z F \\ \partial_t F \end{pmatrix} + \mathcal{O}(\varepsilon),$$

where $\boldsymbol{T}$ is a $d \times (d-1)$-matrix whose columns are given by a basis of tangent vectors on Γ. Moreover, we have

$$\begin{aligned} \operatorname{div} \boldsymbol{f} &= \tfrac{1}{\varepsilon}\partial_z \boldsymbol{F} \cdot \boldsymbol{\nu} + \operatorname{div}_\Gamma \boldsymbol{F} + \mathcal{O}(\varepsilon), \\ \Delta f &= \tfrac{1}{\varepsilon^2}\partial_{zz} F - \tfrac{1}{\varepsilon}\kappa \partial_z F - z|\kappa|^2 \partial_z F + \Delta_\Gamma F + \mathcal{O}(\varepsilon), \end{aligned}$$

where ∇_Γ, $\operatorname{div}_\Gamma$, Δ_Γ are the surface gradient, the surface divergence, and the surface Laplacian on Γ, and κ is the mean curvature, respectively.

For the inner counterpart $((R_{\alpha,\varepsilon})_\alpha, \boldsymbol{V}_\varepsilon, X_\varepsilon, \Phi_\varepsilon)$ of the outer functions $((\rho_{\alpha,\varepsilon})_\alpha, \boldsymbol{v}_\varepsilon, \chi_\varepsilon, \varphi_\varepsilon)$, we assume

$$R_{\alpha,\varepsilon}(t, \boldsymbol{s}, z) = \sum_{i=0}^{\infty} \varepsilon^i R_{\alpha,i}(t, \boldsymbol{s}, z), \;\; \boldsymbol{V}_\varepsilon(t, \boldsymbol{s}, z) = \sum_{i=0}^{\infty} \varepsilon^i \boldsymbol{V}_i(t, \boldsymbol{s}, z), \tag{17.105}$$

$$X_\varepsilon(t,s,z) = \sum_{i=0}^{\infty} \varepsilon^i X_i(t,s,z), \quad \Phi_\varepsilon(t,s,z) = \sum_{i=0}^{\infty} \varepsilon^i \Phi_i(t,s,z). \tag{17.106}$$

Remark 17.2 Due to our definitions of Γ_ε and Γ we cannot expect $X_0(t,s,0) = 0$ but there will be a translational quantity depending on t and s. We could expand the interface position in ε, which would ensure $X_0(t,s,0) = 0$. However, we prefer the definition of Γ as in the orders studied here the translational constant causes no problems as no interfacial mass density appears. This is in contrast to the situation in Dreyer et al. (2012).

17.4.2 Matching Relations

In matched asymptotic techniques, inner and outer quantities are linked via certain matching conditions, see, e.g., Caginalp and Fife (1988). We impose the following asymptotic behavior for a generic quantity f as $z \to \pm\infty$:

$$F_0(t,s,z) - f_0^\pm = o(1/|z|), \tag{17.107}$$

$$F_1(t,s,z) - f_1^\pm - ((\nabla f_0)^\pm \cdot \boldsymbol{\nu}(t,s))z = o(1/|z|), \tag{17.108}$$

where the superscript $\pm$ denotes $\lim_{\varepsilon \searrow 0} f(t, \varrho(t,s) \pm \varepsilon \boldsymbol{\nu}(t,s))$. Moreover, we have

$$\partial_z F_0(t,s,z) = o(1/|z|), \tag{17.109}$$

$$\partial_{zz} F_0(t,s,z) = o(1/|z|), \tag{17.110}$$

$$\partial_z F_1(t,s,z) - (\nabla f_0)^\pm \cdot \boldsymbol{\nu}(t,s) = o(1/|z|), \tag{17.111}$$

$$\nabla_\Gamma F_0(t,s,z) - (\nabla f_0)^\pm + ((\nabla f_0)^\pm \cdot \boldsymbol{\nu}(t,s))\boldsymbol{\nu}(t,s) = o(1/|z|). \tag{17.112}$$

The idea behind this matching method is that the large z-behavior (for small ε) of the inner quantities coincides with the traces of the outer quantities, see, e.g., Lagerstrom (1988). To this end, a formal term-by-term matching of the ε-expansion of the inner quantities to the Taylor polynomials of the outer ones is made, see Caginalp and Fife (1988); Garcke and Stinner (2006).

17.5 Sharp Interface Limit of the Uncoupled Regime

In this section, we establish the sharp interface limit of the uncoupled regime, where the "small" parameter in the electrostatic equations remains independent of the interfacial layer thickness. Our approach relies on the methodology of matched asymptotic expansions. For a comprehensive exposition of this technique, we refer to Lagerstrom (1988); Caginalp and Fife (1988). A simplified version of the model, excluding electrical effects, has been analyzed in Dreyer et al. (2014).

To simplify notation, for any quantity f indexed by α we will write $(f_\alpha)_\alpha$ instead of $(f_\alpha)_{\alpha=1,\dots,N}$.

We begin by defining the outer, inner, and matching solutions. The outer equations are derived by expanding the relevant quantities in terms of δ and substituting them into the scaled system of equations.

Definition 17.1 A tuple

$$((\rho_{\alpha,0})_\alpha, \boldsymbol{v}_0, \chi_0, \chi_1, \varphi_0)$$

with the regularity conditions

$$\begin{aligned}
\rho_{\alpha,0} &\in C^0([0,T_f), C^2(\Omega^\pm, \mathbb{R}_+)) \cap C^1([0,T_f), C^0(\Omega^\pm, \mathbb{R}_+)),\\
\boldsymbol{v}_0 &\in C^0([0,T_f), C^1(\Omega^\pm, \mathbb{R}^3)) \cap C^1([0,T_f), C^0(\Omega^\pm, \mathbb{R}^3)),\\
\chi_0 &\in C^0([0,T_f), C^2(\Omega^\pm, \mathbb{R})),\\
\chi_1 &\in C^0([0,T_f), C^1(\Omega^\pm, \mathbb{R})),\\
\varphi_0 &\in C^0([0,T_f), C^2(\Omega^\pm, \mathbb{R}))
\end{aligned} \tag{17.113}$$

is called an **outer solution of the uncoupled regime** if it satisfies the following system of equations:

$$\begin{aligned}
0 &= \partial_t \rho_{\alpha,0} + \operatorname{div}(\rho_{\alpha,0}\boldsymbol{v}_0) - \operatorname{div}\Big(\sum_{\beta=1}^{N-1} M_{\alpha\beta}\nabla\big(\mu_{\beta,0} - \mu_{N,0} + \Big(\frac{z_\beta}{m_\beta} - \frac{z_N}{m_N}\Big)\varphi_0\big)\Big)\\
&\quad - \sum_{i=1}^{N_R} m_\alpha \gamma_\alpha^i M_{\mathrm{r}}^i \Big(1 - \exp\Big(\sum_{\beta=1}^{N} m_\beta \gamma_\beta^i \mu_{\beta,0}\Big)\Big) \quad \text{for } \alpha = 1,\dots,N-1.
\end{aligned} \tag{17.114}$$

$$0 = \partial_t \rho_0 + \operatorname{div}(\rho_0 \boldsymbol{v}_0), \tag{17.115}$$

$$0 = W'(\chi_0), \text{ in particular, } \nabla(W(\chi_0)) = 0, \tag{17.116}$$

$$\begin{aligned}
0 &= \partial_t(\rho_0\boldsymbol{v}_0) + \operatorname{div}(\rho_0\boldsymbol{v}_0 \otimes \boldsymbol{v}_0) + \nabla\Big(\sum_{\alpha=1}^{N} \rho_{\alpha,0}\mu_{\alpha,0} - \rho f_0\Big) - \nabla\big(W'(\chi_0)\chi_1\big)\\
&\quad + \varepsilon_0 \operatorname{div}\Big((1+s(\chi_0))\Big(\frac{1}{2}|\nabla\varphi_0|^2\mathbf{1} - \nabla\varphi_0 \otimes \nabla\varphi_0\Big)\Big),
\end{aligned} \tag{17.117}$$

$$0 = W''(\chi_0)\chi_1 + \frac{\partial \rho f}{\partial \chi}(\rho_{1,0},\dots,\rho_{N,0},\chi_0) - \frac{\varepsilon_0}{2}s'(\chi_0)|\nabla\varphi_0|^2, \tag{17.118}$$

$$0 = \varepsilon_0 \operatorname{div}((1+s(\chi_0))\nabla\varphi_0) + \sum_{\alpha=1}^{N} \frac{z_\alpha}{m_\alpha}\rho_{\alpha,0}. \tag{17.119}$$

Here, we use the abbreviations

$$\mu_{\alpha,0} = \mu_\alpha(\rho_{1,0}, \ldots, \rho_{N,0}, \chi_0), \quad \rho f_0 = \rho f(\rho_{1,0}, \ldots, \rho_{N,0}, \chi_0). \qquad (17.120)$$

The equations defining the inner solutions are derived by performing a change of variables in the scaled equations and expanding in terms of δ.

Definition 17.2 A tuple

$$((R_{\alpha,0})_\alpha, (R_{\alpha,1})_\alpha, \boldsymbol{V}_0, X_0, X_1, \Phi_0, \Phi_1)$$

with $X_0 \not\equiv 0$ and the regularity conditions

$$\begin{aligned}
R_{\alpha,0} &\in C^0([0,T_f), C^0(U, C^2(\mathbb{R}_+))),\\
R_{\alpha,1} &\in C^0([0,T_f), C^0(U, C^2(\mathbb{R}))),\\
\boldsymbol{V}_0 &\in C^0([0,T_f), C^0(U, C^1(\mathbb{R}^3))),\\
X_0 &\in C^0([0,T_f), C^1(U, C^0(\mathbb{R}))) \cap C^0([0,T_f), C^0(U, C^2(\mathbb{R}))), \qquad (17.121)\\
X_1 &\in C^0([0,T_f), C^0(U, C^2(\mathbb{R}))),\\
\Phi_0 &\in C^0([0,T_f), C^1(U, C^0(\mathbb{R}))) \cap C^0([0,T_f), C^0(U, C^2(\mathbb{R}))),\\
\Phi_1 &\in C^0([0,T_f), C^0(U, C^2(\mathbb{R})))
\end{aligned}$$

is called an **inner solution of the uncoupled regime** with normal velocity w_ν if it satisfies the following system of equations:

$$0 = \left(\frac{1}{2}(1+s(X_0))|\Phi_{0,z}|^2 - (1+s(X_0))|\Phi_{0,z}|^2\right)_z, \qquad (17.122)$$

$$0 = \left(\sum_{\beta=1}^{N-1} M_{\alpha\beta}(\mathcal{M}_{\beta,0} - \mathcal{M}_{N,0})_z\right)_z \text{ for } \alpha = 1, \ldots, N-1, \qquad (17.123)$$

$$0 = W'(X_0) - \gamma X_{0,zz}, \text{ in particular, } 0 = \boldsymbol{\nu}(-W(X_0) + \frac{\gamma}{2}X_{0,z}^2)_z, \qquad (17.124)$$

$$0 = (R_0(\boldsymbol{V}_0 \cdot \boldsymbol{\nu} - w_\nu))_z = (j_0)_z, \qquad (17.125)$$

$$\begin{aligned}
0 = {} & j_0 \boldsymbol{V}_{0,z} + \boldsymbol{\nu}\left(\sum_{\alpha=1}^{N} R_{\alpha,0}\mathcal{M}_{\alpha,0} - RF_0 - W'(X_0)X_1\right)_z \qquad (17.126)\\
& + \gamma\boldsymbol{\nu}(X_{0,z}X_{1,zz} + X_{0,zz}X_{1,z} - \kappa X_{0,z}^2) - \nabla_\Gamma W(X_0) + \gamma X_{0,zz}\nabla_\Gamma(X_0)\\
& + \frac{\varepsilon_0}{2}\left((\Phi_{1,z})^2 + |\nabla_\Gamma\Phi_0|^2\right)(1+s(X_0))_z\boldsymbol{\nu} - \varepsilon_0((1+s(X_0))\Phi_{1,z})_z\left(\Phi_{1,z}\boldsymbol{\nu} + \nabla_\Gamma\Phi_0\right),
\end{aligned}$$

$$\begin{aligned}
0 = {} & \frac{j_0}{\tau}X_{0,z} + W''(X_0)X_1 - \gamma X_{1,zz} + \gamma\kappa X_{0,z} + \frac{\partial RF_0}{\partial\chi} \qquad (17.127)\\
& - \frac{\varepsilon_0}{2}s'(X_0)\left((\Phi_{1,z})^2 + |\nabla_\Gamma\Phi_0|^2\right),
\end{aligned}$$

$$0 = \left((1+s(X_0))\Phi_{1,z}\right)_z, \qquad (17.128)$$

$$0 = (R_{\alpha,0}(\boldsymbol{V}_0 \cdot \boldsymbol{\nu} - w_{\boldsymbol{\nu}}))_z - \sum_{\beta=1}^{N-1} M_{\alpha\beta}\Big((\mathcal{M}_{\beta,1} - \mathcal{M}_{N,1})_{zz} - \kappa(\mathcal{M}_{\beta,0} - \mathcal{M}_{N,0})_z + \left(\frac{z_\beta}{m_\beta} - \frac{z_N}{m_N}\right)\Phi_{1,zz}\Big). \tag{17.129}$$

Here, $\boldsymbol{\nu}$ represents the unit normal vector to the zeroth-order interface, oriented toward the liquid phase. The operator ∇_Γ denotes the surface gradient, while κ refers to the mean curvature of the interface. Additionally, we have used

$$\begin{aligned}
\mathcal{M}_{\beta,0} &:= \mu_\beta(R_{1,0}, \ldots, R_{N,0}, X_0), \ \ RF_0 = \rho f(R_{1,0}, \ldots, R_{N,0}, X_0), \\
\frac{\partial RF_0}{\partial \chi} &:= \frac{\partial \rho f}{\partial \chi}(R_{1,0}, \ldots, R_{N,0}, X_0), \quad j_0 := R_0(\boldsymbol{V}_0 \cdot \boldsymbol{\nu} - w_{\boldsymbol{\nu}}), \\
\mathcal{M}_{\beta,1} &:= \sum_{\alpha=1}^{N} \frac{\partial \mu_\beta}{\partial \rho_\alpha}(R_{1,0}, \ldots, R_{N,0}, X_0) R_{\alpha,1} + \frac{\partial \mu_\beta}{\partial \chi}(R_{1,0}, \ldots, R_{N,0}, X_0) X_1.
\end{aligned} \tag{17.130}$$

Note that we have already simplified (17.123)–(17.129) by using $\Phi_{0,z} = 0$ which follows directly from (17.122), provided the matching condition $\Phi_{0,z}(z) \to 0$ for $z \to \pm\infty$ is satisfied. This seems reasonable as it helps to keep the notation more concise and is justified since we will only consider matching solutions in the following.

Finally, we define matching solutions which consist of compatible outer and inner solutions.

Definition 17.3 A tuple

$$((\rho_{\alpha,0})_\alpha, \boldsymbol{v}_0, \chi_0, \chi_1, \varphi_0, (R_{\alpha,0})_\alpha, (R_{\alpha,1})_\alpha, \boldsymbol{V}_0, X_0, X_1, \Phi_0, \Phi_1)$$

is called a **matching solution of the uncoupled regime** if

$$((\rho_{\alpha,0})_\alpha, \boldsymbol{v}_0, \chi_0, \chi_1, \varphi_0)$$

is an outer- and

$$((R_{\alpha,0})_\alpha, (R_{\alpha,1})_\alpha, \boldsymbol{V}_0, X_0, X_1, \Phi_0, \Phi_1)$$

is an inner solution, with both solutions being connected through the standard matching conditions (see Lagerstrom (1988); Caginalp and Fife (1988)).

Theorem 17.1 *Let*

$$((\rho_{\alpha,0})_\alpha, \boldsymbol{v}_0, \chi_0, \chi_1, \varphi_0, (R_{\alpha,0})_\alpha, (R_{\alpha,1})_\alpha, \boldsymbol{V}_0, X_0, X_1, \Phi_0, \Phi_1)$$

be a **matching solution of the uncoupled regime**. *Then, the following equations hold in the bulk regions:* $\Omega^\pm$*:*

$$\pm 1 = \chi_0, \chi_1 = 0, \tag{17.131}$$

$$\begin{aligned} 0 &= \partial_t \rho_{\alpha,0} + \operatorname{div}(\rho_{\alpha,0}\boldsymbol{v}_0) - \operatorname{div}\Big(\sum_{\beta=1}^{N-1} M_{\alpha\beta}\nabla\Big(\mu_{\beta,0} - \mu_{N,0} + \Big(\frac{z_\beta}{m_\beta} - \frac{z_N}{m_N}\Big)\varphi_0\Big)\Big) \\ &\quad - \sum_{i=1}^{N_R} m_\alpha \gamma_\alpha^i M_{\mathrm{r}}^i \Big(1 - \exp\Big(\sum_{\beta=1}^{N} m_\beta \gamma_\beta^i \mu_{\beta,0}\Big)\Big) \quad \textit{for } \alpha = 1, \dots, N-1, \end{aligned} \tag{17.132}$$

$$0 = \partial_t \rho_0 + \operatorname{div}(\rho_0 \boldsymbol{v}_0), \tag{17.133}$$

$$\begin{aligned} 0 &= \partial_t(\rho_0\boldsymbol{v}_0) + \operatorname{div}(\rho_0\boldsymbol{v}_0 \otimes \boldsymbol{v}_0) + \nabla\left(\sum_{\alpha=1}^{N} \rho_{\alpha,0}\mu_{\alpha,0} - \rho f_0 + \frac{\varepsilon_0}{2}(1 + s(\chi_0))|\nabla\varphi_0|^2\right) \\ &\quad - (\varepsilon_0(1 + s(\chi_0))\nabla\varphi_0 \otimes \nabla\varphi_0), \end{aligned} \tag{17.134}$$

$$0 = \varepsilon_0 \operatorname{div}((1 + s(\chi_0))\nabla\varphi_0) + \sum_{\alpha=1}^{N} \frac{z_\alpha}{m_\alpha}\rho_{\alpha,0}. \tag{17.135}$$

In addition, the following conditions are satisfied at the interface:

$$0 = [\![\mu_{\alpha,0} - \mu_{N,0}]\!], \tag{17.136}$$

$$0 = [\![\rho_0(\boldsymbol{v}_0 \cdot \boldsymbol{\nu} - w_\nu)]\!], \tag{17.137}$$

$$0 = [\![\rho_{\alpha,0}(\boldsymbol{v}_0 \cdot \boldsymbol{\nu} - w_\nu)]\!] - \Big[\!\!\Big[\sum_{\beta=1}^{N-1} M_{\alpha\beta}\nabla\Big(\mu_{\beta,0} - \mu_{N,0} + \Big(\frac{z_\beta}{m_\beta} - \frac{z_N}{m_N}\Big)\varphi_0\Big)\boldsymbol{\nu}\Big]\!\!\Big], \tag{17.138}$$

$$\begin{aligned} 0 &= \Big[\!\!\Big[j_0\boldsymbol{v}_0 + \Big(\sum_{\alpha=1}^{N} \rho_{\alpha,0}\mu_{\alpha,0} - \rho f_0\Big)\boldsymbol{\nu} + \varepsilon_0(1 + s(\chi_0))\Big(\frac{1}{2}|\nabla\varphi_0|^2 - \nabla\varphi_0 \otimes \nabla\varphi_0\Big)\boldsymbol{\nu}\Big]\!\!\Big] \\ &\quad - \gamma\kappa\boldsymbol{\nu}\int_{-\infty}^{\infty} (X_{0,z})^2 \,\mathrm{d}z, \end{aligned} \tag{17.139}$$

$$0 = \Big[\!\!\Big[\frac{j_0^2}{2\rho_0^2} + \mu_{N,0}\Big]\!\!\Big] + \frac{j_0}{\tau}\int_{-\infty}^{\infty} \frac{1}{R_0}(X_{0,z})^2 \,\mathrm{d}z, \tag{17.140}$$

$$0 = [\![(1 + s(\chi_0))\nabla\varphi_0 \cdot \boldsymbol{\nu}]\!], \tag{17.141}$$

$$0 = [\![\varphi_0]\!], \tag{17.142}$$

where $j_0 := \rho_0^{\pm}(\boldsymbol{v}_0^{\pm} \cdot \boldsymbol{\nu} - w_\nu)$ *and (17.136), (17.138) hold for* $\alpha = 1, \ldots, N-1$. *Furthermore, (17.142) implies*

$$[\![\nabla\varphi_0 - (\nabla\varphi_0 \cdot \boldsymbol{\nu})\boldsymbol{\nu}]\!] = 0. \tag{17.143}$$

Remark 17.3 Observe that all the jump conditions presented in Theorem 17.1 have physical significance. Equations (17.136), (17.140), and (17.138) represent generalized Gibbs–Thompson laws, while (17.137) ensures mass conservation. Furthermore, (17.139) corresponds to a dynamic Young–Laplace law, indicating the presence of surface tension of order δ^0.

Equation (17.141) establishes that, under this scaling, the electrical displacement remains continuous across the interface. Similarly, (17.142) enforces the continuity of the electrical potential, which consequently ensures the continuity of the tangential component of the electric field, as expressed in (17.143).

To facilitate the proof of Theorem 17.1, we divide it into a series of lemmas. The first lemma confirms the continuity of the electrical potential across the interface.

Lemma 17.2 *Any function* Φ_0 *satisfying (17.122) and the associated matching conditions fulfills* $\Phi_{0,z} = 0$. *Consequently, both (17.142) and (17.143) hold.*

Proof We begin with the equation $(1 + s(X_0))|\Phi_{0,z}|^2 = k \in \mathbb{R}$. By virtue of the matching condition $k = 0$ and therefore $s \geq 0$ implies $[\![\varphi_0]\!] = 0$. Taking the surface gradient of $[\![\varphi_0]\!] = 0$ leads directly to (17.143). □

Next, we leverage the mass flux continuity across the interface to eliminate the normal velocity from the governing equations.

Remark 17.4 Equation (17.125) is fulfilled if and only if $\boldsymbol{V}_0 \cdot \boldsymbol{\nu} = \frac{j_0}{R_0} + w_\nu$ for some j_0 independent of z.

The subsequent lemma establishes that the bulk consists of pure phases.

Lemma 17.3 *Let* χ_0, χ_1 *be given as in Definition 17.3. Then*

$$\chi_0 \in \{-1, 1\} \quad \text{and} \quad \chi_1 = 0.$$

Moreover, every solution $\Psi \in C^2(\mathbb{R})$ *of the ordinary differential equation*

$$W'(\Psi) - \gamma \partial_{zz}\Psi = 0 \tag{17.144}$$

with $\partial_z\Psi \to 0$, $\Psi \to \pm 1$ *as* $z \to \pm\infty$ *can be expressed as a one-parameter family of solutions:*

$$\Psi(z) = \bar{\Psi}(z - \bar{z}), \quad \bar{z} \in \mathbb{R}, \tag{17.145}$$

where $\bar{\Psi}$ *is the unique monotonically increasing solution of (17.144) satisfying* $\bar{\Psi}(0) = 0$. *Consequently, all* X_0 *satisfying Definition 17.3 can be written by the one-parameter family*

$$X_0(t, s, \cdot) = \bar{\Psi}(\cdot - \bar{z}(t, s)), \quad \bar{z} \in \mathbb{R}.$$

Proof From (17.117), we establish that $\chi_0 \in \pm 1, 0$. Continuity then implies that χ_0 remains constant in $\Omega^\pm$. A phase portrait analysis, as detailed in Benzoni-Gavage et al. (2007), confirms that (17.144) leads to (17.145) and $\chi_0^\pm = \pm 1$. Thus, $\chi_0 = \pm 1$ in $\Omega^\pm$, and we obtain $\frac{\partial \rho f}{\partial \chi}(\rho_{1,0}, \ldots, \rho_{N,0}, \chi_0) = 0 = s'(\chi_0)$ because of (17.84). Given that $W''(\pm 1) \neq 0$ and using (17.118), we conclude that $\chi_1 = 0$. □

We now rewrite some of Eqs. (17.123)–(17.129) to derive a system that allows us to compute $R_{\alpha,0}$ independently of Φ_1 and X_1.

Lemma 17.4 *For X_0 as defined in Lemma 17.3, Eqs. (17.123), (17.126), (17.127), and (17.128) are equivalent to the system consisting of (17.123), (17.126), (17.128), and the following equation:*

$$\frac{j_0}{R_0}\left(\frac{j_0}{R_0}\right)_z + (\mathcal{M}_{\alpha,0})_z = -\frac{j_0}{R_0 \tau}(X_{0,z})^2, \tag{17.146}$$

for all $\alpha = 1, \ldots, N$, where $R_0 := \sum_{\alpha=1}^N R_{\alpha,0}$.

Proof By combining (17.127) with the normal component of (17.126), we obtain

$$\begin{aligned}
&j_0(\boldsymbol{V}_0 \cdot \boldsymbol{\nu})_z + \left(\sum_{\alpha=1}^N R_{\alpha,0}\mathcal{M}_{\alpha,0} - RF_0 - W'(X_0)X_1\right)_z + \gamma X_{0,zz}X_{1,z} \\
&\quad + \frac{\varepsilon_0}{2}\left((\Phi_{1,z})^2 + |\nabla_\Gamma \Phi_0|^2\right)(1 + s(X_0))_z - \varepsilon_0((1 + s(X_0))\Phi_{1,z})_z\Phi_{1,z} \\
&= -\gamma(X_{0,z}X_{1,zz} - \kappa X_{0,z}^2) \\
&= -\frac{j_0}{\tau}(X_{0,z})^2 - W''(X_0)X_{0,z}X_1 - \frac{\partial RF_0}{\partial \chi}X_{0,z} + \frac{\varepsilon_0}{2}s'(X_0)X_{0,z}\left((\Phi_{1,z})^2 + |\nabla_\Gamma \Phi_0|^2\right).
\end{aligned} \tag{17.147}$$

Using (17.128), Eq. (17.147) simplifies to

$$\begin{aligned}
j_0(\boldsymbol{V}_0 \cdot \boldsymbol{\nu})_z + \left(\sum_{\alpha=1}^N R_{\alpha,0}\mathcal{M}_{\alpha,0} - RF_0 - W'(X_0)X_1\right)_z + \gamma X_{0,zz}X_{1,z} \\
= -\frac{j_0}{\tau}(X_{0,z})^2 - W''(X_0)X_{0,z}X_1 - \frac{\partial RF_0}{\partial \chi}X_{0,z}.
\end{aligned} \tag{17.148}$$

Rearranging, we obtain

$$j_0(\boldsymbol{V}_0 \cdot \boldsymbol{\nu})_z + \left(\sum_{\alpha=1}^N R_{\alpha,0}\mathcal{M}_{\alpha,0} - RF_0\right)_z = -\frac{j_0}{\tau}(X_{0,z})^2 - \frac{\partial RF_0}{\partial \chi}X_{0,z} \tag{17.149}$$

making use of (17.124). By applying the definition of μ_α, this can be rewritten as

$$j_0(\boldsymbol{V}_0\cdot\boldsymbol{\nu})_z+\sum_{\alpha=1}^{N}R_{\alpha,0}(\mathcal{M}_{\alpha,0})_z=-\frac{j_0}{\tau}(X_{0,z})^2. \tag{17.150}$$

Finally, using (17.123) and the positivity of R_0, we obtain (17.146), completing the proof. □

Note that Eqs. (17.123) and (17.146) constitute a system of N equations in which the only unknowns are the $R_{\alpha,0}$, as X_0 is already known up to a translational constant. Consequently, we can determine $(R_{\alpha,0})_\alpha$ up to this translational constant. Since these equations do not involve Φ_0 and Φ_1, we can utilize a result from Dreyer et al. (2014).

Lemma 17.5 *Let $\rho_1^-,\dots,\rho_N^->0$ be given and let $j_0\in\mathbb{R}$ with $|j_0|$ be small enough. Define $\rho^-:=\sum_\alpha\rho_\alpha^-$ and $\mu_\alpha^-:=\mu_\alpha(\rho_1^-,\dots,\rho_N^-,-1)$. Then, there exist functions $R_{1,0},\dots,R_{N,0}\in C^0(\mathbb{R},\mathbb{R}_+)$ and values $\rho_1^+,\dots,\rho_N^+>0$ such that (17.136), (17.140) and the following equations hold:*

$$0=\mu_\alpha(R_{1,0}(z),\dots,R_{N,0}(z),X_0(z))-\mu_N(R_{1,0}(z),\dots,R_{N,0}(z),X_0(z))-\mu_\alpha^-+\mu_N^-, \tag{17.151}$$

$$0=\mu_N(R_{1,0}(z),\dots,R_{N,0}(z),X_0(z))+\frac{j_0^2}{2\sum_\alpha R_{\alpha,0}^2(z)}+\frac{j_0}{\tau}\int_{-\infty}^{z}\frac{(X_{0,z}(\tilde z))^2}{\sum_\alpha R_{\alpha,0}(\tilde z)}\,\mathrm{d}\tilde z, \tag{17.152}$$

$$0=\lim_{z\to\pm\infty}R_{\alpha,0}(z)-\rho_\alpha^\pm \tag{17.153}$$

for $\alpha=1,\dots,N$. In particular, the functions $R_{1,0},\dots,R_{N,0}$ solve (17.123) and (17.146).

Thus, only the solvability criteria for Φ_1, X_1, $(R_{\alpha,1})_\alpha$ and the tangential part of $\boldsymbol{V}_0$ are left to be determined.

Lemma 17.6 *For X_0 as in Lemma 17.3, the function Φ_1 from Definition 17.3 satisfies*

$$\Phi_{1,z}=\frac{k}{1+s(X_0)} \tag{17.154}$$

for some $k\in\mathbb{R}$. Such a k exists if and only if (17.141) holds.

Proof Since $s\geq 0$, the equivalence of (17.128) to (17.154) is straightforward. The equivalence to the interface condition (17.141) follows from the matching conditions. □

Lemma 17.7 *The normal part of (17.126) can be expressed as*

$$\mathcal{L}X_1 = \left(\frac{j_0^2}{\sum_{\alpha=1}^N R_{\alpha,0}}\right)_z + (\sum_{\alpha=1}^N R_{\alpha,0}\mathcal{M}_{\alpha,0} - RF_0)_z - \gamma\kappa X_{0,z}^2$$
$$- \frac{\varepsilon_0}{2}((1+s(X_0))(\Phi_{1,z})^2)_z + \frac{\varepsilon_0}{2}|\nabla_\Gamma \Phi_0|^2(1+s(X_0))_z, \tag{17.155}$$

where

$$\mathcal{L} : W^{2,1}(\mathbb{R}) \to L^1(\mathbb{R}), \quad \Psi \mapsto (W'(X_0)\Psi - \gamma X_{0,z}\Psi_z)_z.$$

Equation (17.155) has a solution if and only if the normal part of (17.139) holds.

Proof The equivalence of the normal part of (17.126) and (17.155) is immediate. Thus, we focus on the solvability condition. The homogeneous adjoint problem of (17.155), given by

$$W'(X_0)\Xi_z + \gamma(X_{0,z}\Xi_z)_z = 0, \tag{17.156}$$

only admits in $L^\infty(\mathbb{R})$ solutions of the form $\Xi(z) = k$ for all $z \in \mathbb{R}$, with $k \in \mathbb{R}$, as shown in (Dreyer et al., 2014, Lemma 4.11). Hence, by Fredholm's theorem, (17.155) has a solution if and only if

$$0 = \int_{-\infty}^{\infty} \left(\frac{j_0^2}{\sum_{\alpha=1}^N R_{\alpha,0}}\right)_z + (\sum_{\alpha=1}^N R_{\alpha,0}\mathcal{M}_{\alpha,0} - RF_0)_z - \gamma\kappa X_{0,z}^2$$
$$- \frac{\varepsilon_0}{2}((1+s(X_0))(\Phi_{1,z})^2)_z + \frac{\varepsilon_0}{2}|\nabla_\Gamma \Phi_0|^2(1+s(X_0))_z \,\mathrm{d}z \tag{17.157}$$

is satisfied which corresponds to normal part of (17.139):

$$\left[\!\left[\frac{j_0^2}{\rho_0} + \sum_{\alpha=1}^N \rho_{\alpha,0}\mu_{\alpha,0} - \rho f_0 - \frac{\varepsilon_0}{2}(1+s(\chi_0))\left((\nabla\varphi_0\cdot\boldsymbol{\nu})^2 - (|\nabla\varphi_0|^2 - (\nabla\varphi_0\cdot\boldsymbol{\nu})^2)\right)\right]\!\right]$$
$$= \int_{-\infty}^{\infty} \gamma\kappa X_{0,z}^2 \,\mathrm{d}z. \tag{17.158}$$

Next, we verify that the tangential component of (17.139) determines the existence of the tangential part of $\boldsymbol{V}_0$. By (17.141) and (17.143), the tangential component of (17.139) simplifies to

$$j_0[\![\boldsymbol{v}_0 - (\boldsymbol{v}_0\cdot\boldsymbol{\nu})\boldsymbol{\nu}]\!] = 0. \tag{17.159}$$

Lemma 17.8 *The tangential part of (17.126) admits a solution if and only if (17.159) is satisfied.*

Proof For any vector $\mathbf{t}$ tangent to the zeroth-order interface Γ, multiplying (17.126) by $\mathbf{t}$ yields

$$0 = j_0(\boldsymbol{V}_0 \cdot \mathbf{t})_z - \nabla_\Gamma W(X_0) \cdot \mathbf{t} + \gamma X_{0,zz} \nabla_\Gamma (X_0) \cdot \mathbf{t} - \varepsilon_0 \left((1 + s(X_0))\Phi_{1,z}\right)_z \nabla_\Gamma \Phi_0 \cdot \mathbf{t}. \tag{17.160}$$

Using (17.124) and (17.128), this simplifies to

$$0 = j_0(\boldsymbol{V}_0 \cdot \mathbf{t})_z, \tag{17.161}$$

which has a solution if and only if (17.159) holds. □

Finally, we analyze solvability of (17.129).

Lemma 17.9 *Given* $((R_{\alpha,0})_\alpha, \boldsymbol{V}_0, X_0, X_1, \Phi_1)$, *there exist* $(R_{\alpha,1})_\alpha$ *satisfying (17.129) if and only if (17.138) holds.*

Proof Based on (Dreyer et al., 2014, Lemma 4.8) the map

$$(0, \infty)^N \to \mathbb{R}^N, \qquad (\rho_\alpha)_{\alpha=1,\dots,N} \mapsto (\mu_1((\rho_\alpha)_{\alpha=1,\dots,N}, \chi), \dots, \mu_N((\rho_\alpha)_{\alpha=1,\dots,N}, \chi))^T$$

is a diffeomorphism for any fixed $\chi \in [-1, 1]$ such that for fixed $\chi \in [-1, 1]$, the matrix

$$\left(\frac{\partial \mu_\beta}{\partial \rho_\gamma}((\rho_\alpha)_{\alpha=1,\dots,N}, \chi)\right)_{\beta,\gamma=1,\dots,N}$$

is invertible for any $\rho_1, \dots, \rho_N > 0$.

Thus, instead of studying criteria determining the existence of functions $(R_{\alpha,1})_\alpha$ solving (17.129) we can equivalently seek conditions ensuring the existence of functions $\mathcal{M}_{\alpha,1} : \mathbb{R} \to \mathbb{R}$, $\alpha = 1, \dots, N$ satisfying (17.129). Once the existence of $(\mathcal{M}_{\alpha,1})_\alpha$ is established, the corresponding $(R_{\alpha,1})_\alpha$ can be computed from (17.130). Conversely, if no $(\mathcal{M}_{\alpha,1})_\alpha$ solve (17.129), no solution for $(R_{\alpha,1})_\alpha$ exists. From (17.123) and the matching conditions, we have

$$(\mathcal{M}_{\beta,0} - \mathcal{M}_{N,0})_z = 0 \quad \text{for all } \beta = 1, \dots, N.$$

Therefore, (17.129) can be rewritten as

$$0 = (R_{\alpha,0}(\boldsymbol{V}_0 \cdot \boldsymbol{\nu} - w_\nu))_z - \sum_{\beta=1}^{N-1} M_{\alpha\beta}\Big((\mathcal{M}_{\beta,1} - \mathcal{M}_{N,1})_{zz} + \Big(\frac{z_\beta}{m_\beta} - \frac{z_N}{m_N}\Big)\Phi_{1,zz}\Big) \tag{17.162}$$

for $\alpha = 1, \dots, N-1$.

Since only the differences $\psi_\beta := \mathcal{M}_{\beta,1} - \mathcal{M}_{N,1}$ for $\beta = 1, \dots, N-1$ are relevant in (17.162), we employ the Fredholm alternative theorem to determine solvability conditions for $(\psi_\beta)_\beta$. To this end, we introduce auxiliary functions $\Xi_\beta \in C^\infty(\mathbb{R})$ for $\beta = 1, \dots, N-1$ such that

$$\Xi_\beta(z) = \begin{cases} (\nabla(\mu_{\beta,0} - \mu_{N,0}))^+ \cdot \boldsymbol{\nu} z + (\mu_{\beta,1}^+ - \mu_{N,1}^+) & \text{for } z > 1 \\ (\nabla(\mu_{\beta,0} - \mu_{N,0}))^- \cdot \boldsymbol{\nu} z + (\mu_{\beta,1}^- - \mu_{N,1}^-) & \text{for } z < -1. \end{cases}$$

We will see that the solvability conditions are independent of the chosen auxiliary functions. Defining $\Psi_\beta = \psi_\beta - \Xi_\beta$, we focus on the following auxiliary problem: Find $(\Psi_\beta)_{\beta=1,\dots,N-1} \in L^1(\mathbb{R})$ such that

$$\sum_{\beta=1}^{N-1}(M_{\alpha\beta}(\Psi_\beta)_z)_z = (R_{\alpha,0}(V_0 \cdot \nu - w_\nu))_z - \sum_{\beta=1}^{N-1} M_{\alpha\beta}\Big((\Xi_\beta)_z + \left(\frac{z_\beta}{m_\beta} - \frac{z_N}{m_N}\right)\Phi_{1,z}\Big)_z. \tag{17.163}$$

The solvability conditions for (17.163) are determined by the solutions of the homogeneous adjoint system of equations in $L^\infty(\mathbb{R})$. The homogeneous adjoint system is given by

$$\sum_{\beta=1}^{N-1}(M_{\alpha\beta}(Z_\beta)_z)_z = 0 \quad \text{for } \alpha = 1, \dots, N-1, \quad (Z_\beta)_{\beta=1,\dots,N-1}. \tag{17.164}$$

Since the matrix $M_{\alpha\beta}$ is constant in z and positive definite there are $N-1$ linearly independent solutions of (17.164) in $L^\infty(\mathbb{R})$, i.e., $\alpha = 1, \dots, N-1$, which can be chosen as

$$Z_\beta(z) = \delta_{\alpha\beta}, \qquad z \in \mathbb{R}.$$

By Fredholm alternative theorem, Eq. (17.163) is solvable if and only if

$$\int_{\mathbb{R}} (R_{\alpha,0}(V_0 \cdot \nu - w_\nu))_z - \sum_{\beta=1}^{N-1}\left(M_{\alpha\beta}(\Xi_\beta)_z + \left(\frac{z_\beta}{m_\beta} - \frac{z_N}{m_N}\right)\Phi_{1,z}\right)_z \mathrm{d}\,z = 0 \tag{17.165}$$

for $\alpha = 1, \dots, N-1$. Integrating (17.165) yields (17.138). □

Proof (*Proof of Theorem* 17.1) The interface conditions and properties of χ_0, χ_1 follow by combining the previous lemmas. Regarding the bulk equations, (17.132) corresponds to (17.114), (17.133) to (17.115), (17.135) to (17.119), and (17.134) follows from (17.117) given that $\chi_1 = 0$. □

17.6 Sharp Interface Limit of the Coupled Regime

This section focuses on establishing the sharp interface limit for the coupled regime, where the "small" parameter in the electrostatic equations is proportional to the thickness of the interfacial layer. As usual, we start by defining outer, inner, and matching solutions.

The outer equations are derived by inserting expansions of the relevant quantities in terms of δ into the rescaled system of equations.

Definition 17.4 A tuple

$$((\rho_{\alpha,0})_\alpha, \boldsymbol{v}_0, \chi_0, \chi_1, \varphi_0)$$

with the regularity conditions

$$\begin{aligned}
\rho_{\alpha,0} &\in C^0([0,T_f), C^2(\Omega^\pm, \mathbb{R}_+)) \cap C^1([0,T_f), C^0(\Omega^\pm, \mathbb{R}_+)),\\
\boldsymbol{v}_0 &\in C^0([0,T_f), C^1(\Omega^\pm, \mathbb{R}^3)) \cap C^1([0,T_f), C^0(\Omega^\pm, \mathbb{R}^3)),\\
\chi_0 &\in C^0([0,T_f), C^2(\Omega^\pm, \mathbb{R})),\\
\chi_1 &\in C^0([0,T_f), C^1(\Omega^\pm, \mathbb{R})),\\
\varphi_0 &\in C^0([0,T_f), C^2(\Omega^\pm, \mathbb{R}))
\end{aligned} \tag{17.166}$$

is called an **outer solution of the coupled regime** if it satisfies (17.114), (17.115), (17.116), and

$$0 = \partial_t(\rho_0 \boldsymbol{v}_0) + \operatorname{div}(\rho_0 \boldsymbol{v}_0 \otimes \boldsymbol{v}_0) + \nabla\Big(\sum_{\alpha=1}^N \rho_{\alpha,0}\mu_{\alpha,0} - \rho f_0\Big) - \nabla(W'(\chi_0)\chi_1), \tag{17.167}$$

$$0 = W''(\chi_0)\chi_1 + \frac{\partial \rho f}{\partial \chi}(\rho_{1,0}, \dots, \rho_{N,0}, \chi_0), \tag{17.168}$$

$$0 = \sum_{\alpha=1}^N \frac{z_\alpha}{m_\alpha}\rho_{\alpha,0}\,. \tag{17.169}$$

Here, we use the abbreviations

$$\mu_{\alpha,0} = \mu_\alpha(\rho_{1,0}, \dots, \rho_{N,0}, \chi_0), \quad \rho f_0 = \rho f(\rho_{1,0}, \dots, \rho_{N,0}, \chi_0). \tag{17.170}$$

The equations defining inner solutions are derived by performing a change of variables in the scaled system and expanding the relevant quantities in terms of δ.

Definition 17.5 A tuple

$$((R_{\alpha,0})_\alpha, (R_{\alpha,1})_\alpha, V_0, X_0, X_1, \Phi_0, \Phi_1)$$

with $X_0 \not\equiv 0$ and the regularity conditions

$$
\begin{aligned}
R_{\alpha,0} &\in C^0([0,T_f), C^0(U, C^2(\mathbb{R}_+))),\\
R_{\alpha,1} &\in C^0([0,T_f), C^0(U, C^2(\mathbb{R}))),\\
\boldsymbol{V}_0 &\in C^0([0,T_f), C^0(U, C^1(\mathbb{R}^3))),\\
X_0 &\in C^0([0,T_f), C^1(U, C^0(\mathbb{R}))) \cap C^0([0,T_f), C^0(U, C^2(\mathbb{R}))),\\
X_1 &\in C^0([0,T_f), C^0(U, C^2(\mathbb{R}))),\\
\Phi_0 &\in C^0([0,T_f), C^0(U, C^2(\mathbb{R}))),\\
\Phi_1 &\in C^0([0,T_f), C^0(U, C^2(\mathbb{R})))
\end{aligned}
\tag{17.171}
$$

is called an **inner solution of the coupled regime** with normal velocity w_ν if it satisfies (17.125) and the following system of equations:

$$0 = ((1+s(X_0))\Phi_{0,z})_z, \tag{17.172}$$

$$0 = \left(\sum_{\beta=1}^{N-1} M_{\alpha\beta}(\mathcal{M}_{\beta,0} - \mathcal{M}_{N,0})_z + \left(\frac{z_\beta}{m_\beta} - \frac{z_N}{m_N}\right)\Phi_{0,z}\right)_z, \tag{17.173}$$

$$0 = \boldsymbol{\nu}(-W(X_0) + \gamma\boldsymbol{\nu}X_{0,z}^2 - \frac{\varepsilon_0}{2}(1+s(X_0))(\Phi_{0,z})^2)_z, \tag{17.174}$$

$$0 = W'(X_0) - \gamma X_{0,zz} - \frac{\varepsilon_0}{2}s'(X_0)(\Phi_{0,z})^2, \tag{17.175}$$

$$
\begin{aligned}
0 = j_0\boldsymbol{V}_{0,z} &+ \boldsymbol{\nu}\left(\sum_{\alpha=1}^{N} R_{\alpha,0}\mathcal{M}_{\alpha,0} - RF_0 - W'(X_0)X_1\right)_z\\
&+ \gamma\boldsymbol{\nu}(X_{0,z}X_{1,zz} + X_{0,zz}X_{1,z} - \kappa X_{0,z}^2) - \nabla_\Gamma W(X_0) + \gamma X_{0,zz}\nabla_\Gamma(X_0),
\end{aligned}
\tag{17.176}
$$

$$0 = \frac{j_0}{\tau}X_{0,z} + W''(X_0)X_1 - \gamma X_{1,zz} + \gamma\kappa X_{0,z} + \frac{\partial RF_0}{\partial\chi}, \tag{17.177}$$

$$
\begin{aligned}
0 = (R_{\alpha,0}(\boldsymbol{V}_0\cdot\boldsymbol{\nu} - w_\nu))_z - \sum_{\beta=1}^{N-1} M_{\alpha\beta}\Big((\mathcal{M}_{\beta,1} - \mathcal{M}_{N,1})_{zz} - \kappa(\mathcal{M}_{\beta,0} - \mathcal{M}_{N,0})\\
+ \left(\frac{z_\beta}{m_\beta} - \frac{z_N}{m_N}\right)\Phi_{1,zz} - \kappa\left(\frac{z_\beta}{m_\beta} - \frac{z_N}{m_N}\right)\Phi_{0,z}\Big),
\end{aligned}
\tag{17.178}
$$

$$0 = ((1+s(X_0))\Phi_{1,z})_z + \sum_{\alpha=1}^{N}\frac{z_\alpha}{m_\alpha}R_{\alpha,0}. \tag{17.179}$$

Here, we use the abbreviations from (17.130).

Note that we already simplified (17.176), (17.177), and (17.179) by using $\Phi_{0,z} = 0$ which follows directly from (17.172), provided the matching condition $\Phi_{0,z}(z) \to 0$ for $z \to \pm\infty$ is satisfied. This seems reasonable as it helps to keep the notation more concise and is justified as we will only consider matching solutions in the following.

Before stating our theorem, we introduce one final definition. We define matching solutions as those that ensure compatibility between the outer and inner solutions.

Definition 17.6 A tuple $((\rho_{\alpha,0})_\alpha, \boldsymbol{v}_0, \chi_0, \chi_1, \varphi_0, (R_{\alpha,0})_\alpha, (R_{\alpha,1})_\alpha, \boldsymbol{V}_0, X_0, X_1, \Phi_0, \Phi_1)$ is called a *matching solution of the coupled regime* if $((\rho_{\alpha,0})_\alpha, \boldsymbol{v}_0, \chi_0, \chi_1, \varphi_0)$ is an outer- and $((R_{\alpha,0})_\alpha, (R_{\alpha,1})_\alpha, \boldsymbol{V}_0, X_0, X_1, \Phi_0, \Phi_1)$ is an inner solution and both are connected through the matching conditions.

Theorem 17.2 *If* $((\rho_{\alpha,0})_\alpha, \boldsymbol{v}_0, \chi_0, \chi_1, \varphi_0, (R_{\alpha,0})_\alpha, (R_{\alpha,1})_\alpha, \boldsymbol{V}_0, X_0, X_1, \Phi_0, \Phi_1)$ *is a* **matching solution of the coupled regime**, *then the following equations hold in the bulk regions* $\Omega^\pm$*:*

$$\pm 1 = \chi_0, \chi_1 = 0, \tag{17.180}$$

$$\begin{aligned} 0 = \partial_t \rho_{\alpha,0} + \operatorname{div}(\rho_{\alpha,0}\boldsymbol{v}_0) - \operatorname{div}\Big(\sum_{\beta=1}^{N-1} M_{\alpha\beta}\nabla\Big(\mu_{\beta,0} - \mu_{N,0} + \Big(\frac{z_\beta}{m_\beta} - \frac{z_N}{m_N}\Big)\varphi_0\Big)\Big) \\ - \sum_{i=1}^{N_R} m_\alpha \gamma_\alpha^i M_{\mathrm{r}}^i \Big(1 - \exp\Big(\sum_{\beta=1}^{N} m_\beta \gamma_\beta^i \mu_{\beta,0}\Big)\Big), \end{aligned} \tag{17.181}$$

$$0 = \partial_t \rho_0 + \operatorname{div}(\rho_0 \boldsymbol{v}_0), \tag{17.182}$$

$$0 = \partial_t(\rho_0 \boldsymbol{v}_0) + \operatorname{div}(\rho_0 \boldsymbol{v}_0 \otimes \boldsymbol{v}_0) + \nabla\left(\sum_{\alpha=1}^{N} \rho_{\alpha,0}\mu_{\alpha,0} - \rho f_0\right), \tag{17.183}$$

$$0 = \sum_{\alpha=1}^{N} \frac{z_\alpha}{m_\alpha} \rho_{\alpha,0}. \tag{17.184}$$

In addition, the following conditions are satisfied at the interface:

$$0 = [\![\mu_{\alpha,0} - \mu_{N,0}]\!], \tag{17.185}$$

$$0 = [\![\rho_0(\boldsymbol{v}_0 \cdot \boldsymbol{\nu} - w_\nu)]\!], \tag{17.186}$$

$$0 = [\![\rho_{\alpha,0}(\boldsymbol{v}_0 \cdot \boldsymbol{\nu} - w_\nu)]\!] - \Big[\!\!\Big[\sum_{\beta=1}^{N-1} M_{\alpha\beta}\nabla\Big(\mu_{\beta,0} - \mu_{N,0} + (\frac{z_\beta}{m_\beta} - \frac{z_N}{m_N})\varphi_0\Big) \cdot \boldsymbol{\nu}\Big]\!\!\Big], \tag{17.187}$$

$$0 = \left[\!\!\left[j_0 \boldsymbol{v}_0 + \left(\sum_{\alpha=1}^{N} \rho_{\alpha,0}\mu_{\alpha,0} - \rho f_0\right)\boldsymbol{\nu}\right]\!\!\right] - \gamma\kappa\boldsymbol{\nu}\int_{-\infty}^{\infty} (X_{0,z})^2 \,\mathrm{d}z, \tag{17.188}$$

$$0 = \left[\!\!\left[\frac{j_0^2}{2\rho_0^2} + \mu_{N,0}\right]\!\!\right] + \frac{j_0}{\tau}\int_{-\infty}^{\infty} \frac{1}{R_0}(X_{0,z})^2 \,\mathrm{d}z, \tag{17.189}$$

$$0 = [\![(1 + s(\chi_0))\nabla\varphi_0 \cdot \boldsymbol{\nu}]\!] - \int_{-\infty}^{\infty}\Big(\sum_{\alpha=1}^{N} \frac{z_\alpha}{m_\alpha} R_{\alpha,0}(z)\Big)\,\mathrm{d}z, \tag{17.190}$$

$$0 = [\![\varphi_0]\!], \tag{17.191}$$

where $j_0 := \rho_0^{\pm}(\boldsymbol{v}_0^{\pm} \cdot \boldsymbol{\nu} - w_\nu)$ *and* $\alpha = 1, \ldots, N-1$ *in (17.185) and (17.187). Furthermore, (17.142) implies*

$$[\![\nabla\varphi_0 - (\nabla\varphi_0 \cdot \boldsymbol{\nu})\boldsymbol{\nu}]\!] = 0. \tag{17.192}$$

Remark 17.5 The evolution equations for $(\rho_{\alpha,0})_\alpha$ together with the closure relation (17.184) and the charge conservation condition

$$\sum_\alpha \tfrac{z_\alpha}{m_\alpha} r_\alpha = 0$$

lead to the following elliptic problem for φ_0 :

$$0 = \operatorname{div}\left(\sum_{\alpha,\beta=1}^{N-1} M_{\alpha\beta}\left(\left(\frac{z_\alpha}{m_\alpha} - \frac{z_N}{m_N}\right)\nabla(\mu_\beta - \mu_N) + \left(\frac{z_\alpha}{m_\alpha} - \frac{z_N}{m_N}\right)\left(\frac{z_\beta}{m_\beta} - \frac{z_N}{m_N}\right)\nabla\varphi_0\right)\right). \tag{17.193}$$

Moreover,

$$\sum_{\alpha,\beta=1}^{N-1} M_{\alpha\beta}\left(\frac{z_\alpha}{m_\alpha} - \frac{z_N}{m_N}\right)\left(\frac{z_\beta}{m_\beta} - \frac{z_N}{m_N}\right) > 0,$$

if $\frac{z_\alpha}{m_\alpha} - \frac{z_N}{m_N} \neq 0$ for at least one α, due to the positive definiteness of $M_{\alpha\beta}$.

The proof of Theorem 17.2 follows the same strategy as that of Theorem 17.1, so we will only emphasize the key differences.

Lemma 17.10 *Let* Φ_0 *be given as in Definition 17.6. Then, it follows that* $\Phi_{0,z} = 0$*, which implies* $[\![\varphi_0]\!] = 0$.

Proof By integrating (17.172) and using the matching conditions, we obtain

$$(1 + s(X_0))\Phi_{0,z} = 0.$$

Since $s \geq 0$, this directly implies the claim. □

As a consequence of Lemma 17.10, Eq. (17.175) simplifies to

$$0 = W'(X_0) - \gamma X_{0,zz} \tag{17.194}$$

and Lemma 17.3 applies. Regarding the normal velocity, Remark 17.4 remains valid in this setting. Moreover, Eq. (17.173) simplifies to

$$0 = \left(\sum_{\beta=1}^{N-1} M_{\alpha\beta}(\mathcal{M}_{\beta,0} - \mathcal{M}_{N,0})_z\right)_z. \tag{17.195}$$

Similar to Lemma 17.4, we extract a subsystem of equations that determine $(R_{\alpha,0})_\alpha$.

Lemma 17.11 *For X_0 given as in Lemma 17.3 the system of equations (17.176), (17.177), (17.194), (17.195) is equivalent to (17.176), (17.194), (17.195), and*

$$\frac{j_0}{R_0}\left(\frac{j_0}{R_0}\right)_z + (\mathcal{M}_{\alpha,0})_z = -\frac{j_0}{R_0 T}(X_{0,z})^2, \tag{17.196}$$

for any $\alpha \in \{1, \ldots, N\}$, where $R_0 := \sum_{\alpha=1}^N R_{\alpha,0}$.

Proof The proof follows the same reasoning as Lemma 17.4. □

Since Eqs. (17.195) and (17.196) match (17.123) and (17.146), Lemma 17.5 remains valid in this scaling.

Lemma 17.12 *The normal part of (17.176) can be written as*

$$\mathcal{L}X_1 = \left(\frac{j_0^2}{\sum_{\alpha=1}^N R_{\alpha,0}}\right)_z + \left(\sum_{\alpha=1}^N R_{\alpha,0}\mathcal{M}_{\alpha,0} - RF_0\right)_z - \gamma\kappa X_{0,z}^2, \tag{17.197}$$

where

$$\mathcal{L} : W^{2,1}(\mathbb{R}) \to L^1(\mathbb{R}), \quad \Psi \mapsto (W'(X_0)\Psi - \gamma X_{0,z}\Psi_z)_z.$$

Equation (17.197) has a solution if and only if the normal part of (17.188) is satisfied.

Proof The only solutions of the homogeneous adjoint problem corresponding to (17.197), i.e.:

$$W'(X_0)\Xi_z + \gamma(X_{0,z}\Xi_z)_z = 0 \tag{17.198}$$

that belongs to L^∞ are constants, i.e., $\Xi = k$ for $k \in \mathbb{R}$, see (Dreyer et al., 2014, Lemma 4.11). Thus, the solvability condition for (17.197) is

$$0 = \int_{-\infty}^{\infty}\left(\left(\frac{j_0^2}{\sum_{\alpha=1}^N R_{\alpha,0}}\right)_z + \left(\sum_{\alpha=1}^N R_{\alpha,0}\mathcal{M}_{\alpha,0} - RF_0\right)_z - \gamma\kappa X_{0,z}^2\right) \mathrm{d}z,$$

which is equivalent to (17.188). □

Next, we consider the tangential component of $\boldsymbol{V}_0$. The arguments remain the same as in Lemma 17.8, except that the terms involving φ and Φ are absent. Therefore, we conclude the following:

Lemma 17.13 *The tangential component of (17.176) has a solution if and only if the tangential component of (17.188) holds.*

By integrating (17.179), we obtain the following result.

Lemma 17.14 *Let* $(R_{\alpha,0})_\alpha$ *and* X_0 *be given. Then, a solution* Φ_1 *to (17.179) exists if and only if condition (17.190) is satisfied.*

Finally, similar to Lemma 17.9, we establish the following result.

Lemma 17.15 *Equation (17.178) has a solution if and only if (17.187) is satisfied.*

Proof (*Proof of Theorem* 17.1) We obtain the interface conditions and the properties of χ_0, χ_1 by combining the preceding lemmas. Concerning the bulk equations, (17.181) is (17.114), (17.182) is (17.115), (17.184) is (17.169), and (17.183) follows from (17.167) as $\chi_1 = 0$. □

Acknowledgements The authors express their gratitude to Clemens Guhlke for valuable discussions and insightful advice on various aspects of this work.

References

Abels H (2009) Existence of weak solutions for a diffuse interface model for viscous, incompressible fluids with general densities. Commun Math Phys 289:45–73

Abels H (2012) Strong well-posedness of a diffuse interface model for a viscous, quasi-incompressible two-phase flow. SIAM J Math Anal 44:316–340

Abels H, Depner D, Garcke H (2012) Existence of weak solutions for a diffuse interface model for two-phase flows of incompressible fluids with different densities. J Math Fluid Mech. https://doi.org/10.1007/s00021-012-0118-x

Abels H, Garcke H, Grün G (2012) Thermodynamically consistent frame indifferent diffuse interface models for incompressible two-phase flows with different densities. Math Mod Meth Appl S 22:1150013 (40 p)

Aki GL, Daube J, Dreyer W, Giesselmann J, Kränkel M, Kraus C (2012) A diffuse interface model for quasi-incompressible flows: sharp interface limits and numerics. ESAIM Proc 38:54–77

Aki GL, Dreyer W, Giesselmann J, Kraus C (2014) A quasi-incompressible diffuse interface model with phase transition. Math Mod Meth Appl S 24(5):827–861

Benzoni-Gavage S, Danchin R, Descombes S, Jamet D (2007) Stability issues in the Euler-Korteweg model. Contemp Math, AMS 426:103–127

Blesgen T (1999) A generalization of the Navier-Stokes equations to two-phase flows. J Phys D Appl Phys 32(32):1119–1123

Bothe D, Dreyer W (2015) Rational thermodynamics of chemically reacting multicomponent fluid mixtures. Acta Mech 226:1757–1805

Caginalp G, Fife PC (1988) Dynamics of layered interfaces arising from phase boundaries. SIAM J Appl Math 48:506–518

Dreyer W, Giesselmann J, Kraus C, Rohde C (2012) Asymptotic analysis for Korteweg models. Interfaces Free Bound 14:105–143

Dreyer W, Giesselmann J, Kraus C (2014) A compressible mixture model with phase transition. Phys D 273–274:1–13

Dreyer W, Kraus C (2010) On the van der Waals–Cahn–Hilliard phase-field model and its equilibria conditions in the sharp interface limit. Proc R Soc Edinb, Sect A, Math, 140(6):1161–1186

Feireisl E, Petzeltová H, Rocca E, Schimperna G (2010) Analysis of a phase-field model for two-phase compressible fluids. Math Mod Meth Appl S 20(7):1129–1160

Garcke H, Stinner B (2006) Second order phase field asymptotics for multi-component systems. Interfaces Free Bound 8:131–157

Ghosh T, Bringedal C, Rohde C, Helmig R (2024) A phase-field approach to model evaporation from porous media: modeling and upscaling. arXiv:2112.13104

Gurtin ME, Polignone D, Vinals J (1996) Two-phase binary fluids and immiscible fluids described by an order parameter. Math Mod Meth Appl S 6:815–831

Hantke M, Matern C, Warnecke G, Yaghi H (2023) A new method to discretize a model for isothermal flow with a multi-component equation of state. J Comput Appl Math 422:114876

Hohenberg PC, Halperin BI (1977) Theory of dynamic critical phenomena. Rev Mod Phys 49:435–479

Hutter K, Jöhnk K (2004) Continuum methods of physical modeling. Springer, Berlin

Kotschote M (2012) Strong solutions of the Navier-Stokes equations for a compressible fluid of Allen-Cahn type. Arch Ration Mech Anal 206(2):489–514

Lagerstrom PA (1988) Matched asymptotic expansions: ideas and techniques. Springer, New York

Lowengrub JS, Truskinovsky L (1998) Quasi-incompressible Cahn-Hilliard fluids and topological transitions. R Soc Lond Proc Ser A Math Phys Eng Sci 454:2617–2654

Luckhaus S, Modica L (1989) The Gibbs-Thompson relation within the gradient theory of phase transitions. Arch Ration Mech Anal 107(1):71–83

Müller I (1985) Thermodynamics, vol XVII. Pitman Advanced Publishing Program, Boston - London - Melbourne

Ostrowski L, Massa FC, Rohde C (2020) A phase field approach to compressible droplet impingement. In: Droplet interactions and spray processes. Springer, Berlin, pp 113–126

Owen NC, Rubinstein J, Sternberg P (1990) Minimizers and gradient flows for singularly perturbed bi-stable potentials with a Dirichlet condition. Proc Roy Soc London Ser A 429(1877):505–532

Sternberg P (1988) The effect of a singular perturbation on nonconvex variational problems. Arch Ration Mech Anal 101(3):209–260

Ten Eikelder MFP, Van Der Zee KG, Akkerman I, Schillinger D (2023) A unified framework for Navier-Stokes Cahn-Hilliard models with non-matching densities. Math Models Methods Appl Sci 33:175–221

Truesdell C (1969) Rational thermodynamics. McGraw-Hill Book Co., New York

Truesdell C, Toupin R (1960) The classical field theories. In: Handbuch der physik, Bd. III/1. Springer, Berlin, pp 226–793; Appendix, pp 794–858

Witterstein G (2011) Phase change flows with mass exchange. Adv Math Sci Appl 21(2):559–611

Xu S, Jin X, Chen H, Fan W, Wen H, Wang B (2024) Modelling and simulation on compressible multi-component gas-liquid flows with chemical reaction and phase transition effects. Aerosp Sci Technol 153:109451

Chapter 18
Hyperbolic Relaxation of the Chemical Potential in the Viscous Cahn–Hilliard Equation

Pierluigi Colli and Jürgen Sprekels

In memory of Prof. Dr. Wolfgang Dreyer with admiration, sympathy and friendship.

Abstract In this paper, we study a hyperbolic relaxation of the viscous Cahn–Hilliard system with zero Neumann boundary conditions. In fact, we consider a relaxation term involving the second time derivative of the chemical potential in the first equation of the system. We develop a well-posedness, continuous dependence and regularity theory for the initial-boundary value problem. Moreover, we investigate the asymptotic behavior of the system as the relaxation parameter tends to 0 and prove the convergence to the viscous Cahn–Hilliard system.

Keywords Cahn–Hilliard system · Hyperbolic relaxation · Partial differential equations · Initial-boundary value problem · Well-posedness · Continuous dependence · Regularity · Asymptotic convergence

AMS (MOS) Subject Classification: 35M33 · 35M87 · 35B40 · 37D35

18.1 Introduction

In this paper, we deal with an initial-boundary value problem for a system of partial differential equations of viscous Cahn–Hilliard type, which in particular includes a hyperbolic relaxation term in the first equation.

P. Colli
Dipartimento di Matematica "F. Casorati", Università di Pavia, and Research Associate at the IMATI – C.N.R. Pavia, Pavia, Italy
e-mail: pierluigi.colli@unipv.it

J. Sprekels (✉)
Weierstrass Institute for Applied Analysis and Stochastics, Berlin, Germany
e-mail: juergen.sprekels@wias-berlin.de

J. Fuhrmann et al. (eds.), *Advances in Continuum Physics*, Advanced Structured Materials 238, https://doi.org/10.1007/978-3-031-93918-1_18

The system is stated as follows:

$$\alpha \partial_{tt}\mu + \partial_t\varphi - \Delta\mu = 0 \quad \text{in } Q := \Omega \times (0, T), \tag{18.1}$$

$$\tau\partial_t\varphi - \Delta\varphi + f'(\varphi) = \mu + g \quad \text{in } Q, \tag{18.2}$$

$$\partial_{\mathbf{n}}\mu = \partial_{\mathbf{n}}\varphi = 0 \quad \text{on } \Sigma := \partial\Omega \times (0, T), \tag{18.3}$$

$$\mu(0) = \mu_0, \;\; (\partial_t\mu)(0) = \nu_0, \;\; \varphi(0) = \varphi_0 \quad \text{in } \Omega, \tag{18.4}$$

where $\Omega \subset \mathbb{R}^N$, $N \in \{1, 2, 3\}$, is a bounded and connected domain with smooth (of class C^2, say) boundary $\partial\Omega$ and T denotes some final time. We denote by $\mathbf{n}$ the unit outward normal to $\partial\Omega$, with the associated outward normal derivative $\partial_{\mathbf{n}}$. Note that $\partial_{\mathbf{n}}$ appears in the homogeneous Neumann boundary conditions stated in (18.3) for both the variables μ and φ.

Equations (18.1)–(18.2) constitute a variation of the Cahn–Hilliard system (introduced in Cahn and Hilliard (1958) and first approached mathematically in Elliott and Zheng (1986))

$$\partial_t\varphi - \Delta\mu = 0 \quad \text{in } Q, \tag{18.5}$$

$$-\Delta\varphi + f'(\varphi) = \mu + g \quad \text{in } Q, \tag{18.6}$$

in which a viscosity term $\tau\partial_t\varphi$ has been included in the second equation and where especially the hyperbolic relaxation term $\alpha\partial_{tt}\mu$ has been added in the first equation. The viscous Cahn–Hilliard system

$$\partial_t\varphi - \Delta\mu = 0 \quad \text{in } Q, \tag{18.7}$$

$$\tau\partial_t\varphi - \Delta\varphi + f'(\varphi) = \mu + g \quad \text{in } Q, \tag{18.8}$$

is well known and was already investigated in a number of papers (see Cherfils and Miranville (2023), Colli et al. (2021), Colli et al. (2018a), Colli et al. (2018b), Colli et al. (2019), Colli et al. (2024b), Conti et al. (2024), Gilardi et al. (2023a), Gilardi et al. (2023b), Gilardi and Sprekels (2019), Kagawa and Ôtani (2022) to quote some recent contributions), while to our knowledge an inertial term like $\alpha\partial_{tt}\mu$ in (18.1) is not common, but certainly deserves to be examined. In this class of problems, the unknown functions φ and μ usually stand for the *order parameter*, which can represent a scaled density of one of the involved phases, and the *chemical potential* associated with the phase separation process, respectively.

Moreover, f' denotes the derivative (if it exists) of a double-well potential f, which in general is split into a (possibly nondifferentiable) convex part f_1 and a smooth and concave perturbation f_2. Typical and physically significant examples for f are the so-called *classical regular potential*, the *logarithmic double-well potential*, and the *double obstacle potential*, which are given, in this order, by

$$f_{\text{reg}}(r) := \frac{1}{4}(r^2-1)^2, \quad r \in \mathbb{R}, \tag{18.9}$$

$$f_{\log}(r) := \begin{cases} (1+r)\ln(1+r) + (1-r)\ln(1-r) - c_1 r^2 & \text{if } r \in (-1,1) \\ 2\ln(2) - c_1 & \text{if } r \in \{-1,1\}, \\ +\infty & \text{if } r \notin [-1,1] \end{cases} \tag{18.10}$$

$$f_{2\text{obs}}(r) := \begin{cases} c_2(1-r^2) & \text{if } r \in [-1,1] \\ +\infty & \text{if } r \notin [-1,1] \end{cases}. \tag{18.11}$$

Here, the constants c_i in (18.10) and (18.11) satisfy $c_1 > 1$ and $c_2 > 0$, so that $f_{\log}$ and $f_{2\text{obs}}$ are nonconvex. Notice that for $f = f_{\log}$ the term $f'(\varphi)$ occurring in (18.2) becomes singular as $\varphi \searrow -1$ and $\varphi \nearrow 1$, which forces the order parameter φ to attain its values in the physically meaningful range $(-1, 1)$. In the nonsmooth case (18.11), the convex part f_1 is given by the indicator function of $[-1, 1]$. Accordingly, in such cases, one has to replace the derivative of the convex part by the subdifferential ∂f_1 and, consequently, to interpret (18.2) as a differential inclusion or a variational inequality. We also note that τ is a fixed positive parameter (the viscosity coefficient), while for the positive parameter α we will also discuss the asymptotic convergence to 0. We point out that in (18.2) a known forcing term g is present that may be interpreted as a direct or secondary control term which acts on the system. In this connection, we mention that optimal control problems for viscous Cahn–Hilliard systems with a distributed control term involving g have recently been treated in the paper Colli et al. (2024b).

Some hyperbolic relaxations of the viscous Cahn–Hilliard system have been already considered and studied: let us mention the recent contributions Bonfoh (2021), Chen (2021), Chen et al. (2023), Dor et al. (2024), Savostianov and Zelik (2016), Şen and Khanmamedov (2024). However, the available investigations are concerned with systems where the inertial term involves the phase variable φ. In our case, the system (18.1)–(18.4) couples a wave-type equation for μ combined with a source term given by $-\partial_t\varphi$, with a semilinear parabolic equation in which the source term includes μ.

From the energetic viewpoint, there is a change with respect to the viscous (and nonviscous) Cahn–Hilliard equation. To see this, let us for simplicity argue now on the case $g = 0$. Indeed, for (18.7)–(18.8), as well as for (18.5)–(18.6), the basic energy estimate is obtained by testing (18.7) by μ, (18.8) by $\partial_t\varphi$, and then adding and thus producing a cancellation of the terms containing the product $\mu\partial_t\varphi$. Therefore, one has

$$\int_0^t\int_\Omega |\nabla\mu|^2 + \tau\int_0^t\int_\Omega |\partial_t\varphi|^2 + \int_\Omega\Big(\frac{1}{2}|\nabla\varphi(t)|^2 + f(\varphi(t))\Big) = \text{constant}$$

for $t \in [0, T]$, where the first two terms are dissipative and the energy term is given by the third one. The second term is missing in the case of the Cahn–Hilliard system (18.5)–(18.6), but the energy is the same and there is only one term for dissipation.

On the other hand, it turns out to be more difficult and involved to recover an energy estimate for (18.1)–(18.2): as you will check in the sequel, our main estimate is constructed by testing (18.1) by $\partial_t \mu$ and the time derivative of (18.2) by $\partial_t \varphi$, in order to have a cancellation of the terms containing the product $\partial_t \varphi \, \partial_t \mu$. By integration, we then obtain

$$\begin{aligned} &\int_\Omega \Big(\frac{\alpha}{2} |\partial_t \mu(t)|^2 + \frac{1}{2} |\nabla \mu(t)|^2 + \frac{\tau}{2} |\partial_t \varphi(t)|^2 \Big) \\ &\quad + \int_0^t \int_\Omega |\nabla(\partial_t \varphi)|^2 + \int_0^t \int_\Omega f_1''(\varphi) |\partial_t \varphi|^2 \\ &= \text{constant} - \int_0^t \int_\Omega f_2''(\varphi) |\partial_t \varphi|^2 \quad \text{for } t \in [0, T], \end{aligned}$$

where the energy is now located in the first integral in which neither the nonlinearity f nor any of its derivatives occur. Moreover, note that the last term on the left-hand side induces dissipation (as f_1'' is nonnegative), but on the right-hand side the complementary term may be positive and grow with respect to t, since f_2 is concave, in general. The viscous contribution in (18.2) is important here to control this term on the right-hand side, since the addendum $\frac{\tau}{2} |\partial_t \varphi(t)|^2$ is part in the energy. However, by our estimate we can proceed in the analysis and not only construct a well-posedness theory but also investigate the asymptotic behavior as the parameter α converges to 0.

This paper is dedicated to the memory of Wolfgang Dreyer, who recently passed away. The authors of this paper were fortunate to have benefited from Wolfgang's friendship, as well as from his exceptional expertise and insight in Thermodynamics and Applied Mathematics. He was a brilliant and generous scientist who truly enjoyed engaging in scientific discussions with friends and colleagues. We both feel enriched by having known him and are grateful for the opportunity to have collaborated with him. Together, along with other colleagues, we co-authored the paper Bonetti et al. (2002) that explored the effects of phase separation driven by mechanical actions in tin/lead alloys, and where the corresponding system of partial differential equations already included equations of Cahn–Hilliard type.

The paper is organized as follows. In the following section, we formulate the general assumptions and state the main results concerning the system (18.1)–(18.4). In Sect. 18.3, we then prove the existence of a solution by using a double approximation based on a Yosida regularization of ∂f_1 and on a Faedo–Galerkin scheme. This proof requires the main analytical effort of this paper, since it involves a number of estimates and two passage-to-the-limit processes. In Sect. 18.4, we show the results on continuous dependence with respect to data and on the regularity of the solution: actually, three theorems are proved there. Finally, Sect. 18.5 then brings the asymptotic results of the convergence of the system to the viscous Cahn–Hilliard system as α tends to 0 and an estimate of the difference of solutions in terms of a precise rate of convergence.

We fix some notation. For any Banach space X, we let X^* denote its dual space, and $\|\cdot\|_X$ stands for the norm in X and any power of X. For two Banach spaces X and Y that are both continuously embedded in some topological vector space Z, the linear space $X \cap Y$ is the Banach space equipped with its natural norm $\|v\|_{X\cap Y} := \|v\|_X + \|v\|_Y$ for $v \in X \cap Y$. The standard Lebesgue and Sobolev spaces $L^p(\Omega)$ and $W^{m,p}(\Omega)$ are defined on Ω for $1 \le p \le \infty$ and $m \in \mathbb{N} \cup \{0\}$. For the sake of convenience, we denote the norm of $L^p(\Omega)$ by $\|\cdot\|_p$ for $1 \le p \le \infty$. If $p = 2$, we employ the usual notation $H^m(\Omega) := W^{m,2}(\Omega)$. We also set

$$H := L^2(\Omega), \quad V := H^1(\Omega), \quad W := \left\{v \in H^2(\Omega) : \partial_{\mathbf{n}} v = 0 \text{ on } \partial\Omega\right\}.$$

Moreover, V^* is the dual space of V, and $\langle\cdot\,,\,\cdot\rangle$ stands for the duality pairing between V^* and V. We denote by $(\cdot\,,\,\cdot)$ the natural inner product in H. As usual, H is identified with a subspace of the dual space V^* according to the identity:

$$\langle u, v\rangle = (u, v) \quad \text{for every } u \in H \text{ and } v \in V.$$

Note that $W \subset V \subset H \equiv H^* \subset V^*$ with dense and compact embeddings. In the following, we use the general notation $\overline{v}$ to denote the mean value $\frac{1}{|\Omega|}\int_\Omega v$ of a generic function $v \in L^1(\Omega)$. If v is in V^*, then we can set

$$\overline{v} := \frac{1}{|\Omega|}\langle v, 1\rangle \tag{18.12}$$

as well, noting that the constant function 1 is an element of V. Here, $|\Omega|$ denotes the Lebesgue measure of Ω. Clearly, $\overline{v}$ is the usual mean value of v if $v \in H$.

We also employ the notation

$$Q_t := \Omega \times (0, t), \quad t \in (0, T]; \quad Q := Q_T.$$

About the constants used in the sequel for estimates, we adopt the rule that C denotes any positive constant that depends only on the given data. The value of such generic constants C may change from formula to formula or even within the lines of the same formula. Finally, the notation C_δ indicates a positive constant that additionally depends on the quantity δ.

18.2 Main Results

In this section, we formulate the general assumptions for the data of the system (18.1)–(18.4) and state existence, continuous dependence, and regularity results. First, let us remark that the positive parameter α is not listed in the assumptions below, since it is also involved in the related asymptotic analysis, and, consequently, we let

$$0 < \alpha \leq 1. \tag{18.13}$$

On the other hand, throughout the paper we suppose that

$$\tau > 0 \text{ is a fixed constant.} \tag{18.14}$$

For the nonlinearity f we generally assume that

$$\begin{aligned} &f = f_1 + f_2, \quad \text{where} \\ &f_1 : \mathbb{R} \to [0, +\infty] \text{ is convex and lower semicontinuous with } f_1(0) = 0, \\ &f_2 : \mathbb{R} \to \mathbb{R} \text{ has a Lipschitz continuous first derivative } f_2' \text{ on } \mathbb{R}. \end{aligned} \tag{18.15}$$

A consequence of (18.15) is that

$$\text{the subdifferential } \partial f_1 \text{ is maximal monotone in } \mathbb{R} \times \mathbb{R}, \text{ with } 0 \in \partial f_1(0), \tag{18.16}$$

and an important requirement for the sequel is that

$$\text{the domain } D(\partial f_1) \text{ of } \partial f_1 \text{ has a nonempty interior containing } 0. \tag{18.17}$$

Note that these conditions are fulfilled in each of the cases considered in (18.9), (18.10), (18.11) with the domain $D(\partial f_1)$ given by $\mathbb{R}$, $(-1, 1)$, $[-1, 1]$, respectively. From now onwards we use the symbol $\partial f_1^\circ(r)$ for the element of $\partial f_1(r)$ (with $r \in D(\partial f_1)$) having minimum modulus, and we extend the notations f_1, ∂f_1, $D(\partial f_1)$, and ∂f_1° to the corresponding functionals and the operators induced on L^2 spaces.

Also, we assume for the forcing term g and the initial values μ_0, ν_0, φ_0 that

$$g \in H^1(0, T; H), \tag{18.18}$$

$$\mu_0 \in V, \quad \nu_0 \in H, \tag{18.19}$$

$$\varphi_0 \in W \cap D(\partial f_1) \text{ with } \partial f_1^\circ(\varphi_0) \in H. \tag{18.20}$$

Note that the condition $\varphi_0 \in W$ implies that $\varphi_0 \in C^0(\overline{\Omega})$. Moreover, we require that

$$m_0 := \frac{1}{|\Omega|} \int_\Omega \varphi_0 \text{ lies in the interior of } D(\partial f_1). \tag{18.21}$$

In view of (18.12), note that $m_0 = \overline{\varphi_0}$ represents the mean value of φ_0.

Let us now specify our notion of solution. We state the problem (18.1)–(18.4) in a variational form. We also introduce an additional variable ξ, which plays the role of $f_1'(\varphi)$ in the case when the derivative of f_1 is replaced by a real subdifferential ∂f_1. Namely, the solution is a triple (μ, φ, ξ) satisfying the regularity requirements

$$\mu \in W^{2,\infty}(0, T; V^*) \cap W^{1,\infty}(0, T; H) \cap L^\infty(0, T; V), \tag{18.22}$$

$$\varphi \in W^{1,\infty}(0,T;H) \cap H^1(0,T;V) \cap L^\infty(0,T;W), \tag{18.23}$$

$$\xi \in L^\infty(0,T;H), \tag{18.24}$$

and the following variational equations and initial conditions:

$$\alpha\langle \partial_{tt}\mu(t), v\rangle + (\partial_t\varphi(t), v) + \int_\Omega \nabla\mu(t)\cdot\nabla v = 0$$
$$\text{for a.e. } t \in (0,T) \text{ and every } v \in V, \tag{18.25}$$

$$\tau(\partial_t\varphi(t), v) + \int_\Omega \nabla\varphi(t)\cdot\nabla v + (\xi(t) + f_2'(\varphi(t)), v) = (\mu(t) + g(t), v)$$
$$\text{for a.e. } t \in (0,T) \text{ and every } v \in V, \tag{18.26}$$

$$\xi \in \partial f_1(\varphi) \quad \text{a.e. in } Q, \tag{18.27}$$

$$\mu(0) = \mu_0, \quad (\partial_t\mu)(0) = v_0, \quad \varphi(0) = \varphi_0 \quad \text{a.e. in } \Omega. \tag{18.28}$$

Of course, in view of the regularities in (18.22)–(18.24), the variational equality (18.26) is actually equivalent to an equation (cf. (18.2)) plus the boundary condition for φ which is already encoded in the fact that $\varphi \in L^\infty(0,T;W)$. Therefore, (18.26) can be replaced by

$$\tau\partial_t\varphi - \Delta\varphi + \xi + f_2'(\varphi) = \mu + g \quad \text{a.e. in } Q. \tag{18.29}$$

On the contrary, the analog equivalence for (18.25) (cf. (18.1)) would be true only if μ were more regular (cf. the first and third terms in (18.25)).

Remark 18.1 Note that, owing to the compactness of the embedding $W \subset C^0(\overline{\Omega})$ for $N \le 3$, it follows from Simon (1987, Sect. 8, Cor. 4) and the regularity (18.23) that $\varphi \in C^0(\overline{Q})$. By the same token, we have thanks to (18.22) that $\mu \in C^1([0,T];V^*) \cap C^0([0,T];H)$, and, consequently, $\partial_t\mu$ is at least weakly continuous from $[0,T]$ to H, which gives a meaning to the initial conditions in (18.28).

The next statement yields a well-posedness result for (18.25)–(18.28).

Theorem 18.1 *Assume that* (18.13)–(18.21) *are fulfilled. Then there exists a unique triple* (μ,φ,ξ), *with the regularity as in* (18.22)–(18.24), *that solves problem* (18.25)–(18.28) *and satisfies the estimate*

$$\alpha\|\mu\|_{W^{2,\infty}(0,T;V^*)} + \alpha^{1/2}\|\mu\|_{W^{1,\infty}(0,T;H)} + \|\mu\|_{L^\infty(0,T;V)}$$
$$+ \|\varphi\|_{W^{1,\infty}(0,T;H)\cap H^1(0,T;V)\cap L^\infty(0,T;W)} + \|\xi\|_{L^\infty(0,T;H)} \le K_1 \tag{18.30}$$

for some constant $K_1 > 0$ *that depends only on* Ω, T *and the data in* (18.14)–(18.21), *but is independent of* α.

The uniqueness property stated above is a consequence of the following continuous dependence result: Here, we use the notation

$$1 * v(t) = \int_0^t v(s)ds, \quad \text{for } v \in L^1(0, T; V^*) \text{ at least.}$$

Theorem 18.2 *Under the assumptions* (18.13)–(18.17)*, let* g_i, $\mu_{0,i}$, $\nu_{0,i}$, $\varphi_{0,i}$, $i = 1, 2$, *be two sets of data satisfying* (18.18)–(18.21)*, and let* $(\mu_i, \varphi_i, \xi_i)$, $i = 1, 2$, *denote any corresponding solutions to problem* (18.25)–(18.28) *with the regularity as in* (18.22)–(18.24)*. Then, the estimate*

$$\begin{aligned}
&\alpha^{1/2}\|\mu_1 - \mu_2\|_{L^\infty(0,T;H)} + \|\nabla(1 * (\mu_1 - \mu_2))\|_{L^\infty(0,T;H)} \\
&+ \|\varphi_1 - \varphi_2\|_{L^\infty(0,T;H)\cap L^2(0,T;V)} \\
&\le K_2 \left(\|g_1 - g_2\|_{L^2(0,T;H)} + \alpha^{1/2}\|\mu_{0,1} - \mu_{0,2}\|_H + \alpha^{1/2}\|\nu_{0,1} - \nu_{0,2}\|_H\right) \\
&\quad + K_2\big(1 + \alpha^{-1/2}\big)\|\varphi_{0,1} - \varphi_{0,2}\|_H
\end{aligned} \tag{18.31}$$

holds true with a constant $K_2 > 0$ *that depends only on* Ω, T, τ, *some Lipschitz constant for* f_2', *and is independent of* α.

We observe that, by taking the same data in Theorem 18.2, the estimate (18.31) ensures uniqueness for the solution components μ and φ in the statement of Theorem 18.1, while the uniqueness of ξ results from (18.29) since the other terms in the equality are uniquely determined.

On the basis of the estimates (18.30) and (18.31), we are interested to investigate the asymptotic behavior of the problem (18.25)–(18.28) as $\alpha \searrow 0$. This analysis will be developed in Sect. 18.5, whereas now we discuss some further results that mostly depend on the values of $\alpha > 0$. A regularity result is stated below, under the additional assumption that

$$\mu_0 \in W, \quad \nu_0 \in V. \tag{18.32}$$

Theorem 18.3 *Assume that* (18.13)–(18.21) *and* (18.32) *are fulfilled. Then the unique solution* (μ, φ, ξ) *to problem* (18.25)–(18.28) *satisfies*

$$\mu \in W^{2,\infty}(0, T; H) \cap W^{1,\infty}(0, T; V) \cap L^\infty(0, T; W), \tag{18.33}$$

and there exists a constant K_3*, independent of* α*, such that*

$$\alpha\|\mu\|_{W^{2,\infty}(0,T;H)} + \alpha^{1/2}\|\mu\|_{W^{1,\infty}(0,T;V)} + \|\mu\|_{L^\infty(0,T;W)} \le K_3\big(1 + \alpha^{-1/2}\big). \tag{18.34}$$

Notice that, due to the compactness of the embedding $W \subset C^0(\overline{\Omega})$, it follows from Simon (1987, Sect. 8, Cor. 4) that $\mu \in C^0(\overline{Q})$.

For the next regularity and continuous dependence result, we have to assume further regularity for g, that is,

$$g \in L^\infty(Q), \tag{18.35}$$

and for the nonlinearity f. Namely, we suppose that the effective domain of ∂f_1 is an open interval and that the restriction of f_1 to this interval is a smooth function. More precisely, we assume that

$$\begin{aligned} &D(\partial f_1) = (r_-, r_+), \quad \text{with} \ -\infty \le r_- < 0 < r_+ \le +\infty, \\ &\text{and the restriction of } f_1 \text{ to } (r_-, r_+) \text{ belongs to } C^2(r_-, r_+). \end{aligned} \tag{18.36}$$

Then, for $r \in (r_-, r_+)$, the subdifferential $\partial f_1(r)$ reduces to the singleton $\{f_1'(r)\}$, and we require that

$$\lim_{r \searrow r_-} f_1'(r) = -\infty, \quad \lim_{r \nearrow r_+} f_1'(r) = +\infty. \tag{18.37}$$

Please note that both the potentials f_{reg} and $f_{\log}$ in (18.9) and (18.10) fulfill (18.36)–(18.37) with $(r_-, r_+) = \mathbb{R}$ and $(r_-, r_+) = (-1, 1)$, respectively.

The so-called separation property and a refined continuous dependence result are stated as follows.

Theorem 18.4 *Assume that* (18.13)–(18.21) *and* (18.32)–(18.37) *are fulfilled. There exists two real numbers r_* and r^*, depending on α as well, such that*

$$r_- < r_* \le \varphi(x,t) \le r^* < r_+ \quad \textit{for every } (x,t) \in \overline{Q}. \tag{18.38}$$

Moreover, if for $i = 1, 2$ we let $(g_i, \mu_{0,i}, v_{0,i}, \varphi_{0,i})$ be a set of data and $(\mu_i, \varphi_i, \xi_i)$, with $\xi_i = f_1'(\varphi_i)$, denote the corresponding solution to problem (18.25)–(18.28), *the estimate*

$$\begin{aligned} &\|\mu_1 - \mu_2\|_{H^2(0,T;V^*) \cap W^{1,\infty}(0,T;H) \cap L^\infty(0,T;V)} \\ &+ \|\varphi_1 - \varphi_2\|_{H^1(0,T;H) \cap L^\infty(0,T;V) L^2(0,T;W)} \\ &\le K_4 \Big(\|g_1 - g_2\|_{L^2(0,T;H)} + \|\mu_{0,1} - \mu_{0,2}\|_V + \|v_{0,1} - v_{0,2}\|_H + \|\varphi_{0,1} - \varphi_{0,2}\|_V \Big) \end{aligned} \tag{18.39}$$

holds true for some constant $K_4 > 0$ that depends on α and on the structure of the system.

Remark 18.2 Please note that in the case when the domain $D(\partial f_1)$ is the entire real line, i.e., if $r_- = -\infty$ and $r_+ = +\infty$, then the property (18.38) is directly ensured by the estimate (18.30), since φ is bounded in $C^0(\overline{Q})$; therefore, if $D(\partial f_1) = \mathbb{R}$, then the additional regularity assumptions (18.32) and (18.35) are not needed to prove (18.38).

Remark 18.3 It would be interesting to investigate the system (18.1)–(18.4) from the viewpoint of an optimal control problem, with the distributed control located in the source term g in Eq. (18.2). Thus, in order to discuss differentiability properties and optimality conditions, it could be important to deal with smoother data and a smooth nonlinearity f, and with the control g lying in a control box in $L^\infty(Q)$ (cf. (18.35)). In this framework, stronger stability and continuous dependence estimates can possibly be derived for the system. However, the estimates (18.30), (18.34), and (18.39) are already a good starting point in that direction.

18.3 Existence of Solutions

In this section, we are going to prove the existence result for the problem (18.25)–(18.28), by constructing a solution (μ, φ, ξ) that satisfies (18.22)–(18.24). We adopt two levels of approximation: at first, we replace the subdifferential ∂f_1 in (18.27) by the derivative of the Moreau–Yosida regularization $f_{1,\varepsilon}$ of f_1, depending on a parameter $\varepsilon \in (0, 1)$; then, we apply a Faedo–Galerkin scheme to the resulting approximate system.

To begin with, we consider for every $\varepsilon \in (0, 1)$ the Moreau–Yosida regularization $f_{1,\varepsilon}$ of f_1, that is (see, e.g., Barbu (2010); Brezis (1973)),

$$\begin{aligned} f_{1,\varepsilon}(r) &:= \inf_{s\in\mathbb{R}} \left\{ \frac{1}{2\varepsilon}|r-s|^2 + f_1(s) \right\} \\ &= \frac{1}{2\varepsilon}\left|r - J_\varepsilon(r)\right|^2 + f_1(J_\varepsilon(r)) = \int_0^r f'_{1,\varepsilon}(s)ds, \end{aligned}$$

where $f'_{1,\varepsilon} : \mathbb{R} \to \mathbb{R}$ and the associated resolvent operator J_ε are given by

$$f'_{1,\varepsilon}(r) := \frac{1}{\varepsilon}(r - J_\varepsilon(r)), \quad J_\varepsilon(r) := (I + \varepsilon\partial f_1)^{-1}(r), \quad \text{for all } r \in \mathbb{R},$$

with I denoting the identity operator. Note that the derivative $f'_{1,\varepsilon}$ turns out to be a regularization of the graph ∂f_1. Indeed, $f'_{1,\varepsilon}$ and $f_{1,\varepsilon}$ fulfill, for all $0 < \varepsilon < 1$ (see, e.g., Brezis (1973, pp. 28 and 39)),

$$\begin{aligned} &f'_{1,\varepsilon} : \mathbb{R} \to \mathbb{R} \text{ is monotone and Lipschitz continuous} \\ &\quad \text{with Lipschitz constant } 1/\varepsilon, \text{ and it holds } f'_{1,\varepsilon}(0) = 0, \end{aligned} \tag{18.40}$$

$$|f'_{1,\varepsilon}(r)| \le |\partial f_1^\circ(r)| \quad \text{for every } r \in D(\partial f_1), \tag{18.41}$$

$$0 \le f_{1,\varepsilon}(r) \le f_1(r) \quad \text{for every } r \in \mathbb{R}. \tag{18.42}$$

As for the second approximation, we employ a Faedo–Galerkin discrete scheme using a special basis. To this end, we take the eigenvalues $\{\lambda_j\}_{j\in\mathbb{N}}$ of the eigenvalue problem

$$-\Delta v = \lambda v \quad \text{in } \Omega, \qquad \partial_{\mathbf{n}} v = 0 \quad \text{on } \partial\Omega,$$

and let $\{e_j\}_{j\in\mathbb{N}} \subset W$ be the associated eigenfunctions, normalized by $\|e_j\|_H = 1$, $j \in \mathbb{N}$. Then, we have that

$$0 = \lambda_1 < \lambda_2 \le \dots, \qquad \lim_{j\to\infty} \lambda_j = +\infty,$$
$$\int_\Omega e_j e_k = \int_\Omega \nabla e_j \cdot \nabla e_k = 0 \quad \text{for } j \ne k,$$

and we note that e_1 is just the constant function $|\Omega|^{-1/2}$. We then define the n-dimensional spaces $V_n := \operatorname{span}\{e_1, \dots, e_n\}$ for $n \in \mathbb{N}$, where V_1 is just the space of constant functions on Ω. It is well known that the union of these spaces is dense in both H and V.

The approximating n-dimensional problem is stated as follows: find functions $\mu_{nj}, \varphi_{nj} : [0, \mathrm{T}] \to \mathbb{R}$, $j = 1, \dots, n$, such that

$$\mu_n(x,t) := \sum_{j=1}^n \mu_{nj}(t) e_j(x), \quad \varphi_n(x,t) := \sum_{j=1}^n \varphi_{nj}(t) e_j(x) \tag{18.43}$$

satisfy

$$\alpha(\partial_{tt}\mu_n(t), v) + (\partial_t\varphi_n(t), v) + \int_\Omega \nabla\mu_n(t) \cdot \nabla v = 0$$
$$\text{for all } t \in [0,T] \text{ and every } v \in V_n, \tag{18.44}$$

$$\tau(\partial_t\varphi_n(t), v) + \int_\Omega \nabla\varphi_n(t) \cdot \nabla v + (f'_{1,\varepsilon}(\varphi_n(t)) + f'_2(\varphi_n(t)), v)$$
$$= (\mu_n(t) + g(t), v) \quad \text{for all } t \in [0,T] \text{ and every } v \in V_n, \tag{18.45}$$

$$\mu_n(0) = P_n(\mu_0), \quad (\partial_t\mu_n)(0) = P_n(v_0), \quad \varphi_n(0) = P_n(\varphi_0) \quad \text{a.e. in } \Omega, \tag{18.46}$$

where P_n denotes the H-orthogonal projection onto V_n. Then $P_n(v) = \sum_{j=1}^n (v, e_j)e_j$ for every $v \in H$, and we have (see, e.g., Colli et al. (2024a, formula (3.14)))

$$\|P_n(v)\|_Y \le C_\Omega \|v\|_Y \quad \text{for every } v \in Y, \text{ where } Y \in \{H, V, W\}, \tag{18.47}$$

for some constant $C_\Omega > 0$ depending only on Ω. By comparing (18.44)–(18.46) with (18.25)–(18.28), note that the inclusion (18.27) present in (18.25)–(18.28) is not reproduced in (18.44)–(18.46), since the role of the ξ variable is now played by $f'_{1,\varepsilon}(\varphi_n)$, written as it is, in (18.45).

Next, we take $v = e_k$ in all of Eqs. (18.44)–(18.46), for $k = 1, \dots, n$, obtaining the system

$$\alpha \frac{d^2}{dt^2}\mu_{nk} + \frac{d}{dt}\varphi_{nk} + \lambda_k \mu_{nk} = 0 \quad \text{in } (0,T), \tag{18.48}$$

$$\tau \frac{d}{dt}\varphi_{nk} + \lambda_k \varphi_{nk} + (f'_{1,\varepsilon}(\varphi_n) + f'_2(\varphi_n), e_k) = \mu_{nk} + (g, e_k) \quad \text{in } (0,T), \tag{18.49}$$

$$\mu_{nk}(0) = (\mu_0, e_k), \quad \frac{d}{dt}\mu_{nk}(0) = (v_0, e_k), \quad \varphi_{nk}(0) = (\varphi_0, e_k). \tag{18.50}$$

Then we have to deal with a Cauchy problem for a system of ordinary differential equations, which is of second order in the variables μ_{nk} and of first order in the variables φ_{nk}. This system is set in explicit form and offers Lipschitz continuous nonlinearities and source terms (g, e_k) in $H^1(0,T)$ (cf. (18.18)). By Carathéodory's theorem, the Cauchy problem (18.48)–(18.50) has a unique solution expressed by μ_{nk}, φ_{nk}, with $\mu_{nk} \in H^3(0,T)$ and $\varphi_{nk} \in H^2(0,T)$, for $k = 1, \dots, n$. On account of (18.48)–(18.49) and (18.43), this solution uniquely determines a pair $(\mu_n, \varphi_n) \in H^3(0,T; V_n) \times H^2(0,T; V_n)$ that solves (18.44)–(18.46).

We now derive a series of a priori estimates for the finite-dimensional approximations. In the following, $C > 0$ denotes constants that may depend on the data of the state system, but are independent of $n \in \mathbb{N}$, $\varepsilon \in (0,1)$ and $\alpha \in (0,1]$.

First estimate The aim is taking the time derivative of (18.45) and then testing by $\partial_t \varphi_n$. In addition, we add the resulting equality to (18.44) where we choose $v = \partial_t \mu_n$. By this approach we obtain the cancellation of two terms. Next, we integrate with respect to time and, in order to recover our estimate, we have to control the H-norm of the initial value $\partial_t \varphi_n(0)$. Taking $t = 0$ and $v = \partial_t \varphi_n(0)$ in (18.45), by (18.46) we easily infer that

$$\begin{aligned}
&\tau \|\partial_t \varphi_n(0)\|_H^2 \\
&= (\Delta \varphi_n(0) - f'_{1,\varepsilon}(\varphi_n(0)) - f'_2(\varphi_n(0)) + \mu_n(0) + g(0), \partial_t \varphi_n(0)) \\
&\le \frac{\tau}{2} \|\partial_t \varphi_n(0)\|_H^2 \\
&\quad + \frac{1}{2\tau} \|\Delta P_n(\varphi_0) - f'_{1,\varepsilon}(P_n(\varphi_0)) - f'_2(P_n(\varphi_0)) + P_n(\mu_0) + g(0)\|_H^2,
\end{aligned}$$

thanks to the Schwarz and Young inequalities. By virtue of Assumptions (18.18)–(18.20) on g and the initial data, and of the property (18.47), it turns out that there is a constant C_ε, depending only on the data and on ε, such that

$$
\begin{aligned}
&\tau \|\partial_t \varphi_n(0)\|_H \\
&\le \|\Delta P_n(\varphi_0) - f'_{1,\varepsilon}(P_n(\varphi_0)) - f'_2(P_n(\varphi_0)) + P_n(\mu_0) + g(0)\|_H \le C_\varepsilon . \quad (18.51)
\end{aligned}
$$

The dependence on ε follows from the Lipschitz continuity of $f'_{1,\varepsilon}$ with constant $1/\varepsilon$ (cf. (18.40)), while f'_2 is Lipschitz continuous independently of ε (see (18.15)).

Now, recalling the notation $Q_t := \Omega \times (0,t),\ \ t \in (0,T]$, we can perform the computation described above and deduce that

$$
\begin{aligned}
&\frac{\alpha}{2}\|\partial_t \mu_n(t)\|_H^2 + \frac{1}{2}\|\nabla \mu_n(t)\|_H^2 + \frac{\tau}{2}\|\partial_t \varphi_n(t)\|_H^2 \\
&\quad + \iint_{Q_t} |\nabla(\partial_t \varphi_n)|^2 + \iint_{Q_t} f''_{1,\varepsilon}(\varphi_n)|\partial_t \varphi_n|^2 \\
&\le \frac{\alpha}{2}\|P_n(v_0)\|_H^2 + \frac{1}{2}\|\nabla P_n(\mu_0)\|_H^2 \\
&\quad + \frac{1}{2\tau}\|\Delta P_n(\varphi_0) - f'_{1,\varepsilon}(P_n(\varphi_0)) - f'_2(P_n(\varphi_0)) + P_n(\mu_0) + g(0)\|_H^2 \\
&\quad - \iint_{Q_t} f''_2(\varphi_n))|\partial_t \varphi_n|^2 + \iint_{Q_t} \partial_t g\, \partial_t \varphi_n . \qquad (18.52)
\end{aligned}
$$

By the monotonicity of $f'_{1,\varepsilon}$, the last term on the left-hand side of (18.52) is nonnegative. Moreover, owing to (18.15), we have that $|f''_2(\varphi_n)||\partial_t \varphi_n|^2 \le C|\partial_t \varphi_n|^2$ a.e. in Q_t. Then, in view of (18.18), (18.19), (18.47), Young's inequality, and Gronwall's lemma, it is straightforward to infer from (18.52) that

$$
\begin{aligned}
&\alpha^{1/2}\|\partial_t \mu_n\|_{L^\infty(0,T;H)} + \|\nabla \mu_n\|_{L^\infty(0,T;H)} \\
&+ \|\partial_t \varphi_n\|_{L^\infty(0,T;H)} + \|\nabla(\partial_t \varphi_n)\|_{L^2(0,T;H)} \\
&\le C\big(\|v_0\|_H + \|\mu_0\|_V + \|\partial_t g\|_{L^2(0,T;H)} \\
&\quad + \|\Delta P_n(\varphi_0) - f'_{1,\varepsilon}(P_n(\varphi_0)) - f'_2(P_n(\varphi_0)) + P_n(\mu_0) + g(0)\|_H\big), \quad (18.53)
\end{aligned}
$$

for some constant C depending only on data, as $0 < \alpha < 1$. Therefore, recalling the initial conditions in (18.46), since

$$
\mu_n(t) = P_n(\mu_0) + \int_0^t \partial_t \mu_n(s)ds, \quad \varphi_n(t) = P_n(\varphi_0) + \int_0^t \partial_t \varphi_n(s)ds
$$

for all $t \in [0,T]$, we easily conclude from (18.51) that

$$
\alpha^{1/2}\|\mu_n\|_{W^{1,\infty}(0,T;H)} + \|\nabla \mu_n\|_{L^\infty(0,T;H)} + \|\varphi_n\|_{W^{1,\infty}(0,T;H)\cap H^1(0,T;V)} \le C_\varepsilon . \qquad (18.54)
$$

Complementary estimates Taking now $v \in V$ in (18.44) and using the orthogonality of the projection operator P_n and (18.47), we have that

$$\begin{aligned} \alpha\langle \partial_{tt}\mu_n(t), v\rangle &= \alpha(\partial_{tt}\mu_n(t), P_n(v)) + \alpha(\partial_{tt}\mu_n(t), v - P_n(v)) \\ &\le |\alpha(\partial_{tt}\mu_n(t), P_n(v))| \le \left|(\partial_t\varphi_n(t), P_n(v)) + \int_\Omega \nabla\mu_n(t)\cdot\nabla P_n(v)\right| \\ &\le C\big(\|\partial_t\varphi_n\|_{L^\infty(0,T;H)} + \|\nabla\mu_n\|_{L^\infty(0,T;H)}\big)\|v\|_V \quad \text{for a.e. } t \in (0,T), \end{aligned} \tag{18.55}$$

so that from (18.54) it clearly follows that

$$\alpha\|\partial_{tt}\mu_n\|_{L^\infty(0,T;V^*)} \le C_\varepsilon\,. \tag{18.56}$$

In addition, we can take $v = -\Delta(\varphi_n(t))$ in (18.45) and integrate by parts in some term. With the help of Young's inequality and the Lipschitz continuity of f_2', we obtain

$$\begin{aligned} &\|\Delta\varphi_n(t)\|_H^2 + \int_\Omega f_{1,\varepsilon}''(\varphi_n(t))|\nabla\varphi_n(t)|^2 \\ &= (\tau\partial_t\varphi_n(t) + f_2'(\varphi_n(t)) - g(t), \Delta\varphi_n(t)) + \int_\Omega \nabla\mu_n(t)\cdot\nabla\varphi_n(t) \\ &\le \frac{1}{2}\|\Delta\varphi_n(t)\|_H^2 + C\big(1 + \|\varphi_n\|_{W^{1,\infty}(0,T;H)}^2 + \|g\|_{L^\infty(0,T;H)}^2\big) \\ &\quad + \|\nabla\mu_n\|_{L^\infty(0,T;H)}\|\nabla\varphi_n\|_{L^\infty(0,T;H)} \quad \text{for a.e. } t \in (0,T), \end{aligned} \tag{18.57}$$

where the second term in the first line is nonnegative due to (18.40). Consequently, from (18.54) and the elliptic regularity theory, we find that

$$\|\Delta\varphi_n\|_{L^\infty(0,T;H)} + \|\varphi_n\|_{L^\infty(0,T;W)} \le C_\varepsilon\,. \tag{18.58}$$

Passage to the limit in the Faedo–Galerkin scheme By virtue of the uniform estimates shown above, there exists a pair $(\mu_\varepsilon, \varphi_\varepsilon)$ such that (possibly on a subsequence, which is again labeled by $n \in \mathbb{N}$)

$$\mu_n \to \mu_\varepsilon \quad \text{weakly star in } W^{2,\infty}(0,T;V^*) \cap W^{1,\infty}(0,T;H) \cap L^\infty(0,T;V), \tag{18.59}$$

$$\varphi_n \to \varphi_\varepsilon \quad \text{weakly star in } W^{1,\infty}(0,T;H) \cap H^1(0,T;V) \cap L^\infty(0,T;W). \tag{18.60}$$

By (18.59) and the compact embeddings $V \subset H \subset V^*$, it follows from Simon (1987, Sect. 8, Cor. 4) that

$$\mu_n \to \mu_\varepsilon \quad \text{strongly in } C^1([0,T];V^*) \cap C^0([0,T];H)\,. \tag{18.61}$$

Moreover, owing to the compactness of the embeddings $W \subset C^0(\overline{\Omega})$ and $W \subset V$, it turns out that

$$\varphi_n \to \varphi_\varepsilon \quad \text{strongly in } C^0(\overline{Q}) \cap C^0([0,T];V), \tag{18.62}$$

whence, by the Lipschitz continuity of $f'_{1,\varepsilon}$ and f'_2, we deduce that

$$f'_{1,\varepsilon}(\varphi_n) + f'_2(\varphi_n) \to f'_{1,\varepsilon}(\varphi_\varepsilon) + f'_2(\varphi_\varepsilon) \quad \text{strongly in } C^0(\overline{Q}) \cap C^0([0,T];H), \tag{18.63}$$

at least. Then, taking first $v \in V_k$ with $k \leq n$ in (18.44)–(18.45), and passing to the limit as $n \to \infty$, it is not difficult to infer that

$$\alpha\langle \partial_{tt}\mu_\varepsilon, v\rangle + (\partial_t\varphi_\varepsilon, v) + \int_\Omega \nabla\mu_\varepsilon \cdot \nabla v = 0 \quad \text{a.e. in } (0,T), \tag{18.64}$$

$$\tau(\partial_t\varphi_\varepsilon, v) + \int_\Omega \nabla\varphi_\varepsilon \cdot \nabla v + (f'_{1,\varepsilon}(\varphi_\varepsilon) + f'_2(\varphi_\varepsilon), v) = (\mu_\varepsilon + g, v) \quad \text{a.e. in } (0,T), \tag{18.65}$$

at first for all $v \in \cup_{k\in\mathbb{N}} V_k$, and then, by density, for all $v \in V$. Note that, due to the regularity of φ_ε, the second variational equality can be equivalently rewritten as

$$\tau\partial_t\varphi_\varepsilon - \Delta\varphi_\varepsilon + f'_{1,\varepsilon}(\varphi_\varepsilon) + f'_2(\varphi_\varepsilon) = \mu_\varepsilon + g \quad \text{a.e. in } Q, \tag{18.66}$$

where it is understood that φ_ε satisfies the boundary condition $\partial_{\mathbf{n}}\varphi_\varepsilon = 0$ a.e. on Σ, on account of $\varphi_\varepsilon \in L^\infty(0,T;W)$.

Thanks to (18.61) and (18.62), we can pass to the limit as $n \to \infty$ also in the initial conditions (18.46) and find that

$$\mu_\varepsilon(0) = \mu_0, \quad (\partial_t\mu_\varepsilon)(0) = v_0, \quad \varphi_\varepsilon(0) = \varphi_0, \quad \text{a.e. in } \Omega, \tag{18.67}$$

since (see (18.19)–(18.20))

$$P_n(\mu_0) \to \mu_0, \quad P_n(v_0) \to v_0, \quad P_n(\varphi_0) \to \varphi_0, \quad \text{strongly in } V,\ H,\ W,$$

respectively, and $\partial_t\mu_\varepsilon$ is weakly continuous from $[0,T]$ to H. Also, we can invoke the weak star lower semicontinuity of norms and pass to the limit in (18.53) to derive the inequality:

$$\begin{aligned} &\alpha^{1/2}\|\partial_t\mu_\varepsilon\|_{L^\infty(0,T;H)} + \|\nabla\mu_\varepsilon\|_{L^\infty(0,T;H)} \\ &+ \|\partial_t\varphi_\varepsilon\|_{L^\infty(0,T;H)} + \|\nabla(\partial_t\varphi_\varepsilon)\|_{L^2(0,T;H)} \\ &\leq C\big(\|v_0\|_H + \|\mu_0\|_V + \|\partial_t g\|_{L^2(0,T;H)} \\ &\qquad + \|\Delta\varphi_0 - f'_{1,\varepsilon}(\varphi_0) - f'_2(\varphi_0) + \mu_0 + g(0)\|_H\big). \end{aligned} \tag{18.68}$$

Hence, recalling (18.20) and (18.41), it turns out especially that $\|f'_{1,\varepsilon}(\varphi_0)\|_H$ is bounded independently of ε, which implies that the complete right-hand side of (18.68) is uniformly bounded. Then, arguing as before, we can improve (18.54) for $(\mu_\varepsilon, \varphi_\varepsilon)$ and recover that

$$\alpha^{1/2}\|\mu_\varepsilon\|_{W^{1,\infty}(0,T;H)} + \|\nabla\mu_\varepsilon\|_{L^\infty(0,T;H)} + \|\varphi_\varepsilon\|_{W^{1,\infty}(0,T;H)\cap H^1(0,T;V)} \leq C. \tag{18.69}$$

As a consequence, by repeating the arguments in (18.55) and (18.57) for μ_ε and φ_ε, we easily find out that

$$\alpha\|\partial_{tt}\mu_\varepsilon\|_{L^\infty(0,T;V^*)} + \|\Delta\varphi_\varepsilon\|_{L^\infty(0,T;H)} + \|\varphi_\varepsilon\|_{L^\infty(0,T;W)} \leq C \tag{18.70}$$

for some constant C which is independent of both $\varepsilon \in (0, 1)$ and $\alpha \in (0, 1]$.

Further estimates We insert the constant function $v = 1/|\Omega|$ in (18.64) and, in view of the definition (18.12) of the mean value, deduce that $\overline{\partial_t(\alpha\partial_t\mu_\varepsilon + \varphi_\varepsilon)} = 0$ a.e. in $(0, T)$. Hence, by virtue of (18.67), (18.21) and (18.12) again, it is straightforward to obtain

$$\alpha\,\overline{\partial_t\mu_\varepsilon}(t) + \overline{\varphi_\varepsilon}(t) = \alpha\,\overline{v_0} + m_0 \quad \text{for all } t \in [0, T]. \tag{18.71}$$

Now, we take $v = \varphi_\varepsilon(t) - m_0$ in (18.65) and, without integrating with respect to time, we have that

$$\begin{aligned}&\int_\Omega |\nabla(\varphi_\varepsilon(t) - m_0)|^2 + (f'_{1,\varepsilon}(\varphi_\varepsilon(t)), \varphi_\varepsilon(t) - m_0)\\&= -(\tau\partial_t\varphi_\varepsilon(t) + f'_2(\varphi_\varepsilon(t)) - g(t), \varphi_\varepsilon(t) - m_0) + (\mu_\varepsilon(t), \varphi_\varepsilon(t) - m_0).\end{aligned} \tag{18.72}$$

Now, in view of the properties (18.15)–(18.17) of f_1 and ∂f_1, and on account of (18.21), it turns out that there exist two positive constants δ_0 and C_0, independent of ε, such that

$$f'_{1,\varepsilon}(r)(r - m_0) \geq \delta_0|f'_{1,\varepsilon}(r)| - C_0 \quad \text{for every } r \in \mathbb{R}. \tag{18.73}$$

For this property, we refer to Miranville and Zelik (2004, Appendix, Prop. A.1) and also to the detailed proof given in Gilardi et al. (2009, p. 908). Applying (18.73) to the second term in the left-hand side of (18.72), due to (18.18) on g and to the Lipschitz continuity of f'_2, we infer that

$$\delta_0|f'_{1,\varepsilon}(\varphi_\varepsilon(t))| \leq C\big(1 + \|\varphi_\varepsilon\|^2_{W^{1,\infty}(0,T;H)}\big) + (\mu_\varepsilon(t), \varphi_\varepsilon(t) - m_0). \tag{18.74}$$

As for the last term in (18.74), thanks to (18.71) and the Poincaré–Wirtinger inequality, we can argue as follows:

$$
\begin{aligned}
&(\mu_\varepsilon(t), \varphi_\varepsilon(t) - m_0) \\
&= (\mu_\varepsilon(t), \alpha\, \partial_t \mu_\varepsilon(t) + \varphi_\varepsilon(t) - \alpha\, \overline{v_0} - m_0) + (\mu_\varepsilon(t), \alpha\, \overline{v_0} - \alpha\, \partial_t \mu_\varepsilon(t)) \\
&= (\mu_\varepsilon(t) - \overline{\mu_\varepsilon}(t), \alpha\, \partial_t \mu_\varepsilon(t) + \varphi_\varepsilon(t) - \alpha\, \overline{v_0} - m_0) + \alpha(\mu_\varepsilon(t), \overline{v_0} - \partial_t \mu_\varepsilon(t)) \\
&\le C \|\nabla \mu_\varepsilon\|_{L^\infty(0,T;H)} \big(\alpha^{1/2} \|\partial_t \mu_\varepsilon\|_{L^\infty(0,T;H)} + \|\varphi_\varepsilon\|_{L^\infty(0,T;H)} + 1\big) \\
&\quad + C\big(\alpha \|\mu_\varepsilon\|^2_{W^{1,\infty}(0,T;H)} + 1\big)
\end{aligned}
\tag{18.75}
$$

since $0 < \alpha \le 1$. Hence, by (18.74), (18.75), and (18.69), we can conclude that

$$
\|f'_{1,\varepsilon}(\varphi_\varepsilon)\|_{L^\infty(0,T;L^1(\Omega))} \le C\,. \tag{18.76}
$$

Next, we choose the constant function $v = 1$ in (18.65). We obtain, for a.e. $t \in (0, T)$,

$$
\int_\Omega f'_{1,\varepsilon}(\varphi_\varepsilon(t)) + \int_\Omega (\tau\, \partial_t \varphi_\varepsilon + f'_2(\varphi_\varepsilon(t)) - g(t)) = |\Omega|\, \overline{\mu_\varepsilon}(t). \tag{18.77}
$$

Owing to (18.76) and (18.69), both summands on the left-hand side are bounded in $L^\infty(0, T)$. Then we infer that

$$
\|\overline{\mu_\varepsilon}\|_{L^\infty(0,T)} \le C. \tag{18.78}
$$

Moreover, using again the Poincarè–Wirtinger inequality, we have that

$$
\begin{aligned}
\|\mu_\varepsilon\|_{L^\infty(0,T;H)} &\le \|\mu_\varepsilon - \overline{\mu_\varepsilon}\|_{L^\infty(0,T;H)} + C \|\overline{\mu_\varepsilon}\|_{L^\infty(0,T)} \\
&\le C \|\nabla \mu_\varepsilon\|_{L^\infty(0,T;H)} + C \le C.
\end{aligned}
\tag{18.79}
$$

Now we can go back to (18.65) or, better, to (18.66) and compare the terms in the equation: from (18.69), (18.70), and (18.18) it follows that $\partial_t \varphi_\varepsilon$, $\Delta \varphi_\varepsilon$, $f'_2(\varphi_\varepsilon)$, μ_ε, g are all uniformly bounded in $L^\infty(0, T; H)$, whence

$$
\|f'_{1,\varepsilon}(\varphi_\varepsilon)\|_{L^\infty(0,T;H)} \le C\,. \tag{18.80}
$$

Passage to the limit as $\varepsilon \to 0$ Thanks to the uniform estimates (18.69), (18.70), (18.79), and (18.80), it follows that there is a triple (μ, φ, ξ) such that, for some subsequence ε_k tending to 0, it holds

$$
\mu_{\varepsilon_k} \to \mu \quad \text{weakly star in } W^{2,\infty}(0, T; V^*) \cap W^{1,\infty}(0, T; H) \cap L^\infty(0, T; V), \tag{18.81}
$$

$$
\varphi_{\varepsilon_k} \to \varphi \quad \text{weakly star in } W^{1,\infty}(0, T; H) \cap H^1(0, T; V) \cap L^\infty(0, T; W), \tag{18.82}
$$

$$
f'_{1,\varepsilon_k}(\varphi_{\varepsilon_k}) \to \xi \quad \text{weakly star in } L^\infty(0, T; H). \tag{18.83}
$$

As argued in the previous limit procedure (cf. (18.61)–(18.63)), by compactness, in particular exploiting Simon (1987, Sect. 8, Cor. 4), and by the Lipschitz continuity of f_2', we have that

$$\mu_{\varepsilon_k} \to \mu \quad \text{strongly in } C^1([0,T]; V^*) \cap C^0([0,T]; H), \tag{18.84}$$

$$\varphi_{\varepsilon_k} \to \varphi \quad \text{strongly in } C^0(\overline{Q}) \cap C^0([0,T]; V), \tag{18.85}$$

$$f_2'(\varphi_{\varepsilon_k}) \to f_2'(\varphi) \quad \text{strongly in } C^0(\overline{Q}) \cap C^0([0,T]; H). \tag{18.86}$$

Then we can pass to the limit as $\varepsilon_k \to 0$ in (18.64) and (18.65) by finding (18.25) and (18.26), respectively. Moreover, the initial conditions (18.28) follow from (18.67). It remains to check (18.27): but, since the extension of ∂f_1 to $L^2(0,T; H)$ is a maximal monotone operator and $f_{1,\varepsilon}'$ denotes its Yosida approximation, and since we have that

$$\limsup_{k,n\to\infty} \int_0^T (f_{1,\varepsilon_k}'(\varphi_{\varepsilon_k}(t)) - f_{1,\varepsilon_n}'(\varphi_{\varepsilon_n}(t)), \varphi_{\varepsilon_k}(t) - \varphi_{\varepsilon_n}(t))dt = 0$$

due to the weak convergence of $f_{1,\varepsilon_k}'(\varphi_{\varepsilon_k})$ to ξ in $L^2(0,T; H)$ and the strong convergence of φ_{ε_k} to φ in $L^2(0,T; H)$, we can apply Barbu (2010, Prop. 2.2, p. 38) and recover the inclusion $\xi \in \partial f_1(\varphi)$ in $L^2(0,T; H)$ and almost everywhere in Q.

In conclusion, we note that the triple (μ, φ, ξ) found by the limit procedure is actually the unique solution of the problem (18.25)–(18.28), on account of the continuous dependence result, and that the estimate (18.30) follows easily from the uniform bounds in (18.69), (18.70), (18.79), (18.80) and the weak star lower semicontinuity of norms. Therefore, Theorem 18.1 is completely proved. □

18.4 Continuous Dependence and Regularity

In this section, we report on the proofs of the continuous dependence and regularity results.

Proof of Theorem 18.2 We just have to prove the inequality (18.31) by letting $(\mu_i, \varphi_i, \xi_i)$ be any solution of problem (18.25)–(18.28) with the corresponding data $g_i, \mu_{0,i}, \nu_{0,i}, \varphi_{0,i}$, satisfying (18.18)–(18.21) for $i = 1, 2$. For convenience, within this proof we set

$$g = g_1 - g_2, \quad \mu_0 = \mu_{0,1} - \mu_{0,2}, \quad \nu_0 = \nu_{0,1} - \nu_{0,2}, \quad \varphi_0 = \varphi_{0,1} - \varphi_{0,2},$$

as well as

$$\mu = \mu_1 - \mu_2, \quad \varphi = \varphi_1 - \varphi_2, \quad \xi = \xi_1 - \xi_2.$$

Then, taking the differences of the respective equalities (18.25)–(18.26) and integrating the one resulting from (18.25) with respect to time, we obtain

$$(\alpha\partial_t\mu(t)+\varphi(t),v)+\int_\Omega \nabla(1*\mu)(t)\cdot\nabla v=(\alpha v_0+\varphi_0,v)$$
$$\text{for a.e. } t\in(0,T) \text{ and every } v\in V, \tag{18.87}$$

$$\tau(\partial_t\varphi(t),v)+\int_\Omega \nabla\varphi(t)\cdot\nabla v+(\xi(t),v)=(\mu(t)+g(t)-f_2'(\varphi_1(t))+f_2'(\varphi_2(t)),v)$$
$$\text{for a.e. } t\in(0,T) \text{ and every } v\in V, \tag{18.88}$$

$$\xi_i\in\partial f_1(\varphi_i),\ i=1,2,\quad \text{a.e. in } Q, \tag{18.89}$$
$$\mu(0)=\mu_0\,,\quad \varphi(0)=\varphi_0\quad \text{a.e. in } \Omega\,. \tag{18.90}$$

Next, we take $v=\mu(t)$ in (18.87) and $v=\varphi(t)$ in (18.88), then we add them noting a cancellation of terms and integrate once more with respect to t. With the help of (18.90) and recalling that $\partial_t(1*\mu)=\mu$, we infer that

$$\begin{aligned}
&\frac{\alpha}{2}\|\mu(t)\|_H^2+\frac{1}{2}\|\nabla(1*\mu)(t)\|_H^2+\frac{\tau}{2}\|\varphi(t)\|_H^2+\iint_{Q_t}|\nabla\varphi|^2+\iint_{Q_t}\xi\varphi\\
&\le\frac{\alpha}{2}\|\mu_0\|_H^2+\frac{\tau}{2}\|\varphi_0\|_H^2+\iint_{Q_t}(\alpha v_0+\varphi_0)\mu\\
&\quad+\iint_{Q_t}g\,\varphi-\iint_{Q_t}(f_2'(\varphi_1)-f_2'(\varphi_2))\varphi
\end{aligned} \tag{18.91}$$

for all $t\in(0,T]$. Now, thanks to (18.89) the last term on the left-hand side is nonnegative. On the right-hand side, by the Young inequality, we have that

$$\iint_{Q_t}(\alpha v_0+\varphi_0)\mu\le\alpha\int_0^t\|\mu(s)\|_H^2ds+C\big(\alpha\|v_0\|_H^2+\alpha^{-1}\|\varphi_0\|_H^2\big),$$

and, using also the Lipschitz continuity of f_2', it follows that

$$\iint_{Q_t}g\,\varphi-\iint_{Q_t}(f_2'(\varphi_1)-f_2'(\varphi_2))\varphi\le C\int_0^t\|\varphi(s)\|_H^2ds+\frac{1}{2}\|g\|_{L^2(0,T;H)}^2.$$

Therefore, we can collect these inequalities and apply the Gronwall lemma to the resultant from (18.91) in order to plainly obtain the estimate (18.31). □

Proof of Theorem 18.3 We already know from (18.23) that $\partial_t\varphi$ is in $L^\infty(0,T;H)\cap L^2(0,T;V)$. Then, in view of (18.32), the regularity in (18.33) follows from the variational theory for linear evolution problems of second order in time (see, e.g., Dautray and Lions (1992)). Then, in order to reproduce the estimate in (18.34), let us proceed formally and test Eq. (18.25) by $-\Delta(\partial_t\mu(t))$. By this, we can easily integrate by parts and also with respect to time. With the help of Young's inequality, we obtain

$$\frac{\alpha}{2}\int_\Omega |\nabla(\partial_t\mu(t))|^2 + \frac{1}{2}\int_\Omega |\Delta\mu(t)|^2$$
$$= \frac{\alpha}{2}\int_\Omega |\nabla\nu_0|^2 + \frac{1}{2}\int_\Omega |\Delta\mu_0|^2 - \iint_{Q_t} \nabla(\partial_t\varphi)\cdot\nabla(\partial_t\mu)$$
$$\le \frac{\alpha}{2}\|\nu_0\|_V^2 + \frac{1}{2}\|\Delta\mu_0\|_H^2 + \frac{1}{2\alpha}\iint_{Q_t} |\nabla(\partial_t\varphi)|^2 + \frac{\alpha}{2}\iint_{Q_t} |\nabla(\partial_t\mu)|^2, \quad (18.92)$$

whence the estimate

$$\alpha^{1/2}\|\mu\|_{W^{1,\infty}(0,T;V)} + \|\mu\|_{L^\infty(0,T;W)} \le C\big(1+\alpha^{-1/2}\big) \quad (18.93)$$

follows from an application of Gronwall's lemma, along with (18.32) and (18.30). Having shown (18.93), it is now straightforward to compare the terms in (18.25) and to deduce that $\alpha\|\partial_{tt}\varphi\|_{L^\infty(0,T;H)}$ is bounded by a quantity like the right-hand side of (18.93). Thus, we complete the proof of (18.34). □

Proof of Theorem 18.4 We first show (18.38). It is already known from (18.20) that the initial value of φ, i.e., φ_0, belongs to a compact subset of $D(\partial f_1) = (r_-, r_+)$. By the previous proof, we have checked that μ is bounded in $L^\infty(0, T; W)$, hence in $L^\infty(Q)$, as follows from the above estimate (18.93). Then, let us rewrite Eq. (18.29) as

$$\tau\partial_t\varphi - \Delta\varphi + f_1'(\varphi) = h, \quad \text{with } h = \mu + g - f_2'(\varphi), \quad \text{a.e. in } Q. \quad (18.94)$$

The term ξ in (18.29) has been expressed here as $f_1'(\varphi)$, as it is allowed by Assumption (18.36). Note that the right-hand side h of (18.94) is actually bounded in $L^\infty(Q)$, thanks to (18.35) and the bound for φ ensured by (18.30), along with the Lipschitz continuity of f_2'.

To prove (18.38), it is enough to derive an $L^\infty(Q)$-bound for $f_1'(\varphi)$. Let us outline the argument by proceeding formally and pointing out that just a truncation of the test functions would be needed for a rigorous proof. We take any $p > 2$ and test (18.94) by $|f_1'(\varphi)|^{p-2}f_1'(\varphi)$, a function of φ which is increasing and attains the value 0 at 0 (cf. (18.15)–(18.17)). Then, we integrate from 0 to $t \in (0, T]$, obtaining

$$\tau\int_\Omega\Big(\int_0^{\varphi(t)} |f_1'(s)|^{p-2}f_1'(s)ds\Big)$$
$$+ (p-1)\iint_{Q_t} |f_1'(\varphi)|^{p-2}f_1''(\varphi)|\nabla\varphi|^2 + \iint_{Q_t} |f_1'(\varphi)|^p$$
$$= \tau\int_\Omega\Big(\int_0^{\varphi_0} |f_1'(s)|^{p-2}f_1'(s)ds\Big) + \iint_{Q_t} h|f_1'(\varphi)|^{p-2}f_1'(\varphi). \quad (18.95)$$

Note that the first term on the left-hand side is nonnegative since f_1' is monotone increasing with $f_1'(0) = 0$; moreover, the second term on the left-hand side is nonnegative since the derivative f_1'' is nonnegative everywhere in (r_-, r_+). About the right-hand side we may observe that

$$\tau \int_\Omega \Bigl(\int_0^{\varphi_0} |f_1'(s)|^{p-2} f_1'(s) ds \Bigr) \leq \tau \| f_1'(\varphi_0)\|_\infty^{p-1} \|\varphi_0\|_\infty |\Omega| ,$$

and, with $p' = p/(p-1)$ and the help of the Young inequality, that

$$\begin{aligned} \iint_{Q_t} h |f_1'(\varphi)|^{p-2} f_1'(\varphi) &\leq \|h\|_{L^p(Q_t)} \| \, |f_1'(\varphi)|^{p-1} \|_{L^{p'}(Q_t)} \\ &= \|h\|_{L^p(Q_t)} \|f_1'(\varphi)\|_{L^p(Q_t)}^{p/p'} \leq \frac{1}{p} \|h\|_{L^p(Q_t)}^p + \frac{1}{p'} \|f_1'(\varphi)\|_{L^p(Q_t)}^p . \end{aligned}$$

By rearranging from (18.95), and taking $t = T$, we infer that

$$\begin{aligned} \|f_1'(\varphi)\|_{L^p(Q)} &\leq \Bigl(p\tau \|f_1'(\varphi_0)\|_\infty^{p-1} \|\varphi_0\|_\infty |\Omega| + \|h\|_{L^p(Q)}^p \Bigr)^{1/p} \\ &\leq \bigl(p\tau \|f_1'(\varphi_0)\|_\infty^{p-1} \|\varphi_0\|_\infty |\Omega| \bigr)^{1/p} + \|h\|_{L^p(Q)} . \end{aligned}$$

Then, passing to the limit as $p \to +\infty$ in the above chain of inequalities, we conclude that

$$\|f_1'(\varphi)\|_{L^\infty(Q)} \leq \|f_1'(\varphi_0)\|_\infty + \|h\|_{L^\infty(Q)},$$

which ensures the validity of (18.38), for some constants r_*, r^* as in the statement.

Next, we argue in order to prove the continuous dependence estimate (18.39). We use the same notation as in the proof of Theorem 18.2, so that

$$g = g_1 - g_2, \quad \mu_0 = \mu_{0,1} - \mu_{0,2}, \quad \nu_0 = \nu_{0,1} - \nu_{0,2}, \quad \varphi_0 = \varphi_{0,1} - \varphi_{0,2},$$

and

$$\mu = \mu_1 - \mu_2 , \quad \varphi = \varphi_1 - \varphi_2 ,$$

where $(\mu_i, \varphi_i, \xi_i)$, with $\xi_i = f_1'(\varphi_i)$, is the solution to problem (18.25)–(18.28) corresponding to $g_i, \mu_{0,i}, \nu_{0,i}, \varphi_{0,i}$, $i = 1, 2$, these data satisfying (18.18)–(18.21). Then, taking the differences of the respective equalities (18.25)–(18.26), we obtain

$$\alpha \langle \partial_{tt}\mu(t), v \rangle + \int_\Omega \nabla\mu(t) \cdot \nabla v = -(\partial_t \varphi(t), v) \qquad \text{for a.e. } t \in (0, T) \text{ and every } v \in V, \tag{18.96}$$

$$\tau(\partial_t\varphi(t), v) + \int_\Omega \nabla\varphi(t)\cdot\nabla v = (\mu(t) + g(t) - f'(\varphi_1(t)) + f'(\varphi_2(t)), v)$$
$$\text{for a.e. } t \in (0,T) \text{ and every } v \in V, \tag{18.97}$$

$$\mu(0) = \mu_0\,, \quad (\partial_t\mu)(0) = \nu_0\,, \quad \varphi(0) = \varphi_0 \quad \text{a.e. in } \Omega\,, \tag{18.98}$$

where we have used $f' = f_1' + f_2'$ in Eq. (18.97) noting that the estimate (18.38) and Assumptions (18.36) on f_1 now allow us to take f' as a global Lipschitz continuous function (with Lipschitz constant depending on α). In view of (18.31), we have for the term on the right-hand side of (18.97) that

$$\begin{aligned}
&\|\mu + g - f'(\varphi_1) + f'(\varphi_2)\|_{L^2(0,T;H)}\\
&\le \|\mu\|_{L^2(0,T;H)} + \|g\|_{L^2(0,T;H)} + C_\alpha\|\varphi\|_{L^2(0,T;H)}\\
&\le C_\alpha\Big(\|g\|_{L^2(0,T;H)} + \|\mu_0\|_H + \|\nu_0\|_H + \|\varphi_0\|_H\Big),
\end{aligned} \tag{18.99}$$

where the constants are denoted by C_α since they depend on α as well. Then, using the standard parabolic regularity estimate (see, e.g., Lions (1961) or Dautray and Lions (1992))

$$\begin{aligned}
&\|\varphi\|_{H^1(0,T;H)\cap L^\infty(0,T;V)L^2(0,T;W)}\\
&\le C\big(\|\mu + g - f'(\varphi_1) + f'(\varphi_2)\|_{L^2(0,T;H)} + \|\varphi_0\|_V\big)
\end{aligned}$$

for the φ solution to (18.97) with the respective initial condition, it is straightforward to deduce that

$$\begin{aligned}
&\|\varphi\|_{H^1(0,T;H)\cap L^\infty(0,T;V)\cap L^2(0,T;W)}\\
&\le C_\alpha\big(\|g\|_{L^2(0,T;H)} + \|\mu_0\|_H + \|\nu_0\|_H + \|\varphi_0\|_V\big).
\end{aligned} \tag{18.100}$$

Next, we can choose $v = \partial_t\mu$ in (18.96), integrate with respect to time, and infer that

$$\begin{aligned}
&\frac{\alpha}{2}\|\partial_t\mu(t)\|_H^2 + \frac{1}{2}\|\nabla\mu(t)\|_H^2\\
&\le \frac{\alpha}{2}\|\nu_0\|_H^2 + \frac{1}{2}\|\nabla\mu_0\|_H + \int_0^t \|\partial_t\varphi(s)\|_H\|\partial_t\mu(s)\|_H ds.
\end{aligned} \tag{18.101}$$

Then, first applying Young's inequality to the last term and then Gronwall's lemma, we arrive at the estimate

$$\|\mu\|_{W^{1,\infty}(0,T;H)\cap L^\infty(0,T;V)} \le C_\alpha\big(\|\mu_0\|_V + \|\nu_0\|_H + \|\partial_t\varphi\|_{L^2(0,T;H)}\big),$$

whence from (18.100) it is clear that

$$\|\mu\|_{W^{1,\infty}(0,T;H)\cap L^\infty(0,T;V)} \le C_\alpha\big(\|g\|_{L^2(0,T;H)} + \|\mu_0\|_V + \|\nu_0\|_H + \|\varphi_0\|_V\big).$$

By this estimate, comparison of the terms in (18.96) yields

$$\begin{aligned}\alpha\|\partial_{tt}\mu\|_{L^2(0,T;V^*)} &\le \|\nabla\mu\|_{L^2(0,T;H)} + \|\partial_t\varphi\|_{L^2(0,T;H)}\\ &\le C_\alpha\big(\|g\|_{L^2(0,T;H)} + \|\mu_0\|_V + \|\nu_0\|_H + \|\varphi_0\|_V\big),\end{aligned} \tag{18.102}$$

so that (18.39) is completely proved. □

18.5 Asymptotic Analysis

This section is devoted to the study of the asymptotic behavior of the problem (18.1)–(18.4) as α approaches 0. We allow the initial data for μ and $\partial_t\mu$, as well as the source term g, to depend on α, while we keep fixed φ_0, the initial value of φ, for reasons of simplicity in front of restrictions like (18.20) and (18.21) for φ_0.

Thus, for $0 < \alpha \le 1$, we consider families of data g_α, $\mu_{0,\alpha}$, $\nu_{0,\alpha}$ such that

$$\begin{aligned}&\{g_\alpha\} \text{ is uniformly bounded in } H^1(0,T;H)\\ &\qquad\text{and strongly converges to } g \text{ in } L^2(0,T;H) \text{ as } \alpha \searrow 0,\end{aligned} \tag{18.103}$$

$$\{\mu_{0,\alpha}\} \text{ is uniformly bounded in } V, \tag{18.104}$$

$$\{\nu_{0,\alpha}\} \text{ is uniformly bounded in } H. \tag{18.105}$$

Of course, it follows from (18.103) that $g \in H^1(0,T;H)$ and $\partial_t g_\alpha \to \partial_t g$ weakly in $L^2(0,T;H)$. We can state the following convergence result.

Theorem 18.5 *Assume that* (18.14)–(18.17), (18.20)–(18.21), (18.103)–(18.105) *are fulfilled. For all* $\alpha \in (0,1]$, *let the triple* $(\mu_\alpha, \varphi_\alpha, \xi_\alpha)$, *with*

$$\mu_\alpha \in W^{2,\infty}(0,T;V^*) \cap W^{1,\infty}(0,T;H) \cap L^\infty(0,T;V), \tag{18.106}$$

$$\varphi_\alpha \in W^{1,\infty}(0,T;H) \cap H^1(0,T;V) \cap L^\infty(0,T;W), \tag{18.107}$$

$$\xi_\alpha \in L^\infty(0,T;H), \tag{18.108}$$

be the solution to the initial value problem

$$\begin{aligned}&\alpha\langle\partial_{tt}\mu_\alpha(t), v\rangle + (\partial_t\varphi_\alpha(t), v) + \int_\Omega \nabla\mu_\alpha(t)\cdot\nabla v = 0\\ &\qquad \textit{for a.e. } t \in (0,T) \textit{ and every } v \in V,\end{aligned} \tag{18.109}$$

$$\tau(\partial_t\varphi_\alpha(t), v) + \int_\Omega \nabla\varphi_\alpha(t)\cdot\nabla v + (\xi_\alpha(t) + f_2'(\varphi_\alpha(t)), v)$$
$$= (\mu_\alpha(t) + g_\alpha(t), v) \quad \textit{for a.e. } t \in (0,T) \textit{ and every } v \in V, \tag{18.110}$$

$$\xi_\alpha \in \partial f_1(\varphi_\alpha) \quad \textit{a.e. in } Q, \tag{18.111}$$
$$\mu_\alpha(0) = \mu_{0,\alpha}\,, \quad (\partial_t\mu_\alpha)(0) = \nu_{0,\alpha}\,, \quad \varphi_\alpha(0) = \varphi_0 \quad \textit{a.e. in } \Omega\,. \tag{18.112}$$

Then there exists a triple (μ, φ, ξ) such that, for some subsequence α_k tending to 0, there holds

$$\mu_{\alpha_k} \to \mu \quad \textit{weakly star in } L^\infty(0,T;V), \tag{18.113}$$

$$\alpha_k\mu_{\alpha_k} \to 0 \quad \textit{weakly star in } W^{2,\infty}(0,T;V^*)$$
$$\textit{and strongly in } W^{1,\infty}(0,T;H), \tag{18.114}$$

$$\varphi_{\alpha_k} \to \varphi \quad \textit{weakly star in } W^{1,\infty}(0,T;H)\cap H^1(0,T;V)\cap L^\infty(0,T;W)$$
$$\textit{and strongly in } C^0([0,T];V)\cap C^0(\overline{Q}), \tag{18.115}$$

$$\xi_{\alpha_k} \to \xi \quad \textit{weakly star in } L^\infty(0,T;H). \tag{18.116}$$

Moreover, (μ, φ, ξ) is a solution to the viscous Cahn–Hilliard system

$$(\partial_t\varphi(t), v) + \int_\Omega \nabla\mu(t)\cdot\nabla v = 0$$
$$\textit{for a.e. } t \in (0,T) \textit{ and every } v \in V, \tag{18.117}$$

$$\tau(\partial_t\varphi(t), v) + \int_\Omega \nabla\varphi(t)\cdot\nabla v + (\xi(t) + f_2'(\varphi(t)), v) = (\mu(t) + g(t), v)$$
$$\textit{for a.e. } t \in (0,T) \textit{ and every } v \in V, \tag{18.118}$$

$$\xi \in \partial f_1(\varphi) \quad \textit{a.e. in } Q, \tag{18.119}$$
$$\varphi(0) = \varphi_0 \quad \textit{a.e. in } \Omega\,. \tag{18.120}$$

Proof A closer inspection of the proof of Theorem 18.1 in Sect. 18.3 reveals that the estimate (18.30) still holds under the conditions (18.103)–(18.105). Then, by a standard weak star compactness argument, we deduce the existence of a subsequence $\alpha_k \searrow 0$ and a triple (μ, φ, ξ) such that (18.113)–(18.116) hold. In fact, the strong convergence property in (18.115) is a consequence of the compactness result reported in Simon (1987, Sect. 8, Cor. 4). Moreover, by the Lipschitz continuity of f_2', we also have that

$$f_2'(\varphi_{\alpha_k}) \to f_2'(\varphi) \quad \text{strongly in } C^0(\overline{Q}) \cap C^0([0,T];H). \tag{18.121}$$

Then, we can pass to the limit in (18.109), (18.110), and the third condition in (18.112), all written for α_k, and easily obtain (18.117), (18.118), (18.120). Recovering (18.119) from (18.111) is straightforward, due to the weak convergence of ξ_{α_k} and the strong convergence of φ_{α_k} in $L^2(0,T;H)$, along with the maximal monotonicity of ∂f_1 (see, e.g., Barbu (2010, Cor. 2.4, p. 41)). This concludes the proof. □

Remark 18.4 Note that Theorem 18.5 implicitly yields an existence result for solutions to the viscous Cahn–Hilliard system (18.117)–(18.120). The found solution is already a regular and strong solution. Indeed, the component φ is in $W^{1,\infty}(0,T;H) \cap H^1(0,T;V) \cap L^\infty(0,T;W)$ and therefore also in $C^0([0,T];H^{2-\delta}(\Omega))$ for all $\delta > 0$, by compactness (see, e.g., Simon (1987, Sect. 8, Cor. 4)). As a consequence of the dimension $N \le 3$ of the space and of the Sobolev embedding theorems, it turns out that $\varphi \in C^0(\overline{Q})$. Moreover, from (18.113), a comparison in (18.117), and the elliptic regularity theory, it turns out that $\mu \in L^\infty(0,T;W) \cap L^2(0,T;H^3(\Omega))$ in addition, so that $\mu \in L^\infty(0,T;C^0(\overline{\Omega}))$, in particular. Both the equalities (18.117) and (18.118) can be equivalently rewritten as the equations

$$\partial_t\varphi - \Delta\mu = 0 \quad \text{a.e. in } Q, \tag{18.122}$$

$$\tau\partial_t\varphi - \Delta\varphi + \xi + f_2'(\varphi) = \mu + g \quad \text{a.e. in } Q, \tag{18.123}$$

plus the homogeneous boundary conditions

$$\partial_{\mathbf{n}}\mu = \partial_{\mathbf{n}}\varphi = 0 \quad \text{a.e. on } \Sigma. \tag{18.124}$$

The mentioned regularity for (μ, φ, ξ) is exactly the same as in Colli et al. (2024b, Theorem 2.2), where a slightly more general system is investigated. However, the existence of a less regular solution can also be proved, along with the uniqueness of the component φ of the solution, as it results, for instance, from Colli et al. (2021, Theorem 2.5). Please note that in general uniqueness cannot be expected for ξ and μ unless ∂f_1 is single valued (like the case considered in (18.36)); otherwise, only the difference $\xi - \mu$ is uniquely determined from (18.123).

Remark 18.5 By the uniqueness property for the component φ, which is pointed out in the previous remark, we infer that not only a subsequence $\{\varphi_{\alpha_k}\}$ but the entire family $\{\varphi_\alpha\}_{\alpha\in(0,1]}$ converges to φ in the sense of (18.115) as $\alpha \searrow 0$.

The next result is devoted to an error estimate of the difference $\varphi_\alpha - \varphi$ in certain norms and in terms of the parameter α.

Theorem 18.6 *Under the same assumptions as in Theorem 18.5, we let* $(\mu_\alpha, \varphi_\alpha, \xi_\alpha)$ *denote the solution to* (18.109)–(18.112), *for* $\alpha \in (0, 1]$, *and* (μ, φ, ξ) *be the solution to* (18.117)–(18.120) *found by the asymptotic limit in* (18.113)–(18.116). *Then there is a constant* $K_5 > 0$, *which depends on the structure of the system but is independent of* α, *such that*

$$\begin{aligned}
&\alpha^{1/2}\|\mu_a\|_{L^\infty(0,T;H)} + \|\nabla(1 * (\mu_\alpha - \mu))\|_{L^\infty(0,T;H)} \\
&+ \|\varphi_\alpha - \varphi\|_{L^\infty(0,T;H)\cap L^2(0,T;V)} \\
&\le K_5\left(\alpha^{1/4} + \|g_\alpha - g\|_{L^2(0,T;H)}\right).
\end{aligned} \tag{18.125}$$

Proof We argue similarly as in the proof of Theorem 18.2. We take the difference of (18.109) and (18.117), then we integrate with respect to time with the help of (18.112) and (18.120). We obtain

$$\begin{aligned}
&((\varphi_\alpha - \varphi)(t), v) + \int_\Omega \nabla(1 * (\mu_\alpha - \mu))(t) \cdot \nabla v = (\alpha\, \nu_{0,\alpha} - \alpha\, \partial_t \mu_\alpha(t), v) \\
&\text{for a.e. } t \in (0, T) \text{ and every } v \in V.
\end{aligned} \tag{18.126}$$

At the same time, we subtract (18.118) from (18.110) and have that

$$\begin{aligned}
&\tau(\partial_t(\varphi_\alpha - \varphi)(t), v) + \int_\Omega \nabla(\varphi_\alpha - \varphi)(t) \cdot \nabla v + ((\xi_\alpha - \xi)(t), v) \\
&= ((\mu_\alpha - \mu)(t) + (g_\alpha - g)(t) - f_2'(\varphi_\alpha(t)) + f_2'(\varphi(t)), v) \\
&\text{for a.e. } t \in (0, T) \text{ and every } v \in V\,.
\end{aligned} \tag{18.127}$$

Then we take $v = (\mu_\alpha - \mu)(t)$ in (18.126) and $v = (\varphi_\alpha - \varphi)(t)$ in (18.127), sum up noting that a cancellation occurs, and integrate with respect to t. Since the product $(\xi_\alpha - \xi)(\varphi_\alpha - \varphi)$ is nonnegative due to (18.111), (18.119) and the monotonicity of ∂f_1, we easily derive the inequality

$$\begin{aligned}
&\frac{\alpha}{2}\|\mu_\alpha(t)\|_H^2 + \frac{1}{2}\|\nabla(1 * (\mu_\alpha - \mu))(t)\|_H^2 \\
&\quad + \frac{\tau}{2}\|(\varphi_\alpha - \varphi)(t)\|_H^2 + \iint_{Q_t} |\nabla(\varphi_\alpha - \varphi)|^2 \\
&\le \frac{\alpha}{2}\|\mu_{0,\alpha}\|_H^2 + \iint_{Q_t} \alpha\, \nu_{0,\alpha}(\mu_\alpha - \mu) + \iint_{Q_t} \alpha\, \partial_t \mu_\alpha\, \mu \\
&\quad + \iint_{Q_t} (g_\alpha - g)(\varphi_\alpha - \varphi) - \iint_{Q_t} (f_2'(\varphi_\alpha) - f_2'(\varphi))(\varphi_\alpha - \varphi)
\end{aligned} \tag{18.128}$$

for all $t \in (0, T]$. Now, we recall the boundedness properties (18.104) and (18.105), the estimate (18.30) for $\|\mu_\alpha\|_{L^\infty(0,T;H)}$ and $\alpha^{1/2}\|\partial_t \mu_\alpha\|_{L^\infty(0,T;H)}$, as well as the regularity $\mu \in L^\infty(0, T; H)$, in order to deduce that

$$\frac{\alpha}{2}\|\mu_{0,\alpha}\|_H^2 + \iint_{Q_t} \alpha\, v_{0,\alpha}(\mu_\alpha - \mu) + \iint_{Q_t} \alpha\, \partial_t \mu_\alpha\, \mu$$
$$\leq \alpha \|\mu_{0,\alpha}\|_H^2 + C\alpha \|v_{0,\alpha}\|_H + C\alpha^{1/2} \leq C\alpha^{1/2}.$$

In addition, by virtue of the Lipschitz continuity of f_2' and the Young inequality, we have that

$$\iint_{Q_t} (g_\alpha - g)(\varphi_\alpha - \varphi) - \iint_{Q_t} (f_2'(\varphi_\alpha) - f_2'(\varphi))(\varphi_\alpha - \varphi)$$
$$\leq \|g_\alpha - g\|_{L^2(0,T;H)}^2 + C \int_0^t \|(\varphi_\alpha - \varphi)(s)\|^2 ds.$$

Then, collecting the above computations in (18.128) and applying Gronwall's lemma, the estimate (18.125) follows. □

We finally notice that (18.125) gives an error estimate, in particular, for

$$\|\varphi_\alpha - \varphi\|_{L^\infty(0,T;H)\cap L^2(0,T;V)}$$

of order 1/4, provided that the convergence of $\|g_\alpha - g\|_{L^2(0,T;H)}$ to 0 is at least of this order.

Acknowledgements The authors are grateful to the anonymous referees for a careful reading of the manuscript and for the related suggestions. This research has been supported by the MIUR-PRIN Grant 2020F3NCPX "Mathematics for industry 4.0 (Math4I4)". In addition, PC indicates his affiliation to the GNAMPA (Gruppo Nazionale per l'Analisi Matematica, la Probabilità e le loro Applicazioni) of INdAM (Istituto Nazionale di Alta Matematica). During the preparation of this paper, individual visits to the WIAS in Berlin and to the Dipartimento di Matematica "F. Casorati" in Pavia have been exchanged by the two authors: the warm hospitality of the two institutions is gratefully acknowledged.

References

Barbu V (2010) Nonlinear differential equations of monotone type in Banach spaces. Springer, London, New York

Bonetti E, Colli P, Dreyer W, Gilardi G, Schimperna G, Sprekels J (2002) On a model for phase separation in binary alloys driven by mechanical effects. Phys D 165:48–65

Bonfoh A (2021) Existence and continuity of inertial manifolds for the hyperbolic relaxation of the viscous Cahn-Hilliard equation. Appl Math Optim 84:3339–3416

Brezis H (1973) Opérateurs Maximaux Monotones et Semi-groupes de Contractions dans les Espaces de Hilbert. North-Holland Math Stud 5, North-Holland, Amsterdam

Cahn JW, Hilliard JE (1958) Free energy of a nonuniform system I. J Chem Phys 2:258–267

Chen H (2021) Error estimates for the scalar auxiliary variable (SAV) schemes to the viscous Cahn–Hilliard equation with hyperbolic relaxation. J Math Anal Appl 499, Paper No. 125002, 21 pp

Chen X, Ma L, Yang X (2023) Error analysis of second-order IEQ numerical schemes for the viscous Cahn-Hilliard equation with hyperbolic relaxation. Comput Math Appl 152:112–128

Cherfils L, Miranville A (2023) On the viscous Cahn-Hilliard system with two mobilities. Discrete Contin Dyn Syst Ser S 16:3606–3621

Colli P, Gilardi G, Marinoschi G (2021) Solvability and sliding mode control for the viscous Cahn-Hilliard system with a possibly singular potential. Math Control Relat Fields 11:905–934

Colli P, Gilardi G, Signori A, Sprekels J (2024) Curvature effects in pattern formation: well-posedness and optimal control of a sixth-order Cahn-Hilliard equation. SIAM J Math Anal 56:4928–4969

Colli P, Gilardi G, Sprekels J (2018) On a Cahn–Hilliard system with convection and dynamic boundary conditions. Ann Mat Pura Appl (4) 197:1445–1475

Colli P, Gilardi G, Sprekels J (2018) On the longtime behavior of a viscous Cahn-Hilliard system with convection and dynamic boundary conditions. J Elliptic Parabol Equ 4:327–347

Colli P, Gilardi G, Sprekels J (2019) Recent results on well-posedness and optimal control for a class of generalized fractional Cahn-Hilliard systems. Control Cybernet 48:153–197

Colli P, Sprekels J, Tröltzsch F (2024) Optimality conditions for sparse optimal control of viscous Cahn–Hilliard systems with logarithmic potential. Appl Math Optim 90, Paper No. 47, 48 pp

Conti M, Gatti S, Miranville A (2024) A perturbation of the Cahn-Hilliard equation with logarithmic nonlinearity. J Differ Equ 382:50–76

Dautray R, Lions J-L (1992) Mathematical analysis and numerical methods for science and technology. Vol 5. Evolution problems. I.", with the collaboration of M. Artola, M. Cessenat and H. Lanchon, Springer-Verlag, Berlin

Dor D, Miranville A, Pierre M (2024) Hyperbolic relaxation of the viscous Cahn-Hilliard equation with a symport term for biological applications. Math Methods Appl Sci 47:5999–6035

Elliott CM, Zheng S (1986) On the Cahn-Hilliard equation. Arch Rat Mech Anal 96:339–357

Gilardi G, Miranville A, Schimperna G (2009) On the Cahn-Hilliard equation with irregular potentials and dynamic boundary conditions. Commun Pure Appl Anal 8:881–912

Gilardi G, Rocca E, Signori A (2023) Well-posedness and optimal control for a viscous Cahn-Hilliard-Oono system with dynamic boundary conditions. Discrete Contin Dyn Syst Ser S 16:3573–3605

Gilardi G, Signori A, Sprekels J (2023) Nutrient control for a viscous Cahn-Hilliard-Keller-Segel model with logistic source describing tumor growth. Discrete Contin Dyn Syst Ser S 16:3552–3572

Gilardi G, Sprekels J (2019) Asymptotic limits and optimal control for the Cahn-Hilliard system with convection and dynamic boundary conditions. Nonlinear Anal 178:1–31

Kagawa K, Ôtani M (2022) Asymptotic limits of viscous Cahn–Hilliard equation with homogeneous Dirichlet boundary condition. J Math Anal Appl **512**, Paper No. 126106, 23 pp

Lions JL (1961) "Équations différentielles opérationnelles et problèmes aux limitesâŁž, Die Grundlehren der mathematischen Wissenschaften, vol 111. Springer, Berlin-Göttingen-Heidelberg

Miranville A, Zelik S (2004) Robust exponential attractors for Cahn-Hilliard type equations with singular potentials. Math Methods Appl Sci 27:545–582

Savostianov A, Zelik S (2016) Global well-posedness and attractors for the hyperbolic Cahn-Hilliard-Oono equation in the whole space. Math Models Methods Appl Sci 26:1357–1384

Şen Z, Khanmamedov A (2024) The Cahn–Hilliard/Allen–Cahn equation with inertial and proliferation terms. J Math Anal Appl 530, Paper No. 127736, 20 pp

Simon J (1987) Compact sets in the space $L^p(0, T; B)$. Ann Mat Pura Appl 146(4):65–96

Chapter 19
A New Approach to Thermodynamically Consistent Models for SMAs

Claudio Giorgi and Angelo Morro

Abstract The paper derives a thermodynamic scheme for SMAs. The new idea of the thermodynamic approach is the formulation of the second law where also the entropy production is given by a constitutive equation. Also by analogy with the usual one-dimensional experimental setting, the paper describes SMAs in a thin parallelepiped form. The body is taken to be subjected to the standard balance equations and the material is characterized by constitutive equations where the temperature, the stress, their time derivatives, and the mass fraction of a phase (martensite) are the variables. The thermodynamic analysis leads to a skeleton surface consisting of equilibrium states and to rate equations governing the evolution of the fields, outside the skeleton surface. It emerges that the rate equations allow a whole account of hysteretic loops induced by temperature or by stress.

19.1 Introduction

Shape memory alloys (SMAs) are metallic alloys that exhibit a rich variety of hysteretic behaviors. In essence, SMAs show hysteretic effects induced by temperature and stress. Hysteresis is associated with the shape memory effect whereby the body undergoes a change of spatial form. This is due to the fact that the metallic lattice has two different equilibrium configurations, one is the high-symmetric high-temperature austenite and the other is the low-temperature martensite.

The practical interest arising from the shape memory is accompanied by the concern into the mathematical models of hysteretic phenomena induced by both temperature and stress. A reasonable report on the literature on SMAs is given by

C. Giorgi
Dipartimento di Ingegneria Civile Ambiente Territorio Architettura e Matematica, Università di Brescia, Brescia, Italy
e-mail: claudio.giorgi@unibs.it

A. Morro (✉)
Dipartimento di Informatica, Bioingegneria, Robotica e Ingegneria dei Sistemi, Università di Genova, Genoa, Italy
e-mail: angelo.morro@unige.it

J. Fuhrmann et al. (eds.), *Advances in Continuum Physics*, Advanced Structured Materials 238, https://doi.org/10.1007/978-3-031-93918-1_19

recent review papers on the subject; among them we mention (Kumar and Lagoudas 2008; Xu et al. 2019; Cisse et al. 2016) while further references on the subject are (Brinson 1993; Huo and Müller 1993; Brokate and Sprekels 1996; Seelecke and Müller 2004).

Many works are based on microscopic models with various potential energies of interaction between particles. Other approaches of macroscopic character have been developed where the main technical point is the evolution equation for the mass fraction ξ of a phase (Cisse et al. 2016). Often suitable internal variables are associated with the deformation and then the scheme is completed through corresponding thermodynamic forces (Scalet et al. 2019). Thermodynamic approaches have been developed by modeling the mass fraction as a phase change variable (e.g., Jin et al. (2001)) or an order parameter (e.g., Levitas (2013)) and accordingly introducing a generalized thermodynamic force in the balance equations.

Physically admissible models are required to be thermodynamically consistent. Furthermore, a systematic thermodynamic approach should simplify the description of a material by using a few constitutive functions possibly without introducing additional forces and powers and hence changing the balance equations. With this idea, a recent approach of ours allows the formulation of thermodynamically consistent models of hysteresis in terms of free energy and entropy production (see, e.g., Morro and Giorgi (2023), Giorgi and Morro (2024)).

The new idea of our approach is to regard the entropy production as a constitutive function per se, just as it happens for all of the constitutive functions. It is apparent that this feature results in more general schemes. Technically, this allows the formulation of rate equations where the entropy production is a non-negative-valued function of the variables under consideration (see, e.g., Giorgi and Morro (2024)).

Also in view of the usual one-dimensional experimental setting, in this paper, we describe SMAs in a thin parallelepiped form with the stress applied in the longitudinal direction. The body is taken to obey the standard balance equations and the material is characterized by constitutive equations where the temperature, the stress, their time derivatives, and the mass fraction of a phase (martensite) are the variables. The purpose is to derive from thermodynamics a scheme where both equilibrium states and evolution equations are established. The scheme is expected to describe equilibrium states; evolutions; and, in particular, hysteretic loops induced by temperature or by stress.

19.2 Notation and Balance Equations

The body occupies a time-dependent three-dimensional region Ω; we let Ω be in a parallelepiped form $\Omega = \{I \times S\}$ where I is a segment and S is a square orthogonal to I. The position vector of a point in Ω is denoted by $\mathbf{x}$. The symbol ∇ denotes the gradient while $\nabla\cdot$ is the divergence operator.

The motion is specified by a function $\mathbf{x} = \boldsymbol{\chi}(\mathbf{X}, t)$, where $\mathbf{X}$ is the position vector of the point in a reference configuration R. Hence formally we can write $\Omega = \boldsymbol{\chi}(\mathrm{R}, t)$.

The symbol $\mathbf{F}$ denotes the deformation gradient, in suffix notation $F_{iK} = \partial_{X_K} \chi_i$. The symbol $f(\mathbf{x}, t)$ denotes the value of f at $\mathbf{x}$, at time $t \in \mathbb{R}$. The symbols ∂_t, ∂_{x_i}, ∂_{X_K} denote partial derivatives. Hence $\mathbf{v} = \partial_t \chi(\mathbf{X}, t)$ is the velocity. A superposed dot denotes the total time derivative so that for $f(\mathbf{x}, t)$ it is $\dot{f} = \partial_t f + (\mathbf{v} \cdot \nabla) f$.

Let ρ be the mass density. The conservation of mass is expressed locally in the form:

$$\dot{\rho} + \rho \nabla \cdot \mathbf{v} = 0.$$

If $J = \det \mathbf{F} > 0$ then it follows that

$$\rho J = \rho_R,$$

$\rho_R(\mathbf{X})$ being the mass density in the reference configuration.

The equation of motion is written in the form:

$$\rho \dot{\mathbf{v}} = \nabla \cdot \mathbf{T} + \rho \mathbf{b},$$

where $(\nabla \cdot \mathbf{T})_i = \partial_{x_j} T_{ij}$ and $\mathbf{b}$ is body force density. Also, $\mathbf{T}$ is the Cauchy stress tensor; by the balance of angular momentum we have $\mathbf{T} = \mathbf{T}^T$.

Let $\mathbf{L}$ be the velocity gradient, $L_{ij} = \partial_{x_j} v_i$, and $\mathbf{D}$ the stretching tensor, namely, $\mathbf{D} = \mathrm{sym}\mathbf{L}$. Denote by ε the internal energy density, $\mathbf{q}$ the heat flux vector, and r the energy supply. The balance of energy leads to

$$\rho \dot{\varepsilon} = \mathbf{T} \cdot \mathbf{D} - \nabla \cdot \mathbf{q} + \rho r. \tag{19.1}$$

Let θ be the absolute temperature, η the specific entropy density, $\mathbf{j}$ the entropy flux, and r/θ the entropy supply. The balance of entropy leads to

$$\rho \dot{\eta} + \nabla \cdot \mathbf{j} - \frac{\rho r}{\theta} = \rho \gamma,$$

where γ is the (rate of) entropy production.

A process $\mathcal{P}$ is the set of fields entering the balance equations. As with recent applications of ours, we let γ be given by a constitutive function. Physically admissible sets of constitutive functions are required to satisfy the second law of thermodynamics in the following form.

Postulate. For every process $\mathcal{P}$ admissible in a body, the inequality

$$\rho \dot{\eta} + \nabla \cdot \mathbf{j} - \frac{\rho r}{\theta} = \rho \gamma \geq 0 \tag{19.2}$$

is valid at any internal point.

Henceforth Eq. (19.2) is referred to as Clausius–Duhem (CD) inequality. Let $\mathbf{j} = \mathbf{q}/\theta + \mathbf{k}$. Substitute $\nabla \cdot \mathbf{q} - \rho r$ from (19.1) into (19.2) to obtain

$$-\rho(\dot{\psi}+\eta\dot{\theta})+\mathbf{T}\cdot\mathbf{D}+\theta\nabla\cdot\mathbf{k}-\frac{1}{\theta}\mathbf{q}\cdot\nabla\theta=\rho\theta\gamma, \tag{19.3}$$

where $\psi=\varepsilon-\theta\eta$ is the Helmholtz free energy.

The following scheme holds with $\mathbf{k}=\mathbf{0}$. Nonzero $\mathbf{k}$ occur in connection with nonlocal properties expressed by gradients of the variables. Hereafter it is understood and not written that $\gamma\geq 0$.

19.3 One-Dimensional Thermodynamic Scheme for SMAs

Consider Ω in the parallelepiped form $\Omega=I\times S$ with $I=[0,l(t)]$. We let $x\in[0,\infty)$ be the variable along the axis comprising I and $X\in[0,l(0)]$. The deformation gradient $\mathbf{F}$ is diagonal and we let $F=\partial_X x>0$. Let $L=\partial_x v$ be the velocity gradient in the chosen x-direction; hence $L=\dot{F}F^{-1}$.

Let σ be the traction in the x-direction while the other entries of $\mathbf{T}$ are zero. Hence the surface $I\times\partial S$ is traction free. We let $J=\det\mathbf{F}$. In view of the diagonal structure of $\mathbf{F}$, we can write

$$J=FF_\perp^2,$$

where $F_\perp$ is the common value of $\mathbf{F}$ at the lateral faces. The product $J\rho$ is the referential mass density, say ρ_R. While in general ρ_R can depend only on the position we assume that ρ_R is independent of $\mathbf{X}$ and hence is constant. For simplicity we neglect heat conduction ($\mathbf{q}=\mathbf{0}$) and the CD inequality (3) can be written in the form

$$-\rho(\dot{\psi}+\eta\dot{\theta})+\sigma\dot{F}F^{-1}=\rho\theta\gamma. \tag{19.4}$$

Define $h=\ln F$. Hence it follows

$$-\rho(\dot{\psi}+\eta\dot{\theta})+\sigma\dot{h}=\rho\theta\gamma.$$

The crucial property of SMAs is the transformation between austenite and martensite. In this paper, we confine our attention to positive-oriented (or detwinned) martensite which occurs when the values of both absolute temperature θ and tensile stress σ are not too small. The single-martensite model has often been chosen for a detailed study because of its simplicity and ease of experimental verification (Kumar and Lagoudas 2008). To describe the related properties, we use the (martensitic) phase fraction ξ,

$$\xi=\frac{\rho_M}{\rho_A+\rho_M},$$

ρ_A, ρ_M being the mass densities of austenite and martensite.

Multiply (19.4) by J and let $J\sigma = \tau$ to obtain

$$-\rho_R(\dot{\psi} + \eta\dot{\theta}) + \tau\dot{h} = \rho_R\theta\gamma. \tag{19.5}$$

Let

$$\phi = \psi - \frac{1}{\rho_R}\tau h. \tag{19.6}$$

Hence the CD inequality (19.5) can be written in the form:

$$-\rho_R(\dot{\phi} + \eta\dot{\theta}) - h\dot{\tau} = \rho_R\theta\gamma. \tag{19.7}$$

We append some comments on the modeling of h and η and the definitions of τ and ϕ. By analogy with the literature (e.g., Xu et al. (2019)), we assume h and η are given by the additive decompositions:

$$h = h^e + h^{tr}, \qquad \eta = \eta^e + \eta^{tr}. \tag{19.8}$$

The terms h^e and η^e are then viewed as strain and entropy and are assumed to arise from an energy potential, as with elastic solids, though here h^e and η^e can depend also on the mass fraction ξ. The quantities h^{tr} and η^{tr} (often called transition or transformation functions) are additional terms related to the presence of austenite and martensite and arise from the transformation mechanism between the two phases.

By the Nanson formula (Truesdell and Toupin (1963), p. 20; Eringen (1967), p. 71) the current surface element $\mathbf{n}\,da$ and the corresponding reference surface element $\mathbf{n}_R da_R$ are related by

$$\mathbf{n}\,da = J\mathbf{F}^{-T}\mathbf{n}_R da_R.$$

Since $\mathbf{T} = \sigma\mathbf{n}\otimes\mathbf{n}$, and $\mathbf{F}$ is diagonal, then $\mathbf{n} = \mathbf{n}_R$ and

$$da = JF^{-1}da_R.$$

If we denote by σ_R the reference stress we have $\sigma\,da = \sigma_R da_R$, whence

$$\sigma_R = JF^{-1}\sigma.$$

This relation is consistent with the definition of the first Piola stress tensor (Gurtin et al. (2010), Chap. 24; Morro and Giorgi (2023), p. 94)

$$\mathbf{T}_R = J\mathbf{T}\mathbf{F}^{-T}$$

as the stress in the reference configuration. Instead,

$$\tau = J\sigma$$

is merely the **n**-component of the Kirchhoff stress $\mathbf{T}_K = J\mathbf{T}$.

In connection with (19.6) we observe that, in classical thermodynamics, $\phi = \psi + p/\rho$ is the specific Gibbs free energy (or total free energy if the kinetic energy is included in $\varepsilon = \psi + \theta\eta$ Dreyer et al. (2011)). Since $\tau/\rho_R = \sigma/\rho$ then, relative to the Gibbs free energy, the present function ϕ is a modified Gibbs free energy with the mechanical term $-\sigma h/\rho$ in place of p/ρ. This definition is motivated directly by the balance of entropy (19.4).

19.4 Thermodynamic Restrictions

The expected evolution equation is thought as a relation among $\dot{\xi}, \dot{\tau}, \dot{\theta}$, possibly with $\dot{\xi}$ as a function of $\dot{\tau}, \dot{\theta}$ and θ, τ, ξ. This suggests that we let

$$\theta, \tau, \xi, \dot{\theta}, \dot{\tau}$$

be the variables. Hence ϕ, h, η, γ, and $\dot{\xi}$ are taken to be given by constitutive functions. Indeed h, η, γ, and $\dot{\xi}$ are given by continuous functions while ϕ is continuously differentiable. Upon computation of $\dot{\phi}$ we can write the CD inequality (19.7) in the form:

$$-\rho_R(\partial_\theta\phi + \eta)\dot{\theta} - (\rho_R\partial_\tau\phi + h)\dot{\tau} - \rho_R\partial_\xi\phi\,\dot{\xi} - \rho_R\partial_{\dot{\theta}}\phi\ddot{\theta} - \rho_R\partial_{\dot{\tau}}\phi\ddot{\tau} = \rho_R\theta\gamma. \tag{19.9}$$

The linearity and arbitrariness of $\ddot{\theta}$ and $\ddot{\tau}$ imply

$$\partial_{\dot{\theta}}\phi = 0, \qquad \partial_{\dot{\tau}}\phi = 0.$$

In dealing with a thermoelastic solid we would find that $\eta = -\partial_\theta\phi, h = -\partial_\tau\phi$. By analogy we let

$$\eta^e = -\partial_\theta\phi, \qquad h^e = -\rho_R\partial_\tau\phi. \tag{19.10}$$

Consequently the functions η^{tr}, h^{tr} and $\dot{\xi}$ are required to satisfy the reduced inequality

$$-\rho_R\eta^{tr}\dot{\theta} - h^{tr}\dot{\tau} - \rho_R\partial_\xi\phi\,\dot{\xi} = \rho_R\theta\gamma. \tag{19.11}$$

By the general constitutive assumption, the entropy production γ is a function of $\theta, \tau, \xi, \dot{\theta}, \dot{\tau}$. Instead, it is $\phi = \phi(\theta, \tau, \xi)$. Likewise, and also in connection with experimental data (Wu et al. 2003), we let both η^{tr} and h^{tr} be functions of θ, τ, ξ.

In the following sections, we will see that Eq. (19.11) gives a full description of the one-dimensional martensitic phase transitions. Indeed, upon suitable selections of the functions ϕ, h^{tr}, η^{tr} and γ, Eq. (19.11) determines the hysteretic loops related to the austenite $\leftrightarrows$ martensite transformations.

Since we are considering a single variant of martensite, we confine our attention to the temperature–stress domain

$$\mathcal{D} = \{(\theta, \tau) : \theta \geq \hat{\theta}_M > 0\}.$$

In fact, for temperatures below $\hat{\theta}_M$, the material is always in the martensitic phase (twinned, mixed, or detwinned). The introduction of a phase transition temperature $\hat{\theta}_M$ is not new in the literature (see, e.g., Fig. 1a of Cisse et al. (2016) where M_f^0 stands for $\hat{\theta}_M$). In particular, $\hat{\theta}_M \simeq 200°$ K for NiTi-based single crystals (see, e.g., Timofeeva et al. (2022)).

19.4.1 Isothermal Cyclic Processes

Further consequences follow in connection with closed thermodynamic processes as it happens with hysteretic loops. At isothermal processes ($\dot{\theta} = 0$) Eq. (19.11) simplifies to

$$-\rho_R\dot{\phi} - h\dot{\tau} = \rho_R\theta\gamma, \tag{19.12}$$

while ϕ, h, and γ are given by functions of the form: $\hat{\phi}(\tau, \xi)$, $\hat{h}(\tau, \xi, \dot{\tau})$, $\hat{\gamma}(\tau, \xi, \dot{\tau})$. Let $(\tau(t), \xi(t)) : \mathbb{R} \to \mathbb{R}^2$ be a closed function on $[0, T]$, namely, $\tau(T) = \tau(0), \xi(T) = \xi(0)$. Hence $\phi(t) = \hat{\phi}(\tau(t), \xi(t))$ satisfies $\phi(T) = \phi(0)$. Since $\gamma \geq 0$ then integrating (19.12) on $[0, T]$ we find

$$\int_0^T h\dot{\tau}dt = -\rho_R\theta \int_0^T \gamma(t)dt \leq 0.$$

Denote by C_τ the closed curve $C_\tau : t \mapsto (\tau(t), h(t)), \tau(T) = \tau(0), h(T) = h(0)$. Hence

$$\int_{C_\tau} hd\tau = \int_0^T h\dot{\tau}dt \leq 0.$$

Consequently the closed curve C_τ is oriented in the counterclockwise sense, in the $\tau - h$ plane, and

$$\mathcal{A}_\tau = \rho_R\theta \int_0^T \gamma(t)dt$$

is the positive area bounded by C_τ. However, in the most natural strain–stress plane the overturned closed curve $\tilde{C}_\tau$ is oriented in the clockwise sense as depicted in the schematic Fig. 19.1 inspired by Fig. 1b of Ref. Cisse et al. (2016).

In Fig. 19.1, we implicitly assume $\theta_0 \geq \hat{\theta}_A$, where $\hat{\theta}_A$ represents the final temperature of an austenitic transformation during heating in a stress-free state and corresponds to A_f^0 in Cisse et al. (2016, Fig. 1a). Since the transformation temperatures are sensitive to small alloy composition changes, in commercial NiTi-based alloys $\hat{\theta}_A$ covers a range from approximately 250 to 400° K (see, e.g., Abel et al. (2004)). Moreover, τ_0 and τ_1 denote the starting stresses of a martensitic transformation at

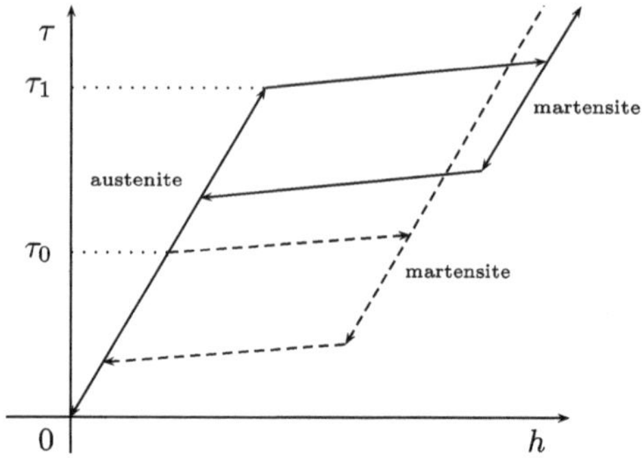

Fig. 19.1 Schematic major hysteresis loops in the strain–stress plane at constant temperature: θ_0 (dashed) and $\theta_1 > \theta_0$ (solid)

temperatures θ_0 and θ_1, respectively. This means that $\tau_0 = \tau_M^s(\theta_0)$ and $\tau_1 = \tau_M^s(\theta_1)$, as can be seen from the subsequent Fig. 19.2.

Likewise look at processes with $\dot{\tau} = 0$. We can then write Eq. (19.11) in the form:

$$-\dot{\phi} - \eta\dot{\theta} = \theta\gamma,$$

where ϕ is a function of θ, ξ while η, γ are functions of $\theta, \xi, \dot{\theta}$. As $t \in [0, T]$ the image $(\theta(t), \eta(t))$ yields the closed curve C_θ. Hence we have

$$\int_{C_\theta} \eta d\theta = \int_0^T \eta\dot{\theta} dt = -\int_0^T \theta\gamma dt \leq 0.$$

Thus

$$\int_{C_\theta} \eta d\theta = -\mathcal{A}_\theta,$$

where $\mathcal{A}_\theta$ is the (positive) area bounded by C_θ. The negative value of the integral denotes that C_θ is run counterclockwise in the $\theta - \eta$ plane. Thus, closed curves in the stress–strain and temperature–entropy planes are oriented counterclockwise. Unfortunately, loops of this type are usually neither studied nor represented in the SMA literature. However, temperature–entropy hysteresis loops are described in a recent paper on first-order transitions in magnetocaloric materials Gottschall et al. (2016).

19.4.2 Mass Fraction Loops

For cycles involving the mass fraction ξ, no direct conclusion follows directly from thermodynamics. In order to develop a simple but complete model, it is however reasonable to assume that both the logarithmic deformation h and the entropy $\rho_R\eta$ linearly depend on ξ.

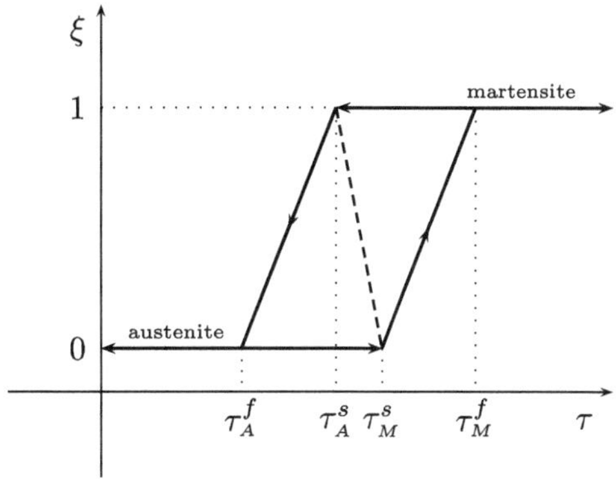

Fig. 19.2 Schematic description of the stress-induced transition. Major hysteresis loop of the martensite concentration ξ as a function of the stress τ at constant temperature $\theta = \theta_0 > \hat{\theta}_A$

If h is assumed in the form

$$h = \kappa(\tau, \theta) + \lambda(\tau)\xi, \tag{19.13}$$

then, along cyclic isothermal processes,

$$0 \geq \int_0^T h\dot{\tau}dt = \int_0^T \hat{\kappa}(\tau)\dot{\tau}dt + \int_0^T \xi\lambda(\tau)\dot{\tau}dt,$$

where $\hat{\kappa}$ denote the function κ at constant temperature. Consider the closed curve $C^\dagger : t \to (\tau(t), \xi(t)), \tau(T) = \tau(0), \xi(T) = \xi(0)$. If $0 < \bar{\lambda} \leq \lambda(\tau)$ then we have

$$\bar{\lambda}\int_{C^\dagger} \xi d\tau \leq \int_{C^\dagger} \xi\lambda(\tau)d\tau \leq 0.$$

Hence the closed curve $C^\dagger$ in the $\tau - \xi$ plane is oriented counterclockwise (see Fig. 19.2), which is consistent with experimental data (see, e.g., Mandolino et al. (2023)). Since $\lambda(\tau) = h_{\xi=1} - h_{\xi=0}$, then $\lambda(\tau)$ equals the variation of h due to a complete phase change at a fixed stress τ.

Likewise, assume the entropy is given in the form

$$\rho_R\eta = \ell(\tau, \theta) - \mu(\theta)\xi, \tag{19.14}$$

and consider the closed curve $C^\ddagger : t \to (\theta(t), \xi(t)), \theta(T) = \theta(0), \xi(T) = \xi(0)$. Along cyclic processes at constant stress, $\tau = \tau_0$, by (19.7) and (19.14) it follows that

$$0 \geq \rho_R \int_0^T \eta\dot{\theta}dt = \int_0^T \hat{\ell}(\theta)\dot{\theta}dt - \int_0^T \xi\mu(\theta)\dot{\theta}dt,$$

where $\hat{\ell}(\theta) = \ell(\tau_0, \theta)$. Let $0 < \mu(\theta) \leq \hat{\mu}$. It follows that

$$\hat{\mu}\int_{C^\ddagger}\xi d\theta \geq \int_{C^\ddagger}\xi\mu(\theta)d\theta \geq 0. \tag{19.15}$$

Hence the closed curve $C^\ddagger$ in the $\theta-\xi$ plane is oriented clockwise (see Fig. 19.3), which is consistent with experimental data (see, e.g., Lagoudas (2008, Chap. 1)) that show the condition $\dot{\theta}\dot{\xi} \leq 0$. Since $\mu(\theta) = \rho_R(\eta_{\xi=0} - \eta_{\xi=1})$, then $\mu(\theta)$ equals the variation of entropy $\rho_R\eta$ due to a complete phase change at a fixed temperature θ.

19.5 Constitutive Assumptions

We now derive a thermodynamically consistent model by choosing appropriate functional expressions for the Gibbs free energy, ϕ, the transition functions, η^{tr}, h^{tr} and the entropy production function, γ.

19.5.1 Gibbs Free Energy

First, we look for a Gibbs free energy ϕ consistent with (19.13) and (19.14). Assume $\rho_R\phi$ in the simple form (quadratic in ξ)

$$\rho_R\phi = \mathcal{G}(\tau,\theta) - \tfrac{1}{2}\alpha\xi^2 + [\mathcal{S}(\theta) - \mathcal{T}(\tau)]\xi, \tag{19.16}$$

where α is a constant and $\mathcal{G}(\tau,\theta)$ represents the Gibbs free energy of the austenite. In view of (19.13) and (19.14), we let

$$\alpha > 0, \quad \mathcal{S}'(\theta) > 0, \quad \mathcal{T}'(\tau) \geq \bar{\lambda} > 0, \quad \mathcal{S}(0) = 0, \quad \mathcal{T}(0) = 0. \tag{19.17}$$

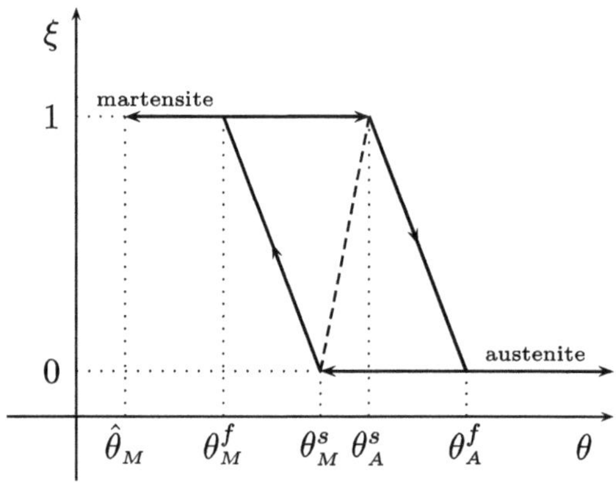

Fig. 19.3 Schematic description of the temperature-induced transition. Major hysteresis loop of the martensite concentration ξ as a function of the temperature θ at constant stress $\tau = \tau_0 > \tau_*$

From (19.16) we have

$$\rho_R \partial_\xi \phi = -\alpha\xi + \mathcal{S}(\theta) - \mathcal{T}(\tau), \tag{19.18}$$

$$\rho_R \partial_\theta \phi = \partial_\theta \mathcal{G}(\tau, \theta) + \mathcal{S}'(\theta)\xi, \quad \rho_R \partial_\tau \phi = \partial_\tau \mathcal{G}(\tau, \theta) - \mathcal{T}'(\tau)\xi.$$

Consequently, from definitions (19.10) it follows

$$\begin{aligned} \rho_R \eta^e &= -\rho_R \partial_\theta \phi = -\partial_\theta \mathcal{G}(\tau, \theta) - \mathcal{S}'(\theta)\xi, \\ h^e &= -\rho_R \partial_\tau \phi = -\partial_\tau \mathcal{G}(\tau, \theta) + \mathcal{T}'(\tau)\xi. \end{aligned} \tag{19.19}$$

19.5.2 Transition Functions

For definiteness in later developments, we assume the following expressions of the transition functions

$$\eta^{tr} = k(\theta)\partial_\xi \phi, \qquad h^{tr} = p(\tau)\rho_R \partial_\xi \phi, \tag{19.20}$$

with the functions $k(\theta)$ and $p(\tau)$ to be determined. By virtue of (19.18), it follows that

$$\begin{aligned} \rho_R \eta^{tr} &= -k(\theta)[\alpha\xi - \mathcal{S}(\theta) + \mathcal{T}(\tau)], \\ h^{tr} &= -p(\tau)[\alpha\xi - \mathcal{S}(\theta) + \mathcal{T}(\tau)]. \end{aligned} \tag{19.21}$$

To determine the connections with κ, λ, ℓ, and μ appearing in (19.13) and (19.14), we observe that

$$\rho_R \eta = \rho_R(\eta^e + \eta^{tr}) = -[\alpha k(\theta) + \mathcal{S}'(\theta)]\xi - \partial_\theta \mathcal{G}(\tau, \theta) + k(\theta)[\mathcal{S}(\theta) - \mathcal{T}(\tau),$$

$$h = h^e + h^{tr} = [\mathcal{T}'(\tau) - \alpha p(\tau)]\xi - \partial_\tau \mathcal{G}(\tau, \theta) + p(\tau)[\mathcal{S}(\theta) - \mathcal{T}(\tau)]. \tag{19.22}$$

Hence, a comparison with (19.14) and (19.13) yields

$$\mu(\theta) = \alpha k(\theta) + \mathcal{S}'(\theta), \qquad \ell(\tau, \theta) = -\partial_\theta \mathcal{G}(\tau, \theta) + k(\theta)[\mathcal{S}(\theta) - \mathcal{T}(\tau)],$$

$$\lambda(\tau) = \mathcal{T}'(\tau) - \alpha p(\tau), \qquad \kappa(\tau, \theta) = -\partial_\tau \mathcal{G}(\tau, \theta) + p(\tau)[\mathcal{S}(\theta) - \mathcal{T}(\tau)].$$

Consistency with the results on the orientation of closed curves obtained in Sect. 19.4.2 holds if

$$\mathcal{T}'(\tau) - \alpha p(\tau) \geq \bar{\lambda} > 0, \quad \mathcal{S}'(\theta) + \alpha k(\theta) > 0.$$

By virtue of (19.17), these conditions are satisfied provided only that $k(\theta) > 0$ and $p(\tau) < 0$. For formal convenience, we then let

$$p(\tau) = -\varkappa(\tau), \quad \varkappa(\tau) > 0$$

and h^{tr} can be rewritten as

$$h^{tr} = \varkappa(\tau)[\alpha\xi - \mathcal{S}(\theta) + \mathcal{T}(\tau)]. \tag{19.23}$$

One might introduce expressions for the free energy ϕ that are more involved than (19.16). For example, quite a general expression for ϕ might be

$$\rho_R\phi = \mathcal{G}(\tau,\theta) - \Xi(\xi) + [\mathcal{S}(\theta) - \mathcal{T}(\tau)]\Xi'(\xi),$$

where $\Xi'(\xi)\xi > 0$. Unfortunately, the corresponding expressions for h and η would no longer be linear in ξ and the term $\alpha\xi$ would be replaced with $\Xi'(\xi)$ in Eq. (19.21) and in the following ones. Hence, for a more direct illustration of the method, hereafter we consider $\rho_R\phi$ in the form (19.16).

19.5.3 Entropy Production

The functions η^{tr}, h^{tr} are represented by (19.21), (19.23) and are required to satisfy the reduced dissipation inequality (19.11). In this connection, we complete the scheme by assuming the entropy production in the form:

$$\theta\gamma = g(\xi)|\partial_\xi\phi|\big[\beta(\theta)|\dot\theta| + \nu(\tau)|\dot\tau|\big], \tag{19.24}$$

where g, β, and ν are positive functions. Hence, by virtue of (19.20), Eq. (19.11) can be written in the form:

$$-\partial_\xi\phi k(\theta)\dot\theta + \partial_\xi\phi\varkappa(\tau)\dot\tau - \partial_\xi\phi\dot\xi = g(\xi)|\partial_\xi\phi|\big[\beta(\theta)|\dot\theta| + \nu(\tau)|\dot\tau|\big]. \tag{19.25}$$

If $\partial_\xi\phi \neq 0$ then (19.25) reduces to a rate-type equation that determines the evolution of $\xi(t)$ induced by $\theta(t)$ and $\tau(t)$. Indeed, by virtue of (19.18),

$$k(\theta)\dot\theta - \varkappa(\tau)\dot\tau + \dot\xi = g(\xi)\,\mathrm{sgn}\big[\alpha\xi - \mathcal{S}(\theta) + \mathcal{T}(\tau)\big]\big[\beta(\theta)|\dot\theta| + \nu(\tau)|\dot\tau|\big]. \tag{19.26}$$

At constant stress, say $\tau = \tau_0$, $\dot\tau = 0$, Eq. (19.26) simplifies to

$$\dot\xi = -k(\theta)\dot\theta + \beta(\theta)g(\xi)\,\mathrm{sgn}[\alpha\xi - \mathcal{S}(\theta) + \mathcal{T}(\tau_0)]|\dot\theta|, \tag{19.27}$$

a hysteretic Duhem model describing temperature-induced transitions. If instead the temperature is constant, say $\theta = \theta_0$, $\dot\theta = 0$, then

$$\dot{\xi} = \varkappa(\tau)\dot{\tau} + \nu(\tau)g(\xi)\ \mathrm{sgn}[\alpha\xi - \mathcal{S}(\theta_0) + \mathcal{T}(\tau)]|\dot{\tau}| \tag{19.28}$$

provides the corresponding hysteretic Duhem model of stress-induced transitions.

It is worth remarking that Eq. (19.26) is the general nonlinear model equation for the one-dimensional evolution of an SMA when nonlocal effects are neglected. The formulation of a three-dimensional model that also describes non-local aspects will be considered in a future contribution.

As we show in the next sections, particular cases of Eqs. (19.27) and (19.28) allow a direct closed-form determination of the time functions $\theta(t)$, $\tau(t)$, $\xi(t)$.

19.6 Evolution Equations and Hysteresis Loops

In this section, we apply the previous thermodynamic results and establish an operative description of the evolution of the temperature, stress, and mass fraction in SMAs. Based on the thermodynamic relation (19.26) we look for an operative though qualitative scheme without fitting data of particular alloys.

19.6.1 Equilibrium Surface and Its Projections

We start by looking for the modeling functions $\mathcal{S}(\theta)$, $\mathcal{T}(\tau)$ so that they are consistent with the observed hysteretic properties of the transitions. If $\partial_\xi \phi \neq 0$ then Eq. (19.25) determines an evolution equation involving $\theta(t)$, $\tau(t)$, $\xi(t)$. If instead $\partial_\xi \phi = 0$ then we have

$$\alpha\xi - \mathcal{S}(\theta) + \mathcal{T}(\tau) = 0, \tag{19.29}$$

which characterizes equilibrium states (θ, τ, ξ) of the alloy. Indeed, from (19.24), it follows $\gamma = 0$. We then refer to (19.29) as the *equilibrium surface* and we can read it from two different points of view. First, for any given pair (θ, τ) Eq. (19.29) determines the equilibrium fraction ξ. Second, for any given martensitic fraction ξ it gives a curve in the $\theta - \tau$ plane representing the temperature–stress equilibrium values. Since the martensite fraction is subject to the constraint $\xi \in [0, 1]$, the projection of the equilibrium surface on the $\theta - \tau$ plane is

$$\mathcal{E} = \{(\theta, \tau) : 0 \le \mathcal{S}(\theta) - \mathcal{T}(\tau) \le \alpha\}.$$

For definiteness hereafter we let $\mathcal{S}(\theta)$ and $\mathcal{T}(\theta)$ be linear in that

$$\mathcal{S}(\theta) = c_\theta\, \theta + d_\theta, \qquad \mathcal{T}(\tau) = c_\tau\, \tau + d_\tau,$$

where c_θ and c_τ are positive constants by virtue of (19.17). Hence Eq. (19.29) is taken in the form

$$\alpha\xi = c_\theta\theta - c_\tau\tau + c_d, \quad c_d = d_\theta - d_\tau, \tag{19.30}$$

and represents the set of all possible equilibrium states (θ, τ, ξ). Letting

$$m = c_\theta/c_\tau > 0, \qquad \tau_d = c_d/c_\tau, \qquad q = \alpha/c_\tau > 0,$$

and accounting for $0 \le \xi \le 1$, the region $\mathcal{E}$ is represented by the strip

$$0 \le m\theta - \tau + \tau_d \le q.$$

We now examine the relationship between the material parameters introduced so far and the parameters characterizing the schematic hysteretic loops depicted in Figs. 19.2 and 19.3. To this end, we consider the projection of the equilibrium surface on the planes $\theta = \theta_0$ or $\tau = \tau_0$.

At constant temperature, $\theta = \theta_0$, the projection of (19.30) results in a segment in the $\tau - \xi$-plane, namely:

$$\alpha\xi = c_\theta\theta_0 - c_\tau\tau + c_d, \quad 0 \le \xi \le 1.$$

In Fig. 19.2, this segment is represented by a dashed line, sometimes referred to as "skeleton curve" of the cycle. In light of this schematic picture, as $\xi = 0$ or $\xi = 1$, we obtain the values

$$\tau^s_M|_{\theta=\theta_0} = m\theta_0 + \tau_d, \qquad \tau^s_A|_{\theta=\theta_0} = m\theta_0 + \tau_d - q.$$

On the other hand, at constant stress, $\tau = \tau_0$, the projection of (19.30) yields (see the dashed line in Fig. 19.3)

$$\alpha\xi = c_\theta\theta - c_\tau\tau_0 + c_d, \quad 0 \le \xi \le 1,$$

whence, at $\xi = 0$ and $\xi = 1$, we have

$$\theta^s_M|_{\tau=\tau_0} = (\tau_0 - \tau_d)/m, \qquad \theta^s_A|_{\tau=\tau_0} = (q + \tau_0 - \tau_d)/m. \tag{19.31}$$

In particular, we let

$$\theta_\dagger := \theta^s_A|_{\tau=0} = (q - \tau_d)/m. \tag{19.32}$$

Accordingly, the upper boundary of the equilibrium strip, namely (see Fig. 19.4)

$$r_u = \{(\theta, \tau) : \tau = m\theta + \tau_d\}$$

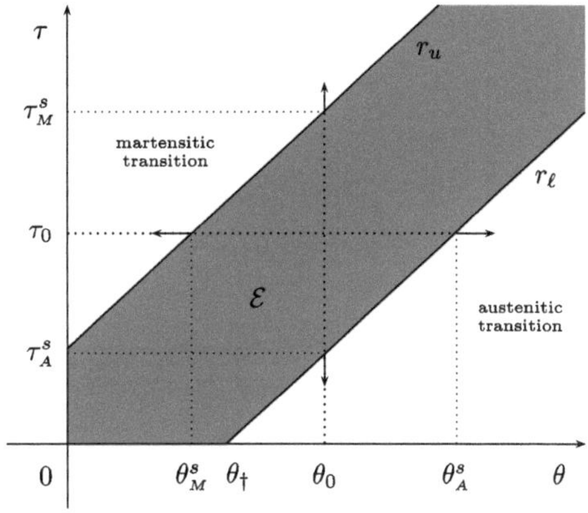

Fig. 19.4 Schematic diagram of the starting points of the martensitic and austenitic transition at constant stress, $\tau = \tau_0$, or constant temperature, $\theta = \theta_0$

can be viewed as the set of temperature–stress pairs at which a martensitic transition starts when the evolution process leaves the equilibrium set $\mathcal{E}$, whereas its lower boundary, namely

$$r_\ell = \{(\theta, \tau) : \tau = m\theta + \tau_d - q\}$$

represents the set of temperature–stress pairs at which an austenitic transition starts.

19.6.2 Schematic Representation of the Phase Transition Diagram

At low stress, however, experimental data suggest that the onset temperature of a martensitic transformation during cooling remains nearly constant (Timofeeva et al. 2022). Let

$$\theta_* = \theta_M^s|_{\tau=0}$$

be the starting temperature of a martensitic transformation during cooling in a stress-free state. The evolution of θ_M^s as the tensile stress increases ($\tau > 0$) is described in Fig. 3 of Timofeeva et al. (2022). With a good approximation we can represent this evolution with the piecewise-linear dashed curve in Fig. 19.5: the value of θ_M^s remains equal to θ_* (M_s^0 in Timofeeva et al. (2022)) until the stress reaches a threshold value τ_* ($\sigma_{cr}(M_s^0)$ in Timofeeva et al. (2022)) and then assumes a linear trend. According to experimental data concerning NiTi-based SMAs, also called Nitinol, we infer that

$$\theta_* \simeq 225°\text{K} > \hat{\theta}_M, \qquad \tau_* \simeq 125\,\text{MPa}$$

and the slope $1/m$ is proportional to the strain of the CVP (corresponding variants pairs) structure formation of the SMA (see Timofeeva et al. (2022) Sect. 4).

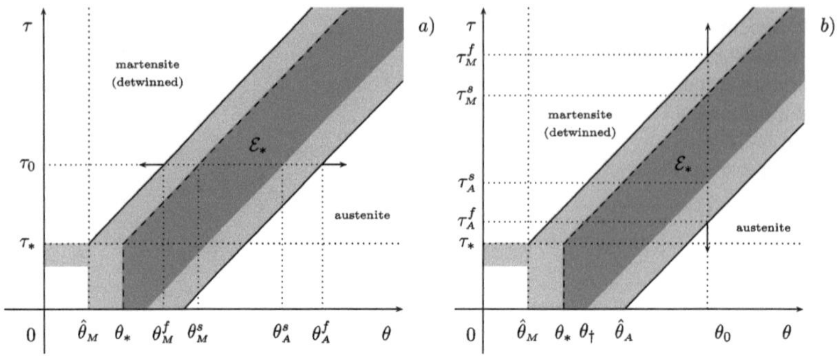

Fig. 19.5 Schematic $\theta - \tau$ phase diagram: **a** temperature-induced transition ($\tau = \tau_0 > \tau_*$); **b** stress-induced transition ($\theta = \theta_0 > \hat\theta_A$)

Accordingly, Fig. 19.5 provides a schematic but complete representation of the phase transition diagram for a SMA (to be compared, e.g., with Fig. 1a of Cisse et al. (2016)). Observe that all possible mixed states (austenite + martensite with $\xi \in (0, 1)$) are allowed in the equilibrium region (see the dark gray strip in Fig. 19.5):

$$\mathcal{E}_* = \{(\theta, \tau) : \theta \geq \theta_*, \ \tau \in [m\theta + \tau_d - q, m\theta + \tau_d]\}.$$

Since it is required that $\tau_* = m\theta_* + \tau_d$ we infer that

$$\tau_d = \tau_* - m\theta_* \tag{19.33}$$

(τ_d is negative in Fig. 19.5). Moreover, by (19.32) we have

$$q = \tau_* + m(\theta_\dagger - \theta_*). \tag{19.34}$$

The two light gray bands in Fig. 19.5, instead, represent transition temperature–stress values in which the martensite concentration ξ lies in a neighborhood of 0 (the lower band) or 1 (the upper band). We now look for determining the values of θ^f_M, θ^f_A and τ^f_M, τ^f_A. Indeed, the upper and lower contours of the above bands consist of the collection of all pairs (θ^f_M, τ^f_M) and (θ^f_A, τ^f_A), respectively.

Consider first the evolution, at constant stress τ_0, i.e., $(\theta(t), \xi(t))$ as $t \in [0, T]$ such that

$$\xi(0) = 0, \ \xi(T) = 1, \quad \theta(0) = \theta^s_M, \ \theta(T) = \theta^f_M.$$

This process describes the temperature-induced transition from austenite ($\xi = 0$) to martensite ($\xi = 1$) as the temperature decreases from θ^s_M to θ^f_M (see Fig. 19.3). Now, at $t = 0$

$$\rho_R \partial_\xi \phi = c_\theta \theta^s_M - c_\tau \tau_0 + c_d = 0.$$

Next ξ increases and θ decreases and then

$$\rho_R \partial_\xi \phi = -\alpha\xi + c_\theta\theta - c_\tau\tau_0 + c_d < 0$$

as $t \in (0, T]$. Thus Eq. (19.27) becomes

$$\dot{\xi} = -k(\theta)\dot{\theta} - \beta(\theta)g(\xi)\,\mathrm{sgn}[\alpha\xi - c_\theta\theta + c_\tau\tau_0 - c_d]\dot{\theta} = -k(\theta)\dot{\theta} - \beta(\theta)g(\xi)\dot{\theta}. \quad (19.35)$$

Integrating (19.35) on $(0, T)$ gives the value of θ_M^f. For simplicity let $k(\theta) = k_0$ and $\beta(\theta)g(\xi) = \beta_0 g_0$. Then (19.35) simplifies to

$$\dot{\xi} = -k_0\dot{\theta} - \beta_0 g_0\dot{\theta}. \quad (19.36)$$

Along the line from θ_M^s to θ_M^f the fraction ξ varies from 0 to 1; integration of (19.36) then gives

$$\theta_M^f = \theta_M^s - \frac{1}{k_0 + \beta_0 g_0}.$$

On the other hand, along the line from θ_A^s to θ_A^f the fraction ξ varies from 1 to 0 and then integration of (19.36) yields

$$\theta_A^f = \theta_A^s + \frac{1}{k_0 + \beta_0 g_0}.$$

In light of (19.31) we can write θ_M^f and θ_A^f in the form:

$$\theta_M^f = \frac{\tau_0 - \tau_d}{m} - \frac{1}{k_0 + \beta_0 g_0}, \qquad \theta_A^f = \frac{\tau_0 - \tau_d + q}{m} + \frac{1}{k_0 + \beta_0 g_0}.$$

Note that in order to obtain the loop as depicted in Fig. 19.3 we need $\theta_M^f(\tau_0) \geq \hat{\theta}_M$ (see also Fig. 19.5a). Hence it is required that

$$\tau_0 \geq \tau_d + \frac{m}{k_0 + \beta_0 g_0} + m\hat{\theta}_M := \tau_*.$$

Remembering (19.33) it follows

$$\frac{1}{k_0 + \beta_0 g_0} = \theta_* - \hat{\theta}_M. \quad (19.37)$$

Likewise we can consider the evolution at constant temperature, $\theta = \theta_0$. Consider the line $(\tau, \xi)(t)$ from $(\tau_M^s, 0)$ to $(\tau_M^f, 1)$ (see Fig. 19.2). Now,

$$\rho_R \partial_\xi \phi = -\alpha\xi + c_\theta\theta - c_\tau\tau + c_d \quad (19.38)$$

vanishes at $\xi = 0, \tau = \tau^s_M, \theta = \theta_0$. Next ξ and τ increase so that $\partial_\xi \phi < 0$ as $t \in (0, T]$. In view of (19.28) with $\varkappa(\tau) = \varkappa_0$ and $\nu(\tau)g(\xi) = \nu_0 g_0$, the evolution equation reads

$$-\varkappa_0 \dot{\tau} + \dot{\xi} = g_0 \nu_0 \dot{\tau}.$$

Integration from $t = 0$ to $t = T$ yields

$$\tau^f_M = \tau^s_M + \frac{1}{\varkappa_0 + \nu_0 g_0}$$

and likewise

$$\tau^f_A = \tau^s_A - \frac{1}{\varkappa_0 + \nu_0 g_0}.$$

Since $\tau^s_M = m\theta_0 + \tau_d$ and $\tau^s_A = m\theta_0 + \tau_d - q$ then we find that

$$\tau^f_M = m\theta_0 + \tau_d + \frac{1}{\varkappa_0 + \nu_0 g_0}, \qquad \tau^f_A = m\theta_0 + \tau_d - q - \frac{1}{\varkappa_0 + \nu_0 g_0},$$

where $\tau_d = \tau_* - m\theta_*$. The loop can occur in the form depicted in Fig. 19.2 provided that $\tau^f_A(\theta_0) \geq 0$. Hence, it is required that (see also Fig. 19.5b)

$$\theta_0 \geq \frac{q - \tau_d}{m} + \frac{1}{m(\varkappa_0 + \nu_0 g_0)} =: \hat{\theta}_A$$

and from (19.32)–(19.33) it follows

$$\frac{1}{\varkappa_0 + \nu_0 g_0} = m(\hat{\theta}_A - \theta_\dagger). \tag{19.39}$$

By construction of the schematic phase diagram in Fig. 19.5, $\hat{\theta}_A - \theta_\dagger = \theta_* - \hat{\theta}_M$ and then from (19.37)–(19.39) it follows $k_0 + \beta_0 g_0 = m(\varkappa_0 + \nu_0 g_0)$.

19.6.3 An Operative Example

The operative character of the present approach is now shown by considering particular linear cases.

19.6.3.1 Temperature-Induced Transition Processes

The austenite–martensite transition, induced by temperature, shows a behavior like that in Fig. 19.3. We now show how transitions are obtained through the model equations derived above. Also for numerical convenience we use relative non-dimensional variables and hence the results hold irrespective of the alloy under consideration.

Let $\vartheta = \theta/\hat{\theta}_M$. In view of (19.27) we consider the evolution equation in the form:

$$\dot{\xi} = -k_0\dot{\vartheta} + \gamma_0 \operatorname{sgn}(\xi - \mathcal{S}(\vartheta) + \mathcal{T}(\varsigma_0))|\dot{\vartheta}|,$$

where $k_0 > 0$, $\gamma_0 := \beta_0 g_0 > 0$, and $\varsigma_0 = \tau_0/\tau_*$, $\tau_0 \geq \tau_*$. An hysteretic loop involves both transitions, from austenite to martensite and vice versa. Note that the saturation property at $\xi = 0$ (during heating) and at $\xi = 1$ (during cooling) is verified only if $\gamma_0 = k_0$.

In Fig. 19.3, the theoretical description of the loop is represented with its clockwise orientation. Figure 19.6, instead, shows the computed major hysteretic loop associated with the data

$$k_0 = 1, \quad \gamma_0 = \begin{cases} 1 \text{ if } 0 \leq \xi \leq 1, \\ 0 \text{ otherwise}, \end{cases} \quad \mathcal{S}(\vartheta) = \vartheta, \quad \mathcal{T}(\varsigma_0) = \tfrac{5}{2}.$$

19.6.3.2 Stress-Induced Transition Processes

We let $\theta = \theta_0 \geq \hat{\theta}_A$ and describe the evolution of $\tau = \tau_*\varsigma$ and ξ via the linear differential equation (19.28) that takes the form:

$$\dot{\xi} = \varkappa_0\dot{\varsigma} + \delta_0 \operatorname{sgn}[\xi - \mathcal{S}(\vartheta_0) + \mathcal{T}(\varsigma)]|\dot{\varsigma}|, \tag{19.40}$$

where $\varkappa_0 > 0$, $\delta_0 := \nu_0 g_0 > 0$, $\vartheta_0 = \theta_0/\hat{\theta}_A$. Note that the saturation property at $\xi = 0$ (during cooling) and at $\xi = 1$ (during heating) is verified only if $\delta_0 = \varkappa_0$.

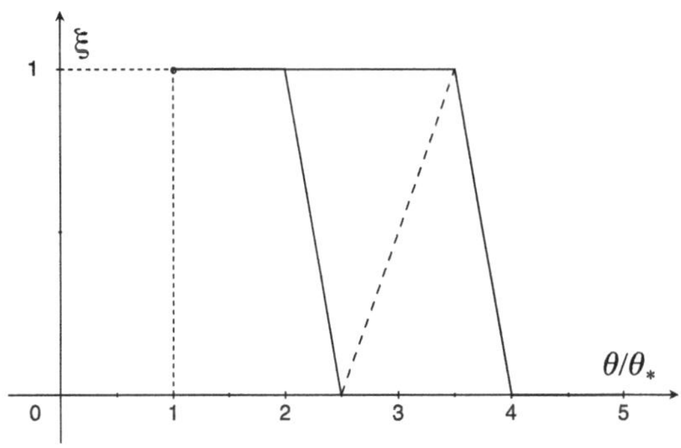

Fig. 19.6 Major loop of the temperature-induced transition

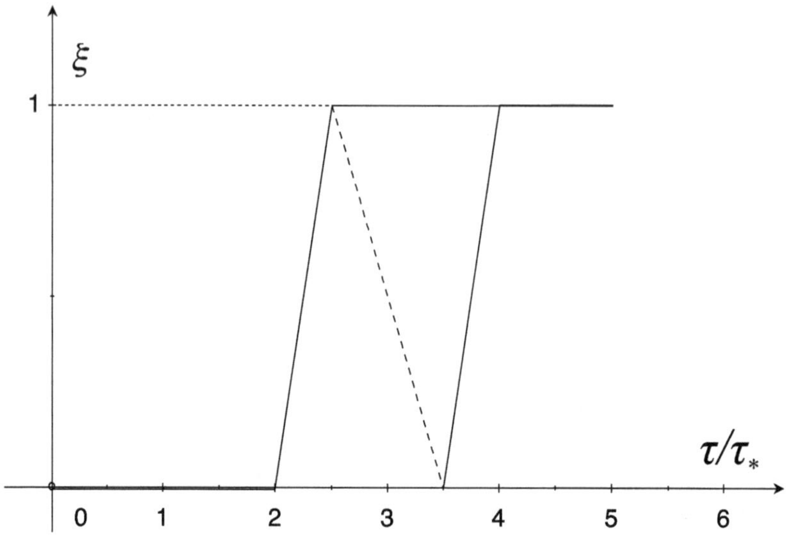

Fig. 19.7 Major loop of the stress-induced transition

While Fig. 19.2 describes the qualitative hysteretic loop oriented in the counter-clockwise sense, Fig. 19.7 shows the numerical simulation obtained in the particular case

$$\varkappa_0 = 1, \quad \delta_0 = \begin{cases} 1 \text{ if } 0 \le \xi \le 1, \\ 0 \text{ otherwise,} \end{cases} \quad \mathcal{T}(\varsigma) = \varsigma, \quad \mathcal{S}(\vartheta_0) = \tfrac{7}{2}.$$

19.6.3.3 Stress–Strain Hysteresis at Constant Temperature (Pseudoelasticity)

Let $\theta = \theta_0$ be the constant temperature. In view of (19.22) we can write the strain variable h in the form:

$$h = [\mathcal{T}'(\tau) + \alpha\varkappa(\tau)]\xi - \partial_\tau \mathcal{G}(\tau, \theta_0) - \varkappa(\tau)[\mathcal{S}(\theta_0) - \mathcal{T}(\tau)].$$

We keep regarding $\mathcal{T}(\tau)$ as linear. Furthermore, we let $\varkappa$ be constant, $\varkappa(\tau) = \varkappa_0$, and $\partial_\tau \mathcal{G}(\tau, \theta_0)$ be linear in τ. We can then write h in the form $h = \Sigma\tau + \Lambda\xi + \Upsilon$, where Σ, Λ, Υ are constants, and then

$$\dot{h} = \Sigma\dot{\tau} + \Lambda\dot{\xi}. \tag{19.41}$$

Using (19.40) with $\varkappa_0 = \delta_0$ we can determine $\dot{\xi}$ in terms of $\dot{\tau}$. Then substitution in (19.41) yields

$$\dot{h} = [\Sigma + \delta_0 \Lambda/\tau_*]\dot{\tau} + \delta_0(\Lambda/\tau_*)\,\mathrm{sgn}[\xi - \mathcal{S}(\theta_0) + \mathcal{T}(\tau)]|\dot{\tau}|.$$

A numerical simulation of this hysteretic Duhem equation provides loops at constant temperature θ_0 in the stress–strain plane. Finally, a direct interchange of axes results in the sought hysteretic loop in the $h - \tau$ plane (see Fig. 19.1).

By (19.23) and (19.38), the linear approximation of the transition function h^{tr} can be written in the form

$$h^{tr} = \varkappa_0(\alpha\xi + c_\tau \tau - c_\theta \theta - c_d)$$

that exhibits a linear dependence of h^{tr} on ξ, τ and θ. Therefore, even in the linear approximation at constant temperature, the current approach represents a generalization with respect to others present in the literature, in which h^{tr} depends (linearly) only on ξ (see, e.g., Xu et al. (2019), Berti et al. (2013)).

19.7 Relation to Other Approaches

Within the literature on models for SMAs there is a variety of approaches with differences and similarities to the present one. To illustrate the conceptual differences, it is natural to consider the pertinent aspects in the corresponding one-dimensional setting.

19.7.1 Dissipation-Consistent Modeling

In McBride et al. (2018), the generalized stress and strain are decomposed additively in energetic and dissipative parts. The Helmholtz free energy (i.e., energy storage potential in McBride et al. (2018)) per unit volume is subject to the dissipation inequality

$$d := -\dot{\Psi} + \tau\dot{h} \geq 0,$$

where d is the dissipation density; in our notation $d = \rho_R \theta \gamma$. Letting $\Psi = \Psi(h, \xi)$ we have

$$-\,\partial_h \Psi \dot{h} + \tau \dot{h} - \partial_\psi \Psi \dot{\xi} \geq 0.$$

The stress τ is decomposed as the sum

$$\tau = \tau^{en} + \tau^{dis},$$

respectively, energetic and dissipative parts. Upon the assumption $\tau^{en} = \partial_h \Psi$ it follows the reduced dissipation inequality

$$d = \tau^{dis}\dot{h} - \partial_\xi \Psi \, \dot{\xi} \ge 0.$$

So far the steps are consistent with the Coleman–Noll and present procedure.

Next X^{en} and X^{dis} are defined by

$$X^{en} := \partial_\xi \Psi, \qquad X^{dis} := -X^{en}$$

and then d can be written in the form:

$$d = \tau^{dis}\dot{h} + X^{dis}\dot{\xi} \ge 0.$$

With this form the pair $\mathbf{A} = (\tau^{dis}, X^{dis})$ is said to be the generalized force and $\mathbf{a} = (\dot{h}, \dot{\xi})$ the corresponding velocity. Next, based on the general assumption of normal dissipativity (see, e.g., Halphen and Quoc Son (1975) and refs. therein), it is assumed that there exists a convex potential $\varphi(\mathbf{A})$ such that

$$\mathbf{a} \in \partial\varphi(\mathbf{A}),$$

where ∂ denotes the sub-differential.

Aside from the convexity requirement, the assumption would amount to considering τ^{dis} and X^{dis} as the variables and $\dot{h}$ and $\dot{\xi}$ as the sought constitutive functions. Even though γ is not allowed to be a constitutive function, we might say that in our approach the derivative $\dot{\xi}$ depends on both the other derivatives (e.g., $\dot{h}$) and the partial derivatives $\partial_h \Psi$, $\partial_\xi \Psi$. This type of dependence is missing in principle within the normal dissipativity.

19.7.2 Modeling Through Internal State Variables

Among the approaches involving internal state of variables we revisit Lagoudas (2008), Chap. 3, in the one-dimensional setting. The infinitesimal strain is split in the form $\epsilon = \epsilon^{el} + \epsilon^{tr}$ and the CD inequality is assumed as

$$\sigma\dot{\epsilon} - \rho(\dot{\psi} + \eta\dot{\epsilon}) \ge 0.$$

Next the free energy $\Phi = (\psi - \sigma\epsilon/\rho)$ is considered so that

$$-\rho(\dot{\Phi} + \eta\dot{\theta}) - \epsilon\,\dot{\sigma} \ge 0,$$

possibly letting ρ be constant. Hence, with the variables $\sigma, \theta, \epsilon^{tr}, \xi$ we have

$$-(\epsilon + \rho\partial_\sigma\Phi)\dot{\sigma} - \rho(\partial_\theta\Phi + \eta)\dot{\theta} - \rho\partial_{\epsilon^{tr}}\Phi\dot{\epsilon}^{tr} - \rho\partial_\xi\Phi\dot{\xi} \ge 0.$$

It follows

$$\epsilon = -\rho\partial_\sigma \Phi, \qquad \eta = -\partial_\theta \Phi$$

and

$$\rho\partial_{\epsilon^{tr}} \Phi\dot{\epsilon}^{tr} + \rho\partial_\xi \Phi\dot{\xi} \le 0.$$

Thanks to the particular free energy

$$\Phi = -\frac{1}{2\rho}\mathcal{S}\sigma^2 - \frac{1}{\rho}\sigma[\alpha(\theta-\theta_0)+\epsilon^{tr}] + c[\theta-\theta_0-\theta\ln(\theta/\theta_0)] - \eta_0\theta + \Phi_0 + \frac{1}{\rho}f(\xi)$$

we have

$$\partial_{\epsilon^{tr}} \Phi = -\frac{1}{\rho}\sigma$$

and then the inequality can be written as

$$\sigma\dot{\epsilon}^{tr} - \rho\partial_\xi \Phi\dot{\xi} \ge 0. \tag{19.42}$$

At constant temperature, $\dot{\theta} = 0$, Eq. (19.42) appears to be similar to (19.11) once we recall that $\gamma \ge 0$. Yet the procedure in Lagoudas (2008) does not determine evolution equations on the basis of (19.42). Rather it is postulated that

$$\dot{\epsilon}^{tr} = \Lambda\dot{\xi}$$

with an ad hoc assumption on Λ and an appropriate parameterization depending on the sign of $\dot{\xi}$.

19.7.3 A Model Involving a Generalized Surface Force

An apparently different approach is developed in Levitas (2013) through a nonlocal scheme. Without entering the details we mention that the balance of energy, and then the CD inequality, is changed by allowing for a surface force density Q conjugate to $\dot{\xi}$ in that the corresponding power density is $Q\dot{\xi}$ though without affecting the balance of linear momentum; in three dimensions, the surface power density is $\mathbf{Q}\dot{\xi} \cdot \mathbf{n}$ and $\nabla \cdot (\mathbf{Q}\dot{\xi})$ is the corresponding power per unit volume. The heat conduction inequality is set aside and then the CD inequality becomes

$$\sigma\dot{\epsilon} - \rho(\dot{\psi} + \eta\dot{\theta}) + \partial_x(Q\dot{\xi}) \ge 0, \tag{19.43}$$

x being the Cartesian coordinate. Accordingly $Q\dot{\xi}$ may be viewed as an extra-entropy flux. To account for nonlocality also $\partial_x\xi$ is a variable in addition to ϵ, θ, ξ. Consequently, the CD inequality (19.43) yields

$$- \rho(\partial_\theta \psi + \eta)\dot{\theta} + \sigma\dot{\epsilon} - \rho\partial_\epsilon \psi \dot{\epsilon} - \rho\partial_{\partial_x \xi}\psi(\partial_x \xi)^{\cdot} + \partial_x(Q\dot{\xi}) \geq 0.$$

The classical relation $\eta = -\partial_\theta \psi$ follows. Now, since $\partial_x(Q\dot{\xi}) = (\partial_x Q)\dot{\xi} + Q\partial_x \dot{\xi}$ and, in the linear approximation,

$$(\partial_x \xi)^{\cdot} = \partial_x \dot{\xi} - \partial_x \xi \dot{\epsilon}.$$

The remaining inequality has the form

$$[\sigma - \rho(\partial_\epsilon \psi + \partial_{\partial_x \xi}\psi\ \partial_x \xi)]\dot{\epsilon} + (\partial_x Q - \rho\partial_\xi \psi)\dot{\xi} + (Q - \rho\partial_{\partial_x \xi}\psi)\partial_x \dot{\xi} \geq 0.$$

Next letting $\sigma = \sigma^e + \sigma^d$, and allowing σ^d depend also on $\dot{\epsilon}$ Ref. Levitas (2013) arrives to the dissipation inequality

$$\rho\delta = \sigma^d \dot{\epsilon} + X\dot{\xi} \geq 0,$$

where

$$X = -\rho\partial_\xi \psi + \partial_x(\rho\partial_{\partial_x \xi}\psi),$$

and possibly split in

$$\sigma^d \dot{\epsilon} \geq 0, \qquad X\dot{\xi} \geq 0.$$

The evolution of ξ is then established by letting

$$\dot{\xi} = L\,X,$$

L being a positive constant.

19.7.4 Models Based on the Ginzburg–Landau Theory

It is worth mentioning Falk's model (Falk (1982), Falk (1983), Brokate and Sprekels (1996), Chap. 5) which is based on the analogy with Landau–Devonshire form of the free energy for hysteretic materials. The free energy is assumed in the form:

$$\psi(\theta, F) = \psi_0(\theta) + \alpha_1(\theta - \theta_0)F^2 - \alpha_2 F^4 + \alpha^3 F^6,$$

where $\alpha_1, \alpha_2, \alpha_3$ are positive constants. The body is thought as a stack of layers and then to account for interfacial energies a term $\frac{1}{2}\nu(\partial_x F)^2$ is comprised in the internal energy. Hence, the total free energy is considered in the form:

$$\psi_{tot} = \psi(\theta, F) + \tfrac{1}{2}\nu(\partial_x F)^2 - \rho b u,$$

while $\frac{1}{2}\rho(\partial_t u)^2$ is the kinetic energy density. Next Hamilton's principle is applied and the equation of motion is obtained in the form:

$$\rho\partial_t^2 u - \partial_x \partial_F \psi(\theta, F) + \nu\partial_x^4 u = \rho b(x,t), \tag{19.44}$$

where $x \in [0, 1]$ and $t > 0$. Among the conceptual differences, (19.44) is a fourth-order equation in the displacement u via the term $\nu\partial_x^4 u$. In addition, the evolution of the temperature is governed by the classical Fourier scheme.

In the framework of a Ginzburg–Landau–Devonshire theory, the minimum (Gibbs) free energy representation has been obtained in Berti et al. (2013) as a quadratic function of the stress σ and the anelastic deformation h^{tr}. The temperature is regarded as a given parameter. This approach involves an order parameter φ whose modulus represents the martensitic mass fraction, namely, $\xi = |\varphi|$. In addition, the anelastic deformation h^{tr} is represented as a quadratic function of φ^2, so that the free energy turns out to be an eighth-order function of φ. This approach allows to highlight the role of the Ginzburg–Landau equation when phase transitions occur in materials with hysteresis such as SMAs.

19.8 Conclusions

Within the wide literature on the modeling of SMAs we can distinguish among microscopic and macroscopic models (see, e.g., Cisse et al. (2016)). Microscopically, a strain measure is considered which is based on interatomic potentials. Macroscopic approaches are based on a free energy function of temperature and strain, the strain being decomposed in elastic and transformation parts $\boldsymbol{\epsilon}^e, \boldsymbol{\epsilon}^{tr}$. Conceptually the macroscopic approaches are characterized by the modeling of the transformation strain $\boldsymbol{\epsilon}^{tr}$; some ways are mentioned in Sect. 6. In essence, we can say that macroscopically the mass fraction ξ of a constituent (e.g., martensite) is considered and the constitutive properties emerge from a free energy potential while the transition strain $\boldsymbol{\epsilon}^{tr}$ is eventually related to the ξ through a properly defined rate equation.

In the present approach, the constitutive equations depend on the temperature θ, the one-dimensional stress τ, the mass fraction ξ, and the derivatives $\dot\theta, \dot\tau$ while $\dot\xi$ is given by a constitutive rate equation. The main novelty of our approach is that the entropy production density (here γ) is itself given by a constitutive function (of $\theta, \tau, \xi, \dot\theta, \dot\tau$). For simplicity the model is one-dimensional and the strain variable is $h = \ln F$, with F the deformation derivative in the longitudinal direction. Within this nonlinear theory the whole set of thermodynamic restrictions is derived. Next a free energy function ϕ is considered in the form (19.16). Hence, we find that

$$0 = \rho_R \partial_\xi \phi = -\alpha\xi + \mathcal{S}(\theta) - \mathcal{T}(\tau),$$

here referred to as skeleton surface, characterizes the equilibrium mixed states (θ, τ, ξ). Meanwhile, upon the selection of the constitutive function (19.24) for the entropy production, the evolution condition for ξ is obtained in the form (19.26).

The operative aspect of the theory is emphasized in Sect. 19.6 by using a linearized version of the equations so that their integration can be done in closed form. Hence, detailed hysteretic loops are derived accordingly as they are induced by temperature or by stress. Note that some characteristic parameters of the model, $k_0, \varkappa_0, \tau_d, q$, are related to experimentally measurable quantities such as $m, \theta_*, \tau_*, \theta_\dagger, \hat{\theta}_M, \hat{\theta}_A$.

A further development along the lines of this paper is the generalization to bodies subject to fields within a three-dimensional setting.

References

Abel E, Hongyan L, Pridham M, Slade A (2004) Issues concerning the measurement of transformation temperatures of NiTi alloys. Smart Mater Struct 13:1110

Berti A, Giorgi C, Vuk E (2013) Free energies and pseudo-elastic transitions for shape memory alloys. Discrete Contin Dyn Syst Ser S 6:293–316

Brinson LC (1993) One dimensional constitutive behavior of shape memory alloys: thermomechanical derivation with nonconstant material functions and redefined martensite internal variable. J Intell Mater Syst Struct 4:229–242

Brokate M, Sprekels J (1996) Hysteresis and phase transitions. Springer, New York

Cisse C, Zaki W, Ben Zineb T (2016) A review of constitutive models and modeling techniques for shape memory alloys. Int J Plasticity 76:244–284

Dreyer W, Guhlke C, Herrmann M (2011) Hysteresis and phase transition in many-particle storage systems Cont. Mech Thermodyn 23:211–231

Eringen AC (1967) Mechanics of Continua. Wiley, New York

Falk F (1982) Landau theory and martensitic phase transitions. J Phys C 12:3–15

Falk F (1983) One-dimensional model of shape memory alloys. Arch Mech 35:63–84

Giorgi C, Morro A (2024) On the second law of thermodynamics in continuum physics. Thermo 4:273–294

Giorgi C, Morro A (2024) Strain-rate and stress-rate models of nonlinear viscoelastic materials. Mathematics 12:3011

Gottschall T et al (2016) On the S(T) diagram of magnetocaloric materials with first-order transition: kinetic and cyclic effects of Heusler alloys. Acta Mater 107:1–8

Gurtin ME, Fried E, Anand L (2010) The mechanics and thermodynamics of Continua. Cambridge University Press

Halphen B, Quoc Son N (1975) Sur les matériaux standards généralisés. J Mécan 14:39-63

Huo Y, Müller I (1993) Nonequilibrium thermodynamics of pseudoelasticity. Cont Mech Thermodyn 5:163–204

Jin Y, Artemev A, Khachaturyan A (2001) Three-dimensional phase-field model of low-symmetry martenitic transformation in polycrystal: simulation of $\zeta_2$martensite in AuCd alloys. Acta Mater 49:2309–2320

Kumar PK, Lagoudas DC (2008) Introduction to shape memory alloys. In: Lagoudas DC (ed) Shape memory alloys: modeling and engineering applications. Springer, New York

Lagoudas DC (ed) (2008) Shape memory alloys modeling and engineering applications. Springer, New York

Levitas VI (2013) Thermodynamically consistent phase field approach to phase transformations with interface stresses. Acta Mater 61:4305–4319

Mandolino MA, Scholtes D, Ferrante F, Rizzello G (2023) A physics-based hybrid dynamical model of hysteresis in polycrystalline shape memory alloy wire transducers. IEEE/ASME Trans Mechatron (TMECH) 28:2529–2540

McBride AT, Reddy BD, Steinmann P (2018) Dissipation-consistent and classification of extended plasticity formulations. J Mech Phys Solids 119:118–139

Morro A, Giorgi C (2023) Techniques for the thermodynamic consistency of constitutive equations. Thermo 3:260–276

Morro A, Giorgi C (2023) Mathematical modelling of continuum physics. Birckhäuser, Cham

Scalet G, Niccoli F, Garion C, Chiggiato P, Maletta C, Auricchio F (2019) A three-dimensional phenomenological model for shape memory alloys including two-way shape memory effect and plasticity. Mech Mater 136:103085

Seelecke S, Müller I (2004) Shape memory alloy actuators in smart structures: modeling and simulation. Appl Mech Rev 57:23–46

Timofeeva EE et al (2022) On the stress-temperature dependences in TiNi-based shape memory alloys. J Alloys Compd 905:164227

Truesdell C, Toupin RA (1960, The classical field theories, in Encyclopedia of Physics, vol. III/1, ed. S. Flügge (Springer, Berlin)

Wu X, Sun G, Wu J (2003) The nonlinear relationship between transformation strain and applied stress for nitinol. Mater Lett 57:1334–1338

Xu L, Baxevanis T, Lagoudas DC (2019) A three-dimensional constitutive model for the martensitic transformation in polycrystalline shape memory alloys under large deformation. Smart Mater Struct 28:074004

Chapter 20
Which Frequency Bands are Responsible for Noise-Induced Triggering of Thermoacoustic Instabilities?

Maria A. Heckl and Sadaf Arabi

Dedication by Maria Heckl
I got to know Wolfgang Dreyer at the Technical University of Berlin during the last two years of my physics studies. At the time, Wolfgang was working on his Ph.D under the guidance of Professor Ingo Müller, whose lecture courses I keenly followed. Wolfgang was not only a passionate scientist, but also a caring person who would stand up for fair play. I will be forever grateful to Wolfgang for his crucial support during the saga that surrounded my Diplomarbeit (the equivalent of a master's thesis). This was supervised and examined by two professors from different research groups. After a dispute I had with them about a joint publication, they threatened to fail my Diplomarbeit and thus derail my academic career, which had only just started. Wolfgang was instrumental in exposing their behaviour and thus helped me to fight back. Now, some 40 years later, I look back over a successful research career, facilitated in no small part by Wolfgang Dreyer. Thank you, Wolfgang!

Abstract This paper presents an analytical study of noise-induced triggering, which is the spontaneous transition of a thermoacoustic system from a linearly stable state to an unstable state. Our aim is to find out whether there are some frequency bands in the noise spectrum that are more effective at triggering than others. We model the complete combustion system as a combination of three elements: a combustion chamber, and within it, a flame and a noise source. The empty combustion chamber is described by its impulse response (also called tailored Green's function). The flame is described by its heat release rate, which has a nonlinear relationship with the acoustic velocity. The noise is narrow-band and described in terms of its frequency

M. A. Heckl (✉)
School of Chemical and Physical Sciences, Keele University, Newcastle-Under-Lyme, Staffordshire, United Kingdom
e-mail: m.a.heckl@keele.ac.uk

S. Arabi
Department of Mathematics, University of York, York, United Kingdom

J. Fuhrmann et al. (eds.), *Advances in Continuum Physics*, Advanced Structured Materials 238, https://doi.org/10.1007/978-3-031-93918-1_20

spectrum; it is filtered white noise and has three variable parameters: noise intensity, bandwidth and centre frequency. We use Green's function approach to combine the three elements; this leads to an integral equation for the time history of the acoustic field in the combustion chamber. The calculated time histories reveal the consequences of varying noise properties. The noise level is an important parameter: if it is sufficiently high, triggering occurs; this manifests itself by a rapid increase in amplitude, which is soon followed by a limit cycle. The bandwidth of the noise also plays a role: if the bandwidth is increased (without changing the centre frequency), triggering occurs more readily. The parameter with the most remarkable behaviour is the centre frequency of a frequency band: if this is in the very low-frequency range, triggering is achieved with a low noise level, but as the frequency increases, substantially higher noise levels are required. Preferential triggering at the resonance frequency of the combustion chamber has not been observed.

20.1 Introduction

Thermoacoustics is the branch of thermodynamics, which studies the interaction between sound and flames. In combustion systems consisting of a combustion chamber with a flame inside, the sound-flame interaction can evolve into a positive feedback loop; this then leads to growing acoustic pressure amplitudes until a limit cycle is reached (see Fig. 20.1). This phenomenon is called thermoacoustic instability. The associated pressure amplitude can be extreme and inflict immense mechanical damage on the affected combustion system (Lieuwen and Yang 2005, Chap. 8). Thermoacoustic instabilities are a serious issue in various combustion technologies, such as stationary gas turbine engines, jet aircraft engines, domestic and industrial boilers, rockets and potentially in any other system where a flame is burning continuously inside a cavity. The fact that thermoacoustic instabilities tend to occur without warning adds to the need to understand them and to the challenge to avoid them.

Intense research efforts in the last five decades have led to a growing understanding of the complexities of thermoacoustic instabilities. Despite all these efforts, the scientific understanding of thermoacoustic instabilities remains incomplete, and it is still not possible to design a practical combustion system, which is safe from thermoacoustic instabilities across its whole operation range. In order to move away from the combustion of fossil fuels, alternative fuels (such as hydrogen, ammonia, biofuels and syngas) are being considered for the development of carbon-neutral combustion technologies. Unfortunately, such new technologies come with an increased risk of thermoacoustic instabilities.

Put simply, combustion systems are both nonlinear and noisy. The nonlinearity is in the sound-flame interaction and leads to effects, such as limit cycles, bifurcations, bistability and hysteresis (Culick 2006; Sujith and Pawar 2021). Noise, which is inherent in all combustors, adds a random element to the dynamics and is an important contributor to the erratic nature of thermoacoustic instabilities. The interplay

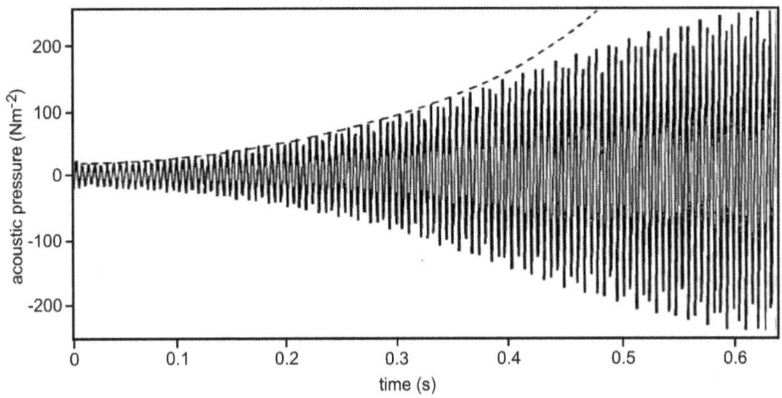

Fig. 20.1 Time history of the acoustic pressure during a thermoacoustic instability, measured at a point near the heat source in a laboratory combustor (Heckl 1985). The dashed line has been added to show the envelope for the hypothetical case of unlimited exponential growth

between nonlinear behaviour and noise can lead to several phenomena, amongst them "noise-induced triggering". This is the spontaneous transition of a combustion system from a linearly stable state to an unstable state, which is triggered if the noise level is sufficiently high. Once the system is unstable, the oscillation amplitude keeps growing until it is limited by some other physical effect. The seemingly random occurrences of triggering compounded by the potentially high amplitudes that are reached in the aftermath make this phenomenon quite dangerous from a practical point of view.

Several researchers have experimentally observed noise-induced triggering on diverse thermoacoustic test rigs. Lieuwen (2002) performed experimental investigations on a laboratory test rig simulating a gas turbine combustor with a lean premixed flame anchored to a bluff body; he observed that the noise inherent in his test rig could trigger instability under certain operating conditions. Kim and Hochgreb (2012) studied a combustor with a lean premixed swirl flame and also investigated the effect of the inherent background noise; their focus was on triggering and the subsequent transient growth of thermoacoustic oscillations. Jegadeesan and Sujith (2013) experimented with a non-premixed flame in a duct, into which they injected white noise via a loudspeaker. They varied the noise level and achieved triggering if the noise level was above a certain threshold; this threshold was surprisingly low and well below the level that would be required according to a nonlinear stability analysis. We also mention the study by Pun et al. (2003) here, although this is not explicitly about noise-induced triggering. Pun et al. forced their thermoacoustic system not with noise, but with single-frequency sound. The effect on the stability behaviour was assessed in terms of the local Rayleigh index (a measure of the spatially resolved energy from the flame), which they determined experimentally. They observed zones within the flame, where the Rayleigh index was positive, indicating that acoustic energy was gained from parts of the flame, thus encouraging instability. These zones were largest for the lowest forcing frequency, which was 22 Hz. However, their size reduced as the frequency increased, and they disappeared altogether for frequencies above 55 Hz.

Noise-induced triggering has also been studied from a theoretical perspective. Waugh and Juniper (2011) analysed a Rijke tube (a straight tube with open ends, which contains a hot wire gauze that acts as the heat source instead of a flame); this is the prototype of a thermoacoustic system. The heat release rate was modelled by an empirical nonlinear function of the time-lagged velocity at the hot gauze, the tube was modelled by a superposition of classical acoustic modes, and the complete system was analysed by a Galerkin approach. Three different types of noise were considered: pink noise (higher amplitudes at low frequencies), white noise (similar amplitudes at all frequencies) and blue noise (higher amplitudes at high frequencies). Waugh et al. found that pink noise was more effective at causing triggering than white noise, which was more effective than blue noise. In a similar study, involving again the Rijke tube, Waugh and Juniper (2011) had a noise term representing pink noise of variable intensity. They predicted that noise-induced triggering depends strongly on the noise level, with increasing noise level enhancing the risk of instability. Arabi and Heckl (2025) modelled a noisy combustion chamber by treating it as an acoustic resonator with two acoustic sources: one was the flame, which they described by a heat release rate that was coupled nonlinearly to the acoustic field; the other source was noise, which was independent of the acoustic field. They considered both white noise and pink noise with varying intensity and calculated the time evolution of the acoustic velocity with Green's function approach. They also observed that pink noise is more effective than white noise in triggering a thermoacoustic instability. This was despite the fact that both noise types had the same energy content in a narrow frequency band around the frequency of the instability.

Most thermoacoustic instabilities have a frequency, which is close to one of the resonance frequencies of the combustion chamber; one might therefore hypothesise that an important role in the triggering is played by those frequency components in the noise spectrum, which are close to the resonance frequency. On the other hand, pink noise is more effective (as mentioned above), and this suggests that the very low frequencies in the noise spectrum are responsible for triggering. The aim of this paper is to shed light on the question of whether some frequency bands in the noise spectrum are more instrumental than others.

This paragraph gives a brief outline of the basic concepts of acoustics for the benefit of the reader unfamiliar with them. Sound propagates as a wave; in the air at room temperature, its speed is about 340 m/s, but in hot air, it is considerably higher, for example, 430 m/s if the temperature is 460K. As the sound wave propagates, the pressure in the fluid changes from the mean pressure $\overline{p}$ to $\overline{p} + p(\mathbf{x}, t)$, and the mass density changes from its mean value $\overline{\rho}$ to $\overline{\rho} + \rho(\mathbf{x}, t)$. Throughout this paper, we denote a mean quantity by an overbar. Mean quantities are constant with respect to position and time; in the open air, $\overline{p}$ is equal to the atmospheric pressure. Perturbation quantities, such as $p(\mathbf{x}, t)$, depend on both position $\mathbf{x}$ and time t. Compared with the mean quantities, the perturbations are weak, with amplitudes being orders of magnitude smaller than the mean quantities, i.e. $|p/\overline{p}| << 1$ and $|\rho/\overline{\rho}| << 1$. For example, $|p/\overline{p}| = 10^{-3}$ for a sound at the pain threshold of hearing (Dowling and Ffowcs Williams 1983). As a consequence, sound is regarded as a linear motion, and it is safe to assume that sound waves superimpose linearly. Viscous effects are usually negligible in a sound wave because at practically relevant frequencies, the pressure represents a far greater stress field than that induced by viscosity (Dowling and Ffowcs Williams 1983). The fluid particles carrying a sound wave perform an oscillating movement. This results in a velocity field $\mathbf{u}(\mathbf{x}, t)$; we call this the *acoustic velocity*. In this paper, we consider only 1-D configurations, in particular sound waves in a straight tube. Then the vector field $\mathbf{u}(\mathbf{x}, t)$ reduces to $u(x, t)$, where x is the position along the tube axis. The corresponding velocity potential ϕ is given in terms of its gradient by $u(x, t) = \frac{\partial \phi(x,t)}{\partial x}$. In combustion systems, there is another important velocity, and this is the flow velocity of the combustible gases that are burnt as they pass through the flame. We call this the *mean velocity* $\overline{u}$. It is typically in the range of 1–10 m/s and hence much smaller than the speed of sound. For this reason, $\overline{u}$ plays only a minor role in the propagation of sound and can therefore be neglected in the acoustics context. However, it cannot be neglected in the *thermoacoustics* context, because it has a major effect on the flame dynamics. We will therefore neglect $\overline{u}$ in our model for the acoustic wave propagation, but not in our model for the sound-flame interaction. Also, we model this interaction by a *nonlinear* expression—in contrast to the acoustic waves.

Returning to the topic of this paper, we will build on the work by Arabi and Heckl (2025) and consider the tube shown in Fig. 20.2.

The tube has length L and ends at $x = 0$ and $x = L$. One-dimensional acoustic waves are present in the tube and travel forwards and backwards in the axial direction. If a wave impinges on one of the ends, it is reflected back into the tube (typically with a change of phase). The temperature jump from $\overline{T}_1$ to $\overline{T}_2$ at the axial position $x = x_q$ divides the tube into two distinct sections: a cold and a hot section. Associated with the temperature jump is a jump in the mean density, from $\overline{\rho}_1$ to $\overline{\rho}_2$, where $\overline{\rho}_2 < \overline{\rho}_1$,

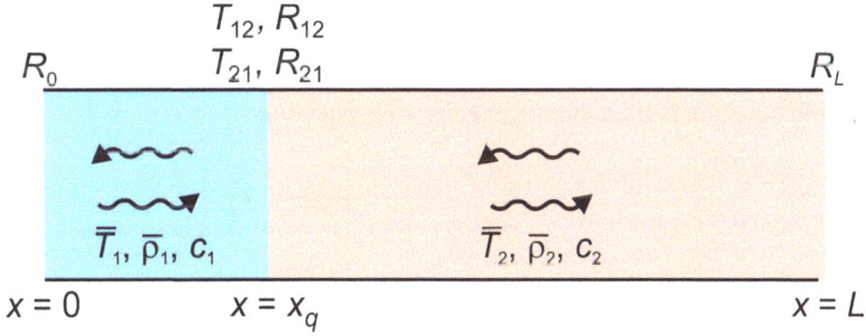

Fig. 20.2 Tube of length L with cold section and hot section, separated by an interface at x_q. $\overline{T}$, $\overline{\rho}$ and c are, respectively, mean temperature, mass density and speed of sound; subscripts 1 refer to the cold section, and subscripts 2 refer to the hot section. R_0, R_L are the reflection coefficients at the tube ends; R_{12}, R_{21} are the reflection coefficients at the interface, and T_{12}, T_{21} are the corresponding transmission coefficients

and a jump in the speed of sound from c_1 to c_2, where $c_2 > c_1$. The product (mean density)x(speed of sound), which is called acoustic impedance, is also discontinuous at the interface, jumping from $\overline{\rho}_1 c_1$ to $\overline{\rho}_2 c_2$, where $\overline{\rho}_2 c_2 < \overline{\rho}_1 c_1$. Waves approaching this discontinuity from either side are partly reflected and partly transmitted; this is described in more detail in Sect. 20.2.1.

Our paper is structured as follows: Sect. 20.2 describes the mathematical model for our combustion system, including the heat source and the noise source. In Sect. 20.3, we present the governing integral equation, which will be solved to obtain predictions for the acoustic field in the combustion chamber. In Sect. 20.4, we will show predictions for the two special cases where either the heat source is present in the combustion chamber and the noise source is absent, or vice versa. Predictions for instability triggering by narrow-band noise will be made in Sect. 20.5. Conclusions are drawn in Sect. 20.6.

20.2 Mathematical Model

Our mathematical model for the complete combustion system has three elements, which are described separately in Sects. 20.2.1, 20.2.2 and 20.2.3 below. Section 20.2.1 is about the empty combustion chamber, i.e. without any acoustic sources. In Sect. 20.2.2, we describe the heat source, and in Sect. 20.2.3 the noise source.

20.2.1 Model for the Combustion Chamber in Terms of the Tailored Green's Function

Green's function is an impulse response, i.e. it is the acoustic field generated by a hypothetical point source located at x', firing an impulsive signal at time t', and measured by an observer at point x and at time t. We denote it by $G(x, x', t, t')$. Its governing equation is the inhomogeneous wave equation

$$\frac{1}{c^2}\frac{\partial^2 G}{\partial t^2} - \frac{\partial^2 G}{\partial x^2} = \delta\left(x - x'\right)\delta\left(t - t'\right). \tag{20.1}$$

We distinguish between the *free-field* Green's function and the *tailored* Green's function. The free-field Green's function is the solution of Eq. (20.1) in a uniform medium of infinite extent. It is given by

$$G(x, t; x', t') = \frac{\delta[(t - t') - |x - x'|/c]}{4\pi\,|x - x'|}, \tag{20.2}$$

which represents two impulses travelling away from the source to $x \to \pm\infty$ (Heckl et al. 2022).

In contrast to the free-field Green's function, the tailored Green's function is the impulse response in a medium *with* boundaries, such as the tube shown in Fig. 20.1. Its name comes from the fact that it is "tailored" to the geometry of the configuration of interest. In addition to Eq. (20.1), it satisfies the boundary conditions of the acoustic field in the tube. The boundary at $x = 0$ is modelled by the pressure reflection coefficient R_0. This means that if a left-travelling pressure wave $p_-(x, t)$ with amplitude p_0,

$$p_-(x, t) = p_0\, \mathrm{e}^{-\mathrm{i}\omega(t+x/c_1)}, \tag{20.3}$$

approaches this end, a reflected wave

$$p_+(x, t) = R_0 p_0\, \mathrm{e}^{-\mathrm{i}\omega(t-x/c_1)}, \tag{20.4}$$

is generated, and this has amplitude $R_0 p_0$. The end at $x = L$ is treated in the same way, with reflection coefficient R_L. Reflection as well as transmission of acoustic waves occurs at the interface at $x = x_q$ between the cold and hot regions. This is modelled similarly in terms of reflection and transmission coefficients. R_{12} and R_{21} are the reflection coefficients for waves coming from the cold side and the hot side, respectively, while T_{12} and T_{21} are the corresponding transmission coefficients. They depend on the acoustic impedances $\overline{\rho}_1 c_1$ and $\overline{\rho}_2 c_2$ as given in the equations below (see Sect. 6.2 in (Kinsler et al. 2000)).

$$R_{12} = \frac{\overline{\rho}_2 c_2 - \overline{\rho}_1 c_1}{\overline{\rho}_1 c_1 + \overline{\rho}_2 c_2}, \quad R_{21} = -\frac{\overline{\rho}_2 c_2 - \overline{\rho}_1 c_1}{\overline{\rho}_1 c_1 + \overline{\rho}_2 c_2}, \tag{20.5a,b}$$

$$T_{12} = \frac{2\overline{\rho}_2 c_2}{\overline{\rho}_1 c_1 + \overline{\rho}_2 c_2}, \quad T_{21} = \frac{2\overline{\rho}_1 c_1}{\overline{\rho}_1 c_1 + \overline{\rho}_2 c_2}. \tag{20.5c,d}$$

The tailored Green's function is effectively the impulse response of an acoustic resonator. It is therefore expected to be a superposition of resonator modes,

$$G(x, x', t, t') = H(t - t')\mathrm{Re} \sum_{n=1}^{\infty} g_n(x, x')\, \mathrm{e}^{-\mathrm{i}\omega_n(t-t')}, \tag{20.6}$$

where $H(t - t')$ stands for the Heaviside function; this ensures that Green's function is causal, i.e. there is no response before the impulse has occurred. g_n and ω_n are, respectively, the amplitude and frequency of mode n. ω_n is identical with the resonator eigenfrequency. It can be determined from an eigenvalue approach (Arabi and Heckl 2025), which leads to the characteristic equation $F(\omega) = 0$, with

$$F(\omega) = e^{-ik_1x_q}e^{-ik_2(L-x_q)} - R_LR_{21}e^{-ik_1x_q}e^{ik_2(L-x_q)} - R_0R_{12}e^{ik_1x_q}e^{-ik_2(L-x_q)} \\ - R_0R_Le^{ik_1x_q}e^{ik_2(L-x_q)} \quad . \tag{20.7}$$

k_1 and k_2 are the wave numbers in the cold and hot regions, respectively, given by

$$k_1 = \omega / c_1 \text{ and } k_2 = \omega / c_2. \tag{20.8}$$

The amplitude g_n of the tailored Green's function differs from the amplitude of the resonator mode n in that it depends not only on the observer position x, but also on the source position x'. The analytical calculation of $g_n(x, x')$requires a generalised function approach and has been performed by Arabi and Heckl (2025). The result is

$$g_n(x, x') = \frac{c_2\, \hat{g}(x, x', \omega_n)}{\omega_n F'(\omega_n)}, \tag{20.9}$$

where

$$\hat{g}(x, x', \omega) = \begin{cases} A(x, \omega)B(x', \omega) \text{ for } 0 < x < x_q \\ B(x', \omega)C(x, \omega) \text{ for } x_q < x < x' \\ C(x', \omega)B(x, \omega) \text{ for } x' < x < L \end{cases} \tag{20.10}$$

and

$$A(x, \omega) = T_{21}\left(R_0e^{ik_1x} + e^{-ik_1x}\right), \tag{20.11}$$

$$B(x, \omega) = R_Le^{ik_2(L-x)} + e^{-ik_2(L-x)}, \tag{20.12}$$

$$C(x, \omega) = e^{ik_2(x-x_q)}\left(R_{21}e^{-ik_1x_q} + R_0e^{ik_1x_q}\right) + e^{-ik_2(x-x_q)}\left(e^{-ik_1x_q} - R_0R_{12}e^{ik_1x_q}\right). \tag{20.13}$$

Green's function depends implicitly on the mean fields of temperature, density and pressure. This can be seen from Eqs. (20.5a,b, 20.5c,d) and (20.8), which show the dependence on the speeds of sound, c_1 and c_2, in the cold and hot section of the combustion chamber. For a gas, the speed of sound c is related to the mean temperature $\overline{T}$ via the formula $c = \sqrt{\gamma R\overline{T}}$, where γ is the specific heat ratio and R is the gas constant. γ and R depend on the type of gas; for air, they are $\gamma = 1.4$ and $R = 287\,\mathrm{J\,kg^{-1}K^{-1}}$. There is also a link between Green's function with the mean pressure and mean density via the perfect gas law, which relates these quantities to $\overline{T}$.

20.2.2 Model for the Heat Source

Our heat source is a flame described by its local heat release rate (i.e. the heat release rate per unit mass), which we denote by $q(x, t)$. The flame is compact and located at the position x_q, so we can write

$$q(x,t) = q(t)\delta(x - x_q). \tag{20.14}$$

The time-dependent part, $q(t)$, is closely related to the global heat release rate, $Q(t)$,

$$Q(t) = S\,\overline{\rho}\,q(t), \tag{20.15}$$

where S is the cross-sectional area of the tube. Our model for $q(t)$ is inspired by Noiray's experiments (Noiray 2007), which measured the global heat release rate for a matrix flame, which he excited with acoustic velocities of different amplitudes. He observed that there was a clear amplitude dependence. Heckl (2015) introduced an analytical approximation for it, which we are going to use here:

$$q(t) = K\big[n_1 u_q(t-\tau) - n_0 u_q(t)\big]. \tag{20.16}$$

u_q is the acoustic velocity at the flame, and K is a measure of the power of the flame. τ (time-lag) and $n_0,\ n_1$ (coupling coefficients) are parameters that depend on the amplitude A of the velocity u_q. We normalise this with the mean velocity $\overline{u}$ and write the amplitude dependence in terms of $A/\overline{u}$. The amplitude dependence of τ is quadratic and given by

$$\tau = \tau_0 + \tau_2\left(\frac{A}{\overline{u}}\right)^2. \tag{20.17}$$

For $n_0,\ n_1$, the amplitude dependence is linear and given by

$$n_0 = \frac{1}{2}\left(g_0 - g_1\frac{A}{\overline{u}} - 1\right), \tag{20.18}$$

$$n_1 = \frac{1}{2}\left(g_0 - g_1\frac{A}{\overline{u}} + 1\right). \tag{20.19}$$

The quantities g_0, g_1, τ_0 and τ_2 in Eqs. (20.17), (20.18) and (20.19) are constants. Their numerical values originate from a curve fitting procedure to optimise the fit of Eq. (20.16) with Noiray's data (2007).

20.2.3 Model for the Noise Source

We denote the noise in the tube by $\xi(x, t)$. For mathematical convenience, we assume that the noise source (like the heat source) is compact and located at the position x_q. This source could be a loudspeaker emitting a stochastic acoustic signal into the tube through a small hole in the tube wall at x_q. In analogy to Eq. (20.14), we can then write

$$\xi(x, t) = \xi(t)\delta(x - x_q). \tag{20.20}$$

$\xi(t)$ is a stochastic quantity and represents the time dependence of a stochastic noise source,

$$\xi(t) = \beta \times \text{(random number)}; \tag{20.21}$$

the parameter β is a measure of the noise level, and the random number is in the range $[-0.1, 0.1]$.

In order to synthesise narrow-band white noise of a specific bandwidth Δf and specific centre frequency f_c, we applied the following procedure: First, we generated a white noise signal in the frequency range [0, 500Hz], using the Matlab function 'randn'. We then transformed this into the frequency domain by fast Fourier transform and applied a narrow Gaussian filter with bandwidth Δf and centre frequency f_c. This result was subsequently transformed back into the time domain and normalised so that the noise signal was within the range [−0.1,0.1]. Figure 20.3 below shows the power spectral density (magnitude) and time history of filtered white noise with $\Delta f = 20$ and $f_c = 20$ Hz. The equivalent curves for unfiltered white noise are shown in Fig. 20.4 for comparison.

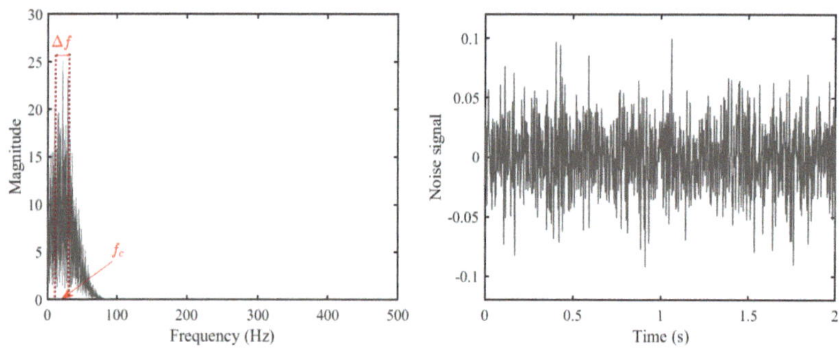

Fig. 20.3 Filtered white noise with $\beta = 1$, $\Delta f = 20$ Hz and $f_c = 20$ Hz. **a** Power spectral density; **b** Typical time history

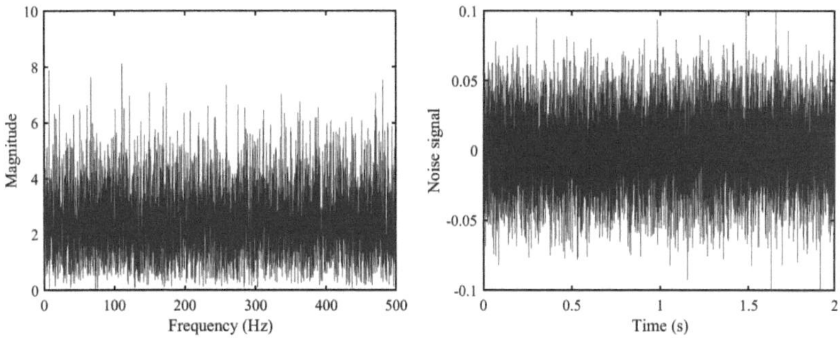

Fig. 20.4 Unfiltered white noise with $\beta = 1$. **a** Power spectral density; **b** Typical time history

20.3 The Integral Governing Equation for the Acoustic Field Driven by the Flame and the Noise Source

We now combine the individual elements described in Sect. 20.2 (combustion chamber, flame and noise source) to get a model for the complete combustion system. The starting point is the inhomogeneous wave equation for the velocity potential ϕ

$$\frac{1}{c^2}\frac{\partial^2\phi}{\partial t^2} - \frac{\partial^2\phi}{\partial x^2} = -\frac{\gamma - 1}{c^2}q(x, t) + \xi(x, t). \tag{20.22}$$

The velocity potential is a convenient field quantity because (i) it satisfies the same boundary conditions as the acoustic pressure, and (ii) its gradient gives the acoustic velocity.

The special case of Eq. (20.22), where the noise is absent ($\xi(x, t) = 0$), is a well-known equation in thermoacoustics, where it is called the "acoustic analogy equation". Its derivation is based on the linearised balance equations for mass and momentum. It is also based on the fact that heat addition changes the functional dependence of the density: if heat is added to a fluid, its density becomes a function of pressure *and* specific enthalpy (the heat supplied to unit mass of fluid, see Sect. 7.3 in (Dowling and Ffowcs Williams 1983).

If noise is present, the inhomogeneous wave Eq. (20.22) has two forcing terms: the first one represents forcing by the fluctuating heat release rate; this is coupled to the acoustic field (see Sect. 20.2.2). The second forcing term, $\xi(x, t)$, represents the noise, which is independent of the acoustic field. We treat both sources as acoustically compact and located at the axial position x_q, as described by Eqs. (20.14) and (20.20). We also assume that the initial conditions act only at the point $x_q$and that they are described by

$$\phi(x, 0) = \varphi_0\delta(x - x_q) \quad \text{and} \quad \frac{\partial\phi(x, 0)}{\partial t} = \dot{\varphi}_0\delta(x - x_q), \tag{20.23a,b}$$

where φ_0 and $\dot{\varphi}_0$ have prescribed values (Bigongiari and Heckl 2016).

Given that Eq. (20.22) for ϕ and Eq. (20.1) for Green's function have the same structure, they can be combined by a series of mathematical manipulations, which are described in (Arabi and Heckl 2025). The outcome is the following integral equation,

$$u_q(t) = \left.\frac{\partial \phi}{\partial x}\right|_{x=x_q} = -\frac{\gamma - 1}{c^2}\int_{t'=0}^{t} \left.\frac{\partial G(x, x', t, t')}{\partial x}\right|_{\substack{x = x_q \\ x' = x_q}} q(t')\mathrm{d}t' + \int_{t'=0}^{t} \left.\frac{\partial G(x, x', t, t')}{\partial x}\right|_{\substack{x = x_q \\ x' = x_q}} \xi(t')\mathrm{d}t' - \frac{\varphi_0}{c^2}\left.\frac{\partial^2 G}{\partial x \partial t'}\right|_{\substack{x=x_q \\ x'=x_q \\ t'=0}} + \frac{\dot{\varphi}_0}{c^2}\left.\frac{\partial G}{\partial x}\right|_{\substack{x=x_q \\ x'=x_q \\ t'=0}}. \tag{20.24}$$

Eq. (20.24) is a Volterra-type integral equation for the velocity $u_q(t)$; $G(x, x', t, t')$ is the tailored Green's function given by Eq. (20.6), while $q(t)$ and $\xi(t)$ are the time-dependent parts of $q(x, t)$ and $\xi(x, t)$ (see Eqs. (20.14) and (20.20)) and given by Eqs. (20.16) and (20.21), respectively. Eq. (20.24) has a clear physical interpretation. This can be seen if the time histories of $q(t)$ and $\xi(t)$ are considered as a series of tightly spaced impulses in the interval $t' = 0, \ \ldots \ t$; each of these impulses leads to a response in terms of an acoustic velocity field, and Eq. (20.24) represents the sum of these velocity fields. The last two terms in Eq. (20.24) are a byproduct of the initial conditions.

The integral equation is solved numerically by a straightforward time-stepping approach. The iteration is started by the initial conditions described by φ_0 and $\dot{\varphi}_0$. We put $\dot{\varphi}_0 = 0$ throughout, and φ_0 we choose a value, which gives the desired initial velocity amplitude. As the iteration progresses, the time history $u_q(t)$ will reach a peak value every so often. We regard the most recent peak value as the current amplitude and calculate from it the amplitude-dependent parameters n_0, n_1 and τ in the heat release law (16). Details about the numerical implementation of the iteration can be found in (Arabi and Heckl 2025).

20.4 Predictions for the Acoustic Field Generated by a Single Source

In this section, we study the effects of the flame and the noise separately. We calculate time histories for $u_q(t)$ by solving the integral equation (20.24) and applying the associated equations given in Sect. 20.2. By putting $\beta = 0$ for the noise level, we get results for the flame without noise (see Sect. 20.4.1), and by putting $K = 0$ for the power of the flame, we get results for noise without the flame (see Sect. 20.4.2).

Table 20.1 Parameters describing the setup

Parameter	Symbol	Numerical value	Units
Reflection coefficient at $x = 0$ (upstream end of tube)	R_0	-1	
Reflection coefficient at $x = L$ (downstream end of tube)	R_L	-1	
Transmission coefficient at $x = x_q$ for wave coming from cold side	T_{12}	0.90	
Transmission coefficient at $x = x_q$ for wave coming from hot side	T_{21}	1.10	
Reflection coefficient at $x = x_q$ for wave coming from cold side	R_{12}	-0.10	
Reflection coefficient at $x = x_q$ for wave coming from hot side	R_{21}	0.10	
Upstream temperature	$\overline{T}_1$	304	K
Downstream temperature	$\overline{T}_2$	460	K
Tube length	L	2	m
Mean flow velocity	$\overline{u}$	1	$\mathrm{m\,s^{-1}}$
Specific heat ratio	γ	1.4	
Gas constant for air	R	287	$\mathrm{J\,kg^{-1}K^{-1}}$
Parameters describing nonlinear flame	g_0	1.4	
	g_1	0.3	
	τ_0	5×10^{-3}	s
	τ_2	4.4×10^{-3}	s

A number of parameters will be kept constant, and their numerical values are listed in Table 20.1.

The setup is a tube with ideal open ends. The mean temperature in the tube is *not* uniform, but jumps at x_q from 304 K to 460 K; the corresponding jump in the speed of sound is from 349 $\mathrm{ms^{-1}}$ to 430 $\mathrm{ms^{-1}}$. As a consequence, waves are reflected and transmitted at the jump, with the coefficients listed in Table 20.1. Only the mode $n = 1$ of the thermoacoustic system will be considered.

20.4.1 *Flame Without Noise*

The thermoacoustic system characterised by the parameters in Table 20.1 has been studied previously by Bigongiari and Heckl (2016), and by Arabi and Heckl (2025) for the case without noise. We summarise their predictions here in order to introduce the relevant concepts from nonlinear dynamics. Also, this case will serve as a baseline against which we will compare our new predictions for a noisy combustion system.

The nonlinear behaviour of mode 1 can be described in terms of an amplitude-dependent stability map. This is shown in Fig. 20.5 with the source position x_q in the role of the control parameter and traversing the whole length of the tube from $x_q = 0$ to L. The power of the flame is $K = 3 \times 10^5 \mathrm{W\,s\,kg}^{-1}$. Unstable regions are shaded in grey, while stable regions are white.

The thermoacoustic system is linearly stable in the x_q–ranges [0, 0.075], [0.9, 0.95] and [1.7, 2]m. If x_q is in a fixed position within one of these ranges, and a large enough initial amplitude (inside the grey region above x_q) is imposed on the acoustic field, the ensuing oscillation is nonlinearly unstable: its amplitude will grow until it reaches the upper edge of the grey region, and at that point a limit cycle is established.

Figure 20.6 shows two typical time histories starting from the blue and red points marked in Fig. 20.5; both are for $x_q = 1.8$ m. The first one, shown in Fig 20.6a, starts with the amplitude $A/\overline{u} = 0.2$; this is a stable oscillation. The second one, shown in Fig. 20.6b, starts with the amplitude $A/\overline{u} = 0.3$; this oscillation is unstable at first until the amplitude reaches the value $A/\overline{u} = 1.2$, which is at the upper edge of the grey region above the red dot.

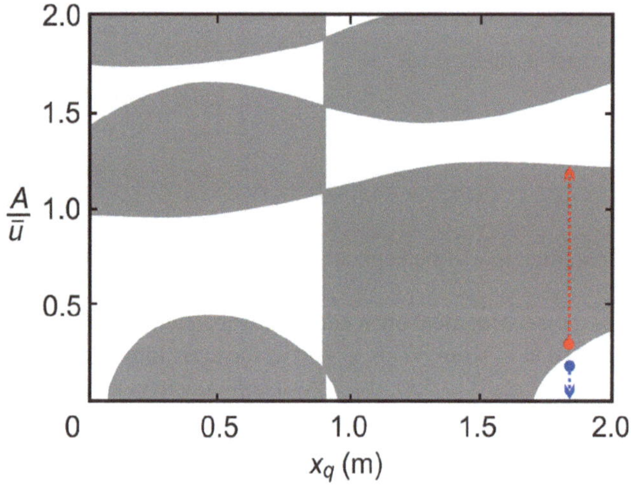

Fig. 20.5 Stability map in the $x_q - A/\overline{u}$ plane, where x_q is the source position, and $A/\overline{u}$ is the velocity amplitude (normalised by the mean velocity $\overline{u}$). White regions indicate stability, while grey regions indicate instability. The blue dot with coordinates $(x_q, A/\overline{u}) = (1, 8\,\mathrm{m}, 0.2)$ is in a stable region; the dotted blue arrow marks its trajectory as time passes. The red dot with coordinates $(x_q, A/\overline{u}) = (1, 8\,\mathrm{m}, 0.3)$ is in an unstable region; the dotted red arrow marks its trajectory as time passes

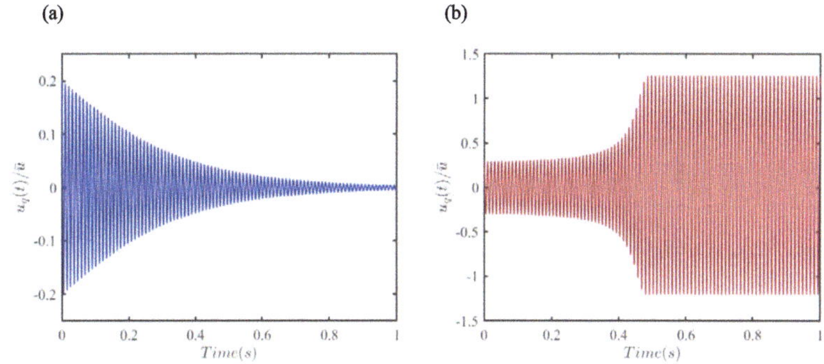

Fig. 20.6 Time histories of the acoustic velocity $u_q(t)/\overline{u}$ for $x_q = 1.8\,\mathrm{m}$. **a** Initial amplitude: $A/\overline{u} = 0.2$; **b** initial amplitude: $A/\overline{u} = 0.3$

20.4.2 Noise Without Flame

We exclude the flame as a source by putting $K = 0$. Our configuration is then simply an acoustic resonator with external forcing by noise. Its resonance frequencies are identical with the eigenfrequencies of the tailored Green's function and can be determined by solving $F(\omega) = 0$, where $F(\omega)$ is given by Eq. (20.7). The result for the first mode is $\omega_1 = 2\pi \times 89\,\mathrm{s}^{-1}$. This is purely real, indicating that mode 1 in our configuration is undamped. If an undamped oscillator is forced with one of its resonance frequencies, it responds with an oscillation whose amplitude grows linearly with time (Boyce and DiPrima 2001, Sect. 3.9). If the forcing is exerted by noise, which contains the resonance frequency amongst many other frequencies, the resulting oscillation is expected to be noise, superimposed by a harmonic oscillation at the resonance frequency with linear amplitude growth. This is indeed the case, as can be seen from Fig. 20.7, which was produced by solving Eq. (20.24).

20.5 Predictions for Noise-Induced Triggering of Thermoacoustic Oscillations

We now consider the complete system, where both flame and noise are present in the tube. In Sect. 20.5.1, we show results to illustrate the influence of broad-band white noise. We contrast these in Sect. 20.5.2 with the influence of narrow-band noise of different bandwidths and centre frequencies.

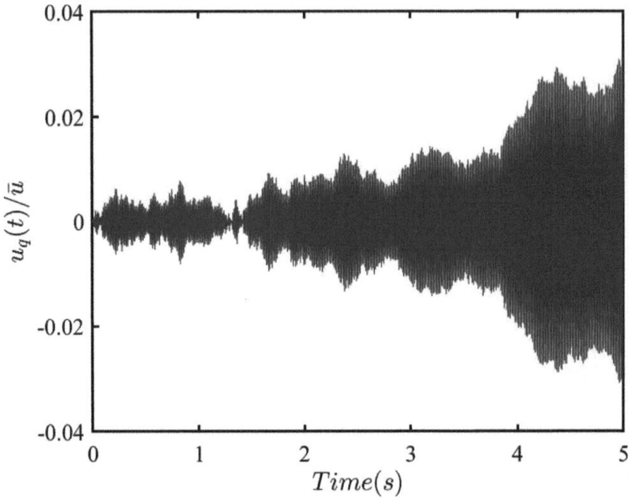

Fig. 20.7 Resonant forcing of mode 1 by white noise with $\beta = 2$

20.5.1 Triggering by Unfiltered White Noise

In this section, we illustrate how white noise (see Fig. 20.4) affects the dynamics of the thermoacoustic system under consideration. We calculate the time history of the acoustic velocity $u_q(t)/\overline{u}$ as described in Sect. 20.3. The point $(x_q, A/\overline{u}) = (1, 8\,\text{m}, 0.2)$ in the stability map is selected; this is the point near the stability boundary, marked in blue in Fig. 20.5. The noise level is varied by gradually increasing β. For low β-values, the system remains stable, i.e. the oscillation amplitude tends to zero as time progresses. Once β reaches the value of 80, triggering is observed.

The stochastic nature of the noise requires us to consider triggering in probabilistic terms. For each β-value, we repeat the time history calculation 20 times. If a transition to limit cycle occurs in 15 or more out of the 20 repeats, we classify it as a triggering event. Figure 20.8 shows a triggering event for $\beta = 80$.

For the first 6 seconds, the time history is dominated by intermittent oscillation bursts; after that, it becomes unstable and soon reaches a limit cycle. The amplitude of that is determined by the upper boundary of the stability region, as was the case for the noiseless example shown in Fig. 20.6b.

20.5.2 Triggering by Narrow-Band White Noise

We now consider *filtered* white noise with three different bandwidths: $\Delta f = 10, 20, 30$ Hz. Also, we vary the centre frequency f_c in steps of 40Hz in the range

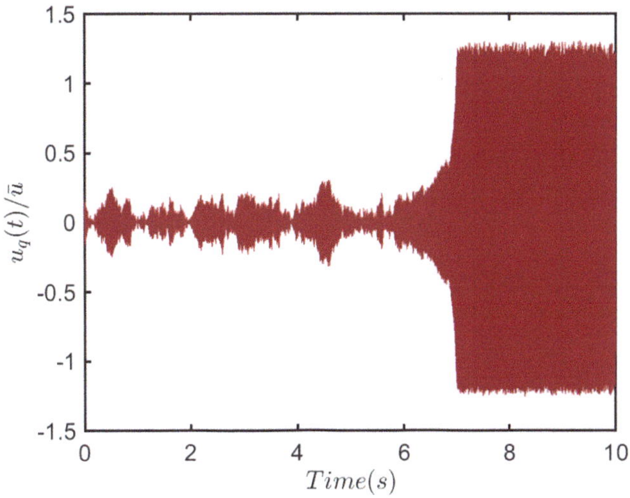

Fig. 20.8 Time history for a case of noise-induced triggering with $\beta = 80$; the starting point $(x_q, A/\overline{u}) = (1, 8\,\text{m}, 0.2)$ is in a stable region of the stability map (blue point in Fig. 20.5)

between $f_c = 20\text{Hz}$ and 300Hz. The power spectral densities for all possible parameter combinations are shown in the three figures below: Fig. 20.9 is for $\Delta f = 10\text{Hz}$; Fig. 20.10 is for $\Delta f = 20\text{Hz}$, and Fig. 20.11 is for $\Delta f = 30\text{Hz}$.

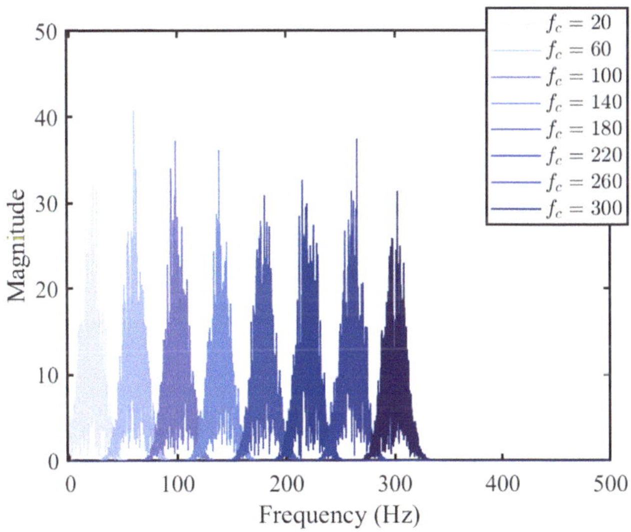

Fig. 20.9 Power spectral density of filtered white noise with bandwidth 10Hz and 8 different centre frequencies

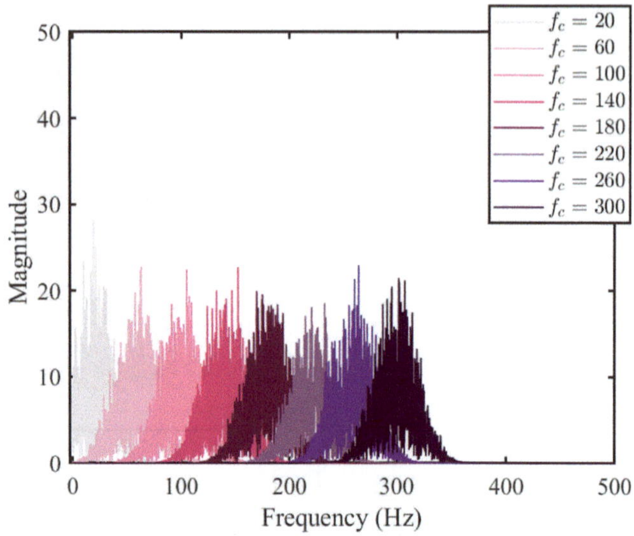

Fig. 20.10 Power spectral density of filtered white noise with bandwidth 20Hz and 8 different centre frequencies

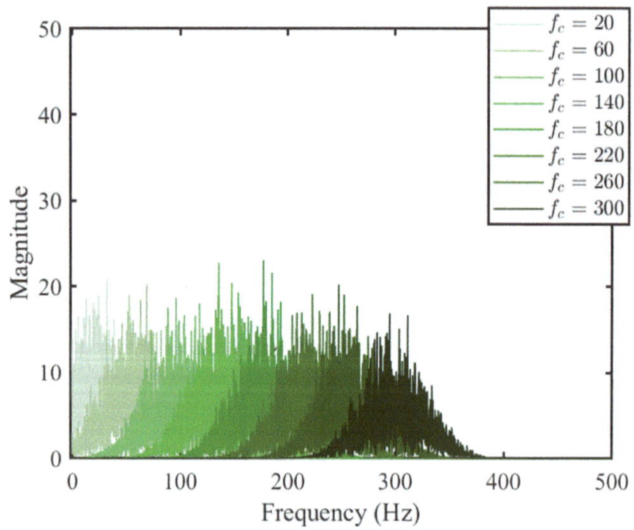

Fig. 20.11 Power spectral density of filtered white noise with bandwidth 30Hz and 8 different centre frequencies

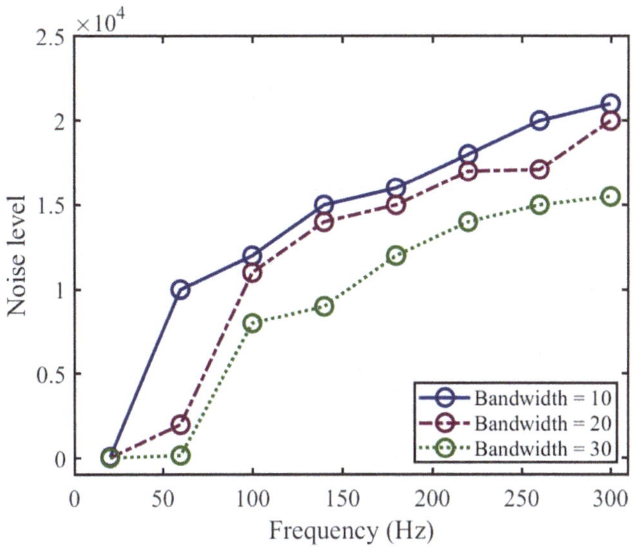

Fig. 20.12 Noise level β required for triggering with narrow-band noise as a function of centre frequency, for $\Delta f = 10, \ 20, \ 30$ Hz

We calculated time histories and identified triggering in the same way as described in Sect. 20.5.1. This gave us the threshold β-values that would lead to triggering. The results are summarised in Fig. 20.12.

It is evident that the centre frequency f_c has a strong influence, and that this influence increases with bandwidth Δf. More specifically, we make the following observations.

- At the lowest centre frequency, $f_c = 20$ Hz, the triggering occurs when $\beta = 100$; this is quite similar to the case of unfiltered white noise (with frequencies evenly distributed in the range [0, 500 Hz]), where we found $\beta = 80$.
- As f_c increases, the β-value required for triggering increases substantially: for $f_c = 300$ Hz and $\Delta f = 10$ Hz, it is $\beta = 21000$.
- For a fixed Δf, the increase with f_c of the required β-value is monotonic. The resonance frequency of the first mode is at 89Hz; however, there is no sign that frequency bands, which include this resonance frequency, might be more effective in that they would trigger at a significantly reduced β-value.

20.6 Conclusions

We have performed a largely analytical study to investigate the intricacies of noise-induced triggering in a basic combustion chamber. The combustion chamber was represented by a tube containing a nonlinear flame and a noise source. The mathematical concept is based on Green's function approach, which features the tailored Green's function, i.e. the impulse response of the combustion chamber. The acoustic field generated by the two sources was calculated from an integral equation, with time as the integration variable. The parameters of interest were noise level, bandwidth and centre frequency. The noise level is crucial in that it needs to be sufficiently high in order to trigger instability. The bandwidth is of moderate importance: an increase in bandwidth requires somewhat lower noise levels for triggering. The most striking observation was made when we varied the centre frequency of the frequency bands. Very low frequencies turned out to be highly effective, but as the frequency is increased, triggering required higher and higher noise levels. For example, the noise level required for a centre frequency of 300Hz is 150 times higher than for 20Hz. We had expected that the frequency band, which contains the resonance frequency of the considered mode, would stimulate triggering more than other bands. However, our results show that this is not the case.

References

Arabi S, Heckl M (2025) Green's function analysis of nonlinear thermoacoustic effects under the influence of noise in a combustion chamber. J Sound Vib 594, paper 118621

Bigongiari A, Heckl MA (2016) A Green's function approach to the rapid prediction of thermoacoustic instabilities in combustors. J Fluid Mech 798:970–996

Boyce WE, DiPrima RC (2001) Elementary differential equations and boundary value problems, 7th edn. John Wiley & Sons, New York

Culick FEC (2006) Unsteady motions in combustion chambers for propulsion systems. Research and Technology Organisation, NATO, RTO-AG-AVT-039

Dowling AP, Ffowcs Williams JE (1983) Sound and sources of sound. Ellis Horwood, Chichester

Heckl MA (1985) Heat sources in acoustic resonators. PhD thesis, University of Cambridge

Heckl M (2015) A new perspective on the flame describing function of a matrix flame. Int J Spray Combust Dyn 7:91–112

Heckl MA, Gopinathan SM, Surendran A (2022) A unified framework for acoustic instabilities based on the tailored Green's function. J Sound Vib 541, paper 117279

Jegadeesan V, Sujith RI (2013) Experimental investigation of noise induced triggering in thermoacoustic systems. Proc Combust Inst 34:3175–3183

Kim KT, Hochgreb S (2012) Measurements of triggering and transient growth in a model lean-premixed gas turbine combustor. Combust Flame 159:1215–1227

Kinsler LE, Frey AR, Coppens AB, Sanders JV (2000) Fundamentals of acoustics. John Wiley & Sons

Lieuwen T (2002) Experimental investigation of limit-cycle oscillations in an unstable gas turbine combustor. J Propuls Power 18:61–67

Lieuwen TC, Yang V (2005) Combustion instabilities in gas turbine engines. American Institute of Aeronautics and Astronautics

Noiray N (2007) Linear and nonlinear analysis of combustion instabilities, application to mulitipoint injection systems and control strategies. PhD thesis, École Centrale Paris, Laboratory EM2C
Pun W, Palm SL, Culick FEC (2003) Combustion dynamics of an acoustically forced flame. Combust Sci Technol 175:499–521
Sujith RI, Pawar SA (2021) Thermoacoustic instability-a complex systems perspective. Springer International Publishing
Waugh IC, Juniper MP (2011) Triggering in a thermoacoustic system with stochastic noise. Int J Spray Combust Dyn 3:225–241
Waugh I, Geuß M, Juniper M (2011) Triggering, bypass transition and the effect of noise on a linearly stable thermoacoustic system. Proc Combust Inst 33:2945–2952

Chapter 21
A Model for Tidal Effects on a Planet—The Quantization of Angular Velocity of Rotation

Ingo Müller

Abstract The planet Mercury rotates 3 times about its axis while it revolves 2 times about the Sun Internet source 2025, so that Mercury offers *opposite* sides to the Sun after each consecutive revolution. This was unexpected by astronomers. Indeed, they thought that a planet as close to the sun as Mercury always ought to offer the *same* side to the Sun because of tidal forces; so that 1 rotation of the planet occurs during 1 revolution. Bouncing off radar waves from Mercury in 1965 taught them the truth, see Asimov (1972). I shall characterize the two possibilities as the 1.5/1-mode and the 1/1-mode respectively.

This paper introduces a model in which the tidal deformation of a planet is simulated by visco-elastic springs subject to the gravitational field of the Sun. It turns out that both modes may occur. The model can be shown to exhibit other modes upon a change of its visco-elastic properties: $\frac{n}{2}/1$ with n being any integer greater than 1. Thus the ratio of the mean angular velocities of rotation and revolution might be said to be "quantized."

21.1 Introduction

The planet Mercury revolves about the Sun in an ellipse with the fairly great excentricity $\varepsilon = 0.21$ meaning that the short axis of the ellipse and the long one have the ratio $\frac{b}{a} = \sqrt{1-\varepsilon^2}$. It is convenient to represent the ellipse in polar coordinates (r, φ) so that

$$r = \frac{a(1-\varepsilon^2)}{1-\varepsilon\cos\varphi} \tag{21.1}$$

I. Müller (✉)
TU Berlin, Berlin, Germany
e-mail: ingo.mueller@alumni.tu-berlin.de

J. Fuhrmann et al. (eds.), *Advances in Continuum Physics*, Advanced Structured Materials 238, https://doi.org/10.1007/978-3-031-93918-1_21

holds, where r is the distance of planet and Sun and a is the long half axis of the ellipse. φ is the angle between the distance vector and the long axis. The angular velocity $\frac{d\varphi}{dt}$ follows from Newton's law of gravitation

$$\frac{d\varphi}{dt} = \sqrt{a\gamma M}\sqrt{1-\varepsilon^2}\frac{1}{r^2}, \quad \text{or by (21.1)}$$

$$\frac{d\varphi}{dt} = \sqrt{\frac{\gamma M}{a^3}}\frac{1}{\sqrt{1-\varepsilon^2}^3}(1-\varepsilon\cos\varphi)^2, \tag{21.2}$$

where γ is the gravitational constant and M is the mass of the Sun.

I introduce dimensionless variables

$$\rho = \frac{r}{a} \quad \text{and} \quad \tau = \sqrt{\frac{\gamma M}{a^3}}t \tag{21.3}$$

and write (21.2) in the form

$$\frac{d\varphi}{d\tau} = \frac{1}{\sqrt{1-\varepsilon^2}^3}(1-\varepsilon\cos\varphi)^2 \tag{21.4}$$

The solution $\varphi(\tau)$ of this differential equation for $\varphi(0) = 0$ is shown in Fig. 21.1, obtained and plotted by Mathematica, like all of the following graphs as well.

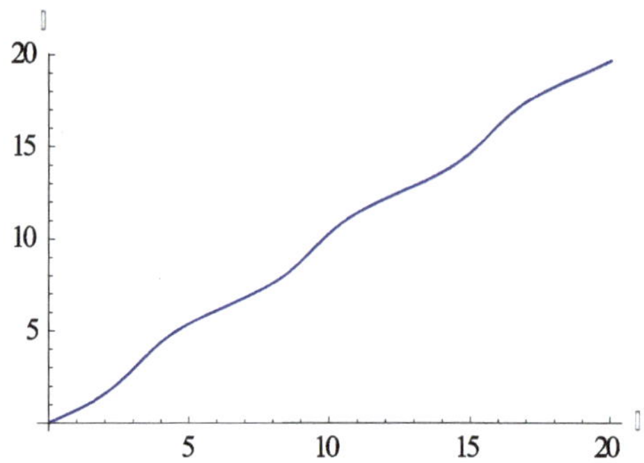

Fig. 21.1 The angle $\varphi(\tau)$ wriggles around the bi-sector of the (φ, τ)-plane as the planet (with $\varepsilon = 0.21$) revolves around the Sun. $\tau = 2\pi$ is the dimensionless time for one revolution

All of this is well known and represented in many books on mechanics, e.g. see (Joos 1956). It is repeated in the present introduction in order to remind the reader and to introduce notation.

21.2 A Model for the Planet

We may model the planet as in Fig. 21.2: By visco-elastic springs of length L on whose end-points half the planetary masses $\frac{m}{2}$ are attached. We introduce notation as shown in the figure and write equations of motion as

$$\begin{aligned} \frac{m}{2}\frac{d^2 \overset{1}{x}_i}{dt^2} &= -D\frac{dL}{dt}\frac{(\overset{1}{x}_i - x_i)}{L} - \lambda(\overset{1}{x}_i - x_i) - \gamma M \frac{m}{2}\frac{1}{\overset{1}{r}{}^3}\overset{1}{x}_i \\ \frac{m}{2}\frac{d^2 \overset{2}{x}_i}{dt^2} &= -D\frac{dL}{dt}\frac{(\overset{2}{x}_i - x_i)}{L} - \lambda(\overset{2}{x}_i - x_i) - \gamma M \frac{m}{2}\frac{1}{\overset{2}{r}{}^3}\overset{2}{x}_i \end{aligned} \qquad (21.5)$$

Given Fig. 21.2 the meaning of these equations is self-explanatory. Let us mention, however, that the terms with λ reflect the elastic forces in the springs, while the terms with D reflect friction forces that tend to slow down the elongations of the springs dissipatively. The gravitational terms in (21.5) represent the tidal forces on the masses $\frac{m}{2}$. $\overset{\alpha}{l}_i$ are the distance vectors of the masses from the point S of attachment of the springs to each other.

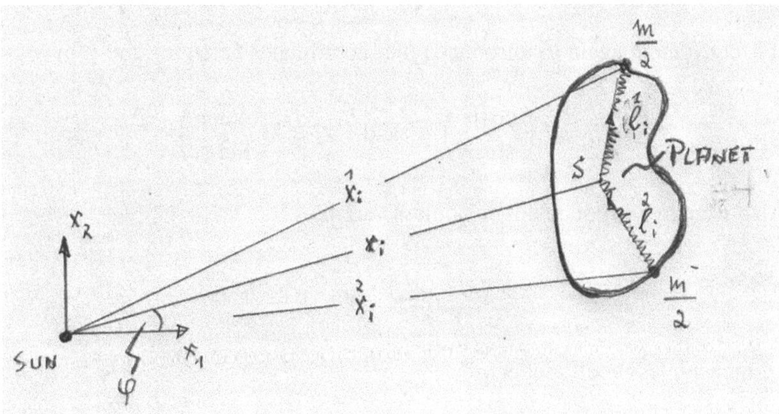

Fig. 21.2 A model of the planet and the configuration of planet and Sun

Given the assumed planar character of the problem the Eq. (21.5) are four in number and they allow for many types of deformation; like folding the two halves and giving them different directions. We may reduce these degrees of freedom by assuming that $\overset{1}{l}_i = \overset{2}{l}_i = l_i$ holds. In that case S may be considered as the center of gravity of the planet. The angle between l_i and the x_1-axis is denoted by α, just as φ is the angle between the distance vector x_i of Sun to planet and the x_1-axis. In this manner the four Eq. (21.5) are reduced to two. As such we choose the difference between (5)1 and (5)2, namely

$$m\frac{d^2 l_i}{dt^2} = -2D\frac{dL}{dt}\frac{l_i}{L} - 2\lambda l_i - \gamma M m\frac{1}{r^3}(l_i - 3\cos(\alpha-\varphi)\frac{L}{r}x_i + \left(\frac{15}{2}\cos^2(\alpha-\varphi) - \frac{3}{2}\right)\frac{L^2}{r^2}l_i \tag{21.6}$$

The gravitational term in (21.6) is approximate, because we have set

$$\frac{1}{\frac{1}{r^3}} = \frac{1}{\sqrt{(x_i+l_i)(x_i+l_i)}^3} \approx \frac{1}{r^3}\left(1 - 3\cos(\alpha-\varphi)\frac{L}{r} + \left(\frac{15}{2}\cos^2(\alpha-\varphi) - \frac{3}{2}\right)\frac{L^2}{r^2}\right) \tag{21.7}$$

assuming that $L << r$. Similar for $\frac{1}{\frac{2}{r^3}}$.

Alternative to (21.6) are the balance equations of energy and spin which result from (21.6) by scalar multiplication with $\dot{l}_i$ and vectorial multiplication with l_i.[1]

$$\begin{aligned} \tfrac{m}{2}\tfrac{d}{dt}(\dot{l}_i\dot{l}_i) &= -2D\dot{L}\tfrac{l_i}{L}\dot{l}_i - 2\lambda l_i\dot{l}_i - \gamma M m\tfrac{1}{r^3}(l_i\dot{l}_i - 3\cos(\alpha-\varphi)\tfrac{L}{r}x_i\dot{l}_i \\ &\quad + (\tfrac{15}{2}\cos^2(\alpha-\varphi) - \tfrac{3}{2})\tfrac{1}{\rho^2}\tfrac{L^2}{a^2}l_i\dot{l}_i), \\ m\tfrac{d}{dt}(\varepsilon_{3ki}l_k\dot{l}_i) &= +\gamma M m\tfrac{1}{r^3}\varepsilon_{3ki}3\cos(\alpha-\varphi)\tfrac{L}{r}l_k x_i \end{aligned} \tag{21.8}$$

It is convenient again to introduce polar coordinates $L,\ \alpha,\ r,$ and φ by

$$l_i = L\begin{pmatrix}\cos\alpha \\ \sin\alpha\end{pmatrix}, \quad \text{and} \quad x_i = r\begin{pmatrix}\cos\varphi \\ \sin\varphi\end{pmatrix} \tag{21.9}$$

Also we may introduce dimensionless variables

$$\tau = \sqrt{\frac{\gamma M}{a^3}}t \quad \text{and} \quad \rho = \frac{r}{a} \tag{21.10}$$

and dimensionless coefficients

$$l = \frac{4\frac{\lambda}{m}}{\frac{\gamma M}{a^3}} \quad \text{and} \quad d = \frac{4\frac{D}{m}}{\sqrt{\frac{\gamma M}{a^3}}} \tag{21.11}$$

for the elasticity and dissipative coefficient in the springs. In this manner we obtain after considerable though straightforward calculation.

[1] For convenience in notation we introduce a dot for the time derivative. It has the same meaning as $\frac{d}{dt}$. Also recall that the model is two dimensional in the plane perpendicular to the x_3 axis so that vectorial multiplication of (21.6) amounts to multiplication by $\varepsilon_{3ki}l_k$.

$$\begin{aligned}
&\frac{d}{d\tau}\left\{\left(\frac{dL}{d\tau}\right)^2 + L^2\left(\frac{d\alpha}{d\tau}\right)^2\right\} = -d\left(\frac{dL}{d\tau}\right)^2 - l\cdot L\frac{dL}{d\tau} + \\
&+\frac{1}{\rho^3}\Big\{L\frac{dL}{d\tau} + 3L\frac{dL}{d\tau}\cos[2(\varphi-\alpha)] + 3L^2\frac{d\alpha}{d\tau}\sin[2(\varphi-\alpha)] \\
&\qquad +\underline{\underline{\left(\frac{15}{2}\cos^2(\alpha-\varphi) - \frac{3}{2}\right)\frac{1}{\rho^2}\frac{L^3\frac{dL}{d\tau}}{a^2}}}\Big\} \\
&\frac{d}{d\tau}\left(L^2\frac{d\alpha}{d\tau}\right) = +\frac{3}{2}\frac{1}{\rho^3}L^2\sin[2(\varphi-\alpha)]
\end{aligned} \tag{21.12}$$

The underlined term represents the second order in the approximation (21.7). We shall neglect that term because $\frac{L}{a}$ is very small in the case of any planet.ρ stands for $\frac{1-\varepsilon^2}{1-\varepsilon\cos\varphi}$ just like in (21.1). Note that—without the underlined 2nd order term—there is no characteristic length in the model to compare L with. In subsequent numerical calculations we consider $\Lambda = L/L(0)$ as the non-dimensional length, made non dimensional by $L(0)$, the initial value of L.

Of course $L = 0$ is the state where the springs are unloaded and their extension is zero, a case which we trust will never occur during a revolution because the tidal forces keep the springs under a tensile load.

The Eq. (21.12) represent equations for the length $\Lambda(\tau)$ and the orientation $\alpha(\tau)$ of the model planet.

21.3 The Modes $\frac{n}{2}/1$ for Integers $2 \leq n \leq 6$

We choose initial conditions

$$\Lambda(0) = 1, \quad \left.\frac{d\Lambda}{d\tau}\right|_0 = 0.01, \quad \alpha(0) = 0, \quad \left.\frac{d\alpha}{d\tau}\right|_0 = 1. \tag{21.13}$$

The initial condition $(13)_1$ is dictated by the definition of Λ. Also we choose pairs (l, d) with decreasing values of the elastic constant l and a damping coefficient d. The latter is chosen so as to give stationary periodic solutions $\Lambda(\tau)$[2] after an initial adjustment to both: the initial conditions (21.13) and the periodic disturbance $\varphi(\tau)$ given by Fig. 21.1. d increase as l drops.

Solutions are shown in Figs. 21.3 through 21.7. The graphs for $\Lambda(\tau)$ are qualitative, because L was made dimensionless with $L(0)$. For each pair (l,d) we list the mean elongation $\overline{\Lambda}$ and the maximal width $\Delta\Lambda$ of the oscillation in Table 21.1, both in the eventual stationary periodic phase.

[2] Meaning that the periodicity of the solution is stationary.

Table 21.1 Properties of $\Lambda(\tau)$ to be read off the Figs. 21.3 through 21.7

l	20	15	10	8.1	6
d	1.360	2.75	6.35	5.8	11.89
$\overline{\Lambda}$	0.435	0.673	0.861	1.020	1.052
$\Delta\Lambda$	0.269	0.484	0.781	0.690	0.770

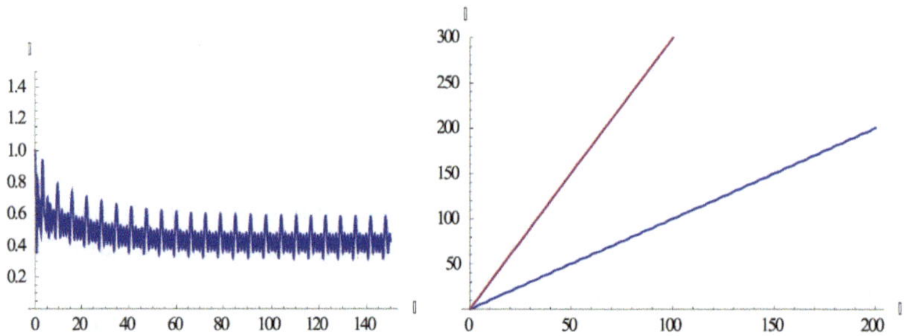

Fig. 21.3 $\Lambda(\tau)$ and $\alpha(\tau)$ for $l = 20$ and $d = 1.36$ and $\varepsilon = 0.21.\Rightarrow \frac{\alpha}{\varphi} \approx 3$

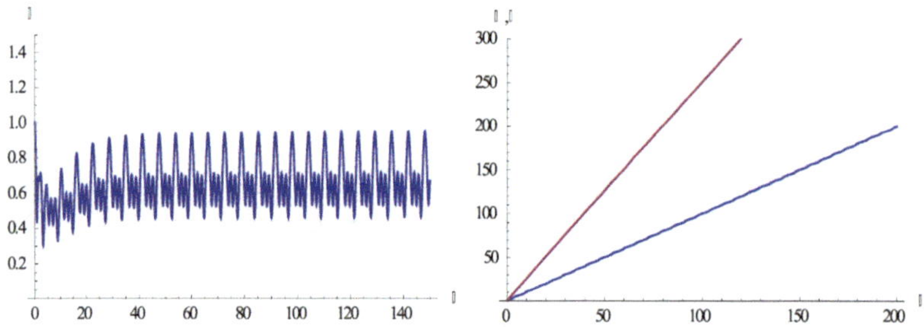

Fig. 21.4 $\Lambda(\tau)$ and $\alpha(\tau)$ for $l = 15$ and $d = 2.75$ and $\varepsilon = 0.21.\Rightarrow \frac{\alpha}{\varphi} \approx 2.5$

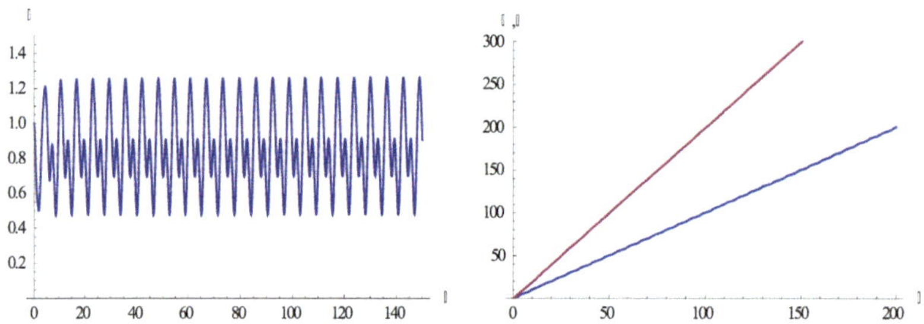

Fig. 21.5 $\Lambda(\tau)$ and $\alpha(\tau)$ for $l = 10$ and $d = 6.35$ and $\varepsilon = 0.21. \Rightarrow \frac{\alpha}{\varphi} \approx 2$

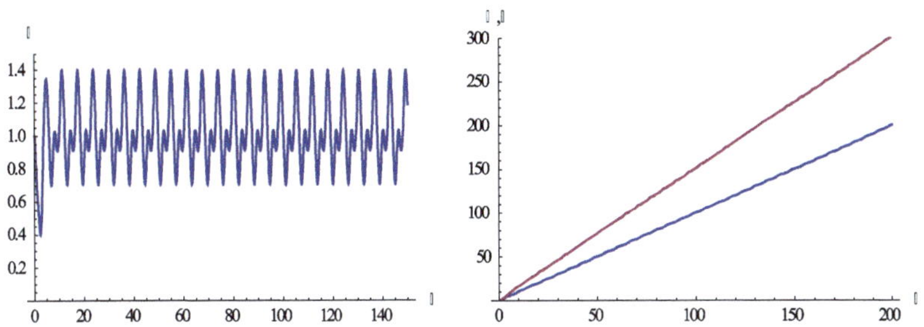

Fig. 21.6 $\Lambda(\tau)$ and $\alpha(\tau)$ for $l = 8.1$ and $d = 5.8$ and $\varepsilon = 0.21. \Rightarrow \frac{\alpha}{\varphi} \approx 1.5$

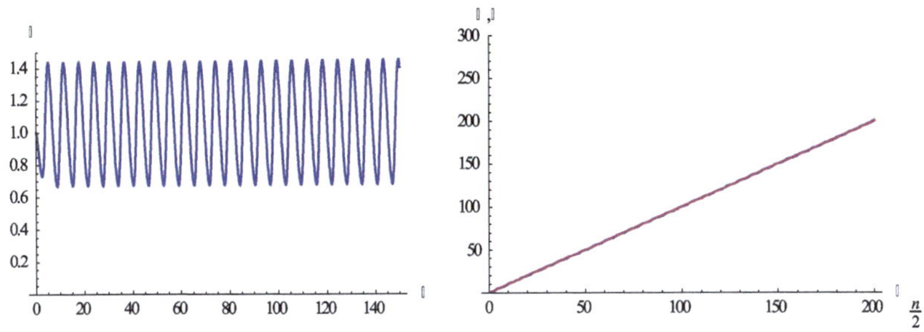

Fig. 21.7 $\Lambda(\tau)$ and $\alpha(\tau)$ for $l = 6$ and $d = 11.89$ and $\varepsilon = 0.21. \Rightarrow \frac{\alpha}{\varphi} \approx 1$

We observe that there is a tendency of $\overline{\Lambda}$ to grow for softer springs and that $\Delta\Lambda$–grossly–also grows for increasing values of the dissipative coefficient d.

What is perfect about the Figs. 21.3 through 21.7 is the observation that the ratio $\frac{\dot{\alpha}}{\dot{\varphi}}$ of angular velocities seems to come down in steps of 0.5 as l increases.[3] It was this phenomenon–or the special case of the possible modes 1.5/1 and 1/1 for Mercury–which has motivated this work. The model simulates this feature.

In Figs. 21.3 through 21.7 we observe an eventual adjustment to a stationary periodic $\Lambda(\tau)$ and–above all–a "quantized" quotient of angles $\frac{\alpha(\tau)}{\varphi(\tau)} \approx \frac{n}{2}$ with $n =$ 6,5 ,4, 3, 2. The transition between these different modes occus quite suddenly. In the next chapter we shall see how and why this behaviour comes about.

21.4 The Modes $\frac{n}{2}$/1 (Again) for Integers N = 6,5,4,3,2

The calculations of Sect. 21.3 have offered no hint as to why $\frac{\alpha}{\varphi}$ is restricted to integers and half integers, i.e. why it is "quantized" by $\frac{n}{2}$. Therefore we appraoch the problem again starting from the original Eq. (21.6). We eliminate $L(\tau)$ and $\alpha(\tau)$ from those

[3] Of course, both α and φ are different "wriggly" graphs. Therefore their ratio is only approximately given by half integers.

by replacing them with $l_1(\tau)$ and $l_2(\tau)$ as the new dependent variables.[4] Thus we obtain the equations

$$\begin{aligned}\frac{d^2 l_1}{d\tau^2} &= -\frac{d}{2}\left(\frac{l_1}{{l_1}^2+{l_2}^2}\frac{dl_1}{d\tau} + \frac{l_2}{{l_1}^2+{l_2}^2}\frac{dl_2}{d\tau}\right)l_1 - \frac{l}{2}l_1 \\ &\quad -\frac{1}{\rho^3}\left(l_1 - 3l_1\cos^2\varphi - 3l_2\sin\varphi\cos\varphi\right) \\ \frac{d^2 l_2}{d\tau^2} &= -\frac{d}{2}\left(\frac{l_1}{{l_1}^2+{l_2}^2}\frac{dl_1}{d\tau} + \frac{l_2}{{l_1}^2+{l_2}^2}\frac{dl_2}{d\tau}\right)l_2 - \frac{l}{2}l_2 \\ &\quad -\frac{1}{\rho^3}\left(l_2 - 3l_1\cos\varphi\sin\varphi - 3l_2\sin^2\varphi\right)\end{aligned} \tag{21.14}$$

We choose initial conditions

$$\lambda_1(0) = 1, \quad \left.\frac{d\lambda_1}{d\tau}\right|_0 = 0, \ \lambda_2(0) = 0, \quad \left.\frac{d\lambda_2}{d\tau}\right|_0 = 0.1 \tag{21.15}$$

and values for (l,d) as in Figs. 21.3 through 21.7. Since the Eq. (21.14) are equivalent to the Eq. (21.12), it is clear that the solutions must be stationary periodic for those parameters and for large values of τ. We plot $\lambda_1(\tau)$ and $\lambda_2(\tau)$ during the revolution between $\tau = 99\pi$ and $\tau = 101\pi$ and observe that–for this rather large time–the two functions are indeed periodic, meaning that $\lambda_1(99\pi)$ and $\lambda_2(99\pi)$ are equal to $\lambda_1(101\pi)$ and $\lambda_2(101\pi)$. For $l = 20$–the case of Fig. 21.3–there are six zero points of $\lambda_1(\tau)$ and $\lambda_2(\tau)$ during a revolution, meaning that there are three rotations in a revolution. In the bottom part of Fig. 21.8 we have plotted the three rotations pertaining to one revolution.

[4] Once again, l_i are made dimensionless by the initial value $L(0)$, so that $\lambda_i = l_i/L(0)$ are the dimensionless variables.

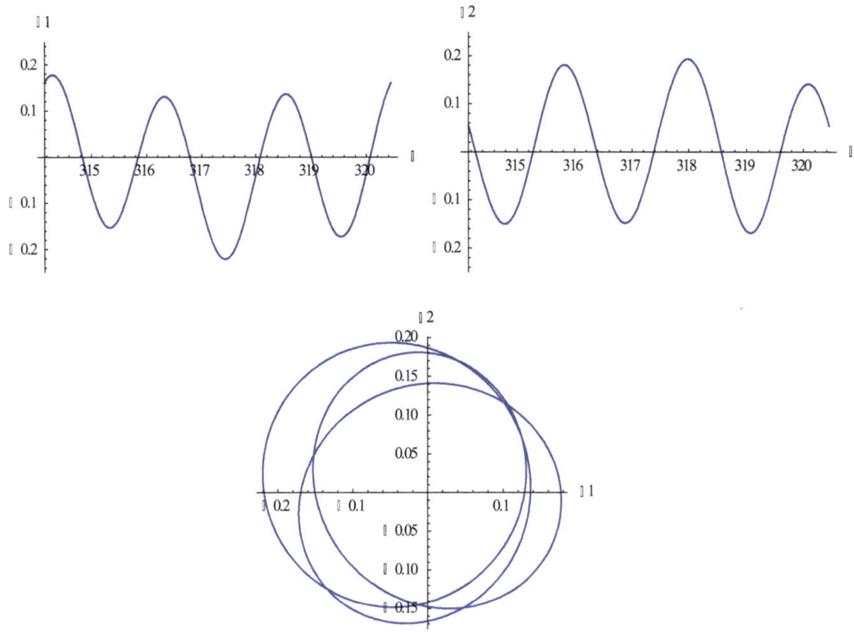

Fig. 21.8 Top: $\lambda_1(\tau)$ and $\lambda_2(\tau)$ during a revolution for $l = 20$, $d = 1.36$. And $\varepsilon = 0.21$ Bottom: three rotations during a revolution in $99\,\pi < \tau < 101\,\pi$

The graphs of $\lambda_1(\tau)$ and $\lambda_2(\tau)$ change when we choose (l,d) different from (20,1.35) and so do their zero points. Those zero points are not spaced equally. However, there must be either an even number of them or an odd number of them. In the latter case we obtain an integer number of rotations **plus** half a rotation. Let us confirm that for the case $n = 5$. According to Fig. 21.4 this is the case $(l,d) = (15, 2.75)$. We obtain the graphs of Fig. 21.9 with 2.5 rotations.

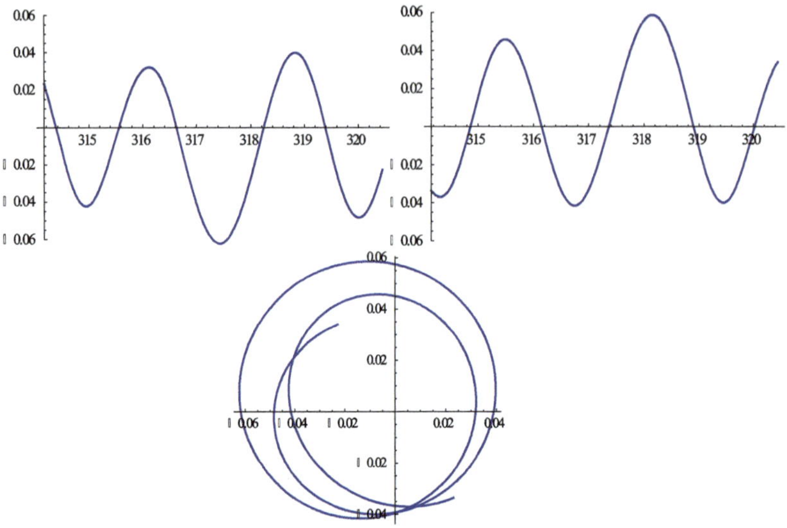

Fig. 21.9 Top: $\lambda_1(\tau)$ and $\lambda_2(\tau)$ during a revolution for $l = 15$, $d = 2.75$. And $\varepsilon = 0.21$ Bottom: two and a half rotations during a revolution in $99\,\pi < \tau < 101\,\pi$

We continue with the next case $n = 4$, the case of two rotations in arevolution. Everything fits as described above, see Fig. 21.10.

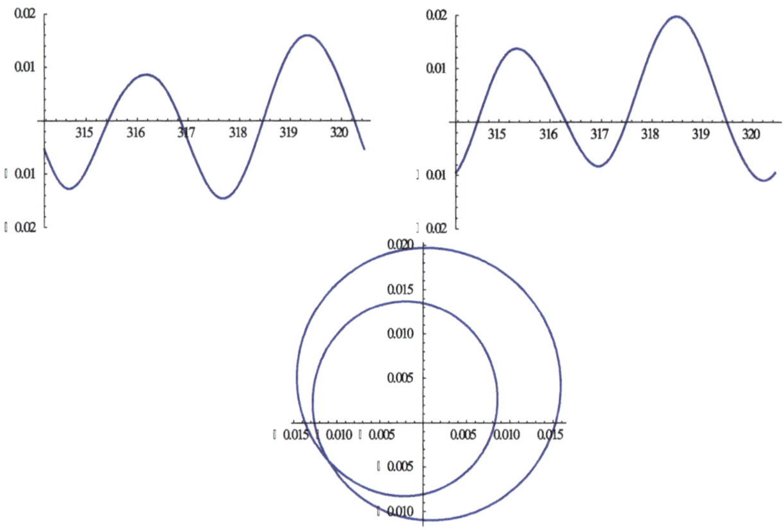

Fig. 21.10 Top: $\lambda_1(\tau)$ and $\lambda_2(\tau)$ during a revolution for $l = 10$, $d = 6.35$. And $\varepsilon = 0.21$ Bottom: two rotations during a revolution in $99\,\pi < \tau < 101\,\pi$

Next is the case $n = 3$. Here we expect and get three zero points of $\lambda_1(\tau)$ and $\lambda_2(\tau)$ and accordingly 1.5 rotations. Qualitatively this is the case of Mercury, see Fig. 21.11.

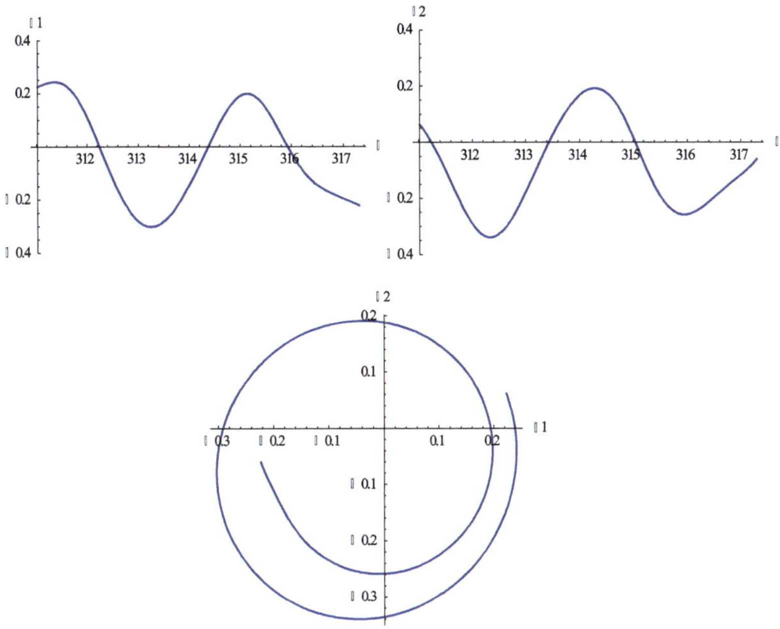

Fig. 21.11 Top:$\lambda_1(\tau)$ and $\lambda_2(\tau)$ during a revolution for $l = 8.1$, $d = 5.8$. And $\varepsilon = 0.21$ Bottom: One and a half rotations during a revolution in $99\,\pi < \tau < 101\,\pi$

There remains the case for $n = 2$–exhibited in Fig. 21.12–with two zero points of $\lambda_1(\tau)$ and $\lambda_2(\tau)$ and accordingly one rotation. This case comes closest to a tidal lock, because it represents a single rotation. But it is not really a tidal lock, because there is a fairly large excentricity, i.e. $\varepsilon = 0.21$.

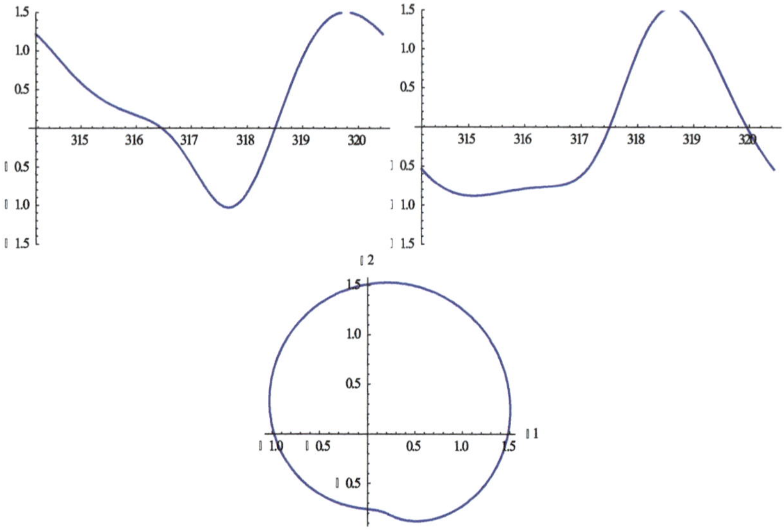

Fig. 21.12 Top:$\lambda_1(\tau)$ and $\lambda_2(\tau)$ during a revolution for $l = 6$, $d = 11.89$ and $\varepsilon = 0.21$. Bottom: one rotation during a revolution in $99\,\pi < \tau < 101\,\pi$

21.5 Dissipative Power and Dissipated Energy

The damping terms in (21.5), (21.6) and (21.8)–the ones with D or d in (21.12), (21.14)–give rise to dissipative power or heating. The dissipative power may be calculated as the scalar product of the dissipative force $2D\dot{L}\frac{l_i}{L}$ and the velocity $\dot{L}\frac{l_i}{L}$.

Thus the dissipative power or heating is given by.[5]

$\dot{Q} = 2D\dot{L}^2$ or non-dimensionally with $\widehat{Q} \equiv 2\frac{\dot{Q}}{m\frac{\gamma M}{a^3}L^2(0)}$

$$\widehat{Q} = d \cdot \left(\frac{d\Lambda}{d\tau}\right)^2. \tag{21.16}$$

The power is plotted in Fig. 21.13 during one revolution for the two modes corresponding to $(l, d) = (20, 1.36)$ (three rotations) and $(l, d) = (6, 11.89)$ (one rotation), both for $\varepsilon = 0.21$. Note that the dissipative power is greater by two magnitudes in the latter case.

[5] $\dot{Q}$ is **not** generally a time derivative. But it represents the canonical notation of heating in thermodynamics. Also recall that L is made dimensionless by $L(0)$.

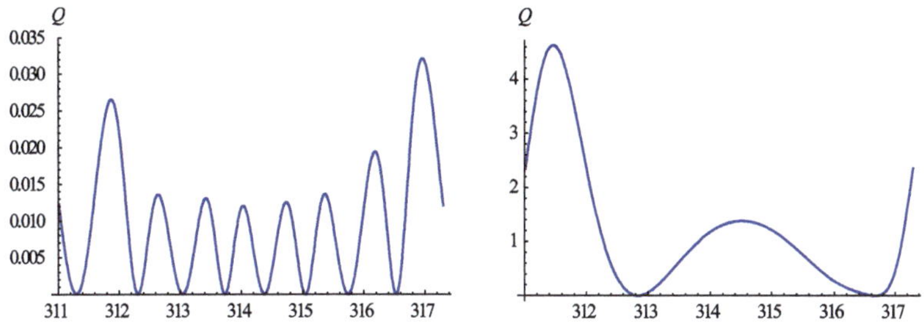

Fig. 21.13 Dissipative power $\widehat{Q}$ for $(l, d) = (20, 1.36)$ and $(l, d) = (6, 11.89)$. And $\varepsilon = 0.21$

Incidentally we may note that, of course, the heating does not lead to a progressive temperature rise in the planet as it might do in a visco-elastic spring in the laboratory. Rather it is radiated off into the universe.

21.6 About Excentricity and the "Tidal Lock"

The visco-elastic properties of the planet are characterized by the pair (l, d) and they represent the elastic property of the planetary substance and its dissipation. It is thus clear that just *one* pair (l, d) characterizes a given planet in its tidal rotation. The values of both coefficients are essential to determine the eventual mean elongation $\overline{L}$ of the springs, e.g. $\overline{L} = 1.052$ for $(l, d) = (6, 11.89)$ according to Table 21.1.

The rotations of all planets in Figs, 21.3 through 21.7 are dissipative in various degrees, e.g. see Fig. 21.13. In the real world–rather than in the world of our model–the dissipated energy during a revolution will tend to reduce the excentricity of the planetary orbit. And eventually it will lead to a circular orbit when the planet is locked rigidly to the Sun; it is then in a state of "dynamic equilibrium" the tidal lock. Obviously it is then in the 1/1- mode. And we might be interested to investigate the process which the planet–initially in the 3/1-mode (say)–undergoes in the transition. This cannot be done with the present model, because itdoes not provide for the interaction of the rotational motion and the revolutionary one. That question is left for a future research; it might well be answered by a similar model to the present one.

However, let us assume that the transition to the tidal lock has occured. We can easily determine the elastic coefficient l for which this happens: Indeed, the centrifugal force $m\dot{\alpha}^2 L$ must then be equal to the sum of the gravitational force, cf. (21.12)

$$-\gamma M m \frac{1}{r^3}(L - 3L - 6\frac{L^2}{a^2}L)$$

and the elastic force $-2\lambda L$. Hence follows

$$\left(\frac{d\alpha}{d\tau}\right)^2 = \frac{1}{\rho^3}\left(2 - 6\frac{L^2}{a^2}\right) - \frac{l}{2}. \tag{21.17}$$

Knowing that ρ and $\left(\frac{d\alpha}{d\tau}\right)^2$ must be equal to 1 in that final state and neglecting $\frac{L^2}{a^2}$ as small, we conclude that l must be equal to 2.

Incidentally, that argument will determine the orbital radius of a planet in the tidal lock, because by (21.11) we have

$$a_{TL} = \left(\frac{\gamma M}{2}\frac{m}{\lambda}\right)^{1/3} \tag{21.18}$$

So, if $l = 6$,–value that we have previously assumed to draw Figs. 21.7 and 21.12–, this means that $a = 1.8a_{TL}$. Such a planet will have to decrease its orbital half-axis by a factor of 1.8 while it changes into tidal lock.

21.7 An Estimate for *l* Appropriate for Mercury

All arguments sofar have a qualitative flavour and, in particular, the various values for the pair (l, d) are arbitrarily chosen. They were chosen so as to provide solutions of the model equations which are stationary periodic.

There are widely different pairs of (l, d) which do have stationary periodic solutions, for instance the pair (100, 0.5). A priori we might therefore be tempted to investigate the neighbourhood of that pair for our investigation.[6] It is therefore expedient to estimate at the outset probable values (l, d) that might relate to Mercury.

Here is a way in which this may be done,–at least for a realistic value of l, the elastic coefficient: Let us assume that the model planet has the same elastic properties as our moon which is in an approximate tidal lock with the earth. Therefore we have, by (21.11)

$$\lambda_{Moon} = \frac{1}{2}\frac{\gamma M_{Earth}}{a_{Moon}^3} m_{Moon}$$

We know the values.

$$a_{Moon} = 3.8 \cdot 10^8 m$$

$$m_{Moon} = 7.3 \cdot 10^{22} kg$$

$$M_{Earth} = 6 \cdot 10^{24} kg \text{ hence } \lambda_{Moon} = 2.66 \cdot 10^{11} \frac{N}{m}$$

[6] It is true that for this pair the model is in the 7/1-mode. Bur this is a secondary concern.

$$\gamma = 6.67 \cdot 10^{-11} \frac{m^3}{s^2 kg}$$

If Mercury has the same value of λ, it follows from (21.11) and from the data

$$a_{Mercury} = 7 \cdot 10^{10} m$$
$$m_{Mercury} = 3 \cdot 10^{23} kg \text{ that } l = 9.1$$
$$M_{Sun} = 2 \cdot 10^{30} kg$$

We recall from Figs. 21.6 and 21.11 that we obtained the 1.5/1-mode of the model for the pair of coefficients $(l,d) = (8.1, 5.8)$. This is not a bad fit with Mercury as far as the value l is concerned.

No such simple argument can be had for the coefficient d which governs dissipation. The value of that coefficient will determine the transition of the planet into the eventual tidal lock; and the model cannot describe that transition.

21.8 Conclusion

The main result of this investigation is the "quantization"–in the mean–of the ratio of angular velocities $\dot{\alpha}$ and $\dot{\varphi}$ by $\frac{n}{2}$. This is due to the natural demand that in the expected stationary periodic state of a planet one periodic motion, the rotation, fit into another one, the revolution. Such a phenomenon is known in quantum mechanics for atoms; it came as a surprise, however, to astronomers who observed the special case $n = 3$ for Mercury.

We wonder whether the phenomenon can be observed in other applications of physics or, perhaps, in biology or sociology.

References

Asimov I (1972) The dance of the sun. In: The solar system and back. Avon Books, NewYork, p 31 ff

Internet source (2025) https://www.sendenberg-schule.de/merkur

Joos G (1956) Lehrbuch der theoretischen Physik, 8th edn. Akademische Verlagsgesellschaft Leipzig, pp 78–80

Chapter 22
Mass Transport in Fokker–Planck Equations with General Potentials

B. Niethammer and J. H. Schröders

To Wolfgang and Michael for all the inspiring discussions.

Abstract We consider Fokker–Planck equations with general nonconvex potentials that have several local minima. In the limit of vanishing diffusion, matched asymptotic expansions predict simple reaction–diffusion dynamics for the masses concentrated in the local minima. We present here a rigorous derivation based on the elementary approach developed in Herrmann and Niethammer (2020) for the case of tilted periodic potentials. The proof relies on suitably defined substitute masses and error bounds that make use of an energy-dissipation relation.

Keywords Fokker–Planck equation · Potential with several minima · Model reduction for multiscale dynamical systems · Asymptotic analysis of singular limits

MSC (2010): 35B2 · 35B40 · 35Q84

22.1 Introduction

We consider the Fokker–Planck equation

$$\tau(\partial_t - \Delta_x)\varrho(t, x, p) = \partial_p(\varepsilon^2\partial_p\varrho(t, x, p) + H'(p)\varrho(t, x, p)), \tag{22.1}$$

for $p \in \mathbb{R}$, $x \in \mathbb{R}^d$ and small parameters τ and ε. Since Eq. (22.1) preserves the mass we impose, for all times t, the normalization

B. Niethammer (✉) · J. H. Schröders
Institut für Angewandte Mathematik, Universität Bonn, Bonn, Germany
e-mail: niethammer@iam.uni-bonn.de

J. H. Schröders
e-mail: s6jdschr@uni-bonn.de

J. Fuhrmann et al. (eds.), *Advances in Continuum Physics*, Advanced Structured Materials 238, https://doi.org/10.1007/978-3-031-93918-1_22

$$\int_{\mathbb{R}^n} \int_{\mathbb{R}} \varrho(t, x, p)\, dp\, dx = 1\,. \tag{22.2}$$

Fokker–Planck equations arise in many fields of mathematics and the sciences, see e.g., Risken (1996) for background information. Equation (22.1) is also known as the Kramers–Smoluchowski equation and arises, for example, in the context of chemical reactions, where H describes a chemical potential and where the diffusion term $\varepsilon^2 \partial_p^2 \varrho$ represents fluctuations due to Brownian motion. Then ϱ is the probability density of a particle that undergoes a random walk influenced by this potential. In this setting, typical potentials H are nonconvex with several local minima corresponding to the metastable configurations of the system.

22.1.1 Kramers Formula

The expected behavior of solutions in the limit $\varepsilon \to 0$ is well known since the early work of Kramers (1940). Because the fluctuation term $\varepsilon^2 \partial_p^2 \varrho$ is small, almost all mass is located in peaks of width $O(\varepsilon^2)$ around the local minima of H. On the other hand, fluctuations still allow that some mass moves from one minimum of H to another, crossing over some local maximum of H. The limit model of Eq. (22.1) in the limit of $\varepsilon \to 0$ can be derived by matched asymptotic expansions, see e.g., (Herrmann et al. (2012), Sect. 2.1). The main point is that around the minima the outer expansion behaves essentially as a stationary solution to (22.1), that is, a solution to

$$\partial_p \left(\varepsilon^2 \partial_p \varrho + H'(p)\, \varrho\right) = 0,$$

while the inner expansion (around the local maxima) is essentially determined by the mass flux $J(t)$ between the local minima, i.e., behaves approximately like a solution of

$$\partial_p \left(\varepsilon^2 \partial_p \varrho + H'(p)\, \varrho\right) = J(t).$$

The matching condition between the inner and outer expansions provides a description of the mass flux J. The time scale τ for which the mass flux is of order one turns out to be exponentially small (cf. (22.4) below and Fig. 22.1 for a brief discussion in the case of an asymmetric double-well potential). This also provides an explicit formula for the rate constants, the Kramers constants, in terms of the second derivatives of H at the local extrema, cf. (22.3). If there are more than two local minima multiple timescales that achieve a non-trivial limit for the mass flux can appear.

Rigorous proofs of this limit model exist for many cases of the potential H, but are mostly restricted to the case with just one relevant timescale. We refer, for example, to Arnrich et al. (2011), Peletier et al. (2010), Peletier and Schlichting (2023), Herrmann et al. (2010) for the case of even double-well potentials and Michel and Zworski (2018) for potentials with several minima and one relevant timescale.

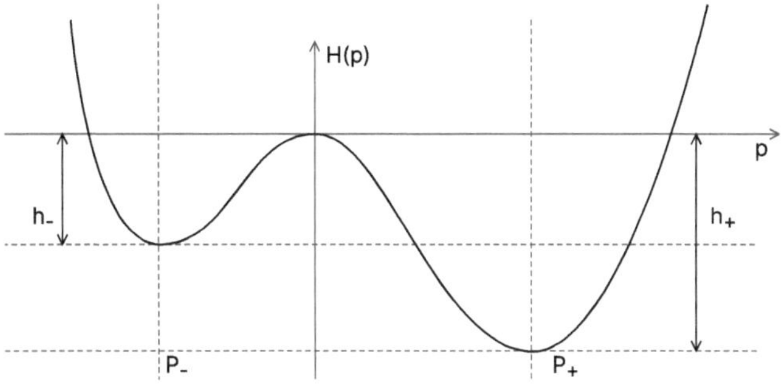

Fig. 22.1 The figure shows a double-well potential with local minima $P_{\pm}$ and local maximum at 0. Since we have $H(P_-) > H(P_+)$, it is more likely for a particle to move from the left to the right due to random fluctuations. Hence, the net mass flux J is positive, the relevant times scale is $\tau = \exp(-h_-/\varepsilon^2)$ and the rate constant turns out to be $C = \sqrt{|H''(x_-)H''(0)|}/(2\pi)$. For the masses m_+, m_- in the peaks around P_+, P_-, one obtains in the limit $\varepsilon \to 0$ the linear dynamics $\dot{m}_- = -Cm_-$ with $m_+ + m_- = 1$

We point out, that while we consider only the case of a one-dimensional potential H, similar results, with one timescale, exist also in higher dimensions, see Bovier et al. (2004), Helffer et al. (2004), and the review Berglund (2013) for a derivation of the timescale in higher dimensions, as well as Evans and Tabrizian (2016), Seo and Tabrizian (2019) and Michel and Zworski (2018) for limit dynamics in higher dimensions.

Our goal is to present a fairly simple, but rigorous, approach, which was originally developed in Herrmann and Niethammer (2020) for tilted periodic potentials, and to extend it to the case of general potentials H with finitely many local minima and multiple relevant timescales. This approach might also be useful in more complicated situations, such as when Eq. (22.1) is coupled with a non-local constraint, as it arises, for example, in the modeling of Lithium-ion batteries. We refer to Dreyer et al. (2011) for the derivation of the corresponding model, to Dreyer et al. (2015), Eberle et al. (2017) for well-posedness results and to Herrmann et al. (2012), Herrmann et al. (2014) for an asymptotic analysis of the corresponding equation.

22.2 Effective Mass Transport for Potentials with Multiple Wells

Throughout this paper, we will make the following assumptions on the potential H (see also Fig. 22.2 for a typical form of H).

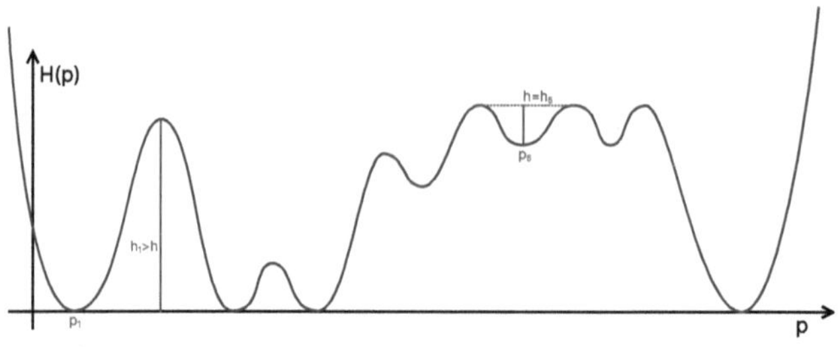

Fig. 22.2 Sketch of some potential H. It will serve as a running example throughout this discussion. Note that, e.g., $h_1 > h$ means, that, according to Theorem 22.4.1, m_1 does not participate in the mass exchange at the timescale determined by h

Assumption 22.2.1 The chemical potential $H : \mathbb{R} \to \mathbb{R}$ is smooth and satisfies

1. $H(p) \geq 0$,
2. H has finitely many local minima and $H(p)$ grows at least quadratically as $|p| \to \infty$.

In this setting, existence and uniqueness of smooth solutions to (22.1) for sufficiently nice initial data is a standard result, see e.g., Friedman (1964). For $\varepsilon > 0$, we denote the solution to (22.1) by $\varrho_\varepsilon = \varrho_\epsilon(t, x, p)$. Let p_i, for $i = 1, \dots, n$, denote the local minimizers of H, while the local maxima between p_i and p_{i+1} are attained at q_i, $i = 1, \dots, n-1$. We define the partial masses near p_i as

$$m_i^\varepsilon(t, x) := \int_{J_i} \varrho_\varepsilon(t, x, p)\, dp,$$

where we set $J_i = (q_{i-1}, q_i)$, with $q_0 = -\infty$ and $q_n = +\infty$ and the height of the energy barrier at q_i, $i = 1 \dots, n-1$, as

$$h_i := \min\{H(q_i) - H(p_i), H(q_i) - H(p_{i+1})\}.$$

In addition, we define Kramers' constants for the mass flux from p_i over the maximum of H at q_j as

$$c_K(i, j) := \frac{\sqrt{|H''(p_i) H''(q_j)|}}{2\pi}, \tag{22.3}$$

and set our timescale

$$\tau := \exp(-h/\varepsilon^2), \tag{22.4}$$

with $h := \min(h_i)$.

Our main result shows that the partial masses m_i^ε satisfy in the limit $\varepsilon \to 0$ the following limit dynamics.

Main result I: The partial masses evolve in the limit $\varepsilon \to 0$ according to

$$(\partial_t - \Delta_x)m_i = a_i m_{i-1} - b_i m_i + c_i m_{i+1}, \tag{22.5}$$

with the convention $m_0 = 0 = m_{n+1}$ and rate constants given by

$$a_i := \begin{cases} c_K(i-1, i-1), & \text{if } H(p_i) \le H(p_{i-1}), \text{ and } h_{i-1} = h \\ 0 & \text{else} \end{cases}$$
$$b_i := \begin{cases} c_K(i,i) + c_K(i, i-1), & \text{if } H(p_i) \ge H(p_{i-1}), H(p_{i+1}), \text{ and } h_i = h_{i-1} = h \\ c_K(i,i), & \text{if } H(p_i) \ge H(p_{i+1}), \text{ and } h_i = h \\ c_K(i, i-1), & \text{if } H(p_i) \ge H(p_{i-1}) \text{ and } h_{i-1} = h \\ 0 & \text{else} \end{cases}$$
$$c_i := \begin{cases} c_K(i+1, i), & H(p_i) \le H(p_{i+1}), \text{ and } h_i = h \\ 0 & \text{else} \end{cases} \tag{22.6}$$

We refer to Sect. 22.4 for the precise formulation of the result and the corresponding proofs.

We proceed by further characterizing the mass exchange between the potential wells on the next slower relevant time scale.

Definition 22.2.1 (Graph structure for H) Let $\mathcal{G}_h(H) = \mathcal{G}_h$ be the directed graph with the p_i as its vertices and directed edges between p_i and p_{i+1} satisfying the rule

1. no edge between p_i and p_{i+1} if $h_i > h$,
2. if $h = h_i$, then a directed edge from p_k to p_l exists, if $H(p_k) \ge H(p_l)$, where $l, k \in \{i, i+1\}$.

With this graph structure available, we can group potential wells accordingly. $\mathcal{G}_h$ captures which potential wells J_i have exchanged mass, and also the direction of this mass transfer, see Fig. 22.3 for an example.

Definition 22.2.2 (Modified wells and masses) Let $(G_k)_{k=1}^m$ be a partition of the set $\{p_1, \ldots, p_n\}$, such that,

1. If $p_i \in G_k$ satisfies $H(p_i) = \min\{H(p),\ p \in G_k\}$, then, for any $p_j \in G_k$, there exists a (directed) path from p_j to p_i.
2. If for $p_i \in \mathcal{G}_k$, $H(p_i) = \min\{H(p),\ p \in G_k\}$, then no path from p_i to any $p_j \notin G_k$ with $H(p_j) \le H(p_i)$ exists.

Define

$$K_k := \bigcup_{p_i \in G_k} J_i,$$

as the modified potential wells and correspondingly the new masses as

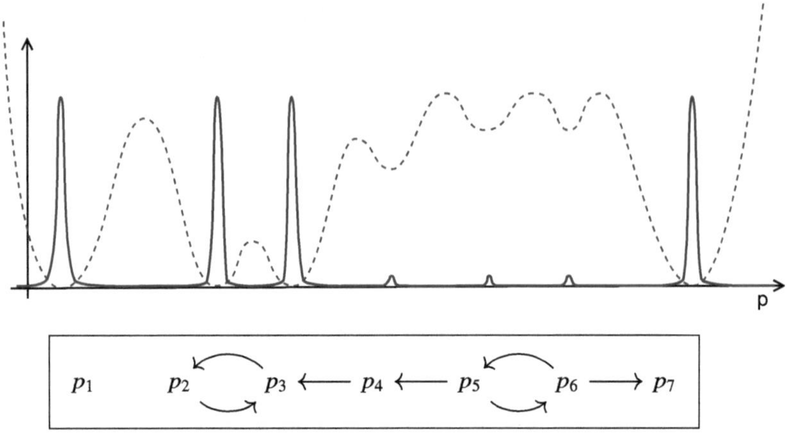

Fig. 22.3 Sketch of the distribution of the mass (solid gray plot) predicted by (22.5) for large times $t \gg 0$, and $\mathcal{G}_h$ for the running example H (dashed). Here we have assumed that all h_i, except h_1, equal h. Notice $\{p_1\} = G_1$, $p_2, p_3, p_4 \in G_2$, and $p_7 \in G_3$, but p_5, p_6 can belong to either G_2 or G_3. In the limit $\varepsilon \to 0$, the mass will be concentrated around p_1, p_2 and p_3, and in p_7, compare Definition 22.2.2. G_h also shows that $H(p_2) = H(p_3)$ and $H(p_5) = H(p_6)$

$$M_k^\varepsilon(t, x) := \int_{K_k} \varrho(t, x, p)\, dp.$$

In addition, we define

$$L_k := \bigcup_{H(p_i)=\min_{G_k} H(p_i)} J_i,$$

which are precisely the intervals in which we expect the masses to be concentrated in the K_k.

One can show that the K_k and L_k are in fact again intervals, i.e., $K_K := (q_{a_k}, q_{b_k})$ with $b_{k-1} = a_k$. This allows us to identify the next slower relevant timescale.

Definition 22.2.3 (Modified timescale) Set

$$H := \min\{H_k : \ p_k \in G_k\}, \qquad \tau' := \exp(-H/\varepsilon^2),$$

where $H_k := \min\{H(q_{a_k-1}),\ H(q_{b_k})\} - \min\{H(p) : \ p \in G_k\} > 0$.

We have to adjust the assumption on the initial data for the new timescale. For that, we introduce the so-called Gibbs function

$$\gamma_\varepsilon(p) := \exp(-H(p)/\varepsilon^2). \tag{22.7}$$

Assumption 22.2.1 ensures that the Gibbs function is integrable, and therefore the spatially homogeneous equation admits the equilibrium solution

$$\overline{\gamma}_\varepsilon := ||\gamma_\varepsilon||_{L^1}^{-1} \cdot \gamma_\varepsilon,$$

with the normalization condition $\int_{\mathbb{R}^n}\int_0^\infty \overline{\gamma}_\varepsilon \, dt\, dx = 1$.

Assumption 22.2.2 We assume that, for all ε

$$0 \le \varrho_\varepsilon(0, x, p)/\overline{\gamma}_\varepsilon(p) \le C,$$

for some constant C that is independent of ε.

Remark 22.2.1 By the maximum principle for $u := \varrho_\varepsilon/\overline{\gamma}$, this assumption will also hold for all times $t > 0$. See also (22.12) for the corresponding equation solved by these relative densities.

This assumption ensures that the configuration of the partial masses are already concentrated in the L_k, on the other hand, it is quite restrictive, in the limit $\varepsilon \to 0$ only, the masses located in the global minima will be non-zero and therefore produce meaningful limit dynamics. Therefore, we restrict our attention now to potentials H, which in addition to Assumption 22.2.1 also satisfy the following.

Assumption 22.2.3 The global minima of H are precisely the local minima which lie in some L_k, i.e.,

$$\operatorname{argmin}\{H(p) : \; p \in \mathbb{R}^d\} = \bigcup_k \bigcup_{p_i \in L_k} p_i.$$

On the timescale τ', we then obtain the analogous limit dynamics.

Main result II: The modified partial masses M_i^ε evolve in the limit $\varepsilon \to 0$ according to

$$(\partial_t - \Delta_x)M_k = A_k M_{k-1} - B_k M_k + C_k M_{k+1} \tag{22.8}$$

with explicit, but more involved, rate constants that can also be expressed in terms of the Kramers constants $C_K(i, j)$. We refer to Sect. 22.5 and especially Theorem 22.5.1 for the precise statements and proofs.

Remark 22.2.2 We emphasize that any subsequent timescale can be treated by the analogous regrouping of thc partial masses.

22.3 Preliminaries About the Fokker–Planck Equation

From now on, we neglect for notational convenience the dependence on ε in the notation of ρ, γ, etc. We first establish asymptotic estimates for the Gibbs function γ as defined in (22.7). For this, we set

$$\mu_i := \int_{J_i} \gamma(p)\, dp, \qquad 1 \le i \le n,$$

$$\eta_j := \int_{q_j}^{q_{j+1}} \frac{1}{\gamma(p)}\,dp, \qquad 1 \le j \le n-1,$$

and the normalizations

$$\gamma_i := \frac{\gamma}{\mu_i}\chi_{J_i}.$$

Laplace's method for asymptotic integrals (cf. Bender and Orszag (1999)) yields

$$\mu_i = \int_{J_i} \exp(-H(p)/\varepsilon^2)\,dp = \exp(-H(p_i)/\varepsilon^2)\frac{\varepsilon\sqrt{2\pi}}{\sqrt{H''(p_i)}}\big(1 + O(\varepsilon^2)\big), \quad (22.9)$$

$$\eta_i = \int_{p_i}^{p_{i+1}} \exp(H(p)/\varepsilon^2)\,dp = \exp(H(q_i)/\varepsilon^2)\frac{\varepsilon\sqrt{2\pi}}{\sqrt{|H''(q_i)|}}\big(1 + O(\varepsilon^2)\big).$$

Equation (22.1) can also be viewed as a Wasserstein gradient flow on the space of probability measures, since it can be written as

$$\tau\partial_t\varrho = (\tau^{1/2}\varepsilon^{-1}\partial_x + \partial_p)(\varrho(\tau^{1/2}\varepsilon^{-1}\partial_x + \partial_p)\partial_\varrho\mathcal{E}).$$

Here, $\mathcal{E} = \mathcal{E}(t) = \mathcal{E}(\varrho, t)$ denotes the free energy of the system and ∂_ϱ denotes the first variation of a functional. This free energy is given by

$$\mathcal{E}(t) := \int_{\mathbb{R}^n}\int_{\mathbb{R}} \varepsilon^2\varrho(t,x,p)\ln(\varrho(t,x,p))\,dp\,dx + \int_{\mathbb{R}^n}\int_{\mathbb{R}} H(p)\varrho(t,x,p)\,dp\,dx. \quad (22.10)$$

This interpretation was introduced in Jordan et al. (1997), see also Jordan et al. (1998), and allows viewing the Fokker–Planck equation (22.1) as a generalized continuity equation. In particular, with this energy, the second law of thermodynamics reads

$$\tau\dot{\mathcal{E}}(t) = -\tau\varepsilon^2\mathcal{C}(t) - \varepsilon^4\mathcal{D}(t), \quad (22.11)$$

where

$$\mathcal{C}(t) := \int_{\mathbb{R}^n}\int_{\mathbb{R}} \frac{(\nabla_x\varrho(t,x,p))^2}{\varrho(t,x,p)}\,dp\,dx$$

is the dissipation in x-space and the dissipation in p-space is given by

$$\mathcal{D}(t) := \int_{\mathbb{R}^n}\int_{\mathbb{R}} \frac{\big(\partial_p\varrho(t,x,p) + \varepsilon^{-2}H'(p)\varrho(t,x,p)\big)^2}{\varrho(t,x,p)}\,dp\,dx.$$

Remark 22.3.1 By Assumption 22.2.1, the energy is bounded from below and its minimizer, in the class of functions with normalized mass, is given by $\overline{\gamma}$. However, Assumption 22.2.1 is in fact not always completely necessary, but the lower bound for $\mathcal{E}$ simplifies some proofs, see Herrmann and Niethammer (2020) for an example with infinitely many wells.

Besides ϱ itself, we will also consider the densities with respect to the invariant measures $\gamma_i(p)\,d\mathcal{L}^1(p)$, supported on J_i. We define the relative densities $u_i = w_i^2$ of ϱ in the potential wells J_i via

$$\varrho(t, x, p) \cdot \chi_{J_i}(p) = w_i^2(t, x, p)\gamma_i(p).$$

On the interval J_i, Eq. (22.1) now reads

$$\tau\gamma_i(\partial_t - \Delta)u_i = \partial_p(\varepsilon^2\gamma_i\partial_p u_i) \tag{22.12}$$

and, due to the convenient introduction of the second power, the p-dissipation is now given by $\mathcal{D} = \sum_{i=1}^n \mathcal{D}_i$, with

$$\mathcal{D}_i(t, x) = \int_{\mathbb{R}^n}\int_{J_i} (\partial_p w_i(t, x, p))^2\gamma_i\,dp\,dx. =: \int_{\mathbb{R}^n} D_i(t, x)\,dx$$

22.4 Effective Mass Transport

This section is devoted to the precise statement and rigorous proof of the limit dynamics (22.5). For that, we will also approximate the partial masses in Sect. 22.4.1 appropriately by some $\tilde{m}_i(t, x)$ with approximation error $O(\varepsilon^2)$, see (22.13) for the precise construction.

Theorem 22.4.1 (Limit dynamics for m_i) *We define for given $\tilde{m}_i(0, x)$ as in* (22.13) *below $\breve{m}_i(t, x)$ as the solution to the linear system*

$$\begin{aligned}(\partial_t - \Delta_x)\breve{m}_i &= a_i\breve{m}_{i-1} - b_i\breve{m}_i + c_i\breve{m}_{i+1},\\ \breve{m}_i(0, x) &= \tilde{m}_i(0, x),\end{aligned}$$

with the convention $\breve{m}_0 = 0 = \breve{m}_{n+1}$ and rate constants given by (22.6).

Then, for any $T > 0$, there exists $C(T)$ such that

$$\sum_{i=1}^n \int_0^T \int_{\mathbb{R}^n} |\breve{m}_i - m_i|\,dx\,dt \le C(T)\,\varepsilon^2 .$$

The strategy of the proof of Theorem 22.4.1 follows the one in Herrmann and Niethammer (2020) and will be to introduce two types of approximations of the partial masses, one of which will be $\tilde{m}_i$, such that the dynamics of one are expressed in terms of the other, and then to provide the error bounds needed to pass to the limit for the m_i.

22.4.1 Substitute Masses

The first substitute masses we define are simply the relative densities evaluated at the p_i, i.e.,

$$\overline{m}_i(t,x) := w_i^2(t,x,p_i).$$

Next, we define another type of substitute masses, such that their derivatives can be expressed in terms of the $\overline{m}_i$. We define weight functions ψ_i, uniquely determined by

$$\psi_i'(p) = \begin{cases} \frac{1}{\eta_i \gamma}, & p \in J_i, \\ 0 & \text{otherwise.} \end{cases}$$
$$\psi_i(p_i) = 0,$$
$$\psi_i(p_{i+1}) = 1.$$

Note that, as can be seen, for example, by Laplace's method, $\psi_{i-1} - \psi_i$ approximate the indicator function of J_i for $i \neq 1, n$, and similarly $1 - \psi_1$ and ψ_{n-1} approximate χ_{J_1} and χ_{J_n}, respectively, see Fig. 22.4. Hence, we set

$$\begin{aligned} \tilde{m}_1(t,x,p) &:= \int_{\mathbb{R}} (1-\psi_1)\varrho(t,x,p)\,dp, \\ \tilde{m}_i(t,x,p) &:= \int_{\mathbb{R}} (\psi_{i-1}-\psi_i)\varrho(t,x,p)\,dp, \qquad 1 < i < n, \\ \tilde{m}_n(t,x,p) &:= \int_{\mathbb{R}} \psi_{n-1}\varrho(t,x,p)\,dp. \end{aligned} \tag{22.13}$$

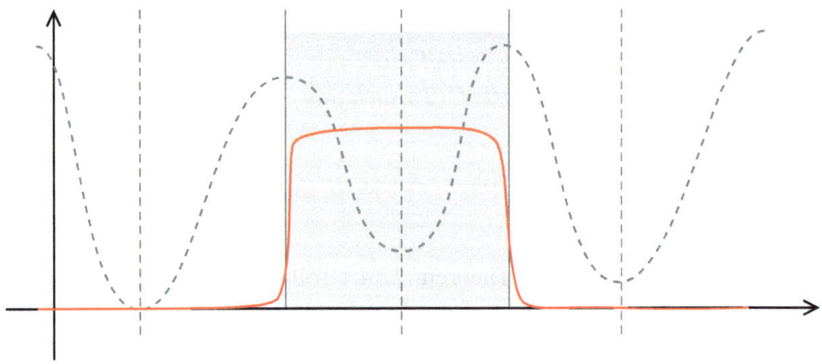

Fig. 22.4 Sketch of some $(\psi_i - \psi_{i+1})$ (solid line) for some potential H (dashed)

Note, that due to

$$(1-\psi_1)+\sum_{i=2}^{n-1}(\psi_{i-1}-\psi_i)+\psi_{n-1}=1,$$

the conservation of mass (22.2) also implies that the total substitute mass $\tilde{m} = \sum_{i=1}^{n} \tilde{m}_i$ is conserved. At the given timescale τ only the smallest energy barriers should permit mass flux. The effective mass flux is also expected to be directed into the deeper potential well. This behavior will be captured by the following constants which we associate to the energy barriers q_i.

$$\begin{aligned} \theta(q_i) &:= \tau\eta_i\tilde{\mu}_i/\varepsilon^2, \qquad 1 \le i \le n-1, \text{ where} \\ \tilde{\mu}_i &:= \min\{\mu_i, \mu_{i+1}\}. \end{aligned}$$

Remark 22.4.1 Recalling (22.9), we note that the $\theta(q_i)$ are of order $\exp((h_i - h)/\varepsilon^2)$. In particular, we verify

$$\lim_{\varepsilon\to 0}\theta(q_i) = \begin{cases} \overline{\theta}_i := \frac{2\pi}{\sqrt{|H''(\tilde{p}_i)H''(q_i)|}}, & \text{if } h_i = h \\ \infty, & \text{else} \end{cases}$$

where $\tilde{p}_i$ denotes the p_j, for $j = i, i+1$, with $\mu_j = \tilde{\mu}_i$. Here, the Kramers constants reappear as inverse of the limit $\overline{\theta}_i$, while $\overline{\theta}_i = \infty$ indicates that no mass flux can happen at this timescale.

Since any m_i can exchange mass across two barriers (at least for $i \neq 1, n$), we also define

$$\theta_i := \min\{\theta(q_{i-1}), \theta(q_i)\} =: \tau\hat{\eta}_i\hat{\mu}_i/\varepsilon^2, \qquad 1 < i < n,$$

and rename $\theta_1 = \theta(q_1)$ and $\theta_n = \theta(q_n)$ for consistency. We now calculate the dynamics of the $\tilde{m}_i$.

Proposition 22.4.1 (Dynamics of the substitute masses) *The substitute masses satisfy the equations*

$$\begin{aligned} \theta_1(\partial_t-\Delta)\tilde{m}_1 &= \tilde{\mu}_1\left(\frac{-\overline{m}_1}{\mu_1}+\frac{\overline{m}_2}{\mu_2}\right), \\ \theta_i(\partial_t-\Delta)\tilde{m}_i &= \frac{\hat{\eta}_i}{\eta_{i-1}}\hat{\mu}_i\left(\frac{\overline{m}_{i-1}}{\mu_{i-1}}-\frac{\overline{m}_i}{\mu_i}\right)+\frac{\hat{\eta}_i}{\eta_i}\hat{\mu}_i\left(\frac{-\overline{m}_i}{\mu_i}+\frac{\overline{m}_{i+1}}{\mu_{i+1}}\right), \qquad 1<i<n, \\ \theta_n(\partial_t-\Delta)\tilde{m}_n &= \tilde{\mu}_{n-1}\left(\frac{\overline{m}_{n-1}}{\mu_{n-1}}-\frac{\overline{m}_n}{\mu_n}\right). \end{aligned}$$

Proof We only consider the case that $1 < i < n$, the cases $i = 1$ and $i = n$ are treated analogously. Using the smoothness of the solution ϱ and integrating by parts, we obtain

$$\begin{aligned}\theta_i(\partial_t + \Delta_x)\tilde{m}_i &= \hat{\eta}_i\hat{\mu}_i \int_{\mathbb{R}} (\psi_{i-1}(p) - \psi_i(p)) \cdot \tau\varepsilon^{-2}(\partial_t + \Delta_x)\varrho(t,x,p)\,dp \\ &= \hat{\eta}_i\hat{\mu}_i \int_{\mathbb{R}} (\psi_{i-1}(p) - \psi_i(p))\partial_p \left(\partial_p\varrho + \varepsilon^{-2}H'(p)\varrho(t,x,p)\right) dp \\ &= -\hat{\eta}_i\hat{\mu}_i \int_{\mathbb{R}} \partial_p(\psi_{i-1}(p) - \psi_i(p)) \left(\partial_p\varrho + \varepsilon^{-2}H'(p)\varrho(t,x,p)\right).\end{aligned}$$

Now consider the (unscaled) density $u := \varrho/\gamma$. On J_i the function u satisfies $\mu_i \cdot u(t,x,p) = u_i(t,x,p)$ and u solves (22.12) even for all $p \in \mathbb{R}$. We verify

$$(\partial_p\varrho + \varepsilon^{-2}H'(p)\varrho(t,x,p) = \gamma\partial_p u(t,x,p).$$

Therefore, after plugging in the definition of the $\partial_p\psi_i$, the fundamental theorem of calculus yields

$$\begin{aligned}\theta_i(\partial_t - \Delta_x)\tilde{m}_i &= -\frac{\hat{\eta}_i}{\eta_{i-1}}\hat{\mu}_i \int_{p_{i-1}}^{p_i} \partial_p u(t,x,p)\,dp + \frac{\hat{\eta}_i}{\eta_i}\hat{\mu}_i \int_{p_i}^{p_{i+1}} \partial_p u(t,x,p)\,dp \\ &= \frac{\hat{\eta}_i}{\eta_{i-1}}\hat{\mu}_i\left(\frac{\overline{m}_{i-1}}{\mu_{i-1}} - \frac{\overline{m}_i}{\mu_i}\right) + \frac{\hat{\eta}_i}{\eta_i}\hat{\mu}_i\left(\frac{-\overline{m}_i}{\mu_i} + \frac{\overline{m}_i}{\mu_{i+1}}\right).\end{aligned}$$

□

Remark 22.4.2 By definition, the coefficients of the form $\frac{\hat{\eta}_i\hat{\mu}_i}{\eta_k\mu_j}$ are bounded by 1 and, in the limit $\varepsilon \to 0$, we obtain

$$\lim_{\varepsilon\to 0}\left(\frac{\hat{\eta}_i\hat{\mu}_i}{\eta_k\mu_j}\right) = \begin{cases} \frac{\sqrt{|H''(p_j)H''(q_k)|}}{\sqrt{|H''(\hat{p}_i)H''(\hat{q}_i)|}} & \text{if } H(p_j) - H(q_k) = H(\hat{p}_i) - H(\hat{q}_i), \\ 0 & \text{else} \end{cases}$$

where $\hat{p}_i$ and $\hat{q}_i$ correspond to $\hat{\mu}_i$ and $\hat{\eta}_i$. This limit can also be written as $c_K(j,k) \cdot \overline{\theta}_i$; hence, we expect the Kramers constants to appear as the rate constants in the limit dynamics.

22.4.2 Passage to the Limit

With the substitute masses set up, we can now reproduce the asymptotic bounds on the approximation error from (Herrmann and Niethammer 2020, Sect. 2.3) for our modified setting. We begin with yet another application of Laplace's method.

Lemma 22.4.1 *There exist intervals $I_i \subset J_i$, such that*

$$\int_{J_i\setminus I_i} \gamma_i(p)\,dp \le C\varepsilon\sqrt{\tau}$$

and

$$\sup_{p\in I_i}(|\psi_i| + |1-\psi_{i-1}|) \le C\varepsilon\sqrt{\tau}.$$

Proof By definition, we have $h \le h_j$, and therefore there exist $\ell_i < p_i < r_i$ such that $H(\ell_i) = H(r_i) = H(p_i) + h/2$. We set $I_i := (\ell_i, r_i)$. The first assertion now follows by Laplace's method, since $\gamma_i = \mu_i^{-1}\gamma$ attains its maximum on $J_i \setminus I_i$ at ℓ_i and r_i. Conversely, ψ_{i-1}' attains it maximum on $\overline{I_i}$ also at ℓ_i, and ψ_i at r_i. Therefore, Laplace's method also provides the second assertion. □

This then allows us to bound the approximation error pointwise.

Proposition 22.4.2 (Bounds on the approximation errors) *With* $m(t,x) = \sum_{i=1}^n m_i(t,x)$, *we have*

$$\sum_{i=1}^{n} |m_i - \tilde{m}_i| + |m_i - \overline{m}_i| \le C\,\tau^{-1/2}\,D(t) + C\,\tau^{1/2}\varepsilon^{-2} m(t,x) + C\,r(t,x),$$

with constants independent of ε *and the term* $\|r(t,\cdot)\|_{L^1(\mathbb{R}^n)} \le C\varepsilon^2$ *for all* $t > 0$.

Proof

Bound for $|m_i - \overline{m}_i|$. Using the quadratic growth of H, Assumption 22.2.1, and that the energy functional $\mathcal{E}$, (22.10), is decreasing, we see that for a constant $C = C(\mathcal{E}(0)) \cdot \frac{1}{\varepsilon} > 0$, we have

$$\int_{\mathbb{R}^n} r(t,x)\,dx := \int_{\mathbb{R}^n} m_i(t,x)\,dx - \int_{\mathbb{R}^n}\int_{J_i\cap(-C,C)} \varrho(t,x,p)\,dx\,dp \le \varepsilon^2.$$

Strictly speaking, this term will only be relevant for the non-compact intervals J_1, J_n, but to avoid case distinctions we will consider it for all J_i. Let $A_i := J_i \cap (-C, C)$. Hölder's inequality yields

$$\begin{aligned}
\left|\int_{A_i} \varrho(t,x,p)\,dp - \overline{m}_i\right| &= C\left|\int_{J_i} w_i^2(p)\gamma_i(p) - w_i^2(p_i)\gamma_i(p)\,dp\right| \\
&\le C\int_{A_i}\left|\int_{p_j}^{p} 2w_i(q)\,\partial_p w_i(q)\,dq\right|\gamma_i(p)\,dp \\
&\le C\int_{A_i}\left(\int_{p_i}^{p}\frac{w_i^2(q)}{\gamma_i(q)}dq\right)^{1/2}\left(\int_{p_i}^{p}(\partial_p w_i(q))^2\gamma_i(q)\,dq\right)^{1/2}\gamma_j(p)\,dp \\
&\le C\int_{A_i}\left(\int_{p_i}^{p} w_i^2(q)\gamma_i(q)\,dq\right)^{1/2}\left(\int_{p_i}^{p}(\partial_p w_i(q))^2\gamma_i(q)\,dq\right)^{1/2}dp,
\end{aligned}$$

where we used, that γ_i^{-1} is monotone increasing for $p > p_i$ and decreasing for $p < p_i$. Using now that A_i is compact,

$$|m_i - \overline{m}_i|\,dx \le \frac{C}{\varepsilon}\sqrt{D_i(t,x)m_i(t,x)} + r(t,x).$$

Bound for$|m_i - \tilde{m}_i|$: With the I_i of Lemma 22.4.1, we write

$$\begin{aligned}\int\limits_{J_i\setminus I_i} w_i^2\gamma_i\,dp &= m_i - \int\limits_{I_i} w_i^2\gamma_i\,dp\\ &= (m_i - \overline{m}_i)\int\limits_{I_i}\gamma_i\,dp + m_j\int\limits_{J_i\setminus I_i}\gamma_i\,dp + \int\limits_{I_i}(w_i^2(p) - w_i^2(p_i))\gamma_i(p)\,dp.\end{aligned}$$

Notice, that the first and the last summands can be bounded by $\frac{C}{\varepsilon}\sqrt{D_i(t,x)m_i(t,x)} + r(t,x)$, due to the calculations of the previous step. Since ψ_i, and $1 - \psi_{i-1}$ are bounded, we use Lemma 22.4.1 to see

$$\begin{aligned}\int_{J_i}\psi_i\,w_i^2\gamma_i\,dp &= \int_{J_i\setminus I_i}\psi_i\,w_i^2\gamma_i\,dp + \int_{I_i}\psi_i\,w_i^2\gamma_i\,dp\\ &\le \frac{C}{\varepsilon}\sqrt{D_i(t,x)m_i(t,x)} + r(t,x) + C\varepsilon\sqrt{\tau}\,m_i.\end{aligned}$$

With the piecewise definition of the ψ_i, we can write

$$\begin{aligned}m_i - \tilde{m}_i = \int_{J_{i-1}}\psi_{i-1}w_{i-1}^2\gamma_{i-1}\,dp &+ \int_{J_i}(1 - \psi_j + \psi_{j-1})w_i^2\gamma_i\,dp\\ &+ \int_{J_{i+1}}(1 - \psi_i)w_{i+1}^2\gamma_{i+1}\,dp.\end{aligned}$$

Consequently,

$$|m_i - \tilde{m}_i| \le C\sum_{|i-j|\le 1}\left(\frac{C}{\varepsilon}\sqrt{D_i(t,x)m_i(t,x)} + r(t,x) + C\,\varepsilon\sqrt{\tau}\,m_j\right).$$

Concluding arguments: Using Cauchy–Schwarz and Young's inequality, we conclude

$$\begin{aligned}\sum_{i=1}^n|m_i - \tilde{m}_i| &\le \sum_{i=1}^n\left(\frac{C}{\varepsilon}\sqrt{D_i(t,x)m_i(t,x)} + r(t,x) + C\varepsilon\sqrt{\tau}\,m_j\right)\\ &\le C\,\varepsilon^{-1}\left(\sum_{i=1}^n D_i\right)^{1/2}\left(\sum_{i=1}^n m_i\right)^{1/2} + C\varepsilon\sqrt{\tau} + C\,r(t,x)\\ &\le C\tau^{-1/2}\,D(t) + C\tau^{1/2}\varepsilon^{-2}m(t,x) + C\,r(t,x).\end{aligned}$$

To apply this error bound, we will need control on the dissipation, which will be given by the following proposition.

Proposition 22.4.3 (Control on Dissipation) *The dissipation D satisfies*

$$\int_0^\infty \mathcal{D}(t)\,dt \le \tau\varepsilon^{-4}C,$$

for some constant $C > 0$.

Proof As noted before, due to Assumption 22.2.1, the free energy is bounded from below, i.e., there is a constant $C > 0$ such that $\mathcal{E}(t) > -C$. We integrate the energy balance (22.11) up to $T > 0$

$$\int_0^T \varepsilon^4\mathcal{D}(t) + \tau\varepsilon^2\mathcal{C}(t)\,dt = -\int_0^T \tau\dot{\mathcal{E}}(t)\,dt \le \tau\left(\mathcal{E}(0) + C\right).$$

Using the positivity of $\mathcal{C}$, we obtain the desired

$$\int_0^\infty \mathcal{D}(t)\,dt \le \tau\varepsilon^{-4}(\mathcal{E}(0) + C).$$

We proceed to prove Theorem 22.4.1. We first reformulate the rate constants (22.6) according to Remarks 22.4.1 and 22.4.2 in terms of the constants μ_i and η_i. Theorem 22.4.1 now reads as follows.

Theorem 22.4.2 (Limit dynamics) *Let $\breve{m}_i(t,x)$ be the solution to the system*

$$\begin{aligned}(\partial_t - \Delta_x)\breve{m}_i &= a_i\breve{m}_{i-1} - b_i\breve{m}_i + c_i\breve{m}_{i+1},\\ \breve{m}_i(0,x) &= \tilde{m}(0,x),\end{aligned}$$

where

$$\begin{aligned}a_i &:= \lim_{\varepsilon\to 0}\left(\theta_i^{-1}\frac{\hat{\eta}_i\hat{\mu}_i}{\eta_{i-1}\mu_{i-1}}\right),\\ b_i &:= \lim_{\varepsilon\to 0}\left(\theta_i^{-1}\frac{\hat{\eta}_i\hat{\mu}_i}{\eta_{i-1}\mu_i} + \theta_i^{-1}\frac{\hat{\eta}_i\hat{\mu}_i}{\eta_i\mu_i}\right),\\ c_i &:= \lim_{\varepsilon\to 0}\left(\theta_i^{-1}\frac{\hat{\eta}_i\hat{\mu}_i}{\eta_i\mu_{i+1}}\right).\end{aligned}$$

Then for all $T > 0$

$$\sum_{i=1}^n \int_0^T \int_{\mathbb{R}^n} |\breve{m}_i - m_i|\,dx\,dt \le C(T)\,\varepsilon^2,$$

for a time-dependent constant $C(T)$.

Remark 22.4.3 If the rate constants a_i, c_i are non-zero, they are indeed the Kramers constants $c_K(i-1, i-1)$ and $c_K(i+1, i)$, respectively. Similarly, we can express the summands of b_i as $c_K(i, i-1)$ and $c_K(i, i)$, if b_i is not zero. Hence, we obtain indeed the rate constants of Theorem 22.4.1.

Proof From Proposition 22.4.1, we obtain

$$\begin{aligned}(\partial_t - \Delta_x)\tilde{m}_i &= \frac{1}{\theta_i}\left(\frac{\hat{\eta}_i}{\eta_{i-1}}\hat{\mu}_i\left(\frac{\tilde{m}_{i-1}}{\mu_{i-1}} - \frac{\tilde{m}_i}{\mu_i}\right) + \frac{\hat{\eta}_i}{\eta_i}\hat{\mu}_i\left(\frac{-\tilde{m}_i}{\mu_i} + \frac{\tilde{m}_i}{\mu_{i+1}}\right)\right)\\ &\quad - \frac{1}{\theta_i}(e_i + f_i + g_i),\end{aligned}$$

where the error terms are given by

$$\begin{aligned}e_i &= \frac{\hat{\eta}_i\hat{\mu}_i}{\eta_{i-1}} \cdot \frac{(\tilde{m}_{i-1} - \overline{m}_{i-1})}{\mu_{i-1}},\\ f_i &= -\left(\frac{\hat{\eta}_i\hat{\mu}_i}{\eta_{i-1}} + \frac{\hat{\eta}_i\hat{\mu}_i}{\eta_i}\right) \cdot \frac{(\tilde{m}_i - \overline{m}_i)}{\mu_i},\\ g_i &= \frac{\hat{\eta}_i\hat{\mu}_i}{\eta_i} \cdot \frac{(\tilde{m}_{i+1} - \overline{m}_{i+1})}{\mu_{i+1}}.\end{aligned}$$

For $i = 1$ or $i = n$, we make the obvious modification and set $e_1 = 0$ and $g_n = 0$, respectively. Next, we want to replace the coefficients by their limits. By Laplace's method, we notice, e.g., for the a_i:

$$a_i - \frac{\hat{\eta}_i\hat{\mu}_i}{\eta_{i-1}} = O(\varepsilon^2),$$

because the fraction $\frac{\hat{\eta}_i\hat{\mu}_i}{\eta_{i-1}}$ is bounded. We rewrite again

$$\begin{aligned}(\partial_t - \Delta_x)\tilde{m}_i &= a_i\tilde{m}_{i-1} - b_i\tilde{m}_i + c_i\tilde{m}_{i+1}\\ &\quad - \frac{1}{\theta_i}(e_i + f_i + g_i) + C\varepsilon^2(\tilde{m}_i + \tilde{m}_{i-1} + \tilde{m}_{i+1}).\end{aligned}$$

By Proposition 22.4.2 and Lemma 22.4.3, we find

$$\begin{aligned}R(t) &:= \sum_{i=1}^{n}\int_0^t\int_{\mathbb{R}^n} e_i + f_i + g_i + C\varepsilon^2(\tilde{m}_i + \tilde{m}_{i-1} + \tilde{m}_{i+1})\,dx\,dt\\ &\leq C\int_0^t\int_{\mathbb{R}^n}\left(\tau^{-1/2}D(t,x) + \tau^{1/2}\varepsilon^{-2}m(t,x)\right)dx\,dt\\ &\quad + \int_0^t\int_{\mathbb{R}^n} C\,\varepsilon^2\tilde{m}(t,x) + C\,r(t,x)\,dx\,dt\\ &\leq C\,(\varepsilon^{-4}\sqrt{\tau} + \varepsilon^2 t).\end{aligned}$$

We recall that $\varepsilon^{-4}\sqrt{\tau}$ decays exponentially, as ε tends to zero. The limit equation preserves the L^1-norm and positivity; hence, we may apply Duhamel's formula to see

$$\sum_{i=1}^{n}\int_{\mathbb{R}^n}\int_0^T |\tilde{m}_i - \breve{m}_i|\, dt\, dx \le \int_0^T R(t)\, dt \le CT^2\varepsilon^2.$$

The claim follows now by integrating the error bound for $|m_i - \tilde{m}_i|$, for the same bound $C(T)\,\varepsilon^2$. □

22.5 Additional Timescales

This section is devoted to precisely stating and proving our second main result (cf. Theorem 22.5.1 below). We begin investigating some properties of the new potential wells. Recalling Definitions 22.2.1 and 22.2.2 we have the following auxiliary results on the modified wells, which will be useful in the following.

Lemma 22.5.1 *On the intervals K_k, the following hold*

1. *Each G_k is of the form $\{p_{a_k}, p_{a_k+1} \ldots, p_{b_k}\}$, for some $a_k < b_k$. Therefore,*

$$K_k = (q_{a_k-1}, q_{b_k}).$$

   Additionally, L_k is also an interval.
2. *There exists and index $a_k \le \ell \le b_k$ such that, for all $p_i, p_{i+1} \in G_k$, we have $H(p_i) \le H(p_{i+1})$ for $i < \ell$ and $H(p_i) \ge H(p_{i+1})$ for $i \ge \ell$.*
3. *For all $a_k - 1 \le \ell \le b_k$, we have $H(q_\ell) \le H(q_{a_k-1})$ or $H(q_\ell) \le H(q_{b_k})$.*
4. *If $H(p_\ell) = H(p_{\ell+1}) = \min(H(p_i),\ p_i \in G_k)$, then $H(q_\ell) = H(p_\ell) + h$, in particular, we have $H(q_\ell) < H(q_{a_k-1})$* **and** *$H(q_\ell) < H(q_{b_k})$.*
5. *Up to relabeling, the L_k are independent of the choices of partition $(G_k)_k$ and of K_k, as are the values $H(q_{a_k-1})$ and $H(q_{b_k})$.*

Proof The first point is an easy consequence of Definition 22.2.1 since edges can only exist between p_i and p_j, if $|i - j| = 1$. The second claim then follows from the first and Definitions 22.2.2 and 22.2.1 by choosing ℓ such that $H(p_\ell) = \min(H(p_i),\ p_i \in G_k)$. The third point again follows from the second and $h_i = h$, for $a_k \le i \le b_k$. For the fourth, note that $H(q_\ell) = H(p_\ell) + h$ and if we assumed, for a contradiction, e.g., $H(q_{b_k}) \le H(q_\ell)$, then we would have $H(p_{b_k}) \le H(q_{b_k}) - h = H(p_\ell)$, and by minimality of h also $H(q_{b_k}) = H(p_\ell) + h$. But then also, $H(p_{b_k+1}) \le H(q_{b_k}) - h$ and this would violate the second point in the definition of the G_k in 22.2.2 because an edge from p_{b_k} to p_{b_k+1} would exist. For the fifth point, note that by definition the L_k are such that along no path the value of H can decrease, while from any possible q_{b_k}, q_{a_k+1} the value of H cannot strictly increase along any path. □

Assumption 22.5.1 After relabeling, we assume $q_{b_k} = q_{a_{k+1}-1}$, i.e., $b_k = a_{k+1} - 1$.

□ We can now define the appropriate timescale for the dynamics of the M_k as claimed in Sect. 22.2.

Definition 22.5.1 (Modified timescale) Set

$$H := \min\{H_k : p_k \in G_k\}, \qquad \tau' := \exp(-H/\varepsilon^2),$$

where $H_k := \min\{H(q_{a_k-1}),\ H(q_{b_k})\} - \min\{H(p) : p \in G_k\} > 0$.

22.5.1 Modifying the Substitute Masses

The goal of this section is to modify the substitute masses and error bounds of Sect. 22.4.

Definition 22.5.2 (Relative densities on K_k) Define the normalized quantities

$$\Gamma_k := \gamma \cdot \Big(\sum_{i=a_k}^{b_k} \mu_i\Big)^{-1} \qquad \text{on } K_k,$$
$$W_k^2 := \varrho / \Gamma_k, \qquad \text{on } K_k.$$

In the K_k, the mass is expected to be concentrated in the local minima $p_i \in L_k$ and, for $H(p_i) < H(p_j)$, we recall $\mu_j \ll \mu_i$. We link this to the density via Assumption 22.2.2. We notice that by this assumption we also have

$$\sup_{\varepsilon}\Big(\sup_{x\in\mathbb{R}^n, p\in\mathbb{R}} W_i^2(0, x, p)\Big) \le C. \tag{22.14}$$

If the relative density is uniformly bounded, the solution ϱ must be exponentially small in the $p_i \in (K_k \setminus L_k)$, where $\Gamma_k \approx \overline{\gamma}$ is exponentially small.

Definition 22.5.3 (Substitute masses) For fixed $\ell \neq 1, k$, we define auxiliary functions on K_ℓ

$$\psi_{i,j,k} := \big(\psi_{i,j} - \psi_{j,k}\big), \qquad i \in L_{\ell-1},\ j \in L_\ell,\ k \in L_{\ell+1},$$

whereas before

$$\psi'_{i,j} := \frac{1}{\gamma \cdot \big(\sum_{r=i}^{j-1} \eta_r\big)},$$
$$\psi_{i,j}(p_i) = 0, \qquad \psi_{i,j}(p_j) = 1.$$

Then, for coefficients $c_{i,j,k}$ to be chosen later, the substitute masses for M_ℓ are

$$\overline{M}_\ell(t,x) := \sum_{p_r \in L_\ell} \overline{m}_r(t,x) = \Big(\sum_{i=a_k}^{b_k} \mu_i\Big) \cdot \sum_{p_r \in L_\ell} W_r^2(t,x,p_r)/\mu_r,$$

$$\tilde{M}_\ell(t,x) := \sum_{i,j,k} c_{i,j,k} \int_{\mathbb{R}} \psi_{i,j,k}(p)\varrho(t,x,p)\,dp, \qquad \text{with} \sum_{i,j,k} c_{i,j,k} = 1.$$

For $\ell = 1$ and $\ell = n$, we make the obvious adjustment, taking $1 - \psi_{j,k}$ and $\psi_{i,j}$, respectively, and dropping the summation over the superfluous index.

Remark 22.5.1 Using Lemma 22.5.1, we note that for $q_r \in L_\ell$, we also have $\eta_r \ll \min(\eta_{a_k-1}, \eta_{b_k})$. Therefore, we see that the derivative $\psi'_{i,j,k}$ is exponentially small on the interval L_ℓ. This shows that every $\psi_{i,j,k} \sim 1$ on L_ℓ, and similarly $\psi'_{i,j,k}$ also approximates 0 on $L_{\ell\pm1}$. The fact that we do not have such information on, e.g., $K_\ell \setminus L_\ell$ makes Assumption 22.2.2 necessary to bound the approximation error of $\tilde{M}_\ell$, see Fig. 22.5.

We proceed by bounding the approximation error.

Proposition 22.5.1 (Asymptotic error bounds) *The approximations of M_ℓ satisfy*

$$\sum_\ell |M_\ell - \overline{M}| + |M_\ell - \overline{M}_\ell| \le C\,\varepsilon^2 \Big(\sqrt{\frac{\tau'}{\tau}}\Big)^{-1} D(t,x) + C\,\varepsilon^{-2}\sqrt{\frac{\tau'}{\tau}} M(t,x) + R(t,x),$$

where again $||R(t,\cdot)||_{L^1(\mathbb{R}^n)} \le C\,\varepsilon^2$, for all $t > 0$.

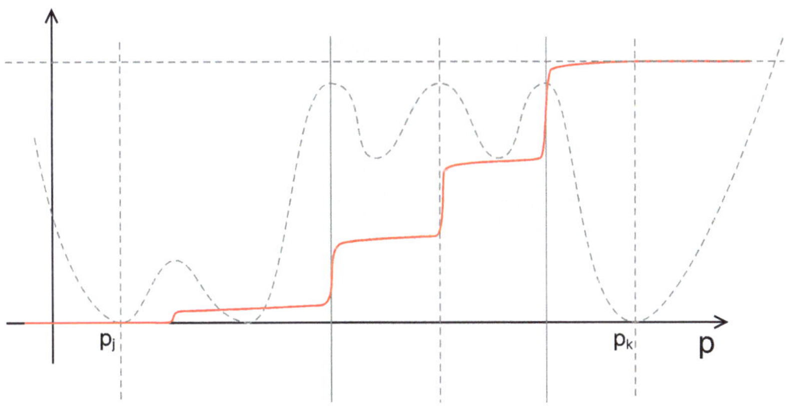

Fig. 22.5 Sketch of some $\psi_{j,k}$ (solid graph) and the corresponding potential H (dashed graph). Inside the gray region $\psi_{j,k}$ does not approximate the indicator function of the corresponding K_{k+1} as $\varepsilon \to 0$, necessitating Assumption 22.2.2

Proof We note that Assumption 22.2.3 simplifies the proof a bit because we can choose the interval $(-C, C)$, to which we restricted in Proposition 22.4.2, independent of ε. We follow now the approach of Proposition 22.4.2.

Bounding$|\overline{M}_\ell - M_\ell|$: Again, by Assumption 22.2.2 and the maximum principle, we conclude that

$$\int\limits_{K_\ell \setminus L_\ell} \varrho(t, x, p)\, dp = \int\limits_{K_\ell \setminus L_\ell} U_\ell(t, x, p)\Gamma_\ell(p)\, dp \le c \int\limits_{K_\ell \setminus L_\ell} \Gamma_\ell(p)\, dp = O(\exp(-C/\varepsilon^2), \tag{22.15}$$

for some constants $c, C > 0$. On L_ℓ, we use the bounds from Proposition 22.4.2, with $A_i = (-C, C) \cap J_i$.

$$\begin{aligned} |\overline{M}_\ell - \int_{L_\ell} \varrho(t, x, p)\, dp| &\le \sum_{p_i \in L_k} \Big| \int_{A_i} u_i(t, x, p)\gamma_i(p)\, dp - \overline{m}_i \Big| + C\, R(t, x) \\ &\le C \sum_{p_i \in L_k} \sqrt{m_i\, D_i} + R(t, x). \end{aligned}$$

Using Cauchy–Schwarz yields

$$|\overline{M}_\ell - M_\ell| \le C \sqrt{M_\ell \sum_{p_i \in K_\ell} D_i} + R.$$

Bounding$|\tilde{M}_\ell - M_\ell|$: We find an interval $\tilde{I}_\ell \subset L_\ell$, that contains all local minima $p_i \in L_\ell$ and satisfies

$$H(x) + \frac{H - h}{2} < H_i, \qquad \text{for all } x \in \tilde{I}_\ell, \text{ for all } i, j, k,$$

due to Lemma 22.5.1. This interval then satisfies the following analogue of Lemma 22.4.1

$$\sup_{p \in \tilde{I}_\ell} |\psi_{j,k}(p)| + |1 - \psi_{i,j}(p)| \le C\varepsilon \sqrt{\frac{\tau'}{\tau}},$$

$$\int_{K_\ell \setminus \tilde{I}_\ell} \Gamma_\ell(p)\, dp \le C\varepsilon \sqrt{\frac{\tau'}{\tau}}.$$

In particular, we can make the error introduced in Eq. (22.15) more explicit as $O(\varepsilon\sqrt{\frac{\tau'}{\tau}})$. Proceeding as in Proposition 22.4.2, we conclude

$$|\tilde{M}_\ell - M_\ell| \le C \sum_{|i-\ell|\le 1} \left(\sqrt{M_\ell \sum_{p_i \in K_\ell} D_i} + \varepsilon \sqrt{\frac{\tau'}{\tau}} \right) + C\, R(t,x).$$

Finally, we sum over all ℓ and obtain, by Cauchy–Schwarz and Young's inequality

$$\sum_\ell |M_\ell - \overline{M}| + |M_\ell - \overline{M}_\ell| \le C\varepsilon^2 \left(\sqrt{\frac{\tau'}{\tau}}\right)^{-1} D(t,x) + C\varepsilon^{-2} \sqrt{\frac{\tau'}{\tau}} M(t,x) + C\, R(t,x).$$

Remark 22.5.2 This is indeed an analogue of Proposition 22.4.2 since by definition $H > h$ and hence $\tau' \ll \tau$. Due to Assumption 22.2.2, these error bounds hold independently of the choice of partition $(G_k)_k$ because L_k and $H(q_{a_k-1})$, $H(q_{b_k})$ are uniquely determined, see assertion 5 of Lemma 22.5.1.

22.5.2 *Substitute and Limit Dynamics*

The dynamics of the substitute masses $\tilde{M}_\ell$ can be expressed in terms of $\overline{m}_i$, $\overline{m}_j$, $\overline{m}_k$, for $i \in L_{\ell-1}$, $j \in L_\ell$, $k \in L_{\ell+1}$, and the coefficients $c_{i,j,k}$, by the same argument as in Proposition 22.4.1. But recovering $\overline{M}_\ell$ in this dynamics imposes a system of additional conditions on the $c_{i,j,k}$ determining them uniquely.

Since the dynamics of M_ℓ will only involve $M_{\ell-1}$ and $M_{\ell+1}$, we fix an index ℓ and remove the subscript ℓ from the extrema of H. We modify the setting by renumbering the minima p_i and maxima q_j of H, such that

$$\begin{aligned} K_{\ell-1} &= (q_1, q_r), & K_\ell &= (q_r, q_s), & K_{\ell+1} &= (q_s, q_v) \\ L_{\ell-1} &= (q_{a-1}, q_b), & L_\ell &= (q_{c-1}, q_d), & L_{\ell+1} &= (q_{e-1}, q_f). \end{aligned}$$

We begin by introducing analogues of the $\theta(q_i)$. Set

$$\Theta(r) := \tau'\, \nu_r \zeta_r \varepsilon^{-2}, \qquad \Theta(s) := \tau'\, \nu_s \zeta_s \varepsilon^{-2},$$

where

$$\begin{aligned} \nu_r &:= \min\left\{ \sum_{i=1}^{r} \mu_i, \sum_{i=r+1}^{s} \mu_i \right\}, & \zeta_r &:= \int_{p_b}^{p_c} 1/\gamma(p)\, dp = \sum_{i=b}^{c-1} \eta_i, \\ \nu_s &:= \min\left\{ \sum_{i=r+1}^{s} \mu_i, \sum_{i=s+1}^{v} \mu_i \right\}, & \zeta_s &:= \int_{p_b}^{p_c} 1/\gamma(p)\, dp = \sum_{i=d}^{e-1} \eta_i. \end{aligned}$$

As the analogue for the θ_i, we set

$$\Theta = \Theta_\ell := \min(\Theta_r, \Theta_s) =: \tau' \hat{\zeta}_\ell \hat{\nu}_\ell.$$

Proposition 22.5.2 (Dynamics of $\tilde{M}_\ell$) *The substitute masses $\tilde{M}_\ell$ satisfies*

$$\Theta(\partial_t - \Delta)\tilde{M}_\ell = \sum_{i,j,k}\Big(c_{i,j,k}\cdot\Big(\Big(\frac{\hat{\zeta}_\ell\hat{\nu}_\ell}{\sum_{r=i}^{j-1}\eta_r}\Big)\Big(\frac{\overline{m}_i}{\mu_i}-\frac{\overline{m}_j}{\mu_j}\Big)+\Big(\frac{\hat{\zeta}_\ell\hat{\nu}_\ell}{\sum_{r=j}^{k-1}\eta_r}\Big)\Big(\frac{\overline{m}_k}{\mu_k}-\frac{\overline{m}_j}{\mu_j}\Big)\Big)\Big).$$

If we choose

$$c_{i,j,k} = \frac{\mu_i\,\mu_j\,\mu_k}{\sum(\mu_i\,\mu_j\,\mu_k)},$$

we obtain

$$\Theta(\partial_t - \Delta)\tilde{M}_\ell = c_{\ell-1}\cdot\Big(\frac{\hat{\zeta}_\ell\hat{\nu}_\ell}{\zeta_r}\Big)\overline{M}_{\ell-1} - c_\ell\cdot\Big(\frac{\hat{\zeta}_\ell\hat{\nu}_\ell}{\zeta_r}+\frac{\hat{\zeta}_\ell\hat{\nu}_\ell}{\zeta_s}\Big)\overline{M}_\ell + c_{\ell+1}\cdot\Big(\frac{\hat{\zeta}_\ell\hat{\nu}_\ell}{\zeta_s}\Big)\overline{M}_{\ell+1} + E,$$

with $c_\ell = \sum_{i,k}(\mu_i\mu_k)$, etc., and error term $E = E(\overline{m}_i, \overline{m}_j, \overline{m}_k)$ exponentially small.

Proof The first equation is derived exactly as in the proof of Proposition 22.4.1. It remains to show the choice of $c_{i,j,k}$. For this, we note that we can write, e.g.,

$$\frac{\hat{\zeta}_\ell}{\sum_{v=i}^{j-1}\eta_v} = \frac{\hat{\zeta}_\ell}{\zeta_r} + e_{i,j},$$

where $e_{i,j} \ll \hat{\zeta}_\ell/\zeta_r$. Next, we require that there are constants $c(i,j)$, $c(i,k)$, $c(j,k)$ such that

$$\begin{aligned}
\sum_i c_{i,j,k}\overline{m}_i/\mu_i &= c(j,k)\sum_i \overline{m}_i = c(i,j)\overline{M}_{\ell-1}\\
\sum_j c_{i,j,k}\overline{m}_j/\mu_j &= c(i,k)\sum_i \overline{m}_j = c(i,k)\overline{M}_\ell\\
\sum_k c_{i,j,k}\overline{m}_k/\mu_k &= c(i,j)\sum_k \overline{m}_k = c(j,k)\overline{M}_{\ell+1}\\
\sum_{i,j,k} c_{i,j,k} &= 1.
\end{aligned}$$

The first equation then implies $c_{i,j,k} = c(i,j)\mu_i$ and so forth. Hence, we obtain the claimed form for the $c_{i,j,k}$, and the term E is then composed of $e(i,j)$, $e(j,k)$. □

Having established the dynamics of the substitute masses, we can now state the rigorous formulation of the second main result, Eq. (22.8).

Theorem 22.5.1 (Limit dynamics for M_ℓ) *Let $\breve{M}_\ell(t,x)$ be the solutions to the system*

$$\begin{aligned}
(\partial_t - \Delta_x)\breve{M}_\ell &= A_\ell\breve{M}_{\ell-1} - B_\ell\breve{M}_\ell + C_\ell\breve{M}_{\ell+1},\\
\breve{M}_\ell(0,x) &= \tilde{M}_\ell(0,x),
\end{aligned}$$

with

$$A_\ell := \lim_{\varepsilon\to 0} \frac{c_{\ell-1}}{\Theta_\ell}\Big(\frac{\hat{\zeta}_\ell \hat{\nu}_\ell}{\zeta_r}\Big), \quad B_\ell := \lim_{\varepsilon\to 0} \frac{c_\ell}{\Theta_\ell}\Big(\frac{\hat{\zeta}_\ell \hat{\nu}_\ell}{\zeta_r} + \frac{\hat{\zeta}_\ell \hat{\nu}_\ell}{\zeta_s}\Big), \quad C_\ell := \lim_{\varepsilon\to 0} \frac{c_{\ell+1}}{\Theta_\ell}\Big(\frac{\hat{\zeta}_\ell \hat{\nu}_\ell}{\zeta_s}\Big).$$

Then for all $T > 0$

$$\sum_{\ell}^{n} \int_0^T \int_{\mathbb{R}^n} |\breve{M}_\ell - M_\ell| \, dx \, dt \le C(T)\, \varepsilon^2,$$

for a time-dependent constant $C(T)$.

Proof The proof is identical to the one of Theorem 22.4.1. □

Remark 22.5.3 This limit dynamics is, up to the more involved rate constants, identical to the limit dynamics at the faster timescale τ. In particular, in the case where $L_\ell, L_{\ell\pm1}$, each only contain one local minimum, we also recover the Kramers constants whenever $A_\ell,\ B_\ell,\ C_\ell$ are non-zero.

References

Arnrich S, Mielke A, Peletier M, Savaré G, Veneroni M (2011) Passing to the limit in a Wasserstein gradient flow: from diffusion to reaction. Siam J Math Anal 44:419–454. https://doi.org/10.1007/s00526-011-0440-9

Bender CM, Orszag SA (1999) Advanced mathematical methods for scientists and engineers: I: Asymptotic methods and perturbation theory. Springer, New York

Berglund N (2013) Kramers' law: validity, derivations and generalisations. Markov Process Related Fields 19(3):459–490

Bovier A, Eckhoff M, Gayrard V, Klein M (2004) Metastability in reversible diffusion processes i: sharp asymptotics for capacities and exit times. J Europ Math Soc 006(4):399–424. https://doi.org/10.4171/JEMS/14

Dreyer W, Guhlke C, Herrmann M (2011) Hysteresis and phase transition in many-particle storage systems. Contin Mech Thermodyn 23(3):211–231. https://doi.org/10.1007/s00161-010-0178-1

Dreyer W, Huth R, Mielke A, Rehberg J, Winkler M (2015) Global existence for a nonlocal and nonlinear Fokker-Planck equation. Z Angew Math Phys 66(2):293–315. https://doi.org/10.1007/s00033-014-0401-1

Eberle S, Niethammer B, Schlichting A (2017) Gradient flow formulation and longtime behaviour of a constrained Fokker Planck equation. Nonlinear Anal 158:142–167. https://doi.org/10.1016/j.na.2017.04.009

Evans LC, Tabrizian P (2016) Asymptotics for scaled kramers-smoluchowski equations. SIAM J Math Anal 48(4):2944–2961. https://doi.org/10.1137/15M1047453

Friedman A (1964) Partial differential equations of parabolic type. Dover books on mathematics. Prentice-Hall, Inc., Englewood Cliffs, N.J., 1964. reprinted 2008 by Dover Publications

Helffer B, Klein M, Nier F (2004) Quantitative analysis of metastability in reversible diffusion processes via a Witten complex approach. Mat Contemp 26:41–85

Herrmann M, Niethammer B (2010) Kramers' formula for chemical reactions in the context of wasserstein gradient flows. Commun Math Sci 9:623–35. https://doi.org/10.4310/CMS.2011.v9.n2.a15

Herrmann M, Niethammer B (2020) Mass transport in fokker-planck equations with tilted periodic potential. Europ J Appl Math 31(4):709–736. https://doi.org/10.1017/S0956792519000251

Herrmann M, Niethammer B, Velázquez J (2012) Kramers and non-kramers phase transitions in many-particle systems with dynamical constraint. Multiscale Model Simul 10(3):818–852. https://doi.org/10.1137/110851882

Herrmann M, Niethammer B, Velázquez JL (2014) Rate-independent dynamics and Kramers-type phase transitions in nonlocal Fokker-Planck equations with dynamical control. Arch Ration Mech Anal 214(3):803–866. https://doi.org/10.1007/s00205-014-0782-5

Jordan R, Kinderlehrer D, Otto F (1997) Free energy and the fokker-planck equation. Physica D: nonlinear Phenomena 107(2):265–271. https://doi.org/10.1016/S0167-2789(97)00093-6. 16th Annual International Conference of the Center for Nonlinear Studies

Jordan R, Kinderlehrer D, Otto F (1998) The variational formulation of the fokker-planck equation. SIAM J Math Anal 29(1):1–17. https://doi.org/10.1137/S0036141096303359

Kramers H (1940) Brownian motion in a field of force and the diffusion model of chemical reactions. Physica 7(4):284–304. https://doi.org/10.1016/S0031-8914(40)90098-2

Michel L, Zworski M (2018) A semiclassical approach to the kramers-smoluchowski equation. SIAM J Math Anal 50(5):5362–5379. https://doi.org/10.1137/17M1124826

Peletier M, Savaré G, Veneroni M (2010) From diffusion to reaction via Γ-convergence. SIAM J Math Anal 42(4):1805–1825. https://doi.org/10.1137/090781474

Peletier MA, Schlichting A (2023) Cosh gradient systems and tilting. Nonlinear Anal 231, Paper No. 113094, 113. https://doi.org/10.1016/j.na.2022.113094

Risken H (1996) The Fokker-Planck Equation: methods of solution and applications. Springer, Berlin Heidelberg, Berlin, Heidelberg

Seo I, Tabrizian P (2019) Asymptotics for scaled kramers-smoluchowski equations in several dimensions with general potentials. Calculus of variations and partial differential equations, 59. https://doi.org/10.1007/s00526-019-1669-y

Chapter 23
Distortion and Residual Stresses in an Additively Manufactured Disk Stack: Continuum Modeling and FEM Simulation for Polyamide 12

Umut Kaya and Alfons Noe

Abstract In this study, an engineering scale model for the simulation of Selective Laser Sintering (SLS) of industry level parts made of Polyamide 12 (PA 12) is introduced. This model is employed to investigate the evolution of temperature, stresses, deformations, and phase transformations during the SLS process. A finite element model is constructed using ABAQUS for a stack of five circular plates. The chemical kinetics, the thermoelastic and the thermoviscoelastic (TVE) material behavior of PA 12 are programmed with Fortran in three ABAQUS subroutines, needed due to phase changes occurring during the process. The computations are performed for both elastic and time-dependent constitutive mechanical laws. The results are compared and assessed.

Keywords Selective laser sintering · Stacked disks · Material modeling · Polyamide 12 · Crystallization kinetics · Thermoviscoelasticity · Warpage · Residual stresses

23.1 Introduction

The objective of this study is to create a finite element model in order to investigate the evolution of stresses, deformations, temperature, and crystallization during Selective Laser Sintering (SLS) of industry level parts made of Polyamide 12. In the first section, SLS process and the characteristics associated with this manufacturing method are introduced. Existing modeling methods for the simulation of SLS are reviewed. In the second section, the theoretical background of kinetics

U. Kaya · A. Noe (✉)
Laboratory of Engineering Mechanics and Simulation, Department of Management and Technology, University of Applied Sciences of South Westphalia, Soest, Germany
e-mail: noe.alfons@fh-swf.de

U. Kaya
e-mail: kaya.umut1@fh-swf.de

J. Fuhrmann et al. (eds.), *Advances in Continuum Physics*, Advanced Structured Materials 238, https://doi.org/10.1007/978-3-031-93918-1_23

of polymer crystallization, the thermoviscoelastic (TVE) material behavior, and the associated tailored thermo-mechanically coupled numerical implementation are carefully detailed. In Sect. 23.3, the finite element model developed for simulations is described. Results for field variables and time histories are presented in Sect. 23.4 and are discussed in Sect. 23.5.

23.1.1 Manufacturing by Selective Laser Sintering (SLS)

Selective Laser Sintering (SLS) is a variant of an Additive Manufacturing (AM) process. Figure 23.1 depicts the process and basic setup of the printing machine. In SLS, the laser is used as a heat source to melt powder particles partially in each layer (Yasa 2021). After repetitive layer deposition, a solid object is obtained.

A SLS process is divided into preheating, building, and cooling phases. During the preheating the powder is heated up to just below the melting temperature for semi-crystalline polymers and close to the glass transition temperature for amorphous polymers. The purposes are to minimize required energy to melt the material, to keep the thermal gradient between the melted material and the powder small, and to minimize the thermal expansion of the material when the laser is applied (Goodridge et al. 2012). After preheating, in the building phase, contours are traced by a laser and the powder at those contours is sintered with the laser power. A roller or re-coater adds a new powder layer from a reservoir after lowering the platform (see Fig. 23.1). The preheating and building phases are repeated until the final layer. The building platform is often maintained at elevated temperatures by heat sources at the bottom, sides, and top, which avoids fast cooling. Finally, the part remains in the chamber until it cools down to room temperature.

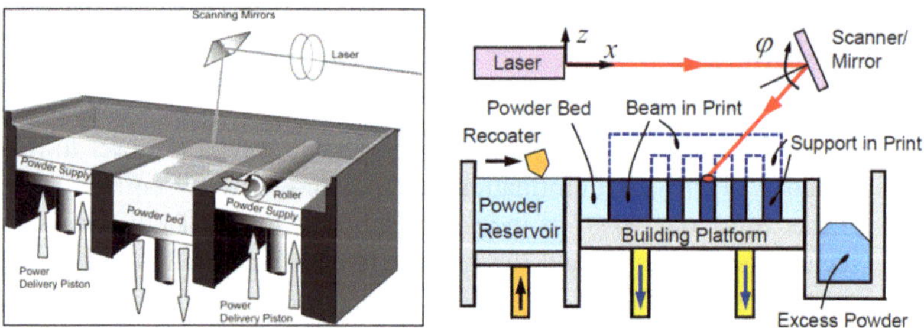

Fig. 23.1 A laser sintering equipment on left (Hopkinson and Dickens 2006) and right (Gladbach et al. 2024)

23.1.2 Modeling Options for the SLS Process

As described above, AM is quite sensitive to a number of parameters. In order to understand the effect of those parameters and to determine optimum values for them, a simulation model becomes a necessity. There are different approaches for the modeling of additive manufacturing processes. These can be classified according to various scales and modeling methods. A first approach classifies between using meshes or meshfree modeling. Meshfree models are known as Optimal Transportation Meshfree (OTM), Smoothed Particle Hydrodynamics (SPH), Monte Carlo Ray trace, Discrete Element Modeling (DEM), and Lattice Boltzmann Method (LBM) (Soundararajan et al. 2021). Mesh-based modeling techniques are Phase Field, Finite Element Method/Analysis (FEM or FEA), and Finite Difference/Finite Volume Methods (FDM/FVM) procedures. Cellular automation FEM and combining DEM with FEM/FDM combine both means.

Another classification can be made based on the dimensional scale and single layer time scale of the models. The models are suitable for the microscopic scale in order to investigate the microstructure evolution, grain growth, and laser beam penetration. The dimensional scales of these models may range from μm to mm, and their single layer time scale may range from μs to ms. Monte Carlo ray trace, cellular automation FEM, and phase field methods are suitable for the microscopic scale (Soundararajan et al. 2021). At the macroscopic level, temperature history, residual stress, and deformations can be studied. The macroscopic models apply from mm to cm scale dimensionally and between ms and s scale for a single layer time. The investigations of Denlinger and Michaleris (2016), Williams et al. (2018), and Heigel et al. (2015) on metal alloys are examples of the FEM at the macroscopic level. Mesoscopic models, as the name suggests, are dimensionally and single layer time-wise between the micro- and the macro-scale and they are used for studying powder spreading, melt pool dynamics, and pore formation (Soundararajan et al. 2021). For example, Shen et al. (2021) conducted a mesoscopic level study and investigated the melt pool as well as the stress and temperature in SLS with PA 12. The study also shows the effect of crystallization and cooling rate on the stress. A macroscopic level study was conducted by Amado (2016) applying SLS for the polymers of PA 12 and icoPP with a FEM model. His study investigated the stress and warpage only in the first layers of the process. The OTM, SPH, DEM, FEM, and FDM/FEA are suitable for both macro- and mesoscopic simulations (Soundararajan et al. 2021). Furthermore, the FEM and FDM/FEA even enable limited applications on microscopic scales. Balemans et al. (2020) used the FEM for the analyses of crystallization kinetics and grain growth of PA 12. Lastly, the Lattice Boltzmann Method (LBM) is applicable for micro- as well as mesoscopic scale simulations (Soundararajan et al. 2021).

In this contribution, an engineering-level configuration made up of PA 12 and built by SLS is analyzed by FEM. The study includes thermoviscoelastic material properties on a macroscopic scale. The configuration consists of multiple parts in a single batch, which is common in industrial applications. Thus, five vertically stacked disk-shaped plates, separated by powder layers, are used as a FEM model. Peters (2020) had previously simulated the chemical kinetics of the stacked disks model. Kaya (2024) modeled the thermoelastic behavior and coupled it with the chemical kinetics. Additional time-dependent coupled behavior is investigated in this work.

There are many challenges to use the FEM for AM modeling and some are addressed in this investigation. They cover phase transformation phenomena and temperature-dependent material properties. Temperature-dependent properties are the source of nonlinearity (Gouge and Michaleris 2018). Phase transformation is a challenge for the convergence of the algorithm. These thermo-chemical features serve as an input to the advanced case of coupled thermo-mechanical modeling and analyses of deformation and residual stress.

23.1.3 Stacked Disks Model for the Manufacturing Scale

In order to study distortion and stress in SLS, a set of five disks in a vertically stacked configuration is chosen. This simple geometry can provide enough complexity for an engineering analysis (see Fig. 23.2). The deflection at the end of cooling is a visible, relevant quantity that can be measured but it can also be calculated, as shown hereafter, where the present simulation allows to record the deformation history easily.

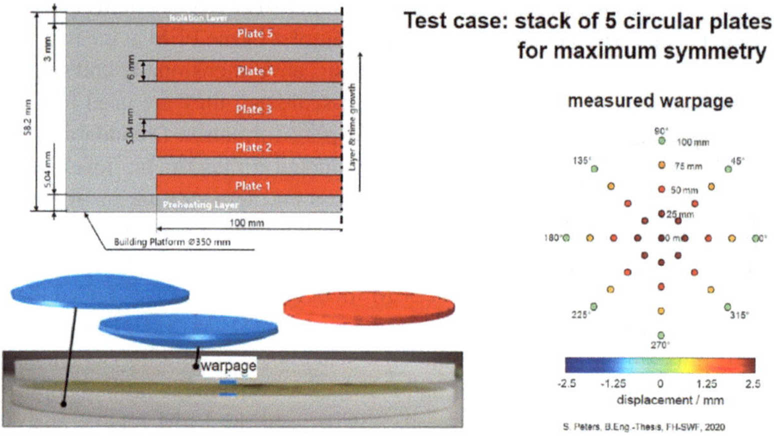

Fig. 23.2 Configuration of five disks in the building chamber on the top left and deformation at the end of the process at the bottom left (Peters 2020). Distortion ranges from concave to convex

23.1.4 Stress and Warpage Due to Constraints and Inherent Strains

The important problems that arise in context with SLS are temperature gradients between layers and volume changes by chemical kinetics of semi-crystalline PA 12. These two reasons cause two different phenomena. The first one is the Temperature Gradient Mechanism (TGM). The TGM is used intentionally in laser bending of sheet metal but it occurs also in SLS and SLM (Mercelis and Kruth 2006; Kruth et al. 2004) although not desired. The effect of this mechanism becomes considerable when printing is continuing on the already solidified layers. A large temperature gradient occurs due to the rapid heating of the upper layer by the laser and relatively slow heat conduction through the layers. The upper layers expand more than the lower layers due to this temperature gradient and bend away from the laser (see Fig. 23.3a). During bending, hot upper layers push the colder lower layers outward. Therefore, the upper layers experience compressive stress and lower layers undergo plastic deformation. When heat input by the laser ceases, the part starts cooling. This time, upper layers shrink more than lower layers because they are hotter than lower layers. This causes bending toward the laser beam shown on the bottom of Fig. 23.3a. During this bending, the colder lower layers resist shrinking of upper layers, causing tensile stress there. Thus, the sheet first bends away from the heat source and then toward it.

Another phenomenon that occurs during SLS is polymer crystallization. Molecular chains of semi-crystalline polymers form regular structures during crystallization, which starts after solidification. During the formation of these regular structures, semi-crystalline polymers shrink. Crystallization is a phenomenon driven by temperature decrease in a small interval, as explained in Sect. 23.2. Therefore, its progress follows the temperature gradient that starts at the colder bottom and advances toward

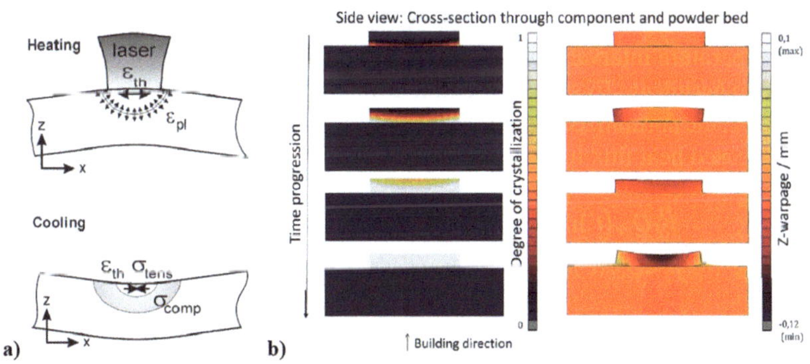

Fig. 23.3 **a** Temperature Gradient Mechanism (TGM): It deforms the part away from the heat source while heating, seen on top, then toward the opposite direction during cooling, seen on the bottom (Kruth et al. 2004). **b** Oscillating bending due to the progress of crystallization strain (shrinkage) from bottom to top (Neugebauer 2017)

the hotter top during printing. Therefore, chemical shrinkage starts at the bottom and advances to the top, resulting in alternating bending behavior similar to TGM (see Fig. 23.3b). The resultant warpage and residual stresses are the combined effects of these two behaviors.

23.2 Continuum Models for of Crystallization, Temperature, Displacement, and Stress

Constitutive, continuum-based modeling for the coupled phenomena of crystallization kinetics, temperature fields emerging from first law of thermodynamics and momentum balance for a thermoviscoelastic (TVE) material is outlined. This computationally cheap TVE approach enables to focus on the history of thermo-chemical–mechanical fields for the polymer PA 12, but this approach leaves out irreversible (plastic) deformations.

23.2.1 Crystallization Kinetics and Shrinkage Strains

Polymers are composed of large molecular chains. In amorphous polymers, these chains form into an irregular structure, whereas in semi-crystalline polymers a fraction of the chains creates regular patterns. The formation of these regular structures is known as crystallization. However, semi-crystalline materials also keep some of their irregular (amorphous) regions during crystallization. For PA 12, the crystallized volume fraction attains 35–45% (Ehrenstein 2001). Crystallization occurs during cooling at a temperature range, which is lower than the melting temperature after heating, and this state is also named undercooled liquid. During the exothermal crystallization, the crystallization enthalpy $\Delta H_{cr} = V^{cr} \Delta h_{cr}$ is mainly released in the small temperature interval $\Delta T^{cr} \leq 10\,\text{K}$ at ca. 161–171 °C. It can be measured in a caloric experiment. To specify the time evolution of the crystallization in a given volume V^{cr}, the normalized quantity $\alpha(t) = \alpha^{\text{cr}}(t), \quad \alpha \in [0, 1]$ is introduced, defined by the released heat flux history:

$$\alpha^{cr} = \frac{\int_0^t \dot{Q}_{cr}(t')\mathrm{d}t'}{\int_0^\infty \dot{Q}_{cr}(t')\mathrm{d}t'} = \frac{\int_0^t \dot{h}_{cr}(t')\mathrm{d}t'}{\int_0^\infty \dot{h}_{cr}(t')\mathrm{d}t'} = \frac{\Delta h_{cr}(t)}{\Delta h_{cr,\max}} \in [0,\ 1]. \tag{23.1}$$

The time evolution of the crystallization can be calculated from the kinetic theory of Avrami, devised for isothermal cases (Avrami 1939)

$$\alpha(t) = \alpha^{cr}(t) = 1 - \exp\left[-k(T_{iso}) \cdot t^n\right] \quad \text{or} \quad \alpha(t) = 1 - \exp\left[-\left(K(T_{iso}) \cdot t\right)^n\right]. \tag{23.2}$$

The parameter n is the Avrami exponent and $K(T_{iso})$ is the crystallization rate at constant temperature. During SLS, the temperature changes with time and position, hence the isothermal approach is not applicable. Therefore, Nakamura et al. (1972) developed a non-isothermal kinetic rate model for crystallization based on Avrami by

$$\frac{\partial\alpha}{\partial t} = n\,K(T)\,(1-\alpha)[-\ell n(1-\alpha)]^{\frac{n-1}{n}}, \quad \alpha = \alpha^{cr}(t(T)). \tag{23.3}$$

The parameter $K(T)$ is related to the Avrami exponent n and to $k(T)$ by the means of

$$K(T) = k(T)^{1/n} = \ell n(2)^{1/n}\left[1/t_{1/2}\right]. \tag{23.4}$$

The term $t_{1/2}$ is the half-crystallization time $t_{1/2} = t\left(\alpha = \alpha^{cr} = 1/2\right)$, which is obtained from the crystal nucleation theory of Hoffman–Lauritzen theory (Hoffman et al. 1976). It reads

$$\left(\frac{1}{t_{1/2}}\right) = K_0 \exp\left(\frac{-U}{R(T-T_\infty)}\right) \exp\left(\frac{-K_G(T+T_0)}{2T^2(T_0-T)}\right), \tag{23.5}$$

where U is the molar activation energy for crystallization, $R = 8.31447\,\mathrm{kJ}/(\mathrm{kmol}\cdot\mathrm{K})$ is the universal gas constant, T_∞ is the limit temperature below which crystallization ceases, and T_0 is the equilibrium melting temperature. The measured parameters K_0 and K_G are the Hoffman–Lauritzen constants. Herein, for the used PA 12 numerical values presented in Table 23.1 have been adopted.

Isothermal crystallization functions $\alpha(t)$ for various temperatures are shown in Fig. 23.4.

Apparently, the growth of crystallization $\alpha(t)$ is faster at lower temperatures, which results from larger under-cooling. For $T = \{150°\mathrm{C},\ 160°\mathrm{C}\}$, the half-times assume $t_{1/2} = t\left(\alpha = 1/2\right) = \{54\mathrm{s},\ 230\mathrm{s}\}$. The half-time corresponds with the peak crystallization rate $\dot{\alpha}_{\max} = \dot{\alpha}(t_{1/2})$. Since crystallization strain ε^{cr} is proportional to crystallization, this strain increases also faster in lower temperatures regimes but it always reaches the same terminal value (Brinson and Brinson 2008).

Table 23.1 Hoffmann–Lauritzen constants used for simulation (Hoffman et al. 1976)

$K_0[1/\mathrm{s}]$	8639	n	2.8
$K_G[\mathrm{K}^2]$	109258	$T_\infty[°\mathrm{C}]$	30
$U[kJ/kmol]$	6270	$T_0[°\mathrm{C}]$	190

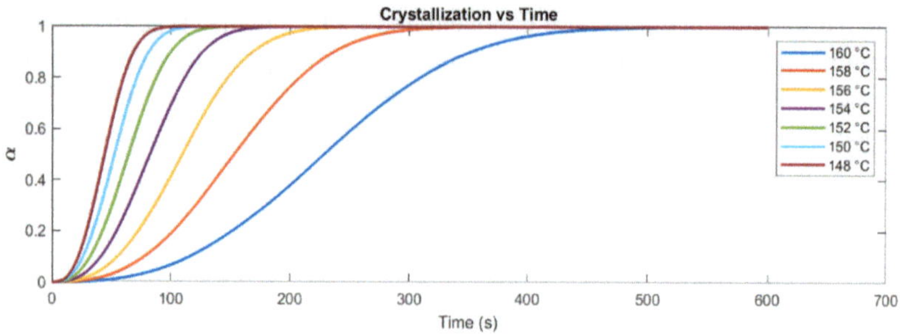

Fig. 23.4 Isothermal crystallization versus time for different temperatures

23.2.2 Energy Balance and Thermal Material Laws for Temperatures

Based on the first law of thermodynamics for a mass element, the rate of internal energy per volume is determined by thermal conduction, the exothermal crystallization, the laser input, and the mechanical power. The balance reads

$$\rho\, c_p \frac{\partial T}{\partial t} = -\nabla \cdot \dot{\mathbf{q}}_{cond} + \rho\, \dot{r}_{cr} + p_{laser} + \dot{w}_{mech}. \tag{23.6}$$

The heat conduction vector is linked to the temperature by Fourier's law and implies

$$\dot{\mathbf{q}}_{cond} = -(k(\alpha) \cdot \operatorname{grad} T) \rightarrow -\operatorname{div} \cdot \dot{\mathbf{q}}_{cond} = \nabla \cdot (k(\alpha) \cdot \nabla T) \tag{23.7}$$

Referring to the definition of $\alpha^{cr}(t)$, the crystallization power per volume is the product of the material-specific crystallization enthalpy and the unknown crystallization rate

$$\dot{r}_{cr} = \Delta\dot{h}_{cr}(t)\ \frac{\partial \Delta h_{cr}}{\partial t} = \frac{\partial \alpha^{cr}}{\partial t} \cdot \Delta h_{\text{cr, max}},\ \Delta h_{cr,\max} = \Delta h_{cr}\big(\alpha^{cr} = 1\big). \tag{23.8}$$

In this investigation, the moving laser, which is a local power source penetrating into the surface layer, is not defined as a boundary condition. Instead, the laser heating is simplified as an initial temperature value on the top surface.

Calculations with typical material parameters show that among the contributing powers the mechanical power formed by stress times strain rates is by magnitude the smallest part and can therefore be neglected (Gladbach et al. 2024). This also justifies the adopted one-way coupling of energy and momentum balance. The PDE for the temperature field $T = T(t, \alpha(T), \mathbf{x})$ is instead strongly coupled with time and temperature depending crystallization function $\alpha = \alpha(T, t)$. Thus, the following partial differential equation

$$\rho(T,\alpha)c_p(T,\alpha)\frac{\partial T}{\partial t} = \nabla\cdot(k(\alpha)\cdot\nabla T) + \rho(T,\alpha)\,\Delta h_{cr}\frac{\partial\alpha}{\partial t} \tag{23.9}$$

for the temperature evolution must be solved, which is coupled with the ODE (23.3) for the crystallization evolution. Regardless of the nonlinearity of the ODE (23.3), this system is nonlinear, because the density $\rho(T,\alpha)$, the specific heat capacity $c_p(T,\alpha)$, and the heat conduction coefficient $k(\alpha)$ are functions of up to two unknowns. Here, the subsequent empirically founded approximations for the three material states have been utilized:

$$\rho(T,\alpha) = \begin{cases} \rho^{pw} & & \text{powder} \\ \rho^{am} & T > 160^\circ\text{C} & \text{amorph} \\ \rho^{cr} & T \leq 160^\circ\text{C} & \text{crystallized,} \end{cases} \tag{23.10}$$

$$c_p(T,\alpha) = \begin{cases} c_p^{pw}(T) = c_p^{cr}(T) & \text{powder} \\ \alpha\, c_p^{cr}(T) + (1-\alpha)\, c_p^{am}(T) & \text{sintered,} \end{cases} \tag{23.11}$$

$$k(\alpha) = \begin{cases} k^{pw} & \text{powder} \\ \alpha k^{cr} + (1-\alpha)k^{am} & \text{sintered} \end{cases} \tag{23.12}$$

23.2.3 *Momentum Balance and Thermoviscoelastic Material Laws*

Since the printing process is very slow and body forces are irrelevant, the static momentum balance

$$\mathbf{0} = \nabla\cdot\boldsymbol{\sigma}\big(\boldsymbol{\varepsilon}(T,\alpha^{cr}), \nabla\mathbf{u}, t\big) \Leftrightarrow \quad \mathbf{0} = \text{div}\boldsymbol{\sigma} \tag{23.13}$$

applies, with a strong impact by the temperature $T(\mathbf{r},t)$ due to the energy balance and $\alpha^{cr}(t,T)$ from crystallization. The reverse impact of stresses and strains is negligible.

23.2.3.1 Thermoelasticity

The simplest constitutive model for the stress and deformation adopts time-independent behavior that is expressed by the thermoelastic Hooke's law

$$\boldsymbol{\sigma} = \mathbf{C}\,\boldsymbol{\varepsilon}^{el} = \mathbf{C}\big(\boldsymbol{\varepsilon} - \boldsymbol{\varepsilon}^{th} - \boldsymbol{\varepsilon}^{cr}\big), \tag{23.14}$$

where $\boldsymbol{\sigma}$ is the stress vector, $\mathbf{C}$ is the stiffness matrix, $\boldsymbol{\varepsilon}$ denotes the total strain vector built by the displacement vector $\mathbf{u}$, and $\boldsymbol{\varepsilon}^{th}$, $\boldsymbol{\varepsilon}^{cr}$ are the normal vectors representing the thermal strains and the crystallization strains. For the axially symmetric disk model at hand, the definitions for stress, displacement, and strain are given by the following expressions:

$$\boldsymbol{\sigma} = \begin{pmatrix} \sigma_{rr} \\ \sigma_{zz} \\ \sigma_{\theta\theta} \\ \sigma_{rz} \end{pmatrix} = \begin{pmatrix} \sigma_{11} \\ \sigma_{22} \\ \sigma_{33} \\ \sigma_{12} \end{pmatrix}, \quad \mathbf{u} = \begin{pmatrix} u_r \\ u_z \end{pmatrix} = \begin{pmatrix} u_1 \\ u_2 \end{pmatrix},$$

$$\boldsymbol{\varepsilon} = \begin{pmatrix} \varepsilon_{rr} \\ \varepsilon_{zz} \\ \varepsilon_{\theta\theta} \\ \varepsilon_{rz} \end{pmatrix} = \begin{pmatrix} \varepsilon_{11} \\ \varepsilon_{22} \\ \varepsilon_{33} \\ \varepsilon_{12} \end{pmatrix}, \tag{23.15}$$

where the displacement–strain relations read

$$\varepsilon_{rr} = \frac{\partial u_r}{\partial r}, \ \varepsilon_{zz} = \frac{\partial u_z}{\partial z}, \ \varepsilon_{\theta\theta} = \frac{u_r}{r}, \ 2\varepsilon_{rz} = \gamma_{rz} = \frac{\partial u_r}{\partial z} + \frac{\partial u_z}{\partial r}. \tag{23.16}$$

The volumetric strain vectors for $\boldsymbol{\varepsilon}^{th}$, $\boldsymbol{\varepsilon}^{cr}$ are determined as

$$\boldsymbol{\varepsilon}^{th} = \Delta\varepsilon^{th}(\Delta T, t) \begin{pmatrix} 1 \\ 1 \\ 1 \\ 0 \end{pmatrix} = \Delta\varepsilon^{th}(\Delta T, t)\,\boldsymbol{\kappa},$$

$$\boldsymbol{\varepsilon}^{cr} = \Delta\varepsilon^{cr}\left(\Delta T, \alpha^{cr}, t\right) \begin{pmatrix} 1 \\ 1 \\ 1 \\ 0 \end{pmatrix} = \Delta\varepsilon^{cr}\boldsymbol{\kappa}. \tag{23.17}$$

The isotropic thermal strain for a constant thermal expansion coefficient α^{th} reads

$$\varepsilon^{th}(T) - \varepsilon^{th}(T_0) = \int_{T'=T_0}^{T'=T} \alpha^{th}\left(T'\right) \mathrm{d}T' = \alpha^{th}(T(t) - T_0(t_0)) = \alpha^{th}\Delta T \tag{23.18}$$

and changes proportionally to the time-variable temperature change $\Delta T(\Delta t)$.

The isotropic crystallization strain $\Delta\varepsilon^{cr}$ is well approximated as a factor of crystallization change, $\Delta\alpha^{cr} = \Delta\alpha^{cr}(\Delta T(\Delta t))$, $\alpha^{cr} \in [0, 1]$ and it is highly nonlinear due to temperature and time dependencies. The negative chemically induced shrinkage strain ε_{vol}^{sh} at the end of the crystallization, $\alpha^{cr} = 1$, is derived from

$$\begin{aligned} \varepsilon^{cr}(T) - \varepsilon^{cr}(T_0) &= \left(\varepsilon_{vol}^{sh}/3\right) \cdot \left[\alpha^{cr}(T(t)) - \alpha^{cr}(T_0(t_0))\right] \\ &= \left(\varepsilon_{vol}^{sh}/3\right) \cdot \Delta\alpha^{cr} = \Delta\varepsilon^{cr}(T) \end{aligned}. \tag{23.19}$$

The shrinkage strain ε_{vol}^{sh} is defined by Bogetti and Gillespie (1992) as

$$V_0 + \alpha^{cr}\Delta V_{\max} = V_0\left(1+\varepsilon^{cr}\right)^3, \quad \varepsilon_{vol}^{sh} = \Delta V_{\max}/V_0 = \text{constant} < 0. \quad (23.20)$$

Since chemically induced shrinkage strain is up to 3.3% (see Table 23.3), a Taylor expansion yields the linear law

$$\varepsilon^{cr} = \left(1+\alpha^{cr}\,\mathrm{e}_{vol}^{sh}\right)^{1/3} - 1, \quad \left|\varepsilon_{vol}^{sh}\right| << 1 \rightarrow \varepsilon^{cr} \approx \alpha^{cr}\varepsilon_{vol}^{sh}/3. \quad (23.21)$$

For further analysis, Hooke's law is limited to elastic isotropy. Polymers are distinctively sensitive to volumetric (=bulk) and deviatoric deformations and therefore Hooke's law is expressed with the bulk modulus K and the shear modulus G, respectively, which requires to decompose the stresses and strains in a bulk and deviatoric part, reading

$$\begin{aligned}
\sigma_{ij} &= \sigma_{ij}^{dev} + \sigma_{ij}^{vol} &&\Leftrightarrow \quad \boldsymbol{\sigma} = \boldsymbol{\sigma}^{dev} + \boldsymbol{\sigma}^{vol}\\
\sigma_{ij}^{dev} &= \sigma_{ij} - \delta_{ij}\sigma_{kk}/3 &&\Leftrightarrow \quad \boldsymbol{\sigma}^{dev} = \boldsymbol{\sigma} - \mathbf{I}\cdot \mathrm{tr}(\boldsymbol{\sigma})/3, \quad \mathrm{tr}\left(\boldsymbol{\sigma}^{dev}\right) = 0\\
\sigma_{ij}^{vol} &= \delta_{ij}\sigma_{kk}/3 &&\Leftrightarrow \quad \boldsymbol{\sigma}^{vol} = \mathbf{I}\cdot \mathrm{tr}(\boldsymbol{\sigma})/3, \quad \sigma^{vol} = (\sigma_{11}+\sigma_{22}+\sigma_{33})/3,
\end{aligned} \quad (23.22)$$

and similar for the strain, thus replacing all $\boldsymbol{\sigma}$ by $\boldsymbol{\varepsilon}$. Then Hook's law becomes

$$\begin{aligned}
\sigma_{ij} &= 2G\left(\varepsilon_{ij} - \delta_{ij}\varepsilon_{kk}/3\right) + K\left(\varepsilon_{kk} - 3\varepsilon^{th} - 3\varepsilon^{cr}\right)\delta_{ij},\\
\varepsilon_{ij}^{\mathrm{dev}} &= \varepsilon_{ij} - \delta_{ij}\varepsilon_{kk}^{\mathrm{vol}}/3.
\end{aligned} \quad (23.23)$$

The moduli K and G can be obtained from Young´s modulus E and Poisson number υ by the formulas

$$\begin{aligned}
G &= \frac{E(T)}{2(1+\upsilon)}, \quad K = \frac{E(T)}{3(1-2\upsilon)} = \frac{G\,E}{9G-3E},\\
E &= \begin{cases} E^{pw}, & \text{powder}\\ \frac{E^{sol}(T)}{E_{\infty}^{sol}(T_{melt})} = \left(\frac{T_{melt}}{T}\right)^M, & \text{sintered}\end{cases}
\end{aligned} \quad (23.24)$$

where temperature dependence of the elasticity constants is characterized by the phase-dependent E-modulus. For sintering at $T \leq T_{melt}$ an empirical power law is convenient (see Table 23.3). Explicit time dependence is excluded, signified by the index infinity in E_{∞}. The thermally induced strains contribute by the means of

$$3\varepsilon^{th}(\Delta T) = 3\alpha^{th}\cdot\Delta T, \qquad 3\varepsilon^{cr}(\Delta T) = \varepsilon_{vol}^{sh}\cdot\Delta\alpha_{(\Delta T)}^{cr}. \quad (23.25)$$

For axial symmetry the time-independent thermoelastic Hooke's law reads

$$\begin{pmatrix} \sigma_{rr} \\ \sigma_{zz} \\ \sigma_{\theta\theta} \\ \sigma_{rz} \end{pmatrix} = \frac{2}{3}G \cdot \begin{pmatrix} +2 & -1 & -1 & 0 \\ -1 & +2 & -1 & 0 \\ -1 & -1 & +2 & 0 \\ 0 & 0 & 0 & 3 \end{pmatrix} \cdot \begin{pmatrix} \varepsilon_{rr} \\ \varepsilon_{zz} \\ \varepsilon_{\theta\theta} \\ \varepsilon_{rz} \end{pmatrix}$$

$$+ K \cdot \begin{pmatrix} 1 \\ 1 \\ 1 \\ 0 \end{pmatrix} \cdot \left[(\varepsilon_{rr} + \varepsilon_{zz} + \varepsilon_{\theta\theta}) - 3\left(\varepsilon^{th} + \varepsilon^{cr}\right) \right]. \tag{23.26}$$

The sum $\varepsilon_{kk} = \varepsilon_{rr} + \varepsilon_{zz} + \varepsilon_{\theta\theta} = \varepsilon_{kk}^{vol} \equiv \varepsilon^{vol} = \mathrm{tr}(\boldsymbol{\varepsilon})$ is the volumetric strain.

23.2.3.2 Thermoviscoelasticity

For polymers, such as PA 12, stress–strain-curves are known to be time dependent, in addition to the temperature dependence, which can be included into thermoelasticity as parameter (Brinson and Brinson 2008; Schwarzl 1990). To treat the long manufacturing times in a practically reasonable simulation time by FEM, e.g., about 20 h for the disk stack, the constitutive modeling by thermoviscoelasticity is good compromise, compared to the refined thermoviscoplasticity, leading to stress equilibrium iteration at each time step.

Thermoviscoelasticity for a strain-controlled simulation can be visualized by a one-dimensional model consisting of a spring and a dashpot in series (Maxwell element), connected in parallel to a permanently elastic spring (elastic Hooke element). This so far viscoelastic super-element is subjected to a strain rate and provides the desired time-dependent stress function. The strain control is indicated by displacements as primary unknowns in FEM. By incorporating a temperature-dependent fluid into the dashpot thermoviscoelasticity is created. To capture a wide range of observed relaxation times, parallel Maxwell elements are used. Figure 23.5 depicts the model, separated into a deviatoric and volumetric part, where the latter accounts for normal, thermal, and crystallization strain, in serial with elastic bulk strain.

For each Maxwell element, the stress is a function of time and expressed by the following first-order Ordinary Differential Equation (ODE)

$$\frac{\mathrm{d}\sigma}{\mathrm{d}t} + \frac{1}{\tau}\sigma(t) = E\frac{\mathrm{d}\varepsilon}{\mathrm{d}t}, \quad \tau = \tau(T) = \frac{\eta(T)}{E(T)}, \quad T = \mathrm{const.}, \quad \frac{\mathrm{d}\varepsilon}{\mathrm{d}t} = \dot{\varepsilon}(t) \tag{23.27}$$

The solution of Eq. (23.27) can be expressed by a hereditary integral. This integral sums up the stress increments due to strain steps initiated at all previous time events. This is known as Boltzmann superposition principle. For a constant temperature, the solution of ODE given in Eq. (23.27) reads

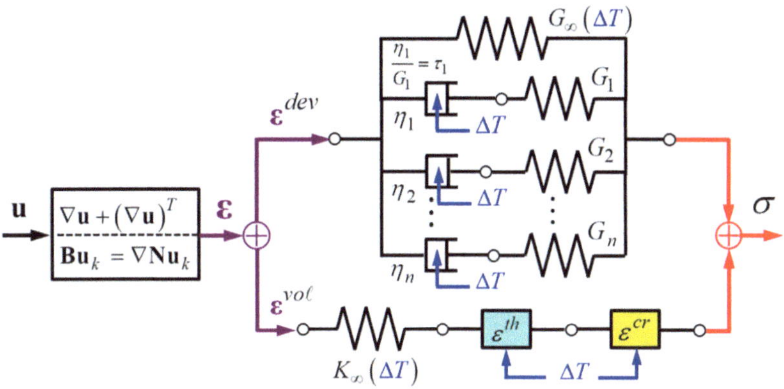

Fig. 23.5 Constitutive model of thermoviscoelasticity

$$\sigma(t) = \sigma(t=0)\, e^{-(t-t_0)/\tau} + \int_{t'=t_0}^{t'=t} \left[E\, e^{-(t-t')/\tau}\right] \frac{\mathrm{d}\varepsilon(t')}{\mathrm{d}t'} dt',$$
$$E\, e^{-(t-t')/\tau} = E(t-t') \tag{23.28}$$

where the term in square brackets is a time-dependent Young's modulus, evaluated at delayed times. This exponentially decaying Young's modulus also arises in a stress relaxation test caused by given constant strain beginning at time $t_0 = 0$, because

$$\sigma(t) = \sigma(0)\cdot e^{-t/\tau} = E\,\varepsilon_0 \cdot e^{-t/\tau} = \left[E\, e^{-t/\tau}\right]\cdot \varepsilon_0$$
$$= E(t)\cdot\varepsilon_0 \quad \Leftrightarrow \quad \sigma(t) = E_{relax}(t)\cdot\varepsilon_0. \tag{23.29}$$

Therefore, the notion of relaxation modulus is common. Note that the relaxation modulus is also the solution to the homogeneous part of the ODE in Eq. (23.27).

When changing to another constant temperature, the decay of the relaxation modulus becomes much faster, namely, $T_2 > T_1 \mapsto \tau_2(T_2) < \tau_1(T_1)$, due to higher mobility of molecules, which would require new experiments to record the modulus. However, for the important class of thermorheologically simple polymers, the relaxation modulus at the temperature time pair $(t_1, T_1) \mapsto E_1(t_1, T_1)$ and the pair $(t_2, T_2) \mapsto E_2(t_2, T_2)$ for the identical modulus $E_1(t_1, T_1) = E_1(t_2, T_2)$ can be calculated, when for a given time the so-called time-shift-factor function $a_T(T)$ is known. It determines the ratio of times like characteristic time constants for a referencing measuring temperatures and an operation temperature. This rule is named correspondence principle. The a_T-function reads

$$a_T(T) = \frac{\tau}{\tau_{ref}} = \frac{\tau(T)}{\tau(T_{ref})}, \ \tau = a_T(T)\tau_{ref}, \ \log\tau - \log\tau_{ref} = \log a_T, \tag{23.30}$$

where in the experimentally motivated log-version the time shift property is obvious. For temperatures above the glass transition temperature, the semi-empirical function

$$\log(a_T(T)) = \log\left(\frac{\tau}{\tau_{ref}}\right) = \frac{-C_{1,ref} \cdot \left(T(t) - T_{ref}\right)}{C_{2,ref} + \left(T(t) - T_{ref}\right)} = \log(a_T(t)), \quad T > T_G \tag{23.31}$$

introduced by Williams, Landel, Ferry, or short WLF equation is most common (Schwarzl 1990), where the constants $\left(C_{1,ref}, C_{2,ref}\right)$ obey the stated transformation rules

$$\begin{aligned} C_{2,ref} &= C_{2,G} + \left(T_{ref} - T_G\right), \text{for } C_1 \cdot C_2 = \text{constant} \\ \mapsto \quad C_{1,ref} &= \frac{C_{1,G} \cdot C_{2,G}}{C_{2,G} + \left(T_{ref} - T_G\right)}. \end{aligned} \tag{23.32}$$

The limit temperature T_G signifies the glass transition temperature. For PA 12, values in Table 23.3 can be used (Tschoegl et al. 2002). Referring to the damper in the Maxwell element and taking into consideration that the E-modulus is much less temperature sensitive than the viscosity, the time ratio $a_T(T)$ is essentially a viscosity ratio $a_T = \tau / \tau_{ref} = \eta / \eta_{ref}$. The constants (C_1, C_2) can be explained by the free volume at glass transition temperature and the sudden increase of the free volume with respect to temperature. The application of the WLF formula is critical, because strictly, phase changes are not admitted. However, due the lack of alternative it is a good pragmatic mean (Schwarzl 1990; Tschoegl et al. 2002).

Applying the WLF equation to the relaxation modulus to get them equal yields

$$\begin{aligned} 1 &= \frac{E_{relax,2}}{E_{relax,ref}} = \frac{e^{-t_2/\tau_2}}{e^{-t_{ref}/\tau_{ref}}} = \exp\left[\left(\frac{t_{ref}}{\tau_{ref}} - \frac{t_2}{a_T(T_2)\tau_{ref}}\right)\right] \\ &= \exp[0] \mapsto \frac{t_2}{t_{ref}} = a_T(T_2). \end{aligned} \tag{23.33}$$

In case of continuously changing temperatures $T = T(t)$, the WLF function changes to a time function $a_T(T(t))$ and the time-dependent scaling holds only for the infinitesimal time $\mathrm{d}t = a_T(t)\,\mathrm{d}t_{ref}$, leading to a significant more complicated hereditary integral as solution of the ODE for time-dependent time constants in the Maxwell model. Based on

$$\frac{\mathrm{d}\sigma}{\mathrm{d}t} + \frac{1}{\tau(t)}\sigma(t) = E\frac{\mathrm{d}\varepsilon}{\mathrm{d}t}, \tau(t) = \tau(T(t)) = \frac{\eta(T(t))}{E(T(t))}, \quad T = T(t), \quad \frac{\mathrm{d}\varepsilon}{\mathrm{d}t} = \dot{\varepsilon}(t) \tag{23.34}$$

the stress function built by the decaying initial stress and the integral generalizes to

$$\sigma(t) = \sigma(t=0)e^{-[\tilde{\tau}(t)-\tilde{\tau}(t_0)]/\tau_{ref}} + \int_{t'=t_0}^{t'=t} E\left(T\left(t'\right)\right)e^{-[\tilde{\tau}(t)-\tilde{\tau}(t')]/\tau_{ref}} \frac{\mathrm{d}\varepsilon\left(t'\right)}{\mathrm{d}t'}\mathrm{d}t' \tag{23.35}$$

where

$$\tilde{\tau}(t) - \tilde{\tau}(t_0) = \int_{t'=t_0}^{t'=t} \frac{\mathrm{d}t'}{a_T(T(t'))}, \ \tilde{\tau}(t) - \tilde{\tau}\left(t'\right) = \int_{t''=t'}^{t''=t} \frac{\mathrm{d}t''}{a_T(T(t''))} \tag{23.36}$$

are accumulated scaled times, implying a nested time integration. Since $a_T(t)$ is a function with an exponential of complicated time dependence, the integrals for the reduced time and subsequently for the stress are usually solved numerically. For a constant temperature and constant $a_T(T)$, the standard exponent is obtained, because

$$\tilde{\tau}(t) - \tilde{\tau}(t_0) = \int_{t''=t'}^{t''=t} \frac{\mathrm{d}t''}{a_T(T)} = \frac{t-t'}{a_T(T)} \Rightarrow \frac{\tilde{\tau}(t) - \tilde{\tau}(t_0)}{\tau_{ref}} = \frac{t-t'}{\tau_{ref}\, a_T(T)} = \frac{t-t'}{\tau(T)}. \tag{23.37}$$

Many WLF functions of $\log a_T(T)$ are fairly linear with $\log a_T(T_G) = 0$, so they can be well approximated by an exponential function through the glass transition temperature or formally centered at $T_{ref} = T_G + 100°\mathrm{C}$, which is 10 °C below the end of crystallization. This linear characteristic is the onset to obtain simpler integrals for the reduced time or even analytic integral formulas. Performing a first-order Taylor expansion provides

$$\begin{aligned} \log\left(\frac{1}{a_T}\right) &= \frac{C_{1,G}\cdot(T-T_G)}{C_{2,G}+(T-T_G)} \approx \frac{C_{1,G}\cdot\left(T_{ref}-T_G\right)}{C_{2,G}+\left(T_{ref}-T_G\right)} \\ &\quad + \frac{C_{1,G}\cdot C_{2,G}}{\left(C_{2,G}+\left(T_{ref}-T_G\right)\right)^2}\left(T-T_{ref}\right) \end{aligned} \tag{23.38}$$

or in a more compact expression

$$\frac{1}{a_T(T(t))} \approx \exp\left[D_1 + D_2\left(T(t)-T_{ref}\right)\right], \begin{cases} D_1 \log e = \dfrac{C_{1,G}\cdot\left(T_{ref}-T_G\right)}{C_{2,G}+\left(T_{ref}-T_G\right)} \\ D_2 \log e = \dfrac{C_{1,G}\cdot C_{2,G}}{\left(C_{2,G}+\left(T_{ref}-T_G\right)\right)^2} \end{cases} \tag{23.39}$$

For the considered AM problem, the highest stresses occur during the 10 °C of crystallization interval with a roughly linear drop of temperature with respect to time, which creates a linear temperature time function in the $a_T(t)$ exponent:

$$
\begin{gathered}
\frac{1}{a_T(T(t))} \approx \exp\left[p_1 - p_2\,(t - t_1)\right], \\
\left\{ \begin{array}{c} T(t) \approx T_1 - \frac{T_1 - T_2}{t_1 - t_2}(t - t_1) \\ p_1 = D_1 + D_2\left(T_1 - T_{ref}\right), \quad p_2 = D_2\,\frac{T_1 - T_2}{t_1 - t_2} \end{array} \right.
\end{gathered} \tag{23.40}
$$

A linearization of a temperature–time function can also be realized for sub-intervals.

A robust interpolation scheme relying on a section-wise exponentially linear time growth is also implemented in the FEM software ABAQUS. Herein, by applying the time linearization (23.40) to the integral for the reduced time difference, we obtain

$$
\begin{aligned}
\tilde{\tau}(t) - \tilde{\tau}\left(t'\right) &= \int_{t''=t'}^{t''=t} \frac{\mathrm{d}t''}{a_T(T(t''))} = \int_{t''=t'}^{t''=t} e^{p_1 - p_2\,(t''-t_1)}\mathrm{d}t'' \\
&= \frac{1}{a_T(t_1)}\,\frac{e^{-p_2\,(t'-t_1)} - e^{-p_2(t-t_1)}}{p_2}
\end{aligned} \tag{23.41}
$$

and this shows that the time difference does not follow a mere time difference typical for a hereditary integral appearing for a constant temperature. If the temperature change remains negligibly small compared to the constant part, the real time difference is recovered up to the constant time magnification factor $\exp(p_1) = 1\big/a_T$, or time constant

$$
p_2(t - t_1) \ll p_1 \rightarrow \tilde{\tau}(t) - \tilde{\tau}\left(t'\right) \approx e^{p_1} \cdot \left(t - t'\right) = \left(t - t'\right)\big/a_T\left(t'\right) \tag{23.42}
$$

Referring to Fig. 23.5, parallel Maxwell elements whose characteristics are outlined above act parallel to the purely elastic component with different weighing factors. Similar to the thermoelastic case, stresses and strains are separated into deviatoric and bulk (volumetric) parts. Both parts are functions of time, similar to the relaxation modulus functions. Since stresses from initial conditions usually decay fast and loads are strain/displacement controlled, they are neglected below. In summary, for the axial symmetric case the thermoviscoelastic material laws takes the subsequent form:

$$
\begin{aligned}
\begin{pmatrix} \sigma_{rr}(t) \\ \sigma_{zz}(t) \\ \sigma_{\theta\theta}(t) \\ \sigma_{rz}(t) \end{pmatrix} &= \frac{2}{3}\begin{pmatrix} +2 & -1 & -1 & 0 \\ -1 & +2 & -1 & 0 \\ -1 & -1 & +2 & 0 \\ 0 & 0 & 0 & 3 \end{pmatrix} \int\limits_{t'=t_0}^{t'=t} \left(G_\infty + G_{TVE}\left(\tilde{\tau}(t) - \tilde{\tau}\left(t'\right)\right)\right)\frac{\mathrm{d}}{\mathrm{d}t'}\begin{pmatrix} \varepsilon_{rr} \\ \varepsilon_{zz} \\ \varepsilon_{\theta\theta} \\ \varepsilon_{rz} \end{pmatrix}\mathrm{d}t' \\
&+ \begin{pmatrix} 1 \\ 1 \\ 1 \\ 0 \end{pmatrix} \int\limits_{t'=t_0}^{t'=t} \left(K_\infty + K_{TVE}\left(\tilde{\tau}(t) - \tilde{\tau}\left(t'\right)\right)\right)\frac{\mathrm{d}\varepsilon^{vol}}{\mathrm{d}t'}\mathrm{d}t' - \begin{pmatrix} 3 \\ 3 \\ 3 \\ 0 \end{pmatrix} K_\infty\left[\varepsilon^{th+cr}(t) - \varepsilon^{th+cr}(t_0)\right]
\end{aligned} \tag{23.43}
$$

where the first and the second rows represent the deviatoric and the bulk stress, respectively.

At constant temperature, the moduli $G(t) = G_\infty + G_{VE}(t)$ and $K(t) = K_\infty + K_{VE}(t)$ are expressed in summation, which is named Prony series, reading

$$G(t) = G_\infty + \sum_{j=1}^{n} G_j e^{-t/\tau_j} = G_\infty + G_0 \sum_{j=1}^{n} g_j e^{-t/\tau_j}, \quad g_j = \frac{G_j}{G_0} \tag{23.44}$$

with a common normalization modulus G_0 according to the following:

$$\left.\begin{array}{l} G(t=0) = G_0 = G_\infty + \sum\limits_{j=1}^{n} G_j = G_\infty + G_0 \sum\limits_{j=1}^{n} g_j \\ G(t=\infty) = G_\infty \end{array}\right\} \rightarrow \quad G_0 = \frac{G_\infty}{1 - \sum g_j}. \tag{23.45}$$

For the bulk modulus, the equivalent summation formula is given by

$$K(t) = K_\infty + \sum_{j=1}^{n} K_j e^{-t/\tau_j} = K_\infty + K_0 \sum_{j=1}^{n} k_j e^{-t/\tau_j}, \quad k_j = \frac{K_j}{K_0}, \tag{23.46}$$

with the simplification from experiments that the numbers of serial elements are the same. The values of the time constants are assumed to be identical, which is critical, because intrinsically different viscoelastic shear and bulk (=volumetric) are required to be related, formally by the Poisson ratio υ. For a time-dependent Poisson ratio, measurements of Young´s modulus $E(t)$ for tension and bending allow to calculate $(G(t), K(t))$. The modulus $E(t)$, represented by a Prony series, and the associated time constants τ_j are obtained by conducting frequency sweep DMTA experiments (dynamic mechanical thermal analysis) at different temperatures. By adopting the thermo-rheological simplicity with its shift function, so-called master curve functions w.r.t. the applied sweep frequency are assembled. They are generated for the storage modulus $E'(\omega)$ and the loss modulus $E''(\omega)$, respectively. The magnitude of both represents the magnitude of modulus, whereas the ratio is a measure for the viscose dissipation. The coefficients of the Prony series are determined from least-square fits to the master curve. For many polymers, such as PA 12, the volumetric (bulk) viscoelasticity is less distinct than the one for shear (deviatoric). Thus, only viscoelastic shear modulus $G(t) = G_\infty + G_{VE}(t)$ is used here. The bulk modulus is assumed as thermoelastic modulus $K(t) \approx K_\infty$, where the time enters through Young's modulus which is temperature and thus time dependent.

The limitation to the shear modulus $G(t) = G_\infty + G_{VE}(t)$ allows obtaining the Prony parameter from DMTA shear tests with sinusoidal shear angle excitation. By excluding initial pre-strains and limiting to the stationary long-term case, the integral

$$\sigma(t) = G_\infty \gamma(t) + G_0 \sum_{j=1}^{n} g_j \int_{t_0}^{t} e^{-(t-t')/\tau_j} \frac{d\gamma}{dt'} dt',$$
$$\gamma(t) = \hat{\gamma} \sin \omega t, \ \omega = 2\pi f \tag{23.47}$$

yields the series

$$\sigma(t) = G_\infty \hat{\gamma} \sin \omega t + \sum_{j=1}^{n} \frac{g_j (\omega \tau_j)^2}{1 + (\omega \tau_j)^2} \cdot G_0 \hat{\gamma} \sin \omega t + \sum_{j=1}^{n} \frac{g_j (\omega \tau_j)}{1 + (\omega \tau_j)^2} \cdot G_0 \hat{\gamma} \cos \omega t, \tag{23.48}$$

where the first summand is the permanent elastic sinusoidal stress part. The second and third summands with frequency-varying amplitudes represent the storage modulus $G'(\omega)$ and the loss modulus $G''(\omega)$, at a fixed temperature, due to $\{G_\infty(T), \ \tau_j = \tau_j(T) / G_j, \ g_j(T)\}$. Typically used 20 Prony elements with 40 unknowns $(g_j, \ \tau_j)$ allow to cover short and very long-term features appearing in AM between the sintering and the cooling phases. For PA 12, results reported by Hoda (2015) have been adopted (see Table 23.4).

The restriction to deviatoric viscoelasticity simplifies the stress integral in (23.43) to

$$\begin{pmatrix} \sigma_{rr}(t) \\ \sigma_{zz}(t) \\ \sigma_{\theta\theta}(t) \\ \sigma_{rz}(t) \end{pmatrix} = \frac{2}{3} \begin{pmatrix} +2 & -1 & -1 & 0 \\ -1 & +2 & -1 & 0 \\ -1 & -1 & +2 & 0 \\ 0 & 0 & 0 & 3 \end{pmatrix} \int_{t_0}^{t} \left(G_\infty + G_{TVE}\left(\tilde{\tau}(t) - \tilde{\tau}(t')\right)\right) \frac{d}{dt'} \begin{pmatrix} \varepsilon_{rr} \\ \varepsilon_{zz} \\ \varepsilon_{\theta\theta} \\ \varepsilon_{rz} \end{pmatrix} dt' + \begin{pmatrix} 1 \\ 1 \\ 1 \\ 0 \end{pmatrix} K_\infty \left[\left(\varepsilon^{vol}(t) - \varepsilon^{vol}(t_0)\right) - 3 \cdot \left(\varepsilon^{th+cr}(t) - \varepsilon^{th+cr}(t_0)\right)\right], \tag{23.49}$$

where $\varepsilon^{vol} = \varepsilon_{rr} + \varepsilon_{zz} + \varepsilon_{\theta\theta} = \mathrm{tr}(\boldsymbol{\varepsilon})$, or in a more compact notation it reads

$$\boldsymbol{\sigma}(t) = \boldsymbol{\Gamma} \cdot \int_{t'=t_0}^{t'=t} \left(G_\infty + G_{TVE}\left(\tilde{\tau}(t) - \tilde{\tau}(t')\right)\right) \frac{d\boldsymbol{\varepsilon}(t')}{dt'} dt' + \boldsymbol{\kappa} \cdot K_\infty \cdot \left\{\left[\mathrm{tr}\left(\boldsymbol{\varepsilon}^{vol}(t)\right) - \mathrm{tr}\left(\boldsymbol{\varepsilon}^{vol}(t_0)\right)\right] - 3 \cdot \left[\varepsilon^{th+cr}(t) - \varepsilon^{th+cr}(t_0)\right]\right\}, \tag{23.50}$$

where the first and the second rows represent the deviatoric and the volumetric (bulk) parts, respectively.

23.2.3.3 Time Discretization of Thermoviscoelasticity

A general applied strain rate, and the non-isothermal AM process through the reduced time, requires the stress integrals to be evaluated numerically, where the accurate scheme by Taylor et al. (1970) is adopted. For a generic deviatoric stress component in the interval $[t_k,\ t_k + \Delta t]$, the numerically explicit procedure becomes

$$\sigma_{k+1}^{dev} = \sigma^{dev}(t_k + \Delta t) = \sigma^{dev}(t_k) + \Delta\sigma^{dev}(t_k) = \sigma_k^{dev} + G_k^{dev}(\varepsilon_{k+1} - \varepsilon_k) \quad (23.51)$$

with the "deviatoric" modulus $G_k^{dev} = G_k^{VE}$. The discrete volumetric (bulk) stress vector

$$\begin{aligned}\boldsymbol{\sigma}_{k+1}^{vol} = \boldsymbol{\sigma}_k^{vol} + \boldsymbol{\kappa}\, K_\infty\Big\{&\big[\mathrm{tr}\left(\boldsymbol{\varepsilon}^{vol}(t_k + \Delta t)\right) - \mathrm{tr}\left(\boldsymbol{\varepsilon}^{vol}(t_k)\right)\big] \\ &- 3\left[\varepsilon^{th+cr}(t_k + \Delta t) - \varepsilon^{th+cr}(t_k)\right]\Big\}\end{aligned} \quad (23.52)$$

adds and yields the total incremental stress vectors, e.g., $\boldsymbol{\sigma}(t_k + \Delta t) = \boldsymbol{\sigma}_{k+1} = \boldsymbol{\sigma}_{k+1}^{dev} + \boldsymbol{\sigma}_{k+1}^{vol}$.

For the sake of this algorithm, any deviatoric stress component $\sigma_k^{dev} = \sigma^{dev}(t_k)$

$$\sigma^{dev}(t_k) = \sigma_k^{dev} = G_\infty(\varepsilon(t_k) - \varepsilon(t_0)) + G_0 \sum_{j=1}^{n} g_j \int_{t_0}^{t_k} e^{-(t_k - t')/\tau_j} \frac{\mathrm{d}\varepsilon}{\mathrm{d}t'} dt' \quad (23.53)$$

is expressed with the function $h_j(t_0 \to t_k) = h_j(t_k)$ and assumes the expression

$$\sigma_k^{dev} = G_\infty(\varepsilon_k - \varepsilon_0) + G_0 \sum_{j=1}^{n} g_j h_j(t_k), \quad h_{j,k} = h_j(t_k) = \int_{t_0}^{t_k} e^{-(t_k - t')/\tau_j} \frac{\mathrm{d}\varepsilon}{\mathrm{d}t'} dt'. \quad (23.54)$$

The next stress $\sigma_{k+1}^{dev} = \sigma^{dev}(t_k + \Delta t)$ is given as

$$\sigma_{k+1}^{dev} = G_\infty(\varepsilon(t_k + \Delta t) - \varepsilon(t_0)) + G_0 \sum_{j=1}^{n} g_j \int_{t_0}^{t_k + \Delta t} e^{-(t_k + \Delta t - t')/\tau_j} \frac{\mathrm{d}\varepsilon}{\mathrm{d}t'} \mathrm{d}t'. \quad (23.55)$$

It is decomposed into the interval $[t_0,\ t_k]$ and the additional short one $\left[t_k,\ t_{k+1}\right]$

$$\sigma_{k+1}^{dev} = G_\infty(\varepsilon_k - \varepsilon_0) + G_0 \sum_{j=1}^{n} g_j \cdot e^{-\Delta t/\tau_j} \int_{t_0}^{t_k} e^{-(t_k - t')/\tau_j} \frac{\mathrm{d}\varepsilon}{\mathrm{d}t'} \mathrm{d}t'$$
$$+ G_\infty(\varepsilon_{k+1} - \varepsilon_k) + G_0 \sum_{j=1}^{n} g_j e^{-(t_k+\Delta t)/\tau_j} \int_{t_k}^{t_k+\Delta t} e^{t'/\tau_j} \frac{\mathrm{d}\varepsilon}{\mathrm{d}t'} \mathrm{d}t', \quad (23.56)$$

where the integral in the first row can be traced back to $h_j(t_k)$. For the stress increment in the second row, strain rate is discretized, and thus the remaining integral can be solved analytically, which explains the higher precision of the Taylor scheme compared to a simple linear approximation for the whole integrand. Inserting the result

$$\frac{\mathrm{d}\varepsilon}{\mathrm{d}t'} \approx \frac{\varepsilon_{k+1} - \varepsilon_k}{\Delta t} \mapsto$$
$$e^{-(t_k+\Delta t)/\tau_j} \int_{t_k}^{t_k+\Delta t} e^{t'/\tau_j} \frac{\mathrm{d}\varepsilon}{\mathrm{d}t'} \mathrm{d}t' = \frac{1 - e^{-\Delta t/\tau_j}}{\Delta t/\tau_j} (\varepsilon_{k+1} - \varepsilon_k) \quad (23.57)$$

into Eq. (23.56) generates the deviatoric stress formula at time $t_k + \Delta t$ as

$$\sigma_{k+1}^{dev} = \left\{ G_\infty(\varepsilon_k - \varepsilon_0) + G_0 \sum_{j=1}^{n} g_j \cdot h_j(t_k) + G_0 \sum_{j=1}^{n} g_j \cdot \frac{1 - e^{-\Delta t/\tau_j}}{\Delta t/\tau_j} \left(\frac{\Delta t}{\tau_j} h_{j,k} \right) \right\}$$
$$+ \left[G_0 \left(1 - \sum_{j=1}^{n} g_j \right) (\varepsilon_{k+1} - \varepsilon_k) + G_0 \sum_{j=1}^{n} g_j \cdot \frac{1 - e^{-\Delta t/\tau_j}}{\Delta t/\tau_j} (\varepsilon_{k+1} - \varepsilon_k) \right], \quad (23.58)$$

where the Prony series property $G_\infty = G_0\big(1 - \sum g_j\big)$ was used. The increment of deviatoric (viscoelastic) stress $\Delta\sigma_k^{dev} = \Delta\sigma^{dev}(t_k) = \Delta\sigma^{VE}(t_k)$ reads

$$\sigma_{k+1}^{dev} = \{\sigma_k^{dev}\} + [\Delta\sigma_k^{dev}]$$
$$\Delta\sigma_k^{dev=VE} = G_0 \left[1 - \sum_{j=1}^{n} g_j \left(1 - \frac{1 - e^{-\Delta t/\tau_j}}{\Delta t/\tau_j} \right) \right] (\varepsilon_{k+1} - \varepsilon_k). \quad (23.59)$$

In addition, the deviatoric viscoelastic modulus related to the stress increment in

$$\sigma_{k+1}^{dev} = \sigma_k^{dev} + \Delta\sigma_k^{dev} = \sigma_k^{dev} + G_k^{dev=VE}(\varepsilon_{k+1} - \varepsilon_k) = \sigma_k^{dev} + \frac{\partial\sigma_k^{dev}}{\partial\varepsilon_k^{dev}}(\varepsilon_{k+1} - \varepsilon_k) \quad (23.60)$$

is determined as derivative of Eqs. (23.58) or (23.59). The result reads

$$G_k^{VE} = G_\infty + G_0 \sum_{j=1}^{n} g_j \frac{1 - e^{-\Delta t/\tau_j}}{\Delta t/\tau_j} = \frac{G_\infty}{1 - \sum g_j}\left[1 - \sum_{j=1}^{n} g_j \left(1 - \frac{1 - e^{-\Delta t/\tau_j}}{\Delta t/\tau_j}\right)\right] \geq G_\infty. \tag{23.61}$$

This formula proves that the shear modulus at any intermediate time supersedes the terminal one G_∞ at time infinity, which is verified by a Taylor expansion. Note that the modulus in Eq. (23.61) is dependent only on time but not current strain or displacement, which maintains the stress–strain law linear.

In addition, the time-independent volumetric (bulk) stress part is reformulated as

$$\sigma_{k+1}^{vol} = \sigma_k^{vol} + K_\infty\left(\varepsilon_{k+1}^{vol} - \varepsilon_k^{vol}\right) - \sigma_k^{th+cr} - 3K_\infty\left(\varepsilon_{k+1}^{th+cr} - \varepsilon_k^{th+cr}\right). \tag{23.62}$$

Inserting (23.58) and (23.62) in (23.43) yields the explicit axial symmetric discrete stress vector

$$\begin{pmatrix} \sigma_{rr,k+1} \\ \sigma_{zz,k+1} \\ \sigma_{\theta\theta,k+1} \\ \sigma_{rz,k+1} \end{pmatrix} = \frac{2}{3}\begin{pmatrix} +2 & -1 & -1 & 0 \\ -1 & +2 & -1 & 0 \\ -1 & -1 & +2 & 0 \\ 0 & 0 & 0 & 3 \end{pmatrix}\left[\begin{pmatrix} \sigma_{rr,k}^{dev} \\ \sigma_{zz,k}^{dev} \\ \sigma_{\theta\theta,k}^{dev} \\ \sigma_{rz,k}^{dev} \end{pmatrix} + G_k^{dev}\left(\begin{pmatrix} \varepsilon_{rr,k+1} \\ \varepsilon_{zz,k+1} \\ \varepsilon_{\theta\theta,k+1} \\ \varepsilon_{rz,k+1} \end{pmatrix} - \begin{pmatrix} \varepsilon_{rr,k} \\ \varepsilon_{zz,k} \\ \varepsilon_{\theta\theta,k} \\ \varepsilon_{rz,k} \end{pmatrix}\right)\right]$$
$$+ \begin{pmatrix} 1 \\ 1 \\ 1 \\ 0 \end{pmatrix}\left[\sigma_k^{vol} + K_\infty\left(\varepsilon_{k+1}^{vol} - \varepsilon_k^{vol}\right) - \sigma_k^{th+cr} - 3K_\infty\left(\varepsilon_{k+1}^{th+cr} - \varepsilon_k^{th+cr}\right)\right], \tag{23.63}$$

and its compact representation reads

$$\boldsymbol{\sigma}_{k+1} = \boldsymbol{\Gamma}\cdot\left[\boldsymbol{\sigma}_k^{dev} + G_k^{dev=VE}(\boldsymbol{\varepsilon}_{k+1} - \boldsymbol{\varepsilon}_k)\right] + \boldsymbol{\kappa}\cdot\left[\sigma_k^{vol} + K_\infty\left(\mathrm{tr}\left(\boldsymbol{\varepsilon}_{k+1}^{vol}\right) - tr\left(\boldsymbol{\varepsilon}_k^{vol}\right)\right) - \sigma_k^{th+cr} - 3K_\infty\left(\varepsilon_{k+1}^{th+cr} - \varepsilon_k^{th+cr}\right)\right]. \tag{23.64}$$

In order to obtain the stiffness matrix the volumetric strain is expressed by the strain vector components, which leads to

$$\begin{pmatrix}\sigma_{rr,k+1}\\ \sigma_{zz,k+1}\\ \sigma_{\theta\theta,k+1}\\ \sigma_{rz,k+1}\end{pmatrix} = \frac{2}{3}\begin{pmatrix}+2 & -1 & -1 & \\ -1 & +2 & -1 & 0\\ -1 & -1 & +2 & 0\\ 0 & 0 & 0 & 3\end{pmatrix}\begin{pmatrix}\sigma^{dev}_{rr,k}\\ \sigma^{dev}_{zz,k}\\ \sigma^{dev}_{\theta\theta,k}\\ \sigma^{dev}_{rz,k}\end{pmatrix} + \begin{pmatrix}1\\1\\1\\0\end{pmatrix}\sigma_k^{vol} - \begin{pmatrix}1\\1\\1\\0\end{pmatrix}\left[\sigma_k^{th+cr} - 3K_\infty\left(\varepsilon_{k+1}^{th+cr} - \varepsilon_k^{th+cr}\right)\right]$$
$$+\left[\frac{2}{3}G_k^{dev=VE}\begin{pmatrix}+2 & -1 & -1 & 0\\ -1 & +2 & -1 & \\ -1 & -1 & +2 & 0\\ 0 & 0 & 0 & 3\end{pmatrix} + K_\infty\begin{pmatrix}1&1&1&0\\1&1&1&0\\1&1&1&0\\0&0&0&0\end{pmatrix}\right]\cdot\left[\begin{pmatrix}\varepsilon_{rr,k+1}\\ \varepsilon_{zz,k+1}\\ \varepsilon_{\theta\theta,k+1}\\ \varepsilon_{rz,k+1}\end{pmatrix} - \begin{pmatrix}\varepsilon_{rr,k}\\ \varepsilon_{zz,k}\\ \varepsilon_{\theta\theta,k}\\ \varepsilon_{rz,k}\end{pmatrix}\right] \tag{23.65}$$

and its corresponding compact matrix vector version

$$\begin{aligned}\boldsymbol{\sigma}_{k+1} &= \left[\boldsymbol{\Gamma}\boldsymbol{\sigma}_k^{dev} + \boldsymbol{\kappa}\sigma_k^{vol}\right] - \boldsymbol{\kappa}\left[\sigma_k^{th+cr} + 3K_\infty\left(\varepsilon_{k+1}^{th+cr} - \varepsilon_k^{th+cr}\right)\right]\\ &\quad + \left[G_k^{dev}\boldsymbol{\Gamma} + K_\infty \mathbf{K}_{tr}^{vol}\right](\boldsymbol{\varepsilon}_{k+1} - \boldsymbol{\varepsilon}_k)\end{aligned} \tag{23.66}$$

provides the only time-dependent (viscoelastic) stiffness matrix

$$\begin{aligned}\mathbf{C}_k &= \mathbf{C}(t_k) = G_k^{dev=VE}\boldsymbol{\Gamma} + K_\infty \mathbf{K}_{tr}^{vol}\\ &\Leftrightarrow\\ &= \frac{2}{3}G_k^{dev=VE}\begin{pmatrix}+2 & -1 & -1 & 0\\ -1 & +2 & -1 & 0\\ -1 & -1 & +2 & 0\\ 0 & 0 & 0 & 3\end{pmatrix} + K_\infty\begin{pmatrix}1&1&1&0\\1&1&1&0\\1&1&1&0\\0&0&0&0\end{pmatrix}.\end{aligned} \tag{23.67}$$

For non-isothermal processes with significantly varying temperatures $T = T(t)$, the time constants τ_j are summed up reduced times. The reformulated WLF Eq. (23.40) yields

$$\begin{aligned}\tilde{\tau}(t_k) - \tilde{\tau}(t_{k-m}) &= \sum_{j=1}^{m} \frac{1}{a_T(t_k - j\,\Delta t)}\left(\frac{1 - e^{-p_{2,jk}\,\Delta t}}{p_{2,jk}}\right)\\ &\approx \sum_{j=1}^{m} \frac{\Delta t}{a_T(T(t_k - j\,\Delta t))}.\end{aligned} \tag{23.68}$$

Moreover, the algebraic time dependence in the (deviatoric) shear modulus $G_\infty(t_k)$ and in the bulk modulus $K_\infty(t_k)$, representing thermoelasticity, must be included.

23.3 Finite Element Analysis of the Multiphysics Model

Hereafter, the generation of a tailored axisymmetric finite element model and the procedures for Additive Manufacturing (AM) using the software ABAQUS is explained. The mesh with initial and boundary conditions is presented. The implementation of nonlinear thermoviscoelasticity and chemical kinetics by the developed subroutines is addressed. The spatial and temporal discretizations of temperatures and stresses based on Galerkin's weak equilibrium method are carried out. Finally, the AM process stages for preheating, sintering, and cooling are explained.

23.3.1 Finite Element Model Generation

The axisymmetric FEM model used in the simulation is shown in Fig. 23.6. Thickness and radius of each disk are 6 and 100 mm, respectively. Each disk is made of 50 layers with a thickness of 0.12 mm and identical with the element thickness. A powder bed of 5.04 mm thickness is deposited under the first disk and between consecutive disks. A powder segment of 3 mm thickness is added on top of the fifth disk for isolation. The print time is 14,708.6 s, while cooling to about 25 °C takes additional 60,000 s.

In FEM modeling of AM, addition of layers is performed by two different methods: The "*Quiet*" method and the "*Dead-Alive*" method (Gouge and Michaleris 2018). In the *Quiet* method, the material values of the layers which are not added yet are taken as very small. Therefore, the equations related to the elements of those layers are in the system matrices from the beginning. In the *Dead-Alive* method, the elements in layers that are not added yet are inactive, implying, when new layers are added, the element equations for those layers are added to the global matrices. The *Dead-Alive*

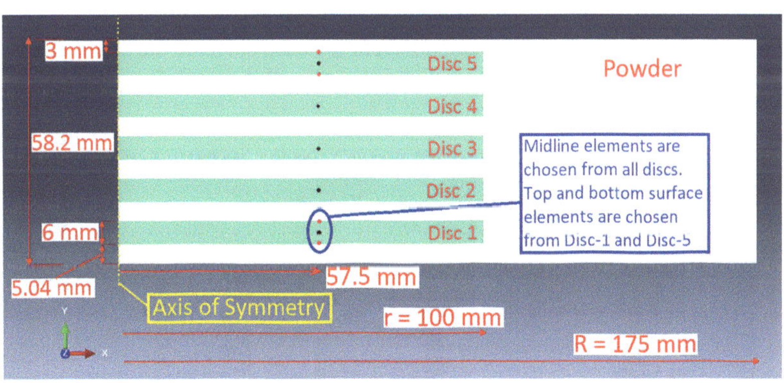

Fig. 23.6 FEM model with disk's (sintered) region in green and powder (non-sintered) region in white color. The locations of nine elements chosen for the history time analyses are also shown

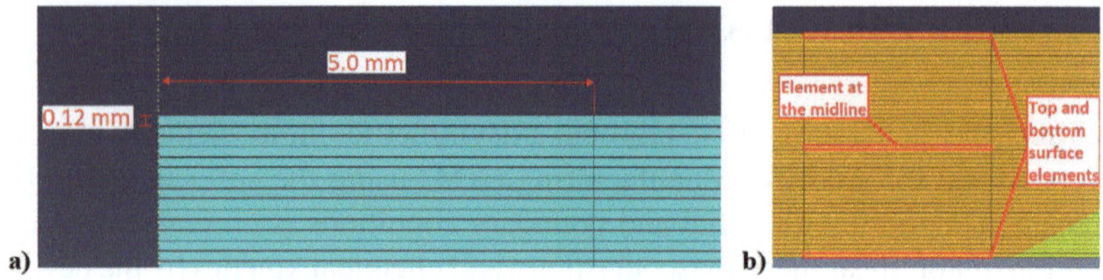

Fig. 23.7 **a** Mesh size and orientation of the elements. The yellow dashed line is the symmetry line. **b** Positions of chosen elements for time history

method is used here. Elements are activated layer by layer in accordance with the single layer time interval of 33.2 s.

Nine elements, 57.5 mm distant from the center, are chosen for recording variables throughout analyses (see Figs. 23.6 and 23.7b). The approximations for the model read

- The powder state is assumed as a continuum, particle interactions are neglected.
- All powder particles in the sintered region melt completely.
- Each new layer is assumed to be instantly deposited across the entire surface.
- Heat input by the laser is replaced by an initial top surface temperature defined as a transient boundary condition.
- The rectangular built chamber is assumed as a cylinder of 350 mm diameter.

The model is meshed with quadratic, full integration, thermo-mechanically coupled, axisymmetric, quadrilateral (CAX8T) elements. The element size is 0.12 mm in height and 5.0 mm in length (see Fig. 23.7a), and it is uniform throughout the model. The implicit, coupled temperature-displacement solver is used.

Elements at the bottom surface are only constrained in vertical direction. Elements at the side surface are unconstrained. Figure 23.8 shows the mechanical boundary conditions.

Thermal boundary conditions are different for each stage of the simulation. They are explained in detail in the stages of simulation in Subsect. 3.5.

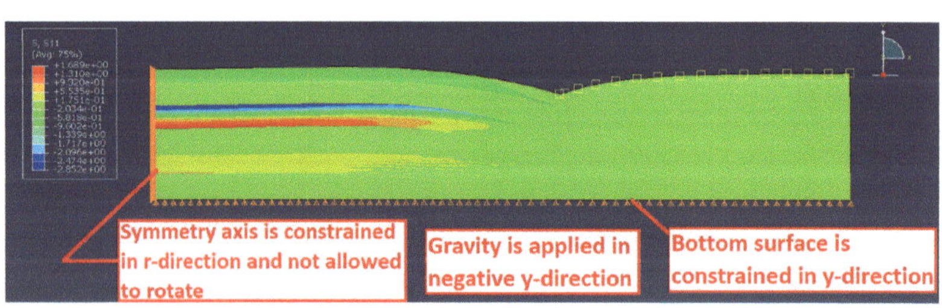

Fig. 23.8 Mechanical boundary conditions

23.3.2 Material Properties

Material data of PA2200 of EOS GmbH is given in Tables 23.2, 23.3, and 23.4. Elements of the FEA model are divided into two regions, the disks and the powder. The powder region does not undergo any phase change so material properties of it are directly entered to ABAQUS GUI. The disk's region undergoes a phase change from powder to melt due to heat input of laser and from melt to solid again due to heat loss. Shortly after it solidifies, and crystallization starts. The material behavior of disks is different in each phase. Therefore, material behavior of disk's region is modeled numerically and submitted to ABAQUS as subroutines. Varying convection coefficients in different temperature regimes cannot be defined through user interface so they also need to be calculated in a subroutine.

Young's modulus is taken as 50 MPa for the liquid phase and the for powder phase of the disk's region. When Young's modulus is taken one MPa for the powder or the melt phase, it is followed up by a sudden jump to about 242 MPa at phase change to solid. This causes convergence problems in the FEM. A very low value of Young's modulus was chosen for the powder region because powder should not

Table 23.2 Thermal material interpolation and data of PA 12 (Neugebauer 2017)

Powder region (no phase change)	Disk's region (for elasticity and viscoelasticity)
$k^{pw} = 0.1\,\mathrm{W}/\mathrm{mK}$	$k^{cr} = 0.3\,W/mK,\ k^{am} = 0.27W/mK$
$c_p^{pw} = c_p^{cr}$	$c_p^{cr} = 0.0112593 \cdot (T - 25°C)/°C + 1.723\,kJ/kgK$
$\rho^{pw} = 0.4\mathrm{g}/\mathrm{cm}^3$	$c_p^{am} = 0.00535 \cdot (T - 200°\mathrm{C})/°\mathrm{C} + 2.806\,kJ/kgK$
	$\rho^{am} = 0.9899 - 5.55 \cdot 10^{-4}(T/°C)\ g/cm^3$
	$\rho^{cr} = 1.0377 - 6.83 \cdot 10^{-4}(T/°C)\ g/cm^3$
	$T_{melt} = 175°\mathrm{C}, \Delta h_c = 45.92\,kJ/kg$

Table 23.3 Mechanical material data of PA 12 (Neugebauer 2017) with WLF constants (Tschoegl et al. 2002)

Powder region (no phase change)	Disk's region (elasticity and viscoelasticity)
$\upsilon = 0.001$ $\alpha_L = 140 \cdot 10^{-6} 1/K$ $E^{pw} = 1.0\ \mathrm{MPa}$	$\alpha_{L,\,solid} = 140 \cdot 10^{-6}\ 1/K$ $\alpha_{L,\,powder} = \alpha_{L,\,melt} = 0\ 1/K$ $E^{melt} = E^{pw} = 50.0\mathrm{MPa}, \upsilon = 0.43$ $V^{cr\,max} = -0.033$
Differences between elastic and viscoelastic models	
Thermoelastic model	**Thermoviscoelastic model**
$E^{solid}(T) = 2.831 \cdot 10^4\,(T/°C)^{-0.922}\mathrm{MPa}$	$E_\infty^{solid}(T) = 2.831 \cdot 10^4\,(T/°\mathrm{C})^{-0.922}\mathrm{MPa}$ and 20 Prony shear terms (see Table 23.4) together with two WLF constants: $T_{ref} = 27°\mathrm{C},\ C_1 = 8.86,\ C_2 = 101.6\ \mathrm{K}$

Table 23.4 Prony series parameters of PA 12 at $T = 27$ °C (Hoda 2015)

Prony series for PA 12 (Nylon) by Hoda (2015)		
n	$\mathbf{g = G_n/G_0}$	τ_n**(seconds)**
1	2.14588984247896E-02	1.38949549437314E+14
2	8.74936724518971E-03	1.93069772888325E+13
3	1.78240333062958E-02	2.68269579527973E+12
4	1.90990430098981E-02	3.72759372031494E+11
5	2.48196983105329E-02	5.17947467923122E+10
6	3.25738158885021E-02	7.19685673001153E+09
7	4.03811725652702E-02	1.00000000000000E+09
8	4.95693882400111E-02	1.38949549437314E+08
9	6.08963940400106E-02	1.93069772888325E+07
10	6.61290823614873E-02	2.68269579527973E+06
11	7.52098116294961E-02	3.72759372031494E+05
12	7.21587240520127E-02	5.17947467923120E+04
13	6.58199570529645E-02	7.19685673001153E+03
14	5.54523158199102E-02	1.00000000000000E+02
15	4.48207984904598E-02	1.38949549437314E+02
16	2.79202973897582E-02	1.93069772888325E+01
17	2.54885990411734E-02	2.68269579527973E+00
18	1.53566767495790E-02	3.72759372031494E-01
19	6.99410508699794E-03	5.17947467923122E-02
20	1.33617192480207E-02	7.19685673001153E-03

restrict the deformation of the disks. A very low value of Poisson's ratio is taken for the powder region, because powder should not undergo strain in transverse directions. The density is also determined based on the phase, which is either powder, crystalline, or amorphous. A constant value is used for the powder regime. For the amorphous and the crystalline phases, two different temperature-dependent functions are utilized based on the approximation that crystallization occurs around 160 °C, as explained in Subsect. 2.2.

23.3.3 Implementation of the Thermoviscoelastic Material Model in the FEA Model

Subroutines of ABAQUS programmed for the simulation are UMAT, UMATHT, and FILM. How these subroutines interact with the finite element formulation is shown in Fig. 23.9. The subroutines perform the shown calculations—material model—for

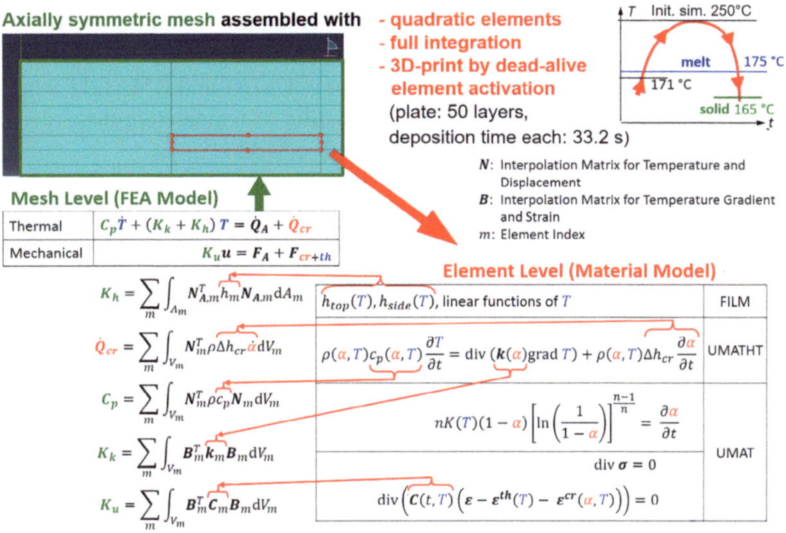

Fig. 23.9 Structure of finite element analysis for additive manufacturing with nonlinear, temperature-, and crystallization-dependent material models and nonlinear convection integrated by user subroutines

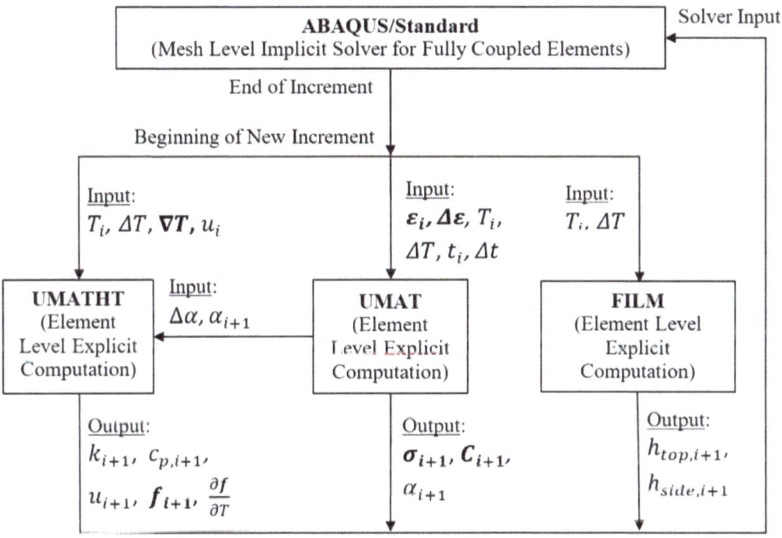

Fig. 23.10 Interactions of the subroutines and the solver in a single time increment, where ABAQUS submits variables to the subroutines, and the subroutines exchange and forward variables internally

each element in each increment. ABAQUS builds the finite element formulation—FEA model—and performs mesh level procedures. Computation of material properties, stress, and thermal energy are carried out in UMAT and UMATHT which are explained in Subsects. 2.2 and 2.3. The calculation of temperature-dependent heat transfer coefficients in FILM is presented in Subsect. 3.5.

Referring to Fig. 23.10, UMAT is utilized for the computation of stress, tangent modulus of the solid phase and crystallization. UMATHT is used for the computation of thermal material properties, thermal energy, and heat flux. FILM is employed for the computation of temperature-dependent heat transfer coefficients, combining convection and radiation, on the top and side surfaces in different temperature regimes.

23.3.4 FEM Discretization

23.3.4.1 Temperature Discretization by Weak Energy Balance

The rigorous energy balance at a volume element yields a "strong" equilibrium. The partial differential equation to determine the temperature field $T = T(t, \alpha(T),\ x_k)$ reads

$$\begin{matrix} \rho\, c_p \dfrac{\partial T}{\partial t} + \dfrac{\partial \dot{q}_k}{\partial x_k} - \rho\, \Delta h_{cr} \dot{\alpha}^{cr}(t) = 0 \\ \rho\, c_p \frac{\partial T}{\partial t} + \nabla \cdot \dot{\mathbf{q}} - \rho\, \Delta h_{cr} \dot{\alpha}^{cr}(t) = 0 \end{matrix}, \left\{ \begin{matrix} T_{A_T} = T\left(x_{k,A_T}\right) \text{ on } A_T \\ \dot{q}_m \cdot x_m = \dot{\mathbf{q}}_n \cdot \mathbf{n} = \dot{q}_n \text{ on } A_h \\ \left(-k \frac{\partial T}{\partial x_m}\right) \cdot x_m = \dot{q}_n = h\left(T_{A_h} - T_{am}\right) \end{matrix} \right. \tag{23.69}$$

constrained by the stated temperature and convective heat flux boundary conditions. The laser energy near the body surface is replaced by a temperature boundary condition.

For the weak form of energy balance, referring to Galerkin, a residual $r = r(x_k)$ of (23.69), which is caused by an approximate solution of the PDE, is replaced by a prescribed zero residual for the product $(\delta T(x_m) \cdot r(x_k))$ accumulated in an enclosed volume, where $\delta T(x_m)$ is a (virtual) test function. The according weak energy equilibrium reads

$$\oint_V \delta T(x_k) r(x_k) \mathrm{d}V = 0 = \oint_V \delta T(x_k) \left(\rho\, c_p \frac{\partial T}{\partial t} + \frac{\partial q_k}{\partial x_k} - \rho\, \Delta h_{cr}\, \dot{\alpha}^{cr} \right) \mathrm{d}V. \tag{23.70}$$

Carrying out integration by parts and along with Gauss's divergence theorem yields

$$\oint_V \delta T(x_m) \frac{\partial \dot{q}_m}{\partial x_m} \mathrm{d}V = \oint_A \delta T(x_m)(\dot{q}_m n_m)\, \mathrm{d}A - \oint_V \frac{\partial \delta T(x_m)}{\partial x_m} \dot{q}_m \, \mathrm{d}V. \tag{23.71}$$

Using (23.71) enables to reformulate the heat flux contribution $\partial \dot{q}_m / \partial x_m$ on the right-hand side of (23.70) and results into the balance (23.72). It contains unknown temperatures in the volume and on the sub-surface A_h subject to convective heat transfer (and/or radiation). Heat fluxes crossing the sub-surface A_h are determined by convection referring to (23.69). The convection law $h\ (T(x_m) - T_{ambient})$ is linear. Nonlinear radiation with $\dot{q} \sim T^4$ can be approximated by convection. On the complementary surface A_T, the surface temperatures T_{A_T} are prescribed. The volumetric temperature gradient obeys Fourier's law of heat conduction. The volume integral on the right-hand side of (23.72) represents the exothermic crystallization power, which inherently also depends on temperature.

Note that (23.72) is independent of FEM, in general, but FEM is the standard tool for an approximate spatial discretization and its solution generates a system of algebraic equations in the static case, or ordinary differential equations (Bathe 2014).

$$\begin{aligned}\oint_V \delta T(x_m)\rho\, c_p \frac{\partial T}{\partial t}\, \mathrm{d}V + \oint_V \frac{\partial \delta T}{\partial x_k}\left(k\frac{\partial T}{\partial x_k}\right) \mathrm{d}V + \int_{A_h} \delta T(x_m)\cdot h \cdot T(x_m)\, \mathrm{d}A \\ = \int_{A_h} \delta T(x_m)\cdot h\cdot T_{am}(x_m)\, \mathrm{d}A \\ + \int_{A_{qn}} \delta T(x_m)(-\dot{q}_k)n_k\, \mathrm{d}A \\ + \oint_V \delta T \rho\, \Delta h_{cr} \dot{\alpha}^{cr}\, \mathrm{d}V. \end{aligned} \tag{23.72}$$

The weak form approach fulfills only the heat flux boundary conditions (BCs) in (23.69). In FEM, this case is readily solved by first having built a (reduced) subsystem with eliminated boundary temperatures (or displacement BCs in the mechanics case), and, second, the subsequent insertion of that solution into the remaining equations.

The formal FEM approach is founded on the product decomposition $T(x_k, t)$ into spatial (polynomial) interpolation functions, passing through element nodal points, and nodal point temperatures as function of time. As the searched temperature in (23.72) is an approximation with respect to the exact temperature in the strong PDE (23.69), a rigorous denomination implies the changed notation $T(x_k, t) \to \overline{T}(x_k, t)$. This reads

$$\begin{aligned} \bar{T}(x_k, t) \approx \sum_p N_p(x_k) T_p(t) \\ \bar{T}(x_k, t) \approx \mathbf{N}(x_k)\mathbf{T}(t) \end{aligned}, \quad \mathbf{N} = [N_1, N_2, \cdots], \quad \mathbf{T}(t) = \begin{bmatrix} T_1 \\ T_2 \\ \vdots \end{bmatrix} \tag{23.73}$$

$$\frac{\partial \bar{T}}{\partial x_k} \approx \frac{\partial \mathbf{N}}{\partial x_k}\mathbf{T}(t) = \mathbf{B}(x_k)\mathbf{T}(t), \quad \frac{\partial \bar{T}}{\partial t} \approx \mathbf{N}(x_k)\dot{\mathbf{T}}(t)$$

where the vector $\mathbf{N}(x_k)$ comprises the interpolation polynomials and $\mathbf{T}(t)$ denotes the unknown vector of nodal point temperatures. It is a main idea of Galerkin to use same interpolation functions for the virtual test temperature δT, which implies

$$\delta T(x_k, t) \approx \mathbf{N}(x_k)\ \delta\mathbf{T}(t), \quad \frac{\partial \delta T}{\partial x_k} \approx \mathbf{B}(x_k)\ \delta\mathbf{T}(t) \tag{23.74}$$

with generally non-zero virtual nodal temperatures $\delta\mathbf{T}(t)$. The insertion of (23.73) and (23.74) into the weak energy balance equilibrium (23.72) generates the scalar product $\delta\mathbf{T}^{\mathrm{T}}[\cdots] = 0$, in which the vector in square brackets must be zero. This provides the following system of inhomogeneous first-order Ordinary Differential Equations (ODEs) for the nodal temperature $\mathbf{T}(t)$ of each finite element, namely:

$$\mathbf{C}_p \frac{\mathrm{d}\mathbf{T}}{\mathrm{d}t} + (\mathbf{K}_T + \mathbf{K}_h)\ \mathbf{T}(t) = \dot{\mathbf{Q}}_{h,am} + \dot{\mathbf{Q}}_{qn} + \dot{\mathbf{Q}}_{cr} \tag{23.75}$$

with the heat capacity matrix $\mathbf{C}_P$, also named thermal mass matrix, given by

$$\mathbf{C}_p(t) = \oint_V \mathbf{N}^{\mathrm{T}} \rho(t)\, c_p(t) \mathbf{N} \cdot \mathrm{d}V, \quad \rho(t) = \rho(t(\mathbf{T})), \quad c_p = c_p(t) \tag{23.76}$$

and the heat conductance matrix $\mathbf{K}_T$, generalized to anisotropy case, which is also obtained from volume integrals

$$\mathbf{K}_T(t) = \oint_V \mathbf{B}^{\mathrm{T}}\ \mathbf{k}(t)\ \mathbf{B}(x_k)\mathrm{d}V, \quad k = k(t) \rightarrow \mathbf{k} = diag(k_1, k_2, k_3). \tag{23.77}$$

The second convection matrix $\mathbf{K}_h$ refers to natural boundary condition in (23.69) valid on the sub-surface A_h. This matrix reads

$$\mathbf{K}_h = \int_{A_h} \mathbf{N}_{A_h}^{\mathrm{T}} h \mathbf{N}_{A_h}\, \mathrm{d}A, \quad \mathbf{N}_{A_h} = \mathbf{N}\big(x_{k,A_h}\big) \tag{23.78}$$

where matrix $\mathbf{N}_{A_h}$ has non-zero entries only for the elements on the surface A_h. The load vector of heat power is composed of, firstly, the convective heat power $\dot{\mathbf{Q}}_{qn,am}$ through the surface A_h and driven by the ambient temperature T_{am}, reading

$$\dot{\mathbf{Q}}_{h,am} = \int_{A_h} \mathbf{N}_{A_h}^{\mathrm{T}}\ hT_{am}\, \mathrm{d}A, \tag{23.79}$$

secondly, the heat power from a known normal component of the heat flux vector $\dot{\mathbf{Q}}_{qn}$ on the surface A_{qn}, or calculated from a FE mesh. For this vector, the surface integral

$$\dot{\mathbf{Q}}_{qn} = \int_{A_{qn}} \mathbf{N}_{A_{qn}}^{\mathrm{T}} (-\dot{q}_k) n_k\, \mathrm{d}A = \int_{A_{qn}} \mathbf{N}_{A_{qn}}^{\mathrm{T}}\ \dot{q}_n\, \mathrm{d}A, \quad -\dot{q}_k n_k = \dot{\mathbf{q}} \cdot \mathbf{n} = \dot{q}_n \tag{23.80}$$

has to be solved. Thirdly, the volumetric power $\dot{\mathbf{Q}}_{cr}$ of the exothermic, time-dependent crystallization enthalpy of PA 12 which is

$$\dot{\mathbf{Q}}_{cr} = \oint_V \mathbf{N}^{\mathrm{T}} \rho \, \Delta h_{cr} \dot{\alpha}^{cr}(t) \, \mathrm{d}V. \tag{23.81}$$

Since the problem is temperature and thus time dependent, the ODEs become nonlinear. In addition, the crystallization rate $\dot{\alpha}^{cr}(t(T))$ is a nonlinear function of the temperature that undergoes strong changes in an interval of 10 °C between 155 °C and 160 °C during the process.

It is common to discretize the ODEs by the implicit backward Euler method. In ABAQUS, Taylor expansions of time are applied to the matrices $(\mathbf{C}_P, \ \mathbf{K}_T + \mathbf{K}_h)$. To present the idea, a slightly simplified nonlinear implicit algorithm

$$\begin{gathered} \mathbf{C}_p(t_{k+1}) \tfrac{\mathbf{T}_{k+1} - \mathbf{T}_k}{\Delta t} = -\big(\mathbf{K}_{T,k+1} + \mathbf{K}_{h,k+1}\big)\mathbf{T}_{k+1} + \dot{\mathbf{Q}}_{k+1} \rightarrow \\ \big[\mathbf{C}_{p,_{k+1}} + \big(\mathbf{K}_{T,k+1} + \mathbf{K}_{h,k+1}\big)\Delta t\big]\mathbf{T}_{k+1} = \mathbf{C}_{p,_{k+1}} \mathbf{T}_k + \Delta\mathbf{Q}_{k+1} \end{gathered} \tag{23.82}$$

is shown. The thermal work increments $\Delta\mathbf{Q}_{k+1} = \Delta\mathbf{Q}_{qn,k+1} + \Delta\mathbf{Q}_{h,am,k+1}$ are given by

$$\Delta\mathbf{Q}_{qn,k+1} + \Delta\mathbf{Q}_{h,am,k+1} = \int_{A_{qn}} \mathbf{N}_{A_{qn}}^{\mathrm{T}} \big(\dot{q}_{n,k} \Delta t\big) \, \mathrm{d}A + \int_{A_h} \mathbf{N}_{A_h}^{\mathrm{T}} \big(h T_{am,k} \Delta t\big) \, \mathrm{d}A. \tag{23.83}$$

For the crystallization part $\Delta\mathbf{Q}_{cr,k+1}$, in (23.84), the rate is approximated at the previous time step $t_k = t_{k+1} - \Delta t$. In the realized code, the crystallization increment is calculated by the fourth-order Runge–Kutta scheme, instead of the shown first-order Euler method.

$$\begin{aligned} \Delta\mathbf{Q}_{cr,k+1} &= \oint_V \mathbf{N}^{\mathrm{T}} \rho \, \Delta h_{cr} \big(\dot{\alpha}^{cr}_{(t_k)} \Delta t\big) \, \mathrm{d}V \\ \big(\dot{\alpha}^{cr}_{(t_k)} \Delta t\big) &= \Big[nK(T_k)(1 - \alpha_k^{cr}) \big[- \ell n(1 - \alpha_k^{cr})\big]^{\frac{n-1}{n}} \Big] \cdot \Delta t = \Delta\alpha_{k+1}^{cr} \end{aligned} \tag{23.84}$$

23.3.4.2 Stresses and Displacements by Weak Momentum Balance

For the mechanical case, the weak equilibrium approach is invoked once more. For the reasoned admissible static momentum balance with negligible body forces in (23.13), namely, $\operatorname{div} \boldsymbol{\sigma} = \mathbf{0} \ \Leftrightarrow \ \sigma_{ij,i} = 0_j$, introducing a virtual displacement vector $\delta v_j \ \Leftrightarrow \ \delta\mathbf{v}$, and adopting Galerkin's approach of a zero weighed residuum for an approximate stress tensor $\overline{\sigma} \ \Leftrightarrow \ \overline{\sigma}_{ij}$ acting in a closed volume, namely

$$\oint_V \big(\delta v_j \cdot \overline{\sigma}_{ij,i}\big) \, \mathrm{d}V = 0 \Leftrightarrow \oint_V \big(\delta\mathbf{v}^{\mathrm{T}} \operatorname{div} \overline{\sigma}\big) \, \mathrm{d}V = 0 \tag{23.85}$$

employing partial integration and Gauss's theorem leads to the weak work equilibrium

$$\begin{aligned}\oint_V \left(\delta v_{j,i}\cdot\overline{\sigma}_{ij}\right)\mathrm{d}V &= \int_{A_u}\delta v_j \bar{s}_j\,\mathrm{d}A + \int_{A_\sigma}\delta v_j\bar{s}_j\,\mathrm{d}A,\quad \bar{s}_j=\overline{\sigma}_{ij}\,n_i\\ \oint_V \left(\nabla\,\delta\mathbf{v}^{\mathrm{T}}\,\overline{\sigma}\right)\mathrm{d}V &= \int_{A_u}\delta\mathbf{v}^{\mathrm{T}}\,\bar{\mathbf{s}}\,\mathrm{d}A + \int_{A_\sigma}\delta\mathbf{v}^{\mathrm{T}}\,\bar{\mathbf{s}}\,\mathrm{d}A,\quad \bar{\mathbf{s}}=\overline{\sigma}\ \cdot\mathbf{n}\end{aligned}\tag{23.86}$$

where $\bar{\mathbf{s}}=\overline{\sigma}\ \cdot\mathbf{n}$ is the surface stress vector associated with the force $d\,\mathbf{F}=\bar{\mathbf{s}}\,dA$ on the surface area dA. Given surface stresses on the sub-surface A_σ provides boundary work, while on the surface A_u displacement vectors are given. Following the usual procedure, zero displacements on A_u are assumed, leading to the reduced boundary value problem. For FEM discretization, displacement shape functions interpolated by element nodal displacement vectors are introduced. Time dependence for nodal displacements is included, because of the viscoelastic constitutive law, regardless of time-varying temperatures. This approach reads

$$\begin{aligned}&\bar{\mathbf{u}}(\mathbf{x},t)\approx\sum_p N_p(\mathbf{x})\cdot\mathbf{u}_p(t)\\ &\bar{\mathbf{u}}(\mathbf{x},t)\approx\mathbf{N}(\mathbf{x})\,\mathbf{u}(t),\quad \mathbf{N}=[N_1,N_2,\cdots]\\ &\varepsilon(\mathbf{x},t)\approx\mathbf{B}(\mathbf{x})\,\mathbf{u}(t)\\ &\text{where}\quad \mathbf{B}_p\approx\frac{\partial\bar{\mathbf{u}}_p}{\partial x_k}=\frac{\partial N_p}{\partial x_k}=\nabla\mathbf{N}_p(x_k)\\ &\delta\mathbf{v}\approx\mathbf{N}(\mathbf{x})\cdot\mathbf{v},\quad \frac{\partial\delta\bar{\mathbf{v}}}{\partial x_k}=\nabla\delta\bar{\mathbf{v}}\approx\mathbf{B}(x_k)\bar{\mathbf{v}}.\end{aligned}\quad,\ \mathbf{u}=\begin{bmatrix}\begin{pmatrix}u_{r1}\\u_{z1}\end{pmatrix}\\ \begin{pmatrix}u_{r2}\\u_{z2}\end{pmatrix}\\ \vdots\end{bmatrix}=\begin{bmatrix}\mathbf{u}_1\\\mathbf{u}_2\\\vdots\end{bmatrix}\tag{23.87}$$

An arbitrary virtual displacement vector $\delta\mathbf{v}=\mathbf{N}\,\mathbf{v}$ provides the reduced system of spatially discrete weak mechanical equilibrium

$$\oint_V \mathbf{B}^{\mathrm{T}}\overline{\sigma}\mathrm{d}V=\int_{A_\sigma}\mathbf{N}_{A_\sigma}^{\mathrm{T}}\bar{\mathbf{s}}\mathrm{d}A=\int_{A_\sigma}\mathbf{N}_{A_\sigma}^{\mathrm{T}}(\overline{\sigma}\cdot\mathbf{n})\mathrm{d}A=\mathbf{F}=\mathbf{F}_{A_\sigma},\bar{\mathbf{s}}=\overline{\sigma}\cdot\mathbf{n}\tag{23.88}$$

for each element (Bathe 2014). The difference of (23.88) for two time steps provides an algorithm for the increments $\Delta\mathbf{u}_k\ \rightarrow\ \mathbf{u}_{k+1}\ =\ \mathbf{u}_k+\Delta\mathbf{u}_k$ and all other next quantities. The onset reads

$$\oint_V \mathbf{B}^{\mathrm{T}}\left(\overline{\sigma}_{k+1}-\overline{\sigma}_k\right)\mathrm{d}V=\mathbf{F}_{k+1}-\mathbf{F}_k\tag{23.89}$$

where the element boundary forces $(\mathbf{F}_k,\ \mathbf{F}_{k+1})$ are assumed to be known at time steps or they are obtained from the standard post-processing of FEM simulation. The stress increment $\boldsymbol{\sigma}_{k+1}-\boldsymbol{\sigma}_k=\Delta\boldsymbol{\sigma}(t_k,\Delta t)=\Delta\boldsymbol{\sigma}_k$ is available from the rearranged stress Eq. (23.66) and can be presented as

$$\boldsymbol{\sigma}_{k+1}-\boldsymbol{\sigma}_k=-\boldsymbol{\kappa}\left[\sigma_k^{th+cr}+3K_\infty\left(\varepsilon_{k+1}^{th+cr}-\varepsilon_k^{th+cr}\right)\right]+\mathbf{C}_k(\boldsymbol{\varepsilon}_{k+1}-\boldsymbol{\varepsilon}_k)$$

$$\Delta\boldsymbol{\sigma}_k = -\left[\boldsymbol{\sigma}_k^{th+cr} + \boldsymbol{\kappa} \cdot 3K_\infty\left(\Delta\varepsilon_k^{th} + \Delta\varepsilon_k^{cr}\right)\right] + \mathbf{C}_k\,\Delta\boldsymbol{\varepsilon}_k \tag{23.90}$$

comprising the accumulated volumetric thermal and crystallization stress in $\boldsymbol{\sigma}_k^{th+cr}$, the volumetric scalar increments of both $\boldsymbol{\kappa} \cdot 3K_\infty\left(\Delta\varepsilon_k^{th} + \Delta\varepsilon_k^{cr}\right)$, and the thermo-viscoelastic growth due to the time (and temperature) varying stiffness matrix $\mathbf{C}_k = \mathbf{C}^{VE}(t_k)$ as factor of the strain increment vector that is related to the nodal displacement vector by the spatial FEM discretization $\Delta\boldsymbol{\varepsilon}_k = \mathbf{B}\,\Delta\mathbf{u}_k$. The attained mechanical stress $\boldsymbol{\sigma}(t_k)$ is

$$\boldsymbol{\sigma}_k = \boldsymbol{\sigma}(t_k) = \boldsymbol{\Gamma}\boldsymbol{\sigma}_k^{dev} + \boldsymbol{\kappa}\sigma_k^{vol} \tag{23.91}$$

and yields the updated approximated stress

$$\overline{\sigma}_{k+1} = \overline{\sigma}_k + \mathbf{C}_k^{VE}\,\mathbf{B}\,\Delta\mathbf{u}_k - \boldsymbol{\kappa}\left[\sigma_k^{th+cr} + 3K_\infty\left(\Delta\varepsilon_k^{th} + \Delta\varepsilon_k^{cr}\right)\right] \tag{23.92}$$

Expressing Eq. (23.89) in terms of displacements and including the thermal and crystallization forces, one obtains

$$\left[\oint_V \mathbf{B}^{\mathrm{T}}\,\mathbf{C}_k^{VE}\,\mathbf{B}dV\right]\,\Delta\mathbf{u}_k = (\mathbf{F}_{k+1} - \mathbf{F}_k) + \mathbf{F}_k^{th+cr} + \Delta\mathbf{F}_k^{th} + \Delta\mathbf{F}_k^{cr}$$
$$\rightarrow\quad \mathbf{u}_{k+1} = \Delta\mathbf{u}_k + \mathbf{u}_k. \tag{23.93}$$

The force vectors are dependent on thermal strain and crystallization fields via volume integrals. They read

$$\mathbf{F}_{k,k+1} = \int_{A_\sigma} \mathbf{N}_{A_\sigma}^{\mathrm{T}}(\mathbf{x})\,\bar{\mathbf{s}}_{k,k+1}\,\mathrm{d}A = \mathbf{F}_{boundaries} \tag{23.94}$$

for the "mechanical" forces on the boundaries at time steps (t_k, t_{k+1}), and furthermore

$$\mathbf{F}_k^{th+cr} = \oint_V \mathbf{B}^{\mathrm{T}}(\mathbf{x})\boldsymbol{\sigma}_k^{th+cr}(\mathbf{x})\,\mathrm{d}V \tag{23.95}$$

$$\Delta\mathbf{F}_k^{th} = 3K_\infty \oint_V \mathbf{B}^{\mathrm{T}}(\mathbf{x}) \cdot \boldsymbol{\kappa}\;\alpha_k^{th}\,\Delta T_k(\mathbf{x})\mathrm{d}V \tag{23.96}$$

$$\Delta\mathbf{F}_k^{cr} = K_\infty\,\varepsilon_{vol}^{shrink}\,\Delta t \cdot \oint_V \mathbf{B}^{\mathrm{T}}(\mathbf{x}) \cdot \boldsymbol{\kappa}\;\dot{\alpha}_k^{cr}(T_k(\mathbf{x}))\mathrm{d}V \tag{23.97}$$

for the incremental volumetric thermal and crystallization loads.

23.3.5 Simulation Stages

As mentioned above, mechanical boundary conditions do not change throughout the simulation. However, thermal boundary conditions differ for each stage of the simulation and they are explained below.

23.3.5.1 Preheating Stage

During real SLS process, the powder is deposited layer by layer and heated with a radiant heater until a quasi-steady state is reached. In the simulation, the powder segment under the first disk is deposited as bulk with the predefined temperature 171 °C and it reaches the steady state with defined boundary conditions. Since there is no sintering at this stage, the simulation is carried out as a 1-second steady-state step in ABAQUS. The temperature boundary condition is defined for the top surface and heat transfer coefficients are defined for the top and side surfaces, shown in Fig. 23.11.

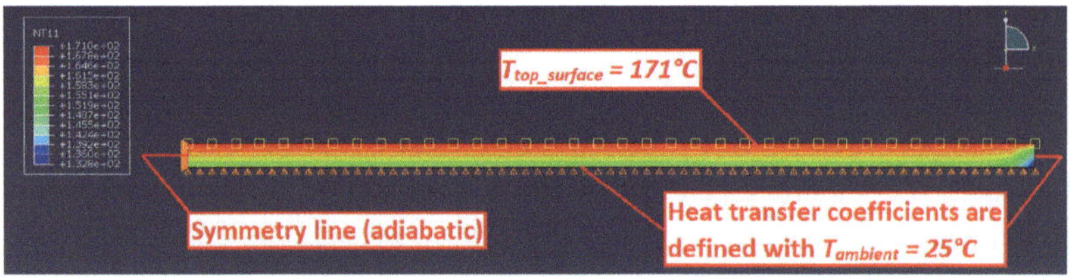

Fig. 23.11 Steady-state preheating stage and thermal boundary conditions of the powder bed

The heat transfer coefficients at the side and the bottom surfaces are defined below (Neugebauer 2017) and they are implemented in ABAQUS through the subroutine FILM.

$$h_{sides} = \begin{cases} 3 \cdot 10^{-6} \frac{\mathrm{W}}{\mathrm{mm}^2 K}, & 20°\mathrm{C} \le T \le 90°\mathrm{C} \\ \left(3 \cdot 10^{-6} + 5.71 \cdot 10^{-8 \frac{T - 90°\mathrm{C}}{°\mathrm{C}}}\right) \frac{W}{mm^2 K}, & 90°\mathrm{C} \le T \le 160°\mathrm{C} \end{cases}$$
$$h_{bottom} = 2.6 \cdot 10^{-6} \frac{W}{mm^2 K} \tag{23.98}$$

23.3.5.2 Sintering Stage

Layer thickness of 0.12 mm, the real process parameter, is adopted for the simulation. All elements are activated at the temperature 171 °C. Heat flux coming from the laser is modeled as initial temperature of 250 °C (Neugebauer 2017) on the top surface of the elements in the sintered area. The laser affects the entire surface of the sintered

a) b)

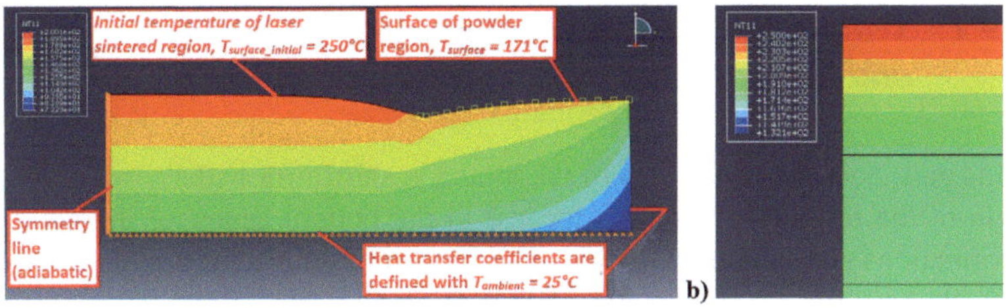

Fig. 23.12 **a** Thermal boundary conditions for the sintering stage. **b** Temperature of the top surface of a sintered element (initially at 250 °C at $t_{step} = 0$ s) and its conduction through the element

portion at once as a macroscopic-scale approach (Neugebauer 2017) (see Fig. 23.12). The same heat transfer coefficients applied during preheating are kept in this stage. The sintering stage lasts 14,707.6 s.

23.3.5.3 Cooling Stage

Cooling follows the deposition of the last layer. The boundary temperature of 171 °C on the top surface is removed and the heat convection coefficient h_{top} (Neugebauer 2017) defined in FILM is activated.

$$h_{top} = \begin{cases} 6 \cdot 10^{-6} \frac{W}{mm^2 K} & 20°\mathrm{C} \leq T \leq 100°\mathrm{C} \\ \left(6 \cdot 10^{-6} + 1.33 \cdot 10^{-8} \frac{T - 100°\mathrm{C}}{°\mathrm{C}}\right) \frac{W}{mm^2 K} & 100°\mathrm{C} \leq T \leq 145°\mathrm{C} \\ \left(5.4 \cdot 10^{-6} - 2 \cdot 10^{-8} \frac{T - 145°\mathrm{C}}{°\mathrm{C}}\right) \frac{W}{mm^2 K} & 145°\mathrm{C} \leq T \leq 200°\mathrm{C} \end{cases} \quad (23.99)$$

At this stage, convective heat loss from the top surface becomes effective in addition to the heat loss from the side and bottom surfaces during sintering. Therefore, the dominant temperature gradient in the model is expected to change.

23.4 Results

Simulation results are presented and discussed with respect to first, the space and time characteristics of the calculated fields and, second, the time history at selected points.

23.4.1 Temperature, Crystallization, Displacement, and Stress Fields

In this part, only the disk's area of the model is shown for better visualization. The origin of the reference system is located in the bottom center of the build space. Deformations are magnified tenfold. The following figures present the temperature fields, the resulting crystallization, and the radial (bending) stress fields. It is seen in Fig. 23.13 that the first disk is bending downward. The first reason for this behavior is the starting location of the crystallization at the bottom right corner. The second reason is Temperature Gradient Mechanism (TGM) explained in Subsect. 23.1. The dominant temperature gradient is in the vertical direction because of the heat input on the upper layers.

The dominant temperature gradient is not changing until cooling starts. This is visible at a later time in Figs. 23.14 and 23.15. Referring to Sect. 23.2, the crystallization occurs faster at lower temperatures, which can be observed on the disks. The crystallization follows the temperature gradient.

At t = 7072 s in Fig. 23.14, Disk-2 and Disk-3 undergo a similar deformation as Disk-1. It can also be seen that Disk-1 is recovered to a great extent due to the crystallization and thermal shrinkage on the top side of it. Note that bending (radial) stresses at the concave sides of the disks are positive as opposed to the bending due to external force.

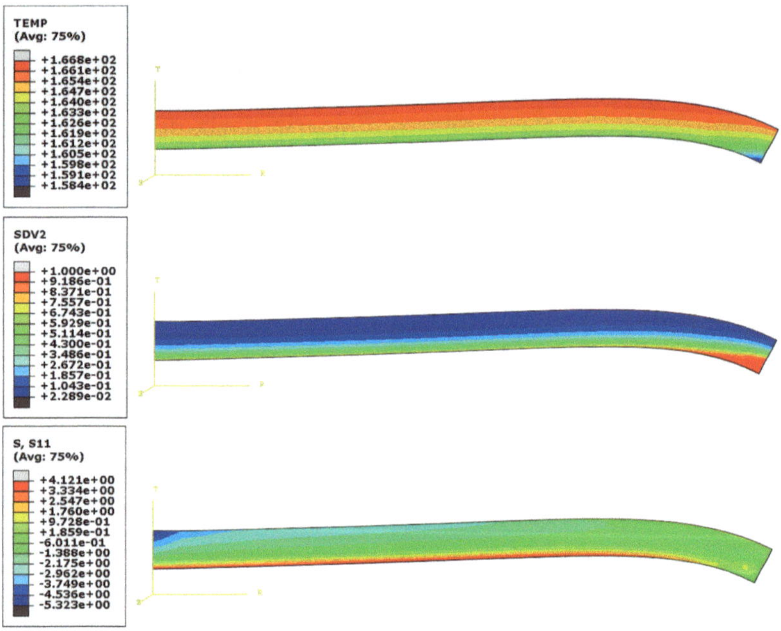

Fig. 23.13 Temperature (top), crystallization (middle), and radial stress of the TVE model at $t = 2656$ s

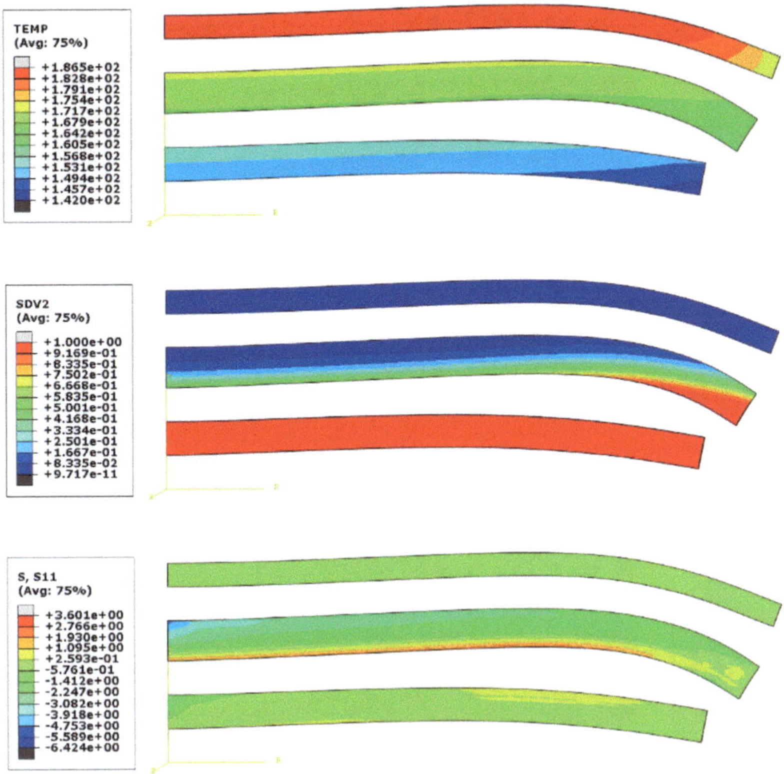

Fig. 23.14 Temperature (top), crystallization (middle), and radial stress of the TVE model at $t = 7072$ s

At t = 14,704 s, just before the cooling phase stars, approximately half of the disks are crystallized. The deformation behavior of all disks is similar so far as seen in Fig. 23.15. The disks at the lower parts of the model have been recovering from downward bending. The top surface is the hottest part of the model. Considering the temperature field and advancement of the crystallization until this point, it is concluded that this pattern is in accordance with Temperature Gradient Mechanism (TGM) and the crystallization strain.

At t = 14,708.6 s printing finishes and cooling starts. As defined in the boundary conditions, the model starts losing heat from top in addition to bottom and sides. Temperature gradient changes in the top portion of the model due to cooling. The hottest portion of the model is moving from the top to the center slowly (see Fig. 23.16).

At t = 15,416 s in Fig. 23.16 Disk-5 behaves similar to Disk-1 because of a similar temperature gradient. However, it behaves in the opposite direction of Disk-1 because it is losing heat faster from top surface. On the top surface of Disk-5 the stress is tensile and on the bottom it is compressive. In Fig. 23.16, Disk-4 had already bent downward but a portion bent upward, like a horizontal "S" shape after cooling started.

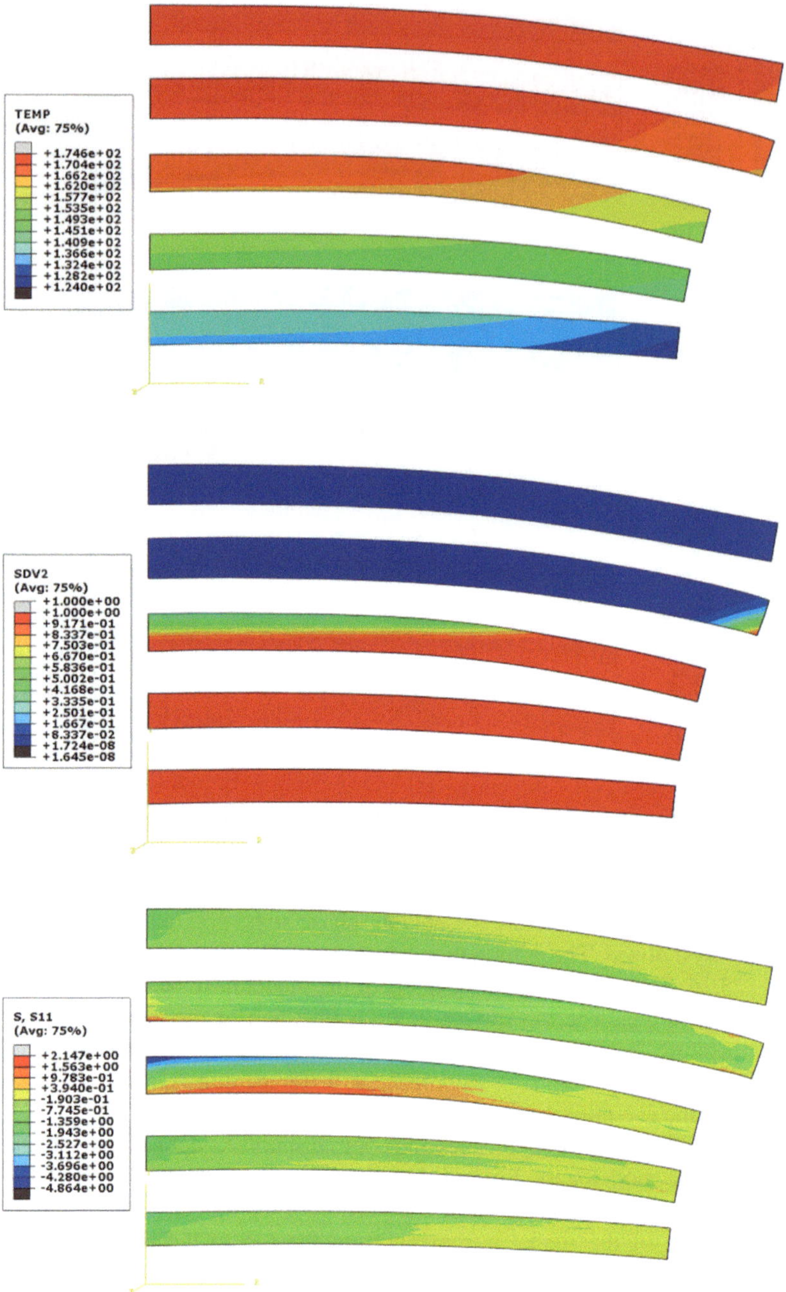

Fig. 23.15 Temperature (top), crystallization (middle), and radial stress of the TVE model at $t = 14{,}704$ s

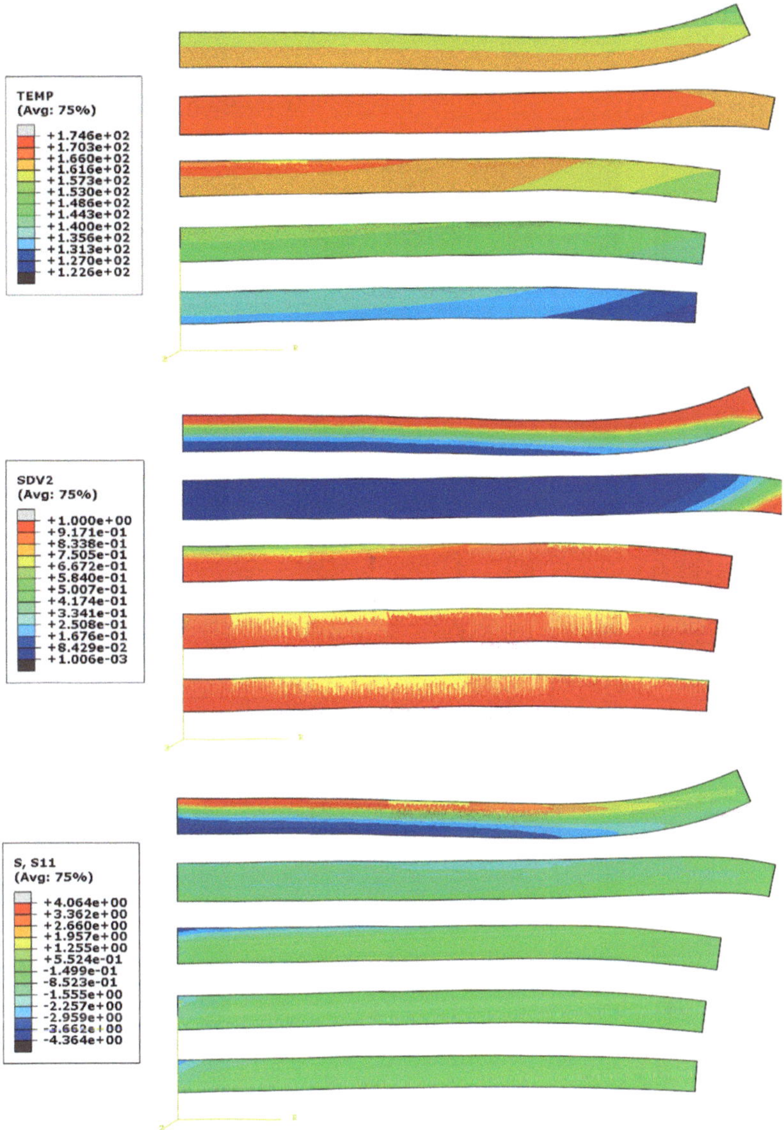

Fig. 23.16 Temperature (top), crystallization (middle), and radial stress of the TVE model at $t = 15{,}416$ s

As the cooling continues, Disk-5 recovers from upward bending also and bends slightly downward because hottest portion of the model moves from top surface to the core (see Fig. 23.17). Disk-1 bends slightly up after recovery due to identical temperature gradients.

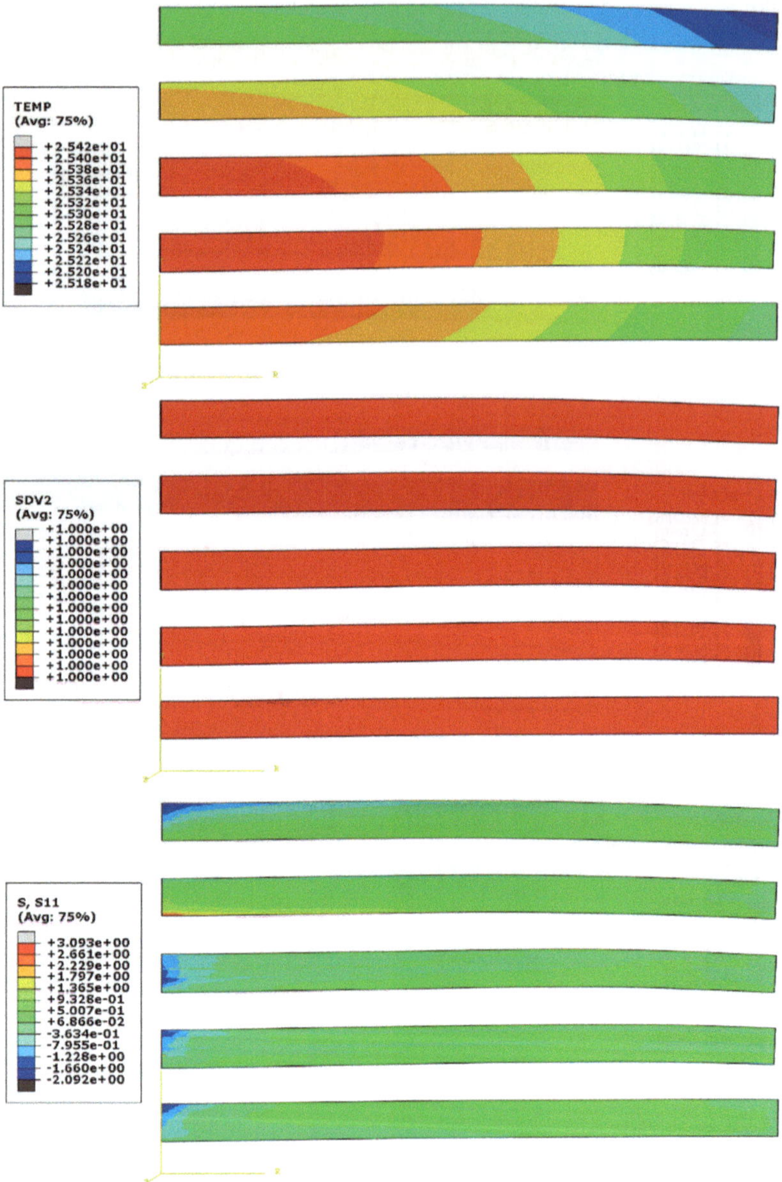

Fig. 23.17 Temperature (top), crystallization (middle), and radial stress of the TVE model at $t = 74{,}708$ s

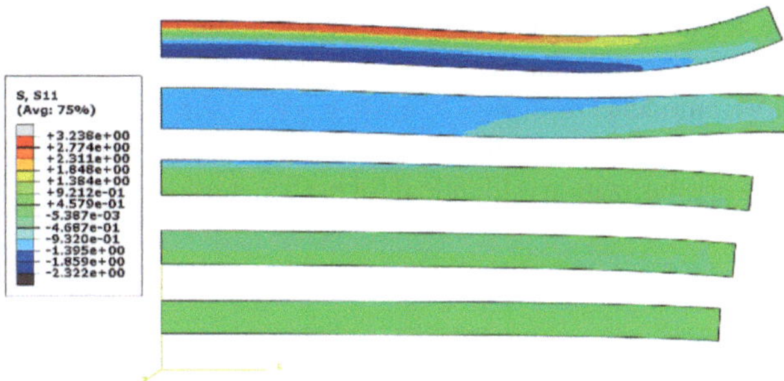

Fig. 23.18 Radial stress and deformation of the elastic model at $t = 15{,}416$ s. Its shape is similar to the viscoelastic model (see Fig. 23.16)

Fig. 23.19 Radial stress and deformation of the elastic model at $t = 74{,}708$ s. Its shape is different from the viscoelastic model (see Fig. 23.17)

Regarding the elastic model, although the deformation is similar with the viscoelastic model at $t = 15{,}416$ s (see Fig. 23.18), it is different at $t = 74{,}708$ s (see Fig. 23.19). In the elastic model, all of the disks are bent upward slightly, while in the viscoelastic model the upper three disks bent downward, and the rest upward. The reason is that the deformation continues in the viscoelastic model even after the effect of temperature becomes negligible. After 74,708 s changes are not notable.

23.4.2 Time History Analyses of Fields at Characteristic Locations

In Fig. 23.20, temperatures of five midline elements (see Fig. 23.6) are presented. The reason for the zig-zag pattern is application of the laser on consecutive layers.

Disk-5 cools faster than the other ones because of the thermal boundary condition on the top surface. The thermal behavior does not differ between elastic and the viscoelastic (TVE) models.

Figure 23.21 shows the crystallization behavior of the midline elements (see Figure 23.6). The crystallization is identical for both materials, because it is a function of temperature.

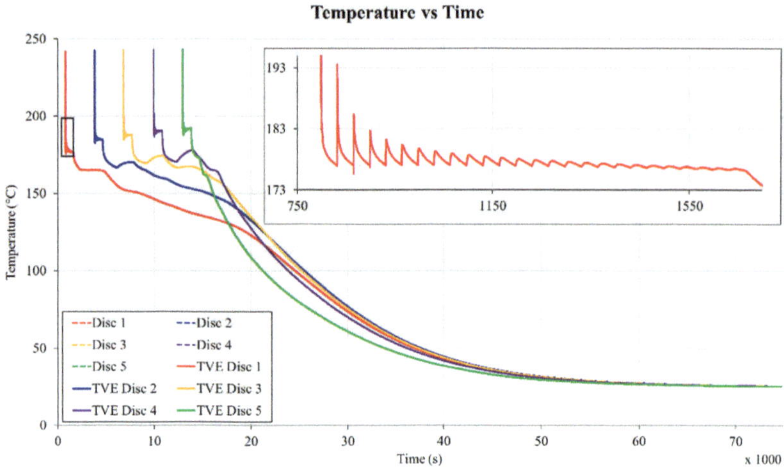

Fig. 23.20 Temperature of the midline elements for the elastic and viscoelastic (TVE) mechanical model

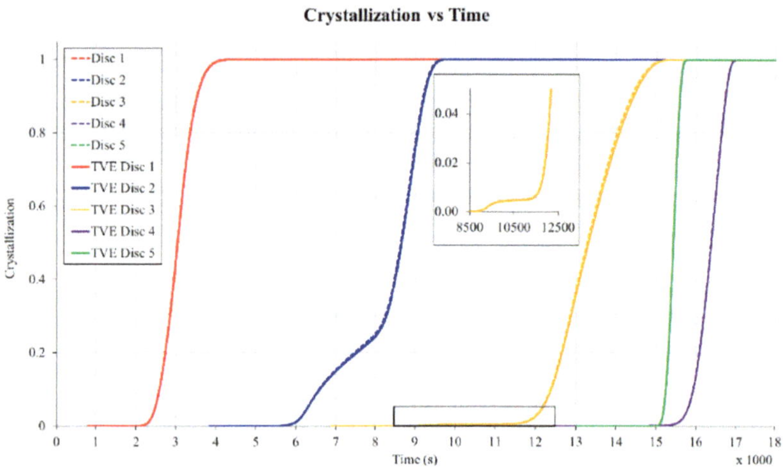

Fig. 23.21 Crystallization of the midline elements of the disks for elasticity and viscoelasticity (TVE)

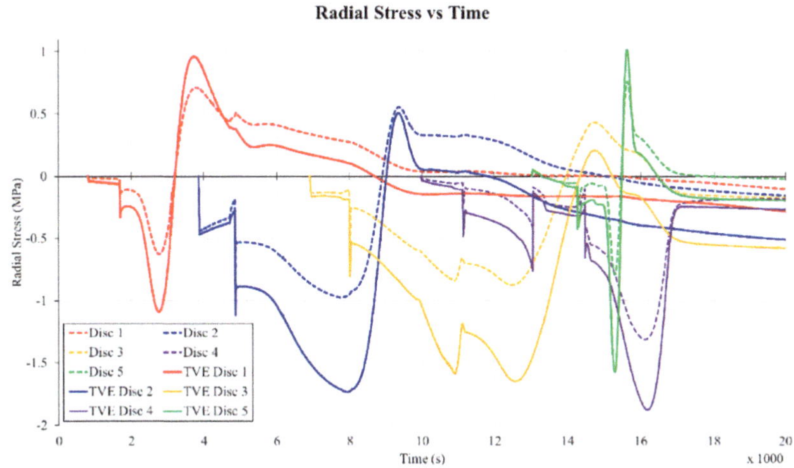

Fig. 23.22 Radial stress in the midline elements until 20,000 s for elasticity and viscoelasticity (TVE)

The oscillating bending behavior can be clearly seen in Fig. 23.22 for the first 20,000 s. The behavior of elastic material model is similar to viscoelastic model. The differences are the maximum stress amplitudes, which are higher for the viscoelastic case due to the Maxwell elements parallel to the elastic ones.

In Fig. 23.23, the evolution of temperature, crystallization, and stress for the elastic model is presented for elements at the top and bottom surfaces of Disk-1 and Disk-5. This covers the whole time interval of printing and the beginning of cooling. The temperature is the driving parameter of the process. In the top plot, it can be seen that the top of Disk-5 cools faster than the bottom. This affects the crystallization speed as shown in the middle plot. Although top of Disk-5 is sintered later than its bottom, the crystallization starts there earlier. It can also be seen that crystallization at the top of Disk-1 is quite slower because heat loss at that location is much smaller due to the thermal resistance of the layers around it and the added heat during the process. It is also seen that if the crystallization takes long time, the stress change shows a wide peak, and vice versa for shorter duration of crystallization. Moreover, the stress is obviously higher for viscoelasticity.

To investigate the results in more detail, additional elements were chosen at the bottom and top sides of Disk-1 and Disk-5. The temperature, crystallization, and stress values at the bottom surface of Disk-1 arc shown in Figs. 23.24, 23.25, and 23.26 for the same time interval. The zig-zag pattern due to heat addition with each new layer can be seen in both temperature and stress. Note also, while the thermal and crystallization behavior are identical for elasticity and viscoelasticity, stress magnitudes are different. Finally, Fig. 23.25 demonstrates drastic stress variations occurring during crystallization, where both the elastic and viscoelastic stresses are clearly visible.

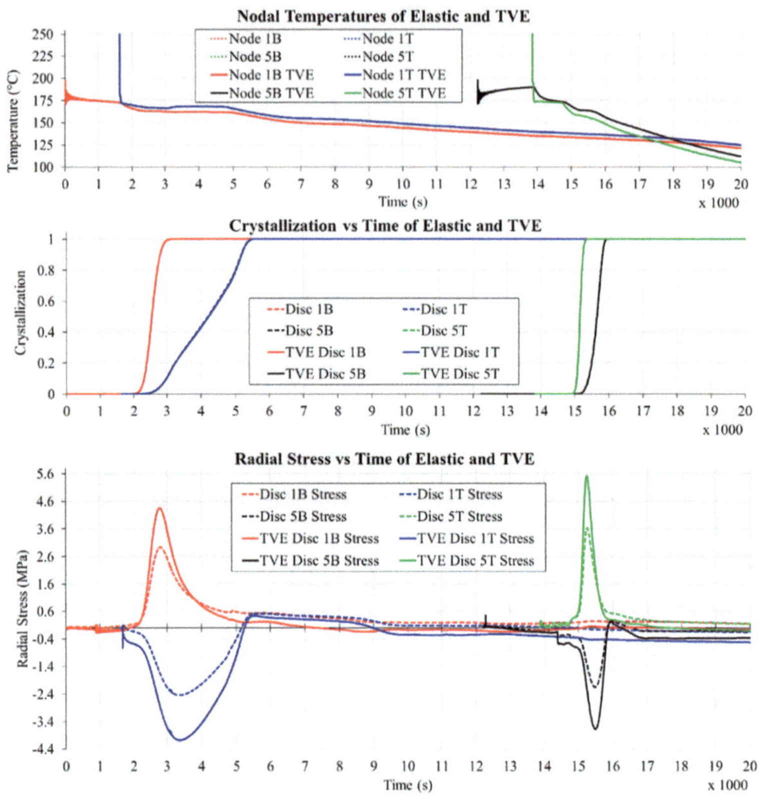

Fig. 23.23 Nodal temperature, crystallization, and stress history at the bottom and top surfaces of Disk-1 and Disk-5 until 20,000 s for the elastic and the viscoelastic (TVE) models

Since the process is temperature driven, variation versus temperature is analyzed. Figures 23.27 and 23.28 depict the crystallization and radial stress versus temperature on the bottom and the top surfaces of Disk-5. The stress peaks during the whole process occur in a short temperature interval and it is evident that they are the result of evolving crystallization strains.

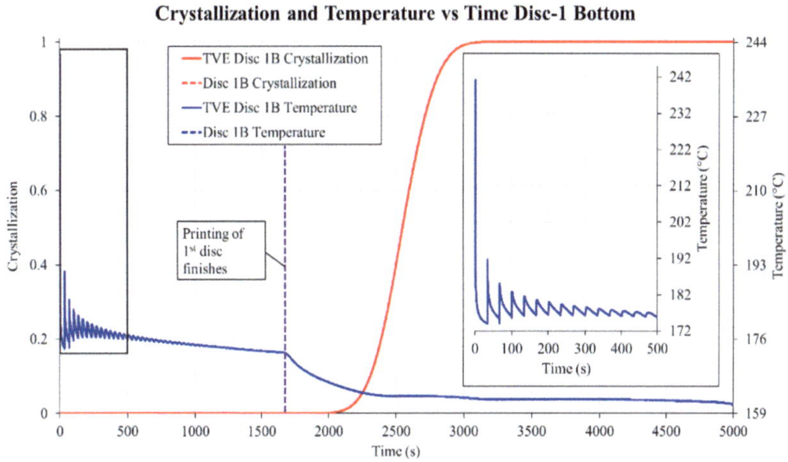

Fig. 23.24 Crystallization and temperature versus time for the element at the bottom surface of Disk-1. Notice the zig-zag pattern due to added hot layers

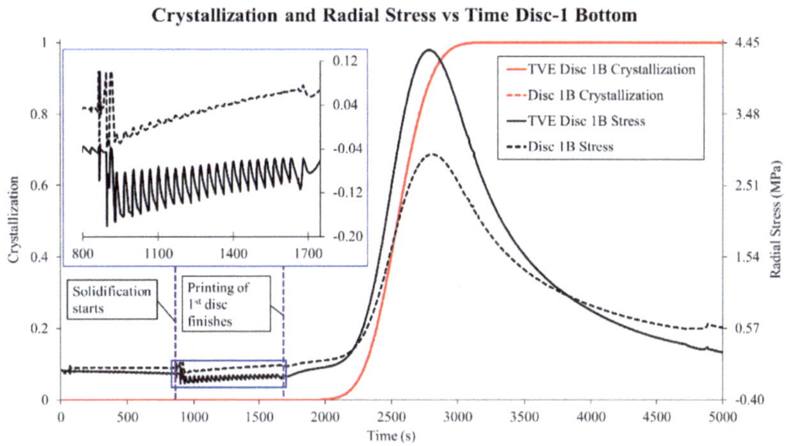

Fig. 23.25 Crystallization and radial stress versus time for the element on the bottom surface of Disk-1

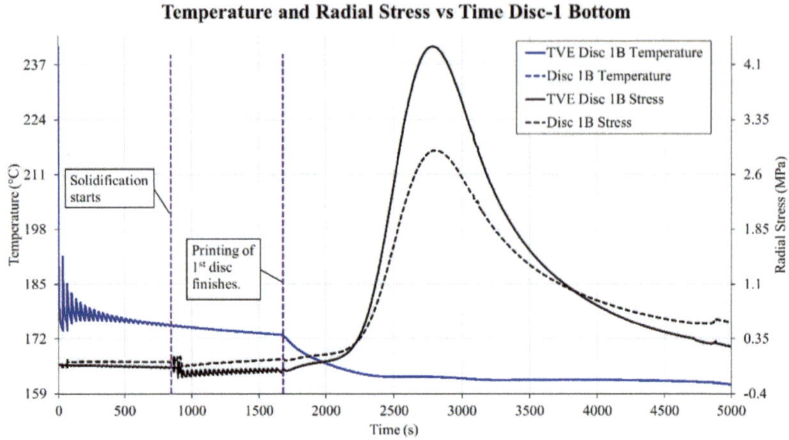

Fig. 23.26 Temperature and radial stress versus time for the element on the bottom surface of Disk-1. The zig-zag pattern in the stress curve starts with solidification

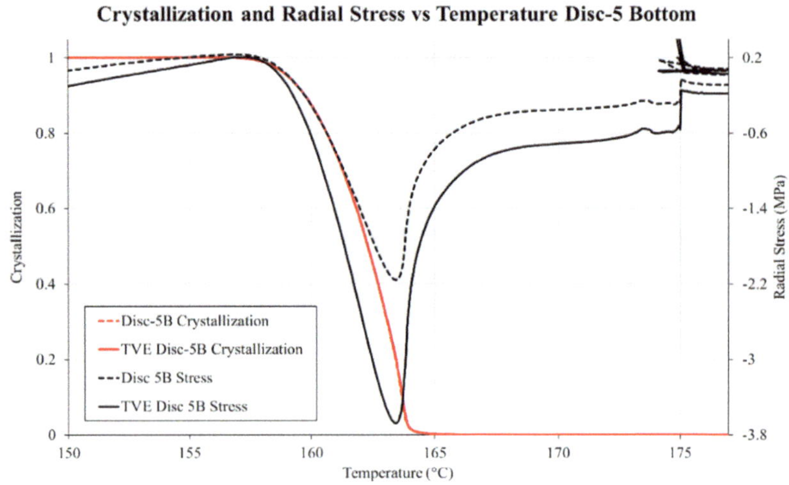

Fig. 23.27 Crystallization and stress versus temperature for the element on the bottom surface of Disk-5

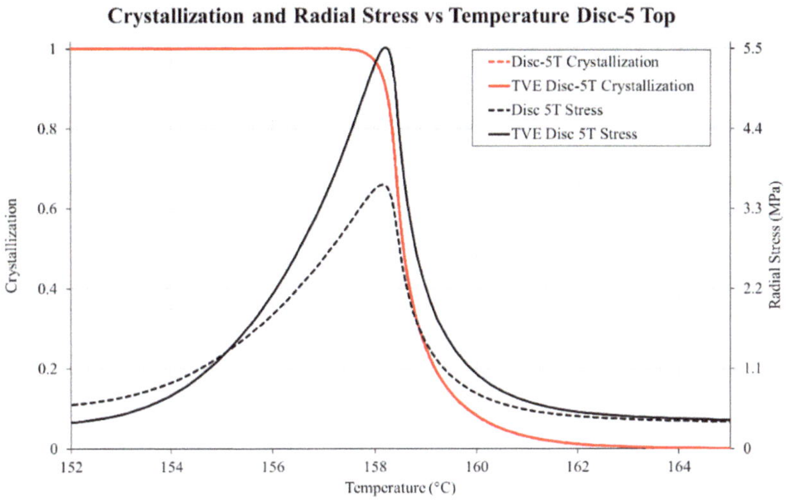

Fig. 23.28 Crystallization and stress versus temperature for the element on the top surface of Disk-5

23.5 Discussion and Conclusion

In this study, a tailored simulation tool for analysis of selective laser sintering (SLS) of industry level parts made of semi-crystalline PA 12 has been developed. Finite element implementation of a thermally, chemically, and mechanically coupled nonlinear system has been realized. Theory of the aforementioned phenomena has been implemented numerically through subroutines of a commercial software package.

Regarding the findings, the oscillating bending behavior has been calculated in agreement with the deformation mechanisms observed in SLS processes. The evolution of crystallization is in good agreement with the theory and it is a major factor in the built-up of residual stress and deformation. This is a considerable difference between AM of semi-crystalline polymers and other materials.

Viscoelasticity causes about 50% larger stresses than elasticity. The final deformation of the thermoviscoelastic disk stack is different from the thermoelastic one because of inherent time dependency of thermoviscoelastic model.

Future application of the developed tool should analyze the variation of material parameters and boundary conditions of the manufacturing machine. Moreover, the impact of different batch configurations such as part orientation and distances should be studied with simulation.

Improvements of the tool capabilities should include the plasticity effect of polymers at elevated temperatures by using thermoviscoplasticity. A more realistic laser power transfer should be modeled as a heat flux instead of an initial surface temperature. However, moving laser source is not a suitable method for an engineering scale part although it is more accurate. Finally, change of viscoelastic Prony parameters with respect to crystallization should be included in the model when the data is available.

Acknowledgements The authors are very grateful for the manufacturing of the disk's stack by the staff of the 3D-Printing Center of Fachhochschule Südwestfalen directed by Prof. Dr.-Ing. Jens Bechthold.

References

Amado A (2016) Characterization and prediction of SLS processability of polymer powders with respect to powder flow and part warpage. Dr. Sc. Dissertation, ETH Zürich

Avrami M (1939) Kinetics of phase change. I General theory. J Chem Phys **7**(12):1103–1112

Balemans C, Looijmans SFSP, Grosso G, Hulsen MA, Anderson PD (2020) Numerical analysis of the crystallization kinetics in SLS. Addit Manuf 33. https://doi.org/10.1016/j.addma.2020.101126

Bathe KJ (2014) Finite element procedures, 2nd edn. Watertown, MA

Bogetti TA, Gillespie JW (1992) Process-induced stress and deformation in thick section thermoset composite laminates. J Comp Mater 26(5):626–660

Brinson HF, Brinson LC (2008) Polymer engineering science and viscoelasticity. Springer

Denlinger ER, Michaleris P (2016) Effect of stress relaxation on distortion in additive manufacturing process modeling. Addit Manuf 12:51–59

Ehrenstein GW (2001) Polymeric materials: structure—properties—applications, Immergut E, Theriault R (transl.). Carl Hanser Verlag, Munich

Gladbach B, Noe A, Rosenhövel T (2024) Warpage reduction in additively manufactured parts based on thermomechanical modeling and a novel simulation strategy for laser scanning. In: Müller WH, Noe A, Ferber F (eds) New achievements in mechanics. advanced structured materials, vol 205. Springer, Cham, pp 301–339. https://doi.org/10.1007/978-3-031-56132-0_13

Goodridge RD, Truck CJ, Hague RJM (2012) Laser sintering of polyamides and other polymers. Prog Mater Sci 57:229–267

Gouge M, Michaleris P (2018) An introduction to additive manufacturing processes and their modeling challenges. In: Gouge M, Michaleris P (eds) Thermo-mechanical modeling of additive manufacturing. Elsevier, pp 10–15

Heigel JC, Michaleris P, Reutzel EW (2015) Thermo-mechanical model development and validation of directed energy deposition additive manufacturing of Ti–6Al–4V. Addit Manuf 5:9–19

Hoda A (2015) Investigating the cyclic performance of laser sintered nylon 12. PhD Dissertation, University of Sheffield

Hoffman J, Davis GT, Lauritzen J (1976) The rate of crystallization of linear polymers with chain folding. In: Hannay NB (eds) Treatise on solid state chemistry. Springer, Boston

Hopkinson N, Dickens P (2006) Emerging rapid manufacturing processes. In: Hopkinson N, Hague RJM, Dickens P (eds) Rapid manufacturing: an industrial revolution for the digital age. John Wiley & Sons, Ltd, West Sussex, p 65

Kaya U (2024) Finite element simulation of the evolution of residual stresses and distortions during selective laser sintering (SLS) of semi-crystalline polyamide 12 sheets. M.S. Thesis, University of Applied Sciences of South Westphalia

Kruth JP, Froyen L, Van Vaerenbergh J, Mercelis P, Rombouts M, Lauwers B (2004) Selective laser melting of iron-based powder. J Mater Process Technol 149:616–622

Mercelis P, Kruth JP (2006) Residual stresses in selective laser sintering and selective laser melting. Rapid Prototyping J 12(5):254–265

Nakamura K, Watanabe T, Katayama K, Amano T (1972) Some aspects of non-isothermal crystallization of polymers. I. Relationship between crystallization temperature, crystallinity, and cooling conditions. J Appl Polym Sci 16(5):1077–1091

Neugebauer F (2017) Material- und Prozessmodellierung zur effizienten FEM-Berechnung der Verzugsentstehung beim Laser-Sintern von teilkristallinen Thermoplasten. Ph.D. Dissertation, Universität Bremen

Peters S (2020) Modellbildung und FEM-Simulation zur orts- und zeitabhängigen Entwicklung der Kristallisation beim Selektiven-Laser-Sintern von Platten aus teilkristallinem Polyamid 12. B.S. Thesis, University of Applied Sciences of South Westphalia

Schwarzl RF (1990) Polymermechanik. Springer

Shen F, Zhu W, Zhou K, Ke L (2021) Modeling the temperature, crystallization, and residual stress for selective laser sintering of polymeric powder. Acta Mech 232:3635–3653

Soundararajan B, Sofia D, Barletta D, Poletto M (2021) Review on modeling techniques for powder bed fusion processes based on physical principles. Addit Manuf 47

Taylor RL, Pister KS, Goudreau GLF (1970) Thermomechanical analysis of viscoelastic solids. Int J Numer Methods Eng 2(1):45–59. https://doi.org/10.1002/nme.1620020106

Tschoegl N, Knauss WG, Emri I (2002) The effect of temperature and pressure on the mechanical properties of thermo- and/or piezorheologically simple polymeric materials in thermodynamic equilibrium—a critical review. Mech Time-Dependent Mater 6:53–99. https://doi.org/10.1023/A:1014421519100

Williams RJ, Davies CM, Hooper PA (2018) A pragmatic part scale model for residual stress and distortion prediction in powder bed fusion. Addit Manuf 22:416–425

Yasa E (2021) Selective laser melting: principles and surface quality. In: Pou J, Riveiro A, Davim JP (eds) Handbooks in advanced manufacturing additive manufacturing. Elsevier, p 77

Chapter 24
Approaches to Conservative Smoothed Particle Hydrodynamics with Entropy

Michal Pavelka, Václav Klika, and Ondřej Kincl

Abstract Smoothed particle hydrodynamics (SPH) is typically used for barotropic fluids, where the pressure depends only on the local mass density. Here, we show how to incorporate the entropy into the SPH, so that the pressure can also depend on the temperature, while keeping the growth of the total entropy, conservation of the total energy, and symplecticity of the reversible part of the SPH equations. The SPH system of ordinary differential equations with entropy is derived by means of the Poisson reduction and the Lagrange → Euler transformation. We present several approaches toward SPH with entropy, which are then illustrated on systems with discontinuities, on the expansion of ideal gas, and on the Rayleigh-Bénard convection without the Boussinesq approximation. Finally, we show how to model hyperbolic heat conduction within the SPH, extending the SPH variables not only with entropy but also with a heat-flux-related vector field.

24.1 Introduction

Smoothed Particle Hydrodynamics (SPH) is a particle-based discretization of the hydrodynamic partial differential equations (Gingold and Monaghan 1977; Lucy 1977). Each particle is typically described by its position and velocity, and the particles interact in a way that mimics the hydrodynamic behavior. Although simulation can be carried out in a purely reversible way (being even symplectic and globally reversible (Kincl and Pavelka 2023), viscous effects reduce the kinetic energy of the

M. Pavelka (✉) · O. Kincl
Mathematical Institute, Faculty of Mathematics and Physics, Charles University, Prague, Czech Republic
e-mail: pavelka@karlin.mff.cuni.cz

V. Klika
Department of Mathematics, FNSPE, Czech Technical University in Prague, Prague 2, Czech Republic

O. Kincl
Department of Civil, Environmental and Mechanical Engineering, University of Trento, Trento, Italy

J. Fuhrmann et al. (eds.), *Advances in Continuum Physics*, Advanced Structured Materials 238, https://doi.org/10.1007/978-3-031-93918-1_24

particles, and the total energy is often not conserved by SPH schemes. The addition of entropy to each particle should restore total energy conservation (Violeau 2012; Monaghan 2005). In this paper, we show several approaches toward SPH with the entropy of each particle while keeping the geometric structure of the reversible part of the SPH equations (symplecticity).

Cleary and Monaghan Cleary and Monaghan (1999) add the internal energy to each particle, the evolution of which compensates the dissipation by Fourier heat conduction so that the total energy is conserved. However, the integration scheme is not symplectic. In Antuono et al. (2010), the energy equation contains a term that violates global energy conservation (the second term on the right-hand side of the energy equation in Eqs. (12) therein). In Antuono et al. (2015), the δ-SPH scheme does not contain energy or entropy of individual particles. The compressional and kinetic energies are not conserved (nor their sum is), so the total energy is not conserved in that scheme. In general, the total energy cannot be conserved if the SPH particles possess no entropy, internal energy, or temperature.

However, even when SPH particles are equipped with entropy or internal energy (Ellero et al. 2023), another source of irreversibility prevails in the non-symplecticity of the underlying numerical scheme. Consequently, the solution of the SPH numerical scheme is irreversible even if dissipative terms (viscosity and heat conductivity) are omitted.

In this paper, we combine symplectic integrators for SPH particles equipped with entropy and conservative dissipative dynamics. Numerical illustrations contain free expansion of ideal gas (purely Hamiltonian expansion and dissipative Joule expansion), thermal convection of stiffened gas within a container heated from below, and hyperbolic heat conduction. The global energy errors are typically of order $10^{-5}\%$ in the non-dissipative case and $10^{-3}\%$ in the dissipative case (using only first-order discretization of the dissipative terms).

Section 24.2 recalls the usual reversible (and symplectic) formulation of SPH without entropy, but it shows how to derive the evolution equations by a reduction of Poisson brackets. In particular, the Lagrangian continuum mechanics is discretized to Lagrangian particles (interpreted as the SPH particles). Section 24.3 shows five approaches to the entropy density in SPH, based on four definitions of particle volumes (mass-based, entropy-based, direct, and implicit), as well as a mixed definition. Section 24.4 contains the reversible (Hamiltonian) part of the five approaches to SPH with entropy, mimicking the derivation of the Poisson bracket from Sect. 24.2. Finally, in Sect. 24.5 we add dissipative (non-Hamiltonian) terms for viscosity, Fourier heat conduction (using gradient dynamics), and hyperbolic heat conduction.

The problem of discontinuities (shocks) in a compressible fluid is not covered in this paper. The most common approach to shocks in SPH is via artificial viscosity (Cullen and Dehnen 2010). When it comes to Riemann solvers, they are difficult to combine with SPH. Nonetheless, this direction was investigated by Inutsuka Inutsuka (2002).

24.2 SPH Without Entropy as a Hamiltonian System

Both the standard SPH and the SPH with entropy are Hamiltonian systems equipped with dissipative terms (viscosity, heat conduction, etc.). Before approaching the derivation of the reversible (Hamiltonian) part of the entropic SPH, let us first recall the Hamiltonian formulation of continuum mechanics, which serves as the starting point of the derivation of the usual SPH without entropy.

24.2.1 Hamiltonian Continuum Mechanics

Hamiltonian continuum mechanics on the Lagrangian (reference) manifold can be derived by means of the principle of stationary action. The Lagrangian is a function of the Lagrange→Euler mapping $\mathbf{x}(t, \mathbf{X})$ and its time-derivative, where t is the time, $\mathbf{X}$ the Lagrangian coordinates of a point, and $\mathbf{x}$ its Eulerian coordinate (current configuration) (Goldstein 2002). The principle of stationary action then gives the Lagrange-Euler equation, which can be rewritten (by the Legendre transformation) to the Hamiltonian setting with the fields of position and momentum density, $x^i(\mathbf{X})$ and $M_i(\mathbf{X})$, as the state variables. These state variables are equipped with the canonical Poisson bracket

$$\{F, G\}^{(\text{Lagrangian})} = \int d\mathbf{X} \left(F_{x^i} G_{M_i} - G_{x^i} F_{M_i} \right), \tag{24.1}$$

and Hamiltonian evolution of any functional of the state variables, $F(\mathbf{x}, \mathbf{M})$, is given by the Poisson bracket of the functional and the energy $E(\mathbf{x}, \mathbf{M})$,

$$\dot{F} = \{F, E\}. \tag{24.2}$$

In particular, if F is chosen as the state variables, we obtain the Hamiltonian evolution equations of the positions and momenta,

$$\dot{\mathbf{x}} = E_{\mathbf{M}} \tag{24.3a}$$

$$\dot{\mathbf{M}} = -E_{\mathbf{x}}, \tag{24.3b}$$

where the subscripts denote derivatives (more precisely, the functional derivatives Pavelka et al. (2018)). Equation (24.3) represent the canonical Hamiltonian equations for the Lagrangian continuum mechanics, which can be used as the dynamics of elastic bodies (Landau et al. 1986; Pavelka et al. 2020).

However, to close the equations, we have to prescribe the energy $E(\mathbf{x}, \mathbf{M})$. If we know the internal energy density per the Eulerian volume $\epsilon(\rho)$, the Lagrangian energy is the sum of the kinetic energy and the internal energy transformed to the Lagrangian manifold,

$$E^{Lagrangian} = \int d\mathbf{X} \left(\frac{\mathbf{M}^2}{2\rho_0} + \det \frac{\partial \mathbf{x}}{\partial \mathbf{X}} \epsilon(\rho(\mathbf{x}(\mathbf{X}))) \right). \tag{24.4}$$

Here, ρ is the Eulerian mass density, which is related to the Lagrangian density $\rho_0(\mathbf{X})$ (mass density per the Lagrangian volume) through

$$\rho(\mathbf{x}) = \rho_0(\mathbf{X}(\mathbf{x})) \det \frac{\partial \mathbf{X}}{\partial \mathbf{x}}. \tag{24.5}$$

Without going into more detail on the Hamiltonian continuum mechanics on the Lagrangian manifold, let us go to the Hamiltonian formulation of the standard SPH method.

24.2.2 Hamiltonian Form of the Standard Weakly Compressible SPH

The standard SPH (Gingold and Monaghan 1977; Violeau 2012) expresses the evolution of the positions and velocities of the smoothed particles, and the reversible part of the evolution (disregarding viscosity) is also a Hamiltonian system. To write down the classical SPH formulation, we need two approximations:

1. smoothed state variables (positions and momenta of the particles)
2. differential operators (appearing in the evolution equations like div or grad)

Different operator discretization may yield different results as clearly observed in the discussion of density evolution (Violeau, 2012, Chap. 5.3.1). Here, we propose a quite different approach, employing the least unnecessary knowledge to derive discrete operators, thus reducing ambiguity in the formulation of SPH.

Instead of discretizing the differential operators, we derive SPH by Poisson reduction from the Hamiltonian form of Lagrangian continuum mechanics, recalled in Sect. 24.2.1. To this end, we first need a mapping projecting the continuum state variables $\mathbf{x}$ and $\mathbf{M}$ to discrete SPH positions and momenta. Let Ω_0 be the Lagrangian manifold, which can be split into mutually disjoint cells (for instance, a grid in 2D, see Fig. 24.1), $\Omega_0 = \bigcup_\alpha \Omega_{0\alpha}$. Each of $\Omega_{0\alpha}$ represents a Lagrangian particle that will become an SPH particle. The volume of the Lagrangian cell with index α is

$$V_{0\alpha} = \int_{\Omega_{0\alpha}} d\mathbf{X}, \tag{24.6a}$$

and we will also need the normalized characteristic functions of the Lagrangian particles

$$\bar{\chi}_\alpha(\mathbf{X}) = \begin{cases} \frac{1}{V_{0\alpha}} & \text{if } \mathbf{X} \in \Omega_{0\alpha} \\ 0 & \text{otherwise} \end{cases}. \tag{24.6b}$$

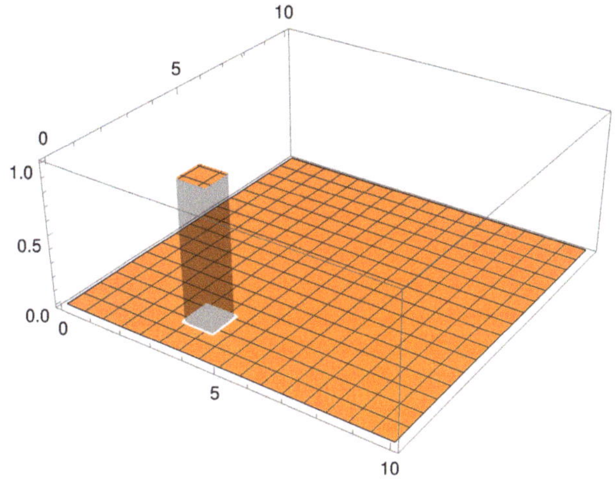

Fig. 24.1 An example of partitioning of the Lagrangian manifold into Lagrangian particles, which then become the SPH particles

The mapping

$$\mathbf{x}_\alpha = \int d\mathbf{X}\bar{\chi}_\alpha(\mathbf{X})\mathbf{x}(\mathbf{X}) \tag{24.7a}$$

$$\mathbf{M}_\alpha = \int d\mathbf{X} V_{0\alpha}\bar{\chi}_\alpha(\mathbf{X})\mathbf{M}(\mathbf{X}) \tag{24.7b}$$

then defines the position and momentum of the SPH particle α, which play the role of the SPH state variables. Later, in Sect. 24.3, we will also add entropy to the state variables, but first we show how to derive the standard SPH by reduction from the Lagrangian continuum mechanics.

In order to reduce the continuum Poisson bracket (24.1) to a Poisson bracket for the SPH state variables ($\mathbf{x}_\alpha$ and $\mathbf{M}_\alpha$), we plug the functionals dependent on the SPH variables into the Poisson bracket. Because the Poisson bracket contains derivatives of the functionals, we need to calculate derivatives of the SPH state variables with respect to the continuum state variables,

$$\frac{\delta x^i_\alpha}{\delta x^j(\mathbf{X})} = \delta^i_j \bar{\chi}_\alpha(\mathbf{X}) \quad \text{and} \quad \frac{\delta M_{\alpha i}}{\delta M_j(\mathbf{X})} = \delta^j_i V_{0\alpha}\bar{\chi}_\alpha(\mathbf{X}) \tag{24.8}$$

while the remaining derivatives are zero. Derivatives of an arbitrary functional of the SPH state variables, $F(\mathbf{x}_\alpha, \mathbf{M}_\alpha)$, are then

$$\frac{\delta F}{\delta x^i(\mathbf{X})} = \sum_\alpha \frac{\partial F}{\partial x^i_\alpha}\bar{\chi}_\alpha(\mathbf{X}) \quad \text{and} \quad \frac{\delta F}{\delta M_i(\mathbf{X})} = \sum_\beta \frac{\partial F}{\partial M_{\beta i}} V_{0\beta}\bar{\chi}_\beta(\mathbf{X}). \tag{24.9}$$

Plugging two arbitrary functionals dependent on the discrete state variables to Poisson bracket (24.1) leads to

$$\{F, G\}^{(\mathrm{SPH})} = \sum_{\alpha}\sum_{\beta}\left(F_{x^i_\alpha} G_{M_{\beta i}} - G_{x^i_\alpha} F_{M_{\beta i}}\right) \int d\mathbf{X} V_{0\beta} \bar{\chi}_\alpha(\mathbf{X}) \bar{\chi}_\beta(\mathbf{X})$$
$$= \sum_{\alpha}\left(F_{x^i_\alpha} G_{M_{\alpha i}} - G_{x^i_\alpha} F_{M_{\alpha i}}\right), \tag{24.10}$$

which is the SPH Poisson bracket expressing the kinematics of the SPH state variables. This Poisson bracket can be seen as the Hamiltonian counterpart of the Lagrangian form of SPH (Violeau 2012).

The reversible Hamiltonian evolution equations implied by this bracket are

$$\dot{\mathbf{x}}_\alpha = E_{\mathbf{M}_\alpha} \tag{24.11a}$$
$$\dot{\mathbf{M}}_\alpha = -E_{\mathbf{x}_\alpha}, \tag{24.11b}$$

which are the Hamilton canonical equations for the SPH state variables. However, to close the equations, we need to express the energy in terms of the SPH state variables.

24.2.3 Energy Functional in the Standard SPH

To find the dependence of the energy on the SPH state variables, we need to approximate the exact energy of the continuous system (24.4), which also depends on the full Lagrange→Euler mapping $\mathbf{x}(\mathbf{X})$. Knowledge of discrete particle positions is not enough for the precise reconstruction of the exact continuum energy, and hence we turn to the standard SPH smoothing using the weighing functions W:

$$\overline{A}_\alpha = \sum_{\beta} A_\beta W(|\mathbf{x}_\alpha - \mathbf{x}_\beta|), \tag{24.12}$$

which is an approximate form of mollification. It can be shown (Violeau 2012) that if the weighing function satisfies a certain normalization property, it provides an approximation precise up to the second order. The normalization requirement can be translated into a condition that W is a homogeneous function of the distance of a degree $-n$, that is, $W(\alpha|\mathbf{x}|) = \alpha^{-n} W(|\mathbf{x}|)$, where n denotes the spatial dimension. From this homogeneous property, one can also see that any scaling of the local volume (as in bulk expansion) is exactly captured by the weighing function W. Finally, as the dimension of W is $1/m^n$, we may define any density variable discretization in Eulerian frame $\overline{A}_\alpha$ from its Lagrangian counterpart A_α via the above expression (24.12). We shall use such smoothing to define the SPH-mass density ρ_α. In particular, ρ_α is the Eulerian mass density of particle α, $m_\alpha = \int d\mathbf{X} \rho_0(\mathbf{X}) \bar{\chi}_\alpha(\mathbf{X})$ is the mass of the Lagrangian particle and hence

$$\rho_\alpha = \sum_\beta m_\beta W_{\alpha\beta}, \tag{24.13}$$

where $W_{\alpha\beta} \stackrel{def}{=} W(|\mathbf{x}_\alpha - \mathbf{x}_\beta|)$.

Now we can close the Hamilton canonical equation (24.11) by supplying an energy functional that depends only on the SPH state variables. This is actually the step that depends on material properties, since the Poisson bracket (24.10) is already fixed by the SPH state variables. The continuum energy (24.4) can be approximated by

$$\begin{aligned} E^{SPH} &= \sum_\alpha V_{0\alpha} \frac{\mathbf{M}_\alpha^2}{2\rho_{0\alpha}} + \sum_\alpha V_{0\alpha} \frac{\rho_{0\alpha}}{\rho_\alpha} \epsilon(\rho_\alpha) \\ &= \sum_\alpha \frac{\mathbf{M}_\alpha^2}{2m_\alpha} + \sum_\alpha V_\alpha \epsilon(\rho_\alpha), \end{aligned} \tag{24.14}$$

where $\rho_{0\alpha} = m_\alpha / V_{0\alpha}$ is its density in the Lagrangian space. This approximate energy function can also be obtained by means of the principle of maximum entropy (MaxEnt), which is shown in Appendix 24.7. We keep the Eulerian volume V_α unspecified (unrelated to Eulerian density, for example), as it is a key step subjected to further in Sect. 24.3, where five possible definitions are shown.

24.2.4 *Standard SPH as a Hamiltonian System*

The purpose of this Section is to collect the above results on the SPH Poisson bracket and SPH energy and to show that they indeed lead to the standard formulation of SPH.

When energy (24.14) is plugged into the Hamilton canonical equation (24.11), we need to take its derivatives with respect to the SPH state variables. In particular, derivative of ρ_α (defined in Eq. (24.13)) with respect to the positions reads

$$\begin{aligned} \frac{\partial \rho_\beta}{\partial x_\alpha^i} &= \sum_\gamma m_\gamma W'(|\mathbf{x}_\beta - \mathbf{x}_\gamma|) \frac{\partial \sqrt{(x_\beta^j - x_\gamma^j)(x_\beta^j - x_\gamma^j)}}{\partial x_\alpha^i} \\ &- \sum_\gamma m_\gamma W'(|\mathbf{x}_\beta - \mathbf{x}_\gamma|) \frac{x_\beta^j - x_\gamma^j}{|\mathbf{x}_\beta - \mathbf{x}_\gamma|} \delta_i^j \left(\delta_{\beta\alpha} - \delta_{\gamma\alpha}\right) \\ &= \sum_\gamma m_\gamma W'_{\beta\gamma} e_{i\beta\gamma} \left(\delta_{\beta\alpha} - \delta_{\gamma\alpha}\right), \end{aligned} \tag{24.15}$$

where $W'_{\beta\gamma} = W'(|\mathbf{x}_\beta - \mathbf{x}_\gamma|)$ is the derivative of the SPH kernel W evaluated on the distance between particles β and γ and where $\mathbf{e}_{\beta\gamma} = \frac{\mathbf{x}_\beta - \mathbf{x}_\gamma}{|\mathbf{x}_\beta - \mathbf{x}_\gamma|}$ is the unit vector[1] pointing from $\mathbf{x}_\gamma$ to $\mathbf{x}_\beta$.

Now, we are finally in the position to write down the resulting ordinary differential equations expressing the Hamiltonian evolution of the SPH state variables,

$$\dot{x}^i_\alpha = \frac{M^i_\alpha}{m_\alpha} \tag{24.16a}$$

$$\begin{aligned}\dot{M}_{\alpha i} &= -\sum_\beta \left(-\frac{m_\beta}{\rho_\beta^2}\epsilon + \frac{m_\beta}{\rho_\beta}\frac{\partial \epsilon}{\partial \rho_\beta}\right)\sum_\gamma m_\gamma W'_{\beta\gamma} e_{i\beta\gamma}(\delta_{\beta\alpha} - \delta_{\gamma\alpha}) \\ &= -\sum_\beta \left(\frac{m_\alpha m_\beta}{\rho_\alpha^2} p_\alpha + \frac{m_\alpha m_\beta}{\rho_\beta^2} p_\beta\right) W'_{\alpha\beta} e_{i\alpha\beta},\end{aligned} \tag{24.16b}$$

where the pressure of particle α is defined by the usual barotropic relation

$$p_\alpha = -\epsilon + \rho_\alpha \frac{\partial \epsilon}{\partial \rho_\alpha}, \tag{24.17}$$

see Callen (1960), Pavelka et al. (2018) or Appendix 24.9. Note that the energy density ϵ is understood as the ratio of energy and volume. When the dependence of the internal energy on the mass density is prescribed, which can be integrated from an equation of state, the system of equations is completely specified. See Violeau (2012) for some choices of internal energy typical in SPH.

Equation (24.16) represents a symplectic SPH discretization of fluid mechanics where the energy depends on the mass density (Violeau 2012). This system of equations forms a Hamiltonian system because it is constructed from a Poisson bracket and energy. Combined with a symplectic time integrator, such as the Verlet scheme, the numerical solution has energy error bounded uniformly with respect to the simulation time t, and the numerical schemes can even be made time reversible by using fixed-point arithmetic (Kincl and Pavelka 2023). For all simulations in this paper, we use our code (Kincl and Pavelka 2025).

However, in the SPH literature (Gingold and Monaghan 1977) it is customary to use the discretized continuity equation instead of the closed expression (24.13). Then, Eq. (24.16) can be generalized, using the definition of ρ_a, to another Hamiltonian system,

$$\dot{x}^i_\alpha = \frac{M^i_\alpha}{m_\alpha} \tag{24.18a}$$

[1] Actually, it is the covector expressing gradient of the distance between the particles while the vector is dual to the covector. But in Euclidean space we do not need to distinguish them Fecko (2006).

$$\dot{\rho}_\alpha = \sum_\beta m_\beta W'_{\alpha\beta} \mathbf{e}_{\alpha\beta} \left(\frac{\mathbf{M}_\alpha}{m_\alpha} - \frac{\mathbf{M}_\beta}{m_\beta} \right) \tag{24.18b}$$

$$\dot{M}_{\alpha i} = -\sum_\beta \left(\frac{m_\alpha m_\beta}{\rho_\alpha^2} p_\alpha + \frac{m_\alpha m_\beta}{\rho_\beta^2} p_\beta \right) W'_{\alpha\beta} e_{i\alpha\beta}, \tag{24.18c}$$

where the pressure is again given by Equation (24.17). Equation (24.18) are the standard SPH equations (Violeau 2012). This system of equations is Hamiltonian because it is generated by a Poisson bracket

$$\{F, G\}^{(\mathrm{SPH},\rho_\alpha)} = \{F, G\}^{(\mathrm{SPH})} + \sum_\alpha \sum_\beta m_\beta W'_{\alpha\beta} e_{i\alpha\beta} \left(F_{\rho_\alpha} (G_{M_{\alpha i}} - G_{M_{\beta i}}) - G_{\rho_\alpha} (F_{M_{\alpha i}} - F_{M_{\beta i}}) \right), \tag{24.19}$$

which can be obtained by projection of the continuum Poisson bracket (24.1) to the state variables $\mathbf{x}_\alpha$, $\mathbf{M}_\alpha$, and ρ_α, see Appendix 24.8 for details (here we dropped s_α from state variables considered in Appendix 24.8).

Poisson bracket (24.19) is different from the bracket (24.10) and leads to Hamiltonian but non-symplectic mechanics (Marsden and Ratiu 1999). From a numerical perspective, it is advantageous to keep the symplecticity (non-degenerate Poisson bracket) because, for instance, the energy error is then capped by a constant decreasing with the time step when a symplectic integrator is used (Hairer et al. 2013; Leimkuhler et al. 2004). Therefore, we make the dynamics symplectic by directly evaluating the density from its definition (24.13) at each time step while keeping the evolution equations for $\mathbf{x}_\alpha$ and $\mathbf{M}_\alpha$. This results, for example, in a globally reversible scheme for SPH (Kincl and Pavelka 2023). On the other hand, this direct evaluation of the density within the boundaries may cause other numerical artifacts (Antuono et al. 2015) if not properly treated (Kincl and Pavelka 2023).

In summary, the standard SPH equation (24.18) can be derived from Lagrangian continuum mechanics by reducing the continuum Poisson bracket (24.1) to the SPH Poisson bracket (24.19). The reduction relies on the projection from the continuous fields of positions and momenta to the SPH positions and momenta (24.7), where the Lagrangian manifold is partitioned into discrete particles. Then, the SPH energy (24.14) is found as an approximation of the continuum energy (24.4), and this approximation can be seen as an application of the principle of the MaxEnt principle (Sect. 24.7). When a concrete dependence of the internal energy on the mass density is chosen (a material is chosen), the SPH equation (24.18) can be solved as long as one chooses an appropriate interpretation of the particle volume.

However, what if the energy also depends on the entropy of the material, which means that the fluid is not barotropic? Then, the formulation of SPH has to be enriched to include also the entropy of the SPH particles, which is the purpose of the following Section.

24.3 Approximations of Particle Volumes, Mass Density, and Entropy Density

The usual formulation of SPH (24.18) does not involve the entropy of the particles, which in particular means that the pressure depends only on the density, and thus the fluid is barotropic. In reality, however, the pressure depends on both the density and the entropy (or temperature) of the fluid, and this dependence is captured by more precise models (see, for instance, the ideal gas or the stiffened gas in Appendix 24.9). We include entropy in SPH as a variable of the density of the volume. However, this highlights the issue of various possible definitions of SPH particle volumes, which we discuss now. For example, even the very definition of SPH energy (24.14) depends on such choice as is immediate from its extension to the case with entropy. For nonbarotropic fluids, the energy density depends on the entropy density, and hence the expression for energy reads as

$$
\begin{aligned}
E^{SPH} &= \sum_{\alpha} V_{0\alpha} \frac{\mathbf{M}_{\alpha}^2}{2\rho_{0\alpha}} + \sum_{\alpha} V_{0\alpha} \frac{V_{\alpha}}{V_{0\alpha}} \epsilon(\rho_{\alpha}, s_{\alpha}) \\
&= \sum_{\alpha} \frac{\mathbf{M}_{\alpha}^2}{2m_{\alpha}} + \sum_{\alpha} V_{\alpha} \epsilon(\rho_{\alpha}, s_{\alpha}),
\end{aligned} \tag{24.20}
$$

where s_α is the Eulerian entropy density of particle α. The energy is expressed as a function of the entropy density rather than the temperature, so the thermodynamic information is complete (forming the fundamental thermodynamic relation Callen (1960)). Temperature can be introduced by taking the derivative of energy with respect to entropy, see Appendix 24.9.

Consider a volume density state variable as is standard continuum mechanics which we discretize to the SPH particles. We denote its Lagrangian variant by $Z_\alpha(X)$ (typically prescribed as an initial condition), and then its Eulerian counterpart $\zeta_\alpha(x)$ is then related via

$$
\zeta_\alpha(x) = \frac{V_{0\alpha}}{V_\alpha} Z_\alpha(X). \tag{24.21}
$$

The Lagrangian particle volume follows directly from the SPH discretization but the Eulerian volume can be estimated in several distinct ways, for example, from the update of particle positions x_α or from the evolution of any volume density state variable. This gives two possibilities of particle volume evolution in the classical SPH formulations (the second one stemming from mass density) but the introduction of entropy density introduces another choice. They all represent an admissible approximation of the particle volume but differ in practice.

Hence, naturally, we start with a discussion of approximation errors.

24.3.1 SPH Discretization Error Estimates

The fundamental idea behind SPH discretization lies in approximative formulas for the Dirac delta function and convolution. In particular, if we take a classical integrable and normalized function, that is, $\int_{\mathbb{R}} W = 1$, and define a rescaled function as $W_h(x) = \frac{1}{h}W(x/h)$, then $\int_{\mathbb{R}} W_h = 1$ and $\lim_{h\to 0} W_h = \delta(x)$. As $f \star \delta = f$, we get $f \approx f \star W_h$. In particular, we have

$$f(x_\alpha) \approx \int_{\mathbb{R}^n} f(y)W_h(x-y) \approx \sum_\beta V_\beta f(x_\beta)W_h(x_\alpha - x_\beta).$$

With particular choices of the approximated function f, we obtain different approximation formulas for volume and different corresponding errors.

For the Wendland kernel, one can show that the relative error in 1D is proportional to the gradient of a logarithm at that point (Violeau, 2012, Eq. (5.115))

$$\frac{f(x) - f(x_\alpha)}{f(x)} = C(h,r)\frac{d \ln f}{dx}, \tag{24.22}$$

where $C(h,r)$ is a constant independent of the function f but dependent of the kernel width h and particle size r.

24.3.2 Particle Volume Definitions

Using the above approximations of convolution with the Dirac delta function, we may propose several distinct definitions of particle volumes. Namely, the choice $f = \rho$ gives

$$\rho(x_\alpha) = \rho_\alpha \approx \sum_\beta V_\beta \rho(x_\beta)W_{\alpha\beta} = \sum_\beta m_\beta W_{\alpha\beta},$$

and hence entails the above mass density update expression Eq. (24.13) or, generally, Eq. (24.12). From this expression, one may define a particle volume (based on mass density) as

$$(V_\alpha^m)^{-1} = \frac{\rho_\alpha}{m_\alpha} \approx \sum_\beta \frac{m_\beta}{m_\alpha} W_{\alpha\beta}. \tag{24.23}$$

However, the introduction of another volume density variable, as is the case with entropy density, gives another possibility

$$(V_\alpha^s)^{-1} = \frac{s_\alpha}{S_\alpha} \approx \sum_\beta \frac{S_\beta}{S_\alpha} W_{\alpha\beta}, \tag{24.24}$$

where we chose $f = s$ and where S_α is the entropy of particle α.

At the same time, the choice $f : x \mapsto 1/V_\alpha, x \in V_\alpha$ gives directly an estimate of the particle volume as

$$(V^d_\alpha)^{-1} \approx \sum_\beta W_{\alpha\beta}. \tag{24.25}$$

Finally, the choice $f = \chi_\Omega$ being the characteristic function of a given domain Ω yields an implicit relation for particle volumes

$$1 \approx \sum_\beta \tilde{V}^I_\beta W_{\alpha\beta},$$

where $\tilde{V}^I_\alpha$ is the volume of the part of the particle that is within the domain Ω. This expression can be rearranged into

$$\tilde{V}^I_\alpha = \sum_\beta W^{-1}_{\alpha\beta}, \tag{24.26}$$

where $W^{-1}_{\alpha\beta}$ denotes the inverse to the matrix $W_{\alpha\beta}$.

Let us make a few observations on the errors and convergence. All four definitions of volumes are based on Eulerian quantities whose evolution we aim to calculate. Hence, the error upper bound is only a rough guidance to the suitable volume definition selection as it is dependent on the systems evolution. Nevertheless, we expect from Eq. (24.22) that the mass-based volume should give a good particle volume representations in systems, where the spatial gradient of density is small and entropy-based volume should similarly yield a well-represented system with shallow entropy gradients. The direct definition of particle volume, V^d_α, has errors proportional to the spatial variation of particle volumes and hence is dependent on the heterogeneity of discretization itself. Finally, the implicit definition of volume, V^I_α, can be expected to have the smallest discretization error in the bulk, however, it is expected to introduce errors in the boundary particles where there is a jump in the discretized function—the characteristic function of the domain Ω. In addition, the determination of implicit volume V^I_α comes at a greater computational cost as it involves inverting the large matrix $W_{\alpha\beta}$.

The precision of the implicit particle volumes V^I_α for the bulk integration can be perceived from the opposite perspective: we are looking for particle volumes such that the integral is exactly represented by the sum. Namely, the particle volumes V^I_α can be equivalently defined as weights of a Gaussian quadrature rule

$$\int f \mathrm{d}\mathbf{x} \approx Q[f] = \sum_\alpha V^I_\alpha f(\mathbf{x}_\alpha),$$

which is exact for all functions in the form

$$f(\mathbf{x}) = \sum_\beta f_\beta W(\mathbf{x} - \mathbf{x}_\beta) \tag{24.27}$$

(where $f_\beta \in \mathbb{R}$ are arbitrary coefficients). Indeed, from the definition (24.26):

$$Q[f] = \sum_\alpha V^I_\alpha f(\mathbf{x}_\alpha) = \sum_\alpha \sum_\beta V^I_\alpha f_\beta W_{\alpha\beta} =$$
$$= \sum_\beta f_\beta = \sum_\beta f_\beta \int W(\mathbf{x} - \mathbf{x}_\beta)\mathrm{d}\mathbf{x} = \int f(\mathbf{x})\mathrm{d}\mathbf{x}.$$

To obtain the reverse implication, substitute $f(\mathbf{x}) = W(\mathbf{x} - \mathbf{x}_\alpha)$ into $\int f \mathrm{d}\mathbf{x} = Q[f]$. Note that again the issue with boundary elements appear, here in the requirement of the function form (24.27), which effectively requires Ω to be exactly discretized into SPH particles.

Let us, moreover, comment on the convergence of all the four above definitions of volume. The discretization of the convolution integral and Dirac delta function is improving, and hence the error is being reduced if the discretization step is decreased. At the same time, it is known that there is a trade-off between the particle size and the width of the SPH kernel, reaching an optimum and preventing the truncation error to be arbitrarily small, Quinlan et al. (2006).

Yet another possibility is to define both mass and entropy densities as the SPH averages,

$$\rho_\alpha = \sum_\beta m_\beta W_{\alpha\beta} \tag{24.28a}$$

$$s_\alpha = \sum_\beta S_\beta W_{\alpha\beta}. \tag{24.28b}$$

However, this choice, called the mixed-volume approach, does not specify how a single-particle volume is defined and instead needs two-particle volumes at once. Despite this loss of elegance, it shows good results in the presence of discontinuities, see Sect. 24.3.5.

24.3.3 Particle Volumes and Conservation Laws

We shall now turn to the discussion of the suitability of the particle volume definitions from a different perspective the validity of discretized conservation laws.

Using the mass-based volume V^m_α, Eq. (24.23), has the advantage of conserving the total mass, $M = \sum_\alpha V^m_\alpha \rho_\alpha = \sum_\alpha m_\alpha$, as particle masses m_α are constant. Similarly, the total entropy, $S = \sum_\alpha V^m_\alpha s_\alpha = \sum_\alpha S_\alpha$, is conserved by the reversible part of the evolution (the Hamiltonian part, which does not affect S_α).

The expression for total energy, Eq. (24.20), however, differs according to the choice of the particle volume definition. In the mass-based volume case, it reads

$$E^{SPH} = \sum_\alpha \frac{\mathbf{M}_\alpha^2}{2m_\alpha} + \sum_\alpha \frac{m_\alpha}{\rho_\alpha}\epsilon(\rho_\alpha, s_\alpha).$$

If we consider the entropy-based particle volumes, the energy discretization takes on a slightly different form:

$$E^{SPH} = \sum_\alpha \frac{\mathbf{M}_\alpha^2}{2m_\alpha} + \sum_\alpha \frac{S_\alpha}{s_\alpha}\epsilon(\rho_\alpha, s_\alpha).$$

In the rest of this paper, we shall explore these four choices of particle volume definitions both in the reversible and irreversible processes and discuss the conservation of energy.

24.3.4 Particle Volumes and Quadratures

As a purely mathematical comparison, let us investigate the accuracy of numerical integration

$$\int f_1(\mathbf{x})\,\mathrm{d}\mathbf{x} \approx \sum_a V_a f(\mathbf{x}_a) \tag{24.29}$$

for the different definitions of particle volume V_a. To this end, we consider functions

$$\begin{aligned} f_1 &= e^{-x-y} \\ f_2 &= 1 + \cos 8x + \sin 8x \\ f_3 &= 1 - xy + 4x^2y^2 \end{aligned} \tag{24.30}$$

in a square $[0, 1] \times [0, 1]$. The exact integral can be easily evaluated. We approximate this value by N particles arranged in a hexagonal pattern. Let us define the spatial resolution of the problem as

$$\mathrm{d}r = \frac{1}{\sqrt{N}}. \tag{24.31}$$

The particle masses m_α and entropies S_α were all set equal to 1 and Wendland's quintic kernel with support radius $h = 3\mathrm{d}r$ was used. The result is shown in Fig. 24.2. We found that the implicit volume is more precise in most cases (at the expense of a longer computation time). Therefore, we suggest that the implicit-volume quadrature should be used in a situation where an integral has to be computed from data in a post-process.

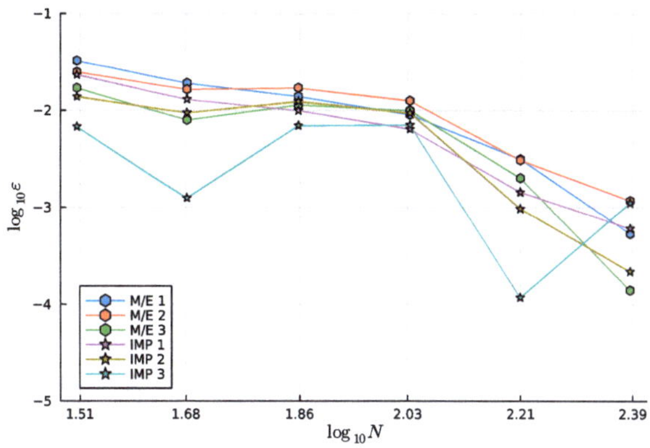

Fig. 24.2 The relative accuracy of numerical quadrature ϵ for explicit integration using mass volume or entropic volume (M/E) and implicit volume (IMP) for various values of N and functions f_1, f_2, f_3 from (24.30). (Note that in this example, mass volume and entropic volume yield identical results.) Dummy particles were used to prevent large error near boundary

24.3.5 Particle Volumes in the Presence of Discontinuities

Let us now illustrate the difference between the volumes in the presence of a discontinuity in the mass density while keeping the entropy density constant in space. We consider a one-dimensional system of particles and make the particles twice denser in the right part of the domain (interval $x \in (0, 1)$ split into two halves). All particles have the same mass, the interparticle spacing is $1.0 \cdot 10^{-04}$ in the left part of the interval while $0.5 \cdot 10^{-04}$ in the right part. Moreover, we set the SPH kernel size as $h = 0.001$ (using the Wendland 1D kernels). Therefore, the density of the particles in the left part should be approximately 1.0 while 2.0 in the right part, experiencing a jump in the middle. The entropies of the particles in the left part are set equal to the respective interparticle spacings so that the entropy is approximately constant over the whole domain.

Figure 24.3 shows the entropy density and the mass density evaluated using the mass-volume method. While the mass density shows a smooth monotone profile, the entropy density oscillates considerably. On the other hand, when we use the entropic-volume discretization, the entropy gets a bump in the middle (caused by oversampling the particles to the right of the jump), as well as the mass density; see Fig. 24.4. The implicit volume shows oscillations in both the mass density and the entropy density, see Fig. 24.5. The direct-volume approach is shown in Fig. 24.7. Finally, Fig. 24.6 shows the behavior of the mixed volume, which gives the best results.

In summary, it may be advantageous to replace the standard mass volume with the entropic volume when the entropy has a smoother behavior than the mass density. In the following Section, we show the reversible part of the SPH evolution equations for each of those SPH variants.

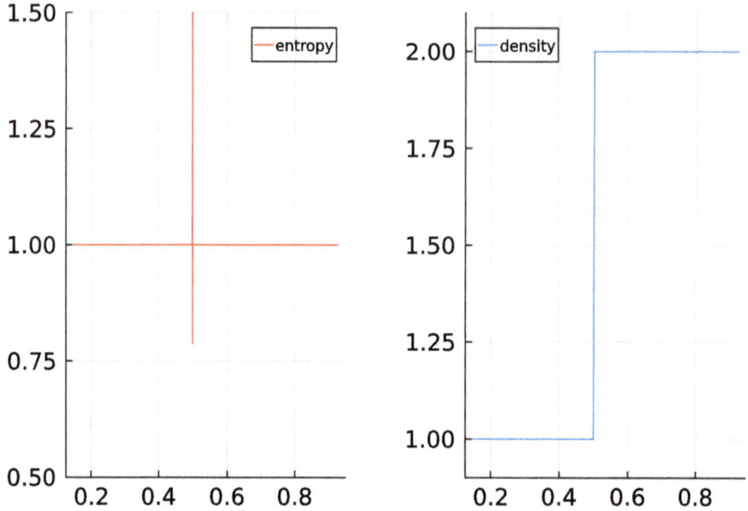

Fig. 24.3 Entropy density and mass density in the presence of the mass-density jump evaluated using the mass volume

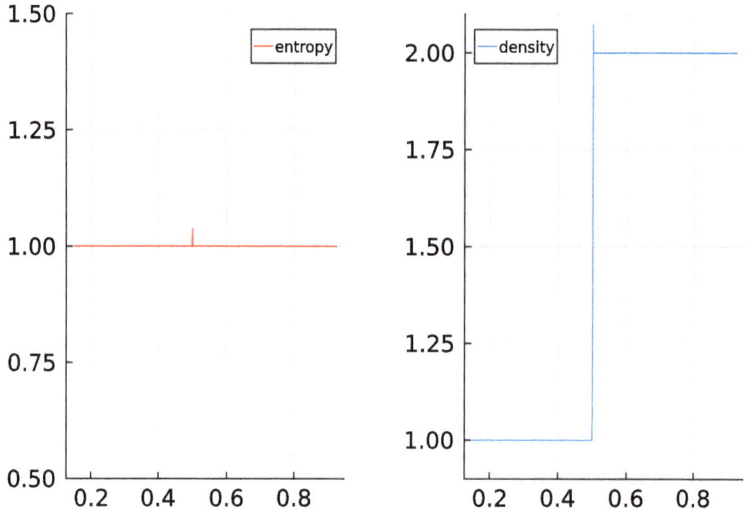

Fig. 24.4 Entropy density and mass density in the presence of the mass-density jump evaluated using the entropic volume

24.4 Hamiltonian Part of the SPH Evolution with Entropy

Since the reversible part of the SPH evolution can be seen as a discretization of the Lagrangian continuum mechanics, it is generated by a Poisson bracket. However, the five forms of SPH with entropy discussed in this paper have different Poisson

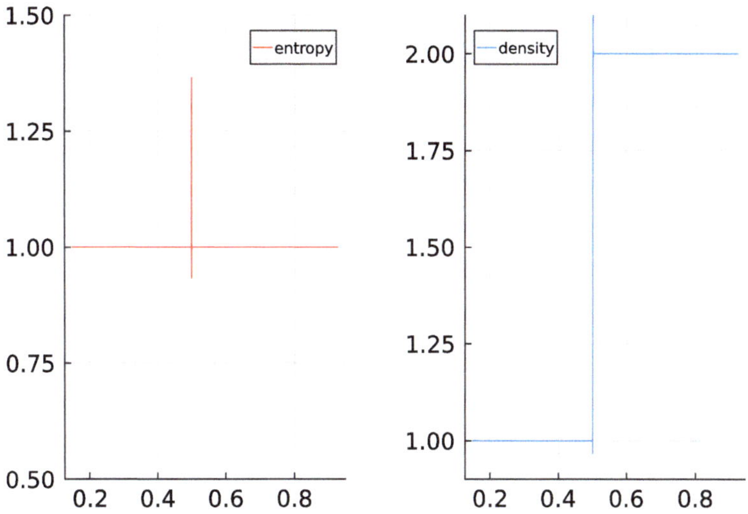

Fig. 24.5 Entropy density and mass density in the presence of the density jump evaluated using the implicit volume

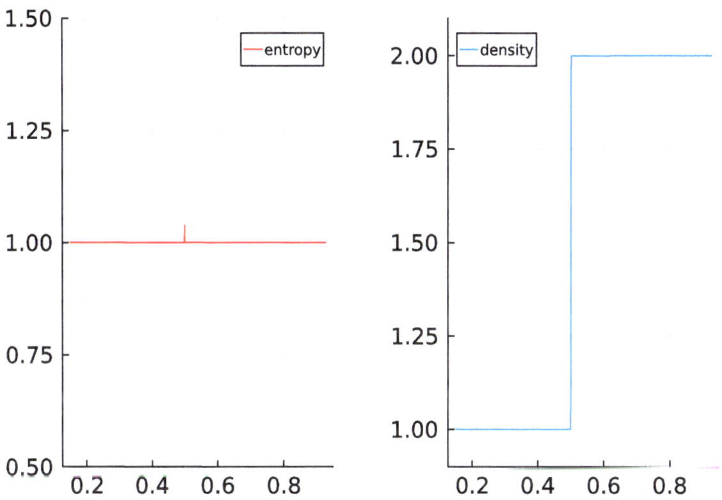

Fig. 24.6 Entropy density and mass density in the presence of the density jump evaluated using the mixed volume. This method shows the best behavior, capturing the mass-density jump well while having a relatively small oscillation in the entropy density

brackets and different reversible parts of the evolution equations. As in the case of standard barotropic SPH, Poisson brackets are derived by projection from the Lagrangian Poisson bracket (24.1). The projection consists of the mapping from the Lagrangian state variables $\mathbf{x}(\mathbf{X})$ and $\mathbf{M}(\mathbf{X})$ to the SPH position $\mathbf{x}_\alpha$ and $\mathbf{M}_\alpha$ by

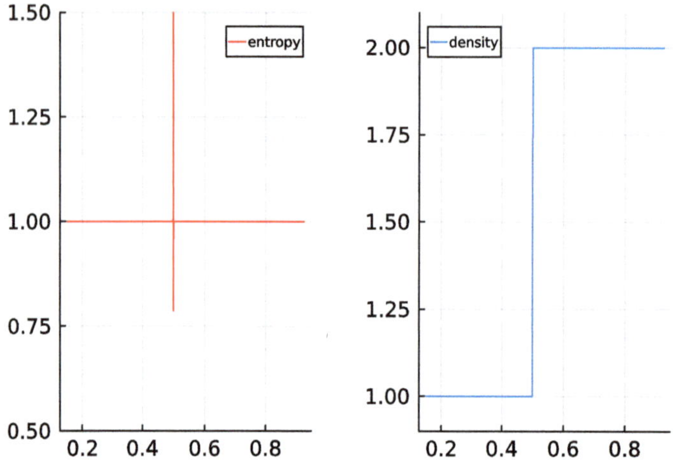

Fig. 24.7 Entropy density and mass density in the presence of the density jump evaluated using the direct volume

Eq. (24.7), and of the respective definitions of the mass density and entropy density based on the choice of the definition of the discrete volume.

24.4.1 Mass-Based Volume Approach

The Poisson bracket governing kinematics of $\mathbf{x}_\alpha$, $\mathbf{M}_\alpha$ (given by Eq. (24.7)), mass density ρ_α, and the volumetric entropy density s_α (given by Eq. (24.21)), that is

$$\rho_\alpha = \sum_\beta m_\beta W_{\alpha\beta} \tag{24.32a}$$

$$s_\alpha = \frac{S_\alpha \rho_\alpha}{m_\alpha}, \tag{24.32b}$$

is obtained by plugging these projections into the Lagrangian Poisson bracket (24.1). Note that S_α represents the entropy of the SPH particle α. Appendix 24.8.1 contains details of the calculation. The resulting Hamiltonian evolution equations are

$$\dot{\mathbf{x}}_\alpha = E_{\mathbf{M}_\alpha} \tag{24.33a}$$

$$\dot{\mathbf{M}}_\alpha = -E_{\mathbf{x}_\alpha} - \sum_\beta m_\alpha m_\beta \left(\frac{E_{\rho_\alpha}}{m_\alpha} + \frac{E_{\rho_\beta}}{m_\beta} + S_\alpha \frac{1}{m_\alpha^2} E_{s_\alpha} + S_\beta \frac{1}{m_\beta^2} E_{s_\beta} \right) W'_{\alpha\beta} \mathbf{e}_{\alpha\beta} \tag{24.33b}$$

$$\dot{\rho}_\alpha = \sum_\beta m_\beta W'_{\alpha\beta} \mathbf{e}_{\alpha\beta} \cdot (E_{\mathbf{M}_\alpha} - E_{\mathbf{M}_\beta}) \tag{24.33c}$$

$$\dot{s}_\alpha = \frac{S_\alpha}{m_\alpha} \sum_\beta m_\beta W'_{\alpha\beta} \mathbf{e}_{\alpha\beta} \cdot (E_{\mathbf{M}_\alpha} - E_{\mathbf{M}_\beta}) \tag{24.33d}$$

These equations are an extension of the standard SPH equation (24.18) with an explicit evolution of ρ_α of s_α. Note that $\dot{\rho}_\alpha$ in (24.33c) is directly compatible with the "update" relation for ρ_α (Equation (24.13)) as one can check by differentiating and using (24.33a), and if the update relation is used instead of the evolution equation for ρ_α, the system of evolution equations becomes symplectic.

To close Eq. (24.33), we need to supply an energy functional as a function of the state variables $\mathbf{x}_\alpha$, $\mathbf{M}_\alpha$, ρ_α, and s_α, which in particular can also depend on the entropy. Equation (24.33) then turn to

$$\dot{x}^i_\alpha = \frac{M^i_\alpha}{m_\alpha} \tag{24.34a}$$

$$\dot{\rho}_\alpha = \sum_\beta m_\beta W'_{\alpha\beta} \mathbf{e}_{\alpha\beta} \cdot \left(\frac{\mathbf{M}_\alpha}{m_\alpha} - \frac{\mathbf{M}_\beta}{m_\beta} \right) \tag{24.34b}$$

$$\dot{M}_{\alpha i} = -\sum_\beta m_\alpha m_\beta \left(\frac{p_\alpha}{\rho_\alpha^2} + \frac{p_\beta}{\rho_\beta^2} \right) W'_{\alpha\beta} e_{\alpha\beta i} \tag{24.34c}$$

$$\dot{s}_\alpha = \frac{S_\alpha}{m_\alpha} \sum_\beta m_\beta W'_{\alpha\beta} \mathbf{e}_{\alpha\beta} \cdot \left(\frac{\mathbf{M}_\alpha}{m_\alpha} - \frac{\mathbf{M}_\beta}{m_\beta} \right), \tag{24.34d}$$

where pressure and temperature are defined as

$$p_\alpha = -\epsilon_\alpha + \rho_\alpha \frac{\partial \epsilon}{\partial \rho_\alpha} + s_\alpha \frac{\partial \epsilon}{\partial s_\alpha} \quad \text{and} \quad T_\alpha = \frac{\partial \epsilon}{\partial s_\alpha}, \tag{24.35}$$

and where $\mathbf{v}_\alpha = \frac{\mathbf{M}_\alpha}{m_\alpha}$ is the particle velocity. The formula for pressure comes from the thermodynamic relation (24.93).

Note that $E_{\mathbf{x}_\alpha} = 0$ for each α for energy (24.20) with $V_\alpha = V^m_\alpha$ and that the evolution equations for position and momentum are identical while the evolution equations for particle density and entropy are equivalent to their updated version. As noted above, the total mass and entropy are conserved, since the Hamiltonian evolution does not change the particle entropies, S_α.

24.4.2 Entropic-Based Volume Approach

Also with volume defined via the ratio of the entropy density and entropy of each particle V^s_α, the SPH evolution equations can be derived by reduction from the Poisson bracket of Lagrangian continuum mechanics (24.1). The resulting Hamiltonian evolution equations are analogical to Eq. (24.33) with the roles of ρ_α and m_α swapped with s_α and S_α, that is

$$s_\alpha = \sum_\beta S_\beta W_{\alpha\beta} \tag{24.36a}$$

$$\rho_\alpha = \frac{m_\alpha s_\alpha}{S_\alpha}, \tag{24.36b}$$

and

$$\dot{\mathbf{x}}_\alpha = E_{\mathbf{M}_\alpha} \tag{24.37a}$$

$$\dot{\mathbf{M}}_\alpha = -E_{\mathbf{x}_\alpha} - \sum_\beta S_\alpha S_\beta \left(\frac{m_\alpha}{S_\alpha^2} E_{\rho_\alpha} + \frac{m_\beta}{S_\beta} E_{\rho_\beta} + \frac{1}{S_\alpha} E_{s_\alpha} + \frac{1}{S_\beta} E_{s_\beta}\right) W'_{\alpha\beta} \mathbf{e}_{\alpha\beta} \tag{24.37b}$$

$$\dot{s}_\alpha = \sum_\beta S_\beta W'_{\alpha\beta} \mathbf{e}_{\alpha\beta} \cdot (E_{\mathbf{M}_\alpha} - E_{\mathbf{M}_\beta}) \tag{24.37c}$$

$$\dot{\rho}_\alpha = \frac{m_\alpha}{S_\alpha} \sum_\beta S_\beta W'_{\alpha\beta} \mathbf{e}_{\alpha\beta} \cdot (E_{\mathbf{M}_\alpha} - E_{\mathbf{M}_\beta}). \tag{24.37d}$$

In addition, these equations can be made symplectic by keeping the evolutions of $\mathbf{x}_\alpha$ and $\mathbf{M}_\alpha$ while evaluating s_α and ρ_α. Taking energy (24.20) with $V_\alpha = V_\alpha^s$, the evolution equations simplify to

$$\dot{x}_\alpha^i = \frac{M_\alpha^i}{m_\alpha} \tag{24.38a}$$

$$\dot{M}_{\alpha i} = -\sum_\beta S_\alpha S_\beta \left(\frac{p_\alpha}{s_\alpha^2} + \frac{p_\beta}{s_\beta^2}\right) W'_{\alpha\beta} e_{\alpha\beta i} \tag{24.38b}$$

$$\dot{s}_\alpha = \sum_\beta S_\beta W'_{\alpha\beta} \mathbf{e}_{\alpha\beta} \cdot (\mathbf{v}_\alpha - \mathbf{v}_\beta) \tag{24.38c}$$

$$\dot{\rho}_\alpha = \frac{m_\alpha}{S_\alpha} \sum_\beta S_\beta W'_{\alpha\beta} \mathbf{e}_{\alpha\beta} \cdot (\mathbf{v}_\alpha - \mathbf{v}_\beta), \tag{24.38d}$$

which can again be made symplectic when both the entropy density and mass density are calculated directly via relations (24.36).

24.4.3 Direct Volume Approach

The definition of the direct volume is

$$V_\alpha^d = \left(\sum_\beta W_{\alpha\beta} \right)^{-1} . \tag{24.39}$$

The density state variables simply follow from the evolution of direct volume V_α^d expression and their Lagrange initial values:

$$\rho_\alpha = m_\alpha / V_\alpha^d, \quad s_\alpha = S_\alpha / V_\alpha^d, \tag{24.40}$$

where ρ_α and s_α are independent state variables (once dissipative evolution is considered).

The corresponding Poisson bracket is the same as in the mass-based volume approach, see Appendix 24.8.1, but where m_β is replaced by m_α. The resulting Hamiltonian evolution equations are

$$\dot{\mathbf{x}}_\alpha = E_{\mathbf{M}_\alpha} \tag{24.41a}$$

$$\dot{\mathbf{M}}_\alpha = -E_{\mathbf{x}_\alpha} - \sum_\beta \left(m_\alpha E_{\rho_\alpha} + m_\beta E_{\rho_\beta} + S_\alpha E_{s_\alpha} + S_\beta E_{s_\beta} \right) W'_{\alpha\beta} \mathbf{e}_{\alpha\beta} \tag{24.41b}$$

$$\dot{\rho}_\alpha = m_\alpha \sum_\beta W'_{\alpha\beta} \mathbf{e}_{\alpha\beta} \cdot (E_{\mathbf{M}_\alpha} - E_{\mathbf{M}_\beta}) \tag{24.41c}$$

$$\dot{s}_\alpha = S_\alpha \sum_\beta W'_{\alpha\beta} \mathbf{e}_{\alpha\beta} \cdot (E_{\mathbf{M}_\alpha} - E_{\mathbf{M}_\beta}) \tag{24.41d}$$

Using the considered form of energy (24.20), Eq. (24.41) then turn to

$$\dot{x}_\alpha^i = \frac{M_\alpha^i}{m_\alpha} \tag{24.42a}$$

$$\dot{\rho}_\alpha = m_\alpha \sum_\beta W'_{\alpha\beta} \mathbf{e}_{\alpha\beta} \cdot \left(\frac{\mathbf{M}_\alpha}{m_\alpha} - \frac{\mathbf{M}_\beta}{m_\beta} \right) \tag{24.42b}$$

$$\dot{M}_{\alpha i} = -\sum_\beta \left[\left(\frac{m_\alpha}{\rho_\alpha} \right)^2 p_\alpha + \left(\frac{m_\beta}{\rho_\beta} \right)^2 p_\beta \right] W'_{\alpha\beta} e_{\alpha\beta i} \tag{24.42c}$$

$$\dot{s}_\alpha = S_\alpha \sum_\beta W'_{\alpha\beta} \mathbf{e}_{\alpha\beta} \cdot \left(\frac{\mathbf{M}_\alpha}{m_\alpha} - \frac{\mathbf{M}_\beta}{m_\beta} \right), \tag{24.42d}$$

where pressure and temperature are defined as in Equation (24.35).

24.4.4 Implicit-Based Volume Approach

The definition of implicit-based volume relates mass and entropy density as

$$V_\alpha^I = \frac{m_\alpha}{\rho_\alpha} = \frac{S_\alpha}{s_\alpha} = \sum_\beta W_{\alpha\beta}^{-1} \tag{24.43}$$

but note that both ρ_α and s_α are independent state variables as we shall consider dissipative evolution as well, where we let S_α to evolve to reflect the dissipation. Nevertheless, the evolution of Eulerian entropic density s_α is such that the above relation Eq. (24.43) for implicit particle volume holds.

The Poisson bracket governing the reversible evolution of $\mathbf{x}_\alpha$, $\mathbf{M}_\alpha$, ρ_α, and s_α can be obtained by plugging functionals dependent on these state variables into the Poisson bracket of Lagrangian continuum mechanics (24.1), see Appendix 24.8.4. We again assume that the Lagrangian particle mass and entropy are independent of the particle positions (although S_α can change over time due to irreversible effects). The reversible part of the evolution equations then reads

$$\dot{\mathbf{x}}_\alpha = E_{\mathbf{M}_\alpha} \tag{24.44a}$$

$$\dot{\rho}_\alpha = \frac{\rho_\alpha^2}{m_\alpha}\sum_\beta \frac{m_\beta}{\rho_\beta}\sum_\delta W_{\alpha\delta}^{-1}W'_{\delta\beta}\mathbf{e}_{\delta\beta}\cdot(E_{\mathbf{M}_\delta} - E_{\mathbf{M}_\beta}) \tag{24.44b}$$

$$\dot{\mathbf{M}}_\alpha = -E_{\mathbf{x}_\alpha} - \sum_\beta\left(\frac{m_\beta}{\rho_\beta}(E_\rho^{-1})_\alpha + \frac{m_\alpha}{\rho_\alpha}(E_\rho^{-1})_\beta + \frac{S_\beta}{s_\beta}(E_s^{-1})_\alpha + \frac{S_\alpha}{s_\alpha}(E_s^{-1})_\beta\right)W'_{\alpha\beta}\mathbf{e}_{\alpha\beta} \tag{24.44c}$$

$$\dot{s}_\alpha = \frac{s_\alpha^2}{S_\alpha}\sum_\beta \frac{S_\beta}{s_\beta}\sum_\delta W_{\alpha\delta}^{-1}W'_{\delta\beta}\mathbf{e}_{\delta\beta}\cdot(E_{\mathbf{M}_\delta} - E_{\mathbf{M}_\beta}), \tag{24.44d}$$

where $(E_\rho^{-1})_\alpha = \sum_\gamma E_{\rho_\gamma}\frac{\rho_\gamma^2}{m_\gamma}W_{\gamma\alpha}^{-1}$ and $(E_s^{-1})_\alpha = \sum_\gamma E_{s_\gamma}\frac{s_\gamma^2}{S_\gamma}W_{\gamma\alpha}^{-1}$. Note that in the case where m_α and S_α are constants (the reversible case), the evolution equations for ρ_α and s_α, Eq. (24.44b, d), can be rewritten in terms V_α^I as expected.

Equation (24.44) represents a Hamiltonian system of ODEs that is non-symplectic. Indeed, the total mass and total entropy are preserved regardless of the choice of energy, so the underlying Poisson bracket has Casimir functionals (mass and entropy) that are conserved regardless of the choice of energy. The system of equations can be made symplectic as before by explicitly evaluating the entropy and mass density by updating the rule (24.21) (while removing the equations $\dot{\rho}_\alpha$ and $\dot{s}_\alpha$ from (24.44)). With energy $E = \sum_\alpha\left(\frac{\mathbf{M}_\alpha^2}{2m_\alpha} + V_\alpha^I\epsilon(\rho_\alpha, s_\alpha)\right)$, symplectified Eq. (24.44) become

$$\dot{\mathbf{x}}_\alpha = E_{\mathbf{M}_\alpha} \tag{24.45a}$$

$$\dot{\mathbf{M}}_\alpha = -\sum_\gamma\sum_\beta p_\gamma(V_\beta W_{\gamma\alpha}^{-1} + V_\alpha W_{\gamma\beta}^{-1})W'_{\alpha\beta}\mathbf{e}_{\alpha\beta}. \tag{24.45b}$$

However, there is a complication in the numerical implementation of Eq. (24.45) caused by the presence of the inverse matrix $W_{\alpha\beta}^{-1}$. Although the matrix $W_{\alpha\beta}$ is typically sparse, its inverse is dense and therefore it is impractical to store it in memory. We can overcome this obstacle by solving the linear system of equations $\sum_\beta W_{\gamma\beta} y_\beta = p_\gamma$ at each time step by conjugate gradients, since the solution gives us the terms on the right-hand side of the momentum equations, $y_\alpha = p_\gamma W^{-1}_{\gamma\alpha}$. Although this approach slows down the simulation (we are using explicit time stepping to keep the symplecticity), the simulation is still viable on standard desktops (for 10^4 particles in two dimensions).

24.4.5 Mixed-Volume Approach

Finally, another possibility is to define the mass and entropy densities using Eq. (24.28), which mixes the mass-based particle volume with the entropy-based volume. Despite leaving the elegance of having a single-particle volume, the mixed approach provides smoothing to both fields, which might be advantageous.

Poisson bracket (24.85), derived in Appendix 24.8.3, leads to the following evolution equations for density and entropy density defined through (24.28),

$$\dot{\mathbf{x}}_\alpha = E_{\mathbf{M}_\alpha} \tag{24.46a}$$

$$\dot{\mathbf{M}}_\alpha = -E_{\mathbf{x}_\alpha} - \sum_\beta (m_\alpha E_{\rho_\beta} + m_\beta E_{\rho_\alpha} + S_\alpha E_{s_\beta} + S_\beta E_{s_\alpha}) W'_{\alpha\beta} \mathbf{e}_{\alpha\beta} \tag{24.46b}$$

$$\dot{\rho}_\alpha = \sum_\beta m_\beta W'_{\alpha\beta} \mathbf{e}_{\alpha\beta} (E_{\mathbf{M}_\alpha} - E_{\mathbf{M}_\beta}) \tag{24.46c}$$

$$\dot{s}_\alpha = \sum_\beta S_\beta W'_{\alpha\beta} \mathbf{e}_{\alpha\beta} (E_{\mathbf{M}_\alpha} - E_{\mathbf{M}_\beta}), \tag{24.46d}$$

where energy is yet to be supplied. Using the energy (24.20) with the mass volume, evolution equation (24.46) become

$$\dot{x}^i_\alpha = \frac{M^i_\alpha}{m_\alpha} \tag{24.47a}$$

$$\dot{\rho}_\alpha = \sum_\beta m_\beta W'_{\alpha\beta} \mathbf{e}_{\alpha\beta} \left(\frac{\mathbf{M}_\alpha}{m_\alpha} - \frac{\mathbf{M}_\beta}{m_\beta} \right) \tag{24.47b}$$

$$\begin{aligned} \dot{M}_{\alpha i} = -\sum_\beta \Bigg[& \frac{m_\alpha m_\beta}{\rho_\alpha^2} p_\alpha + \frac{m_\alpha m_\beta}{\rho_\beta^2} p_\beta \\ & + m_\alpha m_\beta \left(\left(\frac{S_\beta}{m_\beta} - \frac{s_\alpha}{\rho_\alpha} \right) \frac{T_\alpha}{\rho_\alpha} + \left(\frac{S_\alpha}{m_\alpha} - \frac{s_\beta}{\rho_\beta} \right) \frac{T_\beta}{\rho_\beta} \right) \Bigg] W'_{\alpha\beta} \mathbf{e}_{\alpha\beta} \end{aligned} \tag{24.47c}$$

$$\dot{s}_\alpha = \sum_\beta S_\beta W'_{\alpha\beta} \mathbf{e}_{\alpha\beta} \left(\frac{\mathbf{M}_\alpha}{m_\alpha} - \frac{\mathbf{M}_\beta}{m_\beta} \right), \tag{24.47d}$$

where temperature is defined as $T_\alpha = \frac{\partial \epsilon}{s_\alpha}$. Note the extra terms on the right-hand side of the evolution equation for $\mathbf{M}_\alpha$, which depend on the entropies. Although these terms are not present in the usual form of SPH equations, they are necessary to maintain the Hamiltonianity of the equations when the mixed-volume approach is taken, and they follow directly from the projection of Poisson brackets.

24.4.6 Illustration—Free Expansion

In this example, we consider a purely Hamiltonian free expansion of the ideal gas, where the entropy of individual particles S_α does not change. This example can be seen as a baseline numerical experiment, the final state of which coincides with the result of the quasistatic adiabatic expansion (Landau and Lifschitz 1969). We shall use it to discuss the effects of the different choices of particle volumes introduced above.

We consider a double-compartment vessel with only the left half initially filled with an ideal gas; see Appendix 24.9 for the thermodynamics of the ideal gas. The whole vessel is eventually occupied and the temperature of the gas is eventually reduced as in the reversible adiabatic expansion, see Figs. 24.8 and 24.9.

24.5 Entropic SPH with Dissipative Evolution

In this Section, we show how to add dissipative processes to the entropic SPH (to all of its formulations). In particular, we include Fourier heat conduction and viscous dissipation, and finally we add algebraic dissipation in the context of hyperbolic heat conduction.

It should be noted that dissipative terms have to update the particle entropies directly S_α, since entropy densities s_α are calculated from the particle densities and not vice versa.

24.5.1 Fourier Heat Conduction

As we have entropy among state variables, we can capture the transformation of the kinetic energy into the internal energy (dissipation) while keeping the total energy constant. In other words, entropic SPH preserves the total energy also when dissipation (viscosity, heat conduction, etc.) is present.

Here we outline a model for classical Fourier heat conduction, which is based on a dissipation potential with a gradient of conjugate state variable. The SPH evolution can be then seen as a realization of the General Equation for Non-Equilibrium Reversible-Irreversible Coupling (GENERIC) (Grmela and Öttinger 1997; Öttinger and Grmela 1997; Öttinger 2005; Pavelka et al. 2018). In order to discretize the dissipation potential, we need to discretize differential operators first, as is done in the standard SPH.

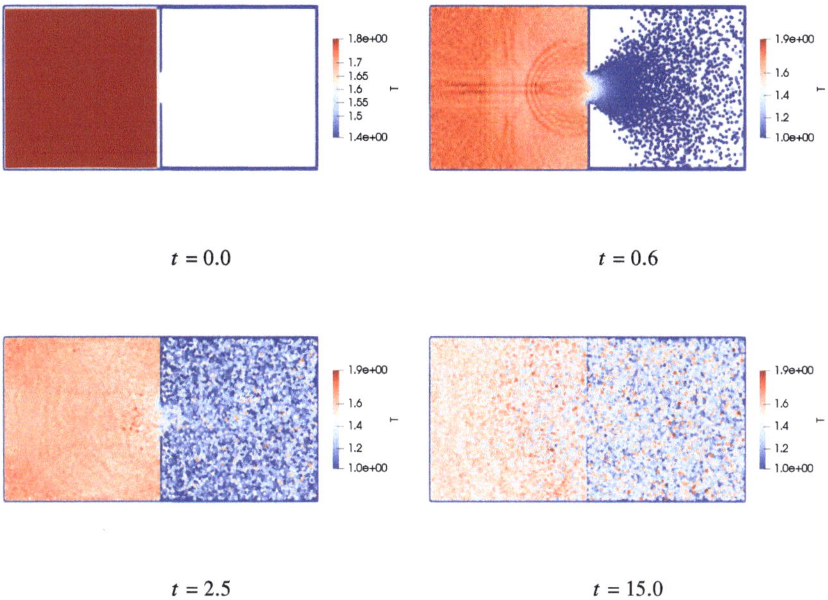

Fig. 24.8 A purely Hamiltonian simulation of free expansion of an ideal gas (with 43017 SPH particles, the Verlet integrator (Verlet 1967), final time 15, time step 1.3E-05, heat capacity $c_V = 1.0 J/kgK$, $\gamma = 1.4$). The color indicates the temperature. The error in the total energy is $10^{-4}\%$ and decreases quadratically with the time step. The simulation is carried out with the mass-based particle volume. Simulations with the entropic, direct, and mixed particle volumes give very similar results. Simulation with the implicit volume leads to negative particle volumes due to boundary effects, so it is not shown

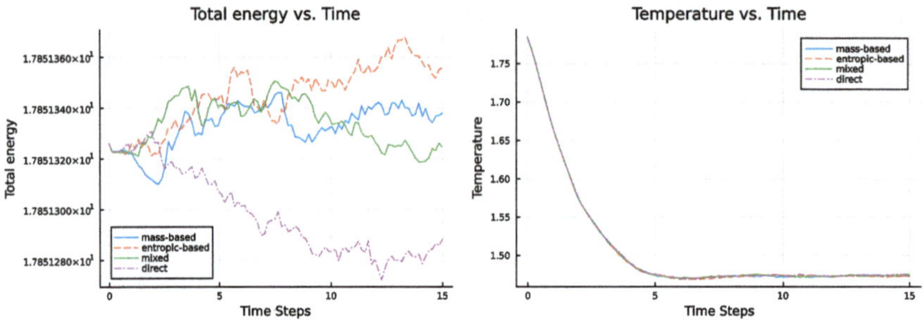

total energy in time average temperature

Fig. 24.9 The total energy error is of order $10^{-4}\%$ and decreases with the time step quadratically for four shown approaches: mass-based volume, entropic volume, direct volume, and mixed volume. The temperature decreases and reaches the value given by laws of the isentropic adiabatic expansion and all the four approaches give the same temperature profiles

Let us now recall the discrete gradient and divergence operators,

$$\tilde{G}^0_\alpha(A_\square) = -\sum_\beta V_\beta(A_\alpha - A_\beta)W'_{\alpha\beta}\mathbf{e}_{\alpha\beta} \tag{24.48a}$$

$$D^0_\alpha(A_\square) = \sum_\beta V_\beta(A_\alpha + A_\beta)W'_{\alpha\beta}\mathbf{e}_{\alpha\beta}, \tag{24.48b}$$

where $A_\square$ is a collection of discrete values $\{A_\alpha\}$ for all the SPH particles, see Violeau (2012), and particle volume V_α is subjected to one of the choices above. Note that these operators satisfy the duality

$$\langle A_\square, \tilde{G}^0_\alpha(B_\square)\rangle = -\langle D^0_\alpha(A_\square), B_\square\rangle, \tag{24.49}$$

where $\langle A_\square, B_\square\rangle = \sum_\alpha V_\alpha A_\alpha B_\alpha$.

Then, the continuous dissipation potential

$$\Xi^{(heat)} = \frac{1}{2}\int d\mathbf{r}\lambda(T)(\nabla\epsilon^*)^2, \tag{24.50}$$

where the conjugate internal energy $\epsilon^* = \frac{\delta S}{\delta \epsilon}$ can be interpreted as the inverse temperature T^{-1}, can be discretized as

$$\begin{aligned}\Xi^{(SPH-heat)} =& \frac{1}{2}\sum_\alpha V_\alpha\lambda(T_\alpha)\tilde{G}^0_\alpha(\epsilon^*_\square)\tilde{G}^0_\alpha(\epsilon^*_\square)\\ =& \frac{1}{2}\sum_\alpha V_\alpha\lambda(T_\alpha)\sum_\beta V_\beta(\epsilon^*_\alpha - \epsilon^*_\beta)W'_{\alpha\beta}\mathbf{e}_{\alpha\beta}\sum_\gamma V_\gamma(\epsilon^*_\alpha - \epsilon^*_\gamma)W'_{\alpha\gamma}\mathbf{e}_{\alpha\gamma},\end{aligned} \tag{24.51}$$

where $\epsilon^*_\alpha = T_\alpha^{-1}$ is the SPH inverse temperature of particle α. Irreversible evolution of the discrete energy density ϵ_α is then obtained as the functional derivative of the dissipation potential with respect to the conjugate variable ϵ^*,[2] which can be calculated from the variation of the dissipation potential,

$$\begin{aligned}\Xi^{(SPH-heat)}(\epsilon^* + \delta\epsilon^*) =& \frac{1}{2}\sum_\alpha V_\alpha \lambda(T_\alpha)\tilde{G}^0_\alpha(\epsilon^*)\tilde{G}^0_\alpha(\epsilon^*)\\ &+ \sum_\alpha V_\alpha \lambda(T_\alpha)\tilde{G}^0_\alpha(\delta\epsilon^*)\tilde{G}^0_\alpha(\epsilon^*) + \mathcal{O}(\delta\epsilon^*)^2\\ =& \Xi^{(SPH-heat)}(\epsilon^*)\\ &- \sum_\alpha V_\alpha D^0_\alpha(\lambda(T_\Box)\tilde{G}^0_\Box(\epsilon^*_\Box))\delta\epsilon^*_\alpha + \mathcal{O}(\delta\epsilon^*)^2, \end{aligned} \tag{24.52}$$

as

$$\dot{\epsilon}_\alpha = \frac{1}{V_\alpha}\frac{\partial \Xi}{\partial \epsilon^*_\alpha} = -D^0_\alpha(\mathbf{q}_\Box) = -\sum_\beta V_\beta(\mathbf{q}_\alpha + \mathbf{q}_\beta)\cdot W'_{\alpha\beta}\mathbf{e}_{\alpha\beta} \tag{24.53}$$

where

$$\begin{aligned}\mathbf{q}_\alpha = \lambda(T_\alpha)\tilde{G}^0_\alpha(\epsilon^*_\Box) &= -\sum_\beta V_\beta \lambda(T_\alpha)(\epsilon^*_\alpha - \epsilon^*_\beta)W'_{\alpha\beta}\mathbf{e}_{\alpha\beta}\\ &= -\sum_\beta V_\beta \lambda_F(T_\alpha - T_\alpha^2/T_\beta)W'_{\alpha\beta}\mathbf{e}_{\alpha\beta}\end{aligned} \tag{24.54}$$

is the discrete heat flux, $\lambda(T_\alpha) = \lambda_F T_\alpha^2$ for λ_F being the Fourier heat conductivity, and s_α being the particle entropy density. The result is similar to the usual approach to heat conduction in SPH (Monaghan 2005; Jeong et al. 2003), except that $T_\alpha - T_\beta$ is replaced by $T_\alpha - T_\alpha^2/T_\beta$.

Discrete heat equation (24.53) satisfies the second law of thermodynamics, which can be shown as

$$\dot{S} = \sum_\alpha V_\alpha \epsilon^*_\alpha \dot{\epsilon}_\alpha = -\langle \epsilon^*_\alpha, D^0_\alpha \mathbf{q}\rangle = \langle \tilde{G}^0_\alpha \epsilon^*, \mathbf{q}_\alpha\rangle = \left\langle \frac{\mathbf{q}_\alpha}{\lambda(T_\alpha)}, \mathbf{q}_\alpha \right\rangle \geq 0. \tag{24.55}$$

However, this could already be inferred from the convexity of the continuous dissipation potential (24.50) because the dissipation potential is convex (Pavelka et al. 2020) and composition of a convex and affine function is convex as well.[3]

[2] The $1/V_\alpha$ prefactor is present as energy *density* is on the left-hand side and the right-hand side has to have the same property. In continuum thermodynamics, the functional derivative produces the density-like behavior.

[3] The second derivative of composition $\Xi \circ f(y)$, where f is affine, is $\frac{\partial^2 \Xi}{\partial f^2}\left(\frac{\partial f}{\partial y}\right)^2 \geq 0$.

Mass-based volume approach.
In this case, we consider the mass-based volume, that is, $V_\alpha = V_\alpha^m = \frac{m_\alpha}{\rho_\alpha}$. Then, as mentioned above, the dissipative processes are included via an update in the discrete Lagrangian quantities. In the case of Fourier heat conduction, we need to update the entropy S_α. This is done by taking

$$\dot{S}_\alpha = V_\alpha^m \frac{\partial s_\alpha}{\partial \epsilon_\alpha}\dot{\epsilon}_\alpha = -\epsilon_\alpha^* \sum_\beta V_\alpha^m V_\beta^m (\mathbf{q}_\alpha + \mathbf{q}_\beta) \cdot W'_{\alpha\beta}\mathbf{e}_{\alpha\beta}, \tag{24.56}$$

where $\mathbf{q}_\alpha$ is given by Equation (24.54) with $V_\alpha = V_\alpha^m$. This dissipative evolution is then added to the Hamiltonian evolution equations according to the GENERIC framework.
Entropic-based volume and mixed-based volume approach.
As in the mass-based approach, we need to convert the Eulerian change of entropy $\dot{s}_\alpha$ to the Lagrangian one, $\dot{S}_\alpha$. This is achieved again by taking

$$\dot{S}_\alpha = V_\alpha^s \epsilon_\alpha^* \dot{\epsilon}_\alpha = -\epsilon_\alpha^* \sum_\beta V_\alpha^s V_\beta^s (\mathbf{q}_\alpha + \mathbf{q}_\beta) \cdot W'_{\alpha\beta}\mathbf{e}_{\alpha\beta}, \tag{24.57}$$

where $\mathbf{q}_\alpha$ is given by Equation (24.54) and $V_\alpha = V_\alpha^s = \frac{S_\alpha}{s_\alpha}$. The same relation is also used in the mixed-volume-based approach.
Implicit-based volume approach
We again need to provide a way to update the Lagrangian particle entropy S_α due to dissipation while respecting the definition of the implicit volume V_α^I. Taking advantage of the assumption made earlier that the dissipative dynamics does not alter particle positions, that is $\frac{\delta S_\alpha}{\delta x_\beta} = 0$, differential of the definition of the entropy-based volume gives

$$0 = \sum_\beta \left(\frac{dS_\beta}{s_\beta} - \frac{S_\beta}{s_\beta^2} ds_\beta \right) W_{\alpha\beta}. \tag{24.58}$$

This is satisfied when the particle entropy is updated as

$$\dot{S}_\alpha = \frac{S_\alpha}{s_\alpha}\dot{s}_\alpha. \tag{24.59}$$

Analogical update rules for the Lagrangian entropies S_α are used also in the presence of other dissipation sources, as for instance the viscous dissipation in the following Section.

24.5.2 Viscous Dissipation

Although viscous terms can also be formulated as gradient dynamics on the continuous level, see Pavelka et al. (2018), its discretization is not as straightforward as in the case of heat conduction. The problem lies in the Lagrangian nature of the momentum in SPH, in contrast to the Eulerian character of the entropy s_α (or density ρ_α). In other words, an SPH discretization of the continuous dissipation potential (Pavelka et al. 2018) would produce a discrete evolution for the Eulerian momentum density $\mathbf{m}_\alpha = \sum_\beta W_{\alpha\beta} M_\beta$, not for $\mathbf{M}_\alpha$ itself. Therefore, we opt out of adapting a usual SPH discretization of the viscous terms (Monaghan 1992),

$$(\dot{\mathbf{M}}_\alpha)_{diss} = \sum_\beta 2(n+2)\mu V_\alpha V_\beta (\mathbf{v}_\alpha - \mathbf{v}_\beta) \cdot \mathbf{e}_{\alpha\beta} \frac{W'_{\alpha\beta}}{r_{\alpha\beta}} \mathbf{e}_{\alpha\beta}, \tag{24.60}$$

where n is the geometric dimension of the system.

Viscous dissipation also changes the kinetic energy of particle α as

$$\dot{e}^{kin}_\alpha = \frac{d}{dt}\left(\frac{1}{2}\mathbf{v}_\alpha \mathbf{M}_\alpha\right) = \mathbf{v}_\alpha \dot{\mathbf{M}}_\alpha = 2(n+2)\mu \mathbf{v}_\alpha \cdot \sum_\beta \mathbf{e}_{\alpha\beta} V_\alpha V_\beta (\mathbf{v}_\alpha - \mathbf{v}_\beta) \cdot \mathbf{e}_{\alpha\beta} \frac{W'_{\alpha\beta}}{r_{\alpha\beta}}, \tag{24.61}$$

which eventually reduces the total kinetic energy

$$\dot{e}^{kin} = \sum_\alpha \dot{e}^{kin}_\alpha = (n+2)\mu \sum_\alpha \sum_\beta ((\mathbf{v}_\alpha - \mathbf{v}_\beta) \cdot \mathbf{e}_{\alpha\beta})^2 V_\alpha V_\beta \frac{W'_{\alpha\beta}}{r_{\alpha\beta}} \leq 0. \tag{24.62}$$

The sign of this inequality is caused by the assumed monotonicity of the smoothing kernel $W(r)$.

While the kinetic energy is reduced, the internal energy grows so that the total energy is conserved. Internal energy is a function of mass density and entropy density, and the growth of internal energy is caused by the entropy production

$$(\dot{S}_\alpha)_{diss} = -\frac{1}{T_\alpha} \sum_\beta (n+2)\mu V_\alpha V_\beta ((\mathbf{v}_\alpha - \mathbf{v}_\beta) \cdot \mathbf{e}_{\alpha\beta})^2 \frac{W'_{\alpha\beta}}{r_{\alpha\beta}} \geq 0. \tag{24.63}$$

Then the overall entropy $S = \sum_\alpha S_\alpha$ is produced, $\dot{S} > 0$, and the second law of thermodynamics is satisfied. The overall energy $E = E^{kin} + E^{int}$ is not changed, at least as far as the continuous form of the entropic SPH, before the temporal discretization, is considered. Note that, as a consequence, we shall use Eq. 24.63 as an update for the Lagrangian particle entropy S_α

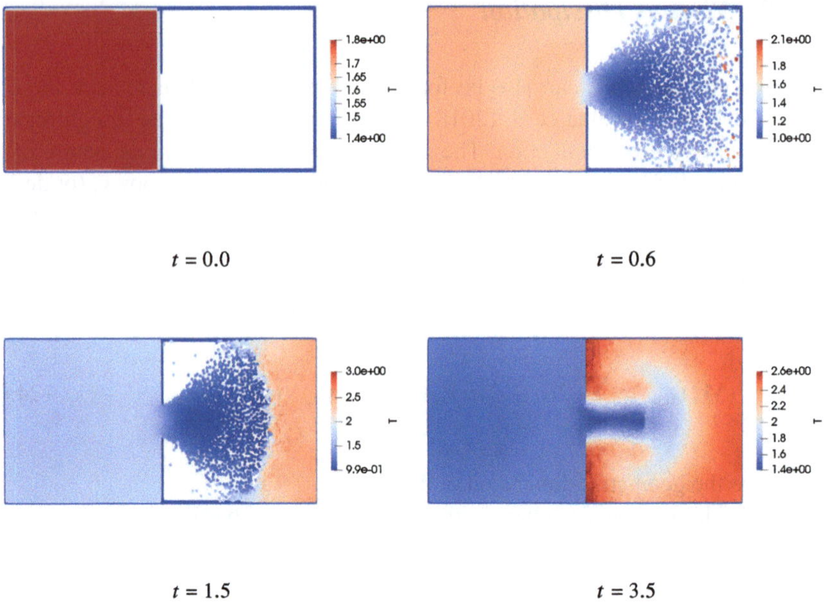

Fig. 24.10 Joule expansion (43017 particles). All particles are initially in the left compartment only and gradually occupy the whole volume. Their temperature drops as in the adiabatic expansion, but then increases due to viscous ($\mu = 0.001$) and Fourier dissipation ($\lambda = 0.1$). The heat flux was pointing to the left almost everywhere in the domain. The simulation was carried out with the mass-volume approach, and the entropic and mixed-volume approaches give quite similar results. When we raise the heat conductivity to $\lambda = 10.0$, the temperature difference at time 15 is approximately 0.2%, so the temperatures approach a common value

24.5.3 *Illustration—Joule Expansion*

Let us now illustrate the Fourier and viscous dissipative effects on the free expansion of the ideal gas, where the dissipative effects raise the entropy. Figure 24.10 shows the results for the mass-volume method (the entropic and mixed methods give practically the same results). Instead of a temperature drop as in the purely Hamiltonian case, we observe that the temperature returns to its initial value because of the dissipative effects, as expected in Joule expansion (Landau and Lifschitz 1969).

Figure 24.11 shows that the energy increases slightly due to the numerical errors in the first-order time discretization of the dissipative terms. The error decreases with the time step and could be further decreased, for instance, by the Runge-Kutta method (Runge 1895; Kutta 1901; Kincl 2023).

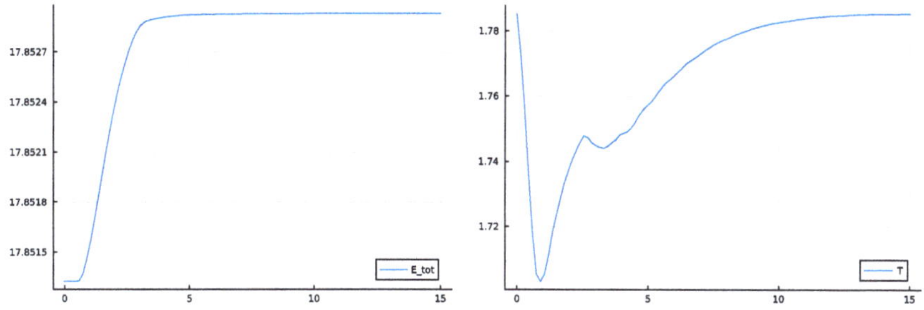

total energy in time — average temperature

Fig. 24.11 Total energy in time gradually increases due to numerical errors in the dissipation (by 0.006%). However, the error decreases with the time step (linearly, as it is dominated by the first-order dissipative terms). The temperature initially drops due to Hamiltonian effects (expansion while not raising entropy fast enough) and then recovers its initial value due to the dissipative effects

24.5.4 Illustration—Rayleigh-Bénard Convection

Our final example is a vessel with a fluid described by the stiffened gas equation of state (see Appendix 24.9 or Dumbser et al. (2016)) so that it has only limited but non-zero compressibility. The fluid is in a gravitational field and is subject to heating from the bottom boundary as in the Rayleigh-Bénard convection (Bénard 1900; Rayleigh 1916; Lebon and Jou 2008). Note that we do not use the usual Boussinesq approximation (Boussinesq 1897) as the fluid is treated as compressible. We use the mass-based volume approach (11k particles, viscosity $\mu = 8.4 \cdot 10^{-4}$, and heat conductivity $\lambda_F = 1.0 \cdot 10^{-4}$).

After the initial equilibration of the gravitational energy, we observe mushroom-like structures caused by the buoyancy due to lower density of the heated parts of the fluid, see Fig. 24.12.

24.5.5 Hyperbolic Heat Conduction—Lagrangian Approach

Although classical Fourier heat conduction is usually sufficient for the description of heat transfer, very fast heat pulses require hyperbolic heat conduction (McNelly et al. 1970; Kovács and Ván 2015). Fourier heat conduction can be seen to be an approximation of hyperbolic heat conduction for sufficiently long times (Jou et al. 2010). There are two routes toward evolution equations for hyperbolic heat conduction. The first route starts with the kinetic theory of phonons (quasiparticles present in a crystal lattice) (Peierls 1955). The kinetic theory can be reduced to a less detailed theory, taking into account only a few moments of the distribution function (Dreyer and Struchtrup 1993), or the Eulerian entropy density and phonon momentum density. This route leads to the Eulerian hyperbolic heat conduction studied in the subsequent Section.

In this Section, we follow the second route that leads to a Lagrangian version of hyperbolic heat conduction. We commence by introducing a scalar field $\psi(\mathbf{X})$

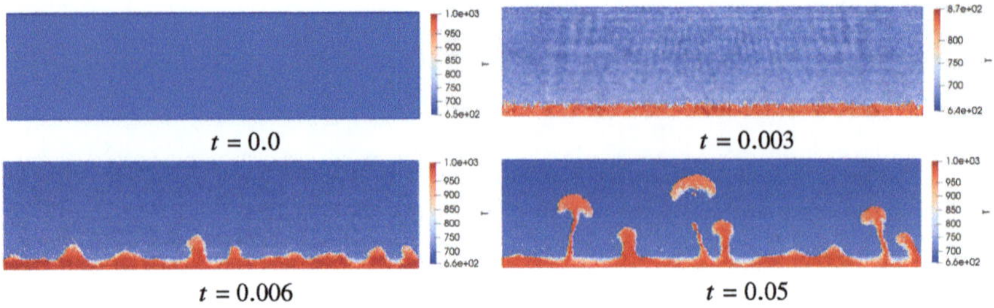

Fig. 24.12 Temperature in a fluid with heated bottom boundary and cooled upper boundary. The first left-top figure shows the initial condition, the right-top figure shows the initial hydrostatic equilibration, the left-bottom figure shows the onset of instability, and the right-bottom figure contains the mushroom-like plumes. The instability is caused by buoyancy due to the lower density of the fluid near the bottom boundary, leading to the formation of mushroom-like plumes. Both viscous dissipation and Fourier heat conduction were switched on

canonically coupled to the field of Lagrangian entropy density, $s_0(\mathbf{X})$, via Poisson bracket

$$\{F, G\}^{(s_0-\psi)} = \int d\mathbf{X} \left(F_{s_0} G_\psi - G_{s_0} F_\psi \right). \tag{24.64}$$

Instead of the scalar field ψ, only its gradient $W_I = \frac{\partial \psi}{\partial X^I}$ is considered, which will represent a vectorial quantity related to the heat flux (Peshkov et al. 2018; Godunov et al. 1996; Malyshev and Romenskii 1986). Finally, Lagrangian quantity W_I is projected to its Eulerian counterpart

$$w_i = \frac{\partial X^I}{\partial x^i} W_I = \frac{\partial \psi(\mathbf{X}(\mathbf{x}))}{\partial x^i}, \tag{24.65}$$

called the conjugate entropy flux, as the derivative of energy with respect to $\mathbf{w}$ turns out to be the entropy flux. In SPH, we shall follow the second route, as it leads to similar results as the first route (based on kinetic theory Sýkora et al. (2023)) while being more suitable for the SPH discretization.

Within SPH, hyperbolic heat conduction has already been implemented in Jiang and Sousa (2006) by discretization of the resulting Cattaneo equation (a telegraphist's equation for temperature). Here, we formulate another model within SPH that is Hamiltonian and symplectic (when disregarding the dissipation) and thus suitable for symplectic integrators. Based on the above geometric construction of the continuous hyperbolic heat conduction, we introduce the SPH variables for hyperbolic heat conduction as

$$S_\alpha = \int d\mathbf{X} \chi_\alpha(\mathbf{X}) s_0(\mathbf{X}) \tag{24.66a}$$

$$\mathbf{w}_\alpha = \tilde{G}^0_\alpha(\psi_\Box) \tag{24.66b}$$

where $\psi_\alpha = \int d\mathbf{X}\bar{\chi}_\alpha(\mathbf{X})\psi(\mathbf{X})$. The Poisson bracket for variables S_α (entropy of particle α) and $\mathbf{w}_\alpha$ (conjugate entropy flux of particle α) is calculated from the canonical Poisson bracket for s_0 and ψ,

$$\{F, G\}^{(S_\alpha, \mathbf{w}_\alpha)} = -\sum_\alpha V_\alpha F_{S_\alpha} D^0_\alpha \left(\frac{G_{\mathbf{w}_\Box}}{V_\Box}\right) - \sum_\alpha \tilde{G}^0_\alpha (G_{S_\Box}) F_{\mathbf{w}_\alpha}. \tag{24.67}$$

The reversible evolution equations implied by this bracket are

$$(\dot{S}_\alpha)_{rev} = -\sum_\beta V_\alpha V_\beta \left(\frac{E_{\mathbf{w}_\alpha}}{V_\alpha} + \frac{E_{\mathbf{w}_\beta}}{V_\beta}\right) \cdot W'_{\alpha\beta}\mathbf{e}_{\alpha\beta} \tag{24.68a}$$

$$(\dot{\mathbf{w}}_\alpha)_{rev} = \sum_\beta V_\beta (E_{S_\alpha} - E_{S_\beta}) W'_{\alpha\beta}\mathbf{e}_{\alpha\beta}. \tag{24.68b}$$

These equations automatically conserve the total entropy, $\dot{S} = \sum_\alpha \dot{S}_\alpha = 0$, as well as the total energy $E(S_\Box, \mathbf{w}_\Box)$. It can be seen from the equation for entropy that the derivative of energy with respect to the field $\mathbf{w}$ is the entropy flux, and thus $\mathbf{w}_\alpha$ is called the conjugate entropy flux at the particle α.

Apart from the reversible evolution, the conjugate entropy flux $\mathbf{w}_\alpha$ has also irreversible evolution that drives it toward an equilibrium value,

$$(\dot{\mathbf{w}}_\alpha)_{irr} = -\frac{1}{\tau} E_{\mathbf{w}_\alpha}. \tag{24.69a}$$

This dissipative evolution, which represents collisions of phonons with crystal impurities (Callaway 1959; Dreyer and Struchtrup 1993), is accompanied with entropy production

$$(\dot{S}_\alpha)_{irr} = \frac{1}{T_\alpha \tau}(E_{\mathbf{w}_\alpha})^2 \geq 0, \tag{24.69b}$$

so that the total energy is conserved. Note that $\tau > 0$ is a relaxation parameter that characterizes collisions.

Equations (24.68) and (24.69) are then summed to the final evolution equations for the SPH hyperbolic heat conduction,

$$\dot{S}_\alpha = -\sum_\beta V_\alpha V_\beta \left(\frac{E_{\mathbf{w}_\alpha}}{V_\alpha} + \frac{E_{\mathbf{w}_\beta}}{V_\beta}\right) \cdot W'_{\alpha\beta}\mathbf{e}_{\alpha\beta} + \frac{1}{T_\alpha \tau}(E_{\mathbf{w}_\alpha})^2 \tag{24.70a}$$

$$\dot{\mathbf{w}}_\alpha = \sum_\beta V_\beta (E_{S_\alpha} - E_{S_\beta}) W'_{\alpha\beta}\mathbf{e}_{\alpha\beta} - \frac{1}{\tau} E_{\mathbf{w}_\alpha}. \tag{24.70b}$$

Figure 24.13 shows four snapshots of the temperature in a rectangular domain with an initial horizontal temperature gradient. Figure 24.14 shows that the total energy error was less than $10^{-4}\%$ and that the second law of thermodynamics was satisfied (entropy growing).

Reduction to Fourier heat conduction. Finally, Eq. (24.70) can be reduced to an equation for S_α when assuming that the equation for $\mathbf{w}_\alpha$ relaxes quickly to its equilibrium, $\dot{\mathbf{w}}_\alpha \approx 0$,

$$\frac{\mathbf{q}_\alpha}{T_\alpha} = E_{\mathbf{w}_\alpha} = \tau \sum_\beta V_\beta (E_{S_\alpha} - E_{S_\beta}) W'_{\alpha\beta} \mathbf{e}_{\alpha\beta} = -\tau \tilde{G}^0_\alpha(T_\square). \tag{24.71}$$

Note that the entropy flux is equal to the heat flux divided by temperature, $E_{\mathbf{w}_\alpha} = \mathbf{q}_\alpha / T_\alpha$. When we plug this value of $E_{\mathbf{w}_\alpha}$ into the equation for S_α (as in the Dynamic MaxEnt method Klika et al. (2019)), we obtain

$$\begin{aligned} \dot{S}_\alpha &= -\sum_\beta V_\alpha V_\beta \left(\frac{\mathbf{q}_\alpha}{T_\alpha V_\alpha} + \frac{\mathbf{q}_\beta}{T_\beta V_\beta} \right) \cdot W'_{\alpha\beta} \mathbf{e}_{\alpha\beta} + \frac{1}{T_\alpha \tau} \left(\frac{\mathbf{q}_\alpha}{T_\alpha} \right)^2 \\ &= -V_\alpha D^0_\alpha \left(\frac{\mathbf{q}_\square}{T_\square V_\square} \right) + \frac{1}{T_\alpha \tau} (\tilde{G}^0_\alpha)^2. \end{aligned} \tag{24.72}$$

Heat flux (24.71) closer to the usual formula for heat flux (Jeong et al. 2003), a difference being in the presence of particle volumes in the denominator. The overall entropy grows due to the last term,

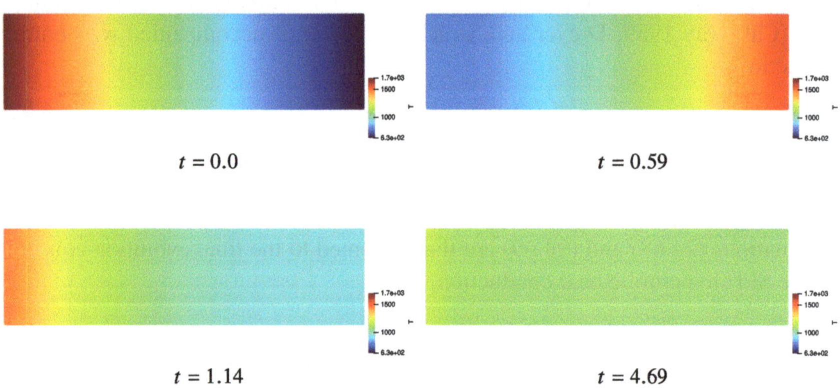

Fig. 24.13 Four snapshots of the rectangular domain with hyperbolic heat propagation. While the heat waves travel through the domain and reflect, they are also being damped by dissipation. The energy was chosen as the sum of the kinetic energy of the heat flux $\sum_\alpha 0.05 \frac{m_\alpha}{\rho_\alpha} \mathbf{w}^2_\alpha$ and the internal energy of the stiffened gas ($c_V = 1.0$, $p_0 = 10$, $\rho_0 = 10$, $\gamma = 1.6$), see Appendix 24.9.2. The relaxation parameter was chosen as $\tau = 10^{-5}$

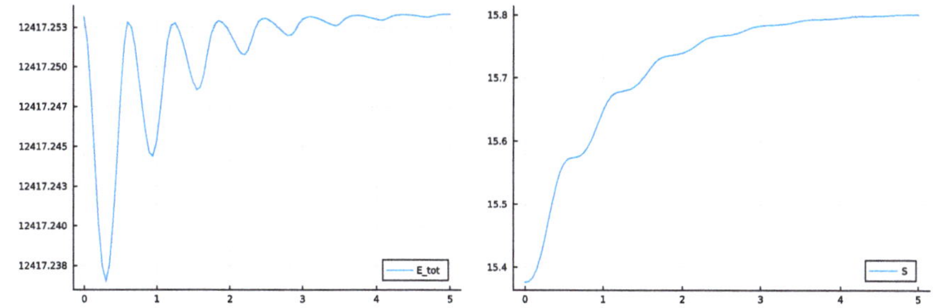

total energy in time total entropy in time

Fig. 24.14 While the total energy is approximately conserved, the overall entropy is raised as expected (Jou et al. 2010)

$$\begin{aligned}\frac{d}{dt}\sum_\alpha S_\alpha &= -\left\langle 1, D^0_\alpha\left(\frac{\mathbf{q}_\Box}{T_\Box V_\Box}\right)\right\rangle + \frac{1}{T_\alpha \tau}\sum_\alpha (\tilde{G}^0_\alpha)^2 \\ &= \underbrace{\left\langle \tilde{G}^0_\alpha(1), \frac{\mathbf{q}_\Box}{T_\Box V_\Box}\right\rangle}_{=0} + \frac{1}{T_\alpha \tau}\sum_\alpha (\tilde{G}^0_\alpha)^2 \geq 0,\end{aligned} \tag{24.73}$$

and the overall energy is conserved as well,

$$\begin{aligned}\frac{dE}{dt} &= \sum_\alpha T_\alpha \dot{S}_\alpha = -\left\langle T_\alpha, D^0_\alpha\left(\frac{\mathbf{q}_\Box}{T_\Box V_\Box}\right)\right\rangle + \frac{1}{\tau}\sum_\alpha \frac{T_\alpha}{T_\alpha \tau}\left(\frac{\mathbf{q}_\alpha}{T_\alpha}\right)^2 \\ &= \underbrace{\left\langle \tilde{G}^0_\alpha(T_\Box), \frac{\mathbf{q}_\alpha}{T_\alpha V_\alpha}\right\rangle}_{=-\frac{1}{\tau}\sum_\alpha \frac{(\mathbf{q}_\alpha)^2}{T_\alpha^2}} + \frac{1}{\tau}\sum_\alpha \frac{T_\alpha}{\tau}\left(\frac{\mathbf{q}_\alpha}{T_\alpha}\right)^2 = 0.\end{aligned} \tag{24.74}$$

and the overall energy is conserved as well. In summary, the model for hyperbolic heat conduction can be reduced to a model for Fourier heat conduction, where the relaxation parameter plays the role of heat conductivity. In the following Section, we derive an Eulerian approach to hyperbolic heat conduction model, which does not need any SPH operators.

24.5.6 Hyperbolic Heat Conduction—Eulerian Approach

The model for hyperbolic heat conduction in the preceding Section required the discrete SPH gradient in the definition of the Lagrangian conjugate entropy flux, $\mathbf{w}_\alpha$. However, there are several versions of the SPH gradients that we might have chosen (Violeau 2012). Here, we proceed in an Eulerian way that does not need any beforehand specification of the SPH gradient.

We add to the SPH state variables the Eulerian phonon momentum density attached to an SPH particle, $\boldsymbol{\pi}_\alpha$, which is related to the conjugate entropy flux by $\boldsymbol{\pi}_\alpha = \mathbf{w}_\alpha S_\alpha$. The aim is to pair it with entropy via a modification of the Poisson bracket, where we take inspiration from the above knowledge of coupling particle momenta with particle entropy, in particular from the Poisson bracket (24.82) with the mass-based volume. We consider the natural coupling between particle mass density and momentum as described in (24.82) and we add a coupling between the auxiliary flux $\boldsymbol{\pi}_\alpha$ and entropy s_α as in the phonon Poisson bracket (Pavelka et al. 2018):

$$
\begin{aligned}
\{F, G\}^{(\mathrm{SPH-m},\pi)} &= \{F, G\}^{(\mathrm{SPH})} \\
&+ \sum_\alpha \sum_\beta m_\beta W'_{\alpha\beta} e_{i\alpha\beta} \left(F_{\rho_\alpha}(G_{M_{\alpha i}} - G_{M_{\beta i}}) - G_{\rho_\alpha}(F_{M_{\alpha i}} - F_{M_{\beta i}})\right) \\
&+ \sum_\alpha \sum_\beta S_\alpha \frac{m_\beta}{m_\alpha} W'_{\alpha\beta} e_{i\alpha\beta} \left(F_{s_\alpha}(G_{\pi_{\alpha i}} - G_{\pi_{\beta i}}) - G_{s_\alpha}(F_{\pi_{\alpha i}} - F_{\pi_{\beta i}})\right).
\end{aligned}
\tag{24.75}
$$

After the transformation to the total momentum $\bar{\mathbf{M}}_\alpha = \mathbf{M}_\alpha + \boldsymbol{\pi}_\alpha$, the bracket turns to

$$
\begin{aligned}
\{F, G\}^{(\mathrm{SPH-}\bar{\mathrm{M}},\pi)} &= \{F, G\}^{(\mathrm{SPH-mass})} \\
&+ \sum_\alpha \sum_\beta S_\alpha \frac{m_\beta}{m_\alpha} W'_{\alpha\beta} e_{i\alpha\beta} \left(F_{s_\alpha}(G_{\pi_{\alpha i}} - G_{\pi_{\beta i}}) - G_{s_\alpha}(F_{\pi_{\alpha i}} - F_{\pi_{\beta i}})\right),
\end{aligned}
\tag{24.76}
$$

that leads to an extension of evolution equation (24.33),

$$\dot{\mathbf{x}}_\alpha = E_{\bar{\mathbf{M}}_\alpha} \tag{24.77a}$$

$$\dot{\bar{\mathbf{M}}}_\alpha = -E_{\mathbf{x}_\alpha} - \sum_\beta m_\alpha m_\beta \left(\frac{E_{\rho_\alpha}}{m_\alpha} + \frac{E_{\rho_\beta}}{m_\beta} + S_\alpha \frac{1}{m_\alpha^2} E_{s_\alpha} + S_\beta \frac{1}{m_\beta^2} E_{s_\beta}\right) W'_{\alpha\beta} \mathbf{e}_{\alpha\beta} \tag{24.77b}$$

$$\dot{\rho}_\alpha = \sum_\beta m_\beta W'_{\alpha\beta} \mathbf{e}_{\alpha\beta} \cdot (E_{\bar{\mathbf{M}}_\alpha} - E_{\bar{\mathbf{M}}_\beta}) \tag{24.77c}$$

$$\dot{s}_\alpha = \frac{S_\alpha}{m_\alpha} \sum_\beta m_\beta W'_{\alpha\beta} \mathbf{e}_{\alpha\beta} \cdot (E_{\bar{\mathbf{M}}_\alpha} - E_{\bar{\mathbf{M}}_\beta}) + \sum_\beta S_\alpha \frac{m_\beta}{m_\alpha} \left(E_{\boldsymbol{\pi}_\alpha} - E_{\boldsymbol{\pi}_\beta}\right) \cdot W'_{\alpha\beta} \mathbf{e}_{\alpha\beta} \tag{24.77d}$$

$$\dot{\boldsymbol{\pi}}_\alpha = -\sum_\beta \left(S_\alpha \frac{m_\beta}{m_\alpha} E_{s_\alpha} + S_\beta \frac{m_\alpha}{m_\beta} E_{s_\beta}\right) W'_{\alpha\beta} \mathbf{e}_{\alpha\beta}. \tag{24.77e}$$

This set of evolution equations can be used to model a moving fluid with hyperbolic heat conduction.

24.6 Conclusion

In this paper, we compare five approaches toward SPH for nonbarotropic fluids, that is, for SPH with entropy. There are four ways to define the discrete particle volume (based on the mass, entropy, a direct definition, or an implicit definition), and five ways to introduce the entropy (corresponding to the various particle volumes and to a mixed version). SPH with entropy conserves the total energy (dumping the kinetic energy into the internal energy). We formulate all five approaches as the sum of a symplectic part (reversible) and a conservative dissipative part (irreversible, raising the total entropy).

The approach toward entropy using the mass-based particle volume, used, for instance, in the SDPD method (Español and Revenga 2003), is perhaps the easiest to use because it does not change the usual SPH evolution equations, except for letting the pressure depend on the entropy. However, in cases with large inhomogeneities in entropy, the method does not provide sufficient detail.

The approach with the entropic volume improves the resolution of entropy density. However, it loses some details of the mass-density profile. Moreover, it is sensitive to negative particle entropies, which cause negative particle volumes.

The mixed approach combines the advantages of both the mass-based and the entropy-based particle volumes. On the other hand, the use of two-particle volumes makes the approach ambiguous, and one should pay attention to the extra entropic terms in the evolution equations.

Finally, the implicit definition of the mass density and entropy density should be the most precise for interpolation. However, the price to pay is that the matrix inversion has to be carried out at each time step, which causes problems, especially at the boundaries.

These various approaches toward entropy in SPH are illustrated in the free expansion of an ideal gas (purely Hamiltonian evolution and Joule expansion), Rayleigh-Bénard convection without the Boussinesq approximation, and hyperbolic heat conduction.

In the future, we would like to apply these approaches toward entropy in numerical simulations of superfluid helium-4, where temperature waves (referred to as the second sound) are important when compared with experiments (Barenghi et al. 2014; Van Sciver 2012; Mongioví et al. 2018).

Acknowledgements OK, MP, and VK were supported by the Czech Grant Agency (grant number 23-05736S). MP is a member of the Nečas Centre for Mathematical Modeling.

24.7 MaxEnt Derivation of the SPH Energy

The SPH energy (24.20) is an approximation of the exact continuum energy (24.4), and here we show that it can be obtained by means of the principle of maximum entropy (MaxEnt) (Jaynes 1967). The former depends on the SPH state variables, which are obtained by projection (24.7) from the state variables of the Lagrangian continuum mechanics. Therefore, the SPH energy expresses less information than the continuum energy.

With a definition of energy on a more detailed level, one can identify the least biased estimate of the energy on the lower (less detailed) level by maximizing the detailed entropy constrained by the knowledge of the less detailed state variables (Shannon 1948; Jaynes 1967). To this end, we use the analogue of the MaxEnt principle, the principle of minimum energy, which is equivalent to MaxEnt for concave entropies (Callen 1960). The energy on the less detailed level of description (here SPH) follows from a double usage of the Legendre transform or, alternatively, from one Legendre transform and the use of the connection between the two levels (also called a projection) (Grmela et al. 2015). Thence, we seek minimizers of the

$$\phi(\mathbf{x}, \mathbf{M}; \mathbf{x}^\dagger_\alpha, \mathbf{M}^\dagger_\alpha) = -E^{Lagrangian}(\mathbf{x}, \mathbf{M}) \\ + \langle(\pi_1(\mathbf{x}, \mathbf{M}), \pi_2(\mathbf{x}, \mathbf{M})), (\mathbf{x}^\dagger_\alpha, \mathbf{M}^\dagger_\alpha)\rangle. \quad (24.78)$$

The solution $\mathbf{M}(\mathbf{M}^\dagger_\alpha)$ to

$$\frac{\delta\phi}{\delta M_i} = 0$$

reads $M_i = \rho_0 V_{0\alpha}\bar{\chi}_\alpha M^\dagger_{\alpha,i}$ while the solution $\mathbf{x}(\mathbf{x}^\dagger_\alpha, \mathbf{M}^\dagger_\alpha)$ to

$$\frac{\delta\phi}{\delta x_i} = 0$$

is given by a condition

$$\frac{\partial\epsilon}{\partial x_i} = \det\left(\frac{\partial\mathbf{x}}{\partial\mathbf{X}}\right)\bar{\chi}_\alpha x^\dagger_{\alpha,i}.$$

The reduced conjugate energy is

$$E^{SPH,\dagger} = \phi(\mathbf{x}(\mathbf{x}^\dagger_\alpha, \mathbf{M}^\dagger_\alpha), \mathbf{M}(\mathbf{x}^\dagger_\alpha, \mathbf{M}^\dagger_\alpha); \mathbf{x}^\dagger_\alpha, \mathbf{M}^\dagger_\alpha). \quad (24.79)$$

For the particle-momentum contribution to the energy at the SPH level, we need to find the relation $\mathbf{M}_\alpha(\mathbf{M}^\dagger_\alpha)$ to identify the reduced energy that then follows from the contribution of the momentum to the conjugate SPH energy $\sum_\alpha = \frac{m_\alpha}{2}(\mathbf{M}^\dagger_\alpha)^2$. Then, from the projection and $\mathbf{M}(\mathbf{M}^\dagger_\alpha)$, we get that the momentum contribution to the SPH energy is $\sum_\alpha \frac{1}{m_\alpha}(\mathbf{M}_\alpha)^2$. The internal energy contribution to ϕ is approximated as follows

$$\int d\mathbf{X}x^{\dagger}_{\alpha,i}\left(\frac{\mathbf{x(X)}}{V_{0\alpha}} - \left(\frac{\partial\epsilon}{\partial x_i}\right)^{-1}\epsilon\right) = \frac{1}{V_{0\alpha}}\int_{V_{0\alpha}} d\mathbf{X}x^{\dagger}_{\alpha,i}\mathbf{x(X)} - V_\alpha\epsilon$$
$$= x^{\dagger}_{\alpha,i}x_{\alpha,i} - \frac{1}{V_{0\alpha}}\int_{V_{0\alpha}} d\mathbf{X}V_\alpha\epsilon \approx x^{\dagger}_{\alpha,i}x_{\alpha,i} - V_\alpha\epsilon_\alpha, \quad (24.80)$$

where we introduced $\epsilon_\alpha(\rho_\alpha, s_\alpha) = \frac{1}{V_{0\alpha}}\int_{V_{0\alpha}} d\mathbf{X}\epsilon(\rho(\mathbf{X}), s(\mathbf{X})$. Therefore we have

$$E^{SPH} = \sum_\alpha \frac{\mathbf{M}_\alpha^2}{2m_\alpha} + \sum_\alpha V_\alpha\epsilon_\alpha(\rho_\alpha, s_\alpha) \quad (24.81)$$

as the Legendre transform of the conjugate SPH energy which removes $x^{\dagger}_{\alpha,i}x_{\alpha,i}$ and flips the sign. If the internal energy does not explicitly depend on the position of the particle, then this energy reduces to the SPH energy (24.20). In other words, the SPH energy (24.14) can be seen as the MaxEnt estimate of the energy subject to knowledge of the SPH state variables.

24.8 Derivation of the Poisson Bracket for the Entropic SPH

This section contains details of the derivation of the Poisson brackets for the mass-volume, entropic-volume, mixed-volume, and implicit-volume approaches.

24.8.1 Mass-Based Volume Poisson Bracket

Consider an arbitrary functional $F(\mathbf{x}_\alpha, \mathbf{M}_\alpha, \rho_\alpha, s_\alpha)$ of particle positions, momenta, densities, and entropy densities as defined in the main text, Eq. (24.21) with the particle volume $V_\alpha = V_\alpha^m$ as defined in Eq. (24.23). By plugging Equation (24.7) and (24.21) for particle mass and entropy density into bracket (24.1), we obtain the Poisson bracket for the mass-based volume approach,

$$\{F, G\}^{(\text{SPH}-\text{mass})} - \{F, G\}^{(\text{SPH})} \quad (24.82)$$
$$+\sum_\alpha\sum_\beta m_\beta W'_{\alpha\beta}e_{i\alpha\beta}\left(F_{\rho_\alpha}(G_{M_{\alpha i}} - G_{M_{\beta i}}) - G_{\rho_\alpha}(F_{M_{\alpha i}} - F_{M_{\beta i}})\right)$$
$$+\sum_\alpha\sum_\beta s_\alpha\frac{m_\beta}{m_\alpha}W'_{\alpha\beta}e_{i\alpha\beta}\left(F_{s_\alpha}(G_{M_{\alpha i}} - G_{M_{\beta i}}) - G_{s_\alpha}(F_{M_{\alpha i}} - F_{M_{\beta i}})\right).$$

This Poisson bracket leads to the Hamiltonian equation (24.33).

24.8.2 Entropic-Based Volume Poisson Bracket

Consider an arbitrary functional $F(\mathbf{x}_\alpha, \mathbf{M}_\alpha, \rho_\alpha, s_\alpha)$ of particle positions, momenta, densities, and entropy densities as defined in the main text, Eq. (24.21) with the particle volume $V_\alpha = V_\alpha^s$ as defined in Eq. (24.24). By plugging Eq. (24.7) and (24.21) for particle mass and entropy density into bracket (24.1), we obtain the Poisson bracket for the entropic-based volume approach,

$$\begin{aligned}\{F, G\}^{(\mathrm{SPH-entropic})} &= \{F, G\}^{(\mathrm{SPH})} \\ &+ \sum_\alpha \sum_\beta S_\beta W'_{\alpha\beta} e_{i\alpha\beta} \left(F_{s_\alpha}(G_{M_{\alpha i}} - G_{M_{\beta i}}) - G_{s_\alpha}(F_{M_{\alpha i}} - F_{M_{\beta i}})\right) \\ &+ \sum_\alpha \sum_\beta m_\alpha \frac{S_\beta}{S_\alpha} W'_{\alpha\beta} e_{i\alpha\beta} \left(F_{\rho_\alpha}(G_{M_{\alpha i}} - G_{M_{\beta i}}) - G_{\rho_\alpha}(F_{M_{\alpha i}} - F_{M_{\beta i}})\right).\end{aligned} \tag{24.83}$$

24.8.3 Mixed-Volume Poisson Bracket

Consider now an arbitrary functional $F(\mathbf{x}_\alpha, \mathbf{M}_\alpha, \rho_\alpha, s_\alpha)$ of positions, momenta, densities, and entropy densities with the latter defined by Eq. (24.28). Before plugging such functionals into the Lagrangian Poisson bracket (24.1), let us compute the derivatives of the functional. The derivative with respect to the position becomes

$$\begin{aligned}\frac{\delta F}{\delta x^i(\mathbf{X})} &= \sum_\alpha \frac{\partial F}{\partial x^i_\alpha} \bar{\chi}_\alpha(\mathbf{X}) + \sum_\alpha \sum_\gamma \frac{\partial F}{\partial \rho_\alpha} \frac{\partial \rho_\alpha}{\partial x^j_\gamma} \frac{\delta x^j_\gamma}{\delta x^i(\mathbf{X})} \\ &+ \sum_\alpha \sum_\gamma \frac{\partial F}{\partial s_\alpha} \frac{\partial s_\alpha}{\partial x^j_\gamma} \frac{\delta x^j_\gamma}{\delta x^i(\mathbf{X})} \\ &= \sum_\alpha \frac{\partial F}{\partial x^i_\alpha} \bar{\chi}_\alpha(\mathbf{X}) + \sum_\alpha \frac{\partial F}{\partial \rho_\alpha} \sum_\beta \sum_\gamma m_\beta W'_{\alpha\beta} e_{i\alpha\beta} (\delta_{\alpha\gamma} - \delta_{\beta\gamma}) \bar{\chi}_\gamma(\mathbf{X}) \\ &+ \sum_\alpha \frac{\partial F}{\partial s_\alpha} \sum_\beta \sum_\gamma S_\beta W'_{\alpha\beta} e_{i\alpha\beta} (\delta_{\alpha\gamma} - \delta_{\beta\gamma}) \bar{\chi}_\gamma(\mathbf{X}).\end{aligned} \tag{24.84}$$

Plugging such two functionals into Poisson bracket (24.1) then leads to

$$\begin{aligned}\{F, G\}^{(\mathrm{SPH-mixed})} &= \{F, G\}^{(SPH)} \\ &+ \sum_\alpha \sum_\beta \sum_\gamma \sum_\delta F_{\rho_\alpha} m_\beta W'_{\alpha\beta} e_{i\alpha\beta} (\delta_{\alpha\gamma} - \delta_{\beta\gamma}) G_{M_{\delta i}} \int d\mathbf{X} V_{0\delta} \bar{\chi}_\gamma(\mathbf{X}) \bar{\chi}_\delta(\mathbf{X}) \\ &- \overset{F\leftrightarrow G}{\ldots}\end{aligned}$$

$$
\begin{aligned}
&+\sum_\alpha\sum_\beta\sum_\gamma\sum_\delta F_{s_\alpha} S_\beta W'_{\alpha\beta} e_{i\alpha\beta}(\delta_{\alpha\gamma}-\delta_{\beta\gamma}) G_{M_{\delta i}} \int d\mathbf{X} V_{0\delta}\bar{\chi}_\gamma(\mathbf{X})\bar{\chi}_\delta(\mathbf{X}) \\
&-\overset{F\leftrightarrow G}{\cdots} \\
&=\{F,G\}^{(SPH)} \\
&+\sum_\alpha\sum_\beta\sum_\gamma F_{\rho_\alpha} m_\beta W'_{\alpha\beta} e_{i\alpha\beta}(\delta_{\alpha\gamma}-\delta_{\beta\gamma}) G_{M_{\gamma i}} - \overset{F\leftrightarrow G}{\cdots} \\
&+\sum_\alpha\sum_\beta\sum_\gamma F_{s_\alpha} S_\beta W'_{\alpha\beta} e_{i\alpha\beta}(\delta_{\alpha\gamma}-\delta_{\beta\gamma}) G_{M_{\gamma i}} - \overset{F\leftrightarrow G}{\cdots} \\
&=\{F,G\}^{(SPH)} \\
&+\sum_\alpha\sum_\beta m_\beta W'_{\alpha\beta} e_{i\alpha\beta}\left(F_{\rho_\alpha}(G_{M_{\alpha i}}-G_{M_{\beta i}}-G_{\rho_\alpha}(F_{M_{\alpha i}}-F_{M_{\beta i}})\right) \\
&+\sum_\alpha\sum_\beta S_\beta W'_{\alpha\beta} e_{i\alpha\beta}\left(F_{s_\alpha}(G_{M_{\alpha i}}-G_{M_{\beta i}})-G_{s_\alpha}(F_{M_{\alpha i}}-F_{M_{\beta i}})\right), \qquad (24.85)
\end{aligned}
$$

which is the Poisson bracket for the approach toward the entropic SPH with the mixed volume. This Poisson bracket then leads to the reversible evolution equation (24.46).

24.8.4 Implicit-Volume Poisson Bracket

Consider now an arbitrary functional $F(\mathbf{x}_\alpha, \mathbf{M}_\alpha, \rho_\alpha, s_\alpha)$ of particle positions, momenta, densities, and entropy densities as defined in the main text, Eq. (24.21) with the particle volume $V_\alpha = \tilde{V}^I_\alpha$ as defined in Eq. (24.26). Before plugging such functionals into the Lagrangian Poisson bracket (24.1), let us compute the derivatives of the functional. The derivative with respect to the position becomes

$$
\begin{aligned}
\frac{\delta F}{\delta x^i(\mathbf{X})} &= \sum_\alpha \frac{\partial F}{\partial x^i_\alpha}\bar{\chi}_\alpha(\mathbf{X}) + \sum_\alpha \frac{\partial F}{\partial \rho_\alpha}\frac{\delta \rho_\alpha}{\delta x^i(\mathbf{X})} + \sum_\alpha \frac{\partial F}{\partial s_\alpha}\frac{\delta s_\alpha}{\delta x^i(\mathbf{X})} \\
&= \sum_\alpha \frac{\partial F}{\partial x^i_\alpha}\bar{\chi}_\alpha(\mathbf{X}) + \sum_\alpha \frac{\partial F}{\partial \rho_\alpha}\sum_\beta\sum_\gamma \frac{m_\beta}{\rho_\beta}\frac{\rho_\alpha^2}{m_\alpha} W^{-1}_{\alpha\gamma} W'_{\gamma\beta} e_{i\delta\gamma}(\bar{\chi}_\gamma - \bar{\chi}_\beta) \\
&+ \sum_\alpha \frac{\partial F}{\partial s_\alpha}\sum_\beta\sum_\gamma \frac{S_\beta}{s_\beta}\frac{s_\alpha^2}{S_\alpha} W^{-1}_{\alpha\gamma} W'_{\gamma\beta} e_{i\delta\gamma}(\bar{\chi}_\gamma - \bar{\chi}_\beta), \qquad (24.86)
\end{aligned}
$$

where the partial derivatives $\frac{\partial \rho_\alpha}{\delta x^i(\mathbf{X})}$ follow from the implicit definition of volume $1 = \sum_\beta \tilde{V}^I_\alpha W_{\alpha\beta}$ noting that $\tilde{V}^I_\alpha = \frac{m_\alpha}{\rho_\alpha} = \frac{S_\alpha}{s_\alpha}$ for all α. Hence

$$
0 = -\frac{m_\delta}{\rho_\delta^2}\frac{\delta \rho_\delta}{\delta x^i(\mathbf{X})} + \sum_\alpha\sum_\beta\sum_\gamma \frac{m_\beta}{\rho_\beta} W^{-1}_{\delta\alpha} W'_{\alpha\beta} e_{i\alpha\beta}(\delta_{\alpha\gamma}-\delta_{\beta\gamma})\bar{\chi}_\gamma(\mathbf{X})
$$

and thus

$$\frac{\delta \rho_\alpha}{\delta x^i(\mathbf{X})} = \sum_\beta \sum_\gamma \frac{m_\beta}{\rho_\beta} \frac{\rho_\alpha^2}{m_\alpha} W^{-1}_{\alpha\gamma} W'_{\gamma\beta} e_{i\delta\gamma} (\bar{\chi}_\gamma - \bar{\chi}_\beta).$$

Similarly, one obtains the expression used for $\frac{\delta s_\alpha}{\delta x^i(\mathbf{X})}$ in the implicit-based volume case. Note that we again assumed that the Lagrangian particle mass and entropy are independent of the particle positions (although S_α can change in time due to irreversible effects).

Plugging such two functionals into the Poisson bracket (24.1) then leads to the Poisson bracket for the implicit-based volume approach toward the entropic SPH

$$\begin{aligned}\{F, G\}^{(\text{SPH-implicit})} = \{F, G\}^{(SPH)} \\ +\sum_\alpha \sum_\beta \sum_\delta F_{\rho_\alpha} \frac{m_\beta \rho_\alpha^2}{\rho_\beta m_\alpha} W^{-1}_{\alpha\delta} W'_{\delta\beta} e_{i\delta\beta} (G_{M_i^\delta} - G_{M_i^\beta}) \\ +\sum_\alpha \sum_\beta \sum_\delta F_{s_\alpha} \frac{S_\beta s_\alpha^2}{s_\beta S_\alpha} W^{-1}_{\alpha\delta} W'_{\delta\beta} e_{i\delta\beta} (G_{M_i^\delta} - G_{M_i^\beta}) \\ -\sum_\alpha \sum_\beta \sum_\delta G_{\rho_\alpha} \frac{m_\beta \rho_\alpha^2}{\rho_\beta m_\alpha} W^{-1}_{\alpha\delta} W'_{\delta\beta} e_{i\delta\beta} (F_{M_i^\delta} - F_{M_i^\beta}) \\ -\sum_\alpha \sum_\beta \sum_\delta G_{s_\alpha} \frac{S_\beta s_\alpha^2}{s_\beta S_\alpha} W^{-1}_{\alpha\delta} W'_{\delta\beta} e_{i\delta\beta} (F_{M_i^\delta} - F_{M_i^\beta}).\end{aligned} \tag{24.87}$$

24.9 Thermodynamics

In this Section, we recall relations between the ideal gas model and the model of stiffened gas (Callen 1960; Dumbser et al. 2016).

As we are using the volumetric densities of energy and entropy, let us also show a useful identity used in the main body of this manuscript. Temperature, pressure, and chemical potential are defined as the corresponding derivatives of energy with respect to the total entropy, volume, and mass,

$$\mathrm{d}E = T\mathrm{d}S - p\mathrm{d}V + \mu\mathrm{d}M. \tag{24.88}$$

Then, due to the extensivity, the Euler 1-homogeneous theorem gives that

$$E = TS - pV + \mu M, \tag{24.89}$$

and when this equation is divided by the volume, we get

$$\epsilon = Ts - p + \mu\rho. \tag{24.90}$$

Be differentiation of Equation (24.89), we obtain the Gibbs-Duhem formula

$$0 = s\mathrm{d}T - \mathrm{d}p + \rho\mathrm{d}\mu. \tag{24.91}$$

By differentiation of Equation (24.90), we get that

$$\mathrm{d}\epsilon = T\mathrm{d}s + \mu\mathrm{d}\rho. \tag{24.92}$$

Equation (24.90) can be then also rewritten as

$$p = -\epsilon + \frac{\partial \epsilon}{\partial s}s + \frac{\partial \epsilon}{\partial \rho}\rho. \tag{24.93}$$

24.9.1 Ideal Gas

The fundamental thermodynamic relation, which gives the equilibrium energy as a function of entropy, volume, and for an ideal gas reads

$$\epsilon = \frac{\rho^\gamma}{\gamma - 1}e^{\frac{s}{\rho c_V}} \tag{24.94}$$

where ϵ is the volumetric energy density, s is the volumetric entropy density, and $\gamma = \frac{c_P}{c_V}$ is the ratio of heat capacities. From this equation, it follows that

$$p = \rho^\gamma e^{\frac{s}{c_V \rho}} = (\gamma - 1)c_V \rho T \tag{24.95a}$$

$$T = \frac{\rho^{\gamma-1}}{c_V(\gamma - 1)}e^{\frac{s}{c_V \rho}} \tag{24.95b}$$

$$\mu = \frac{\gamma\rho^{\gamma-1}}{\gamma - 1}e^{\frac{s}{c_V \rho}} - \frac{\rho^{\gamma-2}}{\gamma - 1}e^{\frac{s}{c_V \rho}}\frac{s}{c_V} \tag{24.95c}$$

and $\epsilon = \rho c_V T$. In particular, by comparing the formulas for pressure and temperature, the equation of state $pV = nRT$ recovered, where n is the number of moles, $R = M_{mol}(c_P - c_V)$ is the gas constant, M_{mol} is the molar mass, and T is the temperature.

24.9.2 Stiffened Gas

The fundamental thermodynamic relationship of the stiffened gas is a simplification of the Grüneisen equation of state (Mie 1903; Grüneisen 1912), which generalizes the equation of state of the ideal gas to liquids (for example water) and solids (certain metals) (Harlow and Amsden 1971; Menikoff and Plohr 1989). The fundamental thermodynamic relation reads

$$\epsilon = \rho \left(\frac{c_0^2}{\gamma(\gamma-1)} \left(\frac{\rho}{\rho_0}\right)^{\gamma-1} e^{\frac{s}{c_V \rho}} + \frac{\rho_0 c_0^2 - \gamma p_0}{\gamma \rho} \right) \tag{24.96}$$

where c_0 is a reference speed of sound, ρ_0 is a reference density, and p_0 is a reference pressure. The pressure, temperature, and chemical potential are then calculated as the corresponding derivatives of energy with respect to the total entropy, volume, and mass density,

$$p =(\gamma-1)\epsilon - (\rho_0 c_0^2 - \gamma p_0) \tag{24.97a}$$

$$T =\frac{c_0^2}{c_V \gamma(\gamma-1)} \left(\frac{\rho}{\rho_0}\right)^{\gamma-1} e^{\frac{s}{c_V \rho}} \tag{24.97b}$$

$$\mu =\frac{c_0^2}{\gamma-1} \left(\frac{\rho}{\rho_0}\right)^{\gamma-1} e^{\frac{s}{\rho c_V}} - \rho \frac{c_0^2}{\gamma(\gamma-1)} \left(\frac{\rho}{\rho_0}\right)^{\gamma-1} \frac{s}{\rho^2 c_V} \tag{24.97c}$$

In particular, if we set $\rho = \rho_0$ and $p = p_0$, we get $s = 0$ and that the speed of sound becomes c_0. Moreover, the equation of state reads

$$p = (\gamma-1)\rho c_V T - \frac{\rho_0 c_0^2 - \gamma p_0}{\gamma}. \tag{24.98}$$

References

Antuono M, Colagrossi A, Marrone S, Molteni D (2010) Comput Phys Commun 181(3):532, https://doi.org/10.1016/j.cpc.2009.11.002. https://www.sciencedirect.com/science/article/pii/S0010465509003506

Antuono M, Marrone S, Colagrossi A, Bouscasse B (2015) Comput Methods Appl Mech Eng 289:209, https://doi.org/10.1016/j.cma.2015.02.004. https://www.sciencedirect.com/science/article/pii/S0045782515000456

Barenghi CF, Skrbek L, Sreenivasan KR (2014) Proc Natl Acad Sci USA 111:4647

Bénard H (1900) Revue Générale des Sciences Pures et Appliquées (in French) 11:1261

Boussinesq J (1897) Théorie de l'écoulement tourbillonnant et tumultueux des liquides dans les lits rectilignes a grande section. No. díl 1 in Théorie de l'écoulement tourbillonnant et tumultueux des liquides dans les lits rectilignes a grande section (Gauthier-Villars et fils). https://books.google.cz/books?id=sbpAAQAAMAAJ

Callaway J (1959) Phys Rev, 113

Callen H (1960) Thermodynamics: an introduction to the physical theories of equilibrium thermostatics and irreversible thermodynamics. Wiley, http://books.google.cz/books?id=mf5QAAAAMAAJ

Cleary PW, Monaghan JJ (1999) J Comput Phys 148(1):227, https://doi.org/10.1006/jcph.1998.6118. https://www.sciencedirect.com/science/article/pii/S0021999198961186

Cullen L, Dehnen W (2010) Monthly Notices Royal Astron Soc 408(2):669

Dreyer W, Struchtrup H (1993) Continuum Mech Thermodyn 5:3

Dumbser M, Peshkov I, Romenski E, Zanotti O (2016) J Comput Phys 314:824

Ellero M, Español P, Flekkoy EG (2003) Phys Rev E 68(041504)

Español P, Revenga M (2003) Phys Rev E 67:026705

Fecko M (2006) Differential geometry and lie groups for physicists. Cambridge University Press

Gingold R, Monaghan J (1977) Mon Not R Astron Soc 181(3):375

Godunov S, Mikhailova T, Romenskii E (1996) Siberian Math J 37(4):690

Goldstein H (2002) Classical mechanics. Pearson Education

Grmela M, Klika V, Pavelka M (2015) Phys Rev E 92(032111)

Grmela M, Öttinger HC (1997) Phys Rev E 56:6620, https://doi.org/10.1103/PhysRevE.56.6620

Grüneisen E (1912) Annalen der Physik 344(12):257

Hairer E, Lubich C, Wanner G (2013) Geometric numerical integration: structure-preserving algorithms for ordinary differential equations. Springer Series in Computational Mathematics (Springer Berlin Heidelberg), https://books.google.cz/books?id=cPTxCAAAQBAJ

Harlow F, Amsden AA (1971) Fluid dynamics, vol. Monograph LA-4700 (Los Alamos Scientific Laboratory, Los Alamos, N. Mex.)

Inutsuka SI (2002) J Comput Phys 179(1):238

Jaynes ET (1967) Delaware Seminar in the foundation of physics, M. Bunge ed. (Springer New York), chap. Foundations of probability theory and statistical mechanics

Jeong J, Jhon M, Halow J, van Osdol J (2003) Comput Phys Commun 153(1):71, https://doi.org/10.1016/S0010-4655(03)00155-3. https://www.sciencedirect.com/science/article/pii/S0010465503001553

Jiang F, Sousa A (2006) Numer Heat Transfer Part B 50:499. https://doi.org/10.1080/10407790600646677

Jou D, Casas-Vázquez J, Lebon G (2010) Extended irreversible thermodynamics, 4th edn. Springer, New York

Kincl O, Pavelka M (2023) Comput Phys Commun 284:108593. https://doi.org/10.1016/j.cpc.2022.108593. https://www.sciencedirect.com/science/article/pii/S0010465522003125

Kincl O, Pavelka M (2025) Smoothedparticles.jl, https://github.com/OndrejKincl/SmoothedParticles.jl. Please cite this software using these metadata

Kincl O, Peshkov I, Pavelka M, Klika V (2023) Appl Math Comput 439:127579, https://doi.org/10.1016/j.amc.2022.127579. https://www.sciencedirect.com/science/article/pii/S0096300322006531

Klika V, Pavelka M, Vágner P, Grmela M (2019) Entropy 21(715). https://doi.org/10.3390/e21070715

Kovács R, Ván P (2015) Int J Heat Mass Transfer 83:613, https://doi.org/10.1016/j.ijheatmasstransfer.2014.12.045. http://www.sciencedirect.com/science/article/pii/S0017931014011533

Kutta W (1901) Zeitschrift für Mathematik und Physik 46:435

Landau L, Lifschitz E (1969) Statistical physics. No. pt. 1 in Course of theoretical physics. Pergamon Press

Landau L, Lifshitz E, Kosevich A, Pitaevskii L (1986) Theory of elasticity. Course of theoretical physics (Butterworth-Heinemann)

Lebon G, Jou D, Vázquez J (2008) Understanding non-equilibrium thermodynamics: foundations, applications. Springer e-Books (Springer, London, Limited, Frontiers. SpringerLink

Leimkuhler B, Reich S, Press CU (2004) Simulating hamiltonian dynamics. Cambridge Monographs on Applie (Cambridge University Press). https://books.google.cz/books?id=tpb-tnsZi5YC

Lucy LB (1977) Astron J 82:1013. https://doi.org/10.1086/112164

Malyshev A, Romenskii E (1986) Sib Math J 27(5):734

Marsden J, Ratiu TS (1999) Introduction to mechanics and symmetry, texts in applied mathematics 17, vol, 2nd edn. Springer, New York

McNelly T, Rogers S, Channin D, Rollefson R, Goubau W, Schmidt G, Krumhansl J, Pohl R (1970) Phys Rev Lett 24:100

Menikoff R, Plohr BJ (1989) Rev Mod Phys 61:75, https://doi.org/10.1103/RevModPhys.61.75

Mie G (1903) Annalen der Physik 316(8):657

Monaghan JJ (2005) Reports Progress Phys 68(8):1703, https://doi.org/10.1088/0034-4885/68/8/R01

Monaghan J (1992) Ann Rev Astr Astroph 30:543

Mongioví MS, Jou D, Sciacca M (2018) Phys Reports 726:1, https://doi.org/10.1016/j.physrep.2017.10.004. https://www.sciencedirect.com/science/article/pii/S0370157317303472

Öttinger H (2005) Beyond equilibrium thermodynamics. Wiley, New York

Öttinger HC, Grmela M (1997) Phys Rev E 56:6633. https://doi.org/10.1103/PhysRevE.56.6633

Pavelka M, Klika V, Grmela M (2020) J Statist Phys 181(1):19

Pavelka M, Klika V, Grmela M (2018) Multiscale thermo-dynamics. de Gruyter, Berlin

Pavelka M, Peshkov I, Klika V (2020) Physica D: Nonlinear Phenomena 408(132510)

Peierls R (1955) Quantum theory of solids. Oxford University Press

Peshkov I, Pavelka M, Romenski E, Grmela M (2018) Continuum Mech Thermodyn 30(6):1343

Quinlan NJ, Basa M, Lastiwka M (2006) Int J Numer Methods Eng 66(13):2064, https://doi.org/10.1002/nme.1617

Rayleigh L (1916) Philos Mag, 6th series 32(192):529

Runge C (1895) Mathematische Annalen 46(2):167. https://doi.org/10.1007/BF01446807

Shannon CE (1948) Bell Syst Techn J 27:379

Sýkora M, Pavelka M, Restuccia L, Jou D (2023) Physica Scripta 98(10):105234, https://doi.org/10.1088/1402-4896/acf418

Van Sciver S (2012) Helium cryogenics. International cryogenics monograph series. Springer, New York

Verlet L (1967) Phys Rev 159:98, https://doi.org/10.1103/PhysRev.159.98

Violeau D (2012) Fluid mechanics and the SPH method: theory and applications. Oxford University Press, Oxford, UK

Chapter 25
Martensitic Nucleation—A Phonon Thermodynamics Approach

Oliver Kastner, Nizan Mizrahi, and Roni Z. Shneck

In memory of Wolfgang Dreyer—Brilliant thermodynamicist and fiery academic critic

Abstract We investigate the stability of a nonlinear atomistic model for austenite and martensite lattices in 2D. In the harmonic limit, where the atomitic interaction forces are linearized so as to decouple the atomic Equations of Motion, the model incorporates four independent branches of the dispersion relation due to longitudinal and transversial waves in both optical and accoustic vibration modes. Most part of their frequency spectra have real values, but the transverse acoustic wave mode also includes imaginary frequencies. These indicate unstable shear waves, known as soft modes, which lead to the formation of martensitic microtwins. Switching to the quantum mechanical picture, the harmonic spectra are interpreted as a Bose gas, known as phonons. We consider this gas as a mixture of two species: stable phonons associated with real frequencies and soft mode phonons associated with imaginary frequencies. Calorical equations of state for these species can be defined for (local) thermal equilibria according to the Bose-Einstein distribution. These allow us to evaluate the phase equilibrium temperature T_E between austenite and martensite by comparing the respective free energy densities of these lattice types, as is customary. We then provide a preliminary argument for the stability of the *soft mode* phonon gas in an austenitic lattice. We show that the density of states functions for the austenitic

O. Kastner (✉)
Elastokalorik Ltd, Berlin, Germany
e-mail: kastner@elastokalorik.com

N. Mizrahi · R. Z. Shneck
Department of Materials Engineering, Ben Gurion University of the Negev, Negev, Israel
e-mail: nizan.mizrahi@gmail.com

R. Z. Shneck
e-mail: roni@bgu.ac.il

J. Fuhrmann et al. (eds.), *Advances in Continuum Physics*, Advanced Structured Materials 238, https://doi.org/10.1007/978-3-031-93918-1_25

phonons species result in free energy densities that may coincide at the temperature $M_s < T_E$. Stable phonons are energetically preferred above M_s, while unstable soft mode phonons are favored below. Accordingly, we interpret M_s as the stability limit for martenite nucleation within the austenite.

Keywords Martensitic transformation · Martensite nucleation · Lattice stability · Phonon thermodynamics

25.1 Introduction

Martensitic transformations are first-order solid-state phase transitions between distinct crystal lattices. They are characterized by coordinated atomic displacements and lead to the formation of regularly twinned structures through nucleation and growth (Otsuka and Wayman 1998; Bhattacharya 2003). This work focuses on the nucleation process of martensite (M) in an ideally homogeneous and infinite (single-crystalline) austenite (A). We refer to martensite as the product phase and austenite as the parent phase of the transformation.

The thermodynamic phase equilibrium is determined by the specific free enthalpies, $g_{(A,M)} = h_{(A,M)} - Ts_{(A,M)}$ of the two phases (A,M). These functions combine the respective enthalpies $h_{(A,M)}$ and entropies $s_{(A,M)}$, with the entropy terms weighted by the temperature T (Müller and Seelecke 2001). These quantities refer to the *isolated* phases. At the phase equilibrium temperature T_E, the two enthalpy functions $g_{A,M}$ coincide, as illustrated schematically in Fig. 25.1b.

In the groundstate $T \to 0$, the specific enthalpy of the austenite is higher than that of the martensite, making the latter energetically favorable. However, the specific entropy—-represented by the the slopes of the individual graphs—is larger for austenite. As a result, the two free enthalpy functions may intersect at the phase equilibrium temperature T_E. Below this value, $g_M(T)$ is minimal, meaning martensite is the stable phase. Above T_E, the situation is reversed, and austenite becomes the stable phase. This behaviour is well known as *entropic stabilisation* of the austinitic lattice upon heating (Müller and Weiss 2005).

Laboratory experiments indicate that the thermodynamic phase equilibrium temperature should be understood as a theoretical reference value. In cyclic cooling/heating experiments conducted in calorimetry devices, martensite nucleates upon cooling at the martensite-start temperature $M_S < T_E$, while the austenite nucleates upon heating at the austenite-start temperature $A_S > T_E$, as illustrated in Fig. 25.1a.

Martensitic transformations occur as dynamical processes in situations of global lattice instability. The relevant length and time scales pertain to the atomic microlevel, which hampers in-situ observations with today's state-of-the-art laboratory equipment. Therefore, our understanding of the transformation dynamics relies on dynamical simulations of atomic model lattices. Specifically, in this work, we employ a simple yet comprehensive binary lattice model that has been shown to capture the kinematic and dynamic characteristics of martensitic transformations in real

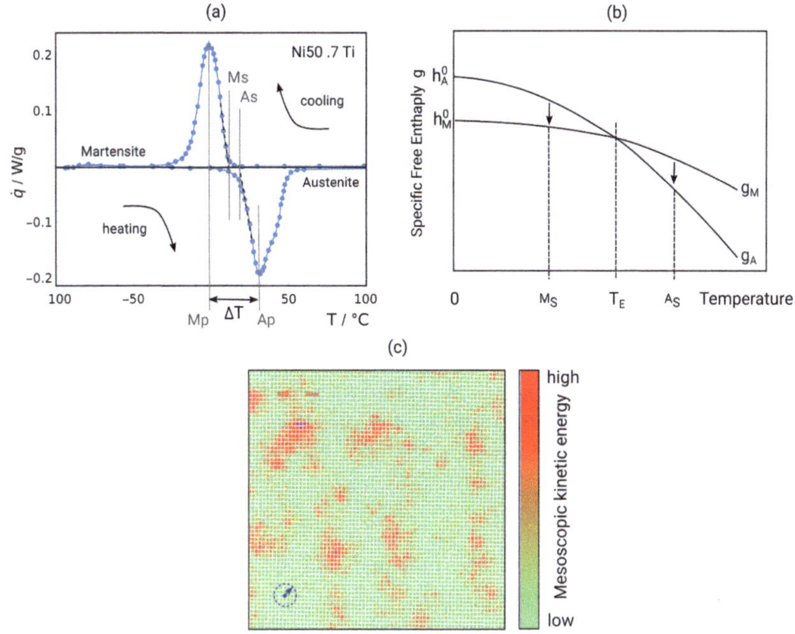

Fig. 25.1 **a** Example of differential scanning calorimetry (DSC) data indicating the nucleation temperatures of martensite M_s upon cooling and austenite A_s upon heating (Data Burow (2010)). **b** Schematic representation of the specific free enthalpy of martensite M and austenite A over the temperature T for constant external load. The specific potential groundstate enthalpies $h^0_M < h^0_A$ of the two phases are indicated. The specific entropies $s_{A,M}$ are defined by the respective slope magnitudes; $s_A > s_M$. **c** MD simulation Snapshot of mesoscale kinetic energy fluctuations in undercooled model austenite. The size of an atomic interaction neighborhood is indicated by the dashed circle. The colour code indicates spots of elevated atomic coordination which indicate nucleation embryos that may develop into stable nuclei

materials qualitatively, while conceptually reducing the geometric complexity to 2D (Kastner 2003, 2006; Kastner and Eggeler 2009; Kastner et al. 2011; Kastner 2012). The model lattice is shown in Fig. 25.2. It incorporates two generic atom species, A and B, to construct the lattices. A square lattice structure is used as a model for the austenite. The square lattice may transform by shear and shuffle deformation into hexagonal lattice structures, modeling the martensite. Due to the presence of various shear and shuffle modes, the model also incorporates martensite twinning. The atomic interaction forces are described by central pair potentials of the Lennard-Jones type, which involve only six constant interaction parameters. The specific model parameter set used here is taken from (Kastner and Shneck 2015). Using this lattice model, the lattice stability can be investigated through molecular dynamics (MD) simulations, where macroscopic variables (temperature T, externally induced stress or strain) can be controlled using numerical thermostats and barostats. Thus, situations of global

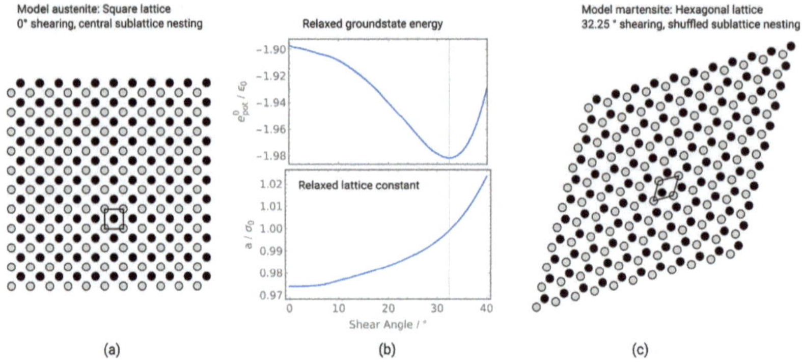

Fig. 25.2 A binary lattice model in 2D with Lennard-Jones interactions comprising six interaction parameters. These parameters may be fixed at values that result in the groundstate energy shown in (**b**). The transformation from **a** to **c** involves shear of unit cells and shuffle of the sublattices as described (Kastner 2012)

lattice instability can be numerically modeled, and the nucleation dynamics can be studied in atomic detail Kastner (2025).

Our dynamical lattice model reveals how the nucleation of a martensitic lattice transformation proceeds: Atoms within the lattice cannot transfer individually from the parent phase to the product phase, as they are strongly interacting with their lattice neighbors. The interaction forces are confined to an effective interaction range of only a few lattice constant lengths; for our 2D model material, approximately 50 atoms lie within this range, making such interaction neighborhoods mesoscopic entities. Successful nucleation of a martensitic particle requires momentarily coordinated motions among neighboring atoms to open the transformation path that interconnects the parent and product domains in the phase space. MD simulations show that this coordination occurs dynamically *before* the product lattice is formed Kastner and Shneck (2015). We can detect such events in MD simulations, since they are accompanied by local build-ups of kinetic energy at the mesoscopic scale of the involved interaction neighborhoods, as shown in Fig. 25.1c. Mesoscopic regions with elevated coordinated activity may be interpreted as nucleation embryos, which have the potential to develop into stable martensite nuclei. However, the dynamical simulations reveal that most of these nucleation embryos dissolve back into the matrix, while only a few may develop into stable martensite plates. Dissolving embryos are regarded as unstable nuclei, while the developing ones are considered stable nuclei.

This contribution proposes a thermodynamic criterion for the formation of stable martensite nuclei. We show that, within the harmonic approximation, the dispersion relation of the austenitic model lattice incorporates two types of vibrational modes: real and imaginary frequency modes. Real frequencies correspond to stable harmonic waves, while imaginary frequencies indicate lattice instability. These unstable modes can be identified as the *soft modes*, which drive the transformation. In this context, the lattice instability of the austenite is associated with the presence of soft mode lattice waves.

The paper proceeds by adopting the quantum mechanical approach and interprets the frequency spectrum of the lattices in the harmonic limit as "phonons", consisting of discrete energy quanta $\hbar\omega(\mathbf{k})$ and corresponding momenta $\hbar\mathbf{k}$, with the wave vector domain confined to the Brillouin zone of the lattice structure under consideration. In this context, we distinguish between two species of phonon gases: stable harmonic modes (associated with real frequencies) and soft harmonic modes (associated with imaginary frequencies). For both species, we assume thermal equilibrium, where the phase space densities follow the Bose-Einstein distribution as a function of temperature. This enables us to compute the caloric equations of state for the two phonon gas species as moments of the phase space densities, based on their respective dispersion relations, as functions of temperature. We propose that for the stable phonon gas, the phase stability of austenite/martensite can be evaluated based on the free energy densities of these lattice phases, yielding the phase equilibrium temperature T_E. We then separately introduce a preliminary criterion for the phase stability of the soft mode phonons in the phonon gas mixture of the austenitic lattice. This phonon gas phase stability criterion allows the evaluation of the soft mode stability as a function of temperature, with the stability limit interpreted as the M_s temperature, at which nucleation occurs.

25.2 Harmonic Lattice Dynamics

In the harmonic approximation, the atomic interaction forces are linearized about the ground-state lattice positions to decouple the differential equation system of the atomic displacements $u_i^\alpha(t)$, cf. Appendix 25.5.1. The result is given by

$$m_\alpha \ddot{u}_i^\alpha = \sum_{\gamma=1}^{N} A_{ij}^{\alpha\gamma} u_j^\gamma(t) \quad . \tag{25.1}$$

Here, $A_{ij}^{\alpha\gamma}$ are the interaction coefficients, which can be interpreted as the spring constants of virtual elastic springs fixed between individual atom pairs in the ground-state configuration. For the square lattice structure (model austenite), these coefficients may have different signs, depending on whether the respective pairing experiences attractive or repulsive interaction forces. Analytical solutions of the linearized Equations of Motion (25.1) are obtained by employing harmonic wave ansatz functions for the atomic displacements $\mathbf{u}^\gamma(t)$:

$$u_j^\gamma(t) = \bar{u}_j^\gamma(\mathbf{k}) \exp\left[i(k_l X_l^\gamma - \omega t)\right] \quad ,$$

where $\bar{u}_j^\gamma$ denote the (complex) amplitudes of the atomic displacements about the equilibrium positions X_l^γ. k_l are the components of the wave vector and ω is the angular frequency. This leads to a characteristic equation for the eigenvalues ω^2 of the

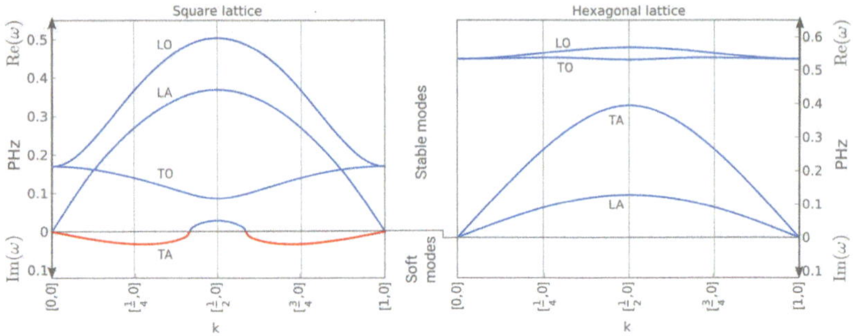

Fig. 25.3 The four modes of the dispersion relation $\omega_P(\mathbf{k})$ for the square lattice (model austenite) and the hexagonal lattice (model martensite). Note that the vertical axis of the diagram is split into an upper part with axes directed upwards, indicating real dispersions, and a lower part with axes directed downwards, indicating imaginary dispersions (red colour line). The wave vector direction $\mathbf{k} = [0...1, 0]$ points along a selected edge of the Brillouin zone; cf. Figs. 25.8 and 25.9 in the Appendix for contour plots of the entire Brillouin zone

dynamic matrix $D_{ij}^{\alpha\gamma}(\mathbf{k}) = A_{ij}^{\alpha\gamma} \exp(ik_l(X_l^\gamma - X_l^\alpha))$. By considering the geometric periodicity of crystal lattices in 2D, the dimension of this matrix can be effectively reduced to two times the number of atoms per primitive cell, which in our case is also two. This results in the manifold of $P = 1 \ldots 4$ eigenvalue modes $\omega_P^2(\mathbf{k})$. Their positive roots are known as the dispersion relations $\omega_P(\mathbf{k})$. Figure 25.3 shows the four modes for the square lattice (model austenite) and the hexagonal lattice (model martensite). The wave vector direction $\mathbf{k} = [0...1, 0]$ points along selected edges of the respective Brillouin zones. The two diagrams are characteristic and indicate the four branches of the dispersion relations, which are known as the transverse and longitudinal optical (TO, LO) and transverse and longitudinal acoustic (TA, LA) frequencies. For completeness, Appendix Sect. 25.5.3 Figs. 25.8 and 25.9 (pages 19, 20) show the respective contour plots of the dispersion relations for the entire Brillouin zones.

Note that in the diagrams of Fig. 25.3, the vertical diagram axes are split into an upper part with upward axes (indicating real dispersions) and a lower part with downward axes (indicating imaginary dispersions). We observe that the model austenite exhibits imaginary frequencies in parts of the transverse-acoustic frequency modes $\omega_{\mathrm{TA}}(\mathbf{k})$. The related branches are indicated by a red line color. For the model martensite, a corresponding signature is absent.

According to the wave ansatz, imaginary frequencies cause exponential amplitude growth with time, leading to the instability of the square lattice. These imaginary frequencies arise from negative eigenvalues of $\omega_{\mathrm{TA}}^2(\mathbf{k})$. For the square lattice, such eigenvalues can exist because the ground-state energy e_{pot}^0 of this structure is concave, as shown in the energy plot of Fig. 25.2b for zero shear angle. This indicates that some pair-force coefficients $A_{ij}^{\alpha\gamma}$ are negative.

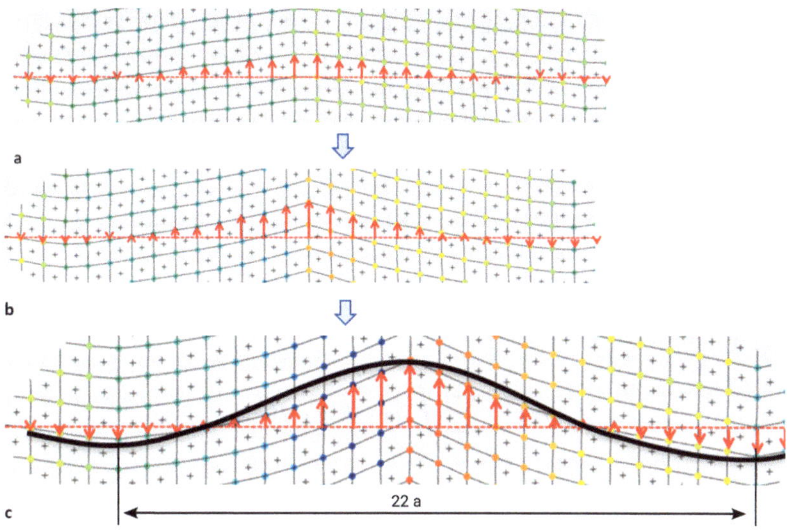

Fig. 25.4 Example of three consecutive MD simulation snapshots a-b-c as observed during martensite nucleation

With the current parameterization of the underlying Lennard-Jones model, this is not the case for the hexagonal lattice. Accordingly, martensite exhibits a convex ground-state energy e^0_{pot}, as shown in Fig. 25.2b.

Figure 25.4 shows three lattice snapshots during martensite nucleation obtained from MD simulation. The nucleation process is indicated by shear deformation of the austenitic lattice, which develops symmetrically on either side of a selected invariant line. The deformation progresses in a wave-like manner, with a wavelength of $\lambda = 22\,a$ as indicated, and it propagates in the lateral direction. Eventually, the wave forms a stable martensitic microtwin lamella.

The observed shear transformation wave approximates a harmonic TA wave with a wave vector $\mathbf{k} \approx [\pm 0.03, 0]$. Inspection of the dispersion relation for the square lattice in Fig. 25.3 confirms that the corresponding TA branch lies in the imaginary domain. We conclude that the imaginary TA modes of the linearized model may correspond to soft mode waves capable of inducing martensite nucleation.

25.3 Phonon Gas Thermodynamics

25.3.1 Macroscopc Fields

Now, we may switch from classical mechanics to quantum mechanics. Accordingly, we consider the frequency spectra of the dispersion relations of the linearized model as quasi-particle "phonons". An individual phonon is defined by its vibrational frequency according to the dispersion relation $\omega_P(\mathbf{k})$ of mode P. The quantum of an energy state, $\hbar\omega_P(\mathbf{k})$, is associated with an elementary lattice excitation. For large crystals, the wave vectors populate the respective Brillouin zones sufficiently densely, allowing the application of quantum statistical arguments for the calculation of macroscopic thermodynamic quantities (Solyom 2007; Gurevich 1986; Srivastava 1990).

We consider the densities of the $(\mathbf{x}, \mathbf{k})$-phase space for the dispersion mode P, denoted by $f_P(\mathbf{x}, \mathbf{k})$. The moments of these density functions define macroscopic quantities. We restrict our attention to the following fields:

- the phonon gas energy density,

$$e = \sum_P \int \hbar\omega_P(\mathbf{k}) f_P \, d\mathbf{k} \quad , \tag{25.2}$$

(omitting the constant groundstate term "1/2")

- the phonon gas entropy density

$$s = -k_B \sum_P \int \left[\log \frac{f_P}{Y} - \left(1 + \frac{Y}{f_P}\right) \log\left(1 + \frac{f_P}{Y}\right) \right] f_P \, d\mathbf{k} \quad . \tag{25.3}$$

The entropy density of equation (25.3) makes use of the corrected quantum statistics. Y denotes the total number of quantized microstates available within a phase space element ($d\mathbf{x}d\mathbf{k}$). The ratio f_P/Y hence represents the probability density of finding a phonon in such a microstate. At thermal equilibrium "E", this quantity takes the form of a Bose-Einstein gas,

$$\frac{f_P^E}{Y} = \frac{1}{\exp\left(\frac{\hbar\omega_P(\mathbf{k})}{k_B T}\right) - 1} \quad . \tag{25.4}$$

It may be used to compute the phonon gas energy density from equation (25.2) and the phonon gas entropy density from equation (25.3) at thermal equilibrium. We consider this equilibrium state as "local" equilibrium. The resulting energy and entropy densities are denoted by $e(T)$ and $s(T)$. These functions read

$$e(T) = \sum_P \int \hbar\omega_\lambda(\mathbf{k}) \frac{Y}{\exp\left(\frac{\hbar\omega^P(\mathbf{k})}{k_B T}\right) - 1} \, d\mathbf{k} \quad , \tag{25.5}$$

$$s(T) = -k_B \sum_P \int \left[\log\left(\frac{1}{\exp\left(\frac{\hbar\omega^P(\mathbf{k})}{k_B T}\right) - 1} \right) - \exp\left(\frac{\hbar\omega^P(\mathbf{k})}{k_B T}\right) \log\left(1 + \frac{1}{\exp\left(\frac{\hbar\omega^P(\mathbf{k})}{k_B T}\right) - 1}\right) \right] \frac{Y}{\exp\left(\frac{\hbar\omega^P(\mathbf{k})}{k_B T}\right) - 1} \, d\mathbf{k} \quad . \tag{25.6}$$

With the dispersion relations $\omega^P(\mathbf{k})$ being computed from the linearized Equations of Motion, the thermodynamic fields $e(T)$ and $s(T)$ can be calculated from these equations. The integrations rely on numerical discretizations of the phase space. Technically, it is convenient to transform the integration domain from **k**-space into ω-space for this purpose, which involves related density-of-states functions, see Appendix 25.5.2.

For the model austenite, both real and imaginary branches of the dispersion relation exist. Accordingly, we consider a phonon gas mixture consisting of two phonon species: one phonon gas of stable phonons with real frequencies (subscript "ST"), and another of unstable phonons with imaginary frequencies (subscript "SM"). Each species is associated with its respective thermodynamic fields as functions of temperature: the phonon-gas energy densities, $e_{\mathrm{ST}}(T)$ and $e_{\mathrm{SM}}(T)$, and their entropies, $s_{\mathrm{ST}}(T)$ and $s_{\mathrm{SM}}(T)$.

For the model martensite, there are only real branches of the dispersion relation, and therefore only stable phonons. The resulting fields are shown in Fig. 25.11 in Appendix Sect. 25.5.3 (page 21) for both phonon-gas species and both lattice structures.

25.3.2 Austenite/Martensite Phase Equilibrium

We proceed to calculate the thermodynamic equilibrium temperature T_E based on the phononic equations of state. For this purpose, we consider the stable phonons of the square and hexagonal lattices and calculate their respective caloric density functions.

The energy densities of austenite (denoted as $e_A(T)$) and martensite (denoted as $e_M(T)$) must incorporate the corresponding groundstate values $e_{\mathrm{pot}}^{0,A}$ and $e_{\mathrm{pot}}^{0,M}$. The atom-specific value of the groundstate energy e_{pot}^{0} is plotted in Fig. 25.2 as a function of the shear angle; the relevant values for the two phases can be obtained from the diagram for shear angles of 0° and 32°, respectively. We consider two atoms per primitive lattice cell of size a^2. Thus, $e_{\mathrm{pot}}^{0,A} = \frac{2e_{\mathrm{pot}}^{0}(0^\circ)}{a(0^\circ)^2}$ and $e_{\mathrm{pot}}^{0,M} = \frac{2e_{\mathrm{pot}}^{0}(32^\circ)}{a(32^\circ)^2}$ are the respective groundstate densities. With these values, the energy densities are given by

$$\begin{aligned} e_A(T) &= e_{\mathrm{pot}}^{0,A} + e_{\mathrm{ST}}^{A}(T) \\ e_M(T) &= e_{\mathrm{pot}}^{0,M} + e_{\mathrm{ST}}^{M}(T) \quad , \end{aligned} \tag{25.7}$$

where $e_{\mathrm{ST}}^{A}(T)$ and $e_{\mathrm{ST}}^{M}(T)$ represent the caloric contributions, which are functions of temperature T and are shown as black solid lines in Figs. 25.11 (a) and (b), respectively.

The free energy densities for austenite and martensite are then given by

$$\begin{aligned} f_A(T) &= e_A(T) - T s_{\mathrm{ST}}^{A}(T) \\ f_M(T) &= e_M(T) - T s_{\mathrm{ST}}^{M}(T) \quad , \end{aligned} \tag{25.8}$$

where $s_{\mathrm{ST}}^{A}(T)$ and $s_{\mathrm{ST}}^{M}(T)$ are the entropy densities, as represented by the black lines in Fig. 25.11(c) and (d), respectively.

We consider a phase mixture of austenite and martensite with partial volumes V_A and V_M, such that $V = V_A + V_M$ is the total volume of the crystal, assumed to be constant; external loads are not considered. Hence, the partial volumes depend on one another, and we may select V_M, say, as a variable for the characterization of the phase mixture. The total free energy of the mixture is given by

$$F(T, V_M) = (V - V_M) f_A(T) + V_M f_M(T) \quad . \tag{25.9}$$

Equilibrium states are indicated by a minimum in the total free energy. This incorporates the following necessary condition for phase equilibrium between austenite and martensite:

$$\frac{\partial F(T, V_M)}{\partial V_M} = 0 \quad \rightarrow \quad f_A(T) = f_M(T) \tag{25.10}$$

Regarding the phase-specific free energy densities of equation (25.8), this condition can be evaluated, see Fig. 25.5. We see that the groundstate energy density of martensite is lower than that of austenite. From the slopes of the graphs, we observe that the phononic entropy density of austenite has a larger magnitude than that of martensite, as expected. Hence, the two functions may coincide at the thermodynamic

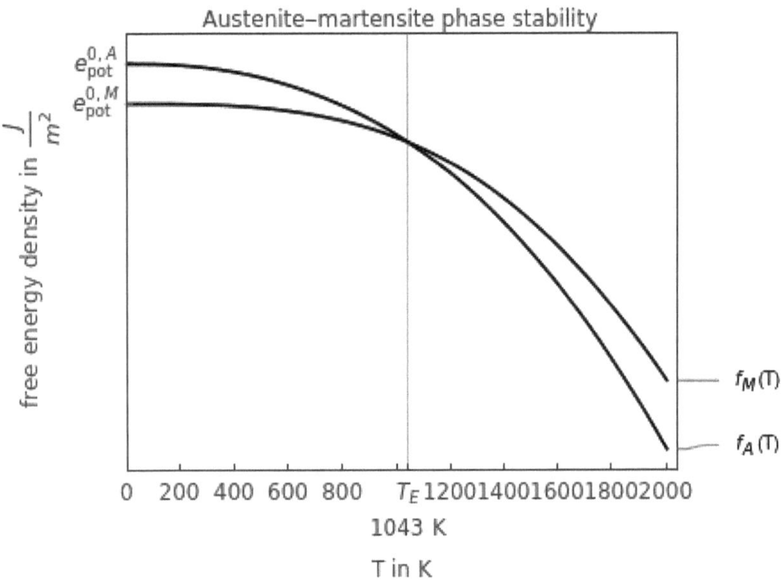

Fig. 25.5 Free energy densities of the model austenite and the model martensite as functions of temperature, based on the groundstate energies of these phases and the caloric state functions due to the stable phonon dispersion branches of the square and the hexagonal lattice structures, respectively. The thermodynamic phase equilibrium temperature is indicated

A/M phase equilibrium temperature T_E. For the model material, the numerical value is read off the diagram at $T_E = 1043$ K. This value agrees acceptably with alternative (approximate) calculations, see e.g. (Kastner et al. 2011).[1]

25.3.3 Phonon Gas Phase Equilibrium of the Model Austenite

In order to evaluate the stability of soft mode phonons in the square lattice as a function of temperature T, we suggest a criterion similar to the one of austenite/martensite phase equilibrium. This is motivated by the fundamental idea that the phonon gas of the square lattice is constituted as a phonon gas phase *mixture* consisting of both stable phonons and soft mode phonons, which reside in the same Brillouin Zone. We suggest that the vibrational pattern of the crystal may be decided by that phonon gas species, which allows minimizing the thermal free energy of the phonon gas mixture. We proceed to formulate a related equilibrium condition for the phonon gas mixture. In a leap of imagination, we may assume that the two phonon gas species reside in the partial volumes V_{ST} and V_{SM} of the austenitic crystal, such that $V = V_{\mathrm{ST}} + V_{\mathrm{SM}}$ is its total volume, which is assumed constant.

The thermal energy densities of the two species are $e_{\mathrm{ST}}(T)$ and $e_{\mathrm{SM}}(T)$, and the entropy densities are $s_{\mathrm{ST}}(T)$ and $s_{\mathrm{SM}}(T)$, respectively. Their combinations yield the free energy densities of stable and soft-mode phonon gas species, according to

$$\begin{aligned} f_{\mathrm{ST}}(T) &= e_{\mathrm{ST}}(T) - T s_{\mathrm{ST}}(T) \\ f_{\mathrm{SM}}(T) &= e_{\mathrm{SM}}(T) - T s_{\mathrm{SM}}(T) \end{aligned} \tag{25.11}$$

We regard the groundstate energies of the stable and the soft modes as identical because they refer to the same (square) lattice structure. Therefore, they are neglected here. The total free energy of the phonon gas mixture hence reads

$$F = (V - V_{\mathrm{SM}})\, f_{\mathrm{ST}}(T) + V_{\mathrm{SM}} f_{\mathrm{SM}}(T) \quad . \tag{25.12}$$

An equilibrium state is indicated by minimal total free energy, hence the condition for phonon gas stability implies

$$\frac{\partial F}{\partial V_{\mathrm{SM}}} = 0 \quad \rightarrow \quad f_{\mathrm{ST}}(T) = f_{\mathrm{SM}}(T) \quad . \tag{25.13}$$

[1] Referring to Fig. 25.7: Transformation temperature in this reference, using the value $\varepsilon_{AB} = 0.17$ for the Lennard-Jones interaction parameter modeling the strength of atom species A-B interactions used here.

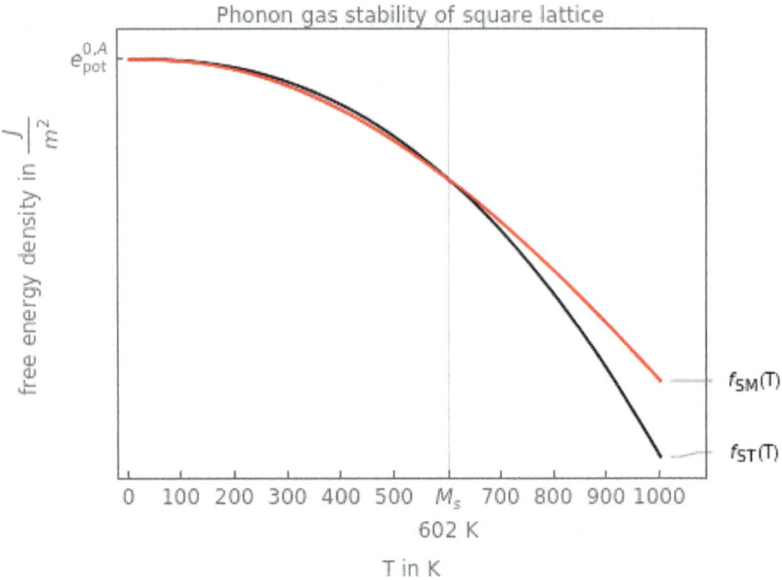

Fig. 25.6 Free energy densities of the stable phonon gas (black line) and the imaginary soft phonon gas (red line)

We may evaluate this condition by plotting the two functions $f_{ST}(T)$ and $f_{SM}(T)$ into the same diagram, see Fig. 25.6. We see that at elevated temperature $T > M_S$, the stable phonon gas exhibits smaller free energy density than the soft mode phonon gas. In this regime, the soft mode phonons are unstable, and accordingly, no martensite nuclei can be formed. At $T = M_S$, the two free energy densities coincide; at this temperature, the *phonon gas phase equilibrium* between stable phonons and soft mode phonons is established. Hence, soft mode waves may progress to full extent so as to form stable martensite nuclei. In this regard, it is plausible to interpret the phonon gas phase equilibrium temperature as the martensite-start temperature M_s. Below M_S, the soft mode free energy provides minimal free energy density and accordingly, the lattice is fully unstable. Existing martensite nuclei may grow, and new nuclei may form. Eventually, the crystal is fully transformed into martensite.

We may note that the temperature of $M_s = 602$ K thus found indeed approximates the nucleation temperature observed during MD simulations of cooling experiments well. More importantly, we see that this temperature is also *below* the austenite/martensite phase equilibrium temperature, $M_S < T_E$, which resembles the need for undercooling with respect to T_E in order to nucleate the martensite.

25.4 Discussion

In 1947, Zener (1947) suggested that a displacive transformation is set by an anomalous decrease of the shear modulus of the parent (austenitic) phase when the temperature decreases. Such softening was indeed found experimentally in many materials (e.g. (Venkataraman 1979; Nakanishi et al. 1982), as well as an anomalous increase in the dielectric constants of many oxides, which also indicated softening of the internal restoring forces near a critical transition temperature, usually designated as M_s (Cowley and Bruce 1980; Cowley 1980). Neutron scattering experiments directly showed that the frequency of certain phonons sharply decreases near M_s, where they become "soft modes". Molecular dynamics simulations reveal that austenite has higher lattice entropy than martensite as a result of geometrically broadened atomic interaction potentials. Broader potential energy translates into broader atomic mobilities about the mean lattice sites and consequently, the occupied phase space domains are more extended for the austenite than for the martensite at identical temperature. This effect is resembled by the phononic free energies of the stable modes presented in Sect. 25.3.2. Here, the atomic mobility is reflected by the vibrational density of states, with richer vibrational patterns in the model austenite than in the martensite.

The dynamic stabilization of low-T stable modes at elevated temperatures is often assumed to be a result of anharmonicity. There is significant experimental evidence for anharmonic effects that become significant at high temperatures, e.g. (Petry et al. 1991; Trubitsin and Dolgusheva 2007; Trubitsin and Ostanin 2004; Li et al. 2015). Harmonic wave modes do not interact, but this is not generally true for waves that may result from the anharmonic terms present in the real interaction forces, cf. Eq. (25.19) in the Appendix. These anharmonic terms may affect wave-wave interactions, such that individual wave modes may both be dynamically created or annihilated. Because the related wave vectors must always reside in the Brillouin Zone before and after the interaction, two types of nonlinear wave-wave interactions are distinguished in the literature, known as "normal" and "umklapp" processes in quantum mechanics (Solyom 2007; Gurevich 1986; Srivastava 1990). It is plausible to assume that interactions of this type are responsible for the annihilation of soft mode waves at elevated temperatures. Still, even above M_s, the soft mode waves represent mechanically allowed states and therefore, they are always attempted to form. However, for temperatures above M_s, their mean lifetime must be shorter than 2π over their frequency, because during this time period, a soft mode wave may extend to its full wavelength and become effective. Mizrahi (2022) investigated the dispersion relation of the (non-linear) model lattice used in the present work by means of time-windowed Fourier analysis of MD simulation data. Using a methodology developed earlier by de Koker (2009), he was able to identify soft phonons above M_s as stable oscillatory modes with short free paths.

Our approach is motivated by the picture of a cylinder which encloses an ideal gas by a piston against a pre-loaded snap spring. The spring tends to relax but is kept

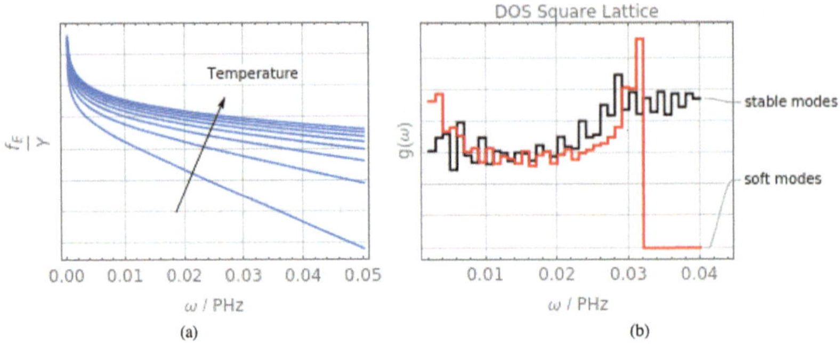

Fig. 25.7 **a** Bose-Einstein distribution. **b** Progression of the density of state populations of stable and soft mode phonons in the square lattice for small frequencies

in place by the ideal gas pressure as a function of temperature. Once the temperature falls below a threshold, the spring may snap in, which represents the nucleation. Putting this picture forward towards phonon gases, we arrive at the stability condition for a phonon gas mixture of stable and soft phonons suggested in this work. Since the criterion is based on linearized equations of motion, it indeed hides the specific character of anharmonic mode-mode interactions; however, it convincingly incorporates the competition between stable and soft modes for the lattice-dominating wave form. This competition, in turn, is modeled as a competition between the free energy densities of the phonon gas species as a function of temperature.

To appreciate this competition, it is necessary to recall that the energy quantum states of a phonon are always populated from low to high energy states. The Bose-Einstein distribution reflects this behavior, as it represents a monotonously decreasing function in the frequency domain. This general behavior is shifted with temperature towards higher frequencies, cf. Figure 25.7a. This augments the low-frequency behavior of the model austenite at low temperature, while the medium- and high-frequency behavior is augmented at elevated temperature. The density of states function of the square lattice indeed exhibits slightly more soft phonons than stable phonons in the low-frequency range, see enlargement in Fig. 25.7b, but also it shows a missing signature in the medium- and high-frequency range, see the full spectrum in Fig. 25.10a. Accordingly, both the energy and the entropy of stable and soft phonons of the model austenite show similar progressions in the low temperature regime, with the soft phonon entropy being slightly higher than the stable phonon one. In the high-T regime, in contrast, only stable phonons can offer frequencies to be activated, and accordingly, the caloric equations divert from one another, see Fig. 25.11a, c.

Combining these functions into the free energies, it is this difference in the density of state functions of stable- and soft-mode phonons, that decides the phonon gas stability: At elevated temperature, the stable modes can activate a frequency spectrum, which does not exist for the soft modes. This affects more available microstates for the stable phonon gas than for the soft phonon gas; more microstates affect higher entropy according to Boltzmann's famous formula. Accordingly, the free energy density of the stable modes is more favorable than the one of the soft modes, favoring the stable modes. Consequently, the square lattice is entropically stabilized *upon heating*. In the low-temperature regime, in contrast, the low-frequency parts of the two density of state functions dominate. Here, the soft-mode phonons offer slightly more available microstates than the stable-mode phonons, which favors the soft-mode entropy over the stable-mode entropy, and in turn, the soft-mode free energy. Therefore, we may say that the soft-mode phonons are entropically stabilized *upon cooling*. Since martensitic nucleation is induced by soft mode waves of the austenitic lattice, we may interpret the stability limit of the soft-mode phonon gas indicated in Fig. 25.6 as the martensite start temperature M_s.

25.5 Appendix

25.5.1 Lattice Dynamics with Central Interaction Forces

In classical mechanics, individual atoms are represented by mass points m_α which move along their trajectories $x_i^\alpha(t)$ under the effect of a force field f_i^α. Since this paper deals with symmetric, square lattices in 2D, we may represent vectors in a Cartesian frame, where the Equations of Motion read

$$m_\alpha \ddot{x}_i^\alpha = f_i^\alpha \quad . \tag{25.14}$$

For an N-atomic assembly in 2 spatial dimensions, these represent an ODE system of $2N$ coupled equations. f_i^α denotes the spatial vector components i of the force field f which acts on atoms $\alpha = 1 \ldots N$[2]. Omitting external forces here, this force field results from the internal interactions between the atoms. We assume that the force acting on an atom α may be represented by the superposition of pairwise interaction forces $f_i^{\alpha\beta}(r_{\beta\alpha})$ which are functions of the pair separations $r_{\beta\alpha}$ between atoms α and β, the magnitude of the related separation vector $r_i^{\beta\alpha} = x_i^\beta - x_i^\alpha$: The pair interaction forces, in turn, may be represented as gradient fields of the central pair interaction potential $\Phi_{\alpha\beta}(r_{\beta\alpha})$,

$$f_i^{\alpha\beta} = -\frac{\partial \Phi_{\alpha\beta}}{\partial x_i^\alpha} = \Phi'_{\alpha\beta} \frac{r_i^{\beta\alpha}}{r_{\beta\alpha}} (1 - \delta^{\alpha\beta}) \quad . \tag{25.15}$$

[2] The summation convention over the spatial vector components is intended.

In this work, we employ central pair potentials of the Lennard-Jones type, see (Kastner 2003, 2012). $\phi'_{\alpha\beta} = \frac{\partial\phi_{\alpha\beta}}{\partial r_{\beta\alpha}}$ denotes the magnitude of the pair interaction force between atoms α and β, which acts into the direction of the separation unit vector $r_i^{\beta\alpha}/r_{\beta\alpha}$. $\delta^{\beta\alpha}$ denotes the generalized Kronecker symbols defined by

$$\delta^{\alpha\beta} = \begin{cases} 1, & \text{if } \alpha = \beta \\ 0, & \text{else} \end{cases} \qquad \alpha, \beta = 1 \ldots N \quad . \tag{25.16}$$

The resulting atomic interaction forces hence read

$$f_i^\alpha = \sum_{\beta=1,\beta\neq\alpha}^{N} f_i^{\alpha\beta} = \sum_{\beta=1,\beta\neq\alpha}^{N} \phi'_{\alpha\beta}\frac{r_i^{\beta\alpha}}{r_{\beta\alpha}} \quad . \tag{25.17}$$

The term with $\delta^{\beta\alpha}$ vanishes upon summation over index β, since self-interactions $\alpha = \beta$ must be excluded.

The force field f_i^α depends non-linearly on the positions of all atoms within the interaction range in the lattice. A Taylor-series expansion of the interaction forces about the atomic ground-state positions X_i^μ, which follow from static relaxation of the interaction forces,

$$f_i^\alpha(x_k^\mu) = 0 \quad \Rightarrow \quad x_k^\mu = X_k^\mu \quad , \mu = 1 \ldots N \quad , \tag{25.18}$$

yields

$$f_i^\alpha \approx \underbrace{f_i^\alpha(X_k^\mu)}_{=0} + \overbrace{\sum_{\gamma=1}^{N} \underbrace{\left(\frac{\partial f_i^\alpha}{\partial x_j^\gamma}\right)_{X_k^\mu}}_{\equiv A_{ij}^{\alpha\gamma}} u_j^\gamma(t)}^{\text{Harmonic approximation}} + \overbrace{\sum_{\gamma=1}^{N}\sum_{\delta=1}^{N} \left(\frac{\partial f_i^\alpha}{\partial x_j^\gamma x_k^\delta}\right)_{X_k^\mu} u_j^\gamma u_k^\delta + \ldots}^{\text{Anharmonic terms}} \tag{25.19}$$

$u_j^\gamma(t)$ denotes the atomic displacement vectors relative to the ground-state positions,

$$u_j^\gamma(t) = x_j^\gamma(t) - X_j^\gamma \quad . \tag{25.20}$$

$A_{ij}^{\alpha\gamma}$ denotes the coefficients of the linear term.

An analytical route toward a solution of the equations of motion (25.14) can be given only for the linear expansion of the forces, which is also known as the harmonic approximation. In this case, the interactions are reduced to the coefficients $A_{ij}^{\alpha\gamma}$. For central pair potentials $\Phi_{\alpha\beta}(r_{\beta\alpha})$, these result in

$$A_{ij}^{\alpha\gamma} = \sum_{\beta=1,\,\beta\neq\alpha}^{N} \left(\left(\phi''_{\alpha\beta} - \frac{\phi'_{\alpha\beta}}{r_{\beta\alpha}} \right) \frac{r_i^{\beta\alpha} r_j^{\beta\alpha}}{r_{\beta\alpha}^2} + \frac{\phi'_{\alpha\beta}}{r_{\beta\alpha}} \delta_{ij} \right)_{X_k^{\mu}} \left(\delta^{\beta\gamma} - \delta^{\alpha\gamma} \right) \quad . \qquad (25.21)$$

We see that these coefficients are due to the superposition of the first and second derivatives of the pair potential, which may be interpreted as elastic constants of virtual springs fixed between individual atom pairs in the ground-state X_k^{μ}.

The linear approximation of the interaction forces decouples the ODE. Solutions for the atomic displacements represent planar harmonic waves according to the ansatz

$$u_j^{\gamma}(t) = \bar{u}_j^{\gamma}(\mathbf{k}) \exp\left[i \left(k_l X_l^{\gamma} - \omega t \right) \right] \quad . \qquad (25.22)$$

$\bar{u}_j^{\gamma}$ denote the (complex) amplitudes of the atomic displacements about their equilibrium positions X_l^{γ}. k_l are the components of the wave vector, such that the expression $k_l X_l^{\gamma}$ represents the scalar product between these two spatial component vectors in a Cartesian frame.

Applying the ansatz (25.22) to the linearised equations of motion yields

$$-\omega^2 \bar{u}_i^{\alpha}(\mathbf{k}) = \sum_{\gamma=1}^{N} \underbrace{\frac{1}{m_\alpha} A_{ij}^{\alpha\gamma} \exp\left[k_l \left(X_l^{\gamma} - X_l^{\alpha} \right) \right]}_{\equiv D_{ij}^{\alpha\gamma}(\mathbf{k})} \bar{u}_j^{\gamma}(\mathbf{k}) \quad . \qquad (25.23)$$

The underbraced quantity is known as the dynamical matrix, denoted by $D_{ij}^{\alpha\gamma}(\mathbf{k})$. It comprises complex exponentials due to the separation vectors $R_l^{\gamma\alpha} = (X_l^{\gamma} - X_l^{\alpha})$ of position pairs in the ground state and the wave vector k_l.

Excluding the trivial solution $\bar{u}_j^{\gamma} = 0$, Equation (25.23) establishes an eigenvalue problem for the eigenvalues ω^2 of the dynamical matrix $D_{ij}^{\alpha\gamma}$. These eigenvalues are obtained from the condition that the matrix $\left(D_{ij}^{\alpha\gamma}(\mathbf{k}) + \omega^2 \delta_{ij}^{\alpha\gamma} \right)$ must be singular, or

$$\mathrm{Det}\left(D_{ij}^{\alpha\gamma}(\mathbf{k}) + \omega^2 \delta_{ij}^{\alpha\gamma} \right) = 0 \quad , \qquad (25.24)$$

which is known as the characteristic equation for the eigenvalues ω^2, and it must be solved for a given wave vector $\mathbf{k}$.

Note that $D_{ij}^{\alpha\gamma}(\mathbf{k})$ shares the dimensions of the coefficient matrix $A_{ij}^{\alpha\gamma}$, which is $2N \times 2N$ for our 2D lattice model. However, evaluating the translational lattice periodicty, the dimension of this matrix effectively reduces to two times the number of atoms per primitive cell, which, in our case, is also two. This results in a manifold of $P = 1 \ldots 4$ eigenvalue modes $\omega_P^2(\mathbf{k})$. Their positive roots are known as the dispersion relations $\omega_P(\mathbf{k})$.

25.5.2 K-Space Discretization, Density of States Function

The moments of the phase density, Eqs. (25.2)–(25.3), incorporate integrations over all vibrational modes present in the crystal lattice. Due to the lattice's translational periodicity, all these vibrational modes may be represented as points located in the first Brillouin Zone (BZ) of the reciprocal lattice. The base vectors of the reciprocal lattice, $\mathbf{e}_i$, are related to the base vectors $\mathbf{E}_i$ of the direct lattice through the operation $\mathbf{E}_i \cdot \mathbf{e}_j = 2\pi\delta_{ij}$. In 2D, the volume of the Brillouin zone reduces to the area size, denoted by A_{BZ}. This can be related to the area size of a primitive cell of the direct lattice, denoted by A_{PC}, as

$$A_{BZ} = \frac{4\pi^2}{A_{PC}} \quad . \tag{25.25}$$

The primitive cell of the model austenite has the size $A_{PC} = a^2$, with a being the lattice constant. Hence, the size of the Brillouin zone is given by $A_{BZ} = \frac{4\pi^2}{a^2}$.

When calculating moments of the phase space density, the calculus can be eased by the use of the density of states function, denoted by $g(\omega)$. To calculate the density of states, the dispersion relation $\omega_P(\mathbf{k})$ is evaluated and binned over incremental $\mathbf{k}$-intervals of the Brillouin Zone according to

$$g(\omega) = \frac{A_{BZ}}{Y_k} \sum_P \sum_{\mathbf{k} \in BZ} \delta\left(\omega - \omega_P(\mathbf{k})\right) \quad , \tag{25.26}$$

where δ denotes the Kronecker delta function, used to count the phonon states with the specific frequency $\omega_P(\mathbf{k})$. The density of states is then normalized by the volume of the Brillouin zone and the number Y_k of $\mathbf{k}$ microstates sampled.

The density of states function hence provides the number of quantum states or phonon modes at a specific vibrational frequency ω. This allows discretization of the phase space prior to integration. With the $\mathbf{k}$-space thus discretized, the integration of a generic function $h(\omega(\mathbf{k}))$ over $\mathbf{k} \in BZ$ transforms into integration over the frequency domain:

$$\int_{BZ} h(\omega(\mathbf{k}))\, d\mathbf{k} = \int_0^\infty h(\omega) g(\omega)\, d\omega \quad . \tag{25.27}$$

25.5.3 Figures

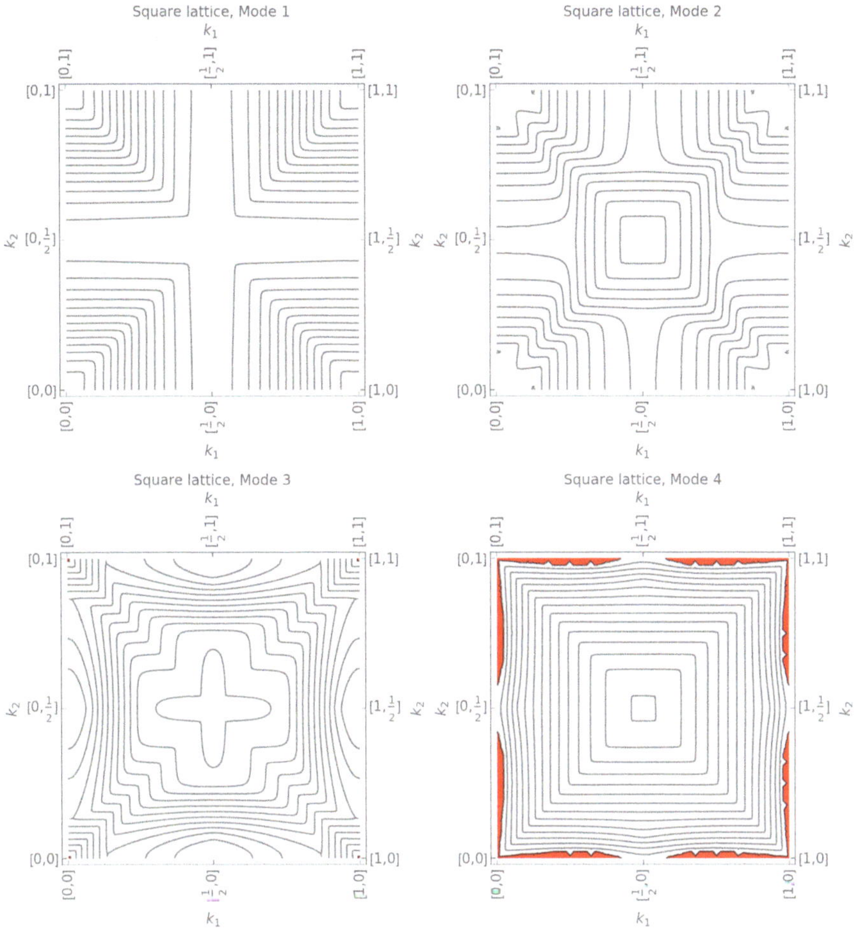

Fig. 25.8 Square lattice (model austenite): Contour plots of the four modes of the dispersion relation $\omega_P(\mathbf{k})$ for wave vectors covering the entire Brillouin Zone, $\mathbf{k} = [0 \ldots 1, 0 \ldots 1]$. Red contour shadings indicate regions where the frequencies are imaginary. Inspection shows that this is the case for mode 4, where imaginary frequencies appear in narrow regions near the edges of the Brillouin zone

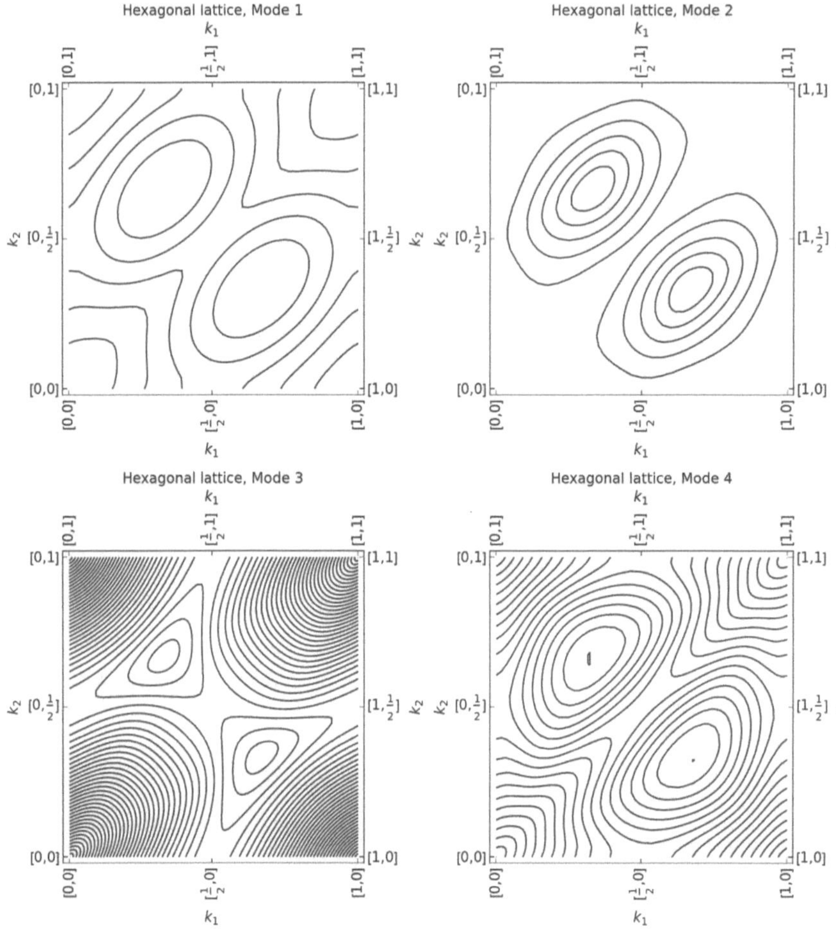

Fig. 25.9 Hexagonal lattice (model martensite): Contour plots of the four modes of the dispersion relation $\omega_P(\mathbf{k})$ for wave vectors covering the entire Brillouin Zone, $\mathbf{k} = [0 \ldots 1, 0 \ldots 1]$. For the current parameterization of the underlying Lennard-Jones interaction potentials, all modes exhibit entirely real frequencies

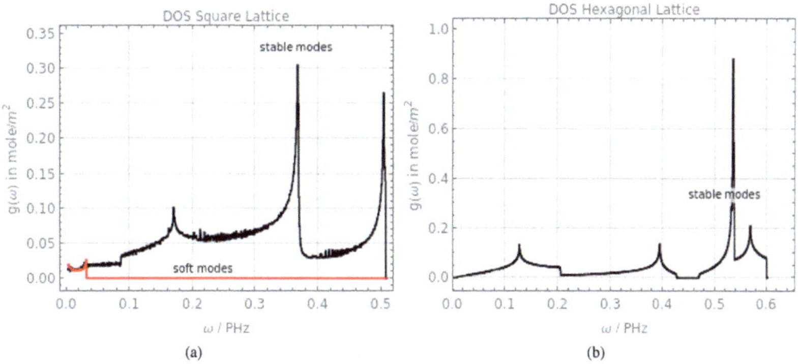

Fig. 25.10 Density of state (DOS) functions for the square and hexagonal lattices. For the square lattice, two functions are plotted: one for the DOS of stable phonons with real frequencies (black line) and the other for the DOS of soft phonons with imaginary frequencies (red line). There are no soft mode vibrations in the dispersion relation of the hexagonal lattice

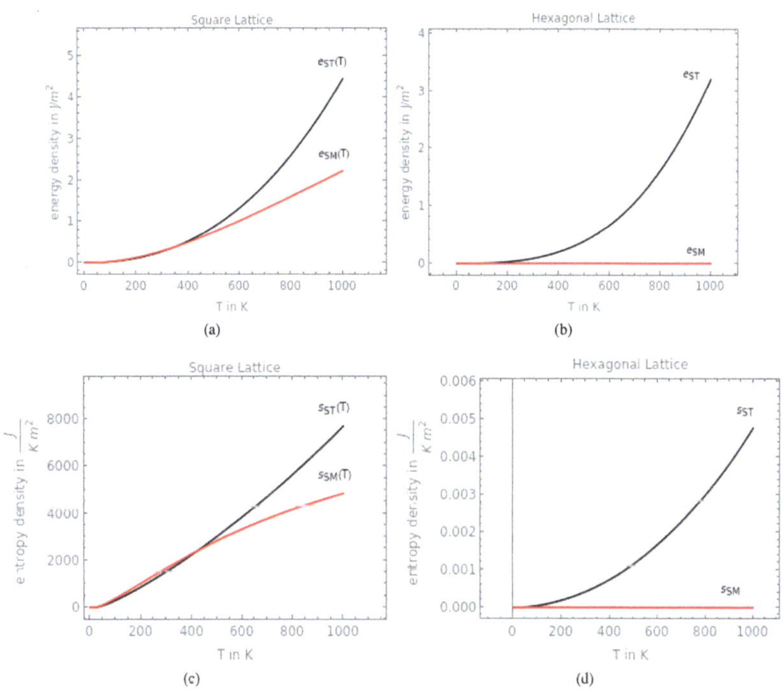

Fig. 25.11 The phonon-gas energy densities, $e_{\mathrm{ST}}(T)$, $e_{\mathrm{SM}}(T)$, and entropies, $s_{\mathrm{ST}}(T)$, $s_{\mathrm{SM}}(T)$ for both phonon-gas species (ST: black, SM: red line color) and both lattice structures (austenite: left column, martensite: right column)

References

Bhattacharya K (2003) Microstructure of martensite—why it forms and how it gives rise to the shape-memory effect. Oxford University Press, Oxford (etc)

Burow J (2010) Herstellung, Eigenschaften und Mikrostruktur von ultrafeinkörnigen NiTi-Formgedächtnislegierungen. PhD thesis, Ruhr-University Bochum

Cowley R (1980) Structural phase transitions I. landau theory. Adv Phys 29:1–110

Cowley RA, Bruce AD (1980) Critical dynamics and quasi-elastic scattering. Adv Phys 29

de Koker N (2009) Thermal conductivity of mgo periclase from equilibrium first principles molecular dynamics. PRL 103:125902

Gurevich VL (1986) Transport in phonon systems. Modern problems in condensed matter sciences. Elsevier Science Publishers

Kastner O (2003) Molecular dynamics of a 2D model of the shape memory effect. Part I: model and simulations. Continuum Mech Therm 15(5):487–502

Kastner O (2006) Molecular dynamics of a 2D model of the shape memory effect. Part II: Thermodynamics of a small system. Continuum Mech Therm 18(1–2):63–81

Kastner O (2012) First principles modelling of shape memory alloys, volume 163 of Springer Series in Materials Science. Springer

Kastner O, Shneck RZ (2015) On the entropic nucleation barrier in a martensitic transformation. Philosoph Mag 95:1282–1308

Kastner O, Eggeler G, Weiss W, Ackland GJ (2011) Molecular dynamics simulation study of microstructure evolution during cyclic martensitic transformations. JMPS 9:1888–1908

Kastner O, Eggeler G (2010) Molecular dynamics simulations of the shape memory effect in a chain of Lennard-Jones crystals. MMMS 6(1)

Kastner O, Web archive of molecular dynamics simulations. https://olli-kastner.de/pub/md

Li CW, Smith HL, Lan T, Niedziela JL, Munoz JA, Keith JB, Mauger L, Abernathy DL, Fultz B (2015) Phonon anharmonicity of monoclinic zirconia and yttrium-stabilized zirconia. Phys Revi B 91(14):144302

Mizrahi N (2022) Atomic level simulation of martensitic nucleation. PhD thesis, Ben-Gurion University of the Negev, Faculty of Natural Sciences, Department of Computer Science, Sept 2022

Müller I, Seelecke S (2001) Thermodynamic aspects of shape memory alloys. Math Comput Model 34:1307–1355

Müller I, Weiss W (2005) Entropy and energy. Interaction of mechanics and mathematics. Springer, Berlin

Nakanishi N, Nagasawa A, Murakami Y (1982) Lattice stability and soft modes. Le J De Physique Colloquies C-43:35–55

Otsuka K, Wayman CM (1998) Shape memory materials. Cambridge University Press

Petry W, Heiming A, Trampenau J, Alba M, Herzig C, Schober HR, Vogl G (1991) Phonon dispersion of the bcc phase of group-iv metals. I. BCC Titanium. Phys Rev B 43:10933

Solyom J (2007) Fundamentals of the physics of solids, volume 1 — Structure and dynamics. Springer

Srivastava GP (1990) The physics of phonons, Adam Hilger, Bristol

Trubitsin VY, Dolgusheva EB (2007) Anharmonic effects and vibrational spectrum of bcc zr under pressure studied by molecular dynamics simulations. Phys Rev B 76(2):024308

Trubitsin VY, Ostanin S (2004) Lattice dynamics of an anharmonic crystal: evidence for interactions between atomic vibrations at high temperatures. Phys Rev Lett 93(15):155503

Venkataraman G (1979) Soft modes and structural phase transitions. Bull Mater Sci 1:129–170

Zener C (1947) Contributions to the theory of beta-phase alloys. Phys Rev 7:846

Chapter 26
The Maximum Entropy Principle in Nonequilibrium Thermodynamics: A Brief History and the Contributions of Wolfgang Dreyer

Takashi Arima and Tommaso Ruggeri

Dedication: This paper is dedicated to the memory of Wolfgang Dreyer. In particular, one of the authors, T.R., was a close friend and colleague of Wolfgang and always greatly appreciated his outstanding contributions as a scientist.

Abstract We present a brief history of how the famous Maximum Entropy Principle was used as closure of moments of the Boltzmann equation. In particular, we want to remark on the important role of two fundamental papers by Wolfgang Dreyer, one in the classical framework and one in a relativistic context, to use this principle and to compare the result with the macroscopic theory of Rational Extended Thermodynamics.

26.1 Introduction

The *Maximum Entropy Principle* (MEP), rooted in probability theory and statistical mechanics, was formalized in its modern form by Jaynes (1957a, b). Jaynes developed the principle to provide a general approach for statistical inference, extending beyond physics to other fields where systems might be out of equilibrium (see e.g., (Kapur 1989)). The MEP has served as both a philosophical and practical foundation for making unbiased, informed predictions in the presence of incomplete data. In image reconstruction, it helps create images that align with known data without introducing unwarranted assumptions. In modern AI, the MEP continues to influence techniques for uncertainty management, probabilistic inference, and decision-making,

T. Arima
Department of Engineering for Innovation, National Institute of Technology, Tomakomai College, Tomakomai, Japan
e-mail: arima@tomakomai-ct.ac.jp

T. Ruggeri (✉)
Department of Mathematics, University of Bologna, and Accademia dei Lincei, Rome, Italy
e-mail: tommaso.ruggeri@unibo.it

J. Fuhrmann et al. (eds.), *Advances in Continuum Physics*, Advanced Structured Materials 238, https://doi.org/10.1007/978-3-031-93918-1_26

ensuring that models remain general and data-driven without making unnecessary assumptions.

In the context of nonequilibrium thermodynamics, the MEP has become a powerful tool. It aids in analyzing and predicting the behavior of systems far from equilibrium, where conventional equilibrium assumptions do not apply. By maximizing entropy under given constraints, MEP offers a systematic and unbiased method to model complex, evolving, and fluctuating systems, making it widely applicable in physical, biological, and socio-economic domains.

Regarding the closure of moments associated with the Boltzmann equation, the MEP is widely known as the Levermore procedure (Levermore 1996). However, significant work in this area predates this paper, notably conducted by Kogan (1969), Dreyer (1987a, 1987b), and Müller and Ruggeri (1993). The aim of this paper is to provide a brief history of the contributions made on this subject, particularly emphasizing the crucial role that Wolfgang Dreyer played in both classical and relativistic frameworks. In particular, we also highlight Dreyer's significant contributions in elucidating the equivalence between the MEP and the nonequilibrium thermodynamics framework of Rational Extended Thermodynamics (RET) (Müller and Ruggeri 1993, 1998; Ruggeri and Sugiyama 2015, 2021).

The paper is organized as follows: In Sect. 26.2, we present the history of MEP in the classical (non-relativistic) context, and in Sect. 26.3, we examine it in the relativistic framework.

26.2 MEP in the Kinetic Theory of Gases and Connection with Rational Extended Thermodynamics: Classical Framework

In this section, we present a concise history of the MEP as a closure procedure for the moment equations derived from the classical Boltzmann equation, as documented in the literature available to the authors. This discussion expands upon the mathematical details provided in the recent book by Ruggeri and Sugiyama (2021).

The initial applications of MEP were focused on monatomic gases, establishing the foundational framework.

26.2.1 *Moment Equations and Problem of Closure*

The kinetic theory of gases describes the state of a rarefied monatomic gas using the velocity distribution function $f(t, \mathbf{x}, \boldsymbol{\xi})$. Here, $f(t, \mathbf{x}, \boldsymbol{\xi})\, d\mathbf{x}\, d\boldsymbol{\xi}$ denotes the number of particles within the phase-space volume element $d\mathbf{x}\, d\boldsymbol{\xi}$ at position $(\mathbf{x}, \boldsymbol{\xi}) \in \mathbb{R}^3 \times \mathbb{R}^3$. In the absence of external forces, the evolution of f is governed by the Boltzmann equation:

$$\partial_t f + \xi_i \partial_i f = Q(f), \tag{26.1}$$

where the collision integral $Q(f)$ encapsulates the rate of change in f due to molecular collisions. Here, ∂_t and ∂_j denote partial derivatives with respect to time t and spatial coordinate x_j $(j = 1, 2, 3)$, respectively. Summation over repeated indices follows Einstein's convention unless explicitly stated otherwise.

The distribution function allows us to construct macroscopic quantities defined as velocity-space moments of f:

$$F = m \int_{\mathbb{R}^3} f \, d\boldsymbol{\xi}, \quad F_{i_1 \dots i_n} = m \int_{\mathbb{R}^3} f \, \xi_{i_1} \cdots \xi_{i_n} \, d\boldsymbol{\xi}, \quad i_k \in \{1, 2, 3\}. \tag{26.2}$$

From the Boltzmann equation, the governing equations for these moments are expressed as a set of infinite balance laws[1]:

$$\partial_t F_{i_1 \dots i_n} + \partial_j F_{i_1 \dots i_n j} = P_{i_1 \dots i_n}, \qquad n = 0, 1, \dots \tag{26.3}$$

where

$$F_{i_1 \dots i_n j} = m \int_{\mathbb{R}^3} f \, \xi_j \xi_{i_1} \cdots \xi_{i_n} \, d\boldsymbol{\xi}, \qquad P_{i_1 \dots i_n} = m \int_{\mathbb{R}^3} Q(f) \, \xi_{i_1} \cdots \xi_{i_n} \, d\boldsymbol{\xi}.$$

Since the first five equations for F, F_i, and F_{ll} represent the conservation laws of mass, momentum, and energy, by comparing these laws of continuum mechanics, we can deduce the physical meanings of the first 13 moments: F represents the mass density, F_i the momentum density, F_{ll} twice the total energy density, F_{ij} the momentum flux, and F_{lli} twice the energy flux. In particular, the first five production terms, P, P_i, and P_{ll}, vanish due to the conservation laws.

To truncate this hierarchy at tensorial order N, we need to know the expression of the last flux $F_{i_1 i_2 \dots i_{N+1}}$ and of the production terms $P_{i_1 \dots i_n}$ with $n = 2, \dots N$ and $P_{ll} = 0$. The truncated balance laws construct the following system

$$\partial_t F_A + \partial_i F_{iA} = P_A, \qquad 0 \leqslant A \leqslant N, \tag{26.4}$$

where the following notations are introduced for the sake of compactness:

$$F_A = \begin{cases} F & \text{for } A = 0 \\ F_{i_1 \cdots i_A} & \text{for } 1 \leqslant A \leqslant N \end{cases}, \qquad F_{iA} = \begin{cases} F_i & \text{for } A = 0 \\ F_{i\, i_1 \cdots i_A} & \text{for } 1 \leqslant A \leqslant N \end{cases}, \tag{26.5}$$

$$P_A = \begin{cases} 0 & \text{for } A = 0, 1 \\ P_{i_1 \cdots i_A} & \text{for } 2 \leqslant A \leqslant N \quad (\text{with } P_{ll} = 0). \end{cases}$$

With these symbols we have:

$$F_A = m \int_{\mathbb{R}^3} f \, \xi_A \, d\boldsymbol{\xi}, \quad F_{iA} = m \int_{\mathbb{R}^3} f \, \xi_i \xi_A \, d\boldsymbol{\xi}, \quad P_A = m \int_{\mathbb{R}^3} Q \, \xi_A \, d\boldsymbol{\xi},$$

[1] For $n = 0$, we assume $F_{i_1 i_2 \dots i_n} = F$, $F_{i_1 i_2 \dots i_n j} = F_j$ and $P_{i_1 i_2 \dots i_n} = P = 0$.

where

$$\xi_A = \begin{cases} 1 & \text{for } A = 0 \\ \xi_{i_1} \cdots \xi_{i_A} & \text{for } 1 \leqslant A \leqslant N. \end{cases}$$

26.2.2 Kogan's MEP for Non-degenerate Gases (1967)

In 1967, Mikhail N. Kogan from the Academy of Sciences of the USSR, Moscow, authored a book in Russian, which was translated in 1969 under the title *Rarefied Gas Dynamics* Kogan (1969). This book offers an elegant presentation of the Boltzmann equations. In Chapter III, *General Methods of Solution of the Boltzmann Equation*, Kogan introduces the method of moments. Notably, in Sect. 3.16, page 267, there is a section titled *The Maximum Probability Principle*. Here, he begins by noting that, in equilibrium, the Maxwellian distribution maximizes the probability under the constraint that the first five moments, i.e., F, F_i, and F_{ll}, are prescribed. He then writes:

> It is natural to try to find the distribution function as the most probable function also for nonequilibrium processes.

For this aim, he considered non-degenerate gas, of which entropy density is given by

$$h = -k_B \int_{\mathbb{R}^3} f \ln \frac{f}{y} \, d\boldsymbol{\xi},$$

being k_B the Boltzmann constant and y is a constant with the same dimension of f that is defined as

$$y = e(2\bar{s} + 1)\frac{m^3}{h^3}$$

where h is Planck's constant, and $\bar{s}$ is the spin quantum number of a particle.[2]

Then, Kogan required that the entropy is maximal under the constraints that the moments (26.2) are prescribed until the tensorial index N: $n = 0, \ldots, N$. Therefore, he constructed the functional

$$\mathcal{L}(f) = -k_B \int_{\mathbb{R}^3} f \ln \frac{f}{y} \, d\boldsymbol{\xi} + \Lambda_A \left(F_A - m \int_{\mathbb{R}^3} f \, \xi_A \, d\boldsymbol{\xi} \right),$$

where $\Lambda_A \equiv \Lambda_A(t, \mathbf{x})$, defined in a similar way to F_A in (26.5), are the Lagrange multipliers, and we omit the summation over A from 0 to N. Requiring the constraint maximum, he found immediately that the distribution function such that $\delta\mathcal{L}(f) = 0$, $\delta^2\mathcal{L}(f) < 0$, is

$$f_N = y \exp\left(-1 - \frac{m}{k_B}\chi\right), \qquad \chi = \Lambda_A \xi_A. \tag{26.6}$$

[2] In Kogan's book (Kogan 1969), the value $y = 1$ is chosen. In fact, in many references, $y = 1$ or $y = e$ is often used.

Therefore, Kogan was the first to derive the expression for the distribution function that maximizes entropy for non-degenerate gases under the constraint that a specified number of moments are prescribed.

By inserting (26.6) into the last flux and into the production terms, he was able to obtain a closed system having as unknown the Lagrange multipliers. Kogan was also the first to observe that this approach encounters issues related to the integrability of moments with f_N. He wrote:

> However, formal use of the method immediately encounters several well-known difficulties.
>
> The first difficulty is of a mathematical nature. Suppose that we examine a problem in which the highest moment determining the problem is an odd moment, for example, the heat flux q_i. Then, ... we obtain divergent integrals ... and the problem has no solution.

Therefore, again, Kogan was the first to realize that in the nonlinear closure, it is necessary (but not sufficient) that N is even such that the moments are integrable.

In particular, the 13-moment model which corresponds to a specific instance of $N = 3$ and only the trace part in the third order tensor is taken into account, has no significance in the MEP.

Additionally, he was also the first to consider the approximation near equilibrium. He wrote on page 271[3]:

> Therefore, we may speak of the distribution function as most probable only for flows with small Knudsen numbers, i.e., for flows close to local equilibrium. In that case, all the moments, with the exception of the hydrodynamic ones, may be considered small, and we may linearize expression (26.6) relative to the equilibrium distribution function, i.e., we may write it in the form

$$f_N = f_E\left(1 - \frac{m}{k_B}\tilde{\chi}\right), \qquad \tilde{\chi} = \tilde{\Lambda}_A \xi_A. \tag{26.7}$$

Here, the $\tilde{\Lambda}$'s represent the difference between the nonequilibrium and equilibrium Lagrange multipliers, and f_E is the Maxwellian distribution. As the Maxwellian is given by a convergent exponential that dominates any polynomial, the moments are always integrable for any N.

Kogan, then, considered 13 moments by using the approximated distribution function (26.7) and demonstrated its equivalence to the Grad distribution function Grad (1958). Therefore, the Grad system also maximizes entropy, but only near the equilibrium state. On the other hand, Grad's procedure is a perturbative method relative to equilibrium.

[3] Here, we have modified the equation number (originally Eq. (16.5) in Kogan (1969)) and the mathematical expressions in (26.7) to correspond to the context and notation used in this paper.

26.2.3 *Dreyer's Work and Comparison Between MEP and RET (1987)*

Twenty years after Kogan's works, Wolfgang Dreyer who was previously a student of Ingo Müller, first presented fundamental results concerning MEP in both the classical Dreyer (1987a) and relativistic frameworks Dreyer (1987b). Here, we exhibit his contribution to the classical context. Later, in Sect. 26.3, we discuss his contribution to the relativistic framework.

Dreyer generalized Kogan's result for the first time, presenting the MEP not only for non-degenerate gases but also for degenerate ones, whose entropy is given by

$$h = -k_B \int_{\mathbb{R}^3} \left\{ \left(s^2 - 1 + \ln \frac{f}{Y} \right) + s \frac{Y}{f} \left(1 - s \frac{f}{Y} \right) \ln \left(1 - s \frac{f}{Y} \right) \right\} f \, d\boldsymbol{\xi}, \quad (26.8)$$

where $Y = y/e$. The constant s is 0 for non-degenerate gases and $s = 1$ for Fermions, and -1 for Bosons. For this entropy, the f_N that maximizes the entropy is:

$$f_N = \frac{Y}{e^{\frac{m}{k_B}\chi} + s}. \quad (26.9)$$

For $s = 0$, (26.9) reduce to the Kogan's one (26.6), while for degenerate gases becomes:

$$f_N = \frac{Y}{e^{\frac{m}{k_B}\chi} \mp 1}, \quad (26.10)$$

where χ is given by $(26.6)_2$, and the upper sign is for Bosons, and the lower sign is for Fermions.

Dreyer's closure with f_N given in (26.10) *is fully nonlinear, as it does not rely on an expansion near equilibrium, and for non-degenerate gases, he was the first to use MEP to close the system of moments.*

Nevertheless, the main aim of Dreyer (1987a) was to prove the equivalence between the closure using MEP and the one at the phenomenological level of RET for rarefied gases, particularly in the 13-moment case. Since RET was initially closed only near equilibrium, Dreyer considered a similar expansion to (26.7).

Before discussing the details, it is necessary to exhibit a brief survey about the RET theory.

26.2.3.1 Brief Introduction of Rational Extended Thermodynamics

The first hyperbolic phenomenological model for a viscous, heat-conducting gas was introduced by Ingo Müller in his doctoral thesis Müller (1966). It was based on modifying the Gibbs equation, recognizing that the hypothesis of so-called *local equilibrium* is overly restrictive for phenomena far from equilibrium.

However, Ruggeri (1983) criticized this approach for several mathematical reasons, particularly because the differential system is not, a priori, in the form of a balance laws. This limitation implies that it is not possible to define weak solutions, making the study of shock waves impossible. Ruggeri, using the theory of hyperbolic systems compatible with an entropy law that can be symmetrized, proposed a model that is symmetric hyperbolic for any fields. It satisfies both the entropy principle and the convexity condition of entropy. However, the system's structure differs from that of moments. Consequently, Müller, along with Liu, revised Extended Thermodynamics, adopting the balance law structure dictated by the moment theory associated with the Boltzmann equation as a starting point Liu and Müller (1983). In a relativistic context, a similar theory was developed by Liu, Müller, and Ruggeri (1986).

This approach became known as *Rational Extended Thermodynamics* and the main results were summarized in the books by Müller and Ruggeri (1993, 1998).

The central idea of the RET approach is to use the structure of balance laws for moments truncated at order N (see (26.4)). However, it then departs from viewing the (F_A, F_{iA}, P_A) as moments of a distribution function solution of the Boltzmann equation. Instead, the system is treated as a phenomenological model typical of continuum mechanics. As is common in such cases, the system involves more fields than equations, requiring constitutive equations to close the system.

In this context, the fields F_A, representing densities, are treated as the unknowns, while the last flux term and production terms serve as the constitutive equations. The requirement is that these constitutive equations be local in nature:

$$F_{N+1} \equiv F_{N+1}(F_B), \quad P_A \equiv P_A(F_B), \quad (A, B = 0, \dots, N).$$

To determine the closure procedure, RET adopts the following universal principles:

(i) *Objectivity and relativity principles*: Constitutive equations must be objective, and balance laws must be invariant under Galilean transformations (or Lorentz transformations for relativistic gases).

(ii) *Entropy principle*: This principle requires that any solutions of the balance laws (26.3) satisfy an additional balance law for an additive, objective scalar h, called entropy:

$$\partial_t h + \partial_i h^i = \Sigma, \tag{26.11}$$

where h_i and Σ are the entropy flux and entropy production, respectively. These are assumed to be local functions of the densities field:

$$h = h(F_A), \quad h^i = h^i(F_A), \quad \Sigma = \Sigma(F_A).$$

Additionally, the entropy production must be non-negative: $\Sigma \geqslant 0$.

The requirement that a system of balance laws

$$\partial_t \mathbf{u} + \partial_i \mathbf{F}^i = \mathbf{P} \tag{26.12}$$

(such as the moment equations (26.4)) is compatible with an additional entropy balance law (26.11) falls under the well-known problem of hyperbolic systems. Starting with Boillat's paper Boillat (1974), Ruggeri and Strumia (1981) proved that if the entropy density h is a convex function of $\mathbf{u}$, there exists a privileged field called the *main field* $\mathbf{u}'$ and four potentials

$$h' = \mathbf{u}' \cdot \mathbf{u} - h, \quad h'^i = \mathbf{u}' \cdot \mathbf{F}^i - h^i, \tag{26.13}$$

such that

$$\mathbf{u} = \frac{\partial h'}{\partial \mathbf{u}'}, \qquad \mathbf{F}^i = \frac{\partial h'^i}{\partial \mathbf{u}'}, \tag{26.14}$$

and inserting (26.14) into the system (26.12) assumes the structure of a special symmetric hyperbolic system, often referred to as Godunov type[4]:

$$\frac{\partial^2 h'}{\partial \mathbf{u}' \partial \mathbf{u}'} \partial_t \mathbf{u}' + \frac{\partial^2 h'^i}{\partial \mathbf{u}' \partial \mathbf{u}'} \partial_i \mathbf{u}' = \mathbf{P}. \tag{26.15}$$

The matrix before the time derivative is positive-definite as h' is the Legendre transformation of h and therefore a convex function of the dual field $\mathbf{u}'$.

The main field is a particular set of Lagrange multipliers proposed by Liu (1972), satisfying:

$$\partial_t h + \partial_i h^i - \Sigma = \mathbf{u}' \cdot \left(\partial_t \mathbf{u} + \partial_i \mathbf{F}^i - \mathbf{P}\right).$$

For more details, readers can refer also to the recent book by Ruggeri and Sugiyama (2021).

26.2.3.2 Equivalence of Closure Using the RET Principles and the MEP for Moment Systems

Dreyer was the first to prove in Dreyer (1987a) *the equivalence between the Lagrange multipliers of MEP and the main field of the phenomenological theory compatible with the entropy principle.*

Specifically, in the 13-moment case near equilibrium, Dreyer demonstrated that MEP provides the same closure as phenomenological RET, thereby showing that three different closure methods-Grad's, RET's, and MEP's-yield the same system.

[4] Godunov (1961) was the first to prove that the Euler fluid equations and systems derived from a variational principle can be written in the symmetric form (26.15).

This result was significant since the three methods are fundamentally different. Grad's method uses a perturbation of the Maxwellian with Hermite polynomials, MEP uses maximum entropy at the kinetic level, while RET is a continuum theory that shares with kinetic theory only the structure of balance laws, where the next density is the previous flux, and the first five equations are the conservation laws for mass, momentum, and energy.

26.2.4 Müller and Ruggeri's Higher-Order Moment Systems and Symmetric Hyperbolic Systems (1993)

In 1993, Clifford Truesdell invited Ingo Müller and Tommaso Ruggeri to publish a book Müller and Ruggeri (1993) on the recent developments in the theory of *Extended Thermodynamics* as part of the series he edited, *Springer Tracts in Natural Philosophy*. This emerging theory also aimed to resolve the longstanding debate that arose in the literature following Müller's landmark paper Müller (1972), where he demonstrated that the Fourier and Navier-Stokes "constitutive" equations violate the principle of objectivity (frame indifference).

At the time, Müller believed his findings suggested that the objectivity principle was not universally valid. This sparked extensive debate, with scholars dividing into proponents and opponents of the objectivity principle.

Independent contributions by Bressan (1982) and Ruggeri (1982) offered a different perspective. They proposed that Müller's result did not undermine the objectivity principle itself but instead indicated that the Fourier and Navier-Stokes equations were not genuine constitutive relations.

The precise and compelling resolution came from RET. In RET, in fact, the Navier-Stokes and Fourier laws are shown to be approximations derived from the balance laws of 13 moments, valid only when certain relaxation times are small. Consequently, these laws are not true constitutive equations and do not need to adhere to the objectivity principle at the same level as fundamental balance laws like momentum and energy, which must only satisfy Galilean invariance.[5]

Returning to the historical contribution on the MEP in Chap. 9 of the first edition of their book in 1993 Müller and Ruggeri (1993), Müller and Ruggeri, considering non-degenerate gases, highlighted the clear equivalence between MEP and RET closure. Moreover they showed that the moments (26.2), derived from the distribution function (26.6) for the truncated system, satisfy the following symmetric relations (see p. 171 of Müller and Ruggeri (1993)):

$$\frac{\partial F_A}{\partial \Lambda_B} = \frac{\partial F_B}{\partial \Lambda_A}, \quad \frac{\partial F_{iA}}{\partial \Lambda_B} = \frac{\partial F_{iB}}{\partial \Lambda_A}.$$

[5] Some authors hold a different opinion and continue to consider simplified hyperbolized formulations of Fourier's law and Navier-Stokes equations, such as the Cattaneo equation or Maxwell-type viscoelasticity models, as constitutive equations. To restore the objectivity, they introduce additional derivatives into these models.

These relations imply the existence of potentials (26.13), enabling the truncated moment system to be formulated as a symmetric hyperbolic system of the Godunov type (26.15).

Therefore, they were the first to prove that the closed system using fully nonlinear MEP is symmetric hyperbolic if we use the Lagrange multipliers as field variables.

Although the book did not address the integrability of the moments, their approach was fully nonlinear, which they termed *Molecular Extended Thermodynamics*. They noted the inherent challenge in the nonlinear closure. On p. 171, they wrote regarding the MEP and RET approaches:

> Both theories have one difficulty in common: The inversion of the transformation $\mathbf{\Lambda} = \mathbf{\Lambda}(\mathbf{u})$. Indeed, inspection of (1.10), shows that neither $\mathbf{u} = \mathbf{u}(\mathbf{\Lambda})$ is easy to calculate, nor is that relation easy to invert. In this situation we are forced to linearize just as in extended thermodynamics.

Consequently, they examined processes near equilibrium and linearized the distribution function using Kogan's approximation (26.7). As concrete examples of closure derivations, they explored expansions near equilibrium, deriving maximized nonequilibrium distribution functions for the 20-moment case. They also developed linear field equations for the 14, 20, 21, 26, and 35-moment cases. The 14-moment field equations correspond to Kremer's phenomenological equations Kremer (1986).

Müller and Ruggeri recognized that this linearization sacrifices the elegant property of symmetric hyperbolicity far from equilibrium. They were the first to define, on p. 146, Sect. 3.2 in Müller and Ruggeri (1993), *the so-called "hyperbolicity region" in the space of nonequilibrium variables for which inside the domain the system is hyperbolic, calculating it explicitly for the 13-field case.* This concept was later revisited by other authors. Notably, Brini and Ruggeri (2020) demonstrated that considering higher-order polynomial expansions in (26.7) slightly enlarges the hyperbolicity domain.

Despite the limitations of the hyperbolicity region, the book Müller and Ruggeri (1993) highlights important results from Weiss (1990), which demonstrated that, with a sufficiently large number of moments, experimental data align closely with RET predictions. This alignment was particularly evident in high-frequency sound wave analysis and light scattering problems. These findings represented a significant success for the theory and contributed to the release of the book's expanded second edition in 1998, where the title was updated from Extended Thermodynamics to Rational Extended Thermodynamics Müller and Ruggeri (1998).

26.2.5 Levermore's Closure (1996)

Three years after the publication of the first edition of Müller and Ruggeri's book, Levermore published a paper in 1996 Levermore (1996) that rigorously reformulates the mathematical properties of the MEP, which had previously been noted by Kogan.

Unfortunately, Levermore appears to overlook Kogan's earlier results, which are cited in Dreyer's paper.

A novel aspect of Levermore's analysis is his explicit focus on the concept of *admissibility* and the emphasis he places on its foundational significance. His so-called "Condition III" closely mirrors Kogan's earlier observation that the integrals involved are not always convergent unless the closure number N is even, making this a necessary condition (see Sect. 26.2.2).

Levermore critiques the work of Dreyer and Müller-Ruggeri, writing:

> Some recent works have employed exponentially based closures of the form (4.10). Dreyer did so within the context of extended thermodynamics. However, he treated the exponential formally, never imposing a condition like (III), and proceeded to retain only the quadratic terms in the exponent while expanding the rest as a polynomial. Both the entropy and hyperbolic structure are generally lost in the resulting moment equations.

However, this characterization seems to miss critical aspects of Dreyer's contributions. Dreyer explicitly enforces the even-degree polynomial constraint to ensure admissibility and assumes the non-negativity of the distribution function as a fundamental premise. Levermore's critiques for linear expansions are confined to the appendix of Dreyer's paper. In the Müller-Ruggeri book, it is explained that they serve solely as a tool for inverting the Lagrange multipliers in terms of the physical variables.

Additionally, contrary to what is stated in the literature, the symmetric form of Godunov-type equations, assuming the integrability of moments, was first explicitly demonstrated in the context of fully nonlinear closure in Müller and Ruggeri (1993). The fact that hyperbolicity is now restricted to a neighborhood of the equilibrium state was also noted before in Müller and Ruggeri (1993), where the concept of the hyperbolicity region was defined, as previously discussed in Sect. 26.2.4.

From the above discussion, the truly novel aspect of Levermore's paper, regarding the MEP, in our view, lies in its rigorous mathematical treatment and in the fully nonlinear closure of the ten-moment system. He considered the following 10 independent fields: $\left(F = \rho,\, F_i = \rho v_i,\, F_{ij} = \rho v_i v_j + p_{ij}\right)$, where $\mathbf{p} \equiv (p_{ij})$ is the pressure tensor (the stress tensor with a sign change) of which trace part is related to the pressure $p_{ll} = 3p$:

$$p_{ij} = m \int_{\mathbb{R}^3} f\, C_i C_j \, d\boldsymbol{\xi}, \qquad C_i = \xi_i - v_i .$$

In this case, the distribution function without expansion is explicitly obtained as the following Gaussian form:

$$f_{10} = \frac{\rho}{m\,(2\pi)^{3/2}\left[\det\left(\mathbf{p}/\rho\right)\right]^{1/2}} \exp\left\{-\frac{1}{2}\left(\frac{\mathbf{p}}{\rho}\right)^{-1}_{ij} C_i C_j\right\}. \tag{26.16}$$

26.2.6 Boillat and Ruggeri's Analysis of Convergence for N Large and General Entropy Functional (1997)

Boillat and Ruggeri, in a paper in 1997 Boillat and Ruggeri (1997), presented new results on this subject that can be summarized in three parts:

- They first considered the MEP closure with a generic entropy functional.[6]

$$h = \int_{\mathbb{R}^3} \psi(f) d\boldsymbol{\xi}, \tag{26.17}$$

  and proved that the MEP closure coincides with the requirement that the truncated moment system satisfies an entropy principle and has a symmetric Godunov form.
- They studied the convergence of the distribution function that maximizes the entropy as $N \to \infty$.
- They demonstrated the existence of a lower bound for the maximal characteristic velocity in equilibrium of the hyperbolic system, showing not only that it increases with N but that it is unbounded in the limit as N tends to infinity.

In the present case the functional $\mathcal{L}$:

$$\mathcal{L} = \int_{\mathbb{R}^3} \psi(f) d\boldsymbol{\xi} + \Lambda_A \left(F_A - m \int \xi_A f d\boldsymbol{\xi} \right).$$

We impose the extremum condition, and we have

$$\frac{d\psi(f)}{df} = m\chi, \tag{26.18}$$

hence it follows that $f(t, \mathbf{x}, \boldsymbol{\xi})$ is a function of a single variable $f \equiv f(\chi)$, with $\chi = \Lambda_A(t, \mathbf{x})\xi_A$. Then (26.18) give:

$$\psi(f) = m \left(\chi \mathcal{F}'(\chi) - \mathcal{F}(\chi) \right),$$

where $\mathcal{F}$ is the partition function such that $\mathcal{F}'(\chi) = f(\chi)$ and a prime on a quantity indicates the derivative with respect χ. As the extremum needs to be a maximum, we have the inequality $\mathcal{F}'' < 0$.

Taking into account (26.13), it is easy to verify that potentials h' and h'^i are expressed as:

$$h' = \int_{\mathbb{R}^3} \mathcal{F}(\chi) d\boldsymbol{\xi}, \quad h'^i = \int_{\mathbb{R}^3} \xi_i \mathcal{F}(\chi) d\boldsymbol{\xi}. \tag{26.19}$$

[6] In the paper, the authors used h as the entropy, changed in sign as is customary in the mathematical community, which prefers defining convex entropy functions instead of concave ones. Consequently, the signs of the Lagrange multipliers and the main field are opposite to those presented here.

Therefore, if the truncated system is compatible with an entropy principle (26.11), then (26.14) are true, and it is simple to prove that the Lagrange multipliers field $\boldsymbol{\Lambda} \equiv (\Lambda_A)$ coincide with the main field $\mathbf{u}'$ and the two closure of MEP and RET at molecular level are equal also for a generic case.

Inserting (26.19) into (26.15), we obtain the closed system for any N is symmetric hyperbolic on the form:

$$\left(m^2 \int_{\mathbb{R}^3} \mathcal{F}'' \xi_A \xi_B d\boldsymbol{\xi}\right) \partial_t \Lambda^B + \left(m^2 \int_{\mathbb{R}^3} \mathcal{F}'' \xi_i \xi_A \xi_B d\boldsymbol{\xi}\right) \partial_i \Lambda^B = P_A.$$

Concerning the convergence for large N and the lower bound for the maximum characteristic velocity, Boillat and Ruggeri had the idea to map the $(k+1)(k+2)/2$ components of the main field (Lagrange multipliers in the MEP) of order k:

$$u'_{i_1 i_2 \dots i_k}, \qquad i_1 \leqslant i_2 \leqslant \dots \leqslant i_k$$

in the corresponding variables:

$$u'_{pqr}, \qquad p+q+r=k.$$

In fact if we collect all powers of ξ_1, ξ_2, ξ_3 in χ, as expressed earlier by $(26.6)_2$, it can be rewritten using this notation as:

$$\chi = \sum_{p,q,r} u'_{pqr} \xi_1^p \xi_2^q \xi_3^r, \qquad 0 \leqslant p+q+r \leqslant N.$$

Then, they proved in Boillat and Ruggeri (1997) the theorem that asserts that, for any N, we have the following lower bound condition for the maximum characteristic velocity evaluated in equilibrium, $\lambda_{\max}$:

$$\frac{\lambda_{\max}}{c_0} \geqslant \sqrt{\frac{6}{5}\left(N - \frac{1}{2}\right)}, \tag{26.20}$$

where c_0 is the sound velocity. Therefore, $\lambda_{\max}$ increases with N (as was noticed numerically by Weiss (1990)) and becomes unbounded when $N \to \infty$.

We discussed before the problem of the convergence of the moments, and according to Kogan, the index of truncation N must be even.

Moreover, if the conjecture—the distribution function f_N, when $N \to \infty$, tends to the distribution function f that satisfies the Boltzmann equation—is true, we need another convergence requirement.

Since

$$\left| \sum_{p,q,r} u'_{pqr} \xi_1^p \xi_2^q \xi_3^r \right| \leqslant a_N |\boldsymbol{\xi}|^N$$

with

$$a_N = \max_{|\mathbf{t}|=1} \nu_N(\mathbf{t}), \qquad \nu_N(\mathbf{t}) = \sum_{p,q,r} u'_{pqr} t_1^p t_2^q t_3^r,$$

the series is absolutely convergent, when $p + q + r = N \to \infty$, for any $\boldsymbol{\xi}$ provided that

$$u'_{pqr} \to 0, \qquad \frac{a_{N+1}}{a_N} \to 0.$$

Hence, the components of the main field become smaller and smaller when N increases. This justifies the truncation of the system. On the other hand, when N is finite, the integrals of moments must also be convergent (Levermore's admissibility discussed in Sect. 26.2.5). When $|\boldsymbol{\xi}|$ is large, $\chi \simeq |\boldsymbol{\xi}|^N \nu_N$. Therefore, it is easy to see, by using the spherical coordinates, that the integrals of moments converge provided that $\nu_N(\mathbf{t}) < 0$ for any unit vector $\mathbf{t}$. But, since $\nu_N(-\mathbf{t}) = (-1)^N \nu_N(\mathbf{t})$, we can conclude that N must be even and $\max_{|\mathbf{t}|=1} \nu_N(\mathbf{t}) < 0$, and we recover in another way the Kogan's observation and the Levermore's admissibility.

26.2.7 Another Problematic of MEP Closure: Junk Paper (1998)

Unfortunately, the elegant form derived in the fully nonlinear case has significant challenges. In addition to the previously mentioned issues-namely, concerns about convergence and the difficulty of inverting the mapping between Lagrange multipliers and physical variables-a more fundamental problem was brought to light by Junk (1998, 2000). Specifically, the complete closure achieved through the MEP introduces a critical limitation: certain physically admissible states correspond to singularities. In other words, for these states, the mapping between Lagrange multipliers and physical variables cannot be inverted.

However, when studying nonequilibrium phenomena near a local equilibrium state, this issue, often referred to as the Junk problem, does not arise. Consequently, for many practical problems, the approximate distribution function proposed by Kogan, as given in (26.7), is employed despite its associated challenges in defining hyperbolicity regions. Nevertheless, as discussed in Sect. 26.2.4, the theory remains in strong agreement with experimental results.

We conclude this section by citing two other highly interesting papers by Dreyer. The first, co-authored with Kunik and published in 1998 Dreyer and Kunik (1998), considers a modification of the MEP to study two extreme cases in kinetic theory: the dominance of particle interactions and free flight. The second, co-authored with Herrmann and Kunik (Dreyer et al. 2004) in 2004, explores another revision of the MEP and discusses Junk's problem in the context of the Boltzmann-Peierls equation, which governs heat conduction in crystalline solids.

26.2.8 Extensions of RET (2011) and MEP (2013) to Rarefied Polyatomic Gases

For a long time, there have been several attempts to generalize both RET and Kinetic Theory from the case of monatomic gases to the more interesting case of polyatomic gases. Regarding RET, a new approach for rarefied polyatomic gases was developed by Arima et al. (2011). This theory adopts two parallel hierarchies (binary hierarchy) for the independent fields: mass density, velocity, internal energy, shear stress, dynamic pressure, and heat flux. One hierarchy consists of balance equations for the mass density F, momentum density F_i, and momentum flux $F_{i_1 i_2}$ (*momentum-like* hierarchy), while the other consists of balance equations for twice the energy density G_{ll} and twice the energy flux G_{lli} (*energy-like* hierarchy);

$$\begin{aligned}
&\partial_t F + \partial_i F_i = 0, \\
&\partial_t F_{i_1} + \partial_i F_{i i_1} = 0, \\
&\partial_t F_{i_1 i_2} + \partial_i F_{i i_1 i_2} = P_{i_1 i_2}, \qquad \partial_t G_{ll} + \partial_i G_{lli} = 0, \\
&\qquad\qquad\qquad\qquad\qquad\qquad\quad \partial_t G_{lli_1} + \partial_i G_{llii_1} = Q_{lli_1},
\end{aligned} \tag{26.21}$$

where $F_{ii_1i_2}$ is the flux of $F_{i_1i_2}$, G_{llii_1} is the flux of G_{lli_1}, and $P_{i_1i_2}$ and Q_{lli_1} are the production term associated with $F_{i_1i_2}$ and G_{lli}, respectively. These hierarchies cannot merge with each other, in contrast to the case of rarefied monatomic gases, because the specific internal energy (the intrinsic part of the energy density) is no longer simply related to the pressure (an intrinsic part of the momentum flux). Moreover, for polyatomic gases, the equations are 14 because among the unknowns there is also the nonequilibrium dynamical pressure Π, which is identically zero for monatomic case.

Regarding the kinetic counterpart, a crucial step in developing the theory of rarefied polyatomic gases was made by Borgnakke and Larsen (1975). In their model, the distribution function is assumed to depend on an additional continuous variable representing the energy of the internal modes of a molecule. This allows the model to account for the exchange of energy (beyond translational energy) during binary collisions. Initially, this model was used for Monte Carlo simulations of polyatomic gases and was later applied to derive the generalized Boltzmann equation by Bourgat et al. (1994).

As a result of introducing an additional parameter I that has the meaning of microscopic internal energy, the velocity distribution function $f(t, \mathbf{x}, \boldsymbol{\xi}, I)$ is defined over an extended domain $[0, \infty) \times \mathbb{R}^3 \times \mathbb{R}^3 \times [0, \infty)$. The quantity $f(t, \mathbf{x}, \boldsymbol{\xi}, I)\varphi(I)\, d\mathbf{x}\, d\boldsymbol{\xi}\, dI$ represents the number of molecules in the 7-dimensional phase space around a point $(\mathbf{x}, \boldsymbol{\xi}, I)$ at time t. Its rate of change is governed by the Boltzmann equation, which retains the same form as that for monatomic gases (26.1), but the collision integral $Q(f)$ incorporates the effects of internal degrees of freedom through the collisional cross-section.

Pavić et al. (2013) demonstrated that the structure of (26.21) can be derived from the Boltzmann equation by identifying the following moments:

$$\begin{pmatrix} F \\ F_{i_1} \\ F_{i_1 i_2} \end{pmatrix} = \int_{\mathbb{R}^3}\int_0^\infty m \begin{pmatrix} 1 \\ \xi_{i_1} \\ \xi_{i_1}\xi_{i_2} \end{pmatrix} f(t, \mathbf{x}, \boldsymbol{\xi}, I)\,\varphi(I)\,dI\,d\boldsymbol{\xi},$$

$$\begin{pmatrix} G_{ll} \\ G_{lli_1} \end{pmatrix} = \int_{\mathbb{R}^3}\int_0^\infty m \begin{pmatrix} \xi^2 + 2\frac{I}{m} \\ (\xi^2 + 2\frac{I}{m})\,\xi_{i_1} \end{pmatrix} f(t, \mathbf{x}, \boldsymbol{\xi}, I)\,\varphi(I)\,dI\,d\boldsymbol{\xi}.$$

The weighting function $\varphi(I)$ is determined in such a way that it recovers the caloric equation of state in equilibrium for polyatomic gases. Then, they used the MEP, in the case of 14 moments, to yield the appropriate macroscopic balance laws under the assumptions of processes near equilibrium. The structure of the distribution function that satisfies the MEP is similar to the one of Kogan (26.6) with a more complex χ:

$$\chi = \Lambda + \Lambda_i \xi_i + \Lambda_{ij}\xi_i\xi_j + \Omega\left(\xi^2 + 2\frac{I}{m}\right) + \Omega_i\left(\xi^2 + 2\frac{I}{m}\right)\xi_i$$

that takes into account the Lagrange multipliers Λ's of the F's hierarchy and the Ω's of the G hierarchy. They showed complete agreement with the RET procedure given in Arima et al. (2011). Later, using the MEP, Ruggeri studied the closure in the most complicated case of non-polytropic gas, in which the internal energy is a nonlinear function of the temperature Ruggeri (2021).[7]

Therefore, even for rarefied polyatomic gases, the closure procedures of RET and MEP yield the same result for any non-polytropic gas. More details, including recent results on polyatomic gases, can be found in the book Ruggeri and Sugiyama (2021) and the references therein.

Although the theory is developed in the neighborhood of equilibrium, one of its most remarkable successes is that, unlike in the case of monatomic gases, for certain gases with large bulk viscosity, a small number of moments (6 or 14) are sufficient to describe the experimental data on shock structure Taniguchi et al. (2014a), Taniguchi et al. (2014b) and of sound waves Arima et al. (2013). Furthermore, numerical simulations of shock waves conducted by Kosuge et al. (2016, 2018, 2019) using the Boltzmann equation for polyatomic gases are indistinguishable from those obtained using the RET theory.

Finally, we remark that the model for monatomic gases can be derived from that of polyatomic gases as its singular limit (see Ruggeri and Sugiyama (2021) and references therein).

[7] In Pavić et al. (2013), there are typos that were correct in Ruggeri (2021) and in the book Ruggeri and Sugiyama (2021).

26.2.9 *Fully Nonlinear MEP Closure for Polyatomic Gases*

In Sect. 26.2.7, we discussed the challenges of the nonlinear MEP closure, which was criticized by Henning Struchtrup in his book Struchtrup (2005). In Sect. 6.6.2, on page 106, he wrote:

> The elegant mathematical formulation of the theory of maximization of entropy is outweighed by far by the problems arising with the method, and its use cannot be recommended.

While this criticism raises important points, this statement is not entirely acceptable. In the case of a theory near equilibrium, MEP still provides a valid tool for closing the system. The fact that hyperbolicity is not global but only in a specific domain is not unusual in nonlinear physical theories. Indeed, global hyperbolicity is an excessively stringent requirement, and only ideal cases (e.g., Euler fluid with the ideal gas assumption) satisfy this requirement for all possible values of the field variables. The lack of hyperbolicity could also be associated with physical effects (phase transition, instability, etc.), as, for example, in van der Waals fluids, in Born-Infeld nonlinear electrodynamics, in nonlinear elasticity, and in the Boltzmann-Vlasov equation (see Brini and Ruggeri (2020) and the references therein). The system, within the hyperbolic domain, has all the beneficial properties of qualitative analysis due to its nature as a symmetric hyperbolic system that satisfies the so-called K-condition. In particular, there exist global smooth solutions for initial data sufficiently small (for further clarification, interested readers can refer to Ruggeri and Sugiyama (2021) and references therein).

Moreover, there are examples where nonlinear closure is possible, and it is also feasible to invert the Lagrange multipliers as functions of the physical variables. Beyond the discussed case of the 10-moment closure by Levermore mentioned before in Sect. 26.2.5, there are, in the polyatomic case, some examples of nonlinear closure that we now mention here.

26.2.9.1 MEP of Polyatomic Gas with Six Moments

The RET with 14 fields, (26.21), provides a satisfactory phenomenological model, but its differential system is rather complex and is valid only near equilibrium. For this reason, a simplified theory with 6 fields (RET_6) Arima et al. (2012), Arima et al. (2015)-the mass density ρ, the velocity v_i, the temperature T, and the dynamic pressure Π-is constructed.

This simplified theory preserves the main physical properties of the more complex RET theory with 14 variables, particularly when the bulk viscosity plays a more important role than the shear viscosity and heat conductivity. This situation is observed in many gases, such as rarefied hydrogen gas and carbon dioxide gas at certain temperature ranges. In this case, in (26.21), we consider only the trace equation of $F_{i_1 i_2}$ and disregard the equation for the heat flux for G_{lli_1}.

The nonlinear version of RET was developed in Arima et al. (2015), while the nonlinear closure via MEP for this model was provided in Ruggeri (2016) for polytropic gases and in Bisi et al. (2017) for non-polytropic ones. In the present case, the 6 moments are:

$$\begin{pmatrix} F \\ F_i \\ F_{ll} \end{pmatrix} = \begin{pmatrix} \rho \\ \rho v_i \\ \rho v^2 + 3(p+\Pi) \end{pmatrix} = \int_{\mathbb{R}^3}\int_0^\infty m \begin{pmatrix} 1 \\ \xi_i \\ \xi^2 \end{pmatrix} f\varphi(I)\, dI\, d\boldsymbol{\xi}, \qquad (26.22)$$

and

$$G_{ll} = \rho v^2 + 2\rho\varepsilon = \int_{\mathbb{R}^3}\int_0^\infty m\left(\xi^2 + 2\frac{I}{m}\right) f\varphi(I)\, dI\, d\boldsymbol{\xi}.$$

Here, the internal energy ε is the sum of the kinetic part due to translation and the internal motion due to rotation and vibration of a molecule:

$$\begin{aligned} &\varepsilon = \varepsilon^K + \varepsilon^I, \quad \text{with} \\ &\varepsilon^K = \int_{\mathbb{R}^3}\int_0^\infty \frac{1}{2} m C^2 f\varphi(I) dI d\boldsymbol{\xi}, \quad \varepsilon^I = \int_{\mathbb{R}^3}\int_0^\infty I f\varphi(I) dI d\boldsymbol{\xi}. \end{aligned} \qquad (26.23)$$

In equilibrium, we have the caloric equations of state

$$\varepsilon_E^K = \frac{3}{2}\frac{k_B}{m}T, \quad \varepsilon_E^I = \varepsilon_E^I(T),$$

where the index E indicates the same quantities in equilibrium. In the present simple case, it is possible to invert the Lagrange multipliers in terms of physical variables far from equilibrium. In Bisi et al. (2017), the nonequilibrium distribution function that satisfies the MEP was determined as follows:

$$f = \frac{\rho}{mA(T^I)}\left(\frac{m}{2\pi k_B T}\frac{p}{p+\Pi}\right)^{3/2} \exp\left(-\frac{mC^2}{2k_B T}\frac{p}{p+\Pi} - \frac{I}{k_B T^I}\right), \qquad (26.24)$$

where the nonequilibrium temperature T^I is defined implicitly through the caloric equation of state as follows:

$$\varepsilon_E^I(T^I) = \varepsilon^I,$$

where ε^I is given by $(26.23)_2$. While considering $\varepsilon = \varepsilon^K + \varepsilon^I = \varepsilon_E^K + \varepsilon_E^I$ and using $2\rho\varepsilon_E^K = 3p$ and $2\rho\varepsilon^K = 3(p+\Pi)$, where the latter relation is derived from (26.22) and $(26.23)_1$, T^I is related to the dynamic pressure Π through the relation:

$$\frac{\varepsilon_E^I(T) - \varepsilon_E^I(T^I)}{\varepsilon_E^K(T)} = \frac{\Pi}{p}.$$

The normalization factor $A\left(T^I\right)$ is given by

$$A\left(T^I\right) = A_0 \exp\left(\frac{m}{k_B}\int_{T_*}^{T^I} \frac{\varepsilon_E^I(x)}{x^2}dx\right),$$

where A_0 is an inessential constant and T_* is a reference temperature. All the moments are convergent, and the bounded solutions satisfy the inequalities:

$$-1 < \frac{\Pi}{p} < \frac{\varepsilon_E^I(T)}{\varepsilon_E^K(T)}.$$

The equilibrium distribution function for a polyatomic, non-polytropic gas, which was derived in two different ways in Bourgat et al. (1994) and Pavić et al. (2013), is obtained from the nonequilibrium distribution function (26.24) by setting $\Pi = 0$.

26.2.9.2 Ellipsoidal Gaussian Distribution: MEP for Polyatomic Non-polytropic Gas with Eleven Moments

By considering up to the second-order moments, similar to the 6-moment case, but including all components, a system formed by the 11-moment; (F, F_i, F_{ij}, G_{ll}), where $F_{ij} = \rho v_i v_j + p_{ij}$ can be constructed. Here, the pressure tensor $\mathbf{p} \equiv (p_{ij})$ for polyatomic gases is defined by

$$p_{ij} = m\int_{\mathbb{R}^3}\int_0^\infty f\, C_i C_j \varphi(I) dI d\boldsymbol{\xi}.$$

being $p_{ll} = 3(p + \Pi)$.

As a result of MEP, the distribution function of the 11-moment system is given by Arima et al. (2024)

$$f = \frac{\rho}{m\,(2\pi)^{3/2}\left[\det\left(\mathbf{p}/\rho\right)\right]^{1/2} A\left(T^I\right)} \exp\left\{-\frac{1}{2}\left(\frac{\mathbf{p}}{\rho}\right)^{-1}_{ij} C_i C_j - \frac{I}{k_B T^I}\right\}.$$

This distribution function is a natural extension of (26.16) to polyatomic gases.

26.2.10 *MEP Incorporating Relaxation Processes of Molecular Rotational and Vibrational Degrees of Freedom*

As observed in RET_6, the RET for polyatomic gases described so far treats the internal energy ε^I as a single entity representing the molecular internal degrees of freedom. However, for gases such as carbon dioxide gases, where slow relaxation of molecular vibrational modes is observed, it becomes necessary to separately describe the nonequilibrium relaxation processes of rotational and vibrational modes.

To address this, in Arima et al. (2017), a kinetic model of gas molecules was proposed, in which the microscopic energy parameter I is divided into parameters representing vibrational and rotational modes. In this case, since ε^I is separated into rotational and vibrational energies, a 7-field theory was constructed as a result of the nonlinear MEP corresponding to RET_6.

Furthermore, although using an expansion near equilibrium, the 15-field MEP corresponding to a 14-field theory has also been studied Arima et al. (2018).

26.3 MEP in Relativistic Kinetic Theory of Gases

In 1987, the emergence of the new approach of Extended Thermodynamics garnered significant interest. The ISIMM (International Society for the Interaction of Mechanics and Mathematics) invited Müller and Ruggeri to organize a symposium, which took place in Bologna. The proceedings were published by a local publishing house, Pitagora Editrice (Bologna).

In this volume, there is a contribution by Dreyer (1987b) on the first use of the MEP in the context of a relativistic fluid. This result was never subsequently published in a journal and thus fell into relative obscurity. One of us, T.R., despite being an editor of the volume, had completely forgotten about Dreyer's work. In fact, even when applying the MEP method to the case of a relativistic polyatomic gas Pennisi and Ruggeri (2017) or in the recent book co-authored with Sugiyama (Ruggeri and Sugiyama 2021), Dreyer's results were never cited. Some time ago in 2022, T.R. received an email from Stefano Boccelli, working at that time as a postdoc in Canada under the supervision of James McDonald, requesting a copy of Dreyer's paper. This gave us the opportunity to revisit it, which we briefly summarize here.

26.3.1 *Relativistic Kinetic Theory and Moment Equations*

In relativity for Minkowski space, the Boltzmann–Chernikov relativistic equation read Cercignani and Kremer (2002), Chernikov (1965), Synge (1957):

$$p^\alpha \partial_\alpha f = Q, \tag{26.25}$$

in which the distribution function f depends on (x^α, p^β), where x^α are the space-time coordinates, p^α is the four-momentum, $\partial_\alpha = \partial/\partial x^\alpha$, Q is the collisional term, and $\alpha, \beta = 0, 1, 2, 3$. For monatomic gases, the relativistic moment equations associated with (26.25), truncated at tensorial index $N + 1$ are:

$$\partial_\alpha A^{\alpha\alpha_1\cdots\alpha_n} = I^{\alpha_1\cdots\alpha_n} \quad \text{with} \quad n = 0, \cdots, N \tag{26.26}$$

with

$$\begin{aligned} A^{\alpha\alpha_1\cdots\alpha_n} &= \frac{c}{m^{n-1}} \int_{\mathbb{R}^3} f\, p^\alpha p^{\alpha_1} \cdots p^{\alpha_n}\, d\boldsymbol{P}, \\ I^{\alpha_1\cdots\alpha_n} &= \frac{c}{m^{n-1}} \int_{\mathbb{R}^3} Q\, p^{\alpha_1} \cdots p^{\alpha_n}\, d\boldsymbol{P}, \end{aligned} \tag{26.27}$$

where c denotes the light velocity, m is the particle mass in the rest frame, and

$$d\boldsymbol{P} = \frac{dp^1\, dp^2\, dp^3}{p^0}.$$

If $n = 0$, the tensor reduces to A^α; moreover, the production tensor in the right side of (26.26) is zero for $n = 0, 1$ because the first 5 equations represent the conservation laws of the particle number and of the energy-momentum, respectively.

When $N = 1$, we have the relativistic Euler system

$$\partial_\alpha A^\alpha = 0, \quad \partial_\alpha A^{\alpha\beta} = 0,$$

where, also in the following, $A^\alpha \equiv V^\alpha$ and $A^{\alpha\beta} \equiv T^{\alpha\beta}$ have the physical meaning, respectively, of the particle number vector and the energy-momentum tensor;

$$V^\alpha = \rho U^\alpha, \quad T^{\alpha\beta} - p h^{\alpha\beta} + \frac{e}{c^2} U^\alpha U^\beta,$$

where $\rho = nm$, with n being the particle number density, U^α is the four-velocity $(U^\alpha U_\alpha = c^2)$, p is the pressure, e is the energy, and $h^{\alpha\beta}$ is the projector:

$$h^{\alpha\beta} = -g^{\alpha\beta} + \frac{1}{c^2} U^\alpha U^\beta,$$

being $g^{\alpha\beta} = \text{diag}(1, -1, -1, -1)$ the metric tensor. In this equilibrium case, the distribution function reduces to the Jüttner distribution function that has the following expression for any kind of gas:

$$f_J = \frac{Y}{\exp\left(\frac{m}{k_B}\alpha + \frac{U_\alpha p^\alpha}{k_B T}\right) + s},$$

where α denotes the fugacity and is expressed as $\alpha = -g/T$ being g the chemical potential.

When $N = 2$, we have the system:

$$\partial_\alpha A^\alpha = 0, \quad \partial_\alpha A^{\alpha\beta} = 0, \quad \partial_\alpha A^{\alpha\beta\gamma} = I^{\beta\gamma}, \tag{26.28}$$

that is formed of 14 equations taking into account that $A^{\alpha\beta}{}_\beta = m^2 c^2 A^\alpha$ and $I^\beta{}_\beta = 0$. In this case, the particle number vector and the energy-momentum tensor are decomposed as

$$V^\alpha = \rho U^\alpha, \quad T^{\alpha\beta} = \frac{e}{c^2} U^\alpha U^\beta + (p + \Pi)\, h^{\alpha\beta} + \frac{1}{c^2}(U^\alpha q^\beta + U^\beta q^\alpha) + t^{\langle\alpha\beta\rangle_3},$$

with 14 physical variables; ρ, T, U^α, Π, q^α, $t^{\langle\alpha\beta\rangle_3}$, where Π is the dynamic pressure, $q^\alpha = -h^\alpha_\mu U_\nu T^{\mu\nu}$ is the heat flux, and $t^{\langle\alpha\beta\rangle_3} = T^{\mu\nu}\left(h^\alpha_\mu h^\beta_\nu - \frac{1}{3} h^{\alpha\beta} h_{\mu\nu}\right)$ is the deviatoric shear viscous stress tensor. We also recall the constraints:

$$U^\alpha U_\alpha = c^2, \quad q^\alpha U_\alpha = 0, \quad t^{\langle\alpha\beta\rangle_3} U_\alpha = 0, \quad t^{\langle\alpha}{}_{\alpha\rangle_3} = 0.$$

As it is well known, the pioneering papers by Müller (1966) and Israel (1976) are the first tentative to obtain a causal relativistic phenomenological theory with a system of equations of a hyperbolic type such that the wave speeds are finite consistently with the relativity principle. For the same reason as the classical case, Liu, Müller and Ruggeri (LMR) Liu et al. (1986) (see also Müller and Ruggeri (1998)) explored the possibility of having a new theory that starts with a few natural assumptions and that uses only universal principles. Therefore, they considered the structure of the 14 moments (26.28) and closed the system (26.28) using the RET procedure at the phenomenological level.

26.3.2 MEP for Monatomic Relativistic Gas: Dreyer (1987)

The closure of MEP of the system (26.28) was the object of the important and complex paper of Dreyer (1987b). He required, under the constraints that the temporal parts $V^\alpha U_\alpha$ and $T^{\alpha\beta} U_\beta$ are prescribed, to find the appropriate distribution function $f \equiv f(x^\alpha, p^\alpha)$, which maximizes the entropy density $h = h^\alpha U_\alpha$ where h^α is the four entropy described by

$$h^\alpha = -k_B\, c\, U_\alpha \int_{\mathbb{R}^3} \left\{ \left(s^2 - 1 + \ln\frac{f}{Y}\right) + s\frac{Y}{f}\left(1 - s\frac{f}{Y}\right) \ln\left(1 - s\frac{f}{Y}\right) \right\} f\, d\boldsymbol{P},$$

which is the relativistic counterpart of (26.8).

The distribution function that maximizes the entropy retains the same form as the classical ones given in (26.9). However, in the relativistic theory, even in near equilibrium, the complexity of Dreyer's paper lies in the inversion between the Lagrange multipliers and the field variables.

Despite the groundbreaking nature of Dreyer's work Dreyer (1987b) on the closure for the relativistic 14-moment system for degenerate gases, it has largely been overlooked, as it appears only in lecture notes Dreyer (1987b), and the present authors are not aware of it being published elsewhere.[8]

Following Dreyer let us close the system (26.28) in which $A^{\alpha\beta\gamma}$ remains to determine in terms of the 14 physical fields. As usual, to obtain a specific expression of the closed field equations, we consider the processes near equilibrium and therefore expand around an equilibrium state as follows[9]:

$$\begin{aligned} f &\simeq f_J\left(1 - \frac{1}{k_B}\tilde{\chi}\right), \\ \tilde{\chi} &= m\,(\lambda - \lambda_E) + (\lambda_\mu - \lambda_{\mu_E})\,p^\mu + \frac{1}{m}\,\lambda_{\mu\nu}\,p^\mu p^\nu . \end{aligned} \tag{26.29}$$

Inserting the distribution function (26.29) into the moments (26.27), we obtain the algebraic system for the Lagrange multipliers in terms of the physical variables.

The calculations are very cumbersome due to the appearance of the integrals:

$$J_{\mu,\nu}(\alpha,\gamma) = \int_0^\infty \frac{\sinh^\mu \rho \cosh^\nu \rho}{\exp\left(\frac{m}{k_B}\alpha + \gamma\cosh\rho\right) \mp 1}\, d\rho .$$

The $J_{\mu,\nu}$ are functions of the fugacity α and the relativistic dimensionless coldness γ defined by

$$\gamma = \frac{mc^2}{k_B T}.$$

First, the expressions for the equilibrium variables were obtained as follows:

$$n = 4\pi Y m^3 c^3 J_{2,1}, \quad e = 4\pi Y m^4 c^5 J_{2,2}, \quad p = \frac{4}{3}\pi Y m^4 c^5 J_{4,0},$$

then, Dreyer derived the following closure expression for the triple tensor in terms of physical variables:

[8] As we cannot provide all details here, readers interested in Dreyer's paper Dreyer (1987b) may contact one of the present authors, T.R., via email. T.R. will be happy to provide a copy of the relevant pages from the volume containing the paper.

[9] In Dreyer's paper Dreyer (1987b), the case of zero rest-mass is also studied. For simplicity, here we excluded this case.

$$
\begin{aligned}
A^{\alpha\beta\gamma} &= (C_0^1 + C_\pi^1 \Pi) U^\alpha U^\beta U^\gamma + \frac{c^2}{6}\left(nm - C_0^1 - C_\pi^1 \Pi\right)\left(g^{\alpha\beta}U^\gamma + g^{\alpha\gamma}U^\beta + g^{\beta\gamma}U^\alpha\right) \\
&+ C_0^3\left(g^{\alpha\beta}q^\gamma + g^{\alpha\gamma}q^\beta + g^{\beta\gamma}q^\alpha\right) - \frac{6}{c^2}C_0^3\left(U^\alpha U^\beta q^\gamma + U^\alpha U^\gamma q^\beta + U^\beta U^\gamma q^\alpha\right) \\
&+ C_0^5\left(t^{\langle\alpha\beta\rangle_3}U^\gamma + t^{\langle\alpha\gamma\rangle_3}U^\beta + t^{\langle\beta\gamma\rangle_3}U^\alpha\right),
\end{aligned}
$$

where

$$
C_0^1 = mn + 8\pi Y m^4 c^3 J_{4,1}, \quad C_\pi^1 = 6\frac{1}{c^2}\frac{\begin{vmatrix} J'_{2,1} & J'_{2,2} & J'_{2,3} \\ J'_{2,2} & J'_{2,3} & J'_{2,4} \\ J'_{4,1} & J'_{4,2} & J'_{4,3} \end{vmatrix}}{\begin{vmatrix} J'_{2,1} & J'_{2,2} & J'_{2,3} \\ J'_{2,2} & J'_{2,3} & J'_{2,4} \\ J'_{4,0} & J'_{4,1} & J'_{4,2} \end{vmatrix}}, \quad C_0^3 = -\frac{1}{5}\frac{\begin{vmatrix} J'_{4,0} & J'_{4,1} \\ J'_{4,2} & J'_{4,3} \end{vmatrix}}{\begin{vmatrix} J'_{4,0} & J'_{4,1} \\ J'_{4,1} & J'_{4,2} \end{vmatrix}},
$$

$$
C_0^5 = \frac{J'_{6,1}}{J'_{6,0}},
$$

with

$$
J'_{\mu,\nu} = \frac{\partial J_{\mu,\nu}(\alpha,\gamma)}{\partial \alpha}.
$$

In Dreyer (1987b), it is noticed that the integrals $J_{\mu,\nu}$ and $J'_{\mu,\nu}$ are related to the following integrals:

$$
I_\nu(\alpha,\gamma) = \int_0^\infty \frac{\cosh \nu\rho}{\exp\left(\frac{m}{k_B}\alpha + \gamma\cosh\rho\right) \mp 1} d\rho,
$$

in particular, the following relations are shown

$$
\begin{aligned}
& J_{2,1} = \frac{1}{4}(I_3 - I_1), \quad J_{2,2} = \frac{1}{8}(I_4 - I_0), \quad J_{4,0} = \frac{1}{8}(I_4 - 4I_2 + 3I_0), \\
& J'_{2,1} = -\frac{m}{k_B\gamma}I_2, \quad J_{4,1} = \frac{1}{16}(I_5 - 3I_3 + 2I_1), \quad J'_{2,2} = -\frac{m}{4k_B\gamma}(3I_3 + I_1), \\
& J'_{2,3} = -\frac{m}{2k_B\gamma}(I_4 + I_2), \quad J'_{2,4} = -\frac{m}{16k_B\gamma}(5I_5 + 9I_3 + 2I_1), \\
& J'_{4,0} = -\frac{3m}{4k_B\gamma}(I_3 - I_1), \quad J'_{4,1} = -\frac{m}{2k_B\gamma}(I_4 - I_2), \\
& J'_{4,2} = -\frac{m}{16k_B\gamma}(5I_5 - 3I_3 - 2I_1), \quad J'_{4,3} = -\frac{3m}{16k_B\gamma}(I_6 - I_2), \\
& J'_{6,0} = -\frac{5m}{16k_B\gamma}(I_5 - 3I_3 + 2I_1), \quad J'_{6,1} = -\frac{m}{32k_B\gamma}(6I_6 - 16I_4 + 10I_2).
\end{aligned}
$$

Table 26.1 Expressions for $I_\nu(\alpha, \gamma)$ for non-zero rest mass. $\rho_F = \text{arcosh}\,|\alpha/(k_B\gamma)|$ which characterize Fermi degeneracy, and $K_\nu(\gamma)$ is the modified Bessel function: $K_\nu(\gamma) = \int_0^\infty \cosh(\nu s)\, e^{-\gamma \cosh s}\, ds$

Gases	Expression for $I_\nu(\alpha, \gamma)$
Non-degenerate	$K_\nu(\gamma) \exp\left(-\dfrac{m\alpha}{k_B}\right)$
Strongly degenerate (Fermions)	$\dfrac{\sinh \nu\rho_F}{\nu}(1 + X + \cdots)$ with $X = \dfrac{\pi^2 \nu}{6\gamma^2} \dfrac{\nu \sinh \nu\rho_F \sinh \rho_F - \cosh \nu\rho_F \cosh \rho_F}{\sinh \nu\rho_F \sinh^3 \rho_F}$
Completely degenerate (Fermions)	$\dfrac{1}{\nu} \sinh \nu\rho_F$
Completely degenerate (Bosons)	$\sum_{r=1}^{\infty} K_\nu(r\gamma) \exp(r\gamma)$

The evaluation of these integrals depends on the specific properties of the gas. The value of $\frac{m\alpha}{k_B\gamma}$ determines the physical regime. For instance, the system corresponds to a non-degenerate gas when $\frac{m\alpha}{k_B} \to \infty$, while for strongly degenerate Fermions, it lies within the range $-\infty < \frac{m\alpha}{k_B\gamma} < -1$. In the case of completely degenerate Fermions, the condition is $\frac{m\alpha}{k_B\gamma} \to -\infty$. Finally, for completely degenerate Bosons, the range is given by $\frac{m\alpha}{k_B\gamma} = -1$. Table 26.1 highlights how the evaluation of $I_\nu(\alpha, \gamma)$ varies across different gas regimes, providing insight into the transition from non-degenerate to strongly degenerate conditions. In the following table is reported the Dreyer results for I_ν.

Dreyer also proved that the results obtained using the closure of the MEP are consistent with the phenomenological LMR closure Liu et al. (1986) and coincide with the Marle closure Marle (1965), which is the relativistic counterpart of the classical Grad method.

26.3.3 Boillat and Ruggeri's Extension to General Entropy Functional (1997)

As in the non-relativistic gas case (see (26.17)), the application of the MEP approach to relativistic expectations for a general entropy functional was carried out by Boillat and Ruggeri (1999a, 1999b), and in this case, the authors proved an upper bound limit for the maximum characteristic velocity similar to the corresponding classical one (26.20) but as we aspect they proved that for $N \to \infty$ the limit is the light velocity according with the expectation of relativistic Boltzmann-Chernikov equation given by Cercignani (1983). Again, this is another indication of the conjecture (never proved) that the f_N obtained by the MEP converges for $N \to \infty$ to the solution of (26.25).

26.3.4 MEP for Polyatomic Relativistic Gas (2017)

Pennisi and Ruggeri first constructed a relativistic RET theory for polyatomic gases with (26.26) in the case of $N = 2$ Pennisi and Ruggeri (2017), and generalized to many moment case Pennisi and Ruggeri (2020). After that, from a Pennisi idea, a more physically reliable model was presented with the following moments Arima et al. (2022a):

$$A^{\alpha\alpha_1\cdots\alpha_n} = \left(\frac{1}{mc}\right)^{2n-1} \int_{\mathbb{R}^3}\int_0^{+\infty} f\, p^{\alpha} p^{\alpha_1}\cdots p^{\alpha_n} \left(mc^2 + I\right)^n \phi(I)\, dI\, d\boldsymbol{P},$$
$$I^{\alpha_1\cdots\alpha_n} = \left(\frac{1}{mc}\right)^{2n-1} \int_{\mathbb{R}^3}\int_0^{+\infty} Q\, p^{\alpha_1}\cdots p^{\alpha_n} \left(mc^2 + I\right)^n \phi(I)\, dI\, d\boldsymbol{P}. \tag{26.30}$$

where the distribution function $f(x^\alpha, p^\beta, I)$ depends on the extra variable I, similar to the classical one (see Ruggeri and Strumia (1981) and references therein), that has the physical meaning of the molecular internal energy of internal modes in order to take into account the exchange of energy due to the rotation and vibration of a molecule, and $\phi(I)$ is the state density of the internal mode.

In the case of $N = 2$, the moments (26.30) provides 15 moments. For such a case, in Arima et al. (2022a), the closed field equations near equilibrium were obtained by adopting MEP.

26.3.5 Classical Limit of MEP

As usual, relativistic theories have compactness properties that are lost in the classical limit, and many properties of classical theories can be understood by considering them as the limit of the corresponding relativistic theories. It is noteworthy that in the relativistic case, there is no issue with the integrability of moments, unlike in the classical case. The reason for this is physically evident, as particle velocities are limited by the speed of light. From a mathematical perspective, this can also be demonstrated (see, for instance, Carrisi et al. (2019)).

Moreover, by taking the classical limit in the monatomic case and truncating the relativistic moment system at tensorial order N, Pennisi and Ruggeri (2020) proved that, in the classical limit, there exists a precise hierarchy of moments (the only admissible one). In this hierarchy, some indices are free while others are repeated (trace part), such that the closure tensorial index of the classical limit is $N_{\text{class}} = 2N$. This value is an even number and thus represents a necessary condition for the integrability of moments. Additionally, not all classical numbers of moments are admissible. Specifically, for a monatomic gas, if the relativistic system is (26.26), then the total number of classical moments, $\mathcal{N}$, is given by the following formula:

$$\mathcal{N} = \frac{1}{6}(N+1)(N+2)(2N+3).$$

This result agrees with the findings of Dreyer and Weiss (1986), who showed that for $N = 2$, the LMR theory converges in the classical limit to Kremer's theory with $\mathcal{N} = 14$ moments Kremer (1986), rather than to Grad's 13-moment theory. A similar result has been obtained for the polyatomic case Arima et al. (2022b), Ruggeri and Sugiyama (2021).

26.4 Conclusions

Our aim here has been to provide a brief chronological account of how the MEP has been applied in nonequilibrium thermodynamics based on kinetic theory. While our narrative is necessarily limited to selected works, we observe that perhaps the most significant contributions to this subject were made, in temporal order, by Kogan, Dreyer, and, later, in the first edition of the book by Müller and Ruggeri.

In particular, Dreyer was not only the first to consider, in the classical context, the case of a general gas-including Fermions and Bosons-but also the first to apply the MEP closure to a relativistic fluid. This challenging work remains largely unknown, as it was published only in the proceedings of a conference by a local publisher. Without diminishing the contributions of Levermore-who deserves credit for rigorously promoting the topic at an international level-the authors modestly believe that the method for closing moment equations should be referred to as the *Kogan-Dreyer method*.

Acknowledgements The work of T. A. was partially supported by JST, PRESTO Grant Number JPMJPR23O1, Japan, and by JSPS KAKENHI Grant Numbers JP22K03912. The work of T. R. was carried out in the framework of the activities of the Italian National Group for Mathematical Physics [Gruppo Nazionale per la Fisica Matematica (GNFM/INdAM)].

References

Arima T, Taniguchi S, Ruggeri T, Sugiyama M (2011) Extended thermodynamics of dense gases. Continuum Mech Thermodyn 24:271

Arima T, Taniguchi S, Ruggeri T, Sugiyama M (2013) Dispersion relation for sound in rarefied polyatomic gases based on extended thermodynamics. Continuum Mech Thermodyn 25:727–737

Arima T, Taniguchi S, Ruggeri T, Sugiyama M (2012) Extended thermodynamics of real gases with dynamic pressure: An extension of Meixner's theory. Phys Lett A 376:2799–2803

Arima T, Ruggeri T, Sugiyama M, Taniguchi S (2015) Nonlinear extended thermodynamics of real gases with 6 fields. Int J Non-Linear Mech 72:6–15

Arima T, Ruggeri T, Sugiyama M (2017) Rational extended thermodynamics of a rarefied polyatomic gas with molecular relaxation processes. Phys Rev E 96:042143

Arima T, Ruggeri T, Sugiyama M (2018) Extended thermodynamics of rarefied polyatomic gases: 15-Field theory incorporating relaxation processes of molecular rotation and vibration. Entropy 20:301

Arima T, Carrisi MC, Pennisi S, Ruggeri T (2022a) Relativistic rational extended thermodynamics of polyatomic gases with a new hierarchy of moments. Entropy 24:43

Arima T, Carrisi MC, Pennisi S, Ruggeri T (2022b) Relativistic kinetic theory of polyatomic gases: classical limit of a new hierarchy of moments and qualitative analysis. Partial Diff Equ Appl 3:39

Arima T, Mentrelli A, Ruggeri T (2024) A novel ES-BGK model for non-polytropic gases with internal state density independent of the temperature. J Stat Phys 191:95

Bisi M, Ruggeri T, Spiga G (2017) Dynamical pressure in a polyatomic gas: interplay between kinetic theory and extended thermodynamic. Kinet Relat Mod 11:71

Boillat G (1974) Sur l'existence et la recherche d'équations de conservation supplémentaires pour les systémes hyperboliques. C R Acad Sci Paris A 278:909

Boillat G, Ruggeri T (1997) Moment equations in the kinetic theory of gases and wave velocities. Continuum Mech Thermodyn 9:205

Boillat G, Ruggeri T (1999a) Maximum wave velocity in the moments system of a relativistic gas. Continuum Mech Thermodyn 11:107

Boillat G, Ruggeri T (1999b) Relativistic gas: moment equations and maximum wave velocity. J Math Phys 40:6399

Borgnakke C, Larsen PS (1975) Statistical collision model for Monte Carlo simulation of polyatomic gas mixture. J Comput Phys 18:405

Bourgat J-F, Desvillettes L, Le Tallec P, Perthame B (1994) Microreversible collisions for polyatomic gases. Eur J Mech B/Fluids 13:237

Bressan A (1982) On relativistic heat conduction in the stationary and nonstationary cases, the objectivity principle and piezoelasticity. Lett Nuovo Cimento 33:108

Brini F, Ruggeri T (2020) Second-order approximation of extended thermodynamics of a monatomic gas and hyperbolicity region. Continuum Mech Thermodyn 32:23

Carrisi MC, Pennisi S, Ruggeri T (2019) Integrability properties for relativistic extended thermodynamics of polyatomic gas. Ricerche mat 68:57–73

Cercignani C (1983) Speed of propagation of infinitesimal disturbances in a relativistic gas. Phys Rev Lett 50:1122

Cercignani C, Kremer GM (2002) The relativistic Boltzmann equation: theory and applications. Birkhäuser, Basel

Chernikov NA (1965) Microscopic foundation of relativistic hydrodynamics. Acta Phys Polonica 27:465

Dreyer W (1987a) Maximization of the entropy in non-equilibrium. J Phys A: Math Gen 20:6505

Dreyer W (1987b) Maximization of entropy for relativistic and degenerate gases in non-equilibrium in ISIMM symposium on kinetic theory and extended thermodynamics, Müller I, Ruggeri T (eds). Pitagora Editrice Bologna, pp 107–123

Dreyer W, Kunik M (1998) Maximum entropy principle revisited. Continuum Mech Thermodyn 10:331

Dreyer W, Herrmann M, Kunik M (2004) Kinetic solutions of the Boltzmann-Peierls equation and its moment systems. Continuum Mech Thermodyn 16:453

Dreyer W, Weiss W (1986) The classical limit of relativistic extended thermodynamics. Ann Inst H Poincaré, Sect A 45:401

Godunov SK (1961) An interesting class of quasilinear systems. Sov Math 2:947

Grad H (1958) Principles of the kinetic theory of gases, Handbuch der Physik 12, 205–294. Springer, Heidelberg

Israel W (1976) Nonstationary irreversible thermodynamics: a causal relativistic theory. Ann Phys 100:310

Jaynes ET (1957) Information theory and statistical mechanics. Phys Rev 106:620

Jaynes ET (1957) Information theory and statistical mechanics II. Phys Rev 108:171

Junk M (1998) Domain of definition of Levermore's five-moment system. J Stat Phys 93:1143

Junk M (2000) Maximum entropy for reduced moment problems. Math Models Methods Appl Sci 10:1001
Kapur JN (1989) Maximum entropy models in science and engineering. Wiley, New York
Kogan MN (1969) Rarefied gas dynamics: translate from Russia. Plenum Press New York (1969)
Kosuge S, Aoki K (2018) Shock-wave structure for a polyatomic gas with large bulk viscosity. Phys Rev Fluids 3:023401
Kosuge S, Aoki K, Goto T (2016) Shock wave structure in polyatomic gases: numerical analysis using a model Boltzmann equation. AIP Conf Proc 1786:180004
Kosuge S, Kuo H-W, Aoki K (2019) A kinetic model for a polyatomic gas with temperature-dependent specific heats and its application to shock-wave structure. J Stat Phys 177:209
Kremer GM (1986) Extended thermodynamics of ideal gases with 14 fields. Ann I H P Phys Theor t 45:419
Levermore CD (1996) Moment closure hierarchies for kinetic theories. J Stat Phys 83:1021
Liu I-S (1972) Method of Lagrange multipliers for exploitation of the entropy principle. Arch Rational Mech Anal 46:131
Liu I-S, Müller I (1983) Extended thermodynamics of classical and degenerate ideal gases. Arch Rat Mech Anal 83:285
Liu I-S, Müller I, Ruggeri T (1986) Relativistic thermodynamics of gases. Ann Phys 169:191
Marle C (1965) Modèle cinétique pour l'ètablissement des lois de la conduction de la chaleur et de la viscositè en thèorie de la relativitè. Comptes Rendus de l'Academie des Sciences, Serie I (Mathematique) 260:6539
Müller I (1966) Zur Ausbreitungsgeschwindigkeit von Störungen in kontinuierlichen Medien, PhD thesis, Aachen
Müller I (1972) On the frame dependence of stress and heat flux. Arch Rational Mech Anal 45:241
Müller I, Ruggeri T (1993) Extended thermodynamics. Springer Tracts in Natural Philosophy, vol 37. Springer, New York
Müller I, Ruggeri T (1998) Rational extended thermodynamics. Springer Tracts in Natural Philosophy, vol 37. Springer, New York, NY, USA
Pavić M, Ruggeri T, Simić S (2013) Maximum entropy principle for rarefied polyatomic gases. Physica A 392:1302
Pennisi S, Ruggeri T (2017) Relativistic extended thermodynamics of rarefied polyatomic gas. Ann Phys 377:415–445
Pennisi S, Ruggeri T (2020) Classical limit of relativistic moments associated with Boltzmann-Chernikov equation: optimal choice of moments in classical theory. J Stat Phys 179:231–246
Ruggeri T (1982) Generators of hyperbolic heat equation in nonlinear thermoelasticity. Rend Sem Mat Padova 68:79
Ruggeri T (1983) Symmetric-hyperbolic system of conservative equations for a viscous heat conducting fluid. Acta Mech 47:167
Ruggeri T (2016) Non-linear maximum entropy principle for a polyatomic gas subject to the dynamic pressure. Bull Inst Math Acad Sin (New Ser) 11(1):1–22
Ruggeri T (2021) Maximum entropy principle closure for 14-moment system for a non-polytropic gas. Ricerche mat 70:207–222
Ruggeri T, Sugiyama M (2015) Rational extended thermodynamics beyond the monatomic gas. Springer, Cham, Heidelberg, New York, Dordrecht, London
Ruggeri T, Sugiyama M (2021) Classical and relativistic rational extended thermodynamics of gases. Springer, Cham
Ruggeri T, Strumia A (1981) Main field and convex covariant density for quasi-linear hyperbolic systems: relativistic fluid dynamics. Ann Inst H Poincaré, Sect A 34:65 (1981)
Struchtrup H (2005) Macroscopic transport equations for rarefied gas flows. Springer, Berlin, Heidelberg
Synge JL (1957) The relativistic gas. North Holland, Amsterdam
Taniguchi S, Arima T, Ruggeri T, Sugiyama M (2014) Thermodynamic theory of the shock wave structure in a rarefied polyatomic gas: beyond the Bethe-Teller theory. Phys Rev E 89:013025

Taniguchi S, Arima T, Ruggeri T, Sugiyama M (2014) Effect of dynamic pressure on the shock wave structure in a rarefied polyatomic gas. Phys Fluids 26:016103
Weiss W (1990) Zur Hierarchie der Erweiterten Thermodynamik, Dissertation, TU Berlin

Chapter 27
Phonon Hydrodynamics Revisited

Henning Struchtrup

Abstract The equations of phonon hydrodynamics are re-considered in the framework of expansions of the phonon kinetic equation for large frequency of momentum conserving N-processes. Specifically, the Chapman-Enskog expansion provides a sound derivation of non-linear transport equations that fulfill all necessary compatibility requirements. The celebrated Guyer-Krumhansl equations can only be derived under the more restrictive condition of slow processes, and are affected by compatibility problems. The impact of different energy-temperature relations on the structure of equations is discussed.

27.1 Introduction

This paper concerns heat transfer in solids at low temperature as described through kinetic theory of phonons, see the review papers by Dreyer and Struchtrup (1993) and Guo and Wang (2015) for detailed discussions of the history, various theories, experiments, and an exhaustive list of related references.

Phonons undergo two types of interaction, namely so-called normal (N) processes, where phonon energy and quasi-momentum are conserved, and resistive (R) processes, where energy is conserved, but phonon momentum is not conserved. R-processes include phonon-phonon Umklapp processes, as well as phonon interactions with dislocations, grain boundaries, etc.

For general processes involving a phonon gas one must consider the full phonon-Boltzmann equation. Phonon hydrodynamics arises as a reduced description for processes in which momentum conserving N-processes dominate, such that the respective collision frequency $1/\tau_N$, i.e., the inverse of the mean free time τ_N, is considerably larger than relevant process frequencies Ω, that is for $\Omega\tau_N \ll 1$ (Dreyer and Struchtrup 1993; Guo and Wang 2015). Phonon hydrodynamics is a macroscopic

H. Struchtrup (✉)
University of Victoria, Victoria, BC, Canada
e-mail: struchtr@uvic.ca

J. Fuhrmann et al. (eds.), *Advances in Continuum Physics*, Advanced Structured Materials 238, https://doi.org/10.1007/978-3-031-93918-1_27

theory, with the phonon energy density $e\,(x_k, t)$ and momentum density $p_i\,(x_k, t)$ as the only variables that describe the local state of the phonon gas; here (x_k, t) denote location and time.

Often the equations for phonon hydrodynamics are derived for a simplified phonon model, as outlined in Sect. 27.2 below. The best known model of this type was presented in 1966 by Guyer and Krumhansl (1966), who found the celebrated Guyer-Krumhansl (GK) equations as, in our notation,

$$\frac{\partial e}{\partial t} + c^2 \frac{\partial p_k}{\partial x_k} = 0\,, \tag{27.1}$$

$$\frac{\partial p_i}{\partial t} + \frac{1}{3}\frac{\partial e}{\partial x_i} - \frac{1}{5}\tau_N c^2 \left[\frac{\partial^2 p_i}{\partial x_k \partial x_k} + 2\frac{\partial^2 p_k}{\partial x_i \partial x_k} \right] = -\frac{p_i}{\tau_R}\,, \tag{27.2}$$

where c denotes the velocity of dispersion-free phonons, and $1/\tau_R$ is the collision frequency of R-processes. A similar system was presented three years prior by Sussmann and Thellung (1963), who, however, ignored R-processes ($\tau_R \to \infty$).

While Sussmann and Thellung used the language of classical kinetic theory, Guyer and Krumhansl derived their equations from an operator formalism, with both pairs of authors agreeing on the factor '2' in front of the derivative $\frac{\partial^2 p_k}{\partial x_i \partial x_k}$ in (27.2). In the simple phonon model used, the dissipative stress tensor must be trace-free, which demands a factor '1/3' instead. Over the years this discrepancy was pointed out by several authors, already in 1974 by Hardy and Albers (1974) based on the methods of Guyer and Krumhansl, by Dreyer and Struchtrup through the moment method of kinetic theory (1993), by Banach and Larecki through the Chapman-Enskog method of kinetic theory (2008), by Guo and Wang from a multiscale expansion (2015), and just recently by Xu, again based on the original method of Guyer and Krumhansl (2024). From these references is it clear that the original derivations by Sussmann and Thellung and by Guyer and Krumhansl are flawed. Indeed, in the Guyer-Krumhansl case Hardy and Albers (1974) and Xu (2024) point to an oversimplifying assumption in the original derivation leading to the factor '2', whereas the corrected procedure gives '1/3'.

In the following, we revisit the derivation of equations of phonon hydrodynamics by expansions of the phonon kinetic equation in the Callaway model (1959) with relaxation times independent of phonon energy and momentum.

The Chapman-Enskog (CE) expansion provides a fully compatible and problem free derivation, which yields the non-linear extension to the above set of transport equations, with factor '1/3'. Due to the simplicity of the kinetic phonon model employed, the application of the CE method can be simplified considerably by relying on the known moments of the distribution functions involved. Thus, we find a faster procedure in comparison to the work of Banach and Larecki (2008), which is fully equivalent, as are the resulting equations.

The Chapman-Enskog expansion relies on the assumption that $\Omega\tau_N = \varepsilon$, where $\varepsilon \ll 1$ is a suitable smallness parameter. Re-reading Sussmann and Thellung's contribution (1963), we noticed that they explicitly consider *slow time variation*, which we interpret as increased scale splitting such that $\Omega\tau_N = \varepsilon^2$. This, indeed, leads to a set of equations similar to (27.1), (27.2), where, however, the energy density e is replaced by an equilibrium energy $e_E(T)$. Moreover, a closer look shows that the slow time expansion has compatibility problems, which are explicitly avoided in the CE method.

In summary, it will become clear that the GK equations *at best* are valid for extremely slow processes, while their modified form—with factor '1/3'—is valid for a considerably wider range of processes.

The remainder of this contribution is structured as follows: The phonon model and its kinetic equation are introduced in Sect. 27.2. The Chapman-Enskog expansion and its results are outlined in Sect. 27.3, with equations of phonon hydrodynamics presented in fully non-linear, quadratic, and linearized CE forms, with the factor '1/3'. Section 27.4 recalls the Guyer-Krumhansl equations and shows that even a splitting of energy density into equilibrium and non-equilibrium contributions cannot make them compatible with the CE based equations. The revised scaling for *slow time variation* is performed and discussed in Sect. 27.5. Here it will be seen that this scaling indeed yields the GK equations—and not those found by Sussmann and Thellung—but the expansion is accompanied by compatibility issues.

27.2 Kinetic Theory of Phonons

27.2.1 Phonon Model

We consider the often used simplified phonon model as discussed e.g., by Dreyer and Struchtrup (1993), Banach and Larecki (2008), Guo and Wang (2015), where longitudinal and two branches of transversal phonons are not distinguished and phonon dispersion is ignored. Then, phonons are described by the wave vector k_i and the Debye velocity c, such that phonon energy and momentum are given as $\hbar ck$ and $\hbar k_i$, respectively, where $\hbar = h/2\pi$ is the reduced Planck constant.

The phase density of phonons f is defined such that

$$f(x_i, t, k_i)\, d\mathbf{x} d\mathbf{k} \tag{27.3}$$

gives the number of phonons at time t at locations in the interval $d\mathbf{x}$ around $\mathbf{x}$ with wave vectors in the interval $d\mathbf{k}$ around $\mathbf{k}$. In the following it is assumed that the phase density decays to zero within the Brillouin zone, so that the integration over the wave vector can be extended to infinity, which greatly simplifies computations.

Moments of the phase density are defined as local averages over the wave vector. Specifically, energy and momentum density of the phonon gas are given by

$$e(x_i, t) = \int \hbar c k f d\mathbf{k} \quad , \quad p_i(x_i, t) = \int \hbar k_i f d\mathbf{k} \, . \tag{27.4}$$

Higher moments that will be considered are the anisotropic phonon stress tensor

$$N_{\langle ij\rangle} = \int \hbar c \frac{k_{\langle i}k_{i\rangle}}{k} f d\mathbf{k} \tag{27.5}$$

and the trace-free third moment

$$M_{\langle ijk\rangle} = \int \hbar c^2 \frac{k_{\langle i}k_j k_{k\rangle}}{k^2} f d\mathbf{k} \, , \tag{27.6}$$

with indices in angle brackets indicating symmetric and trace-free tensors, as

$$k_{\langle i}k_{i\rangle} = k_i k_j - \frac{1}{3} k^2 \delta_{ij} \quad , \quad k_{\langle i}k_j k_{k\rangle} = k_i k_j k_k - \frac{1}{5}\left(k_i \delta_{jk} + k_j \delta_{ik} + k_k \delta_{ij}\right) \, . \tag{27.7}$$

Note that for the full moments traces are given by energy and momentum density, so that, with obvious notation,

$$N_{ij} = \frac{e}{3}\delta_{ij} + N_{\langle ij\rangle} \quad , \quad M_{ijk} = \frac{1}{5}\left(p_i \delta_{jk} + p_j \delta_{ik} + p_k \delta_{ij}\right) + M_{\langle ijk\rangle} \, . \tag{27.8}$$

27.2.2 Callaway Kinetic Equation

The space time evolution of the phase density is given through the Callaway approximation (Callaway 1959), where the complicated phonon collision term is replaced by two relaxation terms that describe resistive and normal phonon interaction processes as

$$\frac{\partial f}{\partial t} + c\frac{k_k}{k}\frac{\partial f}{\partial x_k} = -\frac{1}{\tau_R}\left(f - f_R\right) - \frac{1}{\tau_N}\left(f - f_N\right) \, , \tag{27.9}$$

with relaxation times τ_R and τ_N. In the simplified model considered throughout, dependence of relaxation times on wave vector k_i is ignored, so that they depend only on the local state of the material through temperature $T(x_i, t)$ or, alternatively, energy density $e(x_i, t)$.

In R-processes, energy is conserved but phonon momentum is not, hence

$$P^R = \int \hbar c k \frac{1}{\tau_R}\left(f - f_R\right) d\mathbf{k} = 0 \, , \tag{27.10}$$

while in N-processes energy and momentum are conserved,

$$P^N = \int \hbar ck \frac{1}{\tau_N} (f - f_N) \, d\mathbf{k} = 0 \quad , \quad P_i^N = \int \hbar k_i \frac{1}{\tau_N} (f - f_N) \, d\mathbf{k} = 0 \, . \tag{27.11}$$

With the relaxation times independent of the wave vector, the conservation conditions imply

$$e_R = \int \hbar ck f_R d\mathbf{k} = e \, , \tag{27.12}$$

$$e_N = \int \hbar ck f_N d\mathbf{k} = e \quad , \quad p_i^N = \hbar \int k_i f_N d\mathbf{k} = p_i \, . \tag{27.13}$$

The equilibrium distribution functions f_R and f_N are obtained from maximizing phonon entropy (Dreyer 1987) under constraints of prescribed energy e (for R-processes) and prescribed energy e and momentum p_i (for N-processes), which yields (Dreyer and Struchtrup 1993)

$$f_R = \frac{y}{\exp\left[k\lambda_R\right] - 1} \, , \tag{27.14}$$

$$f_N = \frac{y}{\exp\left[k\left(\lambda + \Lambda_i \frac{k_i}{k}\right)\right] - 1} \, , \tag{27.15}$$

where $y = \frac{3}{8\pi^3}$, and λ_R, λ, Λ_i are Lagrange multipliers that must be determined from (27.12), (27.13). Here, f_R is the equilibrium distribution of the gas which is isotropic in the rest frame of the crystal, and f_N is the distribution of a drifting phonon gas.

27.2.3 R-Processes

Specifically, we obtain from (27.12), (27.14)

$$e = \frac{\pi^2}{10} \frac{\hbar c}{\lambda_R^4} \, , \tag{27.16}$$

which implies

$$\lambda_R = \left(\frac{\pi^2}{10} \frac{\hbar c}{e} \right)^{1/4} \, . \tag{27.17}$$

In thermal equilibrium at temperature T the Lagrange multiplier is identified as (Dreyer and Struchtrup 1993)

$$\lambda_R = \frac{\hbar c}{k_B T} \quad \text{so that} \quad e = \frac{\pi^2}{10} \frac{k_B^4 T^4}{\hbar^3 c^3} \, , \tag{27.18}$$

where the relation between energy and temperature is the well-known Debye law; k_B denotes the Boltzmann constant .

Due to isotropy of the distribution all relevant higher moments of f_R vanish,

$$p_i^R = \hbar \int k_i f_R d\mathbf{k} = 0 \,, \tag{27.19}$$

$$N_{\langle ij\rangle}^R = \int \hbar c \frac{k_{\langle i}k_{j\rangle}}{k} f_R d\mathbf{k} = 0 \,, \tag{27.20}$$

$$M_{\langle ijk\rangle}^R = \int \hbar c^2 \frac{k_{\langle i}k_j k_{k\rangle}}{k^2} f_R d\mathbf{k} = 0 \,. \tag{27.21}$$

27.2.4 N-Processes

We denote the general irreducible moments of the drift distribution f_N as

$$u_{\langle i_1\cdots i_n\rangle}^N = \int \hbar c^{n-1} \frac{k_{\langle i_1}\cdots k_{i_n\rangle}}{k^{n-1}} f_N d\mathbf{k} = (-1)^n \alpha_n \frac{\Lambda_{\langle i_1}\cdots \Lambda_{i_n\rangle}}{\Lambda^n} \,, \tag{27.22}$$

where the tensor structure of the right hand side results from the fact that f_N depends on the single vector Λ_i. Specifically, with (27.13) we identify

$$e = e_N = c^2 u^N \quad , \quad p_i = p_i^N = u_i^N \quad , \quad N_{\langle ij\rangle}^N = u_{\langle ij\rangle}^N \quad , \quad M_{\langle ijk\rangle}^N = u_{\langle ijk\rangle}^N \,. \tag{27.23}$$

For the determination of the α_n, we introduce the direction vector

$$n_i = \frac{k_i}{k} = \{\cos\phi \sin\theta, \sin\phi\sin\theta, \cos\theta\} \tag{27.24}$$

so that with (27.15)

$$\int_0^{2\pi}\int_0^{\pi}\int_0^{\infty} \hbar c^{n-1} y \frac{n_{\langle i_1}\cdots n_{i_n\rangle}}{\exp\left[k\left(\lambda + \Lambda_j n_j\right)\right]-1} k^3 dk \sin\theta d\theta d\phi = (-1)^n \alpha_n \frac{\Lambda_{\langle i_1}\cdots \Lambda_{i_n\rangle}}{\Lambda^n} \tag{27.25}$$

and performing the ϕ and k integrations, we find (recall $y = \frac{3}{8\pi^3}$)

$$\hbar c^{n-1} \frac{\pi^2}{20} \int_0^{\pi} \frac{n_{\langle i_1}\cdots n_{i_n\rangle}}{\left(\lambda + \Lambda_j n_j\right)^4} \sin\theta d\theta = (-1)^n \alpha_n \frac{\Lambda_{\langle i_1}\cdots \Lambda_{i_n\rangle}}{\Lambda^n} \,. \tag{27.26}$$

The last integral is best solved in a frame in which

$$\Lambda_i = \{0, 0, \Lambda\} \,, \tag{27.27}$$

where we must consider only the $3\cdots3$ component of the tensor equation to determine α_n:

$$\hbar c^{n-1}\frac{\pi^2}{20}\int_0^\pi \frac{n_{\langle 3}\cdots n_{3\rangle}}{(\lambda+\Lambda n_3)^4}\sin\theta d\theta = (-1)^n\,\alpha_n\frac{\Lambda_{\langle 3}\cdots\Lambda_{3\rangle}}{\Lambda^n}\,. \tag{27.28}$$

Setting $\cos\theta = \Theta$ and integrating we identify

$$\alpha_0 = \hbar c^{n-1}\frac{\pi^2}{20}\int_{-1}^1 \frac{1}{(\lambda+\Lambda\Theta)^4}d\Theta = \frac{\hbar}{c}\frac{\pi^2}{30}\frac{3\lambda^2+\Lambda^2}{\left(\lambda^2-\Lambda^2\right)^3} \tag{27.29}$$

$$\alpha_1 = \hbar c^{n-1}\frac{\pi^2}{20}\int_{-1}^1 \frac{\Theta}{(\lambda+\Lambda\Theta)^4}d\Theta = \hbar\frac{2\pi^2}{15}\frac{\lambda\Lambda}{\left(\lambda^2-\Lambda^2\right)^3} \tag{27.30}$$

$$\alpha_2 = \frac{3}{2}\hbar c^{n-1}\frac{\pi^2}{20}\int_{-1}^1 \frac{\Theta^2-\frac{1}{3}}{(\lambda+\Lambda\Theta)^4}d\Theta = \hbar c\frac{2\pi^2}{15}\frac{\Lambda^2}{\left(\lambda^2-\Lambda^2\right)^3} \tag{27.31}$$

$$\begin{aligned}\alpha_3 &= \frac{5}{2}\hbar c^{n-1}\frac{\pi^2}{20}\int_{-1}^1 \frac{\Theta^3-\frac{3}{5}\Theta}{(\lambda+\Lambda\Theta)^4}d\Theta \\ &= \hbar c^2\frac{\pi^2}{4}\frac{1}{\lambda\Lambda^3}\left[\frac{1-\frac{8}{3}\left(\frac{\Lambda}{\lambda}\right)^2+\frac{11}{5}\left(\frac{\Lambda}{\lambda}\right)^4}{\left[1-\left(\frac{\Lambda}{\lambda}\right)^2\right]^3}+\frac{1}{2}\ln\frac{1-\frac{\Lambda}{\lambda}}{1-\frac{\Lambda}{\lambda}}\right],\end{aligned} \tag{27.32}$$

where it was used that $\Lambda_{\langle 3}\Lambda_{3\rangle} = \frac{2}{3}\Lambda^2$, $\Lambda_{\langle 3}\Lambda_3\Lambda_{3\rangle} = \frac{2}{5}\Lambda^3$, which follow from (27.7).

The two Lagrange multipliers λ, Λ_i are obtained from energy and momentum densities,

$$e = \hbar c\frac{\pi^2}{30}\frac{3\lambda^2+\Lambda^2}{\left(\lambda^2-\Lambda^2\right)^3}\quad,\quad p_i = -\hbar\frac{2\pi^2}{15}\frac{\lambda\Lambda}{\left(\lambda^2-\Lambda^2\right)^3}\frac{\Lambda_i}{\Lambda}\,, \tag{27.33}$$

which imply

$$\frac{c^2p^2}{e^2} = \frac{16}{9}\frac{\lambda^2\Lambda^2}{\left(\lambda^2+\frac{1}{3}\Lambda^2\right)^2} \tag{27.34}$$

as well as

$$e\sqrt{1-\frac{3}{4}\frac{c^2p^2}{e^2}} = \hbar c\frac{\pi^2}{30}\frac{3\lambda^2-\Lambda^2}{\left(\lambda^2-\Lambda^2\right)^3} \tag{27.35}$$

and

$$\frac{\Lambda^2}{\lambda^2} = 8\frac{e^2}{c^2p^2}\left[1-\frac{3}{8}\frac{c^2p^2}{e^2}-\sqrt{1-\frac{3}{4}\frac{c^2p^2}{e^2}}\right]. \tag{27.36}$$

Combining the above, the second moment is found as the well-known expression (Larecki 1992; Müller and Ruggeri 1998)

$$N^N_{\langle ij\rangle} = 2e\left(1-\sqrt{1-\frac{3}{4}\frac{c^2p^2}{e^2}}\right)\frac{p_{\langle i}p_{j\rangle}}{p^2}\,. \tag{27.37}$$

The third moment is found as

$$M^N_{\langle ijk\rangle} = -\alpha_3\frac{\Lambda_{\langle i}\Lambda_j\Lambda_{k\rangle}}{\Lambda^3} = \frac{\alpha_3}{p^3}p_{\langle i}p_jp_{k\rangle}\,, \tag{27.38}$$

where the expression for $\alpha_3\,(e, p_i)$ must be obtained from (27.32) and (27.36) to replace (λ, Λ_i) by (e, p_i). Strong non-linearity is normally not expected, hence it suffices to consider a Taylor expansion in $\frac{c^2p^2}{e^2}$, which yields, to 7th order in p_i,

$$M^N_{\langle ijk\rangle} = \frac{27}{56}\frac{c^4}{e^2}\left[1+\frac{7}{16}\frac{c^2p^2}{e^2}+\frac{327}{1408}\frac{c^4p^4}{e^4}\right]p_{\langle i}p_jp_{k\rangle}\,. \tag{27.39}$$

27.2.5 Transport Equations and Closure Problem

The energy balance arises by multiplying the Callaway equation (27.9) with phonon energy $\hbar ck$ and integration over $\mathbf{k}$ as

$$\frac{\partial e}{\partial t} + c^2\frac{\partial p_k}{\partial x_k} = 0\,, \tag{27.40}$$

with the energy flux $Q_k = c^2p_k$ proportional to the momentum density. Energy conservation is ensured by (27.12), (27.13).

The transport equation for momentum density is obtained from multiplication of the Callaway equation with momentum $\hbar k_i$ and integration as

$$\frac{\partial p_i}{\partial t} + \frac{1}{3}\frac{\partial e}{\partial x_i} + \frac{\partial N_{\langle ik\rangle}}{\partial x_k} = -\frac{p_i}{\tau_R}\,, \tag{27.41}$$

where phonon pressure is identified as $\frac{1}{3}e$ and $N_{\langle ik\rangle}$ is the phonon stress tensor. The phonon momentum is damped due to R-processes, while the N-processes do not contribute to due to (27.13).

The hydrodynamic description of the phonon gas requires a constitutive relation between the phonon stress $N_{\langle ik\rangle}$ and the hydrodynamic variables energy e and momentum p_i, and their gradients.

The goal in the following is to determine constitutive approximations for the phonon stress $N_{\langle ik\rangle}$ from the Callaway equation by means of the Chapman-Enskog expansion (Sect. 27.3), as well as a revised expansion (Sect. 27.5).

27.3 Chapman-Enskog Method

27.3.1 Scaling and Expansion

The Chapman-Enskog expansion relies on the assumption that N-processes are frequent so that the relaxation time τ_N is smaller than the relevant time scale of the processes. Expressing the time scale through a frequency Ω, we have $\tau_N \Omega \ll 1$. We consider this through a formal smallness parameter ε in the N-process term,

$$\frac{\partial f}{\partial t} + c\frac{k_i}{k}\frac{\partial f}{\partial x_i} = -\frac{1}{\tau_R}\left(f - f_R\right) - \frac{1}{\varepsilon}\frac{1}{\tau_N}\left(f - f_N\right) . \tag{27.42}$$

The smallness parameter plays the role of the Knudsen number that would appear in a dimensionless formulation of the equation. We will expand the dimensional Callaway equation in ε and then set ε to unity at the end of the evaluation, which is tantamount to making the equation dimensionless, expand in the Knudsen number, and than reintroduce dimensional quantities. Note that there are no restrictions on the frequency of R-processes.

We follow the procedure of the Chapman-Enskog expansion as outlined in Struchtrup (2005), Kremer (2010). The distribution function is approximated as a series in the smallness parameter

$$f = f^{(0)} + \varepsilon f^{(1)} + \varepsilon^2 f^{(2)} + \cdots \tag{27.43}$$

The expansion explicitly demands that the higher order terms do not affect the base variables, which is formulated in the compatibility requirement that, for all $\alpha \geq 1$,

$$e = \hbar c \int k f^{(0)} d\mathbf{k} \quad , \quad p_i = \hbar \int k_i f^{(0)} d\mathbf{k} \tag{27.44}$$

$$0 = \hbar c \int k f^{(\alpha)} d\mathbf{k} \quad , \quad 0 = \hbar \int k_i f^{(\alpha)} d\mathbf{k} \tag{27.45}$$

The second moment, however is expanded with the distribution function,

$$N_{\langle ij\rangle} = N_{\langle ij\rangle}^{(0)} + \varepsilon N_{\langle ij\rangle}^{(1)} + \varepsilon^2 N_{\langle ij\rangle}^{(2)} + \cdots \tag{27.46}$$

Inserting this into the momentum balance

$$\frac{\partial p_i}{\partial t} = -\frac{1}{3}\frac{\partial e}{\partial x_i} - \frac{p_i}{\tau_R} - \frac{\partial N^{(0)}_{\langle ik\rangle}}{\partial x_k} - \varepsilon\frac{\partial N^{(1)}_{\langle ik\rangle}}{\partial x_k} - \varepsilon^2\frac{\partial N^{(2)}_{\langle ik\rangle}}{\partial x_k}\cdots \tag{27.47}$$

it is recognized that the time derivative can be considered as an expansion in ε

$$\frac{\partial p_i}{\partial t} = \frac{\partial_0 p_i}{\partial t} + \varepsilon\frac{\partial_1 p_i}{\partial t} + \varepsilon^2\frac{\partial_2 p_i}{\partial t} + \cdots \tag{27.48}$$

where the scaled time derivatives are defined as

$$\frac{\partial_0 p_i}{\partial t} = -\frac{1}{3}\frac{\partial e}{\partial x_i} - \frac{p_i}{\tau_R} - \frac{\partial N^{(0)}_{\langle ik\rangle}}{\partial x_k} \quad , \quad \frac{\partial_\alpha p_i}{\partial t} = -\frac{\partial N^{(\alpha)}_{\langle ik\rangle}}{\partial x_k} \, . \tag{27.49}$$

Note that the energy balance contains only the variables (e, p_i), hence it is not further expanded, that is

$$\frac{\partial_0 e}{\partial t} = \frac{\partial e}{\partial t} = -c^2\frac{\partial p_k}{\partial x_k} \quad , \quad \frac{\partial_\alpha e}{\partial t} = 0 \, . \tag{27.50}$$

We insert the expansion for f into the scaled Callaway equation (27.42) to find with terms up to order ε :

$$\frac{\partial}{\partial t}\left(f^{(0)} + \varepsilon f^{(1)}\right) + c\frac{k_k}{k}\frac{\partial}{\partial x_k}\left(f^{(0)} + \varepsilon f^{(1)}\right) =$$
$$-\frac{1}{\tau_R}\left(f^{(0)} - f_R + \varepsilon f^{(1)}\right) - \frac{1}{\tau_N}\left(\frac{f^{(0)} - f_N}{\varepsilon} + f^{(1)} + \varepsilon f^{(2)}\right) \tag{27.51}$$

From this, the expansion terms $f^{(\alpha)}$ are identified one by one starting at the lowest order where we identify the zeroth order approximation as the drift distribution,

$$f^{(0)} = f_N \, ; \tag{27.52}$$

this gives the leading contribution to phonon stress as

$$N^{(0)}_{\langle ij\rangle} = \hbar c \int \frac{k_{\langle i}k_{j\rangle}}{k} f_N d\mathbf{k} = N^N_{\langle ij\rangle} \, . \tag{27.53}$$

Since $f^{(0)} = f_N$ depends on space and time only through the variables (e, p_i), its time derivative $\frac{\partial f_N}{\partial t}$ must be expanded with the moments, as given in (27.48), (27.50). With this, the remaining equation (27.51) becomes, to order ε,

$$\frac{\partial_0 f_N}{\partial t}+\varepsilon\frac{\partial_1 f_N}{\partial t}+\frac{\partial \varepsilon f^{(1)}}{\partial t}+c\frac{k_k}{k}\frac{\partial f_N+\varepsilon f^{(1)}}{\partial x_k}$$
$$=-\frac{1}{\tau_R}\left(f_N-f_R+\varepsilon f^{(1)}\right)-\frac{1}{\tau_N}\left(f^{(1)}+\varepsilon f^{(2)}\right). \tag{27.54}$$

Equating terms at order ε^0 we find the first correction to the distribution function as

$$f^{(1)}=-\tau_N\left[\frac{\partial_0 f_N}{\partial t}+c\frac{k_k}{k}\frac{\partial f_N}{\partial x_k}+\frac{1}{\tau_R}\left(f_N-f_R\right)\right] \tag{27.55}$$

Before proceeding with the evaluation for phonon stress, we check this expression for compatibility (27.45). With the relaxation times independent of wave vector, this can be done without having to take any explicit derivatives of the distribution f_N. The compatibility condition (27.45) for energy becomes, based on the properties of $f_{R,N}$, and with (27.50)

$$\hbar c\int k f^{(1)} d\mathbf{k}=-\tau_N\left[\frac{\partial_0 e}{\partial t}+c^2\frac{\partial p_k}{\partial x_k}+\frac{1}{\tau_R}\left(e-e\right)\right]=0 \tag{27.56}$$

that is energy compatibility is fulfilled. For momentum we find immediately that, with (27.49) and (27.53),

$$\hbar c^2\int k_i f^{(1)} d\mathbf{k}=-\tau_N\left[\frac{\partial_0 p_i}{\partial t}+\frac{1}{3}\frac{\partial e}{\partial x_i}+\frac{\partial N^N_{\langle ik\rangle}}{\partial x_k}+\frac{p_i}{\tau_R}\right]=0 \tag{27.57}$$

that is momentum compatibility is fulfilled.

With this the compatibility requirements are fulfilled automatically, with no further efforts required. This corresponds to the application of the CE expansion to the BGK equation of classical kinetic theory, while in CE expansion of the full Boltzmann equation additional work is required to ensure compatibility (Struchtrup 2005).

The corresponding contribution to stress is

$$N^{(1)}_{\langle ij\rangle}=\hbar c\int\frac{k_{\langle i}k_{j\rangle}}{k}f^{(1)} d\mathbf{k}$$
$$=-\tau_N\hbar c\int\frac{k_{\langle i}k_{j\rangle}}{k}\left[\frac{\partial_0 f_N}{\partial t}+c\frac{k_k}{k}\frac{\partial f_N}{\partial x_k}+\frac{1}{\tau_R}\left(f_N-f_R\right)\right]d\mathbf{k} \tag{27.58}$$

As for the compatibility checks, time and space derivatives can be pulled out of the integrals, so that

$$N_{\langle ij\rangle}^{(1)} = -\tau_N \frac{\partial_0}{\partial t} \int \hbar c \frac{k_{\langle i}k_{j\rangle}}{k} f_N d\mathbf{k} - \tau_N \frac{\partial}{\partial x_k} \int \hbar c^2 \frac{k_{\langle i}k_{j\rangle}}{k} \frac{k_k}{k} f_N d\mathbf{k}$$
$$-\frac{\tau_N}{\tau_R} \hbar c \int \frac{k_{\langle i}k_{j\rangle}}{k} \left(f_N - f_R\right) d\mathbf{k} \qquad (27.59)$$

To proceed, we must rewrite the argument in the middle term as sum of irreducible tensors. With the definitions of trace-free tensors (27.7)

$$\frac{k_{\langle i}k_{j\rangle}k_k}{k^2} = \frac{k_i k_j k_k}{k^2} - \frac{1}{3} k_k \delta_{ij} = \frac{k_{\langle i}k_j k_{k\rangle}}{k^2} + \frac{2}{5} k_{\langle i}\delta_{j\rangle k}\,, \qquad (27.60)$$

so that after integration the first correction to stress is determined in terms of the moments of f_N and their derivatives,

$$N_{\langle ij\rangle}^{(1)} = -\tau_N \frac{\partial_0 N_{\langle ij\rangle}^N}{\partial t} - \tau_N \frac{\partial M_{\langle ijk\rangle}^N}{\partial x_k} - \frac{2}{5}\tau_N c^2 \frac{\partial p_{\langle i}}{\partial x_{j\rangle}} - \frac{\tau_N}{\tau_R} N_{\langle ij\rangle}^N\,. \qquad (27.61)$$

As a last step, the zeroth-order time derivative is replaced by means of (27.49), (27.50),

$$\frac{\partial_0 N_{\langle ij\rangle}^N}{\partial t} = \frac{\partial N_{\langle ij\rangle}^N}{\partial e} \frac{\partial_0 e}{\partial t} + \frac{\partial N_{\langle ij\rangle}^N}{\partial p_r} \frac{\partial_0 p_r}{\partial t}$$
$$= -\frac{\partial N_{\langle ij\rangle}^N}{\partial e} \left[c^2 \frac{\partial p_k}{\partial x_k}\right] - \frac{\partial N_{\langle ij\rangle}^N}{\partial p_r} \left[\frac{1}{3} \frac{\partial e}{\partial x_r} + \frac{p_r}{\tau_R} + \frac{\partial N_{\langle ik\rangle}^N}{\partial x_k}\right] \qquad (27.62)$$

so that finally

$$N_{\langle ij\rangle}^{(1)} = -\frac{2}{5} c^2 \tau_N \frac{\partial p_{\langle i}}{\partial x_{j\rangle}} - \frac{\tau_N}{\tau_R} N_{\langle ij\rangle}^N + \tau_N c^2 \frac{\partial N_{\langle ij\rangle}^N}{\partial e} \frac{\partial p_k}{\partial x_k} +$$
$$+\tau_N \frac{\partial N_{\langle ij\rangle}^N}{\partial p_r} \left[\frac{1}{3} \frac{\partial e}{\partial x_r} + \frac{\partial N_{\langle ik\rangle}^N}{\partial x_k} - \frac{p_r}{\tau_R}\right] - \tau_N \frac{\partial M_{\langle ijk\rangle}^N}{\partial x_k}\,. \qquad (27.63)$$

We will not consider higher order contributions in the expansion. It is well known that for ideal gases higher order Chapman-Enskog expansions tend to give unstable equations (Burnett, super-Burnett), see, e.g., Struchtrup (2005). Extended moment methods as discussed in Struchtrup (2005) offer an alternative approach with stable equations, see (Dreyer and Struchtrup 1993; Fryer and Struchtrup 2014) for applications to phonons.

The CE expansion of the Callaway equation was first presented by Banach and Larecki (2008), who, however, went through the more cumbersome route of taking the derivatives of f_N with respect to (e, p_i) to determine $f^{(1)}$ explicitly. The approach presented here takes advantage of the simple structure of the Callaway equation with

k_i-independent relaxation times that allows to exchange the order of taking derivatives and integration.

27.3.2 Phonon Hydrodynamics from CE Expansion

We collect the equations in fully non-linear form, and their quadratic and linear approximations, where we now set the scaling parameter $\varepsilon = 1$.

In any case, the transport equations for energy and momentum are

$$\frac{\partial e}{\partial t} + c^2 \frac{\partial p_k}{\partial x_k} = 0 \,, \tag{27.64}$$

$$\frac{\partial p_i}{\partial t} + \frac{1}{3}\frac{\partial e}{\partial x_i} + \frac{\partial N_{\langle ik\rangle}}{\partial x_k} = -\frac{p_i}{\tau_R} \,, \tag{27.65}$$

with differences only in the expressions for the phonon stress

$$N_{\langle ik\rangle} = N^N_{\langle ik\rangle} + N^{(1)}_{\langle ik\rangle} \tag{27.66}$$

with the zeroth and first order contributions given in (27.53), (27.63).

27.3.3 General Non-linear Equations

In the most general non-linear setting, the first order Chapman-Enskog expansion yields

$$\begin{aligned} N_{\langle ik\rangle} = & -\frac{2}{5}\tau_N \frac{\partial c^2 p_{\langle i}}{\partial x_{j\rangle}} + \left(1 - \frac{\tau_N}{\tau_R}\right) N^N_{\langle ij\rangle} + \tau_N c^2 \frac{\partial N^N_{\langle ij\rangle}}{\partial e}\frac{\partial p_k}{\partial x_k} \\ & + \tau_N \frac{\partial N^N_{\langle ij\rangle}}{\partial p_r}\left[\frac{1}{3}\frac{\partial e}{\partial x_r} + \frac{\partial N^N_{\langle ik\rangle}}{\partial x_k} - \frac{p_r}{\tau_R}\right] - \tau_N \frac{\partial M^N_{\langle ijk\rangle}}{\partial x_k} \end{aligned} \tag{27.67}$$

with the non-linear expressions

$$N^N_{\langle ij\rangle} = 2e\left(1 - \sqrt{1 - \frac{3}{4}\frac{c^2 p^2}{e^2}}\right)\frac{p_{\langle i}p_{j\rangle}}{p^2} \,, \tag{27.68}$$

$$M^N_{\langle ijk\rangle} = \frac{27}{56}\frac{c^4}{e^2}\left[1 + \frac{7}{16}\frac{c^2 p^2}{e^2} + \frac{327}{1408}\frac{c^4 p^4}{e^4}\right] p_{\langle i}p_j p_{k\rangle} \,, \tag{27.69}$$

Note that the relaxation times τ_R, τ_N generally are not constants but depend on the local state which implies additional nonlinearities in the momentum balance introduced through the divergence of $N_{\langle ik\rangle}$. These equations are equivalent to those found by Banach and Larecki (2008), who, however, use not the phonon momentum as a variable, but the drift velocity; moreover, they consider the relaxation times as true constants.

27.3.4 Quadratic Non-linearity

Most applications in the literature are linear, hence already the leading non-linear contributions will be of interest. Thus, we consider up to quadratic terms in p_i and the gradients of p_i and e only, that is the approximations

$$N^N_{\langle ij\rangle} \simeq \frac{3}{4}\frac{c^2}{e} p_{\langle i} p_{j\rangle} \quad , \quad M^N_{\langle ijk\rangle} = \beta\left(e, p^2\right) p_{\langle i} p_j p_{k\rangle} \simeq 0 \tag{27.70}$$

and leave only terms up to second order in the expression for stress, which thus reduces to

$$\text{quadratic:} \quad N_{\langle ij\rangle} = -\frac{2}{5}\tau_N c^2 \frac{\partial p_{\langle i}}{\partial x_{j\rangle}} + \frac{1}{2}\tau_N c^2 p_{\langle i} \frac{\partial \ln e}{\partial x_{j\rangle}} + \left(1 - 3\frac{\tau_N}{\tau_R}\right)\frac{3}{4}\frac{c^2}{e} p_{\langle i} p_{j\rangle} \,. \tag{27.71}$$

27.3.5 Linear Theory

Finally, removing all non-linear contributions we recover the linear transport equations, for small deviations from an equilibrium groundstate, where

$$\text{linear:} \quad N_{\langle ik\rangle} = -\frac{2}{5}\tau_N \frac{\partial c^2 p_{\langle i}}{\partial x_{j\rangle}} \tag{27.72}$$

so that

$$\frac{\partial e}{\partial t} + c^2 \frac{\partial p_k}{\partial x_k} = 0 \tag{27.73}$$

$$\frac{\partial p_i}{\partial t} + \frac{1}{3}\frac{\partial e}{\partial x_i} - \frac{1}{5}\tau_N c^2 \left[\frac{\partial^2 p_i}{\partial x_k \partial x_k} + \frac{1}{3}\frac{\partial^2 p_k}{\partial x_i \partial x_k}\right] = -\frac{p_i}{\tau_R} \tag{27.74}$$

Note that on this level of linearization the relaxation times appear at their groundstate values. As discussed in the Introduction, these equations differ from the Guyer-Krumhansl equations (27.1), (27.2) in the factor '1/3' instead of '2'.

27.3.6 *Scaling of τ_R*

The Chapman-Enskog expansion relies on high frequency of N-processes in comparison to the process frequency, $\Omega \ll \frac{1}{\tau_N}$, with no further restrictions on the scaling of the collision frequency for R-processes $1/\tau_R$. Thus, the Chapman-Enskog based equations (27.73), (27.74) are valid for all τ_R.

Famously, the equations of phonon hydrodynamics were used to identify the window condition for second sound as processes with dominant frequencies Ω between the collision frequencies (Guyer and Krumhansl 1964),

$$\text{second sound:} \quad \frac{1}{\tau_R} \ll \Omega \ll \frac{1}{\tau_N} . \tag{27.75}$$

This suggests to scale

$$\text{second sound scaling:} \quad \frac{\varepsilon}{\tau_R} \quad , \quad \Omega \quad , \quad \frac{1}{\varepsilon \tau_N} , \tag{27.76}$$

for which the momentum balance (27.74) with explicit scaling factor reads

$$\frac{\partial p_i}{\partial t} + \frac{1}{3}\frac{\partial e}{\partial x_i} - \varepsilon \frac{1}{5}\tau_N c^2 \left[\frac{\partial^2 p_i}{\partial x_k \partial x_k} + \frac{1}{3}\frac{\partial^2 p_k}{\partial x_i \partial x_k} \right] = -\varepsilon \frac{p_i}{\tau_R} . \tag{27.77}$$

Thus, second sound scaling is compatible with the CE expansion to first order in ε.

Diffusion of phonons is the dominant transport mechanism in case that R-processes are frequent, where

$$\text{diffusion:} \quad \Omega \ll \frac{1}{\tau_R} \tag{27.78}$$

which suggests the scaling

$$\text{diffusion scaling:} \quad \Omega \quad , \quad \frac{1}{\varepsilon \tau_R} \quad , \quad \frac{1}{\varepsilon \tau_N} . \tag{27.79}$$

In this case, the right hand side of the momentum balance becomes $-\frac{p_i}{\varepsilon \tau_R}$, so that CE expansion of the momentum balance irrespective of τ_N gives Fourier's law

$$p_i = -\varepsilon \frac{\tau_R}{3} \frac{\partial e}{\partial x_i} . \tag{27.80}$$

27.3.7 A Remark on Temperature

Temperature is well defined as a measurable property in equilibrium where the Debye law (27.18) holds, and the state of the phonon gas is described by only one property, i.e., either temperature T or energy e. The nonequilibrium states of the drifting phonon gas as described by the closed transport equations are given through energy and momentum (e, p_i), and one might ask how temperature could—or should—be defined in this case. This question is answered differently by different authors.

The compatibility requirement for the Chapman-Enskog expansion ensures that energy has no contributions from higher order terms in the expansion. From that is appears quite natural to extend the equilibrium definition of temperature into the nonequilibrium realm, that is temperature in nonequilibrium is chosen such that

$$e = e_N = e_R = \frac{\pi^2}{10} \frac{k_B^4}{\hbar^3 c^3} T^4 \, . \tag{27.81}$$

This definition was used, e.g., in Dreyer and Struchtrup (1993), and corresponds to the usual definition of nonequilibrium temperature in the kinetic theory of gases by extending the validity of the caloric equation of state into nonequilibrium (Struchtrup 2005).

As will be discussed in the next section in some detail, other expansions lead to contributions to energy from higher order terms. For instance Sussmann and Thellung (1963) find a relation that for the phonon model considered here becomes

$$e = e_E (T) - \tau_N c^2 \frac{\partial p_i}{\partial x_i} \, , \tag{27.82}$$

where

$$e_E (T) = \frac{\pi^2}{10} \frac{k_B^4}{\hbar^3 c^3} T^4 \tag{27.83}$$

is the Debye law as the equilibrium contribution to energy, and $-\tau_N c^2 \frac{\partial p_i}{\partial x_i}$ is a nonequilibrium correction.

Other definitions are possible. Cimmelli et al. (2010), working in the realm of phenomenological nonequilibrium thermodynamics, introduce and evaluate a dynamical nonequilibrium temperature; a further discussion of this is outside the scope of our considerations.

27.4 Guyer-Krumhansl Equations

27.4.1 The Equations

Phonon hydrodynamics (27.73), (27.74) was famously presented by Guyer and Krumhansl (1966), who, however, do not write their equations in terms of the energy and phonon momentum, but in terms of phonon temperature T and phonon drift velocity v_i; the GK equations read (Guyer and Krumhansl 1966)

$$\frac{1}{T_0}\frac{\partial T}{\partial t}+\frac{1}{3}\frac{\partial v_k}{\partial x_k}=0 \tag{27.84}$$

$$\frac{\partial v_i}{\partial t}+\frac{c^2}{T_0}\frac{\partial T}{\partial x_i}-\frac{1}{5}\tau_N c^2\left[\frac{\partial^2 v_i}{\partial x_k \partial x_k}+2\frac{\partial^2 v_k}{\partial x_i \partial x_k}\right]=-\frac{v_i}{\tau_R} \tag{27.85}$$

where T_0 is the temperature of the equilibrium groundstate. To translate into our notation, we use the relations

$$e_E=\frac{\pi^2}{10}\frac{k_B^4 T^4}{\hbar^3 c^3}\quad , \quad v_i=\frac{3}{4}\frac{c^2 p_i}{e}\ , \tag{27.86}$$

where $e_E\,(T)$ denotes the temperature dependent energy density as in (27.83). The relation for drift velocity results from comparing notation with Guyer and Krumhansl (1966) to identify $v_i=-\frac{c\Lambda_i}{\lambda}$, and then using the linearization of (27.33).

The GK equations are valid in the linear regime, where for small deviations from a rest state at T_0, $v_i^0=0$ specific heat is

$$c_v^0=\frac{\partial e_E}{\partial T}_{|0}=4\frac{\pi^2}{10}\frac{k_B^4 T_0^3}{\hbar^3 c^3}=4\frac{e_E^0}{T_0} \tag{27.87}$$

Hence, multiplication of (27.84) with $\frac{4}{3}\frac{e_E^0}{c^2}=\frac{1}{3}c_v^0\frac{T_0}{c^2}$ yields, with $c_v^0\frac{\partial T}{\partial x_i}=\frac{\partial e_E}{\partial x_i}$,

$$\frac{\partial e_E}{\partial t}+c^2\frac{\partial p_k}{\partial x_k}=0\ , \tag{27.88}$$

$$\frac{\partial p_i}{\partial t}+\frac{1}{3}\frac{\partial e_E}{\partial x_i}-\frac{1}{5}\tau_N c^2\left[\frac{\partial^2 p_i}{\partial x_k \partial x_k}+2\frac{\partial^2 p_k}{\partial x_i \partial x_k}\right]=-\frac{p_i}{\tau_R}\ . \tag{27.89}$$

While the energy balance agrees in principle with (27.73), in the momentum balance the expression for divergence of stress has the factor '2', instead of the factor '1/3' that is found in (27.74).

27.4.2 Nonequilibrium Energy Relation

For dispersion-free phonons as considered throughout, energy density and full phonon stress are obtained from the distribution function as

$$e = \int \hbar c k f d\mathbf{k} \quad , \quad N_{ij} = \int \hbar c \frac{k_i k_j}{k} f d\mathbf{k} \, , \tag{27.90}$$

which implies for the the flux of momentum, that is the second moment,

$$N_{ij} = \frac{e}{3}\delta_{ij} + N_{\langle ij \rangle} \, . \tag{27.91}$$

This splitting into energy e as the trace and the trace-free stress $N_{\langle ij \rangle}$ is strictly obeyed in the equations from CE expansion, including the linearized momentum balance (27.74), but appears to be violated in the GK equation (27.89). Since by definition the stress tensor must be trace-free, the GK momentum equation only has the required form (27.74) if one sets

$$e = e_E(T) - \tau_N c^2 \frac{\partial p_i}{\partial x_i} \, , \tag{27.92}$$

that is energy is split into the temperature dependent equilibrium part $e_E(T)$ and a nonequilibrium contribution with the divergence of momentum.

In Dreyer and Struchtrup (1993) we recognized this contradiction between the GK energy and momentum balances, since in the energy balance e_E appears as the sole contribution to energy, while in the momentum balance the nonequilibrium contribution appears.

Sussmann and Thellung, who completely ignore R-processes (that is, $\tau_R \to \infty$), derive a similar splitting which appears in their energy *and* momentum balances (Sussmann and Thellung 1963).

To investigate the link between the GK equations and those from CE, we insert the energy in GK form (27.92) into the linearized balance laws (27.73), (27.74). In the spirit of the CE expansion we are interested only in the leading terms in τ_N (or, $\varepsilon\tau_N$ when scaled), hence we remove all terms that are quadratic in τ_N to obtain

$$\frac{\partial e_E}{\partial t} + c^2 \frac{\partial p_k}{\partial x_k} - \tau_N c^2 \frac{\partial}{\partial x_k}\left[\frac{\partial p_k}{\partial t}\right] = 0 \tag{27.93}$$

$$\frac{\partial p_i}{\partial t} + \frac{1}{3}\frac{\partial e_E}{\partial x_i} - \frac{1}{5}\tau_N c^2 \left[\frac{\partial^2 p_i}{\partial x_k \partial x_k} + 2\frac{\partial^2 p_k}{\partial x_i \partial x_k}\right] = -\frac{p_i}{\tau_R} \tag{27.94}$$

By construction of (27.92) the replacement yields the momentum balance of the GK equations. However, in the energy balance we find the time derivative of the divergence of momentum, which can be replaced by means of the momentum balance and, again, removing higher order terms in τ_N, to give

$$\frac{\partial e_E}{\partial t} + c^2 \left(1 + \frac{\tau_N}{\tau_R}\right) \frac{\partial p_k}{\partial x_k} + \frac{1}{3} \tau_N c^2 \frac{\partial^2 e_E}{\partial x_k \partial x_k} = 0 \tag{27.95}$$

$$\frac{\partial p_i}{\partial t} + \frac{1}{3} \frac{\partial e_E}{\partial x_i} - \frac{1}{5} \tau_N c^2 \left[\frac{\partial^2 p_i}{\partial x_k \partial x_k} + 2 \frac{\partial^2 p_k}{\partial x_i \partial x_k} \right] = -\frac{p_i}{\tau_R} \tag{27.96}$$

The momentum balance remains unchanged and agrees with the GK equations (Guyer and Krumhansl 1966), which however do not have the extra diffusion term $\frac{1}{3}\tau_N c^2 \frac{\partial^2 e_E}{\partial x_k \partial x_k}$ in the energy balance. Sussman and Thellung (1963) assume that no R-processes occur, that is $\tau_R \to \infty$; they find the diffusion term $\frac{1}{3}\tau_N c^2 \frac{\partial^2 e_E}{\partial x_k \partial x_k}$ with the opposite (!) sign.

Stability analysis of both sets of equations shown in this section reveals that these equations are unstable, in the sense that small wavelength disturbances will blow up over time, similar to instabilities in the Burnett equations of rarefied gas dynamics (Struchtrup 2005). The ST equations, with the inverted sign, are unconditionally stable.

27.4.3 Nonequilibrium Energy and Momentum

We must conclude that the shift in energy from e to e_E in (27.92) is not permitted, since it leads to unstable equations. To remove the instability, in addition to the revised energy, we introduce the revised phonon momentum

$$p_i = p_i^E - \frac{\tau_N}{3} \frac{\partial e_E}{\partial x_i} \tag{27.97}$$

into the conservation laws, and once more remove all terms that are quadratic in τ_N. This leads to cancellation of the term causing the instability, and when the energy balance is used to remove the time derivative in the momentum equation ultimately results in

$$\frac{\partial e_E}{\partial t} + c^2 \left(1 + \frac{\tau_N}{\tau_R}\right) \frac{\partial p_k^E}{\partial x_k} = 0\,, \tag{27.98}$$

$$\frac{\partial p_i^E}{\partial t} + \frac{1}{3} \left(1 - \frac{\tau_N}{\tau_R}\right) \frac{\partial e_E}{\partial x_i} - \frac{1}{5} \tau_N c^2 \left[\frac{\partial^2 p_i^E}{\partial x_k \partial x_k} + \frac{1}{3} \frac{\partial^2 p_k^E}{\partial x_i \partial x_k} \right] = -\frac{p_i^E}{\tau_R} \tag{27.99}$$

Hence, redefining energy and momentum leaves the structure of the hydrodynamic equations (27.73), (27.74) intact, with the factor 1/3. Note that in the scaling used $\left(1 + \frac{\tau_N}{\tau_R}\right) = \left(1 - \frac{\tau_N}{\tau_R}\right)^{-1}$, hence the above appear as energy and momentum balance with a revised energy $\left(1 - \frac{\tau_N}{\tau_R}\right) e_E$ or, alternatively, a revised phonon momentum

$\left(1+\frac{\tau_N}{\tau_R}\right) p_k^E$. We mention only in passing that similar factors appear in the equations of Hardy and Albers (1974), however with different sign in the momentum balance.

Sussmann and Thellung (1963) explicitly state both shifts, for energy and momentum (27.92), (27.97), but not the equations above. In the next section it will become clear that the difference is due to their scaling assumptions.

27.5 Slow Time Variation

27.5.1 Revised Scaling and Expansion

The previous section showed how splitting of energy and momentum into contributions at different orders in τ_N changes the transport equations to alternatives. In the Chapman-Enskog expansion this splitting does not occur, since the compatibility requirement ensures that energy and momentum are not altered by terms from higher order expansion.

We ask for possible arguments, or derivations, that lead to such a splitting.

Sussmann and Thellung (1963), before their equations (7-10), explicitly state that they assume *slow time variation*. While these authors use this assumption for an ad-hoc reduction of their approximative distribution function, we consider a formal approach, by rescaling the Callaway equation as

$$\varepsilon\frac{\partial f}{\partial t}+c\frac{k_i}{k}\frac{\partial f}{\partial x_i}=-\frac{1}{\tau_R}\left(f-f_R\right)-\frac{1}{\varepsilon}\frac{1}{\tau_N}\left(f-f_N\right) , \tag{27.100}$$

that is the time derivative is multiplied by the scaling parameter ε to account for slow time variation. With the typical process frequency Ω, the Chapman-Enskog scaling is such that $\tau_N\Omega=\varepsilon\ll 1$, while with slow time scaling $\tau_N\Omega=\varepsilon^2\ll 1$.

The scaling of the balance laws for energy and momentum follows from the rescaled Callaway equation as

$$\varepsilon\frac{\partial e}{\partial t}+c^2\frac{\partial p_k}{\partial x_k}=0 , \tag{27.101}$$

$$\varepsilon\frac{\partial p_i}{\partial t}+\frac{1}{3}\frac{\partial e}{\partial x_i}+\frac{\partial N_{\langle ik\rangle}}{\partial x_k}=-\frac{p_i}{\tau_R} , \tag{27.102}$$

where conservation rules of R- and N-processes for energy and momentum were already used.

As we proceed, we implicitly consider f_R and f_N in their original form (27.14), (27.15), that is written with Lagrange-multipliers λ_R, λ, Λ_i which are yet to be determined.

We are only interested in the first two terms in the expansion, that is the distribution is expanded to first order as

$$f = f^{[0]} + \varepsilon f^{[1]} \tag{27.103}$$

so that the scaled transport equation (27.100) becomes

$$\varepsilon \frac{\partial \left(f^{[0]} + \varepsilon f^{[1]}\right)}{\partial t} + c \frac{k_i}{k} \frac{\partial \left(f^{[0]} + \varepsilon f^{[1]}\right)}{\partial x_i} =$$
$$= -\frac{1}{\tau_R} \left(f^{[0]} + \varepsilon f^{[1]} - f_R\right) - \frac{1}{\varepsilon} \frac{1}{\tau_N} \left(f^{[0]} + \varepsilon f^{[1]} - f_N\right) \tag{27.104}$$

which immediately implies

$$f^{[0]} = f_N \,, \tag{27.105}$$
$$f^{[1]} = -\frac{\tau_N}{\tau_R} \left(f_N - f_R\right) - \tau_N c \frac{k_i}{k} \frac{\partial f_N}{\partial x_i} \,. \tag{27.106}$$

The leading order contribution is the drift distribution f_N as in the CE expansion. The first order contribution, however, is different from the Chapman-Enskog result (27.55), notably the time derivative does not appear. To proceed, we do not evoke compatibility as in the Chapman-Enskog expansion, but determine energy, momentum, and stress from their definitions and the approximated distribution function as

$$e\left(x_i, t\right) = \int \hbar c k \left(f^{[0]} + \varepsilon f^{[1]}\right) d\mathbf{k}$$
$$= e_N - \varepsilon \frac{\tau_N}{\tau_R} \left(e_N - e_R\right) - \varepsilon \tau_N c^2 \frac{\partial p_i^N}{\partial x_i} \tag{27.107}$$

$$p_i\left(x_i, t\right) = \int \hbar k_i \left(f^{[0]} + \varepsilon f^{[1]}\right) d\mathbf{k}$$
$$= \left(1 - \varepsilon \frac{\tau_N}{\tau_R}\right) p_i^N - \varepsilon \tau_N \frac{\partial}{\partial x_j} \left(\frac{e_N}{3} \delta_{ij} + N_{\langle ij\rangle}^N\right) \tag{27.108}$$

$$N_{\langle ij\rangle} = \int \hbar c \frac{k_{\langle i} k_{i\rangle}}{k} \left(f^{[0]} + \varepsilon f^{[1]}\right) d\mathbf{k}$$
$$= \left(1 - \varepsilon \frac{\tau_N}{\tau_R}\right) N_{\langle ij\rangle}^N - \varepsilon \tau_N \frac{\partial}{\partial x_k} \left(M_{\langle ijk\rangle}^N + \frac{2}{5} c^2 p_{\langle i}^N \delta_{j\rangle k}\right) \tag{27.109}$$

Here, e_R, e_N, p_i^N are given through the Lagrange multipliers in the distribution functions f_R, f_N. To simplify, and aiming for agreement with the GK equations, we consider only the linear regime, where $N_{\langle ij\rangle}^N = 0$, $M_{\langle ijk\rangle}^N = 0$. Moreover, continuing with previous notation we evoke

$$e_N - \varepsilon \frac{\tau_N}{\tau_R}(e_N - e_R) = e_E(T) = \frac{\pi^2}{10} \frac{k_B^4 T^4}{\hbar^3 c^3}, \tag{27.110}$$

so that we find expressions that are rather similar to (27.92), (27.97),

$$e(x_i, t) = e_E - \varepsilon \tau_N c^2 \frac{\partial p_i^N}{\partial x_i} \tag{27.111}$$

$$p_i(x_i, t) = \left(1 - \varepsilon \frac{\tau_N}{\tau_R}\right) p_i^N - \varepsilon \frac{\tau_N}{3} \frac{\partial e_N}{\partial x_i} \tag{27.112}$$

$$N_{\langle ij \rangle} = -\varepsilon \frac{2}{5} c^2 \tau_N \frac{\partial p_{\langle i}^N}{\partial x_{j\rangle}} \tag{27.113}$$

Since we are interested only in accuracy up to order ε, we can use the middle equation to write energy and stress tensor with the full momentum, as

$$e(x_i, t) = e_E - \varepsilon \tau_N c^2 \frac{\partial p_i}{\partial x_i} + \mathcal{O}(\varepsilon^2) \tag{27.114}$$

$$N_{\langle ij \rangle} = -\varepsilon \frac{2}{5} c^2 \tau_N \frac{\partial p_{\langle i}}{\partial x_{j\rangle}} + \mathcal{O}(\varepsilon^2) \tag{27.115}$$

These expressions for e and $N_{\langle ij \rangle}$ are now inserted into the time-scaled balance laws, but momentum is not replaced. To order ε this results in

$$\varepsilon \frac{\partial e_E}{\partial t} + c^2 \frac{\partial p_k}{\partial x_k} = 0, \tag{27.116}$$

$$\varepsilon \frac{\partial p_i}{\partial t} + \frac{1}{3} \frac{\partial e_E}{\partial x_i} - \varepsilon \frac{1}{5} \tau_N c^2 \left[\frac{\partial^2 p_i}{\partial x_k \partial x_k} + 2 \frac{\partial^2 p_k}{\partial x_i \partial x_k} \right] = -\frac{p_i}{\tau_R}. \tag{27.117}$$

Upon removal of the formal scaling parameter ε, these are just the Guyer-Krumhansl equations (27.88), (27.89), which thus appear from the revised time scaling.

27.5.2 Further Reduction

With their scaling assumptions, Sussmann and Thellung (1963) should have arrived at the equations (27.116), (27.117) with $\tau_R \to \infty$, but instead have the diffusion term $-\frac{1}{3}\varepsilon \tau_N c^2 \frac{\partial^2 e_E}{\partial x_k \partial x_k}$ in their energy balance which arises from the momentum shift (27.112). With the momentum balance, and considering only linear terms in τ_N, this expression can be rewritten as (with $\tau_R \to \infty$)

$$-\frac{1}{3}\varepsilon \tau_N c^2 \frac{\partial^2 e_E}{\partial x_k \partial x_k} = -\frac{1}{3}\varepsilon \tau_N c^2 \frac{\partial}{\partial x_k} \left[-\varepsilon \frac{\partial p_k}{\partial t} + \frac{p_k}{\tau_R} \right] = \mathcal{O}(\varepsilon^2) \simeq 0. \tag{27.118}$$

If this is taken into account, Sussmann and Thellung agree with Guyer and Krumhansl (1966).

However, just as the term above is identified as being of second order in ε in the scaling for slow time variation, this is true as well for the term with the divergence of momentum in (27.117), which can be written with the energy balance (27.116) as

$$-\varepsilon\frac{1}{5}\tau_N c^2\left[2\frac{\partial^2 p_k}{\partial x_i \partial x_k}\right] = -\varepsilon\frac{1}{5}\tau_N c^2\left[-2\frac{\varepsilon}{c^2}\frac{\partial^2 e_E}{\partial x_i \partial t}\right] = \mathcal{O}\left(\varepsilon^2\right) \simeq 0\,. \qquad (27.119)$$

with that, for slow time variation, energy and momentum balance further reduce to

$$\varepsilon\frac{\partial e_E}{\partial t} + c^2\frac{\partial p_k}{\partial x_k} = 0\,, \qquad (27.120)$$

$$\varepsilon\frac{\partial p_i}{\partial t} + \frac{1}{3}\frac{\partial e_E}{\partial x_i} - \varepsilon\frac{1}{5}\tau_N c^2\frac{\partial^2 p_i}{\partial x_k \partial x_k} = -\frac{p_i}{\tau_R}\,. \qquad (27.121)$$

This implies that in slow time scaling the GK term with factor '2' is not of importance as long as the equations are solved for processes that agree with the scaling assumption in (27.100), which we can indicate as

$$\text{slow time variation scaling:} \quad \varepsilon\Omega \quad , \quad \frac{1}{\varepsilon\tau_N}\,. \qquad (27.122)$$

This two-orders-of-magnitude ratio between process frequency Ω and frequency of N-processes $1/\tau_N$ is not met in experimental examination for second sound, in which the window (27.75) is rather narrow.

We must conclude that the GK equations are not valid for the conditions of second sound, for which they were derived and applied.

27.5.3 Incompatibility

In the Chapman Enskog expansion compatibility is ensured through the conditions (27.45), which were found valid in (27.56), (27.57). Notably, compatibility ensures the proper conservation properties of R- and N-processes as expressed in (27.12), (27.13).

For the slow time scaling that leads to the GK equations (27.116), (27.117), compatibility demands that the production terms retain their values $(0, -\frac{p_i}{\tau_R})$.

Evaluating the collision terms (27.10), (27.11) for R- and N-processes with the distribution (27.103) yields with (27.110), (27.111), (27.112)

$$P^R = -\int \hbar c k \frac{1}{\tau_R}\left(f - f_R\right) d\mathbf{k} = -\frac{1}{\tau_R}\left(e - e_R\right)$$

$$= -\frac{e_N - e_R}{\tau_R} + \varepsilon \frac{\tau_N}{\tau_R}\left[\frac{e_N - e_R}{\tau_R} + c^2 \frac{\partial p_i^N}{\partial x_i}\right], \tag{27.123}$$

$$P_i^R = -\int \hbar k_i \frac{1}{\tau_R}(f - f_R)\, d\mathbf{k} = -\frac{p_i}{\tau_R}, \tag{27.124}$$

$$\begin{aligned} P^N &= -\int \hbar c k \frac{1}{\varepsilon \tau_N}(f - f_N)\, d\mathbf{k} \\ &= -\frac{1}{\varepsilon \tau_N}(e - e_N) = \frac{e_N - e_R}{\tau_R} + c^2 \frac{\partial p_i^N}{\partial x_i}, \end{aligned} \tag{27.125}$$

$$\begin{aligned} P_i^N &= -\int \hbar k_i \frac{1}{\varepsilon \tau_N}(f - f_N)\, d\mathbf{k} \\ &= -\frac{1}{\varepsilon \tau_N}\left(p_i - p_i^N\right) = \frac{1}{\tau_R} p_i^N + \frac{1}{3}\frac{\partial e_N}{\partial x_i}. \end{aligned} \tag{27.126}$$

While P_i^R has the required value, the other terms P^R, P^N, P_i^N do not vanish as they should. Demanding compatibility of the approximated distribution (27.103) one must set

$$P^R = P^N = P_i^N = 0, \tag{27.127}$$

which, with (27.110), (27.111), (27.112), are equivalent to

$$e_N = e_R = e_E(T) \quad , \quad \frac{\partial p_i^N}{\partial x_i} = \frac{\partial p_i}{\partial x_i} = 0 \quad , \quad p_i^N = p_i = -\frac{\tau_R}{3}\frac{\partial e_E}{\partial x_i}. \tag{27.128}$$

These rather strong conditions are simply the leading order terms of the scaled Guyer-Krumhansl equations (27.116), (27.117), that is they result from removing terms of order ε.

Thus, compatibility of the distribution function for slow time scale (27.103) with the Callaway equation is only found for the zeroth order approximation, but not for the first order approximation—which are the GK equations (27.116), (27.117). With that, we must conclude that the GK equations result from an improper scaling, and should not be used.

27.5.4 Dispersion Relations Compared

It is worthwhile to compare the GK and CE equations for propagation of plane harmonic waves. For compact notation we write energy and momentum balances for waves propagating in x-direction as

$$\frac{\partial e}{\partial t} + c^2 \frac{\partial p}{\partial x} = 0 \tag{27.129}$$

$$\frac{\partial p}{\partial t} + \frac{1}{3}\frac{\partial e_E}{\partial x} - \frac{1+\gamma}{5}\tau_N c^2 \frac{\partial^2 p}{\partial x^2} = -\frac{p}{\tau_R} \tag{27.130}$$

where the two sets of equations differ only by the value of the coefficient γ, which is

$$\gamma = \frac{1}{3} \text{ for CE} \quad , \quad \gamma = 2 \text{ for GK} . \tag{27.131}$$

We consider plane harmonic waves of frequency Ω and wave number K, so that energy and momentum deviations from equilibrium are of the form (Struchtrup 2024)

$$u = \tilde{u} \exp\left[i\left(\Omega t - Kx\right)\right] . \tag{27.132}$$

Insertion into the transport equations yields a homogeneous system for the wave amplitudes $\tilde{e}$, $\tilde{p}$,

$$\begin{bmatrix} i\hat{\Omega} & -i\hat{K} \\ -i\frac{1}{3}\hat{K} & i\hat{\Omega} + \frac{1+\gamma}{5}\hat{K}^2 + \frac{\tau_N}{\tau_R} \end{bmatrix} \begin{bmatrix} \tilde{e} \\ c\tilde{p} \end{bmatrix} = \begin{bmatrix} 0 \\ 0 \end{bmatrix} \tag{27.133}$$

where

$$\hat{\Omega} = \Omega\tau_N \quad , \quad \hat{K} = K\tau_N c \tag{27.134}$$

are dimensionless frequency and wave number. Typical heat pulse experiments (Dreyer and Struchtrup 1993) correspond to boundary value problems, hence we ask for the dispersion relation $\hat{K}\left(\hat{\Omega}\right)$ for real values of the frequency $\hat{\Omega}$, which is obtained from vanishing determinant of the coefficient matrix. Dimensionless wave speed $\hat{v}_{ph}$ and damping coefficient $\hat{\alpha}$ are

$$\hat{v}_{ph} = \frac{\hat{\Omega}}{\mathrm{Re}\left[\hat{K}\left(\hat{\Omega}\right)\right]} = \frac{\Omega}{c\,\mathrm{Re}\left[K\right]} \quad , \quad \hat{\alpha} = -\,\mathrm{Im}\left[\hat{K}\right] = -\tau_N c\,\mathrm{Im}\left[K\right] \tag{27.135}$$

The only free parameter in the dimensionless formulation is the ratio of relaxation times $\frac{\tau_N}{\tau_K}$. For conditions that allow for second sound, dimensionless frequencies should be in the window

$$\text{second sound:} \quad \frac{\tau_N}{\tau_R} < \hat{\Omega} < 1 . \tag{27.136}$$

Figure 27.1 shows phase speed and damping for the case $\frac{\tau_N}{\tau_R} = 0.01$ in double logarithmic plots to enhance the features of the curves. Both sets of equations give identical results for rather small frequencies $\hat{\Omega} < 0.05$ which reflects the limited

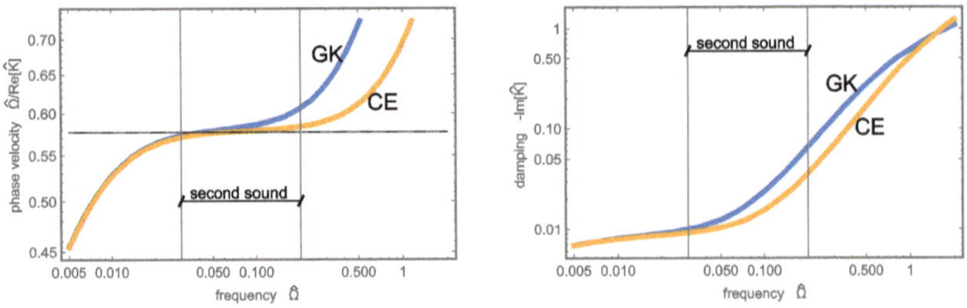

Fig. 27.1 Phase velocity (left) and damping (right) for the GK equations (blue) and the equations from CE expansion (orange) with $\frac{\tau_N}{\tau_R} = 0.01$. In the second sound window, the phase velocity is close to the speed of second sound $v_{\mathrm{II}} = 1/\sqrt{3}$

validity of the GK equations under slow time scaling. For larger frequencies, the two sets shows similar results, but with visible differences.

The phase speed for undamped second sound is $v_{\mathrm{II}} = 1/\sqrt{3}$, which is indicated as the dashed line. For frequencies in $0.04 < \hat{\Omega} < 0.15$ the CE phase velocity exhibits a plateau of waves with this speed and little dispersion—this defines the window for second sound as indicated. The plateau is less pronounced for GK, which also predicts stronger damping of waves.

The results for even higher frequencies do not fulfill the requirement $\hat{\Omega} = \Omega\tau_N \ll 1$ of the CE expansion, and are just shown out of general interest. For higher frequencies transport is effected by ballistic phonons which can be described by extended moment systems as in Dreyer and Struchtrup (1993), Fryer and Struchtrup (2014).

27.6 Conclusions

We have shown that the application of the Chapman-Enskog expansion to the Callaway model for phonon transport provides a clear and problem-free route to the equations of phonon hydrodynamics for phonon energy and momentum densities. The CE expansion requires comparatively high frequencies of momentum conserving N-processes, and the important compatibility conditions are automatically fulfilled. The resulting equations differ from the classical Guyer-Krumhansl equations in a factor ('1/3' instead of '2').

An alternative expansion that demands particularly slow processes yields the Guyer-Krumhansl equations (with factor '2'), but is accompanied by problems with compatibility.

The dispersion relations for both equations agree for low frequencies, that is slow processes, in agreement with the underlying assumption of the expansions. In the interesting range for second sound, however, the predictions differ, and we must conclude that the GK equations—which are derived under more restrictive assumptions—are not valid in this interesting regime.

With their wider range of validity due to less restrictive assumptions in the derivation, the equations from CE expansion, which are also available for strong non-linear deviation from equilibrium, should be used for description of phonon processes in the hydrodynamic limit.

Acknowledgements Wolfgang Dreyer had a profound influence on my development as researcher and instructor. As a student I was drawn to the Institute of Thermodynamics at TU Berlin first by the inspiring and exciting lectures on all aspects of Thermodynamics by Wolfgang and the institute's head, Prof. Ingo Müller. Soon, I could not imagine to study anywhere else, or another topic than Thermodynamics, and thus proceeded to write my bachelor's and diploma theses under Wolfgang's guidance. To this day I vividly remember our discussions, as well as reading and correcting the theses together, until the thoughts, and the writing, were clear. His passionate and clear lectures not only taught me the topic at hand—of course, Thermodynamics and Kinetic Theory!—but as well were exemplary for a style that directly engages the audience. These formative experiences shape my way of thinking and approaching science and teaching to this day. Support from the Natural Sciences and Engineering Research Council of Canada (NSERC)through Discovery Grant RGPIN-2022-03188 is gratefully acknowledged.

References

Banach Z, Larecki W (2008) Chapman-Enskog method for a phonon gas with finite heat flux. J Phys A: Math Theor 41:375502. https://doi.org/10.1088/1751-8113/41/37/375502

Callaway J (1959) Model for lattice thermal conductivity at low temperatures. Phys Rev 113:1046–1051

Cimmelli VA, Sellitto A, Jou D (2010) Nonlinear evolution and stability of the heat flow in nanosystems: beyond linear phonon hydrodynamics. Phys Rev B 82:184302. https://doi.org/10.1103/PhysRevB.82.184302

Dreyer W (1987) Maximisation of the entropy in non-equilibrium. J Phys A: Math Gen 20:6505–6517. https://doi.org/10.1088/0305-4470/20/18/047

Dreyer W, Struchtrup H (1993) Heat pulse experiments revisited. Cont Mech Thermodyn 5:3–50. https://doi.org/10.1007/BF01135371

Fryer MJ, Struchtrup H (2014) Moment model and boundary conditions for energy transport in the phonon gas. Continuum Mech Thermodyn 26:593–618. https://doi.org/10.1007/s00161-013-0320-y

Guo Y, Wang M (2015) Phonon hydrodynamics and its applications in nanoscale heat transport. Phys Rep 595:1–44. https://doi.org/10.1016/j.physrep.2015.07.003

Guyer RA, Krumhansl JA (1964) Dispersion relation for second sound in solids. Phys Rev 133:A1411–A1417. https://doi.org/10.1103/PhysRev.133.A1411

Guyer RA, Krumhansl JA (1966) Solution of the linearized Phonon Boltzmann equation. Phys Rev 148:766–778. https://doi.org/10.1103/PhysRev.148.766

Hardy RL, Albers DL (1974) Hydrodynamic approximation to the phonon Boltzmann equation. Phys Rev B 10:3546–3551. https://doi.org/10.1103/PhysRevB.10.3546

Kremer GM (2010) An introduction to the Boltzmann equation and transport processes in gases. Springer, Berlin, Heidelberg. https://doi.org/10.1007/978-3-642-11696-4

Larecki W (1992) Symmetric conservative form of low-temperature phonon gas hydrodynamics II.—Equations of heat transport and thermal waves in the case of linear isotropic approximation of phonon frequency spectrum, Il Nuovo Cimento 14:141–176. https://doi.org/10.1007/BF02457349

Müller I, Ruggeri T (1998) Rational extended thermodynamics. Springer Springer, New York. https://doi.org/10.1007/978-1-4612-2210-1

Struchtrup H (2024) A thermodynamic introduction to transport phenomena. Springer, Cham, Switzerland. https://doi.org/10.1007/978-3-031-61868--0

Struchtrup H (2005) Macroscopic transport equations for rarefied gas flows. Springer, Berlin. https://doi.org/10.1007/3-540-32386-4

Sussmann JA, Thellung A (1963) Thermal conductivity of perfect dielectric crystals in the absence of Umklapp processes. Proc Phys Soc 81:1122–1130. https://doi.org/10.1088/0370-1328/81/6/318

Xu M, The modified Guyer-Krumhansl equations derived from the Linear Boltzmann transport equation. J Heat Mass Trans 146

Chapter 28
On the Visual Explanation of the Isotherm Dilute Atmosphere

Wolf Weiss

In memory of my friend and colleague Wolfgang Dreyer.

Abstract A very rarefied atmosphere is considered in which all particles fly freely without collisions. The speed of all particles decreases in the vertical direction due to gravity. Because the temperature depends on the speed of the particles, one could assume that the temperature also decreases with altitude. However, if there is an equilibrium distribution on the ground, the temperature remains constant and is independent of altitude. This fact is clearly explained.

28.1 Introduction

In order to calculate the circumstances in an atmosphere in thermodynamic equilibrium we consider the balance equations for mass, momentum and internal energy

$$\frac{\partial \varrho}{\partial t} + \frac{\partial \varrho v_i}{\partial x_i} = 0, \tag{28.1}$$

$$\frac{\partial \varrho v_i}{\partial t} + \frac{\partial}{\partial x_j}\left(\varrho v_j v_i - t_{ij}\right) = \varrho F_i, \tag{28.2}$$

$$\frac{\partial \varrho \varepsilon}{\partial t} + \frac{\partial}{\partial x_j}\left(\varrho \varepsilon v_j + q_j\right) = 0. \tag{28.3}$$

This system connects the mass density ϱ, the macroscopic velocity v_i, the stress tensor t_{ij}, the internal energy ε, and the heat flux q_i as functions of space x_i and time t. An external force is denoted by F_i.

W. Weiss (✉)
Weierstrass Institute for Applied Analysis and Stochastics, Berlin, Germany
e-mail: wolf.weiss@alumni.tu-berlin.de

J. Fuhrmann et al. (eds.), *Advances in Continuum Physics*, Advanced Structured Materials 238, https://doi.org/10.1007/978-3-031-93918-1_28

The stress tensor t_{ij} is decomposed into

$$t_{ij} = -p\delta_{ij} + \sigma_{ij}, \tag{28.4}$$

with the pressure p and the shear stress tensor σ_{ij}. In order to close the system we may choose the classical Navier–Stokes constitutive equation,

$$\sigma_{ij} = \mu \left(\frac{\partial v_i}{\partial x_j} + \frac{\partial v_j}{\partial x_i} \right) + \lambda \frac{\partial v_k}{\partial x_k} \delta_{ij}, \tag{28.5}$$

and for the heat-flux q_i the Fourier law,

$$q_i = -\kappa \frac{\partial T}{\partial x_i}, \tag{28.6}$$

where μ denotes the sheer viscosity, λ the bulk viscosity and κ the thermal conductivity.

If we consider an atmosphere which is at rest, $v_i = 0$, and in thermodynamic equilibrium, we have no dissipation, therefore both non-equilibrium quantities σ_{ij} and q_i vanish. Especially from Fourier's law it follows that

$$T = \text{const.} \tag{28.7}$$

Further the mass balance (28.1) and the balance for the internal energy (28.3) are identically fulfilled and from the balance of momentum (28.2) it follows with $F_i = (0, 0, -g)$, where g is the gravitational acceleration

$$\frac{dp}{dz} = -\varrho g. \tag{28.8}$$

For the pressure we have the thermal equation of state for ideal gases

$$p = \varrho \frac{k_B}{m} T, \tag{28.9}$$

where m denotes the mass of a particle and k_B is the Boltzmann constant. Insertion of (28.9) in to (28.8) leads to the so called *barometric formulas*,

$$p(z) = p_0 \exp\left(-\frac{mg}{k_B T} z \right), \tag{28.10}$$

and

$$\varrho(z) = \varrho_0 \exp\left(-\frac{mg}{k_B T} z \right), \tag{28.11}$$

where p_0 and ϱ_0 are the pressure and the density at the bottom at $z = 0$, respectively.

So far the calculation of temperature, pressure and density in an equilibrium atmosphere. The uniform temperature follows especially from the requirement that the heat flux, given by formula (28.6), vanishes in equilibrium.[1]

All the results were formally derived. However, how can we understand the homogeneous temperature intuitively? We know that temperature is related to the random motion of gas particles. Therefore, the vertical velocity component should decrease with height in the gravitational field. Thus we have a seemingly contradiction to constant temperature. In the following we want to examine this alleged contradiction.

28.2 Short Historical Tour

In the 19th century it was not really clear which behavior the temperature shows in a equilibrium atmosphere. At that time it was thought that the atmosphere was in equilibrium.[2] However, measurements showed that the temperature decreased with altitude. Therefore, it is not surprising that a theory was sought to explain the nonuniform temperature in an equilibrium atmosphere.

As we have seen, the temperature distribution depends strongly on the expression for the heat flow. In 1867, Maxwell published a contribution on the kinetic theory of gases (Maxwell 1867) in which an expression for heat flow was also derived.

However, the first version of this work contained a calculation error.[3] Maxwell derived the following expression for the heat flux in one dimension (direction z),

$$q_z = -\beta \left(\frac{3p^2}{\varrho T} \frac{dT}{dz} - \frac{2p}{\varrho} \frac{dp}{dz} \right), \tag{28.12}$$

where β is a material dependent constant. In equilibrium the heat flow vanishes and with (28.8) and (28.9) a simple differential equation follows,

$$0 = 3k_B \frac{dT}{dz} + 2gm. \tag{28.13}$$

The solution shows a temperature decreasing with altitude

$$T(z) = T_0 - \frac{2}{3} \frac{mg}{k_B} z. \tag{28.14}$$

[1] If relativistic effects are taken into account, an inhomogeneous temperature can exist in an atmosphere in thermodynamic equilibrium, (Eckart 1940; Tolman and Ehrenfest 1930).

[2] Today we know that the atmosphere is not in equilibrium due to radiation, flow and heat conduction.

[3] Strictly speaking, Maxwell made two errors. One of them leads to a temperature that increases with altitude and the other to a temperature that decreases with altitude. Maxwell himself comments on his errors in Maxwell (1867) on page 86.

Maxwell immediately corrected his mistake and the improved theory yielded the well known result[4] (28.6) and thus T = constant.

For Maxwell and Boltzmann, the atmosphere in equilibrium has therefore a constant temperature. Loschmidt, however, believed in a temperature stratification in equilibrium. This difference led to a long dispute[5] between Loschmidt and Boltzmann published in the *Annalen der Physik*. The dispute has been summarized by Dreyer et al. in the paper *Tales of Thermodynamics and Obscure Applications of the Second Law* (Dreyer et al. 2000). Here only a few selected sentences between the opponents are shown:

Boltzmann wrote (Boltzmann 1876, p. 513):

... Wir haben somit einen directen Beweis geliefert, dass die ...[Gleichgewichts-] Zustandsvertheilung durch den Einfluss der Schwere auf die Bewegung der Gasmolecüle nicht gestört wird.[6]

Only slightly later Loschmidt wrote (Loschmidt 1877, p 292):

Bis heute ist übrigens für die behauptete Temperaturgleichheit der verschiedenen Schichten einer verticalen Luftsäule noch in keinem Falle ein stichhältiger Beweis erbracht, ...[7]

Away from formal calculations there is an apparent contradiction. A bullet fired upwards becomes slower and slower as it travels upwards. This should also apply to microscopic particles. Because the temperature is a measure of the average speed of the particles, the temperature should also decrease upwards.

This contradiction has been dealt with by Trupp (1999) and Velasco et al. (1996), among others.

Experiments where made by Graeff (2002, 2007, 2015), Liao (Liao 2009) and Jeong and Park (Jeong and Park 2022), in which some measurements show, for whatever reasons,[8] a negative temperature gradient. However, in this work we follow Maxwell and Boltzmann, which leads to an isothermal atmosphere.

Intuitive explanations are given in Coombes and Laue (1985) for gases without collisions and in Zhang and Zhang (2016) for gases with collisions. But in this work we want to give a more "*visual*" explanation.

---

[4] However, this does not mean that Fourier's expression is universally valid. In general the heat flow depends on the material and the circumstances. For example, the heat flow in a dilute gas in non-equilibrium also depends on the gradient of the pressure tensor (Reitebuch and Weiss 1999; Müller and Ruggeri 2004; Struchtrup and Taheri 2011).

[5] This dispute differs from another dispute between Boltzmann and Loschmidt concerning the contradiction between reversible microscopic and irreversible macroscopic equations.

[6] ... We have thus provided a direct proof that the ... [equilibrium] distribution of states is not affected by the influence of gravity on the motion of the gas molecules.

[7] By the way, until today there has been no sound proof for the equality of temperatures, as claimed, in different layers of a vertical column of air, ...

[8] In Jeong and Park (2022), for example, the authors summarize, among other things: "*The negative temperature gradient of the air column is generated under a gravitational field, and this temperature gradient changes sensitively according to the ambient temperature, even if it is practically insulated.*".

28.3 The Free-Flying Particle

For a single particle we have Newton's law,

$$m\frac{d^2z(t)}{dt^2} = -mg. \tag{28.15}$$

With $z(0) = 0$ and $\frac{dz}{dt}\Big|_{t=0} = C_z(0)$ it follows that

$$z(t) = C_z(0)t - \frac{gt^2}{2}, \tag{28.16}$$

$$C_z(t) = C_z(0) - gt. \tag{28.17}$$

If we replace the time t in (28.17) by (28.16) we get

$$C_z(z) = \sqrt{C_z(0)^2 - 2gz}. \tag{28.18}$$

The maximum height of a particle follows for $C_z(z) = 0$ then to

$$z_{\max}\left(C_z(0)\right) = \frac{C_z(0)^2}{2g}. \tag{28.19}$$

All particles become slower and at a certain height, depending on $C_z(0)$, the particles fall down again.

One could now argue that a gas particle in an atmosphere due to collisions with other particles is not comparable with a free flight shot bullet. Therefore let us consider a very rarefied gas of one atomic particles, also called a *ballistic gas*. In such a gas collisions among the particles are neglected.

This case is very similar to the shooting of bullets upwards. One could perhaps assume that therefore the temperature also decreases with altitude. The appropriate theory to calculate this is the kinetic theory of gases.

28.4 Kinetic Theory

The temperature T in the kinetic theory is proportional to the mean value of the kinetic energy $\langle E\rangle$, which is proportional of the mean value $\left\langle C_i^2\right\rangle$ of the thermal velocity C_i of the particles. It holds (Chapman and Cowling 1990)

$$\langle E\rangle = \frac{1}{2}m\left\langle C_i^2\right\rangle = \frac{3}{2}k_B T. \tag{28.20}$$

The absolute velocity c_i of the particles is given by $c_i = v_i + C_i$, where v_i is again the macroscopic velocity of the gas. Mean values can calculated with the help of the distribution function f, which is the central quantity in the kinetic theory. f describes the distribution of velocities of the particle density at space x_i and time t. Therefore, the whole particle density is given by

$$n = \int_{-\infty}^{\infty} f d\mathbf{C}, \tag{28.21}$$

and for the mass density we have $\varrho = m\,n$. For the temperature follows with (28.20) and the definition of mean values

$$T = \frac{m}{3k_B}\left\langle C_i^2\right\rangle = \frac{m}{3k_B}\frac{\int_{-\infty}^{\infty} C_i^2 f d\mathbf{C}}{\int_{-\infty}^{\infty} f d\mathbf{C}}. \tag{28.22}$$

In thermal equilibrium the distribution function is given by the Maxwell-Boltzmann distribution,

$$f_{\mathrm{eq}} = n\sqrt{\frac{m}{2\pi k_B T}}^{3} \exp\left\{-\frac{m}{2k_B T}C_i^2\right\}. \tag{28.23}$$

Insertion of this function in (28.22) leads to an identity.

To calculate the distribution function in general we may use the Boltzmann equation

$$\frac{\partial f}{\partial t} + c_i\frac{\partial f}{\partial x_i} + F_i\frac{\partial f}{\partial c_i} = \wp(f). \tag{28.24}$$

Here F_i denotes again an external force and $\wp(f)$ is the collision production.

The right hand side of the Boltzmann equation may vanish for two different reasons, thermodynamic equilibrium or extreme rarefied gas. Here we consider a collision free rarefied gas at rest. Further we assume that all quantities depend only on the vertical z coordinate and get

$$C_z\frac{\partial f}{\partial z} - g\frac{\partial f}{\partial C_z} = 0. \tag{28.25}$$

As boundary condition we assume the equilibrium distribution $f_{\mathrm{eq}}(C_z)$ given by (28.23) with the Temperature $T = T_0$ and the number density $n = n_0$ at the bottom $z = 0$.

The equation (28.25) is a simple equation of the Burger type and may be solved by the method of characteristics (Ebert and Reissig 2018). Using this method the partial differential equation (28.25) is replaced by an equivalent system of three ordinary differential equations

$$\frac{dz}{d\tau} = C_z, \quad \frac{dC_z}{d\tau} = -g, \quad \frac{df}{d\tau} = 0, \tag{28.26}$$

which describes f along the so called *characteristic curve* with curve parameter τ. Also the boundary condition can be replaced by a parameterized form with the curve parameter σ

$$z\,(\sigma, \tau = 0) = 0, \quad C_z\,(\sigma, \tau = 0) = \sigma, \quad f\,(\sigma, \tau = 0) = f_{\mathrm{eq}}\,(\sigma)\,. \tag{28.27}$$

Integration of (28.26) leads to

$$z = C_z\tau + K_1, \quad C_z = -g\tau + K_2, \quad f = K_3. \tag{28.28}$$

Determination of the integration constants K_1, K_2 and K_3 with (28.27) leads to the solution in parametric form

$$z\,(\sigma, \tau) = \sigma\tau - \frac{g\tau^2}{2}, \quad C_z\,(\sigma, \tau) = \sigma - g\tau, \quad f\,(\sigma, \tau) = f_{eq}\,(\sigma)\,. \tag{28.29}$$

The two equations $(28.29)_{1,2}$ can be solved for the two curve parameters,

$$\sigma = \pm\sqrt{C_z^2 + 2gz}, \quad \tau = \frac{-C_z \pm \sqrt{C_z^2 + 2gz}}{g}, \tag{28.30}$$

and an explicit solution for f follows,

$$f = f_{\mathrm{eq}}(\sigma) = f_{\mathrm{eq}}\left(\pm\sqrt{C_z^2 + 2gz}\right), \tag{28.31}$$

$$f = n_0\sqrt{\frac{m}{2\pi k_B T_0}}^3 \exp\left\{-\frac{m}{2k_B T_0}\left(C_x^2 + C_y^2 + \sqrt{C_z^2 + 2gz}^2\right)\right\}, \tag{28.32}$$

$$f = f_{\mathrm{eq}} \exp\left(-\frac{mg}{k_B T_0} z\right). \tag{28.33}$$

With this solution and (28.21) and (28.22) we may calculate the number density,

$$n(z) = n_0 \exp\left(-\frac{mg}{k_B T_0} z\right), \tag{28.34}$$

and the temperature,

$$T(z) = T_0. \tag{28.35}$$

Hence, here we get also the formal result that in an atmosphere of ballistic particles (little free flight bullets) the temperature is constant if the particles at the bottom satisfy a equilibrium distribution.[9] However, does this calculation help to understand why the temperature is constant even if the vertical velocity of all particles moves upwards decrease? The answer is 'yes' by inspection of the characteristic curves.

In order to visualize this we introduce some non-dimensional quantities

$$\hat{z} = \frac{mg}{k_B T_0} z, \quad \hat{C}_i = \sqrt{\frac{m}{k_B T_0}} C_i, \quad \hat{f} = \frac{1}{n_0} \sqrt{\frac{k_B T_0}{m}}^{-3}, \tag{28.36}$$

and

$$\hat{\sigma} = \sqrt{\frac{m}{k_B T_0}} \sigma, \quad \hat{\tau} = \frac{mg}{k_B T_0} \tau. \tag{28.37}$$

The characteristic curves are given by $(28.29)_{1,2}$. Along these curves the solution is constant because in this case the collision production vanishes. In Fig. 28.1 some characteristic curves are shown for several values of the parameter σ. The parameter may be written by $(28.27)_2$ as $\sigma = C_z\,(z = 0)$.

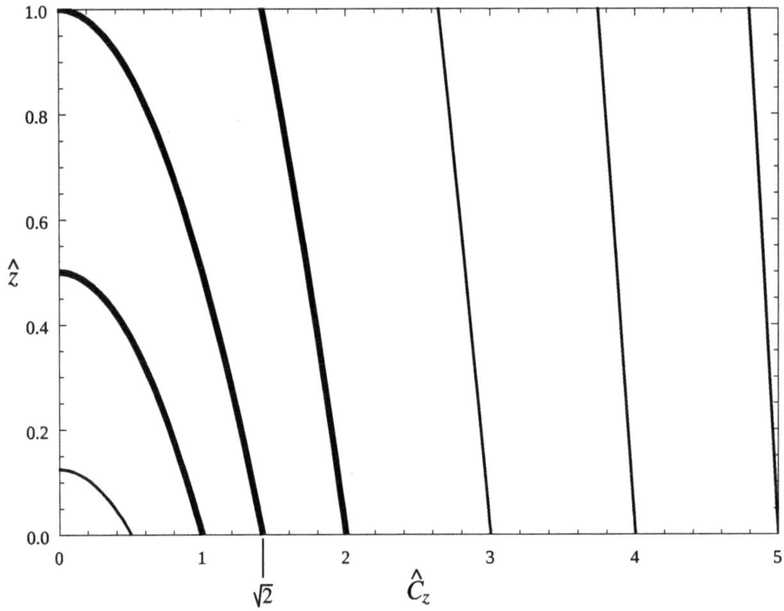

Fig. 28.1 Some characteristic curves for different values of σ

[9] Due to the fact that the collision production in (28.24) vanishes also in equilibrium, the result $T(z) = T_0$ is also valid for gases with collisions. But to illustrate the problem we will stick to a collision-free ballistic gas.

We may ask at what speed a particle must fly upwards at $\hat{z} = 0$ so that at the height $\hat{z} = 1$ its speed disappears. The answer follows from $(28.30)_1$ to $\hat{\sigma} = \hat{C}_z(\hat{z} = 0) = \sqrt{2}$. All particles with velocity $\hat{C}_z(\hat{z} = 0) < \sqrt{2}$ do not reach the height $\hat{z} = 1$ and all particles with $\hat{C}_z(\hat{z} = 0) \geqq \sqrt{2}$ reach $\hat{z} = 1$ or fly beyond it. This behavior may seen in Fig. 28.1 especially for the three bold characteristic curves with $\hat{\sigma}$ equal to 1, $\sqrt{2}$ and 2.

The characteristic curves $(28.29)_{1,2}$ are identical to the trajectories of a free flying particle given by (28.16) and (28.17) if we replace t by τ and $C_z(0)$ by σ. Indeed, in a ballistic gas the upward velocity component C_z of all particles decreases with the height. However, not all particles that start at the bottom at $z = 0$ reach a certain height $z = h$. A certain number falls back and therefore does not contribute to the internal energy or temperature. This is illustrated in Fig. 28.2. This figure shows the boundary condition $\hat{f}\left(\hat{C}_z, \hat{z} = 0\right)$, the explicit solution $\hat{f}\left(\hat{C}_z, \hat{z} = 1\right)$, a characteristic curve for $\hat{\sigma} = 1$, and two implicit solutions, $\hat{f}\left(\hat{\sigma} = \sqrt{2}, \hat{\tau}\right)$ and $\hat{f}\left(\hat{\sigma} = 2, \hat{\tau}\right)$, above their characteristic curves. For a better overview, the solution above the characteristic curve $\hat{\sigma} = 1$ is not shown.

The special property of the Maxwell function is that it deforms with height in such a way that, although fewer and fewer particles are involved, the average kinetic energy remains constant.

One might now think that the integral

$$I_0 = \int_{\sqrt{2}}^{\infty} \hat{f}\left(\hat{z} = 0, \hat{C}_z\right) d\hat{C}_z \tag{28.38}$$

would be equal to the integral

$$I_1 = \int_{0}^{\infty} \hat{f}\left(\hat{z} = 1, \hat{C}_z\right) d\hat{C}_z, \tag{28.39}$$

but this is not the case. Certainly all particles with $\hat{C}_z \geqq \sqrt{2}$ that fly up at $\hat{z} = 0$ arrive at $\hat{z} = 1$, but not at the same time. Particles are constantly flying upwards at $\hat{z} = 0$. Therefore the integral I_1 is also generated by the particles that were released at other times. Furthermore, on average, for every particle that flies upwards one falls down, so that the gas is macroscopically at rest.

By insertion of the general solution (28.31) into (28.22) it may be written, especially for the z−component of the velocity, in dimensionless quantities,

$$\int_{-\infty}^{\infty} \left(\left\langle \hat{C}_z^2 \right\rangle - \hat{C}_z^2\right) f\left(\sqrt{\hat{C}_z^2 + 2\hat{z}}\right) d\hat{C}_z = 0. \tag{28.40}$$

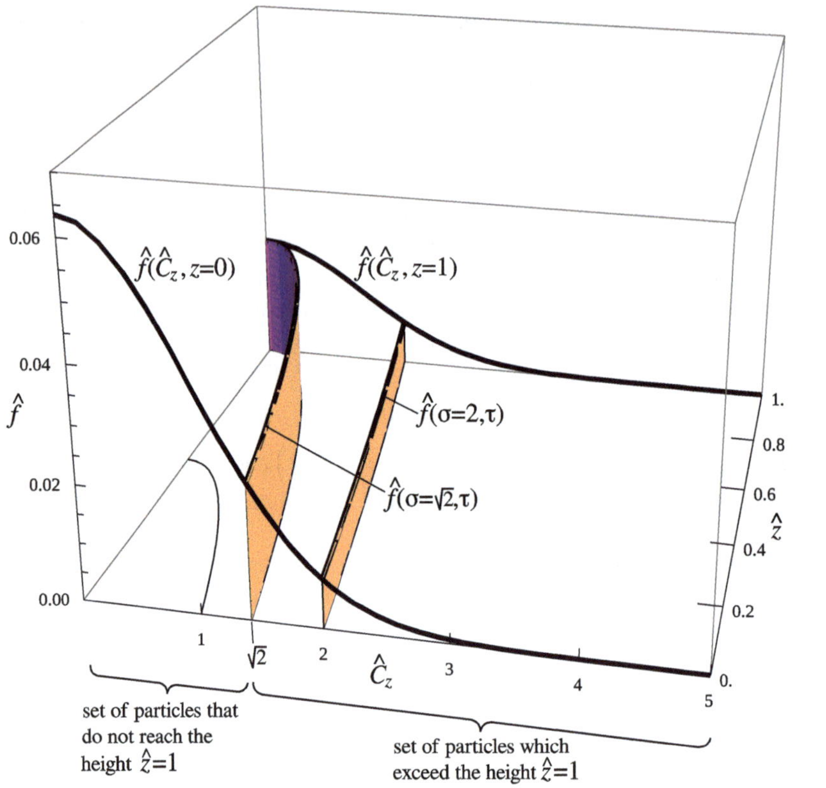

Fig. 28.2 The solution $\hat{f}$ for two characteristic curves

For a given constant $\left\langle \hat{C}_z^2 \right\rangle$ this may be interpreted as an integral equation[10] for the unknown distribution function f . To solve the equation we can expand f into a series of one-dimensional Hermite polynomials $H(i, \hat{C}_z)$ with the weight function $\exp\left(-\hat{C}_z^2/2\right)$ by

$$\hat{f} = \exp\left(-\frac{1}{2}\hat{C}_z^2\right)\sum_{i=0}^{\infty} a_i H(i, \hat{C}_z). \tag{28.41}$$

Insertion of this expression in (28.40) and calculation of the integrals leads to the conclusion that only the expansion coefficient a_0 is different form zero. All other coefficients must disappear if they should not depend on z. Because the expansion in

[10] By introducing the variable $\xi = \sqrt{C_z^2 + 2gz}$ the integral equation may be written as $\int_{\sqrt{2gz}}^{\infty} K(z, \xi)\, f(\xi)\, d\xi = 0$, which is a homogeneous Volterra integral equation of the first kind.

Hermite polynomials is complete, the Maxwell-Boltzmann distribution is the only solution of Eq. (28.40).

28.5 Conclusions

The Maxwell Boltzmann distribution has a very special property. If the particles with this distribution are shot upwards from the ground at $z = 0$, then the vertical velocity component of all particles decreases. But not all particles reach a certain height $z = h$. The slower particles fall down before and therefore no longer contribute to the kinetic energy. The faster particles slow down by exactly the amount that the average kinetic energy remains constant. The deformation of the Maxwell Boltzmann distribution with height is shown visually by solving the Boltzmann equation using the method of characteristics.

References

Boltzmann L (1876) Über die Aufstellung und Integration der Gleichungen, welche die Molekularbewegung in Gasen bestimmen. Sitzungsberichte der Kgl. Akademie der Wissenschaften in Wien (74):503–552

Chapman S, Cowling TG (1990) The mathematical theory of non-uniform gases: an account of the kinetic theory of viscosity, thermal conduction and diffusion in gases. Cambridge University Press

Coombes CA, Laue H (1985) A paradox concerning the temperature distribution of a gas in a gravitational field. Am J Phys 53(3):272–273

Dreyer W, Müller WH, Weiss W (2000) Tales of thermodynamics and obscure applications of the second law. Contin Mech Thermodyn 12:151–184

Ebert MR, Reissig M (2018) Methods for partial differential equations: qualitative properties of solutions, phase space analysis, semilinear models. Birkhäuser

Eckart C (1940) The thermodynamics of irreversible processes. III. Relativistic theory of the simple fluid. Phys Rev 58(10):919

Graeff RW (2002) Measuring the temperature distribution in gas columns. In: AIP conference proceedings, vol 643. American Institute of Physics, pp 225–230

Graeff RW (2007) Viewing the controversy Loschmidt-Boltzmann/Maxwell through macroscopic measurements of the temperature gradients in vertical columns of water. Preprint, Additional results are on the web page

Graeff RW (2015) Present second law statements have to be adjusted due to measured temperature gradients in xenon gas. https://www.andreastrupp.com/Graeff280315.pdf. [Online; Accessed 01 Sept 2024]

Jeong HM, Park S (2022) Temperature gradient of vertical air column in gravitational field. Sci Rep 12(1):6756

Liao C (2009) Temperature gradient caused by gravitation. Int J Mod Phys B: Cond Matter Phys; Stat Phys; Appl Phys 23(22)

Loschmidt J (1877) Über den Zustand des Wärmegleichgewichtes eines Systems von Körpern mit Rücksicht auf die Schwerkraft III. Sitzungsberichte der Kgl. Akademie der Wissenschaften in Wien (75):287

Maxwell JC (1867) Iv. on the dynamical theory of gases. Philos Trans R Soc Lond (157):49–88

Müller I, Ruggeri T (2004) Stationary heat conduction in radially, symmetric situations-an application of extended thermodynamics. J non-Newton Fluid Mech 119(1–3):139–143
Reitebuch D, Weiss W (1999) Application of high moment theory to the plane couette flow. Contin Mech Thermodyn 11:217–225
Struchtrup H, Taheri P (2011) Macroscopic transport models for rarefied gas flows: a brief review. IMA J Appl Math 76(5):672–697
Tolman RC, Ehrenfest P (1930) Temperature equilibrium in a static gravitational field. Phys Rev 36(12):1791
Trupp A (1999) Energy, entropy: on the occasion of the 100th anniversary of Josef Loschmidt's death in 1895: is Loschmidt's greatest discovery still waiting for its discovery? Phys Essays 12(4):614
Velasco S, Román F, White J (1996) On a paradox concerning the temperature distribution of an ideal gas in a gravitational field. Eur J Phys 17(1):43
Zhang K, Zhang YJ (2016) Principle of detailed balance and a dilute gas in gravitational field. arXiv:1607.06692

MIX
Papier aus verantwortungsvollen Quellen
Paper from responsible sources
FSC® C105338
FSC
www.fsc.org

If you have any concerns about our products,
you can contact us on
ProductSafety@springernature.com

In case Publisher is established outside the EU,
the EU authorized representative is:
Springer Nature Customer Service Center GmbH
Europaplatz 3, 69115 Heidelberg, Germany

Printed by Libri Plureos GmbH
in Hamburg, Germany